Standard Designations for Devices

Designation	Device
A	Accelerating contactor or relay
ABE	Alarm or annunciator bell
ABU	Alarm or annunciator buzzer
AH	Alarm or annunciator horn
AM	Ammeter
AT	Autotransformer
B	Brake relay
CAP	Capacitor
CB	Circuit breaker
CH	Chassis or frame (not necessarily grounded)
CI	Circuit interrupter
CON	Contactor
COS	Cable-operated (emergency) switch
CR	Control relay
CRA	Control relay, automatic
CRE	Control relay, emergency
CRH	Control relay, manual
CRL	Control relay, latch
CRM	Control relay, master
CRU	Control relay, unlatch
CS	Cam switch
CT	Current transformer
CTR	Counter
D	Diode
DB	Dynamic braking contactor or relay
DISC	Disconnect switch
F	Forward
FA	Field accelerating contactor or relay
FB	Fuse block
FD	Field decelerating contactor or relay
FF	Full-field contactor or relay
FL	Field-loss contactor or relay
FLD	Field
FLS	Flow switch
FS	Float switch
FTB	Fusible terminal block
FTS	Foot switch
FU	Fuse
FW	Field weakening
GRD	Ground
HTR	Heating element
INST	Instrument
IOL	Instantaneous overload
LO	Lock-out coil (in plugging switch)

Designation	Device
LS	Limit switch
LT	Pilot light
M	Motor starter
MB	Magnetic brake
MC	Magnetic clutch
MCS	Motor-circuit switch
MF	Motor starter, forward
MR	Motor starter, reverse
MSH	Meter shunt
MTR	Motor
NLT	Neon light
OL	Overload relay
PB	Pushbutton
PL	Plug
PLS	Plugging switch
POT	Poentiometer
PRS	Proximity switch
PS	Pressure switch
PSC	Photosensitive cell
R	Reverse
REC	Rectifier
RECP	Receptacle
RES	Resistor
RH	Rheostat
RSS	Rotary selector switch
S	Switch
SCR	Silicon-controlled rectifier
SOC	Socket
SOL	Solenoid
SS	Selector switch
T	Transformer
TACH	Tachometer
TAS	Temperature-actuated switch
TB	Terminal block
T/C	Thermocouple
TCS	Thermocouple switch
TGS	Toggle switch
TR	Time-delay relay
VAT	Variable autotransformer
VM	Voltmeter
WLT	Work light
WM	Wattmeter
X	Reactor

MODERN INDUSTRIAL ELECTRONICS

MODERN INDUSTRIAL ELECTRONICS

TIMOTHY J. MALONEY

Monroe County Community College
Monroe, Michigan

Upper Saddle River, New Jersey
Columbus, Ohio

Library of Congress Cataloging in Publication Data

Maloney, Timothy J.
Modern industrial electronics / Timothy J. Maloney.—4th ed.
p. cm.
Includes index.
ISBN 0-13-015676-0
1. Industrial electronics. I. Title.

TK7881 .M343 2001
629.8′9—dc21

00-049141
CIP

Editor in Chief: Stephen Helba
Assistant Vice President and Publisher: Charles E. Stewart, Jr.
Assistant Editor: Delia K. Uherec
Production Editor: Alexandrina Benedicto Wolf
Production Coordination: York Production Services
Design Coordinator: Robin G. Chukes
Cover Designer: Jeff Vanik
Cover Image: © Uniphoto
Production Manager: Matthew Ottenweller

This book was set in Times and Gill Sans by York Graphic Services, Inc. It was printed and bound by R.R. Donnelley & Sons Company. The cover was printed by Phoenix Color Corp.

10 9 8 7 6 5 4 3 2 1
ISBN 0-13-015676-0

PREFACE

Modern Industrial Electronics, Fourth Edition, provides a total-system view of the world of manufacturing and automated production for students of electrical and electronics technology. It maintains the original commitment, intact since the first edition, of showing how the devices of modern electronics are applied in realistic industrial applications.

In this edition, coverage has been expanded in the following areas:

Chapter 3, PLCs

- Additional instructions are described and demonstrated.
- An explicit standard solution is put forward to a particular program logic situation that is commonly encountered.

Chapter 8, Op Amps

Thorough coverage has been provided for these op amp ideas:

- Open-loop voltage gain
- Virtual zero differential input voltage
- Virtual ground at the nongrounded input terminal
- Virtual zero input signal current
- Inverting amplifier operating characteristics, including closed-loop voltage gain, closed-loop input resistance, and closed-loop output resistance
- Noninverting amplifier operating characteristics
- Output offset, cause, and correction method
- Voltage comparator, inverting and noninverting
- Weighted summing circuit
- Operation from a single-polarity power supply

Chapter 10, Input Transducers

Various flow-measurement devices are introduced and explained, including

- Ultrasonic (Doppler effect)
- Turbine
- Pressure-drop (Venturi)
- Nutating disk

Chapter 19, Industrial Robots

- Expanded coverage of end-of-arm tooling and proximity sensing of objects

A WORD TO STUDENTS

The capabilities of industrial manufacturing systems have expanded at a startling rate since the first edition of *Modern Industrial Electronics* was published in 1979. Part of the new capability has to do with more precise control over machines and processes, and part has to do with our greater ability to measure and make records of production variables. This expansion has two direct effects on you. First, it makes your work more demanding. Second, it gives you the opportunity for even greater satisfaction and personal reward, because anyone who can learn and master today's high-technology industrial controls is sought after by employers. As a technician or engineering technologist working in modern industry, you are a member of a select group, indispensable to your company's productivity and profitability. In fact, your work contribution has obvious impact on our entire society's productivity and economic security. What a compliment to you that you are entrusted with that responsibility.

In this fourth edition, as in the previous three editions that your predecessors used to launch their careers, I have made every effort to help you reach the skill level needed for carrying out your job responsibilities. Toward that goal, this edition features a "Troubleshooting on the Job" exercise at the end of every chapter. These troubleshooting exercises require you to apply the knowledge that you have gained from that chapter to solve a problem. By carrying out the troubleshooting exercises individually or as a team, you will find yourself exercising your technical understanding, thinking imaginatively, and solving realistic problems—in other words, making the transition from classroom student to on-the-job technician or technologist in the industrial arena.

My best wishes for your working career.

FEATURES OF THE TEXT

Chapter-Opening Photograph

Each chapter begins with a photograph and explanatory caption that depicts some modern industrial practice. Figure A shows the opening pages of Chapter 5. Use these presentations to get a feel for some of the interesting opportunities and work responsibilities in the field of industrial electronics. Descriptive captions and credits for the photographs are listed on page x.

Objectives

The first edition, published in 1979, was the original college technology textbook that explicitly stated the learning objectives at the beginning of each chapter. That precedent is continued, naturally, in this fourth edition. As you are reading and studying, try to perform the task that each objective calls for. If you can perform these tasks, then you are learning what the book and the course have to offer. If you find that you cannot satisfy the objectives, ask further questions in class or consult in private with your instructor.

FIGURE A
Opening photograph for Chapter 5: X-ray inspection system for detecting hidden flaws in manufactured parts.

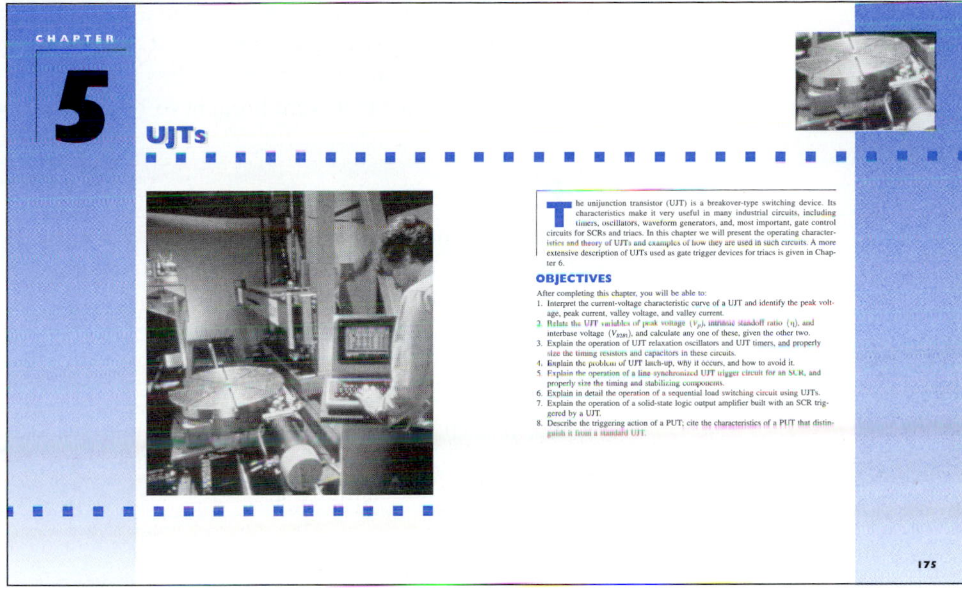

CHAPTER

5 UJTs

The unijunction transistor (UJT) is a breakover-type switching device. Its characteristics make it very useful in many industrial circuits, including timers, oscillators, waveform generators, and, most important, gate control circuits for SCRs and triacs. In this chapter we will present the operating characteristics and theory of UJTs and examples of how they are used in such circuits. A more extensive description of UJTs used as gate trigger devices for triacs is given in Chapter 6.

OBJECTIVES

After completing this chapter, you will be able to:

1. Interpret the current-voltage characteristic curve of a UJT and identify the peak voltage, peak current, valley voltage, and valley current.
2. Relate the UJT variables of peak voltage (V_p), intrinsic standoff ratio (η), and interbase voltage (V_{B2B1}), and calculate any one of these, given the other two.
3. Explain the operation of UJT relaxation oscillators and UJT timers, and properly size the timing resistors and capacitors in these circuits.
4. Explain the problem of UJT latch-up, why it occurs, and how to avoid it.
5. Explain the operation of a line synchronized UJT trigger circuit for an SCR, and properly size the timing and stabilizing components.
6. Explain in detail the operation of a sequential load switching circuit using UJTs.
7. Explain the operation of a solid-state logic output amplifier built with an SCR triggered by a UJT.
8. Describe the triggering action of a PUT; cite the characteristics of a PUT that distinguish it from a standard UJT.

175

Troubleshooting on the Job

The final section of each chapter gives a "Troubleshooting on the Job" exercise that is representative of the duties that you will be performing when working as a technician or engineering aide. Figure B shows the Troubleshooting on the Job in Chapter 13, from pages 596–597, which requires you to analyze a stepper motor test circuit. These assignments invariably require you to use the knowledge that you have learned from that chapter in an imaginative way. Your instructor may ask for your written or drawn solution to

FIGURE B
Troubleshooting on the Job feature in Chapter 13: On-the-job exercises, accompanied by illustrations and photos, challenge you to perform real-life duties.

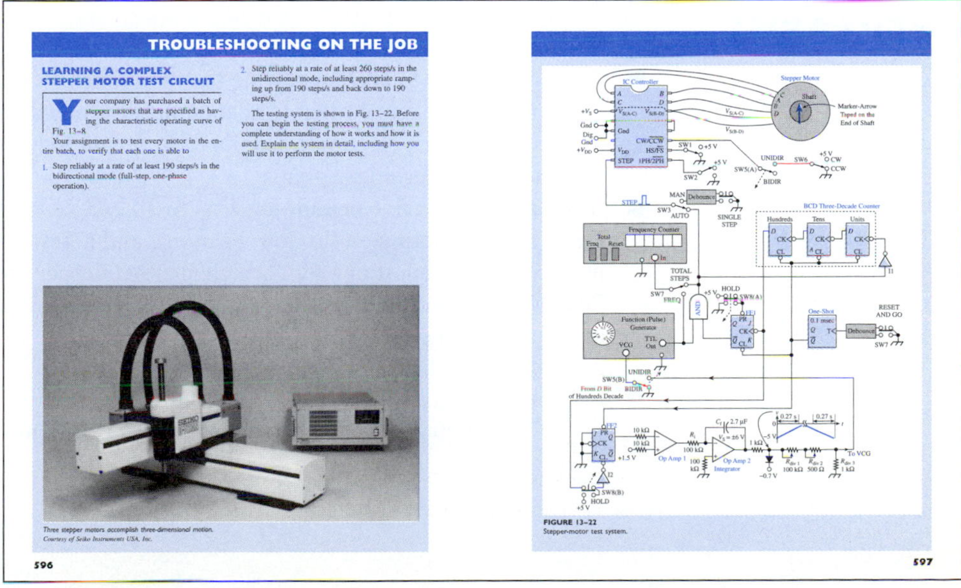

TROUBLESHOOTING ON THE JOB

LEARNING A COMPLEX STEPPER MOTOR TEST CIRCUIT

Your company has purchased a batch of stepper motors that are specified as having the characteristic operating curve of Fig. 13–8.

Your assignment is to test every motor in the entire batch, to verify that each one is able to

1. Step reliably at a rate of at least 190 steps/s in the bidirectional mode (full-step, one-phase operation).
2. Step reliably at a rate of at least 260 steps/s in the unidirectional mode, including appropriate ramping up from 190 steps/s and back down to 190 steps/s.

The testing system is shown in Fig. 13–22. Before you can begin the testing process, you must have a complete understanding of how it works and how it is used. Explain the system in detail, including how you will use it to perform the motor tests.

Three stepper motors accomplish three-dimensional motion.
Courtesy of Seiko Instruments USA, Inc.

FIGURE 13–22
Stepper-motor test system.

596 597

be done individually, or you may be placed in a two- or three-person team to work on the problem. Various solutions are possible in most cases; therefore you and others in the class will present your solutions to the entire class so that everyone can share the differing thoughts and approaches that were brought to bear on the problem.

Examples

When trying to understand new ideas, especially the use of new mathematical formulas, all people are helped by examples. In this text, examples are provided for all situations where numerical calculations are required.

Summary

At the end of each chapter, there is a list of the main ideas that were developed within that chapter. The chapter's mathematical formulas, if any, are also collected for your ready reference in solving the homework.

Questions and Problems

Numerous questions and problems, organized by chapter section, are provided to sharpen your understanding and exercise your problem-solving skills. Your instructor will assign some of them for homework. You may wish to tackle additional problems for your own satisfaction. The more you practice, the more you learn.

Glossary

Definitions of hundreds of terms used in industrial electronics are listed in the glossary. Most of these terms were introduced in this text, but some come from earlier course work in electricity and electronics. Use the glossary to refresh your memory or to verify your understanding of a word's meaning.

ANCILLARIES

- The Lab Manual to accompany the text, by James R. Davis (ISBN 0-13-032332-2), contains experiments written to accommodate students enrolled in industrial electronics courses for electronics engineering technology programs or industrial electronics apprenticeship programs.
- Instructor's Manual: Contains answers to all end-of-chapter questions; solutions to the "Troubleshooting on the Job" sections; and a Test Item File, which contains 20 multiple choice questions for each chapter. Also packaged with each IM are PowerPoint Slides (ISBN 0-13-031557-5). Figures from the text are designed to help instructors with classroom/lecture presentations. The slides are contained on a CD packaged with the Instructor's Manual.
- PH Test Manager (ISBN 0-13-032863-4): Test generator software provides instructors with test questions that can be modified or used just as they are to create student tests.

ACKNOWLEDGMENTS

My thanks to all the people who contributed to this revision. The prodding and encouragement of Charles Stewart and Delia Uherec were quite a psychological help. James Davis of Muskingum Area Technical College in Zanesville, Ohio, prepared the lab manual that accompanies this text. The fine work of manuscript editing and book production were coordinated and accomplished by Alex Wolf and Tonia Grubb.

The comments and opinions of reviewers are essential to an effective textbook revision. I wish to extend my thanks to the following reviewers of this edition for their valuable suggestions: Bob L. Bixler, Austin Community College; Louis E. Frenzel, Austin Community College; William Maxwell, Nashville State Technical Institute; Louis S. Otruba, Pittsburgh Technical Institute; and Frank Riffle.

—T. J. M.

PHOTO CAPTIONS AND CREDITS

Page xxii	Automated machine tooling accomplishes precision cutting of the large gear teeth. *(Courtesy of Society of Manufacturing Engineers)*
Page 34	On the production floor, parts usage is electronically counted and periodically compared to predicted draw-down. A restocking order is then automatically issued to produce just-in-time inventory replenishment. *(Courtesy of General Electric Company)*
Page 74	I/O rack of a large PLC system. *(Courtesy of General Electric Company)*
Page 148	Magnetic levitation vehicle cruising at 400 km/hr. *(Courtesy of Railway Technical Research Institute of Japan)*
Page 174	This Advanced Computed Tomography Inspection System (ACTIS) uses X-rays to make a detailed internal examination of manufactured parts. It can thereby discover internal flaws and stress fractures. *(Courtesy of NASA)*
Page 206	Chemical distillation towers often have their heating and condensing processes controlled by high-power thyristors. *(© Brownie Harris)*
Page 240	Welding robot. *(Courtesy of Society of Manufacturing Engineers)*
Page 282	The MagLev vehicle on its approach track, where it is riding on its mechanical suspension, but propelled by sidewall electromagnets as usual. *(Courtesy of Railway Technical Research Institute of Japan)*
Page 314	When a compact disk is played, the laser pickup head starts close to the center, then moves radially outward to the outside edge. If the disk's spinning speed were kept constant, the outside tracks would move faster across the pickup head than the inside tracks; this cannot be tolerated. CD players use an electronic feedback system to adjust the disk's rotational spinning speed so that the timing bits that are recorded beside the music code bits are detected at a constant rate. Thus, the recorded music is picked up and played at a constant speed. *(© Steve Dunwell for General Electric Company)*
Page 368	Testing apparatus for Hall-Effect proximity detectors used on a magnetic levitation track. *(Courtesy of Railway Technical Research Institute of Japan)*
Page 448	High-frequency test instrumentation. *(Courtesy of Tektronix, Inc.)*
Page 498	State-of-the-art dc motors have overall shaft efficiencies in excess of 95%. *(© Joe McNally, General Electric Company)*
Page 566	Small robotic assembly systems often use stepper motors or position-triggered brushless dc motors for making their motions. *(Courtesy of Seiko Instruments USA, Inc.)*
Page 602	Variable-frequency ac motor drive system. *(Courtesy of Tektronix, Inc.)*
Page 656	Unlike a simple stamping press machine, this large forming press machine uses a proportional controller to regulate its applied force. *(© Brownie Harris)*
Page 710	Machining and polishing of these wheels is done under precisely controlled conditions of rotational speed. *(Courtesy of Railway Technical Research Institute of Japan)*
Page 772	Telemetry receiving station for space satellites. *(Courtesy of Tektronix, Inc.)*
Page 816	Microcomputer software bugs are discovered by using a Logic Analyzer, which provides a CRT display of instruction codes, or a display of waveform graphs like an oscilloscope. *(Courtesy of Tektronix, Inc.)*
Page 856	Cylindrical-configuration robot. *(Courtesy of Seiko Instruments USA, Inc.)*
Page 904	Fire is one of the safety hazards of industrial circuits. *(Courtesy of NASA)*

BRIEF CONTENTS

CONTENTS

CHAPTER

THE TRANSISTOR SWITCH AS A DECISION-MAKER

In any industrial system, the control circuits constantly receive and process information about the conditions in the system. This information represents such things as the mechanical positions of movable parts; temperatures at various locations; pressures existing in pipes, ducts, and chambers; fluid flow rates; forces exerted on various detecting devices; speeds of movements; etc. The control circuitry must take all this empirical information and combine it with input from the human operators. Human operator input usually has the form of selector switch settings and/or potentiometer dial settings. Such operator input represents the desired system response or, in other words, the production results expected from the system.

Based on the comparison between system information and human input, the control circuitry *makes decisions*. These decisions concern the next action of the system itself, such as whether to start or stop a motor, whether to speed up or slow down a mechanical motion, whether to open or close a control valve, or even whether to shut down the system entirely because of an unsafe condition.

Obviously, no real thought goes into the decision making done by the control circuits. Control circuits merely reflect the wishes of the circuit designer, who foresaw all the possible input conditions and designed in the appropriate circuit responses. However, because the control circuits mimic the thoughts of the circuit designer, they are often called *decision-making circuits* or, more commonly, *logic circuits*.

OBJECTIVES

After completing this chapter you will be able to:

1. Name the three parts of an industrial control circuit and describe the general function of each part.
2. Describe how relays can be used to make decisions.
3. Distinguish between normally open and normally closed relay contacts.
4. Describe in detail the operation of a part-classifying system using relay logic.
5. Describe in detail the operation of a part-classifying system using solid-state logic.
6. Name and explain the operation of the various circuits used for input signal conditioning in solid-state logic.
7. Explain the purpose and operation of output amplifiers used with solid-state logic.
8. Discuss the relative advantages and disadvantages of solid-state logic and relay logic.
9. Describe in detail the operation of three real-life solid-state logic systems: a machine tool routing system, a first fault annunciator, and a machine tool drilling system.

1-1 ■ SYSTEMS CONTAINING LOGIC CIRCUITS

An electrical control circuit for controlling an industrial system can be broken down into three distinct parts. These parts or sections are (1) input, (2) logic, and (3) output.

The *input section,* sometimes called the *information-gathering section* in this book, consists of all the devices which supply system information and human operator settings to the circuits. Some of the common input devices are pushbuttons, mechanical limit switches, pressure switches, and photocells.

The *logic section,* sometimes called the *decision-making section* in this book, is that part of the circuit which acts upon the information provided by the input section. It makes decisions based on the information received and sends orders to the output section. The logic section's circuits are usually built with magnetic relays, discrete transistor circuits, or integrated transistor circuits. Fluidic devices can also be used for logic, but they are much less common than electromagnetic and electronic methods. We will not discuss fluidic devices. The essential ideas of logic circuits are universal, no matter what actual devices are used to build them.

The *output section,* sometimes called the *actuating device section* in this book, consists of the devices which take the output signals from the logic section and convert or amplify these signals into useful form. The most common actuating devices are motor starters and contactors, solenoid coils, and indicating lamps.

The relationship among these three parts of the control circuit is illustrated in Fig. 1–1.

1-2 ■ LOGIC CIRCUITS USING MAGNETIC RELAYS

For many years, industrial logic functions were performed almost exclusively with mechanically operated relays, and relay logic still enjoys wide popularity today. In this method of construction, a relay coil is energized when the circuit leading up to the coil is completed by closing certain switches or contacts. Figure 1–2 shows that relay *A* (R*A*) is energized if limit switch 1 (LS1) and pressure switch 4 (PS4) are closed.

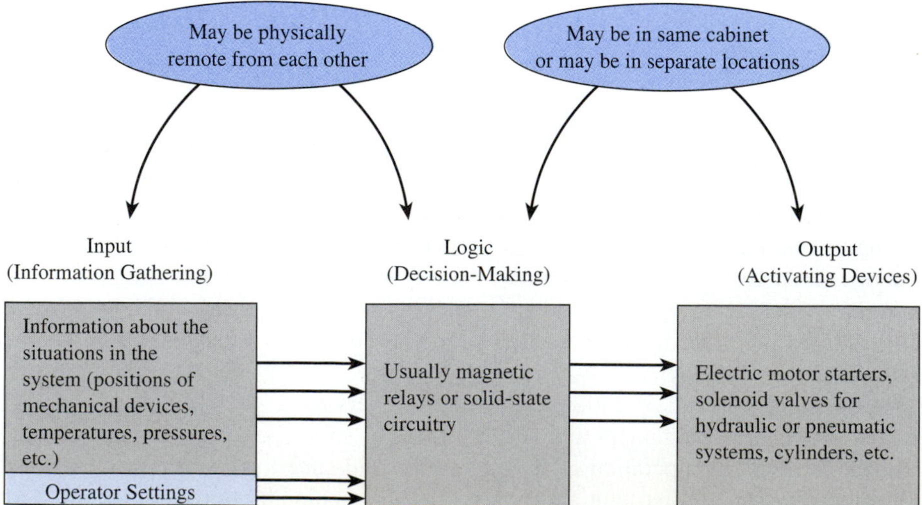

FIGURE 1–1
The relationship among the three parts of an industrial control system.

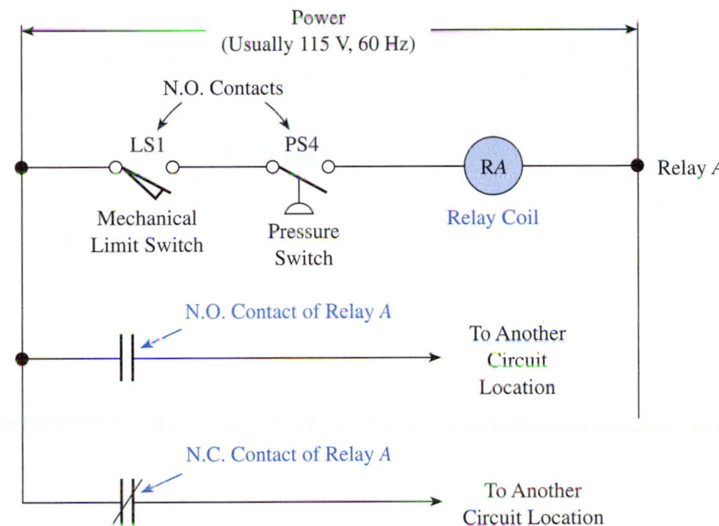

FIGURE 1–2 A relay logic circuit in which the relay coil is controlled by input devices, namely a limit switch and a pressure switch.

The design of the circuit in Fig. 1–2 calls for relay *A* to be energized if a certain combination of events occurs in the system. The necessary combination is the closing of LS1 by whatever apparatus operates LS1 and, at the same time, the closing of PS4 by whatever liquid or gas affects PS4. If both of these things occur at the same time, relay *A* will energize. The terms *picked up* or just *picked* are often used to mean energized, and these terms will sometimes be employed in this book.

If either or both of the switches is open, then R*A* will be deenergized. The terms *dropped out* or *dropped* are often used to mean deenergized, and these terms will also be employed occasionally in this book.

If R*A* is deenergized, the contacts controlled by R*A* revert to their normal state. That is, the *normally closed* (N.C.) contacts are closed and the *normally open* (N.O.) contacts are open. On the other hand, if R*A* is energized, all the contacts associated with R*A* change state. The N.C. contacts go open, and the N.O. contacts go closed. Figure 1–2 shows only one of each kind of contact. Real industrial relays often have several contacts of each kind (several N.C. and several N.O. contacts).

Although this circuit is quite simple, it illustrates the two foundation ideas of relay logic circuits, and for that matter of all logic circuits:

1. A positive result (in this case picking up the relay) is conditional on several other events. The exact conditions needed depend on how the feed-in switch contacts are connected. In Fig. 1–2, *both* LS1 and PS4 must be closed because the contacts are connected in series. If the contacts were connected in parallel, *either* switch being closed would pick up the relay.
2. Once a positive result occurs, the result can branch out to many other circuit locations. It can thereby make its effects felt in several places throughout the control circuits. Figure 1–2 shows R*A* having one N.O. contact and one N.C. contact with each contact leading off to some other destination in the overall circuit. Therefore the action of R*A* would be felt at both of these other circuit destinations.

Expanding on these ideas concerning circuit logic, Fig. 1–3 shows how contacts which feed in to a relay coil sometimes are controlled by *other relays* instead of mechanical limit switches and other independent switches. In Fig. 1–3, the limit switch is

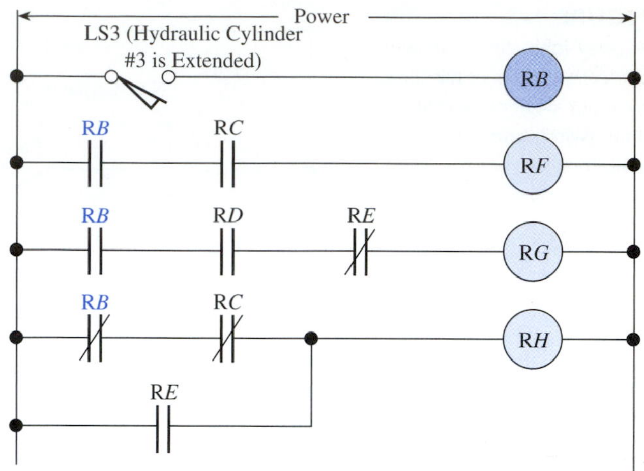

FIGURE 1–3
A relay logic circuit in which relay coils are controlled by the contacts of other relays.

mechanically actuated when hydraulic cylinder 3 is fully extended. Hydraulic cylinder 3 would be located somewhere in the mechanical apparatus of the industrial system and would have some sort of cam attached to it for actuating LS3. When the N.O. contact of LS3 goes closed, R*B* picks up. The "branching-out" idea is illustrated here because R*B* has three contacts, each one leading to a separate circuit location. The action of R*B* therefore affects the results at three other circuit locations, in this case, R*F*, R*G*, and R*H*. This branching-out idea is often referred to as *fan-out*.

To appreciate the decision-making ability of such circuits, consider R*G* carefully. Imagine that R*G* has control over a solenoid valve which can pass or block the flow of water through a certain pipe. Therefore water will flow if the following conditions are met:

1. R*B* is picked, and
2. R*D* is picked, and
3. R*E* is dropped.

We have already seen that R*B* is controlled by hydraulic cylinder 3 through LS3. Relays R*D* and R*E*, though not described in Fig. 1–3, represent conditions in the system, human input, or a combination of both. For concreteness, imagine that R*D* will be picked if adequate water pressure is available and that R*E* will be picked if a certain type of contamination is detected in the water.

What happens here is that R*G* *makes a decision* on whether or not to permit water to flow. It makes its decision by consideration of three conditions:

1. R*B* (N.O.): Hydraulic cylinder 3 must be extended.
2. R*D* (N.O.): Adequate pressure must be available in the system.
3. R*E* (N.C.): The water must not be contaminated.

This therefore is a simple example of how relays are used to build a logic circuit.

1-3 ■ RELAY LOGIC CIRCUIT FOR A CONVEYOR/CLASSIFYING SYSTEM

To cement what we have learned about general logic systems, let us consider the logic for a specific system. The layout is drawn schematically in Fig. 1–4(a).

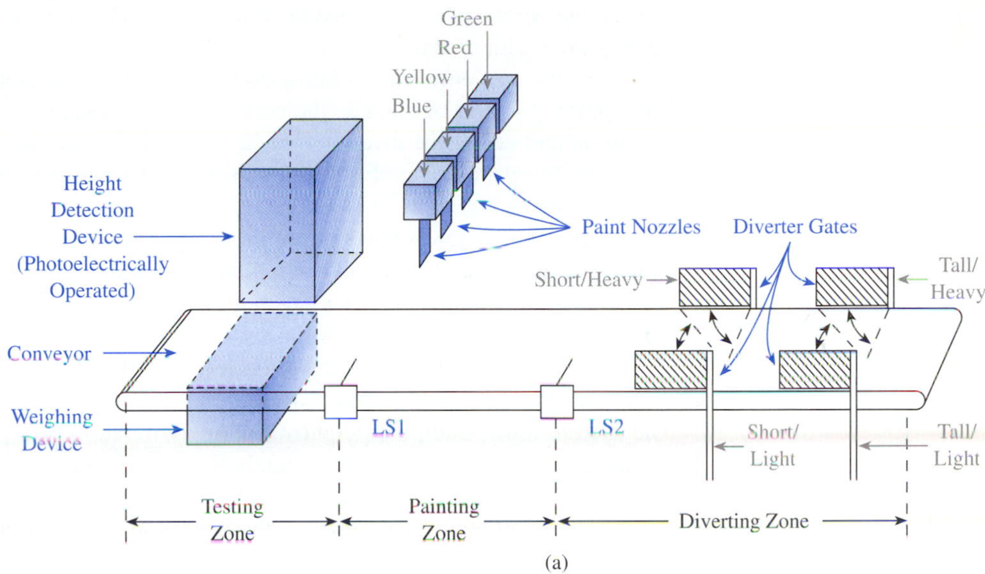

(a)

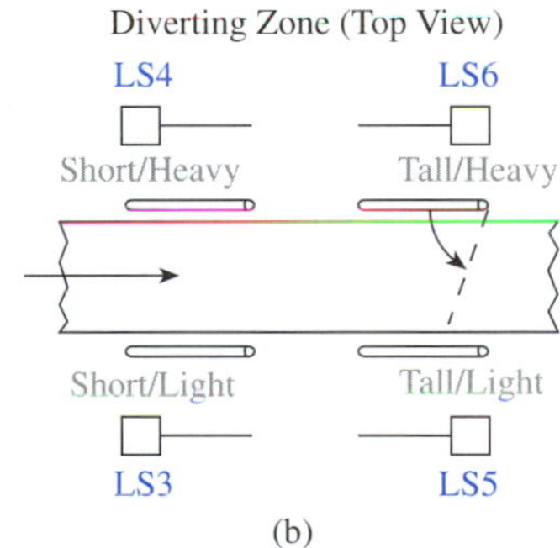

(b)

FIGURE 1–4
(a) Physical layout of a conveyor/classifying system.
(b) Top view of the diverting zone, showing the positions of the four diverting gates and the four chute limit switches.

Manufactured parts of varying height and weight come down the conveyor, moving to the right. A height detector measurers the height of each part and classifies it as either short or tall, depending on whether the part is below or above a certain prescribed height. Likewise, a weighing device classifies it as either light or heavy depending on whether it is below or above a certain prescribed weight. Each part can therefore be put into one of four overall classifications. It is (1) short/light, (2) short/heavy, (3) tall/light, or (4) tall/heavy.

The system then color-codes each part by spraying on a paint stripe of the proper color. After it has been painted, the part is sorted into the proper chute depending on its classification. There are four chutes, one for each classification. This sorting is done by having a diverter gate swing out to direct the part off the conveyor into the proper chute. Each chute has its own gate.

Referring to Fig. 1–4(a), we see that the system is divided into three zones.

A part is measured for height and weight in the testing zone. As it leaves the testing zone and enters the painting zone, the part actuates LS1. LS1 is a limit switch with a cat-whisker extension. Such switches are used when the actuating body does not have an exact repeatable position; parts moving on a conveyor belt are a good example. The part may be offset to either the left or the right side of the conveyor. To detect the passage of a part, the detecting switch must be able to respond to a body located anywhere on a line across the width of the conveyor.

As the part enters the painting zone, one of the four paint solenoid valves is opened up, applying a strip of paint as the part moves underneath it. As the part leaves the painting zone and enters the diverter zone, it strikes LS2, another cat-whisker limit switch. At this time the paint valve closes and one of the four diverter gates swings out. When the part hits the diverter gate, it is guided off the belt and into the proper chute. Figure 1–4(b) indicates how the gate swings out to block the path of the moving part. As the part slides down one of the chutes, it strikes the limit switch mounted in that chute; LS3, LS4, LS5, or LS6. At this time, the diverter gate returns to its normal position, and the system is ready to receive another part in the testing zone.

The parts must be handled in such a way that a new part cannot enter the testing zone until the previous part has cleared one of the chute limit switches. This is because the system must retain the height/weight classification of a part until that part has completely cleared the system. It must retain the classification because it must hold the proper diverter gate out until the part has left the belt.

The relay logic to accomplish the operation is shown in Fig. 1–5. The operation of the logic circuitry will now be explained. An equivalent solid-state logic circuit will be presented and explained in Sec. 1-6. This way, you can become acquainted with a complete practical logic circuit using relay construction. After an understanding of the system itself is gained, we will go on to study the same system using a more modern method of construction.

Let us start on line 9 of Fig. 1–5. The N.C. RCLR* contact is closed at the time a part enters the testing zone. While a part is in the testing zone, the height detector closes its contact if the part is tall but leaves the contact open if the part is short. This will pick RTAL if the part is tall or leave RTAL dropped if the part is short. If RTAL picks, it seals itself with the N.O. RTAL contact on line 10. This is necessary because the height detector contact will return to the N.O. condition after the part has left the testing zone, but the system must retain the information about its tallness or shortness until the part has completely gone.

The actual operation of the height detector is of no concern to us at this time, since we are concentrating on the system logic.

The weight detector on line 11 does the same thing. If the weight of the part is above the preset weight, the contact closes and picks RHVY, which then seals up with the N.O. contact on line 12. If the part is below the preset weight, the weight detector contact remains open, and RHVY remains dropped.

The circuitry between lines 13 and 16 picks the proper relay to indicate the classification of the part. If the part is short, the N.C. RTAL contact on line 13 will remain closed, applying power to the left side of the two RHVY contacts on lines 13 and 14. Then, depending on whether the part is light or heavy, either RSL (short/light) will pick or RSH (short/heavy) will pick.

The same circuit configuration is repeated on lines 15 and 16 through an N.O. contact of RTAL. If the part is tall, the N.O. RTAL contact will go closed, causing RTL

---

*Relays are often named in accordance with the function they perform in the logic circuit. The name of a relay represents an abbreviation of its function. An example is RCLR, where the letters CLR are an abbreviation of the word *cleared*. The preceding R used in all relay names stands for the word *relay*. A more complete description of the relay's function is usually written alongside the coil, as an aid to understanding circuit operation. This helpful practice is followed in Fig. 1–5.

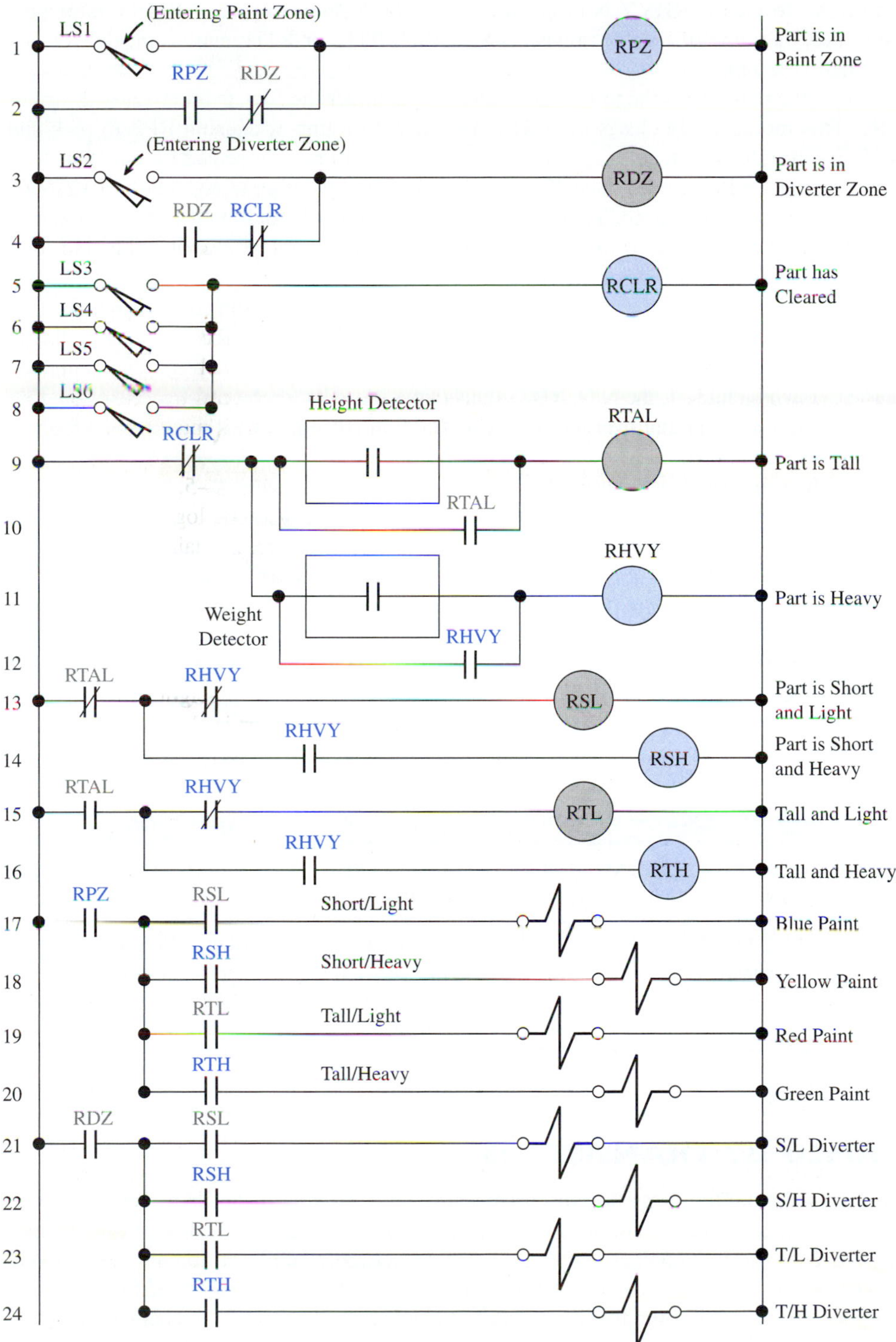

FIGURE 1–5
Control circuitry for the conveyor/classifying system, with the logic performed by magnetic relays.

(tall/light) to pick if RHVY is dropped or RTH (tall/heavy) to pick if RHVY is picked. Note that only one of the four relays, RSL, RSH, RTL, or RTH, can be picked for any part that is tested.

As the part leaves the testing zone and passes under the four paint nozzles, it strikes LS1. This momentarily closes the LS1 N.O. contact on line 1, causing RPZ to pick and seal up through its own N.O. contact on line 2. RPZ will remain sealed up until the N.C. RDZ contact on line 2 goes open. The part is now in the painting zone, and the RPZ N.O. contact on line 17 is closed. Therefore one of the paint solenoids will energize, causing the pr oper color of paint to flow onto the moving part. The paint solenoids and their controlling contacts appear on lines 17–20.

As the part leaves the painting zone, it strikes LS2 and momentarily closes the LS2 contact on line 3. This picks RDZ, which seals up through the N.O. RDZ contact on line 4. RDZ also breaks the seal on RPZ when the N.C. contact on line 2 opens, as mentioned earlier. Down on line 21, the N.O. RDZ contact goes closed, thereby causing one of the four diverter gates to swing out over the conveyor. The four solenoids which operate the four diverter gates are shown on lines 21–24.

When the part has been guided off the conveyor and down one of the chutes, one of the four chute limit switches will close momentarily. These switches are LS3, LS4, LS5, and LS6, and they are all wired in parallel on lines 5–8. Therefore when any one of them closes, RCLR picks momentarily. The RCLR N.C. contact on line 4 breaks the seal on RDZ, indicating that the part has left the diverting zone. Also, the RCLR N.C. contact on line 9 breaks the seal on RTAL and RHVY if either one was sealed up. The operation sequence of the system is now complete, and it is ready to receive a new part into the testing zone.

It was stated in Sec. 1-1 that control circuits are divided into three parts: input, logic, and output. The devices in Fig. 1–5 would be categorized as shown in Table 1–1.

TABLE 1–1
Categories of devices in Fig. 1–5.

INFORMATION GATHERING (INPUT)	DECISION MAKING (LOGIC)	ACTUATING DEVICES (OUTPUT)
LS1, LS2, LS3, LS4, LS5, LS6, height detector, weighing device	Relays RPZ, RDZ, RCLR, RTAL, RHVY, RSL, RSH, RTL, RTH, and their associated contacts	Blue, yellow, red, and green solenoids; S/L, S/H, T/L, and T/H solenoids

1-4 ■ LOGIC PERFORMED BY TRANSISTORS

We can see from the foregoing discussion how relay circuits make decisions. In simple terms, when two contacts are wired in series, the circuit function is called an AND function because the first contact *and* the second contact must be closed to energize the load (pick the relay). When two contacts are wired in parallel, the circuit function is an OR function because either the first *or* the second contact must be closed to energize the load. These two basic relay circuit configurations are illustrated in Fig. 1–6, along with two solid-state circuits for implementing the same functions.

In solid-state logic, instead of contacts being open or closed, input lines are LO or HI. Therefore in the solid-state circuits of Fig. 1–6, the X line going HI (going to +5 V) is equivalent to closing the RX contact in the relay circuit. The X line being LO (being at

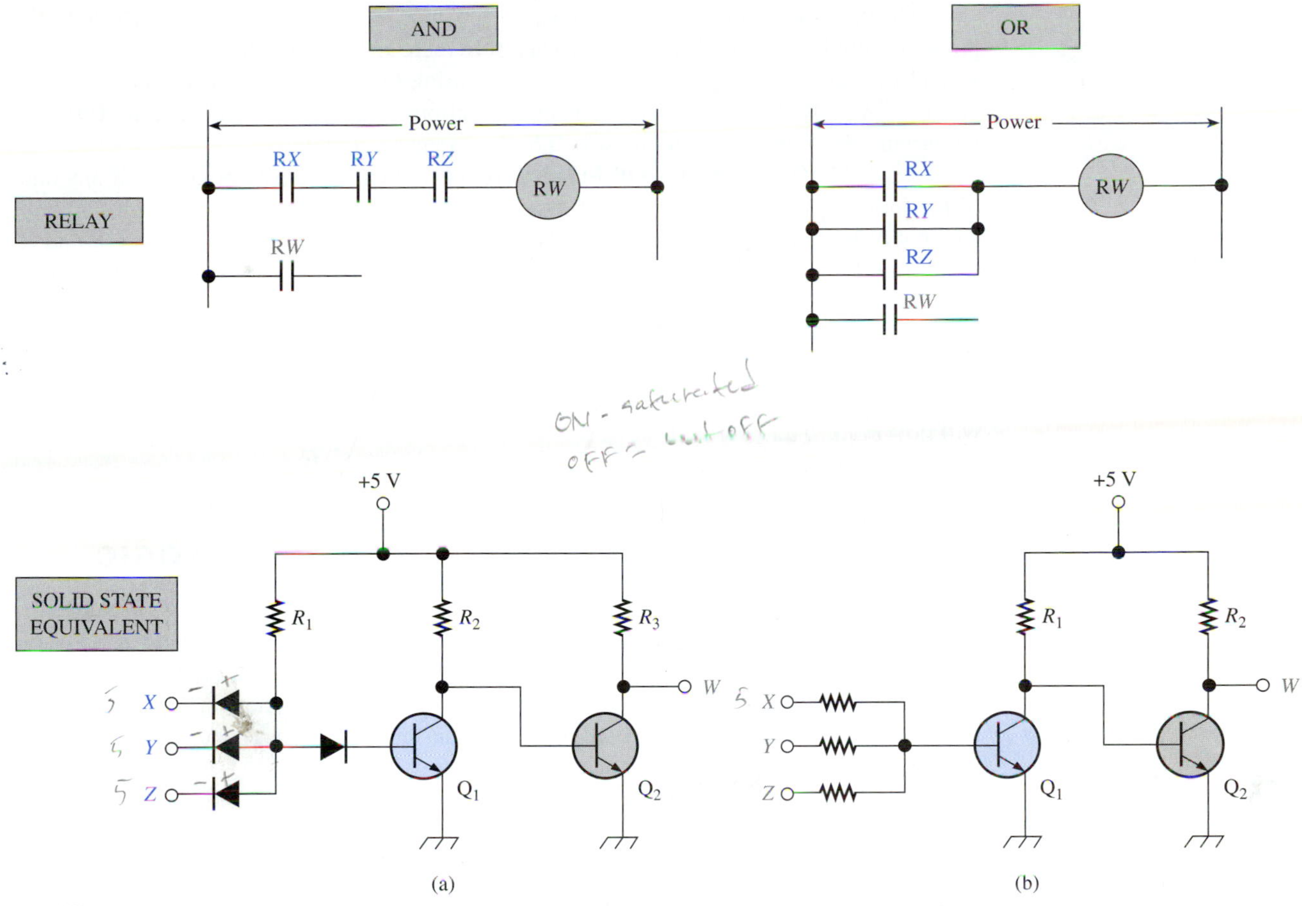

FIGURE 1–6
(a) The AND logic function performed by relay circuitry and by solid-state circuitry. (b) The OR function performed by relay circuitry and by solid-state circuitry.

0 V or ground potential) is equivalent to having the R*X* contact open. The same holds true for the *Y* and *Z* lines.

As for the circuit *result,* in a relay circuit the result is considered to be the energizing of a relay coil and the consequent switching of the contacts controlled by that relay. In a solid-state circuit the result is simply the output line going to a HI state.

With these equivalencies in mind, study the circuit of Fig. 1–6(a). If any one of the inputs is LO (ground voltage), the diode connected to that input will be forward biased. The bias current will come from the +5-V supply, through R_1, through the diode, out the cathode lead of the diode into ground. If a diode is forward biased, its anode can be no more than 0.7 V above the cathode potential. Therefore the anode tie point in Fig. 1–6(a) will be at +0.7 V relative to ground if any one of the *X*, *Y*, or *Z* inputs is LO. With only +0.7 V at the tie point, Q_1* will be cut OFF, due to the hold-off diode in its base lead. Therefore the collector of Q_1 will deliver current into the base of Q_2, turning it ON. With Q_2 saturated, its collector is at approximately 0 V, so the output of the circuit is LO.

*Transistors in electronic schematics can be identified by the letter *Q* or by the letter *T*. We will use *Q* in most situations. The letter *T* is preferred only when *Q* is used for other purposes in the schematic drawing.

On the other hand, if all the inputs X, Y, and Z are HI (+5 V), then the anode tie point will not be pulled down to 0.7 V. Therefore there will be a current flow path through R_1 and into the base of Q_1. Q_1 will saturate, shutting OFF Q_2 and allowing the output to go to +5 V, a HI level. The solid-state circuit action is equivalent to the action of the relay circuit above it. All inputs must be present to get an output.

Figure 1–6(b) shows the OR function. In the solid-state transistor circuit, if any one of the inputs goes HI, Q_1 will turn ON (the resistors are sized to allow this), and its collector will be pulled down to ground. Therefore no base current will flow in Q_2, and it will turn OFF, allowing the output, W, to go to the HI level. Again, the action of the solid-state transistor circuit duplicates that of the relay circuit above it. If any one of the inputs is present, an output will be produced.

For both the relay OR circuit and the solid-state OR circuit, if all inputs are removed (all contacts open in the relay circuit, all inputs LO in the solid-state circuit), the circuit will not produce an output. That is, the relay circuit will fail to energize relay W, and the solid-state circuit will cause a LO signal to appear at output W.

1-5 ■ LOGIC GATES—THE BUILDING BLOCKS OF SOLID-STATE LOGIC

In Sec. 1-4 we showed that solid-state circuits can perform logic functions. It would be very cumbersome and confusing to show every transistor, diode, and resistor in a solid-state logic diagram. Instead, we have invented symbols which stand for the logic function being performed by individual circuits. We then construct complex logic circuits by connecting together many simple individual logic circuits, such as the AND circuit of Fig. 1–6(a).

The simple individual logic circuits thus constitute the *building blocks* of an extensive logic circuit, with each building block indicated by a special symbol. These building blocks are popularly called *logic gates,* or just *gates.*

Review your digital circuits textbook to make sure you have a firm grasp of each of the five basic logic gates—AND, OR, NOT, NAND, and NOR. When you are presented with the schematic symbol of any of these five gates, you should be able to tell at a glance what the output will be for any given combination of inputs. Figure 1–7 shows the symbols for the five basic gates.

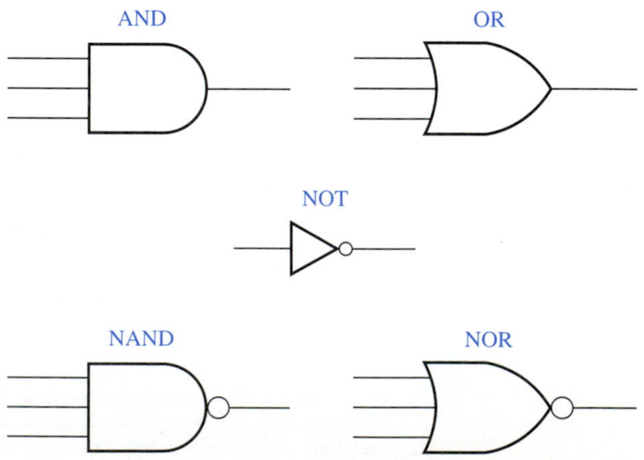

FIGURE 1–7 Schematic symbols of the five basic logic gates. A NOT gate is often called an inverter.

Also review the following topics regarding logic gates:

1. The advantages of inverting gates over noninverting gates (faster operating speed, lower power consumption, lower transistor count in the IC).
2. Current-sinking logic families compared to current-sourcing families.
3. Fan-out capabilities of various logic families: the idea that exceeding the fan-out specification for a current-sinking family jeopardizes the LO output level, while exceeding the fan-out spec for a current-sourcing family jeopardizes the HI output level.
4. Wire-ANDing of gate outputs: the idea that wire-ANDing of outputs is usually permitted if the output transistor has a large-value collector resistor, but is not permitted for totem-pole output circuits, including CMOS.
5. Hanging (unconnected) inputs: the idea that hanging inputs are interpreted as LO by current-sourcing families, but are interpreted as HI by current-sinking families, the noise risk associated with any hanging input, and the disallowance of hanging inputs for all MOS transistors.
6. IC packages and pin identifications (dual-in-line, flat-pack, metal can).
7. Relative noise immunity, operating speed (propagation delay), power consumption, and manufacturing density of various logic families.
8. Positive logic (more-positive voltage level = 1, less-positive level = 0) versus negative logic.

1-6 ■ SOLID-STATE LOGIC CIRCUIT FOR THE CONVEYOR/CLASSIFYING SYSTEM

A solid-state version of the logic for controlling the classifying system of Fig. 1–4 will now be presented and explained.

In Fig. 1–8 the HI logic level is +5 V. As the part passes through the testing zone, the height and weight detectors close their contacts if the height and/or weight are above preset values. Concentrating on the height detector, if the contact closes, a HI will be applied to input 1 of OR3. This causes a HI at the output of OR3, which is then fed back to input 1 of AND3. This causes OR3 to seal up, just as a relay seals up. This happens because input 2 of AND3 is also HI at this time, causing AND3 to be enabled (the output goes HI), which puts a HI on input 2 of OR3. In this circuit arrangement, OR3 is sealed up even after the part leaves the testing zone and the height contact goes back to being open. The only way to break the seal of OR3 is to remove the HI at input 2 of AND3.

It was stated earlier that input 2 of AND3 is HI while the part is in the testing zone. This is so because of the situation at I2, which feeds input 2 of AND3. The input of I2 receives information from limit switches LS3–LS6, all wired in parallel. Since all four of these limit switches are released while the part is in the testing zone, there is no +5-V input applied to I2 at this time. Neither is there a 0-V signal applied to the I2 input. However, the presence of the 1-kΩ resistor connected between the input and ground causes the inverter to treat this situation as if it were a LO input.

Therefore, with the I2 input LO, the output is inverted to a HI, which applies the HI to AND3. The output of OR3 will be maintained HI until the part actuates one of the chute limit switches. At that instant the I2 output will go LO, disabling AND3 and taking away the HI input to OR3. This will break the seal and allow the OR3 output to return to its LO state.

All the foregoing discussion assumed that the height detector contact actually did close, which indicated that the part was tall. Naturally, if the part were short, the contact would fail to close, and OR3 would remain off all through the cycle.

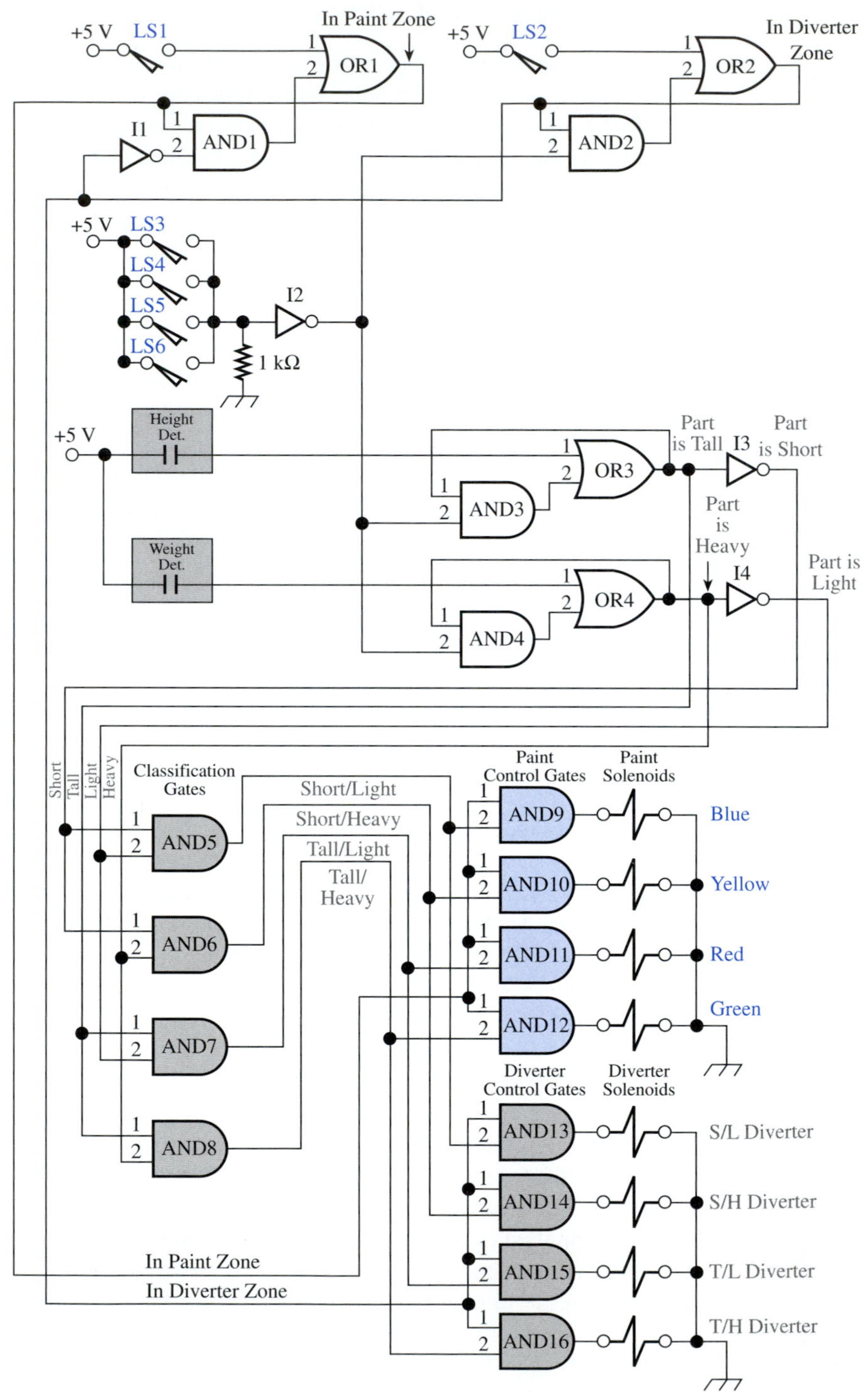

FIGURE 1–8
Control circuitry for the conveyor/classifying system of Fig. 1–4, with the logic performed by solid-state logic gates.

The note on the output line of OR3 describes the meaning of that line going HI. Thus if the output of OR3 is HI, we can conclude that the part is tall. On the other hand, if the output of OR3 is LO, the output of I3 will go HI, meaning that the part is short. The note on the output line of I3 conveys this meaning.

The weighing circuitry, comprised of the weight detector, OR4, AND4, and I4, is an exact duplicate of the height circuitry. Trace through the operation of these gates to be sure that you understand how they work.

AND gates 5–8 can be considered the classification gates. The input signals to this group of gates come from the outputs of the height and weight detection circuitry. Each one of the AND gates has two inputs, representing a certain combination of height and weight outcomes. For example, the two input lines to AND5 are the two lines which indicate (1) the part is short and (2) the part is light. Therefore if the part is short and light, AND5 will be enabled. If the part is short and heavy, AND6 will be enabled, and so on.

The outputs of the classification AND gates feed into two other groups of AND gates. First, they feed AND gates 9, 10, 11, and 12, which control the paint solenoid valves. Second, they feed AND gates 13, 14, 15, and 16, which control the diverters.

AND gates 9, 10, 11, and 12 have input 1 in common with each other. Input 1 of all these paint control gates is driven by the line marked In Paint Zone. This means that when the part enters the paint zone all the number 1 inputs of gates 9–12 will be driven HI. Then, depending on which classification gate is turned on, one of the four paint control gates will be enabled. This in turn will energize the proper solenoid valve. For example, if the tall/light classification gate is turned on (AND7), it will put a HI on input 2 of AND11. When the part enters the paint zone, and the In Paint Zone line goes HI, AND11 will be enabled. This will energize the red paint solenoid. The solenoid will remain energized until the In Paint Zone line goes back LO, disabling AND11.

The diverter control gates, ANDs 13, 14, 15, and 16, work the same way. Their number 1 inputs are tied in parallel and are driven by the In Diverter Zone line. When this line goes HI, one of the four diverter control gates will be enabled, which will energize the proper diverter solenoid. For example, if the tall/light classification gate (AND7) is turned on, it will apply a HI to the number 2 input of AND15. When the In Diverter Zone line goes HI, it will put a HI on input 1 of AND15. The output of AND15 will then go HI, energizing the tall/light diverter solenoid and causing the tall/light diverter in Fig. 1–4(b) to swing out over the conveyor. The diverter solenoid will remain energized until the "in diverter zone" line returns to LO, disabling AND15.

The circuits at the top of Fig. 1–8 furnish the signals which tell the location of the part as it rides the conveyor, namely the In Paint Zone and In Diverter Zone signals.

As a part enters the paint zone it actuates LS1, which applies a +5-V HI to the number 1 input of OR1. The OR1 output goes HI and seals itself in by feeding back to AND1. Thereafter, as long as input 2 of AND1 is HI, is the AND gate will stay enabled, and OR1 will remain on by virtue of its number 2 input. As shown in the diagram, the OR1 output is nothing other than the In Paint Zone line.

When the part leaves the paint zone and enters the diverter zone, it actuates LS2. This applies a HI to input 1 of OR2, which causes the OR2 output to go HI. The OR2 output does several things. First, it puts a HI at the input of I1, which causes a LO at input 2 of AND1. This disables AND1 and breaks the seal on OR1. The In Paint Zone line goes back LO, and the paint solenoid valve shuts off. Second, the output of OR2 feeds into AND2. Since the number 2 input of AND2 is also HI at this time, AND2 turns on and seals up OR2. Third, the OR2 output is the In Diverter Zone signal, which goes down to the bottom of Fig. 1–8 and drives the diverter control gates, as discussed earlier.

When the part is guided off the belt and down a chute, one of the chute limit switches will be actuated, applying a HI to I2. The I2 output goes LO and applies LO signals to AND2, AND3, and AND4. The LO on AND2 breaks the seal on OR2, allowing the In Diverter Zone signal to go back LO. Whichever diverter was swung out thus returns to its normal position. The LOs at AND3 and AND4 disable those gates, applying LOs to the number 2 inputs of OR3 and OR4. This breaks the seals on OR3 and OR4, if they were sealed. Therefore the height and weight circuits are reset and are prepared to test the next part on the conveyor.

1-7 ■ INPUT DEVICES FOR SOLID-STATE LOGIC

The circuit of Fig. 1–8 shows direct-switched connections between the HI logic supply voltage and the gate inputs. For example, LS1 makes a direct connection between the +5-V dc supply line and input 1 or OR1. While this switching arrangement is theoretically acceptable, there are some practical reasons why it is a bad idea.

The primary reason is that mechanical switches never make a clean contact closure. The contact surfaces always "bounce" against each other several times before they make permanent closure. This phenomenon is called *contact bounce* and is illustrated in Fig. 1–9.

In Fig. 1–9(a), when the mechanical switch closes to connect resistor R across dc supply V, the voltage waveform across R looks like Fig. 1–9(b). The elapsed time between initial contact and permanent closure ($t_2 - t_1$ in the waveform) is usually rather short, on the order of a few milliseconds or less. Although the bouncing is quite fast, logic gates respond quite fast, so it is possible for a gate to turn on and off every time a bounce takes place. The unwarranted turning on and off can cause serious malfunctions in logic circuitry.

1-7-1 Capacitive Switch Filters

The solution to this problem is to install some type of filter device between the switch and the logic gate. The filter device must take the bouncing input and turn it into a smooth output. One straightforward method of doing this is shown in Fig. 1–10(a).

When the limit switch first closes, capacitor C starts to charge through the Thevenin resistance of $R_1 \| R_2$. Because the limit switch contacts stay closed only a very short time on the first bounce, the charge buildup on C is not great enough to affect the gate input. The same holds true for all subsequent bounces—the switch never stays closed long enough to trip the gate because of the necessity to charge C. When permanent closure finally occurs, C can charge to the threshold voltage of the gate and turn it on. The filter

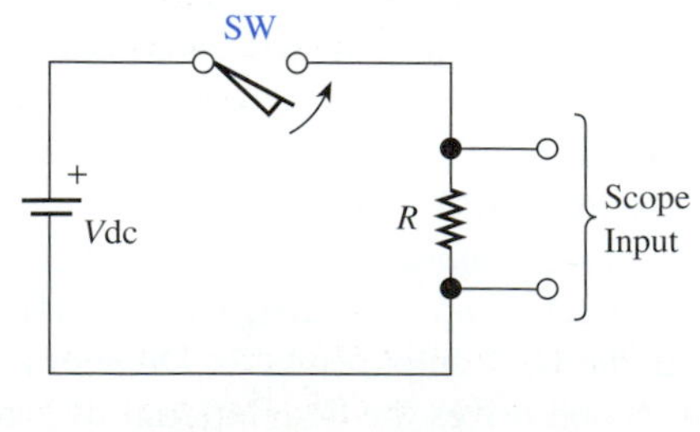

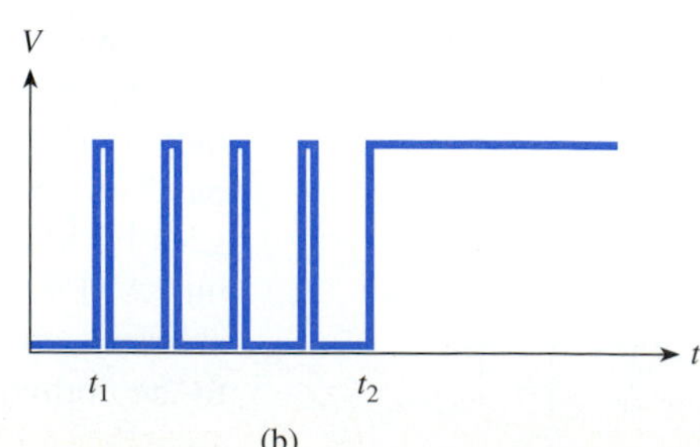

FIGURE 1–9 The problem of contact bounce.

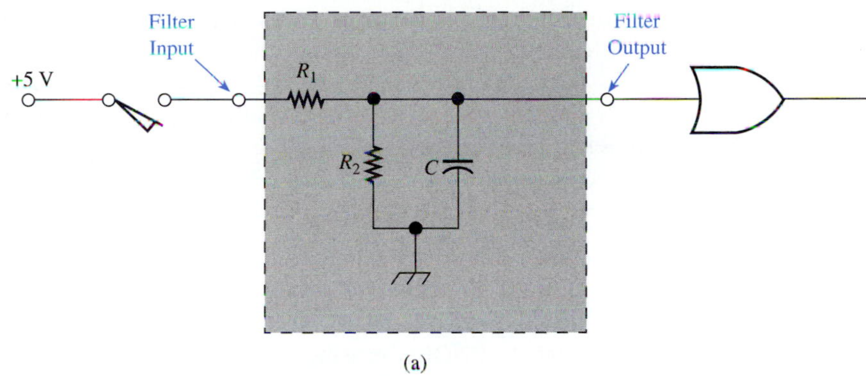

(a)

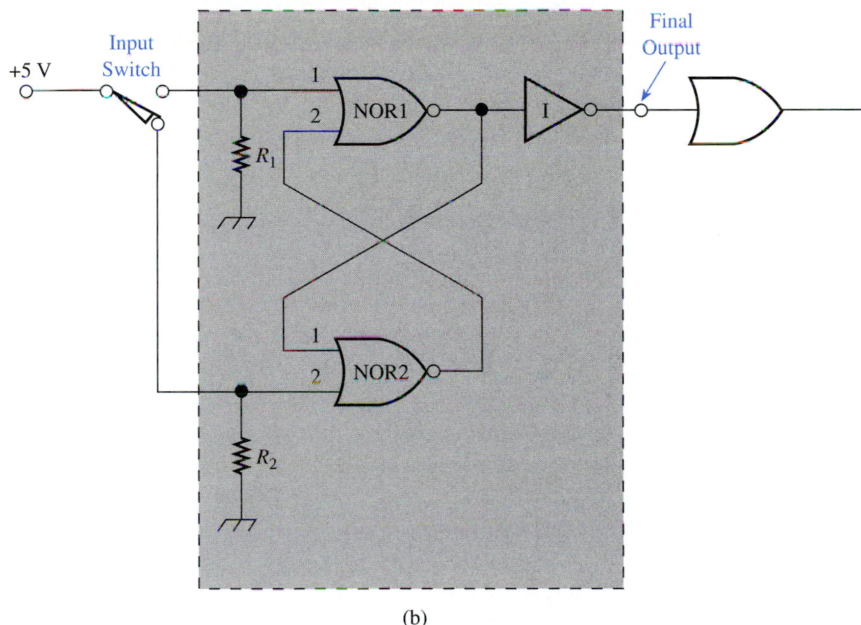

(b)

FIGURE 1–10
(a) *RC* switch filter to eliminate the effects of contact bounce. (b) Bounce eliminator built with solid-state gates.

of Fig. 1–10(a) also serves to reject noise signals from external sources. That is, if a high-speed noise pulse should occur on the lead coming from the switch, it will be rejected by the low-pass filter and will not appear at the gate input.

Of course, when the capacitor does charge up, it cannot charge to the full supply voltage level. It can only charge to the Thevenin voltage of the R_1-R_2 voltage divider. This is normally not a problem, since sold-state gates operate reliably with an input voltage less than full supply voltage.

1-7-2 Bounce Eliminators

Another method of eliminating contact bounce is shown in Fig. 1–10(b). This approach differs from the one in Fig. 1–10(a) in that it triggers on the *first* contact bounce instead of waiting for the final closure. After it turns on, it ignores all subsequent bounces. A disadvantage of this circuit is that it requires a double-throw switch instead of a single N.O. contact. Here is how it works.

With the limit switch released, the N.C. contact is closed and a HI level is applied to R_2 and to input 2 of NOR2. The output of NOR2 is therefore LO, causing input 2 of NOR1 to be LO. Input 1 of NOR1 is also LO because R_1 pulls it down to ground. With both inputs of NOR1 LO, its output is HI; inverter I then makes the final output LO.

During the switching process, this is the sequence of events:

1. The N.C. contact opens first (break-before-make switch), causing the number 2 input of NOR2 to go LO. NOR2 does not change state because its number 1 input is still HI.
2. The N.O. contact closes momentarily on the first closure. This puts a momentary HI on input 1 of NOR1, causing its output to go LO. The inverter drives the final output HI. The NOR1 output feeds input 1 of NOR2, so NOR2 now has two LO inputs. Its output therefore goes HI, applying a HI to input 2 of NOR1. NOR1 has two HI inputs at this point in time.
3. The N.O. contact bounces open on the first rebound. This causes a LO at input 1 of NOR1, but input 2 maintains its HI level. Therefore NOR1 does not change states, and the final output remains HI.
4. There are several more bounces, each one changing the logic level of input 1 of NOR1. However, because the limit switch N.C. contact remains open, a HI remains on input 2 of NOR1, holding NOR1 steady.

When the limit switch is released at some later time, the bounce eliminator does the same thing in reverse, causing a jitter-free transition to the LO level at the final output. You should trace through the operation of the circuit as this happens.

1-7-3 Signal Converters

The capacitive filter and the bounce eliminator that we have considered both assume that the input device is switching a *logic level* voltage (+5 V in Fig. 1–10). Since virtually all industrial logic gates use a supply voltage of 20 V or less, the input devices must operate reliably under relatively low-voltage and -current conditions in order to allow direct switching of this type. This is sometimes feasible, but there are many situations in which it is not feasible. Sometimes the information-gathering devices cannot give reliable operation under low-voltage conditions.

There are two major reasons for this unreliability. First, the input devices may be physically remote from the decision-making logic. Therefore the wire-runs between the input devices and the logic circuits are lengthy and necessarily have higher resistance than if they were shorter. Higher resistance causes a higher *IR* voltage drop along the run. If the beginning voltage is already small, large *IR* voltage drops in the wires cannot be tolerated because the logic might confuse a HI level with a LO level. It is better to start with a large voltage so the system can afford to suffer some voltage loss in the connecting wires.

Second, the contact surfaces of input devices tend to accumulate airborne dust and debris; oxides and other chemical coatings can also form on the surfaces. These things cause the contact resistance to increase, sometimes making it impossible for a small voltage to overcome the resistance. A high voltage level is needed to ensure that the increased resistance can be overcome.

Besides, the very act of switching a high voltage creates arcs between the contact halves. These arcs burn away oxides and residue and keep the surfaces clean.

Under many industrial circumstances, therefore, it is absolutely necessary to use high voltages to drive the input devices. When this is done, there must be an interface device added to convert the high-voltage input signal to a low-voltage logic signal. Such devices are called *signal converters, original input converters, logic input interfaces,* and other names. We will use the term *signal converter* in this book. A schematic symbol of a signal converter is shown in Fig. 1–11(a). A schematic diagram containing three signal converters is shown in Fig. 1–11(b).

In most industrial schematic diagrams, signal converters are drawn with two wires as shown in Fig. 1–11(b), although an actual signal converter usually has four wires attached to it. The schematic representation is simple and uncluttered, yet it suggests the action of a signal converter, namely that a low-voltage logic 1 appears at the output when a high-voltage input signal is applied by the closing of the contact of the input device.

Figure 1–12 shows the internal construction of two typical signal converters for converting a 115-V ac input to a +5-V dc logic level.

Figure 1–12(a) is the familiar full-wave power supply with a center-tapped transformer. The input device switches 115 V ac to the primary winding, and the rectifier and filter circuits convert the secondary voltage to 5 V dc. Note that this type of signal converter has four wire connections even though the schematic symbol is drawn with only two wires.

This signal converter provides electrical isolation between high-voltage input circuits and low-voltage logic circuits by virtue of the magnetic coupling between the transformer windings.

Electrical isolation between the two circuits is desirable because it tends to prevent electromagnetic or electrostatic noise generated by the input circuitry from passing to the

FIGURE 1–11
Signal converters for converting high-voltage input signals to low-voltage logic signals.

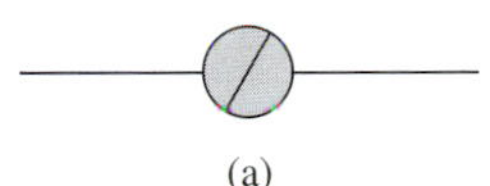

(a)

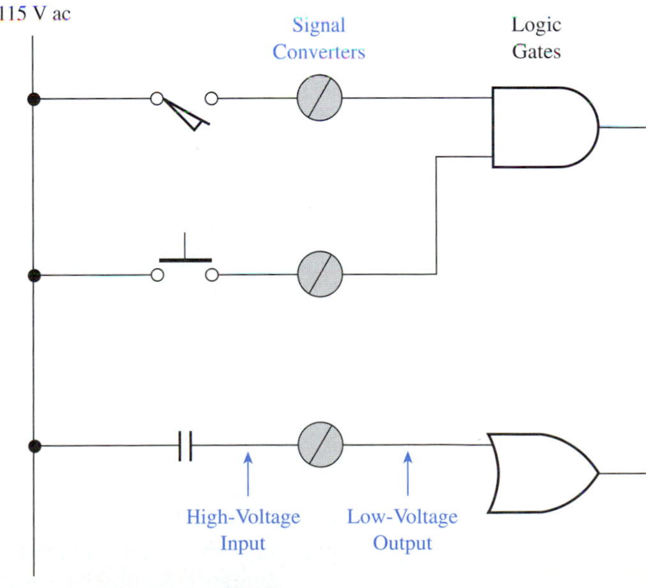

(b)

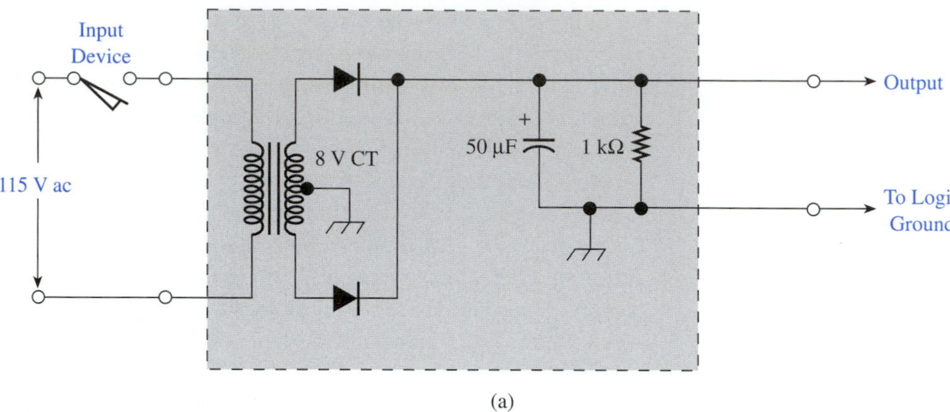

(a)

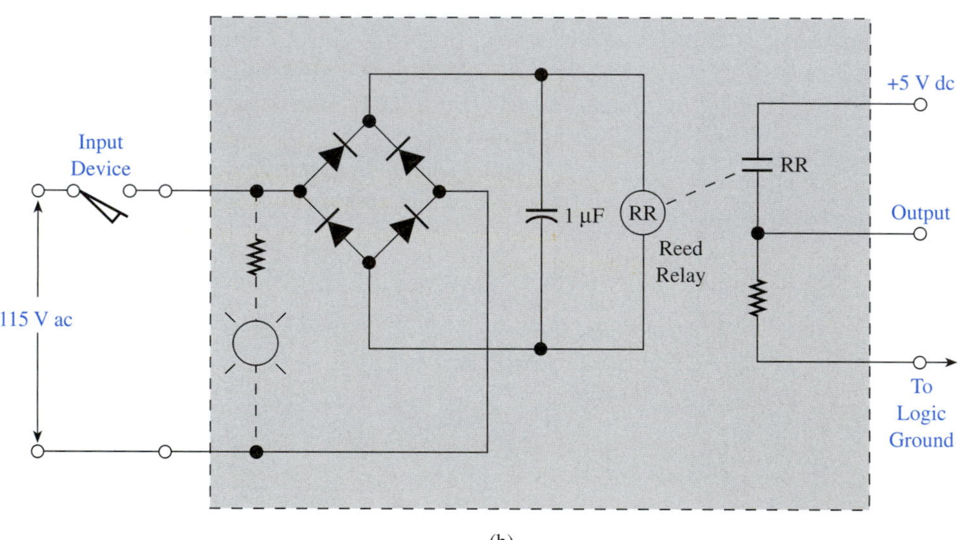

(b)

FIGURE 1–12 (a) Output converter, which uses a transformer to isolate the logic circuit from the input circuit. (b) Signal converter, which uses a reed relay to isolate the logic circuit from the input circuit.

logic circuitry. In an industrial logic system, noise pickup in the input device circuit is very often a problem. This is due to the long wire-runs between the logic panel and the input devices and the tendency to carry the wires in conduit running close to power wires. Power wires driving motors and switchgear are inherently noisy and can easily induce unwanted electrical noise in the connecting wires between input devices and logic.

The signal converter illustrated in Fig. 1–12(b) uses a reed relay. The output of the full-wave bridge energizes the relay coil, and the relay contact switches the logic supply voltage onto the signal converter's output line. The logic circuitry is isolated from the input circuitry through the relay. This type of signal converter does not produce its own logic signal voltage but must have the logic supply brought in from outside. Therefore it has five wire connections. It would still be drawn schematically as shown in Fig. 1–11(b).

Both the signal converters illustrated in Fig. 1–12 contain capacitors which serve to filter out high-frequency noise and switch bounce. Therefore they do not normally require any other filter circuit or bounce eliminator connected to their outputs.

A pilot light indicator can be wired into a signal converter as shown by the dashed lines in Fig. 1–12(b). This is a troubleshooting aid for maintenance personnel. The condition of the input can be seen at a glance; it is then not necessary to apply a voltmeter to tell the state of the input.

Occasionally the input devices in an industrial system are driven by a high-voltage dc supply instead of the usual 115 V ac. A large dc voltage creates more arcing across switch contacts than the same value of ac voltage. Therefore a dc voltage is even more effective at burning away deposits and residue that collects on the contact surfaces. In such cases a dc-to-dc signal converter is used. The circuit of Fig. 1–12(b) would work in such an application.

In recent years optically coupled signal converters have become very popular. Their popularity is due to their light weight, excellent reliability, and low cost. They require no transformer or relay for electrical isolation between input and logic circuits, and their isolating ability is very good. They will be discussed when photoelectric devices are covered in Chapter 10.

1-8 ■ OUTPUT DEVICES FOR SOLID-STATE LOGIC

The solid-state logic diagram of the conveyor/classifying system (Fig. 1–8) shows the paint solenoids and diverter solenoids driven directly by AND gates. While it is possible to drive actuating devices (solenoids, motor starters, etc.) directly from logic gates, this is not the usual practice. Rather, an *output amplifier* is inserted between the logic circuit and the actuating device. The purpose of the output amplifier is to increase the low-voltage/low-current logic power to higher-voltage/higher-current output power.

The symbol for an output amplifier (sometimes called a *driver* or a *buffer*) is shown in Fig. 1–13(a). Output amplifiers as they would appear in an industrial schematic drawing are shown in Fig. 1–13(b).

The OA in the output amplifier symbol is often omitted or replaced by a D, standing for driver. As with signal converters, output amplifiers are shown schematically with only two wires, an input and an output. When the input line goes to a logic HI, the output line energizes the actuating device. In actual construction, most output amplifiers have four wires attached to them.

Most output amplifiers are designed to drive a 115-V ac load, since most industrial solenoid valves, motor starter coils, horns, etc., are designed for use with 115 V ac. This situation is represented in Fig. 1–13(b), with the common power line labeled 115 V ac.

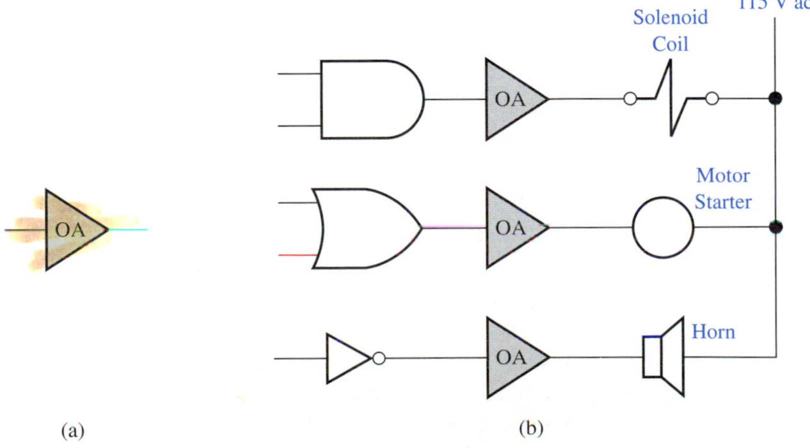

FIGURE 1–13
Output amplifiers for amplifying low-power logic signals into high-power output signals.

Other output amplifiers obtain their operating power from a separate high-voltage dc supply instead of the 115-V ac line. Such amplifiers are used with actuating devices designed to operate on that particular dc voltage. Popular dc voltage levels for driving dc actuating devices are 24, 48, and 115 V dc. An example of the construction of a dc output amplifier is given in Fig. 1–14(a).

The dc output amplifier consists of a power transistor driven by a small signal transistor with an emitter resistor. The load is connected in series with the collector lead of the power transistor and is powered by the 24-V supply. The 24-V supply is referenced to the logic ground supply line by a ground connection somewhere in the control circuit cabinet. This is all shown in Fig. 1–14(a).

The logic supply voltage is brought into the output amplifier as collector supply for Q_1. When the input terminal goes HI, Q_1 turns ON, raising the R_3 voltage high enough to

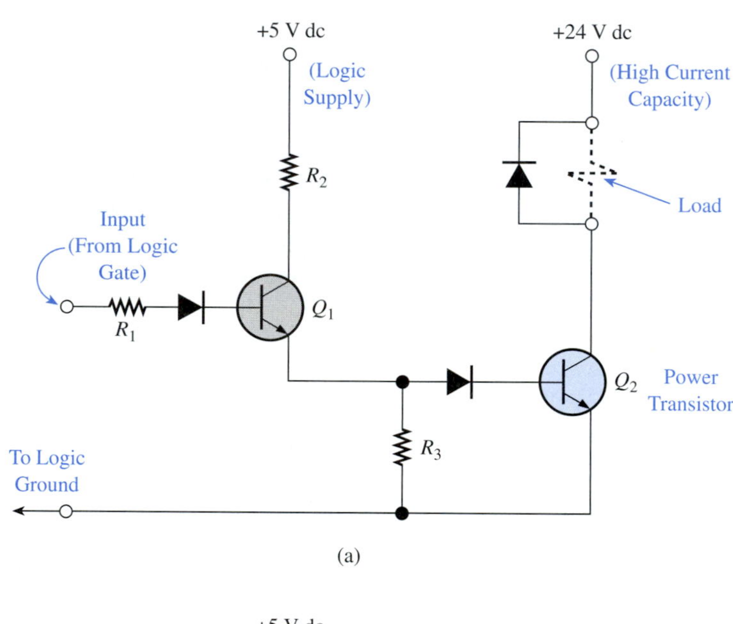

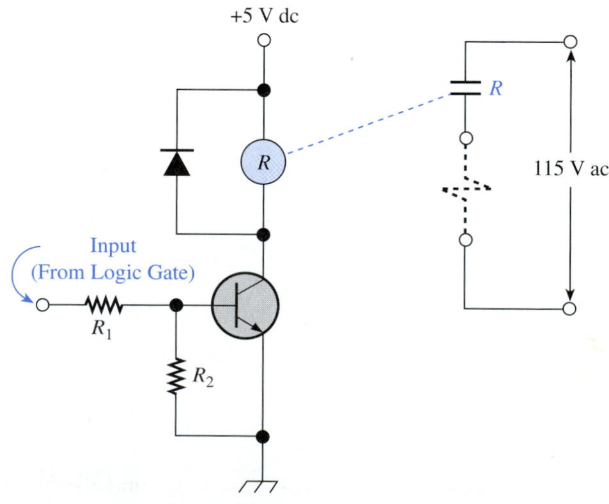

FIGURE 1–14 (a) Output amplifier using a power transistor to control the current through the output device. (b) Output amplifier using a relay contact to control the current through the output device.

bias the power transistor ON. Thereafter most of the Q_1 emitter current flows into the base of the power transistor. The collector current of the power transistor energizes the actuating device. The diode in parallel with the load is placed there to suppress inductive kickback from the load when it is deenergized.

Because of the common ground, there is not complete electrical isolation between logic circuits and output circuits with the output amplifier of Fig. 1–14(a). Careful precautions must therefore be taken with wire routing to avoid injection of noise into the logic.

Figure 1–14(b) is an example of an output amplifier using a reed relay. When the amplifier's input terminal goes HI, it turns ON the transistor and picks the reed relay. The relay contact then connects the load across the 115-V ac lines. This arrangement does provide electrical isolation between logic and output circuits.

More up-to-date ac output amplifiers use solid-state devices instead of reed relays. Such output amplifiers usually have an SCR at their heart, with the SCR often triggered by a unijunction transistor (UJT). A typical design of such a solid-state ac output amplifier will be presented in Chapter 5.

1-9 ■ SOLID-STATE LOGIC COMPARED TO RELAY LOGIC

Magnetic relays have been handling most of the logic requirements of twentieth-century industry for many years, and they will continue to be widely used. Because of improved materials of construction and improved design, relays now are capable of several million trouble-free operations under normal conditions. However, under certain conditions and in certain settings, solid-state logic is demonstrably superior to relay logic. We will discuss the conditions under which solid-state logic is preferred and try to point out some of the important considerations used to decide between the two types of logic.

Reliability. In most industrial situations the most important consideration when selecting logic circuitry is reliable maintenance-free operation. Relays have moving mechanical linkages and contacts, which are subject to wear. Also, their coils must allow fairly large inrush currents in order to create the force necessary to move the linkages. This puts stress on the coil wire and insulation. These are the reasons the life expectancy of relays is limited to a few million operations, as mentioned earlier. This may seem like a remarkable life span, as indeed it is, but consider how long a relay would last if it cycled twice per minute. Two operations per minute figures out to 2880 operations per day, or over 1 million operations per year. At this rate, a relay with a life expectancy of 2 million operations would last only about 2 years. The rate of two operations per minute for 24 hours a day is not unusual in an industrial circuit. Many relays must operate more often than that, with a corresponding reduction of trouble-free life.

Solid-state gates, on the other hand, have an unlimited life expectancy. They have no moving parts and no appreciable inrush current. Barring unexpected thermal shock or overcurrents, a solid-state device will last forever. This is an obvious advantage of solid-state logic over relay logic.

Relay components are exposed to the atmosphere. Dirt particles can therefore work their way into the mechanical apparatus and interfere with proper motion. Chemicals and dust in the atmosphere can attack the surfaces of the contacts, causing them to pit. When contact surfaces are not smooth they may weld together. Coil insulation can also be damaged by chemical action.

By contrast, solid-state gates can be and usually are sealed in packages which are impervious to the atmosphere. Chemicals and airborne particles cannot interfere with their proper operation.

Explosive environments. The fact that relays are exposed to the atmosphere has another important consequence. Relay contacts create sparks when they operate, due to metal clash and load kickback. If there are explosive gases in the atmosphere, sparks are unacceptable. Under these conditions relays can be used only in expensive airtight enclosures.

Solid-state gates turn on and off without sparking, making them inherently safe in explosive environments.

Space requirements. Considering physical size and weight, solid-state logic is clearly more compact. This is not usually an important factor in industrial circuits, but it occasionally can be. An example might be a situation in which a new system was being installed in the space previously occupied by an old system and space was at a premium. If the control circuit was extensive, the space conserved by using solid-state logic could be an important consideration.

Operating speed. Concerning operating speed, it is strictly no contest between relays and logic gates. Relays operate in milliseconds, whereas most solid-state devices operate in microseconds or nanoseconds. Roughly speaking, a solid-state is at least 1000 times as fast as a relay. Again, this high speed is often not an important factor in industrial logic, but it might be. Operating speed becomes important if mathematical computation is required in the decision-making process.

Cost. For an extensive logic circuit containing hundreds of decision-making elements, solid-state logic is cheaper to build and operate than an equivalent relay logic circuit. This is because the low per-gate cost overrides the extra expenses associated with solid-state logic. These extra expenses include the cost of dc power supplies, signal converters and output amplifiers, and special mounting hardware for the printed circuit boards.

Solid-state logic gates consume only a small fraction of the power consumed by relays. Therefore in a large circuit the energy savings can be considerable.

Advantages of relay logic. On the plus side for relays are several assets not possessed by solid-state circuits. First, as implied earlier, relay logic is cheaper to build if the circuit is small. This is because relays require no separate power supply, require no interfacing at the information-gathering (input) end or at the actuating (output) end, and mount very easily on a panel.

Second, relays are not subject to noise pickup. They cannot be fooled by an extraneous noise signal; solid-state gates can be fooled by such noise signals.

Third, relays work well at the high ambient temperatures found in industrial environments. Solid-state logic must often be fan-cooled or air conditioned when used in a hot environment. This negates some of the advantages of energy conservation and reliability, since air conditioning requires energy to run, and the logic is only as reliable as the air conditioner.

Fourth, and often of critical concern, is that many maintenance personnel are thoroughly familiar with relay logic but much less familiar with solid-state logic. Given this situation, down-time may be longer for a system malfunction when solid-state logic is used.

1-10 ■ A SOLID-STATE LOGIC CIRCUIT FOR A MACHINE TOOL ROUTING CYCLE

We will now explore some more examples of circuits using logic gates. The circuit presented in this section is a simple cycling circuit using noninverting gates, ANDs and ORs. Logic circuits using noninverting gates are easier to explain and to understand than circuits using inverting gates.

The circuit presented in Sec. 1-11 is a fairly uncomplicated logic circuit using inverting gates, NANDs.

Finally, in Sec. 1-12 we will explore a more complex circuit using NAND gates. Circuits using NANDs and NORs are more confusing because of the constant necessity to invert the thinking process, but it is necessary to learn to deal with such circuits. They are popular in industrial control for the reasons given in Sec. 2-5; they are cheaper and faster, and draw less current than ANDs and ORs.

Consider the machining application illustrated in Fig. 1–15. The purpose is to route two channels into the top of the workpiece, both running in the east-west direction. The first channel is toward the north edge of the piece, and the second channel is toward the south edge. This is accomplished by loading the workpiece onto a stationary table between two square bars that prevent it from sliding in the east-west direction but permit motion in the north-south direction. The piece is placed on the table so its north side is snug against the face block, which touches the north edge of the table. The face block is loaded with powerful springs so it will not move back from the north edge of the table unless forced back by a hydraulic cylinder. Cylinder B must extend and push the workpiece against the face block to displace the piece a few inches to the north.

The router is mounted on a movable frame that can move east and west. When cylinder A extends, the router frame moves east. When cylinder A retracts, the router frame moves west.

The machining cycle proceeds as follows:

1. When the workpiece is properly positioned between the square bars and snug against the face block, the operator presses the Start button. Cylinder A extends to the east and routes the north channel.
2. When the cylinder A cam hits LS2, indicating that the first channel is complete, cylinder B extends and moves the workpiece to the north. When cylinder B reaches its fully extended position, its cam actuates LS3.
3. Cylinder A retracts to the west and routes the south channel in the top of the piece. It stops when its cam actuates LS1.
4. Cylinder B retracts to the south, allowing the springs to return the workpiece to its original position. This completes the cycle.

Refer to Fig. 1–15(b) for the control schematic. Here is how the circuit works. When the workpiece is properly located between the square bars and up against the face, the "in position" contacts leading into signal converter SC4 go closed. When the Start button is pressed the output of SC4 goes HI, enabling OR1. The OR1 output enables OA1, which energizes the cylinder A solenoid. The cylinder A hydraulic valve shifts, stroking cylinder A to the east. The first channel is routed.

OR1 seals up by putting a HI on input 1 of AND1. This causes the AND1 output to go HI, since input 2 was already HI. That is so because LS3 is released at this time, causing a LO input to I1 and a corresponding HI output from the inverter.

When cylinder A completes its stroke and the router bit has cleared the workpiece, a cam actuates LS2, causing a HI input to OR2 from SC2. The OR2 output goes HI, energizing the cylinder B solenoid through OA2. The cylinder B solenoid valve shifts and extends cylinder B. Meanwhile OR2 has sealed up through AND2. This happens because OR2 supplies a HI to input 1 of AND2 and input 2 of AND2 is already HI. The HI on input 2 comes from I2, whose input is LO due to the LS1 contact being open.

When cylinder B has fully extended, placing the workpiece in position for the second cut, LS3 is actuated. The LS3 N.O. contact closes, applying a 115-V ac input to SC3.

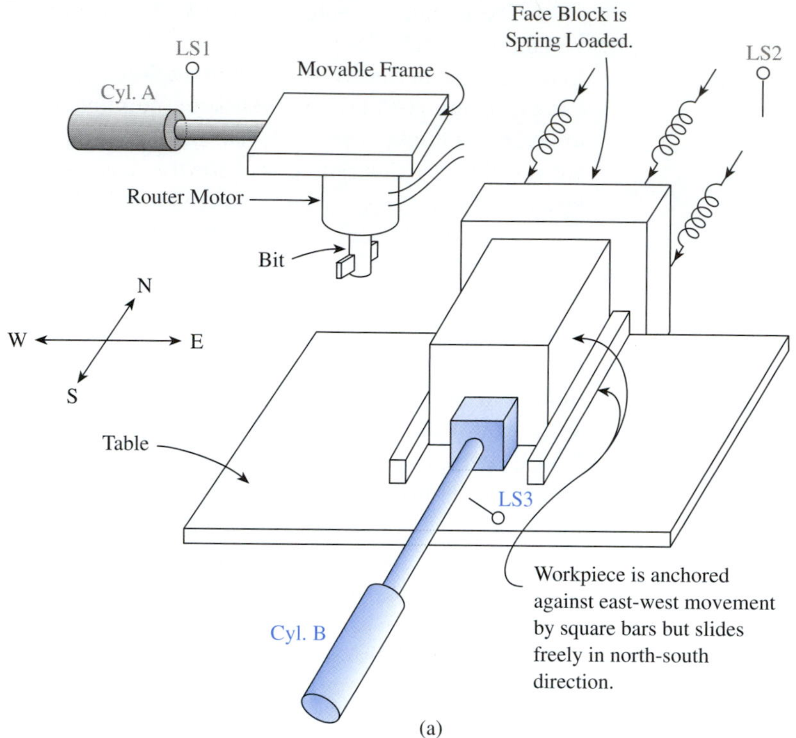

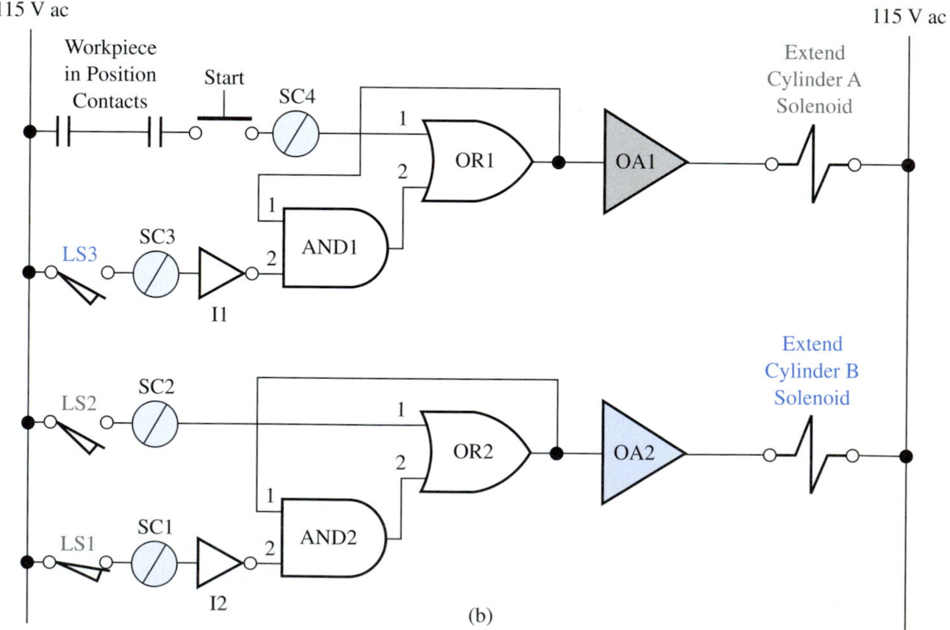

FIGURE 1–15
(a) Physical layout of a machine tool router. (b) Control circuit of the machine tool router.

I1 therefore has a HI input, resulting in a LO applied to input 2 of AND1. This breaks the seal on the OR1 gate, shutting off OA1 and deenergizing the cylinder A solenoid. The hydraulic valve shifts back to its normal position, causing oil to flow into the rod end of cylinder A and causing cylinder A to retract to the west. As the router frame moves to the west, the router cuts the second channel.

When cylinder A has fully retracted it hits LS1. This applies a 115-V ac input to SC1, resulting in a LO output from I2. This LO is applied to the number 2 input of AND2, causing OR2 to lose its seal. When the OR2 output goes LO, OA2 deenergizes the cylinder B solenoid. Cylinder B retracts to the south, allowing the face block springs to shove the workpiece back into the starting position. The machining cycle is now complete, and the operator removes the piece and inserts a new one.

1-11 ■ LOGIC CIRCUIT FOR A FIRST FAILURE ANNUNCIATOR

Figure 1–16 shows the schematic of a *first failure annunciator*. A first failure annunciator is a circuit that tells the system operators which input device gave the warning signal that caused the system to shut down. By way of background, many industrial systems have input devices which constantly monitor conditions in the system, making sure no unsafe condition exists. If an unsafe condition should occur, these devices shut the system down to eliminate the dangerous condition and blow a horn to inform the operators. Unfortunately, by the time the operators arrive on the scene, the unsafe condition may have already corrected itself, or the act of shutting down the system may have made it impossible to tell exactly *which* unsafe condition caused the problem. In such a situation, what is needed is a circuit that can record which input device gave the initial warning and ignore any subsequent warning signals that only occurred because of the shutdown process. This is the purpose of a first failure annunciator.

As a specific system to have in mind, consider an industrial air/gas heated furnace. Three unsafe conditions which could possibly occur in such a system are that (1) the natural gas supply pressure is too high, (2) the combustion air pressure is too low to adequately burn all the gas, or (3) the temperature has exceeded the maximum safe value for this particular furnace; the maximum safe temperature is commonly called the *temperature high-limit*, or just the *high-limit*.

Any one of these conditions is considered unsafe enough to warrant immediate shutdown of the furnace. By the time the human operators come over to see what's wrong, the condition may have corrected itself. The gas pressure may have temporarily surged up and then returned to normal. The air pressure may have momentarily dropped and then recovered, etc. The human operators will have no clue as to the nature of the problem, and no corrective action by them will be possible. Therefore a first failure annunciator is needed. It should be understood that Fig. 1–16 does not show the actual circuitry by which the furnace is shut down; it shows only the first failure annunciator circuitry.

As the circuit monitors a correctly operating furnace, the situation is this. The high gas pressure switch is open because the gas pressure is below the set point of the pressure switch. Therefore SC1 has no 115-V ac input signal and consequently applies a logic LO to input 2 of NAND1. The N.C. low air pressure switch is open because the air pressure is above the set point of the pressure switch. Therefore SC2 has no 115-V ac input, and it applies a LO to input 2 of NAND2. Likewise, the temperature high-limit switch is open because the furnace temperature is below the maximum limit, so input 2 of NAND3 is also LO. NAND gates 1, 2, and 3 all have HI outputs because of the LOs on their number 2 inputs.

Now concentrate on NAND5. Its number 2 input is HI because NAND1 has a HI output. Its number 1 input is also HI due to the action of the Reset line at some time in the past. Here is what happened the last time the circuit was reset: When the Reset button was pushed, the output of the bounce eliminator went HI. Therefore the output of I2 (the Reset line) went LO, applying a LO to input 1 of NAND4. The LO at this input guaranteed a HI at the output of NAND4. This HI fed back around to the number 1 input of

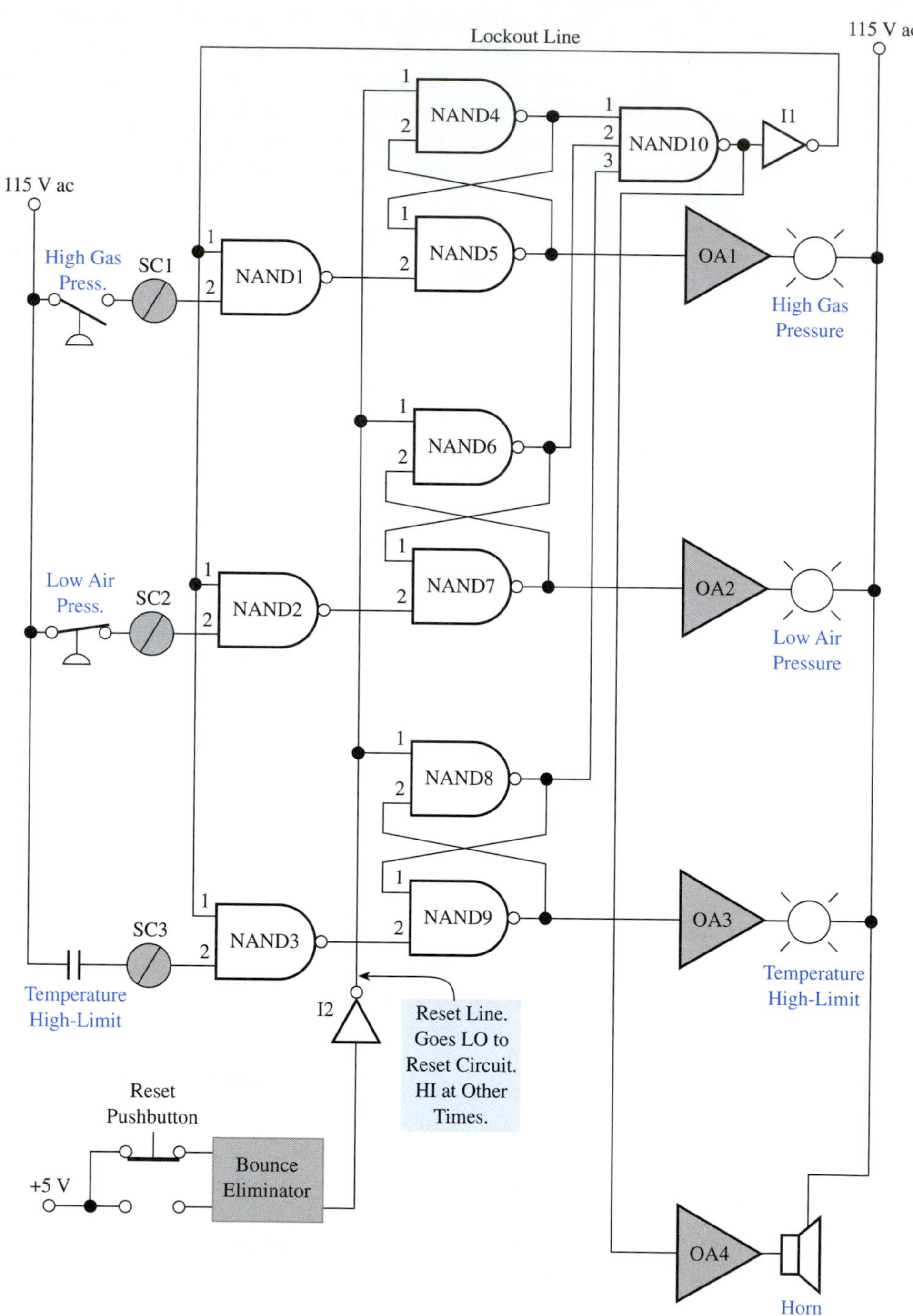

FIGURE 1–16
First failure annunciator, which indicates the initial cause of the failure.

NAND5. With both inputs HI, the output of NAND5 went LO, applying a LO to input 2 of NAND4. Then when the Reset pushbutton was released, allowing the Reset line to return to HI, the state of NAND4 did not change. Its output remained HI because of the LO at input 2.

The preceding paragraph shows that while the circuit is monitoring a correctly operating furnace, NAND4 has a HI output and NAND5 has a LO output. The exact same argument would apply to NAND6 and NAND7 and also to NAND8 and NAND9. Therefore NANDs 6 and 8 have HI outputs and NANDs 7 and 9 have LO outputs. Since this is true, all three output amplifiers have LO inputs from NANDs 5, 7, and 9. Therefore all three indicating lamps are turned off.

NAND10 has all inputs HI from NANDs 4, 6, and 8. Its output is therefore LO, causing a LO to OA4. The horn is deenergized. I1 has a HI output, which is applied to the number 1 inputs of NANDs 1, 2, and 3. This is the complete situation under normal furnace conditions.

Now suppose that there is a failure of the gas pressure in that it temporarily surges up too high. This causes a high-voltage input to SC1, which delivers a logic HI to input 2 of NAND1. There are now two HIs into NAND1, so its output goes LO. This causes a LO at input 2 of NAND5, which drives the output of NAND5 HI. Therefore OA1 has a HI input, and it turns on the High Gas Pressure lamp. The NAND5 output also applies a HI to input 2 of NAND4, driving the output LO. The output goes LO because the number 1 input of NAND4 was already HI. The LO from NAND4 feeds back around to input 1 of NAND5, which seals the NAND5 output HI. That is, as long as the output of NAND4 is LO, the output of NAND5 is held HI, and as long as the output of NAND5 is HI, the output of NAND4 will be LO. The only way to break this seal is by pushing the Reset button and driving the number 1 input of NAND4 LO.

We have seen how NAND4 and NAND5 seal up and how they turn on the proper indicating light. Now look at NAND10. When its number 1 input goes LO, its output goes HI. This HI is fed down to OA4, which causes the alarm horn to blow. Meanwhile the I1 output (the Lockout line) goes LO and applies a LO to input 1 of NANDs 1, 2, and 3. The LO inputs to NAND2 and NAND3 lock these gates in their starting condition, namely, outputs HI. It does not matter what SC2 or SC3 does thereafter, because the outputs of NAND2 and NAND3 are locked HI by the LOs at their number 1 inputs. Therefore the NAND6-NAND7 combination cannot change states and the NAND8-NAND9 combination is also frozen. Thus it is impossible for any *other* indicating lights to turn on once the first one turns on.

Even if the high gas pressure switch returns to the open condition (which it would certainly do when the furnace was shut down), the high gas pressure indicating lamp will continue to glow because of the seal of the NAND4-NAND5 combination.

The foregoing explanation was based on gas pressure being the first failure, but of course the circuit action would be the same if the air pressure or temperature high-limit was the first failure. Air pressure failure would seal up the NAND6-NAND7 combination and lock NAND1 and NAND3 in starting condition with outputs HI. Temperature limit failure would seal up the NAND8-NAND9 combination and lock NANDs 1 and 2 outputs HI when the LOCKOUT line went LO.

When the reset pushbutton is pressed, the RESET line goes LO and breaks whatever seal has occurred. This returns the entire circuit to starting condition.

1-12 ■ LOGIC CIRCUIT FOR A MACHINE TOOL DRILLING CYCLE

Figure 1–17(a) shows a rough drawing of a drilling apparatus. The workpiece is brought into position and held steady by clamps. Two holes are to be drilled in the piece. One hole is vertical, one hole is horizontal, and both must pass through the same internal point. Therefore they cannot both be drilled at the same time. The logic circuit implements the following cycle:

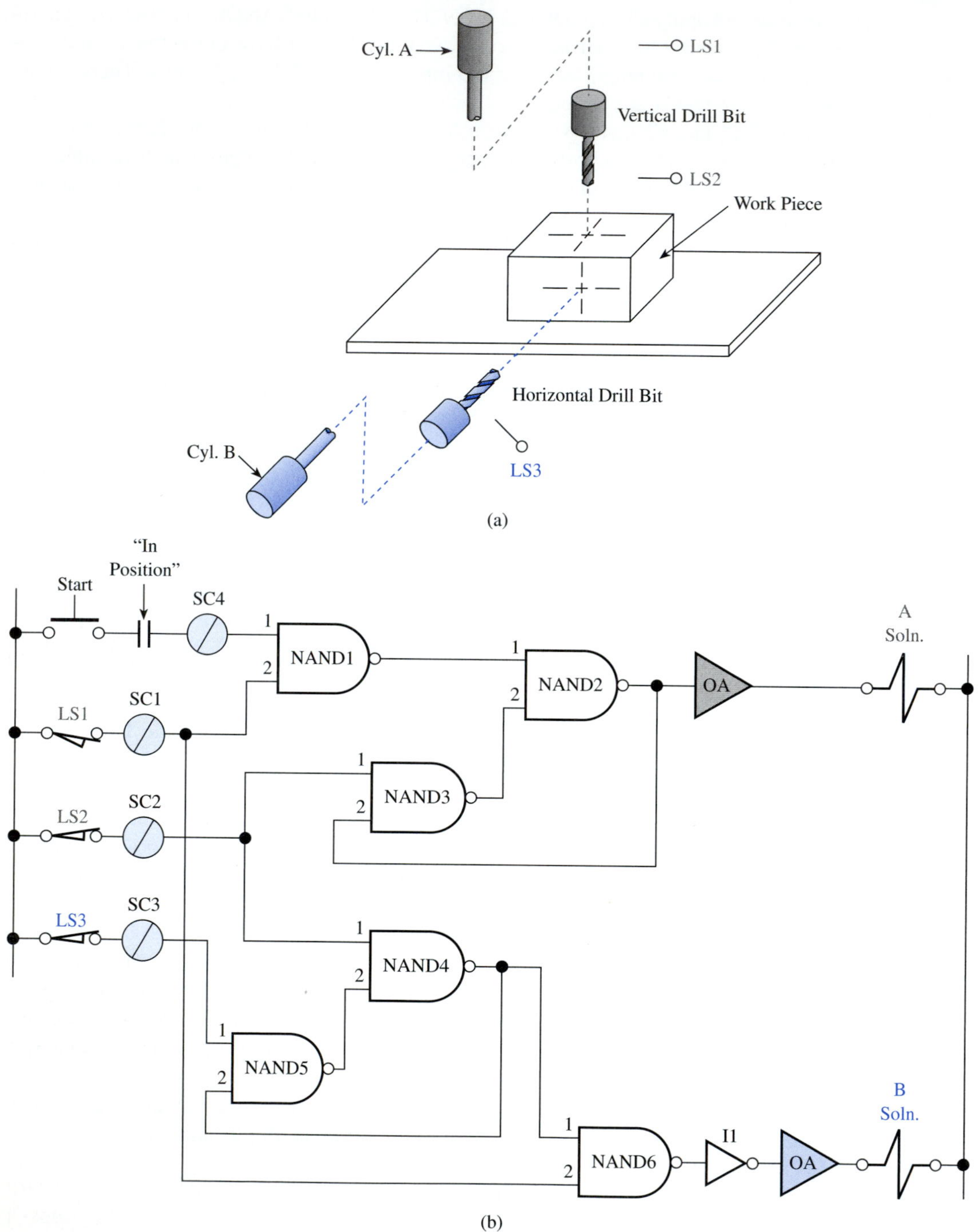

FIGURE 1–17
(a) Physical layout of a machine tool drilling operation. (b) Control circuitry of the machine tool drilling operation.

1. When the workpiece is clamped in position, the operator presses the Start button, causing cylinder A to extend. The rotating drill bit descends and drills the vertical hole.
2. When cylinder A is fully down, LS2 is contacted, causing cylinder A to retract and the drill bit to withdraw from the piece.
3. When cylinder A returns to fully up, LSl is actuated, causing cylinder B to extend and drill the horizontal hole.
4. When cylinder B is fully extended, LS3 is contacted, causing cylinder B to retract and withdraw the drill bit.

The control circuit is shown in Fig. 1–17(b). The description of this circuit is more complex than the descriptions of Figs. 1–15(b) or 1–16. Here is how it works.

When the piece is clamped in position, the "In Position" contact closes and the operator presses the Start button. This delivers a high-voltage input to SC4, which brings input 1 of NAND1 HI. Since LS1 is contacted at this time (cylinder A is retracted at the beginning of the cycle), the number 2 input of NAND1 is also HI. Therefore the output of NAND1 goes LO and applies a LO to input 1 of NAND2. The NAND2 output goes HI, turning on OA1 and energizing the cylinder A solenoid. Therefore the hydraulic valve shifts, and cylinder A starts to descend.

Meanwhile, NAND2 has been sealed HI because of the feedback through NAND3. The NAND2 output puts a HI on input 2 of NAND3. Since LS2 is released at this time, the N.C. contact is closed, and SC2 has a HI output. Therefore input 1 of NAND3 is also HI at this time. With both inputs HI, the NAND3 output goes LO, making input 2 of NAND2 LO. Therefore it doesn't matter what happens to input 1 of NAND2, since the LO at input 2 will guarantee a HI output.

When cylinder A is fully extended, it actuates LS2 and opens the N.C. LS2 contact. When the 115-V ac input is removed from SC2, its output goes LO. This LO does two things:

1. It breaks the seal on NAND2. When input 1 of NAND3 goes LO, its output goes HI. The HI is applied to input 2 of NAND2. Input 1 of NAND2 is already HI because of LSI being released (and the Start button being released). With two HI inputs, NAND2 goes LO at its output and deenergizes the cylinder A solenoid. Cylinder A therefore begins its retraction stroke.
2. The LO from SC2 also arrives at input 1 of NAND4. The output of NAND4 goes HI and seals back through NAND5. This is because LS3 is released at this time, applying a HI to the number 1 input of NAND5. With two HI inputs, the NAND5 output goes LO and delivers a LO to input 2 of NAND4. This seals the NAND4 output HI, no matter what the number 1 input does. Therefore NAND4 will maintain a HI output even after LS2 is released.

The events described in these two paragraphs all take place at the instant LS2 is contacted. Since the cylinder A solenoid has been deenergized, cylinder A immediately starts retracting, and it releases LS2. This cause HIs to reappear at input 1 of NAND3 and input 1 of NAND4, but these HIs have no effect on those gates. When cylinder A is all the way back it contacts LS1, and the output of SC1 goes HI. This HI appears at input 2 of NAND6. Since input 1 of NAND6 is already HI at this time (NAND4 is sealed HI), the output of NAND6 goes LO. This LO is inverted by I1, allowing OA2 to energize the cylinder B solenoid. Therefore cylinder B begins its extend stroke to drill the horizontal hole.

When cylinder B is fully extended it actuates LS3 and opens the LS3 N.C. contact. The SC3 output goes LO, causing the NAND5 output to go HI. This breaks the seal on

TROUBLESHOOTING ON THE JOB

EXPANDING THE MACHINE TOOL DRILLING SYSTEM

It has been decided to expand the machine tool drilling system of Fig. 1–17 so that *three* holes can be drilled into the workpiece. The new drill will be horizontal, from the right side of Fig. 1–17 (a). It will be extended and retracted by new cylinder C, controlled by the C solenoid valve.

Your supervisor has given you the assignment of expanding the logic control circuit of Fig. 1–17(b) to accommodate the third solenoid. This is not so difficult as it might at first seem, since you can make your C solenoid circuitry have the same relationship to the B solenoid circuitry that the B circuitry already has to the A circuitry in that figure.

Thus, to control the C solenoid, you should plan on using three NAND gates, numbered 7, 8, and 9; another inverter, I2; and an output amplifier.

Relying on your understanding of the logic circuit in Fig. 1–17(b), you must decide which additional limit switches will be necessary to accomplish control of the third cylinder solenoid.

Make a complete schematic diagram of your expanded circuit, and be prepared to explain to your supervisor how it functions.

Using computer simulation to test the design of a new control logic circuit.
Courtesy of Hewlett-Packard Company.

NAND4, since NAND4 has two HI inputs at this time. The NAND4 output returns to LO and puts a LO on input 1 of NAND6. The NAND6 output goes HI, so I1 delivers a LO to OA2. The cylinder B solenoid therefore deenergizes and cylinder B retracts, withdrawing the horizontal drill bit.

As the retraction stars, LS3 is again released, causing the N.C. contact to go closed. This reapplies a HI to input 1 of NAND5, but NAND5 is not affected because its number 2 input is LO at this time.

This completes the drilling cycle and the operator removes the workpiece.

■ SUMMARY

- You can think of an industrial control system as having three sections: (1) input, or information gathering, (2) logic, or decision making, and (3) output, or actuating.
- Logic can be performed by electromagnetic relays or by transistors.

- The switched input signals to transistor logic must be processed to eliminate contact bounce and high-frequency noise.
- Switched 120-V ac inputs are converted to transistor-compatible low-voltage dc by signal converters.
- Low-voltage transistor logic signals are converted to 120-V ac by output amplifiers.
- Solid-state logic circuits have many advantages over electromagnetic relay logic circuits, including (1) greater reliability and lifetime, (2) completely sealed, and therefore sparkless, (3) reduced volume and weight, (4) greater speed, (5) lower initial cost, and (6) lower power consumption.

QUESTIONS AND PROBLEMS

Sections 1-1 and 1-2

1. Explain the purpose of each one of the three sections of an industrial logic control circuit.
2. Name some common information-gathering (input) devices used in industrial logic systems.
3. Name some common actuating (output) devices used in industrial logic systems.
4. What two general types of information do input devices supply to an industrial logic circuit?
5. What is the most common voltage used for input circuits for industrial control in the United States?
6. Explain the difference between a normally open limit-switch contact and a normally closed limit-switch contact. Draw the schematic symbol for each.
7. Repeat Question 6 for relay contacts.
8. Explain why series-connected contacts constitute an AND circuit.
9. Explain why parallel-connected contacts constitute an OR circuit.

Sections 1-4 and 1-5

10. In words, explain the operation of an AND gate. Draw the truth table for a two-input AND gate. Then draw the truth table for a four-input AND gate. How many different combinations of inputs are possible when there are four inputs?
11. Repeat Question 10 for an OR gate, a NAND gate, and a NOR gate.
12. In Fig. 1–6(a), draw in all the current flow paths if lines X and Y are at +5 V and line Z is at 0 V.
13. Repeat Question 12 for all three inputs at +5 V.
14. In Fig. 1–6(b), draw in all the current flow paths if lines X and Y are at +5 V and line Z is at 0 V.
15. Repeat Question 14 for all three inputs at 0 V.
16. Draw the logic circuit to implement the following conditions: The solenoid is energized if LS1 and LS2 are both actuated or if LS3 is actuated. Draw the circuit using relay logic and also using solid-state logic.
17. Repeat Question 16 for these conditions: The solenoid energizes if LS1, LS2, and LS3 are all actuated or if LS1 is not actuated.
18. Repeat Question 16 for these conditions: The lamp turns on if LS1 and LS2 are both not actuated or if LS2 is actuated while LS3 is not actuated.
19. Repeat Question 16 for these conditions: The motor starter energizes if either LS1 or LS2 is actuated at the same time that either LS3 is actuated or LS4 is not actuated.

20. Explain the meaning of the term *fan-out* as applied to logic circuits in general and logic gates in particular.
21. Explain the meaning of the term *fan-in* as applied to logic circuits in general and logic gates in particular.
22. Describe the difference between a positive logic system and a negative logic system.
23. Explain the difference between a discrete circuit and an integrated circuit.

Section 1-6

Questions 24–26 apply to the solid-state conveyor/classifying system illustrated in Figs. 1–4 and 1–8.

24. What conditions are necessary to enable AND9?
25. What conditions are necessary to enable AND14?
26. Explain how the momentary closing of one of the chute limit switches breaks the seals on OR2, OR3, and OR4 if they were sealed.

Section 1-7

27. Why are capacitive switch filters sometimes necessary? Describe what they do, and name some of the benefits that arise from using them.
28. Repeat Question 27 for bounce eliminators.
29. What is a logic signal converter, and what does it do?
30. If a commercially built signal converter is used to interface between an input device and solid-state logic circuitry, is a bounce eliminator or switch filter also needed under normal circumstances? Why?
31. Will the signal converter of Fig. 1–12(a) work on dc as well as ac? Why?
32. Repeat Question 31 for the signal converter of Fig. 1–12(b).
33. Why is it a good idea to provide electrical isolation between input circuits and solid-state logic circuits?

Section 1-8

34. What is the purpose of an output amplifier for use with solid-state logic?
35. What is the purpose of the diode in the collector circuit of Q_2 in Fig. 1–14(a)?
36. What is the purpose of the diode in the base lead of Q_2 in Fig. 1–14(a)?
37. Does the output amplifier of Fig. 1–14(b) furnish electrical isolation between the logic circuit and the output circuit? Explain carefully.

Section 1-9

38. Name some conditions under which solid-state logic is preferred to relay logic.
39. Name some conditions under which relay logic might be preferred to solid-state logic.

Section 1-10

Questions 40–42 apply to the machine tool routing system illustrated in Fig. 1–15.

40. If LS3 got stuck in the closed position, what would happen when the operator pressed the Start button?
41. If LS3 failed to close when cylinder B extended, what would happen?

42. If SC3 malfunctioned so that it could not deliver a HI output, what would happen during the machine cycle?

Section 1-11

Question 43–45 apply to the first failure annunciator illustrated in Fig. 1–16.

43. Explain how the *first* failure causes the circuit to ignore all subsequent failures.
44. Explain how the circuit remembers which failure occurred even if the failure corrects itself.
45. Try to explain what would happen if two failures occurred at exactly the same instant. (This would be a fantastic coincidence.)

Section 1-12

Questions 46–52 apply to the machine tool drilling system illustrated in Fig. 1–17.

46. The LS1 contact is normally open, but it is *drawn* in the closed position. Why is this?
47. When the Start button is pressed at the beginning of the cycle, what does NAND1 do? Explain.
48. When the Start button is pressed, explain how NAND2 seals up (output HI).
49. What action of the system breaks the seal of NAND2? Explain.
50. What action of the system causes NAND4 to seal up (output HI)? Explain how NAND4 seals up.
51. What two conditions are necessary to cause the output of NAND6 to go LO? Explain.
52. What action of the system breaks the seal of NAND4? Explain.

CHAPTER

TRANSISTOR SWITCHES IN MEMORY AND COUNTING APPLICATIONS

Besides their usefulness in the construction of decision-making logic gates, transistor switches can also be used to build a circuit that has rudimentary memory—the well-known flip-flop. In turn, flip-flops can be combined with logic gates to build counting circuits. In this chapter we will explore some industrial applications of flip-flops, counters, and related circuits.

OBJECTIVES

After completing this chapter you will be able to:

1. Describe the operation of flip-flops as memory devices in the control circuits presented as examples.
2. Describe how a shift register keeps track of digital data regarding a part moving on a conveyor system.
3. Describe in detail the operation of a carton routing and palletizing system using cascaded decade counters and 1-of-10 decoders.
4. Describe in detail the operation of an automatic tank-filling system using one-shots, flip-flops, a decade counter, and a free running clock.
5. Describe the operation of time-delay relays including the four different types of time-delay contacts.
6. Explain the operation of a solid-state timer based on a series *RC* charging circuit.
7. Describe in detail the operation of a traveling hopper material-supply system using a down-counter, an encoder, and solid-state timers.

2-1 ■ WELDER CONTROL CIRCUIT USING *RS* FLIP-FLOPS

There are two transistor switches at the heart of every flip-flop, as Fig. 2–1(a) illustrates. From your digital circuits textbook, review the operation of this basic flip-flop circuit. The black-box schematic symbol that we will use for an *RS* flip-flop is given in Fig. 2–1(b).

Imagine a situation in which two automatic welders are powered from the same supply bus. The supply bus can deliver enough current to drive one welder, but it cannot power two welders at the same time because of the large current draw. Therefore if an automated system signals for a second welder to begin welding when the first welder is already involved in a weld, the initiation of the second weld must be postponed. When the first welder is finished, then the signal for the second welder will be honored.

To accomplish this, a circuit is needed which knows if a weld is currently being performed and can receive and *remember* input requests for a second weld. Since the circuit must remember something, it will contain flip-flops. A circuit to accomplish this action is shown in Fig. 2–2.

Here is how it works. If a weld is called for by the closing of one of the welder Start contacts, then the appropriate flip-flop turns ON (its *Q* output goes HI). That is, either FF1 or FF3 will turn ON because a HI will appear on its S input. For purposes of illustration, suppose that the A welder Start contact closes, causing the FF1 *Q* output to go HI. This will apply a HI to input 1 of AND1. If the B welder is *not* welding at this time, the number 2 input of AND1 will also be HI. This is explained in the next paragraph. In that case the output of AND1 will go HI and apply a HI to the *S* input of FF2. The *Q* output of FF2 goes HI, causing the output amplifier to energize the A welder contactor. This contactor connects the A welding transformer to the power bus and produces a weld. The preceding description applies equally well if a weld is requested on the B welder when the A welder is turned off. FF3 would turn ON, enabling AND2, which would turn ON FF4.

Therefore if a weld is called for by the system control, it takes place immediately as long as the other welder is not welding at that time. On the other hand, consider what would happen if the A welder Start contact closed while welder B was welding. In that case the number 1 input of the NOR gate would be HI because it is connected to the *Q* output of FF4. The output of the NOR goes LO. This LO is applied to input 2 of AND1,

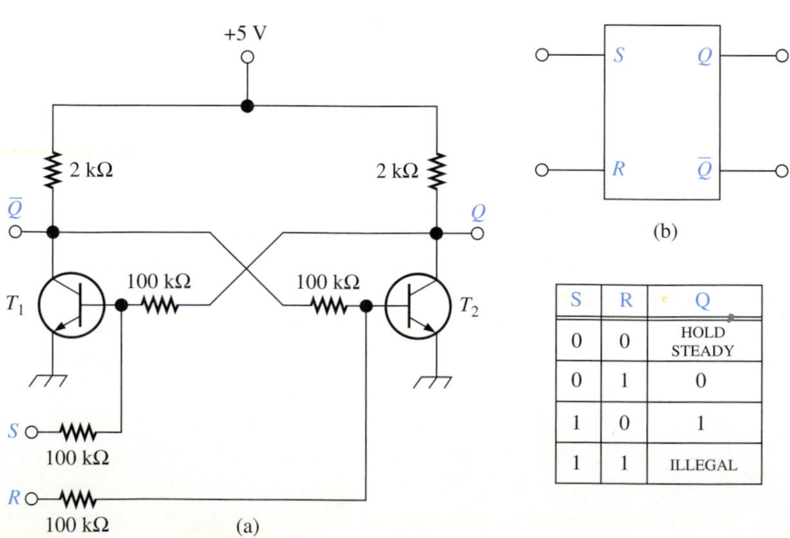

S	R	Q
0	0	HOLD STEADY
0	1	0
1	0	1
1	1	ILLEGAL

FIGURE 2–1
(a) Schematic diagram of an *RS* flip-flop, showing the forcing inputs. The letters *S* and *R* stand for *set* and *reset*. The $\overline{Q}$ output (pronounced "*Q* not") is the digital opposite, or *complement*, of the *Q* output. (b) Black-box symbol of the *RS* flip-flop.

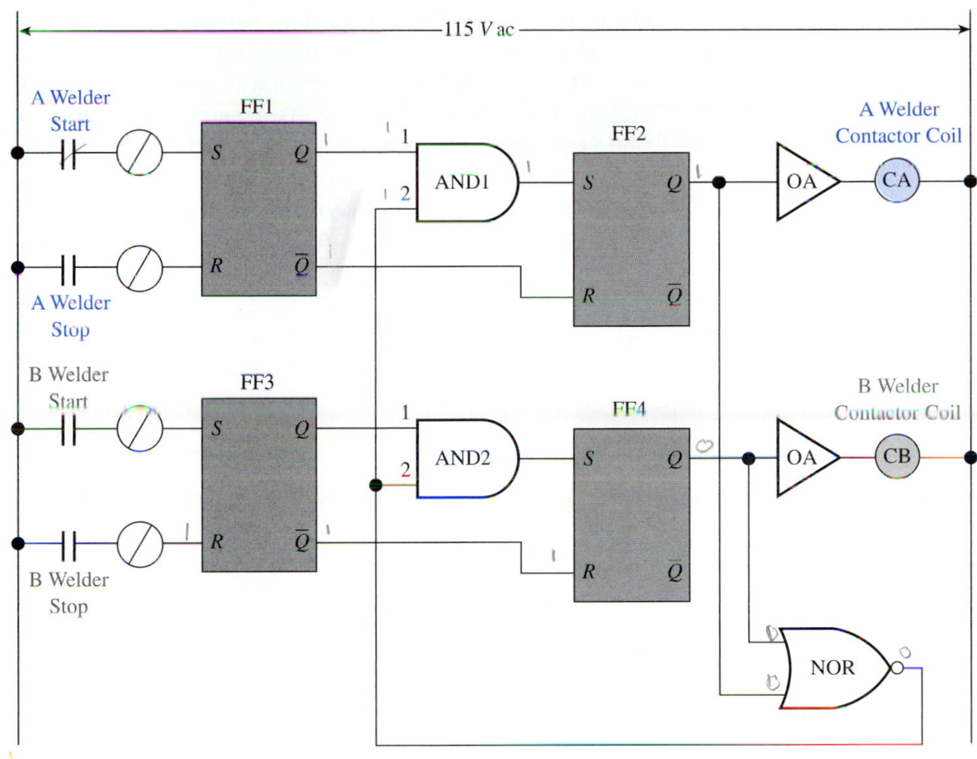

FIGURE 2–2
Welder control circuit illustrating the memory ability of *RS* flip-flops.

guaranteeing a LO output from the AND1 gate and preventing FF2 from turning ON. Thus the A welder cannot start.

As soon as the B welder is finished, the B welder Stop contact will close, applying a HI to *R* of FF3. This turns OFF FF3 and drives the $\overline{Q}$ output HI. This HI appears at the *R* input of FF4, turning that flip-flop OFF. Therefore the *Q* output of FF4 goes LO, removing the HI from the NOR gate input at the same time it deenergizes the B contactor. The output of the NOR gate goes back HI, enabling AND1. At that time, a HI gets through to *S* of FF2, which turns on the A welder.

Thus a weld request is postponed if the other welder is presently operating. However, the circuit of Fig. 2–2 remembers the request and acts upon it when the other welder is free.

2-2 ■ OSCILLATING MACHINING TABLE USING CLOCKED *RS* FLIP-FLOPS

A *clocked RS flip-flop* is one which responds to its *S* and *R* inputs only at the instant when its clock terminal makes a transition. A *positive edge-triggered flip-flop* responds to its static inputs (*S* and *R*) when the clock line makes a positive-going transition, from LO to HI. A *negative edge-triggered flip-flop* responds to its static inputs when its clock makes a negative-going transition, from HI to LO. To avoid confusion, we will assume throughout this book that all clocked flip-flops are negative edge-triggered. The black-box schematic symbol that we will use is shown in Fig. 2–3. In that figure, and in all digital device symbols generally, the small triangle drawn inside the box indicates that the device is an edge-triggered, or clocked device. The small circle outside the box is the general digital symbol for distinguishing a negative edge-triggered clocked device from a positive edge-triggered clocked device. Among static (nonclocked) digital circuits, the same

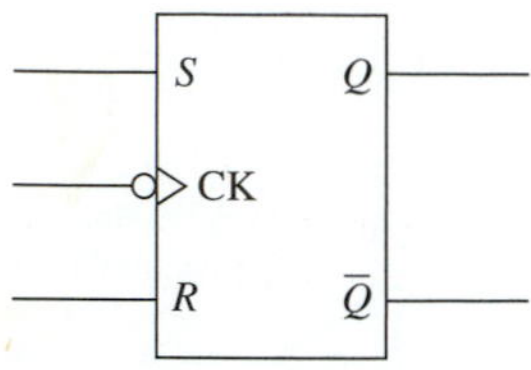

CK	S	R	Q_n
→	X	X	HOLD STEADY
↓	0	0	HOLD STEADY (Q_{n-1})
↓	0	1	0
↓	1	0	1
↓	1	1	ILLEGAL

FIGURE 2–3
Schematic symbol for a negative edge-triggered clocked *RS* flip-flop.

small circle is used to distinguish an active-LO input from an active-HI input. These ideas are explained in detail in your digital electronics text.

Imagine a machining operation in which a table is moved back and forth by a reversing motor. This might arise in a planing operation in which the planing tool remained stationary and the workpiece was mounted on an oscillating table. Figure 2–4(a) shows such a layout. When the motor spins in one direction, the rack and pinion assembly moves the table to the right; when the motor spins in the other direction, the rack and pinion move the table to the left. When the table has moved to the extreme right, it contacts Right LS,

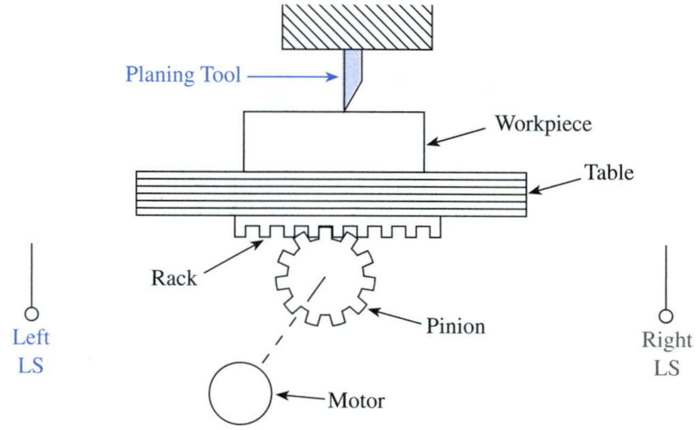

(a)

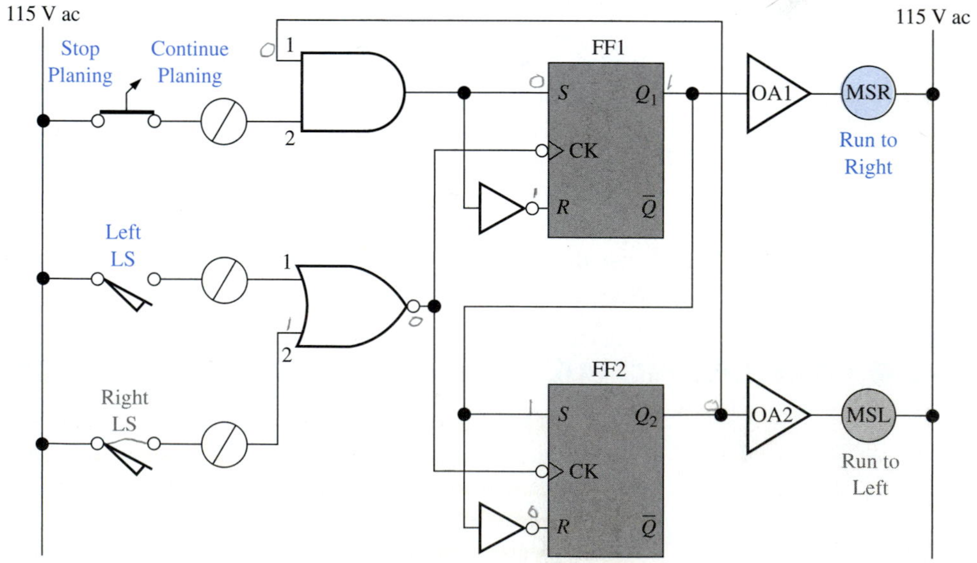

(b)

FIGURE 2–4
(a) Physical appearance of the oscillating planer. (b) Control circuit of the oscillating planing system, illustrating the application of clocked *RS* flip-flops.

which signals the control circuit to reverse the motor and run to the left; when it moves to the extreme left it contacts Left LS, which signals the circuit to run the motor to the right. This action continues as long as necessary to complete the planing operation.

When the operator is satisfied that the planing is complete, he throws a selector switch into the Stop Planing position. The table then continues in motion until it has returned to its extreme left position.

Here is how the circuit works. Assume that the table is running to the right and the two-position selector switch is in the Continue Planing position. The fact that the switch is drawn closed in Fig. 2–4(b) means that the contact is closed when the operator selects that position. Conversely, the contact is open when the operator selects the Stop Planing position.

If the table is running to the right, it is because the Run to Right motor starter is energized, which implies that FF1 is ON. Also, since the Run to Left motor starter coil is necessarily deenergized, it follows that FF2 is OFF. Therefore the situation is this: The HI on Q_1 puts a HI on S of FF2 and a LO on R of FF2 through I2; the LO on Q_2 puts a LO on input 1 of the AND gate, which applies a LO to S of FF1 and a HI to R of FF1.

When the Right LS is contacted, it turns on its signal converter, which applies a HI on input 2 of the NOR gate. The NOR gate output goes LO and delivers a negative edge to both CK inputs. Since the FF1 inputs are ordering FF1 to turn OFF, it does just that: The Run to Right starter coil deenergizes. The FF2 inputs at the instant the edge arrives are $S = 1, R = 0$, so FF2 turns ON. When Q_2 goes HI it enables its output amplifier and energizes the Run to Left starter coil. The table therefore reverses direction and moves to the left.

When the table reaches the extreme left, the Left LS is actuated. At this time the opposite conditions hold true. Q_2 is HI and Q_1 is LO, so FF1 has $S = 1, R = 0$, and FF2 has $S = 0, R = 1$. This is so as long as the SS (selector switch) is in the Continue Planing position and is applying a HI to the number 2 input of the AND gate. When Left LS closes and the clock input terminals receive their negative edges from the NOR, FF1 turns ON and FF2 turns OFF. The motor reverses again and begins moving the table to the right.

Now suppose that the operator decides to terminate the planing operation. At some point in the travel, he throws the SS to the Stop Planing position. This removes the HI from input 2 of the AND gate, forcing the output to go LO. Therefore FF1 has a LO on S and a HI on R, no matter what the state of the number 1 input of the AND. The next time the table contacts Left LS, both FF1 and FF2 will be turned OFF because both flip-flops will have $S = 0$ and $R = 1$. This will be true of FF1 because of the AND gate output being LO; it will be true of FF2 because Q_1 will be LO while the table is moving to the left. With both flip-flops OFF, Q_1 and Q_2 are both LO, and the two motor starters are deenergized. The motor stops, leaving the table in the left position.

If the table had been running to the right at the time the operator changed the SS, the reversing of the motor would take place as usual when the table hit Right LS, because FF2 is free to turn ON regardless of the condition of the AND gate. The table will always stop in the left position.

You may wonder how the cycle is ever started once a new workpiece is placed on the table. This problem has been left as an exercise at the end of the chapter.

When trying to understand the action of clocked *RS* flip-flops in a circuit, it is important to focus on the conditions at the S and R inputs *at the exact instant the clock edge arrives*. In many instances, the very act of triggering a flip-flop causes an almost instantaneous change in the state of the inputs. This is the case in Fig. 2–4(b). Do not pay any attention to the fact that the inputs change state immediately after delivery of the clock edge. The only concern of a flip-flop is the state of the inputs at the exact instant the edge appears.*

To keep this idea clear, it is convenient to think of the clock negative edge as being infinitely fast. That is, it goes from HI to LO in absolutely zero time. If this were true, then any change at the inputs due to triggering the flip-flop would occur too late, since the negative edge is already over and done with by the time the change happens.

Of course, no real-life clock edge can have a fall time of absolutely zero, but this notion helps us to explain and understand the behavior of clocked flip-flops. It prevents confusion in those situations where the inputs change when the flip-flop is triggered.

2-3 ■ *JK* FLIP-FLOPS

The most widely used flip-flop is the *JK flip-flop*. It has two inputs, just like the *RS* flip-flop, but the inputs are referred to as *J* and *K*. The action of a *JK* flip-flop is quite similar to that of a clocked *RS* flip-flop, the only difference being that the *JK* flip-flop has what is called a *toggling mode*.

Many *JK* flip-flops have preset (PR) and clear (CL) static inputs, which override the clocked inputs *J* and *K*. We will use the black-box schematic symbol shown in Fig. 2–5 to represent a full-function *JK* flip-flop. Review your digital circuits text to refresh your memory regarding such *JK* flip-flops.

FIGURE 2–5
JK flip-flop with active-LO preset and clear inputs.

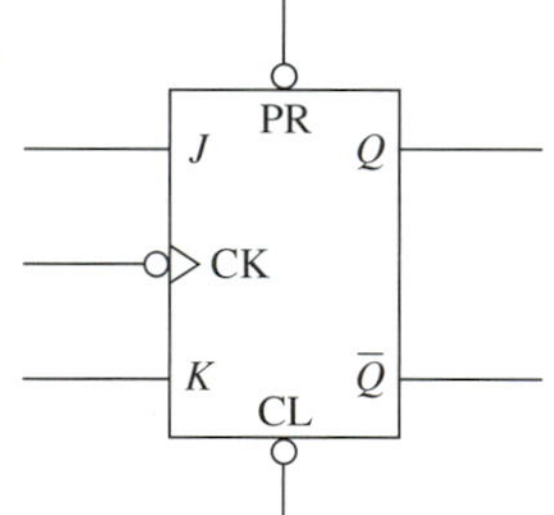

CK	J	K	Q_n
→	X	X	HOLD STEADY
↓	0	0	HOLD STEADY (Q_{n-1})
↓	0	1	0
↓	1	0	1
↓	1	1	TOGGLE TO OPPOSITE ($\overline{Q}_{n-1}$)

2-4 ■ SHIFT REGISTERS

A shift register is a string of flip-flops which transfer their contents from one to another. The best way to understand the operation of a shift register is to look at its schematic diagram and observe how it works.

2-4-1 Shift Registers Constructed of *JK* Flip-Flops

Figure 2–6 shows four *JK* flip-flops connected together so that the outputs of one flip-flop drive the inputs of the next one. That is, Q_1 and $\overline{Q}_1$ are connected to *J* and *K* of FF2, Q_2 and $\overline{Q}_2$ are connected to *J* and *K* of FF3, and so on. This circuit is a 4-bit shift register; it is called *4-bit* because it has four memory elements (flip-flops) and can therefore store four pieces of binary information, or *bits*.

*This idea is valid only for flip-flops built on the so-called *master-slave* principle. In this book we will assume that all our flip-flops are built this way.

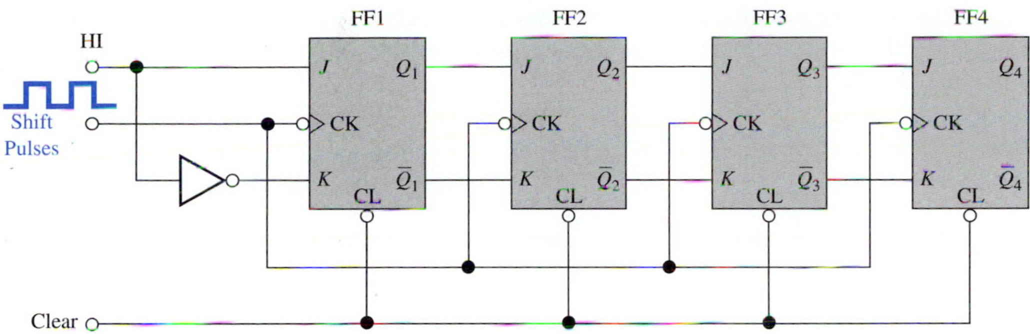

FIGURE 2–6
Shift register built with *JK* flip-flops.

When a negative edge appears on the CK line, it is applied to the CK terminals of all four flip-flops simultaneously. At this instant, all the flip-flops respond to the input levels at their *J* and *K* inputs. However, since the *J* and *K* inputs of one flip-flop are just the Q and $\overline{Q}$ outputs of the neighboring flip-flop, the result is that all information is transferred, or shifted, one place to the right. Therefore if FF1 is ON at the instant the negative edge hits the CK terminals, FF2 will be turned ON. If FF2 is OFF at the instant the negative edge hits, FF3 will be turned OFF. The only flip-flop which doesn't respond this way is FF1, which must have signals applied to its *J* and *K* inputs from some external circuit.

As a specific example, suppose that the Clear line in Fig. 2–6 goes LO to initialize all flip-flops at the OFF state. Assume also that *J* of FF1 is wired to a 1 and *K* is wired to a 0, as shown in that drawing. Now let us see what happens as the pulses start arriving on the Shift line.

As the very first negative edge hits the register, FF4 is told to turn OFF because it has $J = 0, K = 1$. This is so because FF3 is already OFF, making $\overline{Q}_3 = 0$ and $Q_3 = 1$. Since FF4 is already OFF, the signal to turn OFF does not affect it; it just stays OFF.

FF3 is signaled to turn OFF via Q_2 and $\overline{Q}_2$, and it also stands pat. The same is true for FF2, which is signaled by Q_1 and $\overline{Q}_1$. It remains OFF also. FF1, however, turns ON because of the 1 on *J* and 0 on *K*. Therefore at the completion of the first shift pulse, the state of the shift register, reading from left to right, is

1000

Now consider what happens when the second negative edge hits the CKs. FF4 is told to turn OFF by FF3 because FF3 is OFF at this instant. FF3 is likewise told to turn OFF by FF2. FF2, though, is told to turn ON because its *J* input is held HI by Q_1 and its *K* input is held LO by $\overline{Q}_1$. FF2 turns ON at this instant. FF1 still has $J = 1$ and $K = 0$ from outside, so it turns ON again, or in other words, it maintains its ON state. The state of the register is now

1100

What is happening here is that all the stored information in the flip-flops is shifted one place to the right whenever the shift command occurs. Meanwhile an external circuit keeps feeding 1s into the leading flip-flop.

After the third negative edge, the condition would be

1110

and after the fourth shift command the state would be

1111

Any subsequent shift commands will have no effect on the contents of the shift register, since they will merely cause one 1 to be lost on the far right (FF4) while another 1 comes in from the far left (into FF1).

2-4-2 Conveyor/Inspection System Using a Shift Register

The shift register finds widespread use in industrial applications involving conveyor systems, where each flip-flop in the shift register represents one zone on the conveyor system. The state of a particular flip-flop, ON or OFF, stands for some characteristic of the piece that is in that particular zone. The characteristic must be a digital characteristic, one that can be represented by a binary 1 or 0. The most obvious example is pass/fail; either the part passes inspection and is routed to the next production location or it fails inspection and is rejected.

Think of a conveyor which is broken up, mentally at least, into four physical zones. Each time a part moves from one zone to the next, it causes a shift command to be delivered to the shift register. Thus, the binary characteristic of the piece moves to the next flip-flop as the piece itself moves to the next zone.

When the piece leaves the fourth conveyor zone, the binary bit leaves the fourth flip-flop of the shift register. When a new piece enters the first conveyor zone, a new binary bit is fed into the first flip-flop of the shift register. The shift register thereby keeps track of information about the pieces on the conveyor.

In most situations, as information is shifted from one flip-flop to another, it will reach a certain flip-flop where it is acted upon by a detecting circuit. The detecting circuit reads the binary bit at a certain flip-flop and causes some action to be performed at that zone in the industrial system.

Here is a specific example. Suppose we have a production setup in which evenly spaced parts come down a conveyor and are inspected by a person. We shall call the location where the inspection takes place zone 1. Further work is to be done on the parts by other workers in zones 2 and 3 farther down the conveyor. However, if the parts do not meet inspection standards in zone 1, it is useless to waste effort by performing more work in zones 2 and 3. This is why they are inspected in zone 1; if they fail inspection in zone 1, they are not worked on as they pass through zones 2 and 3.

But, because of certain physical constraints, parts that fail the inspection cannot be removed from the conveyor and placed in the reject bin directly from zone 1. Instead, they continue down the conveyor just like good parts, until they reach zone 4. In zone 4, a diverter swings out and diverts bad parts into the reject bin. Good parts leave zone 4 in the normal manner and continue on their way.

The inspector decides if a part passes or fails inspection. If a part fails, he presses a Reject button while the part is still in his zone; he also marks the part for the benefit of the other workers in zones 2 and 3. This might be done by dabbing some paint on it with a brush, by tipping it over, or whatever. The mark signals the workers in zones 2 and 3 not to do any work on the part because it is to be rejected.

When a part leaves zone 1, the shift register keeps track of whether it passed inspection or failed. As the part proceeds through the zones, the pass/fail information keeps pace in the shift register. When the part enters zone 4, the shift register signals the diverter whether or not to swing out to dump it into the reject bin.

Such a layout is shown in Fig. 2–7(a), and the control circuit is shown in Fig. 2–7(b).

The circuit of Fig. 2–7(b) is really quite simple. If the Reject button is pressed while the part is in zone 1, the output of I1 goes LO and pulls the preset input of FF1 down

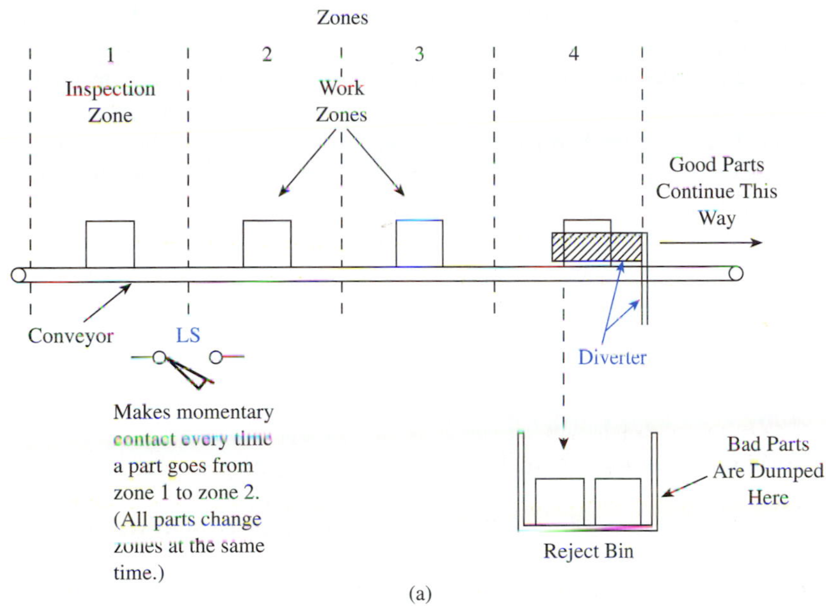

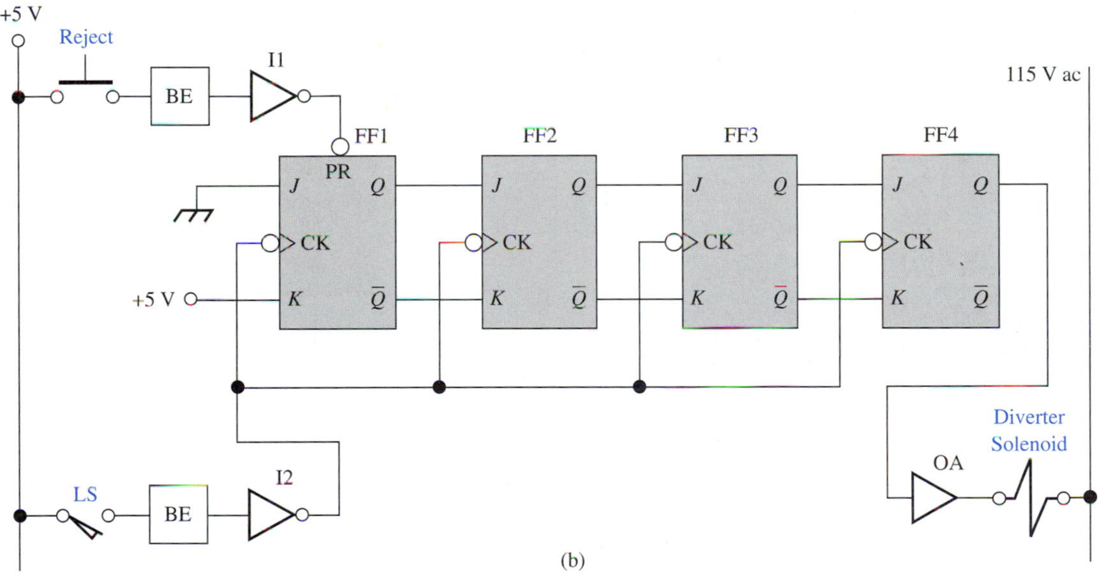

FIGURE 2–7
(a) Physical layout of a conveyor/inspection system. (b) Control circuit of the conveyor/inspection system, illustrating the use of a shift register to keep track of the progress of parts through the system.

LO. This presets FF1 into the ON state (Q_1 goes HI). Remember, we are assuming that flip-flops respond to a LO preset signal.

As the bad part leaves zone 1 and enters zone 2, the limit switch is momentarily contacted. This causes the output of I2 to go LO, delivering a negative edge to all the CK terminals. FF2 turns ON at this instant because Q_1 is applying a 1 to J and $\overline{Q}_1$ is applying a 0 to K. Therefore as the bad part enters zone 2, the information about its badness enters flip-flop 2. A bad part in a zone is indicated by that flip-flop being ON.

FF1 turns back OFF when the negative edge hits the CK terminals because of the hard-wired LO at *J* and HI at *K*.

Since the parts are evenly spaced, every part on the conveyor moves into a new zone at the time the limit-switch contact closes in response to a part passing from zone 1 to zone 2. Thus as the bad part enters zone 3, LS closes again because the following part is entering zone 2. This causes another clock edge to appear, which turns ON FF3. As the bad part enters zone 4, LS causes another clock edge, which turns ON FF4. When Q_4 goes HI, it energizes the diverter solenoid and swings the diverter out. As the conveyor continues to move, the bad part is guided into the reject bin by the diverter.

As the following part enters zone 4, FF4 turns back OFF if the part is good. The diverter immediately swings back into normal position before the part can run into it.

2-4-3 Prepackaged Shift Registers

So far, our illustrations of shift registers have shown several flip-flops serially tied together. To be sure, this is exactly how prepackaged shift registers are internally built, but they are not always illustrated this way. A prepackaged shift register is usually shown as a box having a clock input (CK), a clear input (CL), preset inputs for each bit (PR_n), shift inputs for the first bit (*J* and *K*), and outputs for each bit (Q_n and $\overline{Q}_n$). This symbol is shown in Fig. 2–8(a) for a 4-bit shift register.

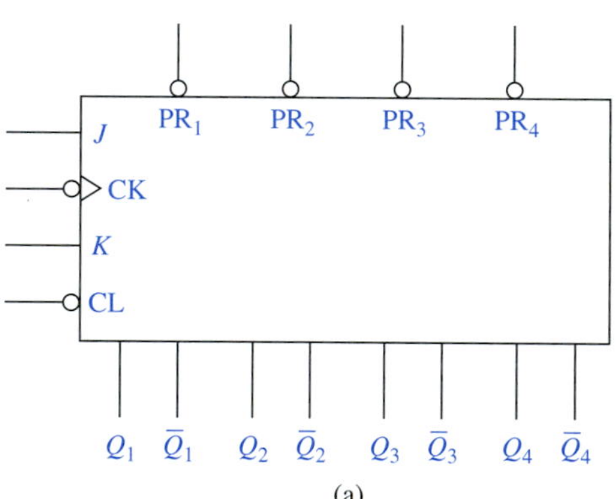

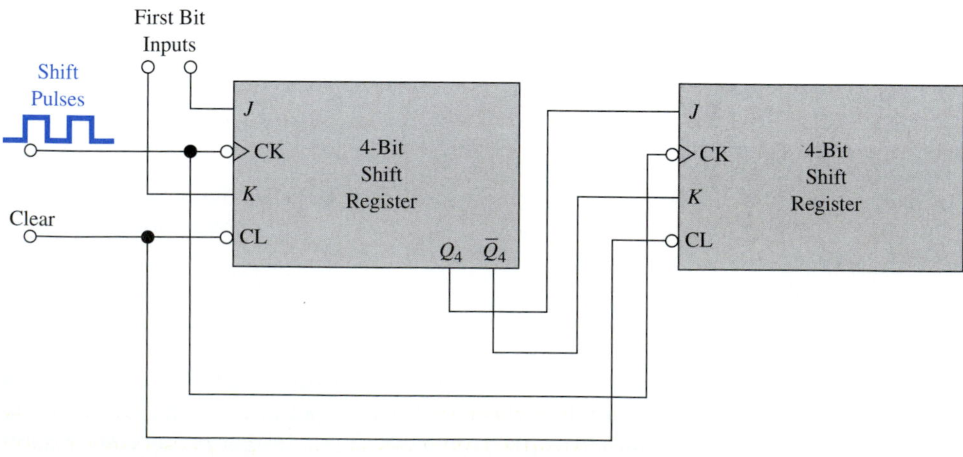

FIGURE 2–8
(a) Black-box symbol of a 4-bit shift register with a common clear input and individual preset inputs.
(b) Tying two shift registers together (cascading). In this schematic diagram the *J* and *K* inputs have been shown separately. It is actually more common for the *J* inputs to be labeled *D* (for Data fit) and for an inverter to be connected internally to *K*. Then the *K* inputs are not brought out to external terminals, as shown here.

PHOTO 2-1
Automated testing and counting of electronic component production.

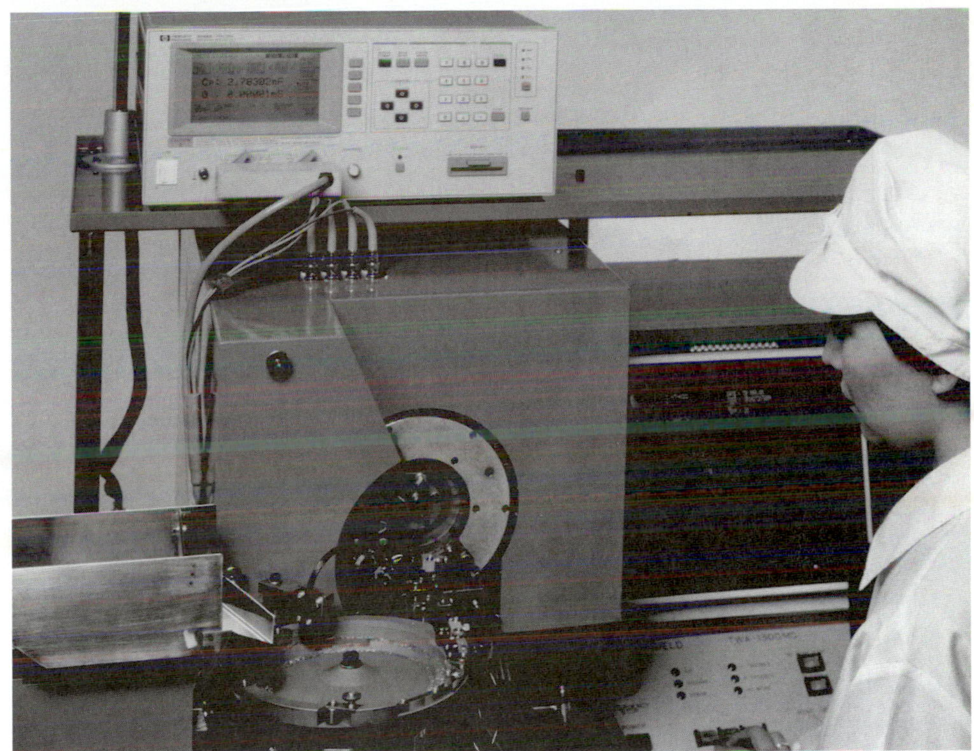

Courtesy of Hewlett-Packard Company.

The most common lengths for prepackaged shift registers are 4, 5, and 8 bits. If a longer shift register is needed, two or more small ones can be cascaded as shown in Fig. 2–8(b). In that figure, two 4-bit shift registers are cascaded to make an 8-bit shift register. As the drawing shows, this is done by tying the CK inputs together, tying the CL inputs together, and connecting the outputs of the last bit to the inputs of the first bit of the next register.

There are many different types of shift registers. They all exhibit the same basic behavior, that of shifting binary bits from one location to the next. Their secondary features differ from one another and from what we have discussed. For example, some shift registers can shift either to the right or the left. Naturally such shift registers have more input terminals than are shown in Fig. 2–8(a) because they must be instructed as to the direction of the shift. Some shift registers have a special LOAD input terminal for signaling when bits are to be preset or "loaded" into the register. To avoid confusion, we will stay with only one type, the type illustrated in Fig. 2–8(a).

2-5 ■ COUNTERS

A *digital counter* is a circuit that counts and remembers the number of input pulses that have occurred. Every time another input pulse is delivered to the CK terminal of a counter, the number stored in the circuit advances by one.

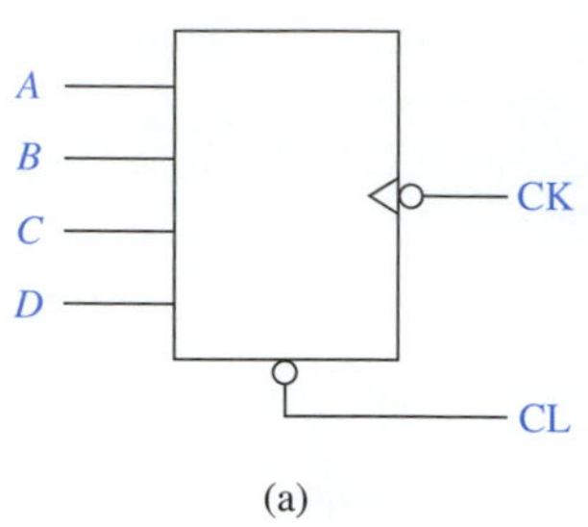

(a)

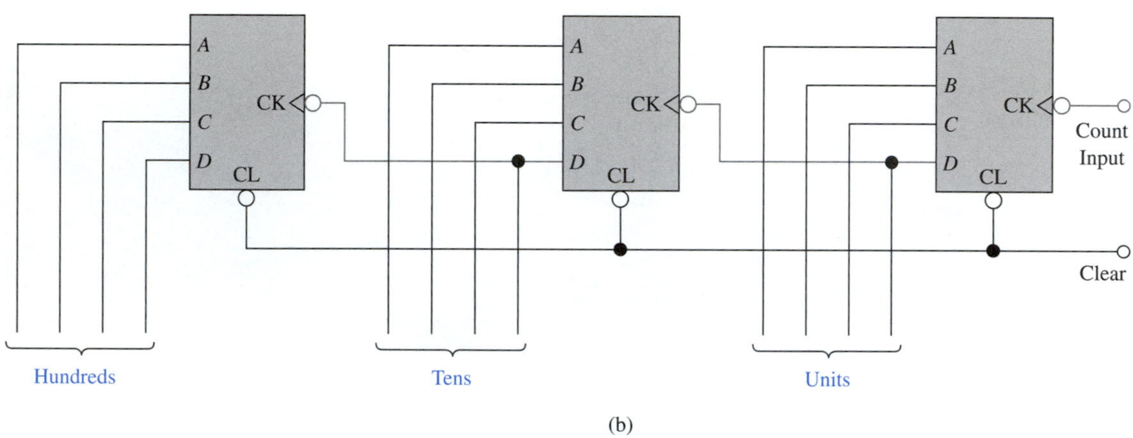

(b)

FIGURE 2–9
(a) Black-box symbol of a decade counter. (b) Cascading decade counters together to count higher than 9.

Of course, since digital counters are built out of flip-flops and logic gates, they must operate in the binary number system. Use your digital circuits text to review the following topics regarding digital counters and the binary number system:

1. Counting in binary
2. Binary ripple counters built with *JK* flip-flops
3. Binary-coded-decimal (BCD) numbers
4. Decade up-counters built by combining *JK* flip-flops with logic gates
5. Cascading decade up-counters

Our schematic symbol for a decade up-counter is shown in Fig. 2–9(a). The output bits are symbolized *D*, *C*, *B*, and *A*, with corresponding numerical values of 8, 4, 2, and 1. All four output bits are cleared to 0 when the counter's CL terminal is pulled to its active-LO state.

As a decade up-counter overflows from 9 to 0, its *D* output bit makes a negative-going transition. Therefore the *D* output terminal can be connected directly to the CK terminal of the next more significant decade counter when two or more counters are cascaded. This up-counter interconnection is shown in Fig. 2–9(b).

2-6 ■ DECODING

In many industrial applications using decade counters, the system operators set a 10-position selector switch to "watch" the counter and take some sort of action when the state of the counter matches the setting on the switch. This idea is illustrated in Fig. 2–10.

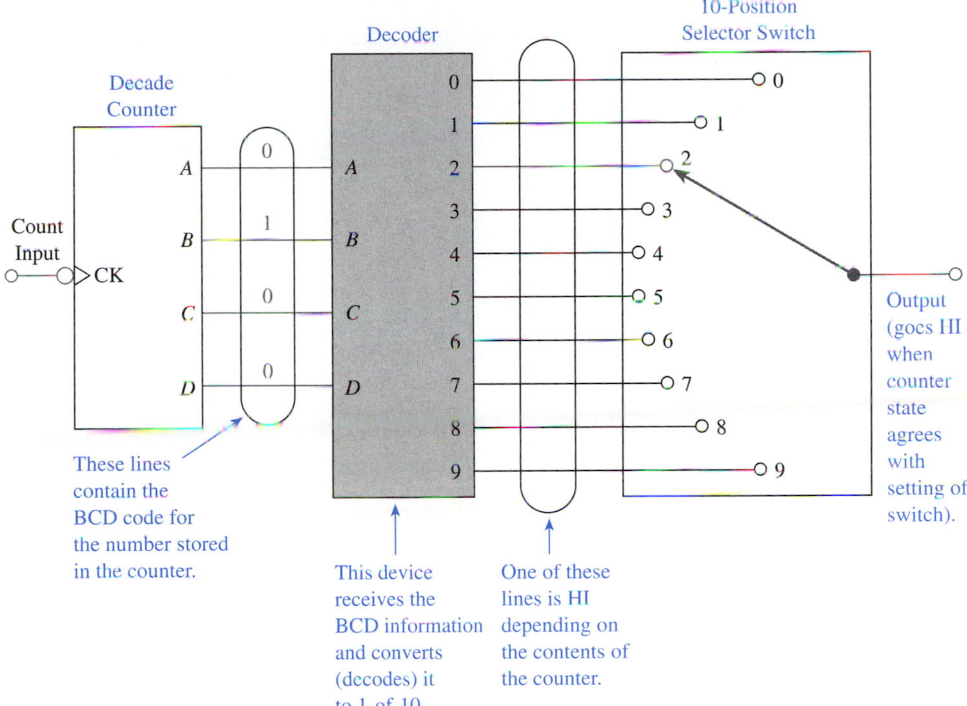

FIGURE 2–10
Combination of a decade counter, a 1-of-10 decoder, and a 10-position selector switch. This combination is often seen in industrial cycle-control circuits.

The decade counter has four output lines, *D*, *C*, *B*, and *A*, which contain the binary code for the number in the counter. The box in between the counter and the selector switch in Fig. 2–10 is called a *decoder* because it takes binary-coded information and converts it to decimal information that human beings can understand. That is, if the binary information represents the decimal digit 2 ($DCBA = 0010$), the decoder drives the 2 output line HI. If the binary information represents the decimal digit 3 ($DCBA = 0011$), the decoder drives the 3 output line HI, and so on. Since it converts coded numbers to uncoded decimal numbers, it is called a decoder.

In Fig. 2–10, if the output of the decoder is the same as the setting of the selector switch, the common terminal of the switch will go HI. Overall, the output of the circuit goes HI when the counter reaches the setting of the 10-position selector switch. The HI on the output could then be used to accomplish some action in the system. This is how a manually operated selector switch can "watch" a counter and take action when it reaches a certain count.

The most straightforward way of constructing a decoder is shown in Fig. 2–11. Look first at Fig. 2–11(a). It shows 10 four-input AND gates, each gate with a different combination of inputs. Each combination of inputs represents one of the possible states of the decade counter. Therefore, for any one of the 10 possible output states, one of the AND gates will be enabled. For example, if the state of the counter is $DCBA = 0101$ (decimal 5), then all four inputs to AND gate number 5 will go HI. Figure 2–11(a) shows that the inputs to that gate are $\overline{D}$, C, $\overline{B}$, and A. If the counter state is $DCBA = 0101$, then $\overline{D}C\overline{B}A = 1111$; when all four inputs are HI, the output goes HI. Thus if the counter has counted to 5, the number 5 output of the decoder goes HI. You should verify that the decoder works properly for the other counter output states.

Figure 2–11(b) shows the same circuit as Fig. 2–11(a). The only difference is that all the interconnections are shown. This schematic drawing drives home the point that there are really only four inputs to the decoder and that these four are decoded to a decimal number from 0 to 9.

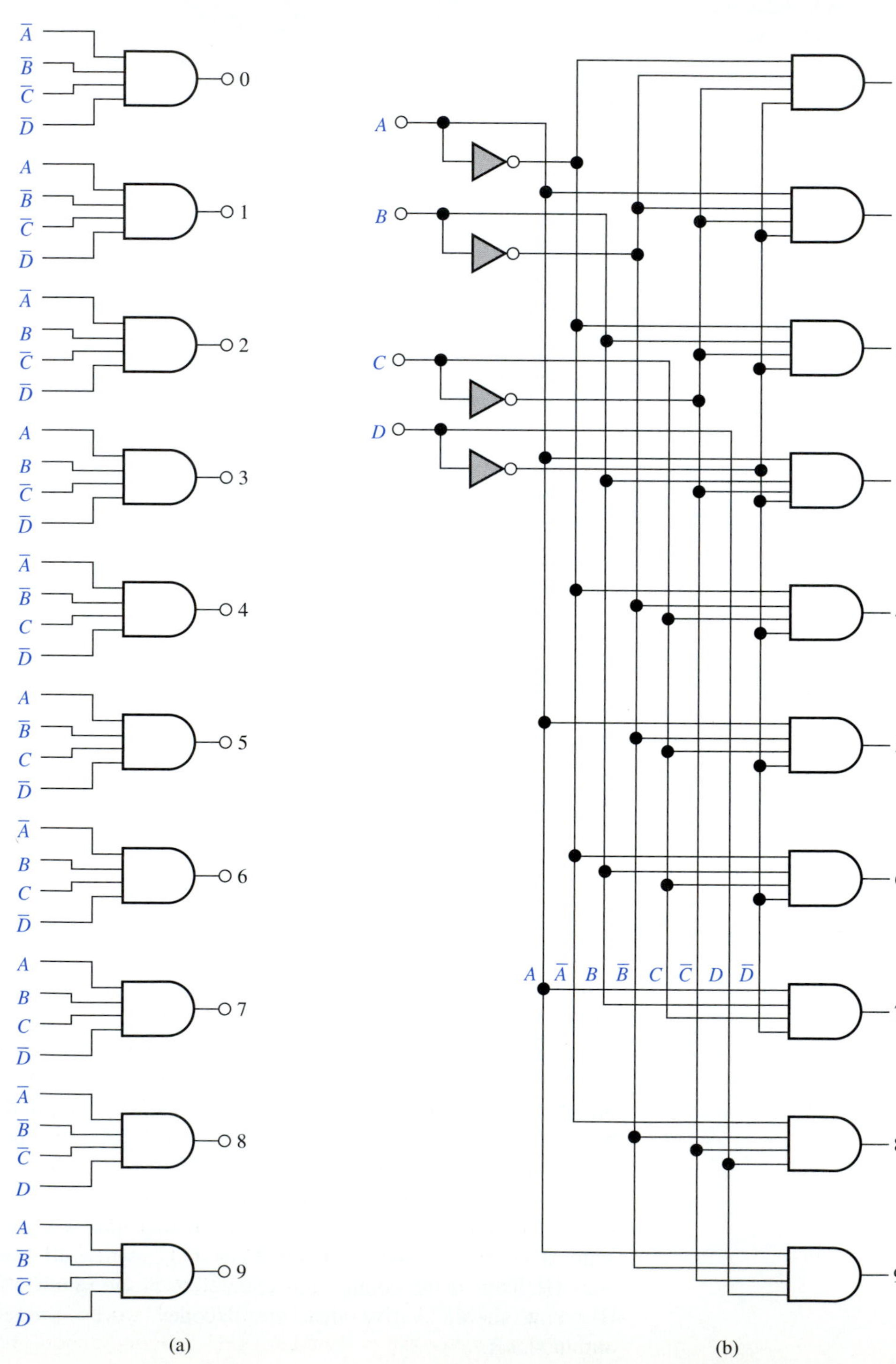

FIGURE 2–11
(a) Construction of a 1-of-10 decoder, showing the inputs of each decoding gate. (b) 1-of-10 decoder showing the actual wire connections.

The decoder of Fig. 2–11 is called a *BCD-to-decimal decoder* or a *BCD-to-1-of-10 decoder*. There are other types of packaged decoders available (BCD-to-seven-segment, Gray-code-to-decimal, excess-three-code-to-decimal, etc.), but for our purposes the term *decoder* will refer to a BCD-to-decimal decoder unless stated otherwise.

2-7 ■ PALLETIZING SYSTEM USING DECADE COUNTERS AND DECODERS

Consider a situation in which cartons are sealed by a carton-sealing machine and then taken away by conveyor to one of two palletizers. Palletizers are machines which stack cartons in an orderly manner on pallets. When one pallet is completely loaded, a diverter swings over and begins routing cartons to the other palletizer. While the second pallet is being loaded, the first loaded pallet is removed and an empty pallet brought into its place.

Because the system handles cartons of different sizes, the number of cartons per pallet will vary. Therefore the operators must be able to easily change the per-pallet carton count. The general layout is depicted in Fig. 2–12(a), and the control circuitry is shown in Fig. 2–12(b).

As can be seen from the drawing, LSA is actuated just prior to a carton being loaded by palletizer A, and LSB is actuated just prior to a carton being loaded by palletizer B. When the preset number of cartons have been loaded onto either palletizer, the diverter swings into the opposite position. The cartons that follow are then routed to the opposite palletizer.

The control circuit works as follows. Assume that the diverter is routing cartons to palletizer A. This implies that A DIVERT SOLENOID is energized, which means that the *JK* flip-flop is OFF. As the cartons pass by LSA, they momentarily close the N.O. contact, causing the I1 input to go HI. As the I1 output goes LO, it delivers a negative edge to the units decade counter, which advances its count by one. The selector switches as drawn are set to 8 on the units switch and to 2 on the tens switch. Therefore the pallet will be loaded with 28 cartons. As the twenty-eighth carton passes LSA, the decade counters go to the states

Tens	*Units*
0010	1000

At this instant the tens decoder is holding the 2 output HI and the units decoder is holding the 8 output HI. Therefore both SS common terminals go HI, which causes both NAND inputs to go HI. As the NAND output goes LO, it delivers a negative edge to the flip-flop. With J and K both HI the flip-flop toggles to ON. The $\overline{Q}$ output goes LO, de-energizing the A DIVERT SOLENOID, and the Q output goes HI, energizing the B DIVERT SOLENOID. This swings the diverter into the dotted position of Fig. 2–12(a), so succeeding cartons are routed to palletizer B.

Meanwhile, the NAND output has pulled the CL terminals down LO on both the units and the tens counter. This immediately resets both counters to 0000, in preparation for beginning the count of cartons to palletizer B. As the counters are cleared, the NAND inputs go back LO. The output of the NAND goes HI, which removes the LO clear signal, putting the counters right back into counting condition.

When 28 cartons have been loaded onto pallet B, the flip-flop toggles back to OFF and the counters are cleared again. The system then begins all over again, loading pallet A.

Whenever a different-sized carton is to be run, the operators just set the selector switches to a different number. Any number of cartons from 0 to 99 can be selected.

FIGURE 2–12
(a) Top view of the palletizing system. (b) Control circuit of the palletizing system, showing the operation of decade counters, 1-of-10 decoders, and 10-position selector switches.

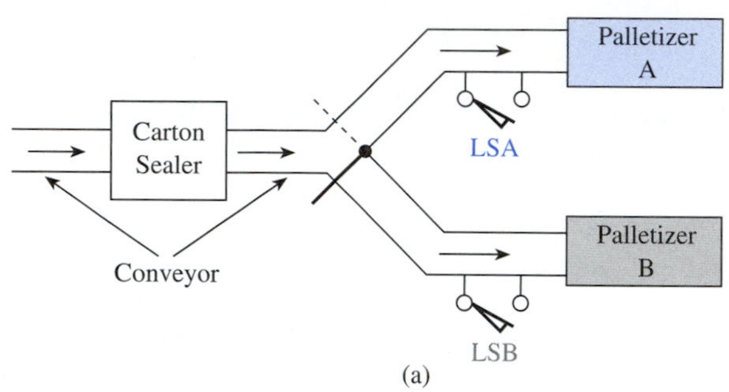

(a)

(b)

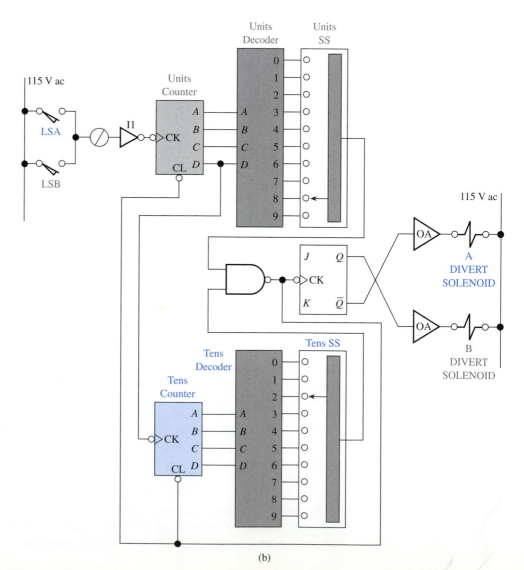

2-8 ■ ONE-SHOTS

The *one-shot* (formal name: monostable multivibrator) is a very useful circuit in digital industrial controls. Its output temporarily goes HI when the circuit is triggered; then it returns to LO after a certain fixed time. A one-shot is used whenever the situation calls for a certain line (the output) to go HI *for a short while* if another line (the input) changes state.

Figure 2–13 illustrates the action of a one-shot. We will assume that one-shots are triggered by a negative edge at the T (trigger) terminal. Actually, some one-shots are triggered by a positive edge at T, but for simplicity we will consider only negative edge-triggered one-shots.

The black-box schematic symbol for a one-shot is shown in Fig. 2–13(a). Notice that it has two outputs, Q and $\overline{Q}$. The $\overline{Q}$ output is the complement of the Q output, just like the outputs of a flip-flop. When the one-shot is triggered, the Q output goes HI while the $\overline{Q}$ output goes LO. After a period of time has elapsed (called the *firing time*, t_f), the Q output returns to LO and the $\overline{Q}$ output returns to HI.

The waveforms in Fig. 2–13(b) show how a one-shot behaves when it is triggered by a short pulse. When the negative-going edge of the short pulse occurs, the one-shot triggers, or fires. The Q output quickly goes HI and remains HI for a length of time equal to t_f. The firing time t_f is usually adjustable by adjusting a resistor or capacitor in the circuit.

When driven by a short pulse as in Fig. 2–13(b), a one-shot may be acting as a pulse stretcher; that is, a short-duration input pulse is converted into a longer-duration output pulse. Or it may be acting as a delay device; that is, when a negative edge occurs at T, another negative edge appears at Q, but delayed by a time t_f. Or it may just be used to clean up a ragged input pulse; that is, the output pulse of a one-shot is well shaped in that it has steep positive-going and negative-going edges, regardless of the condition of the input pulse.

When a one-shot is driven by a long-term level change as shown in Fig. 2–13(c), it is acting rather like a pulse shrinker. One-shots are frequently used in this mode to clear a counter (or a flip-flop) when a certain line changes levels. For example, it is often necessary to begin counting in a counter shortly after a certain level change takes place, but it is necessary for the count to begin at zero. If the level change persists after the next counting sequence is supposed to begin, then we cannot use the level change *itself* to clear the counter. This is because the changed level would hold the counter in the reset state. What is needed is a circuit which can *temporarily* apply a reset signal to the counter when the level change takes place. The reset signal must then go away in time for the next round of counting to begin. A one-shot performs this action perfectly.

We are assuming in this book that counters and flip-flops are reset by a LO level applied to the clear terminal. The fact that the waveforms in Fig. 2–13 show a HI-level output pulse may therefore be a matter of concern. However, we have seen that one-shots also have a $\overline{Q}$ output, which delivers a LO-level pulse during the time the Q output is delivering a HI level. The $\overline{Q}$ output would be used to reset a counter in a situation such as that described in the preceding paragraph.

There are many ways to build discrete one-shots. Fig. 2–13(d) shows one popular way. It works as follows. When the circuit is at rest, T_2 is turned ON and saturated. Its base current is supplied through R_{B2}. The collector of T_2 is virtually at ground potential, so the Q output is LO. The base of T_2 is only 0.6 V above ground potential because of the forward-biased base-emitter junction.

T_1 is cut OFF because it has no base drive. Its base resistor, R_{B1}, is connected to the T_2 collector, which is at 0 V. Therefore R_{C1} is completely disconnected from the grounded emitter to T_1 and is free to carry current to charge capacitor C. Since C is

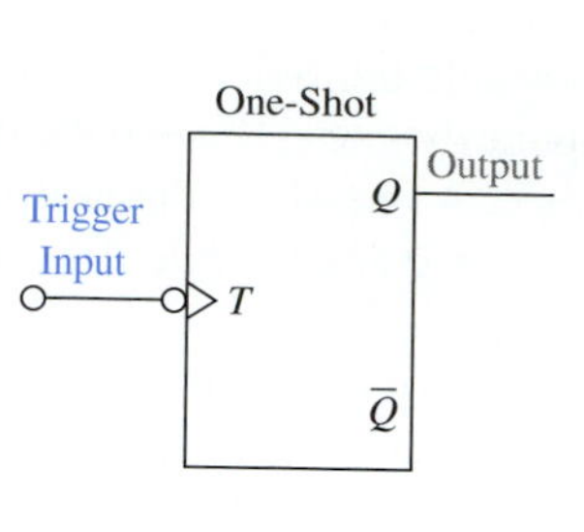

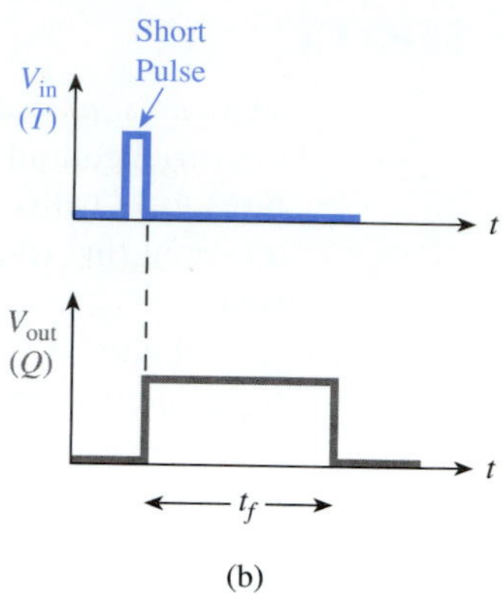

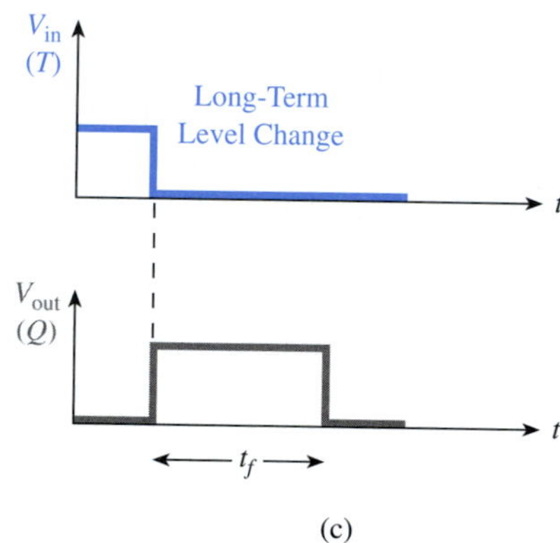

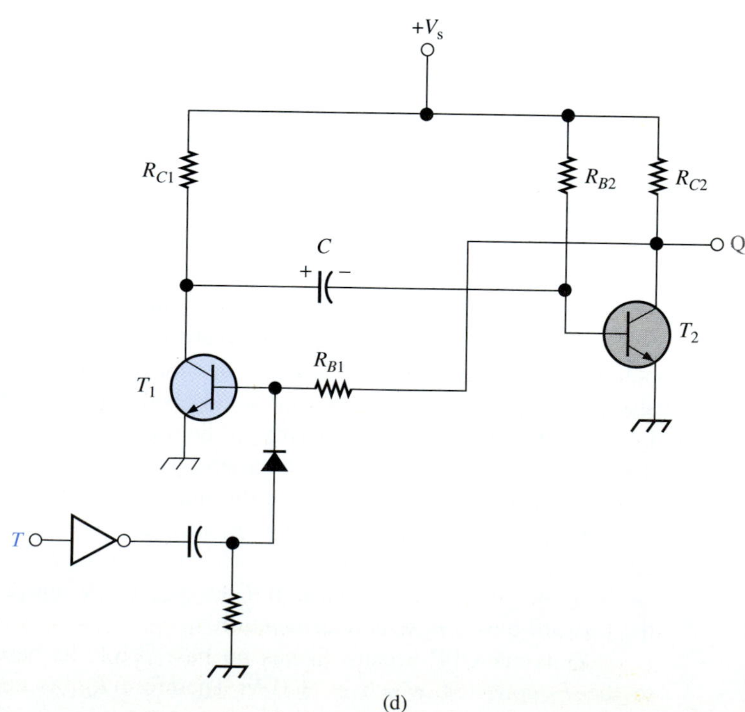

FIGURE 2–13 (a) Black-box symbol of a one-shot. (b) V_{in} and V_{out} waveforms when the one-shot is triggered by a short pulse. (c) Input-output waveforms when the one-shot is triggered by a long-term level change of a logic signal. (d) Schematic diagram of a one-shot.

connected to the base of T_2, which is close to ground potential, it will charge up almost to the supply voltage V_s. The polarity of the charge on C is plus (+) on the left and minus (−) on the right, as shown.

Now, let a negative edge appear at T. The inverter causes a HI to be applied to the RC differentiator, which applies a positive spike to the base of T_1. This turns T_1 ON and pulls the collector of T_1 down to ground. Since the charge on C cannot disappear instantly, the voltage across the capacitor plates is maintained. With the + side of the capacitor pulled down to 0 V by T_1, the − side goes to a voltage far below ground potential. This applies a negative voltage to the base of T_2, shutting it OFF. The collector of T_2 rises toward V_s and is now capable of supplying base current to T_1. Therefore T_1 remains turned ON even after the positive spike from the differentiator goes away. Q is now HI, and Q is LO.

As time goes by, charging current flows onto the plates of C. The flow path is down through R_{B2}, through C, and through collector to emitter of T_1 into ground. As can be seen, this path seeks to charge C to the opposite polarity; what happens is that the voltage across C gets smaller. When the capacitor voltage crosses through zero and reaches 0.6 V in the opposite polarity, it bleeds a small amount of current into the base of T_2. This small base current causes collector current to flow in T_2, lowering the collector voltage. The reduced collector voltage causes a reduction in base current to T_1. This in turn causes a reduction in T_1 collector current. The T_1 collector voltage rises slightly, thereby raising the base of T_2 higher yet. This action is regenerative; once it begins, it avalanches. In the end, T_2 is saturated once again, and T_1 is cut OFF. Q is LO, and $\overline{Q}$ is HI, and the circuit has returned to its original state.

One-shots are usually prepackaged integrated circuits, having the schematic symbol given in Fig. 2–13(a). They generally have provision for the user to connect an external resistor and/or capacitor to set the firing time. The manufacturers of packaged one-shots furnish graphs which show the relationship between t_f and the size of external resistor and capacitor.

One-shots are classified as either *retriggerable* or *nonretriggerable*. Retriggerable means that if a *second* negative edge occurs during the firing time of the one-shot, the output pulse resulting from the first negative edge will be extended beyond its normal duration. We will assume that our one-shots are nonretriggerable; they ignore triggering edges occurring during an output pulse. Several examples of one-shots in industrial controls will be presented in Secs. 2-10 and 2-13.

2-9 ■ CLOCKS

Often in industrial digital circuits it is necessary to keep various digital devices synchronized with each other. In other situations, a continuous train of pulses is needed to supply count pulses to a counter if the system does not generate count pulses naturally as it performs its functions. In either case, what is required is a circuit which supplies a continuous stream of square-edged pulses. Such circuits are referred to as *clocks*.

A black-box symbol of a clock is shown in Fig. 2–14(a). The output waveform shown is a square wave. It can also be thought of as a pulse train with a 50% duty cycle. Many clocks have just such an output; some clocks have duty cycles other than 50%.

The output frequency (pulse repetition rate) of a clock is set by resistor, capacitor, or inductor sizes internal to the circuit. In the case of a crystal-controlled clock, the frequency is determined by the cut of the crystal; crystal-controlled clocks are very frequency-stable. Some clocks have frequency dividers connected to their outputs. A

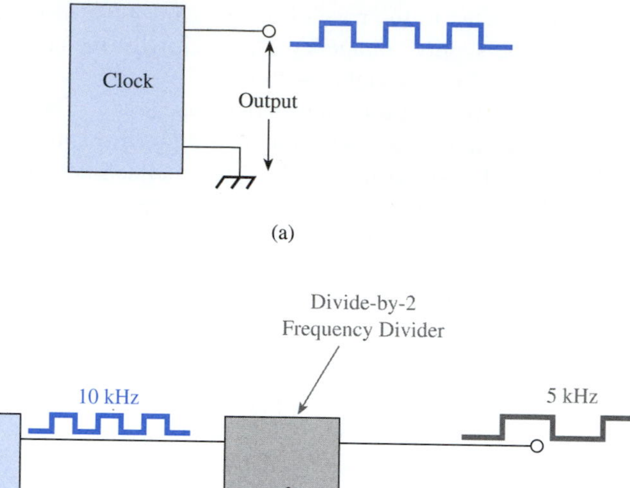

FIGURE 2–14
(a) Black-box symbol of a clock. (b) A clock combined with a frequency divider to obtain a signal at a different frequency.

frequency divider takes the clock pulse frequency, divides it by some integer number, and produces an output pulse train at the lower frequency. Some systems need two or more clock signals of different frequencies in order to synchronize events properly.

Circuit schematics for clocks are presented in many digital electronics books. If not referred to in the index by the name *clock,* the circuits may be indexed by the names *astable multivibrator* or *free-running multivibrator*.

2-10 ■ AUTOMATIC TANK FILLER USING A CLOCK AND ONE-SHOTS

Consider the system illustrated in Fig. 2–15. The four tanks are refilled from a main tank when their liquid levels drop below a certain setting. That is, if the level in tank 2 drops below its low setting, valve 2 will automatically open and refill tank 2 until the liquid level reaches its high setting. Due to certain system restrictions, it is important that only one tank be refilling at one time. The circuit to control this system utilizes a clock and several one-shots, and is shown in Fig. 2–15(b). The abbreviation OS is used for one-shot.

Here is how it works. Each tank has two limit switches, one that closes on low liquid level and one that closes on high liquid level. If all tank levels are satisfactory, FF5 is OFF. Its $\overline{Q}$ output is HI, so input 1 of AND5 is HI. The clock is delivering square-wave pulses to the AND gate, so the AND gate output is also a square wave at the same frequency as the clock. Therefore the decade counter is counting merrily along. As the counter proceeds through its various count states, the decoder steps through *its* output states. That is, first the 1 output goes HI, then the 2 output goes HI while 1 returns to LO, then 3 goes HI while 2 returns to LO, and so on. However, when the counter gets to 5, the 5 output of the decoder causes a negative edge to be delivered to the *T* terminal of OS5 via the inverter. The one-shot fires for a few microseconds, applying a LO to the CL terminal of

FIGURE 2–15
(a) Physical layout of an automatic tank-filling system.
(b) Control circuit of the automatic tank filler, showing the use of one-shots and a free-running clock.

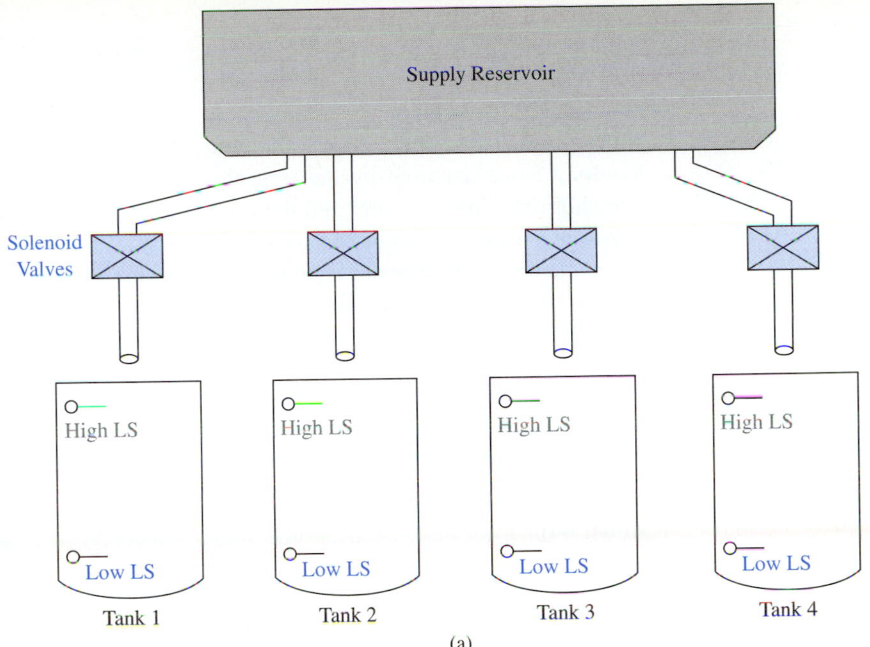

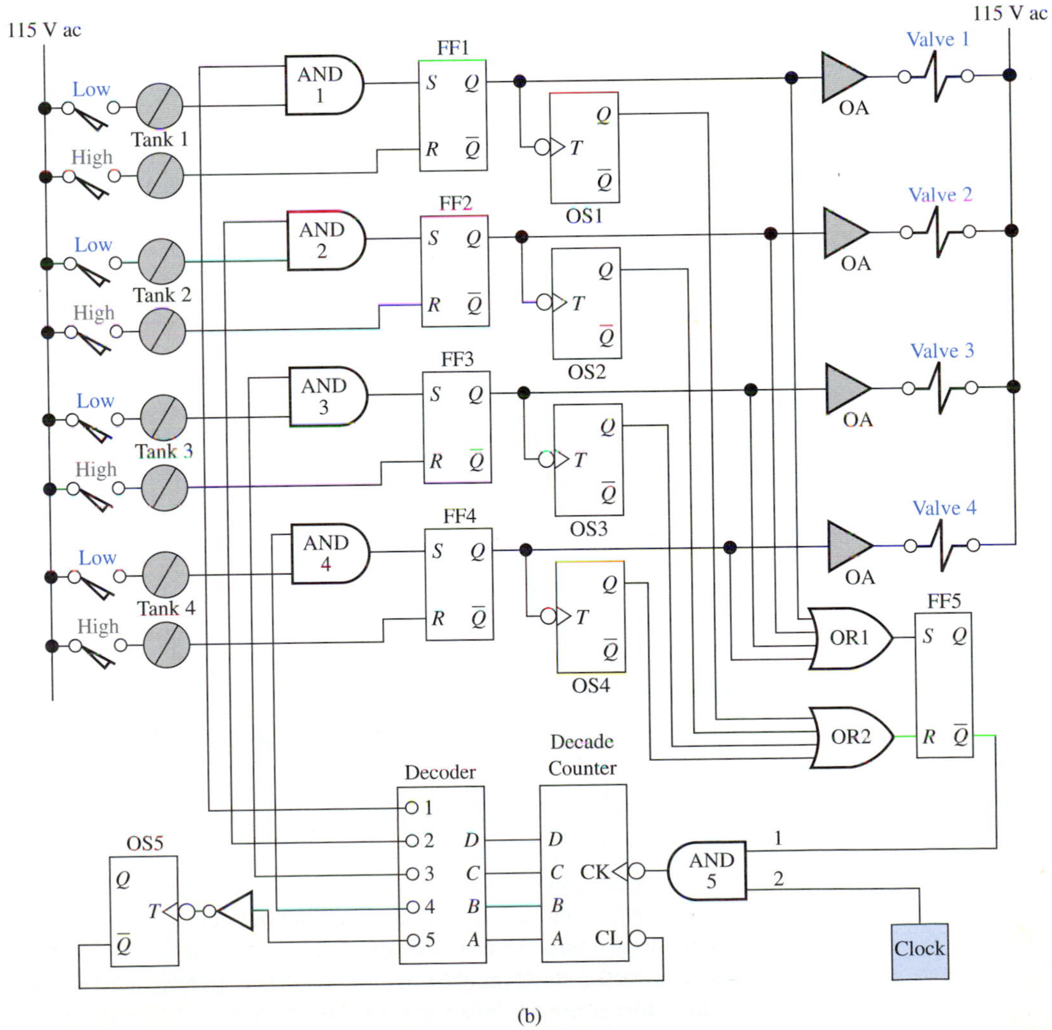

(a)

(b)

the counter. The counter is immediately reset to zero when this occurs. The very next count pulse from AND5 causes the counter to count from 0 to 1, since the clear signal has long since departed by the time the count pulse arrives. This is an illustration of a one-shot resetting a counter and then removing the clear signal in time for the next count; this application was suggested in Sec. 2-8.

Therefore the counter is continually counting through states 0–4; when it reaches 5 it stays in that state only long enough for the clear signal to reset it to 0.

The 1 output of the decoder partially enables AND1. The 2 output of the decoder partially enables AND2, and so on. AND gates 1–4 are partially enabled in succession as the decoder goes through its output states.

If a low liquid level limit switch goes closed, the AND gate it controls will be fully enabled. For example, suppose the low-level limit switch in tank 3 closes. Then, as soon as decoder output 3 goes HI, AND3 will go HI. This applies a HI to the S input of FF3, which turns FF3 ON. The Q output of FF3 signals OA3, which opens valve 3 to refill the tank.

Meanwhile the Q output of FF3 has applied a HI to input 3 of OR gate 1. This causes the OR gate to apply a HI to S of FF5, turning ON that flip-flop. When $\overline{Q}$ of FF5 goes LO, AND5 is disabled and the counter no longer receives count pulses. The counter therefore freezes in its present state.

As the liquid level in tank 3 rises, the low limit switch opens, disabling AND3 and removing the HI from S of FF3. The flip-flop remains ON because of its memory ability. Tank 3 continues refilling until the high liquid level limit switch closes. This applies a HI to the R input of FF3, causing it to turn OFF. When the Q output goes LO, it shuts off valve 3 and disables OR1. The HI is therefore removed from S of FF5. Also, as Q of FF3 goes LO, it delivers a negative edge to the trigger input of OS3, causing that one-shot to fire. The Q output of OS3 goes HI for a few microseconds, enabling OR2. The output of OR2 goes HI temporarily and applies a HI to R of FF5. The flip-flop turns OFF, and its $\overline{Q}$ output returns HI. When this happens, the clock pulses are once again gated into the counter, and counting picks up where it left off.

If you consider the problem of resetting FF5 when the filling operation is finished, you will see why one-shots 1–4 are necessary. OR2, which resets FF5, cannot be driven directly by the $\overline{Q}$ outputs of flip-flops 1–4. With this design, even if one of the flip-flops turned ON, the other three $\overline{Q}$ outputs would hold R of FF5 HI. This would prevent FF5 from ever turning ON, so it would not work. Instead, it is necessary to *temporarily* apply a HI to the R input of FF5 when any one of flip-flops 1–4 turns OFF. One-shots are the best means of doing this.

2-11 ■ DOWN-COUNTERS AND ENCODERS

2-11-1 Decade Down-Counters

The counters discussed in the preceding sections have all counted in the *up* direction. That is, whenever a count pulse was delivered, the count increased by one. In many cases in industrial control, it is very useful to have a counter which counts in the *down* direction. That is, every time a count pulse is delivered, the number stored in the counter *decreases* by one. This type of counting is especially desirable when it is necessary to produce an output signal after a certain presettable number of counts, and also to produce another output signal a fixed number of counts *earlier*. In Sec. 2-13 we will see an example of a down-counter used in just such a manner.

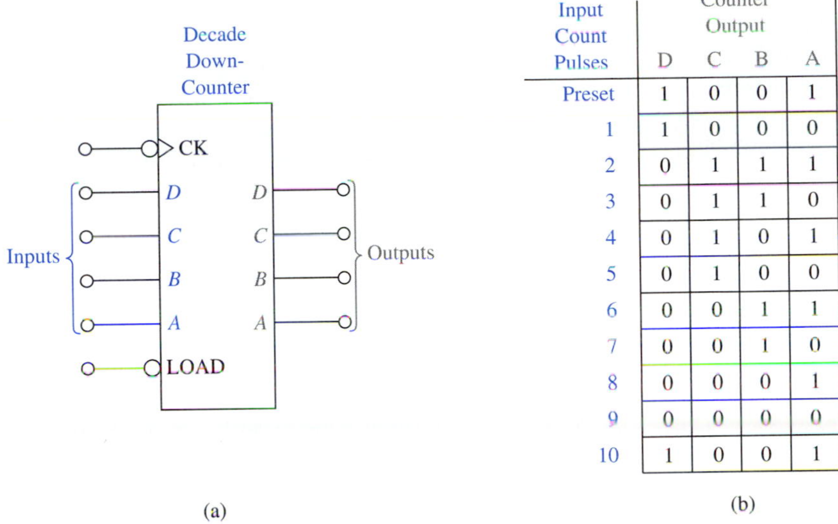

Input Count Pulses	Counter Output D	C	B	A
Preset	1	0	0	1
1	1	0	0	0
2	0	1	1	1
3	0	1	1	0
4	0	1	0	1
5	0	1	0	0
6	0	0	1	1
7	0	0	1	0
8	0	0	0	1
9	0	0	0	0
10	1	0	0	1

FIGURE 2–16
(a) Black-box symbol of a decade down-counter.
(b) The state of the down-counter after each input pulse.

A decade down-counter is shown schematically in Fig. 2–16(a). It is similar in operation to a decade up-counter except that it counts in the down direction. When its contents are zero, the next count input pulse places it in the 9 state ($DCBA$ = 1001).

The down-counter has A, B, C, and D *inputs* as well as outputs in order to preset a number into the counter. When the LOAD input terminal goes LO, the BCD number which appears at the A, B, C, and D inputs is preset or loaded into the counter. During loading, any count pulses appearing at CK are ignored. When the LOAD terminal returns to HI, the A, B, C, and D inputs become disabled, and the count pulses at CK begin stepping the counter.

Figure 2–16(b) shows the output states of the down-counter for 10 successive input count pulses, assuming the counter was preset to 9. If the counter were preset to a lower number, naturally it would reach 0 in a smaller number of counts. When it reaches 0, the next count pulse sends it back to 9.

Some counters can be made to count either up or down. They are called *up/down-counters* and have a special control input to tell them in which direction they are supposed to count.

Down-counters can be cascaded just like up-counters. A cascadable down-counter normally has a special output, which tells the neighboring counter when it is going from 0000 to 1001. This way the tens counter, for instance, can count down by one every time the units counter goes through a complete range of values and returns to 9.

2-11-2 Decimal-to-BCD Encoders

A down-counter often has an *encoder* associated with it, so encoders will be discussed now. An encoder is a device which takes in a decimal number and puts out a binary number. It is the reverse of a decoder.

There are various types of encoders available, but we will concentrate on the type that converts a 1-of-10 decimal input to BCD output. Such an encoder is shown schematically in Fig. 2–17(a), and its truth table is given in Fig. 2–17(b).

FIGURE 2–17
(a) Black-box symbol of a 1-of-10 encoder. (b) The truth table of the encoder, showing the output state for every legal combination of inputs.

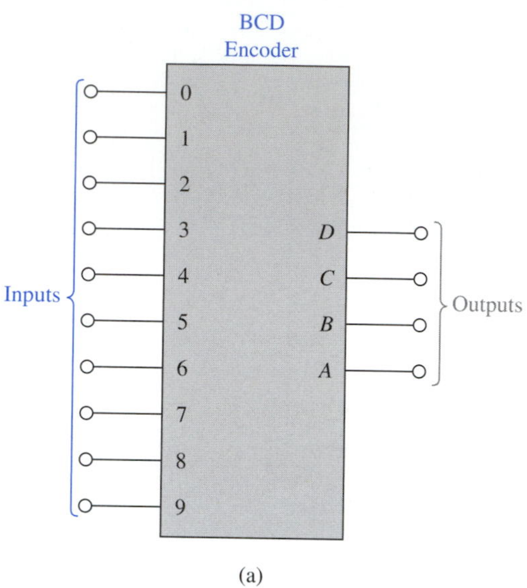

(a)

Truth Table

Inputs 0	1	2	3	4	5	6	7	8	9	Outputs D	C	B	A
1	0	0	0	0	0	0	0	0	0	0	0	0	0
0	1	0	0	0	0	0	0	0	0	0	0	0	1
0	0	1	0	0	0	0	0	0	0	0	0	1	0
0	0	0	1	0	0	0	0	0	0	0	0	1	1
0	0	0	0	1	0	0	0	0	0	0	1	0	0
0	0	0	0	0	1	0	0	0	0	0	1	0	1
0	0	0	0	0	0	1	0	0	0	0	1	1	0
0	0	0	0	0	0	0	1	0	0	0	1	1	1
0	0	0	0	0	0	0	0	1	0	1	0	0	0
0	0	0	0	0	0	0	0	0	1	1	0	0	1

(b)

As can be seen from the truth table, the output is the binary equivalent of the decimal input. The truth table as given implies that there are never two inputs HI at the same time. It is the responsibility of the control circuit designer to make sure this is so.

There is always the possibility that two or more inputs might go HI at the same time, through some malfunction in the input circuitry to the encoder. If it is important to know what the encoder will do in such a case, the manufacturer's specification sheet will explain. Most packaged encoders obey the higher-number input if that problem ever occurs.

Encoders very often receive their input from a 10-position selector switch. The SS is manually set by the system operator, and the number selected appears at the output of the encoder in BCD form. The output of the encoder can then be wired to the input of a

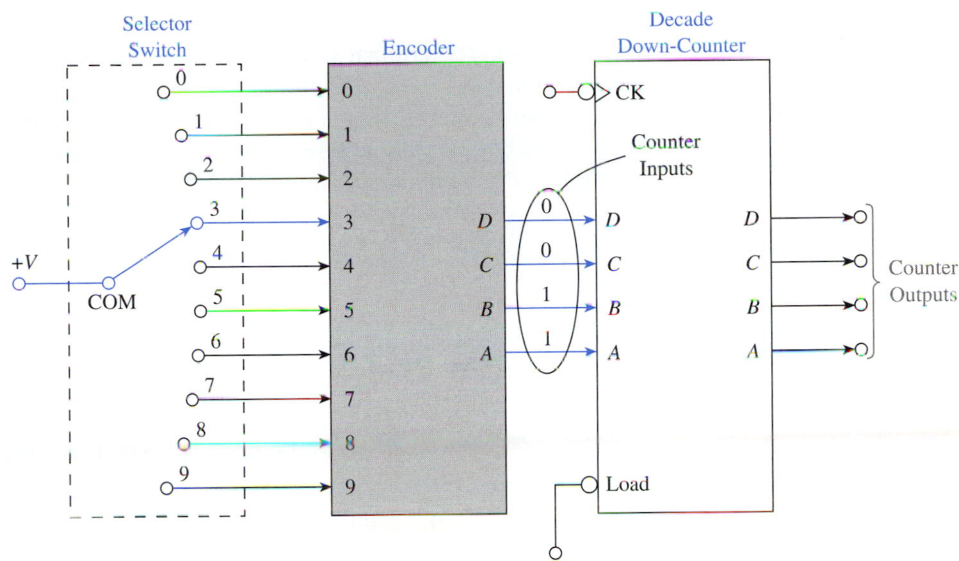

FIGURE 2–18 Combination of a 10-position selector switch, a decimal-to-BCD encoder, and a decade down-counter. This combination is often seen in industrial system control. Outputs *D*, *C*, *B*, and *A* are often symbolized as Q_3, Q_2, Q_1, and Q_0, respectively.

down-counter for presetting the down-counter. When the LOAD terminal of the down-counter goes LO, the SS setting is preset into the counter. This arrangement is illustrated in Fig. 2–18.

Care must be taken with the input lines to an encoder. Some logic families interpret a hanging input to be HI, as we know. If a particular encoder belongs to such a logic family, the simple input method of Fig. 2–18 will not work because all the disconnected inputs will be considered as HI. In such logic families, the manufacturer often gets around this problem by building the encoder to respond to a LO input level instead of a HI input level. That is, whichever one of the ten input lines that is pulled LO is considered to be the desired input number. To simplify our discussion from now on, we will assume that all our packaged encoders respond to a LO input level, and we will use the small circles drawn outside the box to remind ourselves of that fact. Therefore the truth table in Fig. 2–17(b) should be visualized with all the input 0s being 1s and all the input 1s being 0s. The logic level applied to the common terminal of the selector switch then becomes a LO (a ground connection), rather than the HI level $(+V)$ that is indicated in Fig. 2–18.

2-12 ■ TIMERS

In industrial control, it is often necessary to inject a time delay between the occurrence of two events. For example, consider a situation in which two large motors are to be started at about the same time. If both motors are powered from the same supply bus, it is bad practice to switch them both across the lines at the same instant, because large motors draw quite large inrush currents at the instant of starting and continue to draw current far in excess of their normal rated current for several seconds after starting. The motor current drops to its normal rated value only when the motor armature has accelerated up to normal running speed. During the time that the motor is drawing this excessive current, the current capability of the supply bus may be strained. Such a time is no time to require the supply bus to start *another* large motor. Fuses or circuit breakers in the supply lines may open, disconnecting the entire bus. Even if that does not happen, the combination of two starting currents may very well cause excessive voltage drop along the supply lines, resulting in a lower terminal voltage applied to the motors. This prolongs the acceleration period and may cause overheating of the motor windings themselves.

As can be seen from the preceding argument, when two large motors are powered by the same bus, there should be a *time delay* between their starting instants. This can be accomplished with *time-delay relays,* as shown in Fig. 2–19.

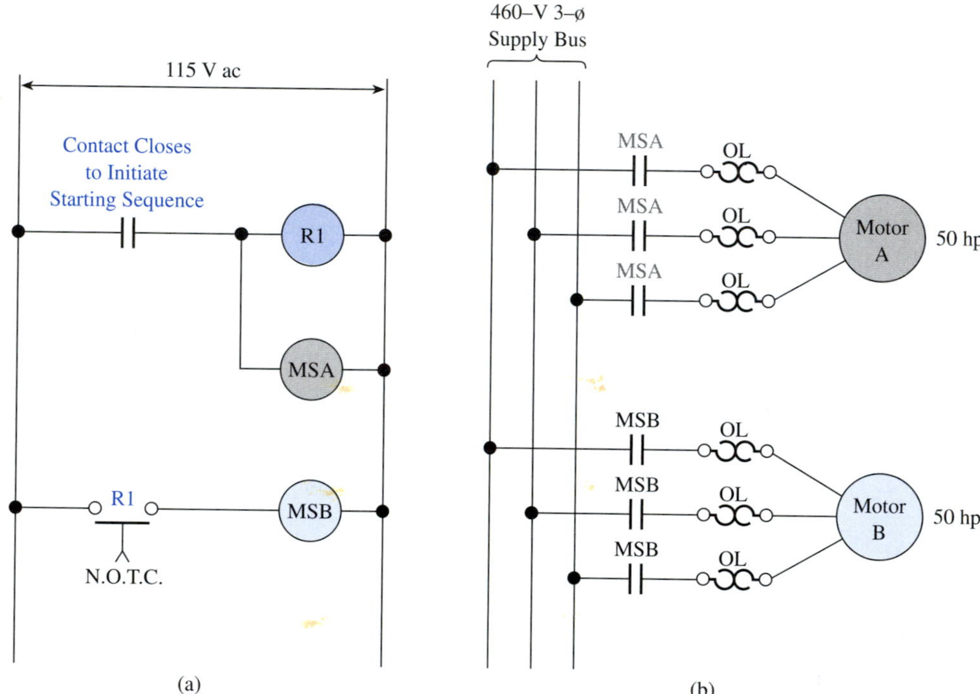

FIGURE 2–19 (a) Simple relay circuit with a time-delay contact. (b) Motor power circuit associated with the relay control circuit in part (a).

2-12-1 Time Delay in Relay Circuits

In Fig. 2–19(b), two large three-phase ac induction motors are driven by a common 460-V supply bus. The contacts that switch the motor windings across the lines are controlled by motor starter A (MSA) and motor starter B (MSB). The control situation calls for motor A and motor B to start at approximately the same time, but it is not necessary that they start at exactly the same time.

When the initiating contact in Fig. 2–19(a) closes, it energizes the coil of MSA and also energizes the coil of relay 1 (R1). The MSA contacts in the high-voltage supply circuit start motor A. Motor A proceeds to draw a large inrush current, perhaps as much as 1000% of rated full-load current. The contact controlled by relay R1 in Fig. 2–19(a) does not close immediately. It delays closing until a certain amount of time has elapsed. By the time it does close to energize MSB, motor A has reached full speed and has relaxed its current demand.

The delayed closing of the relay contact can be accomplished by several methods. The most popular method has been the use of a pneumatic dashpot attached to the moving member of the relay. When the relay coil energizes, a spring exerts a force on the moving member, attempting to close the contact, but a pneumatic (air-filled) dashpot prevents the movement from taking place. As the trapped air bleeds past a needle valve out of the dashpot, the necessary movement occurs and the contacts close. Thus the normally open contacts will not close instantly when the relay is picked. They close after a certain time delay, which is adjustable by adjusting the needle valve. The abbreviation N.O.T.C. in Fig. 2–19(a) stands for "normally open timed closing." The unusual symbol in that figure is the accepted Joint Industry Conference symbol for an N.O.T.C. contact.

Other types of timed contacts are also commonly used. Table 2–1 gives the names, symbols, and brief explanations of each type of contact. The top two types are sometimes called *on-delay* contacts, and relays that have such contacts are called *on-delay* relays, because the delayed action takes place as the relay energizes. The bottom two contacts and the relays that contain them are sometimes described as *off-delay* because the delay action takes place as the relay deenergizes.

TABLE 2–1
The four types of time-delay relay contacts.

	NAME	ABBREVIATION	SYMBOL	DESCRIPTION
Delay Upon Energization (On-Delay)	Normally open timed closing	N.O.T.C.		When the relay energizes, the N.O. contact delays before it closes. When the delay deenergizes, the contact opens instantly.
	Normally closed timed opening	N.C.T.O.	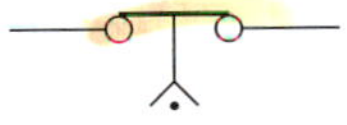	When the relay energizes, the N.C. contact delays before it opens up. When the relay deenergizes, the contact closes instantly.
Delay Upon Deenergization (Off-Delay)	Normally open timed opening	N.O.T.O.	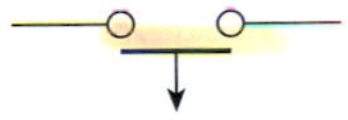	When the relay energizes, the N.O. contact closes instantly. When the relay deenergizes, the contact delays before it returns to the open condition.
	Normally closed timed closing	N.C.T.C.	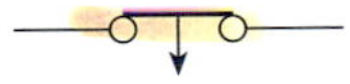	When the relay energizes, the N.C. contact opens instantly. When the relay deenergizes, the contact delays before it returns to the closed condition.

Note that a time-delay contact always delays in one direction only. In the other direction, it acts virtually instantaneously, just like a normal relay contact.

An example of the use of an N.C.T.C. contact is illustrated in Fig. 2–20. In Fig. 2–20(a), a wagon is to be filled with powder from an overhead hopper. The wagon is moved underneath the hopper outlet pipe; then the solenoid is energized to open a valve. When the wagon is sufficiently full, the solenoid is closed, and the wagon is moved away. However, there will be some powder remaining in the fill pipe for a few seconds after the solenoid valve closes. To give this powder a chance to drain out into the wagon, the movement of the wagon is delayed for a few seconds after the valve goes closed. A relay circuit to accomplish this is given in Fig. 2–20(b). When the solenoid deenergizes, RA drops. A little while later, the N.C. contact of RA returns to its closed position; this energizes MSW, which starts a motor to move the wagon away.

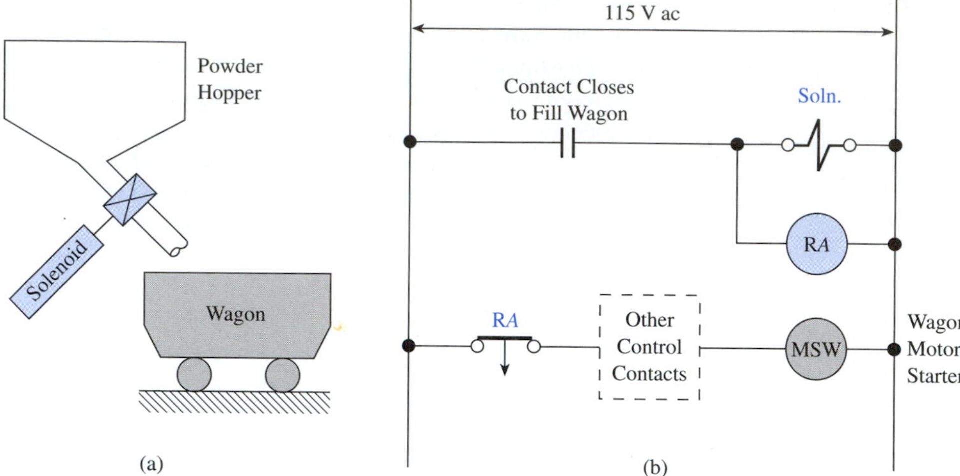

FIGURE 2–20 (a) Physical layout of a wagon being filled from a hopper. (b) Simple relay control circuit, illustrating the use of a time-delay contact to allow the powder to drain out of the supply tube into the wagon before the wagon is moved.

2-12-2 Series Resistor-Capacitor Circuits: Time Constants

The previous examples have shown time delay injected into a control circuit by the action of the contacts of a relay. It is also possible to delay the energization or deenergization of the relay itself. This is usually done by taking advantage of the fact that a certain time must elapse to charge a capacitor through a resistor.

Recall that when a capacitor is charged by a dc source through a series resistor, the charging action is described by the universal time constant curve. Briefly, the rate of charge buildup (voltage buildup) is rapid when the charge on the capacitor is small, but the charging rate decreases as the charge (voltage) on the capacitor gets larger. The behavior of any resistor-capacitor series circuit can be conveniently described in terms of how many *time constants* have elapsed. A time constant for a series *RC* circuit is defined by the formula

$$\tau = RC \tag{2-1}$$

where τ stands for the time constant, measured in seconds; R stands for the resistance in ohms; and C stands for the capacitance, measured in farads, the basic capacitance unit.

Once the idea of a time constant is accepted, the behavior of *all* series *RC* circuits can be described by the universal time constant curve and by certain well-known rules. The most widely used rules are

1. A time equal to five time constants is necessary to charge a capacitor to 99.3% of full supply voltage (99.3% is generally agreed to represent a full charge).
2. In one time constant, a capacitor will charge to 63% of full supply voltage.

The meaning of these rules is graphically illustrated by the universal time constant charging curve in Fig. 2–21.

In our discussion of solid-state logic timers, references will be made to the rules given here for series *RC* circuits. These rules will also prove useful when we discuss other time-related circuit action in later chapters.

2-12-3 Solid-State Timers

In a solid-state control system, the action of time-delay relays is duplicated by solid-state timers. The black-box symbol for a solid-state timer, with its input-output waveform, is shown in Fig. 2–22(a). Also shown in Fig. 2–22(b), (c), and (d) are methods of altering the waveforms to duplicate the actions of the various types of time-delay relay contacts.

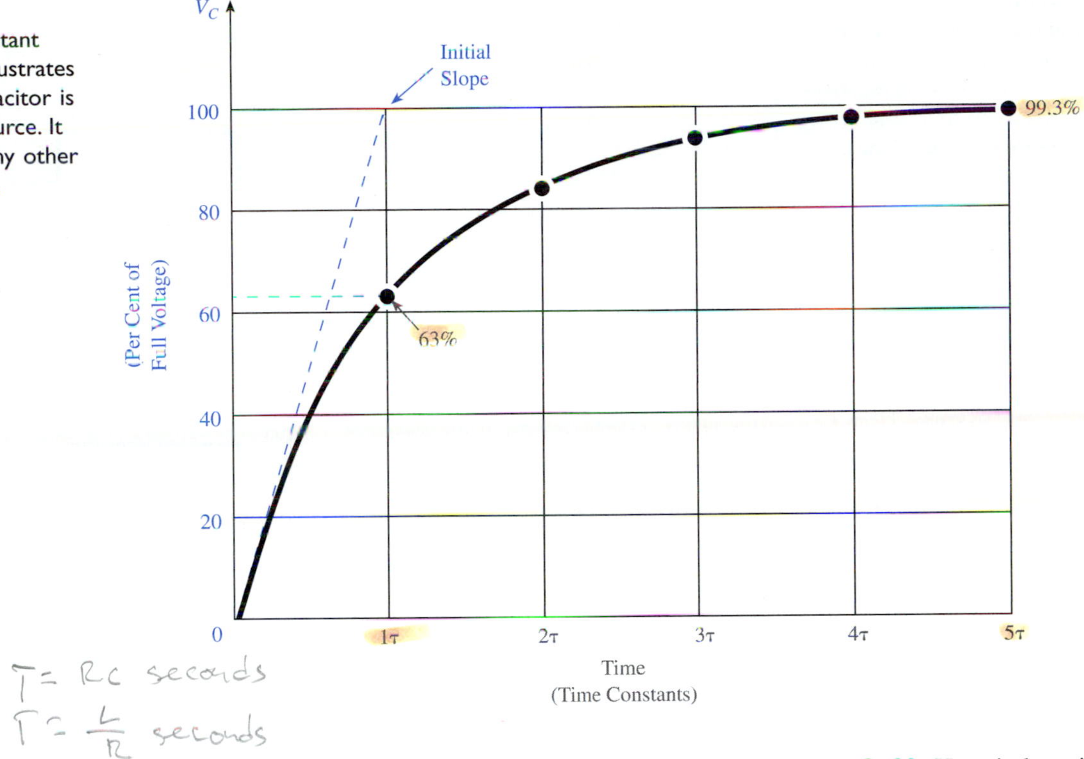

FIGURE 2–21
Universal time constant curve. This curve illustrates in detail how a capacitor is charged by a dc source. It also represents many other natural phenomena.

One method of building a solid-state timer is shown in Fig. 2–23. Here is how it works. When the input is LO, there is no current into the base of T_1, so T_1 is turned OFF. Its collector is near V_s, causing it to turn ON T_2 and T_4 through R_3 and R_{10}. With T_4 ON, its collector is LO, so the output of the overall circuit is virtually 0 V. T_2 comprises a transistor switch which is shorted to ground at this time. It discharges any charge on C_t through D_1. Therefore the voltage at the top of C_1 is virtually 0 V, ensuring that zener diode D_2 is an open circuit. No current can flow into the base of T_3 through R_7 because of the zener diode. No current flows into the base of T_3 through R_6 either, because R_6 is connected to 0 V. Therefore T_3 is OFF, and its collector voltage is near V_s. The T_3 collector delivers current to the base of T_4 through R_4, comprising a second source of base current to hold T_4 ON.

When the input goes HI, it drives the collector of T_1 down to ground. This turns OFF T_2 and also removes one of the sources of base current to T_4. T_4 remains ON because it continues to receive base current through R_9. When T_2 turns OFF, it opens the transistor switch which was preventing timing capacitor C_t from charging up. Therefore C_t begins to charge with a time constant equal to $(R_f + R_t)C_t$. The subscript f on R_f is chosen because it is a fixed resistor. The subscript t on R_t is chosen because it is a timing-adjustment resistor.

As C_t continues to charge, it will eventually reach a voltage which can break down the zener diode. If the reverse breakdown voltage of the zener diode is symbolized by V_z, the voltage necessary to force current through the zener diode is 0.6 V greater than V_z, because any current through zener diode D_2 must go to ground through the base-emitter junction of T_3.

When the top of C_t reaches the necessary voltage, it begins to bleed a little bit of current into T_3 through D_2 and R_7. This causes T_3 to begin carrying a bit of collector current, causing the collector voltage to fall a little bit. This reduces the base current through

FIGURE 2–22
Solid-state timers and their input-output waveforms. This figure shows the equivalence between the four timer configurations and the four types of time-delay relay contacts.

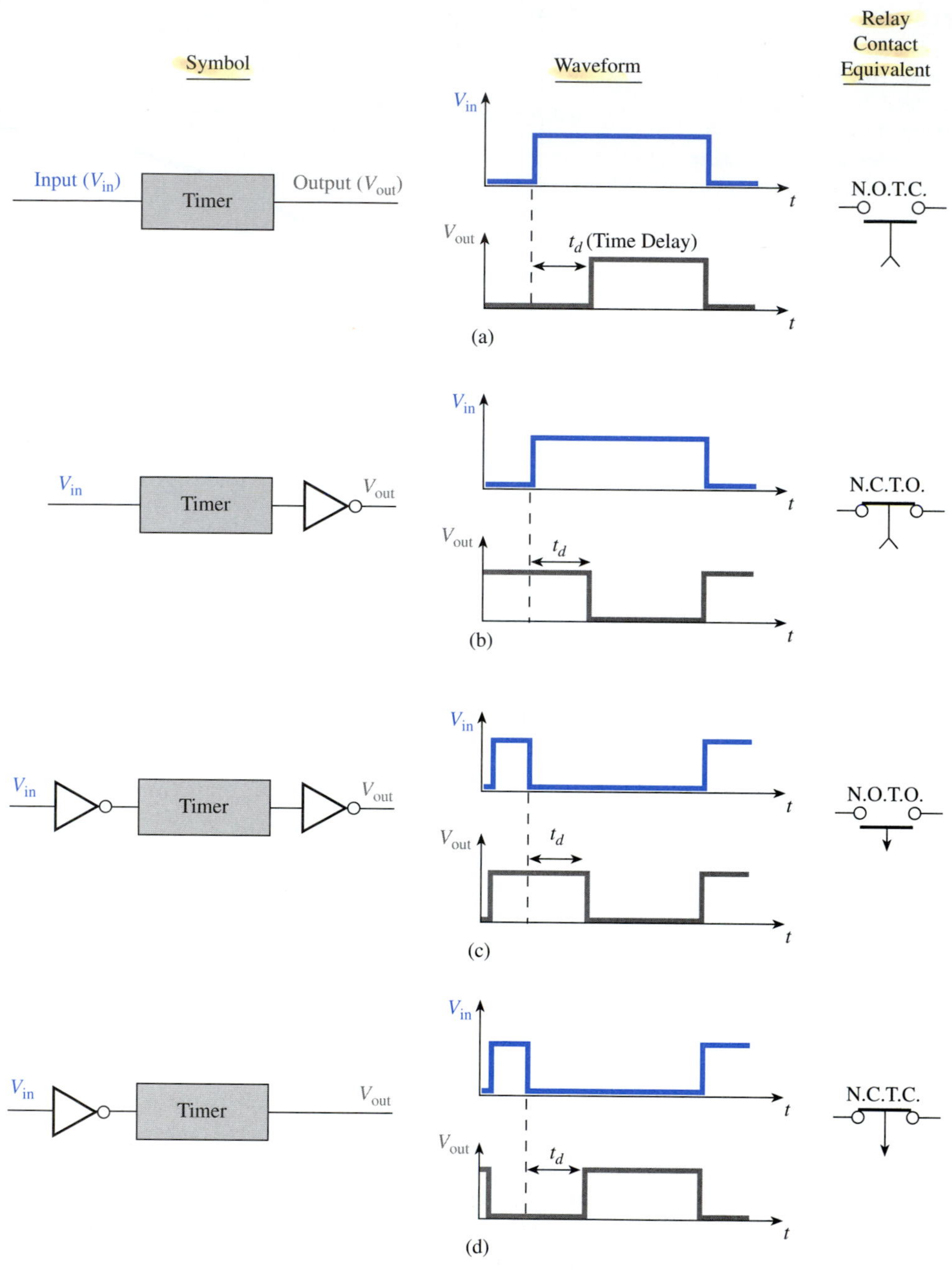

R_9, causing the collector of T_4 to rise a little bit. The rise in T_4 collector voltage reinforces the original base current into T_3, causing it to turn on harder. Therefore the action is self-reinforcing, and it avalanches. Regenerative switching action drives the output level HI in very short time, so the positive-going edge of the output waveform is steep. Thus the output goes HI a certain time after the input goes HI. The elapsed time depends on how long it takes C_t to reach the zener's breakdown point. This time depends on the charging time constant, which is adjusted by potentiometer R_t.

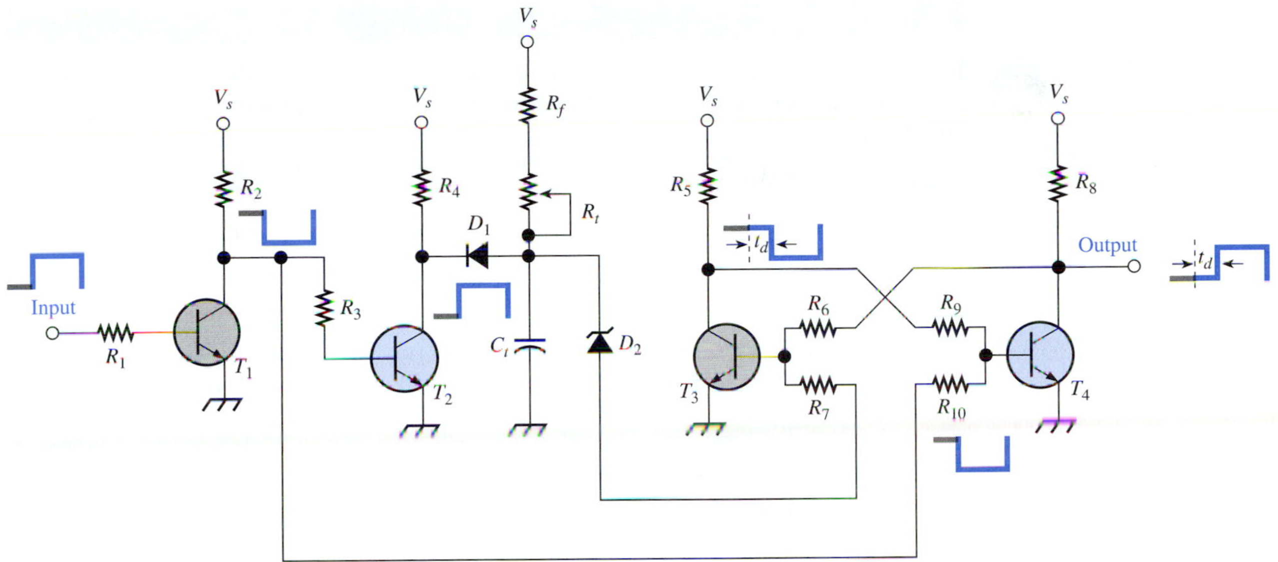

FIGURE 2–23
Schematic diagram showing one way of building a solid-state timer based on an *RC* charging circuit.

When the input returns to LO, T_1 turns OFF, causing its collector voltage to rise. This immediately turns ON T_4 through R_{10}, so the output level immediately goes LO. The T_1 collector also turns ON T_2, closing the switch across the C_t-D_1 combination. When the T_2 switch closes, C_t immediately dumps its positive charge through D_1, through T_2 to ground. This removes the current source for R_7. The current source for R_6 is already removed because the output has gone LO. Therefore T_3 turns OFF, and everything is back to starting condition.

■ EXAMPLE 2-1

In Fig. 2–23, $V_s = 20$ V, zener voltage $V_z = 12$ V, $C_t = 50$ μF, $R_f = 10$ kΩ, and R_t is a 100-kΩ pot. What range of time delays is possible?

Solution. To bleed current through D_2 into the base of T_3, the capacitor voltage must reach 12.6 V. This is given by

$$V_c = V_z + 0.6 \text{ V} = 12.0 + 0.6 = 12.6 \text{ V}$$

12.6 V is exactly 63% of the full capacitor voltage of 20 V. According to rule 2 in Sec. 2-12-2, it takes one time constant for a capacitor to charge to 63% of full voltage. Therefore it will take one time constant to turn ON T_3 after the input goes HI. The time delay is equal to one time constant. τ is given by

$$\tau = (R_f + R_t)C_t$$

The minimum time constant occurs when R_t is dialed completely out. In that case,

$$\tau_{\text{min}} = (10 \text{ k}\Omega + 0)(50\ \mu\text{F}) = 0.5 \text{ s}$$

The maximum time constant occurs when R_t is dialed completely in. In that case,

$$\tau_{\text{max}} = (10 \text{ k}\Omega + 100 \text{ k}\Omega)(50\ \mu\text{F}) = 5.5 \text{ s}$$

The range of possible delay times is therefore from **0.5 to 5.5 s.**

■ EXAMPLE 2-2

In the timer of Fig. 2–23, suppose that a different type of zener diode has been substituted, this one having $V_z = 16$ V. To what value should R_t be adjusted to give a time delay of 8 s?

Solution. In this case, the C_t voltage must reach 16.6 V to turn ON T_3. On a percent basis this is

$$\frac{16.6 \text{ V}}{20.0 \text{ V}} = 83\%$$

of full supply voltage. From the universal time constant curve in Fig. 2–21, it can be seen that about 1.8 time constants are required to charge a capacitor to 83% of full voltage. Therefore

$$(1.8)(\tau) = 8 \text{ s}$$

$$\tau = \frac{8 \text{ s}}{1.8} = 4.44 \text{ s}$$

Since it will require a 4.44-s time constant in order to produce a delay time of 8 s, R_t can be found by

$$\tau = (R_f + R_t)(C_t)$$

$$R_t = \frac{\tau}{C_t} - R_f = \frac{4.44}{50 \times 10^{-6}} - 10 \times 10^3$$

$$= 88.8 \times 10^3 - 10 \times 10^3 = 78 \times 10^3$$

$$= \mathbf{78\ k\Omega}$$

■

2-13 ■ BIN-FILLING SYSTEM USING A DOWN-COUNTER, AN ENCODER, AND TIMERS

The system illustrated in Fig. 2–24 is an efficient method of keeping many material bins full. In this example nine material bins are refilled by a traveling hopper which moves on overhead rails. The traveling hopper is itself loaded from the supply tubes in the home position. The operator then sends it to whichever material bin is in need of replenishing. When it has dumped its material into that bin, the traveling hopper automatically returns home for another load.

The operation must be performed quickly in order to keep the system efficient. Therefore there is a two-speed motor driving the wheels of the traveling hopper. When the hopper leaves the home position with a full load of material, it starts off at the slow speed. After its inertia has been overcome, it changes to high speed. It travels at high speed until it is one position away from its destination. At that time, it switches back to slow speed for the final approach. When it reaches its destination, it stops and opens its dumping doors to dump the material into the bin.

The dumping doors remain open for a certain preset time and then close up. The lightened hopper then returns home at high speed.

The control circuit for this cycle must generate two electrical outputs. One output must occur when the hopper reaches its destination, to cause the motor to stop running. The other output must occur a *fixed distance earlier,* to cause the motor to slow down. This

FIGURE 2–24
(a) Physical layout of the bin-filling system with a traveling hopper. (b) Control circuit for the bin-filling system, illustrating the use of a 10-position selector switch, a decimal-to-BCD encoder, a down-counter, and timers.

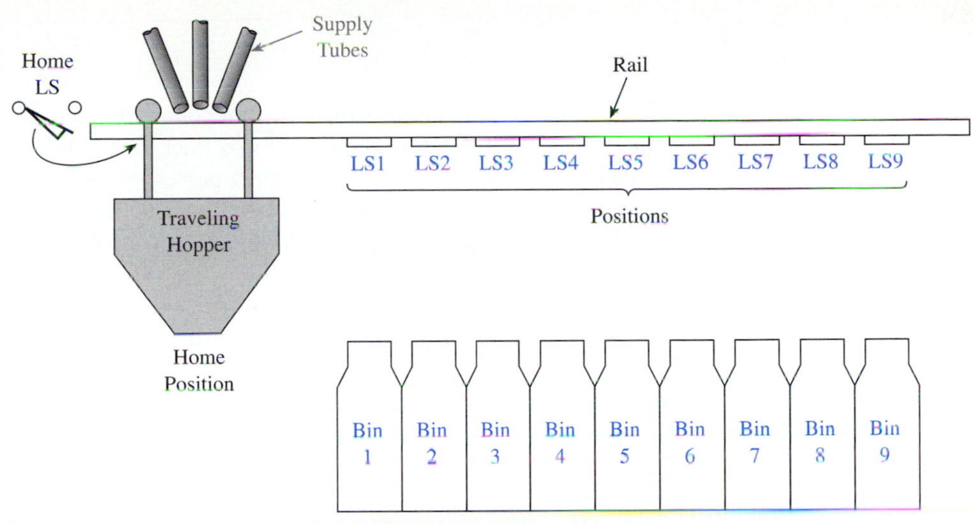

(a)

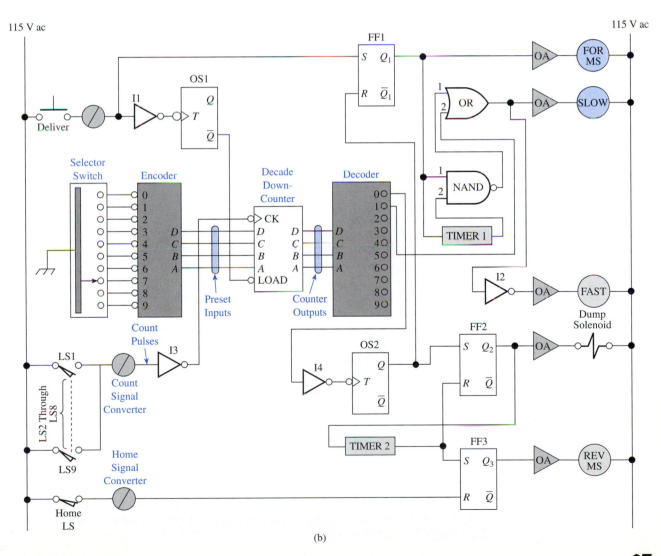

115 V ac
115 V ac
FF1
S Q_1
R $\bar{Q}_1$
OA
FOR MS
OS1
Q
T
$\bar{Q}$
I1
Deliver
OR
SLOW
NAND
TIMER 1
Selector Switch
Encoder
Decade Down-Counter
Decoder
CK
D C B A
LOAD
Preset Inputs
Counter Outputs
0 1 2 3 4 5 6 7 8 9
I2
FAST
Dump Solenoid
Count Pulses
I3
LS1
Count Signal Converter
LS2 Through LS8
LS9
FF2
S Q_2
R $\bar{Q}$
OS2
I4
TIMER 2
FF3
S Q_3
R $\bar{Q}$
REV MS
Home Signal Converter
Home LS

(b)

is the type of situation in which a down-counter is so useful, as mentioned in Sec. 2-11. In Fig. 2–24(b) the down-counter keeps track of the location of the traveling hopper by counting pulses generated as the hopper passes through the nine filling positions. There is an actuating cam mounted on the hopper. As the hopper moves to the right, it actuates a limit switch every time it passes through a new position.

Here is how the control circuit works. The operator receives a signal that a certain bin needs material; the signaling method is not shown. He then causes the hopper to be filled with the proper material from the supply tubes. This mechanism is not shown either. When the hopper is loaded, the operator dials in the destination on the 10-position selector switch. For example, if bin 7 is the one that needs replenishing, he sets the SS to 7. After that, he presses the Deliver pushbutton, and the control circuit takes over.

The output of the Deliver signal converter goes HI, causing FF1 to turn ON. Q_1 goes HI, energizing the forward motor starter coil, labeled FOR MS. This applies power to the motor, which drives the traveling hopper in the forward direction, to the right in Fig. 2–24(a). When the motor is running in the forward direction, its speed depends on which of the two contactors, SLOW or FAST, is energized. When the OR output is HI, the SLOW contactor is energized, and the motor runs slow. When the OR output is LO, the I2 inverter causes the FAST contactor to be energized, and the motor runs fast.

When the Q_1 output initially goes HI, it applies a HI to input 1 of the NAND gate. The output of TIMER 1 remains LO for a certain time setting, so input 2 of the NAND remains LO for a while. The NAND output delivers a HI to the OR gate, causing the OR output to be HI. The motor therefore starts in the slow speed. After a few seconds, whatever time is set on the timer, the output of TIMER 1 goes HI, causing the NAND output to go LO. This removes the HI from input 1 of the OR gate. Input 2 of the OR is probably LO at this time also; we will consider this more carefully later. With both OR inputs LO, the output goes LO, and the motor switches to high speed.

Meanwhile, back at I1, its output goes LO when the operator presses the Deliver button. This causes a negative edge to appear at T of OS1. When OS1 fires, its $\overline{Q}$ output goes LO, applying a LO to the LOAD input of the down-counter. The BCD number which appears at the output of the encoder is thereby loaded into the down-counter. When the output pulse from OS1 goes away, the LOAD input goes back HI, and the down-counter is ready to begin counting when pulses arrive at its CK input. All this takes place in a fraction of a millisecond, so there is absolutely no chance that the down-counter will miss any count pulses generated as the traveling hopper actuates the various Count limit switches, LS1–LS9.

As the motor accelerates the traveling hopper, it approaches LS1. When it hits LS1, the Count signal converter delivers a positive pulse. I3 converts this to a negative edge, and the down-counter counts once. Let us assume that the preset number was 7 (0111). After the traveling hopper contacts LS1, the contents of the down-counter are 0110 or 6.

As the hopper continues to move to the right at high speed, it delivers another negative edge to the down-counter every time it contacts another limit switch. The counter is therefore counting backward toward zero. As the hopper passes through position 5 and actuates LS5, the fifth counter pulse is delivered to the down-counter. This causes its contents to become 0010 (2), since it started at 7 and has received five count pulses. The hopper continues moving to the right at high speed until it hits LS6. The sixth count pulse causes the counter to step into the state $DCBA = 0001$. The decoder immediately recognizes this as the binary code for 1 and accordingly sends its 1 output HI. This HI appears at input 2 of the OR gate and drives the OR output HI. The motor therefore drops into slow speed.

As the hopper proceeds at slow speed, it arrives at its destination above bin 7. It contacts LS7 and delivers the seventh pulse to the down-counter. The counter steps into the state $DCBA$ = 0000. The decoder recognizes this as 0, so it sends its 0 output HI. I4 inverts this HI and fires OS2. The Q output of OS2 goes HI and appears at R of FF1. The flip-flop turns OFF, thereby deenergizing the forward motor starter and stopping the motor. The heavily loaded hopper has low momentum since it was traveling slow, so it does not coast very far. It comes to a halt in position above bin 7.

The Q output from OS2 appears at input S of FF2, turning it ON. Q_2 goes HI, energizing the dump solenoid and starting TIMER 2. The dumping doors of the hopper open and allow the material to fall into bin 7. After a sufficient time has elapsed to get rid of all the material in the hopper, TIMER 2 times out, and its output goes HI. This HI appears at R of FF2 and S of FF3. FF2 turns OFF, closing the dumping doors, and FF3 turns ON. Q_3 goes HI and energizes the reverse motor starter, REV MS. This causes the motor to run in the reverse direction at high speed. Therefore the traveling hopper turns around and heads back toward home. When it contacts the Home LS, the Home signal converter applies a HI to R of FF3. The flip-flop turns OFF and drops out REV MS, so the hopper comes to a stop in the home position.

We said we would return to carefully consider the status of OR input 2 as the hopper is getting under way. We assumed earlier that it would be LO at that time. This assumption is correct as long as the destination is one of the bins 2–9. If the destination is any one of those bins, the number loaded into the down-counter will not be 1 (0001). Therefore as the hopper is starting out, the decoder is not receiving an input of 1, so decoder output 1 will not be HI; it will be LO. Therefore OR input 2 is LO.

However, if bin 1 *is* the destination, then the down-counter was preset to $DCBA$ = 0001, and the decoder output 1 will be HI as the hopper starts out. Under this condition, the motor never does change into fast speed; it makes the entire journey to bin 1 at slow speed. Trace out the circuit behavior and verify this for yourself.

TROUBLESHOOTING ON THE JOB

EXPANDING THE OSCILLATING PLANING CIRCUIT

A prototype of the oscillating planing circuit of Fig. 2–4 has been built and tested. It works perfectly well. Your supervisor now has given you the job of expanding the control circuit so that the operator can select a preset number of planing cycles, up to 999. Making use of the ideas presented in this chapter, draw a detailed schematic diagram of a circuit to accomplish this. Feel free to assume that you can obtain flip-flops that have direct clear and/or direct preset capability and decade counters that can be directly cleared.

The circuit must be designed so that the operator can preset any three-digit decimal number via three 10-position selector switches. The operator presses the momentary Start pushbutton switch. The work table, which waits in the far left position (Left LS actuated) immediately begins a run to the right. The operator can then simply walk away, giving no further human supervision. After the preset number of planing cycles have been accomplished, the control circuit automatically stops the motor with the planing table in the far left position.

Make sure that your circuit design avoids racing problems. For example, do not depend on a lucky outcome in a race between (1) delivering a clock edge to FF1, and (2) disabling the AND gate in Fig. 2–4. Any circuit that is susceptible to the outcome of a circuit race is not an acceptable design.

SUMMARY

- A clocked flip-flop responds to the logic signals that are present on its synchronous input terminals (*R* and *S* or *J* and *K*) at the moment that the CK terminal receives an active transition, or edge.
- A shift register can be used to track a binary characteristic of a part as that part moves from one zone to another in an industrial system.
- The combination of a decade counter, 1-of-10 decoder, and 10-position selector switch is useful for detecting when a certain preset number of events has occurred.
- A one-shot is useful for delivering a fixed-duration pulse when a triggering event occurs.
- The combination of a 10-position selector switch, a decimal-to-BCD encoder, and a down-counter is useful for instructing a control circuit how many events to permit.
- Timers are used to set a fixed amount of time duration between an initiating event and a resulting event.

FORMULA

$\tau = RC$ for a resistor-capacitor series circuit **(Equation 2-1)**

QUESTIONS AND PROBLEMS

Section 2-1

1. Explain why a flip-flop will maintain its present state forever unless ordered to change states by an outside signal.
2. In Fig. 2–2, explain why the B welder contactor cannot be energized if the A welder contactor is already energized.
3. Carefully explain the difference between an *RS* flip-flop and a *clocked RS* flip-flop.
4. Explain the difference between a positive edge-triggered flip-flop and a negative edge-triggered flip-flop.
5. What input combination is not legal for a clocked *RS* flip-flop? Why is it illegal?

Section 2-2

6. In Fig. 2–4(b), is it ever possible for both *S* and *R* of FF1 to be HI at the same time? Why?
7. Explain why the oscillating table of Fig. 2–4 always stops in the extreme left position, never in the extreme right position.
8. Make the necessary additions to Fig. 2–4 to allow the operator to get the oscillations started after he has installed a workpiece.

Section 2-3

9. What is the chief difference between a *JK* flip-flop and a clocked *RS* flip-flop?
10. For a flip-flop, does a direct clear signal (CL) override a signal to turn ON from the clocked inputs?

Section 2-4

11. If the shift register of Fig. 2–6 starts in the state 0000 and two shift pulses are applied with the FF1 input terminal HI, what is the new state of the shift register?
12. Show how the shift register of Fig. 2–6 could be made to *circulate* its information; that is, the information in FF4 is not lost when a shift pulse arrives, but is recycled.
13. In Fig. 2–7 why does FF1 always contain a 0 unless the inspector presets a 1 into it?
14. If it was desired to build a 10-bit shift register, how many packaged 4-bit shift registers are needed? Draw a schematic showing all the interconnections between the packages.

Section 2-6

15. Explain in words the action of a BCD-to-decimal decoder.
16. Referring to Fig. 2–11(a), what decoder output line will go HI if the input situation is $\overline{D}C\overline{B}A = 1111$. Repeat for $D\overline{C}\overline{B}A = 1111$.

Section 2-7

Questions 17–19 refer to the palletizing system of Fig. 2–12.

17. What assures that the decade counters start counting from zero when a new pallet is begun?
18. Suppose the cartons were being loaded six layers high, eight cartons to a layer. What would be the settings of the selector switches? What would be the BCD output of the two counters which would cause the diverter to change positions?
19. Repeat Question 18 for cartons loaded 12 to a layer, 7 layers high.

Section 2-8

20. In words, explain what a one-shot does.
21. One-shots are sometimes called *delay elements*. Why do you think they are called this?
22. What means are utilized to adjust the firing time of a prepackaged one-shot?
23. Suppose we have two one-shots, one of them retriggerable and the other nonretriggerable, both with a firing time of 10 ms. A fast pulse is applied to both trigger inputs at the same instant. Seven milliseconds later, another fast pulse is applied to both trigger inputs. Make a sketch showing the output waveforms of both one-shots.

Section 2-12

24. In very general terms, what is the purpose of an industrial timer?
25. In words, explain the behavior of each of the four types of time-delay relay contacts; N.O.T.C., N.O.T.O., N.C.T.C., and N.C.T.O.
26. What are the standard symbols for each of the four contacts in Question 25?
27. Referring to Fig. 2–23 suppose $(R_f + R_t)C_t = \tau = 0.2$ s. If $V_s = 30$ V and $V_z = 15$ V, what is the time delay of the timer? Use the universal time constant curve of Fig. 2–21.
28. Repeat Question 27 for $V_z = 24$ V.

29. Explain each of the time-delay circuits shown in Fig. 2–22. That is, explain why each circuit has the V_{in}-V_{out} waveform shown.
30. Why is it necessary to have D_1 in Fig. 2–23?

Section 2-13

Questions 31–36 refer to the bin-filling system of Fig. 2–24.

31. The common terminal of the 10-position selector switch is tied to ground. Explain why this is correct (as opposed to tying it to the logic dc supply).
32. Why is the LOAD terminal of the down-counter connected to the $\overline{Q}$ output of OS1 instead of the Q output?
33. Describe the entire process of presetting the down-counter to the proper number.
34. What causes FOR MS to deenergize? Explain the process by which the circuit deenergizes this motor starter.
35. Why is it necessary to slow down the traveling hopper before it arrives at its destination?
36. Explain how the circuit slows down the hopper before it arrives at the destination.

CHAPTER

3 PROGRAMMABLE LOGIC CONTROLLERS

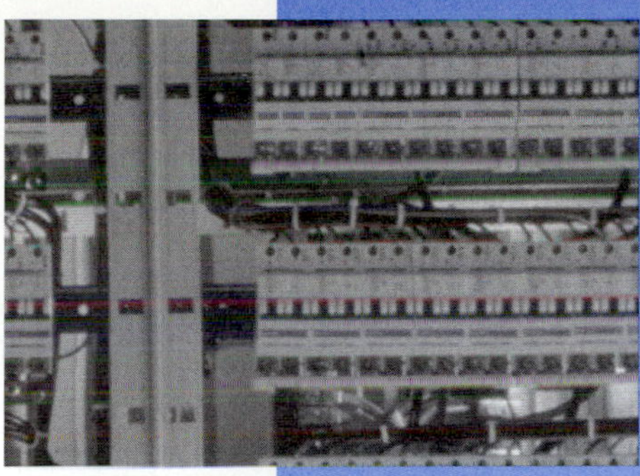

The transistor-based logic systems described in Chapters 1 and 2 possess all the usual advantages of solid-state electronic circuits: they are safe, reliable, small, fast, and cheap. Their only fault, from an industrial user's viewpoint, is that they are not easily modifiable. If modifications need to be made, we must change the actual wire or copper-track connections between the logic devices, or change the devices themselves. Such *hardware* changes are undesirable because they are difficult and time-consuming.

Today, a fundamentally different approach to the construction of industrial logic systems has become popular. In this new approach, the system's decision making is carried out by coded instructions which are stored in a memory chip and executed by a microprocessor. Now if the control system must be modified, only the coded instructions need to be changed. Such changes are called *software* changes, and they are quickly and easily implemented just by typing on a keyboard. This new approach is sometimes referred to as *flexible* automation, to distinguish it from standard *dedicated* automation.

When this flexible approach is used, the entire sequence of coded instructions that controls the system's performance is called a *program*. Therefore we refer to such systems as programmable systems. If all the necessary control components are assembled and sold as a complete unit, which is the most common practice, the complete unit is known as a *programmable logic controller*. That is the subject of this chapter.

OBJECTIVES

After you have completed this chapter, you will be able to:

1. Contrast the software logic of a programmable logic controller to the logic of a hard-wired circuit.
2. Name the three parts of a programmable logic controller and describe each part's function.
3. Define the following terms associated with the input/output function of a programmable logic controller: I/O rack, I/O group, slot, module, and terminal.
4. List the sequence of events in a programmable logic controller's scan cycle and cite approximate time durations for each event.
5. Define the following terms associated with the processor function of a programmable logic controller: user-program, instruction-rung, input image file, output image file, and central processing unit.

6. Give a detailed description of the procedure by which the central processing unit executes one instruction-rung.
7. Explain the operation of the three basic relay-type instructions that are available with a programmable logic controller, namely: examine-On, examine-Off, and output-energize.
8. Discuss the difference between an output-energize instruction that affects a load device and an output-energize instruction that is used solely for internal logic.
9. Describe the following capabilities of a programmable logic controller: timing, counting, value comparison, and arithmetic.
10. Discuss each of the three operating modes of a programmable logic controller: PROGRAM, TEST, and RUN.
11. Given a ladder-logic representation of a user-program, enter that program into memory by typing on the programming terminal's keyboard.
12. Use the program-editing functions that are on the programming terminal's keyboard.
13. Given a memory map of the processor and the arrangement of the input/output section, choose appropriate addresses for input devices, output devices, internal-logic instructions, timers, counters, and data files.
14. Write the program instructions for bringing a measured analog value into the user-program by reading an Analog Input Module.
15. Write the program instructions for performing calculations with a measured analog value and for making program decisions based on the calculated results.

3-1 ■ THE PARTS OF A PROGRAMMABLE LOGIC CONTROLLER

Programmable logic controllers (PLCs) can be considered to have three parts: the input/output section, the processor, and the programming device, or terminal. We'll take each part in turn.

3-1-1 Input/Output Section

The input/output (I/O) section of a programmable logic controller handles the job of interfacing high-power industrial devices to the low-power electronic circuitry that stores and executes the control program. The control program will be called by us the *user-program.*

The I/O section contains input and output modules. Think of each input module as a printed circuit board containing 16 signal converters, discussed in Sec. 1-7. Each one of the module's 16 terminals receives a high-power signal (switched 120 V ac, usually) from an input device and converts it into a low-power digital signal compatible with the electronic circuitry of the processor. All modern PLC input modules use optical signal converters to accomplish electrically isolated coupling between the input circuits and the processor electronics. Optical couplers will be described in Chapter 10.

Each input switching device is wired to a particular input terminal on a module's terminal strip, as illustrated in Fig. 3–1(a). Thus, if the topmost pushbutton switch is closed, 120 V ac appears on input terminal ØØ of that rack.* Input signal converter ØØ,

*The digit zero is commonly written with a slash when a number system other than the decimal system is being used. For us, some PLC numeric values will be expressed in the octal (base 8) number system, so zeros will have slashes, as shown here.

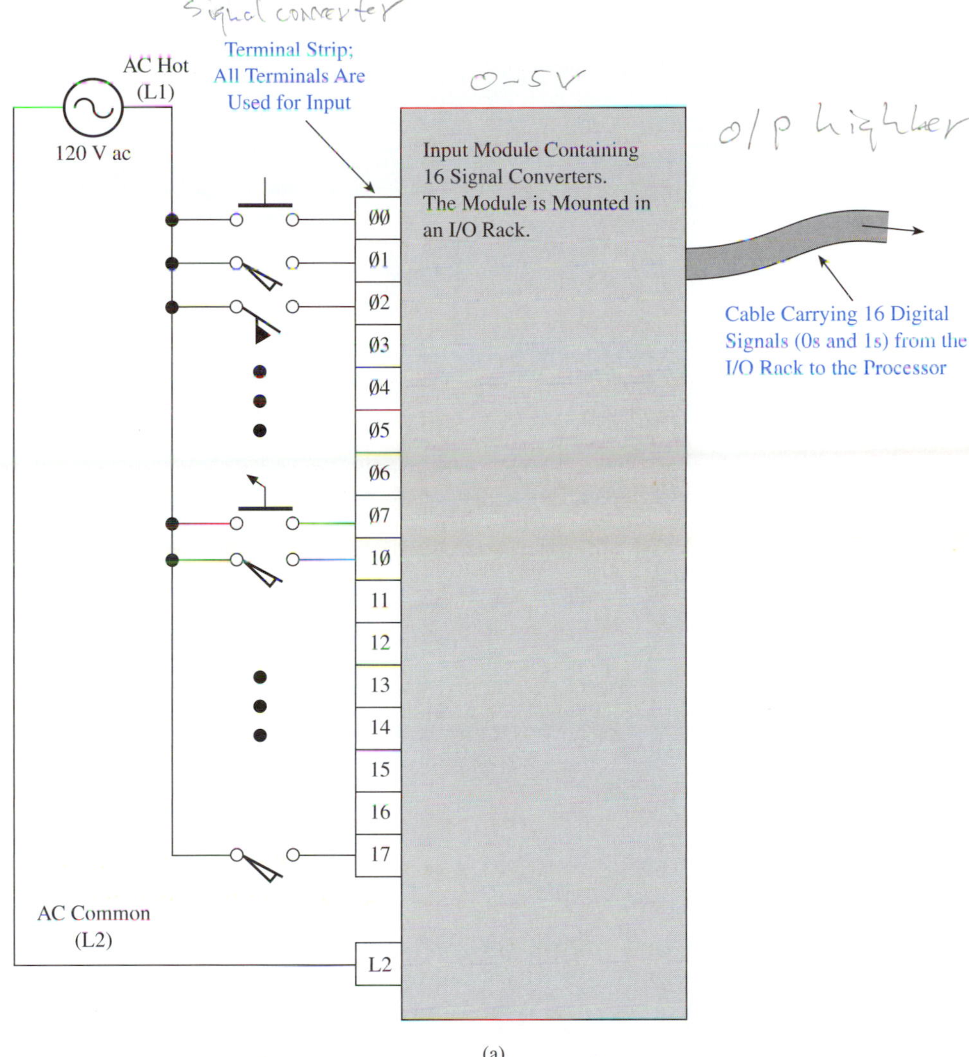

FIGURE 3–1
An I/O rack is a mechanical enclosure with slots for holding printed circuit boards (modules) that contain either 16 input signal converters or 16 output amplifiers. (a) One slot of the I/O rack holds an input module designed for 120-V ac input. Note that terminal numbering starts with Ø, not 1. The numbers are octal, not decimal. (b) Another slot of the I/O rack holds an output module. This output module contains 16 optically coupled output amplifiers designed to drive 120-V ac loads.

which is contained in the module, converts this ac voltage to a digital 1 and sends it to the processor via the connector cable. Conversely, if the topmost pushbutton switch is open, no ac voltage appears on input terminal ØØ. Input signal converter ØØ will respond to this condition by sending a digital 0 to the processor. The other 15 input terminal converters behave identically.

Think of each output module as a printed circuit board containing 16 output amplifiers, discussed in Sec. 1-8. Each output amplifier receives a low-power digital signal from the processor and converts it into a high-power signal capable of driving an industrial load. A modern PLC output module has optically isolated amplifiers which use a triac as the series-connected load-controlling device. Triacs will be discussed in Chapter 6.

Each output load device is wired to a particular terminal on an output module's terminal strip, as illustrated in Fig.3–1(b). Thus, for example, if output amplifier Ø2 receives a digital 1 from the processor, it responds to that digital 1 by applying 120 V ac to output module terminal Ø2, thereby illuminating the lamp. Conversely, if the processor sends a digital 0 to output amplifier Ø2, the amplifier applies no power to module terminal Ø2, and the lamp is extinguished.

FIGURE 3–1
(continued)

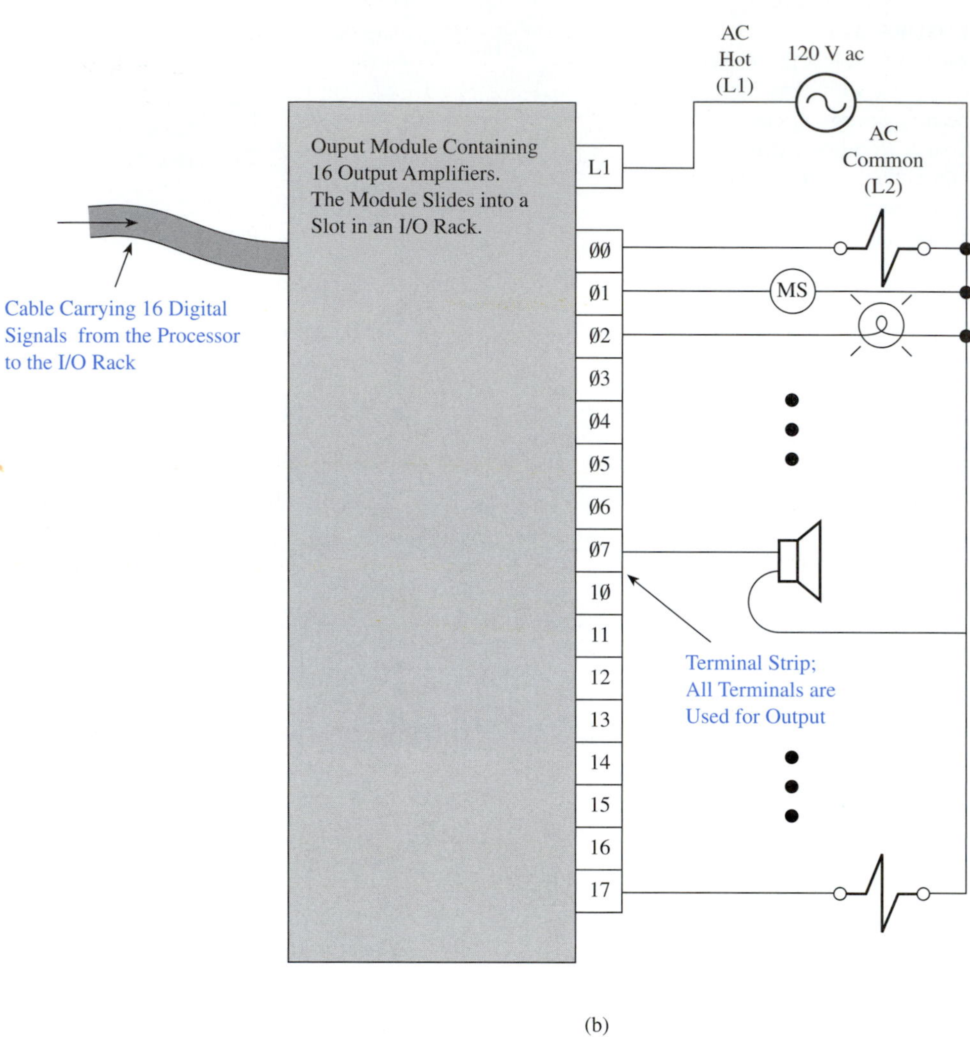

(b)

Besides 120 V ac, I/O modules also are available for interfacing to other industrial levels, including 5 V dc (TTL devices), 24 V dc, etc.

3-1-2 The Processor

The *processor* of a PLC holds and executes the user program. To carry out this job, the processor must store the most up-to-date input and output conditions.

Input image file. The input conditions are stored in the *input image file,* which is a portion of the processor's memory.* That is, every single input module terminal in the I/O section has assigned to it a particular location within the input image file. That particular location is dedicated solely to the task of keeping track of the latest condition of

*This memory is the read-write type of memory, popularly called random access memory, or RAM.

its input terminal. As mentioned in Sec. 3-1-1, if the input terminal has 120-V ac power fed to it by its input device, the location within the input image file contains a binary 1 (HI); if the input module has no 120-V ac power fed to it, the location contains a binary 0 (LO).

The processor needs to know the latest input conditions because the user-program instructions are contingent upon those conditions. In other words, an individual instruction may have one outcome if a particular input is HI and a different outcome if that input is LO.

Output image file. The output conditions are stored in the *output image file,* which is another portion of the processor's memory. The output image file bears the same relation to the output terminals of the I/O section that the input image file bears to the input terminals. That is, every single output terminal has assigned to it a particular memory location within the output image file. That particular location is dedicated solely to the task of keeping track of the latest condition of its output terminal.

Of course, the output situation differs from the input situation with regard to the direction of information flow. In the output situation, the information flow is from the output image file to the output module, while in the input situation the information flow is from the input module to the input image file. These relationships are portrayed by the processor block diagram of Fig. 3–2.

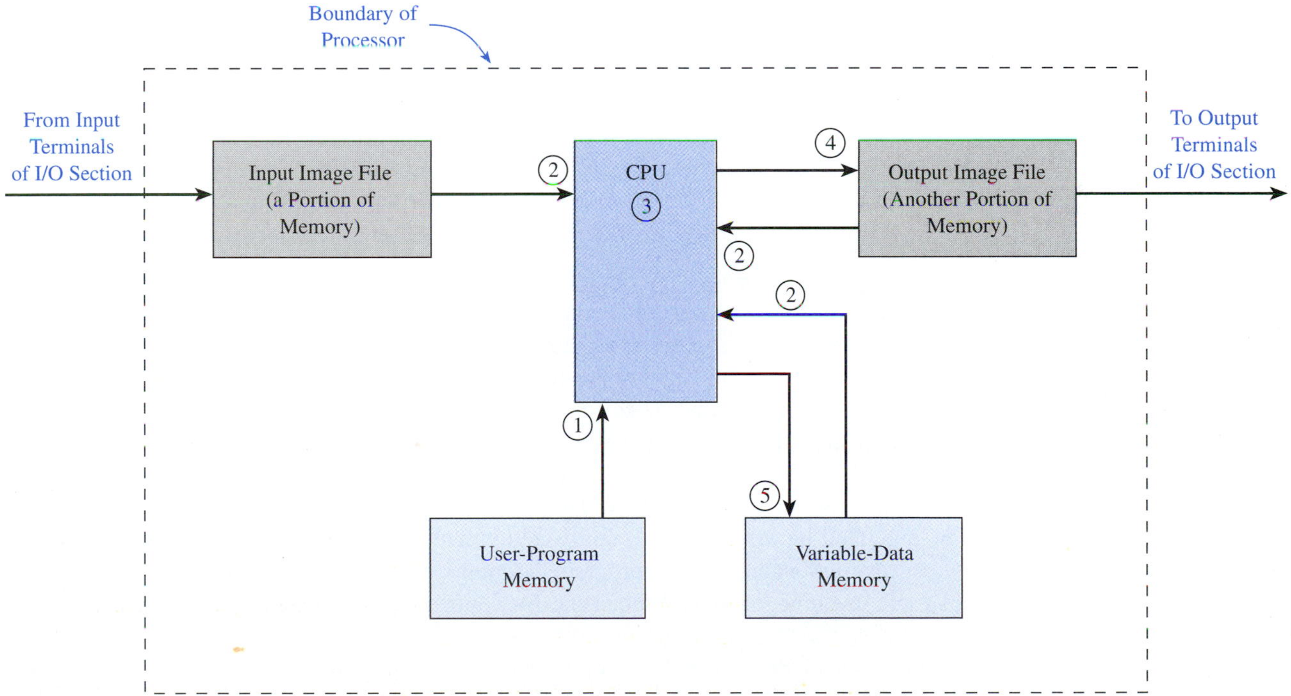

FIGURE 3–2
The processor. The duties of the processor are ① to fetch instructions from user-program memory into the CPU, ② to fetch I/O information from the image files and numerical data from variable-data memory, and ③ to execute the instructions. Execution of the instructions involves ④ making logical decisions regarding the proper states of the outputs and causing those proper states to appear in the output image file, and ⑤ calculating the values of variable data and storing those values in variable-data memory.

The locations within the input and output image files are identified by *addresses*. Each location has its own unique address. For example, a particular memory location within the input image file might have the address I:ØØ1/Ø6, and a particular location within the output image file might be addressed as O:ØØ3/17. The various PLC manufacturers all have their own methods for assigning addresses. In Sec. 3-2 we will study the addressing method used by one prominent manufacturer, the Allen-Bradley Company.

Central processing unit. The subsection of the processor that actually performs the program execution will be called the *central processing unit* (CPU) in this book. The CPU subsection is pointed out in the processor block diagram of Fig. 3–2.

As the CPU executes the user-program, it is continually and immediately updating the output image file. In other words, if an instruction execution calls for a change at one of the output image file locations, that change is effected immediately, before the processor proceeds to the next instruction. This immediate updating is necessary because output conditions often affect later instructions in the program.

For instance, suppose a certain instruction causes output address O:Ø14/17 to change from LO to HI. A later instruction may say, in effect, "If input I:Ø13/Ø6 and output O:Ø14/17 are both HI, then bring output O:Ø15/Ø2 HI." For this later instruction to be carried out correctly, the processor must recognize that output O:Ø14/17 is presently HI by virtue of the earlier instruction.

Thus, we see that the output image file has a dual nature: Its first function is to receive immediate information from the CPU and pass it on (a little later) to the output modules of the I/O section. Second, it also must be capable of passing output information "backward" to the CPU, when the user-program instruction that the CPU is working on calls for an item of output information.

The input image file does not have this dual nature. Its single mission is to acquire information from the input terminals and pass that information "forward" to the CPU when the instruction that the CPU is working on calls for an item of input information.

The information-flow arrows in Fig. 3–2 illustrate these ideas.

User-program memory. A particular portion of the processor's memory is used for storing the user-program instructions. We will use the name *user-program memory* to refer to this processor subsection, as shown in Fig. 3–2.

Before a PLC can begin controlling an industrial system, a human user must enter the coded instructions that make up the user program. This procedure, called *programming* the PLC, will be demonstrated in Sec. 3-1-3.

As the user enters instructions, they are automatically stored at sequential locations within the user-program memory. This sequential placement of program instructions is self-regulated by the PLC, with no discretion needed by the human user. The total number of instructions in the user-program can range from a half-dozen or so, for controlling a simple machine, to several thousand, for controlling a complex machine or process.

After the programming procedure is complete, the human user manually switches the PLC out of PROGRAM mode into RUN mode, which causes the CPU to start executing the program from beginning to end repeatedly.

For organizing and editing programs, we find it convenient to group instructions into instruction-rungs, often just called *rungs*. The word *rung* is taken from the fact that these groups of instructions resemble the rungs of a ladder when the user-program is represented in *ladder-logic* format. Figure 1–5, showing the relay-logic circuit for a conveyor/classifying system, is an example of ladder-logic format.

Let us focus our attention on lines 1 and 2 of Fig 1–5. We will use those lines as a concrete example for demonstrating the correspondence between ladder-logic user-program format and a hardware relay-logic circuit. Those two lines are reproduced in Fig. 3–3(a).

Figure 3–3(b) is a ladder-logic representation of an instruction-rung that can duplicate the action of the hard-wired relay circuit. Think of this instruction-rung as a portion of the complete user-program that is stored in the user-program memory section of the processor. The rung is drawn unannotated in Fig. 3–3(b) to show its actual appearance as

FIGURE 3–3 Correspondence of the ladder-logic representation of a PLC program to a relay ladder-logic diagram. (a) Relay ladder-logic diagram. (b) Corresponding PLC ladder-logic program representation, shown unannotated as it would appear on a CRT screen. (c) PLC ladder-logic diagram, shown annotated.

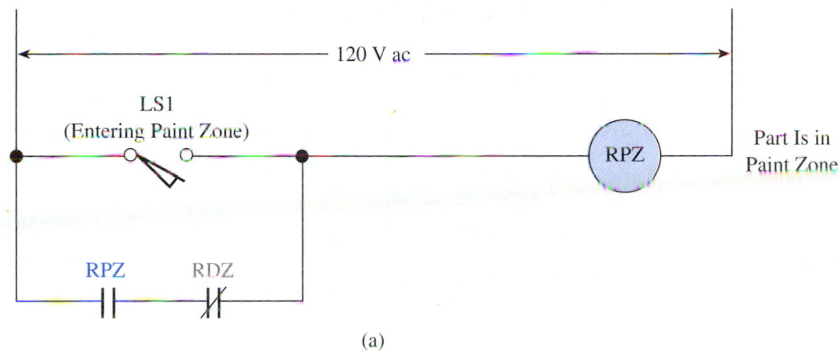

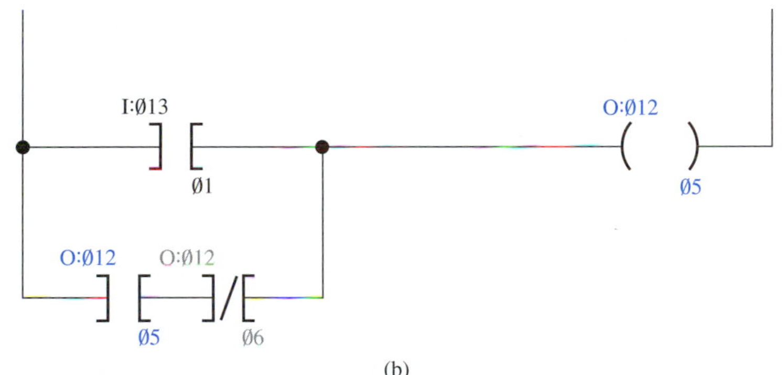

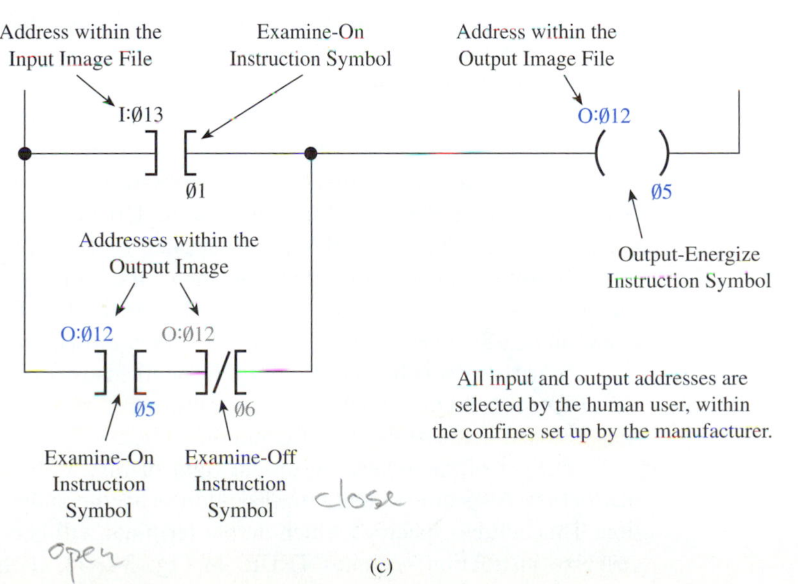

it would be displayed on the CRT screen of the programming terminal, which is usually just a desktop or laptop PC running the programming software published by the PLC manufacturer.

The same instruction-rung is presented in annotated ladder-logic format in Fig. 3–3(c). As that figure shows, the rung consists of four instructions, represented by the three contact-like symbols on the left and the one coil-like symbol on the right. Each one of these symbols corresponds to an identical symbol shown on the CRT screen, along with the F-number of the keyboard's function key that enters that particular symbol. For example, the] [symbol is shown near the bottom of the CRT screen directly above an image of the [F1] key. By pressing the [F1] key, you cause the] [symbol to appear in the instruction-rung's screen display. At the same time, you cause the coded instruction that the symbol represents to be entered into the user-program memory, and likewise for the]/[symbol by pressing the [F2] key, and the -( )- symbol by pressing the [F3] key.

The symbol in the upper left of Fig. 3–3(c) stands for an *examine-On instruction*. An examine-On instruction works as follows: If the input terminal associated with the instruction has 120-V power applied to it, then the overall instruction-rung regards the instruction as producing logic continuity, like a closed electrical contact. However, if the input module has no 120-V power applied to it, the overall instruction-rung regards the instruction as producing logic discontinuity, like an open electrical contact.

The upper-left examine-On instruction in Fig. 3–3(c) has the address I:Ø13/Ø1 displayed with it. This address specifies which input terminal is associated with the instruction. That is, it specifies that input terminal Ø13/Ø1, which is wired to address I:Ø13/Ø1 in the input image file, will be examined for the presence or absence of power when this instruction is executed. Of course, to duplicate the performance of the relay circuit of Fig. 3–3(a), we must physically wire limit switch 1 (LS1) to input terminal Ø13/Ø1 of the module in the I/O rack.

People who work with PLCs often speak of an examine-On instruction as a "normally open instruction," because it behaves much like a normally open electrical contact.

The foregoing description of an examine-On instruction is given in the context of a system input. However, examine-On instructions can also refer to output terminals (output image file addresses); this is the case at the lower left of Fig. 3–3(c). The address O:Ø12/Ø5 appearing with that instruction symbol refers to a location in the *output* image file (in the Allen-Bradley addressing scheme). Therefore that instruction produces logic continuity if output terminal Ø12/Ø5 is powered up (output image file address O:Ø12/Ø5 contains a logic 1) but logic discontinuity if output terminal Ø12/Ø5 is not powered up (output image file address O:Ø12/Ø5 contains a logic 0).

An examine-Off instruction is represented by the symbol containing the slash mark at the lower right of Fig. 3–3(c). An examine-Off instruction works as follows: If the associated I/O terminal has 120-V power applied to it, then the instruction contributes logic discontinuity to the overall instruction rung, like an open electrical contact. However, if the I/O terminal has no 120-V power applied to it, then the instruction contributes logic continuity to the overall instruction-rung, like a closed electrical contact.

Note that the behavior of an examine-Off instruction is the opposite of an examine-On instruction. As you would expect, examine-Off instructions are sometimes called "normally closed instructions" in PLC jargon.

The set of parentheses at the far right of Fig. 3–3(c) represents an output-energize instruction. A specific output terminal address accompanies every output-energize instruction. This address specifies which output terminal will become powered up if the output-energize instruction becomes TRUE. In Fig. 3–3(c), if the output-energize instruction becomes TRUE, execution will cause a digital 1 to be stored at address O:Ø12/Ø5 in the output image file, which in turn will cause output terminal O:Ø12/Ø5 to be powered up.

For an output-energize instruction to become TRUE, the examine-On and examine-Off instructions to its left must produce overall logic continuity through the rung (between the left edge of the rung and the opening parenthesis of the output-energize instruction). If the examine-On and the examine-Off instructions fail to produce such logic continuity, we say that the output-energize instruction is FALSE, or that the rung conditions are FALSE. In this event, execution causes a digital 0 to be stored at the specified address in the output image file, which results in removal of power from the associated output terminal in the I/O rack.

Notice the similarity of these ideas to a hard-wired relay-logic circuit; the output-energize instruction corresponds to a relay coil, and the examine-On and examine-Off instructions correspond to normally open and normally closed contacts, respectively.

We can summarize the behavior of the Fig. 3–3(b) and (c) instruction-rung as follows: Output terminal Ø12/Ø5 will be powered up if either one of two conditions is satisfied:

1. Input terminal Ø13/Ø1 is powered up, or
2. Output terminal Ø12/Ø5 is *already* powered up and output terminal Ø12/Ø6 is not powered up.

This instruction rung thus duplicates the behavior of the Fig. 3–3(a) relay circuit, which calls for relay RPZ to be energized if either one of two conditions is satisfied:

1. LS1 is actuated, or
2. RPZ is already energized and relay RDZ is deenergized.

Now that we have demonstrated the equivalence of an instruction-rung in a PLC user-program to a hard-wired relay-logic circuit, we can put forth an initial definition of an instruction-rung. This definition is quite restricted, but it will serve to help us gain an understanding of the execution process for a PLC user-program. Our definition goes like this:

> *An instruction-rung is a group of instructions that affects a single output terminal, based on the statuses of certain input terminals and output terminals.*

In this definition, the phrase "affects a single output terminal" refers to the fact that the rung contains a single output-energize instruction, as in Fig. 3–3(b) and (c). The phrase "based on the statuses of certain input terminals and output terminals" refers to the collection of examine-On and examine-Off instructions which produce either TRUE rung conditions (logic continuity) or FALSE rung conditions (logic discontinuity).

To execute the user-program, the CPU handles one instruction-rung at a time. Figure 3–4 depicts the events involved in the execution of one instruction-rung.

Part (a) gives a block diagram view of the processor during an instruction-rung execution, and part (b) is a flowchart of the execution process. The circled numbers show event-correspondence between the two diagrams. We will refer to both diagrams to explain the execution process of an instruction-rung.

1. The CPU, which always keeps track of its precise location in the user-program, fetches the next sequential instruction from the user-program memory. This is illustrated in part (a) of Fig. 3–4 by the arrow indicating transfer from user-program memory to the CPU.
2. The instruction that the CPU has just obtained is bound to be an examine-type instruction because our definition of an instruction-rung calls for each rung to begin

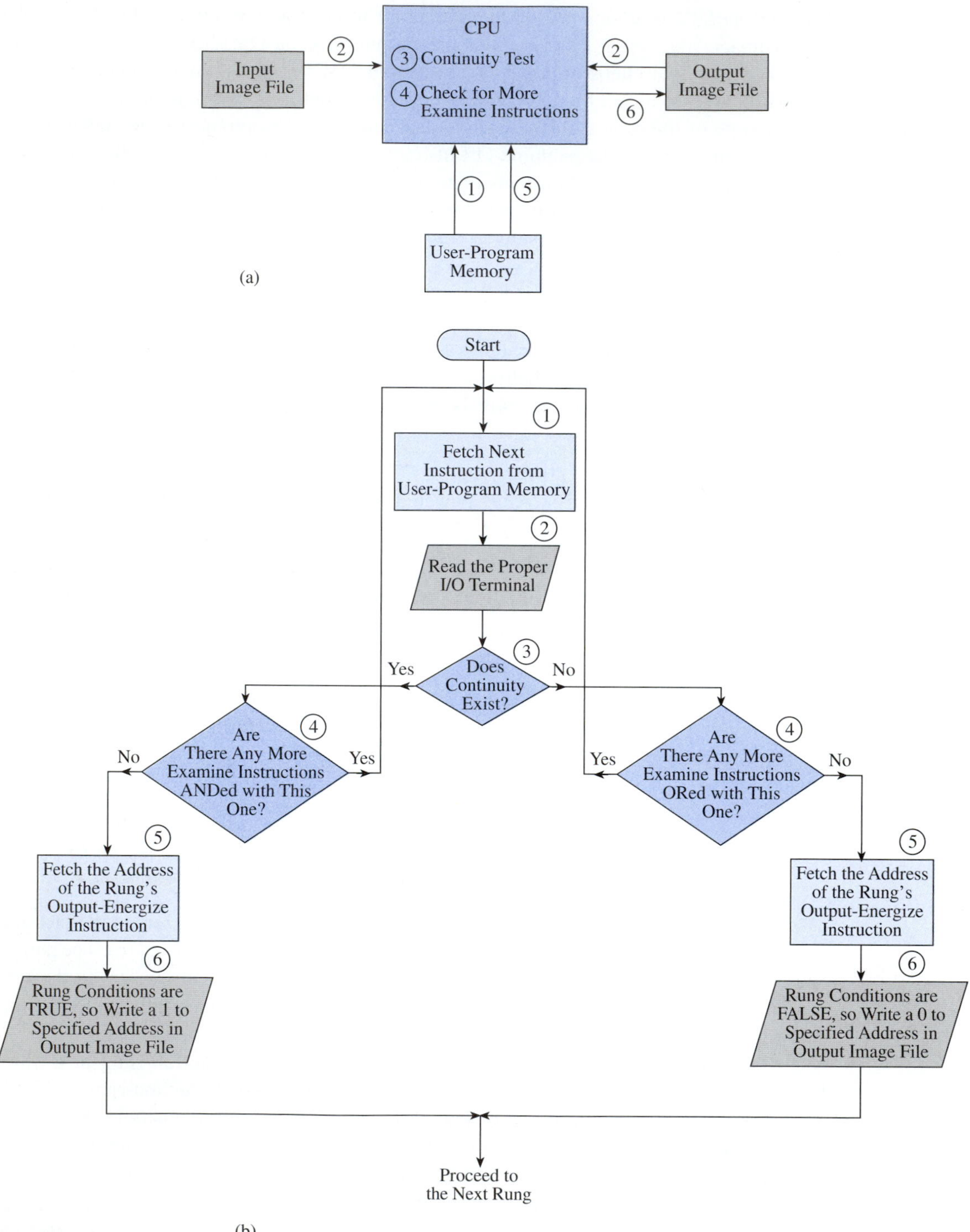

FIGURE 3–4
Execution of a single relay-type instruction-rung. A relay-type instruction-rung is one which contains no variable data, only examine instructions and an output-energize instruction: (a) block diagram; (b) flowchart. The numerical labels in this diagram do not correspond to the numerical labels in Fig. 3–2.

with an examine-type instruction. The CPU brings in the required information from the input or output image file in order to evaluate the instruction. This step is represented in Fig. 3–4(a) by the arrows indicating transfer from the image files to the CPU.

3. The CPU carries out an internal test by combining the instruction from step 1 with the I/O information from step 2. This test determines whether the instruction yields logic continuity or discontinuity. The test is represented in the flowchart of part (b) by the diamond-shaped decision box.
4. The CPU looks ahead at the user-program memory to see whether the next instruction is another examine-type instruction or an output-energize instruction. If it's an examine-type instruction, the CPU notes whether it is logically ANDed or logically ORed with the previous instruction. If it is logically ANDed (in series on the ladder-logic representation), then both instructions must produce continuity for the rung to maintain continuity so far. If the next instruction is logically ORed with the previous one (they appear in parallel paths on the ladder-logic representation), then it is sufficient for *either* instruction to produce continuity in order for the rung to maintain continuity so far.

   It may happen that the CPU can make its decision right now regarding the TRUEness or FALSEness of the rung conditions. A right-now decision is expressed by either of the "no" branches headed toward the outside from the decision boxes labeled ④ in the flowchart. Those branches lead to step 5, which brings in the address of the last instruction of the rung, the output-energize instruction.

   On the other hand, it may happen that the CPU cannot make its TRUE-or-FALSE decision right now, but must fetch the next examine-type instruction for further continuity testing. This situation is expressed by the two "yes" branches headed toward the inside from the decision boxes labeled ④. Those branches return to step 1 in the flowchart; they cause the CPU to repeat steps 1 through 4.
5. Eventually the CPU will progress through the rung to the point where it can decide whether the overall rung conditions are TRUE or FALSE. It then fetches the output-energize instruction from the user-program memory, so it can know which address to affect. This action is expressed by the transfer arrow labeled ⑤ in Fig. 3–4(a).
6. The CPU now knows the rung condition and the correct output address, so it sends the proper digital signal to the output image file, which then passes it to the associated output terminal. This act is represented by the arrow labeled ⑥ in the block diagram. Refer to the descriptions in the parallelogram-shaped I/O boxes labeled ⑥ in the flowchart.

When the processor has finished executing one instruction-rung, it moves to the next sequential location in user-program memory, fetches the next instruction (the first instruction of the next rung), and repeats steps 1 through 6. It continues in this fashion until every instruction has been executed. At that point the user-program has been fully executed one time.

The complete scan cycle. As long as the PLC is left in RUN mode, the processor executes the user-program over and over again. Figure 3–5 depicts the entire repetitive series of events. Beginning at the top of the circle representing the scan cycle, the first operation is the *input scan*. During the input scan, the current status of every input terminal is stored in the input image file, bringing it up to date. Like all PLC operations, the input scan is quite fast. The elapsed time depends on the number of input modules and terminals in the I/O section, the clock speed of the CPU, and other technical features of the

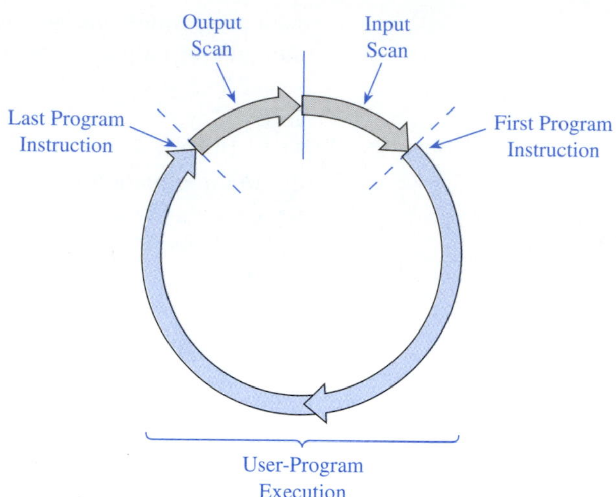

FIGURE 3–5
A PLC's entire scan cycle can be visualized as having three distinct parts: input scan, program execution, and output scan.

CPU. Speaking approximately, a system containing 10 to 20 input terminals would have an input scan time on the order of a few hundred microseconds.

Following the input scan, the processor enters its user-program execution, sometimes called "program scan," as represented in Fig. 3–5. The execution involves starting at the program's first instruction-rung, carrying out the six-step execution sequence described earlier, then moving on to the second rung, carrying out its execution sequence, and so on to the last program rung. The program execution time will depend on the length of the program, the complexity of the instruction-rungs, and the technical specifications of the CPU. Speaking approximately, we can say that a user-program of 20 to 30 instruction-rungs would probably have an execution time of several milliseconds.

As with all digital electronic devices, PLC scan speeds are continually increasing as their CPU clock frequencies keep rising and instructions are redesigned to be more efficient. Throughout the user-program execution, the processor continually keeps its output image file up to date, as stated earlier. However, the output terminals themselves are not kept continually up to date. Instead, the entire output image file is transferred to the output terminals during the *output scan* following the program execution. This is made clear in Fig. 3–5. The output scan time for 10 to 20 output modules would usually be in the neighborhood of a few hundred microseconds, similar to the input scan.

It is perfectly reasonable that the output terminals are updated all together during the output scan, rather than on an immediate individual basis during the user-program execution. This is because, in general, the load devices themselves are hopelessly slow compared to the scan cycle of the PLC. Consider a typical example. A real solenoid might require two or three oscillations of the ac line to become magnetically fluxed and pull in its armature (the movable part of the solenoid-operated mechanism). Two or three ac line oscillations take between 30 and 50 ms, which is enough time for the PLC to pass through its entire scan cycle many times. In other words, if the PLC on one pass through its scan cycle signals the solenoid to energize, it has to keep sending the same signal many times before the solenoid can respond. Under this circumstance, why should we bother with delaying the program execution to pass the output signal to the output device immediately? Waiting for the output scan is soon enough in almost all industrial control situations.

On rare occasions it may be necessary to update an output terminal immediately during user-program execution. Advanced PLCs have provisions for accomplishing this. Their instruction set (list of legal instructions) contains a special *immediate output* instruction which temporarily suspends the normal business of the program, updates the output terminal, and then returns to the program. This capability is portrayed in Fig. 3–6(a).

Some powerful PLCs also contain special *immediate input* instructions which can be used to update a particular location in the input image file just prior to executing an

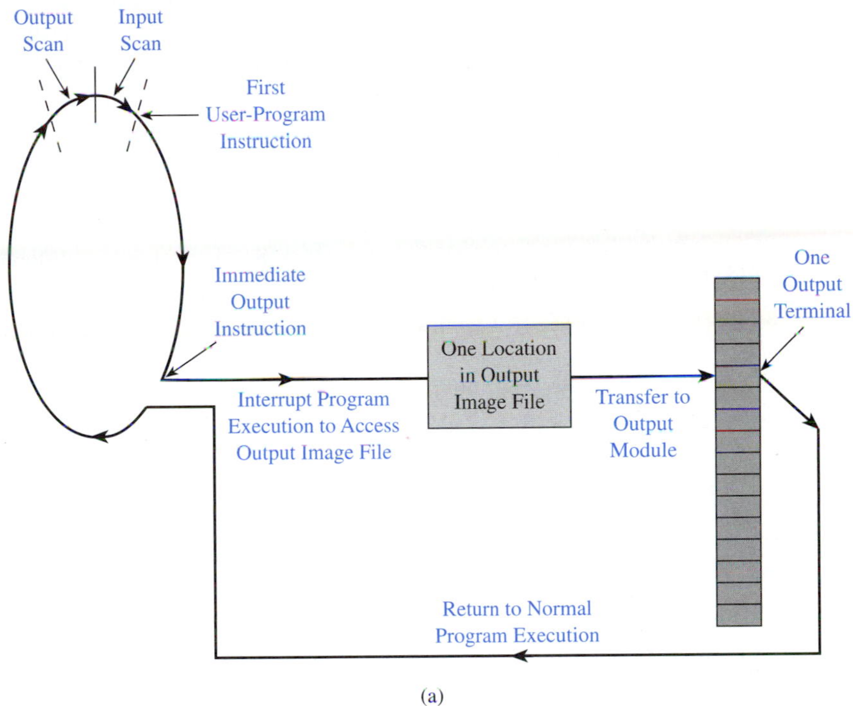

(a)

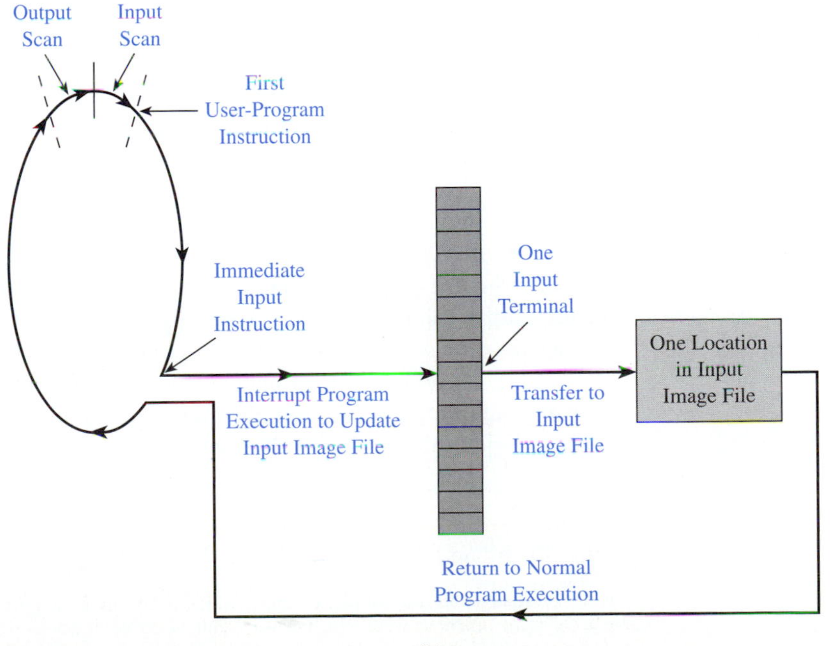

(b)

FIGURE 3–6 Visualizing immediate I/O functions: (a) immediate output; (b) immediate input.

instruction which uses that input. To justify going to this trouble, the control situation must be so exacting that it really matters if the input has changed during the few milliseconds that may have elapsed between the last input scan and the point in the user-program where the critical instruction is encountered. The immediate input capability is portrayed in Fig. 3–6(b).

Variable-data memory. Up until this point we have seen only three instructions, namely, examine-On, examine-Off, and output-energize. These three are classified as *relay-type* instructions because they duplicate the actions of relay contacts and coils. PLCs possess other instructions besides the relay types. Generally, a standard PLC possesses additional instructions that give it the following capabilities:

1. It can introduce time delay into a control scheme. That is, the PLC has internal timers* that duplicate the actions of the timers discussed in Sec. 2-12.
2. It can count events, with the events represented by switch closures. That is, the PLC contains internal counters, like the up-counters and down-counters discussed in Secs. 2-5 and 2-11.
3. A PLC is a computer, after all. Therefore it can perform arithmetic on the numeric data that resides in its memory.
4. It can perform numeric comparisons (greater than, less than, etc.).

All four of these capabilities imply that the PLC can store and work with numbers. Naturally the numbers can change from one scan cycle to the next (events occur and are counted, time passes, etc.). Therefore the PLC must have a section of its memory set aside for keeping track of variable numbers, or data, that are involved with the user-program. This section of memory we will call *variable-data memory,* as indicated in Fig. 3–2.

Many types of numeric data can be present in the variable-data memory. Six types that are important to understand are

1. *The preset value of a timer:* This is the number of seconds that the timer must remain energized in order to give a "timed-out" signal.
2. *The accumulated value of a timer:* This is the current number of seconds that have elapsed since the timer was energized.**
3. *The preset value of a counter:* This is the number that an up-counter must count up to in order to give a "count-complete" signal. For a down-counter it is the starting number that the counter counts down from.
4. *The accumulated value of a counter:* This is the current number of counts that have been recorded by an up-counter. For a down-counter it is the current number of counts remaining before the counter reaches zero.
5. *The value of a physical variable in the controlled process:* Such values are obtained by measuring the physical variable with a transducer and converting the transducer's analog output voltage (or current) into digital form with an A-to-D converter (ADC).
6. *The value of an output signal sent to a controlling device in the controlled process:* Such values are obtained by a mathematical calculation performed by the PLC. The user must tell the PLC just how the mathematical calculation is to be performed; this is done during the entering of the user-program from the programming terminal.

---

*We say *timers,* but we really mean timer *instructions*. Likewise with counters—we really mean counter instructions.

**These descriptions are for *On-delay* timers, which delay upon energization (see Table 2–1). PLCs usually contain *Off-delay* timers as well, which cause a time delay upon deenergization. By changing the words *energized* to *deenergized* the descriptions of preset value and accumulated value would apply to Off-delay timers.

Calculated output values are digital within the PLC and are usually converted to analog by a D-to-A converter (DAC) before being sent out to the controlling device.

When the CPU is executing an instruction for which a certain data value must be known, that data value is brought in from variable-data memory. When the CPU executes an instruction that produces a numeric result, that result is put out into variable-data memory. Thus the CPU can *read from* or *write to* the variable-data memory. This two-way interaction is pointed out in Fig. 3–2. Understand that this relationship is different from the relationship between the CPU and the user-program memory. When the user-program is executing, the CPU can only read from the user-program memory, never write to it.

3-1-3 The Programming Device

The third essential part of a PLC is the *programming device,* which is also called the programming terminal, or just the programmer. Some PLCs are equipped with a dedicated programming device manufactured by the same company that makes the PLC, but in many installations, the programming device is a desktop or laptop computer with a communication interface card installed in an expansion slot. A serial communication cable plugs into the interface card, joining it to the PLC's processor, as shown in Fig. 3–7. With special software installed on the computer's hard disk, keystrokes on the computer's keyboard represent user-program instructions which are converted into appropriate code by the interface card. From there they are passed through the communication cable to the processor. The PLC manufacturer's software gives prompts on the CRT screen to help the human programmer. It also displays several rungs of the ladder-logic user-program on the screen, as suggested in Fig. 3–7. This enables the user to observe the program's rung-by-rung development. Later, when the program is actually functioning, the screen display aids in troubleshooting and editing (changing) it.

Program entry. With the processor switched into PROGRAM mode (out of RUN mode), the human user enters the user-program into the processor's user-program memory by typing on the keyboard. The *function* keys (F keys) in the keyboard's top row are used to enter specific PLC instructions, in accordance with the F-key code shown in Fig. 3–8. It is not necessary to memorize this F-key code because the PLC software makes this very image visible on the CRT screen whenever an instruction is about to be entered. As Fig. 3–8 makes clear, only nine PLC instructions are actually available in the Basic Instruction Menu. These are the nine most frequently used instructions out of the scores that are possessed by a modern PLC. Less frequently used instructions are available through the [F10] key, [All Others]. Pressing the [F10] key brings up a different menu on the screen, which guides you to the class of instructions that you are seeking (instructions for doing mathematical computations, making comparisons of values, reading and writing analog data from/to the I/O rack, and many others). In general the F keys enable you, the user, to navigate through the various menus that the PLC manufacturer's software can place on the screen.

The keyboard's number keys, 0 through 9, are used for entering the addresses that go with instructions. Number keys are also used for entering the initial values of variable data. All program information is displayed in ladder-logic format on the CRT screen as it is entered through the keyboard.

There is a prescribed order in which the program information must be entered. This prescribed order and the exact keystroke details differ from one PLC manufacturer to the next and from one model to the next within a single manufacturer's model lineup. Since we intend to show some actual keystroke examples for entering rungs of a user-program,

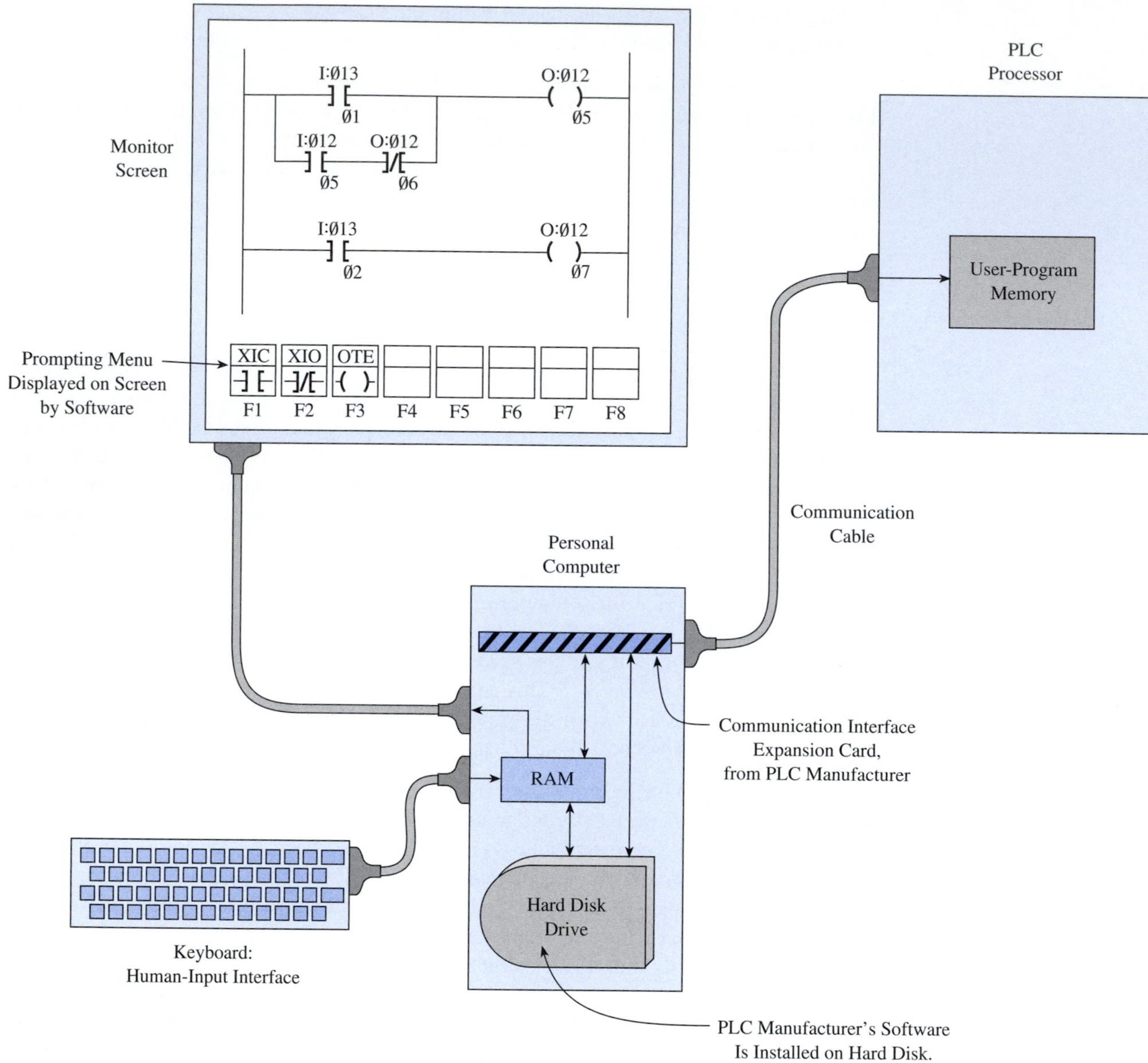

FIGURE 3–7
In most modern PLC systems the programming device is a desktop PC with its peripherals—keyboard and display monitor.

we must choose a particular manufacturer and model for our examples. Let us choose the Allen-Bradley company's model PLC 5/12. This is an advanced model of programmable logic controller with 70 instructions in its instruction set, and it is capable of handling a large number of input and output signals (up to 512 total signals). It is not a top-of-the-line PLC, but it is up to handling most industrial control applications.

The primary rule of user-program entry is that all information regarding one instruction-rung must be entered before the next instruction rung is begun. Within one instruction-rung, the required order is demonstrated in the following nine steps. For con-

XIC	XIO	OTE	OTL	OTU	TON	CTU	CPT	CMP	ALL
-] [-	-]/[-	-( )-	-(L)-	-(U)-					OTHERS
F1	F2	F3	F4	F5	F6	F7	F8	F9	F10

FIGURE 3–8

The Basic Instruction Menu. This menu appears along the bottom 2 in. of the screen when the programming device is expecting one of the common instructions to be entered. The PLC manufacturer's software automatically places this menu on the screen. Once an instruction has been entered by pressing an F key, the software gives an on-screen prompt, asking the user to enter the address that is to be associated with that instruction.

The software automatically jumps from one menu to another at the appropriate time in the program-entry process. Some menus are much more extensive than this basic example and may occupy the entire screen.

creteness, we will relate the program-entry keystrokes to the instruction-rung of Fig. 3–9(a), which is a detailed-description reproduction of Fig. 3–3(b).

By following the instructions given in the Allen-Bradley instruction manual, you must arrange for the appropriate menu to be on the screen, the Edit Rung Menu.

1. Use the [↓] key (down-arrow) to place the cursor on the End-Of-File line (the bottom line) of the ladder-logic screen display.
2. Press function key [F4], which is for [Insert Rung], as the current menu will indicate. The Allen-Bradley (AB) software will automatically jump to a new menu, called the Edit Instruction Menu, which will appear in the bottom 2-in. section of the screen.
3. Press function key [F4] again; in this new Edit Instruction Menu the function of [F4] will have changed to [Insert Introduction], as will be shown on the screen display.

   At this point the AB software will jump to the Basic Instruction Menu shown in Fig. 3–8.
4. Press [F1], which is for [-] [-], an examine-On instruction. On the screen the -] [- instruction will be color-reversed (dark background becomes white, light schematic symbol becomes dark); a blinking cursor appears after the message "Enter Bit Address." This is your prompt to enter the address.
5. Enter the address associated with the instruction. Here the address is I:Ø13/Ø1, so

   Keystrokes—[I] [:] [Ø] [1] [3] [/] [Ø] [1] [Enter]

   If there had been another series instruction in the Conditional Instruction Area (Fig. 3–9(a) on the first line of the rung, you would repeat steps 4 and 5. You continue in this manner until the Conditional Instruction Area of the first (top) line of the rung is completely specified. In a complex program, there may be several series instructions (logically ANDed instructions) in this area.
6. Enter the output-type instruction for this rung. There are many output-type instructions, of which output-energize -( )- is the simplest. There is always one and only one output-type instruction per rung. For Fig. 3–9(a), and referring to the Basic Instruction Menu of Fig. 3–8,

   Keystrokes—[F3] then [O] [:] [Ø] [1] [2] [/] [Ø] [5] [Enter]

   This finishes the rung's top line.
7. Now you must program the parallel branch of Fig. 3–9(a). Use the [←] key to move the cursor to the examine-On instruction on the left of the top line, as called for in Fig. 3–9(b). Press the Escape key [ESC] in the extreme upper left of the keyboard to return to a more fundamental menu. By pressing it twice, you will return to the Edit

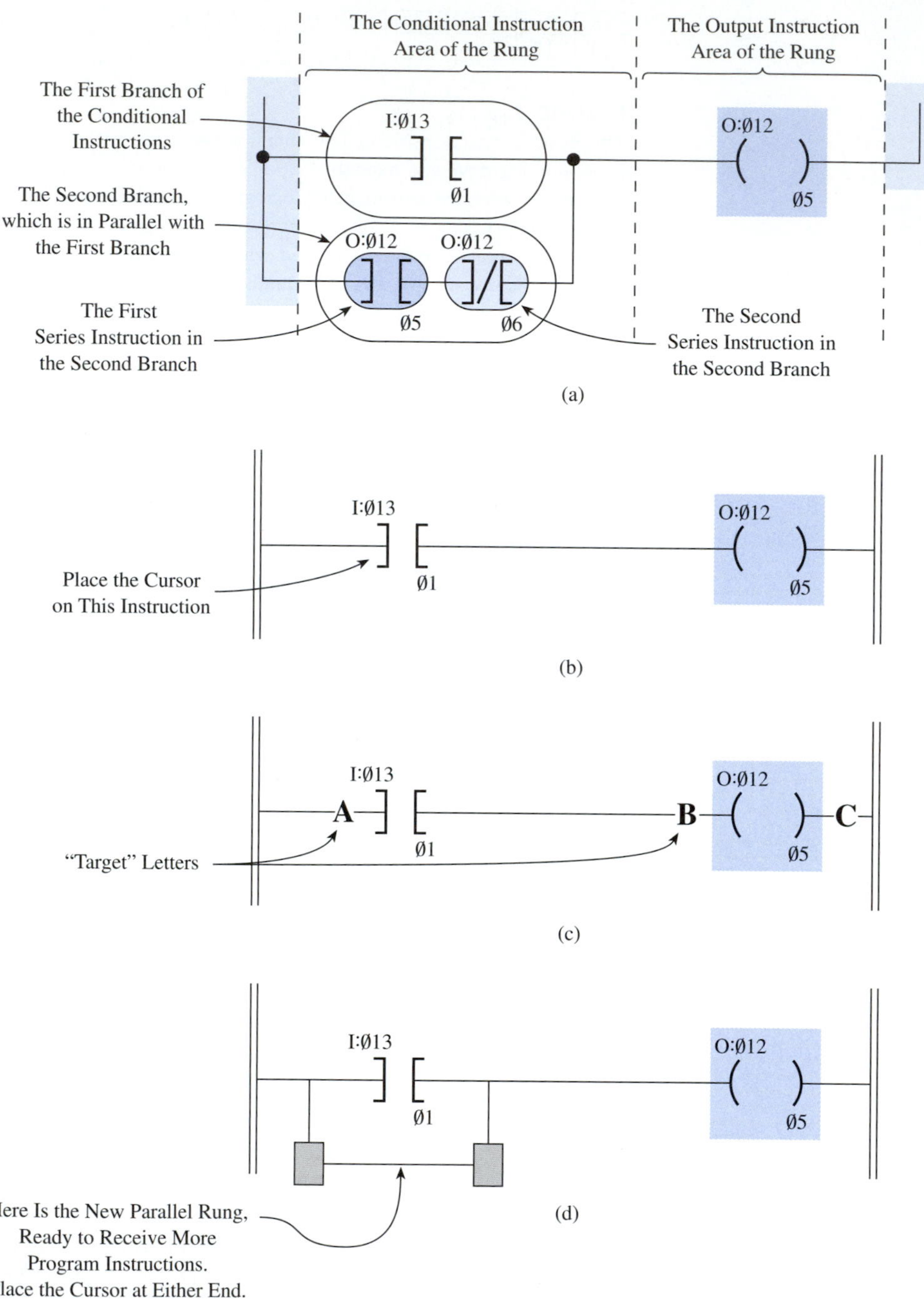

FIGURE 3–9 Practice Rung. (a) Reproduction of the conveyor/classifying system's top rung, identifying the details of the PLC rung's structure. (b) Preparing to program the parallel branch (step 7) after the top line has been completely programmed. (c) Target letters appear after the user issues the Insert Branch command from the Edit Branch Menu. (d) After point B is selected as the finish target, an empty parallel branch appears between logic points A and B. The user now enters instructions into this empty branch.

Rung Menu, which is two steps earlier in the menu hierarchy. This menu has [Modify Rung] as the function of the [F5] key. (In general, going "backward" to an earlier menu is usually accomplished by the [ESC] key, but sometimes it is accomplished by a particular F key, which will be labeled [Return to Menu].)

Press the [F5] key, [Modify Rung]. The AB software will jump to a new menu with [Branch *] as the function of the [F1] key. Press that key. The software will jump to the Edit Branch Menu, which has [Insert Branch] as the function of the [F4] key.

Pressing F4 causes a new parallel branch to start on the *left* side of the cursored instruction in Fig. 3–9(b). (Pressing F3, Append Branch, would cause a new parallel branch to start on the *right* side of the cursored instruction.)

After the F4 key is pressed, the software will place "target letters" at various locations on the ladder-logic line containing the cursored instruction. In Fig. 3–9(c), the only letters that will appear are A, B and C. The letter A will be on the left of the examine-On instruction, and B will be far to the right of it, next to the output instruction. The letter C is on the right rail. (In a more complex logic line with several conditional instructions, the Allen-Bradley software would put a letter of the alphabet at every possible place where the upcoming parallel branch could legally finish.) Each letter will be associated with an F key in the display at the bottom of the screen; A will be associated with F1 and B with F2. The place where you want the upcoming parallel branch to finish (the only place where it can reasonably finish in Fig. 3–9c) is at point B, to the right of the examine-On. Therefore press F2 for point B.

The PLC programming software immediately responds by drawing an empty branch in parallel with the examine-On instruction, as shown in Fig. 3–9(d).

8. Program the conditional instructions that belong on this parallel branch. Do so by using the keyboard's four arrow keys (also called cursor-control keys) to place the cursor at either end of the blank parallel rung in Fig. 3–9(d). If you are on the left end, press F4 for Append Instruction; if your cursor is on the right end of the branch, press F4 for Insert Instruction. In PLC 5 software, *Append* always means do something toward the right; *insert* always means do something toward the left.

   Enter the instruction-data by repeating steps 4 and 5. The AB software will be presenting the Basic Instruction Menu.

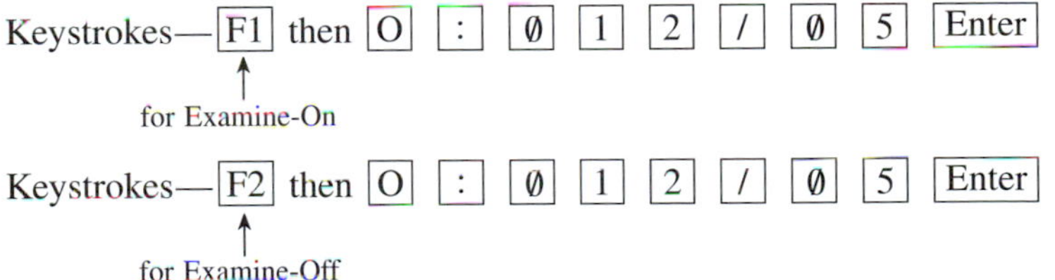

   With this model of PLC, there is no specific instruction that must be entered to indicate the completion of the parallel branch. This is unlike many other PLC models, in which the user must enter a Branch-End instruction to signify that the branch is now to be tied to the line above, logically ORing it with that line. The model PLC 5/12 software has *already* received its branch target letter in step 7.

9. Tell the programming device that the rung is now complete. Do this by using the ESC key to return to an earlier menu that has the F10 key labeled Accept Rung. Press F10 to have the rung officially blessed by the AB software as a complete rung. When this is done, the program code for the rung is copied from the computer's RAM work-file into the processor's user-program memory via the communication interface card.

To begin entering the next instruction-rung of the user-program, cursor down to the End-Of-File line on the ladder-logic screen display. Repeat the entire procedure for the second rung.

Editing the program. As you can imagine, it is unlikely that a program will work perfectly on its very first run. There are so many chances for conceptual error in the design of the program, and so many chances for typographical errors in the entering of the program, that it is almost certain that the program will require some debugging before it's truly ready to operate. With this in mind, the PLC manufacturers have provided editing capabilities in the programming software. Editing capabilities allow us to alter a program in a variety of ways. For example, we can insert or remove individual instructions, insert or remove entire instruction-rungs, change addresses, and change initial values of variable data. Many other kinds of changes are also possible. Editing is done with the processor switched into PROGRAM mode. The editing functions of the Allen-Bradley PLC 5/12 software are invoked by reaching the Ladder Editor Main Menu, which is the menu that has the [F10] key labeled [Edit]. Then press the [F10] key.

When editing a program, it is necessary to position the cursor in the proper position, generally on the instruction or instruction-rung that is to be altered. The cursor can be moved one line or one rung at a time with the up- and down-arrow keys, or it is possible to send the cursor immediately to a specified point in the program, even a point that is many rungs away from its current position on the screen display. This is accomplished in the AB software by using the [Search] function, which is the [F6] key on the Ladder Editor Main Menu. In general, consult the manufacturer's software manual to learn the procedures for searching and editing a program.

Testing the program. Because of the unlikelihood of a program working satisfactorily on its first try, PLC manufacturers provide a third mode of processor operation besides PROGRAM and RUN. This is the TEST mode, in which the processor executes the program without actually powering up the output terminals in the I/O section. Instead, a small indicator LED for each output terminal is illuminated when that output terminal would have been powered up if the processor had been in RUN mode. In this way we are able to simulate the operation of the industrial system without actually energizing the load devices. When we don't have absolute confidence in a newly written program, it's a big relief to be able to watch what the machinery would have done, compared to gritting our teeth and watching what it actually does.

For instance, suppose the industrial system contains two hydraulic cylinders whose rod extensions intersect. It is imperative that the program never allow both cylinders to be extended at the same time, because the one that arrives later will crash into the one that arrived earlier. If we made a logic error in the design of the user-program, however, or if we made a typographical error in the keyboard entry, executing the flawed program with the machinery actually operating might result in such a collision. By first executing the program in the TEST mode we have an opportunity to spot any such flaws. In this example, if we see that both indicator lamps are simultaneously lighted on the two cylinder-controlling output terminals, we will realize there is trouble with the program and we can do something about fixing it.

Once the program is completely debugged, a trial execution in TEST mode will show all outputs operating as planned. Then the processor can be switched to RUN mode with confidence.

To carry out a program test we must have a method of artificially controlling the inputs to make them provide the input signals that would occur naturally if the system were actually operating. For instance, in Figs. 3–3 and 3–9(a), the first rung of the program for the conveyor/classifying system contains an instruction that examines input I:013/Ø1. For that input terminal to receive 120-V power naturally, LS1 must be actuated; but LS1 cannot be actuated because we do not have any parts moving down the conveyor—in TEST mode the conveyor can't even move. So what should we do, send somebody out to push LS1 with a stick? No. That sort of practice is dangerous. Even if we were working with a relay control panel we wouldn't do that; instead, we would run a jumper wire from the hot side of the ac line to the LS1 wire-terminal in the control panel.

With a PLC we don't have to worry about such inconveniences because the programming software provides us with *forcing functions*. The Force-On [F2] and Force-Off [F1] functions are available in the Forcing Menu, which is reached by pressing [F9] for [Force] from the Ladder Editor Main Menu. The Force-On function permits us to place a digital 1 at a particular address in the input image file, regardless of the actual state of its corresponding input terminal. Therefore we can make the processor think that power is present at the input terminal even though it's really absent. This is far better than going out and pushing LS1 with a stick.

The Force-Off key produces a digital 0 in the input image file, regardless of the actual state of the input terminal. It enables us to make the processor think that power is absent at the input terminal when actually it's present.

The procedures for using the forcing functions are explained in the manufacturer's software instruction manual.

Forcing functions can be applied to outputs too. In the TEST mode the output-forcing capability is useful for finding out what the program would do if a particular combination of input and output conditions were to occur.

In the RUN mode an output-forcing function actually affects the output module. This enables us to energize and deenergize the system's load devices at will, which is useful for checking their mechanical performance, making adjustments, etc.

Various PLC manufacturers. When studying technology, the question sometimes arises whether it is better to concentrate on one particular manufacturer's organizational format or to experience a smattering of many manufacturers' nomenclatures and numbering schemes. We have adopted the first view, that concentrating on a single PLC manufacturer's format is a more effective way to learn.

With this approach, you can concentrate on understanding the underlying concepts of PLC structure and functioning while being able to practice with specific program examples that demonstrate those concepts. The alternative, presentation of many different model specifications, distracts a learner's mental resources. So much mental effort goes into distinguishing among the different memory organizations, addressing schemes, symbols, and display formats that it becomes difficult to focus on the essential *ideas*. Of course, the essential ideas are the same for all PLCs, regardless of manufacturer.

If you find yourself working in a PLC environment other than the Allen-Bradley PLC5-series environment that we have chosen, it is a relatively easy matter to then learn the specifics of that environment by relating that model's specs to your knowledge of AB PLC5.

3-2 ■ PROGRAMMING A PLC TO CONTROL THE CONVEYOR/CLASSIFYING SYSTEM

Let us develop a complete user-program for implementing the conveyor/classifying control system of Sec. 1-3. This exercise will give us some introductory practice in program design and will provide further insight into the functional equivalence of PLC software and hard-wired relay logic. We will use the relay logic circuit of Fig. 1–5 as our starting point.

3-2-1 Assigning I/O Addresses

The first thing we must do is select the input and output addresses that we intend to use for the input and output devices in the system. As mentioned earlier, we must choose these

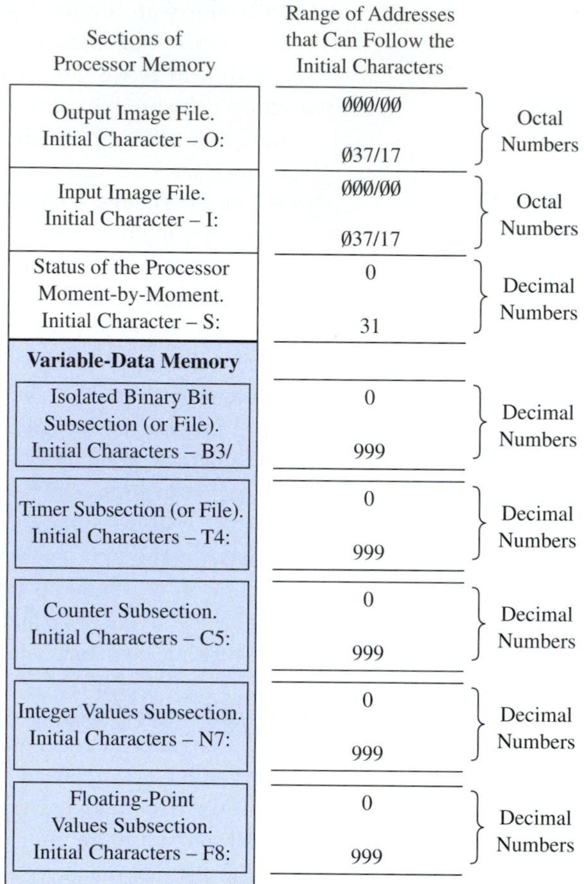

FIGURE 3–10
Memory table for an Allen-Bradley model PLC 5/12. In the I/O image files at the top of this figure, addresses are given in octal. In the Status of the Processor file, and in all the variable-data memory files (or subsections), addresses are given in decimal.

addresses within the constraints of the manufacturer's addressing scheme. The PLC's instruction manual must explain the addressing rules for input and output. The explanation is often given by reference to a *memory table* or *memory map*. A simplified memory table for an Allen-Bradley PLC 5/12 is shown in Fig. 3–10.

Within the I/O image files, the user must normally specify the *complete* address, which identifies one particular individual bit in the memory. To specify a complete address right down to the bit, two pieces of information are needed: the word-address, and the bit-number within that word. An input or output word-address has three octal digits; the I/O bit-number has two octal digits.

Think of each word as containing 16 bits, numbered from octal ØØ on the far right up to octal 17 on the left. This word structure is pictured in Fig. 3–11. Notice the octal numbering sequence for the bits; when the bit-number reaches Ø7, which is the highest value expressible with a single digit (leading zero), the next higher bit-number spills over into the second digit, as 1Ø.

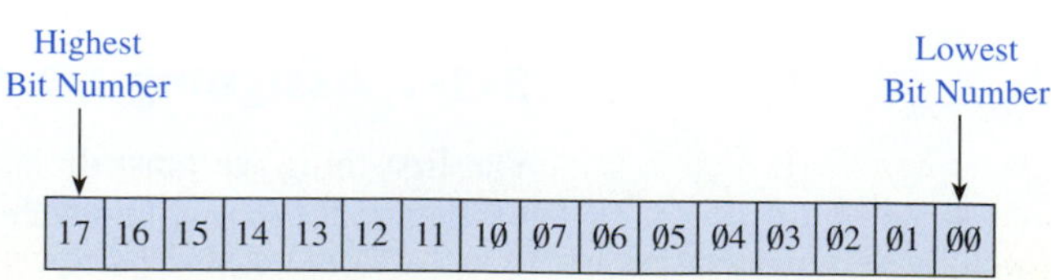

FIGURE 3–11
One word is a group of 16 bits numbered from **ØØ** to octal 17.

We often use the subscript 8 in parentheses to distinguish an octal number from a decimal number. Thus we could write $17_{(8)}$ to refer to the highest-numbered bit in a 16-bit word in the output or input memory sections. Subscripts aren't necessary in Fig. 3–11 because the counting-sequence context makes it clear that the numbers are octal.

Now that the numbering system has been clarified, let's return to Fig. 3–10 and focus our attention on two sections of the processor memory: the output image file, and the input image file. These are the sections that we must deal with first as we develop our user-program for the conveyor/classifying system.

The lowest word-address in the output image file is ØØØ. Because of the manufacturer's wiring that connects the processor to the I/O rack(s) of the PLC, this particular word-address refers to a specific exact physical position in the I/O rack(s). The left two digits refer to a specific rack (of the several racks that might be present in this PLC system), and the rightmost digit refers to a specific slot within that rack. Thus, for example, the word-address Ø13 would refer to rack number Ø1 and slot number 3 within rack Ø1. We, the users, have no say about this matter; it is predetermined by the manufacturer.

Each slot receives one input module or one output module. Each I/O module has 16 wire-connection terminals for connecting the industrial system's I/O devices, as we know from Fig. 3–1(a) and (b).

Slots are numbered from Ø to 7, as shown in Fig. 3–12. Racks are numbered from ØØ to Ø3 for the model PLC 5/12. (Two digits are reserved for rack numbers because larger PLC models can accommodate up to 24 racks.)

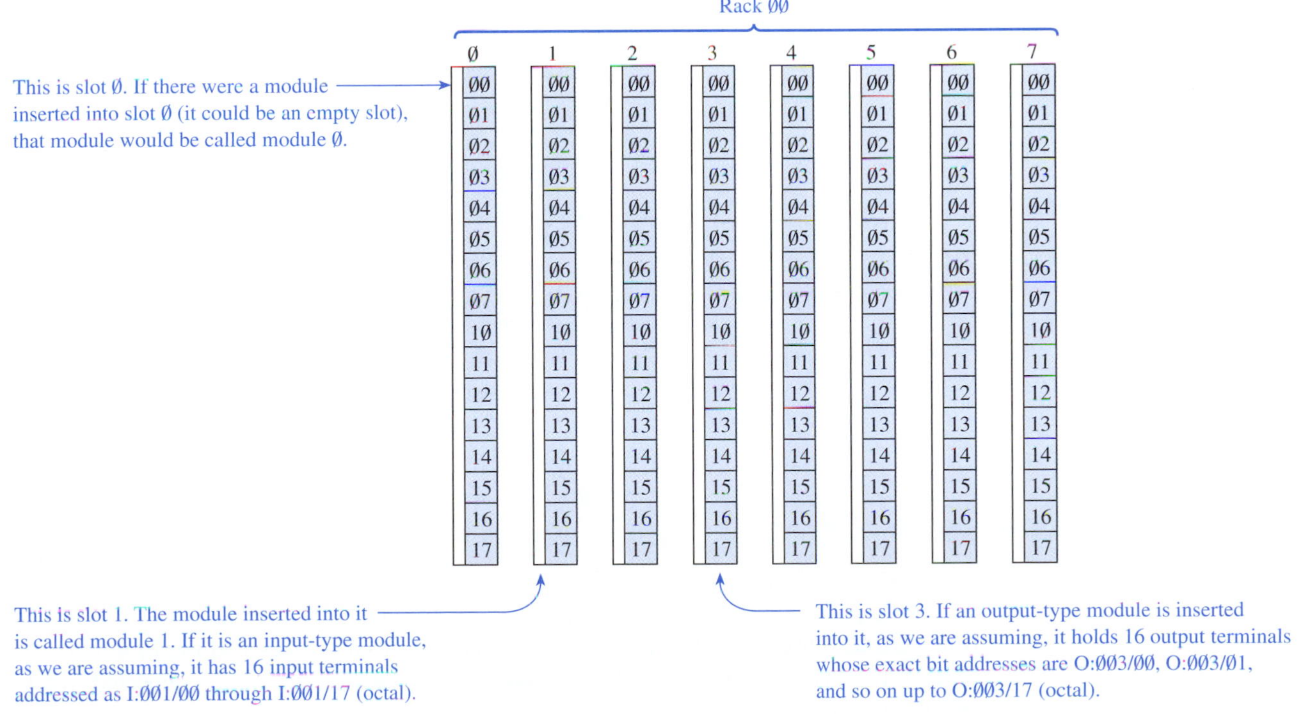
Rack ØØ
Ø 1 2 3 4 5 6 7
ØØ Ø1 Ø2 Ø3 Ø4 Ø5 Ø6 Ø7 1Ø 11 12 13 14 15 16 17
This is slot Ø. If there were a module inserted into slot Ø (it could be an empty slot), that module would be called module Ø.
This is slot 1. The module inserted into it is called module 1. If it is an input-type module, as we are assuming, it has 16 input terminals addressed as I:ØØ1/ØØ through I:ØØ1/17 (octal).
This is slot 3. If an output-type module is inserted into it, as we are assuming, it holds 16 output terminals whose exact bit addresses are O:ØØ3/ØØ, O:ØØ3/Ø1, and so on up to O:ØØ3/17 (octal).
Rack Ø1 Is Here, if Used

FIGURE 3–12
Layout of an eight-slot rack, assuming that each slot accommodates a module with 16 terminals. Each module's data is stored in the image file as one 16-bit word. The bit-identifying numbers (octal) in the memory word correspond to the terminal-identifying numbers on the I/O module.

If we elect to install an input-type module in slot 1 of rack ØØ, then we activate the word-address within the *input* image file that references slot 1 of rack ØØ, but we deactivate the word-address within the *output* image file that references slot 1 of rack ØØ. Referring to the memory table of Fig. 3–10, we activate word-address I:ØØ1 (in the input image file), but we deactivate word address O:ØØ1 (in the output image file). This means that in the user-program we can access the 16 input terminals addressed I:ØØ1/ØØ, I:ØØ1/Ø1, I:ØØ1/Ø2, . . . , I:ØØ1/16, and I:ØØ1/17, but we must not use any of the 16 addresses O:ØØ1/ØØ, O:ØØ1/Ø1, O:ØØ1/Ø2, . . . , O:ØØ1/16, or O:ØØ1/17.

Figure 3–13(a) shows the general syntax of an input or output address for the model PLC 5/12. Figure 3–13(b) shows a valid address for the case where an output-type card is installed in slot Ø of rack Ø1. If an input-type card were installed in slot Ø of rack Ø1, however, that address would be illegal; we could not allow it to appear anywhere in the program.

FIGURE 3–13 Relating the five address digits to I/O rack location. The slot number is often referred to as the module number or the module-group number.

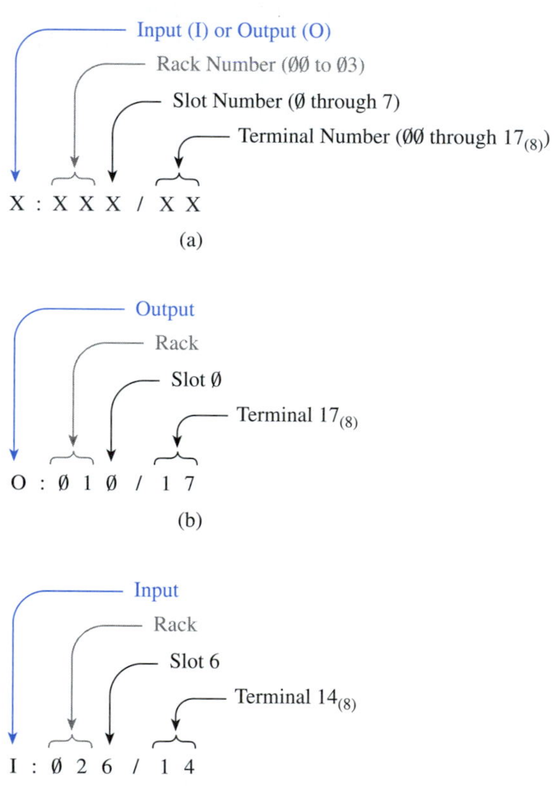

The address in Fig. 3–13(c) is legal only if an input-type module is inserted in slot 6 of rack Ø2. Of course, the wiring between the industrial devices and the PLC must be correct also. That is, there must be an input device, not a load device, connected to terminal 14 of slot 6 of rack Ø2.

Now that we know the rules, we can proceed to assign addresses to the input and output devices of the conveyor/classifying system of Figs. 1–4 and 1–5. There are eight input devices; they are limit switches LS1 through LS6 and the height and weight detectors. Let's say that we have inserted an input card in slot 1 of rack ØØ. Then we can use the input addresses shown in Table 3–1.

TABLE 3–1
Choosing the I/O Rack Locations and Image File Addresses

INPUT DEVICE	ADDRESS	OUTPUT DEVICE	ADDRESS
LS1	I:001/01	Blue paint soln.	O:003/00
LS2	I:001/02	Yellow paint soln.	O:003/01
LS3	I:001/03	Red paint soln.	O:003/02
LS4	I:001/04	Green paint soln.	O:003/03
LS5	I:001/05	S/L diverter soln.	O:003/04
LS6	I:001/06	S/H diverter soln.	O:003/05
Height detector	I:001/07	T/L diverter soln.	O:003/06
Weight detector	I:001/10	T/H diverter soln.	O:003/07

Note that we purposely skipped address 001/00. This way the LS numerals match the last digits of the addresses, which is convenient.

Now move on to the output addresses. An inspection of Fig. 1–5 reveals that there are eight output devices: the four paint solenoids and the four diverter solenoids. Let's say we have inserted an output card into slot 3 of rack 00. Then we can assign output addresses as listed in Table 3–1.

When we installed I/O modules into the rack, we skipped slots 0 and 2, leaving them empty. This way it will be convenient if we need to expand our PLC system to greater input/output capability at a later date. Then we can slip an input module into slot 0 and an output module into slot 2, keeping the input modules adjacent to each other and the output modules adjacent to each other. This arrangement will make it easier to wire the additional industrial devices to the I/O rack; there will be less wire-crossing and tangling. We are reserving slots 4 through 7 for analog measurement inputs and output control signals, which are explained in Secs. 3-6 and 9-11.

3-2-2 Assigning Internal Addresses

There are nine relays in the conveyor/classifying circuit of Fig. 1–5, but they are not truly output devices. Their action is strictly internal to the control circuit. This might cause a misunderstanding, since the placement of the relay coils in the relay ladder-logic diagram corresponds to the placement of the output-energize instructions in the user-program ladder-logic diagram. There are two ways of handling the task of assigning addresses to the user-program's output-type instructions that correspond to internal control-relay coils:

1. Use one of the addresses B3/0 through B3/999 from the Isolated Binary Bit subsection of variable-data memory. Refer to the Memory Table of Fig. 3–10.
2. Use an address from the output image file section of the memory table, but refer to an *empty slot,* one that has no module inserted in it. This way you do not power up a dead-ended output terminal every time one of the user-program's internal output-type instructions becomes TRUE.

Option 1 is preferred. The problem with option 2 is that the rack slot might be empty *now,* but it may not remain empty in the future. If you expand your system and fill that empty slot, this program's internal logic instruction addresses will then need to be changed.

Let us choose option 1 and simply start with the lowest bit address in the B3 subsection, then go in ascending order. Thus for the output-type instructions that correspond to the internal control relays of Fig. 1–5, we can assign addresses as shown in Table 3–2.

TABLE 3–2
Choosing Memory Addresses for the Internal Logic Instructions in the User-Program

RELAY COIL OF FIG. 1–5	ASSIGNED ADDRESS
Part is in paint zone: RPZ	B3/0
Part is in diverter zone: RDZ	B3/1
Part has cleared: RCLR	B3/2
Part is tall: RTAL	B3/3
Part is heavy: RHVY	B3/4
Part is short and light: RSL	B3/5
Part is short and heavy: RSH	B3/6
Part is tall and light: RTL	B3/7
Part is tall and heavy: RTH	B3/8

Note that the address assignments in Tables 3–1 and 3–2 do not coincide with the addresses used for the example instruction-rung in Figs. 3–3 and 3–9(a). We are starting over fresh.

Isolated Bit-type (B3-type) addresses have a difference from all the other address types in the variable-data memory section of Fig. 3–10. The decimal number (0 through 999) following B3/ refers to one single bit in RAM. For all other data types in variable-data memory, the decimal number refers to an entire *word,* consisting of 16 bits.*

3-2-3 Writing the User Program

Figure 3–14(a) shows the ladder-logic representation of the user-program's first instruction-rung, using our agreed-upon addressing schedule. The keystroke sequence is given in Fig. 3–14(b). Verify for yourself that the keystroke sequence coincides with the ladder-logic representation.

The second instruction-rung is similar to the first. It is illustrated in Fig. 3–15.

The third instruction-rung is presented in Fig. 3–16. Trace through that figure and satisfy yourself that it will successfully implement the "part has cleared" circuit on lines 5, 6, 7, and 8 of Fig. 1–5.

The circuit appearing on lines 9, 10, 11, and 12 of Fig. 1–5 contains two relay coils: RTAL and RHVY. The PLC cannot duplicate this circuit straightforwardly because an instruction-rung can contain only one output-type instruction. Therefore we must use two instruction-rungs to implement this logic. The rungs are shown in Fig. 3–17(a). The keystroke sequence for the top rung is shown in Fig. 3–17(b).

The same restriction regarding only one output-type instruction per rung applies to the circuit on lines 13 and 14 of Fig. 1–5. That circuit must be implemented as two rungs. Likewise for lines 15 and 16. The circuitry on lines 17, 18, 19, and 20 takes four rungs in the program, and the same is true for lines 21, 22, 23, and 24. The remainder of the user-program appears in Fig. 3–18. Verify it for yourself.

*Actually, in most cases the word-address that is specified refers to a collection of two or more words. This will become clear when we discuss timer- and counter-instructions in Sec. 3-3. At that time we will also learn how to address one specific bit within one of those words.

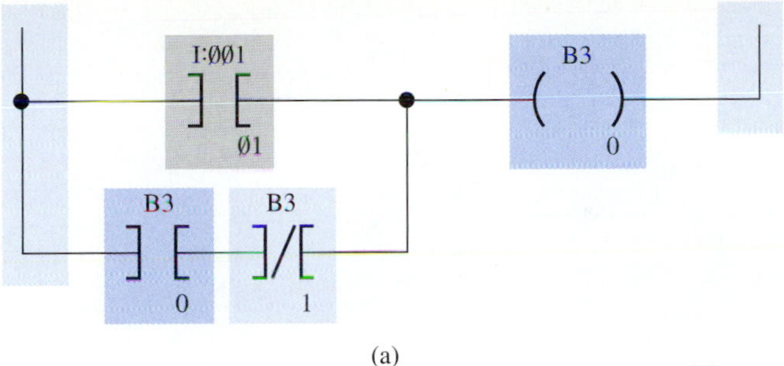

(a)

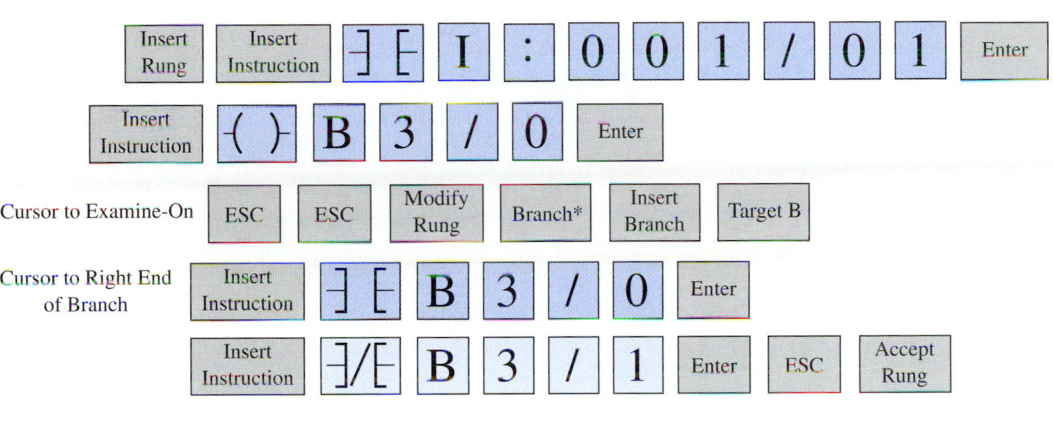

(b)

FIGURE 3–14
First instruction-rung in the conveyor/classifying system's user-program: (a) ladder-logic representation; (b) keystroke sequence.

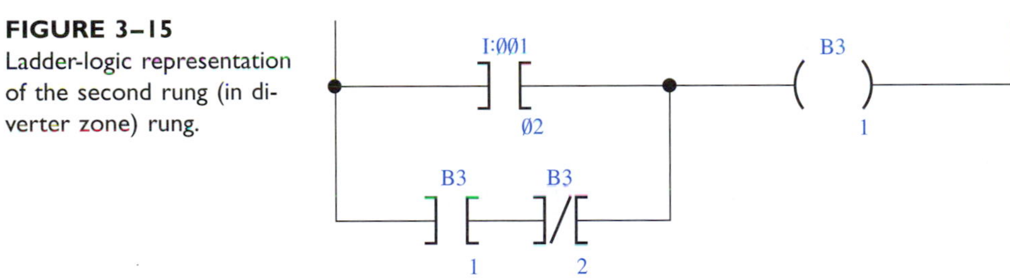

FIGURE 3–15
Ladder-logic representation of the second rung (in diverter zone) rung.

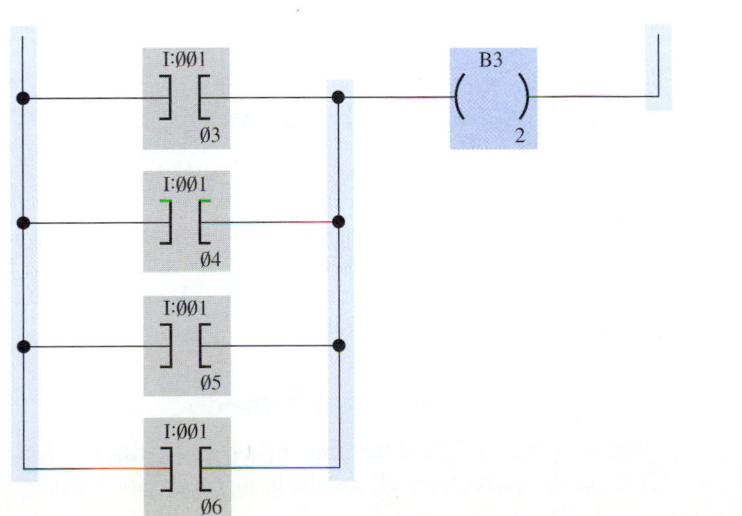

FIGURE 3–16
"Part has cleared" rung ladder-logic representation.

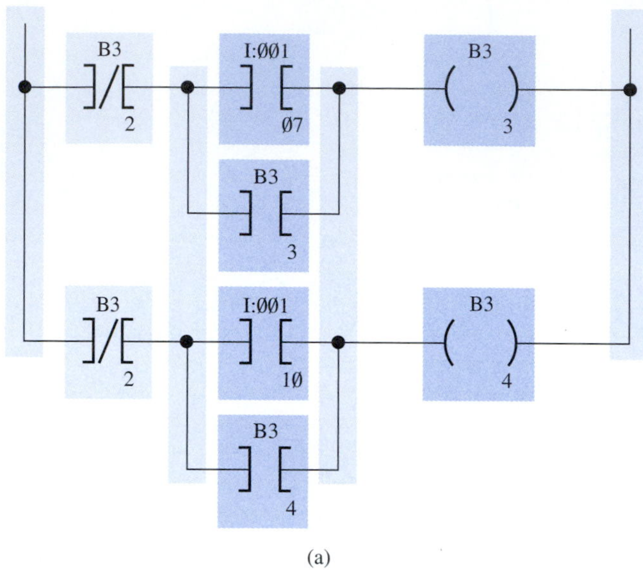

(a)

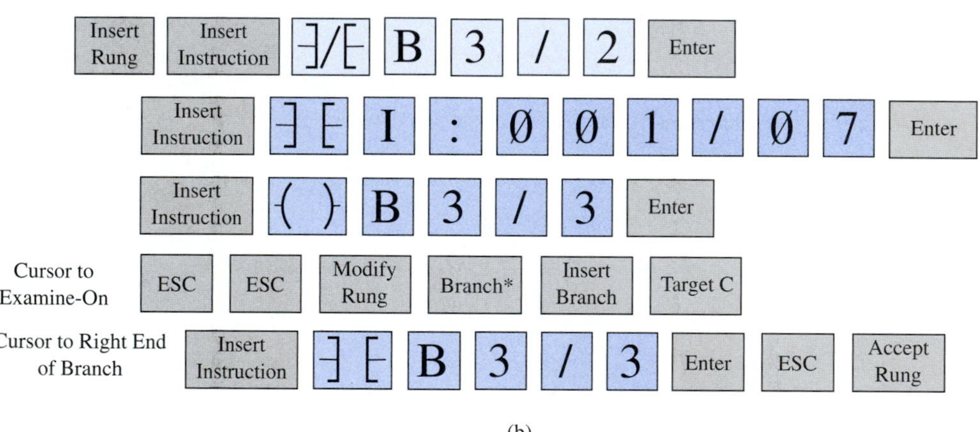

(b)

FIGURE 3–17
Height- and weight-detection rungs: (a) ladder-logic representation; (b) keystroke sequence for the top rung only.

3-3 ■ PROGRAMMING TIMING AND COUNTING FUNCTIONS

As mentioned earlier, PLCs are not limited to relay-type functions. They possess a complete set of other functions, including all the modes of timing (On-delay, Off-delay, retentive), up- and down-counting, comparison (equal to, less than, greater than), flip-flop (latch-unlatch), mathematical functions (basic arithmetic, trigonometry, integration, and differentiation approximations), and more. It is their wide range of capabilities that gives PLCs their industrial versatility.

3-3-1 Programming a Timer

When programming a timer rung, the user specifies the decimal word-address of the timer. The processor reserves all 16 bits in that word for keeping track of the timer's status and

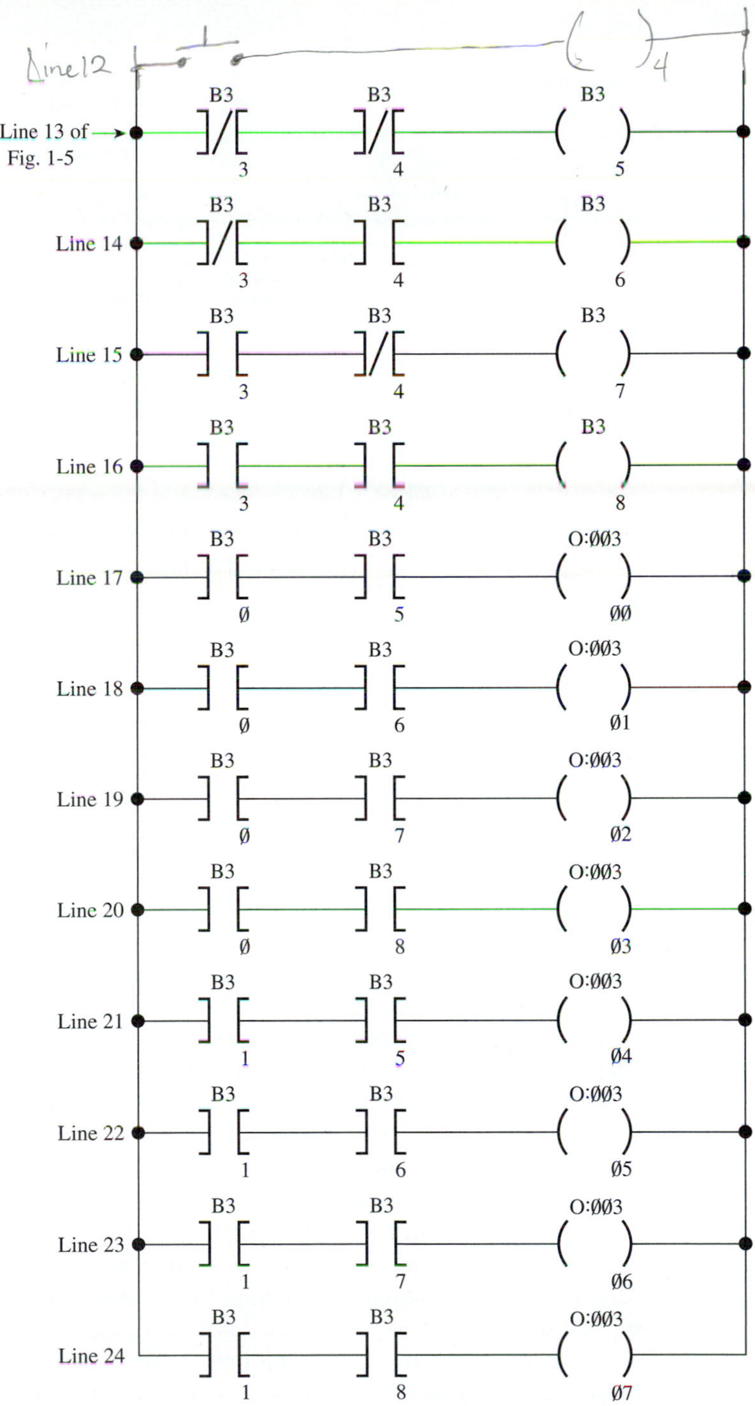

FIGURE 3–18 Ladder-logic representation of the classifying and diverting functions.

progress.* This is unlike the programming of an I/O address, where we, the users, must specify the bit-number in addition to the three-digit word-address.

However, the programming of a timer requires us to furnish two new pieces of information that were not required in the relay-type instructions of the previous section:

*It also reserves all 16 bits in two other associated words that hold the timer's Preset value and Accumulated value. That is, three words of RAM are dedicated to every timer instruction even though just one address is used in the memory table of Fig. 3–10, not three addresses.

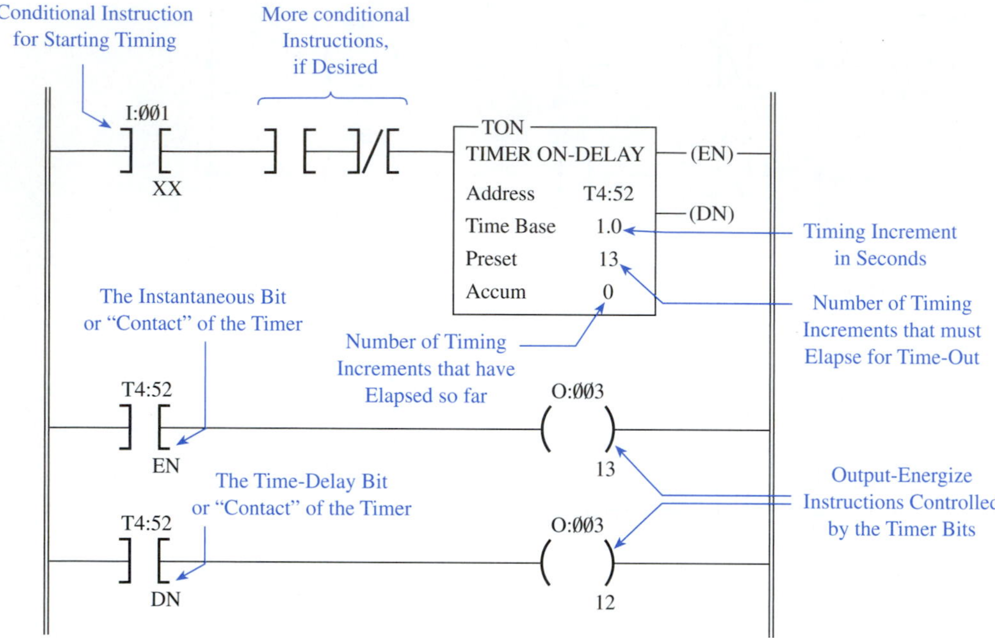

FIGURE 3–19
Ladder-logic relating to an On-delay timer instruction.

1. The timing increment, and
2. The number of timing increments that must elapse for the timer to time out.

As an example, suppose that we wished to establish a time delay of 13 seconds, with a timed interval resolution of 1 second. The instruction-rung would appear on the programming-device display screen as shown in Fig. 3–19, the topmost rung.

In that figure, the left side of the topmost instruction-rung contains the conditional instructions that determine whether or not the timer is running (timing). On the first scan cycle in which the conditional instructions yield logic continuity, the TON output-type instruction becomes TRUE and timing begins. On every subsequent scan cycle, if the rung is maintained in the TRUE condition, the timer continues to accumulate time. Eventually, if the rung's TRUE condition is maintained, enough timing increments will accumulate to match the user-programmed Preset value, and the timer times out.

The top rung in Fig. 3–19 contains the TON output-type instruction, standing for Timer On-Delay. It has been assigned word-address 52 in the T4 subsection of variable-data memory. This assignment is done by the user. The processor automatically reserves the additional 16-bit words needed for holding the timer's Preset and Accumulated values. The user must also key in the timing increment (1.0 second in Fig. 3–19) and the Preset number of increments needed in order for the timer to time out (13 increments in Fig. 3–19). Preset values are expressed in decimal.

In word T4:52, three bits are set aside to be referenced by relay-type instructions. One bit serves as the *time-delay* bit, which is sometimes called the *timed-out* bit. Another bit serves as an *instantaneous* bit.* These time-controlled bits are similar to the contacts of a relay-type timer. This is suggested in the notes accompanying the second and third

*The third bit is similar to a plain instantaneous bit, but it reverts to its original inactive state when the timer times out. Its mnemonic is TT (for Timer is Timing), bit number decimal 14.

rungs of Fig. 3–19. As shown in that diagram, the EN bit (for ENabled) serves as the instantaneous bit, and the DN bit (for DoNe) serves as the time-delay bit. These specific bits are addressed in the relay-type Examine instructions by their mnemonic letters, not by their octal or decimal bit-numbers. Thus on the second rung of Fig. 3–19, the EN bit, which happens to be bit number decimal 15, would be keyed in to the program as

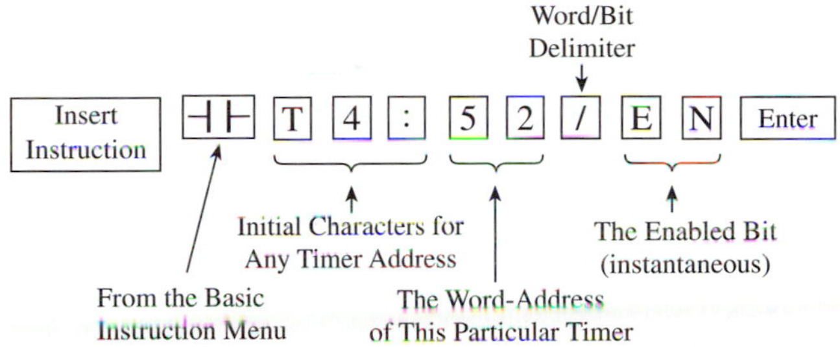

In the third rung of Fig. 3–19, the DN bit, which happens to be bit number decimal 13, serves as a time-delay bit. So when the TON instruction at address T4:52 becomes TRUE because the first rung's conditional instructions give rung continuity, memory bit T4:52/EN immediately changes from a 0 to a 1. Any relay-type instruction later in that scan cycle, and in subsequent scan cycles, will find a 1 in that location. Thus if the first rung of Fig. 3–19 becomes TRUE, the T4:52/EN examine-On instruction in the second rung immediately establishes logic continuity and the O:ØØ3/13 output-energize instruction becomes TRUE for the remainder of that scan cycle.

The time-delay bit provides the actual timing function. In Fig. 3–19, the third rung contains an examine-On instruction referencing time-delay bit T4:52/DN. This instruction shows logic discontinuity for many succeeding scan cycles, until 13 seconds have elapsed. At that time, address T4:52/DN changes from 0 to 1, and the T4:52/DN examine-On instruction shows logic continuity. Thus, after the TON instruction in Fig. 3–19 becomes TRUE, the O:ØØ3/12 output-energize instruction will become TRUE 13 seconds later if logic continuity is maintained to the TON instruction during *every* intervening scan cycle.

The progress of a timer can be watched on the programming device's display screen. Beneath the Preset value appears an Accumulated value, which represents the number of timing increments that have elapsed.

Figure 3–20 shows the keystroke sequence for programming the TON-instruction rung of Fig. 3–19. In this example there is just one conditional instruction for initiating the timing function. In a real program the conditional instructions might be more extensive, as suggested in Fig. 3–19.

After the user chooses the word-address of this timer (52 in our example) and presses the Enter key, the AB software immediately prompts for the timing increment. Only two values are allowed by the PLC 5/12. They are 1.0 second, and 0.01 second. After the increment value of 1.0 has been keyed in and entered, the software prompts for the Preset value. Any decimal value from 1 to 32 767 is legal.

Other types of timers. PLCs usually have the capability to implement Off-delay timing and retentive timing. Briefly, an Off-delay timer begins timing when its conditional instructions produce logic discontinuity; we say that the TOF instruction must become FALSE to start the timer running. The TOF instruction is reached through the F10 Others key from the Basic Instruction Menu of Fig. 3–8.

A retentive timer differs from the On-delay and Off-delay timer functions in that it retains its accumulated value if its conditional instructions stop the timing process by

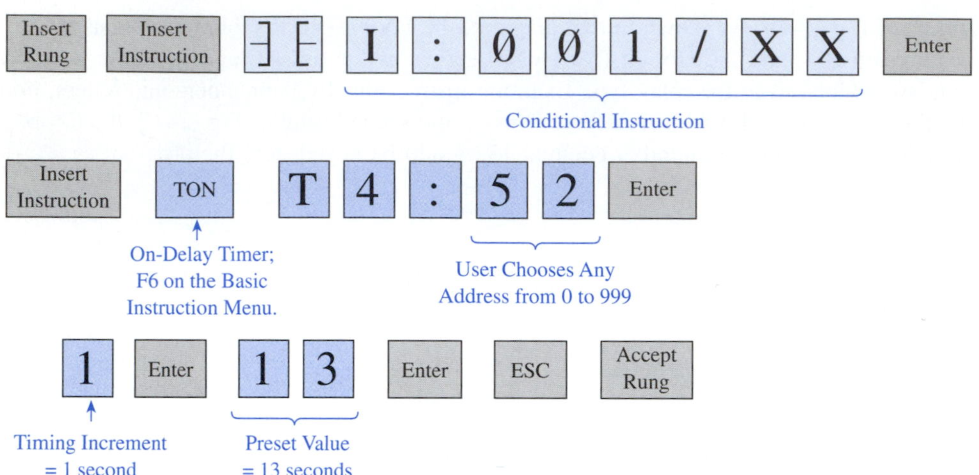

FIGURE 3–20
Keystroke sequence for the top rung of Fig. 3–19.

going FALSE. When the conditional instructions become TRUE again on a later scan cycle, the retentive timer takes up where it left off. In other words, the total time need not be accumulated in an uninterrupted fashion. It can be accumulated "in pieces." An On-delay timer cannot do this because it resets to zero whenever its conditional instructions go FALSE, even if for only a single scan cycle. Similarly, an Off-delay timer resets to zero if its conditional instructions go TRUE, however momentarily.

Since the retentive timing function can't reset simply through a change in its conditional instructions, it must be deliberately reset by a separate instruction in another rung. The retentive timer instruction (RTO for Retentive Timer On-delay) is reached through the Others key from the Basic Instruction Menu. Its Reset instruction (RES) must appear on a different rung of the program; it is linked to the RTO instruction by their matching addresses—T4:345, for example.

3-3-2 Programming a Counter

A counter is very similar to a timer in its programming and addressing. To refer to a counter, we choose a word-address from the C5 subsection of variable-data memory. Any number from 0 to 999 is legal, as the memory table indicates.

The Accumulated value (the number of count events sensed so far) and the Preset value are stored in two other associated words that are automatically handled by the processor. Those two additional words do not appear in the memory table of Fig. 3–10. This is just like a timer.

In the ladder-logic program of Fig. 3–21, we have chosen the word-address 175 to refer to the counter. The three bits that keep track of the counter's current status are stored in this word, addressed by their mnemonics, which are CU (Counting Up) for the counter rung is TRUE, DN for counted out (DoNe), and OV (OVerflow) for the counter has exceeded its maximum capacity and therefore lost its ability to recognize that it has already counted out.

When the counter counts up to its preset value, its *count-complete* bit or *counted-out* bit changes from 0 to 1. The count-complete bit is addressed as DN, as shown by the examine-On C5:175/DN instruction in the second rung of Fig. 3–21.

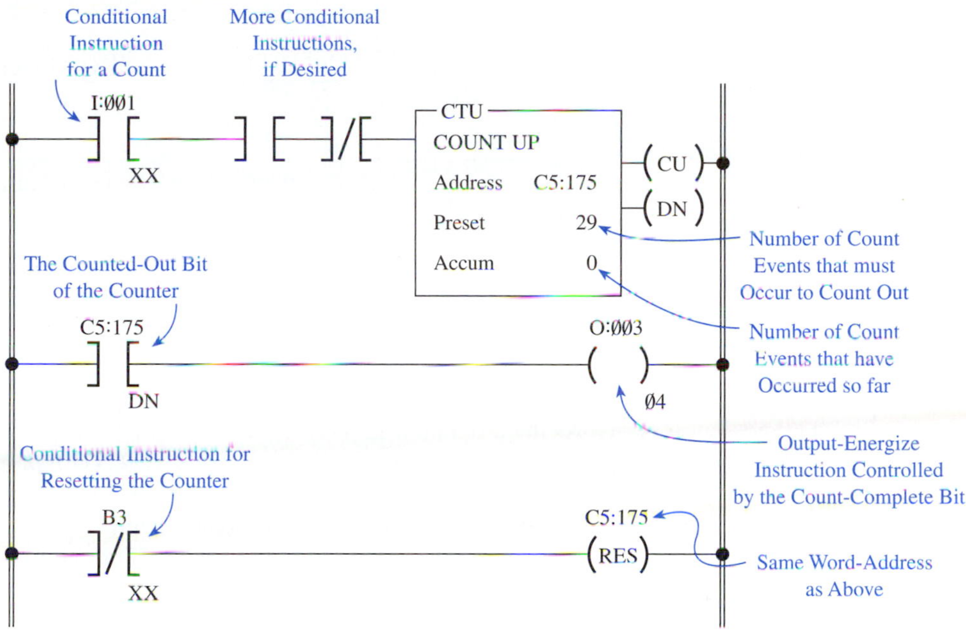

FIGURE 3–21
Ladder-logic relating to an up-counter instruction.

If a counter keeps on counting beyond its preset value, its count-complete bit remains a 1. In the second rung of Fig. 3–21, the O:ØØ3/Ø4 output-energize instruction will be maintained TRUE by the C5:175/DN examine-On instruction as long as the counter's accumulated value is equal to or greater than 29, the Preset value, but not greater than its maximum possible value of 32 767 (decimal).

To increment a counter (advance its count by 1) it is necessary for the CTU rung conditions to be FALSE on one scan cycle and then become TRUE on the next scan cycle. Simply maintaining a TRUE rung condition does not affect a counter. Only a FALSE-to-TRUE transition affects a counter.

A counter must be deliberately reset to zero by a special instruction on a different rung. The reset instruction must have the same word-address as the count instruction. This is illustrated on the third rung of Fig. 3–21, where the counter-reset instruction RES is accompanied by the word-address C5:175.

To enter the count-up instruction into the program of Fig. 3–21, the keystroke sequence is

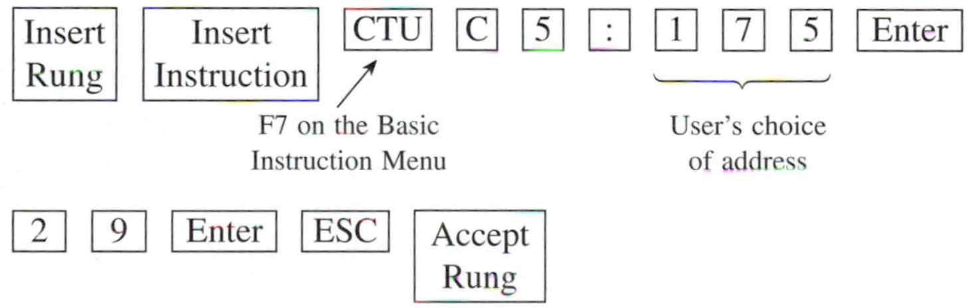

The examine-On instruction for count-complete condition on the second rung is keyed in as

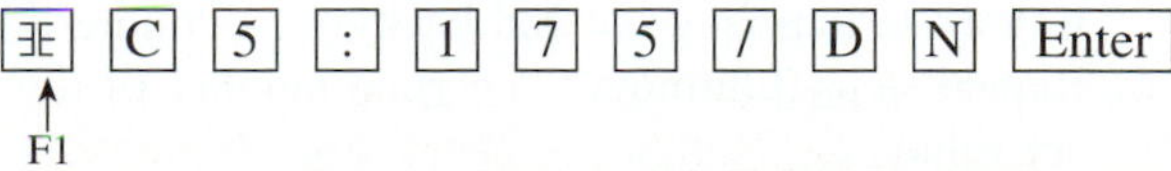

To enter the reset rung, the keystroke sequence is

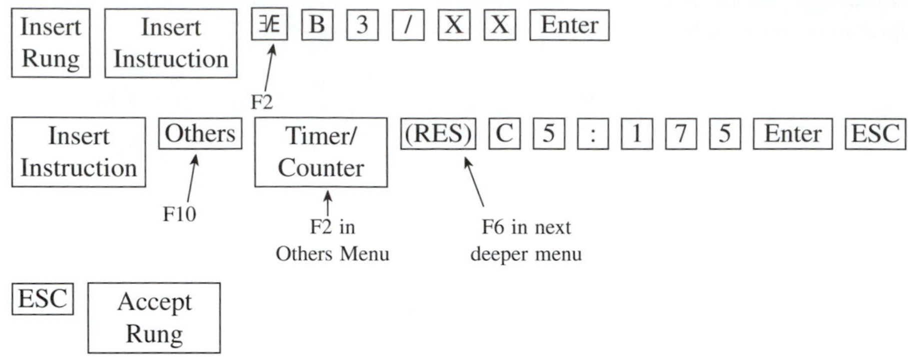

As the program is executing, the counter's actual accumulated value appears on the display screen in the Accum position inside the CTU box of Fig. 3–21.

Down-counter. The foregoing discussion applies to up-counters. Addressing a down-counter is similar. It is programmed by the CTD (count-down) key, reached through the [Others] [F10] key of the Basic Instruction Menu of Fig. 3–8. A down-counter decrements (reduces its value by 1) each time its rung condition changes from FALSE to TRUE. After a count has registered, the rung condition must return to FALSE on a subsequent scan cycle in order to set up the next FALSE-to-TRUE transition that causes another down-count to take place. This is the identical requirement as for an up-counter. The down-counter instruction is usually used in conjunction with an up-counter instruction (the CTU and CTD instructions have the same word-address) to produce an up/down counter.

3-4 ■ A MACHINING APPARATUS UTILIZING TIMING AND COUNTING FUNCTIONS

Figure 3–22 illustrates an apparatus for milling a deep channel into a workpiece. The channel is deepened on each horizontal pass of the milling bit by raising the workpiece slightly each time. Here is the sequence of events:

1. The workpiece is clamped in position on the lift table. This may be done manually, or it may be accomplished by some related piece of automated machinery. Two sensors indicate when the workpiece is properly positioned and clamped.
2. The cylinder extends slowly to the right. During the stroke the high-speed milling bit makes a cut in the workpiece.
3. When the cylinder has completed its left-to-right cutting stroke, it actuates LS2. The slow-speed lift motor then runs for a certain period of time to raise the lift table a slight distance. The cylinder pauses in the extended position for a much longer period of time to allow the milling bit to cool.
4. The cylinder retracts slowly to the left. During this stroke, the milling bit deepens the cut, since the workpiece is now higher than before.
5. When the cylinder has completed its right-to-left cutting stroke, it actuates LS1. The lift motor then raises the workpiece a little farther, and the cylinder again pauses to cool off the milling bit. When LS1 is actuated, the cylinder has completed one cutting cycle (stroking out and back), so a counter increments.
6. Repeat steps 2 through 5 a certain number of times, determined by the counter's preset value.

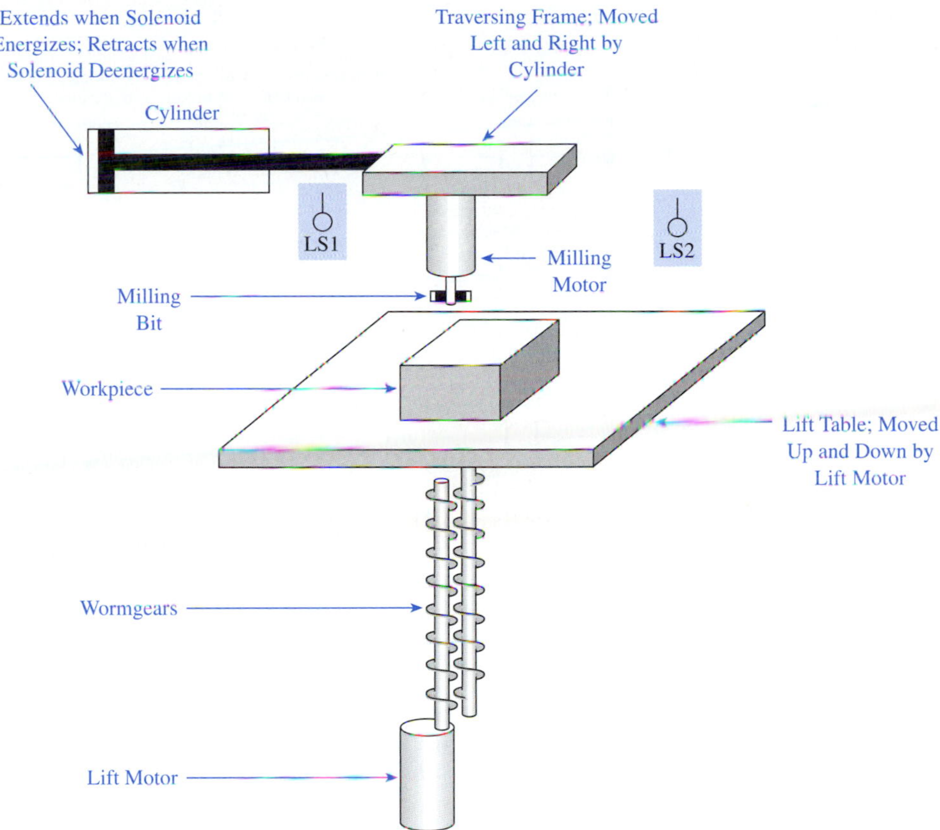

FIGURE 3–22
Physical layout of the channel-milling system.

7. The machining is complete when the counter has counted out. The lift motor then runs in the opposite direction for the appropriate amount of time to return the lift table to its starting height.
8. The workpiece is removed either manually or automatically. This resets the counter and prepares the PLC for a new workpiece.

Suppose that we know from experience that a lift-motor running time of 2.5 seconds is reasonable for the type of material we intend to cut. Also suppose that 15 seconds is a reasonable pause time for cooling the milling bit. Taking into account the known distance that the workpiece will be raised during a 2.5-second run-up time and the final channel depth we wish to achieve, suppose that it will require 18 cylinder cycles (36 cutting strokes) to achieve that depth. We then have all the data we need to program the PLC.

A ladder-logic representation of the program is given in Fig. 3–23. The addressing schedule is shown in Table 3–3. Inputs are listed in part (**a**), outputs in (**b**), internal-logic instructions in (**c**), timers in (**d**), and the lone counter in part (**e**).

Here is how the program works. With the workpiece properly positioned and clamped, examine-On instructions I:ØØ1/Ø3 and I:ØØ1/Ø4 on line 1 have logic continuity. The up-counter is reset at this time (we will see later why this is so); this causes the counted-out bit C5:175/DN to be LO, establishing continuity through the C5:175/DN examine-Off instruction. When the START pushbutton is pressed or an equivalent automatic signal occurs, all the line-1 rung conditions become TRUE and internal-logic instruction B3/4 becomes TRUE.

FIGURE 3–23
Commented user-program for controlling the channel-milling system.

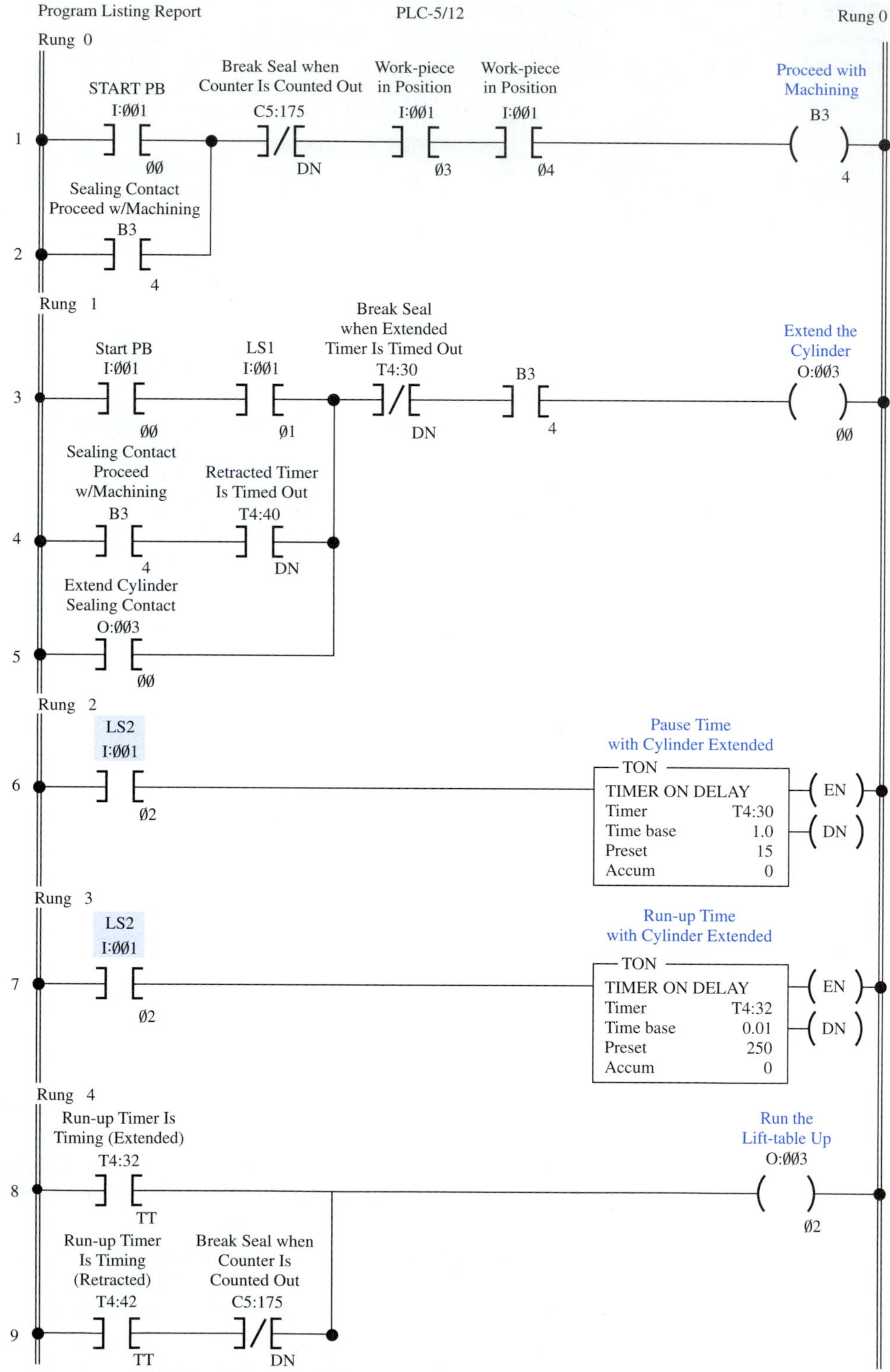

FIGURE 3–23
(*continued*)

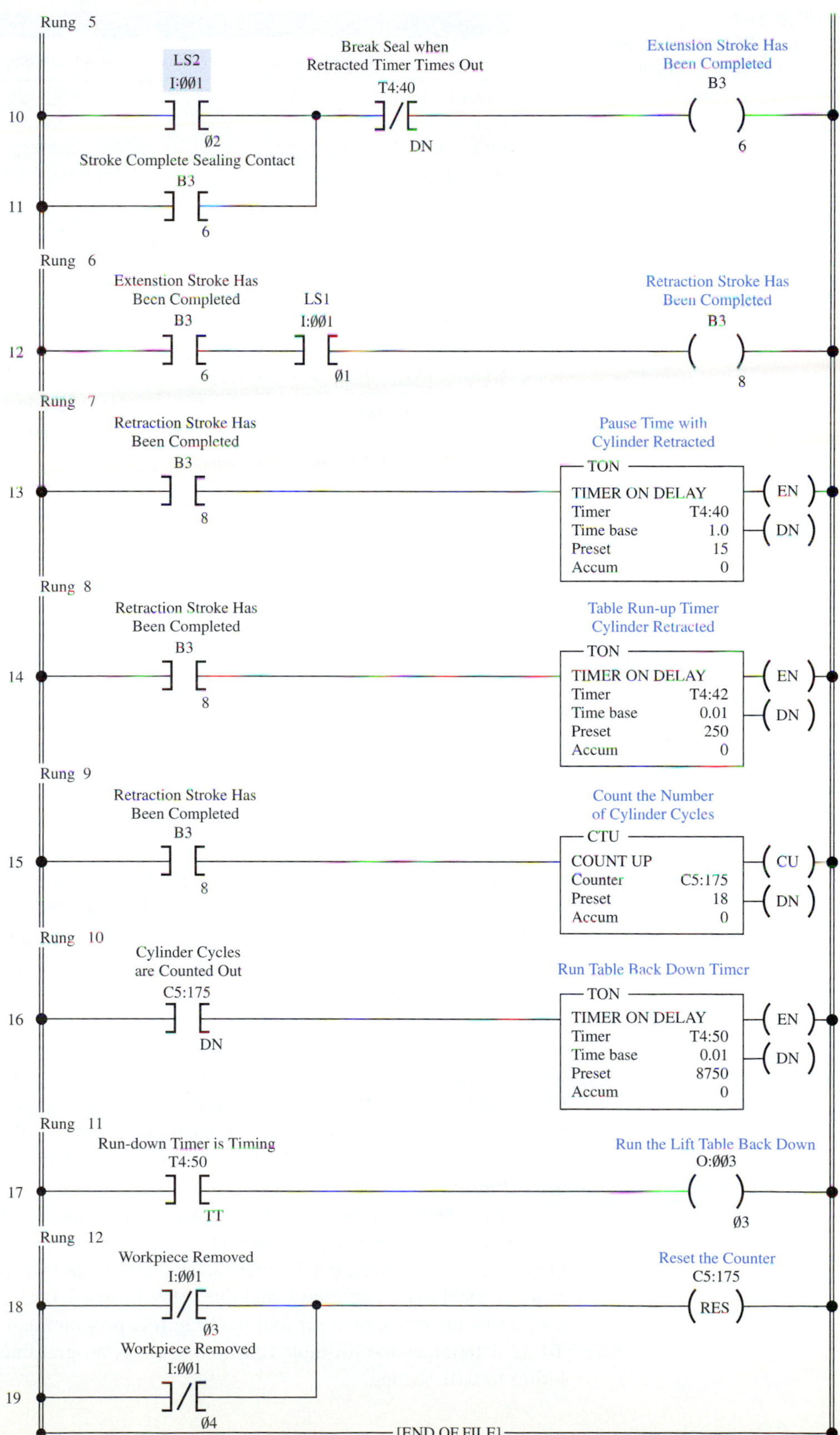
Rung 5
LS2
I:Ø01
Ø2
Break Seal when
Retracted Timer Times Out
T4:40
DN
Extension Stroke Has
Been Completed
B3
6
10
Stroke Complete Sealing Contact
B3
6
11
Rung 6
Extenstion Stroke Has
Been Completed
B3
6
LS1
I:Ø01
Ø1
Retraction Stroke Has
Been Completed
B3
8
12
Rung 7
Retraction Stroke Has
Been Completed
B3
8
Pause Time with
Cylinder Retracted
TON
TIMER ON DELAY
Timer T4:40
Time base 1.0
Preset 15
Accum 0
EN
DN
13
Rung 8
Retraction Stroke Has
Been Completed
B3
8
Table Run-up Timer
Cylinder Retracted
TON
TIMER ON DELAY
Timer T4:42
Time base 0.01
Preset 250
Accum 0
EN
DN
14
Rung 9
Retraction Stroke Has
Been Completed
B3
8
Count the Number
of Cylinder Cycles
CTU
COUNT UP
Counter C5:175
Preset 18
Accum 0
CU
DN
15
Rung 10
Cylinder Cycles
are Counted Out
C5:175
DN
Run Table Back Down Timer
TON
TIMER ON DELAY
Timer T4:50
Time base 0.01
Preset 8750
Accum 0
EN
DN
16
Rung 11
Run-down Timer is Timing
T4:50
TT
Run the Lift Table Back Down
O:Ø03
Ø3
17
Rung 12
Workpiece Removed
I:Ø01
Ø3
Reset the Counter
C5:175
RES
18
Workpiece Removed
I:Ø01
Ø4
19
[END OF FILE]

TABLE 3–3 Choosing Addresses for the Milling System's Control Program

	INPUT	ADDRESS
	START PB	I:001/00
	LS1	I:001/01
(a)	LS2	I:001/02
	Workpiece is in position	I:001/03 and I:001/04
	OUTPUT	**ADDRESS**
	Cylinder-extend solenoid	O:003/00
(b)	Run-up contactor	O:003/02
	Run-down contactor	O:003/03
	INTERNAL-LOGIC INSTRUCTION	**ADDRESS**
	Proceed with machining	B3/4
(c)	Extension stroke has been completed	B3/6
	Retraction stroke has been completed	B3/8
	TIMER	**WORD**
	Hold cylinder extended	T4:30
	Run up while cylinder extended	T4:32
(d)	Hold cylinder retracted	T4:40
	Run up while cylinder retracted	T4:42
	Run down after last cycle	T4:50
	COUNTER	**WORD**
(e)	Count the number of cylinder cycles	C5:175

Instruction B3/4 seals itself in via the instruction on line 2. This seal will hold until the machining process is completed. At that time the counter will count out and the count-complete bit, C5:175/DN, will go HI. Then the examine-Off instruction on line 1 will become FALSE, and the rung's seal will be broken. So we needn't have any further question about bit B3/4—it's HI for the rest of the way.

On line 3, START instruction I:001/00 combines with the T4:30/DN examine-Off instruction to produce overall rung continuity. This is so because On-delay timer T4:30 is now reset, and its timed-out bit at address T4:30/DN is a 0. Output-energize instruction O:003/00 becomes TRUE and seals itself in for the time being, via line 5. Slot 3 holds a genuine output-module group, with the cylinder-control solenoid wired to terminal 00, as stated in Table 3–3(**b**). Therefore the solenoid energizes, the hydraulic valve shifts, and the cylinder gets going.

This state of affairs within the program remains unchanged through many scan cycles, until the slow-moving cylinder has completed its cutting stroke and has actuated LS2. On the scan cycle immediately after that actuation, the I:001/02 input instructions on lines 6 and 7 yield logic continuity, and On-delay timers T4:30 and T4:32 start timing. Timer T4:30 sets the pause time for tool-cooling. It is programmed to run for 15 seconds. Timer T4:32 determines the lift-table run-up time; it is programmed for 2.5 seconds with resolution to 0.01 second.

When timer T4:32 starts timing, instantaneous bit T4:32/TT becomes a 1 and remains that way until the timer times out. Therefore the instruction on line 8 causes output-energize instruction O:ØØ3/Ø2 to become TRUE. This alters the output image file, causing ac power to be applied to output terminal Ø2 on the next output scan. The lift-motor run-up contactor energizes (Table 3–3**b**), and the motor begins to raise the table. After 2.5 seconds of lifting time, the timer times out. Timer is Timing bit T4:32:TT goes LO, causing output image file bit O:ØØ3/Ø2 to go back LO. On the next output scan, ac power is removed from terminal Ø2 of the output module, and the run-up contactor drops out. The table and workpiece freeze at their new elevation.

Meanwhile, down on line 10, limit-switch instruction I:ØØ1/Ø2 has logic continuity, and so does time-delay instruction T4:40/DN, since T4:40, the "pause with cylinder retracted" timer, is now reset. Internal-logic instruction B/6 becomes TRUE and seals itself via line 11 against deactuation of LS2.

All the while, timer T4:30 has been running. After 15 seconds, it times out, causing time-delay bit T4:30/DN to go HI. The examine-Off instruction on line 3 therefore makes that rung FALSE, and O:ØØ3/ØØ returns to 0. On the next output scan the hydraulic valve solenoid deenergizes, and the cylinder begins its right-to-left cutting stroke. As soon as LS2 is released, timers T4:30 and T4:32 are both reset (lines 6 and 7). The line 8 rung remains FALSE due to T4:32/TT remaining LO, but the rung on lines 10 and 11 remains TRUE by virtue of the sealing instruction B3/6.

At the completion of the retraction stroke, LS1 is actuated. Input image file bit I:ØØ1/Ø1 goes HI on the next input scan, so internal logic instruction B3/8 becomes TRUE on line 12. Therefore the three rungs on lines 13, 14, and 15 all become TRUE on that same scan. With the cylinder now retracted, timer T4:40, which produces the cooling pause, and timer T4:42, which produces the lift-table run-up, both start timing. Also, the line 15 rung has just experienced FALSE-to-TRUE transition, so counter C5:175 increments from 0 to 1, representing one cylinder cycle completed.

While timer T4:42 is timing, its instantaneous timer is timing bit is HI. Therefore line 9 has logic continuity through instructions T4:42/TT and C5:175/DN. The C5:175/DN counted-out bit is LO at this time, since the counter is not counted out. Output terminal O:ØØ3/Ø2 receives 120-V ac power on the next output scan, energizing the run-up contactor and raising the workpiece again. After a lift duration of 2.5 seconds, the rung loses continuity through T4:42/TT. Having risen the same distance as on the previous lift, the lift table freezes in this new position.

The cylinder stays retracted, allowing the milling bit to cool, until timer T4:40 times out. The T4:40/DN time-delay bit then becomes a 1, establishing rung continuity via lines 4 and 3, since bit B3/4 is sealed HI until the end of the machining process, and bit T4:30/DN is LO with timer T4:30 reset. The output image file bit O:ØØ3/ØØ immediately becomes a 1, sealing itself through line 5. This seal is needed against loss of continuity through line 4 when bit T4:40/DN goes back to 0, which will happen when timer T4:40 resets later in the program scan.

During the next output scan following the current execution of the program, output module O:ØØ3/ØØ will receive ac power to energize the hydraulic valve solenoid. At that time the cylinder will start another left-to-right cutting stroke, commencing the second cycle.

Later in the current program execution, before the second cylinder cycle begins, the T4:40/DN examine-Off instruction on line 10 causes that rung to go FALSE. Internal-logic bit B3/6 goes LO, breaking the logic continuity on lines 12, 13, 14, and 15. Timers T4:40 and T4:42 are reset, and the C5:175 count-up instruction goes back to FALSE, thereby setting up the next FALSE-to-TRUE transition, which will occur at the completion of the second cylinder cycle.

The system continues cycling in this manner until the eighteenth cycle. When the cylinder completes the right-to-left cutting stroke of the eighteenth cycle, it actuates LS1 and produces the eighteenth FALSE-to-TRUE logic transition on line 15. The accumulated value of the counter then matches its preset value, so counted-out bit C5:175/DN goes HI. This yields logic continuity on line 16, which starts the T4:50 timer. While T4:50 is timing, the rung on line 17 has continuity through the T4:50/TT examine-On instruction. Output module O:003/03 receives ac power, causing the run-down contactor to energize,* as specified in Table 3–3**b**. The T4:50 timer is programmed to keep the run-down contactor energized for 87.5 seconds, which is the same amount of time that the run-up contactor spent in the energized state, since

$$\frac{2.5 \text{ s}}{\text{lift}} \times \frac{2 \text{ lifts}}{\text{cycle}} \times 17.5 \text{ cycles} = 87.5 \text{ s}$$

Therefore timer T4:50 causes the workpiece and lift table to be returned to their original elevation.

Counter C5:175 performs another function besides initiating the running down of the workpiece. On line 1, the C5:175/DN examine-Off instruction breaks the seal on internal-logic instruction B3/4, which has been maintained since the beginning of the machining process. The B3/4 examine-On instruction on line 4 therefore prevents logic continuity when the pause-retracted timer T4:40 times out after 15 seconds. Output instruction O:003/00 does not become TRUE, so the cylinder remains in its retracted position. A nineteenth cycle does not occur.

When the workpiece is removed from its clamped position by manual or automated means, bits I:001/03 and I:001/04 go to 0. By virtue of the examine-Off instructions on lines 18 and 19, either bit has the capability of resetting the counter to zero. This prepares the program for the next machining process, which will begin via line 1 when a new workpiece is clamped into position and the START PB is pressed.

3-5 ■ OTHER PLC RELAY-TYPE FUNCTIONS

The user-program presented in Fig. 3–23 of Sec. 3-4 demonstrates the most common instructions that are utilized in PLC ladder-logic. There are many other instructions also available. Some of them are similar to the basic relay-type instructions already discussed, but many of them have to do with handling analog data, which is a job that relay-type instructions cannot perform. This section examines some of the PLC's additional relay-type instructions. Section 3-6 studies the PLC's ability to read in analog data from the I/O rack, then to manipulate that data to perform numerical calculations and comparisons in order to make decisions affecting the industrial machine or system.

In Chapter 9 we will undertake a study of complete closed-loop process control by a PLC. We will see then that a powerful modern PLC is capable of full three-function (Proportional, Integral, Derivative—PID) control of any analog process.

3-5-1 Latching Functions

A PLC can implement the operation of a *latch-unlatch* relay.** Such a relay has two electromagnetic coils, called the *latch coil* and the *unlatch coil*. When the latch coil is ener-

---

*The motor wiring would be arranged so that the run-down contactor reverses the direction of current through one of the motor's windings.

**Also called a *latch-trip relay*, or simply a *latching relay*.

gized, the relay armature moves and the contacts all change to their nonnormal states. However, unlike a standard relay, when the latch coil is deenergized, the armature does not return the contacts to their normal states. Instead, the armature is mechanically latched, or held in place, so the contacts remain in their nonnormal states. To return a latch-unlatch relay to its normal state, the unlatch coil must be energized after the latch coil is deenergized.

There are some control situations where the latch-unlatch behavior is preferable to standard relay behavior. In such situations we use the *output-latch* and *output-unlatch* instructions rather than the simple output-energize instruction. These instructions are available from the Basic Instruction Menu at key [F4] for [OTL] (OuTput Latch) and at key [F5] for [OTU] (OuTput Unlatch).

When an output-latch instruction becomes TRUE by virtue of logic continuity through its rung, its address bit goes HI as usual, but if logic continuity is lost on subsequent scan cycles, its address bit does not return to LO. Instead, it remains HI until the output-unlatch instruction with the same addressed bit becomes TRUE by virtue of continuity through *its* rung.

For an example of output-latch and output-unlatch performance, refer to Fig. 3–24. On the top rung, if the examine-On I:ØØ1/12 instruction gives continuity on a particular program scan, the Output-Latch B3/37 instruction of that rung stores a 1 at address location B3/37. On subsequent program scans it doesn't matter whether or not address I:ØØ1/12 maintains a HI state. Even if it goes to the LO state, so that the top rung of Fig. 3–24 loses its rung continuity, address location B3/37 retains its 1. The only way to remove the 1 from address B3/37 is to make the Unlatch B3/37 instruction TRUE on the bottom rung. In Fig. 3–24 this can be accomplished only by a HI appearing in the input image file address I:ØØ1/16 on some later scan.

Many people consider latch-unlatch relays desirable because they are not affected by momentary power outages. This advantage, which exists in the electromagnetic relay realm, does not exist in the PLC realm because the output image file and the variable-data memory's Isolated Bit subsection (B3) are not affected by momentary power outages anyway, due to the PLC's automatic battery backup (uninterruptible power supply).

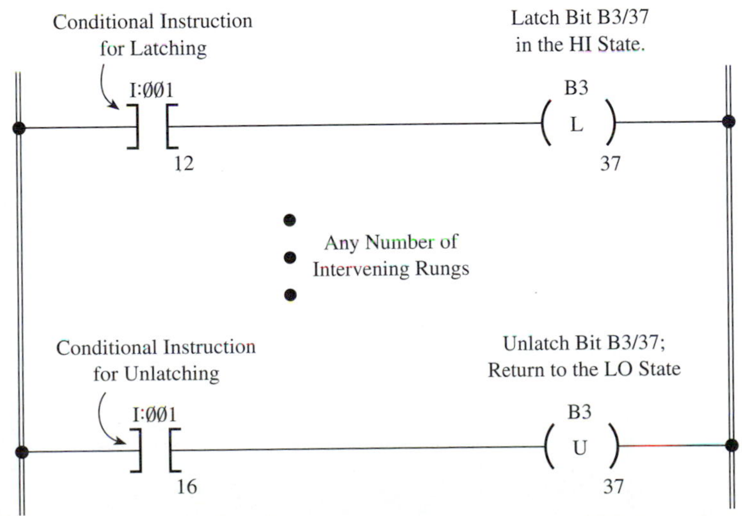

FIGURE 3–24
The OuTput-Latch -(L)- and OuTput-Unlatch -(U)- instructions are used in pairs, with both instructions referring to the same memory address.

3-5-2 Master Control Reset

The master control reset (MCR) function establishes an entire section of the user-program, a zone, in which all the output-type instructions can be disabled (made FALSE). A conditional MCR instruction marks the beginning, or top rung, of the controlled zone; refer to Fig. 3–25.

The processor encounters the beginning MCR instruction-rung on each scan cycle. If this conditional MCR instruction is TRUE, the program functions as it normally would. All the output-type instructions within the controlled zone respond to the rung conditions that exist during this scan. However, if the MCR instruction is FALSE, all the output-type instructions become FALSE regardless of the rung conditions that exist during this scan. A second MCR instruction, which is unconditional, marks the end, or bottom rung, of the program's controlled zone.

In Fig. 3–25, if examine-On instruction I:ØØ1/14 finds a 0 at that address, the conditional MCR instruction on line 1 goes FALSE. Therefore the rungs on lines 2 through 6 all go FALSE regardless of their conditional instructions V:VVV/VV through Z:ZZZ/ZZ. Thus, on line 2 bit address B3/32 goes LO. On line 3 On-delay timer T4:56 resets to 0 on this scan. On line 4 counter C5:63 is FALSE, perhaps setting up a FALSE-to-TRUE

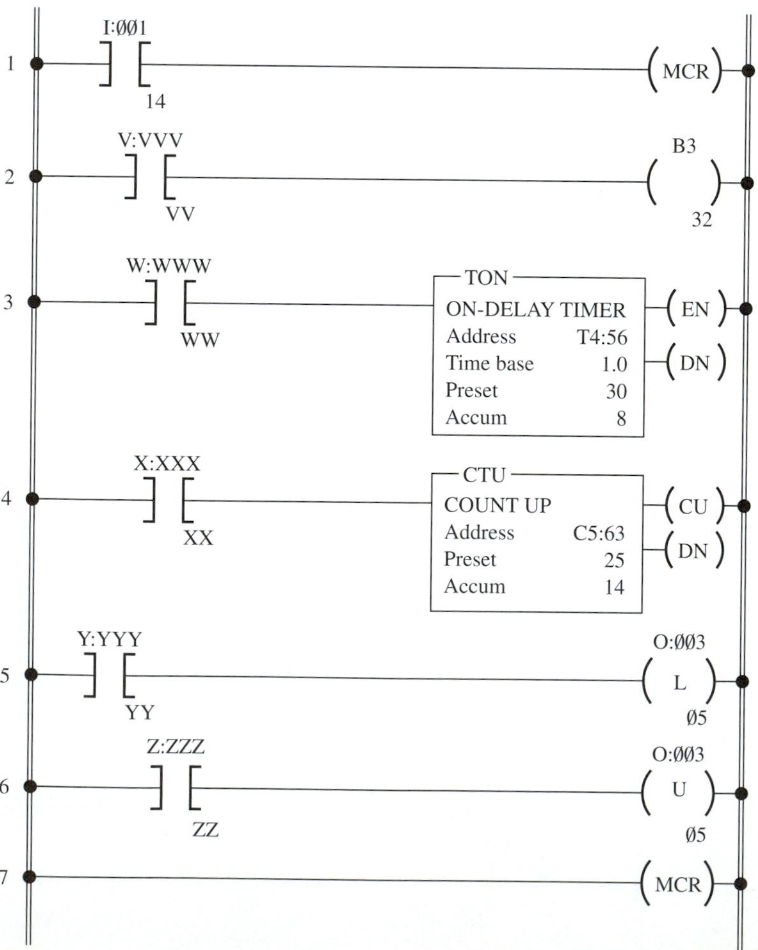

FIGURE 3–25
MCR instructions always appear in pairs: a conditional one to begin the controlled zone, and an unconditional one to mark the end of the controlled zone. If the starting MCR instruction is TRUE, everything functions normally; but if the starting MCR instruction is FALSE, all non-retentive-type outputs are reset to 0 during this scan regardless of their individual rung conditions.

transition on a later scan after MCR has returned to TRUE. However, counter C5:63 maintains its accumulated value of 14; it does not reset to 0. Counters are *retentive*-type instructions, retaining their most recent value unless explicity told to reset by an RES instruction. Simple FALSEness of a counter rung is not sufficient to reset its accumulated value to 0.

Lines 5 and 6 in Fig. 3–25 are the Latch and Unlatch instructions, respectively, for output image file address O:003/05. Since both of these rungs are guaranteed to be FALSE by the FALSEness of the top MCR rung, the MCR has the effect of freezing address O:003/05 in whatever state it was in during the scan just prior to MCR becoming FALSE. For instance, if O:003/05 was a 1 because Y:YYY/YY was a 1 on the immediately preceding scan, it becomes impossible to unlatch O:003/05 during this MCR-FALSE scan because the Unlatch instruction on line 6 cannot be TRUE regardless of the state of Z:ZZZ/ZZ. Likewise, if O:003/05 was a 0 on the immediately preceding scan, it is stuck at 0 during the current scan regardless of the Y:YYY/YY condition on line 5.

The MCR instruction is sometimes used as a safety feature when it is desired to disable all outputs if certain conditions occur. It is reached through the [F10] [All Others] key from the Basic Instruction Menu.

3-5-3 Immediate I/O

As explained in Section 3-1-2, critical input and output transfers can be carried out immediately during user-program execution rather than waiting for the I/O scan that follows program execution. In a model PLC 5/12 an immediate input instruction -(IIN)- cannot be applied to an individual bit of an input module; rather, it must bring in all 16 bits of the addressed input module, thereby immediately updating an entire 16-bit word of the input image file. For example, line **a** of Fig. 3–26 causes temporary suspension of program execution to allow immediate updating of the entire input image file word 001 to reflect the current conditions of all the 16 input terminals at slot 1 of rack 00.

Likewise for the PLC 5/12's immediate output instruction, -(IOT)-. It handles an entire 16-bit output word, not individual bits within a word. If the conditional instructions furnish rung continuity on line **b** of Fig. 3–26, all 16 bits of output image file word 003 are immediately transferred to the output module in slot 3 of rack 00 before program execution picks up where it left off.

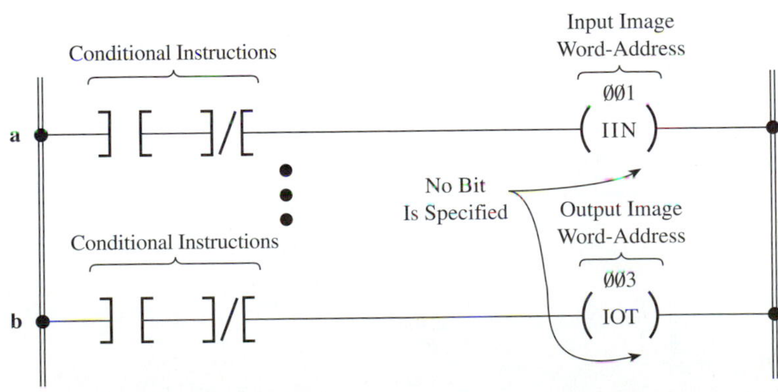

FIGURE 3–26
Immediate input -(INN)- and immediate output -(IOT)- instructions. They deal in 16-bit words, not individual bits.

3-5-4 Off-Delay Timing

In hardware timers, electromechanical and solid-state, Off-delay describes a timing process that begins when a signal is removed, not when a signal is applied. This idea was presented briefly in Secs. 2-12-1 and 2-12-3. Thus, for a relay-type Off-delay timer, the timing process begins when the control coil is deenergized; for electronic Off-delay the timing interval begins when the digital input signal goes LO.

Most PLCs also have an Off-delay timing instruction which begins its timing interval when its rung conditions go False. In the PLC5, this instruction is called Timer Off-delay, acronym TOF. An example is shown in Fig. 3–27. The Off-delay timing instruction is commonly used when the program logic must monitor a condition which is usually true and which, if it should become false, must be corrected back to its true state within a certain amount of time. For example, in Fig. 3–28(a) suppose that the bladder tank is capable by itself of maintaining adequate liquid outflow for 120 seconds. Therefore if the inflow-monitoring pressure switch indicates that the tank's liquid source has failed, the bladder tank can maintain in operation any downstream devices that depend on a flow of cooling liquid, for a maximum of 120 seconds. If the inflow-source pressure switch has not recovered within that time duration, the downstream devices must be deenergized.

3-5-5 Retentive Timing

Standard timers accumulate their time all in one continuous process. They cannot accumulate a portion of their time, then stop accumulating for a while, then resume later from the value at which they left off. This is true for both On-delay and Off-delay timers, for

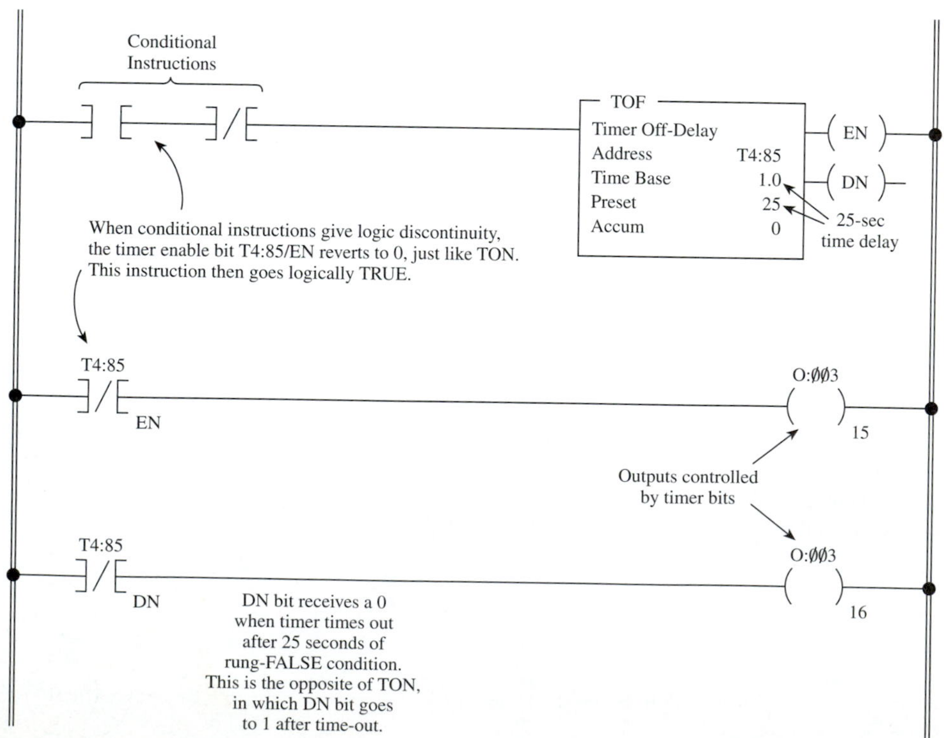

FIGURE 3–27 Ladder logic for the Off-delay timer instruction TOF.

Bladder Tank
compressed air
liquid
Pressure switch
check valve (one-way valve)
source of high-pressure liquid (from a pump)
cooling liquid to downstream devices

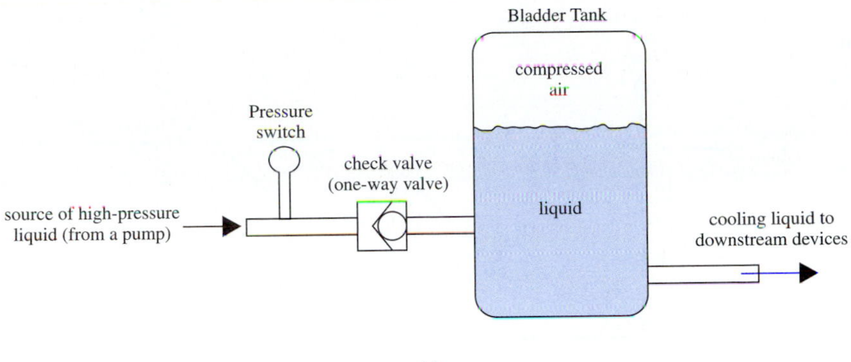

(a)

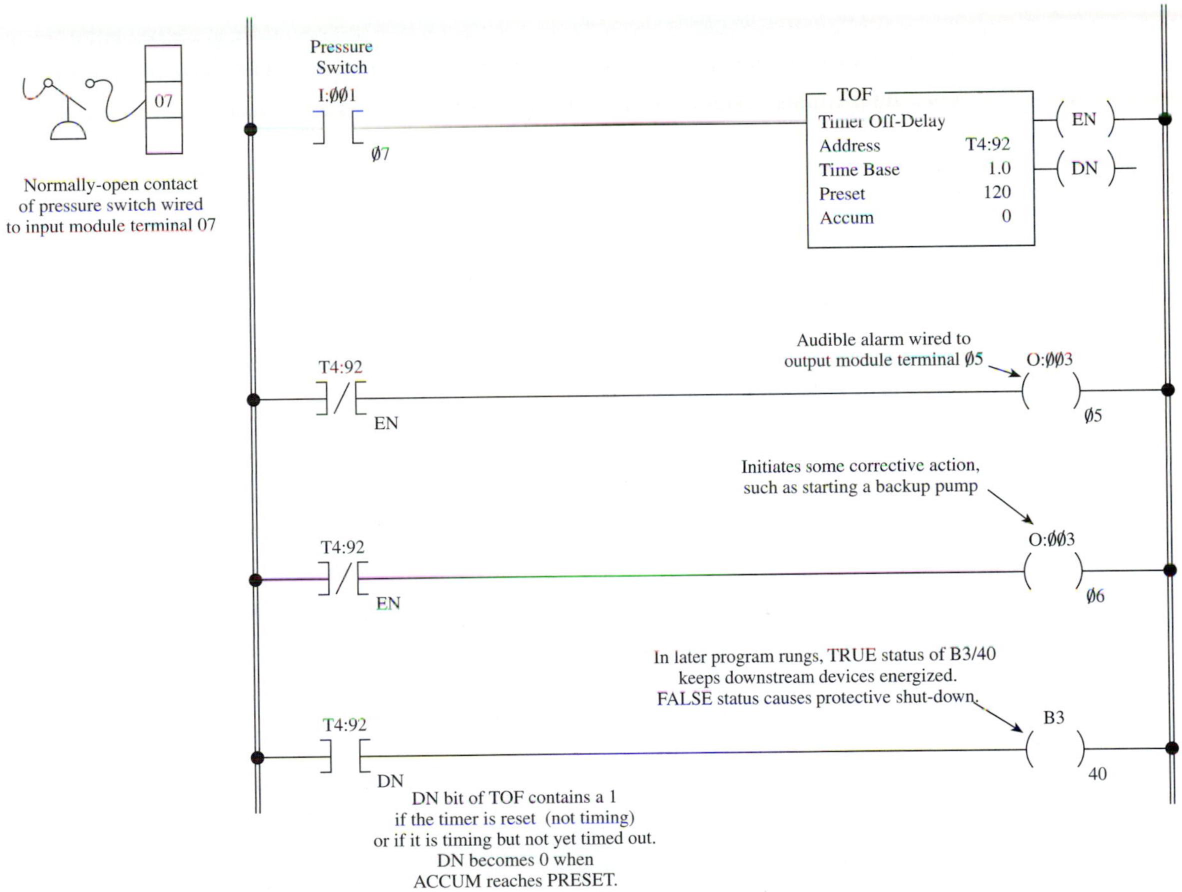

07
Normally-open contact of pressure switch wired to input module terminal 07
Pressure Switch
I:ØØ1
Ø7
TOF
Timer Off-Delay
Address T4:92
Time Base 1.0
Preset 120
Accum 0
EN
DN
Audible alarm wired to output module terminal Ø5
T4:92
EN
O:ØØ3
Ø5
Initiates some corrective action, such as starting a backup pump
T4:92
EN
O:ØØ3
Ø6
In later program rungs, TRUE status of B3/40 keeps downstream devices energized. FALSE status causes protective shut-down.
T4:92
DN
B3
40
DN bit of TOF contains a 1 if the timer is reset (not timing) or if it is timing but not yet timed out. DN becomes 0 when ACCUM reaches PRESET.

(b)

FIGURE 3–28
Example of Off-delay timing for shutting down process devices if a loss of cooling flow is not quickly restored.

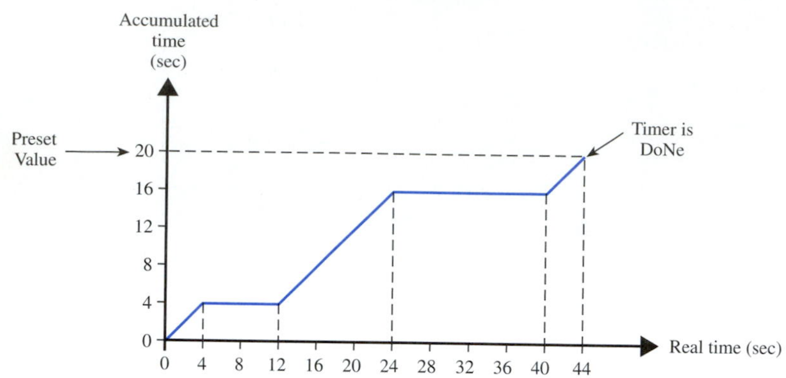

FIGURE 3–29
Accumulating 20 seconds in pieces by a retentive timer. The timing conditions cease to be TRUE after 4 seconds, so the timer holds its 4-second accumulated value until timing resumes. At 12 seconds the timing conditions become TRUE again and continue TRUE until the 24-second moment, so the timer accumulates an additional 12 seconds for a total of 16 seconds. Then it stops again from 24 to 40 seconds, resuming from 40 to 44 seconds.

both hardware and software (PLC) situations. Such standard timers are referred to as *non-retentive,* because they do not retain partially accumulated time. They simply reset to zero when their timing conditions cease to be satisfied.

In some control situations it is desirable to be able to accumulate time "in pieces." That is, we wish for our timer to retain its recently accumulated time value even when it ceases timing; then it begins from that value when it resumes timing in the future. An example of such timing is shown in Fig. 3–29. Such timing capability is necessary whenever the industrial process is unable to give the product its required total exposure all as one event, but must break it into several events. Some processes involving heat-producing chemical reactions have this feature; they require that the catalyst be withdrawn to stop the reaction while the residual heat energy produced by the reaction is dissipated. After the energy has dissipated, the catalyst is reinserted and process timing resumes.

The PLC5 instruction that implements retentive timing is called Retentive Timer On-delay, acronym RTO. As its ladder symbol shows in Fig. 3–30, it requires an additional rung for the purpose of resetting its accumulated value to zero.

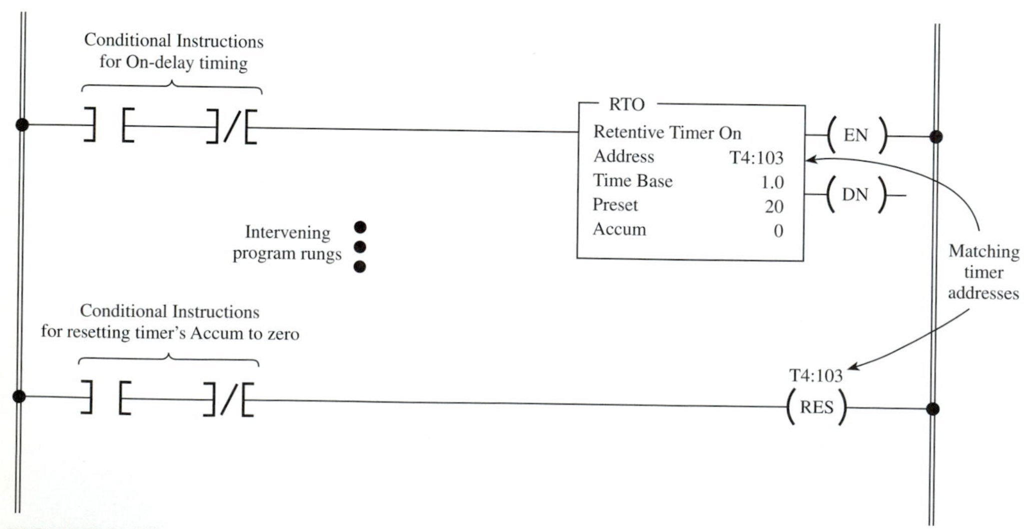

FIGURE 3–30
The RTO Retentive Timer On-delay instruction requires a separate Reset (RES) rung elsewhere in the program.

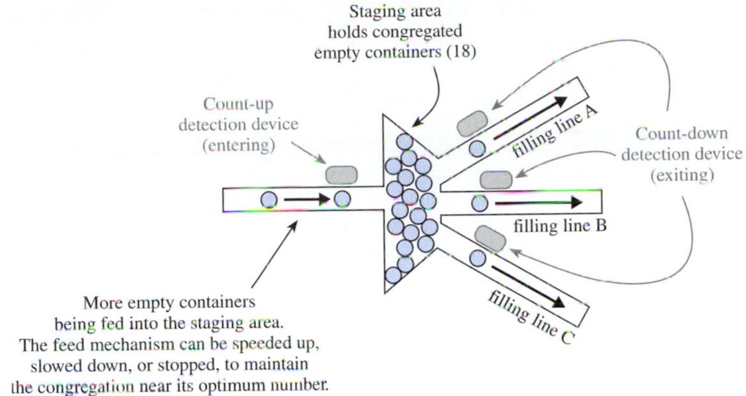

FIGURE 3–31 Automatic container-filing operation using up-down-counting.

3-5-6 Up-Down-Counting

Some industrial operations require that objects be counted when entering and when leaving a certain zone or area. The goal of the control logic is to ensure that there is at least a sufficient number of objects in the zone, but not too great an oversupply. For example, Fig. 3–31 shows an automatic container-filling operation in which empty containers congregate in a staging area. From there they enter one of three filling lines, A, B, or C. The PLC program will attempt to balance the entry of new containers from the left in the figure against the exit of containers from the staging area at the right. An up-counting detector at the left increments the counter's accumulation register, while any one of the three down-counting detectors at the right will decrement the accumulation register.

By continually comparing the actual number of containers with the optimum number, the program is able to adjust the feedrate from the left. Figure 3–32 shows only the counting logic for such a task. It does not show any of the arithmetic comparison instructions and output instructions associated with the control.

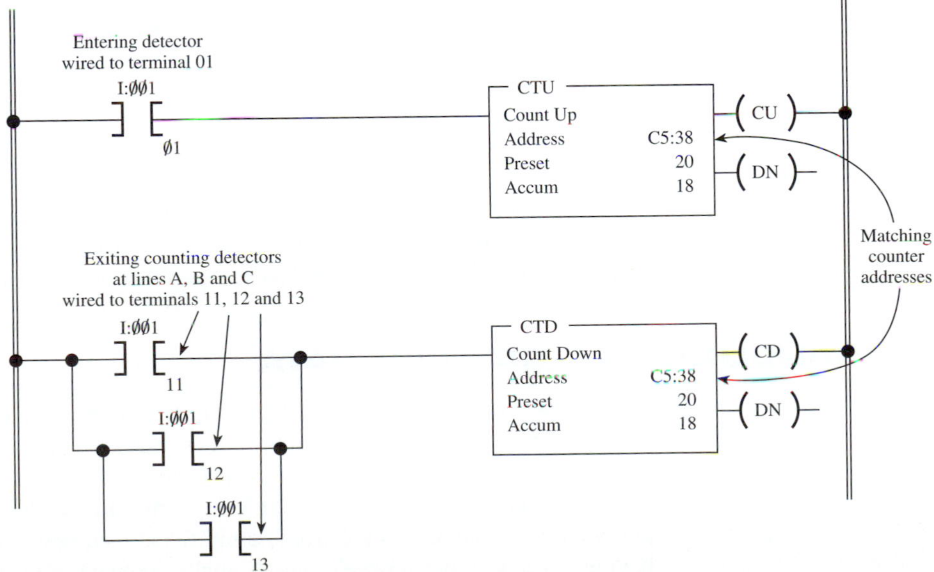

FIGURE 3–32 In the PLC5, up-down-counting involves the use of two separate instructions, CTU and CTD, both referring to a common counter address.

A FALSE-to-TRUE logic transition at the CTU rung will increment the accumulated value of counter C5:38. A FALSE-to-TRUE transition at the CTD rung will decrement C5:38.

3-5-7 The Step-Over Logic Structure

A very common task in machine-control program logic is to begin one action, call it action B, only when the logic is satisfied that a prior action, call it action A, has been successfully completed. There is a standard program-design scheme for accomplishing this. It is called by some people the *step-over* logic function. It involves the use of a "step-over" internal logic bit that becomes true between the two A and B actions. Refer to Fig. 3–33.

Figure 3–33(a) is a replication of the two-cylinder machine-tool drilling apparatus of Fig. 1–17. Recall that the control logic must extend cylinder A to actuate LS2, then retract cylinder A until it actuates LS1. Only then does the logic begin the extension of cylinder B.

The same is true for the PLC program logic shown in Fig. 3–33(b). Extension of cylinder A is accomplished by output-energize instruction O:003/01 on line 1, which is initiated by conditional instructions including examine-On I:001/01 (LS1 actuated). The key thing to realize about the step-over logic is this: The reactuation of LS1 at a later time will initiate the extension of cylinder B (O:003/15) on line 5, but that reactuation of LS1 must be combined with logic proving that cylinder A has first extended and then *retracted* to LS1. Cylinder B must not be energized by simple actuation of LS1 because if it were, both cylinders would stroke simultaneously.

On line 3 of the program segment, extension of cylinder A makes TRUE step-over logic bit B3/100, which latches itself via line 4. On a subsequent program scan the line 3 logic will become discontinuous when cylinder A actuates LS2 to break the seal on O:003/12, thereby beginning its retraction stroke.

At this point the program has prepared itself to initiate the following event, extension of cylinder B. It awaits only the full retraction of cylinder A. When that retraction is complete, the program reactuates LS1, providing complete continuity on line 5. Output O:003/15 goes TRUE and the extension of cylinder B is successfully begun.

On the very next program scan, examine-Off O:003/15 breaks the seal on step-over bit B3/100. The step-over function is restored to its rest condition, ready to handle the next machine cycle.

Cylinder B output O:003/15 seals itself via line 6 to complete the extension stroke. Actuation of LS3 in Fig. 3–33(a) breaks the seal on line 6, causing the B cylinder to retract to its starting position. This completes the logic sequence.

The ideas of the program segment of Fig. 3–33(b) are universal. They can be applied to any pair of machine events in which the later event, B, must begin only after the earlier event, A, has demonstrated its successful completion and return to starting position.

3-6 ■ HANDLING ANALOG INPUT DATA

For the user-program to make its logical decisions based on the measured value of some process variable (such as temperature), it is necessary to install an *Analog Input Module* in the PLC's I/O rack. An Analog Input Module slips into a single slot of the I/O rack, just like a standard 16-terminal, 120-V ac input module. You can think of an Analog Input Module as having 16 electronic amplifiers, each one single-ended (that is, each one having a single input terminal whose analog voltage is measured relative to a common

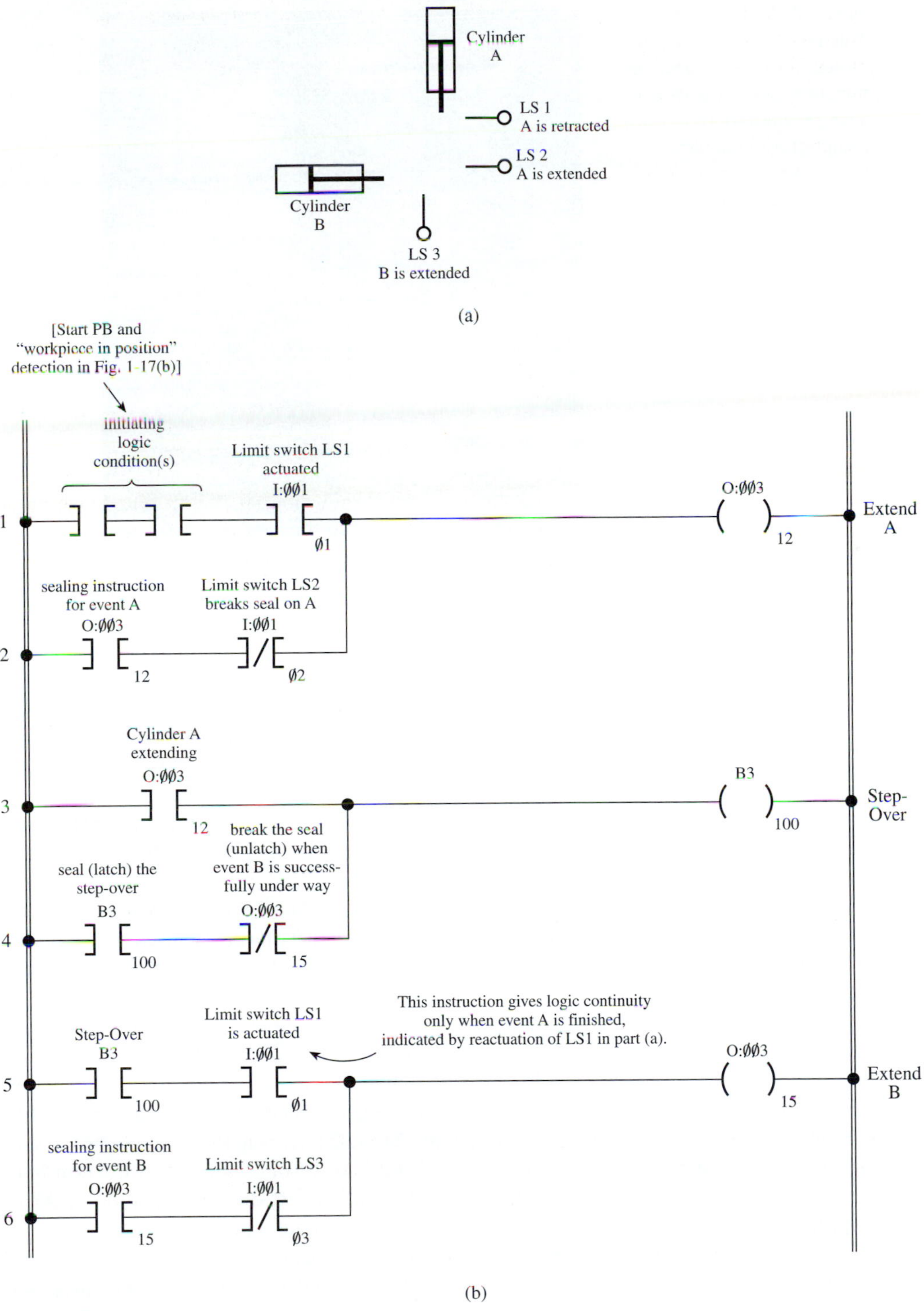

FIGURE 3–33
Step-over logic. (a) Mechanical apparatus. (b) The essential parts of a step-over PLC program segment.

PHOTO 3–1
This test fixture, under the direction of a PLC program, measures several performance parameters of a newly manufactured circuit board. Any out-of-spec parameters are adjusted by the PLC's output tools.

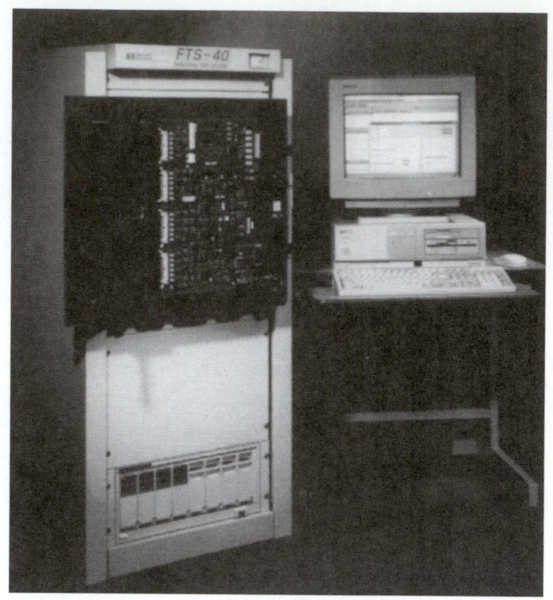

Courtesy of Hewlett-Packard Company.

ground reference point which is shared by all the other 15 amplifiers). Each electronic amplifier receives the analog voltage signal from one input transducer, representing the actual measured value of some physical variable; this is shown in the visualization diagram of Fig. 3–34. You can think of each amplifier as possessing its own analog-to-digital converter (ADC), which converts the analog signal into a 4-digit, binary-coded-decimal digital value.* The collection of an input wiring terminal, amplifier, ADC, and digital holding register is called a *channel*. The channels are numbered 1 through 16 decimal, as indicated in Fig. 3–34.

The digital data from the various channels of an Analog Input Module must be brought into the processor's variable-data memory so that it can be used in the user-program. This is called *reading* the module. It is accomplished by the Block-Transfer-Read (BTR) instruction.

A BTR does not suspend execution of the user-program while the data transfer is occurring. It is unlike an immediate input instruction (IIN) in this respect.** In general, the BTR brings data in from the I/O rack as quickly as it can, while the user-program moves on to the rungs following the rung containing the BTR. At some point in the future the measurement data from the Analog Input Module arrives in the variable-data memory file that you established when you entered the Block Transfer Read instruction into your program. After that event (the transfer of the most recent digital data from the Analog Input Module into the processor's file words) is finished, its completion will be recognized when the BTR instruction is reencountered on some subsequent program scan. From that rung forward, the new up-to-date data becomes available to any program

*Other formats besides 4-digit BCD can be chosen by the user. For example, 12-bit strict binary is another possible format. For a discussion of how analog-to-digital converters work, see D. L. Metzger, *Microcomputer Electronics,* Prentice-Hall, 1989, pp. 310–318.

**We describe this difference by saying that the BTR is asynchronous (not synchronized with) the user-program, whereas the IIN is synchronous with the user-program.

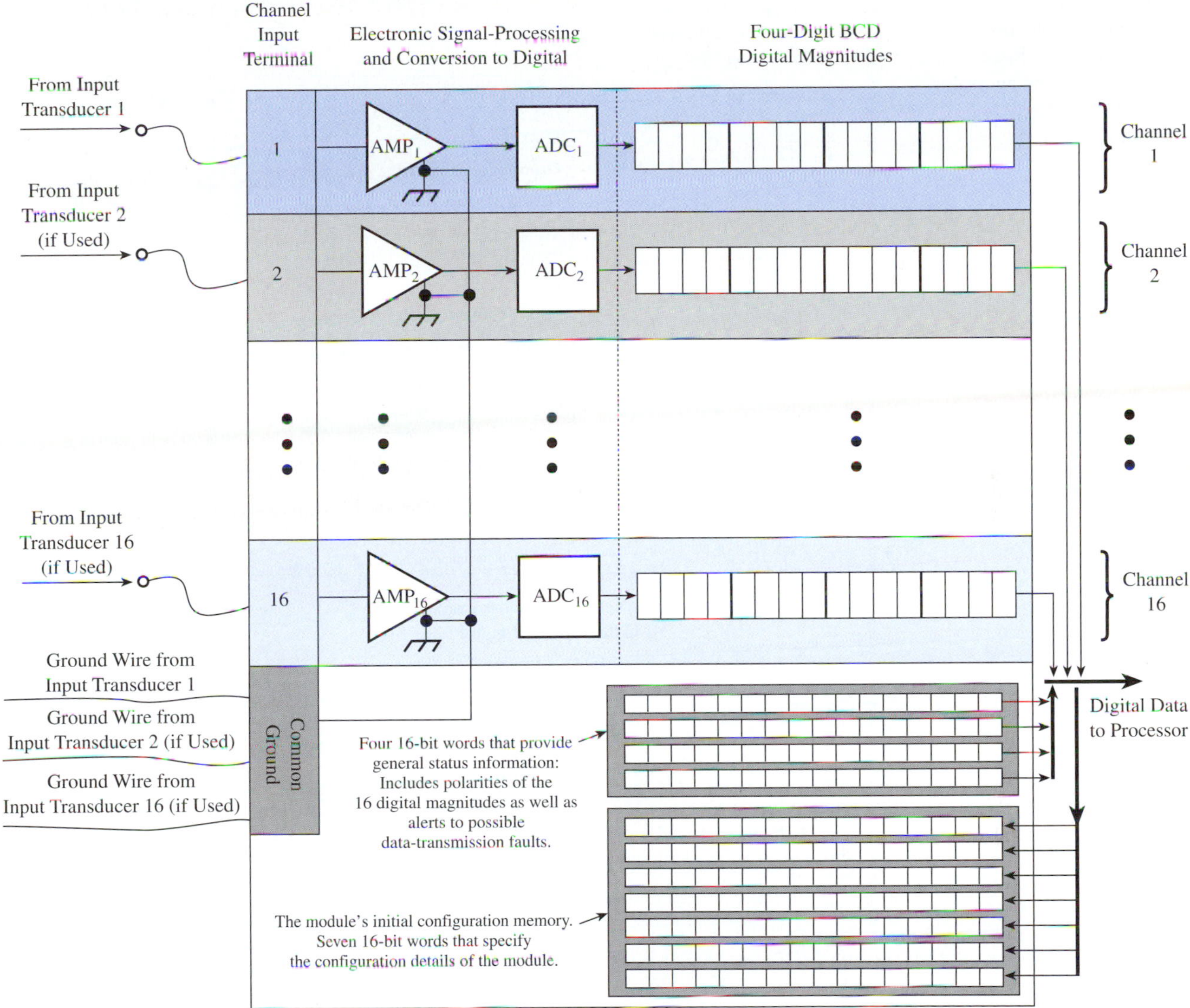

FIGURE 3–34
Visualizing an Analog Input Module. One module has 16 channels, but not all channels are necessarily used. This diagram shows 16 ADCs for conceptual simplicity. In reality, there is only one ADC, which is multiplexed to those channels that are being used.

instructions that refer to those file words. It is very helpful to look at an example in order to understand this action; refer to Fig. 3–35.

The Block-Transfer-Read instruction is an output-type instruction. It is executed only if the conditional instructions on its rung produce rung continuity. On line **a** of Fig. 3–35, if the BTR instruction goes TRUE, a reading of several words from the Analog Input Module begins. The term *block* refers to the fact that we are going to transfer several words from the Analog Input Module, not just a single word (as we do in an IIN instruction, for example). Think of a block as a collection of two or more words which will be stored in contiguous* addresses of the variable-data memory. In Fig. 3–35 we have elaborated the internal la-

*Contiguous means that all the items are right next to each other; there are no "spaces" between them.

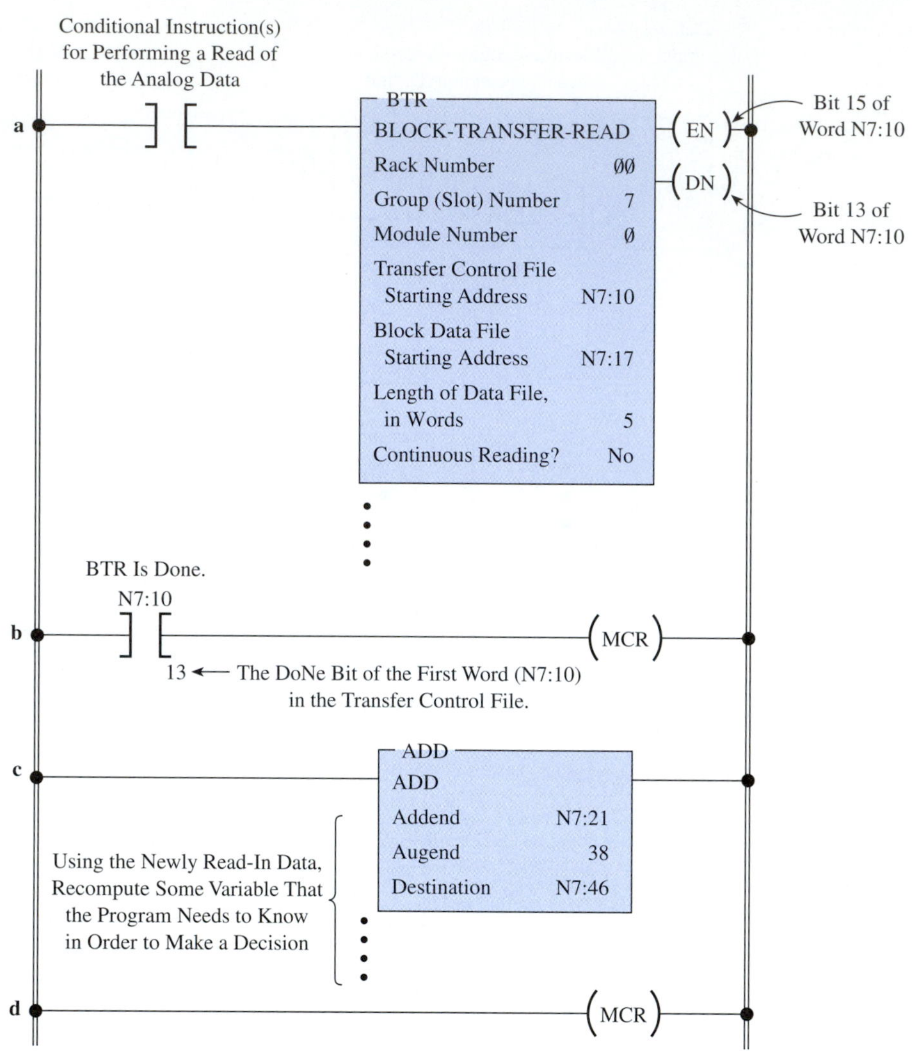

FIGURE 3–35 Understanding the process of Block-Transfer Read (BTR) for bringing an analog value into a PLC user-program.

bels of the BTR box to make their meanings clearer. The actual appearance of those labels will be shown later.

The rack number and group (slot) number tell the processor where the Analog Input Module is located in the I/O rack. For our purposes the module number will always be Ø. It would become nonzero only if an entirely different addressing scheme were being used.

For the processor to manage the details of reading and transferring all the data, it must have five contiguous words of memory reserved solely for that purpose. These memory words should be located in the N7 (integer) subsection of variable-data memory, as shown in Fig. 3–10. When we enter the BTR instruction into our user-program (reached through the [F10] [All Others] key of the Basic Instruction Menu), the Allen-Bradley software will prompt us for the starting address of this five-word block, called the Transfer Control File. In Fig. 3–35 we have chosen to start at word 10 (decimal 10) of the N7 subsection. By making this choice, we have reserved words N7:10, N7:11, N7:12, N7:13, and N7:14 for controlling this particular Block Transfer Read (the same words are reused every time the BTR is executed on future scans). At all other locations in the program where we want to refer to this particular BTR instruction, it will be identified by the starting address of this Read Transfer Control File, namely, N7:10. The words N7:10 through N7:14 cannot be used for any other purpose. They are reserved for the sole use of this BTR. Although *any* word-address from 0 through 999 could have been chosen by us, it is a good practice to always choose a Read Transfer Control File starting address that ends in the digit 0. The reason for this will become clear later.

With N7:14 being the last word in the Transfer Control File, the next word that is available for other purposes in N7:15. Therefore, when the Allen-Bradley software prompts us, we could choose that address as the starting address for the Block Data File, which is the block of words that will store the information coming in from the Analog Input Module. Many programmers prefer to do just that, because there are certain memory-conservation advantages and speed-of-execution advantages to making the Block Data File contiguous with the Transfer Control File. However, for small programs these considerations aren't very important.

For purposes of keeping the address numbers straight in our minds, a better choice is to skip two words and start the Block Data File at the next higher number that ends in the digit 7. This is what we have done in Fig. 3–35 with the Block Data File starting at address N7:17.

Next, the Allen-Bradley software prompts us for the number of contiguous words that we wish to reserve in the Block Data File. In Fig. 3–35 we have chosen five words. Therefore our Block Data File consists of the words at addresses N7:17, N7:18, N7:19, N7:20, and N7:21. At this point the N7: subsection of memory is organized as shown in Fig. 3–36.

When we read the Analog Input Module with the BTR instruction, the first four words that we get are diagnostic and channel-input polarity information. We cannot avoid these words; they must be read into the processor's memory every time the BTR is performed. The processor insists on examining them so that it can know when a malfunction occurs. The first word of actual input data is the fifth word, which is taken from channel 1 in Fig. 3–34. It lands in word-address N7:21 in Fig. 3–36. Thus the last digit of the word-address agrees with the channel number; this resulted from our skipping two words and starting the Block Data File with an address ending in 7.

If there were two channels being used on the Analog Input Module, we would gain access to channel 2 by simply entering a Length value of 6 words in the BTR box of Fig. 3–35 when prompted by the Allen-Bradley software. That would extend the Block Data File of Fig. 3–36 by one additional word, bringing it to N7:22. Then the current value of the channel 2 input signal would appear in binary format in word N7:22. If all 16 channels were in use because 16 separate analog measurements were being made in the industrial control process, the Length entry in Fig. 3–35 would be 20 words (4 unavoidable diagnostic words, plus 16 actual data words) and the Block Data File in the N7 subsection of memory would extend down to N7:36. Agreement is maintained between the address of the word that receives the data and the module channel that provided the data. Thus, for example, word N7:29 receives and stores the data from channel 9; word N7:30 receives and stores the data from channel 10; and so on to word N7:36 storing the data from channel 16.

It is not possible to read a higher-numbered channel or channels while ignoring lower-numbered channels. To get to a higher-numbered channel you must also read all the channels that are on top of it. The phrase "on top" refers to the physical module itself, in which the lower numbers are truly toward the physical top of the module and I/O rack, as suggested in Fig. 3–34. The "on-top" description also applies to the way that we usually sketch the words in the memory, as shown in Fig. 3–36.

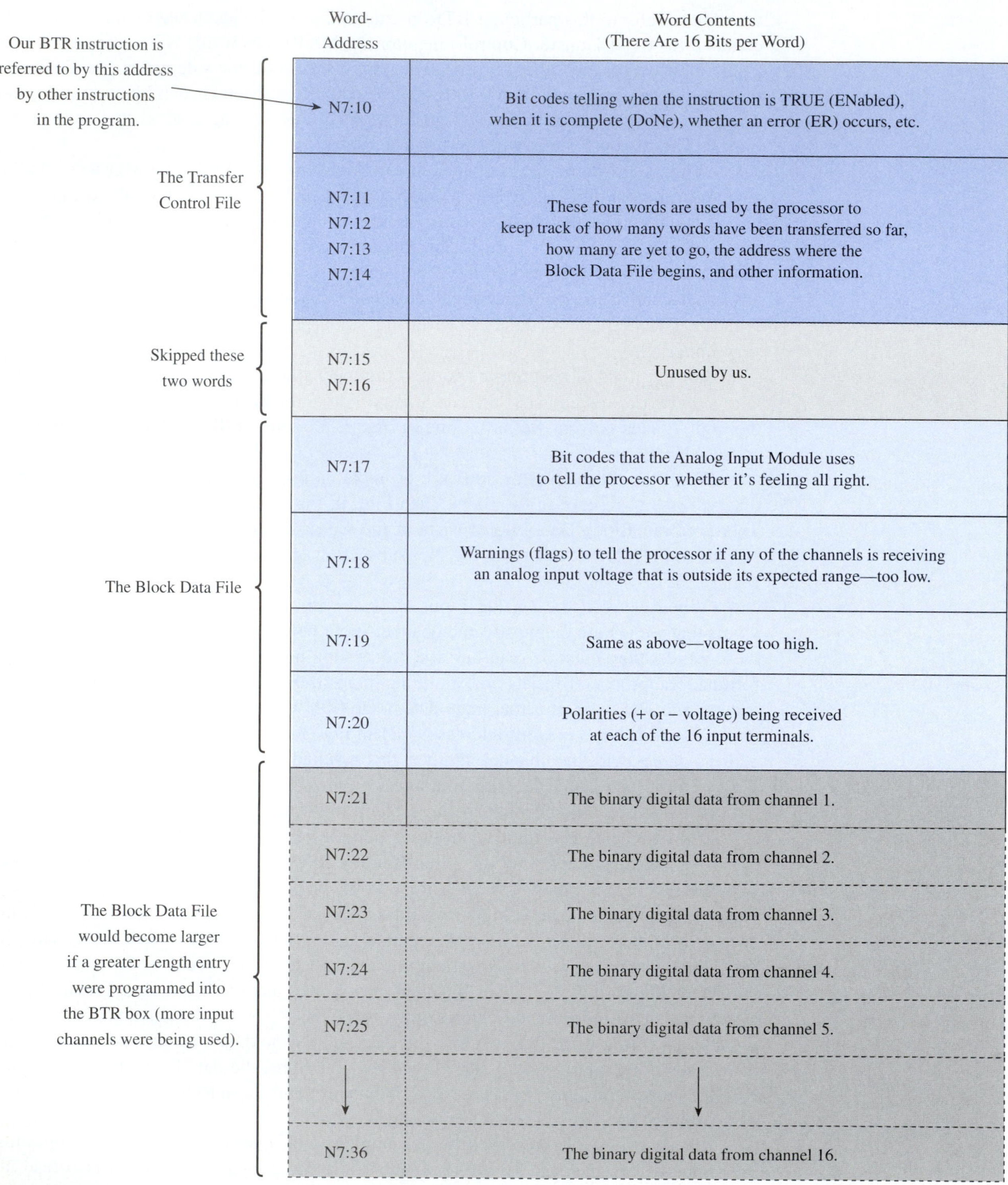

FIGURE 3–36
Visualizing the organization of the Transfer Control File and the Block Data File for a Block-Transfer-Read (BTR) instruction.

In the program example of Fig. 3–35 we have set up a Master Control Reset zone that is conditional on the BTR instruction being successfully completed, with no errors. When that happens, the next program scan of the BTR instruction will place a logic HI at the DN bit (bit-number decimal 13). It doesn't matter whether the conditional instruction(s) on line **a** has maintained rung continuity, since the BTR is a retentive-type instruction. That is, the BTR's EN bit, number 15, is latched HI until the data block transfer is complete and the instruction's DN bit gets set to HI; only then can the EN bit unlatch on a succeeding program scan that finds the line **a** rung with discontinuity. (This description applies to a noncontinuous read instruction: one with No entered on the Continuous Reading? line in the BTR box, which is usual.)

The program of Fig. 3–35 is designed so that the particular scan that causes the BTR's DN bit to go HI also causes the recomputation of some variable value, based on the new data value that is now present in memory word N7:21, having been very recently read in from analog channel 1. The MCR zone that contains all the recomputation instructions is enabled by an examine-On instruction that examines bit N7:10/13 on line **b**.

These are the key things to recognize:

1. When we want to ask about the status of the BTR instruction as a whole, we ask at word N7:10. Such program instructions must be addressed N7:10/XX, as line **b** shows.
2. When we want to do something with a piece of data itself, we must address the specific word in the Block Data File that holds that particular piece of data. For example, the first step in the recomputation process inside the MCR zone is adding the number decimal 38 to the analog value that was recently read in from channel 1. To accomplish this first step the ADD instruction on line **c** is programmed by us to fetch the value from word N7:21, add 38 to that value, and store the result at some destination address elsewhere in N7: memory. The next step in recomputation, on the next rung, would probably involve fetching the value from *that* address (N7:46 in Fig. 3–35) and performing some further computation with it.

These are the ideas involved in bringing analog data into the user-program of a programmable logic controller.

3-6-1 Initial Configuration of an Analog Input Module

We have described the overall process of reading analog data into a user-program, but there are many additional details of the analog input process that have to be dealt with by the programmer. For instance, how does channel 1 know what *range* of input voltages to expect? Should it expect V_{ANALOG} to vary between 0 and +5 V, as suggested in Fig. 3–37? Or perhaps the design of the analog transducer (a topic covered in Chapter 10) is such that the analog input voltage varies from +1 V to +5 V, like the transducer feeding channel 2 in Fig. 3–37. For channel 3, with the op amp powered by ±11 V dc, input transducer No. 3 will produce $V_{\text{ANALOG(3)}}$, varying from about −10 V to +10 V at the wiring terminal. There is nothing in the electronic hardware of a channel's amplifier (Fig. 3–34) that enables that channel to respond appropriately to different V_{ANALOG} ranges. Instead, every channel has to be told by *software* instructions what range it should expect.*

Another major issue that needs to be decided is whether a particular channel is to be scaled or unscaled. An unscaled channel produces a digital output that varies from decimal 0000 to decimal 4095 (the Allen-Bradley ADC has 12-bit resolution) as the analog

*Or by a combination of software instruction and hardware alteration made on the module.

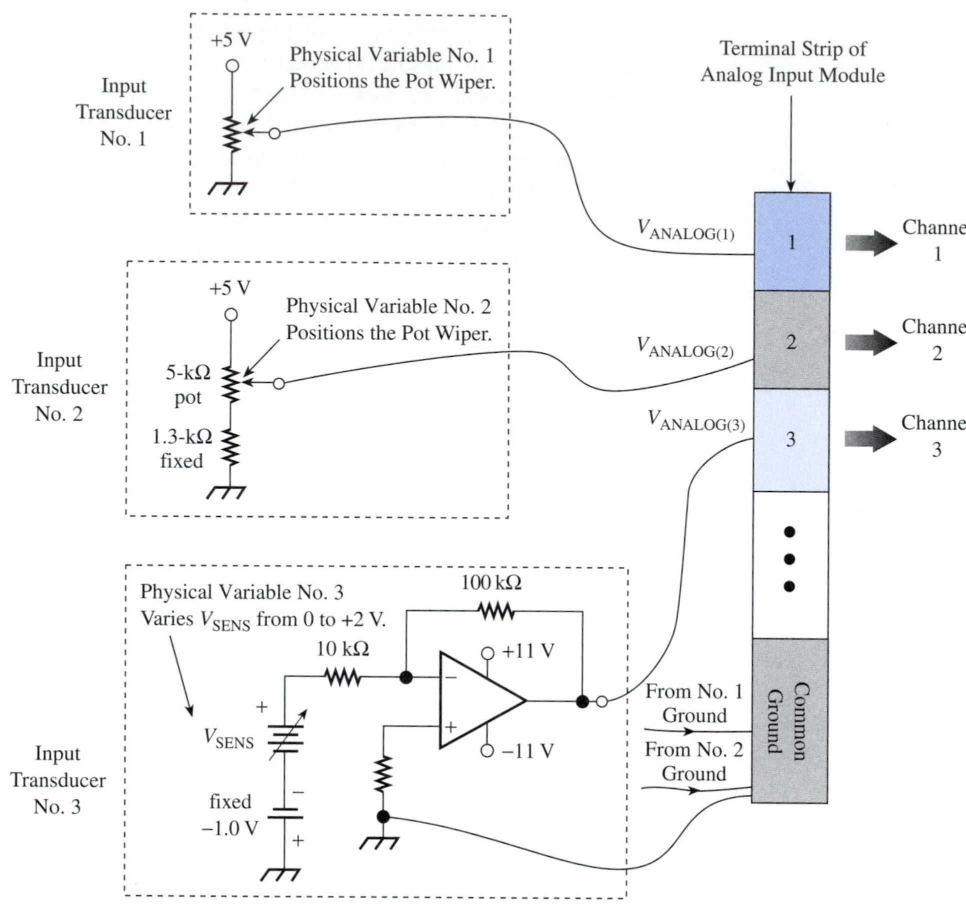

FIGURE 3–37 Industrial input transducers often vary their analog values over the range from 0 to +5 V, or +1 to +5 V, or −10 V, to +10 V, as illustrated here. Other ranges are also used, such as −5 V to +5 V. Furthermore, not all input transducers produce a *voltage* signal. Some produce a *current* signal value, such as the popular 4- to 20-mA range.

input signal varies from its minimum value to its maximum value (for example, 0 to +5 V). But the digital numbers from 0000 to 4095 don't mean anything by themselves. When those digital numbers eventually appear in the user-program, we, the users, need to know in our minds that 0000 really means a temperature of, say, 25°C. We would know this by tracing the circuit all the way back to the sensing element of the transducer. And we would also have to keep it straight mentally that the digital value 4095 appearing in the program really means a temperature of, say, 150°C. Therefore a digital value of 3397, for example, would represent an actual temperature of 128.7°C, since

$$\text{Actual measured temp} = \text{minimum temp} + \left[\frac{\overset{\text{ADC value in program}}{3397 \text{ parts}}}{4095 \text{ total parts}}\right] \times (\text{total range of temperature})$$

$$= \underset{\text{Min temp}}{25^\circ\text{C}} + \left[\frac{3397}{\underbrace{4095}_{2^{12}-1}}\right] \times (\underset{\text{Max temp}}{150^\circ\text{C}} - \underset{\text{Min temp}}{25^\circ\text{C}})$$

$$= 25^\circ\text{C} + (0.8295) \times (125^\circ\text{C}) = 128.7^\circ\text{C}$$

Having to keep all this information straight in our minds when we write program instructions to manipulate the digital data would be very confusing, as you can appreciate. And it wouldn't be just *one* set of facts regarding the minimum and maximum values of the physical variable. It could be as many as 16 sets of facts if all 16 analog channels were being used to measure different physical variables. That would be a nightmare.

Fortunately, the Allen-Bradley software enables us to *scale* the analog inputs. This means that we can program the actual temperature that corresponds to $V_{ANALOG} = 0$ V and the actual temperature that corresponds to $V_{ANALOG} = +5$ V. For the foregoing example, the actual temperature of +25 units would be entered as the minimum scaling value for that channel, and the actual temperature of +150 units would be keyed in as the maximum scaling value for that channel. Under this plan of operation, when the channel sends its digital value to the processor during the Block Transfer Read, the digital number that it sends is automatically expressed in the actual measurement units. In the example that we are discussing, the actual analog input voltage must have been

$$V_{ANALOG(1)} = \left[\frac{3397}{4095}\right] \times (5\text{ V} - 0\text{ V})$$

$$= (0.8295) \times 5\text{ V} = +4.148\text{ V}$$

With channel 1 of the Analog Input Module scaled from +25 units (minimum) to +150 units (maximum), the actual ADC output of 3397 is software-scaled to 129 units (128.7°C rounded to the nearest integer) *before* it is stored in word N7:21 in Fig. 3–36. The channel's digital number that is put into the program has been converted to meaningful units by the PLC software. Therefore we humans don't have to carry the burden of interpreting a nonmeaningful number. This is a huge advantage to us.

The only disadvantage to scaling analog input channels is that it sacrifices resolution because of rounding to the nearest integer. For our example the effective measurement resolution is 1 part in 125, since the software rounds to the nearest degree out of an overall range of 125 degrees. The original ADC resolution was 1 part in 4095, which would have been equivalent to

$$\frac{125°\text{C}}{4095} = 0.03°\text{C}$$

Thus we had to sacrifice our original, terrific measurement resolution of 0.03°C (assuming that the transducer's accuracy justified such a fine resolution) and settle for a final measurement resolution of 1°C.

Getting the Analog Input Module properly set up so that every channel knows what input range to expect and every channel produces meaningful unit-numbers to the program is called *configuring* the module. Configuration must be performed only once, on the very first scan of the program after the processor is placed in RUN mode. It is accomplished by a Block-Transfer-Write (BTW) instruction, which writes the proper codes into the seven words of module memory shown at the bottom of Fig. 3–34.

A BTW instruction is similar to a BTR instruction in terms of the specifications that must be programmed into its schematic box when it is entered into the user-program. Refer to Fig. 3–38 to see the similarity.

The Write instruction is transferring the configuration data to the Analog Input Module located in slot 7 of rack ØØ, so the Rack number and Group/Slot number must identify that physical location. This is shown in Fig. 3–38.

A writing-type transfer of data (from the processor to the I/O rack) requires the same record keeping and error checking as a reading-type transfer (from the I/O rack to the processor). Since the reading-type transfer (BTR) in Fig. 3–36 needed five words to keep track of such things, our writing-type transfer (BTW) also needs five words. The five words that are used for BTW are called the Transfer Control File, the same as they were called for BTR. We could select any contiguous group of five words for this purpose as long as none of their addresses were already being used for some other purpose. (For example, you wouldn't dare choose words N7:30 through N7:34 for the writing Transfer Control File, because those words will be needed for the reading Block Data File if the number of analog input channels rises to 10 or more. Refer to Fig. 3–34 and Fig. 3–36.) A good practice to follow for initial module-configuration writing-transfers is to pick a number ending in the digit 0, which is a little higher than the highest possible address of the module's reading Block Data File. In Fig. 3–38 we have chosen a starting address of N7:50. Therefore the file occupies words N7:50 through N7:54, as Fig. 3–39 shows. The AB software will prompt us for the starting address of the Transfer Control File as we are programming the BTW instruction.

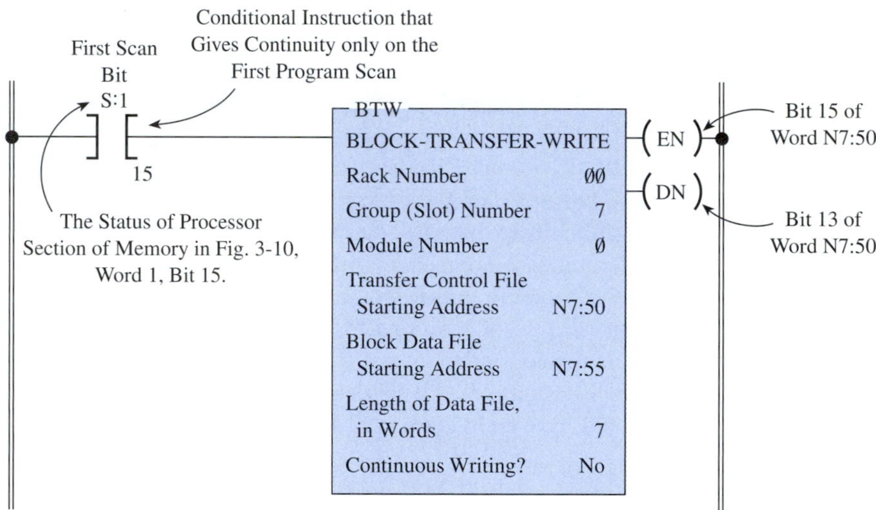

FIGURE 3–38 Understanding the specifications of a Block-Transfer-Write (BTW) instruction.

The Block Data File that is written to the Analog Input Module for configuration has three words that are absolutely essential whether or not any of the channels are scaled. These three words have been located (our choice of starting address, when the software prompts us in the BTW box) as N7:55, N7:56, and N7:57. This is shown in Fig. 3–39.

An additional four words are required if just one channel (channel 1) is to be scaled in meaningful units. These four words must be contiguous with the first three words. In Fig. 3–39, with only channel 1 being used and scaled, the four words are N7:58 through N7:61. Therefore the overall Block Data File length must be seven words. This Length value is entered by us in the BTW box at the time the BTW instruction is entered into the user-program.

If channel 2 of the Analog Input Module were being used in addition to channel 1, it would have to be scaled too.* To do this would require two additional words, N7:62 and N7:63, as suggested in Fig. 3–39. Of course, the BTW instruction would then have

*If you scale any channel of an analog module, you must scale them all. To effectively unscale a channel, if that is what you want, enter a minimum scale value of 0000 and a maximum scale value of 4095. This will make the channel respond as if it were unscaled.

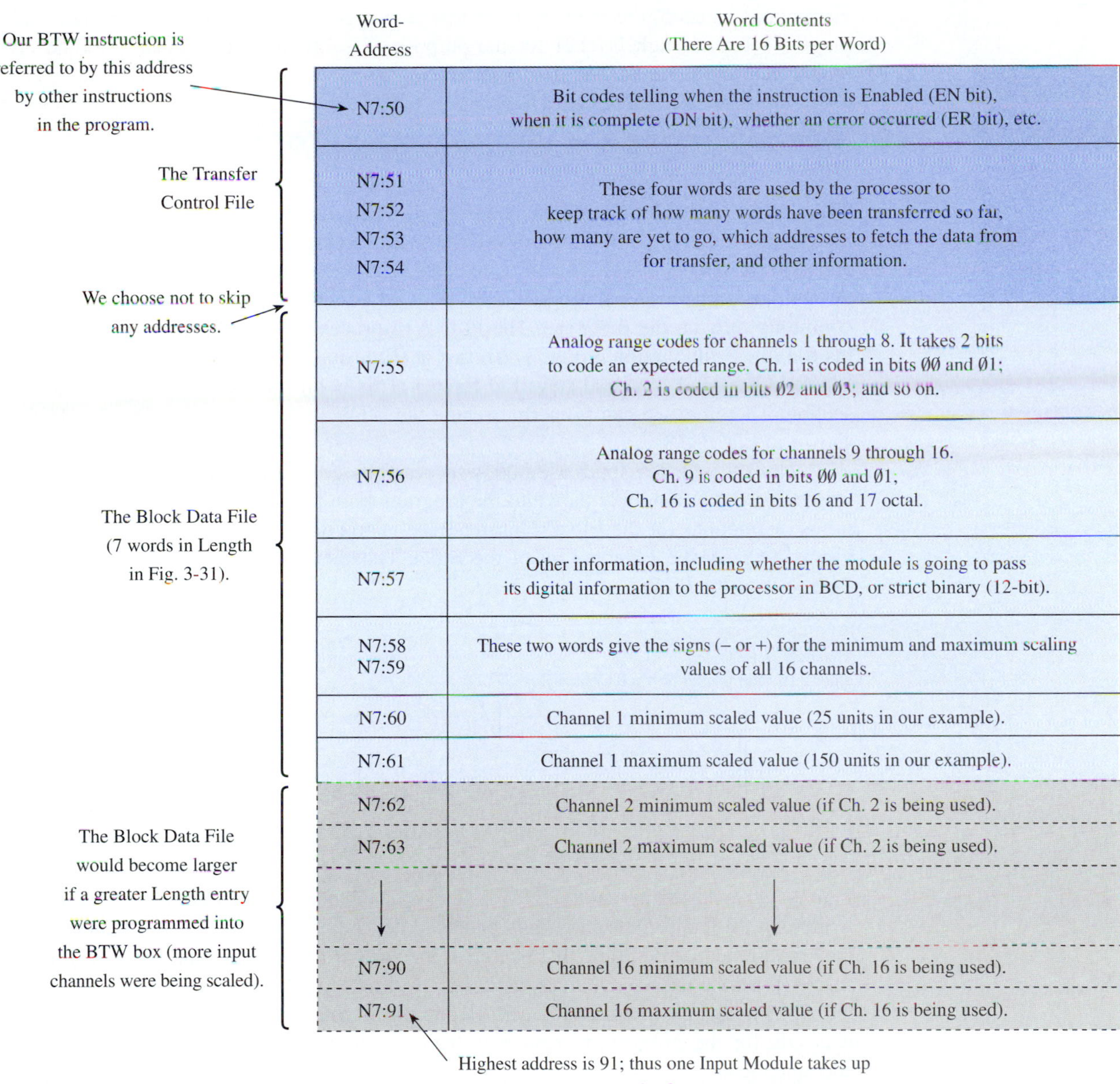

Our BTW instruction is referred to by this address by other instructions in the program.
The Transfer Control File
We choose not to skip any addresses.
The Block Data File (7 words in Length in Fig. 3-31).
The Block Data File would become larger if a greater Length entry were programmed into the BTW box (more input channels were being scaled).
Word-Address
Word Contents (There Are 16 Bits per Word)
N7:50
Bit codes telling when the instruction is Enabled (EN bit), when it is complete (DN bit), whether an error occurred (ER bit), etc.
N7:51
N7:52
N7:53
N7:54
These four words are used by the processor to keep track of how many words have been transferred so far, how many are yet to go, which addresses to fetch the data from for transfer, and other information.
N7:55
Analog range codes for channels 1 through 8. It takes 2 bits to code an expected range. Ch. 1 is coded in bits ØØ and Ø1; Ch. 2 is coded in bits Ø2 and Ø3; and so on.
N7:56
Analog range codes for channels 9 through 16. Ch. 9 is coded in bits ØØ and Ø1; Ch. 16 is coded in bits 16 and 17 octal.
N7:57
Other information, including whether the module is going to pass its digital information to the processor in BCD, or strict binary (12-bit).
N7:58
N7:59
These two words give the signs (– or +) for the minimum and maximum scaling values of all 16 channels.
N7:60
Channel 1 minimum scaled value (25 units in our example).
N7:61
Channel 1 maximum scaled value (150 units in our example).
N7:62
Channel 2 minimum scaled value (if Ch. 2 is being used).
N7:63
Channel 2 maximum scaled value (if Ch. 2 is being used).
N7:90
Channel 16 minimum scaled value (if Ch. 16 is being used).
N7:91
Channel 16 maximum scaled value (if Ch. 16 is being used).
Highest address is 91; thus one Input Module takes up a 100-word chunk of memory, virtually.

FIGURE 3–39
The placement of the Transfer Control File and the Block Data File for a module-configuring (initializing) Write instruction (BTW).

to be programmed with a file Length of 9 in Fig. 3–38. The maximum length of the BTW's Block Data File is 37 words if all 16 input channels are in use. This takes the file to an ending address of N7:91.

It is possible that our PLC has two Analog Input Modules. This might be necessary either because our one industrial process requires a great number of analog measurements (more than 16) or because the PLC is used to control several different industrial processes at different times, with each process requiring a few analog measurements. With two Analog Input Modules present in the I/O rack, our user-program must have two separate BTW instructions to configure them both. In that case a good policy is to reserve the address range N7:150 through N7:191 for the purpose of configuring the second module (100 higher than the range for the first module).

3-6-2 Executing the Block-Transfer-Write Instruction Only Once

The Analog Input Module must be configured when the processor initially starts executing the user-program, which occurs when we take the processor out of PROGRAM mode and place it in RUN mode. Therefore, to enable the BTW instruction on the very first scan through the program but never thereafter, we need a conditional instruction that gives rung continuity *only* on the first scan. The PLC 5/12 processor has a bit in its Processor Status file (the S file section in Fig. 3–10) that is HI during the first program scan, then goes LO on the second scan and stays LO forever. This is bit-number 15 of word 1 in the S section. Therefore we can properly execute the BTW with an examine-On of S:1/15, as shown in Fig. 3–38.

Until the Analog Input Module is successfully configured by the BTW instruction, we don't want to read any data into the program from that module. Therefore the conditional instructions for the Block-Transfer-Read instruction in Fig. 3–35 should contain an examine-Off of the ENabled bit of the Block-Transfer-Write (or an examine-On of the DoNe bit of the BTW), as shown here:

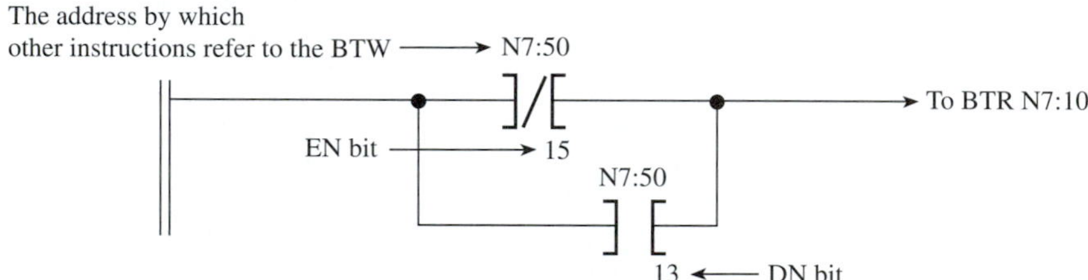

That is, we are proving that the BTW is not still enabled, which would mean that it is still working on its configuration-writing chores. After the writing transfer is complete, which may require an elapsed time that is greater than the program scan time, the next program scan will set the BTW's DN bit (number 13 decimal) to HI. The scan following that scan will reset its EN bit (number 15 decimal) to LO. As soon as either of these events occurs, it is OK for the program to read data from the input module when the BTR rung is encountered.

3-6-3 Entering the Data Values into the Seven Words of the BTW's Block Data File

Programming the BTW instruction as shown in Fig. 3–38 gets the instruction into the user-program, but it doesn't do anything about getting the actual data into the Block Data File. As we are sitting at the keyboard with the processor still in PROGRAM mode, we must key the range codes, digital data format (BCD or strict binary), and channel-scaling minimum and maximum values into the proper words of the Block Data File, N7:55 through N7:61. The most natural way to accomplish this is to place the cursor on the BTW instruction of our ladder-logic screen display while the AB software has the Ladder Editor Main Menu on the bottom of the screen. Press [F10] for [Edit], then press [F2] for I/O Edit. The AB software will jump to the Module Edit Menu. We then cursor to the Data Format line. There will be three choices for the data format: Binary-Coded-Decimal; Strict binary magnitude, with a sign (polarity) bit; and Strict binary, with negative values expressed in 2's complement format. Press the F9 key for Toggle until "2's complement binary" is highlighted—assuming that we actually want a strict binary format rather than BCD. We cannot choose different digital formats for different channels. All the channels of this Analog Input Module must use the same format—2's complement binary, in our example. Press F10 for Accept, then F8 for Yes to confirm the Accept command.

Press F3 for Channel Edit. The AB software will jump to the Channel Edit Menu. Cursor to the Channel line. Then key in the number of the channel whose analog input range and scaling values we wish to enter. In our example, with the temperature transducer wired to channel 1, key 1 Enter. Cursor to the Range line. Press the F7 key for Toggle until the proper analog input range is highlighted. In our example, we will quit toggling when the screen highlights "0 to +5 V." Cursor to the Minimum Scaling Value line. Key in the minimum value of the meaningful units, preceded by a minus sign if the minimum value is negative. In our example, we key in 2 5 Enter. Cursor to the Maximum Scaling Value line. Key in the maximum value, 1 5 0 Enter in our example.

If we wish, we may now cursor to the Unit line and key in the name of the meaningful units that we are working with. This has no effect on the actual execution of the user-program or on the actual control of the industrial process. It is just an explanatory piece of information that is recorded so that another person can look at the display screen and understand what units we had in mind when we designed the program. In our example, key in the letters DEGREESCEL, or CELSIUSDEG, or some 10-letter string that clearly indicates that the meaningful units are degrees Celsius.

Press F10 for Accept, then F8 for Yes to confirm. When we accept and confirm the acceptance, the AB software stores all our configuration information in the Block Data File of the BTW instruction. For our example, the configuration information is placed into words N7:55 through N7:61 of the processor's variable-data memory.

The configuration information is also saved in a database file on the computer's hard disk. The configuration information is software-linked to the name that we assign to our user-program, which we also will save to the hard disk. This way we can remove this user-program from the processor's user-program memory so that the PLC can run some new different user-program. This is one of the great features of a programmable logic controller; it has the capability to perform one job part of the time and a totally different job another part of the time.

It is necessary to save the configuration information into a hard disk database file because the new different user-program might overwrite the words of variable-data memory that hold the information for this program (N7:55 through N7:61). When we restore (download from the hard disk) this user-program into the processor at a future time, however, the Allen-Bradley software has sense enough to also restore the required configuration data into the proper words of variable-data memory. That is, the software automatically fetches the configuration information from the database file.

What we have described here regarding saving of user-programs and related database files on a hard disk also applies to saving on a floppy disk. By saving on a floppy disk, we can create backup copies of all our PLC programs. This is necessary to protect ourselves against hard disk failure. Also, a floppy disk gives us total portability of our program to another programmable logic controller at a different location.

3-7 ■ IMPROVING THE MILLING MACHINE SYSTEM BY MAKING IT TEMPERATURE-SENSITIVE

Section 3-5 presented a milling machining control program that paused for 15 seconds to cool down the milling bit after every stroke; refer to Figs. 3–22 and 3–23. Actually, it is not just the bit itself that needs to be cooled. The milling motor also heats up as it works; so does the workpiece. These parts of the system rely on that 15-second pause too.

Under certain conditions (hard-to-cut workpiece, dull milling bit, warm ambient air temperature) 15 seconds for cooling may not be adequate. A better system design would measure the actual operating temperature of the milling bit or the milling motor and lengthen the cooling time if the temperature rises beyond a certain value.

Let us revise our milling machine program so that we measure the milling motor temperature with a temperature transducer mounted on its surface. The transducer must have flexible wires to allow for its back-and-forth motion. Suppose the transducer is designed so that its output varies from 0 to +5 V as its temperature varies from 25°C to 150°C. No motor could withstand an operating temperature of 150°C (302°F), but that's how our transducer is designed. Our corporate bean-counters, thick-headed clods all, refuse to spring for a replacement transducer with a proper narrow temperature range, so we have to live with this one.

From careful study of the milling process, we have learned that the milling motor and milling bit are not thermally stressed if the motor temperature remains below 54°C (about 129°F), but if the temperature rises above 54°C, it results in shortened life expectancy of the motor and is associated with premature dulling of the bit. Therefore we wish to institute a control program that lengthens the cooling times by 3 seconds for every degree that the motor temperature rises above 53°C. Thus if the motor temperature rises to 54°C, the program lengthens the cooling pauses to 18 seconds. The longer cooling times should allow the motor to cool back down to below 54°C, which would permit the program to reduce the cooling pauses back to 15 seconds. In this way the milling machine will switch between spending part of its time with 15-second cooling pauses and part of its time with 18-second cooling pauses. The motor temperature should then vary between slightly below 54°C and slightly above 54°C.

If working conditions are especially difficult, however, the motor temperature will continue to climb, even with an 18-second cooling pause. If the temperature should reach 55°C, then the program will increase the cooling pause to 21 seconds. And if, 56°C is reached, the cooling pause becomes 24 seconds, and so on. By this means the program should be able to prevent the motor temperature from climbing too far above 54°C.

The motor-temperature transducer is electrically connected to channel 1 of an Analog Input Module that is mounted in slot 7 of the ØØ I/O rack. The revisions to the PLC user-program are shown in Fig. 3–40. Rungs 0 through 13 are inserted *above* the top rung of Fig. 3–23. From rung 14 onward, Fig. 3–40 contains the identical program-rungs as Fig. 3–23. Here is how the temperature-sensitive part of the new program adjusts the cooling pause times.

As the program begins its first scan after entering RUN mode, Status bit S:1/15 is HI. The examine-On instruction on rung 0 of Fig. 3–40 gives rung-continuity, so the BTW instruction is enabled. The processor begins transferring the module-configuration information from Data File N7:55 through N7:61 to the first seven words of the analog module's own memory. While this write-transfer is underway, the processor continues on to the following rungs.

On rung 1, timer T4:1 is not done—it hasn't even begun timing yet. Therefore its DN bit is LO and the examine-Off instruction gives continuity. The timer starts timing toward 30 seconds.

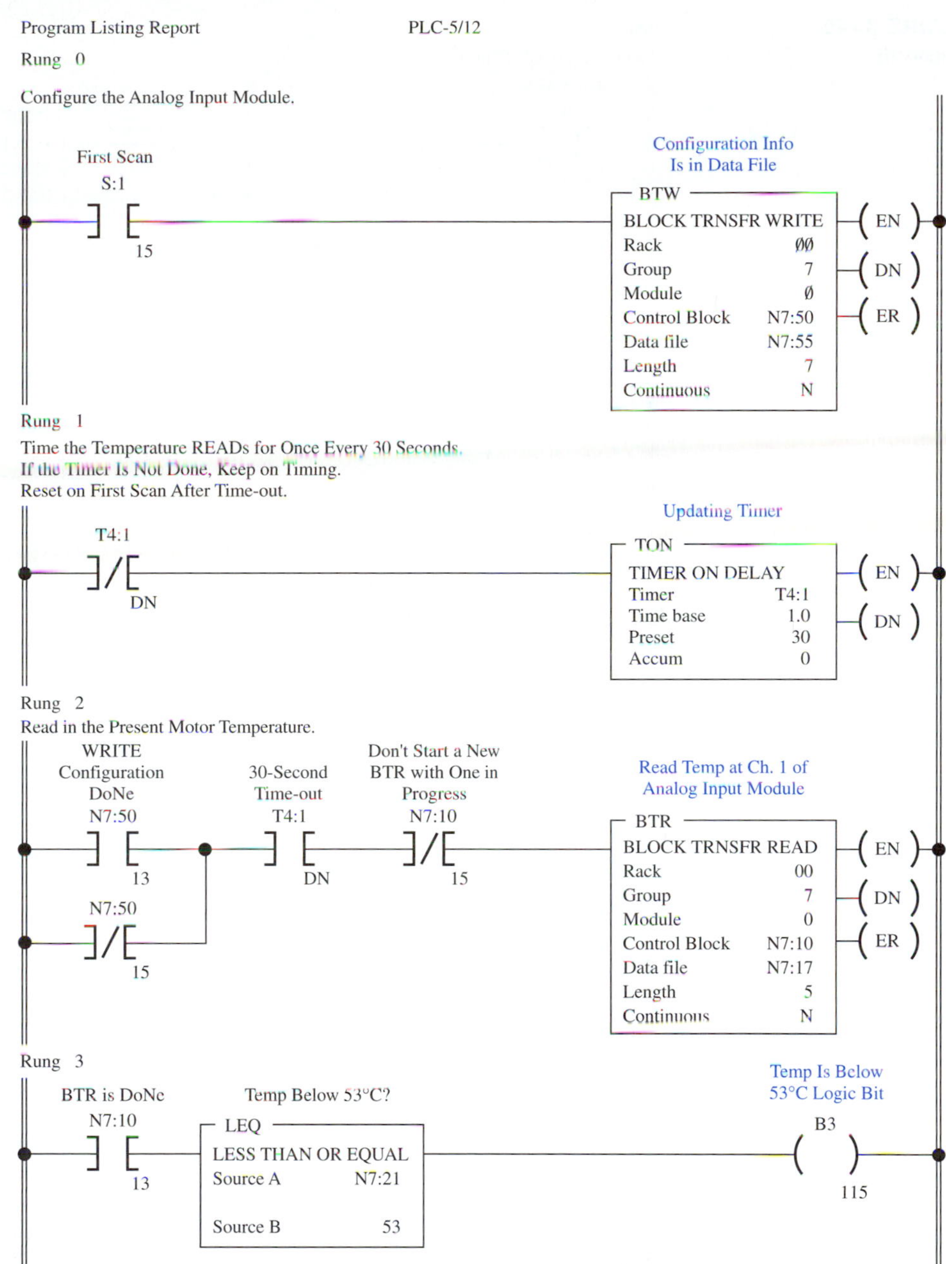

FIGURE 3–40
Rungs that must be added to the original milling machine user-program to make it sensitive to rising motor temperature.

On rung 2, the write configuration done bit N7:50/13 does not provide continuity on the first scan because a writing-type block transfer takes a few milliseconds to complete. Neither does the examine-Off ENabled bit N7:50/15 instruction give continuity; the ENabled bit is a 1 because the BTW is enabled from rung 0. Besides, timer T4:1 isn't DoNe. The BTR remains disabled.

Rung 3 does not have continuity through the BTR DoNe bit, N7:10/13 because the BTR hasn't even begun, let alone finished. Therefore the Output Energize instruction is

FIGURE 3–40 (*continued*)

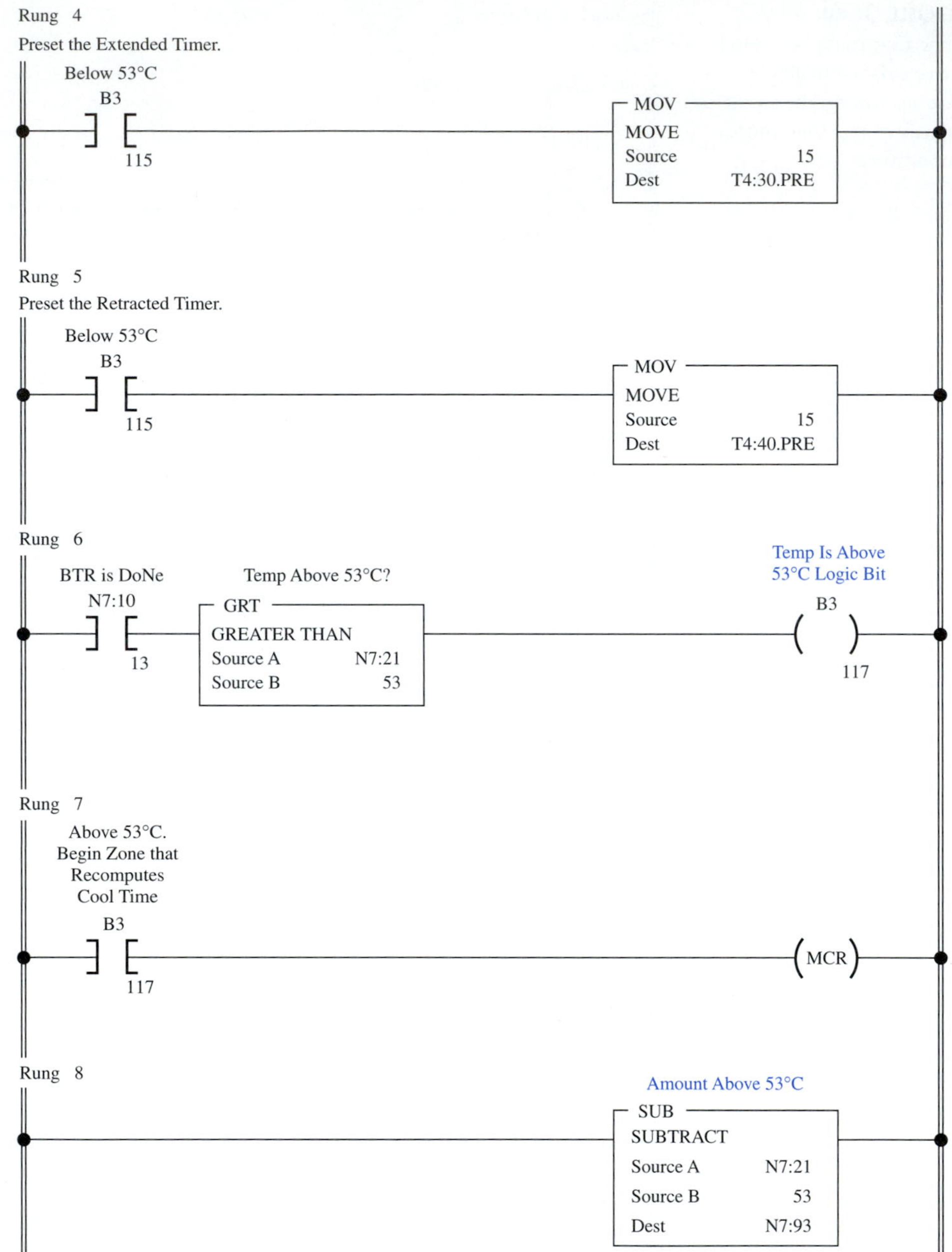

FALSE, sending internal logic bit B3/115 to the 0 state. This breaks the continuity of rungs 4 and 5, so the MOV instructions are FALSE. They are not able to move the source value, which we have programmed as 15 (seconds), into the destination addresses T4:30.PRE and T4:40.PRE. These are the Preset words on cooling timers T4:30 and T4:40. However, these timers were originally Preset to 15 seconds anyway when they were entered into our present user-program (see lines 6 and 13 of Fig. 3–23). We've lost nothing by failing to accomplish the Moves on rungs 4 and 5.

FIGURE 3–40
(continued)

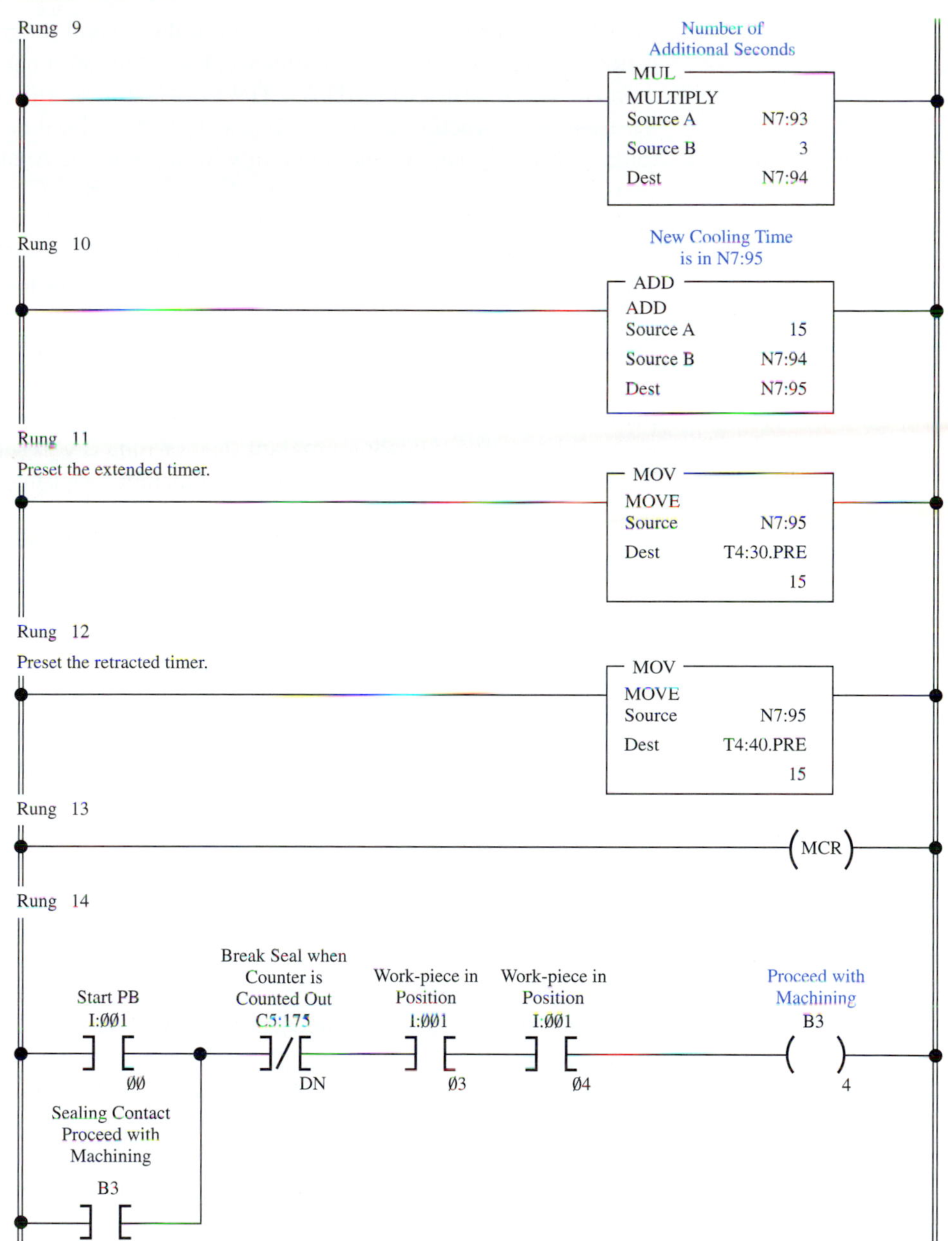

On rung 6, BTR hasn't even begun yet, so examine-On of N7:10/13 DoNe bit gives no continuity. The GRT instruction isn't even executed before the processor puts a 0 at internal logic address B3/117. This causes rung 7 to go FALSE, which disables all the rungs through rung 13, which is the end of the MCR zone. Therefore rung 14 of Fig. 3–33, which is line 1 of Fig. 3-23, begins execution in its usual way, as if the temperature-sensing portion of the program weren't even present.

Eventually the BTW initialization transfer will be finished. It may happen at some moment during the first program scan, when the processor is on one of the rungs below rung 14, or it may happen at some point in the second program scan, or the third. There is no way to predict this with confidence, but whenever it happens, the following program scan of rung 0 will set the BTW's DN bit, N7:50/13. This will give continuity through the leftmost instruction on rung 2, but with the timer still far short of 30 seconds, rung 2 remains FALSE. That is, the program will not read the Analog Input Module during the first 30 seconds of machine operation. The milling machining apparatus simply functions as it did in Sec. 3-4, with a 15-second cooling pause.

When timer T4:1 eventually times out, rung 2 goes TRUE because the BTR ENabled bit N7:10/15 is not *yet* set to 1, so the examine-Off instruction gives continuity. EN will be set to 1 when this rung finishes executing, but by that time the examine-Off instruction becomes irrelevant because the BTR latches itself in the ENabled state. Therefore the BTR instruction begins to read the Analog Input Module's measured motor temperature. This will take a few milliseconds. At some future moment the read transfer will be completed. The following program scan of rung 2 will set a HI in the BTR DoNe bit, N7:10/13. From this point forward the temperature-sensing portion of the user-program is active.

On rung 3, the LEQ conditional instruction performs as follows: It fetches the numeric value from word N7:21, which is its source A. This value is the motor's actual measured temperature, since the channel 1 input signal landed in word 21 of the Read Block Data File. It then fetches Source B, which is the numeric value 53 (a Source can be a word address that contains a numeric value, or a numeric value itself). If the Source A value is less than or equal to the Source B value, the LEQ gives continuity, but if Source A is greater than Source B, it gives discontinuity.

If the motor temperature is all right (not above 53°C), the LEQ instruction makes rung 3 TRUE. Internal logic bit B3/115 goes HI, thereby enabling rungs 4 and 5. Rung 4's MOV instruction takes the value from its Source and Moves that value to the Destination address that is specified by the program. In this case, the destination is the memory word that holds the Preset value for timer T4:30; that memory word is referred to as T4:30.PRE. It is one of the two memory words that are associated with word T4:30 itself, as explained on page 104. Thus the cooling-pause time with the cylinder extended is Preset to 15 seconds. It may be that the timer's Preset value was already 15 seconds. No matter, the program reaffirms that the 15-second value is proper. Rung 5 performs the same duty for T4:40, the cooling-pause timer with the cylinder retracted.

If the LEQ conditional instruction fails, then the GRT conditional instruction on rung 6 must succeed in giving continuity because the measured temperature value in word N7:21 (Source A) must truly be greater than 53 (Source B). Therefore internal logic instruction B3/117 goes HI on rung 6. This in turn activates the MCR zone beginning on rung 7.

The mathematical instructions on rungs 8, 9, and 10 accomplish the calculation

$$\text{new time} = (\text{measured temp} - 53) \times 3 + 15 \tag{3-1}$$

Memory words N7:93, N7:94, and N7:95 are available for use, since the BTW data file can never extend beyond N7:91 (Fig. 3–39). Referring to Eq. 3-1, the quantity (measured temp −53) is the number of degrees by which the present motor temperature exceeds 53°C. That number is calculated and then stored in N7:93 by rung 8. Rung 9 multiplies that number by 3 to obtain the additional seconds of cooling time and stores the result in N7:94. Rung 10 adds this additional cooling time to the 15-second base to calculate the proper cooling time, which is stored in N7:95. Rung 11 places this proper cooling time into the Preset word of T4:30, the cooling timer with cylinder extended. Rung 12 does

TROUBLESHOOTING ON THE JOB

REFINING THE TEMPERATURE-RESPONSE OF THE MILLING PROCESS

In the milling machining process of Fig. 3–22, using the temperature-sensitive control program of Fig 3–40, we add an additional 3 seconds per stroke when the motor temperature rises by 1° above the 53°C baseline. This lengthens the total production time by 105 seconds, because 18 cylinder cycles requires 35 cooling pauses, with

$$\frac{3 \text{ seconds}}{\text{cooling pauses}} \times 35 \text{ cooling pauses} = 105 \text{ seconds}$$

In an effort to reduce manufacturing time, your supervisor has suggested that the motor temperature might be maintained acceptably close to 53°C by reducing the cooling pause increment to only 1 second, but for a reduced temperature rise of only 1/3° Celsius. This way, if the motor temperature does rise by a full degree, we still get our cooling pause increment of 3 seconds, but the increased pause times would kick in sooner perhaps enabling the system to maintain a less fluctuating temperature with less drastic increments in cooling time.

You have been assigned the job of reprogramming the PLC to achieve this finer temperature-response. Then you are to test the system in actual production to find out if it really does reduce the total production time. Show your revised version of the temperature sensitive portion of the PLC user-program to accomplish this.

the same thing for T4:40, the cooling timer with cylinder retracted. Rung 13 contains the unconditional MCR instruction that marks the end of the Master Control Reset zone of the program. From rung 14 to the end, the program is the same as it always was, except for the longer cooling-pause times.

SUMMARY

- A programmable logic controller (PLC) is a computer-based industrial control system that uses program instructions to make the turning-ON and turning-OFF decisions that otherwise would be made by relays or hard-wired logic gates.
- A PLC can be mentally divided into three parts: (1) I/O section, (2) processor, and (3) programming device.
- The I/O section provides the interface for bringing 120-V ac input signals into the computer-like processor (input) and for converting low-voltage processor signals back to 120-V ac industrial signals (output).
- The processor holds and executes the user-program, which is the sequence of instructions the human user has created to control the industrial machine or process.
- The programming device is the keyboard-equipped device through which the human user enters or edits the program to be executed by the processor.
- A PLC's processor can be visualized as a central processing unit (CPU) and four sections of memory: (1) input image file; (2) output image file; (3) user-program memory; and (4) variable-data memory.
- Most PLCs present their program on a CRT screen or on paper in ladder-logic format, which is like a relay ladder-logic schematic.
- Relay-type instructions (examine-On, examine-Off, and output-energize) duplicate the operation of normally open contacts, normally closed contacts, and relay coils.

- An examine-On instruction provides logic continuity in its program rung if its memory address contains a logic 1, which is equivalent to its input rack address terminal being energized (powered) by 120 V ac.
- An examine-Off instruction provides logic continuity in its program rung if its memory address contains a logic 0, which is equivalent to its input rack address terminal being deenergized (no 120-V ac power).
- If the conditional instructions in a program rung provide logic continuity, the rung is said to be TRUE, and its output-energize instruction stores a 1 at its memory address. This results in its output rack address terminal becoming energized by 120 V ac. On the other hand, if there is logic discontinuity, the rung is FALSE, the output-energize instruction stores a 0 in memory, and the output rack terminal receives no 120-V ac power.
- Think of the complete scan cycle as consisting of input scan, followed by program execution, followed by output scan. This sequence is repeated indefinitely, as long as the PLC is left in RUN mode.
- The processor must be placed in PROGRAM mode for entering or editing (changing) the user-program.
- A PLC's memory table or memory map shows the legal ranges of addresses for all the various types of instructions and functions.
- PLCs have instructions that duplicate all the possible functions of timers and counters: On-delay and Off-delay timers, retentive and nonretentive timers, instantaneous and delayed "contacts," up-counting, down-counting, up-/down-counting, and counter overflow detection.
- PLCs have instructions for duplicating the actions of latch/trip relays and master-control relays.
- With an Analog Input Module installed into a slot of the I/O rack, a PLC can receive analog information from the controlled industrial process and use that information in the decision making of the user-program.
- In an Allen-Bradley PLC 5/12, the Block-Transfer-Read (BTR) instruction accomplishes reading the digitally encoded value of an analog variable. Using the BTR instruction requires careful selection of addresses for the Transfer Control File and the Block Data File, as well as careful initial configuration of the Analog Input Module with a Block-Transfer-Write (BTW) instruction.
- Most PLC analog processes have 12-bit resolution (1 part in 4095) if no unit-scaling is used.
- The analog data can have any of the standard ranges (0 to +5 V, for example) used by industrial measurement transducers.

FORMULA

$$\text{New Time} = (\text{measured temp} - 53) \times 3 + 15 \qquad \textbf{(Equation 3–1)}$$

QUESTIONS AND PROBLEMS

Section 3-1

1. What is flexible automation? Explain how it differs from standard industrial automation.
2. Name the three parts of a PLC. Describe the function of each part.

3. Describe a PLC scan cycle. Give approximate time durations for the various events.
4. List the three operating modes of a PLC. Describe the purpose of each mode.
5. What is the direction of information flow between the following pairs of locations?
   a. Input image file and the I/O rack.
   b. Output image file and the I/O rack.
   c. Input image file and the CPU.
   d. Output image file and the CPU.
6. When the CPU completes the execution of an instruction-rung containing an output-energize instruction, it updates the output image file *immediately*. Why is this necessary, since the output modules of the I/O section will not be updated until the output scan following the end of the user-program?
7. True or False: Examine-On and examine-Off instructions always refer to an address in the input image file.
8. Draw the ladder-logic program representation for the following logic conditions: If LS1 and LS2 are both actuated at the same time, solenoid 1 becomes energized. Assume that LS1 is wired to address I:001/01, LS2 is wired to address I:001/02, and solenoid 1 is wired to address O:003/01.
9. Repeat Question 8 for the following logic conditions: If LS1 is actuated while LS2 is not actuated, solenoid 1 becomes energized.
10. Repeat for the following logic conditions: If LS1 is actuated, solenoid 1 becomes energized and seals itself in the energized state until LS2 is actuated.
11. Repeat for the following logic conditions: If LS1 is actuated, solenoid 1 becomes energized and seals itself in the energized state until LS2 is deactuated.
12. True or False: An examine-Off instruction produces logic continuity if power is absent from the associated I/O terminal.
13. True or False: An output-energize instruction causes power to be applied to the associated output terminal only if its instruction-rung has logic continuity.
14. In Fig. 3–4(b), the decision box labeled ③ determines whether the rung has maintained logic continuity *so far*. If the decision is *No*, the next test is whether there are any more ORed instructions (box ④ on the right). Explain why the CPU looks for ORed instructions under this condition rather than ANDed instructions.
15. In Fig. 3–4(b), if the decision from box ③ is *Yes*, the next test is whether there are any more ANDed instructions (box ④ on the left). Explain why this is proper.
16. What practical industrial consideration makes the immediate output instruction seldom necessary?
17. List and describe six types of numeric data that can be stored in a PLC's variable-data memory.
18. What is the direction of information flow between the following pairs of locations?
   a. CPU and user-program memory.
   b. CPU and variable-data memory.

From this point forward assume that all questions refer to the Allen-Bradley model PLC 5/12.

19. True or False: When a personal computer is used as the Programming Device in a PLC system, the user-program is entered into the Processor by passing through the communication interface card.
20. True or False: Most PLC manufacturers' software requires us to use the computer's mouse extensively.
21. In TEST mode, what does the Force-On instruction do to an address in the input image file?

22. In RUN mode, what does the Force-On instruction do to an output terminal in the I/O rack?

Section 3-2

23. If an Input Module were placed in slot 2 of rack Ø1 and a certain limit switch were wired to terminal 15 of that module, what address would be possessed by an instruction that examines the condition of that limit switch?
24. If an Output Module were placed in slot 7 of rack Ø3 and a certain solenoid were wired to terminal Ø6 of that module, what output instruction address would affect the energization of that solenoid?
25. For the situation described in Question 23, what is wrong with entering a user-program that contains an output-energize instruction addressed O:Ø12/Ø7?
26. For the situation described in Question 24, state the range of input addresses that are illegal.
27. True or False: It is a good policy to assign B3/X addresses to all internal logic instructions.
28. What is the alternative to the policy described in Question 27? Why is it not wise?
29. Is this a correct description of the function of the rung of Fig. 3–15? "Logic bit B3/1 will be set HI on the next scan after input terminal I:ØØ1/Ø2 receives ac power, and it will remain in the HI state until the first rung-scan after logic bit B3/2 becomes HI." Explain your answer.
30. Is this a correct description of the function of the rung of Fig. 3–17(a)? "Internal logic bit B3/3 will be set to 1 on the next program scan after input terminal 1:ØØ1/Ø7 receives 120-V ac power, and it will remain in the HI state until the next program scan after 120-V power is removed from input terminal I:ØØ1/Ø7." Explain your answer.

Section 3-3

Questions 31 through 33 refer to the TON program-rungs of Fig. 3–19, with I:ØØ1/XX being the only conditional instruction. Tell which statements are correct and which are incorrect. Explain each answer.

31. On the next program scan after input terminal I:ØØ1/XX receives 120-V ac power the T4:52/EN bit gets set to 1; that causes output terminal O:ØØ3/13 to become energized until the timer times out, after which that terminal becomes deenergized.
32. On any processor scan that finds input terminal I:ØØ1/XX deenergized, output terminal O:ØØ3/12 is bound to be energized.
33. Input terminal I:ØØ1/XX becomes energized and remains energized for 9 seconds. Then it becomes deenergized for 0.5 seconds. Then it becomes reenergized for 7 seconds. At the moment that is about 4 seconds into the 7-second reenergization period, output terminal O:ØØ3/12 becomes energized.

Question 34 refers to the CTU program-rungs of Fig. 3–21, with I:ØØ1/XX being the only conditional instruction. Is the statement correct or incorrect? Explain your answer.

34. If input terminal I:ØØ1/XX is ac-powered on a particular processor scan cycle, it is absolutely certain that the counter will not increment on the following scan cycle.

Section 3-4

Questions 35 through 38 refer to the machine of Fig. 3–22, controlled by the user-program of Fig. 3–23.

35. If the lift-table run-up timers T4:32 and T4:42 are set to 1.5 seconds and the required number of cylinder cycles is 25, what value should be used to preset On-delay timer T4:50?
36. On line 9 of Fig. 3–23, what is the purpose of the C5:175/DN examine-Off instruction?
37. When the 25th retraction stroke is complete and LS1 is actuated, explain why the cylinder does not begin a 26th extension stroke due to O:003/00 becoming TRUE on line 3.
38. After the 25th cylinder cycle has been completed and the workpiece holding table has run back down to its initial position, what would happen if the START PB were pressed again without removing the workpiece?

Section 3-5

39. In Fig. 3–25, if the beginning MCR instruction is FALSE on this scan, which of the following are guaranteed to reset to 0?
    a. On-delay timer T4:56.
    b. Internal logic bit B3/32.
    c. Up-counter C5:63.
    d. Output-energize file bit O:003/05.
40. True or False: The only opportunity that up-counter C5:63 has to increment is during a scan that finds input terminal I:001/14 powered by 120-V ac.

Section 3-6

41. An Analog Input Module contains ________ single-ended amplifiers.
42. True or False: On a given Analog Input Module, every one of the input terminals must be set up to receive the same analog voltage range as every other terminal.
43. True or False: On a given Analog Input Module, every one of the input terminals must be set up to transmit digital data back to the processor's Data File in the same binary format (BCD or strict binary) as every other terminal.
44. An Analog Input Module with a 12-bit ADC has a natural resolution of 1 part in ________.
45. True or False: If a certain channel is scaled and the spread between the minimum and maximum values is less than 4095 (say, minimum = −100 units and maximum = +800 units), the effective measurement resolution of the user-program deteriorates.
46. True or False: If a certain channel is scaled and the spread between the minimum and maximum values is more than 4095 (say, minimum = −2000 units and maximum = +4000 units), the effective measurement resolution of the user-program improves.
47. If a BTR instruction brings in the data from four input channels, its Block Data File will contain ________ words; its Transfer Control File will contain ________ words.
48. If a BTR instruction brings in the data from nine input channels, its Block Data File will contain ________ words; its Transfer Control File will contain ________ words.
49. Why do we prefer to start a Block Data File at a word-address that ends with the digit 7?
50. True or False: If the BTR instruction has more channels to read, that will lengthen the elapsed time between the beginning of the transfer process and the scanning of the following rung of the program.

51. True or False: An Analog Input Module is read from but never written to.
52. When a BTR or BTW instruction is executed, how does the processor know where to read the data from (BTR) or write the data to (BTW)?
53. When a BTW instruction is executed to configure an Analog Input Module, how does the processor know where to find the configuration information codes?
54. When the configuration information for an Analog Input Module is keyed in to the BTW's Block Data File words in the processor's variable-data memory, that information is automatically saved on the computer's hard disk too. Why is this necessary?

Section 3-7

55. In Fig. 3–40, explain how the temperature updating timer is reset after its 30-second time-out. Explain how it becomes reenabled to begin the next 30-second timing duration.
56. In Fig. 3–40, why is it neccessary to include a parallel branch with examine-Off N7:50/EN (bit-number 15) in rung 2? Why isn't it sufficient to simply detect that the BTW has been completed with the examine-On N7:50/DN (bit-number 13) instruction in that rung?
57. Why don't rungs 3 and 6 have the same requirement (parallel branching of the conditional instructions) as rung 2?
58. In the conditional instruction LEQ, which condition gives rung continuity, $A \leq B$ or $B \leq A$?
59. To perform the functions of rungs 10, 11, and 12, could we have just entered the address T4:30.PRE as the destination of the ADD instruction (rung 10), then repeated the ADD instruction with the same sources A and B but a destination of T4:40.PRE, thereby eliminating a rung?
60. What disadvantage would probably accompany the reprogramming suggested in Question 59?

CHAPTER

4

SCRs

Numerous industrial operations require the delivery of a variable and controlled amount of electrical power. Lighting, motor speed control, electric welding, and electric heating are the four most common of these operations. It is always possible to control the amount of electrical power delivered to a load by using a variable transformer to create a variable secondary output voltage. However, in high power ratings, variable transformers are physically large and expensive and need frequent maintenance. So much for variable transformers.

Another method of controlling electrical power to a load is to insert a rheostat in series with the load to limit and control the current. Again, for high power ratings, rheostats are large, expensive, need maintenance, and waste energy to boot. Rheostats are not a desirable alternative to variable transformers in industrial power control.

Since 1960, an electronic device has been available which has none of the faults mentioned above. The SCR is small and relatively inexpensive, needs no maintenance, and wastes very little power. Some modern SCRs can control currents of several hundred amperes in circuits operating at voltages higher than 1000 V. For these reasons, SCRs are very important in the field of modern industrial control. We will investigate SCRs in this chapter.

OBJECTIVES

After studying this chapter and performing the suggested laboratory projects, you will be able to:

1. Explain the operation of an SCR power control circuit for controlling a resistive load.
2. Define firing delay angle and conduction angle, and show how they affect the average load current.
3. Define some of the important electrical parameters associated with SCRs, such as gate trigger current, holding current, forward ON-state voltage, etc., and give the approximate range of values expected for these parameters.
4. Calculate approximate resistor and capacitor sizes for an SCR gate trigger circuit.
5. Explain the operation and advantages of breakover trigger devices used with SCRs.
6. Construct an SCR circuit for use with a 115-V ac supply and measure the gate current and gate voltage necessary to fire the SCR.
7. Construct a zero-point switching circuit and explain the advantages of zero-point switching over conventional switching.

4-1 ■ THEORY AND OPERATION OF SCRS

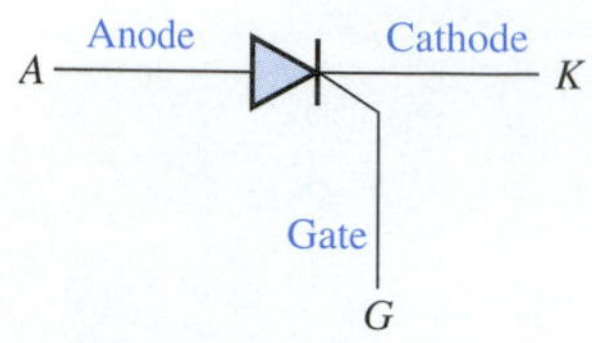

FIGURE 4–1
Schematic symbol and terminal names of an SCR.

A *silicon-controlled rectifier* (SCR) is a three-terminal device used to control rather large currents to a load. The schematic symbol for an SCR is shown in Fig. 4–1, along with the names and letter abbreviations of its terminals.

An SCR acts very much like a switch. When it is turned ON, there is a low-resistance current flow path from anode to cathode; then it acts like a closed switch. When it is turned OFF, no current can flow from anode to cathode; then it acts like an open switch. Because it is a solid-state device, the switching action of an SCR is very fast.

The average current flow to a load can be controlled by placing an SCR in series with the load. This arrangement is shown in Fig. 4–2. The supply voltage in Fig. 4–2 is normally a 60-Hz ac supply, but it may be dc in special circuits.

If the supply voltage is ac, the SCR spends a certain portion of the ac cycle time in the ON state and the remainder of the time in the OFF state. For a 60-Hz ac supply, the cycle time is 16.67 ms, which is divided between the time spent ON and the time spent OFF. The amount of time spent in each state is controlled by the gate. How the gate does this is described later.

If a small portion of the time is spent in the ON state, the average current passed to the load is small, because current can flow from the supply through the SCR to the load only for a relatively short portion of the time. If the gate signal is changed to cause the SCR to be ON for a larger portion of the time, then the average load current will be larger, because current now can flow from the supply through the SCR to the load for a relatively longer time. In this way the current to the load can be varied by adjusting the portion of the cycle time the SCR is switched ON.

As its name suggests, the SCR is a rectifier, so it passes current only during positive half cycles of the ac supply. The positive half cycle is the half cycle in which the anode of the SCR is more positive than the cathode. This means that the SCR in Fig. 4–2 cannot be turned ON more than half the time. During the other half of the cycle time the supply polarity is negative, and this negative polarity causes the SCR to be reverse biased, preventing it from carrying any current to the load.

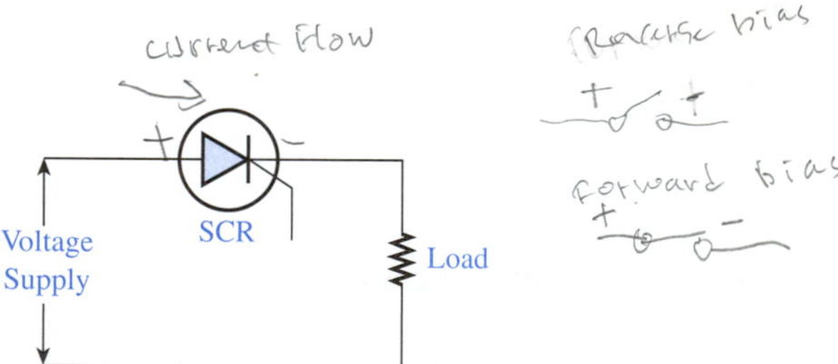

FIGURE 4–2
Circuit relationship among the voltage supply, an SCR, and the load.

4-2 ■ SCR WAVEFORMS

The popular terms used to describe how an SCR is operating are *conduction angle* and *firing delay angle*. Conduction angle is the number of degrees of an ac cycle during which the SCR is turned ON. The firing delay angle is the number of degrees of an ac cycle that elapses *before* the SCR is turned ON. Of course, these terms are based on the notion of total cycle time equaling 360 degrees (360°).

Figure 4–3 shows waveforms for an SCR control circuit for two different firing delay angles. Let us interpret Fig. 4–3(a) now. At the time the ac cycle starts its positive al-

FIGURE 4–3 Ideal waveforms of SCR main terminal voltage (V_{AK}) and load voltage: (a) for a firing delay angle of about 60°, conduction angle of 120°, (b) for a firing delay angle of about 135°, conduction angle of 45°.

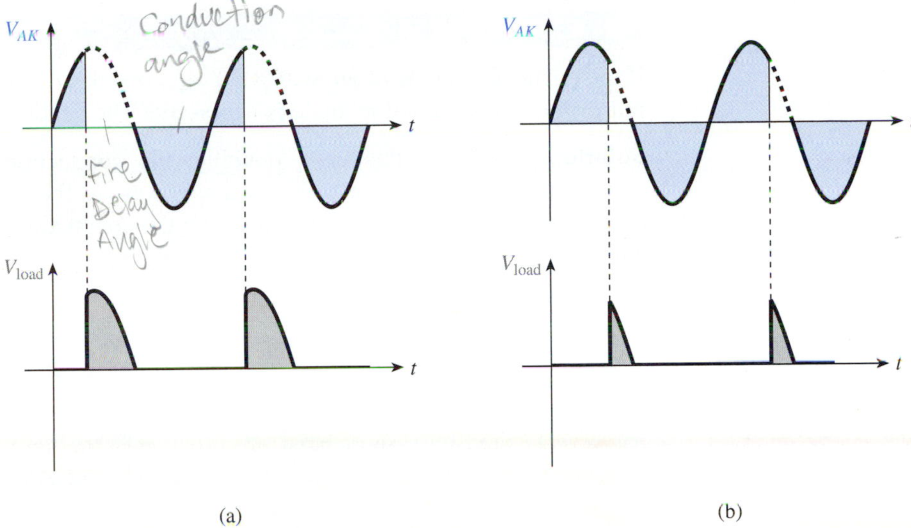

ternation, the SCR is turned OFF. Therefore it has an instantaneous voltage across its anode to cathode terminals equal to the supply voltage. This is just what would be seen if an open switch were put in the circuit in place of the SCR. Since the SCR is dropping the entire supply voltage, the voltage across the load (V_{load}) is zero during this time. The extreme left of the waveforms of Fig. 4–3(a) illustrates these facts. Farther to the right on the horizontal axes, Fig. 4–3(a) shows the anode to cathode voltage (V_{AK}) dropping to zero after about one third of the positive half cycle; this is the 60° point. When V_{AK} drops to zero, the SCR has "fired" or turned ON. Therefore in this case the firing delay angle is 60°. During the next 120° the SCR acts like a closed switch with no voltage across its terminals. The conduction angle is 120°. The firing delay angle and conduction angle always total 180°.

The load voltage waveform in Fig. 4–3(a) shows that when the SCR fires, the supply voltage is applied to the load. The load voltage then follows the supply voltage through the rest of the positive half cycle, until the SCR again turns OFF. Turning OFF occurs as the supply voltage passes through zero.

Overall, these waveforms show that before the SCR fires, the entire supply voltage is dropped across the SCR terminals, and the load sees zero voltage. After the SCR fires, the entire supply voltage is dropped across the load, and the SCR drops zero voltage. The SCR behaves just like a fast-acting switch.

Figure 4–3(b) shows the same waveforms for a different firing delay angle. In these waveforms, the firing delay angle is about 135° and the conduction angle about 45°. The load sees the supply voltage for a much shorter time as compared to Fig. 4–3(a). The average current is smaller as a result.

■ EXAMPLE 4-1

Which condition would cause the larger load current in Fig. 4–2, a firing delay angle of 30° or a firing delay angle of 45°?

Solution. The firing delay angle of 30°, because the SCR would then spend a greater portion of the cycle time in the ON state. The more time spent in the ON state, the greater the average load current.

■ EXAMPLE 4-2

If the conduction angle of an SCR is 90° and it is desired to double the average load current, what new conduction angle is necessary? The supply is an ac sine wave.

Solution. 180°. In this case, doubling the conduction angle doubles the average load current, because the first 90° of a sine wave is the image of the second 90°. However, in general, it is *not* true that doubling the conduction angle will double the average current. ■

4-3 ■ SCR GATE CHARACTERISTICS

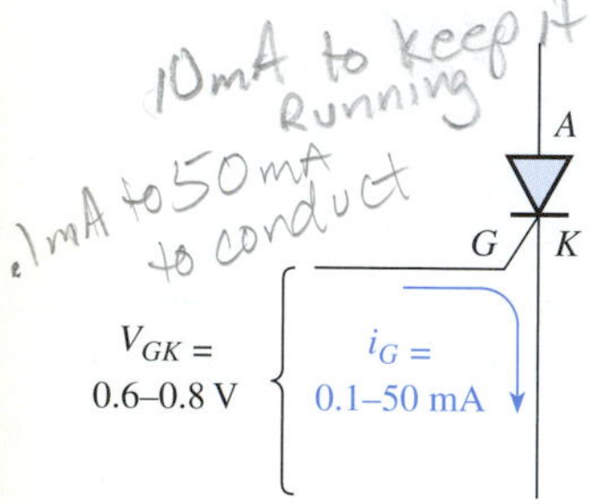

FIGURE 4–4
Gate to cathode voltage (V_{GK}) and gate current (i_G) needed to fire an SCR.

An SCR is fired by a short burst of current into the gate. This gate current (i_G) flows through he junction between the gate and cathode and exits from the SCR on the cathode lead. The amount of gate current needed to fire a particular SCR is symbolized I_{GT}. Most SCRs require a gate current of between 0.1 and 50 mA to fire ($I_{GT} = 0.1 - 50$ mA). Since there is a standard *pn* junction between gate and cathode, the voltage between these terminals (V_{GK}) must be slightly greater than 0.6 V. Figure 4–4 shows the conditions which must exist at the gate for an SCR to fire.

Once an SCR has fired, it is not necessary to continue the flow of gate current. As long as current continues to flow through the main terminals, from anode to cathode, the SCR will remain ON. When the anode to cathode current (i_{AK}) drops below some minimum value, called *holding current,* symbolized I_{HO}, the SCR will shut OFF. This normally occurs as the ac supply voltage passes through zero into its negative region. For most medium-sized SCRs, I_{HO} is around 10 mA.

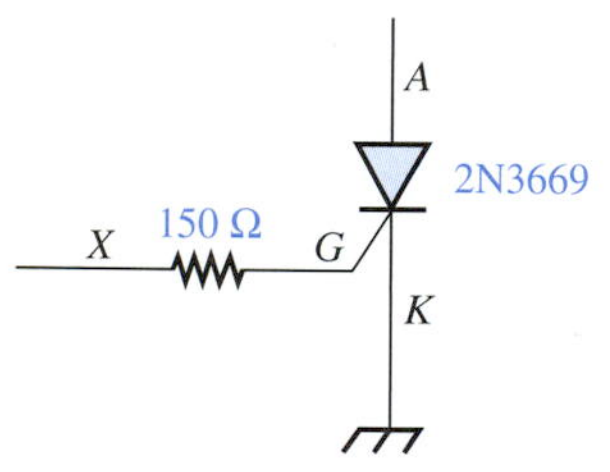

FIGURE 4–5
SCR with a 150-Ω resistor in the gate lead and its cathode terminal connected to circuit ground.

■ EXAMPLE 4-3

For the circuit of Fig. 4–5, what voltage is required at point *X* to fire the SCR? The gate current needed to fire a 2N3669 is 20 mA under normal conditions.

Solution. The voltage between point *X* and the cathode must be sufficient to forward-bias the junction between points *G* and *K* and also to cause 20 mA to flow through 150 Ω. The forward-bias voltage is about 0.7 V. From Ohm's law, $V_{XG} = (20\text{ mA})(150\ \Omega) = 3.0$ V. Therefore total voltage $= 3.0 + 0.7 = 3.7$ V. ■

4-4 ■ TYPICAL GATE CONTROL CIRCUITS

The simplest type of gate control circuit, sometimes called a triggering circuit, is shown in Fig. 4–6. This is an example of using the same voltage supply to power both the gate control circuit and the load. Such sharing is common in SCR circuits. The positions of the SCR and load are reversed from those of Fig. 4–2, but this makes no difference in operation.

In Fig. 4–6, if the supply is ac, operation is as follows. When the switch is open, it is impossible to have current flow into the gate. The SCR can never turn ON, so it is essentially an open circuit in series with the load. The load is therefore deenergized.

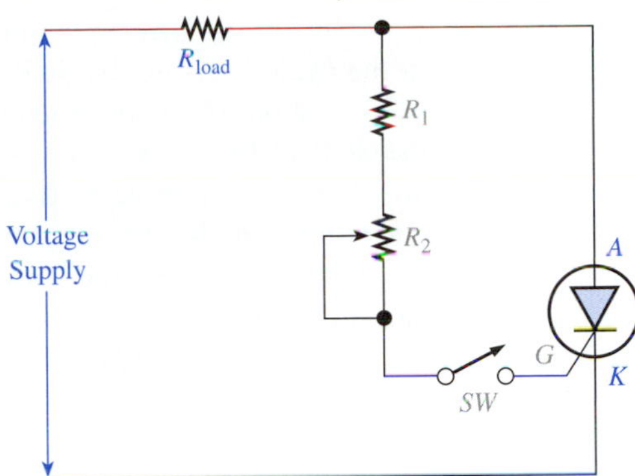

FIGURE 4–6
Very simple triggering circuit for an SCR.

When SW is closed, there will be current into the gate when the supply voltage goes positive. The firing delay angle is determined by the setting of R_2, the variable resistance. If R_2 is low, the gate current will be sufficiently large to fire the SCR when the supply voltage is low. Therefore the firing delay angle will be small, and average load current will be large. If R_2 is high, the supply voltage must climb higher to deliver enough gate current to fire the SCR. This increases the firing delay angle and reduces average load current.

The purpose of R_1 is to maintain some fixed resistance in the gate lead even when R_2 is set to zero. This is necessary to protect the gate from overcurrents. R_1 also determines the minimum firing delay angle. In some cases a diode is inserted in series with the gate to protect the gate-cathode junction against high reverse voltages.

One disadvantage of this simple triggering circuit is that the firing delay angle is adjustable only from about 0° to 90°. This fact can be understood by referring to Fig. 4–7, which shows that the gate current tends to be a sine wave in phase with the voltage across the SCR.

In Fig 4–7(a), i_G barely reaches I_{GT}, the gate current needed to trigger the SCR. Under this circumstance the SCR fires at 90° into the cycle. It can be seen that if i_G were any smaller, the SCR would not fire at all. Therefore, firing delays past 90° are not possible with such a gate control circuit.

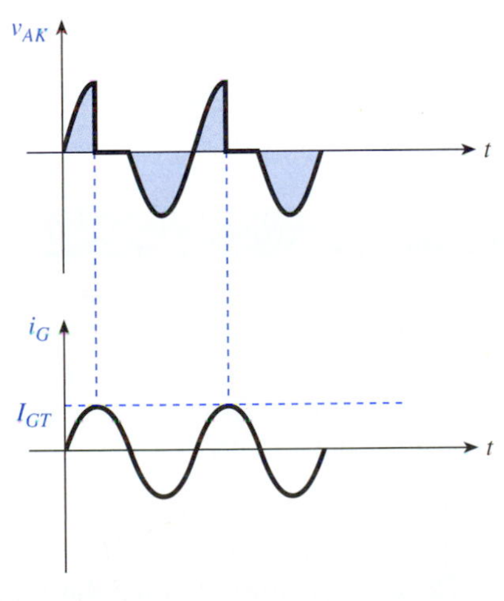

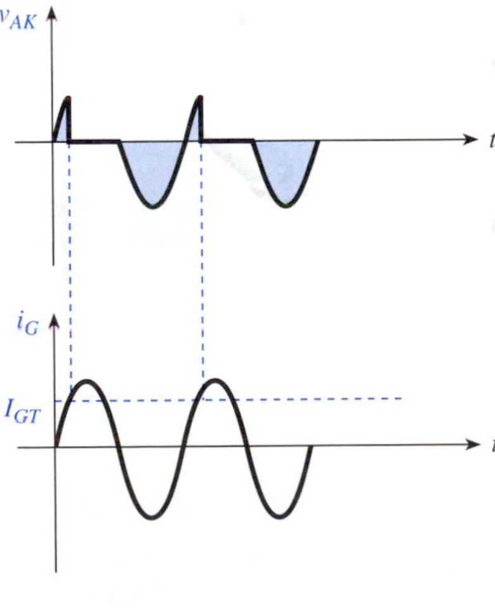

FIGURE 4–7
Ideal waveforms of SCR main terminal voltage and gate current. The dashed line represents the gate current necessary to fire the SCR (I_{GT}). (a) Gate current is low, resulting in a firing delay angle of about 90°. (b) Gate current is greater, resulting in a firing delay angle of nearly 0°.

In Fig. 4–7(b), i_G is quite a bit larger. In this case, i_G reaches I_{GT} relatively early in the cycle, causing the SCR to fire early.

It should be understood that the i_G waveforms of Fig. 4–7 are idealized. As soon as the SCR of Fig. 4–6 fires, the voltage from anode to cathode drops almost to zero (actually 1 to 2 V for most SCRs). Since the gate voltage is derived from the anode to cathode voltage, it also drops virtually to zero, shutting off the gate current. Furthermore, since the gate is reverse biased when the ac supply is negative, there is really no negative gate current as shown in Fig. 4–7. In reality then, the i_G curve is a sine wave in phase with the supply voltage *only* in the region between 0° and the triggering point. At other times i_G is nearly zero.

One more point bears mentioning. Prior to triggering, the v_{AK} waveform is virtually identical to the ac supply waveform, because the voltage dropped across the load in Fig. 4–6 is negligible prior to triggering. The load voltage is so small because the load resistance in such circuits is much lower than the resistance in the gate control circuit. Load resistance is almost always less than 100 Ω and often less than 10 Ω. the fixed resistance in the gate control circuit is typically several thousand ohms. When these two resistances are tied together in series, as they are prior to triggering, the voltage across the small load resistance is naturally very low. This causes almost the entire supply voltage to appear across the SCR terminals.

■ EXAMPLE 4-4

For Fig. 4–6, assume the supply is 115 V rms, $I_{GT} = 15$ mA, and $R_1 = 3$ kΩ. The firing delay is desired to be 90°. To what value should R_2 be adjusted?

Solution. At 90°, the instantaneous supply voltage is

$$(115\text{ V})(1.41) = 162\text{ V}$$

Neglecting the load voltage drop and the 0.7-V drop across the gate-cathode junction (they are both negligible compared to 162 V), the total resistance in the gate lead is given by

$$\frac{162\text{ V}}{15\text{ mA}} = 10.8\text{ k}\Omega$$

Therefore

$$R_2 = 10.8\text{ k}\Omega - 3\text{ k}\Omega = \mathbf{7.8\text{ k}\Omega}$$

■ EXAMPLE 4-5

In Fig. 4–6, if the resistance of the load is 40 Ω and the supply is 115 V rms (103.5 V_{avg}),* how much average power is burned in the SCR when the firing delay angle is 0°? Assume that the forward voltage across the SCR is constant at 1.5 V when it is turned ON and that reverse leakage current through the SCR is so small as to be negligible. (Reverse leakage current is less than 1 mA for most SCRs).

Solution. Since the power burned in the SCR is zero during the negative half cycle (reverse leakage current is negligible), the overall average power is half the average power of the positive half cycle. The average power burned during the positive half cycle equals

*Recall that $V_{avg} = (0.90)V_{rms}$.

the forward voltage, V_T, multiplied by the average forward current during the positive half cycle (I_{Tavg}):

$$P_{(+\text{half})} = (V_T)(I_{Tavg})$$

$$I_{Tavg} = \frac{103.5\text{ V} - 1.5\text{ V}}{40\ \Omega} = 2.55\text{ A}$$

$$P_{(+\text{half})} = (1.5\text{ V})(2.55\text{ A}) = 3.83\text{ W}$$

$$P_{avg} = \tfrac{1}{2}(3.83\text{ W}) = 1.91\text{ W}$$

It can be seen from the example that SCRs are very efficient devices. In Example 4-5 the SCR controlled a load current of several amps while wasting only about 2 W of power. This is much better than a series rheostat, for comparison.

The reason for the remarkable efficiency of SCRs is that when they are OFF, their current is very nearly zero, and when they are ON, their voltage is very low. In either case, the product of current and voltage is very small, resulting in low power dissipation.

It is this low power dissipation which enables the SCR to fit into a physically small package, making it economical. Economy and small size are the two most attractive features of SCRs.

Dc supply operation. Refer to Fig. 4–6 again; if the supply voltage is dc, the circuit operates as follows. When *SW* is closed, the SCR fires. The resistance in the gate lead would be designed so that this will occur. Once fired, the SCR will remain ON and the load will remain energized until the supply voltage is removed. The SCR stays ON even is *SW* is reopened, because it is not necessary to continue the flow of gate current to keep an SCR turned ON.

Although simple, this circuit is very useful in alarm applications. In an industrial alarm application, the *SW* contact could be closed when some malfunction occurs in an industrial process. As a burglar alarm, *SW* could be closed by the opening of a door or window or by interruption of a light beam.

4-5 ■ OTHER GATE CONTROL CIRCUITS

4-5-1 Capacitors Used to Delay Firing

The simplest method of improving gate control is to add a capacitor at the bottom of the gate lead resistance, as shown in Fig. 4–8. The advantage of this circuit is that the firing delay angle can be adjusted past 90°. This can be understood by focusing on the voltage across capacitor *C*. When the ac supply is negative, the reverse voltage across the SCR is applied to the *RC* triggering circuit, charging the capacitor negative on the top plate and positive on the bottom plate. When the supply enters its positive half cycle, the forward voltage across the SCR tends to charge *C* in the opposite direction. However, voltage buildup in the new direction is delayed until the negative charge is removed from the capacitor plates. This delay is applying positive voltage at the gate can be extended past the 90° point. The larger the potentiometer resistance, the longer it takes to charge *C* positive on the top plate and the later the SCR fires.

This idea can be extended by using either of the triggering circuits of Fig. 4–9. In Fig. 4–9(a), a resistor has been inserted into the gate lead, requiring the capacitor to charge higher than 0.6 V to trigger the SCR. With the resistor in place, capacitor voltage must

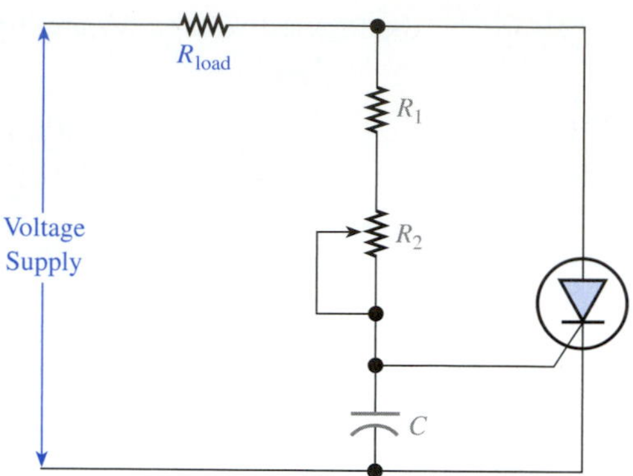

FIGURE 4–8
SCR gate control circuit which is an improvement on the circuit of Fig. 4–6. The capacitor provides a greater range of adjustment of the firing delay angle.

reach a value large enough to force sufficient current (I_{GT}) through the resistor and into the gate terminal. Since C must now charge to a higher voltage, triggering is further delayed.

Figure 4–9(b) shows a double RC network for gate control. In this scheme, the delayed voltage across C_1 is used to charge C_2, resulting in even further delay in buildup of gate voltage. The capacitors in Fig. 4–9 usually fall in the range from 0.01 to 1 μF.

For given capacitor sizes, the minimum firing delay angle (maximum load current) is set by fixed resistors R_1 and R_3, and the maximum firing delay angle (minimum load current) is set mostly by the size of variable resistance R_2.

The manufacturers of SCRs provide detailed curves to help in sizing the resistors and capacitors for the gate control circuits of Fig. 4–9. In general terms, when these gate control circuits are used with a 60-Hz ac supply, the time constant of the RC circuit should fall in the range of 1–30 ms. That is, for the single RC circuit of Fig. 4–9(a), the product $(R_1 + R_2)C$ should fall in the range of 1×10^{-3} to 30×10^{-3}. For the double RC gate circuit of Fig. 4–9(b), $(R_1 + R_2)C_1$ should fall somewhere in that range, and R_3C_2 should also fall in that range.

This approximation method will always cause the firing behavior to be in the right ball park. The exact desired firing behavior can then be experimentally tuned in by varying these approximate component sizes.

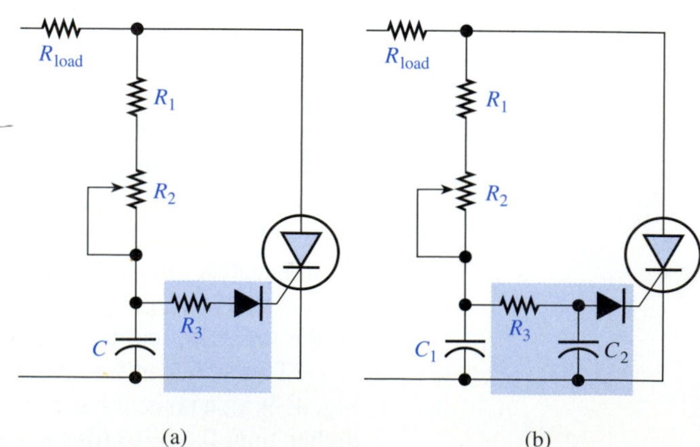

FIGURE 4–9
Improved SCR gate control circuits. Either one of these circuits provides a greater range of adjustment of the firing delay angle than the circuit of Fig. 4–8.

■ EXAMPLE 4-6

Suppose that it has been decided to use $C_1 = 0.068\ \mu F$ and $C_2 = 0.033\ \mu F$ in the gate control circuit of Fig. 4–9(b).

a) Approximate the sizes of R_1, R_2, and R_3 to give a wide range of firing adjustment.

b) If you then built the circuit and discovered that you could not adjust the firing delay angle to less than 40°, what resistor would you experimentally change to allow adjustment below 40°?

Solution. a) The time constant $(R_1 + R_2)C_1$ should fall in the range of about 1×10^{-3} to 30×10^{-3}. To provide a wide range of adjustment, the time constant should be adjustable over a large part of that range. As an estimate, we might try for an adjustment range of 2×10^{-3} to 25×10^{-3}.

The minimum time constant occurs when R_2 is dialed out, so

$$(R_1 + 0)(0.068 \times 10^{-6}) = 2 \times 10^{-3}$$
$$R_1 = 29.4\ \text{k}\Omega$$

The nearest standard size is **27 kΩ.**

The maximum time constant (and maximum firing delay) occurs when R_2 is completely dialed in, so

$$(R_2 + 27 \times 10^3)(0.068 \times 10^{-6}) = 25 \times 10^{-3}$$
$$R_2 = 340\ \text{k}\Omega$$

The nearest standard pot size is **250 kΩ**.

Experience has shown that the second time constant, R_3C_2, should fall somewhere toward the lower end of the suggested range. Let us assume 5 ms. Therefore

$$(R_3)(0.033 \times 10^{-6}) = 5 \times 10^{-3}$$
$$R_3 = \text{about } \mathbf{150\ k\Omega}$$

b) Either R_1 or R_3 should be made smaller to allow lower firing delay angles, because the capacitors will charge faster with smaller resistors (smaller time constants). You would probably try R_3 first. ■

4-5-2 Using a Breakover Device in the Gate Lead

The circuits of Figs. 4–6, 4–8, and 4–9 all share two disadvantages:

1. Temperature dependence
2. Inconsistent firing behavior between SCRs of the same type

Regarding disadvantage 1, an SCR tends to fire at a lower gate current when its temperature is higher (I_{GT} is lowered). Therefore, with any of the triggering circuits discussed so far, a change in temperature causes a change in firing angle and a consequent change in load current. In many industrial situations, this is unacceptable.

The second problem is that SCRs, like transistors, exhibit a wide spread in electrical characteristics within a batch. That is, two SCRs of a given type may show great differences in characteristics. The variation in I_{GT} is the most serious of these differences.

Fig. 4–10 shows how these difficulties can be eliminated. The *four-layer diode* in Fig. 4–10 has a certain breakover voltage. If the voltage across the capacitor is below that

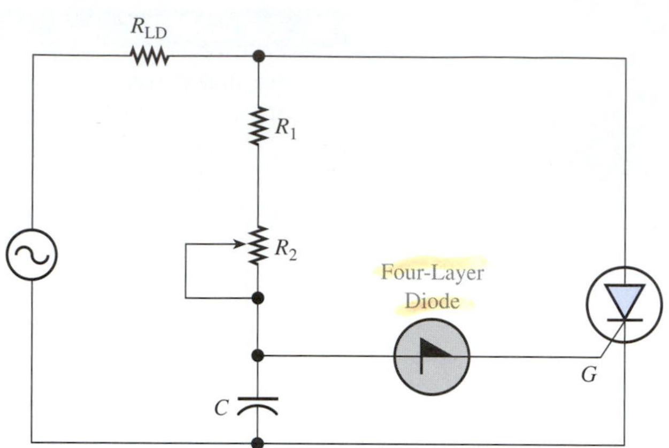

FIGURE 4–10
SCR gate control circuit using a four-layer diode (or any breakover device). The four-layer diode provides consistency of triggering behavior and reduces the temperature dependence of the circuit.

breakover point, the four-layer diode acts like an open switch. When the capacitor voltage rises to the breakover point, the four-layer diode fires and acts like a closed switch. This causes a burst of current into the gate, which provides sure triggering of the SCR.

The advantages of the four-layer diode are that it is relatively independent of temperature and that the breakover voltage can be held consistent from one unit to another. Therefore the imperfections of the SCR are of no importance, since it is the four-layer diode which determines the trigger point.

There are other devices which can be inserted in the gate lead to accomplish the same effect. They all have operating characteristics similar to those of the four-layer diode, and they are all temperature independent and have small spreads in breakover voltage. Some of the common triggering devices are the SUS (silicon unilateral switch), the SBS (silicon bilateral switch), the diac, and the UJT (unijunction transistor). All of these devices will be discussed in detail in Chapters 5 and 6.

4-6 ■ ALTERNATIVE METHODS OF CONNECTING SCRS TO LOADS

4-6-1 Unidirectional Full-Wave Control

Figure 4–11(a) shows how two SCRs can be combined with a center-tapped transformer to accomplish full-wave power control. This circuit behaves much like a full-wave rectifier for a dc power supply. When the secondary winding is in its positive half cycle, positive on top and negative on bottom, SCR_1 can fire. This connects the load across the top half of the transformer secondary. When the secondary winding is in its negative half cycle, SCR_2 can fire, connecting the load across the bottom half of the secondary winding. The current through the load always flows in the same direction, just as in a full-wave dc power supply. Figure 4–11(b) shows waveforms of load voltage and ac line voltage for a firing delay angle of about 45°.

Figure 4–11(a) indicates two separate trigger circuits for the two SCRs. Often these two circuits can be combined into a single circuit designed around one of the triggering devices mentioned in Sec. 4-5. Such a design ensures that the firing delay angle is identical for both half cycles.

4-6-2 Bidirectional Full-Wave Control

Another common SCR configuration is shown in Fig. 4–12(a). In this circuit, SCR_1 can fire during the positive half cycle and SCR_2 during the negative half cycle. The current

FIGURE 4–11
(a) Full-wave rectified power control, using two SCRs and a center-tapped winding. (b) Supply voltage and load voltage waveforms. Both ac half cycles are being used to deliver power, but the load voltage has only one polarity (it is rectified).

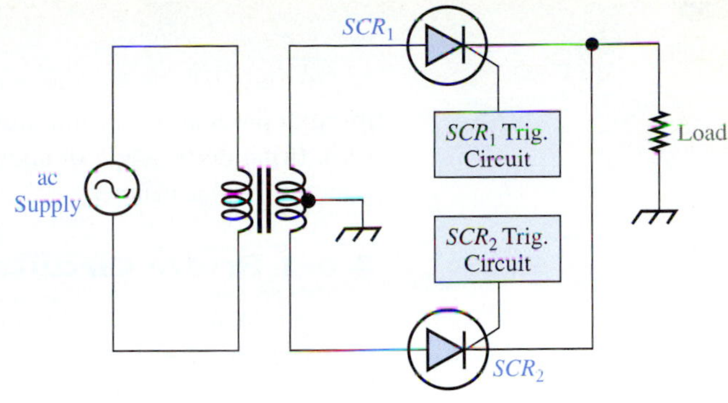

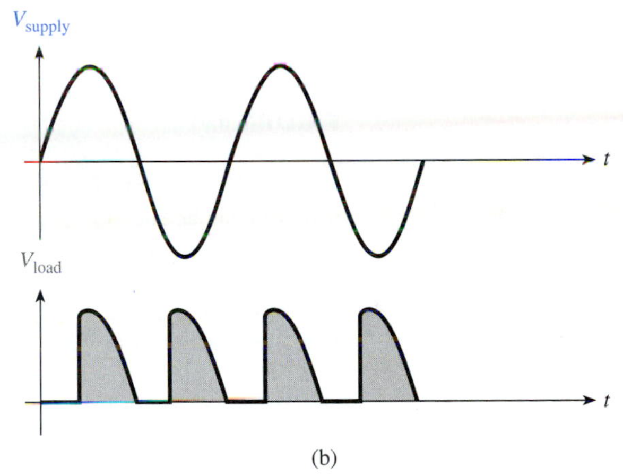

FIGURE 4–12
(a) Full-wave unrectified power control, using two SCRs. (b) The same circuit drawn another way. (c) Load voltage waveform. Both ac half cycles are being used to deliver power, and the load voltage is unrectified.

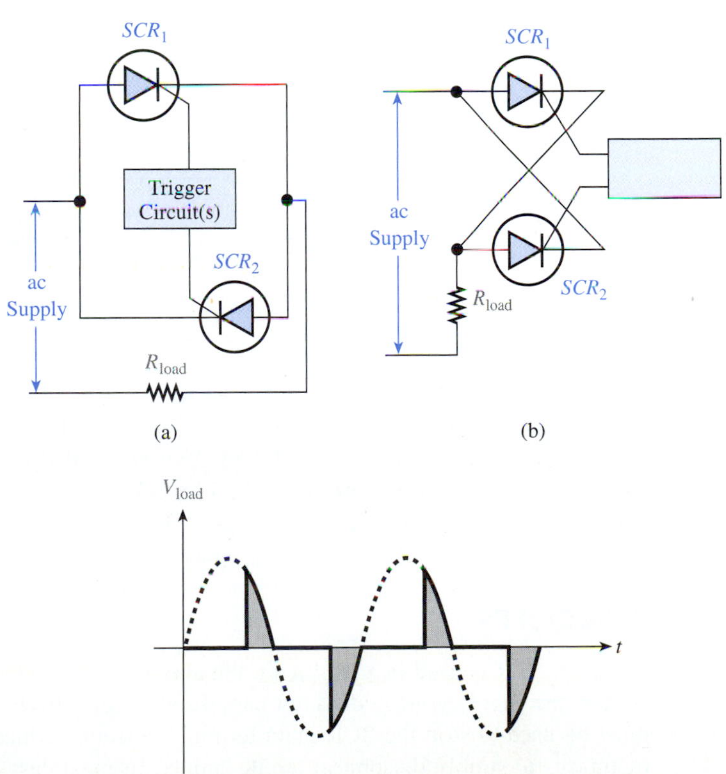

through the load is not unidirectional. Figure 4–12(c) shows a waveform of load voltage for a firing delay angle of about 120°. Figure 4–12(b) shows the same circuit redrawn in a more popular manner.

4-6-3 Bridge Circuits Containing an SCR

A single SCR can control both alternations of an ac supply when connected as shown in Fig. 4–13(a). When the ac line is in its positive half cycle, diodes *A* and *C* are forward biased. When the SCR fires, the line voltage is applied to the load. When the ac line is in its negative half cycle, diodes *B* and *D* are forward biased. Again the ac line voltage is applied to the load when the SCR fires. The load waveform would look like the waveform shown in Fig. 4–12(c).

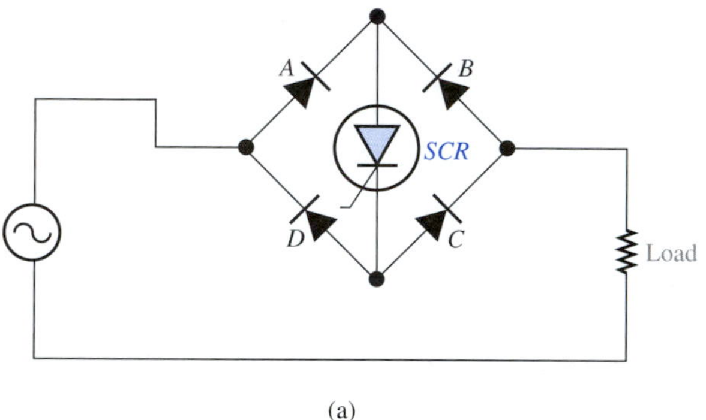

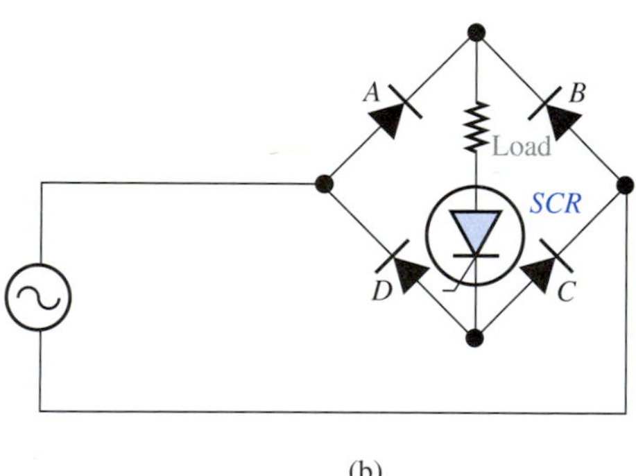

FIGURE 4–13 Full-wave bridge combined with an SCR to control both halves of the ac line. (a) With the load inserted in one of the ac lines leading to the bridge, the load voltage is unrectified, as in Fig. 4–12(b). (b) With the load inserted in series with the SCR itself, the load voltage is rectified, as in Fig. 4–11(b).

Figure 4–13(b) shows a bridge rectifier controlled by a single SCR, this time with the load wired in series with the SCR itself. The load current is unidirectional, with a waveform like that illustrated in Fig. 4–11(b).

4-7 ■ SCRS IN DC CIRCUITS

When an SCR is used in a dc circuit, the automatic turn-OFF does not occur, because, of course, the supply voltage does not pass through zero. In this situation, some other means must be used to stop the SCR main terminal current (reduce it below I_{HO}). One obvious method is to simply disconnect the dc supply. In most cases this is impractical.

Often, main terminal current is stopped by connecting a temporary short circuit from the anode to the cathode. This is illustrated in Fig. 4–14(a), in which a transistor switch is connected across the SCR. When the SCR is to be turned OFF, the trigger circuit pulses the transistor, driving it into saturation. The load current is temporarily shunted into the transistor, causing the SCR main terminal current to drop below I_{HO}. The transistor is held ON just long enough to turn OFF the SCR. This normally takes a few microseconds for a medium-sized SCR. The trigger circuit then removes the base current, shutting the transistor OFF before it can be destroyed by the large load current.

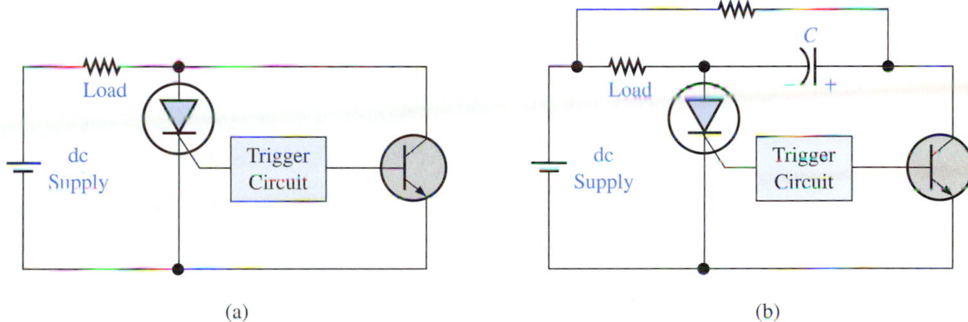

FIGURE 4–14
SCR commutation circuits. (a) The transistor switch shorts out the SCR, thereby turning it OFF. (b) The transistor switch puts a charged capacitor in parallel with the SCR for reverse-bias turn-OFF. Often another SCR is used in place of the transistor.

In this arrangement the trigger circuit is responsible for both turning ON and turning OFF the SCR. Such trigger circuits are naturally more complex than those discussed in Sec. 4-5, which were responsible only for turn-ON.

More reliable turn-OFF can be accomplished by actually reverse-biasing the SCR. This is shown in Fig. 4–14(b). In this circuit the capacitor charges with the polarity indicated when the SCR is turned ON. To turn OFF, the trigger circuit again saturates the transistor, which effectively places the capacitor in parallel with the SCR. Since the capacitor voltage cannot change instantly, it applies a temporary reverse voltage across the SCR, shutting it OFF.

■ EXAMPLE 4-7

In Fig. 4–14(a), imagine that the dc supply is 48 V and the trigger circuit behaves as follows:

1. It delivers a turn-ON pulse to the gate of the SCR.
2. 6.0 ms later it delivers a pulse to the base of the transistor.
3. It repeats this cycle at a frequency of 125 Hz.
   (a) Describe the load waveform. Neglect V_T.
   (b) If the load resistance is 12 Ω, how much average power is delivered to the load?

Solution. a) For a cycle frequency of 125 Hz, the period is

$$T = \frac{1}{f} = \frac{1}{125 \text{ Hz}} = 8 \text{ ms}$$

so the load waveform would be a rectangular wave, 48 volts tall, spending 6 ms up (at 48 V) and 2 ms down (at 0 V).

b) $$P_{\text{ON state}} = \frac{V_{\text{LD}}^2}{R_{\text{LD}}} = \frac{48^2}{12} = 192 \text{ W}$$

$P_{\text{avg}} = (0.75)(P_{\text{ON state}})$, because the SCR is ON for 75% of the total cycle time. Therefore,

$$P_{\text{avg}} = (0.75)(192 \text{ W}) = \mathbf{144 \text{ W}}$$ ■

MAGLEV VEHICLES

Magnetic levitation, MagLev, will become the premier technology for medium- to long-distance overland transportation at some point in the next century, as the earth's petroleum reserves become exhausted. The photo shows a MagLev railway vehicle, which has a cruising speed of 420 km/hr (about 260 mi/hr).

This vehicle contains extremely strong dc electromagnets that require no electric driving source whatsoever. The electromagnets, once energized, are able to achieve their extremely strong magnetic flux densities even in the absence of a power source because their winding coils are made of zero-resistance superconducting material. With $R = 0\ \Omega$, Ohm's law calls for zero voltage to produce large current through the winding coil, which does not overheat because its I^2R power dissipation is also zero. The difficulty is that the electromagnet coils must be held at a very cold temperature, below $-269°C$ ($-452°$ F), to maintain their superconducting ability. This is accomplished by housing them in specially insulated enclosures constantly supplied with high-pressure helium. Compressed liquid helium is carried in the vehicle and must be continually refrigerated. The refrigeration compressor accounts for the only on-board energy consumption.

MagLev vehicle and track, showing sidewall propulsion coils and on-the-ground levitation coils.

Courtesy of Railway Technical Research Institute of Japan.

Magnets made of newly discovered superconductive materials are now under development. They will operate in the $-120°C$ ($-185°F$) temperature range. These new magnets will be even cheaper and easier to refrigerate on-board, using liquid nitrogen rather than helium.

The vehicle's dc supermagnets route flux both vertically and horizontally, through the vehicle's floor and through its sides. The horizontal sideways flux provides the forward propulsion, as explained in Figs. 4–15. The vertical floor flux provides the levitation.

PROPULSION

Figures 4–15 and 4–16 are views from above. They show the railway's sidewall electromagnets that have to do with forward propulsion. These two figuresignore the railway's on-the-ground magnets that have to do with levitation. There are thousands and thousands of railway magnets, spaced close together, as you can tell by a close look at the photograph.

Each sidewall magnet has a pair of electric leads that are controlled by the railway's electronic power circuitry. The circuitry switches each magnet's current off and on, and it controls the current's direction. It therefore is able to switch an individual magnet's polarity, changing it between North and South. Each magnet is electrically in parallel with the one directly across from it on the other sidewall, so these two are always alike in polarity (if they are turned ON).

Figure 4–15 shows the situation at time instant t_1. Figure 4–16 shows the situation a short time later at instant t_2. We have identified four sidewall magnet positions, labeled **a, b, c,** and **d.** At time instant t_1, the magnets at position **d** are turned OFF by the power-control circuitry. The magnets at position **c** are turned ON, polarized South. The magnets at position **b** are turned ON, polarized North. Those at position **a** are

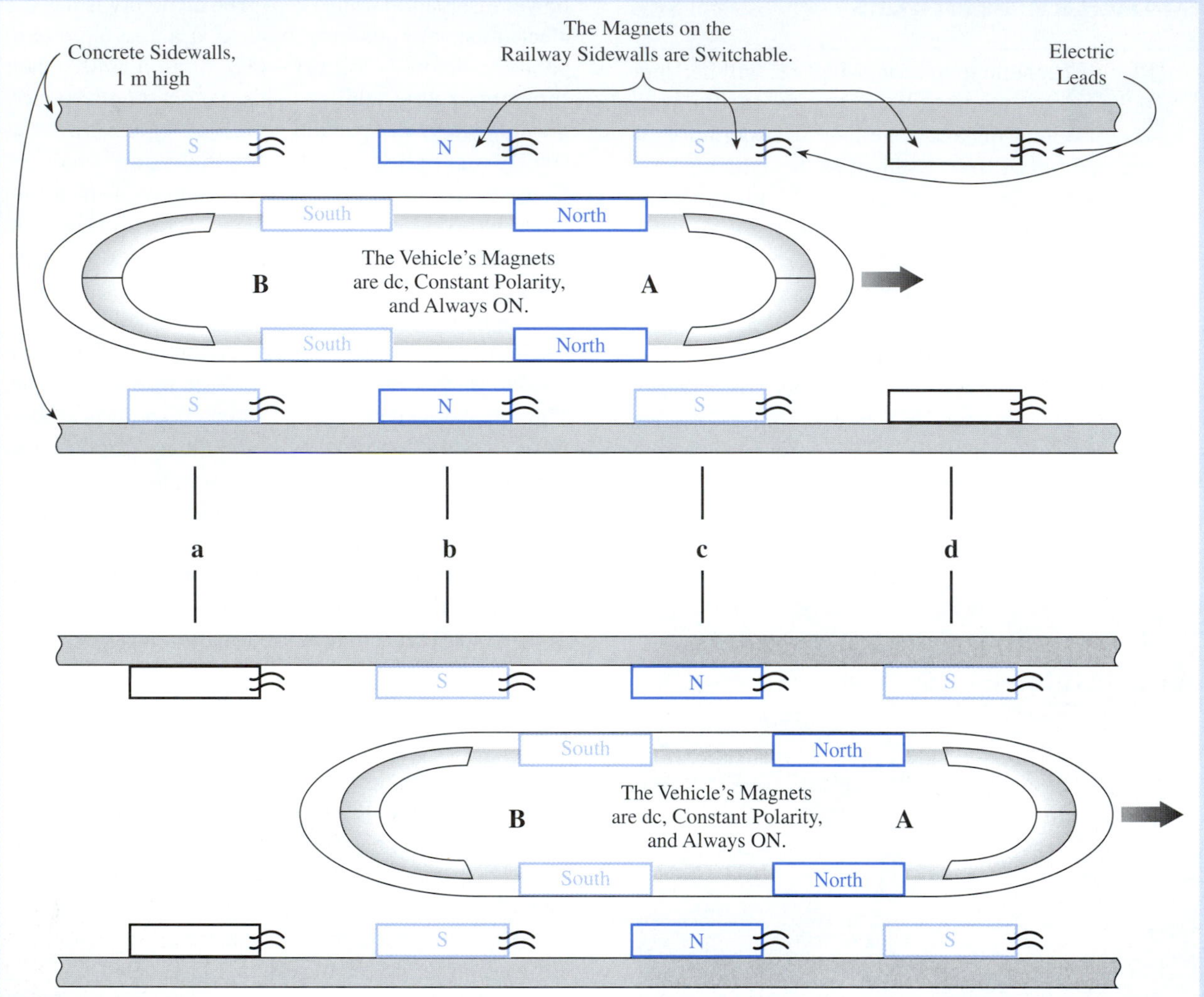

FIGURE 4–15 (Top)
Propulsion events at the t_1 instant.
FIGURE 4–16 (Bottom)
Propulsion events at the t_2 instant.

ON, polarized South. Inside the vehicle, the dc supermagnet positions are labeled **A** and **B.** At this instant the North supermagnets at **A** are attracted to the South sidewall magnets at **c.** This attractive force tends to pull the vehicle forward and to the right. Also, the **A** supermagnets are repelled by the like-polarity (North) sidewall magnets at **b.** This repulsion tends to push the vehicle forward.

Meanwhile, the South supermagnets at **B** are attracted to the unlike-polarity (North) sidewall magnets at **b,** producing more forward propulsion, and the South supermagnets at **B** are repelled away from the like-polarity South sidewall magnets at position **a.**

As the vehicle moves forward, its **A** supermagnets pass by sidewall position **c.** At that moment, the electronic power-control circuitry reverses the current direction through sidewall magnets **b** and **c.** Position **c** becomes North and position **b** becomes South. Figure 4–16 shows this reversal. At the same moment the power control circuitry switches ON the sidewall magnets at **d,** with a South polarity, and it switches OFF the sidewall magnets at **a** since the vehicle is now past that position. All of these conditions are indicated in Fig. 4–16, which shows the position of the vehicle at time t_2, a short time after the switching moment. By checking the magnetic polarities in Fig. 4–16, you can satisfy yourself that all forces are still propelling the vehicle forward to the right.

For simplicity of explanation, these figures show the vehicle with just two magnetic poles (one N and one S) per side. Actually, the 22-m-long vehicle has six magnetic poles per side (three N and three S).

LEVITATION

When the MagLev vehicle starts from a standstill, it has its wheels lowered onto the railbed. Once it reaches a speed of about 100 km/hr, it begins to levitate. Then it retracts its wheels, like an airplane. Figure 4–17 shows how levitation occurs.

Some of the flux from the supermagnets passes through the floor of the vehicle, as mentioned before. As this vertical flux passes at high speed over any one of the on-the-ground magnets, the rapid rate of change of flux induces voltage in the ground magnet's coil, in accordance with Faraday's law and Lenz's law. The induced voltage is maximum at a moment when the leading or trailing edge of a supermagnet is just above it; the voltage is minimum (zero) at a moment when a supermagnet is centered directly above the ground coil. The inductive reactance of the coil causes the coil's current to be phase-shifted by one-quarter cycle. Therefore the ground coil's current is maximum at a moment when it is centered beneath a supermagnet, as shown in Figure 4–17. This maximum current produces the magnetic flux that opposes (points in the opposite direction to) the supermagnet flux. Therefore the ground coil

Testing the MagLev mechanical suspension system; these wheels are retracted when the vehicle is moving fast enough to levitate.
Courtesy of Railway Technical Research Institute of Japan.

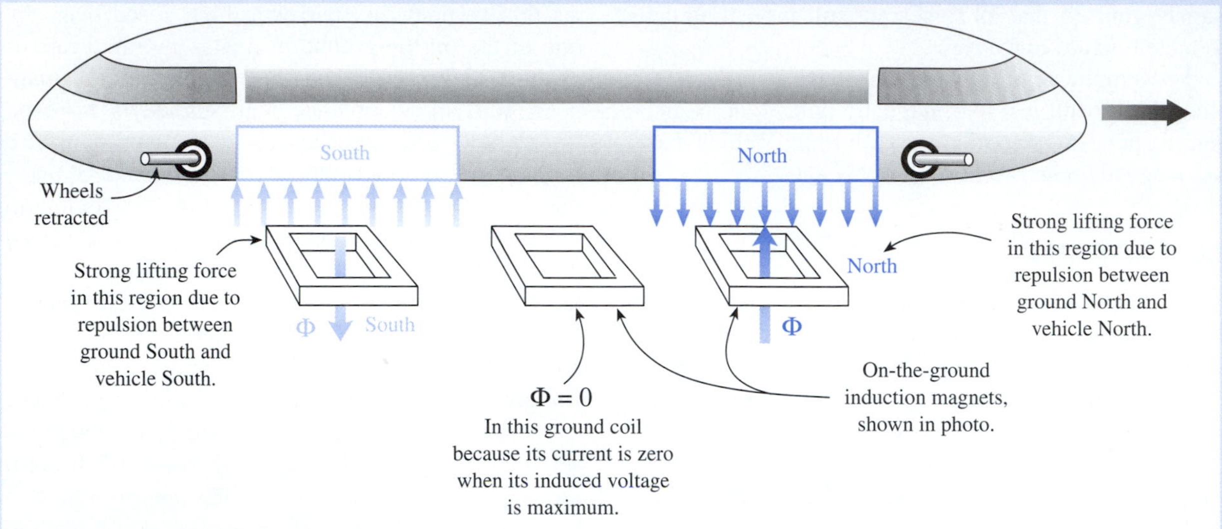

FIGURE 4–17
Levitation.

repels the supermagnet, providing the lifting force on the vehicle.

Figure 4–17 shows only a single ground coil beneath each on-board supermagnet. Actually the ground coils are physically much smaller than the vehicle's supermagnets, as you can see in the photograph. Thus there are really several ground coils interacting with a single supermagnet.

As the vehicle rises, the distance between the two flux surfaces increases. This weakens the magnetic repulsion-lifting force. When the vehicle reaches a height where the lifting force is exactly equal to its weight, it will maintain that levitation. This vehicle is designed to rise 10 cm above the railbed.

The MagLev vehicle has guide-arms, visible in the photograph, that ride in channels in the concrete sidewalls. Rubber wheels on the guide-arms rub against the sidewalls if the vehicle strays too far off center. This rarely happens because there is a natural sideways restoring force exerted on the vehicle by the sidewall magnets. We will not try to explain the origin of this restore-to-center force.

If any system should fail, the vehicle simply settles down with its guide-arms dragging on the bottom surfaces of the channels. Its safety advantage is obvious.

JOB ASSIGNMENT

This day your job assignment is to investigate a certain location on the MagLev track that has been experiencing a propulsion problem. This problem is commonly encountered and usually traceable to one of the following troubles within the propulsion control-module circuit of Fig. 4–18:

1. There is a failure or malfunction in the dc power supply that powers an entire bank of 10 control modules (40 SCRs) that drive an entire bank of 10 pairs of sidewall propulsion coils.
2. One (or more) of the SCRs within the module has failed, either open or shorted.
3. One or more of the turn-OFF (commutating) transistors within the module has failed, either open or shorted.
4. One of the propulsion coils itself has failed, either open or shorted.

Figure 4–18 shows the control-module circuitry for controlling one pair of propulsion coils. The circuit puts the coils into one of three possible states. The first is deenergized, or OFF. The coils carry no current and are not active magnetically. This is their state whenever the train is not passing between them, which is almost all the time, naturally. The second is turned

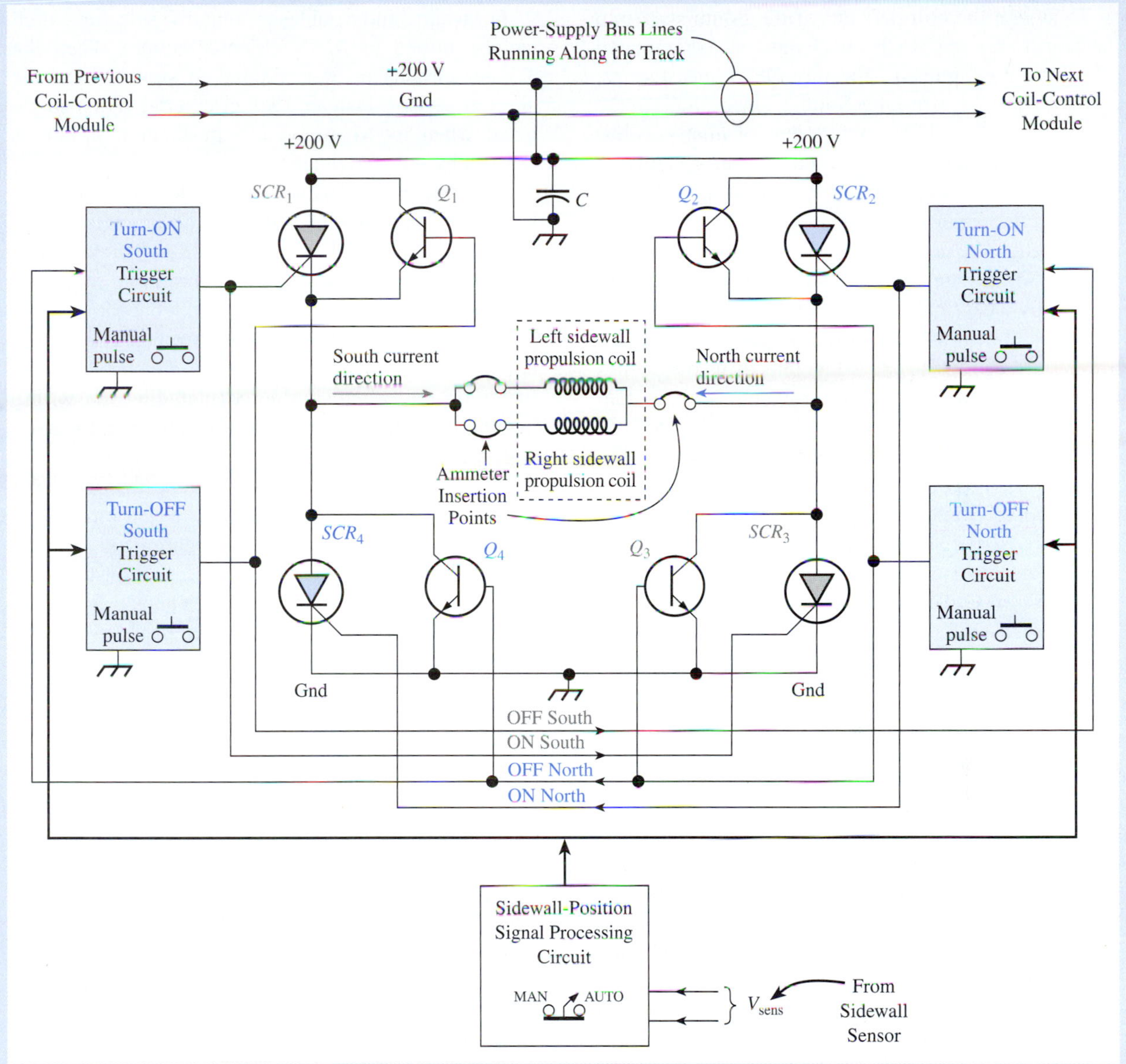

FIGURE 4–18
Schematic diagram of the control module for one pair of propulsion coils, which are directly opposite one another on the concrete sidewalls.

ON, South. This is the state that the coils are switched into when a train is passing, with a North on-board supermagnet approaching at this moment or a South on-board supermagnet receding (Figs. 4–15 and 4–16).

The third is turned ON, North. This is the state that the coils are switched into when a train is passing, with a South on-board supermagnet approaching at this moment or a North on-board supermagnet receding.

To switch the coil-pair into the ON, South condition, SCRs 1 and 3 are simultaneously triggered into the conducting state by a gate pulse from the turn-ON South Trigger Circuit in the upper left of Fig. 4–18.

To switch the coil-pair out of the South state into the North state, SCRs 1 and 3 must be commutated OFF by a base pulse to the turn-OFF transistors Q_1 and Q_3, from the turn-OFF South trigger circuit in the lower left of Fig. 4–18. A few tens of microseconds later, SCRs 2 and 4 are triggered into the conducting state by a gate pulse from the turn-ON North trigger circuit in the upper right of Fig. 4–18.

To switch the coil-pair out of the North state back into the South state, SCRs 2 and 4 must be commutated OFF by a base pulse to turn-OFF transistors Q_2 and Q_4, from the turn-OFF North trigger circuit in the lower right of Fig. 4–18. A few tens of microseconds later, SCRs 1 and 3 are again triggered into the conducting state by another gate pulse from the turn-ON South trigger circuit at the upper left.

To return the coil-pair to the deenergized state after the on-board South supermagnet has receded (Figs. 4–15 and 4–16), SCRs 1 and 3 are commutated OFF by a base pulse to turn-OFF transistors Q_1 and Q_3, from the turn-OFF South Trigger Circuit at the lower left of Fig. 4–18.

This switching sequence would constitute a complete switching sequence for the passage of a two-magnet train, as shown in Figs. 4–15 and 4–16. For a six-magnet train, there is a greater number of magnetic polarity reversals per one-pass switching sequence, naturally.

When the track is operational, all four trigger-control circuits in the control module are under automatic control from the Sidewall-Position-Signal Processing Circuit, at the bottom of Fig. 4–18. A magnetic Hall-effect sensor (described in Sec. 10-13) that is mounted on a sidewall is used to detect the approach of the trains' supermagnets. With the selector switch in AUTO position, the Signal Processing Circuit automatically issues the properly timed commands to the four Trigger circuits.

For testing and troubleshooting, the selector switch can be turned to MAN (Manual mode). Then the Trigger circuits do not respond to signals from the sidewall sensor; instead, each circuit delivers a single pulse when its Manual Pulse pushbutton switch is pressed by the technician.

Also for testing maintenance purposes, the module has provisions for coil isolation and ammeter insertion. As a maintenance technician, one of your test instruments is a dual-polarity, high-current ammeter. It can be inserted in any one of the three locations provided in Fig. 4–18. Those same terminals can be disconnected for ohmmeter and megohmmeter measurements on the propulsion coils (including the wiring leading from the control-module panel out to the sidewall locations). Another of your test instruments is a portable differential-input oscilloscope for observing the pulses emanating from the four trigger circuits.

To begin your checkout procedure, you connect your ammeter into the insertion location on the right side of the Fig. 4–18 schematic. With the Sidewall-Position Signal Processing Circuit disabled (in the Manual mode), you press the Manual Pulse PB in the turn-ON South Trigger Circuit. The ammeter immediately responds with a 20-ampere current indication in the proper left-to-right direction.

1. What components in the system are exonerated from fault by this test? List every component, device, and piece of interconnecting wiring that has been proved to be operating correctly.
2. What is the next logical step in your testing/troubleshooting procedure?
3. Suppose that when you pressed the Manual Pulse PB in the turn-ON South Trigger Circuit, the ammeter did not respond at all. Describe the next logical steps in the troubleshooting process to isolate the fault.

■ SUMMARY

- An SCR is like a fast-acting switch placed in series with the load device that it controls.
- The load's average dc current (and power) is controlled by varying the portion of the cycle time that the SCR is conducting, or turned ON.
- An SCR is triggered into the conducting state—fired—by current into its gate lead. That current, i_G, must reach a certain critical value, called *gate-trigger current,* symbolized I_{GT}.

- An SCR can be fired only when its main terminals are forward biased: Anode A positive, cathode K negative. It cannot be fired when its anode-cathode terminals are reverse biased.
- Once an SCR has been triggered into the ON state by gate current, it latches itself in the ON state until its main-terminal current automatically decreases below the critical value called *holding current,* symbolized I_{HO}. In most SCR applications this occurs at the negative-going zero crossover of the ac line.
- SCRs, like all solid-state electronic devices, exhibit temperature instability and batch instability.
- An SCR's temperature and batch instability can be counteracted by installing a breakover device, such as a four-layer diode, in the gate lead.
- It is possible to obtain full-wave power control by using two SCRs. The control can be unidirectional to the load (dc) or bidirectional (ac).
- An SCR can be used in a dc-supply circuit if special commutation (turn-OFF) circuitry is included.

FORMULA

$$P_{avg} = V_T \times I_{Tavg} \qquad \text{for an SCR circuit}$$

QUESTIONS AND PROBLEMS

Section 4-1

1. The letters SCR stand for silicon-controlled rectifier. Explain the use of the word *rectifier* in the name.

Section 4-3

2. What two things must happen to cause an SCR to fire?
3. Roughly speaking, how much gate current is needed to trigger an medium-power SCR?
4. In words, explain what each of the following symbols stands for:
   a. I_{GT}
   b. I_{Trms}
   c. I_{HO}
   d. V_T
5. After an SCR has fired, what effect does the gate signal have on the SCR?

Section 4-4

6. Roughly speaking, how much voltage appears across the anode-cathode terminals of a medium-power SCR after it has fired?
7. What effect does an increase in anode current have on V_T? Specifically, if anode-cathode current is doubled, does V_T also double?
8. In Fig. 4–6, the supply is 115 V rms, 60 Hz. The SCR has an I_{GT} of 35 mA; $R_1 = 1\ \text{k}\Omega$; what value of R_2 will cause a firing delay of 90°?
9. If R_2 is set to 2.5 kΩ in Question 8, what will be the firing delay angle? What is the conduction angle?
10. Explain why an SCR is superior to a series rheostat for controlling and limiting current through a load.

Section 4-5

11. The gate control circuit of Fig. 4–9(a) is used with a switched dc voltage supply of 60 V. The low-resistance load is connected as shown in Fig. 4–8. $R_1 = 1\ k\Omega$, $R_2 = 2.5\ k\Omega$, $R_3 = 1\ k\Omega$, and $C = 0.5\ \mu F$. The I_{GT} of the SCR is 10 mA. If the dc supply is suddenly switched ON, how much time will elapse before the SCR fires? *Hint:* Use the universal time constant curve in Chapter 2, and use Thevenin's theorem.
12. For the circuit of Question 11, what value of C will cause a time delay of 70 ms between closing the switch and firing the SCR?
13. For the circuit of Fig. 4–8, the supply is 220 V rms, 60 Hz. The load resistance is 16 Ω. Neglect the V_T of the SCR.
    a. How much power is delivered to the load if the firing delay angle = 0°?
    b. How much if the firing delay angle = 90°?
    c. If the firing delay angle = 135°, will the load power be less than one half or more than one half of the amount delivered for a 90° delay angle? Explain.
14. For Fig. 4–9(a), $C = 0.47\ \mu F$. Find appropriate sizes of R_1 and R_2 to give a wide range of adjustment of firing delay angle.
15. In Fig. 4–9(a), if $R_1 = 4.7\ k\Omega$ and $R_2 = 100\ k\Omega$, choose an approximate size of C which will permit the firing delay angle to be adjusted very late.
16. What two important benefits arise from using breakover-type devices to trigger SCRs?
17. Name some of the common breakover-type devices.

Section 4-7

18. Describe the methods used to turn OFF SCRs in dc circuits.

■ SUGGESTED LABORATORY PROJECTS

Project 4-1: SCR Power Control Circuit

Purpose

1. To observe the operation and waveforms of an SCR driving a resistive load.
2. To determine the electrical characteristics of a particular SCR.
3. To observe temperature and batch stabilization by using a breakover device (a diac, which is functionally equivalent to a four-layer diode in this application).

Procedure

Construct the gate control circuit of Fig. 4–9(a), with $R_1 = 2.2\ k\Omega$, $R_2 =$ 25-kΩ pot, $R_3 = 1\ k\Omega$, and $C = 0.68\ \mu F$. Place a diode, rated 200 V or higher, in series with the gate lead and pointing into the gate. The load resistor and ac supply are wired as shown in Fig. 4–8. The ac supply should be 115 V ac, *isolated from earth ground.*

If an earth-isolated 115 V source is not available, there are two ways to proceed: (1) Use an isolation transformer, with a secondary voltage close to 115 V. (2) Check the polarization of the ac line cord and arrange for the SCR cathode to be connected to *ac common* (the white wire which is at nearly earth-ground potential). Then, using a differential scope, connect the scope ground permanently to the SCR cathode and use the differential input to measure the load voltage and the gate resistor voltage.

Use an insensitive-gate medium SCR, RCA type S2800D or similar. The load should be a 100-Ω, 100-W power resistor or a 40- to 60-W light bulb. Insert a 0- to 1-A or 0- to 500-mA analog ammeter in series with the load.

1. Place the oscilloscope across the anode and cathode terminals of the SCR.
   a. Measure and record the minimum firing delay angle and the maximum firing delay angle.
   b. Record the average load current under both conditions. Does this agree with your understanding of the relationship between firing delay angle and load current?
   c. In which direction must you turn the 25-kΩ pot to increase the firing delay angle? Explain why this is so.
   d. Draw the V_{AK} waveform for some intermediate firing delay angle.
   e. Measure the voltage which exists across the SCR after firing (V_T). Is it fairly constant? Is it about as large as you expected?
2. Without disturbing the potentiometer setting of part **d**, connect the oscilloscope across the load resistance.
   a. Draw the load voltage waveform for the same intermediate firing delay angle as before.
   b. Compare the load voltage waveform to the SCR voltage waveform. Does the comparison make sense?
3. Place the oscilloscope across the 1-kΩ gate resistor. The current flowing into the gate terminal can now be found by using Ohm's law for the 1-kΩ resistor. Measure the gate current necessary to fire the SCR (I_{GT}). How much does it change as the firing delay angle is changed? As this what should be expected?
4. Place the oscilloscope across the main terminals of the SCR and adjust to some intermediate firing delay angle. Cool the SCR with freeze-mist and observe the reaction of the firing delay angle. What effect does decreased temperature have on an SCR circuit?
5. Install a diac (about 35-V rating) in series with the 1-kΩ gate resistor. Repeat steps 1 and 4. What important difference do you notice?
6. If several diacs of the same type are available, substitute different ones and repeat step 5. What can you conclude about the batch spread among diacs?

Project 4-2: SCR control with a double *RC* gate trigger circuit

Purpose

1. To observe larger firing delay angles possible with a double *RC* gate control circuit.
2. To observe the nonsinusoidal waveforms that occur when an SCR drives a motor or other inductive load.

Procedure

Construct the gate control circuit of Fig. 4–9(b). The load and ac supply are wired as shown in Fig. 4–8. Again, the 115-V ac source should be isolated from the earth, but if that is not possible, follow the suggestions given in Project 4-1.

Use the following component sizes: $R_1 = 4.7$ kΩ, $R_2 = 100$-kΩ pot, $R_3 = 10$ kΩ, $C_1 = 0.5$ μF, $C_2 = 0.05$ μF. Place a rectifier diode in the gate lead, along with a 1-kΩ resistor to protect the gate and limit the gate current. The SCR should be a 200-V, medium-current SCR, such as a C106B. As a load, use a small universal dc motor, such as a $^1\!/_4$ hp drill motor.

Observe the load voltage waveform by connecting the oscilloscope across the motor terminals. Try to explain why the SCR does not turn OFF exactly when the ac line passes through zero going into its negative region.

Project 4-3: SCR zero-point (soft-start) switch

Zero-point switching is the technique of always switching an SCR ON at the instant when the ac supply voltage is zero. It is desirable for two reasons. (1) It prevents the large inrush of current which occurs when a rather high voltage is suddenly applied to a very low-resistance load. It therefore prevents thermal shock to the load. (2) It eliminates electromagnetic interference which results from the sudden surges of inrush current.

Figure 4–19 shows a zero-point switching circuit. The average load current is controlled by the duty cycle (pulse width) of the rectangular wave from the pulse generator.

Observe the V_{LD} waveform on an oscilloscope. If a dual-trace scope is available, display both V_{LD} and the output wave of the pulse generator on the screen at the same time.

Note that the load voltage always appears in *complete* half cycles but that the number of half cycles spent ON versus the number spent OFF can be varied. This is the essence of zero-point power control.

Bring an inexpensive AM radio up close to the zero-point control circuit. Do you hear any electromagnetic interference on the radio? Repeat this test for any of the circuits built in Project 4-1 or 4-2. Comment on the difference.

Can you explain how this circuit works? *Hint:* The 0.22-μF capacitor charges during the negative half cycle. The capacitor is then the energy source for firing SCR_2 as the ac line passes through zero going positive.

FIGURE 4–19
Zero-point switching power-circuit. Load power is controlled by the variable pulse width.

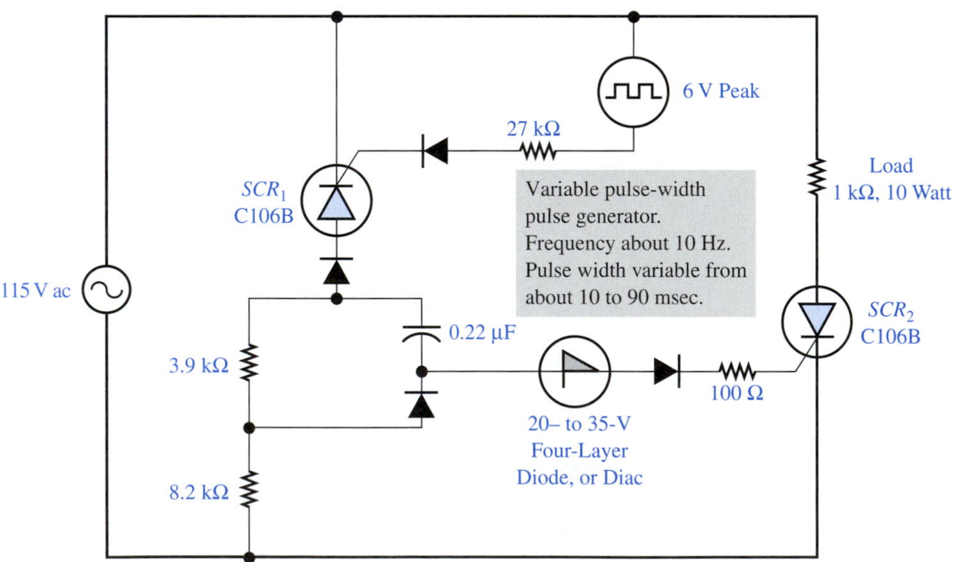

CHAPTER

5

UJTs

The unijunction transistor (UJT) is a breakover-type switching device. Its characteristics make it very useful in many industrial circuits, including timers, oscillators, waveform generators, and, most important, gate control circuits for SCRs and triacs. In this chapter we will present the operating characteristics and theory of UJTs and examples of how they are used in such circuits. A more extensive description of UJTs used as gate trigger devices for triacs is given in Chapter 6.

OBJECTIVES

After completing this chapter, you will be able to:

1. Interpret the current-voltage characteristic curve of a UJT and identify the peak voltage, peak current, valley voltage, and valley current.
2. Relate the UJT variables of peak voltage (V_p), intrinsic standoff ratio (η), and interbase voltage (V_{B2B1}), and calculate any one of these, given the other two.
3. Explain the operation of UJT relaxation oscillators and UJT timers, and properly size the timing resistors and capacitors in these circuits.
4. Explain the problem of UJT latch-up, why it occurs, and how to avoid it.
5. Explain the operation of a line-synchronized UJT trigger circuit for an SCR, and properly size the timing and stabilizing components.
6. Explain in detail the operation of a sequential load switching circuit using UJTs.
7. Explain the operation of a solid-state logic output amplifier built with an SCR triggered by a UJT.
8. Describe the triggering action of a PUT; cite the characteristics of a PUT that distinguish it from a standard UJT.

5-1 ■ THEORY AND OPERATION OF UJTS

5-1-1 Firing a UJT

The UJT is a three-terminal device, the three terminals being labeled emitter, base 1, and base 2. The schematic symbol and terminal locations are shown in Fig. 5–1(a). It is not a good idea to try to mentally relate the terminal names of the UJT to the terminal names of the common bipolar transistor. From a circuit operation viewpoint, there is no resemblance between the emitter of a UJT and the emitter of a bipolar transistor. The same is true for the relationship between the UJT base terminals and the bipolar transistor base terminal. To be sure, these terminal names make sense from an internal viewpoint which considers the action of the charge carriers, but internal charge carrier action is not an important concern to us.

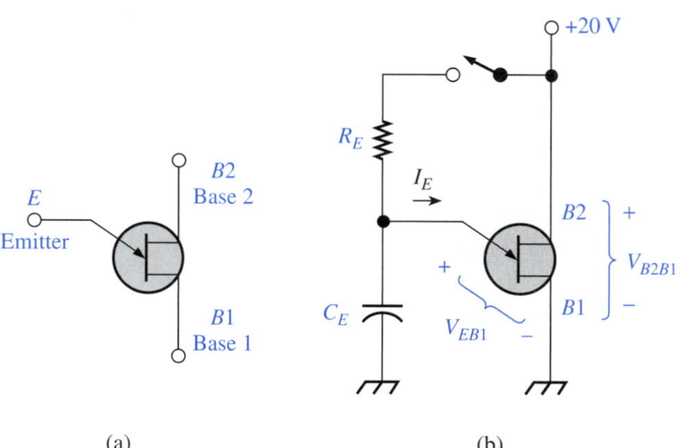

FIGURE 5–1
(a) Schematic symbol and terminal names of a UJT. (b) A UJT connected into a simple circuit. This drawing shows the emitter current (I_E), the emitter-to-base 1 voltage (V_{EB1}), and the base 2-to-base 1 voltage (V_{B2B1}).

In simplest terms, the UJT operates as follows. Refer to Fig. 5–1(b).

1. When the voltage between emitter and base 1, V_{EB1}, is less than a certain value called the *peak voltage*, V_p, the UJT is turned OFF, and no current can flow from E to $B1$ ($I_E = 0$).
2. When V_{EB1} exceeds V_p by a small amount, the UJT fires or turns ON. When this happens, the E to $B1$ circuit becomes almost a short circuit, and current can surge from one terminal to another. In virtually all UJT circuits, the burst of current from E to $B1$ is short-lived, and the UJT quickly reverts back to the OFF condition.

As Fig. 5–1(b) shows, an external dc voltage is applied between $B2$ and $B1$, with $B2$ the more positive terminal. The voltage between the two base terminals is symbolized V_{B2B1}, as indicated. For a given type of UJT, the peak voltage V_p is a certain fixed percent of V_{B2B1}, plus 0.6 V. That fixed percent is called the *intrinsic standoff ratio*, or just the *standoff ratio*, of the UJT, and is symbolized η.

Therefore the peak voltage of a UJT can be written as

$$V_p = \eta V_{B2B1} + 0.6\text{ V} \tag{5-1}$$

where 0.6 V is the forward turn-ON voltage across the silicon *pn* junction which exists between the emitter and base 1.

■ EXAMPLE 5-1

If the UJT in Fig. 5–1(b) has a standoff ratio of $\eta = 0.55$ and an externally applied V_{B2B1} of 20 V, what is the peak voltage?

Solution. From Eq. (5-1),

$$V_p = 0.55(20 \text{ V}) + 0.6 \text{ V} = \mathbf{11.6 \text{ V}}$$

In this case V_{EB1} would have to exceed 11.6 V in order to fire the UJT. ■

Refer again to the circuit of Fig. 5–1(b). The capacitor would begin charging through resistor R_E at the instant the switch was closed. Since the capacitor is connected directly between E and $B1$, when the capacitor voltage reaches 11.6 V the UJT will fire (assuming $\eta = 0.55$ as in Example 5-1). This will allow the charge built up on the plates of C_E to discharge very quickly through the UJT. In most UJT applications, this burst of current from E to $B1$ represents the output of the circuit. The burst of current can be used to trigger a thyristor,* turn ON a transistor, or simply to develop a voltage across a resistor inserted in the base 1 lead.

5-1-2 Current-Voltage Characteristic Curve of a UJT

There is a certain internal resistance existing between the two base terminals $B2$ and $B1$. This resistance is about 5–10 kΩ for most UJTs and is shown as r_{BB} in Fig. 5–2(a). In the physical structure of a UJT, the emitter lead contacts the main body of the UJT somewhere between the $B2$ terminal and the $B1$ terminal. Thus a natural voltage divider is created, because r_{BB} is divided into two parts, r_{B2} and r_{B1}. This construction is suggested by the equivalent circuit in Fig. 5–2(a). The diode in this figure indicates the fact that the emitter is p-type material, while the main body of a UJT is n-type material. Therefore a pn junction is formed between the emitter lead and the body of the UJT.

The total applied voltage, V_{B2B1}, is divided between the two internal resistances r_{B2} and r_{B1}. The portion of the voltage which appears across r_{B1} is given by

$$V_{rB1} = \frac{r_{B1}}{r_{B1} + r_{B2}} V_{B2B1}$$

which is simply the equation for a series voltage divider, applied to the circuit of Fig. 5–2(a).

To fire the UJT, the E to $B1$ voltage must become large enough to forward-bias the diode in Fig. 5–2(a) and bleed a small amount of current into the emitter lead. The value of V_{EB1} required to accomplish this must equal the sum of the diode forward turn-ON voltage plus the voltage drop across r_{B1}, or

$$V_{EB1} = V_D + \frac{r_{B1}}{r_{B1} + r_{B2}} V_{B2B1}$$

*A generic term which includes both SCRs and triacs.

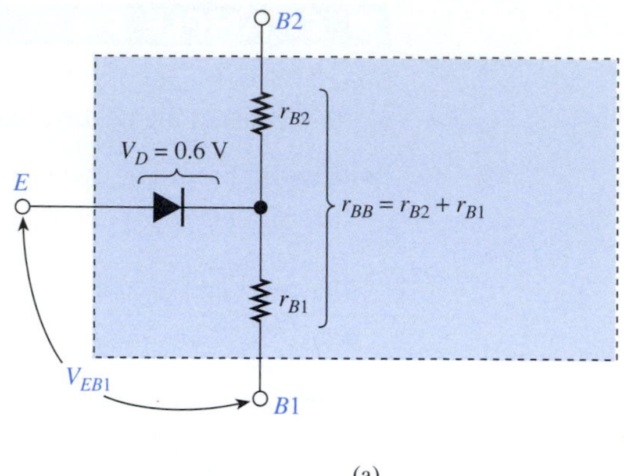

(a)

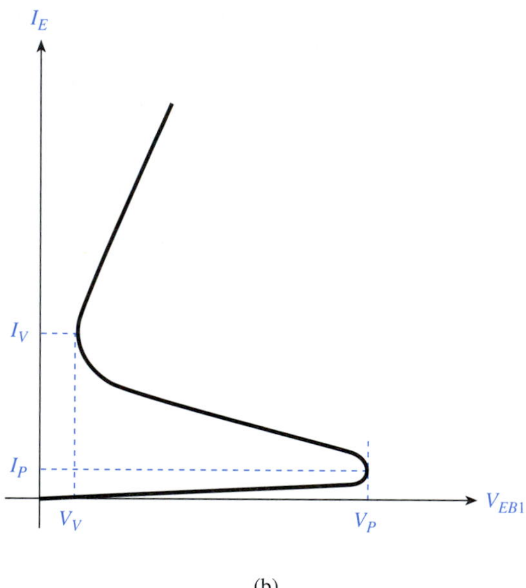

(b)

FIGURE 5–2
(a) Equivalent circuit of a UJT. The total resistance between $B2$ and $B1$ is called r_{BB}. It is divided into two parts, r_{B2} and r_{B1}. The emitter is connected through a diode to the junction of r_{B2} and r_{B1}. (b) Current versus voltage characteristic curve of a UJT (I_E versus V_{EB1}). The four important points on this curve are called peak voltage (V_p) peak current (I_p) valley voltage (V_V) and valley current (I_V).

in order to fire the UJT. Comparing this to Eq. (5-1) shows that the standoff ratio is just the ratio of r_{B1} to the total internal resistance, or

$$\eta = \frac{r_{B1}}{r_{B1} + r_{B2}} = \frac{r_{B1}}{r_{BB}} \tag{5-2}$$

The total internal resistance r_{BB} is called the *interbase resistance*.

■ EXAMPLE 5-2

(a) If the UJT in Fig. 5–1(b) has an r_{B1} of 6.2 kΩ and an r_{B2} of 2.2 kΩ, what is its standoff ratio?
(b) How large is the peak voltage?

Solution. (a) From Eq. (5-2),

$$\eta = \frac{r_{B1}}{r_{B1} + r_{B2}} = \frac{6.2\text{ k}\Omega}{6.2\text{ k}\Omega + 2.2\text{ k}\Omega} = \mathbf{0.74}$$

(b) From Eq. (5-1),

$$V_P = (0.74)(20\text{ V}) + 0.6\text{ V} = \mathbf{15.4\text{ V}}$$

The mechanism by which the UJT fires is suggested in Fig. 5–1(b). When the emitter to base 1 voltage rises to the peak voltage V_p and a small emitter current starts to flow, the UJT "breaks back" to a smaller voltage between the emitter and base 1 terminals. This smaller voltage is called the *valley voltage* and is symbolized V_V in Fig. 5–2(b). This breaking back occurs because of a drastic increase in the number of charge carriers available in the $B1$ region when emitter current starts to trickle into the main body of the device. From the external viewpoint, it appears as though r_{B1} drops to almost zero ohms in a very short time.

It is convenient to think of r_{B1} as a resistance whose value varies drastically, from its original OFF-state value all the way to nearly zero ohms. The resistance of r_{B2}, on the other hand, is fixed at its original OFF-state value. When r_{B1} drops to nearly zero ohms, the emitter to base 1 circuit allows an external capacitor to dump its charge through the device. Because r_{B2} maintains its original high resistance at this time, there is no unmanageable surge of current out of the dc power supply from $B2$ to $B1$.

The external capacitor quickly discharges to the point where it can no longer deliver the minimum current required to keep the UJT turned ON. This minimum required current is called the *valley current* and is symbolized I_V, as shown in Fig. 5–2(b). When the current flow from emitter to base 1 declines to slightly less than the valley current, the UJT reverts to the OFF state. Once it has returned to the OFF state, no current flows from E to $B1$, and V_{EB1} must again climb to V_p in order to fire the device a second time.

5-2 ■ UJT RELAXATION OSCILLATORS

The relaxation oscillator is the heart of most UJT timer and oscillator circuits. It is essentially the same circuit as shown in Fig. 5–1(b), except that resistors are added to the $B1$ and $B2$ leads in order to develop output signals. These external resistors are rather small compared to the internal resistance of the UJT, r_{BB}. The external resistors are usually symbolized R_2 and R_1. Typical component sizes for a relaxation oscillator are given in Fig. 5–3(a).

The oscillator works on the principles discussed in Sec. 5-1. When power is applied, C_E charges through R_E until the capacitor voltage reaches V_P. At this point, the UJT will fire, as long as R_E is not too large. The limitation on R_E occurs because a certain minimum amount of current must be delivered from the dc power supply into the emitter to successfully fire the UJT, even given that V_p is reached. Since this current must arrive at the emitter terminal by way of R_E, the resistance of R_E must be small enough to permit the necessary current to flow. This minimum current is called *peak point current* or *peak current,* symbolized I_p, and is just a few microamps for most UJTs. I_p is shown graphically on the characteristic curve of Fig. 5–2(b).

The equation which gives the maximum allowable value of R_E is easily obtained by applying Ohm's law to the emitter circuit.

$$R_{E\max} = \frac{V_S - V_p}{I_p} \qquad \textbf{(5-3)}$$

In Eq. (5-3), V_S represents the dc source voltage. The quantity $V_S - V_p$ is the voltage available across R_E at the instant of firing.

When the UJT fires, the internal resistance r_{B1} drops to nearly zero, allowing a pulse of current to flow from the top plate of C_E into R_1. This causes a voltage spike to appear at the $B1$ terminal, as shown in Fig. 5–3(b). At the same time that the positive spike appears at $B1$, a negative-going spike occurs at $B2$. This happens because the sudden drop in r_{B1} causes a sudden reduction in total resistance between V_S and ground and a consequent increase in current through R_2. This increase in current causes an increased voltage drop across R_2, creating a negative-going spike at the $B2$ terminal, as illustrated in Fig. 5–3(c).

At the emitter terminal, a sawtooth-shaped wave occurs, shown in Fig. 5–3(d). The sawtooth is not linear on its run-up, because the capacitor does not charge at a constant rate. Also, the bottom of the waveform is not exactly zero volts. There are two reasons for this:

1. The emitter-to-base 1 voltage never reaches 0 V, only V_V, as Fig. 5–2(b) indicates.
2. There is always some voltage drop across R_1 due to the current flow through the main body of the UJT. That is, there is always a complete circuit by which current can flow out of the dc supply terminal, through R_2, through the body of the UJT, through R_1 and into ground.

We saw earlier that in a relaxation oscillator R_E must not be too large, or the UJT will not be able to fire. Likewise, there is a limit as to how *small* R_E may be, to be assured of turning the UJT back OFF after firing. Recall that the reason a UJT turns back OFF is that the capacitor C_E discharges to the point where it cannot deliver an emitter current equal to I_V; the valley current (see Fig. 5–2b). The implication here is that the UJT must not be able to draw enough emitter current through R_E either. Therefore R_E must be large enough to prevent the passage of current equal to I_V. The equation which expresses the minimum value for R_E is

$$R_{E\text{min}} = \frac{V_S - V_V}{I_V} \tag{5-4}$$

which is just Ohm's law applied to the emitter resistor. The quantity $V_S - V_V$ is the approximate voltage across R_E after firing. This is true because upon firing, the emitter to ground voltage drops to about V_V (neglecting the small voltage across R_1).

The frequency of oscillation of a relaxation oscillator of the type shown in Fig. 5–3(a) is given very approximately by

$$f = \frac{1}{T} = \frac{1}{R_E C_E} \tag{5-5}$$

Equation (5-5) is fairly accurate as long as the UJT has an η in the neighborhood of 0.63, which is usually the case. As the η strays above or below 0.63, Eq. (5-5) becomes less accurate.

An intuitive feel for Eq. (5-5) can be gained by remembering that an RC circuit charges through 63% of its total voltage change in one time constant. If $\eta = 0.63$, C_E must charge to about 63% of V_S in order to fire the UJT. This requires a charging time of one time constant, or, in other words,

$$t_{\text{charge}} = R_E C_E \tag{5-6}$$

FIGURE 5–3
(a) Schematic drawing of a relaxation oscillator. For a given UJT (given η), the oscillation frequency depends on R_E and C_E. (b) Base 1-to-ground voltage (V_{B1}) waveform for the relaxation oscillator. (c) Base 2-to-ground voltage (V_{B2}) waveform. (d) Emitter-to-ground voltage (V_E) waveform.

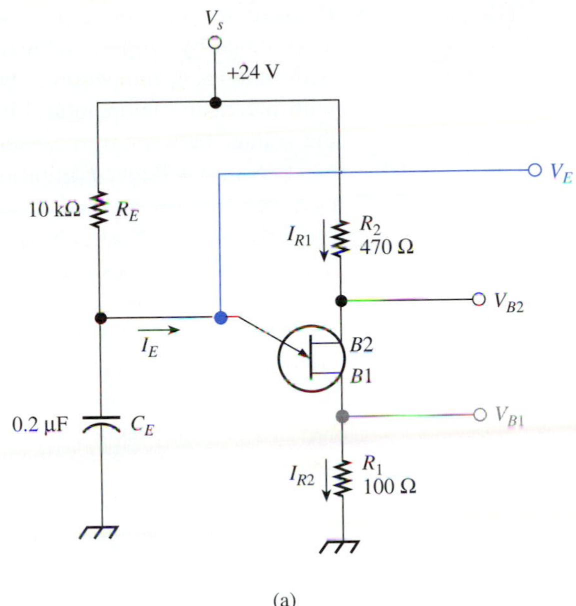

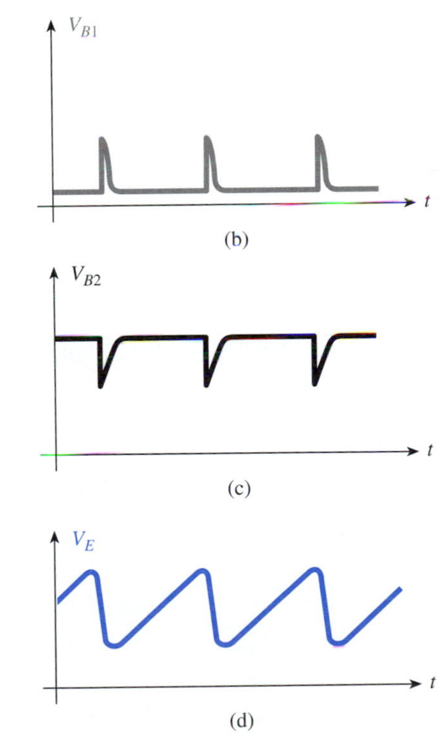

Since the firing and subsequent turn-OFF are both very fast compared to the charging time, the total period of the oscillations is about equal to $R_E C_E$. Frequency equals the reciprocal of period, so Eq. (5-5) is valid.

A UJT's standoff ratio is fairly stable as temperature changes, varying less than 10% over an operating temperature range of $-50°C$ to $+125°C$ for a high-quality UJT.

Relaxation oscillators can be made frequency stable to within 1% over the same temperature range by properly adjusting R_2 in Fig. 5–3(a). The standoff ratio tends to *decrease* with increasing temperature, whereas the total internal resistance, r_{BB}, tends to *increase* with increasing temperature. External resistor R_2 is constant as temperature changes, so the voltage between the base terminals, V_{B2B1}, increases with increasing temperature since r_{BB} becomes a larger portion of the total resistance from V_S to ground. Therefore V_{B2B1} gets larger as η gets smaller. These effects can be made to just cancel each other if R_2 is properly chosen. Under these circumstances V_p is held constant. If V_p is constant, oscillation frequency is also constant, because C_E will always have to charge to the same voltage to fire the UJT, no matter what the temperature.

The batch stability, or variation between UJTs of the same type number, is not so good as the temperature stability. Two supposedly identical UJTs may have standoff ratios which differ by 30% or more. For this reason, UJT relaxation oscillators contain some sort of trim adjustment if a precise oscillation frequency is desired. This is easily done by inserting a potentiometer in series with R_E.

■ EXAMPLE 5-3

Refer to the relaxation oscillator shown in Fig. 5–3. Assume that the UJT has the following characteristics:

$$\eta = 0.63 \qquad r_{BB} = 9.2\ \text{k}\Omega \qquad V_V = 1.5\ \text{V}$$
$$r_{B1} = 5.8\ \text{k}\Omega \qquad I_P = 5\ \mu\text{A}$$
$$r_{B2} = 3.4\ \text{k}\Omega \qquad I_V = 3.5\ \text{mA}$$

(a) Find V_P.
(b) What is the approximate output frequency?
(c) Prove that a 10-kΩ R_E is within the acceptable range. That is, $R_{E\min} < R_E < R_{E\max}$.
(d) Describe the waveform that appears across R_1. How tall are the spikes? What voltage appears across R_1 during the time that the UJT is OFF?

Solution. (a) From Eq. (5-1),

$$V_P = (0.63)(V_{B2B1}) + 0.6\ \text{V}$$

The voltage from base 2 to base 1 can be found by the proportion

$$\frac{V_{B2B1}}{V_S} = \frac{r_{BB}}{R_{\text{total}}} = \frac{r_{BB}}{R_2 + r_{BB} + R_1}$$

$$\frac{V_{B2B1}}{24\ \text{V}} = \frac{9200\ \Omega}{470\ \Omega + 9200\ \Omega + 100\ \Omega}$$

$$V_{B2B1} = 22.6\ \Omega$$

Therefore,

$$V_p = (0.63)(22.6\ \text{V}) + 0.6\ \text{V} = \mathbf{14.8\ V}$$

(b) since $\eta = 0.63$, Eq. (5-5) will predict the oscillator's output frequency quite accurately:

$$f = \frac{1}{(10\ \text{k}\Omega)(0.2\ \mu\text{F})} = \frac{1}{2 \times 10^{-3}} = \mathbf{500\ Hz}$$

(c) From Eq. (5-3)

$$R_{E\max} = \frac{V_S - V_P}{I_P} = \frac{24\text{ V} - 14.8\text{ V}}{5\ \mu\text{A}} = \mathbf{1.84\ M\Omega}$$

From Eq. (5-4),

$$R_{E\min} = \frac{V_S - V_V}{I_V} = \frac{24\text{ V} - 1.5\text{ V}}{3.5\text{ mA}} = \mathbf{6.4\ k\Omega}$$

The actual value of R_E, 10 kΩ, is between 6.4 kΩ and 1.84 kΩ, so it is acceptable. It will allow enough emitter current to flow to fire the UJT but not enough to prevent it from turning back OFF.

(d) The peak value of the spikes across R_1 is given approximately by

$$V_{R1} = V_P - V_V = 14.8\text{ V} - 1.5\text{ V} = \mathbf{13.3\ V}$$

This equation is valid because the capacitor voltage always equals the voltage from emitter to base 1 plus the voltage across R_1. At the instant of firing, the capacitor voltage equals V_p and the emitter-to-base 1 voltage is approximately equal to V_V. Naturally the peak value of V_{R1} occurs at the instant the UJT fires, so it can be calculated as shown in the above equation.

The voltage level to which V_{R1} returns when the UJT is OFF can be calculated by the series circuit voltage-division formula:

$$\frac{V_{R1}}{R_1} = \frac{V_S}{R_{\text{total}}}$$

$$\frac{V_{R1}}{100\ \Omega} = \frac{24\text{ V}}{470\ \Omega + 9200\ \Omega + 100\ \Omega}$$

$$V_{R1} = \mathbf{0.25\ V}$$

The V_{R1} waveform could therefore be described as a rest voltage of 0.25 V with fast spikes rising to 13.3 V, occurring at a frequency of 500 Hz. ■

5-3 ■ UJT TIMING CIRCUITS

5-3-1 UJT Relay Timer

An example of a UJT timing circuit to provide the time delay in picking a relay is shown in Fig. 5–4. In this circuit, power is applied to the load when relay *CR* picks up. This will occur a certain time (adjustable) after SW1 is closed. The time delay is adjusted by adjusting R_E. The circuit works as follows.

When SW1 is closed and 24 V is applied to the top of R_3, a small amount of current starts to flow into the *CR* relay coil. R_3 is sized so that this current is not large enough to *pick* the coil, but is large enough to keep the coil energized once it has already been picked up. This is possible because the holding current for a relay coil is usually only about one half of the pickup current. That is, a relay coil which requires a current of 0.5 A to actually move the armature and switch the contacts might require only 0.25 A to *maintain* contact closure.

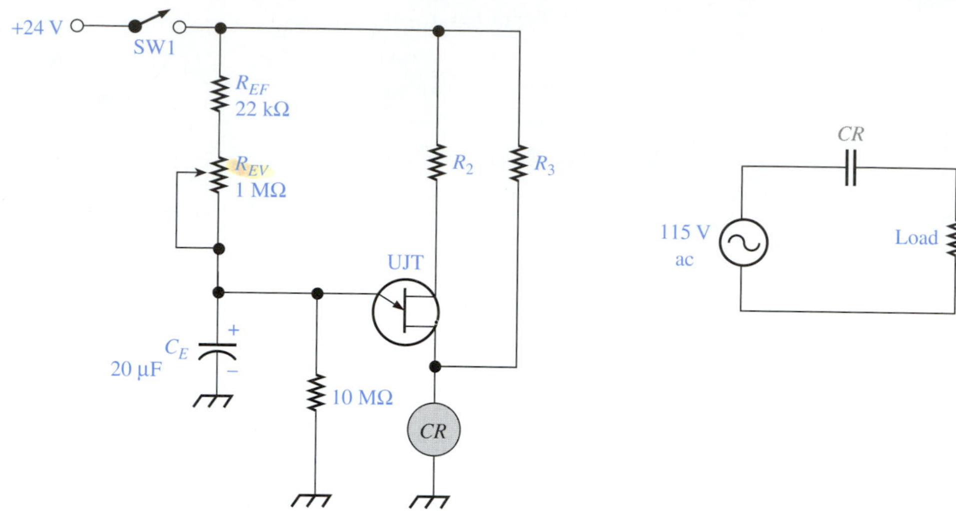

FIGURE 5–4
UJT timing circuit. Relay *CR* energizes a certain time after the switch is closed. The time delay can be varied by potentiometer R_{EV}.

The 20-μF capacitor C_E charges through R_{EF} and the 1-MΩ pot R_{EV}, at a rate determined by the setting of R_{EV}. When C_E reaches a high enough voltage, the UJT fires, dumping the capacitor charge into relay coil *CR*. This is sufficient to energize the coil, picking *CR*. The current pulse into the coil ceases almost immediately, but now the current through R_3 is sufficient to hold the relay energized. The N.O. *CR* contact closes and applies power to the load. The time delay is given by Eq. (5-6):

$$t = (R_{EF} + R_{EV})C_E$$

5-3-2 Improved One-Shot Using a UJT

We have encountered one-shots in Sec. 2-8, and we have seen some of their uses in industrial digital circuitry. A method of constructing a one-shot was presented in Fig. 2–13. This design is adequate in most one-shot applications, but it does have two shortcomings:

1. When the output pulse is complete, the one-shot is not ready to be fired again immediately. It has a certain nonzero *recovery time*. Recovery time is the time that must elapse between the finish of an output pulse and the arrival of the next trigger input pulse.
2. Long firing times are difficult to achieve with this design. Output pulses longer than a few seconds cannot be accomplished.

Refer back to Fig. 2–13(d) and let us see why these problems exist. Here is the reason for problem 1.

At the instant the output pulse is finished, the voltage across *C* is close to zero. Actually it is about 0.6 V, positive on the right, just enough to overcome the base-emitter junction of T_2. At this instant, T_2 turns ON and T_1 turns OFF. When this happens, *C* starts charging through R_{C1}, through the base-emitter junction of T_2, to ground. Until the capacitor can charge fully, the one-shot is not ready to fire again. That is, *C* must charge to a voltage equal to $V_S - 0.6$ V, positive on the left, before the one-shot can be triggered again. If a trigger pulse arrives at the one-shot before the capacitor has charged fully, the resulting output pulse will be too short.

To fully charge C, a time equal to five charging time constants must elapse. Therefore recovery time is given by

$$t_{rec} = 5(R_{C1})C$$

Now here is the reason for problem 2. The output pulse duration (firing time) equals the time it takes to *discharge* C when T_1 turns ON. The discharge path is down from V_S, through R_{B2}, through C, through T_1, to ground. When C is discharged to 0 V and charged just slightly in the opposite direction (about 0.6 V as mentioned) it turns ON T_2. Turning ON T_2 drives the one-shot back into its stable state and terminates the output pulse. Therefore the size of C and the size of R_{B2} determine the duration of the output pulse.

To obtain long-duration output pulses, either C or R_{B2} or both must be made large. However, we saw earlier that the larger C is made, the longer the recovery time becomes. Therefore C must be held to a reasonably small value. As far as R_{B2} is concerned, it cannot be made very large either because it may prevent T_2 from saturating. To pass sufficient base current to saturate T_2, R_{B2} must be held to a reasonably small value. Since C and R_{B2} must both be held small, it is impossible to obtain long firing times.

These two problems can be eliminated by using the improved one-shot shown in Fig. 5–5, which contains a UJT. Here is how it works. In the rest state, T_2 is ON and T_1 is held OFF. The reason T_2 is ON instead of T_1 is that R_{B2} is lower than R_{B1} (10 kΩ compared to 56 kΩ). This ensures that T_2 turns ON and that its 0-V collector holds T_1 OFF. The fact that the T_2 collector is at 0 V means that capacitor C_E is completely discharged.

When a trigger pulse arrives at the trigger terminal, T_1 is driven into the ON state. This forces T_2 to turn OFF because the collector of T_1 falls to 0 V. When T_2 turns OFF, its collector rises quickly and cleanly to almost V_S, thus causing the output pulse to appear at the Q output. When this happens, C_E begins charging. Its charging path is down from V_S, through R_{C2}, through R_E, and into C_E. When V_{CE} climbs to the peak voltage of

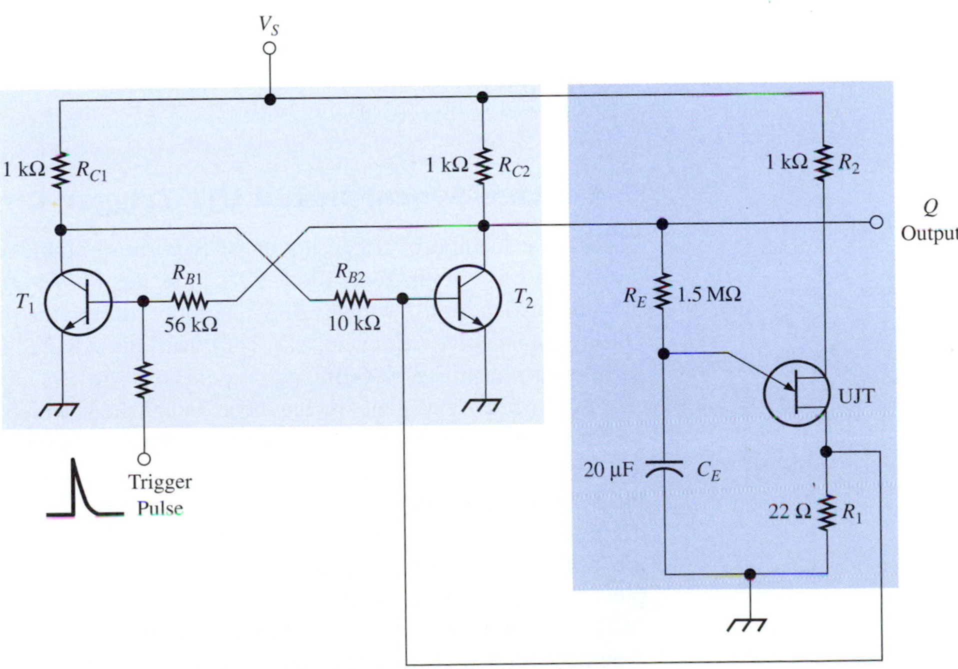

FIGURE 5–5
One-shot built with a UJT. This one-shot is superior to the one shown in Fig. 2–13, because its recovery time is zero and it can deliver very long output pulses.

the UJT, the UJT fires. This creates a positive pulse across R_1 in the base 1 lead of the UJT. This positive pulse is fed back to the base of T_2, thus turning it back ON. The one-shot's output pulse terminates at this instant.

Now let us ask ourselves if this circuit has the same drawbacks as the circuit of Fig. 2–13(d). Is there a recovery time needed before the one-shot can be triggered again? The answer is no, because the only capacitor in the circuit, C_E, is completely discharged and ready to begin charging again whenever required. (C_E discharged all at once through the E to $B1$ circuit of the UJT.)

Is there a limit on the firing time? Again the answer is no, because now the firing time is determined by R_E and C_E. These components can be made very large without any adverse effects on the operation of the rest of the circuit. The R_E and C_E values given in Fig. 5–5 would create a firing time of about 30 s, since the time to reach V_p is about one time constant, or

$$t_f = (1.5\ \text{M}\Omega)(20\ \mu\text{F})$$
$$= (1.5 \times 10^6)(20 \times 10^{-6}) = 30\ \text{s}$$

5-4 ■ UJTS IN SCR TRIGGER CIRCUITS

The UJT is almost ideal as a firing device for SCRs. Most of the UJT triggering principles discussed in this chapter in regard to SCRs apply equally well to triacs, as will be seen in Chapter 6.

There are several reasons for the compatibility between UJTs and SCRs:

1. The UJT produces a pulse-type output, which is excellent for accomplishing sure turn-ON of an SCR without straining the SCR's gate power dissipation capability.
2. The UJT firing point is inherently stable over a wide temperature range. It can be made even more stable with very little extra effort, as explained in Sec 5-3. This nullifies the temperature instability of SCRs.
3. UJT triggering circuits are easily adaptable to feedback control. We will explore this method of control as we proceed.

5-4-1 Line-Synchronized UJT Trigger Circuit for an SCR

The classic method of triggering an SCR with a unijunction transistor is shown in Fig. 5–6(a). In this circuit, zener diode ZD1 clips the V_1 waveform at the zener voltage (usually about 20 V for use with a 120-V ac supply) during the positive half cycle of the acline. During the negative half cycle, ZD1 is forward biased and holds V_S near 0 V. The V_S waveform is shown in Fig. 5–6(b).

Once the dc voltage V_S has been established, which occurs very shortly after the positive-going zero crossing of the ac line, C_E begins charging through R_E. When C_E reaches the peak voltage of the UJT, the UJT fires, creating a voltage pulse across R_1. This fires the SCR, thus allowing current flow through the load for the remainder of the positive half cycle. The V_{R1} waveform and V_{LD} waveform are shown in Fig. 5–6(c) and (d).

This circuit arrangement provides automatic synchronization between the firing pulse of the UJT and the SCR polarity. That is, whenever the UJT delivers a pulse, the SCR is guaranteed to have the correct anode to cathode voltage polarity for turning ON. A simple relaxation oscillator powered by a normal dc supply would not provide such synchronization; the UJT pulses would be as likely to occur during the negative half cy-

FIGURE 5–6
(a) UJT used to trigger an SCR. When the UJT fires, it triggers the SCR. The firing delay angle is adjusted by R_E. (b) V_S waveform. It is almost a perfect square wave. (c) V_{R1} waveform, which is applied to the SCR gate. The V_{R1} resting voltage (the voltage between spikes) must be less than the gate trigger voltage of the SCR. (d) Load voltage waveform, with a firing delay angle of about 60°.

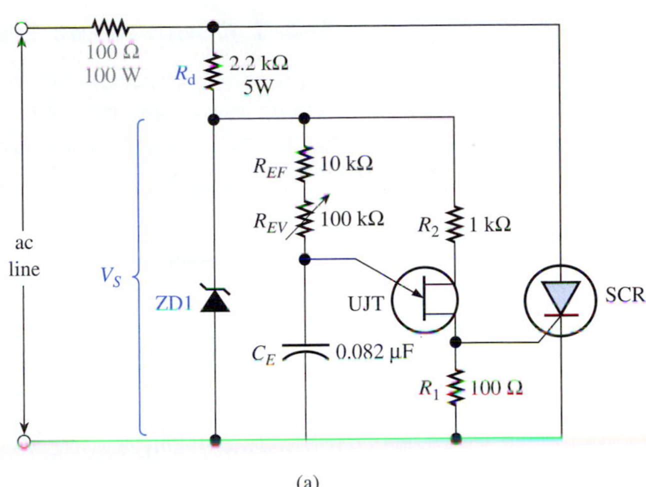

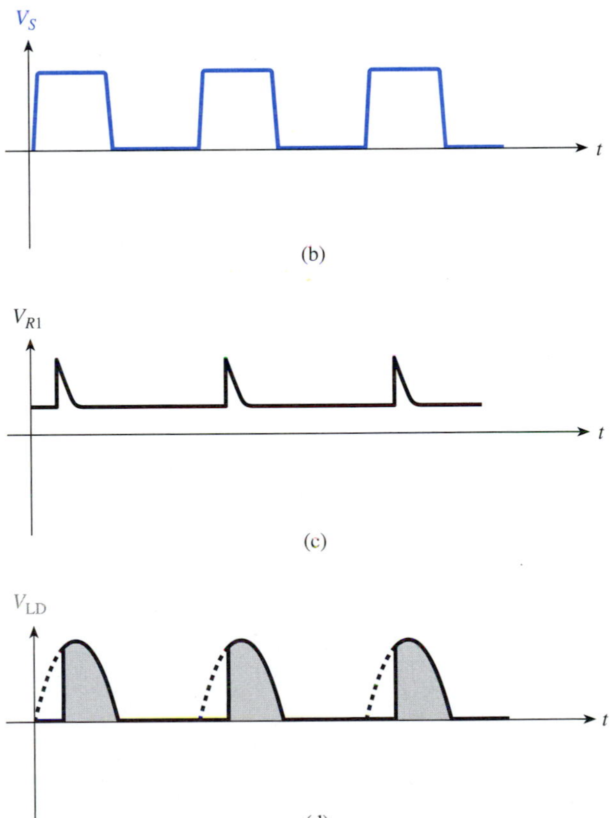

cle as during the positive half cycle. Of course, pulses occurring during the negative half cycle would be worthless.

The load power is controlled by potentiometer R_E. When R_E is low, C_E charges quickly, causing early firing of the UJT and SCR. This results in large average current through the load. When R_E is large, C_E charges more slowly, causing late firing and lower average load current.

5-4-2 Component Sizes for a UJT Trigger Circuit

In Fig. 5–6(a), special care must be taken when selecting R_1. The value of R_1 must be held as low as possible while still being able to generate large enough voltage pulses to fire the SCR reliably. There are two reasons for this:

1. Even before the UJT fires, there is some current flow through R_1, because of the connection through the main body of the UJT to V_S. This current can easily be several milliamps because the OFF-state resistance of the UJT, r_{BB}, is only about 10 kΩ. This is shown by the equation

$$I_{R1} = \frac{V_S}{R_2 + r_{BB} + R_1} \cong \frac{20\text{ V}}{10\text{ K}\Omega} = 2\text{mA}$$

   In this calculation R_1 and R_2 have been neglected, since they are always small compared to r_{BB}. Because of this nonnegligible current, R_1 must be held at a low value so the Ohm's law voltage across its terminals, which is applied to the SCR gate, is also low. Otherwise the SCR may fire inadvertently.
2. With a low value of R_1, there is less chance of an undesirable noise spike falsely triggering the SCR. External sources of noise (dc motor armatures, welders, switchgear, etc.) create unwanted noise signals which might cause this to happen. Small resistors are not as likely to pick up noise signals as large resistors. Specifically, when R_1 is kept small, there is less chance of a noise signal being generated across it which could trigger the SCR.

A method for sizing all the components in Fig. 5–6(a) will be presented now. Assume the UJT to be a 2N4947, which has the following typical characteristics at a supply voltage of 20 V:

$$r_{BB} = 6\text{ k}\Omega \qquad I_V = 4\text{ mA}$$
$$\eta = 0.60 \qquad V_V = 3\text{ V}$$
$$I_P = 2\ \mu\text{A}$$

If ZD1 has a zener breakdown voltage of 20 V, then the current through R_1 before firing is given by

$$I_{R1} = \frac{20\text{ V}}{R_2 + r_{BB} + R_1}$$

Again, neglecting R_2 and R_1 because they are quite a bit smaller than r_{BB}, we can say to a fair approximation that

$$I_{R1} = \frac{20\text{ V}}{r_{BB}} = \frac{20\text{ V}}{6\text{ k}\Omega} = 3.3\text{ m}A$$

Since most SCRs fire at a V_{GK} of about 0.7–1.0 V, it is reasonable to allow V_{R1} to go no higher than about 0.3 V while the UJT is waiting for the signal to fire. This would allow a noise margin of at least 0.4 V (0.7 V − 0.3 V), which is usually adequate. Therefore

$$R_1 = \frac{V_{R1}}{I_{R1}} = \frac{0.3\text{ V}}{3.3\text{ mA}} = 100\ \Omega$$

As explained in Sec. 5–3, R_E must be small enough to allow enough current, I_P, to flow into the emitter to trigger the UJT. Also, R_E must be large enough to prevent the UJT from latching up; that is, R_E must not permit the emitter to carry a current equal to the valley current, I_V, after C_E has discharged. If a current equal to I_V does continue flowing, the UJT cannot turn back OFF and is said to have latched up.

From Eq. (5-4)

$$R_{E\text{min}} = \frac{V_S - V_V}{I_V} = \frac{20\text{ V} - 3\text{ V}}{4\text{ mA}} = 4.25\text{ k}\Omega$$

which means that R_E must be greater than 4.25 kΩ to allow the UJT to turn OFF. Let us choose an R_E value of **10 kΩ.**

It should be pointed out that for the circuit of Fig. 5–6(a), UJT latch-up could not persist for longer than one half cycle, because V_S disappears when the ac line reverses. However, even a one-half-cycle latch-up is undesirable because it would result in continuous gate current to the SCR during the entire conduction angle. This causes the gate power dissipation to increase and could conceivably cause thermal damage to the SCR gate.

Proceeding, we note that V_p is given by Eq. (5-1):

$$V_p = \eta V_{B2B1} + V_D = (0.60)(20\text{ V}) + 0.6\text{ V} + 12.6\text{ V}$$

where V_{B2B1} has been taken as 20 V, which is approximately correct due to the small size of R_2 and R_1.

From Eq. (5-3),

$$R_{E\text{max}} = \frac{V_S - V_p}{I_p} = \frac{20\text{ V} - 12.6\text{ V}}{2\ \mu\text{A}} = 3.7\text{ M}\Omega$$

meaning that R_E must be smaller than 3.7 MΩ in order to deliver enough emitter current to fire the UJT.

To size R_E, there would be nothing wrong with averaging $R_{E\text{min}}$ and $R_{E\text{max}}$, yielding

$$R_E = \frac{4.25\text{ k}\Omega + 3.7\text{ M}\Omega}{2} = 1.85\text{ M}\Omega$$

However, in situations like this where it is desired to find a happy medium between two values that differ by a few orders of magnitude, it is customary to take the *geometric mean,* instead of an average (arithmetic mean). Doing this gives

$$\begin{aligned} R_E &= \sqrt{(R_{E\text{min}})(R_{E\text{max}})} \\ &= \sqrt{(4.25 \times 10^3)(3.7 \times 10^6)} \\ &= 125\text{ k}\Omega \end{aligned}$$

The nearest standard potentiometer value is 100 kΩ, so

$$R_{EV} = \mathbf{100\ k\Omega}$$

To calculate the correct size for C_E, recognize that when all the variable resistance is dialed in, the V_P charging time should be almost one half of the ac line period (the time for a half cycle). This will permit a large delay angle adjustment.

The time to charge to V_P is given approximately by Eq. (5-6). For a 60-Hz ac line, the half cycle time is about 8.3 ms, so

$$R_{E\text{total}}C_E = 8.3 \times 10^{-3}$$

or

$$C_E = \frac{8.3 \times 10^{-3}}{110 \times 10^3} = 0.076\ \mu\text{F}$$

The nearest standard size is C_E = **0.082 μF.**

R_2 is difficult to calculate and is usually determined experimentally or by referring to graphs. For most UJTs, the best temperature stability is realized with an R_2 between 500 Ω and 3 kΩ. Detailed manufacturers' data sheets have graphs which enable the user to choose R_2 for the temperature response desired. In most cases, good stability results when R_2 = **1 kΩ.**

One way of sizing ZD1 and R_d is to proceed as follows. Assume that ZD1 must be no larger than a 1-W zener diode. This is a reasonable condition, since zener regulating characteristics tend to get sloppier at the higher power ratings and the cost goes up considerably.

If ZD1 can dissipate an average power of 1 W, it can dissipate almost 2 W during the positive half cycle because the power burned during the negative half cycle is negligible, due to the low voltage drop when the diode is forward biased ($P = VI$). Therefore, the allowable average current through the zener during the positive half cycle is

$$I = \frac{P_{+\text{ half}}}{V_Z} = \frac{2\text{ W}}{20\text{ V}} = 100\text{ mA}$$

R_d must be sized to allow no more than 100-mA average current during the positive half cycle. To a rough approximation, the average voltage across R_d during the positive half cycle will be 100 V, because

$$V_{\text{line}} - V_Z = 120\text{ V} - 20\text{ V} = 100\text{ V}$$

Therefore,

$$R_d = \frac{100\text{ V}}{100\text{ mA}} = 1\text{ k}\Omega$$

Naturally, R_d should be somewhat larger than this for a safety margin. A power dissipation safety margin of 2 to 1 is considered desirable, so we might choose

$$R_d = \mathbf{2.2\ k\Omega}$$

The power rating of R_d can be determined by assuming a 100-V rms voltage drop across the resistor.

$$P_{Rd} = \frac{V^2}{R_d} = \frac{(100)^2}{2.2\text{ k}\Omega} = 4.5\text{ W}$$

This would call for a 5-W resistor, the nearest standard rating which is greater than 4.5 W. Naturally all such calculations are approximate and would have to be tested experimentally.

5-4-3 Sequential Switching Circuit Using UJTs for Gate Control

An interesting example of the UJT-SCR combination is the sequential switching circuit shown in Fig. 5–7. In this circuit, the three loads are energized in sequence, and each load is energized for a certain length of time. The times are individually variable. That is, it would be possible to have load 1 energized for 5 s, after which load 1 would deenergize

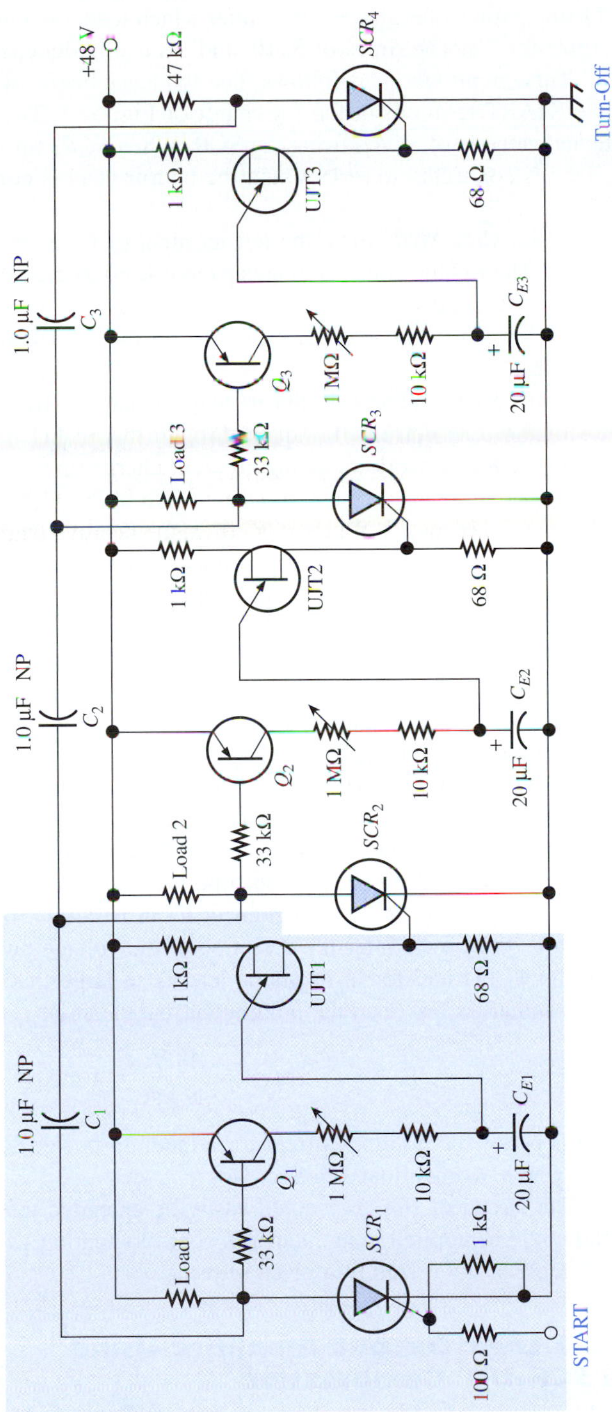

FIGURE 5–7
Sequential switching circuit using UJT-SCR pairs. When a UJT fires, it causes the next SCR to fire. When that SCR fires, it connects a charged commutating capacitor across the main terminals of the preceding SCR, thereby turning it OFF.

and load 2 would energize for 10 s, after which load 2 would deenergize and load 3 would energize for 7 s. The times of 5, 10, and 7 s can be independently adjusted.

This circuit works as follows. The sequence begins when a positive pulse is applied at the START terminal in the lower left of Fig. 5–7. This generates a voltage between gate and cathode of SCR_1, firing the SCR. When SCR_1 fires, load 1 energizes because the top lead is connected to +48 V and the bottom lead is connected to ground through the SCR.

Also, when SCR_1 fires, the left terminal of C_1 is connected to the ground through the SCR. The right terminal of that capacitor is connected through the resistance of load 2 to the +48-V supply line. C_1 quickly charges to 48 V because the load resistance would be fairly low. The charge polarity is plus on the right-hand side and minus on the left-hand side.

While SCR_1 and load 1 are carrying current, *pnp* transistor Q_1 also switches ON because of the base current flow path through the 33-kΩ base resistor, through SCR_1, to ground. The *RC* network including C_{E1} will charge up to the peak voltage of UJT1, causing that UJT to deliver a pulse of current into its 68-Ω base 1 resistor. This in turn fires SCR_2, energizing load 2. When SCR_2 fires, the positive (right) terminal of C_1 is connected to ground through SCR_2. C_1 had previously been charged to 48 V, and since a capacitor cannot discharge instantaneously, the −48 V potential on the left side of C_1 is applied to the anode of SCR_1. This effectively reverse-biases SCR_1 for an instant, shutting it OFF and deenergizing load 1. Transistor Q_1 also switches OFF, so C_{E1} does not charge up again.

This action is repeated in the second stage of the switching circuit, with C_{E2} charging through Q_2 at a rate determined by the 1-MΩ pot in series with C_{E2}. When the proper time has elapsed, UJT2 fires, which fires SCR_3 and connects the C_2 terminals in parallel with SCR_2. C_2 had charged plus on the right and minus on the left during the time load 2 was energized, so it now reverse-biases SCR_2, turning if OFF.

When the load 3 energization time has elapsed, UJT3 fires, causing SCR_4 to fire. The only purpose of SCR_4 is to connect C_3 in parallel with SCR_3, to turn it OFF. SCR_4 turns OFF on its own, after the cessation of the voltage pulse at its gate. This occurs because the 47-kΩ resistor in its anode lead is so large that the current through the SCR_4 main terminals is less than the holding current. That is,

$$I_{AK} = \frac{48\text{ V}}{47\text{ k}\Omega} = 1\text{ mA}$$

which is below the holding current for a medium-power SCR. I_{HO} for a medium SCR is about 10 mA, as mentioned in Sec. 4-3.

The circuit of Fig. 5–7 could easily be extended to any number of stages. Such a circuit could be applied in an industrial control situation whenever there are several loads which must be energized in a given sequence.

5-4-4 Logic Output Amplifier Using an SCR-UJT Combination

In Sec. 1-8, we discussed output amplifiers used to interface between low-voltage logic circuitry and industrial actuating devices. As mentioned then, modern output amplifiers often contain an SCR with a UJT in its gate control circuit. A popular design of such an output amplifier is shown in Fig. 5–8(a). Here is how the output amplifier works. Look first at the right-hand side of Fig. 5–8(a). The load, in this case a solenoid coil, is placed in the ac power line in series with a bridge rectifier which is controlled by a single SCR.

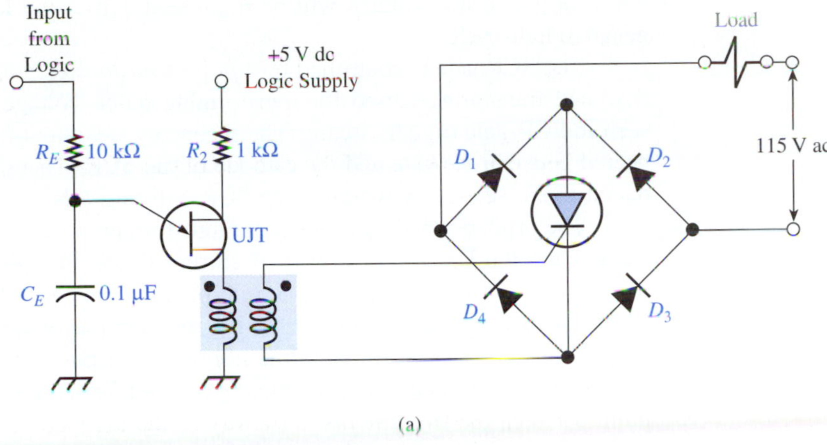

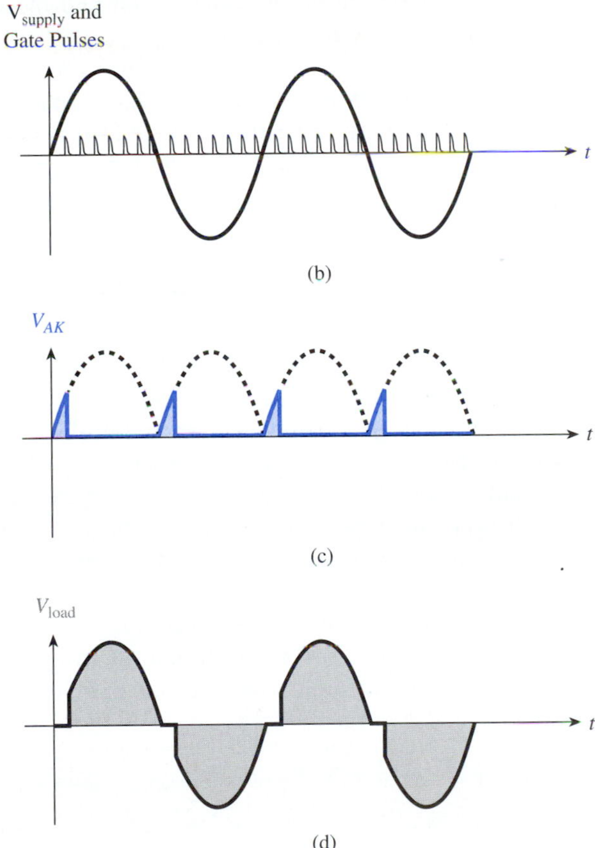

FIGURE 5–8
(a) Schematic of a logic output amplifier using a UJT and an SCR. When the input line goes HI, it causes the relaxation oscillator to start oscillating at a high frequency, delivering a rapid train of gate pulses to the SCR.
(b) The train of gate pulses, shown relative to the 115-V ac supply. (c) V_{AK} waveform, showing that the SCR fires very shortly after the start of a half cycle. (d) Load voltage waveform.

This method of controlling both half cycles of the ac line was presented in Sec. 4-6-3, Fig. 4–13(a). Recall that during the positive half cycle of the ac line, diodes D_1 and D_3 are forward biased, and the SCR is also forward biased and capable of firing. If the SCR does fire, the ac line voltage will be impressed across the load for the rest of the positive half cycle. During the negative half cycle of the ac line, diodes D_2 and D_4 are forward biased, and the SCR is still forward biased and capable of firing. Therefore if it does fire, the negative ac line voltage will be impressed across the load for the remainder of the negative half-cycle.

The SCR gate is controlled by a *pulse transformer*. Pulse transformers are specially designed transformers used for transforming quick voltage pulses. They are frequently seen in SCR gate trigger circuits. The secondary winding of the pulse transformer is connected between the gate and the cathode of the SCR. Therefore if a voltage pulse is produced in the secondary winding, the SCR will turn ON.

The primary winding of the pulse transformer is wired in the base 1 lead of the UJT. Thus when the UJT fires, a burst of current flows through the primary winding of the transformer. This burst creates a current pulse in the secondary winding, which fires the SCR. The arrangement in Fig. 5–8(a) is an example of a situation in which the power supply for the gate control circuit is *not* the same power supply that drives the load. In fact the gate control circuit is completely isolated from the main terminal circuit. The coupling between the two circuits is by way of the magnetic coupling between the primary and secondary windings of the pulse transformer. This provides the usual benefits of electrical isolation between the noisy power circuit and the low-voltage electronic control circuit.

The firing of the UJT is determined as always by R_E, C_E, and the input voltage signal at the top of R_E. If that input voltage is LO, C_E cannot charge, so the UJT never fires. In that case the SCR never fires either, and the load is deenergized.

However, if the input voltage from the logic circuit goes HI (+5 V in this example), the emitter circuit will start charging with a charging time given by Eq. (5-6):

$$\begin{aligned} t_{\text{charge}} &= R_E C_E \\ &= (10\ \text{k}\Omega)(0.1\ \mu\text{F}) = 1\ \text{ms} \end{aligned}$$

As long as the input terminal remains HI, the UJT circuit will behave like a relaxation oscillator, producing output pulses spaced about 1 ms apart. These output pulses continually create gate pulses to the SCR, thereby continually ordering the SCR to fire. The continual arrival of gate trigger pulses is illustrated in Fig. 5–8(b).

With this barrage of trigger pulses arriving at the SCR gate, it can be seen that the SCR is bound to turn ON very early in every half cycle. The latest it can turn ON is 1 ms into the half cycle; in all probability a trigger pulse will arrive before 1 ms has elapsed. There is no synchronization between V_{AK} and the gate control circuit in this case, but none is needed.

The SCR main terminal voltage waveform is drawn in Fig. 5–8(c), assuming about a 1-ms delay between the zero crossover and the firing. The resulting load waveform is shown in Fig. 5–8(d). The overall result is that the load is energized when the input signal goes HI.

5-5 ■ PUTS

A *programmable unijunction transistor* (PUT) has effectively the same operating characteristics as a standard UJT, and is used in similar applications. The schematic symbol and lead identifications of a PUT are shown in Fig. 5–9.

The cathode of a PUT corresponds to base 1 of a UJT: When a PUT fires, a burst of current emerges from the device via the cathode lead, just as a firing burst emerges from the base 1 lead of a UJT. Also, the cathode of PUT, like base 1 of a UJT, is the reference terminal relative to which other voltages are measured.

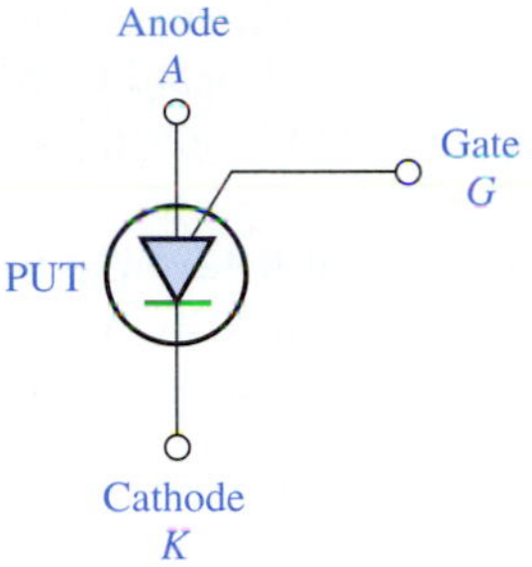

FIGURE 5–9
Schematic symbol and terminal names of a PUT.

The anode of a PUT corresponds to the emitter of a UJT: The PUT's anode voltage rises until it reaches a certain critical value called the peak voltage, V_P, which causes the device to fire.

The gate of a PUT bears a rough correspondence to base 2 of a UJT: For a PUT, the gate receives a voltage from an external circuit, and that voltage sets the peak voltage V_P according to the formula

$$V_P = V_G + 0.6 \text{ V} \tag{5-7}$$

The 0.6 V term in Eq. (5-7) is approximate; it depends mostly on the forward voltage across the anode-gate *pn* junction, which is somewhat temperature-dependent.

Note that a PUT differs from a UJT in that its V_P is determined by external circuitry, rather than an intrinsic standoff ratio associated with the transistor itself. This is what makes the device programmable: By making an adjustment in an external circuit, we can select any desired value of peak voltage.

The characteristic curve of a PUT has the same general shape as the UJT curve of Fig. 5–2(b). For a PUT, the horizontal axis represents anode current, I_A, and the vertical axis represents anode-to-cathode voltage, V_{AK}. As a general rule, the PUT characteristic curve can be regarded as being compressed closer to the origin, compared to the UJT curve. That is, the I_P and I_V values* of the most sensitive PUTs tend to be lower than for the most sensitive UJTs.

A very sensitive PUT may be able to trigger at an I_P value of only 0.1 μA, compared to about 1 to 20 μA required for a standard UJT. Once it has fired, a sensitive PUT may be able to hold itself in the ON state with an anode current of only 50 μA or so (I_V), compared to 1–10 mA of emitter current required for a UJT. In the same spirit, a PUT's valley voltage V_V tends to be lower than that of a UJT; a typical V_V value for a PUT is less than 1 V.

The PUT relaxation oscillator of Fig. 5–10 emphasizes some of the characteristics of the PUT that distinguish it from a standard UJT. First notice that the oscillation fre-

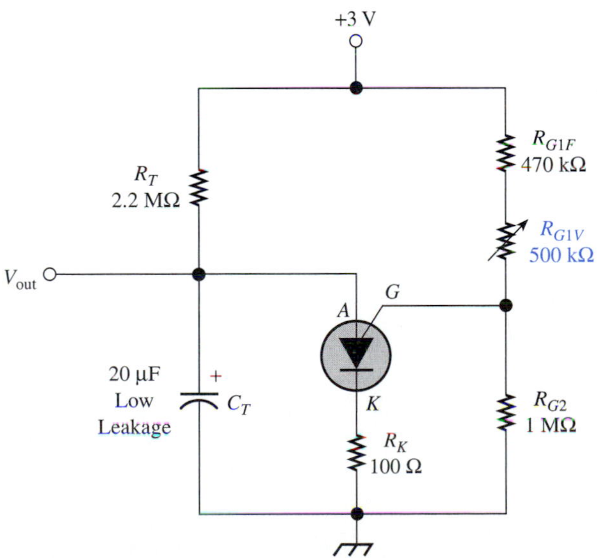

FIGURE 5–10
Schematic drawing of a PUT relaxation oscillator. The oscillator frequency varies with R_{G1V}.

*Actually, the I_P and I_V values of a PUT are themselves programmable to some extent, by selection of resistance values in the gate circuit. For a UJT, these parameters are largely inherent in the transistor itself.

quency is adjusted by varying the dc voltage applied to the gate from the R_{G1}-R_{G2} voltage divider. Contrast this with a UJT oscillator, where the frequency would be adjusted by varying R_T to change the charging rate of timing capacitor C_T. The act of varying V_G can be regarded as programming the PUT.

With cathode resistor R_K present, the ground reference for the circuit is taken to be at its bottom terminal rather than at the cathode terminal itself. This has virtually no effect on V_P, since the voltage across R_K is virtually zero when the PUT is in its OFF state.

With R_{G1V} dialed in, V_G can be calculated as

$$V_G = (3\text{ V})\frac{R_{G2}}{R_{G2} + R_{G1F} + R_{G1V}} = 1.5\text{ V}\frac{1\text{ M}\Omega}{1\text{ k}\Omega + 470\text{ k}\Omega + 500\text{ m}\Omega}$$

V_P is given approximately by

$$V_P = V_G + 0.6\text{ V} = 1.5\text{ V} + 0.6\text{ V} = 2.1\text{ V}$$

The time required for C_T to charge to V_P and fire the PUT is found by

$$\frac{2.1\text{ V}}{3.0\text{ V}} = 0.70\text{ or }70\%$$

From a universal time constant curve it can be seen that 1.2τ are required to charge to 70%.* Therefore,

$$T_{\text{min}} = 1.2\tau = 1.2R_TC_T = 1.2(2.2\text{ M}\Omega)(20\ \mu\text{F}) = 53\text{ s}$$

$$f_{\text{max}} = \frac{1}{T_{\text{min}}} = \frac{1}{53\text{ s}} = 0.019\text{ Hz}$$

With R_{G1V} dialed out,

$$V_G = (3\text{ V})\frac{R_{G2}}{R_{G2} + R_{G1F}} = (3\text{ V})\frac{1\text{ M}\Omega}{1.47\text{ M}\Omega} = 2.0\text{ V}$$

$$V_P \cong 2.0\text{ V} + 0.6\text{ V} = 2.6\text{ V}$$

The time required to charge to V_P is found by

$$\frac{2.6\text{ V}}{3.0\text{ V}} = 0.87\text{ or }87\%$$

It takes about 2.0τ to charge to 87%, so

$$T_{\text{max}} = 2.0(2.2\text{ M}\Omega)(20\ \mu\text{F}) = 88\text{ s}$$

$$f_{\text{min}} = \frac{1}{T_{\text{max}}} = 0.011\text{ Hz}$$

Such slow oscillation is a consequence of the long time constant, which is determined in part by the large value of R_T. A large value of timing resistance implies a small amount of anode current available for firing the PUT. The transistor's peak-point current I_P must be even less than this small amount in order for successful firing to occur. For the circuit of Fig. 5–10, the worst case occurs with $V_P = 2.6$ V. Then,

$$I_{\text{available}} = \frac{3\text{ V} - 2.6\text{ V}}{2.2\text{ M}\Omega} = 0.18\ \mu\text{A}$$

---

*This ignores any initial charge on the capacitor due to incomplete discharge through the PUT the last time it fired.

There are PUTs with I_P ratings below 0.18 μA, as mentioned earlier. A standard UJT could not be used in this situation. In general, PUTs lend themselves to the construction of low-frequency oscillators and long-duration timers better than do standard UJTs.

Notice also that the dc supply voltage in Fig. 5–10 is only 3 V. PUTs, with their valley voltage ratings below 1 V, can operate on such low supply voltages. Most UJTs cannot.

The programmability of the PUT gives it a special usefulness in industrial control applications. Figure 5–11 shows an example. The circuit in that figure is a ramp generator. The output ramps always maintain a constant slope, but the height of the ramps can be adjusted by programming the PUT via gate voltage V_G. Such a circuit could be used to furnish the input signal to an industrial servomechanism which operates in a battering-ram fashion—moving a tool forward, backing out, then moving it forward again, each time a little farther than on the previous stroke. With the voltage ramp furnishing the set-point signal to the servo system, increasing the ramp height would coincide with increasing the mechanism's stroke distance.

Here is how the ramp generator works: *pnp* transistor Q_1, with its supporting resistors R_{B1}, R_{B2} and R_E, forms a constant-current source. The operation of this particular constant-current circuit design is explained in Sec. 6-9-1. With the PUT in the OFF state, the constant current flowing down the Q_1 collector lead causes a constant rate of change of voltage across capacitor *C*, and thus a constant-slope output ramp. When the capacitor voltage (V_{out}) runs up to the PUT's peak voltage value, the PUT fires. This causes capacitor *C* to discharge through the anode-cathode path, terminating the ramp. As soon as the discharging current drops low enough so that I_A becomes less than the I_V rating, the PUT reverts to its blocking state (OFF), and the next ramp begins.

The ramp height is controlled by adjusting the value of V_G, which determines the V_P value according to Eq. (5-7).

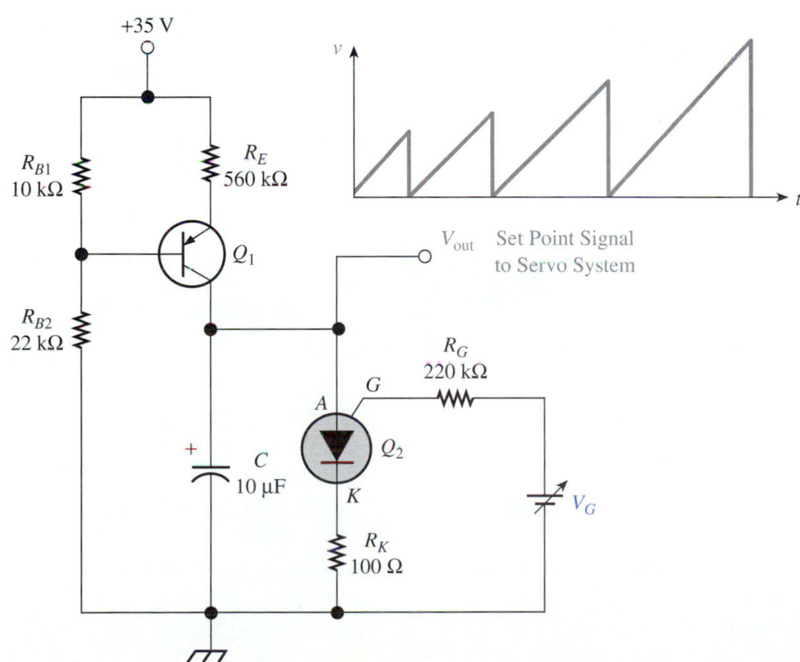

FIGURE 5–11
Constant-slope ramp generator. Ramp height is set by V_G.

TROUBLESHOOTING ON THE JOB

TRIGGERING THE SCRS IN A MAGLEV CONTROL MODULE

Troubleshooting on the Job in chapter 4 had to do with troubleshooting a section of MagLev track. That assignment did not take you inside the four Trigger circuits of Fig. 4–18.

Now suppose that you must investigate the internal operation of one of the Trigger circuits, because your scope measurements have indicated the absence of a trigger pulse, or a faulty pulse. The schematic diagram of the four Trigger circuits is shown in Fig. 5–12. This diagram does not show the detailed digital logic that determines Automatic triggering instants as a high-speed train passes. That issue will be investigated in chapter 8.

In Fig. 5–12, each UJT is controlled by the Q output of a 0.3-ms one-shot. We are using the one-shot's T_2 trigger terminal in Manual mode, the basic troubleshooting mode. The T_1, trigger terminal is for Automatic mode (an actual train passing).

The RC time constant of each UJT timing circuit is $\tau = RC = 1\ \text{k}\Omega \times 0.2\ \mu\text{F} = 0.2\ \text{ms}$. Therefore the UJT will fire about 0.2 ms after the one-shot rises. With the one-shot having an output duration of 0.3 ms, there is not enough time to retrigger the UJT a second time—there is one and only one UJT triggering event for each trigger of the one-shot.

Testing UJT trigger-control circuits with a four-channel menu-driven oscilloscope (only two channels in use here).

Courtesy of Tektronix, Inc.

The two turn-ON Trigger circuits at the top of Fig. 5–12 have dual-secondary pulse transformers. The ON South circuit drives the gate-cathode junctions of SCR_1 and SCR_3. The ON North circuit sends transformer pulses to SCRs 2 and 4.

The two turn-OFF Trigger circuits at the bottom of Fig. 5–12, have triple-secondary pulse transformers. Two of the secondary windings pulse the commutating transistors to short-circuit the conducting SCRs. The third winding drives a rectifying delay circuit which rises to a digital active-HI status about 60 μs after the turn-OFF pulses have been delivered to the commutating transistors. This short-lived digital signal (for example NON, which stands for North go ON, at the lower left in Fig. 5–12) is routed to the digital logic that triggers ON the *opposite* magnetic polarity circuit (look at the NON input at the upper right in

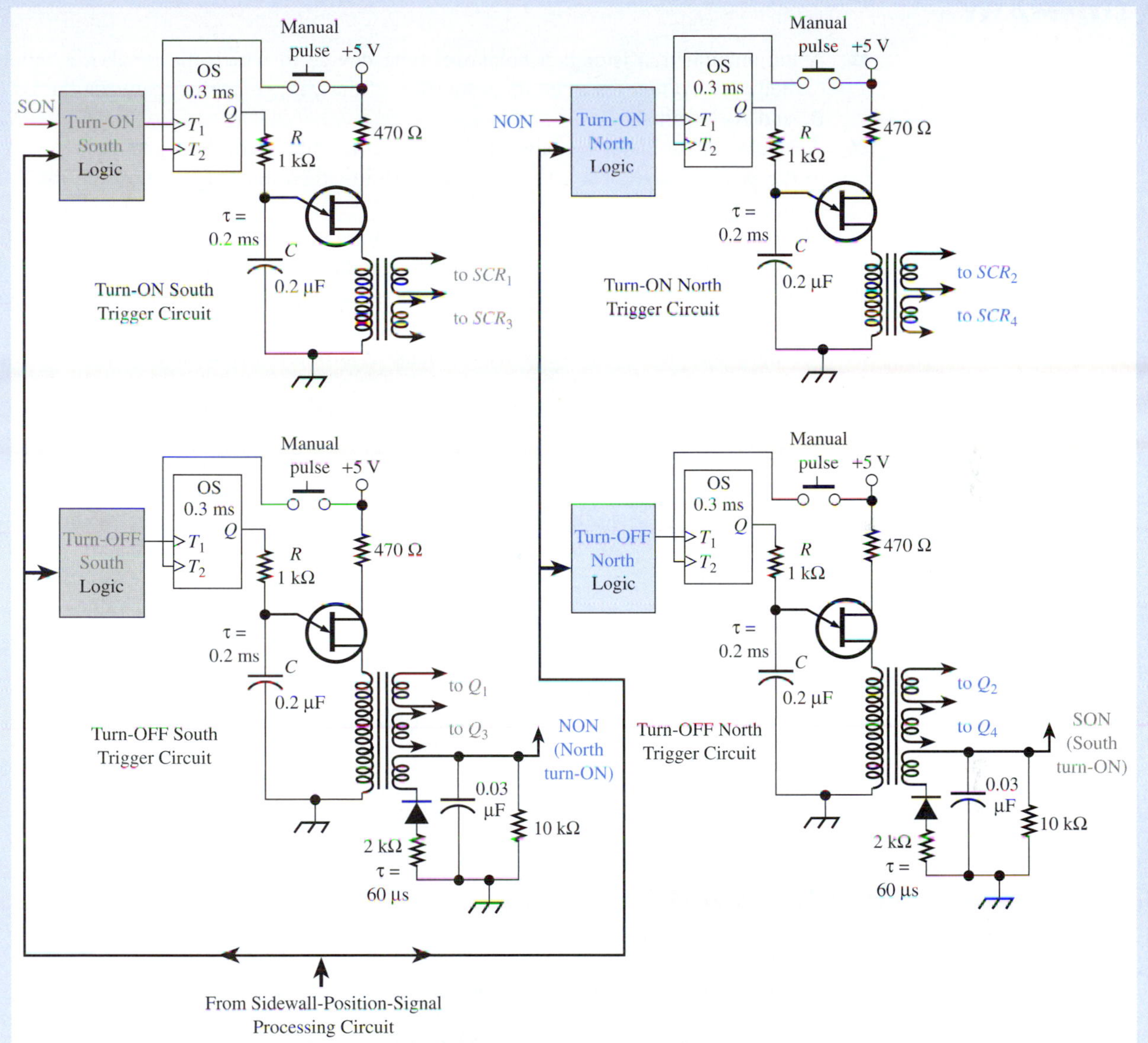

FIGURE 5–12
Schematic of the four Trigger circuits in a propulsion-coil control module.

Fig. 5–12). This is necessary for Automatic operation because at the moment when a passing train turns OFF the sidewall propulsion coils in one magnetic polarity, it must immediately turn the propulsion coils back ON with the opposite polarity. We will call this the Switchover function when the automatic digital logic is examined in chapter 8.

Suppose that in Troubleshooting on the Job in Chapter 4, your scope revealed that the turn-ON South Trigger circuit in Fig. 4–18 was unable to produce proper pulses to either SCR_1, or SCR_3, or both. Referring now to Fig. 5–12, describe an effective troubleshooting procedure to locate the fault in that Trigger circuit.

SUMMARY

- A unijunction transistor is a breakover-type device in which the breakover voltage (called *peak voltage*) is a certain proportion of the voltage applied between the UJT's B_1 and B_2 terminals.
- A relaxation oscillator contains an *RC* time-constant circuit that repetitively charges to the peak voltage of a UJT, with the capacitor then discharging through the E-B_1 path.
- The UJT is useful in the construction of (1) timing circuits; (2) one-shots; and (3) SCR trigger-control circuits, especially those that utilize feedback from the load.
- A pulse transformer is often used to couple a UJT's E-B_1 breakover burst to the gate of an SCR.
- A programmable unijunction transistor behaves similarly to a regular unijunction transistor, except that its peak voltage V_p is easy to vary.

FORMULAS

$$V_p = \eta V_{B2B1} + 0.6 \text{ V} \qquad \text{(for a UJT)} \qquad \text{(Equation 5-1)}$$

$$R_{E\text{max}} = \frac{V_S - V_p}{I_p} \qquad \text{(Equation 5-3)}$$

$$R_{E\text{min}} = \frac{V_S - V_V}{I_V} \qquad \text{(Equation 5-4)}$$

$$f = \frac{1}{R_E C_E} \qquad \text{(for a UJT relaxation oscillator)} \qquad \text{(Equation 5-5)}$$

$$t_{\text{charge}} = R_E \times C_E \qquad \text{(Equation 5-6)}$$

$$V_p = V_G + 0.6 \text{ V} \qquad \text{(for a PUT)} \qquad \text{(Equation 5-7)}$$

QUESTIONS AND PROBLEMS

Section 5-1

1. Is a unijunction transistor a continuously variable device or a switching device? Explain.
2. Roughly speaking, what range of values does η fall into?
3. For the circuit of Fig. 5–1(b), assume that the standoff ratio is 0.70. Calculate the peak point voltage, V_p.

Section 5-2

4. For the relaxation oscillator of Fig. 5–3, what would be the effect on the oscillation frequency of doubling C_E? Of doubling R_1?
5. Explain why there is a maximum limit on the size of the emitter resistor in a UJT circuit.
6. Explain why there is a minimum limit on the size of the emitter resistor in a UJT circuit.
7. Why is it that inserting a resistor in the $B2$ lead of a UJT relaxation oscillator tends to stabilize the oscillation frequency in the face of temperature changes? Explain the two temperature effects which tend to cancel each other.

8. Equation (5-5) for a relcaxation oscillator is only approximate. Identify two reasons it is not exact.
9. For the relaxation oscillator of Fig. 5–3, if $R_E = 10\ \text{k}\Omega$, $C_E = 0.005\ \mu\text{F}$, and $\eta = 0.63$, what is the approximate oscillation frequency? What would be the effect on the frequency if η were larger than 0.63? What would be the effect if it were smaller than 0.63?

Section 5-3

10. In Fig. 5–4, the UJT relay timer, how would the load be deenergized?
11. What is the longest possible time delay in Fig. 5–4? Assume that $\eta = 0.63$.
12. For the one-shot of Fig. 5–5, calculate R_E and C_E to give an output pulse duration of 5 s. Assume $\eta = 0.63$. Is the size of R_{C2} important in determining the firing duration?
13. Explain in detail why the one-shot of Fig. 5–5 can be retriggered immediately after it has fired, whereas the one-shot of Fig. 2–13(d) cannot be retriggered until a certain recovery time has elapsed. Focus your explanation on C of Fig. 2–13(d) and C_E of Fig. 5–5.

Section 5-4

14. When a UJT is being used to trigger a thyristor, as in Fig. 5–6, why is there a limit on the size of R_1?
15. In the circuit of Fig. 5–6, does C_E begin charging again immediately after the UJT has fired during a positive half cycle? Explain.
16. In Fig. 5–7, C_1, C_2, and C_3 are all marked nonpolarized (NP). Why must they be nonpolarized?
17. In Fig. 5–7, SCRs 1–3 are turned OFF by connecting a negatively charged capacitor across their anode to cathode terminals. What is it that turns OFF SCR_4?
18. Why is it necessary for the relaxation oscillator in the gate control circuit in Fig. 5–8 to have such a high frequency?
19. Would the phase relationship between primary and secondary windings of the pulse transformer be important in Fig. 5–8(a)? Explain what would happen if the phasing were reversed.
20. In Fig. 5–7, suppose that the SCRs have a gate trigger voltage of $V_{GT} = 0.7$ V. What is the absolute minimum allowable r_{BB} for the UJTs so that THE SCRs don't fire until ordered to fire? Neglect noise margin considerations.

Section 5-5

21. For the component values in Fig. 5–11, the slope of the output ramps will be 1.85 V/s. If $V_G = 2.5$ V, how much time is required for the ramp to run all the way up? Repeat for $V_G = 5.0$ V.

■ SUGGESTED LABORATORY PROJECTS

Project 5-1: UJT Relaxation Oscillator

Purpose

1. To determine the intrinsic standoff ratio of a UJT
2. To observe and graph the output waveforms of a UJT relaxation oscillator
3. To observe the temperature stability of a UJT relaxation oscillator

Procedure

1. Find the UJT's standoff ratio.
   a. With the ohmmeter, measure the interbase resistance of the UJT, r_{BB}.
   b. Find the two individual emitter to base resistances of the UJT, r_{B1} and r_{B2}. This cannot be done accurately with an ohmmeter. Instead, connect a variable dc supply between the emitter and base 1 with a 10-mA ammeter in the emitter lead. Adjust the dc supply voltage until the ammeter reads 5 mA. Measure V_{EB1}, and subtract 0.6 V for the *pn* junction. The remainder is the voltage actually applied to r_{B1}. Use Ohm's law to calculate r_{B1}:

$$r_{B1} = \frac{V_{EB1} - 0.6\ \text{V}}{5\ \text{mA}}$$

   Why would an ohmmeter give false readings for r_{B1} and r_{B2} but give correct readings for r_{BB}?
   c. Repeat step **b** for the emitter-to-base 2 circuit to find r_{B2}.
   d. Calculate the intrinsic standoff ratio of the UJT from Eq. (5-2).
2. Find the peak voltage of the UJT.
   a. Build the circuit of Fig. 5–3, with $R_E = 100\ \text{k}\Omega$, $C_E = 100\ \mu\text{F}$, and $V_S = 15$ V dc. Let R_1 and R_2 have the values given in that drawing. Install a switch in the V_S supply line. Place a 50-V voltmeter across C_E and a 10-V voltmeter across R_1. Discharge C_E completely; then close the switch and watch the voltmeters. What is the V_P for this circuit? Does it agree with what you would expect from Eq. (5-1)?
   b. Repeat step **a** with a V_S of 10 V.
   c. Measure the time delay before firing. Does it agree with what you would expect from Eq. (5-6)? Does the time delay depend on V_S? Explain this.
3. Open the power switch and change R_E to 22 kΩ and C_E to 0.5 μF. Remove the voltmeters and connect the channel 1 and channel 2 vertical inputs of a dual-trace oscilloscope across C_E and R_1 to observe the waveforms of V_{CE} and V_{R1}. If a dual-trace scope is not available, use a single-trace scope to observe V_{CE} first; then move the input to see V_{R1}.
   a. Using the methods of Example 5–3, predict the appearance of the V_{CE} and V_{R1} waveforms. Sketch what you expect to see for each waveform.
   b. Close the power switch and adjust the scope controls to display several cycles of the waveforms. Sketch the actual waveforms. Do they agree with your predictions? Try to explain any discrepancies.
4. Cool the UJT with freeze spray. Does the oscillation frequency change as the temperature changes? Experiment with different values of R_2 to get the best temperature stability. A frequency counter will come in handy here, but if none is available, you can detect frequency changes by carefully watching the oscillation period on the oscilloscope screen.

Project 5-2: UJT Characteristic Curves

Purpose

1. To observe and graph the current-voltage characteristic curves for a UJT at several different interbase voltages
2. To observe the batch spread among several UJTs of the same type

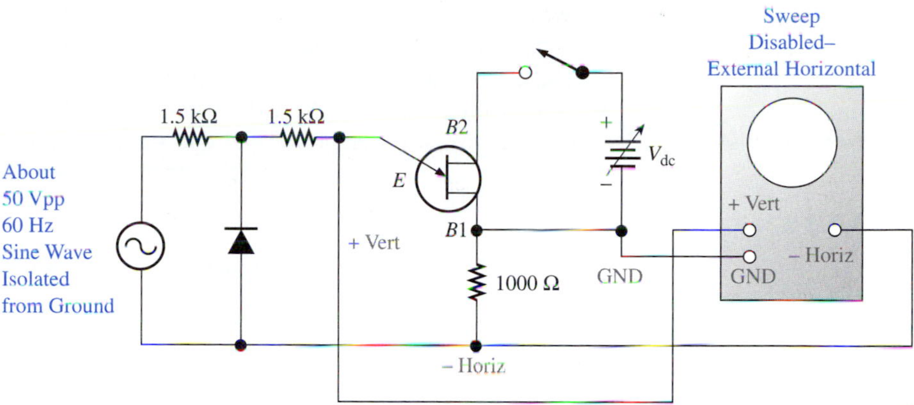

FIGURE 5–13
Circuit for displaying the current versus voltage curve of a UJT. V_{EB1} is applied to the horizontal input of the scope. The signal across the 1000-Ω resistor, which represents I_E, is applied to the vertical input.

Procedure

1. Build the circuit of Fig. 5–13. This circuit will display the I_E versus V_{EB1} curve of the UJT.

   As with Project 4-1, the ac supply should be isolated from the earth. If this is not possible, follow the suggestions given in Project 4-1.

   The horizontal amplifier of the oscilloscope is connected between the emitter terminal and the base 1 terminal in Fig. 5–13, so it will display V_{EB1}.

   The scope vertical amplifier will be connected across the 1000-Ω resistor in the base 1 lead. The only way for current to flow through that resistor is to come from the emitter, because base 2 current will return to the negative terminal of the variable dc supply before passing through the 1000-Ω resistor. Therefore the signal developed across the 1000-Ω resistor represents emitter current by this equation:

$$I_E = \frac{V_{\text{vert}}}{1000\ \Omega}$$

   Set the vertical amplifier gain to 1 volt/cm. With this setting, each centimeter of horizontal deflection represents 1 mA of emitter current, since

$$1\ \text{mA/cm} = \frac{1\ \text{volt/cm}}{1000\ \Omega}$$

   Note that Fig. 5–13 shows the I_E signal applied to the negative vertical input terminal, because the signal developed across the 1000-Ω resistor will be "backward," or 180° out of phase with the actual emitter current. The negative vertical input causes this signal to be reinverted, or put back into phase. If your scope does not have a negative vertical input, use the positive input; then the characteristic curve will be upside-down from that shown in Fig. 5–2(b).
2. With the scope's horizontal amplifier gain set at 2 V/cm, adjust the variable dc supply to 10 V.
   a. Close the switch and observe the *V-I* curve of the UJT. Make an accurate graph of the curve, paying special attention to V_P, V_V, I_P, and I_V. You will have to increase the vertical sensitivity to see I_P accurately. A sensitivity of about 5 mV/cm

will probably work best if your scope can go that low. At a sensitivity of 5 mV/cm, each centimeter represents 5 μA of emitter current.

b. Repeat step **a** with the variable supply set at 15 V(V_{B2B1} = 15 V).

c. Repeat for V_{B2B1} = 20 V, and again for 30 V. Explain what you are seeing.

3. Substitute several different UJTs of that same type, and compare their characteristics for V_{B2B1} = 20 V. How much batch spread is there in V_P? How much in V_V? How much in I_V? How much in η?

Project 5-3: UJT Gate Control Circuit for an SCR

Purpose

To build and observe a UJT gate control circuit for use with an SCR

Procedure

1. Build the UJT-SCR circuit of Fig. 5–6. Use the component sizes calculated in Sec. 5-4-2. If you cannot obtain a 2.2- kΩ, 5-W resistor for R_d, a 6.8-kΩ, 2-W resistor will also work. ZD1 can then be reduced to a ½-W rating.

   As usual the ac supply should be isolated from the earth, or follow the suggestions given in Project 4-1. Use any good UJT (2N4947, for example) and any medium-sized 200-V SCR.

   a. Use your oscilloscope to study the waveforms of V_{CE}, V_{R1}, V_{AK}, and V_{LD}. Graph all these waveforms to the same time reference for a firing delay angle of 90°. Do they all look like what you expected?

   b. What is the minimum possible firing delay angle?

   c. What is the maximum firing delay angle?

2. Readjust the firing delay angle to about 90°. Heat up the SCR with a soldering iron. What happens to the firing delay angle? Does this make sense?

CHAPTER

6 TRIACS AND OTHER THYRISTORS

Triacs behave roughly like SCRs, except that they can carry current in either direction. Both triacs and SCRs are members of the *thyristor* family. The term *thyristor* includes all the semiconductor devices which show *inherent* ON-OFF behavior, as opposed to allowing gradual change in conduction. All thyristors are regenerative switching devices, and they cannot operate in a linear manner. Thus a transistor is not a thyristor because although it *can* operate ON-OFF, that is not its inherent nature; it is possible for a transistor to operate linearly.

Some thyristors can be *gated* into the ON state, as we saw in Chapter 4 for SCRs. Triacs are the same in this respect. Other thyristors cannot be gated ON, but they turn ON when the applied voltage reaches a certain breakover value. Four-layer diodes and diacs are examples of this type of thyristor. Small thyristors which do not switch the main load current are usually referred to as *breakover devices,* and we will use that term in this book. They are useful in the gate triggering circuit of a larger load power-switching thyristor, such as a triac. In this chapter we will deal with the smaller breakover device thyristors as well as with triacs.

OBJECTIVES

After completing this chapter and performing the suggested laboratory projects, you will be able to:

1. Explain the operation of a triac in controlling both alternations of an ac supply driving a resistive load.
2. Define and discuss the important electrical parameters of triacs, such as gate trigger current, holding current, etc.
3. Explain the operation of breakover-type devices in the triggering circuits of triacs, and discuss the advantages of using these devices.
4. Describe the current-voltage behavior of the following breakover devices: diacs, four-layer diodes, silicon bilateral switches (SBSs), and silicon unilateral switches (SUSs).
5. Explain the flash-on effect seen with triacs and why it occurs, and explain how it can be eliminated with an SBS trigger circuit.
6. Explain in detail the operation of resistance feedback for firing a UJT in a triac trigger circuit.
7. Explain in detail the operation of voltage feedback for firing a UJT in a triac trigger circuit.
8. Calculate resistor and capacitor sizes for a triac's UJT trigger circuit employing either resistive feedback or voltage feedback.

9. Construct a triac control circuit for controlling a resistive load, and measure some of the triac's electrical parameters.
10. Construct a circuit which provides an oscilloscope display of the current-voltage characteristic curve of a thyristor.
11. Interpret the characteristic curve of a thyristor, reading the breakover voltages, breakback voltages, and holding currents.

6-1 ■ THEORY AND OPERATION OF TRIACS

A *triac* is a three-terminal device used to control the average current flow to a load. A triac is different from an SCR in that it can conduct current in *either* direction when it is turned ON. The schematic symbol of a triac is shown in Fig. 6–1(a), along with the names and letter abbreviations of its terminals.

When the triac is turned OFF, no current can flow between the main terminals no matter what the polarity of the externally applied voltage. The triac therefore acts like an open switch.

When the triac is turned ON, there is a very low-resistance current flow path from one main terminal to the other, with the direction of flow depending on the polarity of the externally applied voltage. When the voltage is more positive on $MT2$, the current flows from $MT2$ to $MT1$. When the voltage is more positive on $MT1$, the current flows from $MT1$ to $MT2$. In either case the triac is acting like a closed switch.

The circuit relationship among the supply voltage, the triac, and the load is illustrated in Fig. 6–1(b). A triac is placed in series with the load just like an SCR, as this figure shows. The average current delivered to the load can be varied by varying the amount of time per cycle that the triac spends in its ON state. If a small portion of the time is spent in the ON state, the average current flow over many cycles will be low. If a large portion of the cycle time is spent in the ON state, then the average current will be high.

A triac is not limited to 180° of conduction per cycle. With the proper triggering arrangement, it can conduct for a full 360° per cycle. It thus furnishes full-wave power control instead of the half-wave power control possible with an SCR.

Triacs have the same advantages that SCRs and transistors have over mechanical switches. They have no contact bounce, no arcing across partially opened contacts, and they operate much faster than mechanical switches, therefore yielding more precise control of current.

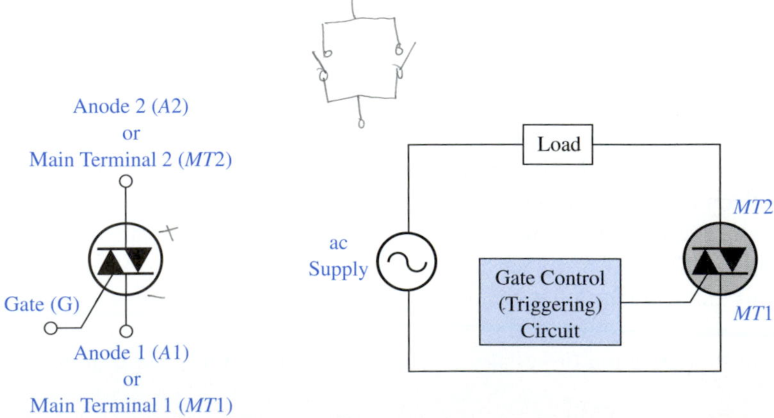

FIGURE 6–1
(a) Schematic symbol and terminal names of a triac. (b) Triac circuit showing how the supply voltage, the load, and the triac are connected.

6-2 ■ TRIAC WAVEFORMS

Triac waveforms are very much like SCR waveforms except that they can fire on the negative half cycle. Figure 6–2 shows waveforms of both load voltage and triac voltage (across the main terminals) for three different conditions.

The Fig. 6–2(a) waveforms show the triac OFF during the first 30° of each half cycle; during this 30° the triac is acting like an open switch. During this time the entire line voltage is dropped across the main terminals of the triac, with no voltage applied to the load. Thus there is no current flow through the triac or the load. The portion of the half cycle during which this situation exists is called the *firing delay angle,* just as it was for an SCR.

Continuing with Fig. 6–2(a), after 30° has elapsed, the triac fires or turns ON, and it becomes like a closed switch. At this instant the triac begins conducting current through

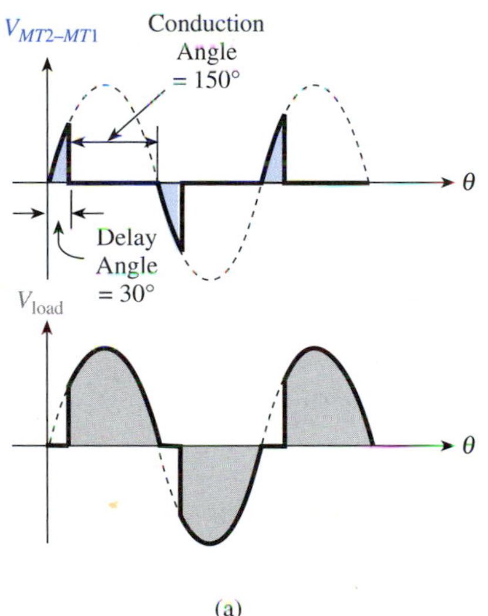

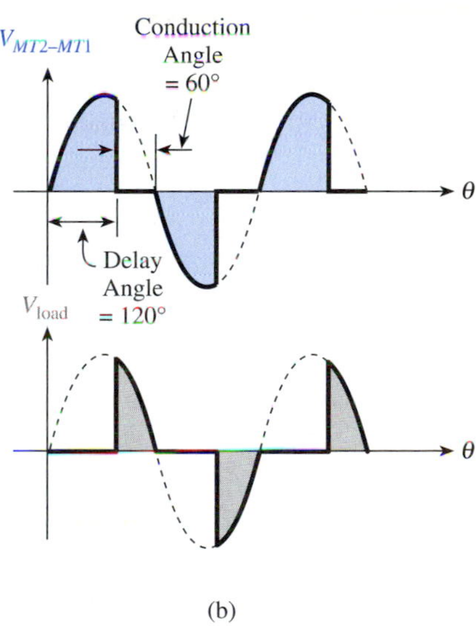

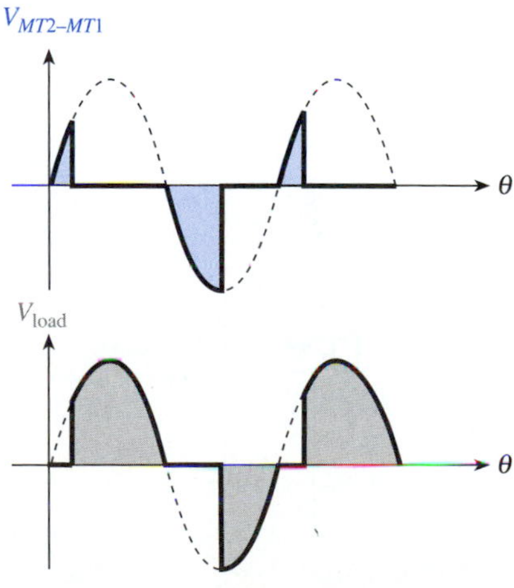

FIGURE 6–2
Waveforms of triac main-terminal voltage and load voltage for three different conditions. (a) Firing delay equals 30° for both the positive half cycle and the negative half cycle. (b) Firing delay equals 120° for both half cycles. (c) Unequal firing delay angles for the positive and negative half cycles. This is usually undesirable.

its main terminals and through the load, and it continues to carry load current for the remainder of the half cycle. The portion of the half cycle during which the triac is turned ON is called its *conduction angle.* The conduction angle in Fig. 6–2(a) is 150°. The waveforms show that during the conduction angle the entire line voltage is applied to the load, with zero voltage appearing across the triac's main terminals.

Figure 6–2(b) shows the same waveforms with a larger firing delay angle. The delay angle is 120° and the conduction angle is 60° in Fig. 6–2(b). Since the current is flowing during a smaller portion of the total cycle in this case, the average current is less than it was for the condition in Fig. 6–2(a). Therefore less power is transferred from source to load.

Triacs, like SCRs and most other semiconductor devices, show notoriously wide variations in their electrical characteristics. This problem is especially evident with triacs because it usually happens that the triggering requirements are different for the two different supply voltage polarities. Figure 6–2(c) shows waveforms which illustrate this problem. The triac waveform in Fig. 6–2(c) shows a smaller delay angle on the positive half cycle than on the negative half cycle due to the tendency of the triac to trigger more easily on the positive half cycle. Another triac of the same type might have a tendency to trigger more easily on the negative half cycle; in that case the negative delay angle would be smaller. Sometimes such inconsistent firing behavior cannot be tolerated. Methods for eliminating unequal firing delays will be discussed in Sec. 6-3.

6-3 ■ ELECTRICAL CHARACTERISTICS OF TRIACS

When a triac is biased with an external voltage more positive on $MT2$ (called *forward* or *positive* main terminal bias), it is usually triggered by current flow from gate to $MT1$. The polarities of the voltages and the directions of the currents in this case are shown in Fig. 6–3(a).

When a triac is biased as shown in Fig. 6–3(a), its triggering is identical to the triggering of an SCR. The G terminal is positive with respect to $MT1$, which causes the trigger current to flow into the device on the gate lead and out of the device on the $MT1$ lead. The gate voltage necessary to trigger a triac is symbolized V_{GT}; the gate current necessary to trigger is symbolized I_{GT}. Most medium-sized triacs have a V_{GT} of about 0.6 to 2.0 V and an I_{GT} of 0.1 to 20 mA. As usual, these characteristics vary quite a bit as temperature changes. Typical variations in electrical characteristics with temperature are graphed on the manufacturer's specification sheets.

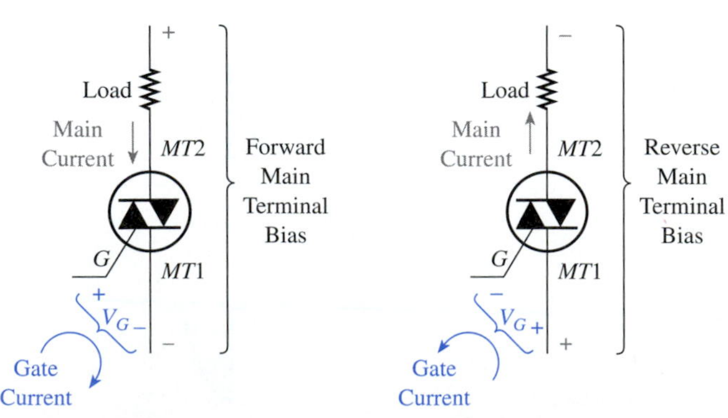

FIGURE 6–3
(a) The situation when a triac has forward main terminal bias. Normally the gate current and gate voltage would have the polarities indicated. (b) The situation at a different point in time when the triac is reverse biased. Normally the gate current and voltage are also reversed.

When the triac is biased more positive on $MT1$ (called *reverse* or *negative* main terminal bias), as shown in Fig. 6–3(b), triggering is usually accomplished by sending gate current into the triac on the $MT1$ lead and out of the triac on the G lead. The gate voltage will be negative with respect to $MT1$ to accomplish this. The voltage polarities and current directions for reverse main terminal bias are illustrated in Fig. 6–3(b).

For a particular individual triac, I_{GT} for the forward main terminal polarity may be quite a bit different from I_{GT} for the reverse main terminal polarity, as mentioned in Sec. 6-2. However, if many triacs of the same type are considered, the I_{GT} for the forward main terminal bias will be equal to the I_{GT} for the reverse main terminal bias.

Like an SCR, a triac does not require continuous gate current once it has been fired. It will remain in the ON state until the main terminal polarity changes or until the main terminal current drops below the holding current, I_{HO}. Most medium-sized triacs have an I_{HO} rating of less than 100 mA.

Other important electrical characteristics which apply to triacs are (1) the maximum allowable main terminal rms current, $I_{T\,\text{rms}}$ and (2) the breakover voltage, V_{DROM}, which is the highest main terminal peak voltage the triac can block in either direction. If the instantaneous voltage applied from $MT2$ to $MT1$ should exceed V_{DROM}, the triac breaks over and begins passing main terminal current. This does not damage the triac, but it does represent a loss of gate control. To prevent breakover, a triac must have a V_{DROM} rating greater than the peak value of the ac voltage driving the circuit. The most popular V_{DROM} ratings for triacs are 100, 200, 400 and 600 V.

For many manufacturers, the sequence of $I_{T\,\text{rms}}$ ratings available is 1, 3, 6, 10, 15, and 25 A. Other similar sequences are also used by triac manufacturers.

One other important electrical rating which is given on manufacturers' specification sheets is V_{TM}, the ON-state voltage across the main terminals. Ideally, the ON-state voltage should be 0 V, but V_{TM} usually falls between 1 and 2 V for real triacs, the same as for SCRs. A low V_{TM} rating is desirable because it means the triac closely duplicates the action of a mechanical switch, applying full supply voltage to the load. It also means that the triac itself burns very little power. The power burned in a triac is given by the product of main terminal current and main terminal voltage. Large power dissipation is undesirable from the point of view of protecting the triac from high temperatures and also from the point of view of economical energy transfer from source to load.

6-4 ■ TRIGGERING METHODS FOR TRIACS

6-4-1 RC Gate Control Circuits

The simplest triac triggering circuit is shown in Fig. 6–4(a). In Fig. 6–4(a), capacitor C charges through R_1 and R_2 during the delay angle portion of each half cycle. During a positive half cycle, $MT2$ is positive with respect to $MT1$, and C charges positive on its top plate. When the voltage at C builds up to a value large enough to deliver sufficient gate current (I_{GT}) through R_3 to trigger the triac, the triac fires.

During a negative half cycle, C charges negative on its top plate. Again, when the voltage across the capacitor gets large enough to deliver sufficient gate current in the reverse direction through R_3 to trigger the triac, the triac fires.

The charging rate of capacitor C is set by the resistance of R_2. For large R_2, the charging rate is slow, causing a long firing delay and small average load current. For small R_2, the charging rate is fast, the firing delay angle is small, and the load current is high.

As was true with SCR triggering circuits, a single RC network cannot delay triac firing much past 90°. To establish a wider range of delay angle adjustment, the double

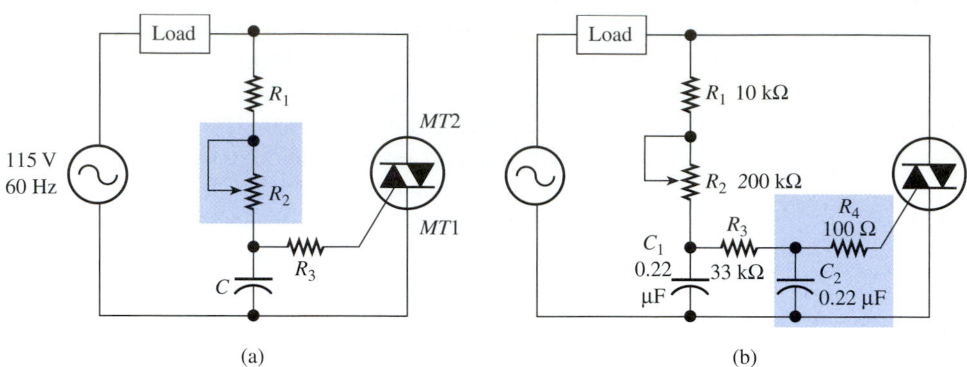

FIGURE 6–4
(a) Simple gate control circuit (triggering circuit) for a triac. Firing delay is adjusted by potentiometer R_2. (b) Improved gate control circuit, which allows a wider range of adjustment of firing delay.

RC network of Fig. 6–4(b) is often used. Typical component sizes are shown for use with a medium-sized triac.

6-4-2 Breakover Devices in Gate Control Circuits of Triacs

The gate control circuits of Fig. 6–4 can be improved by the addition of a breakover device in the gate lead, as shown in Fig. 6–5(a). The breakover device pictured in Fig. 6–5(a) is a *diac,* but there are several other breakover devices which also work well. Use of a breakover device in the gate triggering circuit of a triac offers some important advantages over simple *RC* gate control circuits. These advantages stem from the fact that breakover devices deliver a pulse of gate current rather than a sinusoidal gate current.

The ability of a breakover device to provide a current pulse can be understood by studying Fig. 6–5(b), which shows a typical current-voltage characteristic curve for a diac. (A diac is also known by the names *bidirectional trigger diode* and *symmetrical trigger diode*).

Let us interpret the diac's characteristic curve now. The curve shows that for applied forward voltages less than the *forward breakover voltage* (symbolized $+V_{BO}$) the diac permits virtually no current to flow. Once the forward breakover voltage is reached, however, the diac switches into conduction and the current surges up as the voltage across the terminals declines. Refer to Fig. 6–5(b) to see this. This surge of current on the characteristic curve accounts for the pulsing ability of the diac.

In the negative voltage region, the behavior is identical. When the applied reverse voltage is smaller than the *reverse breakover voltage* (symbolized $-V_{BO}$) the diac permits no current to flow. When the applied voltage reaches $-V_{BO}$, the diac switches into conduction in the opposite direction. This is graphed as negative current in Fig. 6–5(b). Diacs are manufactured to be relatively temperature stable and to have fairly close tolerances on breakover voltages. There is very little difference in magnitude between forward breakover voltage and reverse breakover voltage for a diac. The difference is typically less than 1 V. This enables the trigger circuit to maintain nearly equal firing delay angles for both half cycles of the ac supply.

The operation of the circuit in Fig. 6–5(a) is the same as that of the circuit in Fig. 6–4(a) except that the capacitor voltage must build up to the breakover voltage of the diac in order to deliver gate current to the triac. For a diac, the breakover voltage would be quite a bit higher than the voltage which would be necessary in Fig. 6–4(a). The most popular breakover voltage for diacs is 32 V ($+V_{BO} = +32$ V, $-V_{BO} = -32$ V). This value is convenient for use with a 115-V ac supply. Therefore when the capacitor voltage

FIGURE 6–5
(a) Triac gate control circuit containing a diac (bidirectional trigger diode). This triggering method has several advantages over the methods shown in Fig. 6-4. (b) Current versus voltage characteristic curve of a diac. (c) Another schematic symbol for a diac.

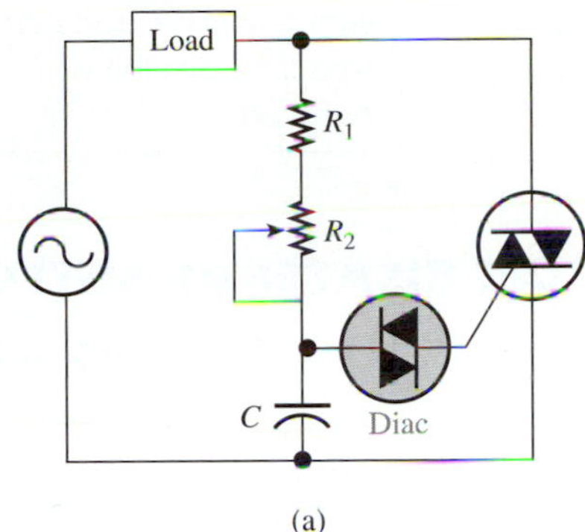

(a)

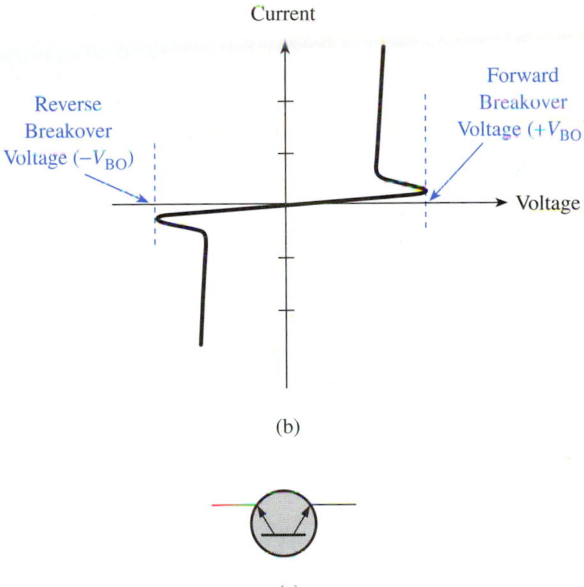

(b)

(c)

reaches 32 V, in either polarity, the diac breaks over, delivering the turn-ON pulse of current to the gate of the triac. Because the capacitor voltage must reach higher values when a diac is used, the charging time constant must be reduced. This means that Fig. 6–5(a) would have smaller component values (resistor and capacitor values) than Fig. 6–4(a).

A second schematic symbol for the diac is presented in Fig. 6–5(c). This symbol is less frequently used, and the diac symbol in Fig. 6–5(a) is preferred.

■ EXAMPLE 6-1

Suppose the circuit of Fig 6–5(a) contains a diac with $V_{BO} = \pm 32$ V. Suppose also that the resistor and capacitor sizes are such that the firing delay angle = 75°. Now if the 32-V diac is removed and replaced by a 28-V diac but nothing else is changed, what will happen to the firing delay angle? Why?

Solution. If the 32-V diac is replaced by a 28-V diac, it means that the capacitor will only have to charge to ±28 V in order to fire the triac, instead of ±32 V. With given sized components, C can certainly charge to 28 V in less time than it can charge to 32 V. Therefore it causes the diac to break over earlier in the half cycle, and the firing delay angle is **reduced.** ■

6-5 ■ SILICON BILATERAL SWITCHES

6-5-1 Theory and Operation of an SBS

There is another breakover device which is capable of triggering triacs. Its name is the *silicon bilateral switch* (SBS), and it is popular in low-voltage trigger control circuits. SBSs have lower breakover voltages than diacs, ±8 V being the most popular rating. The current-voltage characteristic curve of an SBS Is similar to that of the diac, but the SBS has a more pronounced "negative resistance" region. That is, its decline in voltage is more drastic after it enters the conducting state. An SBS is shown schematically in Fig. 6–6(a). Its current-voltage characteristic curve is shown in Fig. 6–6(b). Notice that when the SBS switches into its conducting state the voltage across its anode terminals drops almost to zero (to about 1 V). The SBS is said to have a *breakback voltage* of 7 V, because the voltage between $A2$ and $A1$ decreases by about 7 V when it turns ON.

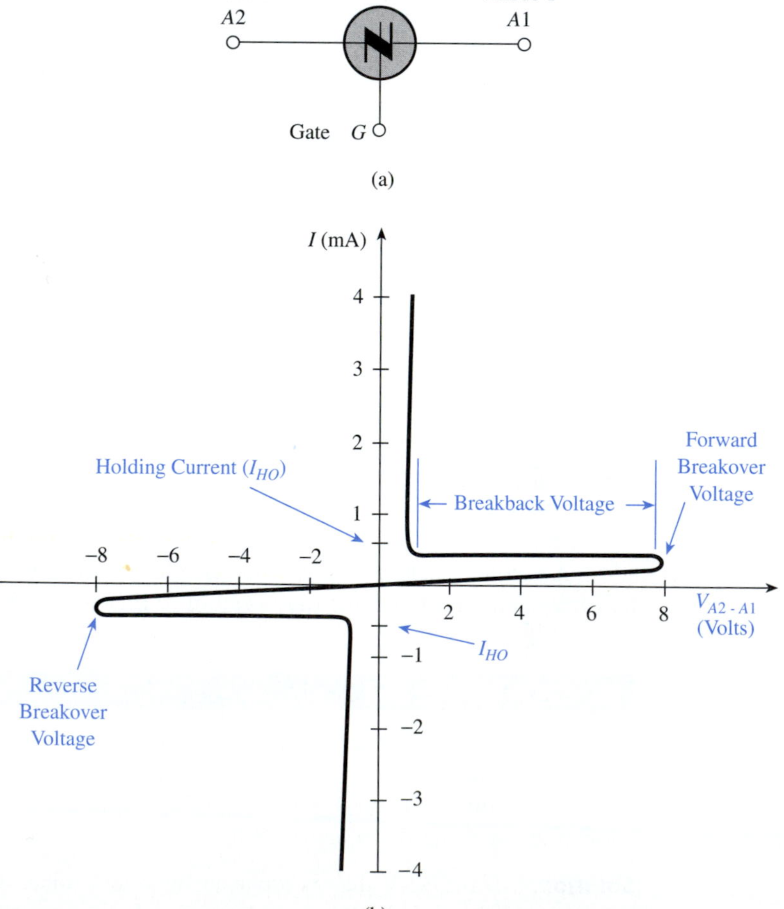

FIGURE 6–6 (a) Schematic symbol and terminal names of an SBS (silicon bilateral switch). (b) Current-voltage characteristic curve of an SBS, with important points indicated.

The characteristic curve of Fig. 6–6(b) is for the gate terminal of the SBS disconnected. The gate terminal can be used to alter the basic current-voltage behavior of an SBS, as we will see shortly. However, the SBS is quite useful even without its gate terminal, just by virtue of the snap-action breakover from $A2$ to $A1$.

To make use of an SBS without its gate terminal, it could be installed in place of the diac in Fig. 6–5(a). Because of the lower V_{BO} of the SBS, the RC timing components would have to be increased in value. You may wonder why we would want to use an SBS in this control circuit instead of a diac. Well, generally speaking, the SBS is a superior device compared to the diac. Not only does the SBS show a more vigorous switching characteristic, as Fig. 6–6(b) indicates, but an SBS is more temperature stable and more symmetrical and has less batch spread than a diac.

To say it with numbers, a modern SBS has a temperature coefficient of about $+0.02\%/°C$. This means that its V_{BO} increases by only 0.02% per degree of temperature change, which figures out to only 0.16 V/100°C, which is very temperature stable indeed.

SBSs are symmetrical to within abut 0.3 V. That is, the difference in magnitude between $+V_{BO}$ and $-V_{BO}$ is less than 0.3 V. This yields virtually identical firing delays for positive and negative half cycles.

The batch spread of SBSs is less than 0.1 V. This means that the difference in V_{BO} among all the SBSs in a batch is less than 0.1 V. The batch spread among diacs is about 4 V, by contrast.

6-5-2 Using the Gate Terminal of an SBS

As mentioned in Sec. 6-5-1, the gate terminal of an SBS can be used to alter its basic breakover behavior. For example, if a zener diode is connected between G and $A1$, as shown in Fig. 6-7(a), the forward breakover voltage ($+V_{BO}$) changes to approximately the V_Z of the zener diode. With a 3.3-V zener diode connected, $+V_{BO}$ would equal 3.3 V + 0.6 V (there is an internal pn junction). This would yield

$$+V_{BO} = 3.9 \text{ V}$$

The reverse breakover voltage would be unaffected and would remain at -8 V. The new current-voltage behavior would be as drawn in Fig. 6–7(b). This behavior would be useful if it were desired to have different firing delay angles for the positive and negative half cycles (which would be unusual).

6-5-3 Eliminating Triac Flash-on (Hysteresis) with a Gated SBS

One of the nicest things about using a gated SBS for the trigger control of a triac is that it can eliminate the *hysteresis* or *flash-on* effect. Let us first come to an understanding of the flash-on problem. The explanation of flash-on is rather involved, so be forewarned.

Refer back to Fig. 6–5(a). Suppose that R_2 is adjusted so that C cannot quite charge up to 32 V in either direction. In this case, the diac would never fire, and the load would be completely deenergized. If the load were a lighting load, it would give no light at all. Since C never discharges any of its built-up charge, it always starts a new half cycle of the ac supply starts, the initial charge on C is minus on the top and plus on the bottom; this charge is left over from the preceding negative half cycle. Likewise, when a negative half cycle of the ac supply line begins, the initial charge on C is plus on the top and minus on the bottom, left over from the preceding positive half cycle. The effect of this initial charge is to mark it more difficult for the capacitor to charge up to the breakover voltage of the diac.

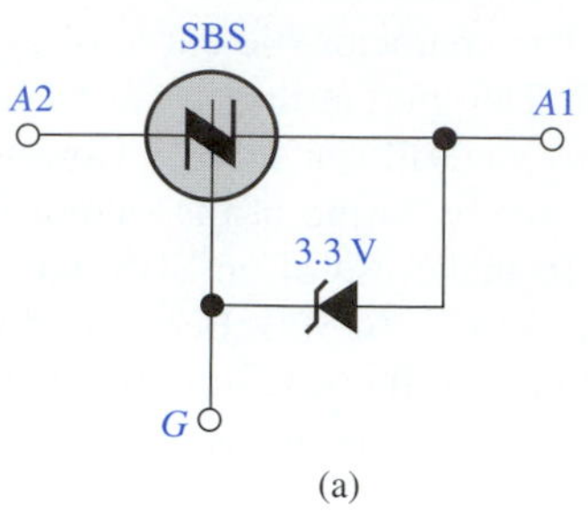

(a)

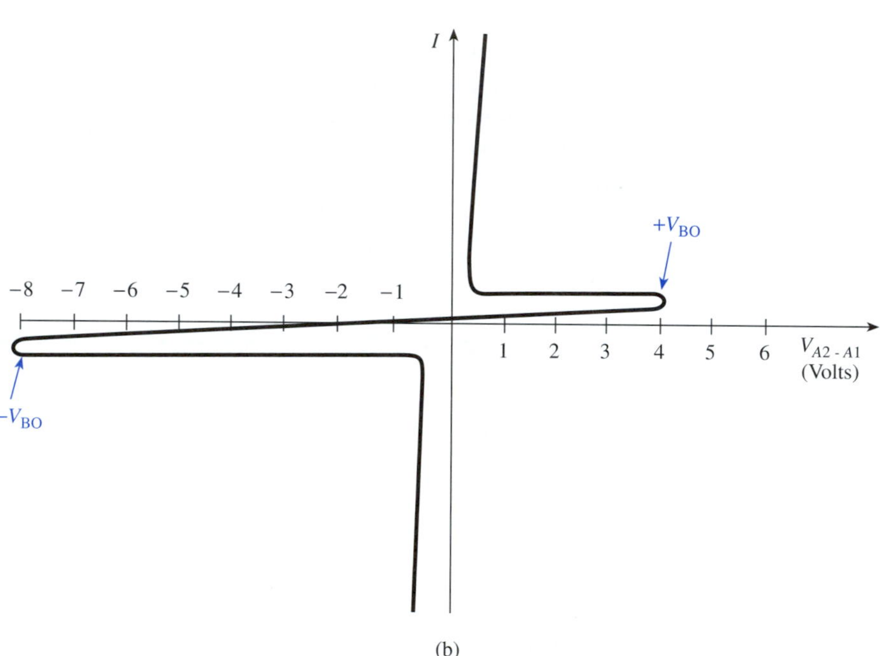

(b)

FIGURE 6–7 (a) SBS combined with a zener diode to alter the breakover point in the forward direction. (b) Characteristic curve of the SBS-zener diode combination. The forward breakover voltage is lower, but the reverse breakover voltage is unchanged.

Now suppose that we slowly decrease R_2 until the capacitor can just barely charge to V_{BO} of the diac. Assume that the first breakover occurs on the positive half cycle (it is just as likely to occur on the negative half cycle as the positive). When the diac breaks over, it discharges part of the + charge which has built up on the top plate of C. the discharge path is through the G to MT_1 circuit of the triac. During the remainder of the positive half cycle, no further charging of C takes place, because the triac shorts out the entire trigger circuit when it turns ON. Therefore when that positive half cycle ends and the next negative half cycle begins, the initial + charge on the top of C is *less* than it was for all the prior negative half cycles. The capacitor has a "head start" this time, as it attempts to charge to $-V_{BO}$.

Because of this head start, C will reach $-V_{BO}$ much earlier in the negative half cycle than it reached $+V_{BO}$ in the preceding positive half cycle. Furthermore, since C will lose some of the − charge on its top plate when the diac breaks over during the negative half cycle, it will begin the next positive half cycle with less initial charge than ever before. Therefore it will fire much earlier in the *next* positive half cycle than it fired in the *first* positive half cycle.

The overall result of this phenomenon is this: You can adjust R_2 to just barely fire the triac, expecting to get very dim light from the lamps, but as soon as first firing takes place, all subsequent firings take place much earlier in the half cycle. It is impossible to smoothly adjust from the completely off condition to the glowing-dimly condition. Instead, the lamps "flash on."

What you *can* do, one the lamps have come on, is to adjust the R_2 resistance back up to a higher value to delay the diac breakover until later in the half cycle. In other words, you must turn the potentiometer shaft back in the direction it just came from in order to create very dim light. You can demonstrate this with almost any commercial lamp dimmer in your home. Unless it is a very good one, it will exhibit flash-on and subsequent reduction in light output as the knob is turned back.

What we have here is a situation in which a single given resistance value of R_2 can cause two completely different circuit results, depending on the *direction* in which R_2 is changing. This phenomenon occurs quite often in electronics and, in fact, in all of nature. Its general name is *hysteresis*. The flash-on of a triac is a specific example of hysteresis.

■ EXAMPLE 6-2

Suppose that it requires an R_2 resistance of 5000 Ω to just barely cause the diac to break over in Fig. 6–5 (a).

(a) If the R_2 resistance is 6000 Ω and we reduce it to 5025 Ω, will any light be created?

(b) If the R_2 resistance is 4700 Ω and we increase it to 5025 Ω, will any light be created?

(c) What word would you use to sum up this behavior?

Solution. (a) When R_2 = 6000 Ω, the diac will not be breaking over because R_2 must decline all the way to 5000 Ω to just barely cause breakover. If we then reduce R_2 to 5025 Ω, the resistance is still a little too high to allow diac breakover, so the triac is not firing and no light is created.

(b) If R_2 is 4700 Ω, this is less than the resistance which just barely causes breakover, so the diac will be breaking over and firing the triac and the lamps will be glowing. If we raise the resistance to 5025 Ω, the diac will still continue breaking over because now the capacitor always starts charging with a smaller opposite charge on its plates than it had in part (a). The smaller charge results from the fact that the capacitor partly discharged on the preceding half cycle. With the diac breaking over, the triac is firing, and the lamps are giving off some light.

(c) The fact that 5025 Ω coming form *above* (from 6000 Ω) caused no light, but 5025 Ω coming from *below* (from 4700 Ω) caused some light means that a given resistance value causes two completely different results, depending on the direction of approach. Therefore we can describe the behavior as showing hysteresis. ■

Triac hysteresis can be almost completely eliminated with the circuit of Fig. 6–8(a). To understand how it works, we must investigate the action of an SBS when a small amount of current flows in its gate lead. Refer to Fig. 6–8(b) and (c).

Figure 6–8(b) shows a resistor R inserted in the gate lead of an SBS and a certain amount of current, I_G, flowing from $A2$ to G. This implies that the voltage applied to the gate resistor is negative relative to $A2$.

FIGURE 6–8
(a) A more complex triac triggering circuit. Triac flash-on can be eliminated with this circuit. (b) Direction of gate current through the SBS as the ac supply approaches its zero crossover. (c) The very low forward breakover voltage when gate current is flowing in the SBS.

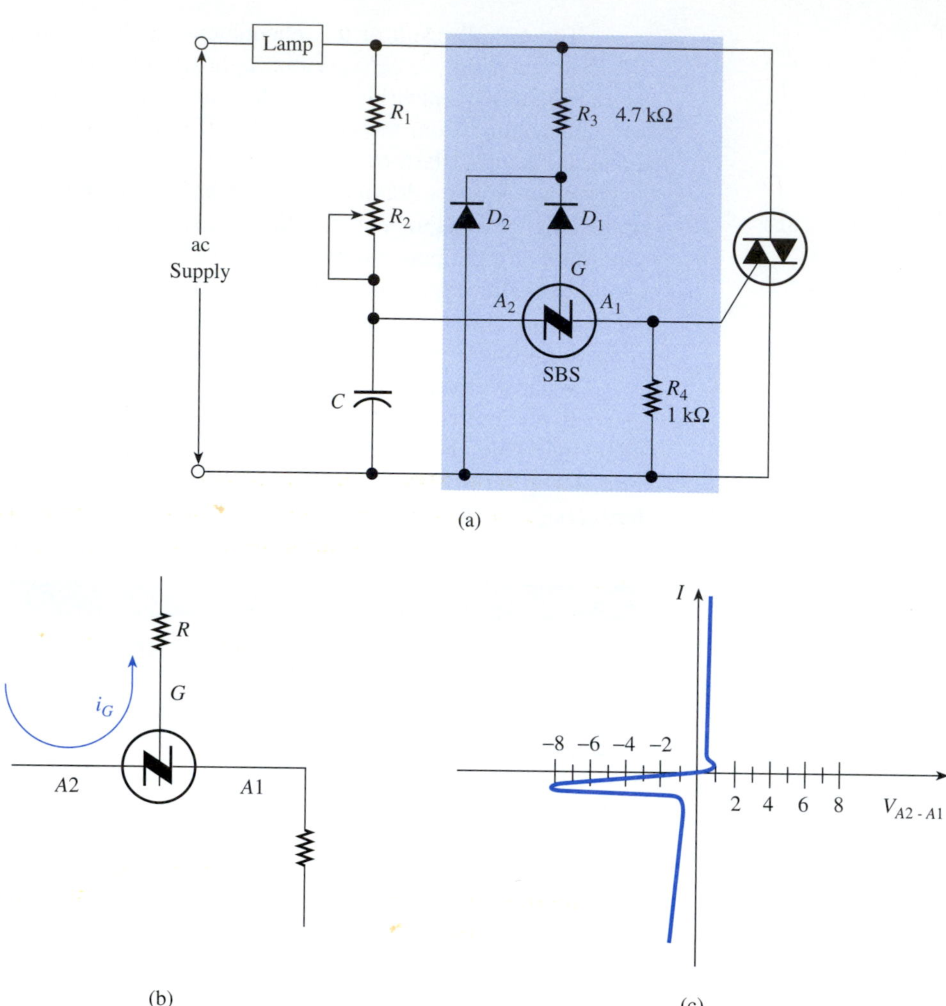

If a small gate current flows between *A*2 and *G*, the forward breakover characteristic is changed drastically. The $+V_{BO}$ voltage drops to about 1 V, as shown in Fig. 6–8(c). This means that the SBS will break over as soon as the *A*2 to *A*1 voltage reaches 1 V. As the curve shows, $-V_{BO}$ is not affected by gate current from *A*2 to *G*.

Now look at Fig. 6–8(a). Suppose that R_2 is set so that the capacitor voltage cannot reach ±8 V to make the SBS break over. The triac will not fire, and the lamp will be extinguished. During the positive half cycle, *C* will charge plus on the top and minus on the bottom. Now look what happens as the ac supply completes the positive half cycle and approaches 0 V. When the top supply line gets near zero relative to the bottom line, it means hat the top of R_3 is near zero volts relative to the bottom of *C*. However, the top of *C* is positive relative to the bottom of *C* at this time because of the capacitor charge. Therefore there is a voltage impressed between *A*2 and the top of R_3; this voltage is plus at *A*2 and minus at the top of R_3. It forward-biases diode D_1 and causes a small amount of SBS gate current to flow. The flow path is into the SBS at *A*2, out of the SBS at *G*, through D_1, and through R_3. With this small I_G, even a very low forward voltage from *A*2 to $A1$ will break over the SBS, as Fig. 6–8(c) shows. There is a small forward voltage between $A2$ and $A1$ at this time, namely the capacitor voltage. As long as it is greater than about 1 volt, the SBS will break over. When the SBS breaks over, it dumps the capacitor charge through R_4. The negative half cycle of the ac supply therefore starts with the capacitor almost completely discharged. The result is that the capacitor starts charging with the same initial charge (about zero) no matter whether the triac is firing or is not firing. Therefore the triac hysteresis is eliminated.

It has been left as a question at the end of the chapter to explain the purpose of diode D_2.

6-6 ■ UNILATERAL BREAKOVER DEVICES

The diac and the SBS are classed as *bilateral* or *bidirectional* breakover devices because they can break over in either direction. There are also breakover devices which break over in only one direction; they are in the class of *unilateral* or *unidirectional* breakover devices. We have already seen one unilateral breakover device, the four-layer diode, in Sec. 4–5. We will now take a detailed look at a more popular modern unidirectional breakover device, the *silicon unilateral switch* (SUS). Although unilateral breakover devices are more frequently seen in SCR triggering circuits, they can also be used in triac triggering circuits if they have some extra supporting circuitry.

The schematic symbols and terminal names of a four-layer diode and an SUS are shown in Figs. 6–9(a) and (b). Their characteristic current-voltage behavior is illustrated in Fig. 6–9(c).

As can be seen, the behavior of four-layer diodes and SUSs is similar to the behavior of an SBS except that only forward breakover is possible. Reverse *breakdown* can happen, but only at a much greater voltage level than $+V_{BO}$. Reverse breakdown may be destructive for the device.

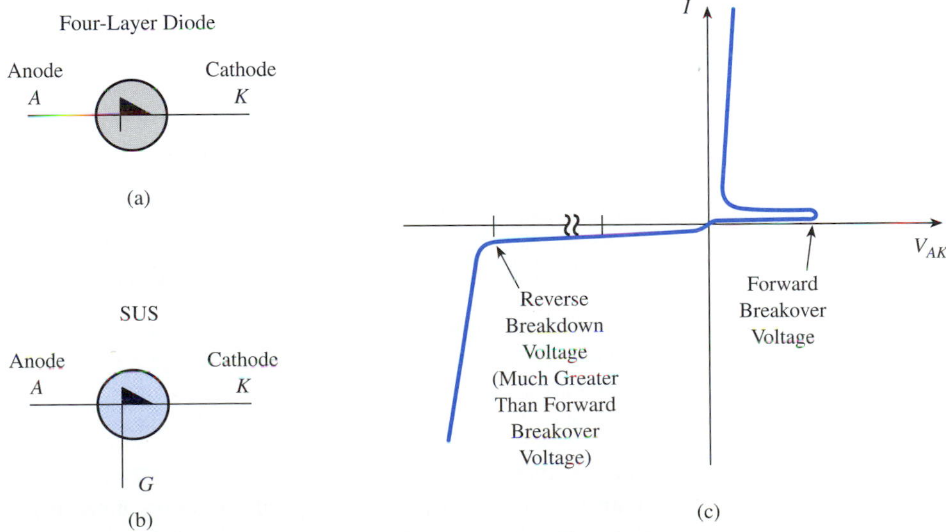

FIGURE 6–9
(a) Schematic symbol of a four-layer diode. (b) Symbol and terminal names of an SUS (silicon unilateral switch). (c) Current-voltage characteristic curve of an SUS.

The SUS, like the SBS, has a gate terminal which can alter the basic breakover behavior shown in Fig. 6–9(c). By connecting a zener diode between the gate and cathode of an SUS, the breakover voltage can be reduced to

$$V_{BO} = V_Z + 0.6 \text{ V}$$

When this is done, the cathode of the zener diode must be connected to the gate of the SUS, and the anode of the zener must be connected to the cathode of the SUS.

The SUS can be fired at a very low anode to cathode voltage (about 1 V) if gate current flows from anode to gate. This is the same kind of control as illustrated in Fig. 6–8 for an SBS.

Silicon unilateral switch are low-voltage, low current devices. Most SUSs have a breakover voltage of 8 V and a current limit of less than 1 A.

6-7 ■ BREAKOVER DEVICE (SUS) USED TO TRIGGER A TRIAC

An example of an SUS and its supporting circuitry in a triac gate control circuit is shown in Fig. 6–10(a). Here is how it works. As the ac supply goes through its positive and negative alternations, the bridge rectifier delivers a full-wave rectified voltage to the *RC* timing network. This is called V_{bridge} in Fig. 6–10(c). The voltage across the capacitor, V_C, tends to follow V_{bridge}, with the amount of lag determined by the setting of R_2. At some point in the half cycle, V_C will reach the breakover voltage of the SUS, shown in Fig. 6–10(d). When that point is reached, the SUS breaks over and allows the capacitor to discharge through the primary winding of the pulse transformer.

The capacitor discharge creates a burst of current in the transformer primary, shown in Fig. 6–10(e). The current burst continues until the capacitor discharges to the point where it cannot supply current equal to the holding current (valley current) of the SUS.

The pulse transformer couples this current pulse into the *G-MT*1 circuit of the triac, thereby firing the triac. The gate current waveform is graphed in Fig. 6–10(f). The pulse transformer is necessary because the *RC* timing circuit must be electrically isolated from the *G-MT*1 circuit, since the *RC* circuit it driven by a bridge rectifier which is itself connected to *MT*1. That is, if the transformer were removed from Fig. 6–10(a) and the trigger circuit were connected directly to *G* and *MT*1, there would be a short circuit across the diode in the lower right-hand corner of the bridge. You should verify this for yourself.

Notice that the I_{sec} pulses all flow in the same direction regardless of the main terminal polarity. Although it has not been mentioned until now, the direction of a triac's gate current does not *have* to agree with its main terminal polarity. That is, a triac with a positive main terminal polarity can be fired by a negative gate current. Likewise, when the triac sees negative main terminal polarity, it can be fired by a positive gate current. SCRs are not like this; reverse gate current cannot trigger an SCR.

To make this clear, the four possible triggering modes of a triac are summarized:

1. Positive main terminal voltage, positive gate current
2. Positive main terminal voltage, negative gate current
3. Negative main terminal voltage, negative gate current
4. Negative main terminal voltage, positive gate current

The positive main terminal voltage means *MT*2 is more positive than *MT*1; the negative main terminal voltage is the opposite. Positive gate current means conventional current flows into the gate and out of main terminal 1; negative gate current is the opposite.

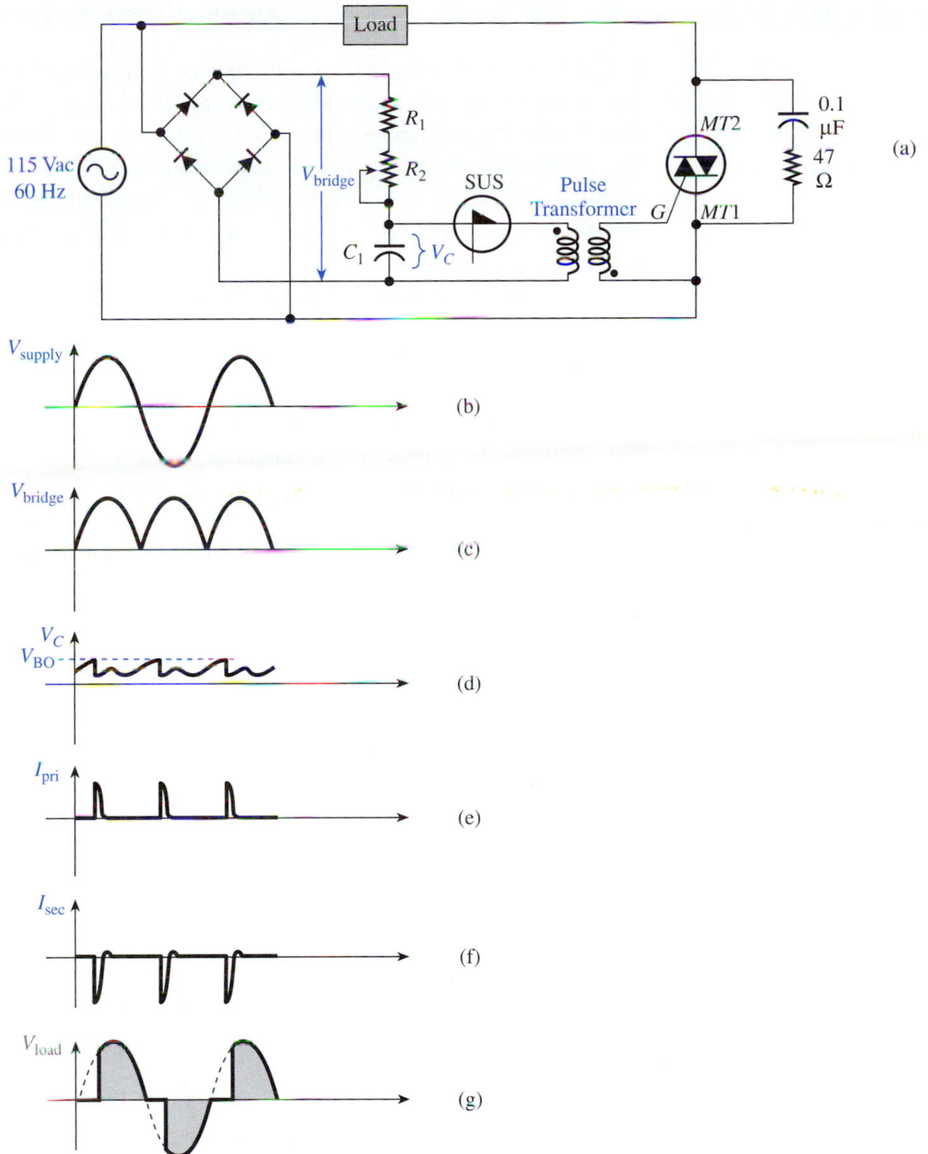

FIGURE 6–10
(a) Complete schematic diagram of a triac control circuit containing an SUS and a pulse transformer. (b) Ac supply voltage waveform. (c) Full-wave rectified voltage (V_{bridge}) which is applied to the triggering circuit. (d) Capacitor voltage waveform, shown reaching the V_{BO} of the SUS. (e) Pulse transformer primary current (f) Inverted pulses of secondary current from the pulse transformer. (g) Load voltage waveform.

Modern triacs trigger very well in modes 1 to 3 but they are more difficult to trigger in mode 4. Therefore mode 4 is avoided. For this reason, whenever both main terminal polarities must be triggered by just one direction of gate current, the direction is *negative*.

A negative gate current pulse can be supplied readily just by inverting the pulse transformer secondary. This has been done in Fig. 6–10(a). Notice the transformer phasing dots indicating that when primary current flows *in* on the top lead, secondary current flows *out* on the bottom lead of the winding. This is the negative direction for gate current, as stated. I_{sec} is shown negative in Fig. 6–10(f).

At the instant the I_{sec} pulse occurs in the gate circuit, the triac fires and causes the supply line voltage to appear across the load. This is drawn in the waveform of Fig. 6–10(g). The delay angle could be reduced or increased by adjusting R_2 to be smaller or larger.

6-8 ■ CRITICAL RATE OF RISE OF OFF-STATE VOLTAGE (*dv/dt*)

In Fig. 6–10, notice the *RC* circuit connected in parallel with the triac. Such *RC* circuits are sometimes installed across triacs in industrial settings. The circuit's purpose is to prevent fast-rising transient signals from appearing across the main terminals of the triac. The reason it is necessary to eliminate fast-rising voltage surges in that all triacs have a certain *dv/dt rating,* which is the maximum rate of rise of the main terminal voltage they can withstand. If this rate of rise is exceeded, the triac may inadvertently turn ON, even when not gate signal occurs.

For most medium-sized triacs, the *dv/dt* rating is around 100 V/μs. As long as any transient surges appearing across the main terminals have slopes less than the *dv/dt* rating, the triac will not turn ON without a gate signal. If a transient surge appears having a voltage vs. time slope in excess of the *dv/dt* rating, the triac may turn ON.

It should be emphasized that this is not the same thing as V_{DROM} breakover. The V_{DROM} rating of a triac is the maximum peak voltage the triac can withstand without breaking over, if that voltage is approached *slowly*. The *dv/dt* rating refers to fast voltage surges, which may have a peak value far less than V_{DROM}. The point is that even though the transient surges may be small in magnitude, their steep slope may cause a triac to fire.

If the ac supply line is guaranteed to be free of transient surges, the *RC* suppression circuit of Fig. 6–10 is not needed. However, in industrial settings the ac lines are usually bristling with transient surges due to switchgear operation, etc. Therefore the *RC* suppression circuit is almost always included.

Surge suppression is actually provided by the *C* part of the *RC* circuit. The 0.1-μF capacitor in Fig. 6–10 tends to short out the triac for high-frequency signals. Thus any fast transients on the ac lines are dropped across the load resistance, since *C* presents a very low impedance to anything fast.

The reason for including *R* is to limit the large capacitive discharge current when the triac turns ON under normal operating conditions. Resistor *R* is of no use in actually suppressing the fast transients; its only purpose is to limit the discharge of capacitor *C* through the main terminals of the triac.

The *R* and *C* sizes shown in Fig. 6–10 are typical. Such *RC* suppression circuits are also used for SCRs, because SCRs have the same problem with fast-rising transients.

6-9 ■ UJTS AS TRIGGER DEVICES FOR TRIACS

So far, the firing delay angle of all our triac and SCR circuits has been set by a potentiometer resistance adjustment. In industrial power control, there are times when the firing point is set by a feedback voltage signal. A feedback voltage signal is a voltage which somehow represents actual conditions at the load. For example, for a lighting load, a voltage proportional to light intensity could be used as a feedback signal to automatically control the triac's firing delay angle and therefore the light which is produced; for a motor load, a voltage proportional to motor shaft speed could be used as a feedback signal to control the firing delay angle and consequently the motor speed. Whenever triac (or SCR) firing control is accomplished via a feedback voltage singal, the UJT is a popular triggering device.

Sometimes a feedback signal takes the form of a varying resistance instead of a varying voltage. In these cases too, the UJT is compatible with the feedback situation.

Figure 6–11 illustrates a frequently seen UJT triggering setup for use with feedback. In Fig. 6–11(a), which shows the complete power control circuit, resistive feedback has been depicted. Resistor R_F is a variable resistance which varies as the load conditions

FIGURE 6–11
(a) Complete schematic diagram of a triac control circuit. The triggering circuit uses a UJT and a constant-current source, which is controlled by resistance feedback. (b) The same triggering circuit as part (a) except that the current source is controlled by voltage feedback. (c) The zener-clipped full-wave voltage which drives the triggering circuit. (d) The capacitor voltage waveform. It rises at a constant slope until it hits V_P of the UJT (15 V in this case). (e) Secondary current pulse from the pulses transformer.

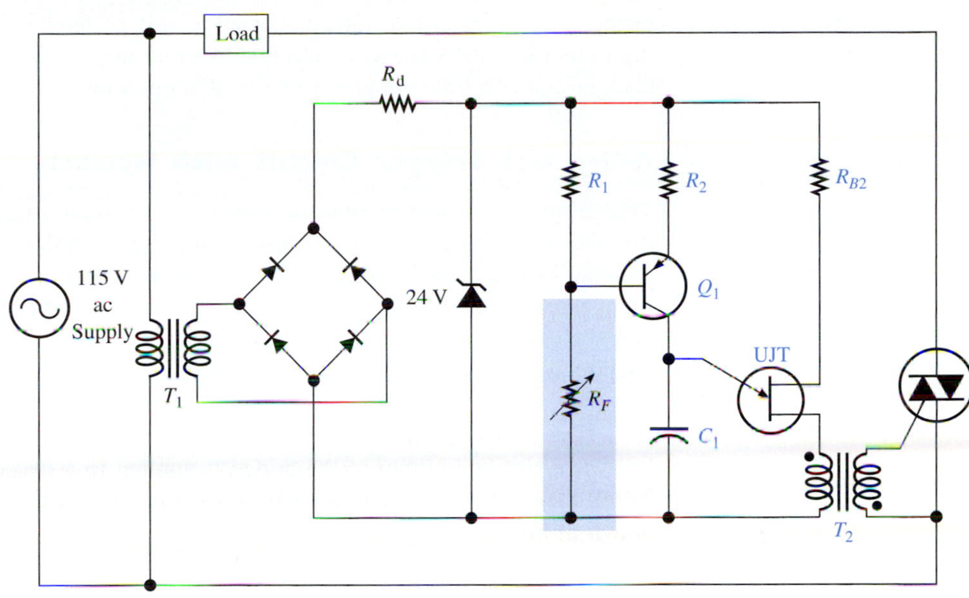

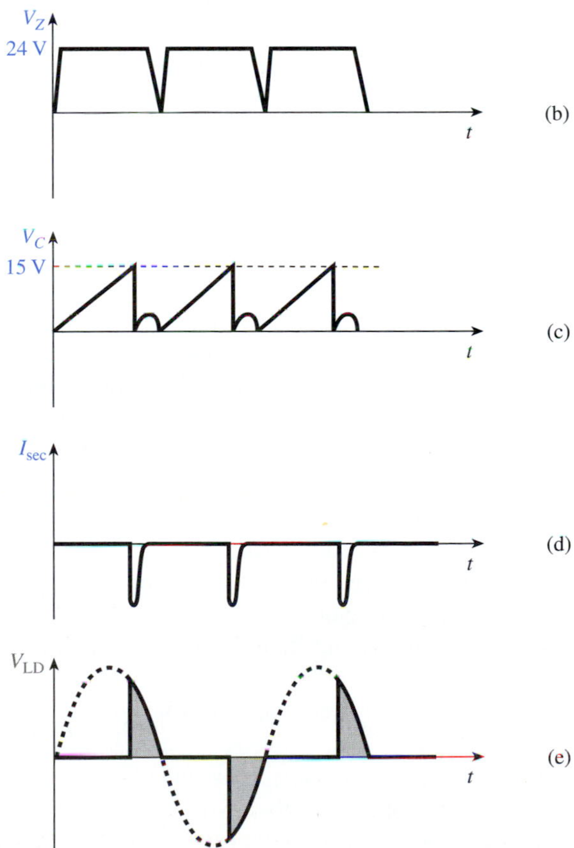

change. The same circuit can be adapted to voltage feedback by removing R_F and inserting in its place the voltage feedback network drawn in Fig. 6–11(b). The resistance feedback circuit is a little simpler, so we will begin with it.

6-9-1 UJT Trigger Circuit with Resistive Feedback

Transformer T_1 is an *isolation transformer*. An isolation transformer has a 1 : 1 turns ratio, and its purpose is to electrically isolate the secondary and primary circuits. In this case the isolation transformer is isolating the ac power circuit from the triggering circuit. Many isolation transformers contain transient supperssion components. When they do contain such components, high-frequency transient signals appearing at the primary are *not* coupled into the secondary winding, thereby helping to keep the secondary circuit noise-free.

The 115-V ac sine wave from the T_1 secondary is applied to a bridge rectifier. The full-wave rectified output of the bridge is applied to a resistor-zener diode combination, which then delivers a 24-V waveform synchronized with the ac line. This waveform is drawn in Fig. 6–11(b).

When the 24-V source is established, C_1 begins charging. When it charges up to V_P of the UJT, the UJT fires and creates a current pulse in the primary winding of pulse transformer T_2. This is coupled into the secondary winding, and the secondary pulse is delivered to the gate of the triac, turning it ON for the rest of the half cycle. The waveforms of capacitor voltage V_{C1}, T_2 secondary current I_{sec}, and load voltage V_{LD}, are drawn in Fig. 6–11(c), (d), and (e). The firing delay angle is about 135° in these waveforms.

The rate at which C_1 charges is determined by the ratio of R_F to R_1, which form a voltage divider. Between them they split up the 24-V dc source which supplies the trigger circuit. If R_F is small relative to R_1, then R_1 will receive a large share of the 24-V supply. This will cause *pnp* transistor Q_1 to conduct hard, because the R_1 voltage is applied to the base-emitter circuit of Q_1. With Q_1 conducting hard, C_1 charges rapidly since C_1 is charged by the Q_1 collector current. Under these conditions the UJT fires early, and the average load current is high.

On the other hand, if R_F is large relative to R_1, then the voltage across R_1 will be smaller than before because of the voltage-division effect. This causes a smaller voltage to appear across the Q_1 base-emitter circuit, reducing the Q_1 drive. With Q_1 conducting less, the charging rate of C_1 is reduced, and it takes longer to build up to the V_P of the UJT. Therefore the UJT and triac fire later in the half cycle, and the average load current is lower than before.

The C_1 charging circuit shown produces a *constant* rate of voltage rise across the capacitor, as Fig. 6–11(c) makes clear. The slope of the voltage waveform is constant because the capacitor charging current is constant. Let us now quantitatively analyze this *constant-current supply* circuit.

First, considering R_1 and R_F to be a series circuit, we can say that

$$V_{R1} = (24\text{ V})\frac{R_1}{R_1 + R_F} \tag{6-1}$$

which expresses the proportionality between voltage and resistance for a series circuit. Of course, strictly speaking, R_1 and R_F are not really in series. The base lead of Q_1 ties into the point between these two resistors; because of that, R_F carries a slight bit of extra current that R_1 does not carry. However, if these resistors are correctly sized, their current draw will far exceed the transistor's base current. With the base current very small by comparison, the percent difference between the R_F current and the R_1 current is negligible. That being so, it is all right to consider R_F and R_1 as a series circuit, and Eq. (6-1) is justified.

V_{R1} appears across resistor R_2 and the base-emitter junction of Q_1. Since R_2 is in the emitter lead of Q_1, we can say that

$$V_{R1} = (I_{E1})R_2 + 0.7 \text{ V}$$

where I_{E1} stands for the Q_1 emitter current. If Q_1 is a high-beta transistor, its collector current is virtually equal to its emitter current, so to a good approximation,

$$V_{R1} = (I_{C1})R_2 + 0.7 \text{ V}$$

where I_{C1} is the collector current in the transistor and also the charging current of the capacitor C_1. Rewriting this equation for I_{C1} and combining with Eq. (6-1), we obtain

$$I_{C1} = \frac{V_{R1} - 0.7 \text{ V}}{R_2} = \frac{1}{R_2}\left[\frac{(24 \text{ V})R_1}{R_1 + R_F} - 0.7 \text{ V}\right] \tag{6-2}$$

Equation (6-2) expresses the fact that the capacitor charging current goes up as R_F goes down, and it also shows that for a given R_F the charging current is constant as time goes by.

Intuitively speaking, this circuit is able to maintain a constant current flow because it reduces the Q_1 collector-to-emitter voltage as the capacitor voltage rises. That is, for every volt that V_{C1} rises, V_{CE} of Q_1 decreases by 1 volt. In this way, the continually rising capacitor voltage cannot retard the flow of current as it normally does in a simple *RC* circuit.

With I_{C1} a constant, the rate of voltage buildup is constant, because for any capacitor,

$$\frac{\Delta v}{\Delta t} = \frac{I_C}{C} \tag{6-3}$$

where $\Delta v/\Delta t$ is the time rate of change of capacitor voltage. Therefore as long as current is constant, the voltage buildup rate is constant, as drawn in Fig. 6–11(c).

■ EXAMPLE 6-3

For the circuit of Fig. 6–11(a), assume the following conditions: $R_1 = 5 \text{ k}\Omega$, $R_F = 8 \text{ k}\Omega$, $R_2 = 2.5 \text{ k}\Omega$, $\beta_1 = 150$, $C_1 = 0.5\ \mu\text{F}$, $\eta = 0.58$.

(a) Find V_{R1}.

(b) Find I_{C1}.

(c) What is the rate of voltage buildup across C_1?

(d) How much time elapses between the beginning of a half cycle and the firing of the triac?

(e) What is the firing delay angle?

(f) What value of R_F would cause a firing delay angle of 120°?

Solution. (a) Assuming that R_1 and R_F are in series, we can use Eq. (6-1):

$$V_{R1} = (24 \text{ V}) = \frac{5 \text{ k}\Omega}{5 \text{ k}\Omega + 8 \text{ k}\Omega} = \mathbf{9.2 \text{ V}}$$

(b) From Eq. (6-2),

$$I_{C1} = \frac{1}{2.5\text{ k}\Omega}\left[\frac{(24\text{ V})(5\text{ k}\Omega)}{5\text{ k}\Omega + 8\text{ k}\Omega} - 0.7\text{ V}\right] = \mathbf{3.4\text{ mA}}$$

(c) From Eq. (6-3),

$$\frac{\Delta V_C}{\Delta t} = \frac{I_C}{C} = \frac{3.4 \times 10^{-3}}{0.5 \times 10^{-6}} = 6.8 \times 10^3\text{ V/s} = \mathbf{6.8\text{ V/ms}}$$

(d) The capacitor must charge to V_P of the UJT, which is given by

$$V_P = \eta V_{B2B1} + 0.6\text{ V} = (0.58)(24\text{ V}) + 0.6\text{ V} = 14.5\text{ V}$$

The time required to charge that far can be found from

$$t = \frac{V_P}{\Delta v/\Delta t} = \frac{14.5\text{ V}}{6.8\text{ V/ms}} = \mathbf{2.1\text{ ms}}$$

(e) Let θ stand for the firing delay angle. Since 360° represents one cycle period and the period of a 60-Hz supply is 16.67 ms, we can set up the proportion

$$\frac{\theta}{2.1\text{ ms}} = \frac{360°}{16.67\text{ ms}}$$

$$\boldsymbol{\theta = 45°}$$

(f) For a firing delay angle of 120°, the time between zero crossover and firing is given by the proportion

$$\frac{t}{120°} = \frac{16.67\text{ ms}}{360°}$$

$$t = 5.55\text{ ms}$$

The peak point of the UJT is still 14.5 V, so to delay the firing for 5.55 ms, the rate of voltage buildup must be

$$\frac{\Delta v}{\Delta t} = \frac{14.5\text{ V}}{5.55\text{ ms}} = 2.6\text{ V/ms}$$

Applying Eq. (6-3), we get a charging current of

$$\frac{I_{C1}}{C_1} = \frac{\Delta v}{\Delta t}$$

$$I_{C1} = \frac{2.6\text{ V}}{1 \times 10^{-3}\text{ s}}(0.5 \times 10^{-6}) = 1.3\text{ mA}$$

From Eq. (6-2), we can find R_F:

$$I_{C1} = 1.3\text{ mA} = \frac{1}{2.5\text{ k}\Omega}\left[\frac{(24\text{ V})(5\text{ k}\Omega)}{(5\text{ k}\Omega + R_F)} - 0.7\text{ V}\right]$$

Manipulating this equation and solving for R_F, we obtain

$$R_F = \mathbf{25\text{ k}\Omega}$$

Therefore if the feedback resistance were increased to 25 kΩ, the firing delay angle would be increased to 120°, and the load current would be reduced accordingly. ■

6-9-2 UJT Trigger Circuit with Voltage Feedback

As stated earlier, UJTs are also compatible with voltage feedback circuits. Mentally remove R_F from Fig. 6–11(a) and replace it with the *npn* transistor circuit shown in Fig. 6–11(f). Now the variable feedback voltage V_F controls the firing delay angle of the triac. Quantitatively, here is how it works. Applying Ohm's law to the base-emitter circuit of Q_2, we obtain

$$V_F = (I_{E2})R_3 + 0.7 \text{ V}$$

in which I_{E2} stands for the emitter current in transistor Q_2. Since the collector current is almost equal to the emitter current for a high-gain transistor, this equation can be written as

$$I_{C2} = \frac{V_F - 0.7 \text{ V}}{R_3}$$

I_{C2}, the Q_2 collector current, is the same as the current through R_1 if we neglect the Q_1 base current. Therefore V_{R1}, which drives Q_1, is determined by I_{C2}. That is,

$$V_{R1} = (I_{C2})R_1 \qquad \textbf{(6-4)}$$

$$V_{R1} = \frac{R_1}{R_3}(V_F - 0.7 \text{ V})$$

From this point on, the circuit action is identical to that in the resistive feedback circuit. The greater V_{R1}, the faster the capacitor charging rate and the earlier the UJT and triac fire. The smaller V_{R1}, the slower the charging rate and the later the UJT and triac fire.

Notice that Q_2 has a rather large resistance in its emitter lead. This provides a high input impedance to the V_F source, resulting in easy loading on the feedback voltage source. Notice also that the feedback voltage source is electrically isolated from the main ac supply lines by virtue of transformers T_1 and T_2, which completely isolate the trigger circuitry.

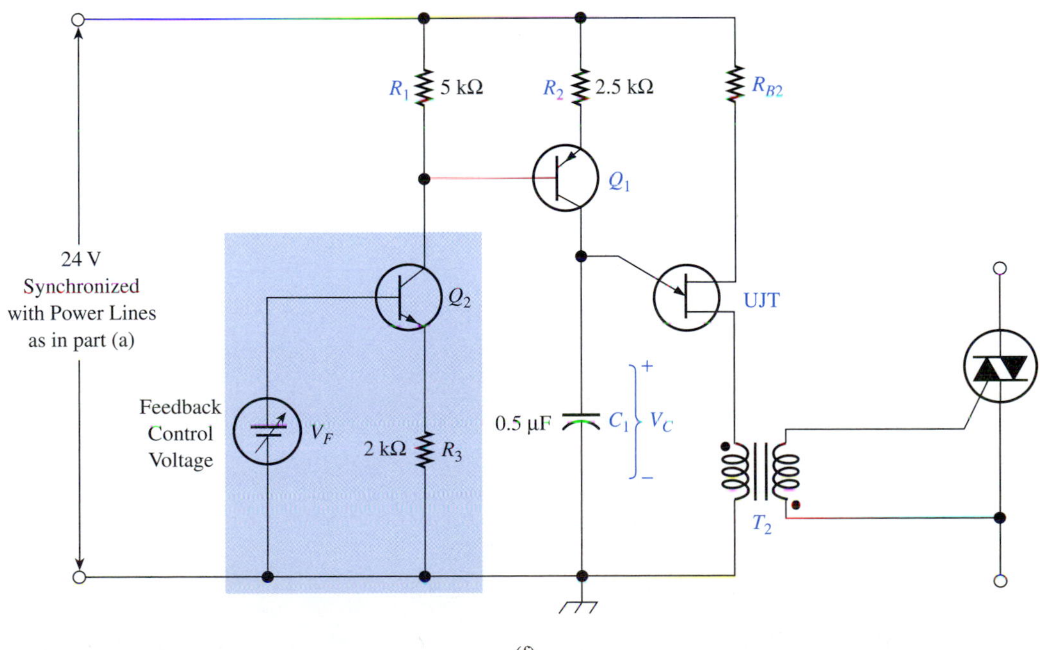

FIGURE 6–11 *(continued)*
(f) Load voltage waveform.

The Q_2 transistor beta and temperature are made unimportant by normal emitter-follower action. Namely, if Q_2 tries to conduct too hard, the voltage developed across R_3 will rise a little and will throttle back the Q_2 base current. This offsets any tendency of the transistor itself to carry a greater than normal collector current. Conversely, if Q_2 starts loafing (not conducting hard enough), the voltage developed across R_3 will drop a little and admit extra base current to offset the tendency of Q_2 to loaf. In the end, the transistor will behave in such a way that Ohm's law, Eq. (6-4) is obeyed.

■ EXAMPLE 6-4

For a voltage feedback situation as shown in Fig. 6–11(f), assume that $\beta_2 = 200$ and $R_3 = 2\ \text{k}\Omega$. All the other trigger circuit component sizes are the same as in Example 6-3.

(a) If the range of feedback voltage is from 2.0 to 7.0 V dc, what is the control range of the firing delay angle?

(b) What is the maximum current drawn from the feedback source?

Solution. (a) For $V_F = 1.8$ V, Eq. (6-4) yields

$$V_{R1} = \frac{5\ \text{k}\Omega}{2\ \text{k}\Omega}(2.0\ \text{V} - 0.7\ \text{V}) = 3.3\ \text{V}$$

Equation (6-2) still applies in the voltage feedback situation, so

$$I_{C1} = \frac{V_{R1} - 0.7\ \text{V}}{R_2} = \frac{3.3\ \text{V} - 0.7\ \text{V}}{2.5\ \text{k}\Omega}$$
$$= 1.0\ \text{mA}$$

Equation (6-3) gives

$$\frac{\Delta v}{\Delta t} = \frac{I_{C1}}{C_1} = \frac{1.0\ \text{mA}}{0.5\ \mu\text{F}} = 2.0\ \text{V/ms}$$

The time to charge to V_P is

$$t = \frac{14.5\ \text{V}}{2.0\ \text{V/ms}} = 7.3\ \text{ms}$$

A 7.3-ms delay figures out to a firing delay angle of

$$\frac{\theta}{7.3\ \text{ms}} = \frac{360°}{16.67\ \text{ms}}$$
$$\theta = 158°$$

This is the latest the triac can fire.

When $V_F = 7.0$ V,

$$V_{R1} = \frac{5\ \text{k}\Omega}{2\ \text{k}\Omega}(7.0\ \text{V} - 0.7\ \text{V}) = 15.8\ \text{V}$$

$$I_{C1} = \frac{V_{R1} - 0.7\ \text{V}}{R_2} = \frac{15.8\ \text{V} - 0.7\ \text{V}}{2.5\ \text{k}\Omega} = 6.0\ \text{mA}$$

The slope of the voltage ramp is then

$$\frac{\Delta v}{\Delta t} = \frac{I_{C1}}{C_1} = \frac{6.0\ \text{mA}}{0.5\ \mu\text{F}} = 12.0\ \text{V/ms}$$

The time to fire is

$$t = \frac{V_P}{\Delta v/\Delta t} = \frac{14.5\ \text{V}}{12.0\ \text{V/ms}} = 1.2\ \text{ms}$$

This figures out to a delay angle of

$$\frac{\theta}{1.2\ \text{ms}} = \frac{360^\circ}{16.67\ \text{ms}}$$

$$\theta = 26^\circ$$

The control range of firing delay angle of the UJT and triac is thus **26° to 158°.**

(b) The maximum current draw from the feedback source will occur when $V_F = 7$ V. It can be found by applying Ohm's law to the input of Q_2. First we must find the input impedance (resistance) of Q_2.

Recall from electronic fundamentals that the input resistance of an emitter-follower is given approximately by

$$R_{\text{in}} = \beta R_E$$

where β is the current gain of the transistor and R_E is the resistance in the emitter lead. In this case,

$$R_{\text{in}} = (200)(2\ \text{k}\Omega) = 400\ \text{k}\Omega$$

Therefore the maximum current draw from the V_F source is given by

$$I_{\text{max}} = \frac{V_{F\text{max}} - 0.7\ \text{V}}{R_{\text{in}}} = \frac{7\ \text{V} - 0.7\ \text{V}}{400\ \text{k}\Omega}$$

$$= \mathbf{16\ \mu A}$$

TROUBLESHOOTING ON THE JOB

REMOVAL OF ASH PARTICULATES FROM A COAL-FIRED ELECTRIC UTILITY STACK

Coal is the most abundant fossil fuel on the earth. It has been estimated that the coal reserves of North America contain perhaps 10 times as much energy as the oil reserves of the Persain Gulf region. The problems with coal are that it's expensive to dig and it's dirtier than oil.

Burning coal for electric power generation affects the environment in several ways: (1) It produces solid ash particulates that do not belong in the atmosphere or on fertile farmland. (2) It produces sulfure dioxide (SO_2) gas, which is the cause of acid rain. (3) It produces carbon dioxide (CO_2) gas, which contributes to the greenhouse effect.

While we're waiting for the nuclear fusion people to get their act together, the coal-burning industry is making big improvements in the first two areas: ash particulates and SO_2 removal. We'll describe the ash process now and talk about SO_2 in the Troubleshooting on the Job assignment of chapter 12.

Solid ash particles are removed by passing the combustion products through an electrostatic precipitator. This is a group of large-area metal collecting plates with charged metal-rods between them. The rods are sometimes called *discharge electrodes.*

All the collecting plates are electrically connected together, and all the rods are also electrically connected together as shown in Fig. 6–12. Very high voltage, about 40-kV dc with 50-kV p-p ac pulses superimposed, is applied between the plates and the rods, negative on the rods in this industry. As the ash particles pass over the rods, they themselves become negatively charged. They are then attracted to and captured by the positive collecting plates. Periodically, the plates must be mechanically rapped to knock the accumulated ash particles off their surface into the collection hopper below. This is accomplished by electromagnetic spring loaded solenoids. Over time, some ash particles also cling to the rods, so they must be rapped occasionally too Figure 6–12 shows one plate rapper and one rod rapper, but really there are dozens of each.

A state of the art "intelligent" precipitator can capture 99.98% of the ash in the exhaust-flow stream. It can do such a good job because it is equipped with advanced electronic controls that sense the buildup of dirt on the plates and automatically adjust the applied voltage. Both the baseline dc value and the magnitude and frequency of the ac pulses are automatically varied as the ash particles accumulate. In general, the control circuit tries to adjust the voltages to the highest possible values without allowing severe arcing between the rods and plates. Prolonged arcing would damage the metal surfaces, as you know.

The electronic controls also adjust the rapping rate. For mild exhaust conditions the rapping rate can be as long as 100 minutes round-trip time. (The rappers are operated sequentially, not all at once.) For very dirty conditions, rapping rate round-trip time can be as quick as 1 minute.

In Fig. 6–12, the variable ac to the high-voltage bridge is produced by triac-switching of the primary winding of a step-up transformer. A simplified version of the circuit is shown in Fig. 6–13(a).

The Trigger Control circuit in that figure is built as shown in Fig. 6–11(f) with V_F, variable from about 6.4 V to about 4.4 V. That variation of V_F changes the triac's firing-delay angle from about 30° to about 45°.

With the triac firing at 30°, the situation is shown in the waveforms of Fig. 6–13(b). The initial secondary voltage bursts have a peak magnitude of about 80 kV, which then becomes the dc baseline voltage between plates and rods in Fig. 6–12.

When the triac firing is delayed to 45°, the situation is shown in the waveforms of Fig. 6–13(c). The later firing angle produces a greater primary transient voltage creating initial secondary bursts of about 40 kV. Thus the bridge rectifier and capacitor filter raise the dc baseline voltage to this higher value.

JOB ASSIGNMENT

The device that monitors the final ash concentration in the final exhaust gases at the top of the smokestack is indicating that too much ash is making its way through the precipitator of Fig. 6–12 and escaping capture. As the technician responsible for maintaining the ash-removal system, you must identify the problem and implement a permanent or temporary repair.

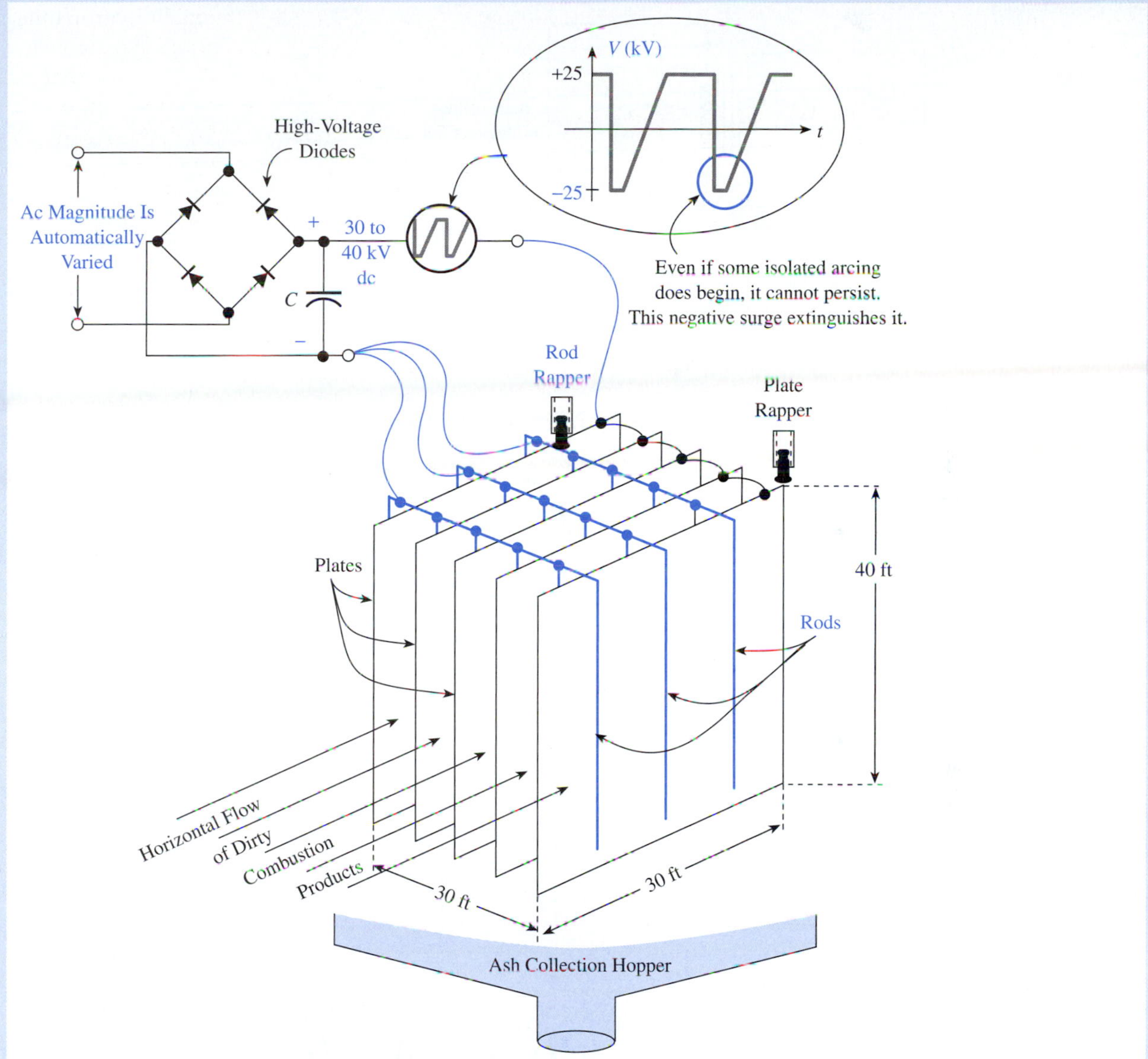

FIGURE 6–12

Electrostatic precipitator for removing ash from electric utility coal combustion.

The actual precipitator has 30 rod-groups and 30 connecting-wires joining them to the negative dc supply terminal not merely the 8 rod-groups shown in Fig. 6–12.

Your test instrumentation includes a very high-voltage dc voltmeter and a very high voltage ac voltmeter both with special safety probes. You also have a 10-A Dc ammeter and dual-trace oscilloscope with a special high-voltage divide by 100 probe.

A properly functioning precipitator of this size is known to have a maximum dc (average) current demand of 25 amperes. You measurement of the dc voltage ahead of the kilovolt ac supply in Fig. 6–12 gives a measurement of only 15 kV. A current measurement in the dc supply line ahead of the kilovolt ac supply causes the 10-A ammeter to be pegged on the right of the scale. Describe the resting/troubleshooting procedure that you will use to isolate the problem.

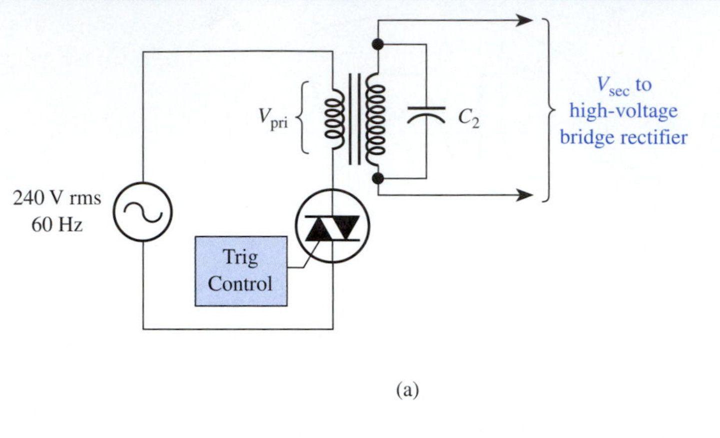

(a)

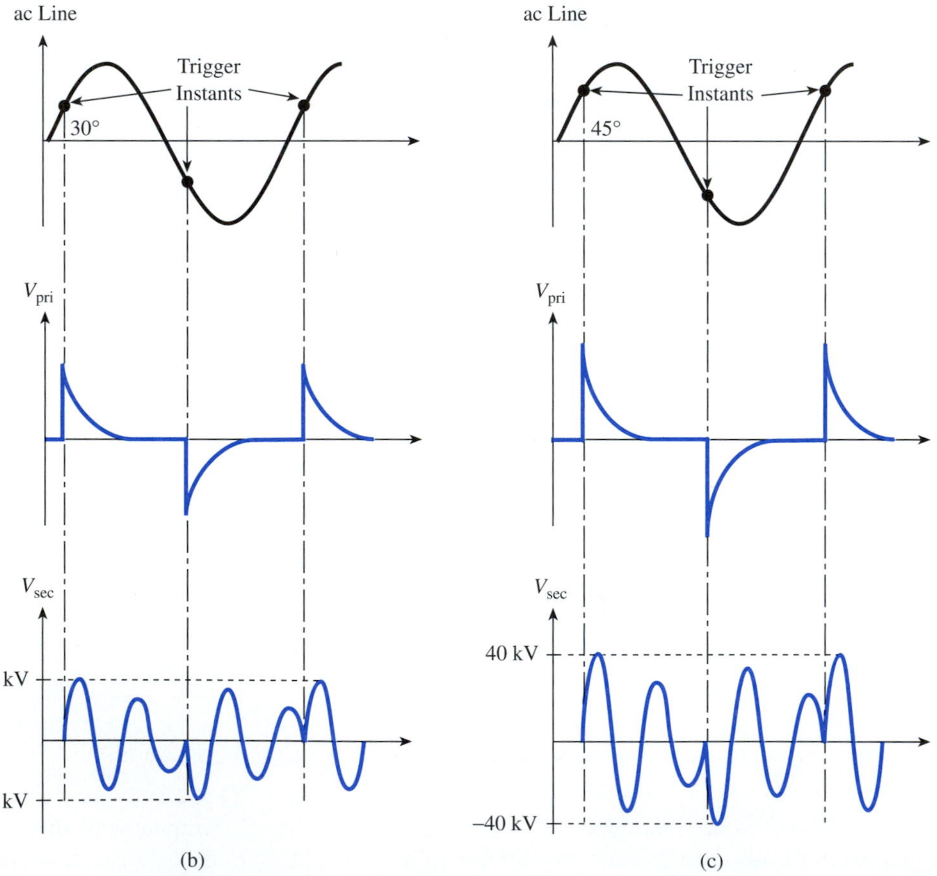

(b) (c)

FIGURE 6–13
Varying the dc kilovolt supply to the precipitator by varying the peak ac voltage to the bridge rectifier.

SUMMARY

- A triac is like a bidirectional SCR. It controls the average ac current to its series-connected load.
- The firing instants at which a triac switches from the blocking to conducting state are controlled by the gate trigger-control circuit, like an SCR.
- In general, triacs are nonsymmetrical between positive and negative half cycles.
- To trigger a triac into the ON state, its gate current must rise to the critical value, I_{GT}.
- A triac's temperature- and batch-instability can be counteracted by installing a bidirectional breakover device, like a diac or SBS, in its gate lead.
- A unilateral breakover device can also be used for triac triggering and is usually accompanied by a pulse transformer to couple the breakover burst to the gate. UJTs are often used for this purpose, especially in circuits with feedback from the load.

FORMULAS

$$V_{BO} = V_Z + 0.6 \text{ V}$$ (for an SUS or SBS with a zener diode at the gate)

$$\frac{\Delta v}{\Delta t} = \frac{I_C}{C}$$ (for a capacitor with constant de current) **(Equation 6-3)**

QUESTIONS AND PROBLEMS

Section 6-1

1. Draw the schematic symbol of a triac and identify all three terminals.
2. What polarity of main terminal voltage is called the *positive* polarity? What is the *negative* polarity?

Section 6-2

3. Does the firing delay angle of a triac during the positive half cycle necessarily equal the firing delay angle during the negative half cycle?

Section 6-3

4. Define V_{GT}. What range of values does it have for medium-sized triacs?
5. Repeat Question 4 for I_{GT}.
6. Repeat Question 4 for I_{HO}.
7. Define V_{DROM}. All else being equal, which would cost more, a triac with high V_{DROM} or a triac with low V_{DROM}?
8. Define V_{TM}. Which is considered better, a high V_{TM} or a low V_{TM}? Why?
9. A triac controls the average ac current through a load. A rheostat can do the same thing. In what ways is a triac superior to a rheostat?

Section 6-4

10. Are triacs inherently temperature stable? What happens to firing delay angle as temperature increases, assuming everything else is constant?

11. Draw the schematic symbol and sketch the characteristic curve of a diac. What other names are used for a diac?
12. Roughly, what is the maximum difference between the magnitudes of $+V_{BO}$ and $-V_{BO}$ for a diac?
13. Compare the two situations of using a 30-V diac in the gate lead of a triac versus triggering the gate directly from the charging circuit. Which of these two situations requires a shorter time constant in the charging circuit? Explain.
14. In Fig. 6–5(a), if the triac is not firing at all, explain why the capacitor starts every half cycle with a residual charge of the "wrong" polarity.
15. For the same circuit as in Question 14, if the triac is firing, does the capacitor still start every half cycle with a residual charge of the wrong polarity? Explain.
16. Try to explain the flash-on effect seen with the circuit of Fig. 6–5(a). Be sure to explain why the load power can be reduced by turning R_2 backward once the triac has just barely started firing.
17. In general, what is hysteresis? Can you think of any good examples of hysteresis in other areas of electricity and magnetism? Can you think of any examples in other nonelectrical areas?

Section 6-5

18. Define *breakback voltage* of a thyristor.
19. How large is the breakback voltage of SBS or an SUS, approximately?
20. Roughly speaking, how symmetrical are SBSs? When symmetry important?
21. What is the purpose of diode D_2 in Fig. 6–8(a)?

Section 6-6

22. If it were desired to alter the basic characteristic of an 8-V SUS so that the breakover voltage was 2.8 V, how could this be done?
23. If it were desired to alter the basic characteristic of an 40-V four-layer diode so that the breakover voltage was 16 V, how could this be done?
24. What is the difference between *breakover* and *breakdown* for an SUS? Is breakdown healthy for an SUS?
25. Which word, *breakover* or *breakdown,* would be appropriate to describe the action of a zener diode?

Section 6-7

26. List the four triggering modes of a triac. Which mode is usually avoided?
27. Why is it that we always deliver *negative* gate current to a triac whenever both main terminal polarities must be triggered by a single gate current polarity?

Section 6-8

28. Explain the meaning of the dv/dt rating of a triac. Do SCRs have a dv/dt rating, too?
29. What simple precaution do we take to protect power thyristors from dv/dt triggering? Give some typical component sizes.

Section 6-9

30. In Fig. 6–11(a), suppose $\eta = 0.55$, $R_1 = 10\ \text{k}\Omega$, $R_2 = 1.5\ \text{k}\Omega$, and $C_1 = 1\ \mu\text{F}$. What value of R_F will cause a firing delay angle of 90°? Make a graph of V_{C1} versus time for this situation.
31. If everything were to stay the same as in Question 30 except that a different UJT, having an $\eta = 0.75$, is substituted, what would be the new firing delay angle?
32. For the circuit of Fig. 6–11(a), suppose $R_1 = 50\ \text{k}\Omega$, $R_F = 100\ \text{k}\Omega$, $R_2 = 1.2\ \text{k}\Omega$, $\beta_1 = 100$, and $C_1 = 1\ \mu\text{F}$.
    - **a.** Prove that this is not a good design because the base current of Q_1 is not negligible compared to the voltage-divider current through R_1 and R_F.
    - **b.** Change the values of R_1 and R_F so that the firing delay is the same but the transistor base current is less than one tenth of the voltage-divider current.
33. For the circuit of Fig. 6–11(b), suppose $R_1 = 10\ \text{k}\Omega$, $R_2 = 1\ \text{k}\Omega$, $R_3 = 15\ \text{k}\Omega$, $C_1 = 0.7\ \mu\text{F}$, and $\eta = 0.60$.
    - **a.** What is the firing delay angle if $V_F = 3.2$ V?
    - **b.** What is the firing delay angle if $V_F = 8.8$ V?
34. For the circuit of Fig. 6–11(f), $R_1 = 18\ \text{k}\Omega$, $R_2 = 2.2\ \text{k}\Omega$, $R_3 = 25\ \text{k}\Omega$, and $\eta = 0.68$. The range of feedback voltages is from 3 to 12 V. Select a size of C_1 to give a firing delay angle range of about 120°, approximately centered on the 90° mark. In other words the firing delay angle is to vary from about 30° to about 150°. It will be impossible to get this exact range. Try to come as close as you can.
35. What essential feature sets thyristors apart from other semiconductors?
36. Elaborate on this statement, explaining in your own words why it is correct: "Although a triac cannot create continuous changes in *instantaneous* current, it can create continuous changes in *average* current."
37. With the tendency of circuit designers to use low-voltage, low-power circuits whenever possible, why are high-power triacs and SCRs even necessary? What is the justification for their existence?

■ SUGGESTED LABORATORY PROJECTS

Project 6-1: Controlling AC Current with a Triac

Procedure

1. Construct the circuit of Fig. 6–4(b). Use an isolated ac supply if possible. If it is not possible, take the steps described in Project 4-1.

   Use the following component sizes and ratings: $R_{\text{LD}} = 100\ \Omega$, 100-W resistor or 40- to 60-W light bulb; triac T2302B, or any triac with a V_{DROM} of at least 200 V, an $I_{T\,\text{rms}}$ of at least 3 A, and with similar gate characteristics. If the gate characteristics (V_{GT} and I_{GT}) are not similar to the T2302B, the triggering circuit components must be changed.

$R_1 = 10\ \text{k}\Omega$	$R_4 = 1\ \text{k}\Omega$
$R_2 = 250\text{-k}\Omega$ pot	$C_1 = 0.22\ \mu\text{F}$
$R_3 = 33\ \text{k}\Omega$	$C_2 = 0.22\ \mu\text{F}$

   - **a.** What is the range of adjustment of the firing delay angle? Are the firing delays equal for both half cycles?

b. Make a sketch showing the waveforms of V_{LD}, $V_{MT2-MT1}$, and V_G, all to the same time reference.
c. Measure I_{GT}, the current necessary to trigger the triac, for both main terminal polarities. This must be done by measuring the voltage across R_4 at the instant of firing and then applying Ohm's law to R_4. The scope must be connected to display the V_{R4} waveform. Compare the measured I_{GT} to the manufacturer's specifications, if you have them.
d. Measure V_{TM}, the voltage across the triac after firing. Compare to the manufacturer's specs.
e. Cool the triac with freeze-spray and note the effect on the firing delay angle. Does this make sense?
f. Investigate the effect of substituting different triacs of the same type number. Explain your results.

2. Insert a diac in the gate lead of the triac. Use a TI43A diac or any equivalent diac with about the same breakover voltage rating (32 V). Change the following component sizes:

$$R_2 = 200\text{-k}\Omega \text{ or } 250\text{-k}\Omega \text{ pot} \qquad C_1 = 0.1\ \mu\text{F}$$
$$R_3 = 4.7\ \text{k}\Omega \qquad C_2 = 0.02\ \mu\text{F}$$

Leave everything else the same.
a. Are the firing delay angles the same for both half cycles? Why?
b. Investigate the effects of cooling the triac. Explain the results.
c. Investigate the effects of substituting different triacs of the same type number. Explain your results.

Project 6-2: Diac and SBS Characteristics

Procedure

Construct the circuit shown in Fig. 6–14. It will furnish a display of the current-voltage characteristic curve of the devices being tested. Adjust the variable ac supply to zero before turning on the power every time a new test is made.

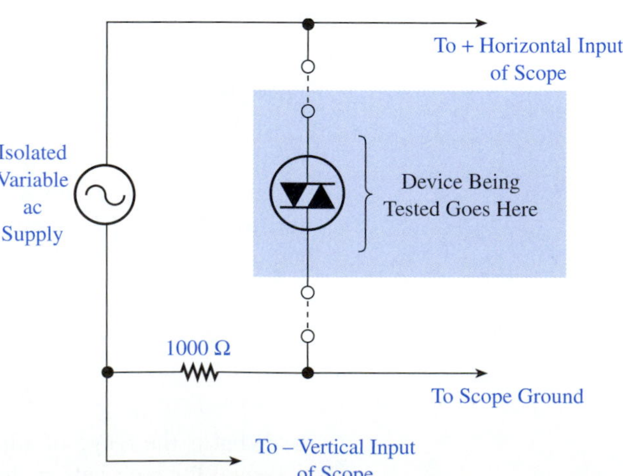

FIGURE 6–14 Circuit for displaying on an oscilloscope the current-voltage characteristic curve of a breakover device.

The horizontal signal applied to the scope represents the voltage applied to the breakover device. The vertical signal represents the device's current, because this signal is developed across the 1000-Ω resistor in series with the device, which carries the same current as the device. The current sensitivity of the oscilloscope is then given by Ohm's law,

$$\text{amperes/cm} = \frac{\text{volts/cm}}{1000} \text{ or 1 mA per 1 V}$$

1. Insert a diac and adjust the ac supply until the diac's breakover points can be seen.
   a. Measure $+V_{BO}$ and $-V_{BO}$. Are they nearly equal in magnitude?
   b. What is the breakback voltage?
   c. What is the holding current (the minimum current necessary to maintain conduction once it has begun)?
   d. Investigate the temperature stability of the diac. Make comments on your findings.
   e. Investigate the batch spread between different diacs of the same type. Comment on your findings.
2. Insert an SBS (MBS 4991 or equivalent) in place of the diac. Adjust the ac supply until the breakover is visible on the scope screen. Repeat steps **a** through **e** of part 1. Make a comparison between the results you obtained for the two devices.
3. Investigate the alteration in SBS characteristics which can be brought about by using the gate.
   a. Insert a low-voltage zener diode (V_Z less than 6 V) as described in Sec. 6-5-2 to alter $+V_{BO}$. Sketch the current-voltage curve obtained. Does this agree with what you expected?
   b. Repeat step a to alter $-V_{BO}$.
   c. Connect two zeners having different zener voltages so that both $+V_{BO}$ and $-V_{BO}$ are altered. Sketch the curve and explain.
   d. With the zeners removed, insert a 2.2-kΩ resistor between G and $A1$. Observe and sketch the new characteristic curve. Does it bear out what you know about SBSs? Repeat with the gate resistor connected to $A2$.

Project 6-3: 100-V rms Voltage Regulator

Note: A true rms voltmeter is needed to perform this project. Without such a voltmeter, the regulation of the circuit cannot be observed.

The circuit of Fig. 6–15 in an rms voltage regulator using resistive feedback. It will maintain the lamp voltage at 100 V rms in the face of the line voltage varying anywere from 110 to 250 V rms.

The amount of light given off by the lamp depends on the true rms value of the voltage across it, since the rms value tells the power-transferring capability of the voltage supply. Therefore if the amount of light given off by the lamp is held constant, the rms voltage across the lamp terminals is being held constant. The circuit of Fig. 6–15 tries to do just that, namely hold the light output constant.

The lamp can be a 100-W incandescent bulb, mounted in a box lined on the inside with aluminum foil. A hole is punched in the box, and a hollow cardboard tube is inserted in the hole. The tube can be the kind that kitchen paper towels are wrapped on. A photocell is mounted at the other end of the tube in such a way that it cannot detect ambient light. It responds only to the light reflected from the aluminum foil down the cardboard tube.

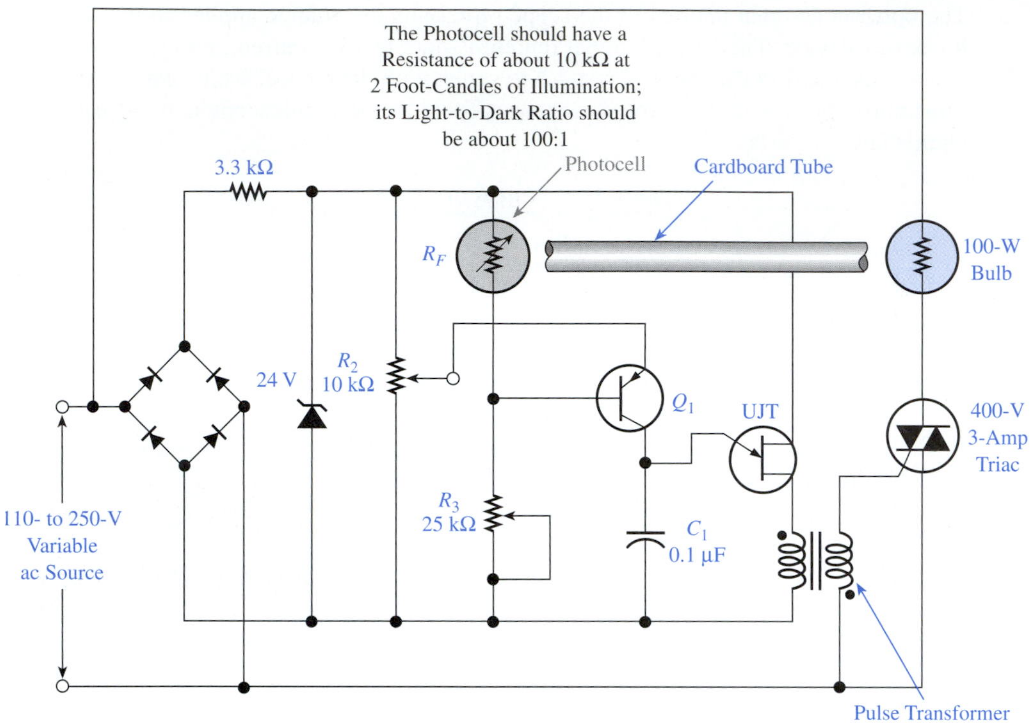

FIGURE 6–15
Circuit that will regulate the rms voltage applied to the lamp. The photocell provides resistive feedback to the triggering circuitry.
Courtesy of Motorola, Inc.

The photocell resistance is the feedback resistance in this circuit, R_F. Its position is different from the position shown in Fig. 6–11(a), which had R_F on the bottom. In this circuit R_F is on top. The proper location depends on whether R_F increases as load power increases or whether R_F decreases as load power increases. In this example, the resistance of a photocell *decreases* as load power increases (as light intensity increases), so R_F must be on top. If the opposite had been true, R_F would have been placed in the bottom position.

Here is how the circuit attempts to maintain a constant light output. If the line voltage increases, tending to increase the light output, more light impinges on the photocell, thereby lowering its resistance. As R_F decreases, the voltage developed across the photocell becomes a smaller portion of the 24-V dc supply voltage. This reduces the drive on transistor Q_1, thereby lowering the C_1 charging rate and causing the triac to fire later. The later firing compensates for the higher line voltage, and the rms lamp voltage increases by only a very small amount.

On the other hand, if the ac line voltage decreases, tending to lower the rms lamp voltage, the reduced light output will cause R_F to increase. This allows R_F to receive a larger share of the 24-V dc supply voltage. The Q_1 drive voltage therefore rises, causing C_1 to charge faster and the triac to fire earlier. The earlier firing cancels the decrease in ac line voltage, and the rms lamp voltage is held almost constant.

To test this circuit, connect a *true rms voltmeter* across the lamp. Adjust potentiometer R_2 so that the load voltage is 100 V rms when the ac line voltage is 120 V rms (line voltage can be measured on a standard peak-detecting voltmeter, since it is a sine wave). Then increase the line voltage to 250 V rms (if possible) and adjust R_3 so that the lamp voltage is still 100 V rms. You may have to go back and forth between these adjustments a few times, since the pots will interact with each other. If the ac line voltage cannot be raised all the way to 250 V rms, take it as high as it can go and then make the R_3 adjustment.

When the adjustments are complete, you should be able to vary the ac line voltage anywhere form 110 to 250 V rms with the lamp voltage maintained at 100 V rms ±2 V.

Do this in equal increments of line voltage and make a table showing ac line voltage, true rms lamp voltage, and firing delay angle (measured on an oscilloscope).

Note that when the firing delay angle is in the 90° neighborhood, it takes only a small change in firing delay to compensate for a given change in line voltage. Yet when the delay angle is far from 90°, either above or below, the circuit will produce a greater change in firing delay to compensate for the same change in line voltage. Can you explain this?

If you are mathematically inclined, you might find it interesting to integrate some of the load voltage waveforms. Look in a good engineering calculus book to see how to integrate these waveforms to solve for the rms values. This integration is not easy because you are not integrating the load voltage itself, but the *square* of the load voltage. The "s" in "rms" stands for *square*.

CHAPTER

7 AN INDUSTRIAL AUTOMATIC WELDING SYSTEM WITH DIGITAL CONTROL

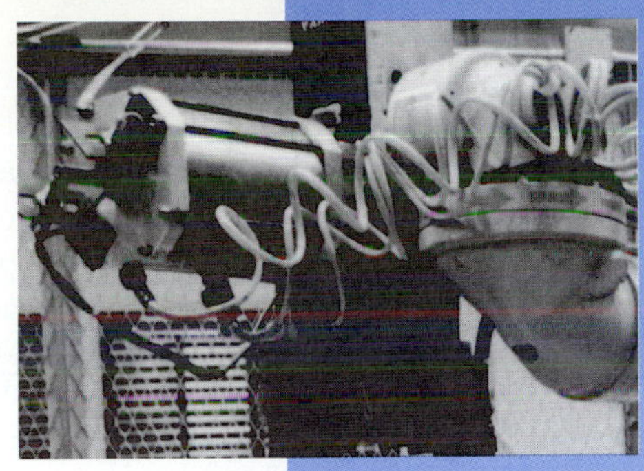

In this chapter we will explain the operation of an automatic welding system. The system presented is a slightly simplified version of a real automobile wheel-welding system capable of a production rate of 600 wheels per hour. Although system operation is explained in terms of wheel welding, the design of the system has much in common with other welding operations utilizing the basic industrial automatic welding sequence of (1) **Squeeze,** (2) **Weld,** (3) **Hold,** (4) **Release,** and (5) **Standby.** In this chapter those five words will be written in boldface whenever they refer to the specific intervals within the automatic welding sequence.

OBJECTIVES

After completing this chapter, you will be able to:

1. Explain in detail how a solid-state logic system receives information from its signal converters about conditions in the apparatus being controlled.
2. Explain how a solid-state logic system exerts control over automated machinery through its output amplifiers.
3. Explain how the system operator can set automatic cycle specifications on selector switches and how those specifications are entered into the memory devices of the logic system.
4. Explain how the logic system keeps track of the progress of the automated cycle, knowing which steps have been completed and which step is to come next.
5. Discuss the *block diagram* approach to complicated industrial systems, and explain the advantages of breaking up a complicated system into small subcircuits.
6. Interpret a block diagram, identifying which blocks interact with each other and which direction the interaction takes.
7. Explain in detail the often-seen industrial practice of shifting, or presetting, selector switch settings into a down-counter.
8. Discuss the common practice of using the ac power line oscillations to "time" the occurrence of events.
9. Explain in detail the action of a decoder circuit in converting a sequence of binary bits into useful form.
10. Explain in detail the action of a diode encoding matrix in converting selector switch settings into a form compatible with down-counters.
11. Explain the use of buffers (drivers) to prevent digital signal degeneration.
12. Name the four variables which are adjusted to produce the best possible weld from an automatic welder, and explain with waveform drawings the meaning of each variable.

13. Describe an ignitron, and point out its advantage over SCRs.
14. Show how an ignitron can be fired by an SCR.
15. Describe how welding current conduction angle is controlled by an ignitron-SCR-UJT control circuit.
16. Discuss the necessity of discharging a UJT's emitter capacitor during the negative half cycle in the control circuit in 15.
17. Describe the saturation problem of welding transformers. Using waveform drawings, show how the saturation problem can be solved.
18. Describe the operation of a delta-connected three-phase welding transformer, and explain why the conduction angle per phase must be limited to 120°.

7-1 ■ PHYSICAL DESCRIPTION OF THE WHEEL-WELDING SYSTEM

The arrangement of the wheel handling and lifting mechanism and the welding electrodes and their associated hydraulic controls is shown in Fig. 7–1. The relationship between the wheel *rim* and the wheel *spider* is shown in Fig. 7–1(a), which is a top view of a spider resting inside a rim. The *rim* of an automobile wheel is the circular outside part that the tire mounts on. The *spider* is the flange-like middle part, which contains the holes for the wheel studs and the center hole for the axle-bearing cap. The spider is welded to the rim to form a complete wheel. The spider is so named because it is welded to the rim in *eight* places. Figure 7–1(a) shows that the spider has four flaps, which are the protuberances that rest on the inside lip of the rim. Each flap is spot-welded to the rim in two places, making eight welds in all.

Figure 7–1(b) shows a side view of a spider-rim combination resting in the *lifting cradle.* The spider could not really be seen through the solid metal of the wheel rim, even though the spider is seen in this figure. The lifting cradle is positioned underneath the hydraulic cylinders which engage and disengage the welding electrodes. The lifting cradle is raised and lowered by the *lift cylinder.* When the lift cylinder extends, it raises the lifting cradle, thus positioning the wheel rim and spider so that the welding electrodes can engage and produce a weld. When the weld is complete and the electrodes have disengaged, the lift cylinder retracts, lowering the lift cradle and wheel.

There is a roof-surface above the welding electrode cylinders which holds the wheel in perfect alignment for the electrode cylinders, but for the sake of simplicity this is not shown in Fig. 7–1(b). Likewise, the two limit switches which detect when the wheel rim is perfectly positioned against the roof-surface have been deleted from Fig. 7–1(b) to keep the drawing uncluttered.

When the lifting cradle has been raised and the wheel is in position, the hydraulic line marked ENGAGE ELECTRODES is pressured up, causing all four electrode cylinders to extend. This brings the electrodes into electrical contact with the wheel rim on the outside and with the spider on the inside. Hydraulic pressure switch PS1 is set to detect when the motion of the cylinders has stopped, which means that the electrodes are pressing against the metal surfaces to be welded. Figure 7–1(b) shows only two pairs of welding electrodes. As stated before, the spider is welded to the rim in eight separate places, so there are really eight pairs of electrodes. Only two pairs are shown in order to keep the drawing simple.

After the welds have been made, the ENGAGE ELECTRODES hydraulic line is relieved and the DISENGAGE ELECTRODES hydraulic line is pressured up. This causes the electrode cylinders to retract, disengaging the electrodes. The lift cylinder then retracts, lowering the finished wheel.

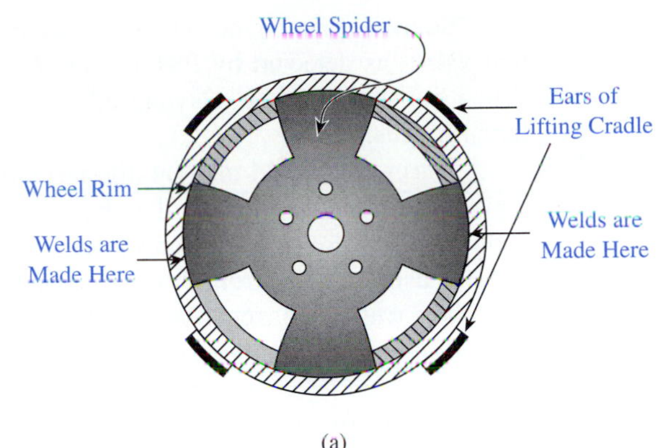

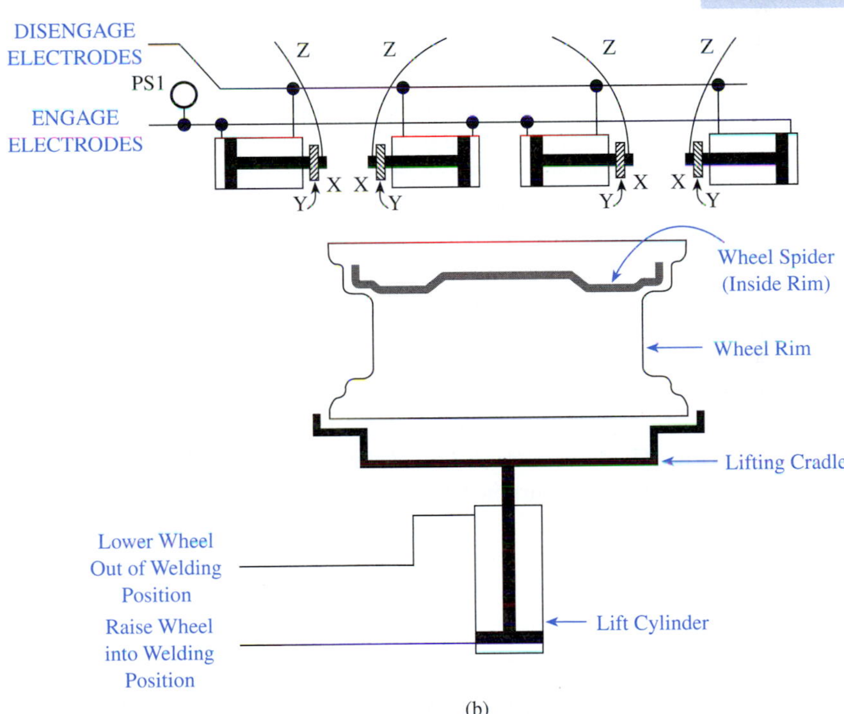

FIGURE 7–1
(a) Top view of a wheel spider resting on the inner lip of a wheel rim. The ears of the lifting cradle restrain the wheel rim from sliding horizontally. (b) Side view showing the physical layout of the welding mechanism. The notes indicate the functions of the various hydraulic lines.

7-2 ■ SEQUENCE OF OPERATIONS IN MAKING A WELD

When the wheel is in position to be welded, the welding electrodes come forward to engage the metal, as explained in Sec. 7-1. Once the electrodes have engaged the metal, they are allowed to press against the surfaces for a short time before the welding current is turned on. This is done to allow the electrodes to conform to the curvature of the surfaces and to make perfect electrical contact. This portion of the overall welding sequence is called the **Squeeze** interval. The time allotted for this interval in the welding sequence is called *squeeze time,* and it can be adjusted by the system operator.

Squeeze time begins when the electrode cylinder hydraulic pressure is up to its normal value, as detected by PS1 in Fig. 7–1(b). Squeeze time is usually about 1 second. When the squeeze time has elapsed, the **Squeeze** interval is complete and the **Weld** interval begins.

During the **Weld** interval, the *welding transformer* (not shown in Fig. 7–1b) is energized. Current flows down the electrode power leads to the electrodes and through the metal-to-metal contact between wheel rim and spider, thereby creating the weld. The **Weld** interval usually takes from 2 to 10 seconds.

The welding current does not flow continuously during the **Weld** interval. It is turned on and off in short bursts, called *pulsations.* The operator sets the number of pulsations which are used to create the weld.

Besides the number of pulsations, the number of cycles of current which flow during a single pulsation is also adjusted by the system operator, as is the number of cycles "missed" between pulsations. Figure 7–2(a) shows a graph of current versus time during the **Weld** interval, assuming that the welding current flows during the entire 180° of an ac half cycle.

In Fig. 7–2(a) it can be seen that welding current flows for three ac cycles. This is followed by an absence of current for two cycles. At the end of these two cycles, the current is turned on for another three cycles. Each time three cycles of current are completed, the system is said to have completed one current *pulsation.*

The example given here shows three cycles of current flow followed by two cycles of current absence. It should be understood that these numbers are adjustable. The operator could select five cycles of current flow followed by three cycles of current absence, or eight cycles of flow followed by two cycles of absence, etc.

The portions of the **Weld** interval during which welding current is flowing are called the **Heat** subintervals. The portions of the **Weld** interval during which the current is absent are called the **Cool** subintervals. The number of cycles in the **Heat** and **Cool** subintervals are adjusted to suit the alloy type and the gauge of the metal.

We have seen that the operator adjusts the number of pulsations in the **Weld** interval, the number of ac cycles in the **Heat** subinterval, and the number of ac cycles in the **Cool** subinterval. In addition to these variables, the number of degrees per half cycle that welding current flows is also adjustable. This number of degrees per half cycle during

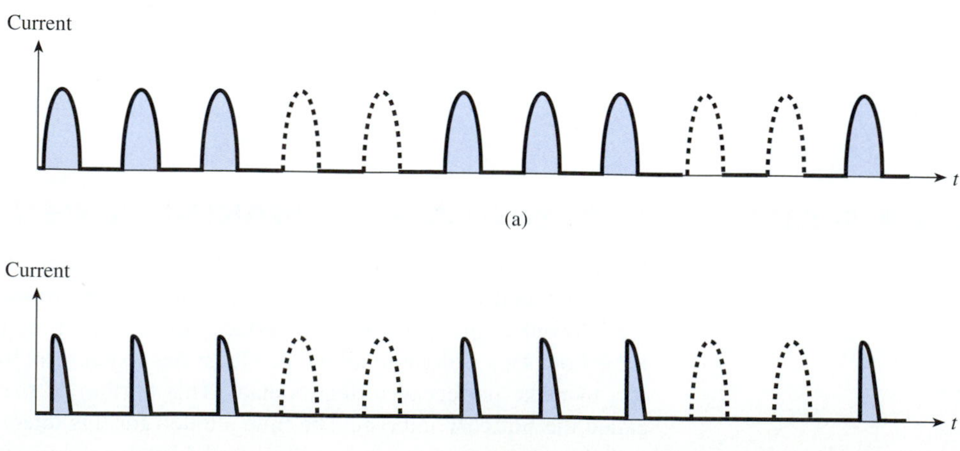

FIGURE 7–2
Waveforms of welding current pulsations. (a) Three cycles of rectified current flow, followed by two cycles of current absence. This means that the **Heat** subinterval is set to three cycles and the **Cool** subinterval is set to two cycles. During the **Heat** subinterval, current flows for the entire 180° of a positive half cycle. (b) The same situation, except that current flows for only 90° of a positive half cycle.

which current actually flows is called the *conduction angle.* Figure 7–2(b) shows a welding current waveform in which the conduction angle is approximately 90°. Alloy type and gauge of the metal and the type of electrode material are taken into account when setting the conduction angle.

Much research has been done to determine the best combination of these four variables for each different welding situation. Each variable has some effect on the final quality of the weld. For every welding situation encountered, the system operators refer to researched tables of settings for each of these four variables (pulsations per **Weld** interval, cycles per **Heat** subinterval, cycles per **Cool** subinterval, and conduction angle). In this way the system produces welds of the highest possible quality. Because all the variables are held perfectly consistent from one weld to another, the weld strength is also perfectly consistent. Of course, consistency is the benefit that arises from any automated machinery.

When the proper number of current pulsations have been delivered, the system leaves the **Weld** interval and enters the **Hold** interval. During the **Hold** interval, electrode pressure is maintained on the metal surfaces, but the welding current is turned off. The purpose of the **Hold** interval is to allow the fused metal of the weld to harden before the mechanical force exerted by the electrodes is removed from the wheel. This prevents any distortion of the wheel while the welded metal is in its molten state.

At the end of the **Hold** interval, which usually lasts about 1 second, the system enters the **Release** interval. During the **Release** interval, the welding electrode cylinders are retracted, releasing the wheel from the electrodes.

When the **Release** interval is complete, the system enters the **Standby** interval, during which the lift cylinder is retracted, lowering the finished wheel from the welding location. Once down, the wheel is removed from the lifting cradle. The system remains in **Standby** until a new wheel rim and spider are loaded onto the lifting cradle, and the lift cylinder is once again given the signal to extend.

To recap and summarize, the entire welding sequence consists of five intervals. In order of occurrence, they are **Standby, Squeeze, Weld, Hold, Release,** and return to **Standby.**

Once entered, the **Standby** interval is maintained until a new wheel is loaded onto the lifting cradle and the signal is given to lift it up.

After the **Standby** interval comes the **Squeeze** interval, during which the welding electrodes make contact with the wheel and press against it. The amount of time spent in this interval is determined by how long it takes the welding electrode cylinders to extend and make firm contact, followed by how much squeeze time the system operator has set. The squeeze time is set by positioning two 10-position selector switches. The circuit details of these selector switches and associated circuits are explained in Sec. 7-6. At this point it is only necessary to realize that the settings of the selector switches determine how many time increments must elapse to complete the **Squeeze** interval. One time increment in this system is the period of the ac line, namely $^1/_{60}$ second (16.67 ms). Therefore the number set on the 10-position selector switches can be thought of as being the number of cycles the ac line must make in order to complete the **Squeeze** interval.

The **Squeeze** interval is followed by the **Weld** interval. The system stays in the **Weld** interval until the proper number of welding current pulsations are delivered to the electrodes. This number of current pulsations is set by positioning two other 10-position selector switches. The number of cycles *per* pulsation (the number of cycles of the **Heat** subinterval) is also set on yet another pair of 10-position selector switches. The same is true for the number of cycles *between* pulsations (the number of cycles of the **Cool** subinterval); this number is also set on a pair of 10-position selector switches.

When the desired number of welding current pulsations has been delivered, the system leaves **Weld** and enters the **Hold** interval. The time spent in the **Hold** interval is again set on a pair of 10-position selector switches. The counting of the ac cycles begins immediately when the **Hold** interval is entered. In the **Squeeze** interval, by contrast, there is a considerable delay before the count of ac cycles begins; this delay occurs as the system waits for the welding electrode cylinders to stroke and for the hydraulic pressure on the ENGAGE ELECTRODES line to reach normal. There is no delay at all in the **Hold** interval.

When the **Hold** interval has counted out (interval has been completed), the system enters the **Release** interval. Timing of the **Release** interval again is set by two 10-position selector switches. The number is set to allow adequate time for the electrode cylinders to retract, freeing the wheel from contact with the welding electrodes. When the **Release** interval has counted out, the system enters **Standby,** which causes the cradle containing the finished wheel to be lowered. This completes one welding sequence. An entire sequence takes from 6 to 15 seconds.

7-3 ■ BLOCK DIAGRAM OF THE SEQUENCE CONTROL CIRCUIT

Figure 7–12 at the end of this chapter shows the complete logic circuit for controlling the wheel-welding sequence. Such large diagrams are difficult to understand all in one chunk. Instead, with complex circuits of this type, it is better to break them up into several small parts (subcircuits). We can then concentrate on these small subcircuits one at a time without being overwhelmed by the entire circuit.

7-3-1 A Complex System Broken Up into Small Subcircuits or Blocks—Explanation of the Block Diagram Approach

In Fig. 7–3, the complete circuitry for controlling the welding sequence has been broken down into subcircuits, with each subcircuit identified by a block. The circuitry contained within a single block has a certain single purpose which contributes to the overall operation of the system. Before studying the specific circuitry of any individual block, it is helpful to understand how each block fits into the overall control scheme. In this section we will try to gain an understanding of the purpose of each block and an understanding of how each block interacts with the other blocks. In the subsequent sections of this chapter we will make a careful study of the specific circuitry of each block.

In further discussion of the welding system, a distinction is made between the terms *welding sequence* and ***Weld** interval.* In fact we have already started making a distinction between them. *Welding sequence* refers to the complete sequence of actions necessary to weld a wheel, from **Standby,** through each interval of the sequence, back into **Standby** again. ***Weld** interval,* on the other hand, refers to that portion (interval) of the sequence during which welding current pulsations are being delivered to the wheel surfaces.

In Fig. 7–3 there are nine blocks. Each block has been assigned a short name which roughly describes its function. Each block has also been assigned a letter, A–I.

The lines between blocks show that there is direct interaction between those blocks or, more precisely, that there is wiring running between the circuitry of those blocks. The arrowhead denotes the direction of flow of information, from signal *sender* to signal *receiver.* In this way the line running out of block A into block B shows that there is interaction between block A and block B, and furthermore that block A is sending the information and block B is receiving the information.

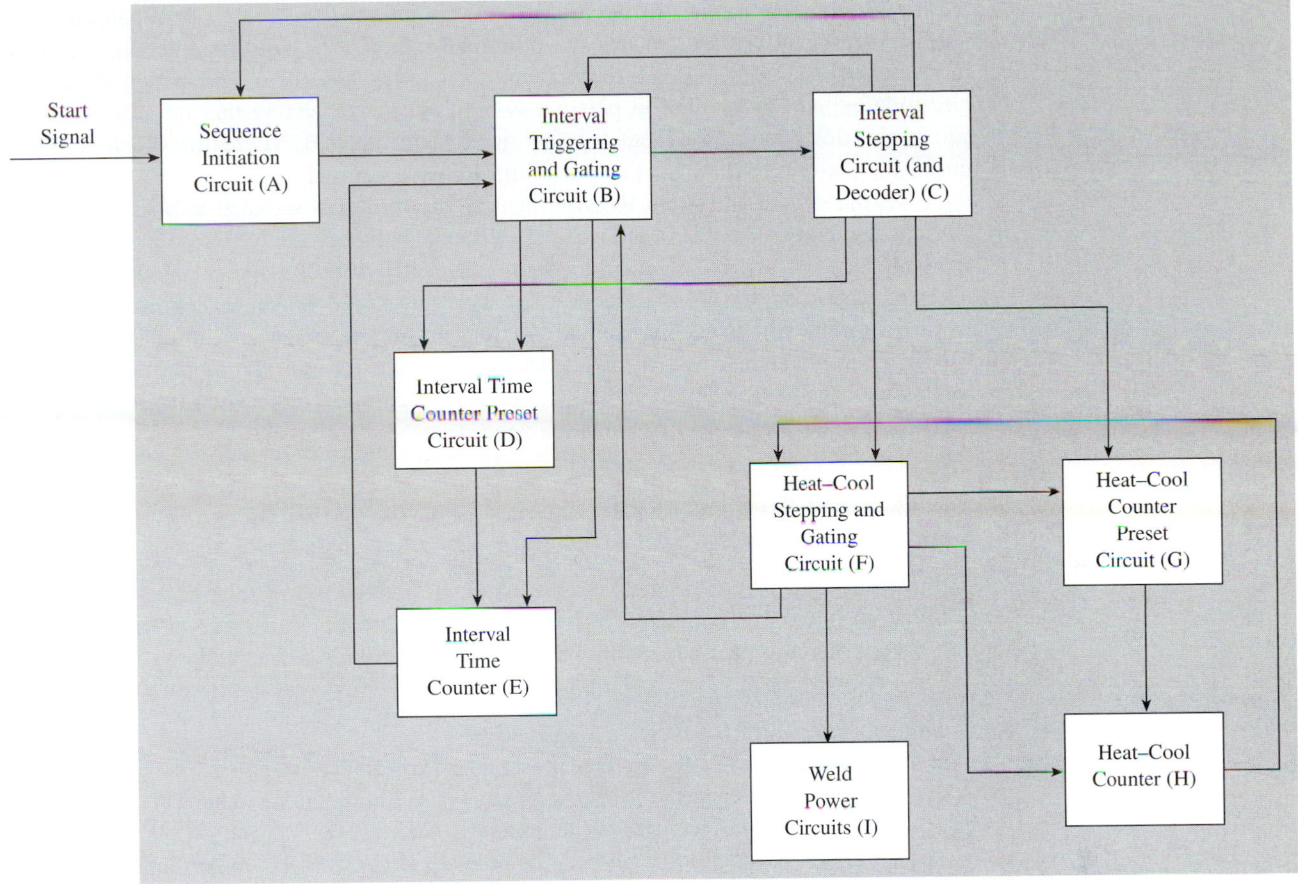

FIGURE 7–3
Block diagram of the entire wheel-welding system. Each block contains a subcircuit which has a certain specific purpose within the overall control scheme.

The fact that Fig. 7–3 shows only one *line* running from one block to another should not be interpreted to mean that there is only one *wire* running between the circuits in the actual wiring. There may be many wires running between those circuits. Figure 7–3 symbolizes the flow of information only; it is not an exact wiring diagram.

In the discussion which follows, a block connection line will be identified by two letters which show the two blocks being connected. The first letter denotes the sending block and the second letter the receiving block. For example, line AB would be the line that goes from block A to block B. Line BE is the line that goes from block B to block E. Line EB is the line that goes from block E to block B. Note that a given block may send information to another block as well as receive information from that block.

7-3-2 How the Sequence Initiation Circuit (Block A) Fits into the Overall System

Block A, the sequence initiation circuit, has the function of lifting a new wheel into position for welding, and detecting when the wheel is properly positioned. It signals this to the interval triggering and gating circuit on line AB, allowing block B to step the system out of **Standby** and into **Squeeze.** Just how block B performs this action is explained in the description of the interval triggering and gating circuit in Sec. 7-3-3. When block A receives the signal via line CA that the system has entered the **Squeeze** interval, it causes the welding electrodes to come forward and engage the wheel rim and spider. When the electrode cylinder hydraulic oil pressure is high enough, meaning that the squeeze time can begin, this condition is detected and signaled to block B. Thereafter block A drops out of the picture until the system enters the **Release** interval.

As the system enters the **Release** interval, that action is signaled to block A on line CA, causing block A to retract the welding electrode cylinders and to release the wheel. After the **Release** interval is over and the system reenters the **Standby** interval, block A receives that information on line CA. At that time it lowers the cradle containing the finished wheel.

In summary, the sequence initiation circuit has the responsibility to raise and lower the wheel and to engage and disengage the welding electrodes at the proper times. It also sends out signals telling the interval triggering and gating circuit when the system is to enter the **Squeeze** interval and when the squeeze time is to begin.

7-3-3 How the Interval Triggering and Gating Circuit (Block B) Fits into the Overall System

The interval triggering and gating circuit, block B, has the job of receiving the information that an interval is complete from the interval time counter. It receives this information on line EB and then acts on that information. It provides these actions:

1. It triggers the interval stepping circuit on line BC, causing that circuit to step into the next interval of the welding sequence.
2. It signals the interval time counter preset circuit on line BD, telling it to preset the interval time counter to the proper numbers. These numbers have been selected on 10-position selector switches, as mentioned in Sec. 7-2.
3. After the first two actions have been completed, it *gates* 60-Hz pulses into the interval time counter to begin the actual timing of the new interval. This is done on line BE. The term *gates* means "allows to pass." The gating is done by a logic gate.

The **Weld** interval is the only exception to action 3. During the **Weld** interval, the count pulses which are passed to the interval time counter do not occur at 60 Hz; instead one count is delivered each time a pulsation of welding current is finished. Note that there is a line running from block F to block B. During the **Weld** interval, line FB delivers a count pulse each time a welding current pulsation is finished. Block B then gates these count pulses into the interval time counter, instead of the usual 60-Hz pulses.

7-3-4 How the Interval Stepping Circuit (Block C) Fits into the Overall System

The interval stepping circuit receives a pulse on line BC each time block B learns that an interval is complete. This pulse causes the circuitry of block C to advance one step. This advancing is done by triggering flip-flops, as will be described in detail in Sec. 7-5. Thus the interval stepping circuit is composed of flip-flops, and the states of the flip-flops indicate which interval the system is in at any point in time. The information about which interval the system is currently in is important to several other subcircuits in the system, as can be seen from Fig. 7–3. The block diagram shows lines leading from block C to blocks A, B, D, F, and G, indicating that information regarding the system's interval is sent to all those blocks. The reason each subcircuit needs to know which interval the system is currently in will become clear when we study the specific circuitry of each block.

The interval stepping circuit is accompanied by a decoder, as the label in block B says. The decoder converts the states of the flip-flops into a useful signal for routing around to the various subcircuit blocks. The decoder has one output line for each interval, five output lines altogether.

For each example, if the flip-flops in the interval stepping circuit are indicating that the system is currently in the **Hold** interval, then the **Hold** output line of the decoder would go HI, while the **Standby, Squeeze, Weld,** and **Release** output lines would all be LO. By detecting which one of the five decoder output lines is HI, the other subcircuits can then tell which of the five intervals the system is currently in.

7-3-5 How the Interval Time Counter Preset Circuit (Block D) Fits into the Overall System

Block D, the interval time counter preset circuit, has the job of setting the correct two-digit number into the interval time counter. It does this immediately after the system enters a new interval. As mentioned in Sec. 7-2, each interval except **Standby** has two 10-position selector switches associated with it, on which the system operator selects the desired number of time increments (ac line cycles) for that interval. The **Weld** interval is different in this regard, as stated before.

The interval time counter preset circuit decides which pair of 10-position selector switches to read, depending on which sequence interval the system has just entered. It knows which interval has been entered by way of connection line CD. The interval time counter preset circuit then shifts the numbers from those selector switches into the interval time counter. It does this via line DE.

7-3-6 How the Interval Time Counter (Block E) Fits into the Overall System

The interval time counter is the circuit which actually counts the time increments during **Squeeze, Hold,** and **Release,** and the welding current pulsations during **Weld.** It is a *down-counter*, counting backward from its preset number to zero. When it reaches zero, it sends a signal on line EB to the interval triggering and gating circuit that the interval is complete and that therefore the system is ready to step into the next interval.

For example, if the preset number shifted into the interval time counter for the **Hold** interval is 45, the counter will count down one unit for each ac line cycle that occurs. After 45 ac line cycles, taking $^{45}/_{60}$ second, the counter will reach zero. When this happens, it signals block B that the system is ready to be triggered into the next interval, the **Release** interval in this example.

7-3-7 How the Heat-Cool Stepping and Gating Circuit (Block F) Fits into the Overall System

The heat-cool stepping and gating circuit has several functions:

1. When the system is in the **Weld** interval, this circuit steps the system back and forth between the **Heat** and **Cool** subintervals.
2. Whenever a new subinterval is entered, it sends a signal on line FG to the heat-cool counter preset circuit, telling that circuit to preset the proper number into the heat-cool counter.
3. Once the presetting has been accomplished, the heat-cool stepping and gating circuit gates 60-Hz pulses down line FH into the heat-cool counter so it can count the

number of ac cycles in the subinterval. When a **Heat** or **Cool** subinterval is completed, the heat-cool stepping and gating circuit receives this information from the heat-cool counter via line HF. This is how it knows when to step to the next subinterval (when to perform function 1).

To summarize the purposes of the heat-cool stepping and gating circuit, it takes care of routing the proper signals to those blocks which have to do with the **Heat** and **Cool** subintervals, blocks G, H, and I. It also sends out a signal whenever a welding current pulsation is finished, so that other subcircuits (blocks B and E) can keep track of the progress of the **Weld** interval itself.

7-3-8 How the Heat-Cool Counter Preset Circuit (Block G) Fits into the Overall System

The heat-cool preset circuit is identical in concept to the interval time counter preset circuit. The heat-cool counter preset circuit shifts a two-digit number from a pair of selector switches into the heat-cool counter. This takes place on line GH. There is a pair of 10-position selector switches which determines the number of ac line cycles in the **Heat** subinterval and another pair of switches which determines the number of cycles in the **Cool** subinterval. The heat-cool counter preset circuit reads the correct pair of switches, depending on which subinterval the system is in. It has access to this information by way of line FG.

7-3-9 How the Heat-Cool Counter (Block H) Fits into the Overall System

Likewise, the heat-cool counter is identical in concept to the interval time counter. It counts backward from the preset number down to zero, making one backward count for each cycle of the ac line. When it reaches zero, it sends out the signal that the subinterval is finished. This signal is sent to block F on line HF, informing block F that it is time to step into the next subinterval.

7-3-10 How the Weld Power Circuit (Block I) Fits into the Overall System

The weld power circuit, block I, receives a signal from block G whenever the system is in the **Heat** subinterval of the **Weld** interval. When the **Heat** signal is received, the weld power circuit energizes the welding transformer and thus delivers current to the welding electrodes. How this is done and how the weld power circuit adjusts the conduction angle will be discussed in detail in Sec. 7-9.

7-4 ■ DETAILED DESCRIPTION OF THE SEQUENCE INITIATION CIRCUIT AND THE INTERVAL TRIGGERING AND GATING CIRCUIT

Starting in this section, and through Sec. 7-9, we will look closely at the details of operation for each of the subcircuits of Fig. 7–3. Before we can do this effectively, we must decide on certain rules of notation. The rules we will use are explained in Sec. 7-4-1.

7-4-1 Notation Used on Schematic Drawings and in the Written Text

Figure 7–4 is a schematic diagram showing the sequence initiation circuit and the interval triggering and gating circuit. Notice that some wires are labeled with capital letters. Each of these letters refers to a note at the bottom of the drawing, which explains the

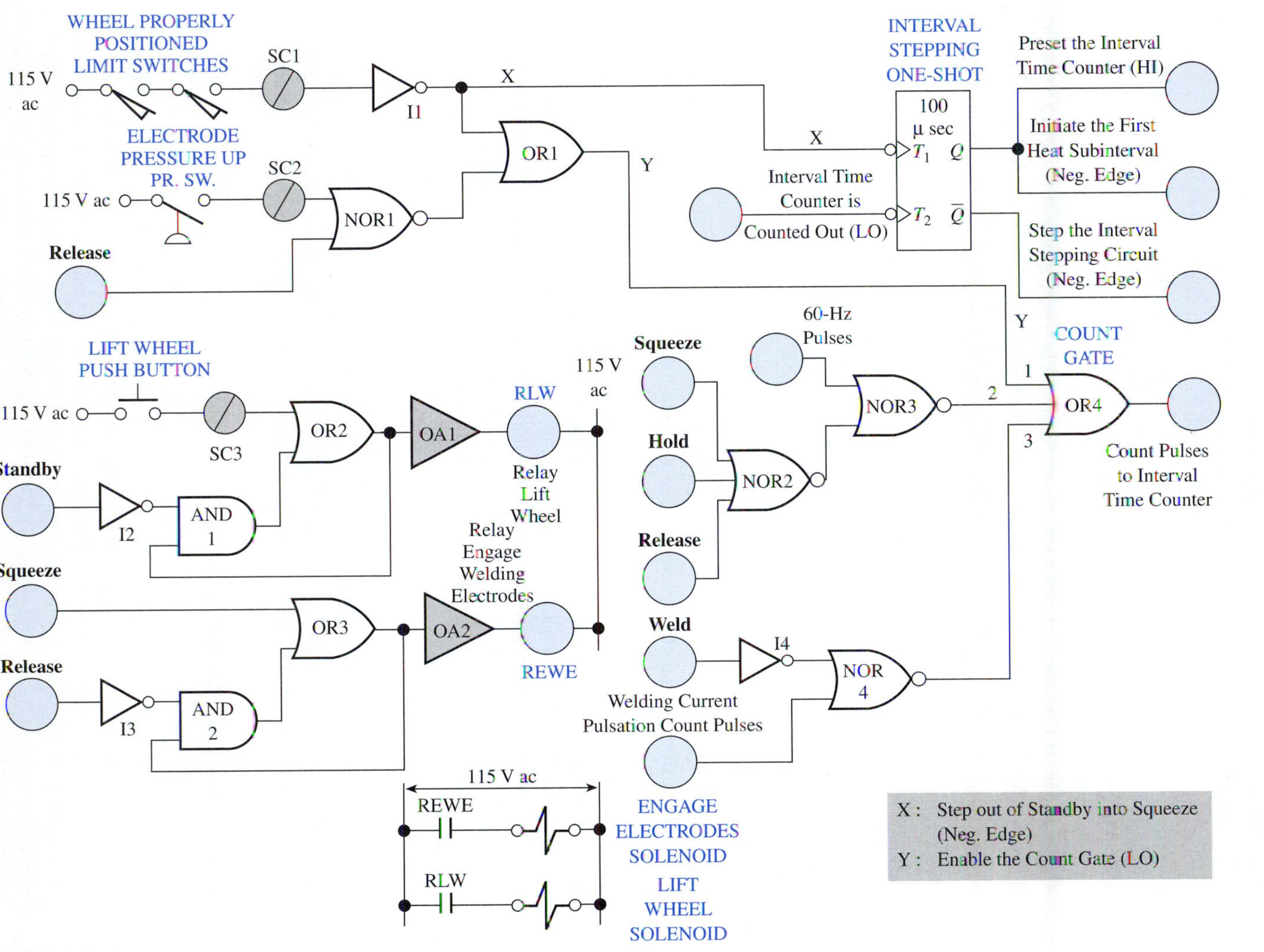

FIGURE 7–4
Schematic diagram of the Sequence Initiation Circuit (block A) and the Interval Triggering and Gating Circuit (block B). The circles represent input and output terminals, which connect to other subcircuits (other blocks).

significance of that wire, telling what that wire accomplishes in the operation of the circuit. In the explanatory text, these wires will be identified by their letter labels.

If there is a set of parentheses following the explanatory note, the condition identified in the parentheses is the condition necessary for the wire to accomplish its purpose. For example, note Y reads "Enable the Count Gate (LO)." This means that the wire labeled Y is the wire that enables the count gate to pass count pulses and that wire Y allows these count pulses to be passed only when it is LO.

When a wire comes into a subcircuit from another subcircuit, its function is identified by a circular terminal having a label which is a word or short phrase. For example, the circular terminal labeled "Release" in Fig. 7–4 indicates that the wire attached to that terminal originally comes from a **Release** terminal in some other subcircuit and that the terminal goes HI when the system enters the **Release** interval.

As another example, the wire attached to the terminal "Interval Time Counter is Counted Out (LO)" comes from another subcircuit. When that terminal goes LO, it means that the interval time counter has counted out (counted down to zero). A precise explanation of the action of that terminal will be given in the written text.

Furthermore, the particular subcircuit being illustrated in a schematic drawing will have *outputs going to* other subcircuits as well as *inputs coming from* other subcircuits. Labeled circular terminals are used to indicate this situation too. When this is done, you can expect to come across that same label on an input terminal in some other subcircuit schematic. For example, the terminal labeled "Preset the Interval Time Counter (HI)" has a wire attached to it which goes off to some other subcircuit. The schematic diagram of that other subcircuit will show an input terminal with exactly the same label.

You cannot confuse input terminals with output terminals because input terminals are always wired to the inputs of solid-state gates, flip-flops, etc., while output terminals are always wired to the outputs of the solid-state circuit devices.

Regarding the identification of circuit parts in the explanatory text, here is the legend we shall follow.

Specific intervals and subintervals of the welding sequence are capitalized and written in boldface, for example, **Release, Standby, Weld, Heat,** and **Cool.**

Particular circuit devices (gates, flip-flops, etc.) appearing in the drawings having specific names are written in capital letters, for example, INTERVAL STEPPING ONE-SHOT, LIFT WHEEL PUSHBUTTON, SIGNAL CONVERTER 2, and NOR3.

Specific subcircuits which are defined in the block diagram of Fig. 7–3 are written capitalized, for example, Interval Time Counter, Interval Stepping Circuit, and Heat-Cool Counter Preset Circuit.

Terminal labels and wire descriptions (notes) are enclosed in quotation marks, for example, "Welding Current Pulsation Count Pulses," "Step the Interval Stepping Circuit (Neg. Edge)," "60-Hz Pulses." An exception is if the terminal label is one of the specific intervals or subintervals of the system. Then it is written in boldface, for example, **Squeeze.**

7-4-2 Circuit Operation

The welding sequence begins when the operator presses and holds the LIFT WHEEL PUSHBUTTON on the left side of Fig. 7–4. In the production process, this would be done when the operator sees that a wheel rim and spider have been properly loaded onto the lift cradle, as described in Sec. 7-1. If the process were completely automated, there would be a solid-state signal or a relay contact instead of the pushbutton switch. At any rate, the application of 115 V ac to the input of SIGNAL CONVERTER 3 will cause a voltage of +5 V dc, a HI, at the input of OR2. The bottom input of OR2 is LO at this time, due to the fact that the **Standby** terminal is HI while the system is in **Standby.** The output of OR2 goes HI, causing output amplifier OA1 to energize relay RLW. The normally open contact of relay RLW in the lower left of Fig. 7–4 closes, applying 115 V ac to the LIFT WHEEL SOLENOID. The energization of this solenoid shifts the hydraulic valve, which causes the lift cylinder to extend. When the lifting cradle has lifted the wheel rim and spider into proper position to be welded, the limit switches in the upper left of Fig. 7–4 close their contacts. This applies 115 V ac to SIGNAL CONVERTER 1, causing a HI at the input of inverter I1. When this HI appears, the output of I1 goes LO, causing a negative edge on wire X. This negative edge appears at the T_1 terminal of the INTERVAL STEPPING ONE-SHOT.

The INTERVAL STEPPING ONE-SHOT has two trigger terminals, T_1 and T_2. It will fire when a negative edge appears at either one of its trigger terminals. Therefore the negative edge at T_1 causes the one-shot to fire, delivering an output pulse 100 μs in duration. As it fires, the $\overline{Q}$ output goes LO, creating a negative edge at the terminal labeled "Step the Interval Stepping Circuit (Neg. Edge)." This causes the Interval Stepping Circuit to step out of **Standby** and into the **Squeeze** interval. This stepping action is discussed in detail in Sec. 7-5.

Meanwhile, the Q output of the INTERVAL STEPPING ONE-SHOT remains HI for 100 μs, far longer than is necessary for the system to step into **Squeeze.** During this 100 μs, the "Preset the Interval Time Counter" terminal is HI. The HI level on this terminal shifts the digits set on the **Squeeze** 10-position selector switches into the Interval Time Counter. This shifting is discussed further in Sec. 7-6.

When the system steps out of **Standby** into **Squeeze,** the **Standby** terminal goes LO, and the **Squeeze** terminal goes HI on the left of Fig. 7–4. Because **Standby** is LO, the output of I2 goes HI. Since the bottom input of AND1 is also HI, the output of AND1 goes HI. This means that the LIFT WHEEL PUSHBUTTON can be released, because the bottom input of OR2 is now HI, eliminating the need for the top input of that gate to be HI. The OR2 gate has sealed itself up, as long as **Standby** remains LO. This keeps RLW energized, keeping the wheel lifted up into the welding position. The OR2-AND1 combination just described is the familiar sealing circuit seen several times in Chapter 1.

As stated earlier, the **Squeeze** terminal on the left of Fig. 7–4 goes HI as the system steps into the **Squeeze** interval. This causes the output of OR3 to go HI. The output feeds back into AND2, sealing up OR3 as long as the **Release** terminal is LO. OR3 drives OA2, which in turn drives relay REWE. This relay energizes the ENGAGE ELECTRODES SOLENOID at the bottom of Fig. 7–4, causing the welding electrode cylinders to extend, bringing the electrodes into contact with the wheel rim and spider. The OR3-AND2 combination is another sealing circuit.

When the pressure on the electrode cylinders is high enough, meaning that the welding electrodes have made firm contact with the metal, the ELECTRODE PRESSURE UP PR. SW. contact closes at the upper left of Fig. 7–4. When the output of SIGNAL CONVERTER 2 goes HI, the output of NOR1 goes LO. At this time both inputs of OR1 are LO, causing wire Y to go LO. This enables the COUNT GATE, OR4, to pass any pulses which show up on its number 2 input. There are 60-Hz pulses existing on input 2 of OR4 at this time, so they feed through OR4 and show up on the "Count Pulses to Interval Time Counter" terminal. The squeeze time has begun, and the Interval Time Counter starts counting down from its preset number.

Let us pause at this point to discuss the action of OR4 in gating the count pulses through to the Interval Time Counter. As long as wire Y was HI, OR4 could not pass count pulses because its output was locked in the HI state by the HI on its number 1 input. Under this condition, pulses appearing at its number 2 input could not get through to the output. Now that input 1 is LO, the OR4 output is able to respond to pulses applied to its number 2 input (assuming that input 3 is LO). This is an example of *gating* pulses to a counter. The gate either passes or blocks the count pulses, in response to the command signal on its number 1 input. Of course, input 3 or OR4 has the same control ability to tell the COUNT GATE to pass or block count pulses.

The number 3 input of OR4 (the COUNT GATE) is LO at this time due to the **Weld** terminal being LO. The I4 output is HI, causing the NOR4 output to go LO, bringing the number 3 input of OR4 to a LO logic level.

It was stated above that 60-Hz count pulses do in fact exist at input 2 of the COUNT GATE at the instant the squeeze time begins. Figure 7–4 shows that such pulses must come from the output of NOR3. Inspection of NOR3 reveals that the pulses appearing on its top input from the "60-Hz Pulses" terminal will be passed to the output of NOR3 only if the bottom input is LO. When they are passed, the count pulses arrive at the output of NOR3 inverted in phase, but that is unimportant in this application. This situation is almost the same as for OR4. If the bottom input were HI, the NOR3 output would be locked in the LO state, and NOR3 would not pass the pulses on its top input. However, its bottom input is LO at this time because the **Squeeze** terminal feeding into NOR2 is HI.

The final result of all this circuit action is that the Interval Time Counter is able to start counting down at the rate of one count per ac line cycle. The **Squeeze** interval is being timed out. When the Interval Time Counter reaches zero, the terminal labeled "Interval Time Counter is Counted Out" goes LO. This supplies a negative edge to terminal T_2 of the INTERVAL STEPPING ONE-SHOT, causing it to fire once again. As before, a negative edge appears at the "Step the Interval Stepping Circuit" terminal. The Interval Stepping Circuit steps into the **Weld** interval. The HI signal on the Q output of the one-shot repeats its function of shifting numbers into the Interval Time Counter. This time it shifts the numbers set on the **Weld** 10-position selector switches.

Since the system is now in the **Weld** interval, all three inputs of NOR2 are LO, causing its output to go HI. This HI is applied to the bottom input of NOR3, disabling that gate by locking its output LO. Thus the 60-Hz pulses are prevented from passing through NOR3 during the **Weld** interval, and they cannot be counted by the Interval Time Counter. However, the **Weld** terminal driving the input of I4 is now HI, which causes a LO level to the top input of NOR4. This LO enables NOR4 to pass any pulses which appear on the "Welding Current Pulsation Count Pulses" terminal.

Remember that the **Weld** interval differs from the **Squeeze, Hold,** and **Release** intervals in that the preset number represents how many welding current pulsations are required to complete the interval, rather than how many ac line cycles. Every time a current pulsation is completed, the Heat-Cool Stepping and Gating Circuit delivers a count pulse to the "Welding Current Pulsation Count Pulses" terminal. From there the pulse is passed through NOR4, through OR4, and eventually into the Interval Time Counter.*

As before, the Interval Time Counter counts backward one bit for each pulse it receives. When it reaches zero, it once again supplies a negative edge to T_2 of the INTERVAL STEPPING ONE-SHOT. The one-shot triggers the Interval Stepping Circuit by means of the negative edge appearing at the "Step the Interval Stepping Circuit" terminal. The system leaves **Weld** and enters **Hold,** and the **Hold** 10-position selector switch

*The word *time* is a little misleading during the **Weld** interval, because the counter is not actually counting time increments but welding current pulsations. During all the other intervals the Interval Time Counter really does count time increments ($^1/_{60}$-second increments).

settings are preset into the Interval Time Counter. The Interval Stepping Circuit takes away the **Weld** signal and sends out the **Hold** interval signal. Therefore the **Weld** terminal in Fig. 7–4 goes LO, disabling NOR4. The **Hold** terminal feeding NOR2 goes HI, causing the bottom input of NOR3 to go back LO. Once again the 60-Hz pulses are routed through NOR3, through OR4, to the Interval Time Counter. The **Hold** interval begins counting out.

When **Hold** is complete, the INTERVAL STEPPING ONE-SHOT receives another negative edge on its T_2 input, and its $\overline{Q}$ output delivers a negative edge to the "Step the Interval Stepping Circuit (Neg. Edge)" terminal. That negative edge steps the system into **Release.** The same actions occur again, resulting in the **Release** selector switch settings being shifted into the Interval Time Counter. NOR3 immediately begins passing the 60-Hz pulses, and the **Release** interval begins counting out.

The **Release** terminal on the lower left of Fig. 7–4 goes HI at this time. This causes the top input of AND2 to go LO, breaking the seal on OR3 for the first time since the system entered the **Squeeze** interval. Output amplifier OA2 goes to a LO level, deenergizing relay REWE. When the ENGAGE ELECTRODES SOLENOID deenergizes, the welding electrode cylinders retract, releasing the wheel. Although the ELECTRODE PRESSURE UP PR. SW. contact opens, allowing SIGNAL CONVERTER 2 to go LO, the output of NOR1 remains LO because its bottom input is now held HI by the **Release** terminal. It is necessary to keep the NOR1 output LO in order to keep wire Y at a LO level, allowing the COUNT GATE, OR4, to continue passing count pulses. When these pulses have driven the Interval Time Counter to zero, the INTERVAL STEPPING ONE-SHOT is triggered once again by the "Interval Time Counter is Counted Out" terminal.

When the system's Interval Stepping Circuit leaves the **Release** interval, it steps into the **Standby** condition. On the far left of Fig. 7–14, the **Standby** terminal goes HI, causing the 12 output to go LO. This LO is applied to the top input of AND1, breaking the seal on OR2. Output amplifier OA1 goes LO, which causes relay RLW to deenergize. This deenergizes the LIFT WHEEL SOLENOID, lowering the finished wheel. The WHEEL PROPERLY POSITIONED LIMIT SWITCHES open, causing SIGNAL CONVERTER 1 to go LO. The output of I1 goes HI, returning wire X to its initial HI state. Wire Y is also HI at this time.

This completes the discussion of circuit action for the Sequence Initiation Circuit and the Interval Triggering and Gating Circuit. In the next section we will deal with the Interval Stepping Circuit and Decoder.

7-5 ■ DETAILED DESCRIPTION OF THE INTERVAL STEPPING CIRCUIT AND DECODER

Figure 7–5 is a schematic diagram of the Interval Stepping Circuit and Decoder. These circuits are not extensive. The Interval Stepping Circuit itself consists of three flip-flops and an AND gate. The Decoder is a diode decoding matrix having six input lines and five output lines. The Decoder also has five output drivers.

7-5-1 Interval Stepping Circuit

In Fig. 7–5 the flip-flop outputs have been identified by the letter name of the individual flip-flop. That is, the outputs of flip-flop A are labeled A and $\overline{A}$ instead of Q and $\overline{Q}$, and the same for flip-flops B and C. Table 7-1 shows the sequence of the flip-flops as pulses

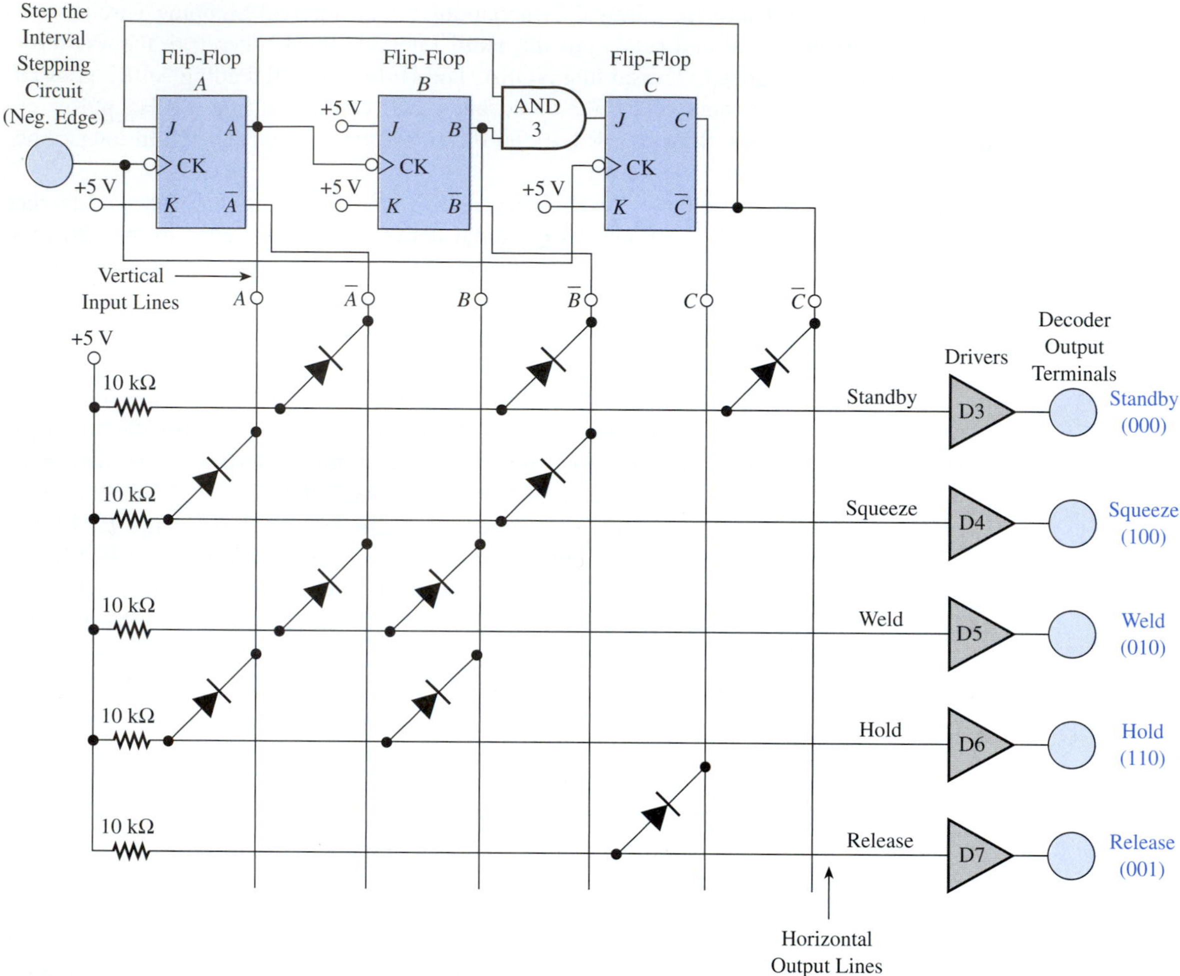

FIGURE 7–5
Schematic diagram of the Interval Stepping Circuit and Decoder (block C). Whenever the system is in a certain interval, the corresponding output terminal goes HI. The digits in parentheses stand for the states of the *A*, *B*, and *C* flip-flops during that interval.

TABLE 7-1
State Table for the Interval Stepping Circuit

NUMBER OF STEPPING PULSES DELIVERED	Flip-Flops			SYSTEM INTERVAL
	A	*B*	*C*	
Start	0	0	0	Standby
1	1	0	0	Squeeze
2	0	1	0	Weld
3	1	1	0	Hold
4	0	0	1	Release
5	0	0	0	Standby

are delivered. The digit 1 in Table 7-1 means that the flip-flop is ON, while a 0 means that the flip-flop is OFF.

All three flip-flops are negative edge-triggered *JK* flip-flops, discussed in Sec. 2-3. To understand the operation of the Interval Stepping Circuit, refer to Table 7-1 and Fig. 7–5.

In the **Standby** condition all flip-flops are OFF. When the first negative edge arrives at the circuit's input terminal, the "Step the Interval Stepping Circuit (Neg. Edge)" terminal on the left of Fig. 7–5, FFA toggles into the ON state because both *J* and *K* are HI. *J* of FFA is HI because *C* is HI with FF*C* in the OFF state.

The negative edge at CK of FF*A* also appears at CK of FF*C*. At this time, though, the output of AND3 is LO, holding *J* of FF*C* LO. FF*C* therefore stays OFF. AND3 is LO because both *A* and *B*, the inputs to AND3, are LO at the instant the negative edge appears. Therefore after the first pulse, the states of the flip-flops are *ABC* = 100.

When the second negative edge appears at the input terminal, *J* of FF*A* is still HI, so FF*A* toggles to the OFF state. The negative edge is delivered to CK of FF*C* also, but the AND3 output is still LO because *B* is LO at this instant. Therefore FF*C* again stays OFF. When the *A* output goes LO it delivers a negative edge to CK of FF*B*. This causes FF*B* to toggle into the ON state. The state of the circuit is *ABC* = 010 after the second pulse.

When the third stepping pulse negative edge arrives at the input terminal, FF*A* toggles into the ON state as before, since *J* of FF*A* is HI. The negative edge appears at CK of FF*C*, but once again *J* of FF*C* is LO. It is LO because the top input of AND3 is LO at the instant the negative edge arrives. After the third pulse, the state of the flip-flop circuit is *ABC* = 110.

On the fourth negative edge, FF*A* toggles into the OFF state because its *J* input is still HI. The negative edge appears at CK of FF*C* also. This time the output of AND3 is HI, so FF*C* toggles into the ON state. The output of AND3 is HI when the edge arrives because both FF*A* and FF*B* are ON at that instant. FF*B* also receives a negative edge at its CK input when *A* goes LO. It therefore toggles into the OFF state. The state of the flip-flop circuit is *ABC* = 001 after the fourth pulse.

When the fifth step pulse negative edge arrives at the input terminal, *J* of FF*A* is LO, because *C* is now LO. Therefore FFA remains in the OFF state. The negative edge appears at CK of FF*C* as usual. This time the *J* input of that flip-flop is LO because both AND gate inputs are LO. FFC therefore turns OFF, making the state of the circuit *ABC* = 000. After five step pulses the Interval Stepping Circuit has returned to its original state.

It can be seen that the Interval Stepping Circuit steps through five separate states, never varying the order. It remains in any given state until it gets a step signal to step into a new state. These features make it an ideal circuit for keeping track of which interval the system is currently in. All that is necessary is to convert the states of the flip-flops, expressed as a sequence of binary bits, into a useful form for the other subcircuits of the system. This is the function of the Decoder.

7-5-2 Decoder

The Decoder of Fig. 7–5 has the same basic purpose as the BCD-to-decimal decoder discussed in Sec. 2-6. It takes in coded information and puts out a logic HI on one of its output terminals. All the other output terminals are held LO while the proper one goes HI.

The way the Decoder does this is by looking at a *portion* of the binary sequence that represents the complete state of the Interval Stepping Circuit. It focuses on that portion that makes one particular state unique. For example, Table 7-1 shows that when the Interval Stepping Circuit is in the **Squeeze** interval, the state is 100. A search of the other entries in Table 7-1 reveals that no other row has $A = 1$ and $B = 0$. Therefore the $AB = 10$ combination makes that state unique, different from all other states. In Fig. 7–5, the horizontal squeeze line has two diodes wired to it, one pointing into the A vertical input line and the other pointing into the B input line. If either of these inputs is LO, the squeeze line will be pulled down LO by one of the two diodes, but if both of these inputs are HI, the squeeze line will be allowed to go HI. An output line will go HI if no diode attached to that line points to a LO potential (0 V). With no diode pointing into a LO, there is no route for current to flow to ground, and therefore there is no voltage drop across the 10-kΩ lead-in resistor. With no voltage drop across the resistor, the output line is at the same potential as the supply voltage, namely +15 V. Thus the squeeze line goes HI whenever both A and B are HI. Of course, B being HI is equivalent to $\overline{B}$ being LO (0). Therefore the squeeze line will go HI whenever both $A = 1$ and $B = 0$. This will cause driver D4 to bring the **Squeeze** terminal HI, which signals to the other subcircuits that the system is in the **Squeeze** interval.

If the system is in **Squeeze,** all four of the other horizontal output lines will be LO, because at least one diode pulls each line down LO. For example, the weld output line is being pulled down by the diode pointing into A. (It is also being pulled down by the diode pointing into B, but one diode is sufficient.)

An output line being pulled down LO means that current is flowing through the 10-kΩ lead-in resistor on the left of that line and then through a diode to ground. The 15 V of the supply is dropped across that 10-kΩ lead-in resistor, leaving only a small voltage on the line itself. Germanium diodes are used in this diode matrix because of their lower forward voltage drop across the *pn* junction (about 0.2 V for germanium versus 0.6 V for a silicon diode).

You should verify for yourself that the other three horizontal output lines, standby, hold, and release, are all pulled down LO when $AB = 10$.

Another example may help to clarify the working of the Decoder. Consider the **Release** interval in Table 7-1. The state of the Interval Stepping Circuit is $ABC = 001$. A search of the rest of the table reveals that no other interval has $C = 1$. Therefore the single bit $C = 1$ distinguishes the **Release** interval from all four other intervals and makes it unique. The Decoder takes advantage of this fact in that it has a single diode pointing from the release output line to the C vertical input line. If C is HI, as it would be during the **Release** interval, the release output lines goes HI. If C goes LO, as it would be during any *other* interval, the release output line is pulled down LO. Therefore the release output line goes HI when the C flip-flop turns ON, and only then. Again, the Decoder is looking at the *unique* portion of the state of the flip-flop circuit and using that portion to control the output line.

The driver attached to each output line has the function of isolating the line from the other subcircuits, so the subcircuits cannot degenerate the quality of the signal level on the output line. Degeneration of signal level (HIs not high enough or LOs not low enough) could occur if the subcircuits were to draw too much current away from the line when it was HI (a current-sourcing logic family) or if they were to dump too much current into the line when it was LO (a current-sinking family).

It will be instructive for you to verify that the Decoder does in fact identify the unique portion of the state of the Interval Stepping Circuit for the other three intervals. Check for yourself that for each interval it brings the proper output terminal HI, leaving all others LO.

7-6 ■ INTERVAL TIME COUNTER AND INTERVAL TIME COUNTER PRESET CIRCUIT

The Interval Time Counter consists of a pair of decade down-counters, one for units and one for tens, and a simple gating circuit to detect when the counter reads zero. It is shown schematically in Fig. 7–6, along with the Interval Time Counter Preset Circuit.

7-6-1 Interval Time Counter

The decade down-counters count down one digit each time a negative clock edge appears at the CK input terminal. The contents of a decade down-counter appear in BCD form at the *DCBA output* terminals ($D = 8, C = 4, B = 2, A = 1$). The *DCBA input*

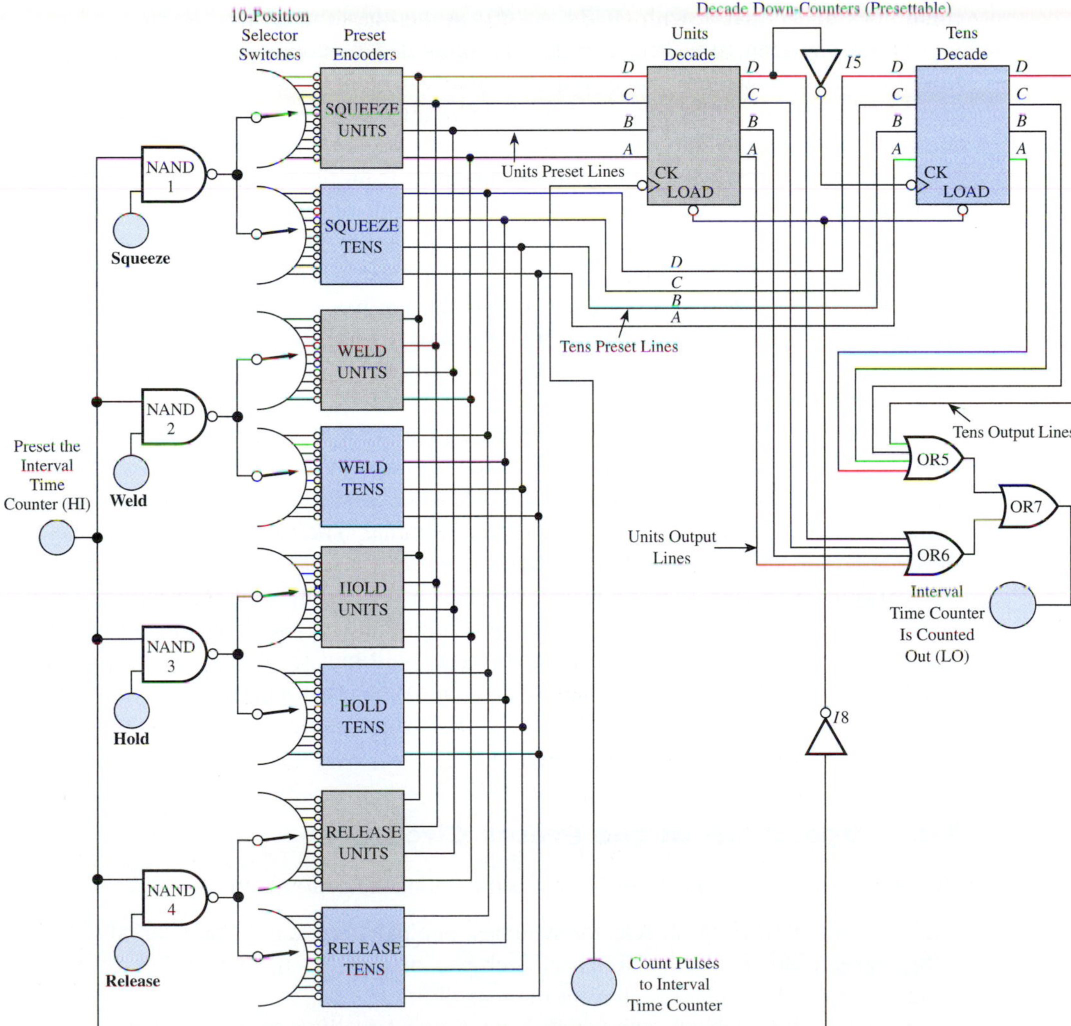

FIGURE 7–6
Schematic of the Interval Time Counter (block E) and the Interval Time Counter Preset Circuit (block D). The length of each interval (except **Standby**) is set on a pair of 10-position selector switches.

terminals are used for presetting a number into the counters before delivering the count pulses. The binary number appearing at the *DCBA* input terminals is shifted into the down-counter when the LOAD terminal goes LO. When the LOAD terminal is HI, the logic levels at the input terminals are ignored by the counter. Down-counter action was discussed in Sec. 2-11.

For example, if the binary data at the input terminals are *DCBA* = 0111, the decimal number 7 would be preset into the counter when the LOAD terminal went LO. The 7 would appear at the output terminals as *DCBA* = 0111. Once the LOAD terminal returns to the HI level, count pulses at the CK terminal can be received by the counter. Any count pulses appearing at CK while the LOAD terminal is LO would be ignored.

As count pulses are received, the counter counts down one digit per pulse. As usual, the actual count transition occurs at the instant the negative-going edge arrives. When the counter gets to zero, all output terminals are LO (*DCBA* = 0000). On the next pulse, the count goes to decimal 9, with *DCBA* = 1001 at the output terminals.

It can be seen from Fig. 7–6 that the units decade counter receives count pulses from the terminal labeled "Count Pulses to Interval Time Counter." This terminal originates in the Interval Triggering and Gating Circuit, as discussed in Sec. 7-4. The tens decade counter receives its count pulses from the units decade counter, since the two counters are cascaded. When the units decade goes from the 0 state into the 9 state, the *D* output line goes HI. This HI is applied to the input of I5, which delivers a negative edge to CK of the tens decade counter. The tens decade moves down one digit at that time. For example, if the number contained in the Interval Time Counter is 40, the units decade has a 0 (*DCBA* = 0000) and the tens decade has a 4 (*DCBA* = 0100). On the next pulse, the units decade goes to a 9 (1001) and the negative edge delivered from the output of 15 steps the tens decade down to a 3 (0011). The number contained in the Interval Time Counter is then 39.

As explained in Sec. 7-2, pulses continue to arrive at the Interval Time Counter until it reaches zero, at which time a negative edge appears at the terminal labeled "Interval Time Counter is Counted Out." OR5, OR6, and OR7 are the gates which detect the zero condition and supply the negative edge to that terminal. The negative edge then triggers the INTERVAL STEPPING ONE-SHOT.

Specifically, when the tens decade contains a 0, the output of OR5 will be LO because every one of its inputs will be LO. When the units decade contains a 0, the OR6 output will be LO because every one of *its* inputs will be LO. When both decades contain 0s, the Interval Time Counter has counted out. When that happens, OR7 will see two LOs on its inputs, and its output will go LO. This supplies the negative edge at the "Interval Time Counter Is Counted Out" terminal.

7-6-2 Operation of the Preset Circuitry

The Interval Time Counter Preset Circuit consists of the following:

1. Four pairs of 10-position selector switches, one pair for each of the intervals **Squeeze, Weld, Hold,** and **Release.** Each pair of switches has one switch for the units digit and a second switch for the tens digit.
2. An encoder for each selector switch—eight encoders altogether.
3. Four NAND gates, one for each of the four intervals which has a pair of switches.

When the "Preset the Interval Time Counter" terminal gets a HI signal from the INTERVAL STEPPING ONE-SHOT in the Interval Triggering and Gating Circuit, two things happen. First, the LOAD terminals of the two decade down-counters are driven LO by

18, permitting the counters to receive their preset numbers. Second, the gates NAND1, NAND2, NAND3, and NAND4 are partially enabled because all their top inputs go HI. Then, depending on which interval has just been entered, one of these four gate outputs will go LO, applying a LO signal to the common terminals of the appropriate pair of selector switches.

For example, if the system has just entered **Hold,** the NAND3 output will go LO, applying 0 V to the common terminals of the **Hold** selector switches. The other three pairs of selector switches will continue to have +5 V on their common terminals. Recall that we said we would expect our encoders to respond to a LO decimal input rather than a HI decimal input. This point was covered in Sec. 2-11-2. Therefore only the selector switches which receive 0 V are able to drive their preset encoders, so the other three pairs drop out of the picture.

In this example, with the system having just entered the **Hold** interval, the HOLD UNITS SWITCH applies its number setting to the units preset lines, and the HOLD TENS SWITCH applies its number setting to the tens preset lines. Thus, BCD data are available on the units preset lines and tens preset lines at the same time that the LOAD terminals of the units and tens decade counters are held LO. In this way the proper selector switch settings are preset into the Interval Time Counter.

Now let us turn our attention to the preset encoders themselves. Decimal-to-BCD encoders are available in packaged ICs, but Fig. 7–7(a) shows how one could be built from scratch. All eight of the preset encoders in Fig. 7–6 are identical and could be built as shown in Fig. 7–7(a).

In Fig. 7–7(a), whichever vertical input line is selected on the switch has 0 V (ground voltage) applied to it from the switch's common terminal. Lowering the potential of an input line to ground level causes the cathode of each diode attached to that line to be at ground potential. When the cathode goes to ground, the horizontal output line that the anode attaches to is also pulled down LO. Current flows through the 1.5-kΩ pull-up resistor at the left end of the output line, down the horizontal line, through the diode, and into the ground. The current flow through the lead-in resistor at the left is sufficient to drop almost the full 5-V supply voltage. Only a few tenths of a volt remain on the horizontal output line itself. The logic LO on an output line is applied to the inverting buffer at the right-hand side of that line, causing the output terminal from that line to go HI. The result is that every horizontal output line that is connected through a diode to the LO input line goes LO itself. Those output terminals of the encoder are then driven HI by their inverting buffers. The inverting buffers serve to isolate, or buffer, the diodes in the encoder from the output terminals and their attached subcircuits. They also serve to invert the internal output lines so that the encoded number appears at the output terminals in positive logic.

Those output lines which are *not* connected through diodes to the 0-V input line have no current flow through their 1.5-kΩ lead-in resistors. Therefore there is no voltage drop across their 1.5-kΩ resistors, leaving these output lines at essentially +5 V (HI). A HI signal existing on a horizontal output line is applied to the inverting buffer at the right-hand end of the line, causing the output terminal to be LO.

As an example, consider that the selector switch is set to 3. This applies ground potential to the number 3 vertical input line. Current will flow through the two diodes that connect the number 3 input line to the *A* output line and to the *B* output line. This current comes through the 1.5-kΩ pull-up resistors at the left end of these two horizontal output lines, causing these two lines to drop almost to 0 V. The *C* output line and the *D* output line have no current flow, so their potential stays essentially at +5 V. Therefore the *A* and *B* inverting buffers receive LO inputs and the *C* and *D* inverting

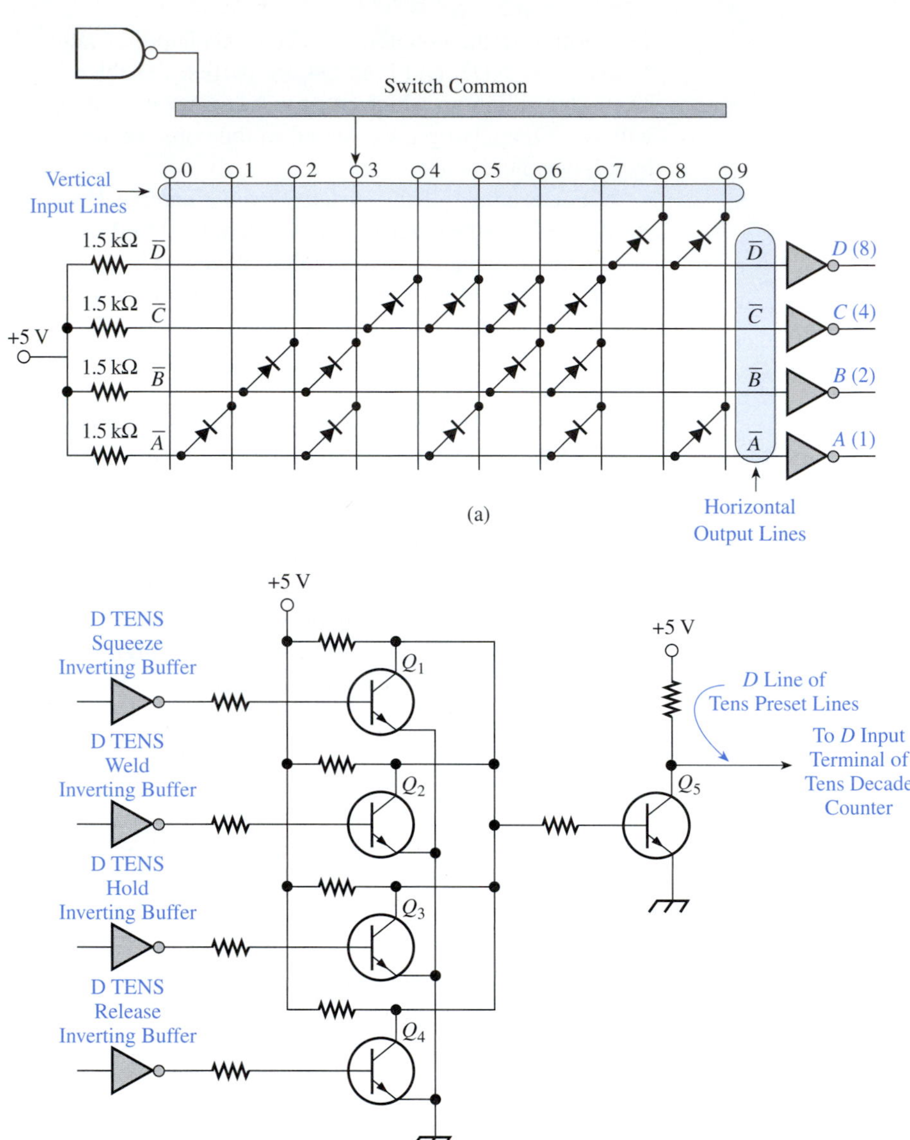

FIGURE 7–7
(a) Detailed schematic of one of the preset encoders. Altogether, there are eight of these preset encoders, as shown in Fig. 7–6. (b) The way in which four output inverting buffers are tied together. During any interval, the three inverting buffers which are not in use all have LO outputs, and they are all isolated by their three transistors. The one inverting buffer which is in use may have either a LO output or a HI output. If it has a LO output, its transistor turns OFF, causing Q_5 to turn ON; this causes the preset line to go LO. On the other hand, if that inverting buffer has a HI output, its transistor turns ON, causing Q_5 to turn OFF; this causes the present line to go HI. The preset line thus follows the state of the one inverting buffer which is in use during that interval.

buffers receive HI inputs. The *A* and *B* output terminals thus go HI, and the *C* and *D* output terminals go LO. The result is that the decimal number 3 has been encoded to $DCBA = 0011$, which is correct. You should analyze the action of the encoder for several other switch settings and prove to yourself that it does correctly encode the decimal number into BCD.

7-6-3 Connecting the Buffers Together

Notice in Fig. 7–6 that the output terminals of the four UNITS ENCODERS are all shown tied together to drive the units preset lines. That is, output terminal *D* of the

SQUEEZE UNITS ENCODER ties to output terminal *D* of the WELD UNITS ENCODER and also to output terminal *D* of the HOLD UNITS ENCODER and output terminal *D* of the RELEASE UNITS ENCODER.* Under these circumstances, "disputes" will arise between the different inverting buffers, with one inverting buffer trying to bring the preset line HI and the other three inverting buffers which attach to that preset line trying to bring it LO. Because of these possible disputes, the inverting buffers cannot be connected directly together. Buffer connection circuits are necessary. The buffer connection circuits are designed so that the inverting buffer trying to bring the line HI "wins the dispute."

Figure 7–7(b) shows how this is accomplished. The buffer connection circuit of Fig. 7–7(b) is drawn for the *D* outputs of the four TENS ENCODERS. However, the entire circuit shown in that figure is repeated eight times. It is repeated for all the *C*, *B*, and *A* outputs of the TENS ENCODERS and again for all the *D*, *C*, *B*, and *A* outputs of the UNITS ENCODERS.

Here is how the buffer connection circuit works. Transistor switches Q_1, Q_2, Q_3, and Q_4 are tied in parallel, with all four collectors connected together. Therefore if the *D* output of the encoder which has been enabled happens to be HI, the collector tie point is pulled down LO by that transistor. The LO on the collector tie point is applied to the base of Q_5, causing the final D output to go HI.

On the other hand, if the *D* output of the encoder which has been enabled happens to be LO, then the collector tie point will go HI. That will occur because none of the transistors Q_1–Q_4 will turn ON. The HI on the collector tie point is applied to the base of Q_5, causing the final *D* output to go LO.

For example, suppose that the HOLD ENCODER has been enabled and that the SQUEEZE, WELD, and RELEASE ENCODERS are disabled. That is, in Fig. 7–6, the NAND3 output is LO, applying 0 V to the common terminals of the **Hold** selector switches, and the NAND1, NAND2, and NAND4 outputs are all HI, applying +5 V to the common terminals of all their selector switches. In this example, the only *D* output terminal in Fig. 7–7(b) which can possibly go HI is the *D* output terminal of the HOLD TENS ENCODER. The other three *D* outputs are guaranteed to be LO because of the HIs at the common terminals of their selector switches. In other words, their encoders are disabled.

If the *D* output of the HOLD TENS ENCODER happens to be HI, Q_3 will turn ON. This will put a LO on the base of Q_5, causing that transistor to turn OFF. Therefore the final *D* output (the *D* line of the tens preset lines) goes HI.

Now consider what will happen if the *D* output terminal of the HOLD TENS ENCODER happens to be LO. In that case Q_3 will turn OFF. Q_1, Q_2, and Q_4 are guaranteed to be OFF at this time because the SQUEEZE, WELD, and RELEASE ENCODERS are all disabled. Therefore the collector tie point will go HI and will turn ON Q_5. The collector of Q_5 goes LO, applying a LO to the *D* line of the tens preset lines.

The overall result is that the *D* line of the tens preset lines obeys the *D* inverting buffer output of the HOLD TENS ENCODER if that encoder is the one that is enabled. Of course, if some other encoder has been enabled, the *D* line would have obeyed *that* encoder's inverting buffer.

---

*The same is true for all four of the *C* output terminals of the four UNITS ENCODERS and also for all four of the *B* and *A* output terminals of the four UNITS ENCODERS. The same situation is also true for the four TENS ENCODERS. Figure 7–6 shows all this.

7-7 ■ HEAT-COOL STEPPING AND GATING CIRCUIT

The Heat-Cool Stepping and Gating Circuit goes into action only during the **Weld** interval. During that interval, its functions are to keep track of the **Heat** and **Cool** subintervals and to control the stepping from one subinterval to the next. The circuit is shown schematically in Fig. 7–8.

The circuit begins operation when the **Weld** terminal goes HI on the far left of Fig. 7–8. This terminal goes HI when the system enters the **Weld** interval. At that time the "Initiate the First **Heat** Subinterval (Neg. Edge)" terminal also goes HI, coming from the Q output of the INTERVAL STEPPING ONE-SHOT. Therefore the output of AND4 goes HI because both of its inputs are HI at this time.

When the INTERVAL STEPPING ONE-SHOT output pulse ends after 100 μs, the "Initiate the First **Heat** Subinterval (Neg. Edge)" terminal goes back LO, driving the top input of AND4 back LO, causing a negative edge to appear at trigger terminal T_1 of the

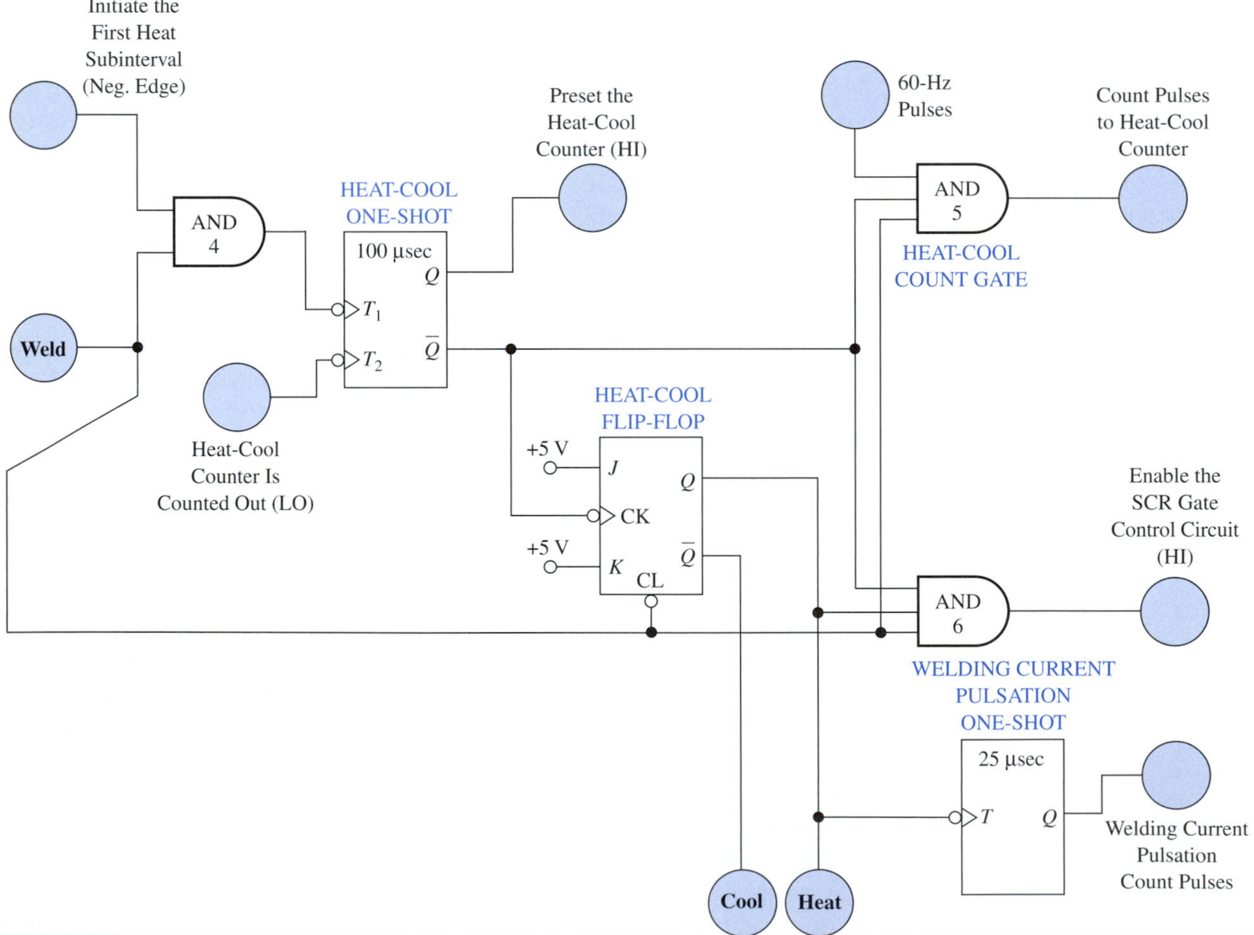

FIGURE 7–8
Schematic diagram of the Heat-Cool Stepping and Gating Circuit (block F). Every time the Heat-Cool Counter is counted out, the HEAT-COOL ONE-SHOT is fired. Its Q output presets the Heat-Cool Counter for the next subinterval, and its Q output causes the HEAT-COOL FLIP-FLOP to toggle into the opposite state.

HEAT-COOL ONE-SHOT. The firing of the HEAT-COOL ONE-SHOT causes its $\overline{Q}$ output to go LO, which causes a negative edge to appear at the clock terminal of the HEAT-COOL FLIP-FLOP. This makes the flip-flop toggle into the ON state. The HEAT-COOL FLIP-FLOP was in the OFF state prior to the system entering the **Weld** interval because its clear input was held LO by the **Weld** terminal. A LO on the CL terminal of a *JK* flip-flop clears the flip-flop, as mentioned in Sec. 2-3.

While the HEAT-COOL ONE-SHOT is still firing, the terminal labeled "Preset the Heat-Cool Counter" is HI. The HI on this terminal is fed to the Heat-Cool Counter Preset Circuit along with the HI signal on the **Heat** terminal. Together, these signals cause the HEAT SELECTOR SWITCH settings to be shifted into the Heat-Cool Counter. This action will be discussed in Sec. 7-8.

When the HEAT-COOL ONE-SHOT pulse ends, the $\overline{Q}$ output of the one-shot goes back to HI. This HI signal is routed to AND5 and AND6. The AND6 gate now has all inputs HI, so it brings the "Enable the SCR Gate Control Circuits" terminal HI, allowing the welding SCRs to commence firing. The welding transformer therefore starts delivering welding current to the wheel rim and spider. This is explained fully in Sec. 7-9.

Meanwhile, AND5 has been enabled to pass 60-Hz pulses to the terminal labeled "Count Pulses to Heat-Cool Counter." Since the welding transformer is carrying 60-Hz current, one pulse is delivered to the Heat-Cool Counter for every cycle of welding current. The Heat-Cool Counter counts backward to zero, just like the Interval Time Counter. When the preset number of cycles of welding current have occurred, the "Heat-Cool Counter Is Counted Out" terminal on the left of Fig. 7–8 goes LO, triggering the HEAT-COOL ONE-SHOT again, this time from terminal T_2. The one-shot fires, putting a HI level on the "Preset the Heat-Cool Counter" terminal once again and causing the HEAT-COOL FLIP-FLOP to toggle into the OFF state. This makes the **Heat** terminal go LO and the **Cool** terminal go HI. The **Weld** interval is now in the **Cool** subinterval.

The negative edge appearing on the *Q* terminal of the HEAT-COOL FLIP-FLOP as it turns OFF is applied to trigger terminal *T* of the WELDING CURRENT PULSATION ONE-SHOT. This one-shot delivers a 25-μs pulse to the "Welding Current Pulsation Count Pulses" terminal, indicating that a welding current pulsation has been completed (a **Heat** subinterval has been completed). This pulse is routed to NOR4 in Fig. 7–4. It is passed through NOR4 and OR4 to the Interval Time Counter as described in Sec. 7–4. Therefore the welding current pulsation that has just finished causes the Interval Time Counter to count down by one digit.

Since the **Weld** interval has just now entered the **Cool** subinterval, the **Cool** terminal is HI and the **Heat** terminal is LO. The output of AND6 goes LO, causing a LO to appear on the terminal labeled "Enable the SCR Gate Control Circuit." This results in shutting off the welding transformer by disabling the SCR gate control circuit. Meanwhile, the "Preset the Heat-Cool Counter" terminal is still HI (it stays HI for 100 μs as the system steps from **Heat** to **Cool**), so the preset numbers on the COOL SELECTOR SWITCHES are shifted into the Heat-Cool Counter. When the HEAT-COOL ONE-SHOT output pulse ends after 100 μs, AND5 is enabled once again. The Heat-Cool Counter again starts counting down as it receives the 60-Hz pulses.

When the Heat-Cool Counter reaches zero, indicating the **Cool** subinterval is complete, it sends another negative edge to the HEAT-COOL ONE-SHOT by way of the "Heat-Cool Counter Is Counted Out" terminal. The one-shot repeats its previous actions, namely toggling the HEAT-COOL FLIP-FLOP into the ON (**Heat**) state, presetting the HEAT SELECTOR SWITCH settings into the Heat-Cool Counter and then reenabling AND5 and AND6 when it finishes firing. Notice that as the system goes from the **Cool** subinterval to the **Heat** subinterval, a positive edge appears at *T* of the WELDING

CURRENT PULSATION ONE-SHOT. This one-shot does not fire on a positive edge, and no pulse appears at the "Welding Current Pulsation Count Pulse" terminal. This is proper because the Interval Time Counter is supposed to count only when a welding current pulsation is complete. No current pulsation has just taken place, so no count pulse is delivered.

This cycle is repeated over and over until the proper number of welding current pulsations have been counted by the Interval Time Counter. At that point the system will step out of **Weld** and into **Hold.** The **Weld** terminal in Fig. 7–8 will go LO, and the entire Heat-Cool Stepping and Gating Circuit will be disabled.

7-8 ■ HEAT-COOL COUNTER AND HEAT-COOL COUNTER PRESET CIRCUIT

The Heat-Cool Counter is identical to the Interval Time Counter. It consists of two decade down-counters cascaded together, and it has the same zero detection circuitry. The arrangement for presetting the Heat-Cool Counter is also similar to the arrangement for the Interval Time Counter. Figure 7–9 shows the Heat-Cool Counter along with the Heat-Cool Counter Preset Circuit.

To preset the Heat-Cool Counter, the LOAD terminals must be driven LO at the same time that LOs are delivered to the common terminals of a pair of selector switches. If the HEAT SELECTOR SWITCH common terminals go LO, the digits selected on those switches are preset into the decade counters. This is what is done at the start of a **Heat** subinterval. If the COOL SELECTOR SWITCH common terminals go LO, digits selected on those two switches are preset into the decade counters. this is what is done at the start of a **Cool** subinterval.

When the HEAT-COOL ONE-SHOT in Fig. 7–8 fires, it temporarily raises the "Preset the Heat-Cool Counter" terminal HI on the left of Fig. 7–9. This terminal applies a HI to *I*6, which drives the LOAD terminals of both decade counters LO, permitting them to accept the data on their preset inputs. Meanwhile, the 100-μs pulse on the "Preset the Heat-Cool Counter" terminal also goes up and partially enables NAND5 and NAND6. If the system has just entered the **Heat** subinterval of the **Weld** interval at this time, then NAND6 will be fully enabled because the **Weld** and **Heat** terminals will both be HI. The LO output of NAND6 will enable the HEAT ENCODERS, thus presetting the number of **Heat** cycles into the Heat-Cool Counter.

On the other hand, if the system has just entered the **Cool** subinterval, the NAND5 output goes LO. This enables the COOL ENCODERS, loading the number of **Cool** cycles into the Heat-Cool Counter.

When the HEAT-COOL ONE-SHOT pulse ends after 100 μs, the presetting operation is complete, and the LOAD terminals go back HI. The Heat-Cool Counter is now ready to begin receiving 60-Hz count pulses at the terminal labeled "Count Pulses to the Heat-Cool Counter." This terminal originates in Fig. 7–8, at the HEAT-COOL COUNT GATE, AND5. This gate is able to pass 60-Hz pulses as soon as the HEAT-COOL ONE-SHOT output pulse is finished.

When both decade counters in Fig. 7–9 reach zero, meaning that the Heat-Cool Counter has counted out, the outputs of OR8 and OR9 both go LO. This causes the output of OR10 to go LO, creating a negative edge at the "Heat-Cool Counter Is Counted Out" terminal. This negative edge is fed back to the HEAT-COOL ONE-SHOT in Fig. 7–8, where it causes the Heat-Cool Stepping and Gating Circuit to step into the next subinterval. This action was described in Sec. 7-7.

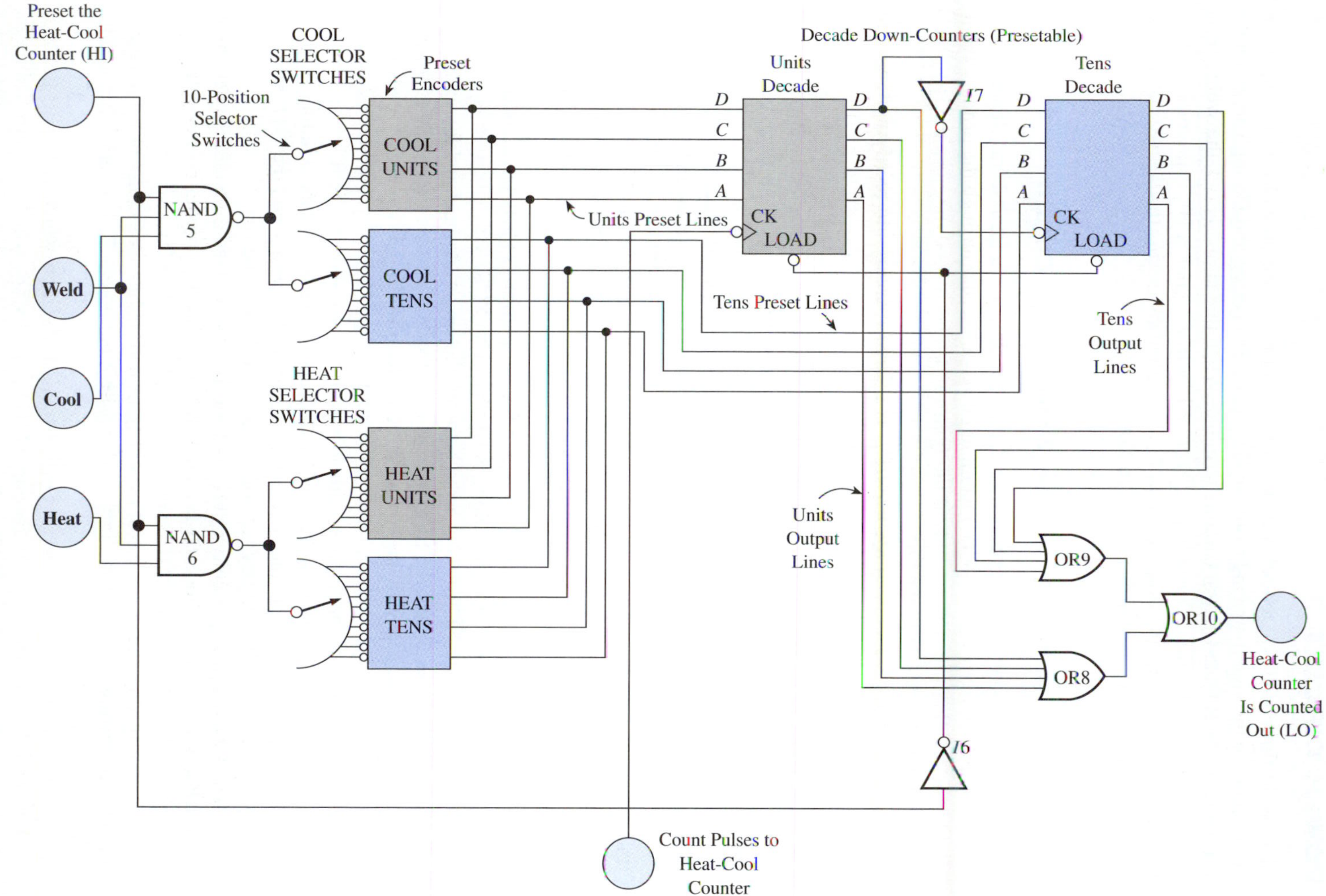

FIGURE 7–9
Schematic of the Heat-Cool Counter (block H) and the Heat-Cool Counter Preset Circuit (block G). The number of ac cycles in the **Heat** subinterval is set by the two **Heat** selector switches, and the number of cycles in the **Cool** subinterval is set by the **Cool** selector switches.

7-9 ■ THE WELD POWER CIRCUIT

The Weld Power Circuit controls the flow of current to the welding electrodes. Its job is to respond to the signal arriving from the Heat-Cool Triggering and Gating Circuit on the "Enable the SCR Gate Control Circuit" terminal. When that signal is LO, the Weld Power Circuit prevents the flow of welding current. When that signal is HI, it permits the flow of welding current. In addition, the Weld Power Circuit maintains the desired conduction angle while welding current is allowed to flow.

7-9-1 Simplified View of the Weld Power Circuit

A simplified schematic of the Weld Power Circuit is shown in Fig. 7–10(a). It shows that current can flow from the 460-V ac incoming power lines, and through the primary winding of the welding transformer, only if the *ignitron* has been fired, because the ignition is in series with the primary winding.

An ignitron (pronounced *igg-nyé-tron*) is a large mercury-arc rectifying tube. The circuit behavior of an ignitron is very much like the circuit behavior of an SCR. It acts either like an open switch in series with the load, or like a closed switch in series with the load. It passes current in only one direction, from anode to cathode. It does not automatically turn ON when the anode to cathode voltage polarity goes positive, but it must be turned ON, or fired, by a third control terminal, called the *ignitor.*

A burst of current into the ignitor lead and out the cathode lead will fire the ignitron, after which it will remain ON until the anode to cathode voltage changes polarity.

The advantage of the ignitron over the SCR is a very simple one: current capacity. In situations where tremendous surges of current must be supplied to a load, the ignitron is often the only device equal to the task. Ignitrons are available which can deliver regular current surges as large as 10 000 amperes. No SCR can come close to that current capacity.

The amount of ignitor current needed to fire an ignitron is fairly large, usually about 25 A. Therefore the ignitor circuit alone is worthy of an SCR. This situation is shown in Fig. 7–10(a), where an SCR is connected between the anode terminal and the ignitor terminal of the ignitron. When the SCR fires, it establishes a current flow path as follows: from the L_1 incoming power line, through the SCR, into the ignitor which leads into a mercury pool inside the ignitron, through the pool of liquid mercury, and out the cathode lead. Thus the firing of the ignitron coincides with the firing of the SCR.

The SCR itself is fired when a pulse appears on the secondary winding of pulse transformer T_2 in its gate circuit. A secondary pulse will appear when the UJT delivers a burst of current into the primary side of T_2, as we have seen before.

The UJT triggering circuity in Fig. 7–10(a) is a fairly standard circuit. When the ac power line goes positive, diode D_1 becomes forward biased, applying a positive half cycle of ac voltage to the R_3-ZD1 combination. Zener diode ZD1 clips the waveform at +15 V shortly after the positive half cycle begins and maintains a constant dc voltage to the UJT triggering circuit for the rest of the positive half cycle. This relationship is illustrated in Fig. 7–10(b) and (c).

Notice transistor Q_2, however, Q_2 is a transistor switch which can short out ZD1 and prevent the UJT from ever triggering. Q_2 is driven by Q_1, which is controlled by the "Enable the SCR Gate Control Circuit" terminal in the lower left of Fig. 7–10(a). This terminal originates in Fig. 7–8, the Heat-Cool Stepping and Gating Circuit. When this terminal is HI, Q_1 is ON, causing a LO to the base of Q_2. Q_2 Turns OFF, thus permitting the dc voltage to be established across the UJT trigger circuit.

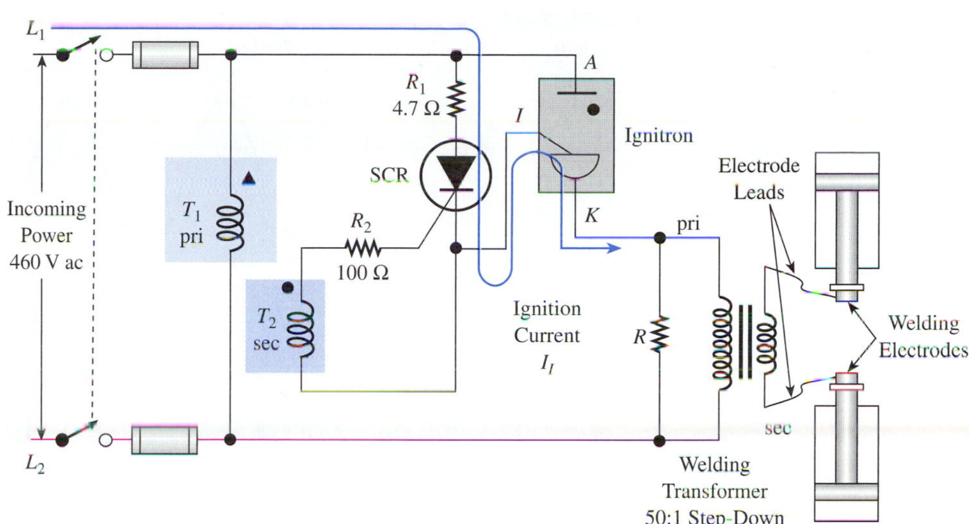

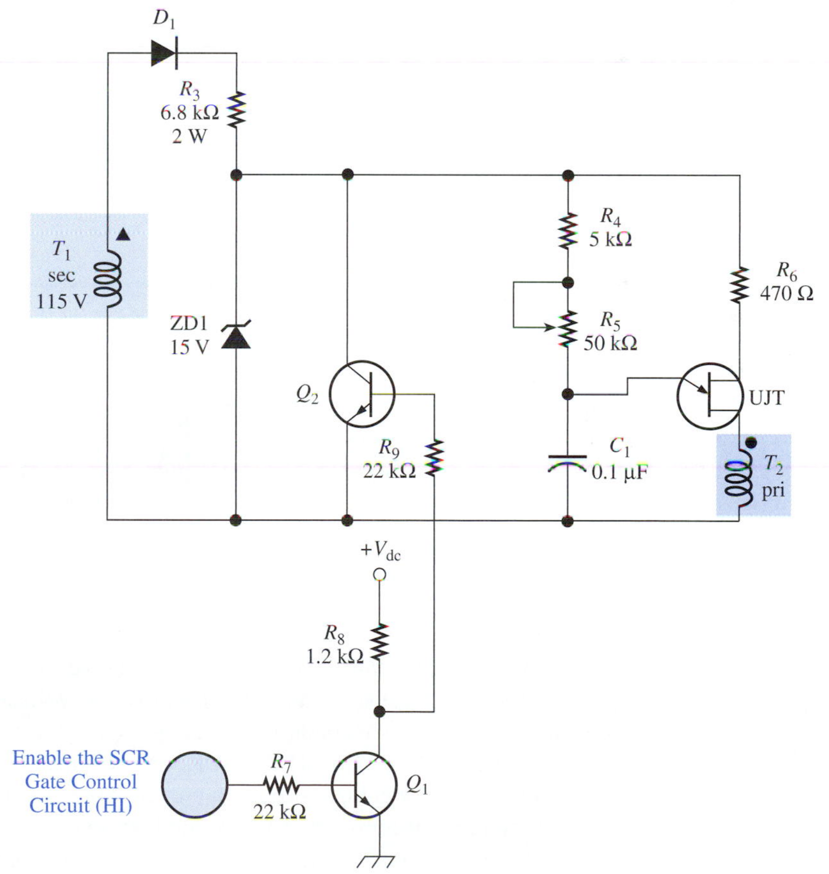

(a)

FIGURE 7–10
(a) Simplified schematic of the Weld Power Circuit (block 1). When the input terminal goes HI, Q_1 turns ON and Q_2 turns OFF. This removes the short circuit across ZD1 and allows C_1 to begin charging when the ac line crosses into its positive half cycle. The pulse transformer triggers and SCR, which in turn fires an ignitron. The ignitron actually carries current to the welding transformer. (b) Waveform of T_1 secondary voltage. (c) Clipped sine wave which powers the UJT timing circuit. (d) Voltage across C_1. When it reaches the V_P of the UJT, the UJT fires; this dumps almost all the charge which is on the top plate of C_1. (e) Pulses of current into the primary side of the pulse transformer. (f) Rectified welding current.

FIGURE 7–10
(continued)

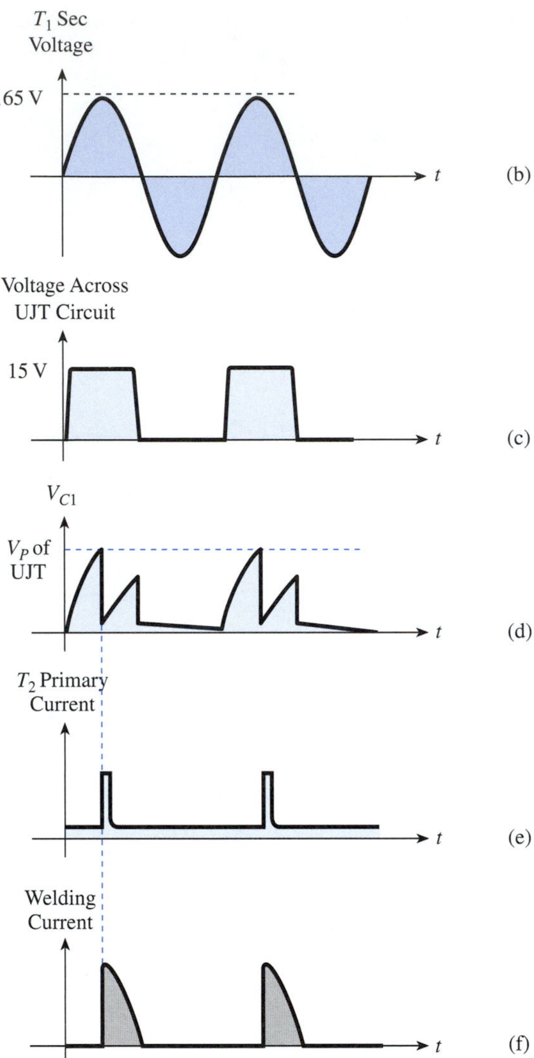

If this control terminal is LO, however, Q_1 turns OFF, causing Q_2 to turn ON. With Q_2 ON, the zener diode is shorted out and no dc voltage can appear across the UJT. In this case the entire T_1 secondary voltage is dropped across R_3.

In this way, the "Enable the SCR Gate Control Circuit" terminal is able to either permit the flow of welding current or prevent the flow of welding current.

If welding current is being permitted, the ZD1 voltage causes capacitor C_1 to start charging when the positive half cycle begins. The charging rate is set by variable resistance R_5. If C_1 charges fast, its voltage reaches the peak voltage of the UJT quickly, and the UJT, SCR, and ignitron fire early in the half cycle. This results in a large conduction angle. If C_1 charges slowly, the UJT, SCR, and ignitron fire late, resulting in a small conduction angle and lower average welding current.

The waveforms of current in the pulse transformer and current through the welding transformer are shown in Fig. 7–10(e) and (f). The welding current waveform is somewhat idealized. It would not really look that clean because of the inductive properties of the transformer windings.

7-9-2 The Real Weld Power Circuit

In the foregoing discussion of the Weld Power Circuit in Sec. 7-9-1, two simplifying changes have been made:

1. We have shown a *single-phase* welding transformer instead of the *three-phase* transformer which is actually used.
2. We have shown only one ignitron-SCR pair, which results in unvarying direction of current flow through the welding transformer and through the wheel metal itself.

In the actual system, there are two ignitron-SCR pairs per phase, which allows the direction of welding current to *reverse* from one current pulsation to the next. This prevents saturation of the welding transformer core. Transformer core saturation can occur due to buildup of residual magnetism if the current always flows in the same direction through the windings.

Figure 7–11(a) shows a three-phase welding transformer, with each phase having two ignitron-SCR pairs. The ignitron-SCR pairs point in opposite directions, which allows the welding current reversal spoken of above earlier.

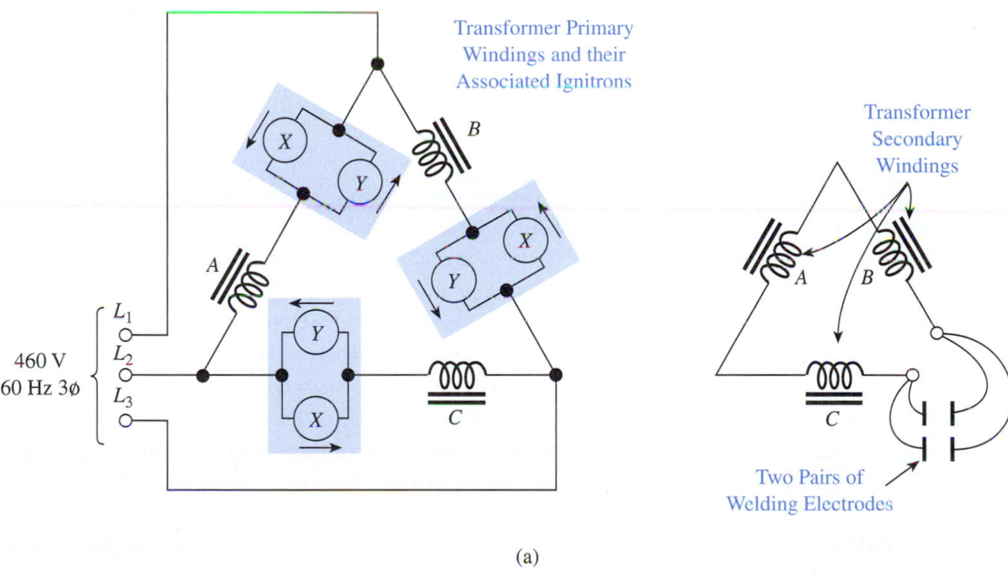

FIGURE 7–11
(a) Three-phase welding transformer with the primary winding connected in delta. During one welding current pulsation the *X* ignitrons fire, and during the next current pulsation the *Y* ignitrons fire. (b) The firing delay angle for the *B* phase must be no less than 60°; this ensures that the *B* winding is not energized before the *A* winding is deenergized. (c) Welding current waveform for a 60° firing delay angle. Only one pulsation is shown. (d) Welding current waveform for a 90° firing delay angle. Two pulsations are shown, illustrating the current reversal from one pulsation to the next.

FIGURE 7–11
(*continued*)

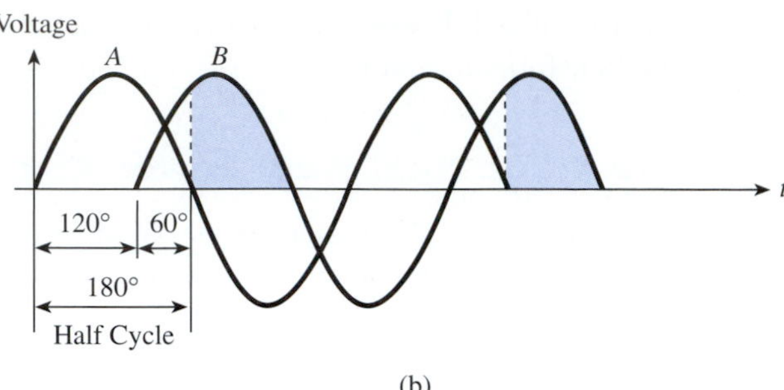

(b)

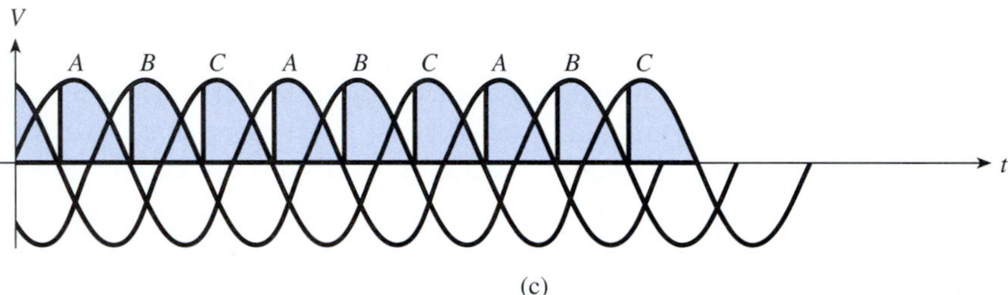

(c)

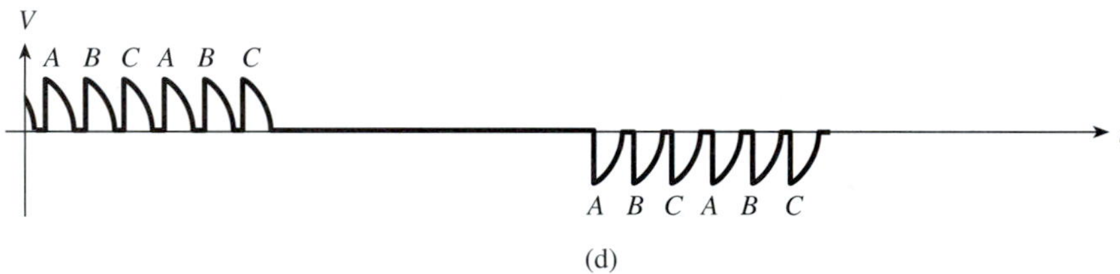

(d)

When the three-phase welding transformer is used, only one phase may be energized at any instant. To understand this, refer to Fig. 7–11(a). If it is desired to energize the *A* phase of the transformer, this can be done by firing the A_X ignitron-SCR pair, allowing current to pass through the *A* primary winding from power line L_1 to L_2, or phase *A* can be energized by firing the A_Y ignitron-SCR pair, allowing current to pass through the *A* primary winding in the opposite direction, from L_2 to L_1. In either case, voltage will be induced in the phase *A* secondary winding, which then delivers welding current to the electrodes. If phase *A* is delivering the welding current, phases *B* and *C* must not interfere. Note that the *B* and *C* secondary windings are connected in series with the phase *A* secondary winding in Fig. 7–11(a). There must be no voltage created in the *B* and *C* secondary windings during the time the *A* secondary is trying to deliver current to the welding electrodes. The *A* secondary must be able to get a clear "shot" at the electrodes. This is why the *B* and *C* primary windings must not be energized while the *A* primary winding is energized. Of course, this argument works the same way when the *B* phase is driving the electrodes, or when the *C* phase is driving the electrodes.

The requirement that only one phase be energized at any instant can be met by designing the SCR gate control circuits so that the ignitrons have a firing delay angle *no less than* 60° (conduction angle no greater than 120°). Figure 7–11(b) shows why this is so. When one transformer phase energizes, it is bound to deenergize when the phase voltage which is driving it crosses into the negative region. This always occurs 60° after the *next* phase voltage crosses into the positive region. This is shown clearly by Fig. 7–11(b). Therefore, if the firing of the ignitrons is delayed by at least 60°, it is impossible for any given transformer phase to begin conducting until the preceding phase has stopped conducting.

This complete idea is illustrated in Fig. 7–11(c), which shows all three phase voltages. In that waveform the phase *A* voltage is shown passing into the negative region 60° after the phase *B* voltage has passed into the positive region. By properly sizing the SCR gate control components, it is possible to prevent ignitron *B* from firing during the first 60° of the cycle of the phase *B* voltage. This ensures that the phase *A* voltage has gone negative by the time the phase *B* ignitron fires, guaranteeing that the *A* phase of the transformer is deenergized before the *B* phase is energized.

The argument given here for the *A*-*B* phase relationship also holds true for the *B*-*C* phase relationship and for the *C*-*A* phase relationship.

In Fig, 7–11(c), the firing delay angle is exactly 60°. It would not have to be exactly 60° of course. The only requirement is that it be not less than 60°.

Reversal of welding current flow direction from one current pulsation to the next is accomplished by alternating between ignitron-SCR pairs. During one welding current pulsation, the ignitron-SCR pairs labeled *X* in Fig. 7–11(a) are sequentially fired. During the next welding current pulsation, the ignitron-SCR pairs labeled *Y* are sequentially fired. That is, during one welding current pulsation, the A_X pair is fired, then the B_X pair is fired, then the C_X pair is fired, and this sequence is repeated as many times as called for by the HEAT SELECTOR SWITCHES. During the next welding current pulsation, the A_Y pair is fired, then the B_Y pair, then the C_Y pair, and this sequence is repeated as many times as the HEAT SELECTOR SWITCHES are set for. The resulting waveform is illustrated in Fig. 7–11(d), this time with a 90° firing delay angle.

Since the actual Weld Power Circuit contains six ignitron-SCR pairs, the control circuit drawn in Fig. 7–10(a) is actually repeated six times. Also, the alternating between the *X* ignitron-SCR pairs and the *Y* ignitron-SCR pairs, which causes welding current reversal, is controlled by the Heat-Cool Stepping and Gating Circuit discussed in Sec. 7-7. Additions would be necessary to that circuit to enable it to alternate between the *X* and *Y* pairs. Those additions will not be shown here. This is not because they are difficult to understand in themselves, but only because they would add further complexity to an already confusing circuit arrangement.

A schematic diagram of the entire welding sequence control circuit is shown in Fig. 7–12. To keep the size of the figure manageable, the Weld Power Circuit is not included in the figure. Instead, the "Enable the SCR Gate Control Circuits (HI)" line is shown leaving Fig. 7–12(d) from the extreme right-hand side. This line signals the Weld Power Circuit, telling it when to start and stop the actual welding operation, as explained in Sec. 7-9-1.

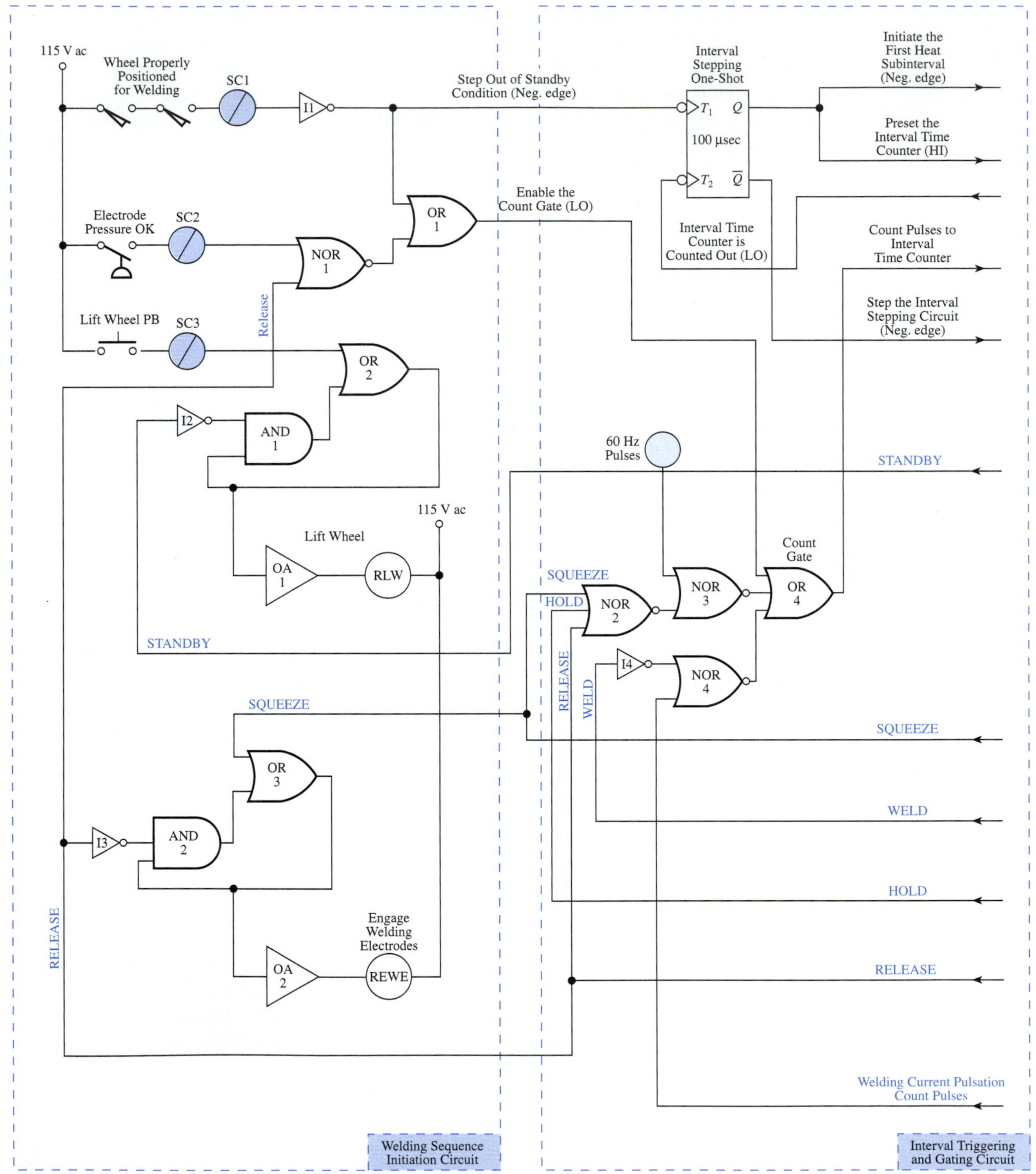

FIGURE 7–12
The welding sequence control circuit in its entirety. Each of the nine subcircuits is enclosed by a dashed line and labeled.

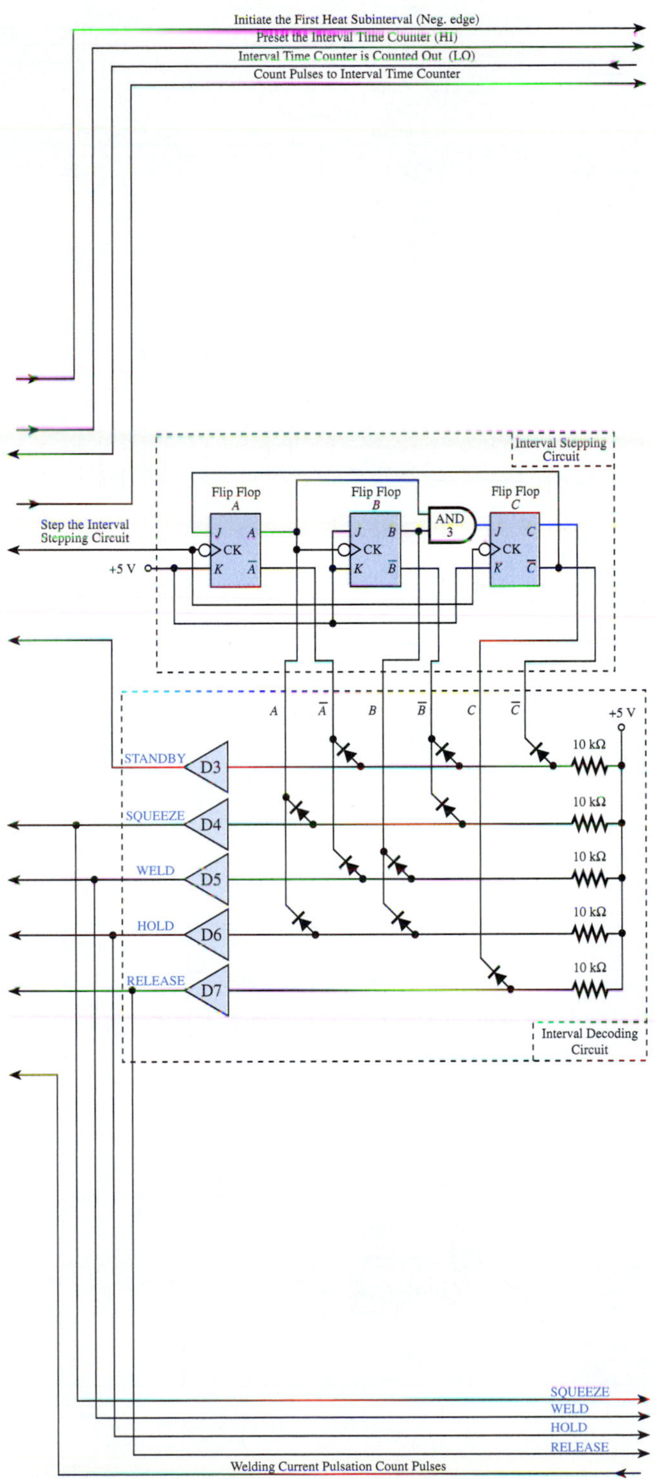

FIGURE 7–12
(continued)

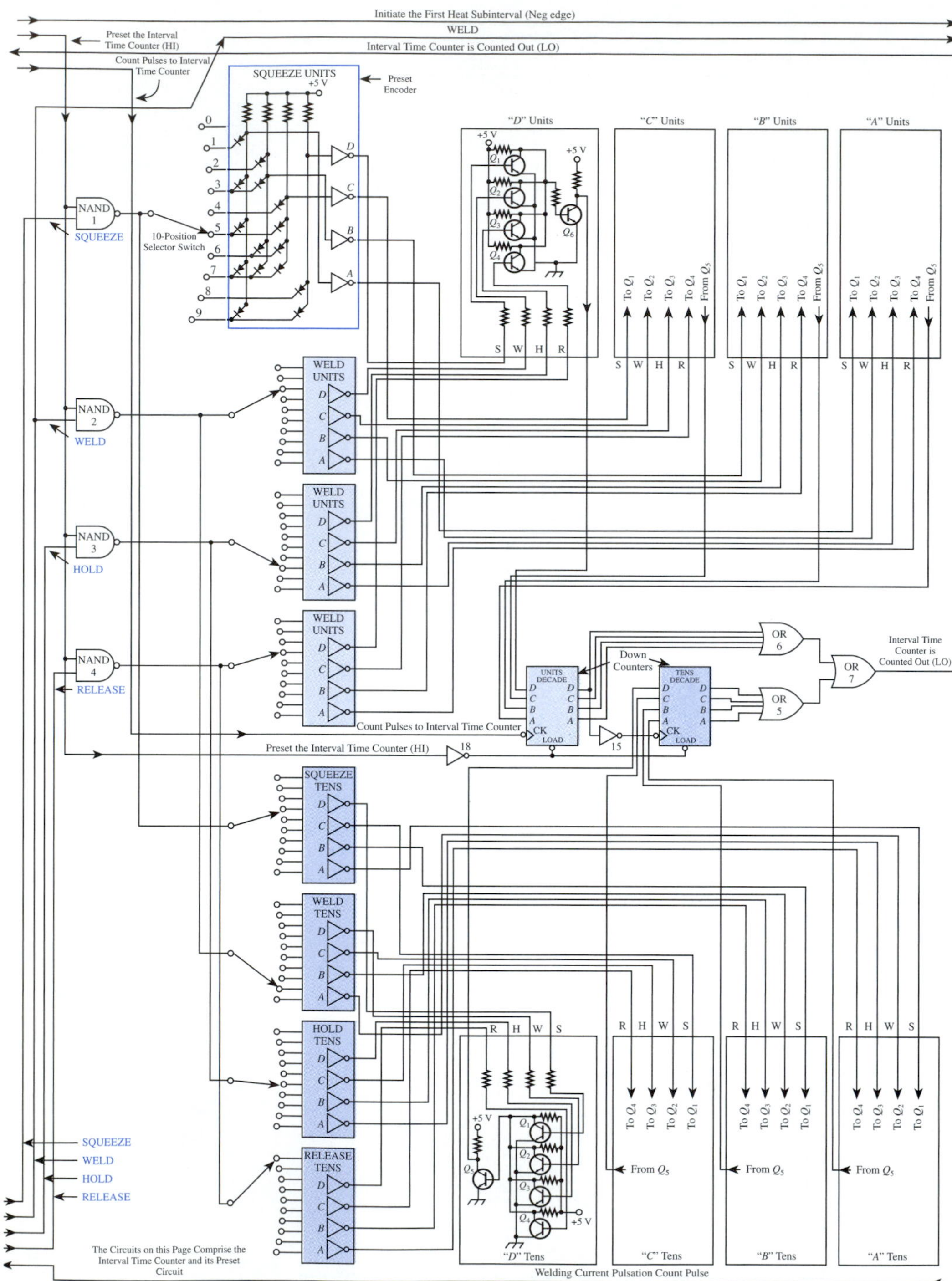

FIGURE 7–12
(continued)

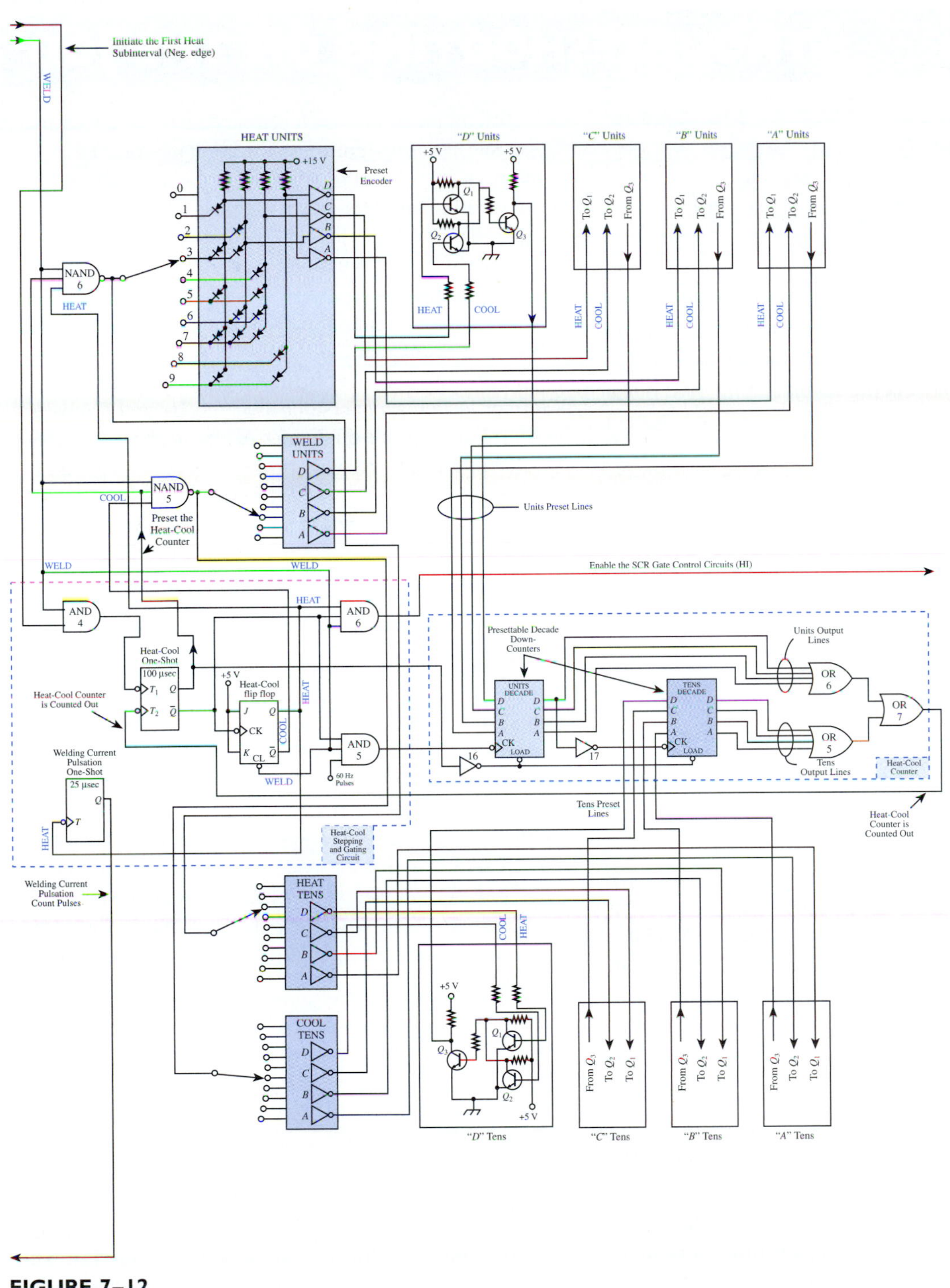

FIGURE 7–12
(continued)

TROUBLESHOOTING ON THE JOB

RE-CREATING THE SCHEMATIC DIAGRAM OF THE WELD POWER CIRCUIT WITH ALTERNATING-POLARITY HEAT SUBINTERVALS

The schematic prints for the Weld Power Circuit (block I) have been lost. Nobody can remember the circuit details, because nobody has looked at the schematic in the last few years. There's been no reason to, since the system has been functioning perfectly.

Now there is trouble in the Weld Power Circuit, apparently due to the inability of the real three-phase welding transformer to alternate the direction of weld current from one **Heat** subinterval to the next, as suggested by Fig. 7–11(d). Your supervisor seems to recall that you were the last technician to work on the Weld Power Circuit, about 3 years ago. Therefore you have been given the assignment of re-creating the exact schematic diagram.

With the system deenergized, it would be possible to remove printed-circuit boards and trace out the circuit's connections one copper track at a time and one interconnecting wire at a time. However, you believe that you can perform the task faster by drawing what you *think* the circuit should look like. Once you have drawn your expected circuit schematic, you intend *then* to trace out the individual copper tracks and bundled interconnecting wires to verify your expectation, or alter it, as the case may turn out.

You do recall that each primary phase is driven by opposite-pointing ignitron-SCR combinations, as suggested in Fig. 7–11(a) and by Fig. 4–12. You also recall clearly that there was a toggling *JK* flip-flop that performed the logic task of switching between the *X* ignitron-SCR combinations and the *Y* ignitron-SCR combinations.

With these remembrances in mind, make your best attempt to re-create the circuit schematic. It is sufficient to show only the A phase of the welding transformer, it being understood that the B- and C-phase control circuits are duplicates of the A phase.

"While I'm at it," you think to yourself, "I'll show the realistic secondary winding situation, with just *one* secondary winding wrapped on the common core of the welding transformer." The original schematic, as you recall it, showed three separate secondary windings interconnected in a delta configuration, which is a normal three phase transformer construction. How- ever, in this particular application, with there never being two primary windings simultaneously energized, a single secondary winding suffices to respond to the magnetic flux variations produced by three primary windings. You recall observing and understanding several years ago that the schematic representation of three secondary windings was a convenient way for the system manufacturer to make the transformer seem familiar, but you resolve to yourself that you will draw the true state of affairs regarding this particular three-phase welding transformer.

■ SUMMARY

- The standard industrial welding sequence in (1) squeeze the electrodes against the work; (2) weld, by energizing the welding transformer; (3) hold the electrodes tight against the work until the welded metal resolidifies; (4) release the electrodes, withdrawing them from the work; (5) stand by until another piece of work is brought into place.
- A welding interval consists of alternating heat and cool subintervals. The welding transformer delivers current to the work for several successive ac line cycles during the heat subinterval; then the transformer deenergizes for several successive line cycles during the cool subinterval.

- Within the heat subinterval, conduction angle may be variable for the SCRs that energize the welding transformer.
- Extremely large welding current surges may be switched On and Off by an ignitron, which is a three-electrode tube containing liquid and gaseous mercury metal.

QUESTIONS AND PROBLEMS

Section 7-1

1. Is it necessary to relieve the DISENGAGE ELECTRODES hydraulic line when pressuring up the ENGAGE ELECTRODES hydraulic line of Fig. 7–1(b)? Explain.

Section 7-2

2. Name the five main intervals of an automatic welding sequence in order. Explain what happens during each interval.
3. Why is the **Hold** interval necessary?
4. Name the two subintervals of the **Weld** interval. Explain what happens during each one.

Section 7-3

Questions 5–8 can be answered by referring solely to Fig. 7–3.

5. When the Interval Time Counter has counted out, which subcircuit does it pass this information to?
6. To which subcircuits does the Interval Stepping Circuit and Decoder send the information about the interval the system is currently in?
7. Which line is used to send count pulses to the Heat-Cool Counter?
8. Which line is used to tell the Interval Stepping Circuit to step into a new interval?

Section 7-4

9. Give an approximate figure for how much time each of the five intervals takes.
10. In this system, what is the longest possible time the **Hold** interval could last?
11. Repeat Question 10 for the **Release** interval.
12. Repeat Question 10 for the **Squeeze** interval.
13. Repeat Question 10 for the **Heat** subinterval.
14. Repeat Question 10 for the **Cool** subinterval.
15. Repeat Question 10 for the **Weld** interval.
16. What conditions are necessary to drive the output of OR1 LO? Express your answer in basic system terms, not in terms of other logic gates. That is, do not say simply that the I1 output must go LO; tell what must happen physically in the system to *make* the I1 output go LO.
17. What conditions are necessary to energize RLW, the LIFT WHEEL relay? Same instructions as for Question 16.
18. What conditions are necessary to energize REWE, the ENGAGE ELECTRODES relay. Same instructions as for Question 16.
19. During which intervals are count pulses delivered to the Interval Time Counter by way of NOR3?

20. What is the purpose of NOR1 and the **Release** terminal connection to NOR1 in Fig. 7–4? Why couldn't we just run the SC2 output into OR1 and eliminate NOR1?
21. Why is the **Release** terminal wired to the I3 input instead of the **Standby** terminal? What would happen during the automatic cycle, if the **Standby** terminal were wired there by mistake?
22. Why is it not necessary to disable OR4 during the firing time of the INTERVAL STEPPING ONE-SHOT? (At first consideration it appears that it *would* be necessary in order to prevent count pulses from entering the counter during the presetting operation.)
23. When is the INTERVAL STEPPING ONE-SHOT triggered from its T_1 terminal? When is it triggered from its T_2 terminal?

Section 7-5

24. During which interval(s) does the output of AND3 go HI?
25. In the Interval Stepping Circuit, FF*A* receives a negative edge at its CK terminal every time the system is about to step into a new interval. When exactly does FF*B* receive a negative edge at its CK terminal? Repeat the question for FF*C*.
26. Explain why only one diode is necessary for the decoding of the **Release** interval state in the Interval Stepping Circuit Decoder. Why not two or three diodes like all the other states?
27. Figure 7–5 shows a diode decoding matrix specially built for this decoding job. Is this special matrix really necessary, or could you make a standard BCD to 1-of-10 decoder work? Explain carefully.

Section 7-6

28. Suppose the system has just entered **Hold** and the settings on the HOLD SELECTOR SWITCHES are being preset into the Interval Time Counter. The selector switches are set to provide a hold time of 47 cycles. Identify the level of each one of the units preset lines, *D*, *C*, *B*, and *A*, and also each one of the tens preset lines, *D*, *C*, *B*, and *A*.
29. Under what conditions does the output of OR5 go LO?
30. Under what conditions does the output of OR6 go LO?
31. Under what conditions does the output of OR7 go LO?
32. The preset number entered into the Interval Time Counter at the beginning of the **Weld** interval does not represent how many ac line cycles are necessary to time the counter out. What does that number represent?

Section 7-7

33. When does the output of AND4 go HI? When does it return to the LO level? Same instructions as for Question 16.
34. When is the HEAT-COOL ONE-SHOT triggered from its T_1 terminal? When is it triggered from its T_2 terminal?
35. Does the WELDING CURRENT PULSATION ONE-SHOT fire as the system *enters* the **Heat** subinterval or as the system *leaves* the **Heat** subinterval?
36. When is the HEAT-COOL FLIP-FLOP held cleared by a LO signal on its CL terminal?

37. What conditions are necessary to drive the output of AND6 HI? Same instructions as for Question 16.
38. At what instant does the counting actually take place when welding pulsations are counted by the Interval Time Counter? Does it take place on the positive-going edge of the WELDING CURRENT PULSATION ONE-SHOT output pulse or on the negative-going edge?

Section 7-8

39. If it were desired to adjust the system controls to deliver 24 current pulsations during the **Weld** interval, with each pulsation consisting of 15 cycles of current flow followed by 36 cycles of no current flow, explain how the operator should adjust the following six selector switches: WELD UNITS, WELD TENS, HEAT UNITS, HEAT TENS, COOL UNITS, and COOL TENS.
40. What conditions are necessary to drive the NAND5 output LO? Same instructions as for Question 16.
41. Explain the function of inverters 15 and 17.

Section 7-9

42. Referring to Fig. 7–10, explain why it is impossible for the UJT to fire when the incoming power line polarity is wrong for firing the ignitron.
43. To increase the average welding current during a current pulsation, should the resistance of R_5 be increased or decreased? Explain.
44. Why is it impossible to fire the UJT when the "Enable the SCR Gate Control Circuit" terminal is LO?
45. Exactly how much time elapses between the T_1 secondary voltage crossing through zero and ZD1 clipping the waveform at +15 V?
46. Exactly why C_1 is in a discharged state at the beginning of every positive half cycle of the T_1 secondary. That is, why doesn't the capacitor start out with a residual charge left over from the previous half cycle?
47. What is the peak current through the collector of Q_2 when it shorts out ZD1?
48. Find the minimum and maximum charging time constants for C_1.
49. Would the circuit of Fig 7–10 work properly if the T_2 secondary winding were reversed? Explain.
50. Figure 7–10(a) is drawn for the simplified situation in which the welding current direction is *not* reversed. Assume this means that the *X* ignitron-SCR pairs are the ones being used. Would anything be different for the *Y* ignitron-SCR pairs? That is, should anything in the schematic be changed for the *Y* pairs? Exactly what?

CHAPTER

8

OP AMPS

The term *operational amplifier* refers to a high-gain dc amplifier with a *differential input* (two input leads, neither of which is grounded). Although discrete operational amplifiers are built, designers of industrial electronic circuitry now use integrated circuit operational amplifiers almost exclusively. We will therefore confine our efforts to IC operational amplifiers, hereafter called *op amps*.

An IC op amp is a complete prepackaged amplifier whose operating characteristics and behavior depend almost entirely on the few *external* components connected to its terminals. That is, the voltage gain, input impedance, output impedance, and frequency bandwidth depend almost solely on stable external resistors and capacitors. This means that the various amplifier characteristics can be tailored to fit a particular application merely by changing a few components, without redesigning the entire amplifier. It is this versatility and ease of adjustment that makes op amps popular in industrial control.

OBJECTIVES

1. Draw the schematic arrangements for the following op-amp circuit applications: inverting amplifier, noninverting amplifier, summing circuit, and voltage comparator.
2. Select resistance values for the feedback loop in an inverting or noninverting amplifier or summing circuit. Relate those values to the amplifier's voltage gain (A_{VCL}), input resistance (R_{in}), and output resistance (R_{out}).
3. Describe the offset problem for an op amp amplifier and show what steps can be taken to correct it.
4. Draw and explain the operation of the op amp differential amplifier and calculate feedback resistor values to produce any desired voltage gain.
5. Draw and explain the operation of the op amp voltage-to-current converter and calculate feedback resistor values to produce any desired conversion factor.
6. Draw and explain the operation of op amp integrators and differentiators and calculate feedback resistor and capacitor values to produce any desired time constant.

8-1 ■ OP AMP IDEAS

An operational amplifier, or op amp, is an integrated circuit amplifier having a voltage gain that is so enormous that it is greater than any application could possibly require. Having more voltage gain than required is a fine luxury because it is a simple matter to reduce the voltage gain to a proper value by connecting external feedback resistors. This is the essential op amp idea.

8-1-1 Schematic Symbol—Input Polarities

An op amp has five required terminals, shown in Fig. 8–1. Usually, a dual-polarity dc power supply with equal-magnitude positive ($+V_{CC}$) and negative ($-V_{EE}$) voltages provides power to the op amp circuit. The power-supply common ground terminal does not connect to the op amp itself, though it does of course connect to the op amp's load.

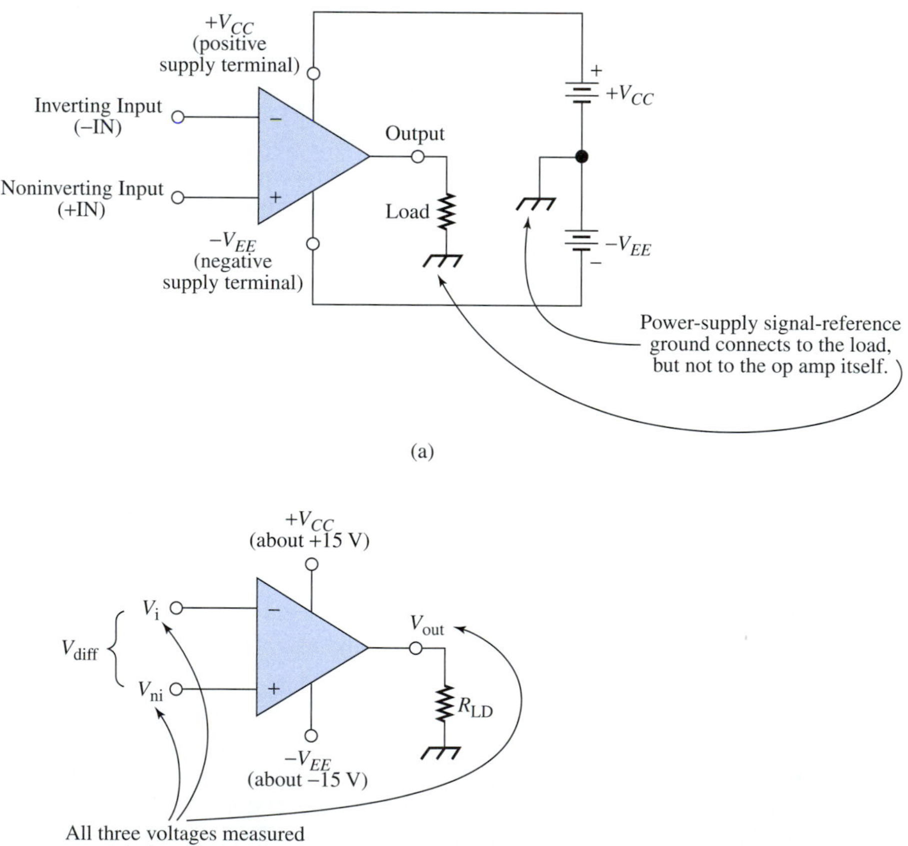

FIGURE 8–1
(a) Op amp terminals (five terminals) connected to a dual-polarity power supply and load device. (b) Symbolizing the signal voltages.

As indicated in Fig. 8–1(a), there are two input terminals labeled – and + . The – input, called the inverting input, is always shown on top in the schematic symbol, and the + input, called the noninverting input, is always shown on the bottom. The names mean what they imply. A signal voltage at the – input tends to produce an output voltage V_{out} that is opposite in polarity to V_i, the voltage at the inverting input. A signal voltage at the +

input tends to produce a V_{out} that is the same polarity as V_{ni}, the voltage at the noninverting input. All voltages are measured with respect to ground, as stated in Fig. 8–1(b).

Think of an op amp like this: When V_i goes positive, V_{out} tends to go negative; when V_{ni} goes positive, V_{out} tends to go positive. The actual input voltage to the op amp itself is the difference between V_i and V_{ni}, symbolized V_{diff} in Fig. 8–1(b). When V_{diff} goes positive on top (more positive at the − terminal) the output V_{out} goes negative. When V_{diff} goes positive on bottom (more positive at the + terminal), V_{out} goes positive.

8-1-2 Open-Loop Gain—Virtual Ground

The voltage gain of a straightforward op amp is huge, as stated earlier. This gain, symbolized A_{VOL} (open-loop amplification of voltage), as distinguished from the voltage gain of a complete amplifier circuit containing external resistors that close a feedback loop, symbolized A_{VCL}, is typically 200 000 : 1, though in some instances it can be greater than 1 000 000 : 1. Thus, an extremely small differential voltage V_{diff} will drive the output into saturation. For $A_{VOL} = 200\,000$, the amount of differential input voltage that is capable of driving the output all the way to saturation is given by

$$V_{diff(sat)} = \frac{V_{out(sat)}}{A_{VOL}} \approx \frac{15\text{ V}}{200\,000}$$
$$\approx 0.07\text{ mV}$$

which is so small as to be almost invisible on a standard oscilloscope with a 2 mV/cm maximum sensitivity.

This smallness of V_{diff} is a crucial fact in understanding the performance of op amp circuits using external feedback components. V_{diff} is so small that we can think of it as virtually zero. If the + input is connected to ground as shown in Fig. 8–2, the − input voltage V_i will be so close to zero (assuming that V_{out} is not saturated) that we can think of it as *virtually at ground* voltage. The − input terminal is called a *virtual ground.*

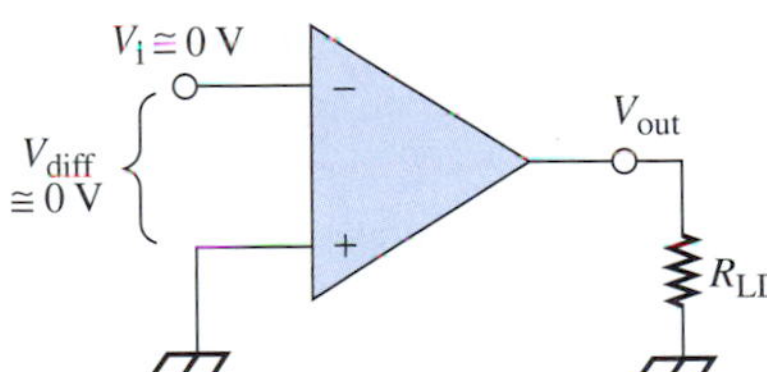

FIGURE 8–2 The voltage gain of an op amp is so great that the differential input voltage V_{diff} must be very small, approximately zero, if it is known that the output is not saturated at either $+V_{CC}$ or $-V_{EE}$. If the noninverting input is connected directly to ground, the inverting input voltage V_i must also be virtually zero.

8-1-3 Open-Loop Input Resistance—Virtually Zero Input Current

The resistance actually present between an op-amp's input terminals is large—typically greater than 1 MΩ. Therefore Ohm's law predicts an input current

$$I_i = \frac{V_{diff}}{R_i} = \frac{\sim 0\text{ V}}{1\text{ M}\Omega} \approx 0$$

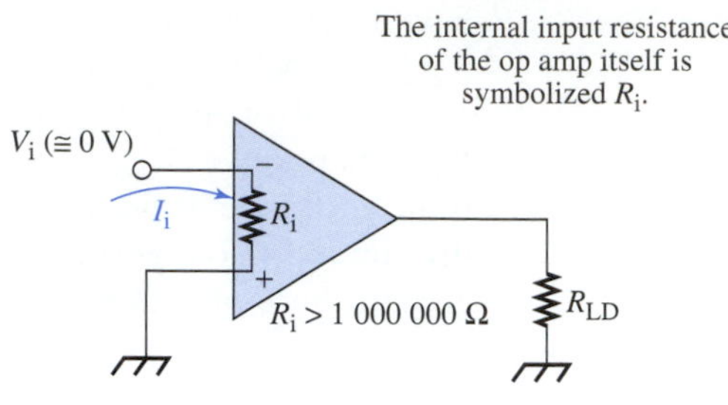
The internal input resistance of the op amp itself is symbolized R_i.
V_i (≅ 0 V)
I_i
R_i
R_i > 1 000 000 Ω
R_{LD}

FIGURE 8–3
With V_i extremely small and R_i rather large, the signal current I_i that enters the inverting input terminal is virtually zero. That is, the op amp exerts essentially zero loading effect on the signal source/feedback network that we intent to connect later.

as indicated in Fig. 8–3. Thus, there is essentially zero current entering and exiting the op amp's input terminals (signal current). This is another important fact because it enables us to say that any current that emerges from the circuit's signal source must all flow through the feedback resistors. None of the signal-source current flows into the op amp itself.

8-2 ■ CLOSED-LOOP CHARACTERISTICS—INVERTING AMPLIFIER

Most op amp applications are accomplished in closed-loop mode, with part of the output voltage fed back to the input in such a way that it tends to decrease the magnitude of input voltage. This idea is called negative feedback. Closing the loop with negative feedback is shown in Fig. 8–4 for an op amp inverting amplifier, one in which output voltage V_{out} has the opposite polarity from input voltage V_S. The feedback resistor R_F combines with resistor R_1 to lower the overall circuit voltage gain. To understand how this works, keep in mind the two conclusions that we drew in Sec. 8-1:

1. Differential input voltage V_{diff} is virtually zero, so the − input terminal voltage V_i is at virtual ground potential.
2. Input current I_i is virtually zero.

In Fig. 8–5, input source voltage $V_S = +0.5$ V has been applied to the circuit. This tends to bring the − inverting input terminal slightly positive, which produces a negative voltage at the output terminal. However, though V_i is slightly positive, it is so slight as to be virtually 0. Therefore the voltage across resistor R_1 is virtually equal to signal source voltage V_S, since

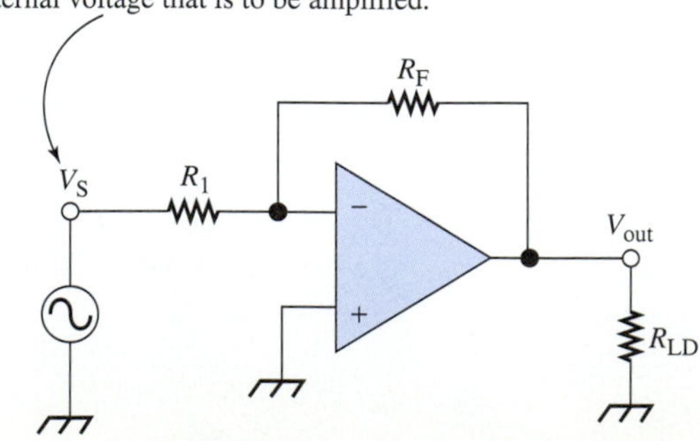
V_S is source (or signal) voltage, the external voltage that is to be amplified.
R_F
V_S
R_1
V_{out}
R_{LD}

FIGURE 8–4
Op amp inverting amplifier. V_{out} and V_S have opposite polarities for dc signals, opposite phases for ac signals. External resistors R_F and R_1 combine to produce negative feedback, making overall closed loop voltage gain A_{VCL} (or just A_V) less than open-loop voltage gain A_{VOL}.

$$
\begin{aligned}
V_{R1} &= \overset{\text{Positive}}{V_S} - \overset{\text{Slightly positive}}{V_i} \\
&= V_S - \sim 0\ \text{V} \\
&\approx V_S \\
&= 0.5\ \text{V}
\end{aligned}
$$

This fact is illustrated in Fig. 8–5(b).

FIGURE 8–5
The situation regarding R_1.

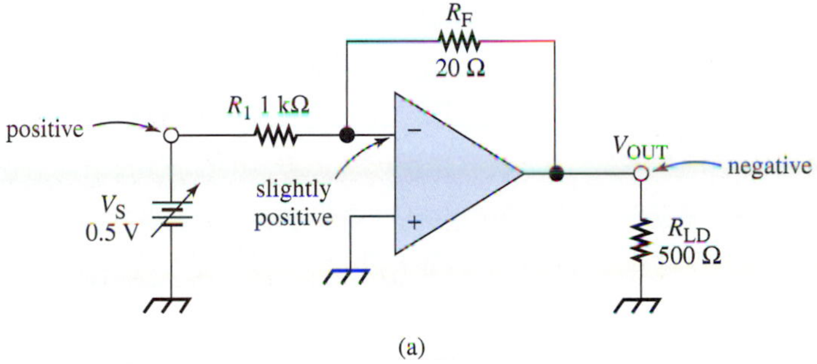

(a)

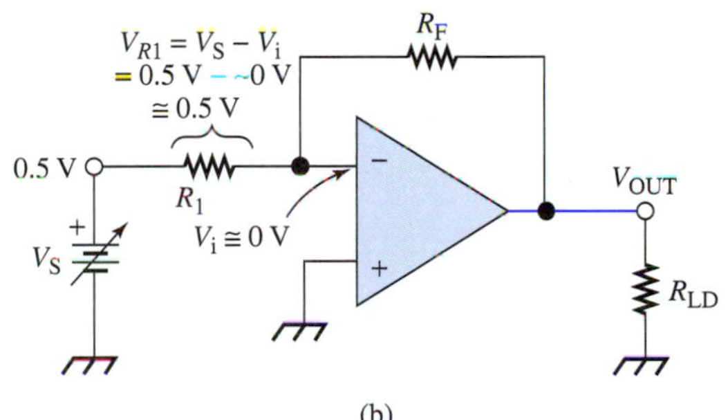

(b)

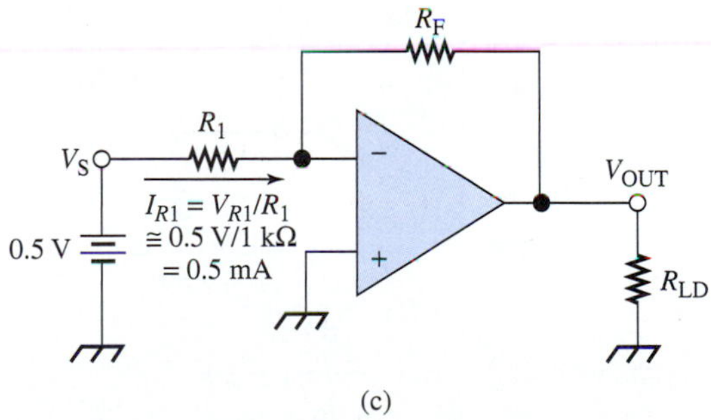

(c)

Current will flow through R_1, in accordance with Ohm's law. As shown in Fig. 8–5(c), the current is given by

$$
\begin{aligned}
I_{R1} &= \frac{V_{R1}}{R_1} \\
&\approx \frac{V_S}{R_1} \\
&= \frac{0.5\ \text{V}}{1\ \text{k}\Omega} = 0.5\ \text{mA}
\end{aligned}
$$

This must be the current value through R_1 because if it were anything different, the V_{R1} voltage drop would be different from 0.5 V, which would cause the voltage V_i at the – input to be different from 0 V. But that is impossible, as established in Sec. 8-1.

The current $I_{R1} = 0.5$ mA arrives at the – input terminal where a split would be theoretically possible—some of the current passing through R_F and some entering the op amp's – input. But we established that virtually zero current enters the – input terminal, so it must be that *all* of the I_{R1} current passes through R_F. This is illustrated in Fig. 8–6(a).

Ohm's law must be satisfied for R_F as well. Therefore the voltage across R_F must be given by

$$
\begin{aligned}
V_{RF} &= (I_{RF})R_F \\
&= (0.5\text{ mA})(20\text{ k}\Omega) \\
&= 10\text{ V}
\end{aligned}
$$

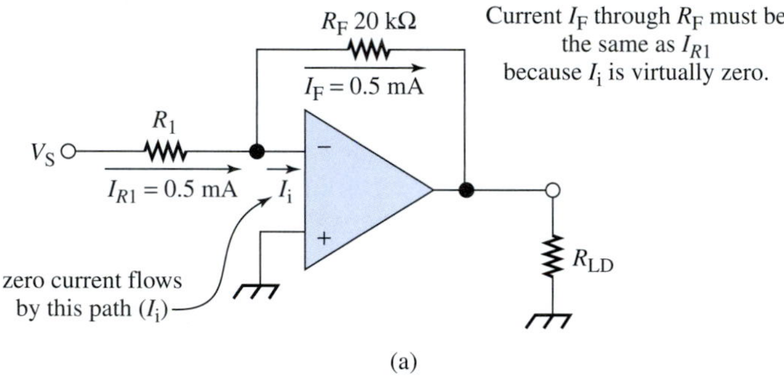

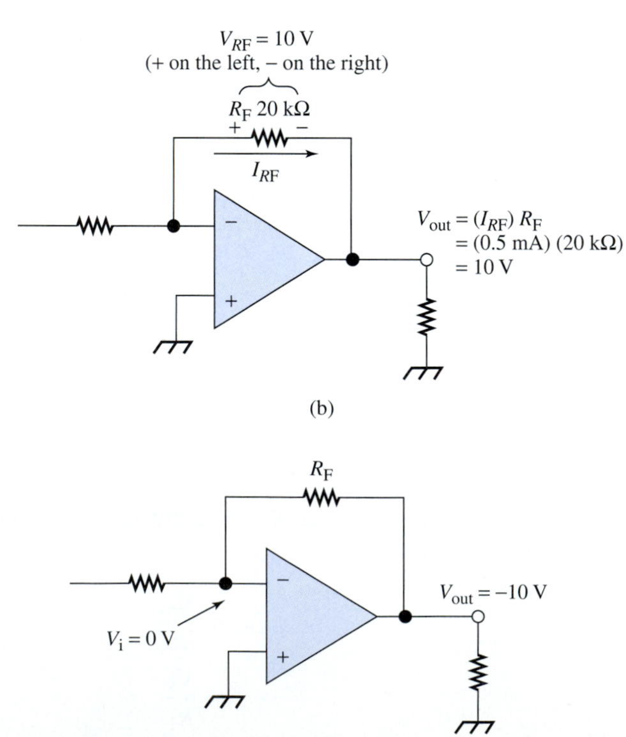

FIGURE 8–6
(a) The current through R_F must be virtually the same as the current through R_1. (b) Because of that fact, the voltage across R_F must be greater than the voltage across R_1 by the factor R_F/R_1. (c) The voltage across R_F is virtually the output voltage V_{out}, since V_i is at virtual ground.

which can happen only if the output terminal goes to the voltage value -10 V, since the voltage on the left side of R_F, which is V_i, is virtually zero. In conclusion, the voltage across R_F must be 20 times greater than the voltage across R_1, because resistance R_F is 20 times greater than R_1 and the two must carry the same current.

The voltage across R_F is virtually equal to output voltage V_{OUT}, just as the voltage across R_1 is virtually equal to input voltage V_S. Therefore the closed-loop circuit's voltage gain, A_{VCL}, is

$$A_{VCL} = \frac{V_{OUT}}{V_S} = \frac{10\text{ V}}{0.5\text{ V}} = 20$$

In general, for an op-amp inverting amplifier of the type shown in Figs. 8–4, 8–5, and 8–6,

Closed-loop voltage gain of the amplifier as a whole

$$A_V = A_{VCL} = \frac{V_{OUT}}{V_S} = \frac{\cancel{(I)}R_F}{\cancel{(I)}R_1}$$

$$A_V = \frac{R_F}{R_1} \qquad \textbf{(8-1)}$$

Input resistance and output resistance. It is immediately apparent that the input resistance R_{in} of an op-amp inverting amplifier is equal to the component resistance R_1, since the right side of R_1 is virtually grounded. That is,

Input resistance of the circuit as a whole

$$R_{in} = R_1 \qquad \textbf{(8-2)}$$

A careful derivation based on the virtual ground idea reveals that the circuit's actual output resistance, R_{out}, is related to the internal output resistance of the op amp itself, R_o in Fig. 8–7, by the equation

Output resistance of the circuit as a whole

$$R_{out} = \frac{R_o}{A_{VOL}/A_{VCL}} \qquad \textbf{(8-3)}$$

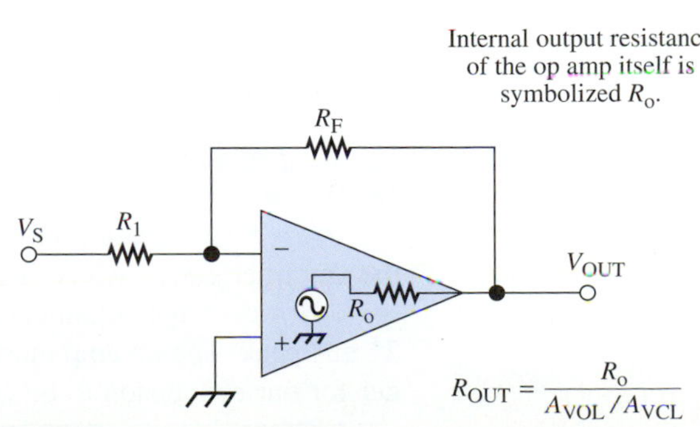

FIGURE 8–7
For an inverting amplifier operating with modest closed-loop voltage gain, the output resistance is extremely low. Of course, this advantage pertains only to loads carrying less than the op amp's maximum rated output current, which can range from about 10 mA maximum to several amperes maximum for heat-sunk audio-power units.

Typical op amp output resistances R_o are less than 100 Ω. So a typical inverting amplifier's R_{out}, assuming a circuit gain of 20 as in Figs. 8–4 through 8–6, would be

$$R_{out} = \frac{100\ \Omega}{200\ 000/20} = \frac{100\ \Omega}{10\ 000} = 0.01\ \Omega$$

which is negligible for almost all applications. That is,

$$R_{out} \approx 0\ \Omega$$

So there is no noticeable loading effect due to increased current demand resulting from load resistance changes within the rated current output range of the op amp.

■ EXAMPLE 8-1

In Fig. 8–4, suppose $R_1 = 2.0\ \text{k}\Omega$, $R_F = 27\ \text{k}\Omega$, and $R_{LD} = 500\ \Omega$. The ac-signal source produces an open-circuit (no-load) voltage of 0.8 V peak, at 400 Hz, driving through its own internal output resistance, $R_{out(source)} = 50\ \Omega$.

(a) Find the amplifier's closed-loop voltage gain from input terminal (V_S) to output terminal (V_{OUT}).

(b) Find the amplifier's input resistance R_{in}.

(c) Taking into account the amplifier's input resistance and the source's 50-Ω output resistance, what value of signal voltage actually reaches the input terminal?

(d) Describe V_{OUT}.

(e) Does the 500-Ω value of load resistance have any appreciable effect on the actual value of V_{OUT}? Explain.

Solution. (a) From Eq. 8-1

$$A_v = \frac{R_F}{R_1} = \frac{27\ \text{k}\Omega}{2.0\ \text{k}\Omega} = \mathbf{13.5}$$

(b) From Eq. 8-2,

$$R_{in} = R_1 = \mathbf{2.0\ k\Omega}$$

(c) The open-circuit signal value of 0.8 V peak is voltage-divided between the amplifier's R_{in} and the source's $R_{out(source)}$.

$$\frac{V_S}{V_{S(oc)}} = \frac{R_{in}}{R_{out(source)} + R_{in}} = \frac{2000\ \Omega}{2000\ \Omega + 50\ \Omega} = 0.976$$

$$V_S = (0.976)(0.8\ \text{V peak}) = \mathbf{0.78\ V\ peak}$$

(d) $V_{OUT} = (A_{VCL})V_S = (13.5)(0.78\ \text{V}) = \mathbf{10.5\ V\ peak}$

V_{out} has $f = 400$ Hz, and it is inverted relative to V_S (phase-shifted by one half cycle, or 180°).

(e) We expected it to have no noticeable effect, since 500 Ω is much greater than the amplifier's R_{out}, which is in the 0.01-Ω neighborhood.

However, the output current is given by Ohm's law as $V_{OUT}/R_{LD} = 10.5\ \text{V}/500\ \Omega =$ 21 mA peak. The op amp must have an output current rating greater than this value in order for our conclusion to be correct. ■

8-3 ■ NONINVERTING AMPLIFIER

Instead of grounding the + input and applying the input signal to the left side of R_1, it is possible to ground the left side of R_1 and apply the input signal to the + input, as shown in Fig. 8–8.

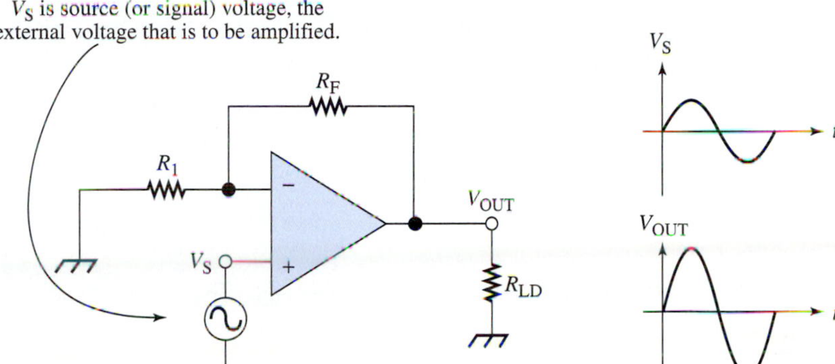

FIGURE 8–8
Op amp noninverting amplifier. (a) Schematic arrangement. (b) Output voltage is in phase with input signal voltage, unlike the inverting amplifier.

Electrical characteristics. From earlier considerations that $V_{diff} \approx 0$ and $I_i \approx 0$, the closed-loop voltage gain formula is derived as

$$V_- = V_S$$

$$I_{R1} = \frac{V_S}{R_1} \qquad I_{RF} = \frac{V_{OUT} - V_S}{R_F}$$

$$I_{RF} = I_{R1} \qquad \frac{V_{OUT} - V_S}{R_F} = \frac{V_S}{R_1}$$

for positive V_S, the voltage across R_F is more positive on the right side, less positive on the left side

$$\frac{V_{OUT}}{R_F} = \frac{V_S}{R_1} + \frac{V_S}{R_F}$$

$$V_{OUT}\left(\frac{1}{R_F}\right) = V_S\left(\frac{1}{R_1} + \frac{1}{R_F}\right)$$

$$\frac{V_{OUT}}{V_S} = R_F\left(\frac{1}{R_1} + \frac{1}{R_F}\right) = \frac{R_F}{R_1} + \frac{R_F}{R_F}$$

Closed-loop voltage gain of a noninverting amplifier

$$A_V = \frac{R_F}{R_1} + 1 \tag{8-4}$$

Thus, for $R_1 = 1\ \text{k}\Omega$ and $R_F = 20\ \text{k}\Omega$ as before, the noninverting amplifier provides voltage gain of

$$A_V = \frac{20\ \text{k}\Omega}{1\ \text{k}\Omega} + 1 = 21$$

Input and output resistances. Input resistance R_{in} is dramatically higher than for the inverting amplifier.

$$R_{in} = R_i\left(\frac{A_{VOL}}{A_{VCL}}\right) \tag{8-5}$$

Input resistance of the amplifier circuit as a whole (R_{in})

Input resistance of the op amp itself (R_i)

The resistance boost can be derived as follows: Refer to Fig. 8–8, imagining internal input resistance R_i connected between the + and − terminals and difference voltage V_{diff} appearing across those terminals. V_{diff} is smaller than V_{out} by a factor A_{VOL}:

$$V_{diff} = \frac{V_{OUT}}{A_{VOL}}$$

The current from the signal source is the current through the op amp's R_i, given by Ohm's law applied to R_i:

$$I_S = \frac{V_{diff}}{R_i}$$

$$I_S = \frac{V_{OUT}}{(A_{VOL})R_i}$$

The output voltage is related to the signal-source voltage by the voltage-gain equation

$$V_{OUT} = (A_{VCL})V_S$$

so

$$I_s = \frac{(A_{VCL})V_S}{(A_{VOL})R_i}$$

and

$$\frac{V_S}{I_s} = R_i\left(\frac{A_{VOL}}{A_{VCL}}\right) = R_{in}$$

(For a noninverting amplifier, R_{in} is much greater than the already large resistance R_i of the op amp itself.)

With the approximate 1-MΩ internal input resistance of the op amp being raised by the factor A_{VOL}/A_{VCL}, it can be seen that the loading effect on a signal source by a noninverting amplifier is negligible.

The output resistance of a noninverting amplifier is very similar to that of an inverting amplifier. For the noninverting amplifier it can be shown that

$$R_{out} = \frac{R_o}{\dfrac{A_{VOL}}{A_{VCL}} + 1} \qquad \textbf{(8-6)}$$

which again produces a negligibly small result for all but very large values of A_{VCL}.

■ EXAMPLE 8-2

In Fig. 8–8, suppose $R_1 = 1.5$ kΩ, $R_F = 22$ kΩ, and $R_{LD} = 300$ Ω. The signal source is the same as in Example 8-1, with open-circuit value $V_{oc} = 0.8$ V peak, through 50-Ω output resistance. The op amp itself has $A_{VOL} = 400\,000$, $R_i = 1.5$ MΩ, $R_o = 40$ Ω and $I_{out(max)} = 50$ mA.

(a) Calculate voltage gain A_{VCL}.
(b) Find the amplifier's R_{in}.
(c) Describe the signal voltage V_S actually appearing at the amplifier's input terminal.
(d) Describe V_{OUT}.
(e) Does the load resistance have any effect on V_{OUT}? Explain.

Solution. (a) From Eq. 8-4,

$$A_V = \frac{R_F}{R_1} + 1 = \frac{22\ \text{k}\Omega}{1.5\ \text{k}\Omega} + 1 = \mathbf{15.7}$$

(b) From Eq. 8-5,

$$R_{in} = R_i\left(\frac{A_{VOL}}{A_{VCL}}\right) = 1.5\ \text{M}\Omega\left(\frac{400\ 000}{15.7}\right) = 1.5\ \text{M}\Omega(25.5 \times 10^3) \approx \mathbf{40\ M\Omega}$$

This is comparable to wire-insulation resistance. The amplifier has effectively infinite input resistance.

(c) With infinite input resistance, all of the no-load signal voltage appears at the amplifier's input terminal—**0.8 V peak.**

(d) $V_{OUT} = (A_V)V_{in} = (15.7)0.8\ \text{V} =$ **12.6 V peak,** 400 Hz, in phase with the input sine wave.

(e) From Eq. 8-6,

$$R_{out} = \frac{R_o}{\frac{A_{VOL}}{A_{VOL}} + 1} = \frac{40\ \Omega}{\frac{400\ 000}{15.7} + 1} \cong \frac{40\ \Omega}{25.5 \times 10^3} \cong 0\ \Omega$$

From Ohm's law,

$$I_{out} = \frac{V_{OUT}}{R_{LD}} = \frac{12.6\ \text{V}}{300\ \Omega} = 42\ \text{mA}$$

R_{out} is negligible and the op amp's 50-mA current limit is not exceeded, so there is no loading effect. ■

8-4 ■ THE OUTPUT OFFSET PROBLEM

The input terminals of an op amp lead to the base terminals of the two front-end transistors in the amplifying circuit. Dc bias current must enter these base terminals via the op amp's − and + input terminals. In either the inverting amplifier, Figs. 8–4 through 8–7, or the noninverting amplifier, Fig. 8–8, dc bias current entering the + terminal has no component resistance to pass through, but current entering the − terminal has the $R_1 - R_F$ combination to pass through.* Refer to Fig. 8–9(a). Thus, one of the bias currents (to the − terminal) produces a voltage drop and the other (to the + terminal) doesn't. This results in a tiny input voltage V_{diff} being applied to the op amp, more negative on the inverting terminal, as shown in Fig. 8–9(b). This voltage is amplified by A_{VOL} to create a positive dc offset voltage at the output. That is, V_{OUT} is some positive dc value even though the signal input voltage is zero.

If an ac signal is then applied, V_{out} oscillates around this positive offset value rather than around zero, as it should. This is shown in Fig. 8–9(c).

This offset problem can be corrected, at least partially, by installing a component resistor in the input lead to the + terminal. This resistance, R_B in Fig. 8–10, should have a value equal to the Thevenin equivalent resistance of the circuit that carries I_{B1}. Looking backward from the − terminal, resistances R_1 and R_F appear in parallel, so

*The signal source's internal output resistance complicates this comparison.

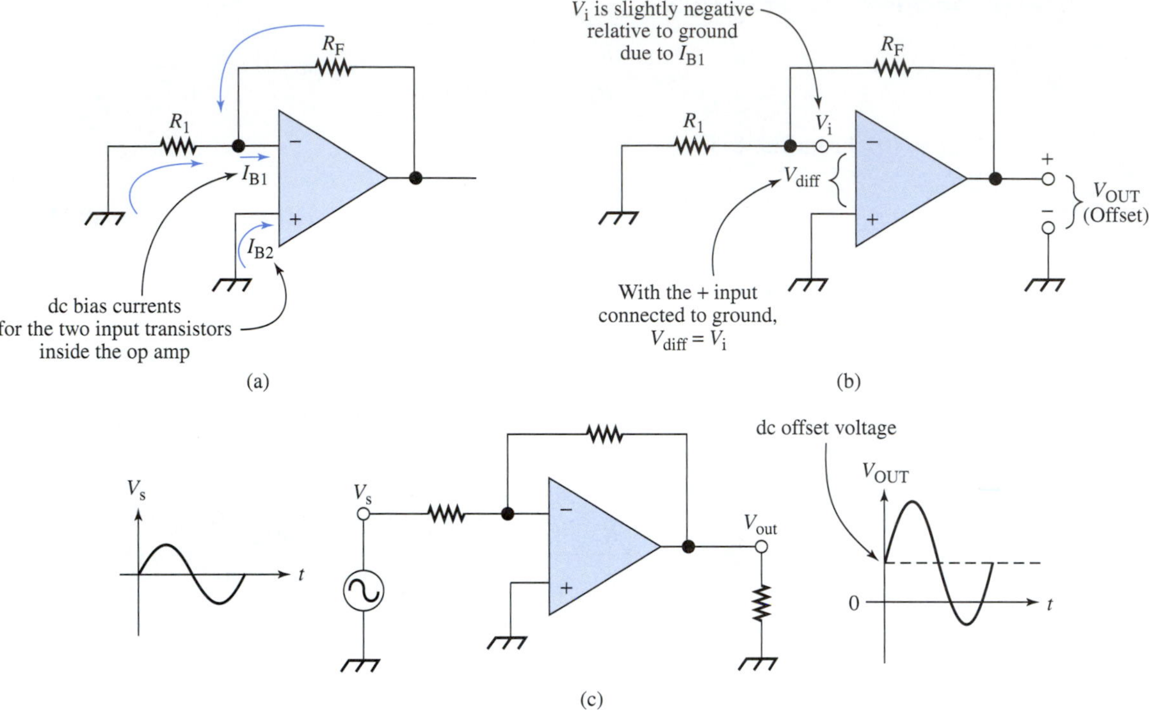

FIGURE 8–9
The output offset problem resulting from input dc bias currents.

$$R_B = \frac{R_1 R_F}{R_1 + R_F} \tag{8-7}$$

For the previously assumed values of 1 kΩ and 20 kΩ,

$$R_B = \frac{(1\text{ k}\Omega)(20\text{ k}\Omega)}{21\text{ k}\Omega} = 952\ \Omega$$

or **953 Ω** standard ±2% value

To the extent that the two bias currents I_{B1} and I_{B2} match each other, the technique of Fig. 8–10(b) will eliminate output offset voltage. But to the extent that the dc bias currents are not identical for the two internal transistors (or their *pn* junction voltages are mismatched), insertion of R_B will not totally eliminate offset voltage. In that case, if output offset cannot be tolerated, an offset nulling potentiometer must be connected to the op amp's offset-correction terminals, as shown in Fig. 8–11.

8-5 ■ OP AMP SUMMING CIRCUIT

A summing circuit is an op-amp inverting amplifier that handles two or more input signals, as shown in Fig. 8–12. With the − input at virtual ground, the voltage across each input resistance is equal to the input voltage applied to its left side. Therefore their currents are

$$I_{R1} = \frac{V_1}{R_1} \qquad \text{and} \qquad I_{R2} = \frac{V_2}{R_2}$$

as shown in Fig. 8–12(b).

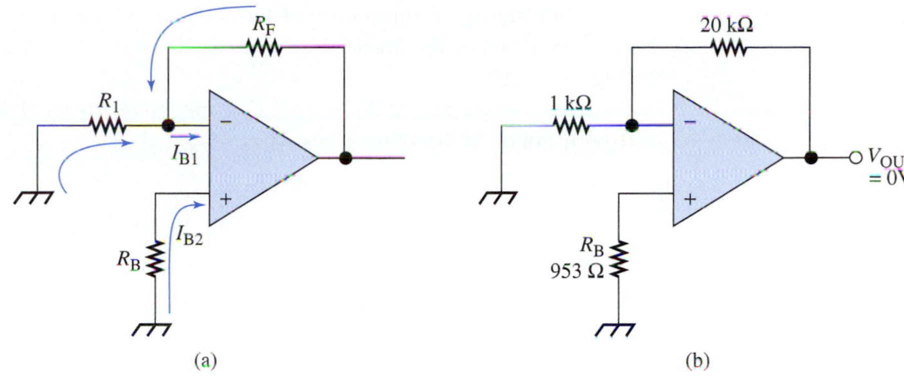

FIGURE 8–10
Correcting the output offset problem. The bias-compensating resistor is symbolized R_B.

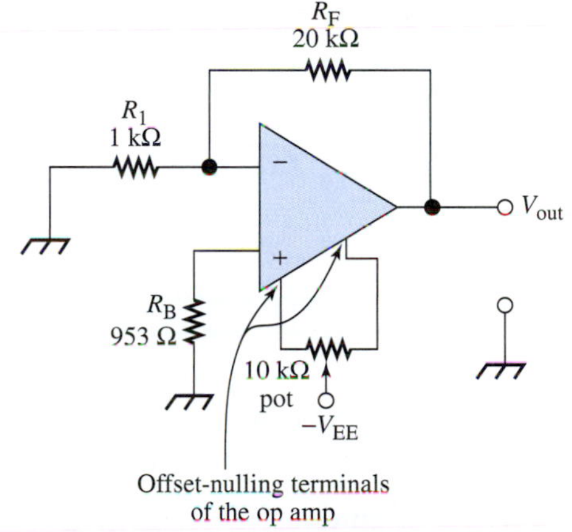

FIGURE 8–11
If compensating resistance R_B is not able to eliminate offset entirely, it may be necessary to install additional correction components.

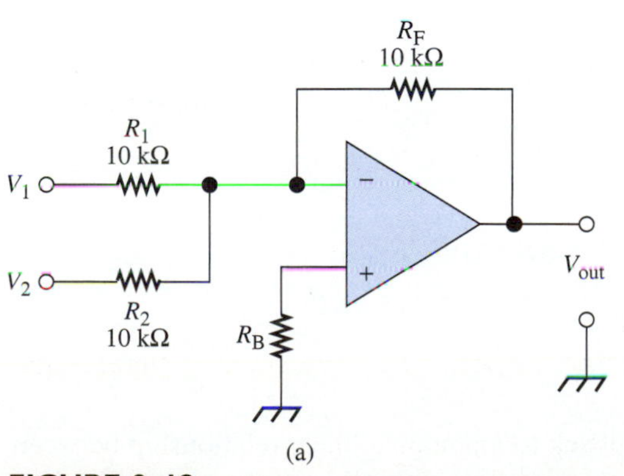

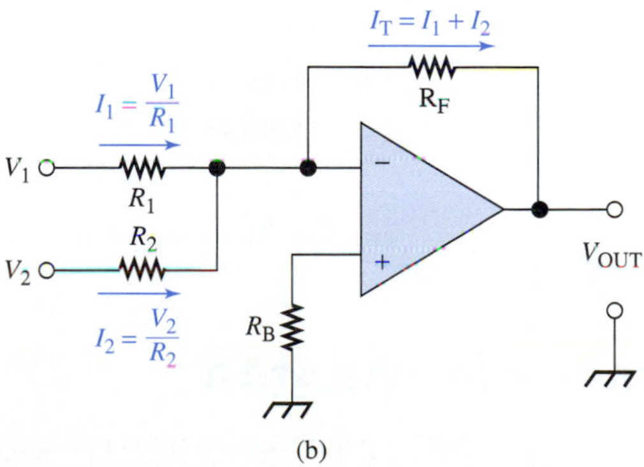

FIGURE 8–12
Op amp summing circuit. (a) Schematic. (b) Current flow for positive input voltage.

With the signal input current virtually zero into the − terminal, the sum of $I_1 + I_2$ must flow through feedback resistance R_F, by Kirchhoff's current law. This is indicated in Fig. 8–12(b).

The voltage across R_F is just V_{out} (negative on its right) because its left side is at virtual ground. So by Ohm's law

$$-V_{OUT} = (I_1 + I_2)R_F$$
$$= \left(\frac{V_1}{R_1} + \frac{V_2}{R_2}\right)R_F$$

If R_1 and R_2 are identical to each other, as shown in Fig. 8–12(a),

$$-V_{OUT} = \left(\frac{V_1}{R_{IN}} + \frac{V_2}{R_{IN}}\right)R_F$$

$$-V_{OUT} = \frac{R_F}{R_{IN}}(V_1 + V_2)$$

If feedback resistance R_F is equal to input resistance R_{IN}, as in Fig. 8–12(a), this reduces to

$$-V_{OUT} = V_1 + V_2 \tag{8-8}$$

with the negative sign representing that V_{out} is inverted to a negative value if V_1 and V_2 are positive voltages. Equation 8-8 vindicates the name *summing circuit.*

For the general case in which R_1 and R_2 are not equal to each other nor to R_F, the circuit equation is

$$-V_{OUT} = \left(\frac{R_F}{R_1}\right)V_1 + \left(\frac{R_F}{R_2}\right)V_2 \tag{8-9}$$

■ EXAMPLE 8-3

In Fig. 8–12, suppose $R_1 = 10\ \text{k}\Omega$, $R_2 = 20\ \text{k}\Omega$, and $R_F = 40\ \text{k}\Omega$. (a) Write the equation for the output voltage as a function of the inputs. (b) Calculate V_{OUT} for $V_1 = +1.2$ V and $V_2 = -1.9$ V. (c) Select the proper value for bias offset resistor R_B.

Solution. (a) From Eq. 8-9,

$$-V_{OUT} = \left(\frac{40\ \text{k}\Omega}{10\ \text{k}\Omega}\right)V_1 + \left(\frac{40\ \text{k}\Omega}{20\ \text{k}\Omega}\right)V_2$$

$$\mathbf{-V_{OUT} = 4V_1 + 2V_2}$$

This describes a weighted summer, since V_1 is assigned greater influence than V_2 in determining V_{out}.

(b) $-V_{OUT} = 4(1.2\ \text{V}) + 2(-1.9\ \text{V})$

$$-V_{OUT} = 4.8\ \text{V} - 3.8\ \text{V} = +1.0\ \text{V}$$

$$\mathbf{V_{OUT} = -1.0\ V}$$

(c) The Thevenin view from the − terminal has $R_1 \| R_2 \| R_F$.

$$R_B = 10\ \text{k}\Omega \| 20\ \text{k}\Omega \| 40\ \text{k}\Omega \cong \mathbf{5.6\ k\Omega}$$

8-6 ■ VOLTAGE COMPARATOR

Most op amp applications use negative feedback to maintain a linear relationship between input(s) and output, which usually prevents the op amp's output from saturating near $+V_{CC}$ or $-V_{EE}$. One application that does not follow this agenda is the *voltage comparator,* also called *comparer.* In a voltage comparator, the output is deliberately intended to be saturated. Positive saturation (near $+V_{CC}$) indicates one result of the comparison operation, and negative saturation (near $-V_{EE}$) indicates the opposite result from the comparison. Refer to Fig. 8–13.

In Fig. 8–13(a) the − input is at ground potential, coupled through R_{B1}. If V_S goes slightly positive relative to ground, V_{diff} must quickly reach a value sufficient to saturate the output. This is certain to occur because the entire open-loop gain A_{VOL} is brought to bear on V_{diff} when there is no feedback resistance R_F. Likewise, if V_S goes slightly negative with respect to ground, the op amp will quickly drive into negative saturation.

The circuit is comparing V_S to 0 V. If it finds V_S above 0 V it produces a V_{OUT} near $+V_{CC}$. If it finds V_S below 0 V it produces a $V_{OUT} \approx -V_{EE}$. These outcomes are displayed in the accompanying waveforms.

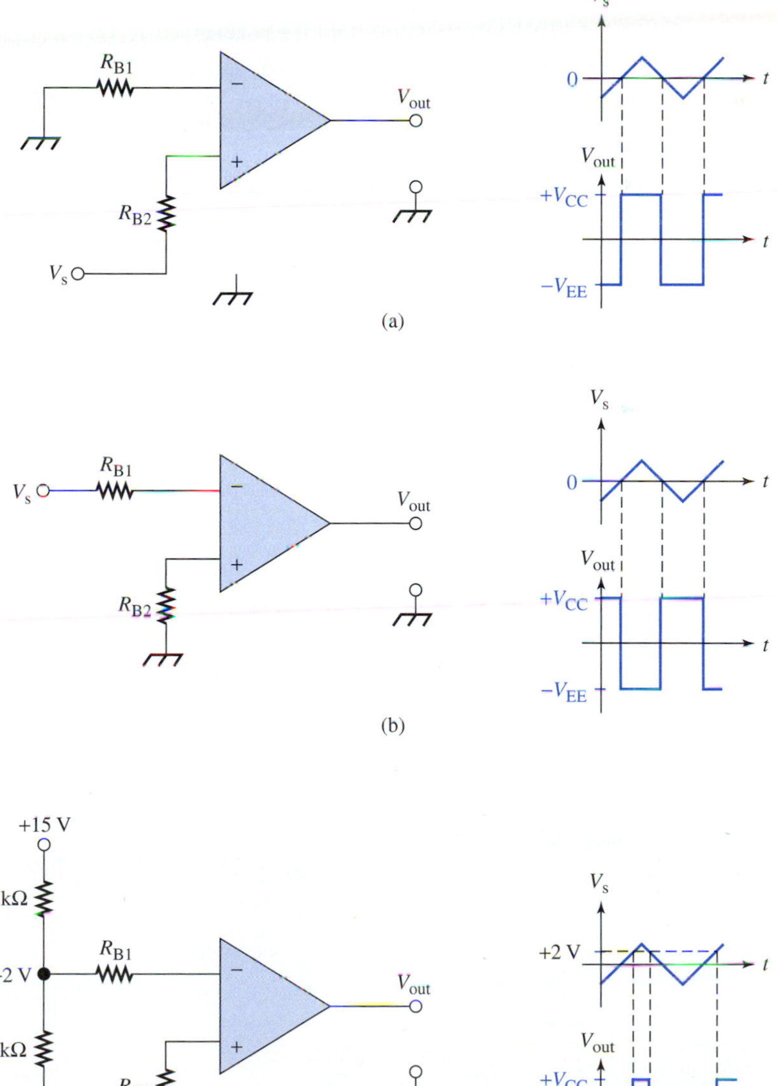

FIGURE 8–13 Voltage comparer. (a) Compare to 0 V, noninverted output. (b) Compare to 0 V, inverted output. (c) Compare to +2 V, noninverted output. (d) Compare to −3 V, noninverted output.

FIGURE 8–13
(continued)

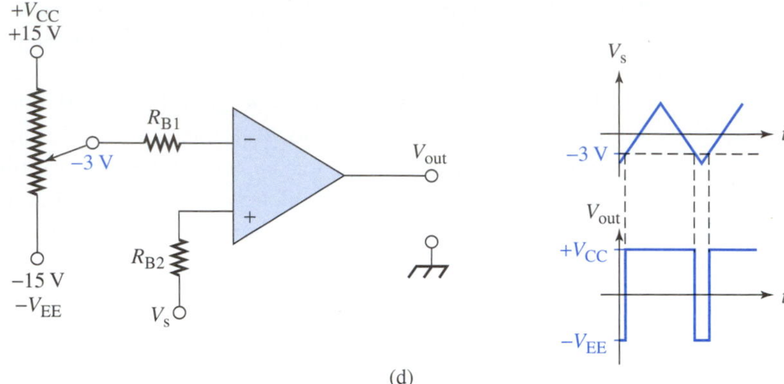

(d)

If the connection polarity is reversed to resemble an inverting amplifier as in Fig. 8–13(b), V_s above 0 V produces $V_{out} \approx -V_{EE}$; and V_s below 0 V produces $V_{out} \approx +V_{CC}$. Noninverting comparison to a positive voltage is represented in Fig. 8–13(c). Comparison to a negative voltage is shown in Fig. 8–13(d).

8-7 ■ OPERATION FROM A SINGLE-POLARITY POWER SUPPLY

There are two reasons for using a dual-polarity power supply for an op amp: (1) The amplifier can handle dc signals of either polarity, with 0-V input producing 0-V output (assuming that offset is corrected). (2) An ac input signal produces an ac output signal that is automatically centered on ground (0 V), so no dc-blocking (ac-coupling) capacitor is required to remove a dc bias component from the output signal.

If these two features are not important for a particular application, there is nothing preventing operation from a single-polarity power supply. Then coupling capacitors are needed on the input side and on the output side, and the noninverting input terminal is just biased at half the dc supply value by a two-resistor equal-value voltage divider. This is shown in Fig. 8–14.

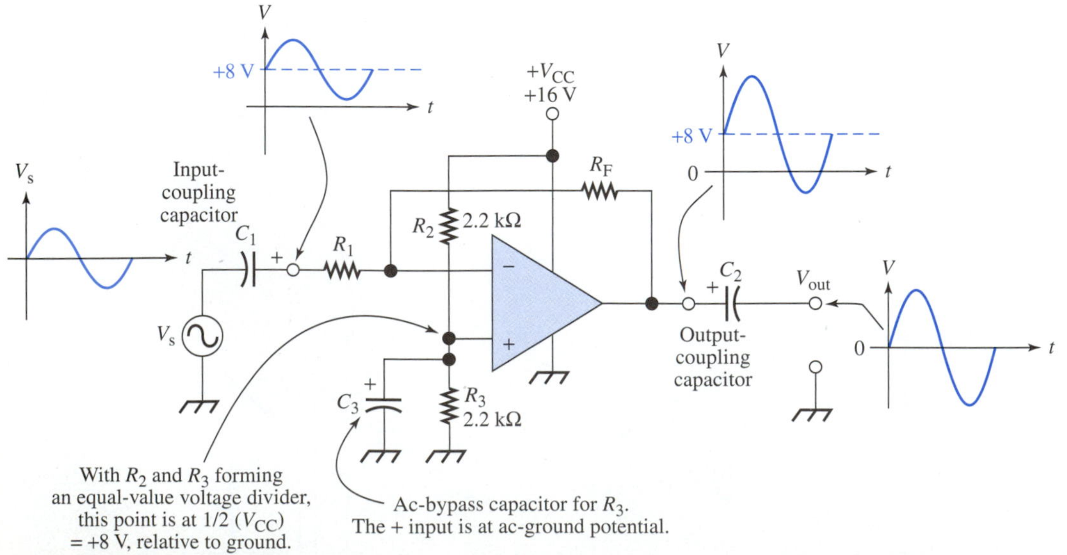

FIGURE 8–14
If centered-on-ground output is not required, an op amp amplifier can combine dc bias with its signal response. This eliminates the need for a negative power supply.

8-8 ■ OP AMP DIFFERENTIAL AMPLIFIER

Sometimes it is necessary to amplify the voltage difference between two input lines, neither of which is grounded. In this case, the amplifier is called a *differential amplifier.* An op amp differential amplifier is shown in Fig. 8–15.

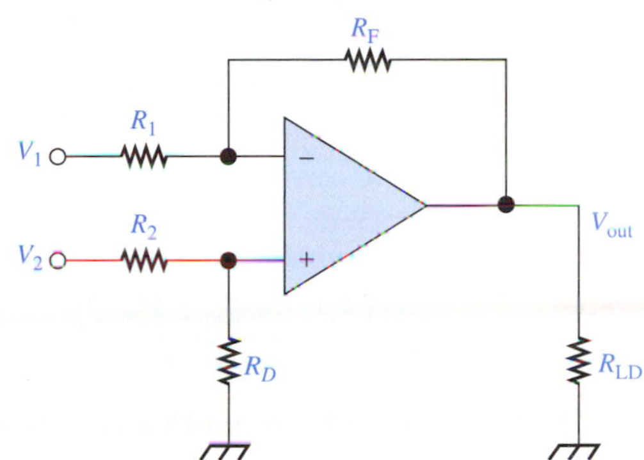

FIGURE 8–15
Op amp differential amplifier. In a differential amplifier, neither input is tied to circuit ground. This can reduce the amount of noise injected into the amplifier, because any noise appears simultaneously on both input terminals. Because the noise is a common-mode signal, the amplifying circuitry rejects it.

Because the differential input current is virtually zero, R_2 and R_D are virtually in series. The voltage-divider equation therefore applies. The voltage at the noninverting input terminal relative to ground is given by

$$V_{ni} = V_2 \frac{R_D}{(R_2 + R_D)} \tag{8-10}$$

The differential input voltage must be virtually zero; this means that the output voltage must assume a value that causes the inverting input terminal voltage to be virtually equal to the noninverting input terminal voltage. That is,

$$V_i = V_{ni} \tag{8-11}$$

Since R_1 and R_F are virtually in series ($I_i = 0$), the voltage drop across R_1 is given by

$$V_{R1} = (V_1 - V_{out}) \frac{R_1}{R_1 + R_F}$$

The voltage at the inverting input terminal is equal to the V_1 input voltage minus the voltage drop across R_1, or

$$\begin{aligned} V_i = V_1 - V_{R1} &= V_1 - (V_1 - V_{out}) \frac{R_1}{R_1 + R_F} \\ &= V_1 \left(1 - \frac{R_1}{R_1 + R_F}\right) + V_{out} \frac{R_1}{R_1 + R_F} \\ &= V_1 \frac{R_1 + R_F - R_1}{R_1 + R_F} + V_{out} \frac{R_1}{R_1 + R_F} \end{aligned}$$

$$V_i = V_1 \frac{R_F}{R_1 + R_F} + V_{out} \frac{R_1}{R_1 + R_F} \tag{8-12}$$

Combining Eqs. (8-10), (8-11), and (8-12) yields

$$V_1 \frac{R_F}{R_1 + R_F} + V_{out} \frac{R_1}{R_1 + R_F} = V_2 \frac{R_D}{R_2 + R_D} \tag{8-13}$$

In most circuits $R_1 = R_2$ and $R_F = R_B$, so Eq. (8-13) becomes

$$V_1 \frac{R_F}{R_1 + R_F} + V_{out} \frac{R_1}{R_1 + R_F} = V_2 \frac{R_F}{R_1 + R_F}$$

$$V_{out} \frac{R_1}{\cancel{R_1 + R_F}} = (V_2 - V_1) \frac{R_F}{\cancel{R_1 + R_F}}$$

$$V_{out} = (V_2 - V_1) \frac{R_F}{R_1} \tag{8-14}$$

Equation (8-14) tells us that the op amp differential amplifier amplifies the *difference* between the two input lines and has a voltage gain dependent solely on the external resistors, as usual.

When an op amp differential amplifier is used, there is a limit to the amount of *common mode* voltage that can be applied to the two inputs. Exceeding this maximum common mode input voltage can damage an op amp. Therefore it is not sufficient to be concerned only with the difference between V_2 and V_1. One must also be concerned with the voltage that V_2 and V_1 have in common. An op amp's data sheet will always specify its maximum common mode input voltage.

8-9 ■ OP AMP VOLTAGE-TO-CURRENT CONVERTER

Occasionally in industrial electronics, it is necessary to deliver a current which is proportional to a certain voltage, even though the load resistance may vary. If the load resistance stayed constant, there would be no problem. The load current would be proportional to the applied voltage just naturally, by Ohm's law. However, if the load resistance varies from unit to unit, or if it varies with temperature or with age, then delivering a current exactly proportional to a certain voltage is not so easy. A circuit which can perform this job is shown in Fig. 8–16. It is called a *voltage-to-current converter.* This circuit is able to convert voltage to current because of the virtual zero across the differential inputs. That is, if V_{in} appears at the + input, then a voltage virtually equal to V_{in} must appear at the − input. The current through R_1 is determined by Ohm's law,

$$I_{R1} = \frac{V_{in}}{R_1}$$

so I_{R1} will never change as long as R_1 itself doesn't change.

Because of the fact that virtually no current flows between the inverting and noninverting inputs, we can say that

$$I_{R1} = I_{load}$$

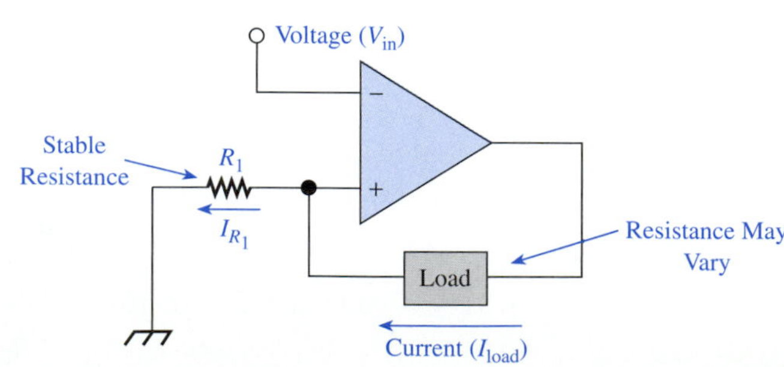

FIGURE 8–16
Op amp voltage-to-current converter. The important idea about the voltage-to-current converter is that the load current is fixed by V_{in} even if there are variations in the load itself.

Therefore

$$I_{\text{load}} = \frac{V_{\text{in}}}{R_1} \tag{8-15}$$

Equation (8-15) holds true irrespective of the load's resistance. The load current is guaranteed to be proportional to the input voltage under *any* conditions of load resistance (within limits).

Another good feature of the op amp voltage-to-current converter is that it can be driven by a voltage source which is not itself capable of supplying the load current called for by Eq. (8-15). This is because the voltage source only has to drive a noninverting amplifier, whose input impedance is very high. The load current itself is supplied by the op amp.

8-10 ■ OP AMP INTEGRATORS AND DIFFERENTIATORS

Besides being able to perform the mathematical operations of addition (summing circuits) and multiplication (amplifiers), op amps can also perform the advanced mathematical operations of integration and differentiation. While these functions are not so commonplace as the basic functions, they are still an important part of the industrial use of op amps.

In simplest terms, a differentiator is a circuit whose output is proportional to *how quickly the input is changing.* An integrator is a circuit whose output is proportional to *how long the input has been present.**

Figure 8–17(a) shows an op amp differentiator. The differentiator can be thought of intuitively this way:

1. If V_{in} is a steady dc voltage, capacitor C will charge to V_{in}, there will be no current through C or R_F, so V_{out} will be 0 V.
2. If V_{in} is a slowly changing voltage, the voltage across capacitor C will always be slightly less than V_{in}, since it has probably had ample opportunity to charge up. This means that only a small current will be flowing in the capacitor leads and through R_F. Therefore V_{out} will be small.
3. If V_{in} is a rapidly changing voltage, then the capacitor voltage will be considerably less than V_{in}, since it probably hasn't had time to charge. This will result in a large current flow through C and R_F and a large V_{out}.

The precise output-input relationship of an op amp differentiator is expressed by the formula beneath the differentiator schematic.

Figure 8–17(b) shows an op amp integrator. An integrator can be thought of intuitively this way:

1. If V_{in} has just very recently appeared at the input terminal, current has not been flowing through R_{in} for very long. Therefore current has not been flowing through C for very long either, and C has not charged very far. The voltage to which C has charged equals the output voltage V_{out}, since one side of C is connected to the output and the other side of C is connected to virtual ground. Since capacitor voltage is small if V_{in} has just recently appeared, V_{out} is also small.
2. If V_{in} has been present for some time, current has been flowing through R_{in} and C for a while. This means that C has had time to charge considerably and therefore has

*This definition is rather an oversimplification of the action of an integrator. It is accurate only if the input voltage is unvarying dc.

FIGURE 8–17
(a) Op amp differentiator. Ideally the output voltage is proportional to the rate of change of the input voltage. (b) Op amp integrator. Ideally, if V_{in} is a dc signal, the output voltage is proportional to the amount of time the input has been present.

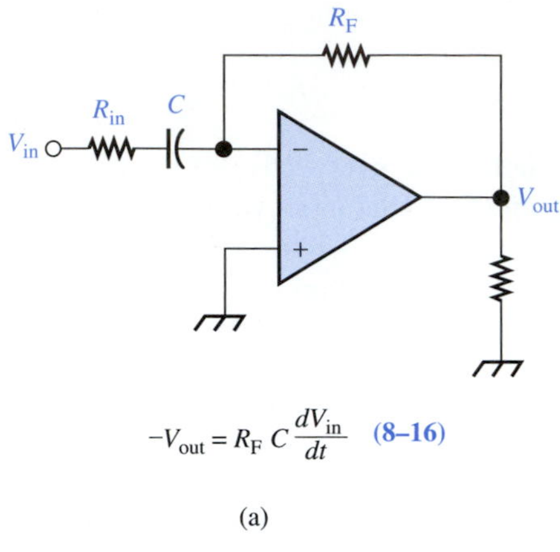

$$-V_{out} = R_F\,C\,\frac{dV_{in}}{dt} \quad \textbf{(8–16)}$$

(a)

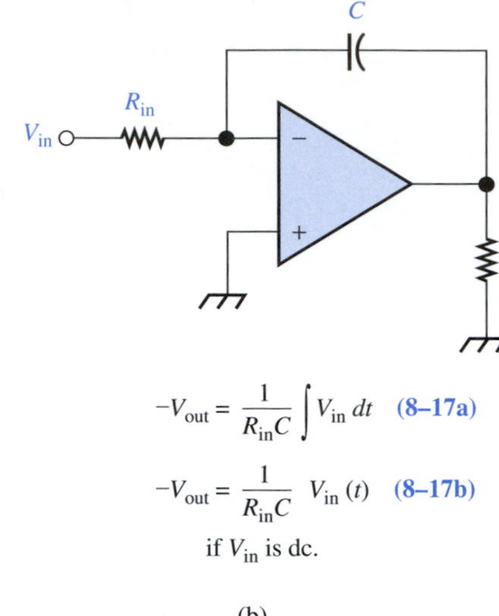

$$-V_{out} = \frac{1}{R_{in}C}\int V_{in}\,dt \quad \textbf{(8–17a)}$$

$$-V_{out} = \frac{1}{R_{in}C}\,V_{in}\,(t) \quad \textbf{(8–17b)}$$

if V_{in} is dc.

(b)

considerable voltage across its plates. Since V_{out} equals capacitor voltage, V_{out} is a considerable voltage (not small).

3. The longer V_{in} persists, the more capacitor C will charge, and the greater will be the output voltage. Thus V_{out} is proportional to *how long* V_{in} has been present.

The general output-input relationship for an op amp integrator is expressed in the first formula beneath the integrator schematic in Fig. 8–17(b). the output-input relationship for the special case of an unvarying dc input voltage is expressed in the second formula.

TROUBLESHOOTING THE MAGLEV SIDEWALL-POSITION-SIGNAL PROCESSING CIRCUIT

The MagLev system of Troubleshooting on the Job in Chapters 4 and 5 depended on a signal from the Sidewall-Position-Signal Processing Circuit (Figs. 4–18 and 5–12); that circuit is shown in Fig. 8–18. To understand how it works, you must work with two additional figures, namely Figs. 8–19 and 8–20. Also refer back to the train diagrams of Figs. 4–16 and 4–17, but mentally substitute six magnets (three N and three S) where those drawings show only two magnets (one N and one S).

DETAILED DESCRIPTION OF THE SIDEWALL-POSITION-SIGNAL PROCESSING CIRCUIT, THE V_{sens} WAVEFORM, AND THE LOGIC CIRCUITS

The Hall-effect sensor is shown at the bottom of Fig. 8–18. It is physically mounted on the sidewall, aligned with the sidewall propulsion coils on the side of the track that houses the electronics. There is one Hall-effect sensor for each pair of propulsion coils, just as there is one Sidewall-Position-Signal Processing Circuit and one coil-control module (Fig 4–18) for each pair of propulsion coils.

The Hall-effect sensor produces a voltage, V_{Hall}, that is proportional to the strength of the magnetic field that it senses from the train's supermagnets. When it is experiencing North flux (a North supermagnet is instantaneously closer than a South supermagnet), V_{Hall} becomes positive. When the sensor has a South supermagnet nearer to it, V_{Hall} becomes negative. An op amp differential amplifier is used to amplify v_{Hall} by a factor of 12 to produce v_{sens}. These components are mounted out on the concrete sidewall with the sensor itself; in that electrically noisy location, the dif amp can reject any common mode noise that is coupled onto the leads of the sensor. The amplified output, v_{sens}, is carried back via shielded cable to the electrically clean location of the control module.

The waveforms of v_{sens} are shown in Fig. 8–19. Part (a) is for the train moving to the right in Figs. 4–15 and 4–16 with a North supermagnet leading. Part (b) is for the train moving to the left with a South supermagnet leading. Let us focus our attention on the train moving to the right; then understanding leftward motion becomes a simple matter of reversing the order of events.

Our explanation will examine what happens to a single pair of propulsion coils, driven by a single control module, receiving a single v_{sens} signal. In understanding the entire six-magnet train, recognize that all these events are simultaneously duplicated at the other five groups of sidewall coil-pairs (three of them with reversed polarity, of course) that are within the train's field of influence.

Train Approaching

As the leading North magnet approaches the Hall sensor, v_{sens} slowly rises in the positive polarity, as

There is an ongoing research and development effort to improve the sidewall and ground magnets for MagLev Transportation.
Courtesy of Railway Technical Research Institute of Japan.

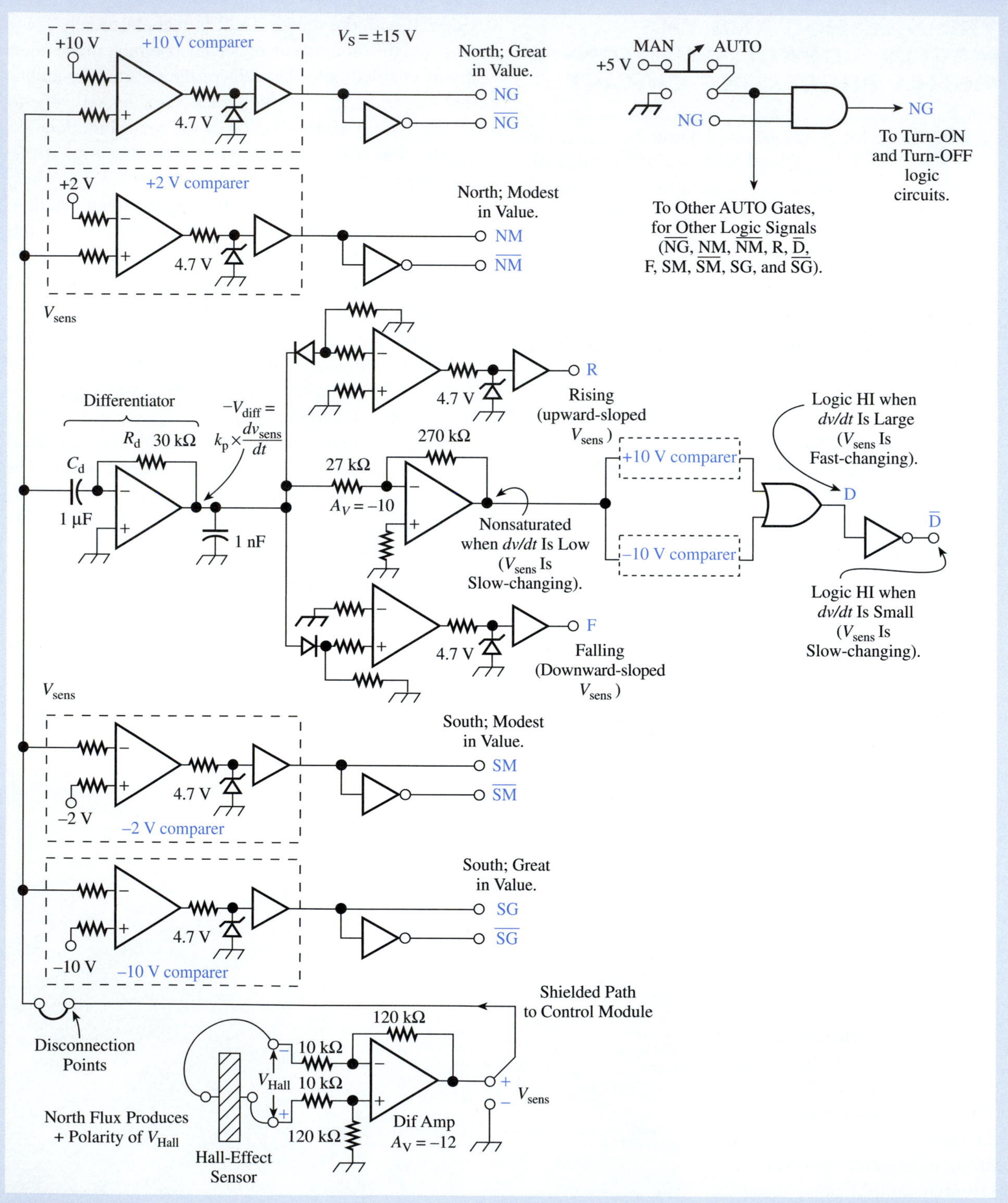

FIGURE 8–18
Schematic of the Sidewall-Position-Signal Processing Circuit and Hall-effect sensor.

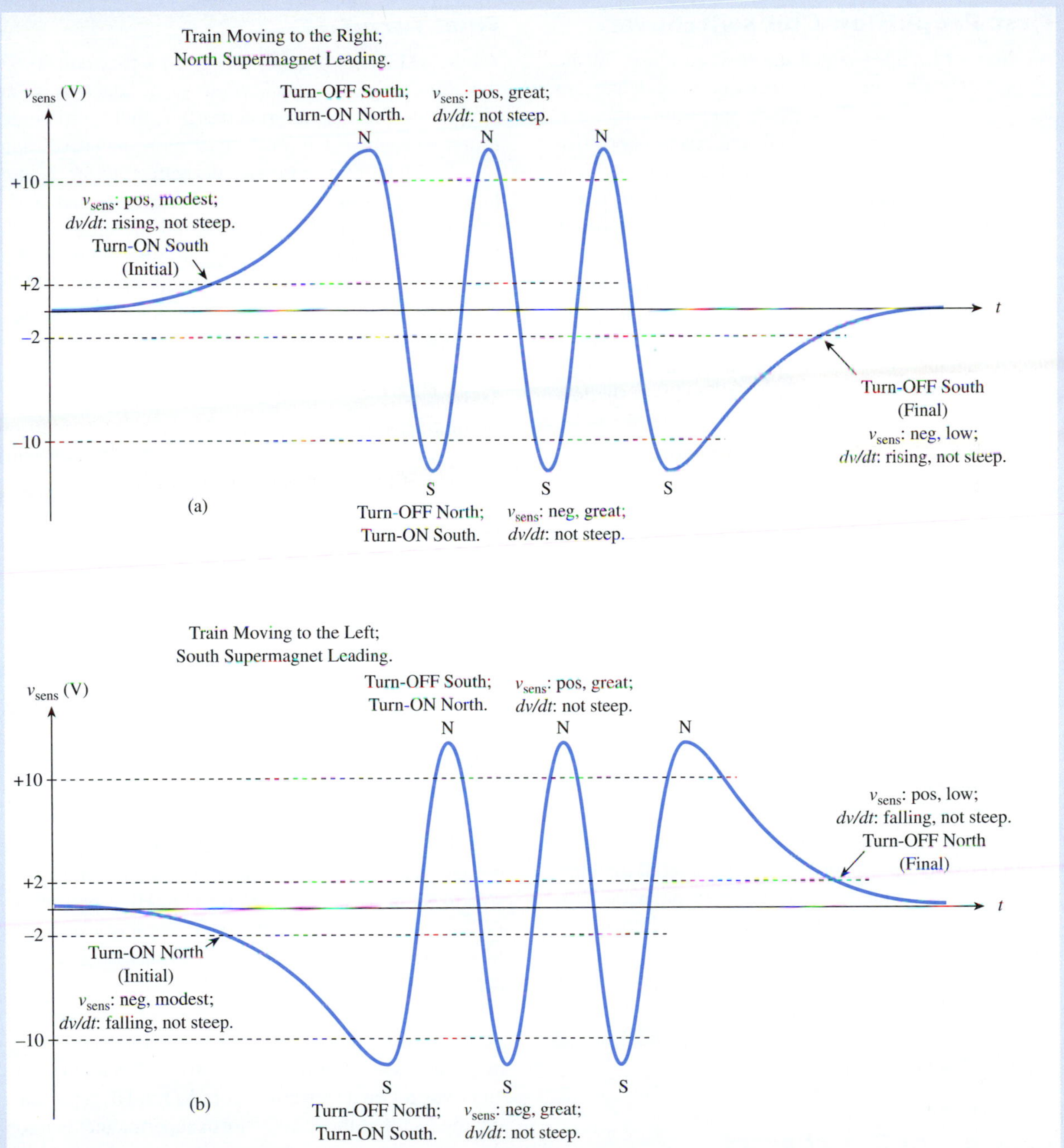

FIGURE 8–19
Waveforms of the v_{sens} signal. (a) Train moving to the right. (b) Train moving to the left.

Fig. 8–19(a) shows. There will come a moment when its magnitude exceeds $+2$ V, which is one of the Processing Circuit's trip values. At this moment, v_{sens} is not changing rapidly; its slope is not steep. Therefore its time derivative, dv/dt, has a low value; dv/dt is positive, since the voltage is rising, not falling.

First Propulsion-Coil Switchover

As the North supermagnet approaches closer to the propulsion coils, the North flux becomes denser, so v_{sens} increases. When the supermagnet is directly upon the coils and their sensor, v_{sens} reaches its peak positive value, which is guaranteed to be at least +11 V. Near the peak of the waveform, the derivative dv/dt is not large: it is small because oscillatory behavior always carries with it the characteristic that large magnitude is accompanied by slow rate of change and small magnitude by fast rate of change. To help in understanding this behavior, imagine tangent lines placed against the oscillating waveform of Fig. 8–19(a). Near the peak any tangent line has a shallow slope; in fact, at the exact peak, the tangent line has a zero slope—it's perfectly flat (horizontal).

Second Propulsion-Coil Switchover

As the train races along, the moment will come when the trailing South supermagnet is the same distance from the propulsion coil as the leading North supermagnet. At that instant the net magnetic flux felt by the sensor is zero; therefore $v_{sens} = 0$. This is the negative-going zero crossover moment on the wave form of Fig. 8–19(a). Notice carefully that dv/dt has a large magnitude (tangent line has a steep slope) when v_{sens} is near a zero crossover.

As the train continues to advance, the approaching South supermagnet becomes more influential than the receding North supermagnet. The net magnetic flux becomes South, so v_{sens} goes negative. When the South magnet is directly upon the Hall sensor, v_{sens} is at its negative peak value, which is guaranteed to be below −11 V (more negative than −11 V). As with the positive peak, the derivative dv/dt is small in the region of the negative peak.

Subsequent Switchovers

At this point the South supermagnet recedes and the train's next North supermagnet approaches. v_{sens} oscillates positive again in Fig. 8–19(a), then South again, giving negative v_{sens}, then North again, giving positive v_{sens}. Then the last supermagnet, which is South, comes upon the Hall sensor; this sends v_{sens} negative for the last time.

Final Turn-OFF

As the last South magnet recedes, its flux density at the sensor gets weaker, so v_{sens} decreases in magnitude. As v_{sens} approaches 0 V, there will come a moment when it is no longer more negative than −2 V. When this happens, derivative dv/dt is *not* large, as it would be if another supermagnet were approaching. Rather, dv/dt is small in value, as can be seen by placing a tangent line against the waveform at the −2-V crossing point on the far right of Fig. 8-19(a). Because v_{sens} is rising, not falling, dv/dt has a positive sign.

v_{sens} Processing Circuit

In the Sidewall-Position-Signal Processing Circuit of Fig. 8-18, there are four voltage comparers along the left side of the schematic.

North Comparers

Starting at the center of the drawing and moving up, the first voltage comparer we encounter is the +2-V comparer. It trips when v_{sens} becomes more positive than +2 V, and it remains tripped for as long as v_{sens} remains more positive than +2 V. The act of tripping involves the op amp changing from negative saturation to positive saturation. As the op amp does this, the voltage at the zener's cathode changes from −0.7 V (zener forward biased by $-V_{SAT}$) to +4.7 V (zener reverse biased by $+V_{SAT}$). The change from −0.7 V to +4.7 V causes the buffer to change its output from digital LO to digital HI. Thus digital terminal NM (North Modest) switches from LO to HI, representing the fact that the net flux is North, and has grown to a Modest value because a North supermagnet is coming near.

At the top of the drawing, the +10-V comparer performs an identical switching action when v_{sens} becomes more positive than +10 V. It switches the NG (North Great) terminal from LO to HI, representing the fact that the train's North supermagnet is upon the sensor and propulsion coils. When the North supermagnet passes by, v_{sens} declines below +10 V, and NG reverts to LO.

South Comparers

Returning to the left center of Fig. 8–18 and moving downward, the op amp in the −2-V comparer saturated negative whenever v_{sens} is more positive than

−2 V. The buffer has a −0.7-V input, so its output is logic LO. But when v_{sens} becomes more negative than −2 V, the op amp's actual input polarity matches its input-terminal polarity marks. That sends the op amp to positive saturation, the buffer gets +4.7 V, and the output goes HI. Digital signal SM (South Modest) goes HI to represent the fact that the net magnetic flux has become South and has grown to a modest value because a South supermagnet is coming near.

Near the bottom of Fig. 8–18, the −10-V comparer performs an identical switching action when v_{sens} becomes more negative than −10 V. It switches the SG (South Great) terminal from LO to HI when a South supermagnet is upon the sensor and coils.

v_{sens} Derivative

Now direct your attention to the center of Fig. 8–18. The far left op amp circuit is a differentiator, or derivative-taker. Its output is proportional to the time rate of change (derivative) of v_{sens}; for the R_d and C_d values in this circuit, the output voltage is given by Eq. (8-18):

$$-v_{diff} = R_d C_d \frac{dv_{sens}}{dt}$$

$$-v_{diff} = 0.03 \frac{dv_{sens}}{dt} \qquad \textbf{(8-18)}$$

The differentiator op amp functions in an inverting mode, so v_{diff} has the opposite polarity from the actual sign of the derivative, dv_{sens}/dt. That is, v_{diff} is instantaneously negative when dv_{sens}/dt is positive (when v_{sens} is rising on the waveform). And v_{diff} is instantaneously positive when dv_{sens}/dt is negative (when v_{sens} is falling on the waveform).

Rising Slope

When v_{diff} goes more negative than about −0.7 V, it trips the op amp comparer which is above it on the right. This causes its buffer to drive the R (Rising) signal to a logic HI state. The R signal being HI represents the fact that v_{sens} is rising (heading upward) on the waveform. Think carefully about this to understand why this is a proper op-amp response and a proper circuit design.

Falling Slope

When v_{diff} goes more positive than about +0.7 V, it trips the op amp comparer which is below it on the right. This causes its buffer to drive the F (Falling) signal to a logic HI state. The F signal being HI represents the fact that v_{sens} is falling (heading downward, or negative-going) on the waveform.

Near-Zero Slope

In Fig. 8–18 the circuitry directly to the right of the differentiator has the following purpose: It detects when v_{diff} is low in magnitude, indicating that v_{sens} is changing only slowly, or not at all. When v_{diff} has a magnitude less than 1 V (it is between −1 V and +1 V), the gain-of-10 inverting amplifier (with R_i = 27 kΩ) will produce an output voltage with a magnitude less than 10 V. Therefore it will not trip either of the parallel comparers. Both the +10-V comparer and the −10-V comparer (identical to the ones at the top and bottom of the schematic drawing) will have logic LO outputs. These are the inputs to the OR gate, so the OR gate is disabled, sending logic signal D to the LO state. The inverter then brings the $\overline{D}$ (Derivative Not large) logic signal to the HI state. When $\overline{D}$ goes HI, it represents the fact that v_{sens} is close to a waveform peak, or near the far left or far right of the waveform. This is an essential piece of information for the Logic circuits of Fig. 5–12. They use the $\overline{D}$ fact to detect when v_{sens} is close to a peak, which is the moment at which the propulsion coils are switched over from one magnetic polarity to the opposite magnetic polarity. The logic circuits also use $\overline{D}$ to distinguish between the slow-changing crossings of the ±2-V lines (at initial approach and final recede of the train) on the one hand, and the fast-changing crossings of the ±2-V lines (when the train is right upon the sensor) on the other hand.

Turn-ON and Turn-OFF Logic

The detailed Logic circuits are shown in Fig. 8–20. As the train approaches from the right, v_{sens} starts up the waveform curve in Fig. 8-19(a). At the moment when it crosses the +2-V line, the digital conditions become correct for enabling the Initial AND gate in the Turn-ON South Logic Circuit of Fig. 8-20. NM is HI because $v_{sens} > +2$ V; NG is LO, and $\overline{NG}$ is HI because $v_{sens} < +10$ V; R is HI because the waveform is Rising. $\overline{D}$ is HI because the Derivative is Not large. Refer to the Signal Processing Circuit of Fig. 8-18 to verify all these conditions.

Turn-ON South (Initial)

When the Initial AND gate's output goes HI, it enables the OR gate. This triggers the T_1 trigger input of the one-shot in the turn-ON South Trigger Circuit of Fig. 5–12. SCRs 1 and 3 are fired, and the propulsion coils become South. An attraction force is developed between the propulsion coils and the leading North supermagnet, thereby pulling the train to the right.

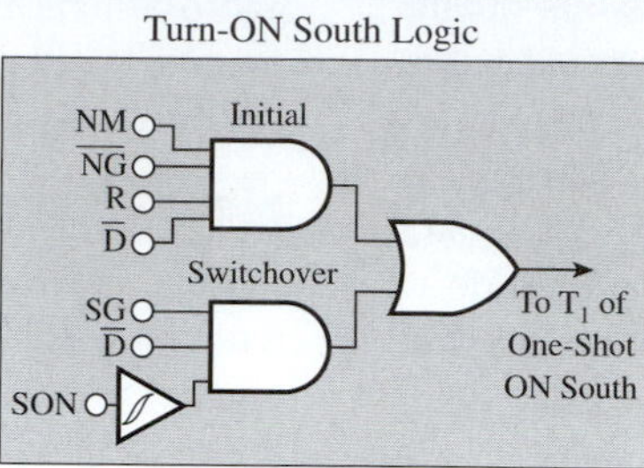

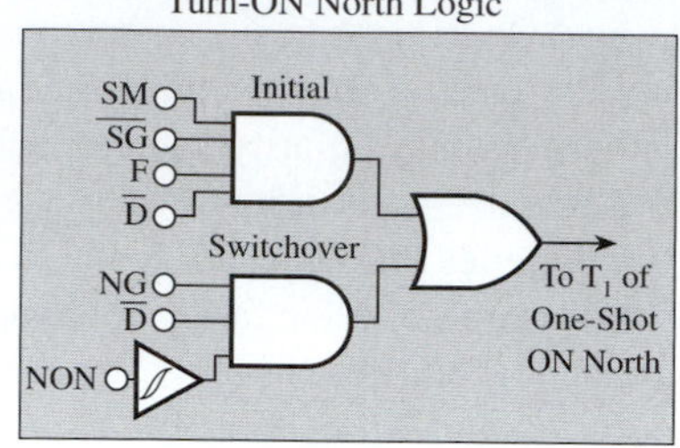

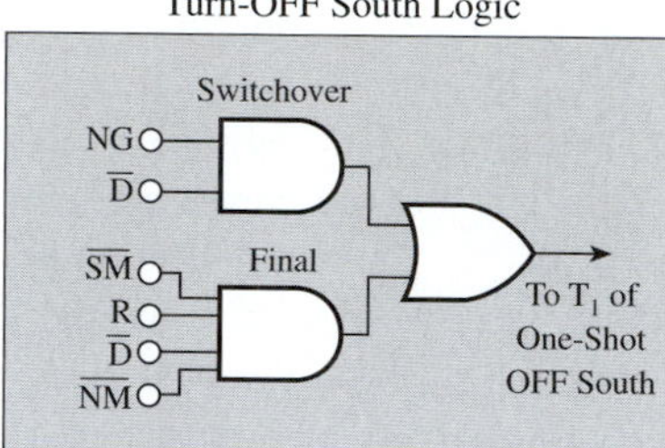

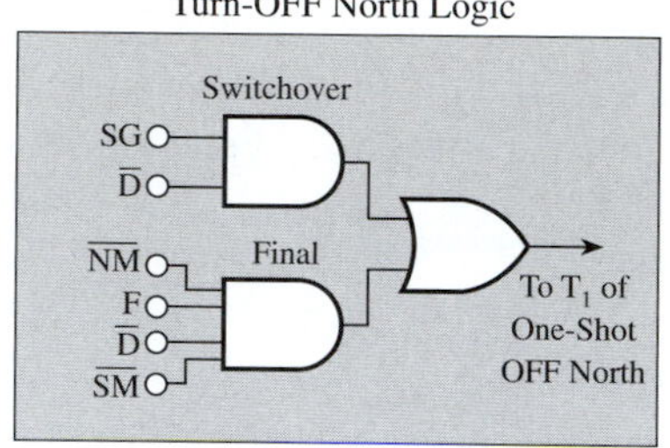

FIGURE 8–20
Logic circuits for turning-ON and turning-OFF propulsion coils. (Refer to Figs. 5–12 and 4–18.)

South-to-North Switchover

When the leading North supermagnet comes directly upon the coils, the coils must switch over from South to North so that they can stop attracting and start repelling the North supermagnet. Refer to Figs. 4–15 and 4–16. In Fig. 8–20 the switchover AND gate in the turn-OFF South Logic Circuit becomes enabled at this moment. NG is HI (North Great), and $\overline{D}$ is HI (Derivative Not large). The OR gate enables, triggering the one-shot in the turn-OFF South Trigger Circuit of Fig. 5–12. In Fig. 4–18, transistors Q_1 and Q_3 are pulsed to commutate SCR_1 and SCR_3. This shuts OFF the flow of South-direction current. In Fig. 5–12 the bottommost secondary winding is building voltage across the 0.03-μF capacitor, with a 60-μs time-constant. That capacitor voltage becomes the logic signal NON (North turn-ON) which is fed to the turn-ON North Logic at the upper right in Fig. 5–12. Referring to the upper right of Fig. 8–20 (turn-ON North Logic), we see that NON feeds a Schmitt-trigger driver gate that connects to the Switchover AND gate. NG and $\overline{D}$ are both HI (they turned OFF the South SCRs just a few tens of microseconds ago), so the Switchover AND gate enables. The OR enables, the turn-ON North one-shot is triggered, and pulses are sent to SCR_2 and SCR_4 from the upper right of Fig. 5–12. In Fig. 4–18 North current powers the propulsion coils.

Remainder of the Propulsion-Coil Cycle

The foregoing description covers the initial turn-ON South and the first switchover from South to North of the propulsion-coil pair. These events have been related to Figs. 4–15, 4–16, 5–12, 8–18, 8–19, and 8–20. You must mentally repeat this explanation for the switchover from North to South, which occurs when the South supermagnet comes directly upon the sensor. Subsequent South-to-North and North-to-South switchovers are exact repeats of the first two switchovers. Then you must mentally trace out the train/circuit actions for the final turn-OFF South event, which occurs when the last South supermagnet recedes to the right of the propulsion coils. On the waveform of Fig. 8–19(a), this is the turn-OFF South (final) point.

Train Moving in the Other Direction

Finally, you must mentally trace out the train/circuit actions for the condition of the train moving to the left in Figs. 4–15 and 4–16, with a South supermagnet leading. The v_{sens} waveform for this direction is shown in Fig. 8–19(b).

Time-Constant Values in the Trigger Circuits

In Fig. 5–12 the UJT-firing time constant is $\tau = 0.2$ ms ($\tau = RC = 1\ \text{k}\Omega \times 0.2\ \mu\text{F} = 0.2$ ms). The switchover delay time constant is $\tau = 60\ \mu\text{s}$

($\tau = RC = 2\ \text{k}\Omega \times 0.03\ \mu\text{F} = 60\ \mu\text{s}$). These time constants are compatible with the train's speed and with the SCRs' commutation times, as we will now show.

From the instant when a Logic circuit signals that a switchover is needed, there is a 0.2-ms delay before the UJT triggers. (Approximately 1 τ must elapse in a UJT circuit like those in Fig. 5–12.) During this time delay, the train travels a distance of

$$d = \text{speed} \times t = (420 \times 10^3\ \text{m}/3600\ \text{s}) \times (0.2 \times 10^{-3}\ \text{s}) \approx 0.02\ \text{m}\ (2\ \text{cm})$$

This is a small enough distance that the accompanying uncertainty in the precise switchover location is tolerable. For example, if the uncertainty in the UJT timing is about $\pm 30\%$, which is a reasonable estimate, the accompanying uncertainty in the exact location of the train is about $\pm 30\%$ of 2 cm, which is ± 0.6 cm. Stated another way, the switchover turning-OFF of the propulsion coils is known to occur within a 1.2-cm-wide span of the sidewall. With the sidewall coils having a width of approximately 0.7 m (70 cm), a 1.2-cm uncertainty is acceptable.

The SCRs used in Fig. 4–18 have commutation turn-OFF times on the order of a few tens of microseconds. That is, once the commutating transistors are turned ON, it may require 20 to 50 μs for the turning-OFF SCRs to achieve blocking status. The turning-ON SCRs must not receive their gate pulse until after this commutation time has elapsed. In Fig. 5–12 the NON and SON switchover-delay circuits require more than 1 τ for their 0.03-μF capacitors to charge to the trigger value of the Schmitt-trigger buffer gates in Fig. 8–20. Thus, the turn-ON South and turn-ON North Logic circuits of Fig. 8–20 cannot go active until at least 60 microseconds *after* the commutating transistors have begun the turning-OFF process. This is how the turning-ON SCRs are prevented from receiving their gate pulse too quickly.

YOUR JOB ASSIGNMENT

Suppose that your efforts in Troubleshooting on the Job in Chapters 4 and 5 revealed no malfunctions of any kind in Manual operating mode. Nevertheless, the train is known to lose speed and levitation height in this section of track.

1. What can you conclude about the origin of the problem?
2. Assume that your portable scope is a dual-trace model; also assume that you have a variable-frequency function generator that can be used to inject an artificial v_{sens} waveform at the break-location in Fig. 8–18. Describe a reasonable troubleshooting procedure to isolate the system malfunction.

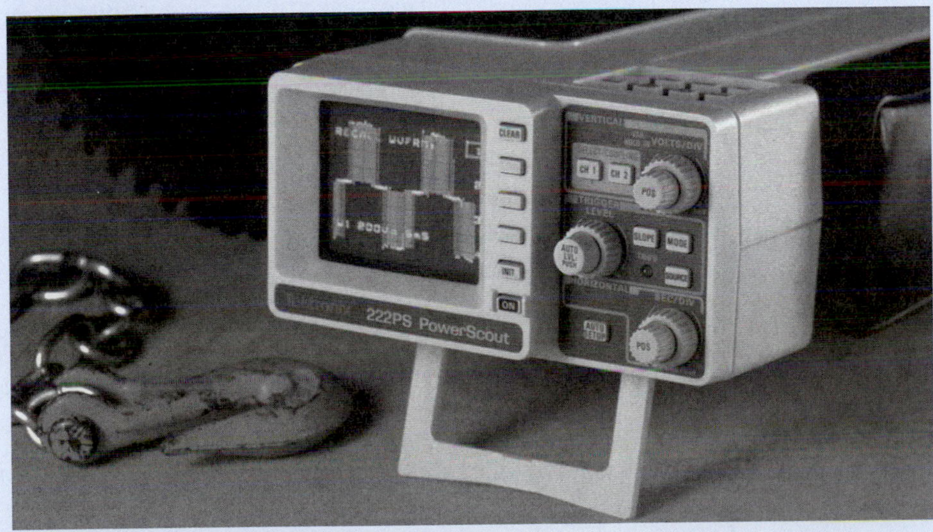

A portable battery-powered oscilloscope is an essential test instrument for the type of field work called for in this troubleshooting assignment.

Courtesy of Tektronix, Inc.

SUMMARY

- A modern op amp is a transistor integrated circuit with very large open-loop voltage gain A_{VOL} and very high input resistance R_{in}.
- Op amps are commonly used to perform the following functions: (1) comparing voltage values–comparator; (2) amplifying single-ended (ground-referenced) voltage and current with polarity inversion—inverting amp; (3) amplifying single–ended voltage and current with no polarity inversion—noninverting amp; (4) amplifying differential voltage and current (neither terminal grounded)—differential amp; (5) performing arithmetic addition of two or more analog signals—summing circuit.
- An op amp voltage-to-current converter delivers a load current that is proportional to its input voltage, regardless of the load's resistance.
- An op amp integrator produces an output voltage V_{OUT} that is proportional to two things: (1) the input voltage V_{IN} and (2) the amount of time that V_{IN} has been present.
- An op amp differentiator produces an output voltage V_{OUT} that is proportional to the time rate of change of input voltage (how rapidly V_{IN} is changing).

FORMULAS

For an inverting amplifier

$$A_V = \frac{R_F}{R_1} \quad \text{(voltage gain)} \qquad \text{(Equation 8-1)}$$

$$R_{in} = R_1 \quad \text{(input resistance)} \qquad \text{(Equation 8-2)}$$

$$R_{out} = \frac{R_o}{A_{VOL}/A_{VCL}} \quad \text{(output resistance)} \qquad \text{(Equation 8-3)}$$

For an noninverting amplifier

$$A_V = \frac{R_F}{R_1} + 1 \quad \text{(voltage gain)} \qquad \text{(Equation 8-4)}$$

$$R_{in} = R_i\left(\frac{A_{VOL}}{A_{VCL}}\right) \quad \text{(input resistance)} \qquad \text{(Equation 8-5)}$$

$$R_{out} = \frac{R_o}{\dfrac{A_{VOL}}{A_{VCL}} + 1} \quad \text{(output resistance)} \qquad \text{(Equation 8-6)}$$

$$R_B = \frac{R_1 R_F}{R_1 + R_F} \quad \text{(for bias-offset correction)} \qquad \text{(Equation 8-7)}$$

$$-V_{OUT} = V_1 + V_2 \quad \text{(for unweighted gain-of-1 summing circuit)} \qquad \text{(Equation 8-8)}$$

$$-V_{OUT} = \left(\frac{R_F}{R_1}\right)V_1 + \left(\frac{R_F}{R_2}\right)V_2 \quad \text{(for general summing circuit)} \qquad \text{(Equation 8-9)}$$

$$V_{out} = (V_2 - V_1)\frac{R_F}{R_1} \quad \text{(for an op amp differential amplifier)} \qquad \text{(Equation 8-14)}$$

$$I_{load} = \frac{V_{in}}{R_1} \quad \text{(for an op amp voltage-to-current converter)} \qquad \text{(Equation 8-15)}$$

$$-v_{OUT} = R_F C\frac{dv_{IN}}{dt} \quad \text{(for an op amp differentiator)} \qquad \text{(Equation 8-16)}$$

$$-v_{OUT} = \frac{1}{R_{IN}C}\int v_{IN}\,dt \qquad \text{(for an op amp integrator)} \qquad \text{(Equation 8-17a)}$$

$$-v_{OUT} = \frac{1}{R_{IN}C}V_{IN}(t) \qquad \text{(for an op amp integrator if } V_{IN} \text{ is dc)} \qquad \text{(Equation 8-17b)}$$

QUESTIONS AND PROBLEMS

Section 8-1

1. Give an approximate value for the voltage gain A_{VOL} of an IC op amp without supporting/feedback resistors.
2. Draw the schematic symbol for an op amp, with its five required terminals. Write the symbols for the voltages at all five terminals.
3. Give an approximate value for the input resistance R_i of an IC op amp.
4. For normal operating circumstances (amplifier not saturated), state approximate values for
   a. The difference voltage V_{diff} between the − and + inputs.
   b. The signal current passing through the op amp between the − and + inputs.

Section 8-2

5. An op amp inverting amplifier like that shown in Fig. 8–6(a) has $R_1 = 1.5\text{ k}\Omega$ and $R_F = 18\text{ K}\Omega$.
   a. Find its voltage gain, A_{VCL}.
   b. What is the value of its input resistance R_{IN}?
6. Suppose that the amplifier of Problem 5 is driven by a source with the following specs: $V_{s(p\text{-}p)} = 1.4\text{ V}$ (open circuit); $R_{out} = 50\ \Omega$
   a. What value of signal voltage actually reaches the amplifier's input terminals?
   b. Describe the output waveform, V_{out}. Mention its time synchronization with the source voltage V_s.

Section 8-3

7. For an op amp inverting amplifier like that shown in Fig. 8–8(a), suppose the component values are $R_1 = 2.0\text{ k}\Omega$, $R_F = 30\text{ k}\Omega$, and the signal source has $V_{s(p\text{-}p)} = 2.2\text{ V}$ (open circuit), $f = 1\text{ kHz}$, output resistance $R_{out} = 600\ \Omega$.
   a. Find the amplifier's voltage gain.
   b. What value of signal voltage actually reaches the amplifier's input terminals?
   c. Describe the output waveform V_{out}. Mention its time synchronization with the source voltage V_s.
   d. For the description of part **c** to be correct, what values of supply voltages V_{CC} and V_{EE} would be required? Explain.
8. For the amplifier of Problem 7, suppose the op amp's maximum output current rating is 70 mA with an output resistance $R_o = 30\ \Omega$.
   a. Can this amplifier successfully drive a 400-Ω load? Explain
   b. Can this amplifier successfully driven a 200-Ω load? Explain.

Section 8-4

9. Describe the output offset problem for an op amp amplifier. What undesirable thing occurs?

10. Describe and draw the basic method for correcting the output offset problem.
11. For an inverting amplifier with $R_1 = 750\ \Omega$, $R_F = 7.5\ k\Omega$, calculate and show the schematic placement of bias correcting resistor R_B.
12. For a noninverting amplifier with $R_1 = 1200\ \Omega$, $R_F = 8.2\ k\Omega$, calculate and show the schematic placement of bias correcting resistor R_B.

Section 8-5

13. In an op amp summing circuit, suppose the component values are $R_1 = 10\ k\Omega$, $R_2 = 20\ k\Omega$, $R_3 = 40\ k\Omega$, $R_F = 40\ k\Omega$. Write the equation for output voltage V_{OUT} as a function of the input voltages.
14. For the summer of Problem 13, find V_{OUT} for $V_1 = -1.5$ V, $V_2 = +0.8$ V, $V_3 = -3.1$ V.
15. Referring to Problem 14, if V_1 and V_3 remain the same, what new value of V_2 will cause $V_{OUT} = +2.0$ V?

Section 8-6

16. Draw the schematic diagram for a voltage comparer that compares the input signal voltage to +2 V, producing noninverting output.
17. For Problem 16, draw the waveform of output voltage, assuming that the input voltage is a triangle wave with peak magnitude = 3 V.
18. Repeat Problems 16 and 17 for this voltage comparer: Compare input voltage to −2 V, producing inverted output.

Section 8-8

19. If the differential amplifier of Fig. 8–15 has a voltage gain of 20 and a difference input of $V_2 - V_1 = 3$ V p-p, describe and sketch the output waveform.

Section 8-9

20. Explain intuitively why the voltage-to-current converter of Fig. 8–16 is able to deliver an unvarying current to a variable resistance load.
21. It is desired to build a voltage-to-current converter to drive a 1200-Ω load (nominal resistance). The proportionality factor is to be 5 mA/V.
    a. Find R_1 to provide this proportionality factor.
    b. If $V_{IN} = 7.2$ V, find I_{LOAD} if $R_{LOAD} = 1200\ \Omega$.
    c. If $V_{IN} = 7.2$ V, find I_{LOAD} if R_{LOAD} changes to 1150 Ω.
    d. Repeat parts **b** and **c** for a V_{IN} of −10.5 V.

Section 8-10

22. If the op amp differentiator of Fig. 8–17(a) had a sawtooth input voltage waveform, what would the output voltage waveform look like?
23. Repeat Question 22 for a square-wave input.
24. If the op amp integrator of Fig 8–17(b) had a square-wave input voltage, what would the output voltage waveform look like?
25. For the differentiator of Fig. 8–17(a), $R_F = 20\ k\Omega$ and $C = 0.075\ \mu F$. V_{in} is a triangle wave centered on ground, with a magnitude of 8 V p-p, at a frequency of 400 Hz. Graph the output voltage waveform, with the voltage and time axes properly scaled and marked.

26. For the integrator of Fig. 8–17(b), $R_{IN} = 1\ M\Omega$ and $C = 2\ \mu F$. $V_{supply} = \pm 15$ V. V_{in} is a step function, 5-V tall. (Imagine a switch closing and causing +5 V to be suddenly applied to the input.) Graph the output voltage waveform, with the voltage and time axes properly scaled and marked.

SUGGESTED LABORATORY PROJECT

Project 8-1: Op Amp Integrator

Construct the op amp integrator of Fig. 8–14(b) and carefully cancel the offset. Put a closed switch in parallel with the capacitor to prevent it from charging. At the instant the switch is opened, C will start charging, and the integration will begin. Use $C = 1\ \mu F$, nonpolarized, and $R_{IN} = 1.5\ M\Omega$.

1. Predict how fast the integrator will integrate. Predict how long it should take to integrate up to 10 V with $V_{IN} = -1$ V.
2. Apply $V_{IN} = -1$ V, open the switch, and measure the time required to integrate to 10 V. Compare to your prediction.
3. Change C to 0.05 μF and R_{IN} to 10 kΩ. Apply a 6-V p-p square wave at 300 Hz to the input. Predict what the output waveform should look like. Then open the switch and observe the output waveform, and compare to your prediction.
4. Change the input voltage waveform to a train of positive pulses, 1-V peak, at 1 kHz. Make sure there are no negative pulses applied to the input. This can be done by using a true pulse generator or by using a square-wave generator with its negative half cycle blocked by a diode.

Open the shorting switch and let the integration begin. You should see a staircase waveform at the output, rising to the saturation voltage of the op amp. It will occur only once per switch opening. Try to explain why this waveform occurs.

CHAPTER

FEEDBACK SYSTEMS AND SERVOMECHANISMS

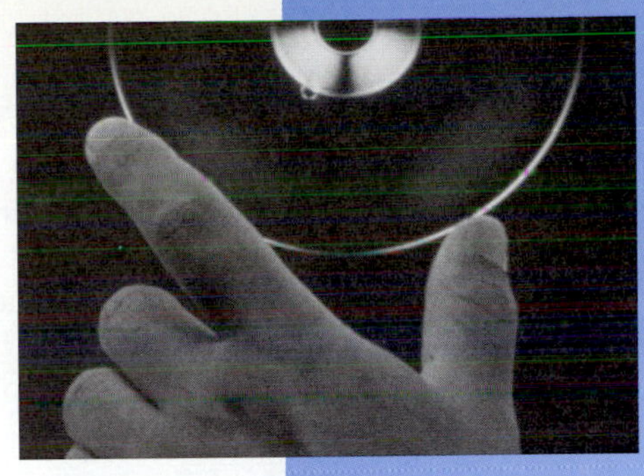

In all the industrial systems we have discussed so far there has been one thing in common: They have not been *self-correcting*. Self-correcting as used here refers to the ability of a system to monitor or "check" a certain variable in the industrial process and *automatically,* without the intervention of a human, correct it if it is not acceptable. Systems which can perform such self-correcting action are called *feedback systems* or *closed-loop systems*.

When the variable which is being monitored and corrected is an object's physical position, the feedback system is assigned a special name; it is called a *servo system*. The basic characteristics and components of closed-loop systems are the subjects of this chapter.

OBJECTIVES

After completing this chapter, you will be able to:

1. Explain the generalized closed-loop block diagram, and state the purpose of each one of the blocks.
2. State the characteristics which are used to judge a closed-loop control system; that is, describe what makes the difference between a good system and a bad system.
3. List the five general closed-loop control modes and explain how each one acts to correct the system error.
4. Cite the reasons why the On-Off control mode is the most popular.
5. Define the term *proportional band,* and solve problems involving proportional band, full scale range of the controller, and the control limits.
6. Discuss the problem of *offset* in proportional control, and show why it cannot be eliminated in a strict proportional controller.
7. Explain why the proportional plus integral control mode overcomes the offset problem.
8. Describe the effects of changing the integral time constant (reset rate) in a proportional plus integral controller.
9. Explain the advantage of the proportional plus integral plus derivative control mode over simpler control modes. State the process conditions which require the use of this mode.
10. Describe the effects of changing the derivative time constant (rate time) in a proportional plus integral plus derivative controller.
11. Define and give examples of the three different types of delay exhibited by industrial processes, namely, time constant delay, transfer lag, and transportation lag.

12. Interpret a table which relates the characteristics of an industrial process to the proper control mode for use with that process.
13. Interpret a table which relates the characteristics of an industrial process to the correct settings of proportional band and reset rate.
14. Write a PLC user-program segment for controlling an industrial process with the PID (Proportional-Integral-Derivative) instruction.

9-1 ■ OPEN-LOOP VERSUS CLOSED-LOOP SYSTEMS

Let us start by considering the essential difference between an open-loop system (not self-correcting) and a closed-loop system (self-correcting). Suppose that it is desired to maintain a given constant liquid level in the tank in Fig. 9–1(a). Liquid enters the tank at the top and flows out via the exit pipe at the bottom.

One way to attempt to maintain the proper level is for a human being to adjust the manual valve so that the rate of liquid flow into the tank exactly balances the rate of liquid flow out of the tank when the liquid is at the proper level. It might require a bit of hunting for the correct valve opening, but eventually the human could find the proper position. If he then stands and watches the system for a while and sees that the liquid level stays constant, he may conclude that all is well, that the proper valve opening has been set to maintain the correct level. Actually, as long as the operating conditions remain precisely the same, he is right.

The problem is that in real life, operating conditions don't remain the same. Numerous subtle changes could occur that would upset the balance he has worked to achieve. For example, the supply pressure on the upstream side of the manual valve might increase for some reason. This would increase the input flow rate with no corresponding increase in output flow rate. The liquid level would start to rise, and the tank would soon overflow. (Actually, there would be some increase in output flow rate because of the increased pressure at the bottom of the tank when the level rises, but it would be a chance in a million that this would exactly balance the new input flow rate.)

An increase in supply pressure is just one example of a change that would upset the manual adjustment. Any temperature change would change the liquid viscosity and thereby

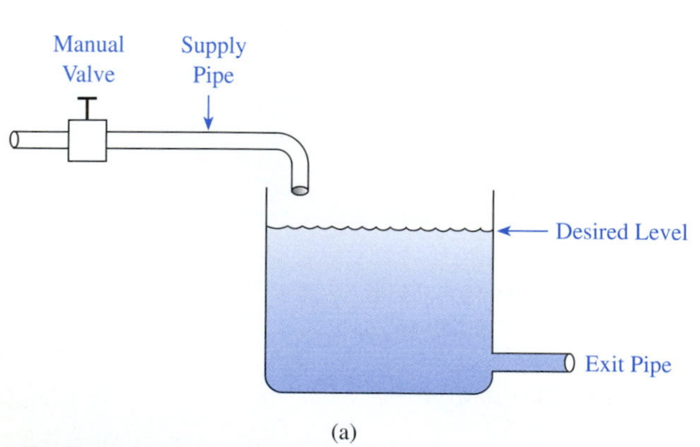

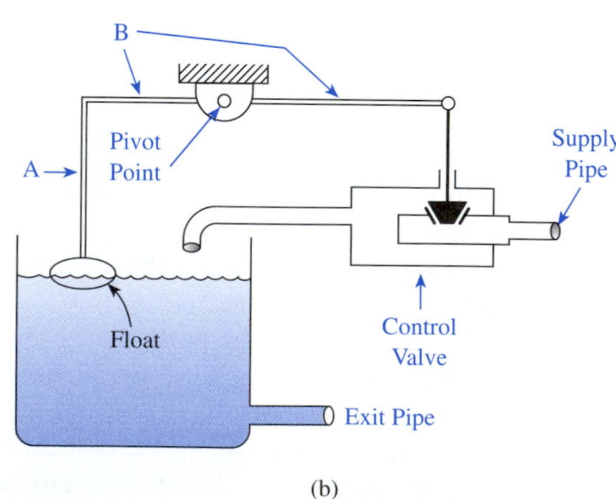

FIGURE 9–1
System for maintaining the proper liquid level in a tank. (a) An open-loop system; it has no feedback and is not self-correcting. (b) A closed-loop system; it has feedback and is self-correcting.

change the flow rates; a change in a system restriction downstream of the exit pipe would change the output flow rate, etc.

Now consider the setup in Fig. 9–1(b). If the liquid level falls a little too low, the float moves down, thereby opening the tapered valve to admit more inflow of liquid. If the liquid level rises a little too high, the float moves up, and the tapered valve closes a little to reduce the inflow of liquid. By proper construction and sizing of the valve and the mechanical linkage between float and valve, it would be possible to control the liquid level very close to the desired point. (There would have to be some slight deviation from desired liquid level to cause the valve opening to change.) With this system the operating conditions can change all they want. No matter which direction the liquid level tries to stray from the desired point and no matter what the reason for the straying, the system will tend to restore it to that desired point.

Our discussion to this point has dealt with the specific problem of controlling the liquid level in a tank. However, in the general view, many different industrial control systems have certain things in common. No matter what the exact system, there are certain relationships between the controlling mechanisms and the controlled variable that never differ. We try to illustrate these cause-effect relationships by drawing block diagrams of our industrial systems. Because of the general "sameness" among different systems, we are able to devise *generalized* block diagrams that apply to *all* systems. Such a generalized block diagram of an open-loop system is shown in Fig. 9–2(a).

Now let us try to relate the blocks in Fig. 9–2(a) to the physical parts in the manual control valve system of Fig. 9–1(a). Figure 9–2(a) shows that a *controller* (in our example the manual valve) affects the overall *process* (in our example the pipes that carry the liquid and the tank containing the liquid). The arrow leading from the controller box to the process box just means that the controller somehow "sends signals to" or "influences or affects" the process. The controller box has an arrow pointing into it called the *Setting*. This means that the human operator must somehow supply some information to the controller (at least once) that indicates what the controller is supposed to do. In our example the setting would be the position of the valve handle. The process box has an arrow pointing into it called *Disturbances*. This means that external conditions can upset the process and affect its outcome. In our example the disturbances are the changing conditions mentioned earlier, such as pressure changes, viscosity changes, etc. The *controlled variable* arrow stands for that variable in the process that the system is supposed to

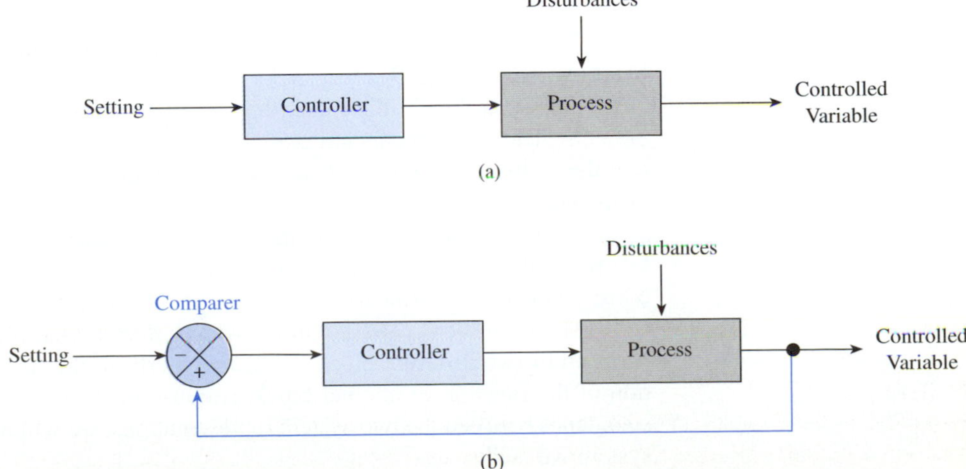

FIGURE 9–2 Block diagrams which show the cause-effect relationships between the different parts of the system: (a) for an open-loop system; (b) for a closed-loop system.

monitor and correct when it needs correcting. In our example the controlled variable is the liquid level in the tank.

The block diagram is basically only a cause-effect indicator, but it shows rather clearly that for a given setting the value of the controlled variable cannot be reliably known. Disturbances that happen to the process make their effects felt in the output of the process, namely in the value of the controlled variable. Because the block diagram of Fig. 9–2(a) does not show any lines *coming back around* to make a circle, or to "close the loop," such a system is called an *open-loop system*. All open-loop systems are characterized by the inability to compare the actual value of the controlled variable to the desired value and to take action based on that comparison.

On the other hand, the system containing the float and tapered valve of Fig. 9–1(b) is represented in block diagram form in Fig. 9–2(b). In this diagram the setting and the value of the controlled variable are compared to each other in a *comparer*. The output of the comparer represents the *difference* between the two values. The difference signal then feeds into the controller, allowing the controller to affect the process. The fact that the controlled variable comes back around to be compared with the setting makes the block diagram look like a "closed loop." A system that has this feature is called a *closed-loop system.* All closed-loop systems are characterized by the ability to compare the actual value of the controlled variable to its desired value and automatically take action based on that comparison.

For our example of the float level control in Fig. 9–1(b), the *setting* represents the location of the float in the tank. That is, the human operator selects the level that he desires by locating the float at a certain height above the bottom of the tank. This setting could be altered by changing the length of rod A, which connects the float to horizontal member B of the linkage in Fig. 9–1(b).

The comparer in the block diagram is the float itself in our example. The float is constantly aware of the actual liquid level, because it moves up or down according to that level. It is also aware of the setting, which is the desired liquid level, as explained earlier. If these two are not in agreement, the float sends out a signal which depends on the magnitude and the polarity of the difference between them. That is, if the level is too low, the float causes horizontal member B in Fig. 9–1(b) to be displaced (rotated) counterclockwise; the amount of counterclockwise displacement of B depends on how low the liquid is.

If the liquid level is too high, the float causes member B to be displaced clockwise. Again, the amount of displacement depends on the difference between the setting and the controlled variable; in this case the difference means how much higher the liquid is than the desired level.

Thus the float in the mechanical drawing corresponds to the comparer block in the block diagram of Fig. 9–2(b).

The controller in the block diagram is the tapered valve in the actual mechanical drawing. The valve opens and closes to raise or lower the liquid level, in the same way that the controller in Fig. 9–2(b) sends an output signal to the process to affect the value of the controlled variable.

In our particular example, there is a fairly clear correspondence between the physical parts of the actual system and the blocks in the block diagram. In some systems the correspondence is not nearly this clear-cut. It may be difficult or impossible to say exactly which physical parts comprise which blocks. One physical part may perform the function of two different blocks, or it may perform the function of one block and a portion of the function of another block. Because of the difficulty in stating an exact correspondence between the two system representations, we will not always attempt it for every system we study.

The main point to be realized here is that when the block diagram shows the value of the controlled variable being fed back and compared to the setting, the system is called a closed-loop system. As stated before, such systems have the ability to automatically take action to correct any difference between actual value and desired value, no matter why the difference occurred.

9-2 ■ CLOSED-LOOP SYSTEM DIAGRAMS AND NOMENCLATURE

9-2-1 The General Closed-Loop Block Diagram

A more detailed general block diagram which adequately describes most closed-loop systems is shown in Fig. 9–3.

FIGURE 9–3
The classic block diagram of a closed-loop system. This generalized drawing gives the names of the main parts (blocks) of a closed-loop system. It also gives the names of the signals which are sent between the various blocks.

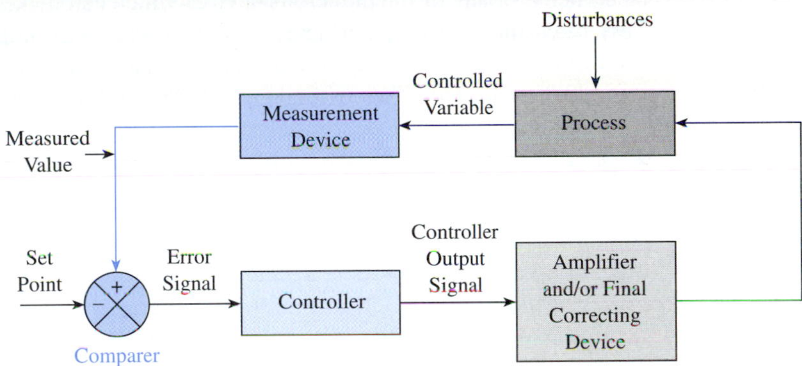

The ideas embodied in this general system block diagram are as follows: A certain process variable which is being controlled (temperature, pressure, fluid flow rate, chemical concentration, humidity, viscosity, mechanical position, mechanical speed, etc.) is measured and fed into a comparer. The comparer, which may be mechanical, electrical, or pneumatic, makes a comparison between the measured value of the variable and the set point, which represents the desired value of the variable. The comparer then generates an error signal, which represents the difference between measured value and desired value. The error signal is considered to be equal to the measured value minus the desired value, so if the measured value is too large, the error signal is positive, and if the measured value is too small, the error signal is of negative polarity. This is expressed in the equation

$$\text{error} = \text{measured value} - \text{set point}$$

The controller, which also may be either electrical, mechanical, or pneumatic, receives the error signal and generates an output signal. The relationship between the controller output signal and the error signal depends on the design and adjustment of the controller and is a detailed subject in its own right. All closed-loop controllers can be classified into five classes or *modes* of control. Within the modes, there are certain variations, but these variations do not constitute an essential control difference. Mode of control has nothing to do with whether the controller is electrical, mechanical, or pneumatic; it depends only on *how drastically* and in *what manner* the controller reacts to an error signal. More precisely, it depends on the mathematical relationship between the controller's output and its input (its input is the error signal). We will take a close look at the five control modes in Secs. 9-4 through 9-8.

Figure 9–3 shows that the controller output is fed to a final correcting device; amplification may be necessary if the controller output signal is not powerful enough to operate the final correcting device. The final correcting device is very often an electric motor, which may be used to open or close a valve, move some mechanical object in one direction or the other, or any similar function. The final correcting device might also be a solenoid valve, a pneumatically actuated valve or damper, or an SCR or triac to control the load power in an all-electric system.

An example of the need for amplification would be a situation in which the controller output is a low-voltage, low-current signal (as would be obtained from a Wheatstone bridge) and the final correcting device is a 2-hp motor. Obviously, the low-power signal from a Wheatstone bridge cannot drive a 2-hp motor directly, so an electric power amplifier would be inserted between them. We will explore and study final correcting devices in Chapter 11.

The measurement device in Fig. 9–3 might be a thermocouple, a strain gage, a tachometer, or any of the numerous devices which can make a measurement of a variable. Very often the measurement device gives an *electrical* output signal (usually a voltage) even though it is measuring a nonelectric quantity. Measurement devices which do this are called *transducers*. We will take a detailed look at several transducers in Chapter 10.

9-2-2 Nomenclature Used in Closed-Loop Systems

Unfortunately, the terms used to describe what's happening in a closed-loop control system are not universally agreed upon. We will now look at the various words and phrases that are used by different people, and we will decide on which ones we will use.

Starting from the left in Fig. 9–3, the term *set point* is also called the "set value," "setting," "desired value," "ideal value," "command," and "reference signal." We will use the term *set point* in this book.

The *comparer* is also called a "comparator," "error detector," "difference detector," etc. We will use the term *comparer*.

The *error signal* also goes by the names of "difference signal," "deviation," and "system deviation." We will use *error signal*.

The *controller output signal* is sometimes referred to as the "control value" (CV), especially when it is produced by a digital calculation performed by a programmable logic controller.

The *final correcting device* is also called the "correcting element" and the "motor element." We will use *final correcting device*.

The *controlled variable* is sometimes called the "controlled condition," "output variable," "output condition," etc. We will use *controlled variable*.

The *measurement device* is also referred to as the "detecting device," "detector," or "transducer." We will use the term *measurement device* in most cases, but when we wish to emphasize the ability of the measurement device to convert a nonelectrical signal into an electrical signal, we will use *transducer*.

The *measured value* is sometimes called the "actual value" or the "process variable" (PV). We will use *measured value* except when referring to the measurement input signal to a programmable logic controller that performs a digital calculation of the controller output signal. Then we will use PV.

9-2-3 Characteristics of a Good Closed-Loop System

It may seem obvious that the measure of a "good" closed-loop control system is its ability to bring the measured value into close agreement with the set point. In other words, a good system reduces the error signal to zero, or almost to zero. The final difference between measured value and set point that the system allows (that it cannot correct) is usually called *offset*. Therefore a good system has low offset. We are now using the word *offset* to have a different meaning than it has when applied to op amps.

There are other features of a closed-loop system that are also important, in some cases even more important than low offset. One of them is speed of response. If conditions occur that drive the measured value out of agreement with the set point, a good system will restore the agreement quickly. The quicker the restoration, the better the system.

It is possible to design systems which have low offset and fast speed of response, but they sometimes tend to be *unstable*. Unstable means that the system causes violent large variations in the value of the controlled variable as it frantically "hunts" for the proper controller output. This occurs because the system overreacts to an error, thereby causing an even larger error in the opposite direction. It then tries to correct the opposite error and overreacts again going in the other direction. When this occurs, the system is said to be oscillating. The oscillations usually eventually die out, and the system settles down at the correct value of controlled variable. In the meantime, though, the process has been effectively out of control, and bad consequences may result.

In certain cases the oscillations may not die out at all. They may continue getting larger and larger until the process is permanently out of control. If the control system is a mechanical positioning system (a servo system), these oscillations may cause the mechanism to shake itself apart and destroy itself.

As can be seen, then, a good system is one which is stable. The less violent the oscillations in the controlled variable, the more stable it is, and the better the system.

9-3 ■ EXAMPLES OF CLOSED-LOOP CONTROL SYSTEMS

It is normally easiest to see the correspondence between the actual physical components and the generalized block diagram of Fig. 9–3 when the system is a servo system. To learn to recognize the block diagram functions of the system components, we will now consider some examples of servo systems.

9-3-1 A Simple Rack-and-Pinion Servomechanism

Figure 9–4 shows a linear positioning system. The pointer is attached to a thin cord which runs over a fixed pulley, around the movable pulley, and over another fixed pulley and attaches to the object which is to be positioned. The object sits on a rack whose pinion is

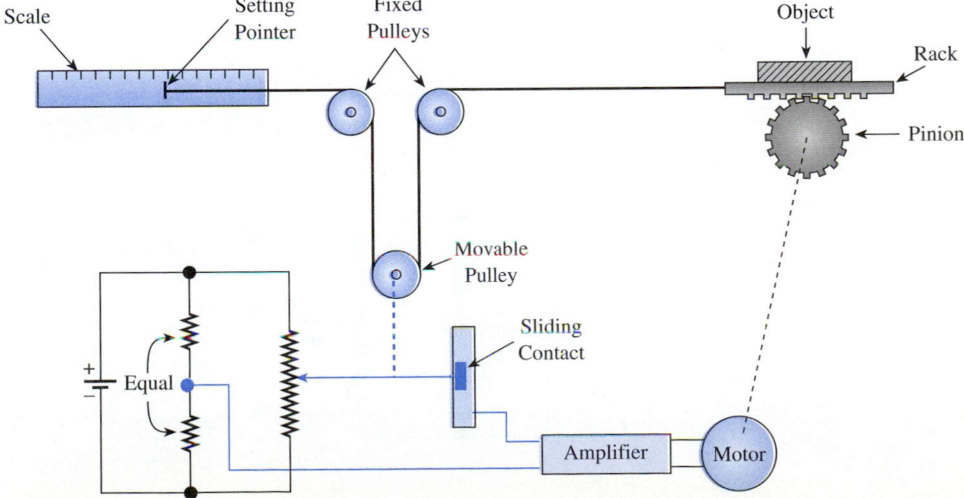

FIGURE 9–4
Mechanical positioning system using a rack and pinion. This is a simple example of a servomechanism.

driven by the motor. If the pointer is moved to the left on the scale, the movable pulley is drawn up by the cord, causing the potentiometer wiper to move up by the same amount. When the potentiometer contact is no longer in the center, the unbalanced bridge circuit delivers an input voltage to the amplifier. The output of the amplifier runs the motor, which drives the object to the left. When the object has moved the same distance as the pointer, the movable pulley returns to its rest position, and the potentiometer contact is centered once again. The bridge goes back into balance, causing zero input voltage to the amplifier, which then stops the motor.

It can be seen that whenever the bridge becomes unbalanced, it will send a low-power signal to the amplifier, which will amplify it to drive the motor. The motor moves the controlled object into such a position that the bridge is put back into balance. Since the bridge is only balanced when the movable pulley is in its rest position, the controlled object always moves exactly the same distance as the pointer, because only by doing so can it return the movable pulley to the rest position.

In this system, the position of the pointer represents the set point. The position of the object represents the controlled variable. The cord and pulley arrangement represents the comparer, with the instantaneous pulley position being the error signal. The bridge circuit is the controller, and the controller output signal is the voltage applied to the input of the amplifier. The motor with rack-and-pinion arrangement represents the final correcting device.

9-3-2 Profile Cutting Machine

The same idea applied to a more elaborate mechanism is illustrated in Fig. 9–5. This system is a profile cutting machine. A pattern piece, or model, is fastened to the mounting support, as is the uncut workpiece. The mounting support is then slowly moved to the left. As it moves, the motor-driven cutting tool cuts an identical profile in the workpiece.

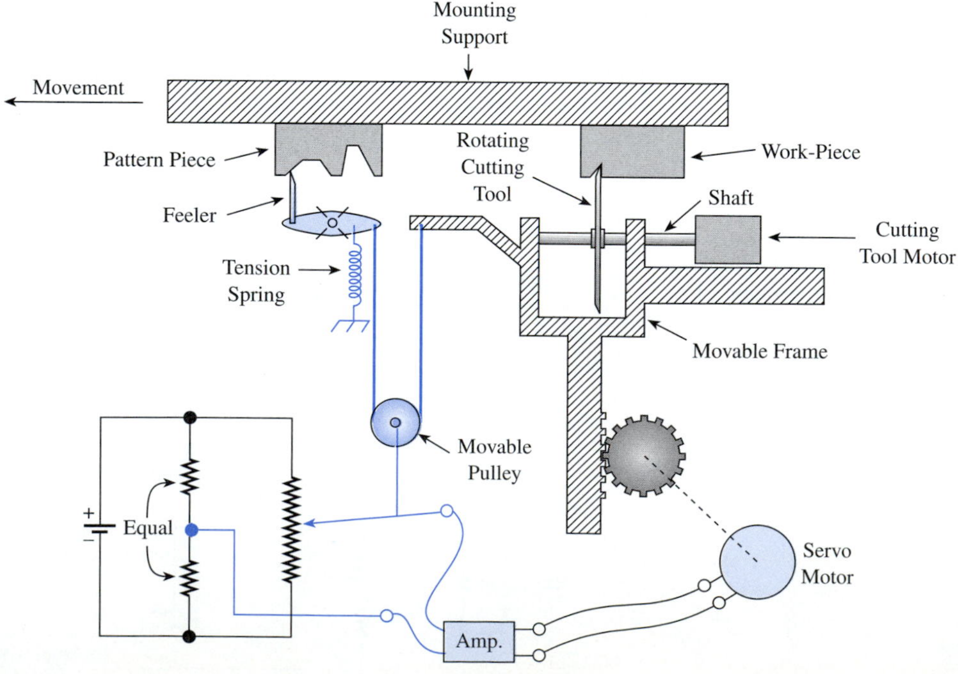

FIGURE 9–5 Profile cutting system. This is an example of a more complex servomechanism.

The system works like this: The rigid feeler is held tight against the pattern by the action of the tension spring on the right-hand side of the pivot. As the feeler moves up and down, its motion is transmitted to the movable pulley through the cord attached to the right-hand side of the feeler arm. This movable pulley is attached to the wiper of a potentiometer, so as the pulley moves away from its centered position, the bridge becomes unbalanced. The unbalanced bridge drives an amplifier, which drives the servo motor. The servo motor causes the movable frame to move up or down the proper amount to return the movable pulley to its centered position. As the frame moves, it causes the rotating cutting tool to cut into the workpiece. As the cutting tool duplicates the position of the feeler, the profile it cuts duplicates that of the pattern.

In this feedback system, the set point is the pattern depth, or feeler position. The controlled variable is the position of the cutting tool, or, equivalently, the position of the movable frame. All other system parts serve the same block diagram functions that they did in Fig. 9–4.

9-3-3 Bimetal Temperature Control System

Figure 9–6 shows a popular method for temperature control which is used for home heating systems and in some industrial systems. The spiral bimetal strip is immersed in the medium whose temperature is being controlled. Because the two component metals have different coefficients of expansion, the spiral either unwinds or winds up tighter as the metal temperature changes. Assume in this example that the spiral strip is constructed with the metal with the greater coefficient of expansion on the inside, so the spiral *unwinds* as the temperature *rises*. Attached to the end of the spiral is a mercury switch, a sealed glass bulb containing liquid mercury and two electrodes. Mercury, although it is a liquid under standard conditions, is a metal and is an excellent electrical conductor. When the mercury

FIGURE 9–6
Closed-loop system for controlling temperature.

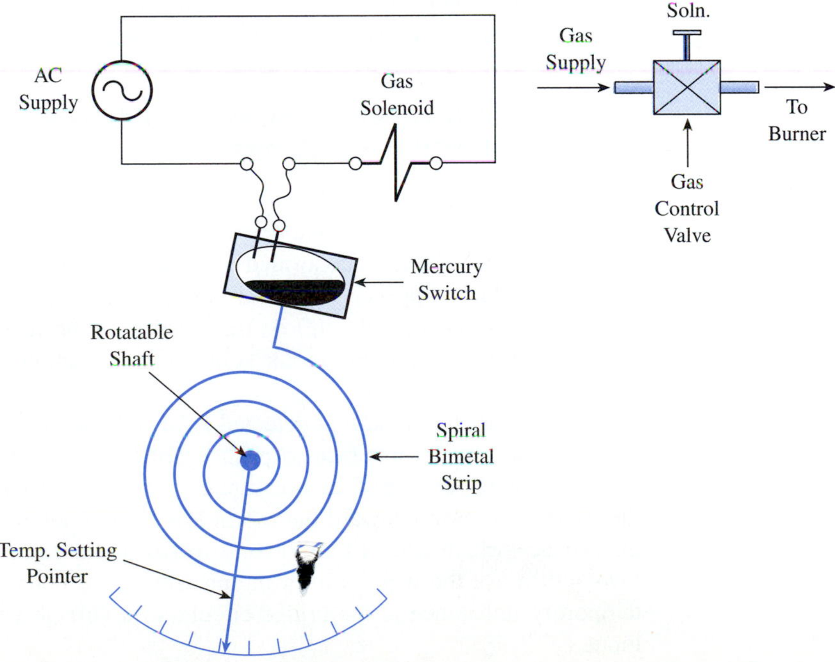

switch tips to the right (rotates clockwise) the mercury drains to the right-hand side of the bulb and breaks the electrical connection between the electrodes. When the switch tips to the left (rotates counterclockwise) the mercury flows to the left-hand side of the bulb and makes an electrical connection between the electrodes.

When the mercury switch is open, the gas solenoid is deenergized and the gas control valve closes, stopping the flow of natural gas through the pipe to the burner. When the mercury switch closes, it energizes the gas solenoid, opening the gas control valve and allowing natural gas to flow to the burner.

The shaft to which the center of the bimetal spiral strip is attached is itself rotatable. The position of this shaft sets the initial position of the spiral strip. The initial position of the spiral strip determines the desired temperature.

Here is how the system works. If the temperature is below the desired control temperature, the bimetal spiral strip will tend to be wound up. This causes the mercury switch to close, energizing the gas solenoid and turning on the burner. As the temperature rises due to the heat released by burning the natural gas, the bimetal strip unwinds. At a certain temperature, the strip will have unwound enough to open the mercury switch. This turns off the burner. With the burner off, the temperature slowly drops until it causes the strip to wind up enough to close the mercury switch. The burner then comes back on to drive the temperature back up. In this manner the system will continue to maintain the actual temperature near the desired temperature.

The rotatable shaft attached to the setting pointer comprises the set point in the generalized block diagram. To raise the set point, the setting pointer is moved to the right. The measured value in the block diagram is the amount that the bimetal spiral strip is wound up. The comparer is the mercury bulb; the position of the mercury could be considered the error signal. The mercury in combination with the electrodes is the controller. The final correcting device is the gas solenoid valve-burner combination.

You may take the point of view that the solenoid valve is part of the controller and that the burner alone represents the final correcting device. This viewpoint is also reasonable and could easily be adopted. This just goes to point out that there may be no definite one-to-one correspondence between the actual system components and the blocks in the generalized block diagram of Fig. 9–3. The correspondence may be quite blurry.

9-3-4 Pressure Control System Using a Motor-Driven Positioner

Consider the control system shown in Fig. 9–7. The pressure at a certain point in a process chamber is to be maintained at a desired value. The method of adjustment is the variable-position damper, which is controlled by a slow-moving positioning motor. If the damper in the inlet duct is opened somewhat, the pressure in the chamber tends to rise. If the damper closes to restrict the inflow, the pressure in the chamber tends to drop. As is commonly done, the positioning motor is driven by an amplifier whose input voltage comes from a Wheatstone bridge.

The process pressure is detected by a *bellows*. As the pressure rises, the bellows expands, causing its left face to push against the compression spring. The bellows is linked to the wiper arm of the pressure error potentiometer so that as pressure increases, the wiper arm moves toward the top in Fig. 9–7. Therefore if the system experiences a disturbance which causes the pressure to rise too far above the desired value, the bellows will move the wiper arm of the pressure error potentiometer up. This will cause a temporary imbalance of the bridge circuit, so a voltage will be applied to the amplifier input.

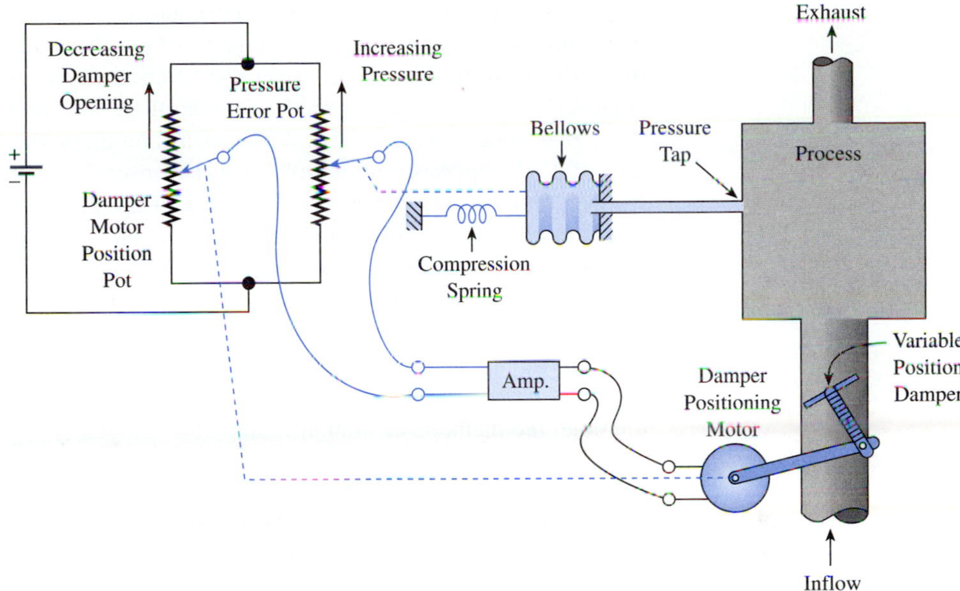

FIGURE 9–7
Closed-loop system for controlling pressure in a process chamber.

The amplifier output will then drive the motor in the correct direction to move the wiper arm of the left-side potentiometer up. As this is taking place the motor linkage is closing off the control damper. When the movement of the left-side potentiometer wiper equals the movement of the right-side potentiometer wiper, the bridge is back in balance and all movement stops. The damper ends up in a more closed position, thereby limiting the pressure increase to a small amount.

In this system the set point is the adjustment screw on the compression spring, which can alter the force that the spring exerts on the face of the bellows. The measurement device is the bellows itself. Low pressure causes the bellows to collapse, moving to the right; high pressure causes the bellows to expand, moving to the left. The comparer is the combination of compression spring, bellows, and potentiometer arm. The position of the potentiometer arm represents the error signal. Assuming that the exact middle position means zero error, then positions toward the top indicate positive errors (measured value greater than set point) and positions toward the bottom indicate negative errors (measured value less than set point).

The combination of Wheatstone bridge, amplifier, and positioner motor can be considered to comprise the controller. The variable position damper is the final correcting element.

9-4 ■ MODES OF CONTROL IN INDUSTRIAL CLOSED-LOOP SYSTEMS

As mentioned in Sec. 9-2, the *manner* in which the controller reacts to an error signal is an indication of the mode of control. It is somewhat difficult to make hard-and-fast classifications of control modes, but it is generally agreed that there are five basic modes:

1. On-Off
2. Proportional
3. Proportional plus integral
4. Proportional plus derivative
5. Proportional plus integral plus derivative

This list of modes is arranged in order of complexity of the mechanisms and circuitry involved. That is, the mode at the top, On-Off, is the simplest to implement; as you move down in the list, the physical construction of the controllers gets more complex. Naturally the more complex modes of control are also more difficult to understand.

In general, the more difficult the control problem, the farther down the list you must go to find the appropriate control mode. However, in many industrial processes the control need not be very precise, or the nature of the process may be such that precise control is easy to accomplish. In these situations, the simpler control modes are completely adequate. In fact, the simplest method, On-Off, is by far the most widely used. It is inexpensive, reliable, and easy to adjust and maintain.

In this book we are concentrating on electrical and electronic industrial control, so specific examples of the various control modes will be electric controllers. The principles involved are the same when discussing pneumatic, hydraulic, or mechanical controllers, although the methods of implementation are altogether different, naturally.

In the succeeding sections of this chapter, Secs. 9-5 through 9-9, we will study each of the five control modes. We shall start with the simplest and work up to the most complex. Each of the five control modes is explained in terms of *temperature* being the controlled variable. Temperature control is easier to visualize than most other variables. However, keep in mind that the principles discussed in this chapter apply just as well to the control of other process variables besides temperature.

9-5 ■ ON-OFF CONTROL

In the *On-Off control mode,* the final correcting device has only two positions or operating states. For this reason, On-Off control is also known as *two-position control* and as *bang-bang control*. If the error signal is positive, the controller sends the final correcting device to one of its two positions. If the error signal is negative, the controller sends the final correcting device to the other position. On-Off control can be conveniently visualized by considering the final correcting device to be a solenoid-actuated valve, as seen in Sec. 9-3-3. When a valve is actuated by a solenoid, it is either fully open or completely closed; there is no middle ground. Therefore a solenoid-actuated valve fits perfectly into an On-Off control system. A graph of final correcting device position (percent valve opening) for ideal On-Off control is given in Fig. 9–8(a). In this figure the controlled variable is considered to be temperature, with the set point equal to 120°F. As can be seen, if the measured value of the temperature is less than 120°F by even a tiny amount, the valve is positioned 100% open. If the measured value of the temperature is greater than 120°F by even a tiny amount, the valve is 0% open, or completely shut off.

Figure 9–8(b) shows a typical graph of measured value of temperature versus time, with valve position plotted against the same time axis. Notice that the actual temperature tends to oscillate around set point. This is a universal characteristic of On-Off control. This particular graph shows a 4°F overshoot in the positive direction and 4°F undershoot in the negative direction. These particular values are picked at random. The actual overshoots depends on the complete nature of the system and may be different in the positive and negative directions (overshoot may be different from undershoot).

Overshoot occurs because the process cannot respond instantly to the change in valve position. When temperature is climbing, it's because the rate of heat input is greater than the rate of heat loss in the process. Quickly shutting off the control valve cannot reverse that trend instantly, because there will be residual heat energy built up in and around the heating device which must diffuse through the process chamber. As this residual heat is distributed, it temporarily continues to raise the temperature.

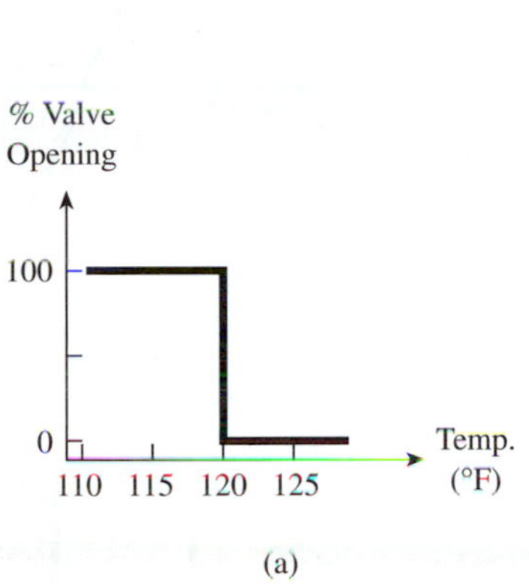

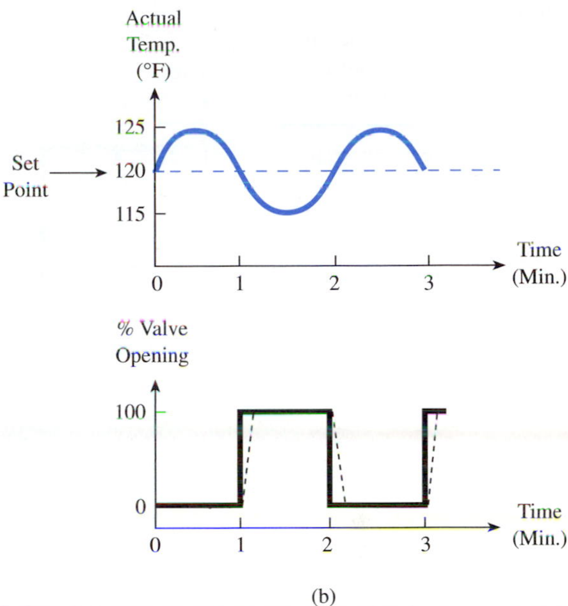

FIGURE 9–8
Graphs pertaining to the On-Off control mode. (a) Valve position versus measured temperature, with a set point of 120°F. (b) Actual measured temperature versus time and valve opening versus time. In the graph of valve opening versus time, the solid line is for a snap-action valve and the dashed line for a slow-acting valve.

In the same way, a temperature downtrend cannot be reversed instantly because it takes a while to distribute new heat energy throughout the process. Until the distribution can occur, the downtrend will continue, causing undershoot.

To be sure, the system can be designed to keep the magnitude of the oscillations small, but this tends to cause more frequent cycling. This aggravates the other disadvantage of On-Off control, namely the wear on the correcting device caused by the frequent operation. In this specific example, the solenoid valve will wear out sooner if its frequency of opening and closing is higher.

The graph of valve position in Fig. 9–8(b) reflects the fact that the valve is wide open when the temperature is below the set point and completely closed when the temperature is above the set point. The dashed lines are for the case in which the valve is not a snap-action valve. This is often encountered when the valve is physically large. Large heavy valves cannot be successfully operated by snap action but must be operated somewhat slowly. A geared-down motor and linkage is the most effective method of actuating such valves.

9-5-1 Differential Gap

No On-Off controller can exhibit the ideal behavior sketched in Fig. 9–8(a) and (b). All On-Off controllers have a small *differential gap,* which is illustrated graphically in Fig. 9–9(a).

The differential gap of an On-Off controller is defined as the smallest range of values the measured value must pass through to cause the correcting device to go from one position to the other. Differential gap is defined specifically for On-Off control; there is no such thing as differential gap in other control modes. It is often expressed as a percent of full scale.

Differential gap is an expression of the fact that the measured value must rise above the set point by a certain small amount (the error signal must reach a certain positive value) to close the valve. Likewise, the measured value must drop below the set point by

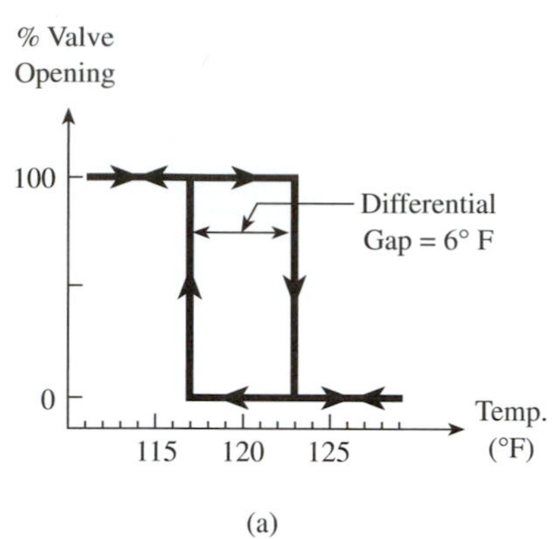

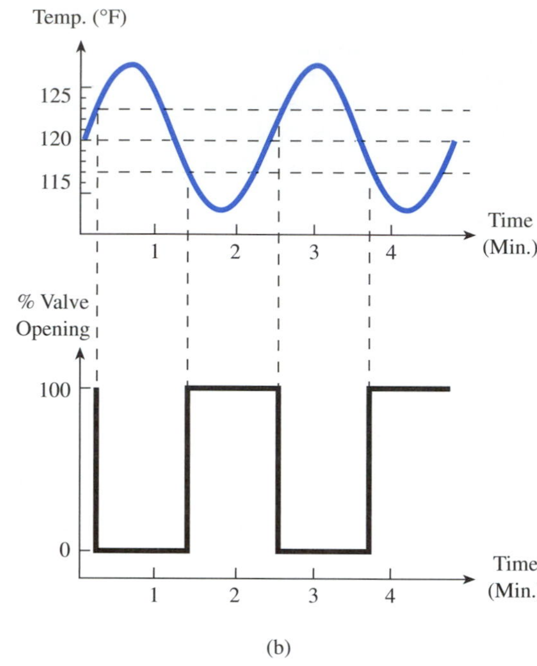

FIGURE 9–9
Graphs illustrating differential gap in On-Off control. (a) Valve opening versus temperature. The set point is 120°F, and the differential gap is 6°F. (b) Actual measured temperature versus time and valve position versus time, with a 6°F differential gap.

a certain small amount (the error signal must reach a certain negative value) in order to open the valve. In the example given in Fig. 9–9, the actual measured temperature must rise above the set point by 3°F to close the valve, and it must drop below the set point by 3°F to open the valve. Therefore the smallest possible temperature change which can drive the valve from open to closed is 6°F. The differential gap is thus 6°F.

Differential gap can also be expressed as a percent of full controller range. If the controller had a range of, say, 60°F to 300°F, then the size of its range would be 240°F (300°F − 60°F). A temperature of 6°F would represent 2.5% of full control range, because

$$\frac{6°\text{F}}{240°\text{F}} = 0.025 = 2.5\%$$

Therefore in this case the differential gap could be expressed as 2.5% instead of 6°F.

The practical effect of differential gap is shown in the time graph of Fig. 9–9(b). As can be seen, the magnitude of oscillation is larger, but the frequency of oscillation is smaller. Differential gap is thus a curse and a blessing. It is a curse because the instantaneous measured value can stray further from the set point, but it is a blessing because wear on the correction device is reduced.

In many On-Off controllers, the differential gap is fixed. If so, it is usually less than 2% of full scale. Some On-Off controllers have adjustable differential gap so that the user can select the amount that suits his application.

If you are familiar with magnetic materials and circuits, you will recognize that differential gap in an On-Off controller has the same effect as hysteresis in a magnetic core. In general, when the switching point of the dependent variable depends not only on the *value* of the independent variable but also on the *direction of approach,* we say that hysteresis exists. Recall that we also saw hysteresis in triac power control in Chapter 6. In magnetism, the dependent variable is flux density (B) and the independent variable is magnetomotive force (H). In the On-Off control mode, the dependent variable is the final correcting device position (valve either open or closed) and the independent variable is the error signal.

A good example of an On-Off controller is the temperature control system utilizing the spiral strip and mercury switch in Sec. 9-3-3. Other On-Off control systems will be shown in Chapter 12.

9-6 ■ PROPORTIONAL CONTROL

In the proportional control mode, the final correcting device is not forced to take an all-or-nothing position. Instead, it has a *continuous* range of possible positions. The exact position that it takes is *proportional* to the error signal. In other words, the output from the controller block is *proportional* to its input.

9-6-1 Proportional Band

Assuming that the final correcting device is a variable-position valve controlled by a slow gear motor and linkage, we can illustrate the effects of proportional control by drawing a graph of percent valve opening versus temperature. This is done in Fig. 9–10(a). To visualize what is happening, imagine that the valve is controlling the flow of fuel to a burner. This setup is illustrated schematically in Fig. 9–10(b). When the valve opening gets larger, more fuel is delivered, and more heat is released into the process. Therefore the process temperature tends to rise. When the valve opening gets smaller, less fuel is delivered to the burner, and the process temperature tends to fall.

Figure 9–10(a) shows the proportional relationship between percent valve opening and error signal. Study this graph carefully. To get started, imagine that the set-point temperature is now 180°F. Furthermore, let us assume that the process temperature is being maintained right on 180°F with the valve 40% open. There would be no way to know that this would be the case, since the percent of valve opening which is needed in order to maintain 180°F will depend on many unpredictable process conditions. Such things as

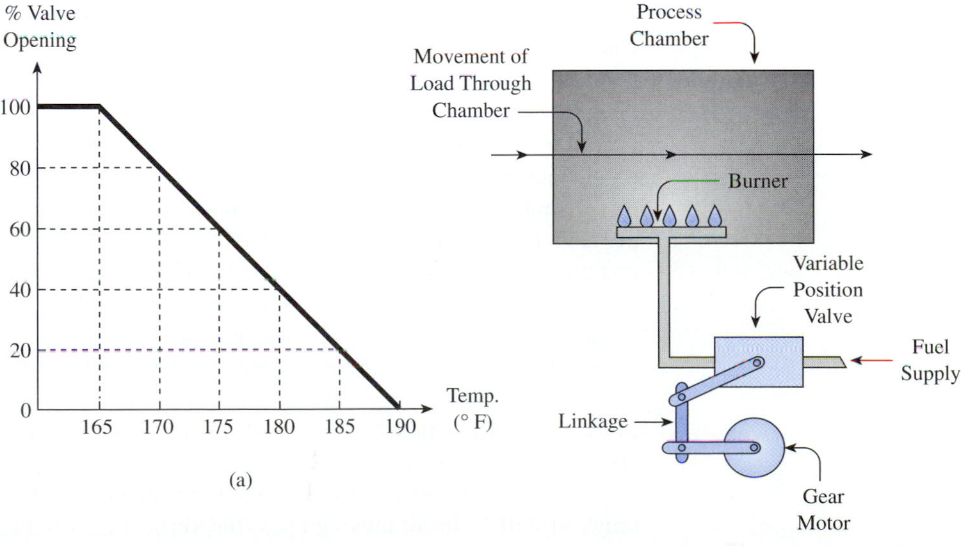

FIGURE 9–10 Proportional control mode. (a) Graph of valve position versus measured temperature. Valve position is *proportional* to the error signal. (b) Layout of the control system.

ambient temperature, rate of heat consumption by the load, fuel supply pressure, heating capacity of the fuel, etc., will all have an effect on what valve opening will be necessary. Therefore, let us just *assume* that a 40% valve opening is correct.

Now if something happens to cause the measured temperature to change, the valve will assume a new position according to the graph of Fig. 9–10(a). If the temperature should somehow drop to 175°F, the valve will open to the 60% point. This will cause the temperature to subsequently rise back toward 180°F. If the original drop in temperature had been more drastic, say, down to 170°F, the valve would have opened 80%. Therefore the controller is responding not only to the fact that the measured temperature is too low; it is responding to the *amount* of the error. The more serious the error, the more drastic the corrective action. This is the essential difference between proportional control and On-Off control.

The word *proportional* is correctly applied in this situation because the amount of correction introduced is in proportion to the amount of the error. When the error is 5°F (measured value equals 175°F), the valve travels from 40% open to 60% open; this means it travels 20% of its full range. If the error is twice as large, namely 10°F (measured value equals 170°F), the valve travels from 40% open to 80% open, or 40% of its full range. Thus the corrective action is also twice as large when the error is twice as large. In general, a given percent change in error brings about a proportional percent change in valve position.

In the example shown in Fig. 9–10(a), a measured temperature of 165°F or less causes the valve to go 100% open, and a measured temperature of 190°F or more causes the valve to go 0% open. The difference between these two points is called the *proportional band of control*. The proportional band in this case is 25°F. Within that band the valve response is proportional to temperature change; outside that band the valve response ceases because it has reached its limit.

Usually, proportional band is expressed as a percentage of the full-scale range of the controller. If the controller set point can be adjusted anywhere between 60°F and 300°F, as assumed earlier, it has an adjustment range of 240°F. The proportional band expressed as a percent would be given by

$$\frac{25°\text{F}}{240°\text{F}} = 0.104 = 10.4\%$$

The formal definition of proportional band is this: The proportional band is the percentage of full controller range by which the measured value must change in order to cause the correcting device to change by 100%. Most proportional controllers have an adjustable proportional band, usually variable from a few percent up to a few hundred percent.

Figure 9–11(a) through (f) gives graphical representation of different proportional band adjustment settings, both on a degrees-Fahrenheit basis and on a percent of full controller range basis. Any two graphs drawn alongside each other are drawn for the same percents proportional band, but the one on the left is plotted versus temperature and the one on the right is plotted versus percent of full controller range. That is, the graph in Fig. 9–11(a) shows percent of valve opening versus *temperature* for a 20% proportional band, while the graph in Fig. 9–11(b) shows percent of valve opening versus *percent of full range of the controller,* also for a 20% proportional band.

The graphs in Fig. 9–11(c) and (d) show the same things, but for a 50% proportional band. The graphs in Fig. 9–11(e) and (f) show the same things again, but for a 100% proportional band.

In all cases, the controller range is assumed to be from 60°F to 300°F, giving a full range of 240°F. In all these graphs, the vertical axis indicates percent valve opening, as stated earlier.

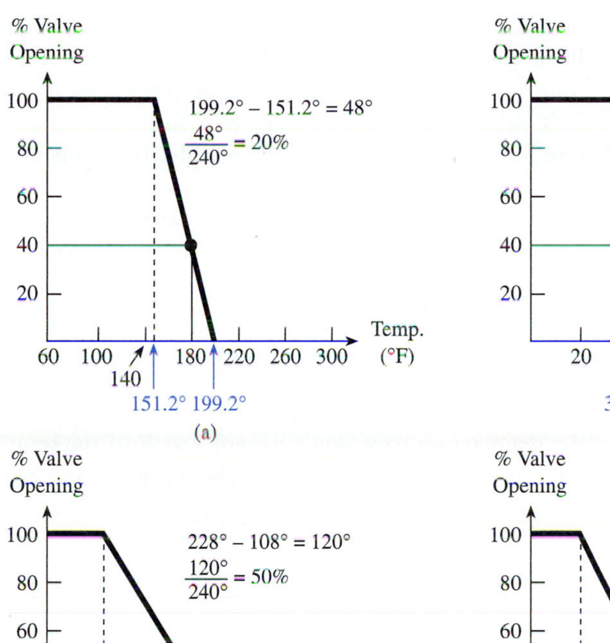

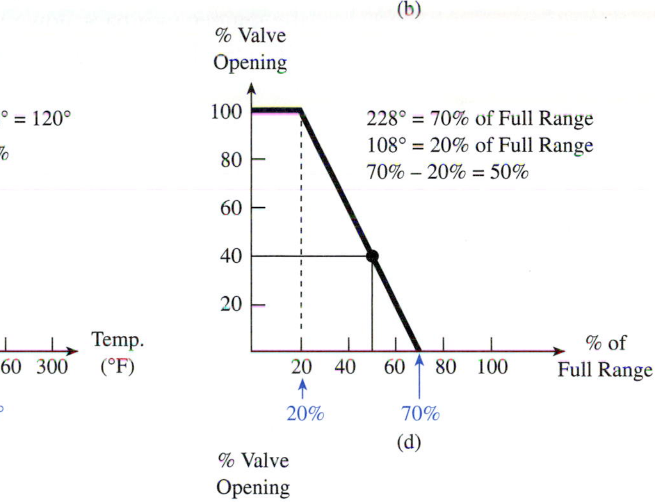

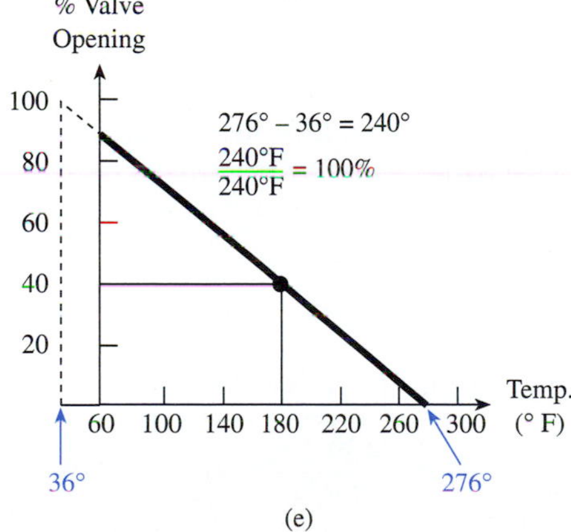

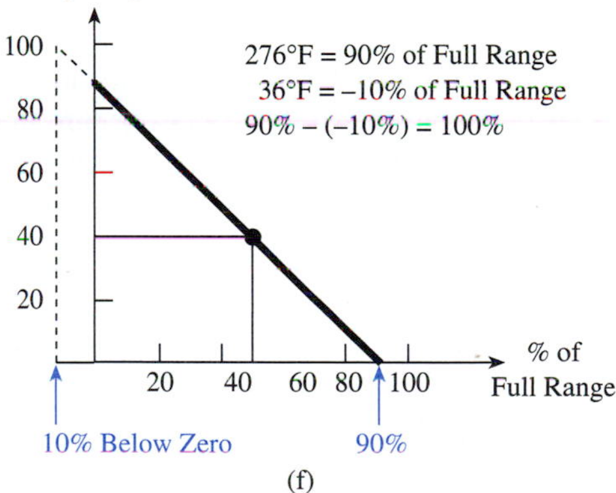

FIGURE 9–11
Graphs of percent valve opening versus temperature and also versus percent of full control range.
(a) Proportional band = 20%. Valve opening plotted versus temperature.
(b) Proportional band = 20%. Valve opening plotted versus percent of full control range.
(c) Proportional band = 50%. Valve opening plotted versus temperature.
(d) Proportional band = 50%. Valve opening plotted versus percent of full control range.
(e) Proportional band = 100%. Valve opening plotted versus temperature.
(f) Proportional band = 100%. Valve opening plotted versus percent of full control range.

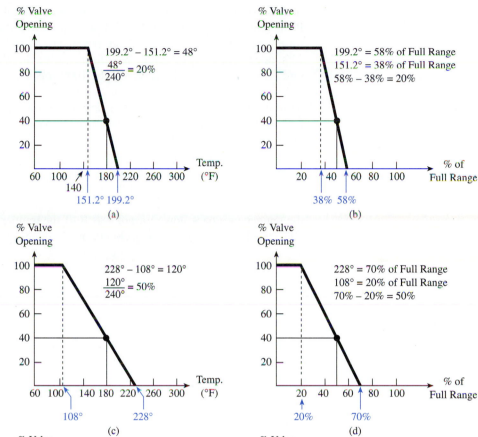

In all the graphs, we have tacitly assumed that the measured temperature is being maintained at the set point of 180°F with a 40% valve opening until a disturbance occurs which upsets the measured temperature.

These graphs deserve careful study. By studying and thoroughly understanding the specific numbers indicated on these graphs, you can get a firm grasp of the meaning of proportional band.

Here is a step-by-step interpretation of the graph in Fig. 9–11(a):

1. If the measured temperature is 180°F, the valve will take a position 40% open.
2. If the temperature drops below 180°F, the valve will start opening further. At 172°F, for example, the valve will be about 57% open. At 164°F, the valve will be about 73% open. These points can be read from the graph. When the measured temperature drops to 151.2°F, the valve will be 100% open. Any further drop in temperature below that point will cause no further corrective action, since the valve has reached its limit. (Of course, if the system is designed properly, the temperature should be able to recover from this level and start rising back up toward 180°F with the valve 100% open.) The temperature which causes a 100% valve opening (151.2°F) is specifically marked on the horizontal temperature axis in Fig. 9–11(a).
3. If the measured temperature should rise above 180°F for any reason, the valve will start closing down to less than 40% open. For example, if the temperature should reach 188°F, the valve would throttle back to about 23% open in an effort to drive the temperature back down toward 180°F. If the temperature should somehow hit 199.2°F, the valve will go fully closed, or 0% open. The exact temperature which causes 0% valve opening is specifically marked on the horizontal temperature axis in Fig. 9–11(a). Beyond this temperature, control is lost because the valve has reached its limit. However, with the valve fully closed and no fuel entering the burner in Fig. 9–10, the temperature is bound to start dropping back down toward 180°F.

To demonstrate that the performance shown by Fig. 9–11(a) constitutes a 20% proportional band, look at the calculations done next to that graph. The range of temperatures which causes the valve to drive from fully open to fully closed is 151.2°F to 199.2°F, which is a span of 48°F. A span of 48°F represents 20% of the total range of the controller, since

$$\frac{48°\text{F}}{240°\text{F}} = 0.2 = 20\%$$

Therefore Fig. 9–11(a) represents a 20% proportional band.

To illustrate how the percent calculations are made in the graphs on the right-hand side, namely Figs. 9–11(b), (d), and (f), here is the derivation of the 58% figure and the 38% figure in Fig. 9–11(b).

The temperature of 199.2°F is 139.2°F higher than the lowest temperature in the range of the controller (60°F). That is,

$$199.2°\text{F} - 60°\text{F} = 139.2°\text{F}$$

To calculate what percent this figure is of the full controller range, we say

$$\frac{139.2°\text{F}}{240°\text{F}} = 0.58 = 58\%$$

Therefore the percent of full controller range which causes the valve to go fully closed is 58%, and this is specifically marked on the horizontal axis of Fig. 9–11(b).

These calculations are then repeated for the fully *open* valve position:

$$151.2°\text{F} - 60°\text{F} = 91.2°\text{F}$$

$$\frac{91.2°\text{F}}{240°\text{F}} = 0.38 = 38\%$$

The exact percent of full controller range that causes the valve to go fully open is therefore 38%, and this is specifically marked on the horizontal axis of Fig. 9–11(b).

To be sure you understand the meaning of proportional band, you should check the calculations done alongside Fig. 9–11(c) for a 50% proportional band. You should then do the appropriate calculations for Fig. 9–11(d) and verify the numbers marked on the horizontal axis of Fig. 9–11(d).

In Fig. 9–11(e) and (f), which are for a 100% proportional band, the lines have been extrapolated below 60°F. The extrapolations are drawn dashed because such temperature measurements are impossible; the controller cannot detect temperatures below 60°F. However, it is convenient to imagine these temperatures anyway, because it makes the calculations for proportional band easier to demonstrate. In a real-life situation, this would mean that the valve could never open up to 100%. The error necessary to open the valve that far is beyond the range of the controller. The maximum valve position in this situation would be 90% open. You should check and verify the calculations presented alongside Fig. 9–11(e) and (f).

Variations in process conditions. In all the graphs of Fig. 9–11 it is assumed that a set-point temperature of 180°F can be accomplished by a 40% valve opening. Remember that this could change drastically as process conditions change. For example, it might require a 65% valve opening to maintain the temperature at 180°F under heavier load conditions; it might even require a 90% valve opening to maintain a temperature of 180°F under very heavy load conditions. If these different load conditions actually existed, the *slopes* of the lines in Fig. 9–11 would remain the same, but their *horizontal locations* on the graphs would change. This idea is illustrated for a 20% proportional band in Fig. 9–12.

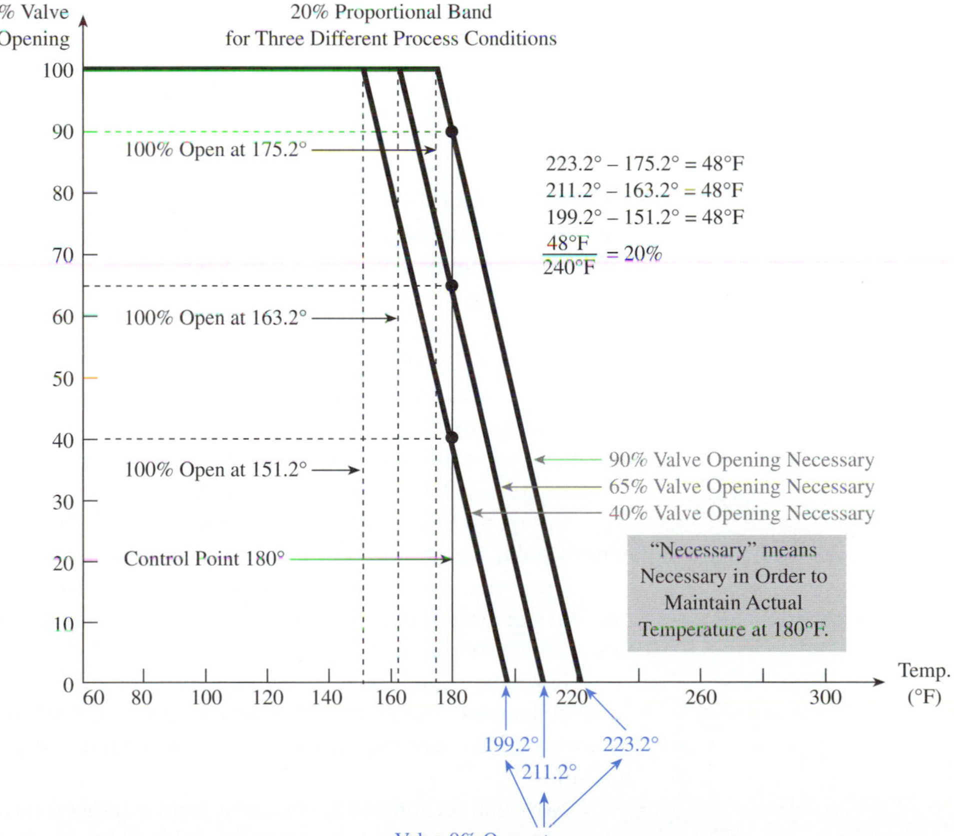

FIGURE 9–12
Graphs of valve opening versus temperature with a 20% proportional band for three different process conditions.

Here is the interpretation of the graphs in Fig. 9–12. The left graph is for a 40% valve opening to produce a measured temperature of 180°F, and that graph is just a repeat of Fig. 9–11(a). The center graph in Fig. 9–12 is for the situation in which process conditions have changed so that a 65% valve opening is necessary to produce a temperature of 180°F. Notice that the 180°F temperature line intersects the center graph at 65% valve opening.

The center graph indicates that the valve will be fully open at 163.2°F and will be fully closed at 211.2°F. The proportional band of temperatures is still 48°F, which is 20% of full controller range. The only thing different between the left graph and the center graph is the horizontal location.

The right-hand graph in Fig. 9–12 is for the situation in which process conditions have changed more drastically, so that a 90% valve opening is necessary to produce an actual temperature of 180°F. Notice that the 180°F temperature line intersects the right-hand graph at 90% valve opening.

In the right-hand graph, the fully open temperature is 175.2°F and the fully closed temperature is 223.2°F. The temperature band is still 48°F, and the proportional band is therefore still 20%.

9-6-2 Effects of Proportional Control

Let us now discuss the control effects of using the proportional control mode. As might be expected, it eliminates the permanent oscillation that always accompanies On-Off control. There may be some temporary oscillation as the controller homes in on the final control temperature, but eventually the oscillations die out if the proportional band is adjusted properly. However, if the proportional band is set too small, oscillations may occur anyway, because a very small proportional band makes proportional control behave almost the same as On-Off control. You should think carefully about the last sentence. If you understand what proportional band means, you will understand why that statement is true.

Thus we can see that the proportional control mode has one important advantage over On-Off control. It eliminates the constant oscillation around the set point. It thereby provides more precise control of the temperature and reduces wear and tear on the valve. A variable-position valve moves only when some sort of process disturbance happens, and even then it moves in a less violent manner than a snap-action valve. Its life expectancy is therefore much longer than that of a solenoid snap-action valve.

Figure 9–13 shows some typical responses of a proportional temperature controller to a load disturbance. In each case in Fig. 9–13, a load disturbance has occurred which tends to drive the temperature down. Figure 9–13(a) shows the response for a narrow proportional band (10%). The approach to the control position is fast, but once it is there, the temperature oscillates awhile before settling down.

In Fig. 9–13(b) a medium proportional band (50%) causes a slower approach to the control point, but it almost eliminates oscillation.

The behavior of a large proportional band (200%) is shown in Fig. 9–13(c). The time for the system to reach the control point is longer, but once it is there, the temperature does not oscillate at all.

If you pay careful attention to the meaning of the graphs in Fig. 9–13, you may become concerned. The graphs of temperature versus time in Fig. 9–13 are showing that after a load disturbance the actual measured temperature *does not return to the original control value*.

Now one reasonable expectation from a temperature controller is that it return the actual temperature to the set point after a load change. Figure 9–13 shows that a propor-

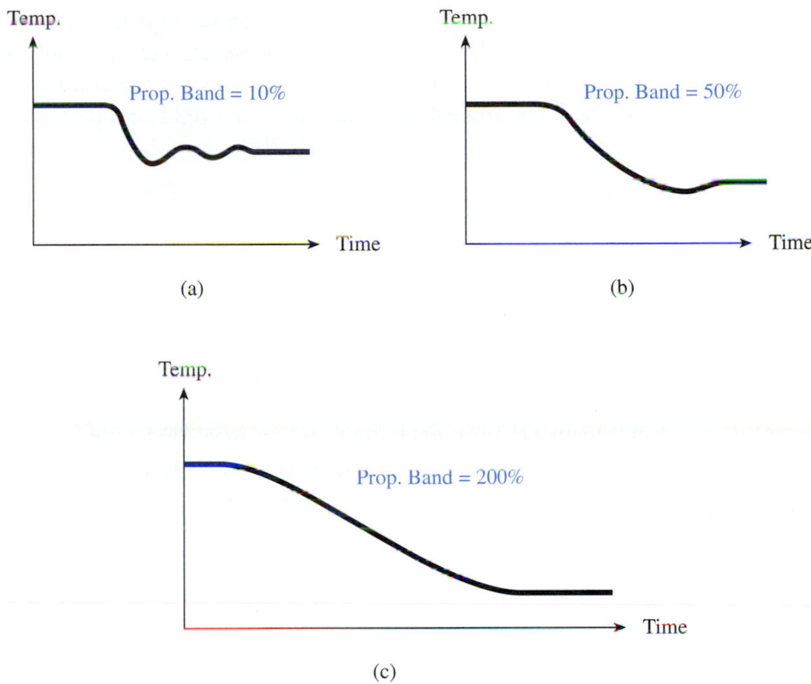

FIGURE 9–13 Graphs of temperature versus time after a load disturbance: (a) narrow proportional band (10%); (b) medium proportional band (50%); (c) wide proportional band (200%).

tional controller does not do this. Furthermore, the wider the proportional band, the greater the difference between the two control values before and after the disturbance.

To understand why this is so, imagine again that a proportional controller is maintaining a control temperature of 180°F with the control valve 40% open. If a disturbance occurs which causes the temperature to drop (an increase in rate of heat loss through the walls of the chamber, for example), the valve will drive open. The increased fuel flow will tend to drive the temperature back up toward 180°F, but it can never recover fully, because the increased fuel flow is now a permanent requirement. The control valve must *permanently remain further open* to meet the increased demand for heat input. Since the percent opening of the valve is proportional to the error signal, a permanently increased valve opening can only happen if there is a permanently increased error.

Looking at it another way, we cannot expect the temperature to recover fully to 180°F, because if it did, the valve would return to 40% open, just as it was before the increased heat demand. If the valve is in the same position as before, how will it satisfy the process demand for more heat input?

Either way we look at it, the temperature cannot recover fully to its predisturbance level. With a narrow proportional band, the recovery is closer, because it takes only a small increase in error to create a large change in valve position. Therefore the increased heat demand can be met with very little permanent error introduced. On the other hand, if the proportional band is large, it takes a relatively large error to produce a given amount of change in valve position. Therefore the increased heat demand can only be met by introducing a large permanent error. These ideas are illustrated in Fig. 9–13; the larger proportional bands create larger permanent errors.

This serious shortcoming of the proportional mode of control means that proportional control is not very useful except in a certain few types of processes. As a general statement, it may be said that proportional control works well only in systems where the process changes are quite small and slow. It helps if the disturbances occur slowly, because then the proportional band can be adjusted quite narrow, since there is not much oscillation produced by a *slow* process change. The only objection to a narrow proportional band is that it may cause oscillation of the controlled temperature. If it is possible to adjust the proportional band nice and tight, the permanent error can be kept small.

9-6-3 Offset in Proportional Control

We have been considering the problem of failure of the actual temperature to return to the original control value after a system disturbance. However, we have purposely avoided the problem of controlled temperature error before a disturbance. In other words, we have not asked the question "Does the actual measured temperature agree with the set point *before* a system disturbance happens?" The answer to this question is "Probably not." There is only one unique set of circumstances under which a proportional controller can ever cause exact agreement between measured temperature and set point. The chance of hitting that set of circumstances is remote. Here is why.

The design of real-life proportional temperature controllers is such that absolutely zero error signal causes a 50% opening in the control valve (this can be altered by the user, but let us consider it to be exactly 50%). The 50% figure is desirable so that the controller has available to it equal maximum corrections in both directions. That is, it has just as much correcting ability for both positive errors and negative errors. Now, under a given set of process conditions, a 50% valve opening will cause a given fixed temperature to actually occur in the process. If the set point happens to be *that particular temperature,* then the controller will hold the valve 50% open when the error reaches zero, and the 50% opening will cause the measured value to exactly agree with the set point.

For example, imagine a set of process conditions that causes the temperature to stabilize at exactly 700°F when the control valve is locked at 50% open. If it so happens that we *want* a set point of exactly 700°F (this would be a fantastic coincidence), then here is what will happen. The temperature will rise toward 700°F from below. With a measured temperature below 700°F, the error will be negative, and the valve will be more than 50% open. The closer the actual measured temperature gets to 700°F, the smaller the error gets, and the closer the valve gets to 50% open. At the point where the measured temperature hits exactly 700°F, the error will be zero, and the valve will be positioned precisely at 50% open. Since a 50% opening happens to be just what's needed to maintain a 700°F temperature, there is no further temperature change, and the system controls at that value.

Keep in mind that this is the *only* possible set point at which the controller could bring about exact agreement. Even if the set point were 705°F, the valve would have to be open more than 50% (say, 50.2%) to attain that temperature. For the valve to be open 50.2%, the error signal must be nonzero. The error would be a very small negative value in this case. Thus the actual measured temperature could never climb to exactly 705°F but would have to stop at about 704.9°F in order to maintain the error necessary to keep the valve open more than 50%.

Of course, with normal luck, the set point we want will be quite a bit different from the stable temperature at 50% valve opening, so the permanent error will be larger than the 0.1°F suggested above. Just as a typical example, if the set point were 950°F, the valve might end up 75% open, with the control point at 944°F. The 6°F permanent difference between set point and control value is called *offset*. The farther the set point is from the 50%-open temperature, the worse the offset will be.

The idea of offset is shown graphically in Fig. 9–14. In Fig. 9–14(a), at the first set point a certain offset exists. When the set point is changed in the same direction as the first offset, the new set point results in a worse offset.

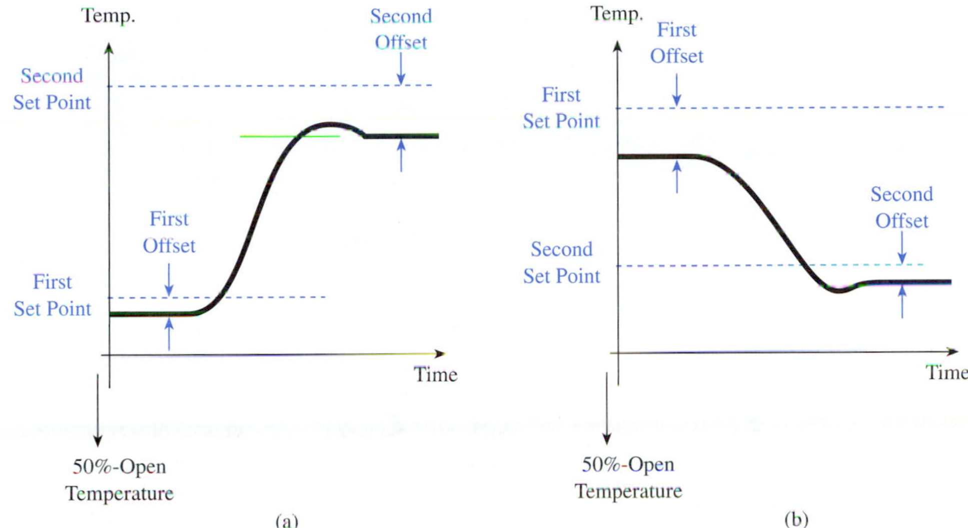

FIGURE 9–14 Graphs of temperature versus time which illustrate the offset problem in proportional control. In both graphs the valve is assumed to be more than 50% open.
(a) The offset gets worse (larger) when the set point is moved further above the 50%-open temperature.
(b) The offset gets better (smaller) when the set point is moved closer to the 50%-open temperature.

In Fig. 9–14(b), when the set point is changed in the opposite direction to the first offset, the new set point causes a better (smaller) offset.

9-6-4 An Electric Proportional Temperature Controller

An example of a proportional temperature controller is illustrated in Fig. 9–15. Two equal potentiometers are arranged in a bridge configuration, with the centers of both pots grounded. The pot on the right is called the *error pot,* and the pot on the left of the bridge is called the *valve position pot.* Assume for a moment that the proportional band adjustment is dialed completely out (shorted out). Then whatever position the wiper of the error pot assumes, the wiper of the valve position pot must assume that same position. If the error pot wiper moves up 200 Ω, for example, the unbalanced bridge will deliver a signal to the electronic amplifier. This will drive the motor in such a direction that the wiper of the valve position pot will also move up. When the valve position pot has moved up the same 200 Ω, the bridge is back in balance, the input is removed from the amplifier, and the motor stops. Thus the electronic amplifier and gear motor will force the valve position pot to follow the error pot.

The wiper of the error pot is positioned by a pressure bellows expanding against a set-point spring. As the process temperature changes, the pressure of the fluid in the filled sensing bulb will change. This pressure change is imparted to the flexible bellows through capillary tubing. Higher temperature causes the bellows to expand to the left, against the set-point spring. Lower temperature causes it to retract to the right, aided by the compressed set-point spring. The motion of the bellows is imparted to the wiper of the error pot. The set point is adjusted by adjusting the compression of the spring. Higher set points require tighter compression, and lower set points mean relaxed compression. When the actual temperature is above the set point (positive error), the error pot wiper moves above midpoint. When the actual measured temperature is below the set point (negative error), the error pot wiper moves below midpoint. The distance the error pot wiper moves from its midpoint is proportional to the magnitude of the error.

It can be seen that for any given amount of error there is a specific position of the control valve which will balance the bridge, and that valve position is proportional to the error.

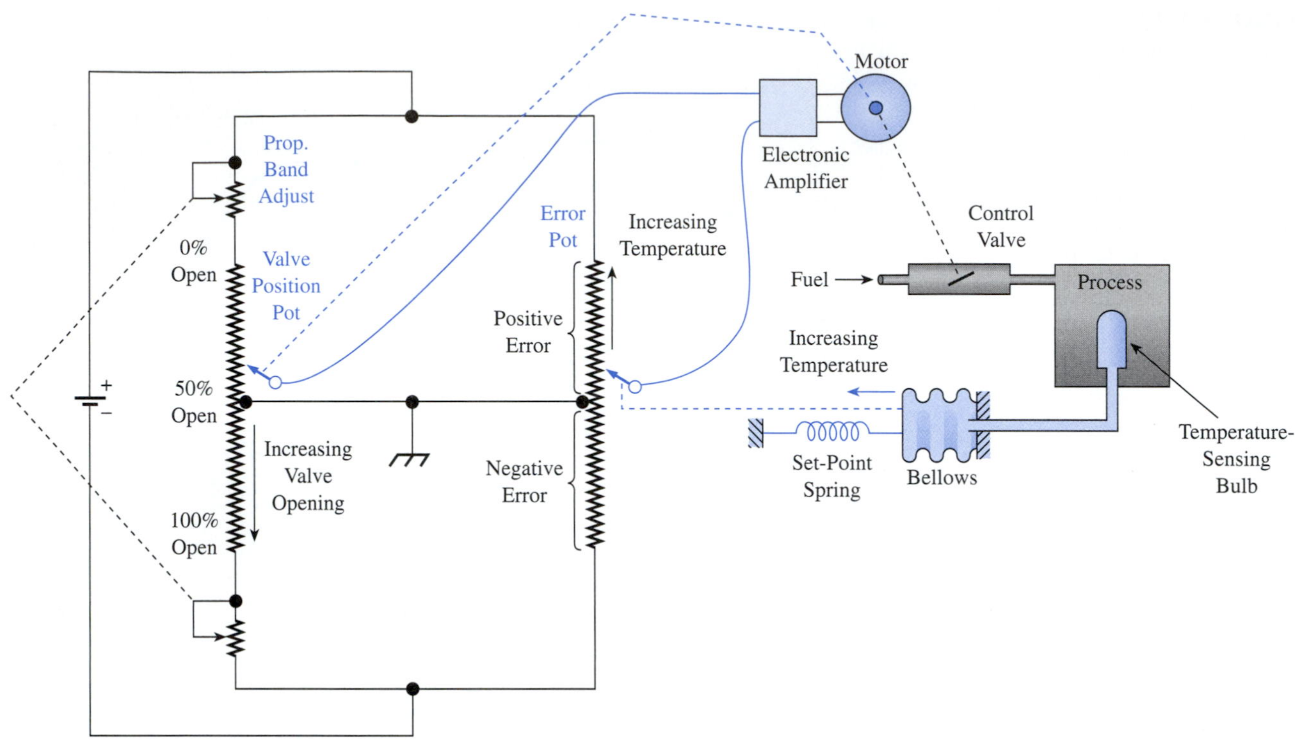

FIGURE 9–15
An electric method of implementing the proportional control mode. Temperature is the controlled variable. The proportional band-adjust pot is actually two pot elements mechanically ganged together. This ensures that the voltage across the valve-position pot is divided equally between its top and bottom halves.

To make the controller more sensitive (decrease its proportional band), we decrease the voltage across the valve position pot. This is accomplished by adjusting the *proportional band adjust pot* on the left of Fig. 9–15 (actually, two pots ganged together). As this pot resistance is raised, the voltage across the valve position pot decreases. When that happens, the wiper of the valve position pot must move *further* to balance a given movement in the error pot wiper. Looking at it another way, it takes less movement of the error pot to produce a given movement in the valve position pot. This means that the proportional band has been reduced.

As an example, suppose the error pot has 20 V applied to it from end to end but that the valve position has only 10 V from end to end due to the setting of the proportional band adjust pot. If the error wiper now moves 100 Ω, the valve position pot wiper must move 200 Ω to create an equal voltage at its wiper terminal and null the amplifier input voltage. Since a 200-Ω change in the valve position pot represents a given valve opening, it can be seen that a given change in valve percent opening has been accomplished with a smaller temperature error than would otherwise be necessary. Thus the proportional band has been reduced.

A careful study of Fig. 9–15 reveals why a permanent offset occurs with this type of electric proportional controller. Suppose we are controlling right on the set point with the valve 50% open. If a load disturbance occurs which drives the temperature up, the error pot will move up a certain distance. The valve position pot must follow it up the same distance because of amplifier-motor action. This causes a reduction in fuel flow, which tends to reduce the temperature. As the temperature starts to fall back toward the set point, the error pot travels back toward its center position, and the valve position pot follows it, thus reopening the valve a little bit. This will continue until a point is found where any further reduction in temperature will cause the valve to be open enough to drive the temperature up once again. When that point is found, the system stabilizes, and all pot movement stops. Unfortunately, this stabilization point must necessarily be a little bit above the set-point temperature. In other words, the error pot never makes it back to its center point. It cannot make it back to the center point because if it did the valve would be 50% open again, and we already know that with the valve 50% open, the temperature rises. That was the beginning premise of the discussion.

Therefore the system stabilizes at a new control point which is a little higher in temperature than the original set point. At stabilization, the error pot is a little above center, indicating a positive error; the valve position pot is also a little above center, indicating an opening of a little less than 50%. Permanent offset has set in.

9-7 ■ PROPORTIONAL PLUS INTEGRAL CONTROL

In Section 9-6 we showed that proportional control eliminates oscillation in the measured variable and reduces wear on the control valve, but it introduces permanent offset in the measured variable. For this reason, it is not too useful in most systems. Strict proportional control can be used only when load changes are small and slow and the variation in set point is small. This point was made in Sec. 9-6. For the more common process situation, in which load changes are large and rapid and set point may be varied considerably, the *proportional plus integral mode* of control is better suited. Proportional plus integral control is also called *proportional plus reset control.*

In proportional plus integral control, the position of the control valve is determined by two things:

1. *The magnitude of the error signal:* This is the proportional part.
2. *The time integral of the error signal:* In other words, the magnitude of the error multiplied by the *time* that it has persisted. This is the integral part.

Since the valve can respond to the time integral of the error, any permanent offset error that results from proportional control alone is eventually corrected as time goes by. It can be thought of this way: The proportional control part positions the valve in proportion to the error that exists. Then the integral control part senses that a small error (offset) is still persisting. As time passes, the integral part moves the valve *further* in the same direction, thus helping to reduce the offset. The longer the error persists, the more additional distance the valve moves. Eventually, the error will be reduced to zero and the valve will stop moving. It stops moving because as more time passes, the time integral of the error does not increase any more because the error is now zero.

To understand the action of the integral part of such a controller, it is helpful to study a schematic diagram which shows how one could be built. Refer to Fig. 9–16. It shows the same controller as Fig. 9–15 except that an integral part has been added to make it into a proportional plus integral controller.

The best way to picture the action of this proportional plus integral controller is to focus on the *RC* circuit attached to the wiper of the valve position pot. Recall that a capacitor never charges instantly and sometimes takes rather a long time to build up any appreciable amount of voltage. This is the case in this circuit, because the *RC* time constant

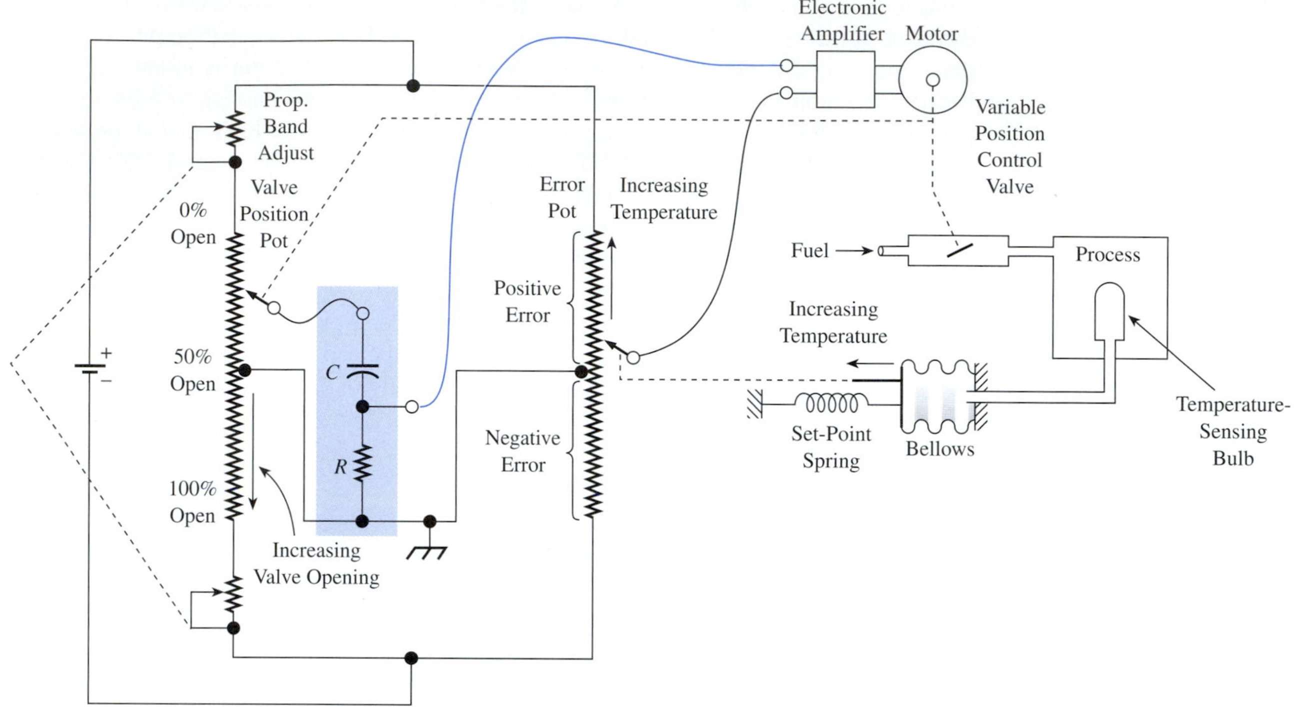

FIGURE 9–16
An electric method of implementing the proportional plus integral control mode.

is rather large. When the valve position pot wiper moves off its center point and applies a voltage to the *RC* circuit, at first the entire wiper voltage appears across *R* because the capacitor *C* has no charge at all. As time passes *C* charges up, thus *reducing* the voltage across *R*. The voltage across *R* is equal to the wiper voltage (potential difference between wiper and ground) minus the capacitor voltage. As capacitor voltage builds up, resistor voltage decreases.

Now imagine that the controller is controlling right on set point (zero error) with the valve 50% open. As in Sec. 9-6, assume that the proportional band adjustment pot is shorted out. If a process disturbance occurs which drives the temperature up, the error pot will move up a certain distance. The valve position pot must follow it up the same distance because of the amplifier/motor action. Thus the percent valve opening is reduced, the temperature is *partially* corrected, and an offset error sets in. The offset error is due to the fact that the error pot must remain off center in order to hold the valve slightly closed, as explained in detail in Sec. 9-6-4.

To give concreteness to our discussion of integral control, assume a specific situation. Let us assume that the voltage on the wiper of the error pot is +1 V relative to ground and that the voltage on the valve position pot is also +1 V relative to ground. Thus the voltage applied to the amplifier is the difference between these, which is 0 V. The motor therefore stops.

As time passes, *C* will start charging + on the top and − on the bottom. This reduces the voltage across *R*, say to 0.75 V. The voltage into the amplifier is now the difference between 1.00 and 0.75, which is 0.25 V. This 0.25 V is amplified and causes the motor to run further in the *same direction* (closing the valve). The valve position pot will move up until the wiper voltage is 1.25 V, which will once again null the amplifier. Therefore the fuel flow is further reduced, and the temperature moves closer to set point. The error pot wiper voltage is now reduced as the temperature error approaches zero.

As time continues to pass, C continues to charge, thereby constantly reducing the voltage across R, which is the signal on one of the amplifier input leads. As long as the error is not zero, the voltage across R can be reduced to *less than* the error pot wiper voltage as time passes; this will continue to drive the valve position pot in the up direction, closing the valve further and further. Eventually, the temperature will be reduced back to set point, causing the error pot to return to the center. This applies 0 V on the amplifier input lead attached to the error pot wiper. At that time, the capacitor will just have reached full charge, and the voltage across R will be zero, causing 0 V on the other amplifier input lead. The valve therefore stops in the correct position to maintain the temperature right at the set point.

The final position of the pots is now quite different from what it would be for strict proportional control. The error pot wiper is centered, and the valve position pot wiper is moved up far enough to establish the proper fuel flow to the process. There is no way to know beforehand at what percent opening it will settle.

It can be seen that the position of the control valve is *initially* determined by the proportional part of the control, but *finally* settles in a position partially determined by the integral part of the control. The relative importance of the proportional and integral parts can be varied by adjusting the resistor R. In most controllers, R is a potentiometer, so the RC time constant can be adjusted. When the time constant is large (large R), the integral part is less effective (slower to make its effects felt). When the time constant is small (small R), the integral part is more effective. Figure 9–17 shows the control effect of changing the integral time constant.

In most industrial controllers, the integral time constant is not used as a reference. Instead, the *reciprocal* of the integral time constant is the variable which is talked about.

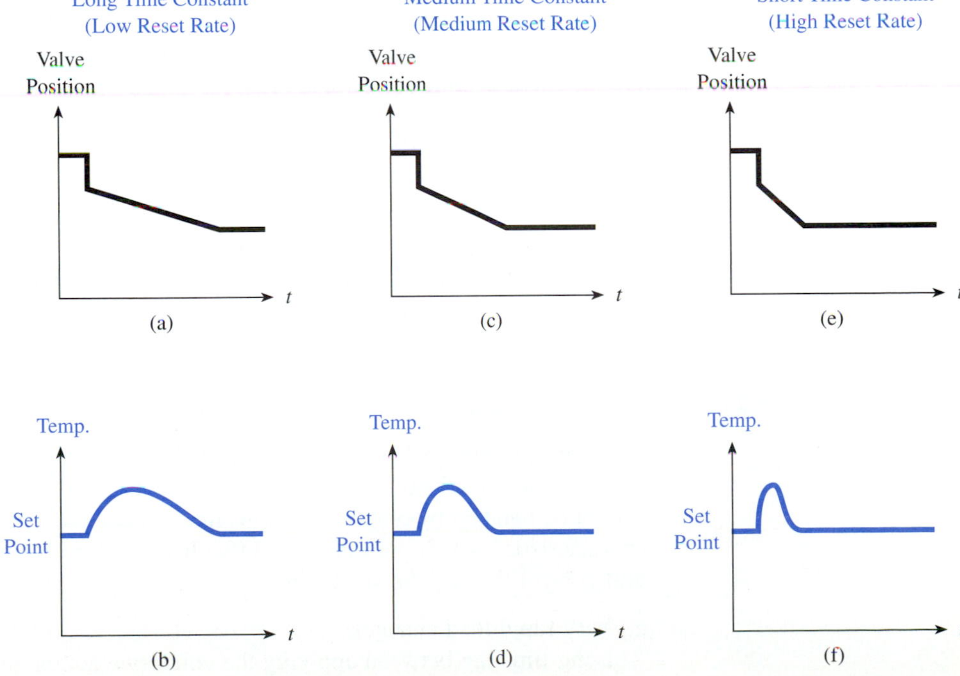

FIGURE 9–17 Graphs of valve position versus time and actual temperature versus time after a disturbance. The control mode is proportional plus integral.
(a) and (b) Long integral time constant.
(c) and (d) Medium integral time constant.
(e) and (f) Short integral time constant.

This variable is termed *reset rate*. The term *reset rate* can be rather confusing if you are used to thinking in terms of time constants. However, just remember that when reset rate is low (large time constant), the integral part is slow to make its effects felt by the process. When reset rate is high (small time constant), the integral part of the control is quick to make its effects felt by the process.

Figure 9–17(a) shows a graph of valve position (percent opening) versus time for a large integral time constant (low reset rate). Farther up on the vertical axis means increased valve opening, and farther down on the vertical axis means reduced valve opening.

Figure 9–17(b) shows actual measured temperature versus time for the large integral time constant. The graphs in Fig. 9–17(a) and (b) should be considered as a pair. The graphs in Fig. 9–17(c) and (d) comprise another pair, for a medium time constant, and those in Fig. 9–17(e) and (f) comprise a third pair, for a small time constant.

Look first at Fig. 9–17(a) and (b) for the large time constant (low reset rate). As can be seen, the valve makes an initial change in position due to the sudden appearance of an error when a load disturbance takes place. The proportional part of the controller provides this initial change. Thereafter, the valve slowly closes further in an effort to correct the resulting offset. Because of the slow reaction of the integral part, the actual temperature is slow to come back to the set point, as shown in Fig. 9–17(b).

In Fig. 9–17(c), the valve reacts more quickly to the offset error because of the medium integral time constant. The temperature therefore is quicker to return to the set point in Fig. 9–17(d).

In Fig. 9–17(e), the valve reacts very quickly to offset error because of the small integral time constant (high reset rate). The temperature quickly returns to the set point in Fig. 9–17(f).

In Fig. 9–17, the valve is shown going further *closed*. This corresponds to an initial disturbance which drove the measured temperature process *higher*. If the initial process disturbance had been in the other direction, tending to drive the actual temperature lower, the valve would have gone further *open,* but the same general behavior would have been shown.

All the graphs in Fig. 9–17 are somewhat idealized. In real life, the temperature would not recover to the set point so smoothly. Instead, it would make some oscillations on the way back to the set point, and it would probably make at least one oscillation around the set point once it had recovered. The graphs in Fig. 9–17 are drawn idealized to more clearly illustrate the effect of varying the reset rate.

There is a limit to how high the reset rate can be adjusted. If it is made too large, the temperature can break into prolonged oscillations after a disturbance.

The proportional plus integral mode of control will handle most process control situations. Large load changes and large variations in set point can be controlled quite well, with no prolonged oscillations, no permanent offset, and quick recovery after a disturbance.

9-8 ■ PROPORTIONAL PLUS INTEGRAL PLUS DERIVATIVE CONTROL

Although proportional plus integral control is adequate for most control situations, it is not adequate for all situations. Some processes present very difficult control problems that cannot be handled by proportional plus integral control. Specifically, here are two process characteristics which present such difficult control problems that proportional plus integral control may not be sufficient:

1. Very rapid load changes
2. Long time lag between applying the corrective action and the appearance of the results of that corrective action in the measurement

In cases where either of these two problems prevails, the solution may be *proportional plus integral plus derivative control.* The term *derivative control* is also called *rate control.** In proportional plus integral plus derivative control the corrective action (the position of the valve) is determined by three things:

1. *The magnitude of the error:* This is the proportional part.
2. *The time integral of the error,* or the magnitude of the error multiplied by the time that it has persisted: This is the integral part.
3. *The time rate of change of the error:* A rapidly changing error causes a greater corrective action than a slowly changing error. This is the derivative part.

In an intuitive sense, the derivative part of the controller attempts to "look ahead" and foresee that the process is in for a bigger change than might be expected based on present measurements. That is, if the measured variable is changing very rapidly, it is a pretty sure bet that it is going to try to change by a large amount. This being the case, the controller tries to "outguess" the process by applying more corrective action than would be called for by proportional plus integral control alone.

As before, to understand what derivative control does, it is helpful to study a schematic diagram of how a derivative controller can be built. To avoid getting mixed up between integral and derivative parts, we will first show a proportional plus derivative controller schematic in Fig. 9–18. The full proportional plus integral plus derivative controller is shown in Fig. 9–19.

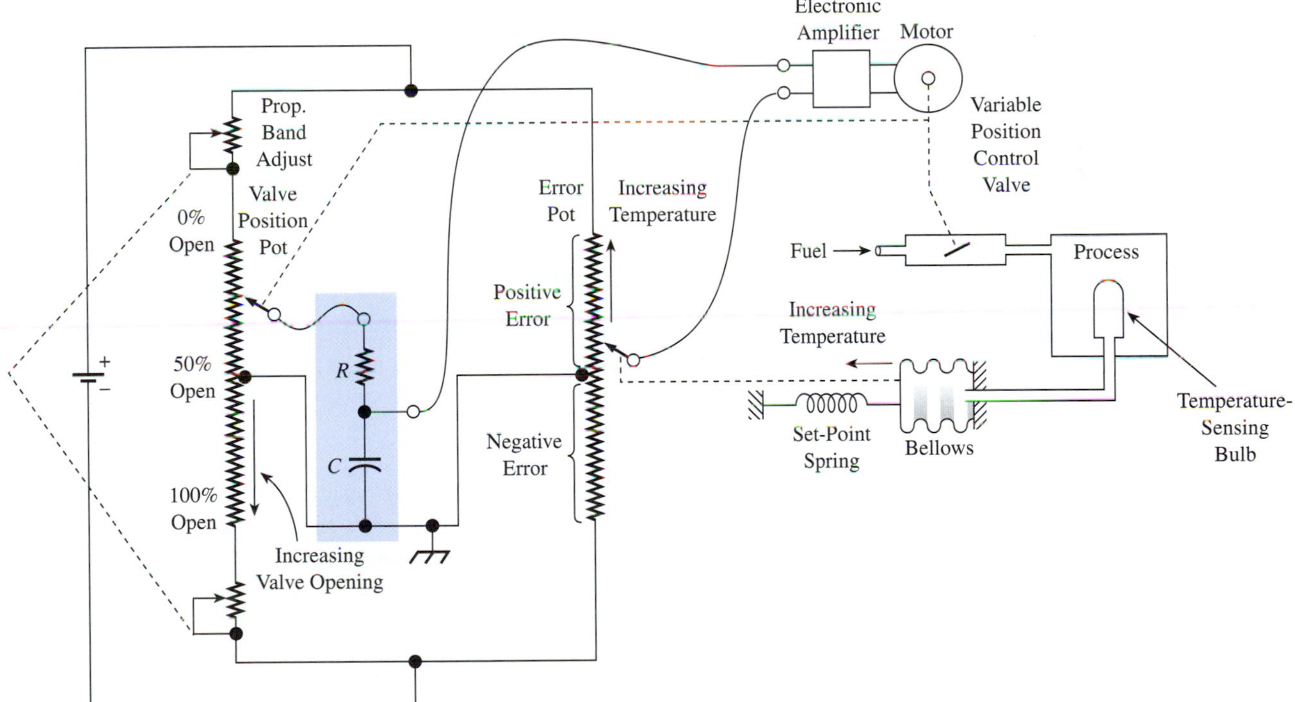

FIGURE 9–18
An electric method of implementing the proportional plus derivative control mode.

*Do not confuse this "rate control" with the phrase "reset rate." *Rate control* refers to control having a derivative part. *Reset rate* refers to the adjustment of the integral time constant in integral control. It is unfortunate that the pioneers in process technology used the same word to convey such different ideas, because now we are stuck with it.

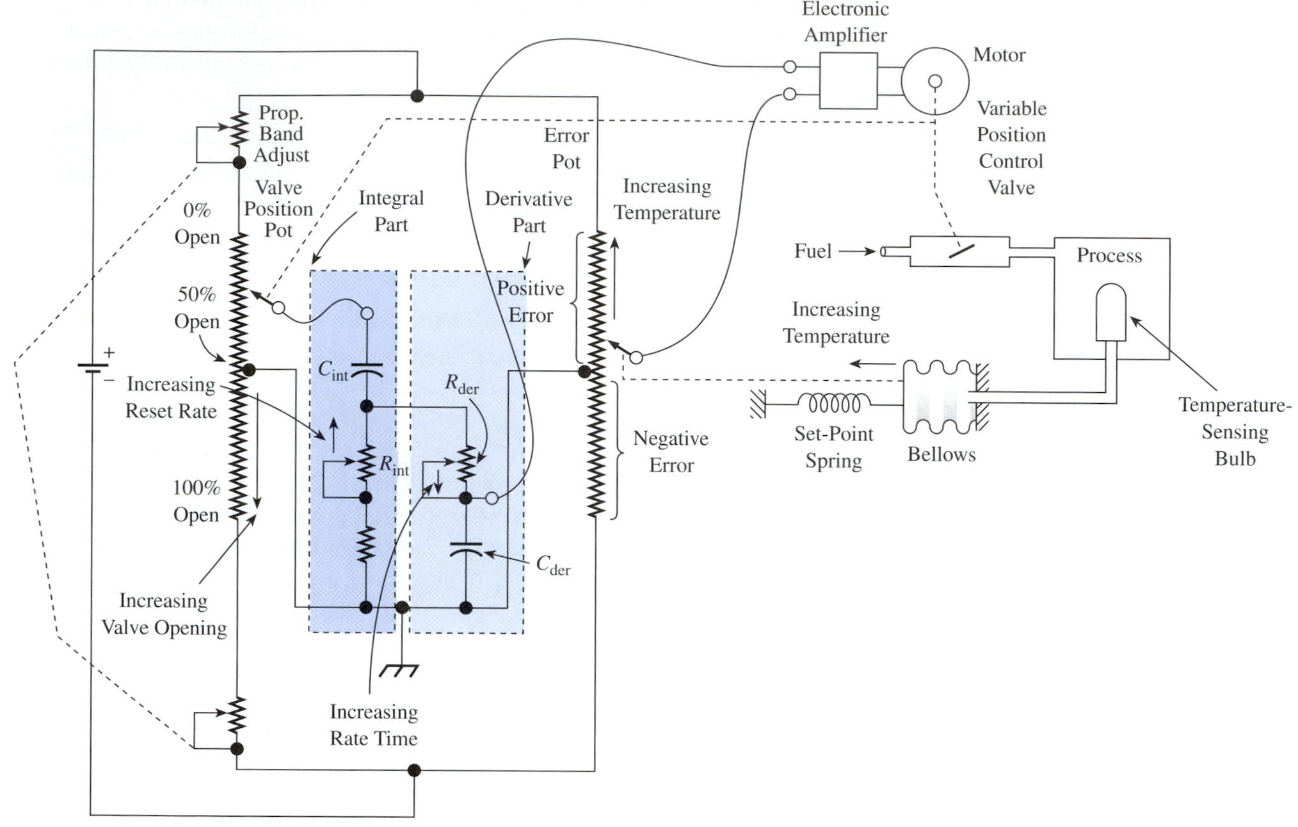

FIGURE 9–19
An electric method of implementing the proportional plus integral plus derivative control mode.

9-8-1 Proportional Plus Derivative Electric Controller

Proportional plus derivative control, as illustrated in Fig. 9–18, is very seldom used in industrial temperature control. It is presented here only to explain the derivative part of a proportional plus integral plus derivative controller. Proportional plus derivative control is popular in industrial servo control systems, however.

Focus on the *RC* circuit connected between the valve position pot wiper and ground. Notice that the positions of resistor and capacitor have been reversed from that of the integral controller in Fig. 9–16. Again, remember the fact that it always takes a certain amount of time to charge a capacitor through a resistor.

If a disturbance occurs that drives the process temperature up, the error pot will move up a certain distance. The valve position pot will attempt to follow it because of the amplifier/motor action. However, to null the amplifier input voltage, the voltage across the capacitor must equal the error pot wiper voltage. Since the voltage across *C* lags behind the valve position pot wiper voltage due to the *RC* time constant delay, the valve position pot must *overcorrect*. That is, it must move up further than it normally would in order to null the amplifier.

Furthermore, the amount by which it overcorrects depends on how rapidly the error is changing. If the error is changing slowly, the position pot wiper will follow slowly, and the capacitor will have time to keep almost fully caught up with the voltage on the position pot wiper. Therefore not much overcorrection is necessary.

On the other hand, if the error is changing rapidly, the position pot wiper will follow rapidly, and the capacitor will lag far behind the position pot wiper voltage. Therefore a large overcorrection is necessary to keep the amplifier nulled (to keep the capacitor voltage equal to the error pot wiper voltage).

In this way, the derivative part of the controller responds to the *rate of change* of the error; it introduces an additional adjustment in valve opening beyond what the proportional control alone would produce. The amount of additional movement depends on the rapidity of the change in error.

In a real industrial controller, the resistor R is a potentiometer, so the derivative time constant can be varied. When the derivative time constant is small (low R), the derivative part of the control is less effective. It introduces only a small overcorrection due to rapid change in error. When the derivative time constant is large (high R), the derivative part becomes more effective. It introduces a large overcorrection when a rapid change in error occurs.

The reference variable which is commonly used when dealing with derivative control is *rate time*. Rate time is a rather complicated variable from the mathematical point of view. Nevertheless, here is its formal definition: Rate time is the amount of time allowed for the measured variable to change through the full controller range, if it is to drive the final correcting device through its full range of adjustment, assuming a 100% proportional band.

Intuitively, rate time is the amount of time by which the controller is "looking ahead" or "seeing into the future." Obviously, this is a very nonrigorous description of rate time, since nothing can really see into the future. It is best to think of rate time as being equal to the derivative time constant multiplied by a numerical constant. The larger the rate time, the greater the amount of overcorrection the controller makes for a rapid change in error.

9-8-2 Proportional Plus Integral Plus Derivative Electric Controller

Figure 9–19 shows a schematic diagram of a complete proportional plus integral plus derivative controller—often called a *PID controller*. Notice that the derivative part is attached to the integral part. The output of the integral RC circuit is the input to the derivative RC circuit.

The direction of adjustment of the integral pot in order to increase the reset rate (to increase the amount of integral part contribution) is shown in the figure. Also, the direction of adjustment of the derivative pot in order to increase the rate time (to increase the amount of derivative part contribution) is shown in Fig. 9–19.

The operation of the controller in Fig. 9–19 can be understood by combining the explanations of the proportional plus integral controller with the proportional plus derivative controller.

The graphs of Fig. 9–20 show the control effect of changing the derivative time constant (which changes the rate time).

Figures 9–20(a) and (b) show the valve position and measured temperature for a large and rapid load change with derivative control removed. As can be seen, the initial error is quite large and consequently takes a long time to be corrected.

In Fig. 9–20(c) and (d), the derivative time constant (rate time) is small, and the initial error is not so large because the initial valve correction is greater. The controller has introduced overcorrection because it recognized that the initial rapid rate of change in measured temperature portended a large total temperature change unless special corrective steps were taken. Because the initial error is smaller, recovery to set point is earlier.

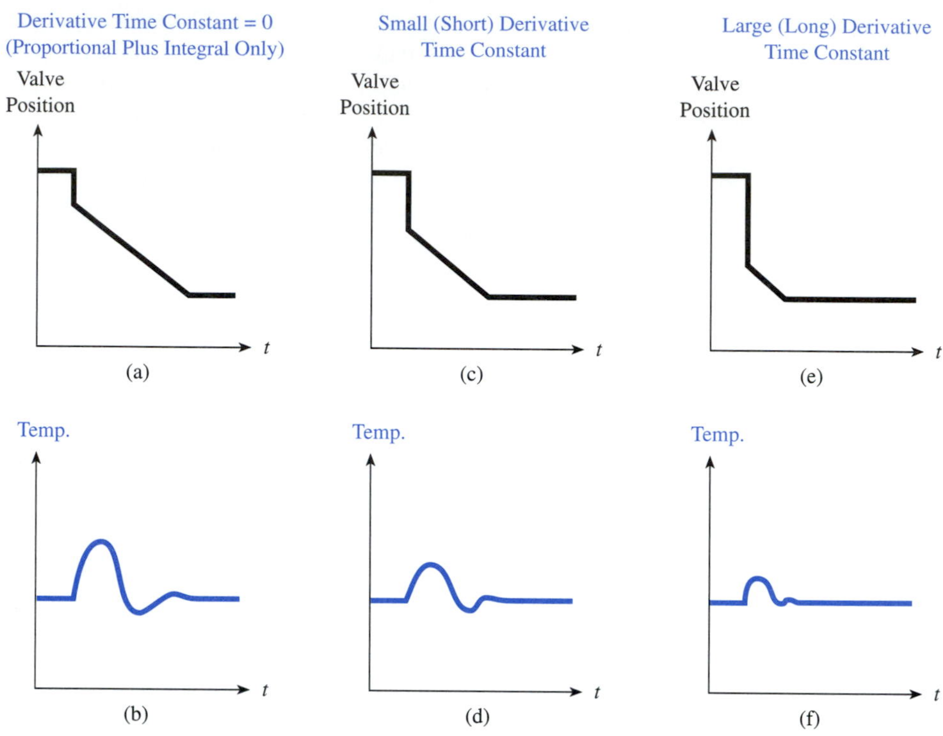

FIGURE 9–20 Graphs of valve position versus time and measured temperature versus time following a disturbance. The control mode is proportional plus integral plus derivative. (a) and (b) Zero derivative time constant. (c) and (d) Short derivative time constant. (e) and (f) Long derivative time constant.

In Fig. 9–20(e) and (f) the derivative contribution has been increased by raising the derivative time constant. Therefore the initial error is even less than before because more initial valve overcorrection is provided. With the reduced initial error, the time to recover and stabilize at set point is reduced even more than before.

Just as there was a limit with reset rate, there is a limit to how far the rate time can be increased. Prolonged oscillations around the set point can occur if too much derivative control is introduced, that is, if rate time is set too high.

9-9 ■ PROCESS RESPONSE

In Secs. 9-4 through 9-8 we have concentrated on the action of the controller block in the generalized block diagram of Fig. 9–3. No matter what particular mode of control is being used, it can in fairness be said that the controller is doing the "thinking" for the entire system. The controller is the component that sends out orders to the final correcting device based on its assessment of the direction and size of the error. We have seen that sophisticated controllers may also consider the length of time that the error has persisted as they decide how to adjust the final correcting device. Some controllers may also consider how rapidly the error is changing as they decide what orders to send to the final correcting device. The controller does all this according to a predetermined plan which existed in the mind of the system designer and also in the mind of the person who made the final adjustments (proportional band, reset rate, etc.).

It should be apparent, however, that the action of the controller does not describe the whole picture. The reaction of the *process itself* to the final correcting device is just as important as the action of the controller in determining the overall behavior of the system. In this section we will discuss the response characteristics of typical industrial processes and show how these characteristics affect overall system response.

9-9-1 Time Constant Delay (Process Reaction Delay) in Industrial Processes

The most obvious characteristic of industrial processes is that they require a certain amount of time to fully respond to a change in input. For example, in the process illustrated in Fig. 9–21(a), a liquid is being heated by a steam heating coil while it is being agitated. The liquid is admitted at the inflow pipe on the lower left of the tank, and it exits at the outflow pipe at the upper right. Assume that the controlled variable is the liquid temperature, and try to imagine what will happen if there is a sudden increase in steam flow through the heating coils (with consequent increase in average temperature).

The liquid temperature will not instantly increase to a new value but will rise more or less according to the curve of Fig. 9–21(b). The reason for the delay is that the tank of liquid has what is called *thermal capacity* and the heat transfer apparatus has what is called *thermal resistance*. The thermal capacity is an expression of the idea that a certain quantity of heat energy (Btu) must be added to the tank before the temperature can rise a given amount. Thermal resistance is an expression of the idea that all mediums have a natural reluctance to carry heat energy from one point to another; in this case the transfer of heat energy is to take place from the hot steam, through the metal walls of the heat coils, and into the surrounding liquid.

Thermal capacity is analogous to electrical capacitance. Both concepts are expressions of the fact that the relevant *quantity* (coulombs of charge in the electrical case, Btu of heat in the thermal case) must be transferred before the relevant *potential* (voltage change in the electrical case, degrees of temperature change in the thermal case) can be built up.

Thermal resistance is analogous to electrical resistance. Both concepts are expressions of the fact that a certain *potential difference* (voltage drop in the electrical case,

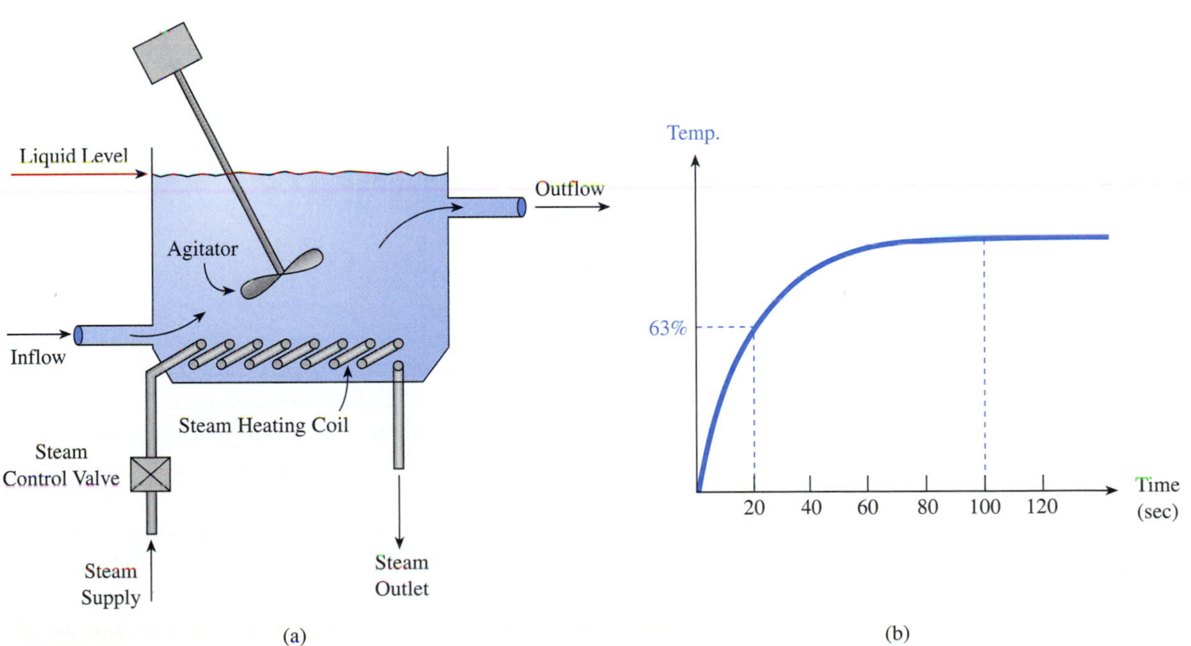

FIGURE 9–21
(a) Temperature control system for understanding process reaction delay. (b) Graph of temperature versus time following a disturbance, illustrating the effects of thermal capacity and thermal resistance.

temperature difference in the thermal case) is necessary to cause a certain *rate of flow* (amperes of current in the electrical case, Btu per second of heat flow in the thermal case) to be established. We are well acquainted with the fact that a certain amount of time is necessary for the voltage across a capacitor to build up to a steady value if a resistor-capacitor circuit is subjected to a sudden change in driving voltage. The exact same situation prevails in the thermal case. A certain amount of time is necessary for the thermal capacity (the tank of liquid) to rise to a steady temperature when a thermal resistance-capacity system is subjected to a sudden change in temperature difference. In the same way that the capacitor will experience 63% of its total voltage change in one electrical time constant, the thermal capacity will experience 63% of its temperature change in one thermal time constant. The larger the thermal resistance, the larger the thermal time constant, and the more time is necessary to reach a final steady temperature value. The same holds true for the thermal capacity; the larger the capacity, the more time is required to bring the temperature up to a steady value. In the example of Fig. 9–21(b), the thermal time constant equals about 20 seconds; about five time constants, or 100 seconds, are required for the temperature to reach the new value.

The thermal time constant depends on the thermal resistance and the thermal capacity, as stated in the preceding paragraph. Thermal resistance depends on the thermal conductivity of the metal in the heating coils, the thickness of the coil walls, and the surface area of the coils. The thermal capacity depends on the size of the tank (amount of liquid present) and the specific heat of the liquid.

The point of this whole discussion is that in a temperature-control process there is a time delay between the application of corrective action and the appearance of the final result of that corrective action.

This delay is called *time constant delay* or *process reaction delay*. We will normally use the term *process reaction delay* unless we wish to specially emphasize its equivalence to *RC* time constant behavior.

Virtually all industrial processes, not only thermal ones, show this type of delay. In many cases the delays are measured in seconds. Some processes have process reaction delays of a few minutes, and some have process reaction delays in the range of 15 to 30 min. Occasionally you may come across industrial processes that have process reaction delays of an hour or more.

9-9-2 Transfer Lag

In some thermal processes there is more than one resistance-capacity combination. An example of such a process is shown in Fig. 9–22(a). Natural gas is burned inside the radiant tube heaters on each side of the furnace. The heat is carried through the tube walls and is transferred to recirculating air passing over the tubes. The fan forces the heated air through distribution nozzles and on to the metal billets which are being heated. In this arrangement the response of the billet temperature to a change in fuel input is even more drastically retarded, as shown in the curve of Fig. 9–22(b). In fact, the response is not even the same shape as the time constant curve of Fig. 9–21(b). The reason for the more retarded temperature response is that there are now two thermal resistance-capacity combinations in series with each other. The first involves the thermal resistance of the radiant tube walls and the capacity of the recirculating air. The second thermal time constant involves the thermal resistance and thermal capacity of the metal itself. The circuit of Fig. 9–22(c) is the electrical equivalent of the thermal process in Fig. 9–22(a). R_1 represents the thermal resistance of the radiant tube walls, and C_1 represents the thermal capacity of the recirculating air. R_2 represents the thermal resistance of the metal comprising the bil-

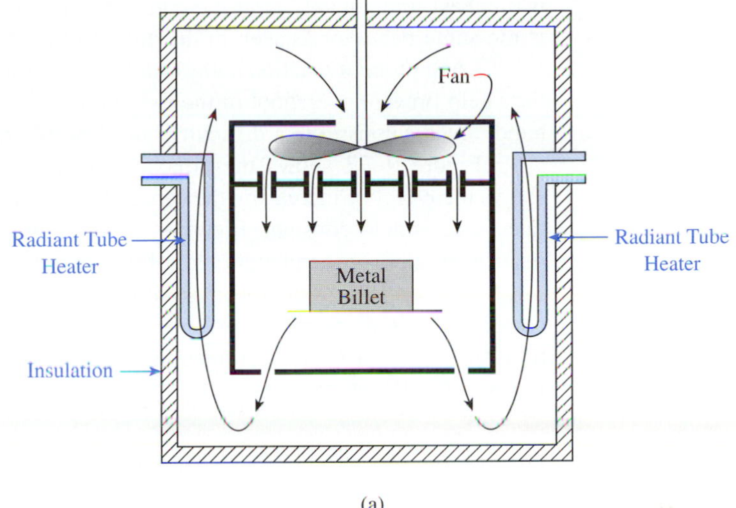

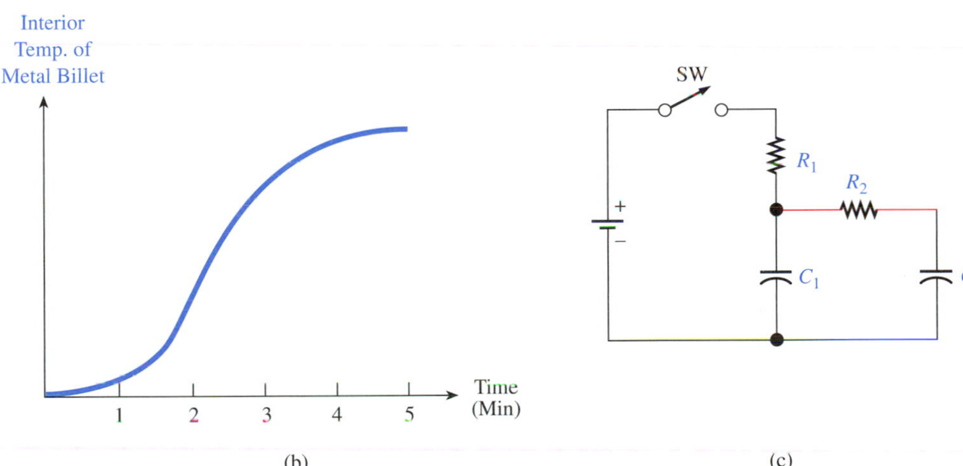

FIGURE 9–22
(a) Temperature control system for understanding transfer lag. (b) Graph of temperature versus time following a disturbance, illustrating the effects of *two* thermal capacities and two thermal resistances. Transfer lag is present. (c) The electrical analogy.

let, while C_2 is the thermal capacity of the billet. It is pretty evident by looking at the electrical circuit that C_1 must charge before C_2 can start to charge. Therefore the charging of C_2 is considerably delayed after the application of the input signal when the switch goes closed. The same problems occur in the thermal process. The billet temperature cannot start to rise until the recirculating air temperature has risen, and of course the recirculating air temperature cannot rise instantly after an increase in the heat input to the radiant tubes. Whenever there are two thermal time constants, the process is referred to as a *two-capacity* process, and the delay is referred to as *transfer lag*.

As a general rule, transfer lag is a much more serious problem than the simple single time constant delay (process reaction delay) seen in the system of Fig. 9–21 because transfer lag causes the measured process temperature to initially respond very slowly to a corrective action. This slow initial response is shown clearly in Fig. 9–22(b), in which the temperature has gone through only about 10% of its total change in the first minute following the correction. Single time constant delay, by contrast, allows the measured temperature to respond quite quickly immediately following a correction. In fact, the response is quickest immediately after the corrective action occurs, as shown in Fig. 9–21(b). This is the same behavior as seen in the universal time constant curve of Fig. 2–21.

Long process reaction delays cannot necessarily be considered a problem at all, since they help prevent overshoot of the measured temperature. Long transfer lags, on the other hand, always constitute a difficult control problem.

Figure 9–23 shows the effects of various process constructions. Figure 9–23(a) shows the effect of increasing the capacity in a one-capacity process, assuming the process resistance is held constant. Figure 9–23(b) shows the effect of increasing the number of resistance-capacity combinations in the process.

Note especially the response of the process temperature immediately following the corrective action (near the zero point on the time axis). In this time area, the effect of transfer lag is very severe compared to the effect of simply increasing the time constant in a single-capacity process.

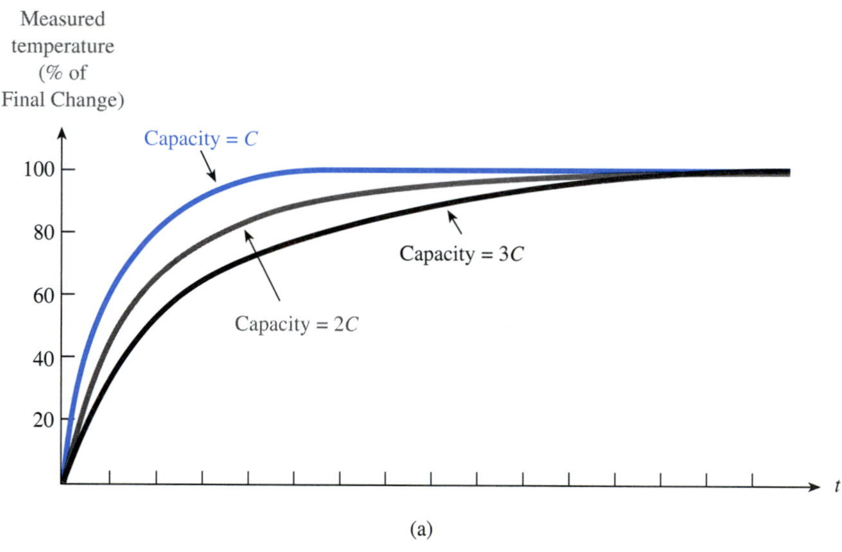

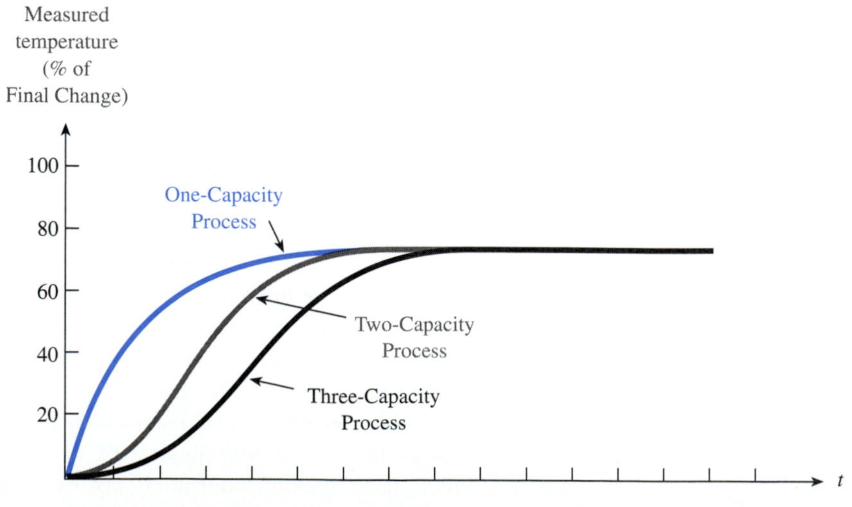

FIGURE 9–23 Graphs of actual temperature versus time following a disturbance for different system constructions, illustrating the severe effect of transfer lag. (a) Effect of increasing the amount of thermal capacity in a one-capacity system. (b) Effect of increasing the number of thermal capacities in the system. The transfer lag problem is worsened when more capacities are present.

The same principles which apply to thermal processes also apply to other types of industrial processes, as we saw in Sec. 9-9-1. Pressure-control processes, liquid level-control processes, and all other industrial processes have their associated resistances and capacitances, and they often suffer from two or more resistance-capacity combinations. Because of this, they are subject to the same transfer lag problems which affect thermal processes. The graphs in Fig. 9–23 could apply to *any* industrial process, no matter what the controlled variable happens to be.

9-9-3 Transportation Lag and Dead Time

When transfer lag is present, the controlled variable takes some time to reach its new steady value after the controller sends an order to the correcting device, but at least *some* partial response is felt immediately. This is clearly shown in Figs. 9–22 and 9–23. A more difficult control problem occurs when absolutely no response whatsoever is felt in the controlled variable for a certain time period after the controller signals the correcting device. This situation usually occurs when the physical location of the correcting device is far away from the physical location of the measuring device. The system shown in Fig. 9–24 is an example of such an arrangement.

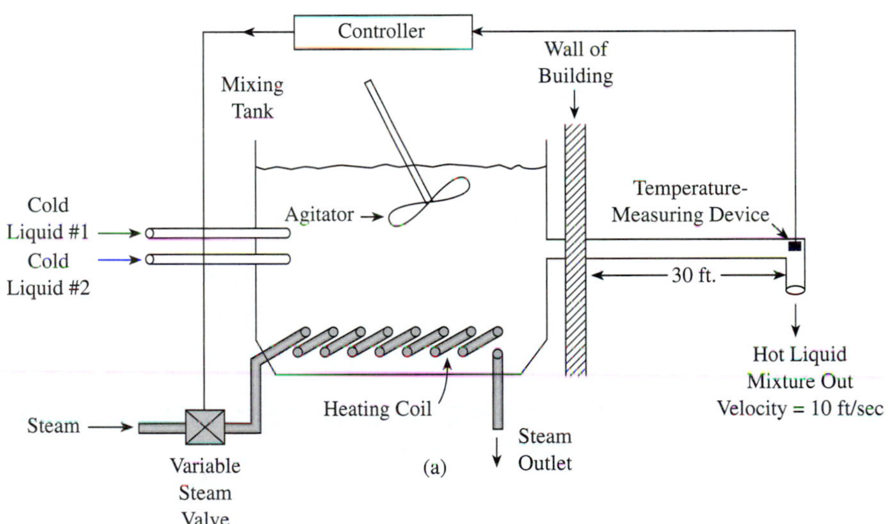

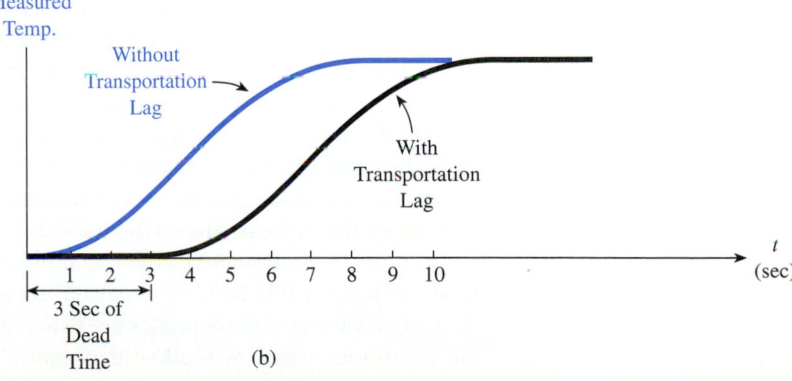

FIGURE 9–24
(a) Temperature control system for understanding transportation lag. (b) Graphs of measured temperature versus time following a disturbance, with and without transportation lag. A dead time of 3 seconds occurs when transportation lag is present. Transfer lag is present in both cases.

Assume that the mixing/heating tank must be located 30 ft away from the point where the hot mixture is to be used. There might be some practical reason for this. For example, it may be that the mixing/heating tank must be located indoors and the discharge opening must be some distance away outdoors. Since there may be some cooling during the long pipe run, the temperature is measured and controlled at the discharge point rather than inside the tank. This allows the controller to eliminate the effect of cooling in the pipe, which may vary widely with outdoor temperature changes.

Since the hot mixture leaves the constant-diameter carrying pipe at a velocity of 10 ft/s, it takes 3 seconds to travel down the 30-ft-long pipe. This being the case, if the temperature in the mixing tank is changed, it will take the temperature-measuring device 3 seconds to realize it. This delay is completely independent of and in addition to any transfer lag that may be present in the tank. A delay of this nature is termed *transportation lag*. Other terms used are transport lag and distance/velocity lag; we shall use the term transportation lag. The actual amount of time that the correcting device change remains undetected is called the *dead time*. Transportation lags are always associated with the controlled medium moving from one physical location to another in the process. The slower the speed of movement, the worse the transportation lag. The farther the distance between the two points, the worse the transportation lag. The effect of transportation lag is shown in Fig. 9–24(b); the dead time is 3 seconds. Transportation lag presents a difficult control problem, even worse than transfer lag.

In industrial control systems, dead time can arise for reasons other than transportation lag. For example, in a position control system, there is always some sort of gearing present. As you may know, all gears are subject to *gear backlash* to some extent. Gear backlash is the problem of gear teeth meshing imperfectly, so that the driving gear must turn through a small initial angle before its teeth make contact with the driven gear. Because of this, the controller in a servo system may cause the servo motor to start turning, but the resultant motion of the load is delayed until the gear teeth make contact. The result is a period of dead time. If the gear train is complex, with several gear combinations between the servo motor and the controlled object, the gear backlash problem is magnified. Dead time can be a serious problem in servo systems of this kind.

9-10 ■ RELATIONSHIP BETWEEN PROCESS CHARACTERISTICS AND PROPER MODE OF CONTROL

Generally speaking, the characteristics of the process being controlled determine which mode of control is best suited to that process. In Sec. 9-9 we discussed three important process characteristics: time constant delay, transfer lag, and transportation lag (dead time). We saw in Secs. 9-6, 9-7, and 9-8 that the size and speed of load disturbances were also rather important characteristics of the process. These five process characteristics determine the nature and difficulty of the control job, and therefore which mode of control is required.

Of course the desired accuracy of control is also a prime determining factor in the choice of control mode; if the measured variable can be allowed to deviate from the set point by wide margins without harm to the product, there is no sense installing a controller capable of keeping the deviation small. In such a case, it doesn't matter very much how nasty the process characteristics are; a simple On-Off controller will suffice.

Table 9–1 summarizes the relationship between process characteristics and control mode. Naturally this table is somewhat rough.

The various control modes are listed in the left column. The other columns describe the conditions which will allow that control mode to be successful. The column entries

TABLE 9–1
The types of processes which can be successfully controlled by each of the five basic control modes

CONTROL MODE	PROCESS REACTION DELAY (MINIMUM)	TRANSFER LAG (MAXIMUM)	DEAD TIME (MAXIMUM)	SIZE OF LOAD DISTURBANCE (MAXIMUM)	SPEED OF LOAD DISTURBANCE (MAXIMUM)
On-Off	Long only (cannot be short)	Very short	Very short	Small	Slow
Proportional only	Long or moderate (cannot be too short)	Moderate	Moderate	Small	Slow
Proportional plus integral	Any	Moderate	Moderate	Any	Slow
Proportional plus derivative	Long or moderate (cannot be too short)	Moderate	Moderate	Small	Any
Proportional plus integral plus derivative	Any	Any	Any	Any	Any

in the four far-right columns of the table describe the *maximum limit* for that particular characteristic. The entry in the "PROCESS REACTION DELAY" column describes the *minimum limit.*

For example, if the entry in the dead time column is "Moderate," it means that control mode will work successfully if dead time is either moderate or short or nonexistent (short or nonexistent dead time is easier to handle than moderate dead time).

However, if the entry in the process reaction delay column is "Long," it means that the process reaction delay must be long, and not short or moderate. (In many cases, a fast reaction causes serious overshoot, as mentioned in Sec. 9-9-2.)

As Table 9–1 shows, On-Off control is generally acceptable only under simple process conditions. It works only when the process reaction delay is fairly long (slow response). A short process reaction delay causes excessive overshoot and undershoot with On-Off control.

Proportional control can tolerate long or moderate process reaction delays because it continually repositions the final correcting device as the controlled variable approaches the set point after a disturbance. Therefore it is not so likely to produce excessive overshoot as is On-Off control. Moderate transfer lag and dead time can be handled by a proportional-only controller. However, long transfer lag and/or long dead time produce sustained cycling. This occurs because the controller does not throttle back on the final correcting element until it is too late. That is, if the lag is too great, by the time the controller realizes that the controlled variable is returning to set point, it has already permitted too much energy to enter the process. The process inertia will then carry the controlled variable beyond the set point in the opposite direction, and the controller cannot do anything about it until the excess energy has dissipated. This sets the stage for sustained cycling.

Proportional plus integral control can handle any process reaction delay and any size of load disturbance. The integral part of the control continually repositions the final correcting element until the set point is attained, no matter how large the load change. Because of this, the proportional band can be adjusted wider, because a narrow

TABLE 9–2
Proper setting of a proportional plus integral controller for different process conditions

Process Characteristics		Controller Adjustments	
PROCESS REACTION RATE	TOTAL LAG	PROPORTIONAL BAND	RESET RATE (RECIPROCAL OF INTEGRAL TIME CONSTANT)
Slow	Short	Narrow	Fast
Slow	Moderate	Medium	Slow
Fast	Short	Medium	Fast
Fast	Moderate	Wide	Slow

proportional band is no longer necessary to keep the offset small. With a wide proportional band, the controller can start to throttle back the correcting device sooner and harder as the controlled variable recovers to set point. This prevents excessive overshoot and possible cycling even if the process reaction rate is fast.

Within the proportional plus integral control mode, we can make finer distinctions among the different control situations. Table 9–2 shows the relative settings of proportional band and reset rate for a proportional plus integral controller. The proportional plus integral controller is assumed to operate under various conditions of process reaction rate and transfer lag/dead time. Transfer lag and dead time (transportation lag) have been lumped together under the term *lag* in Table 9–2.

Process reaction rate in Table 9–2 is the opposite of process reaction delay. That is, a short process reaction delay equates to a fast process reaction rate, and a long process reaction delay equates to a slow process reaction rate.

Returning to Table 9–1, proportional plus derivative control must have moderate or long process reaction delay, because in the absence of an integral part to take care of the offset the proportional band must be made narrow to keep offset small. With a narrow proportional band, a short reaction delay can cause overshoot and cycling. However, with derivative control, rapid load changes are not harmful because the controller overcorrects when it detects a quickly changing error.

When long transfer lag and/or dead time are present in a process, the only successful mode of control is proportional plus integral plus derivative. The proportional band is made quite wide, so only a small portion of the corrective action after a disturbance is due to the proportional part. Most of the *immediate* corrective action is due to the derivative part. This allows a considerable inflow of energy to the process immediately when it is most needed. When the error stops growing and starts shrinking, the corrective action due to the derivative part goes away, leaving only the relatively small corrective action due to the proportional part. By this time, hopefully, enough time has elapsed that dead time is finished and the transfer lag is fairly far progressed. The controller can now sense that the controlled variable is recovering. There is no tendency to overshot, though, because the position of the final correcting device is not drastically different from where it was before the disturbance. This is so because of the wide proportional band.

The integral part of the controller takes over at this point and slowly repositions the correcting device to drive the controlled variable back to set point. The reset rate is usually adjusted to be rather slow.

The effect of having the derivative part present in the controller is shown in Fig. 9–25. Figure 9–25(a) shows the best response that could be obtained in a process having long transfer lag and long dead time (transportation lag) using proportional plus integral control. Figure 9–25(b) shows the response which is possible when a derivative part is added and the controller is properly adjusted.

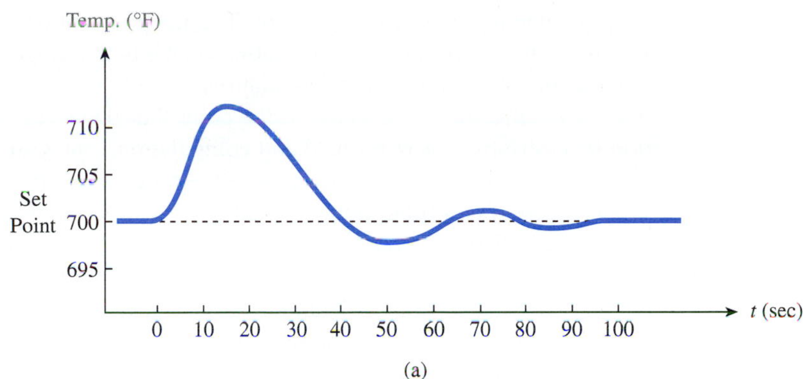

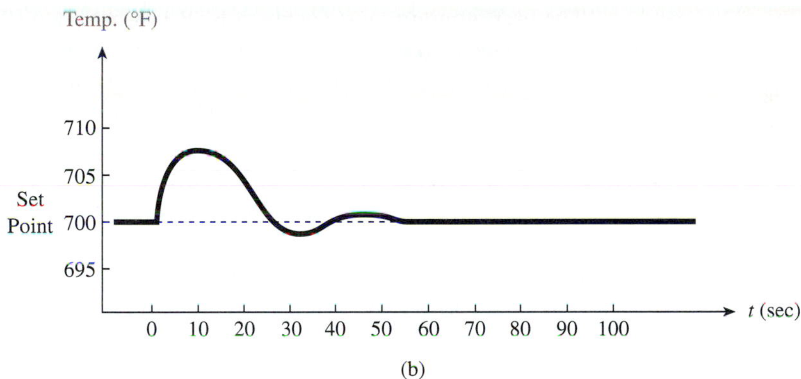

FIGURE 9–25
Graphs of temperature versus time following a disturbance in a system which has both transportation lag and transfer lag.
(a) Without derivative control action, the overshoot and recovery time are great.
(b) When derivative control is added, the overshoot and recovery time are lessened.

9-11 ■ PID PROCESS CONTROL WITH A PROGRAMMABLE LOGIC CONTROLLER

Proportional plus integral plus derivative control is called by the acronym *PID control*. Less commonly, it is also called three-function control. Figures 9–15, 9–16, 9–18, and 9–19 are very useful for understanding the concepts of proportional control and proportional plus integral control, leading up to full PID control in Fig. 9–19. However, these straightforward electric methods of implementing closed-loop control, while understandable, are not practical in industrial settings.

During the decades of the 1950s and 1960s, industrial closed-loop process control in the United States was accomplished by elaborate pneumatic controllers or by analog electronic controller "instruments." These instruments were built around the operational amplifier idea: very high open-loop gain with resistive-capacitive feedback and resistive summing. Refer to Figs. 8–15, 8–17(a) and (b), and 15–4(b). These instruments required very skilled technicians to adjust and maintain them.

Digital computer people demonstrated that a program of computation instructions could successfully set the Controller Output signal that was sent to the final correcting device, but in those years digital computers were very expensive machines that weren't able to exist in an industrial environment. They existed only in air-conditioned, clean, electrically noise-free settings. During the 1970s, as computers became more affordable and reliable, a few applications sprang up that used a "minicomputer" for process control. This trend accelerated in the 1980s as computers became much more affordable, powerful, and reliable. Now the digital computer is the most common method of implementing closed-loop control. The integral and derivative portions of the control signal are found by iterative calculation methods that approximate the mathematical operations of integration and differentiation.

The situation has evolved to the point that powerful PLCs now contain an instruction that performs proportional band computations, integral computations, and derivative computations on the measured value of the process' controlled variable and produces the proper value of controller output signal. This instruction is available on the Allen-Bradley PLC 5/12, called the *PID Instruction*.

To use its PID instruction, the PLC must be equipped with an Analog Input Module *and* an Analog Output Module in its I/O rack; this is shown clearly in Fig. 9–26. The Analog Output Module has the reverse function of the Analog Input Module: It receives the digital value written from the PLC's processor and converts it into an analog output signal for operating the final correcting device.

The user program must contain a BTR instruction, a PID instruction, and a BTW instruction, as indicated in Fig. 9–27. The three instructions are executed in that order, on a fixed time schedule (the update time must be nearly constant from one PID execution to the next). First, the update timer times out. Then BTR instruction reads in the latest value of the measured process variable (PV) from an Analog Input Module. Next, the PID instruction performs its calculation of the new controller output signal (CV). Finally, the BTW instruction writes the new value of CV to the Analog Output Module. The voltage or current analog value that is produced at the module's channel-connection terminal is then sent by wires to the final correction device, which adjusts the energy flow to the industrial process. These three data-processing instructions are reexecuted on a consistent periodic basis.

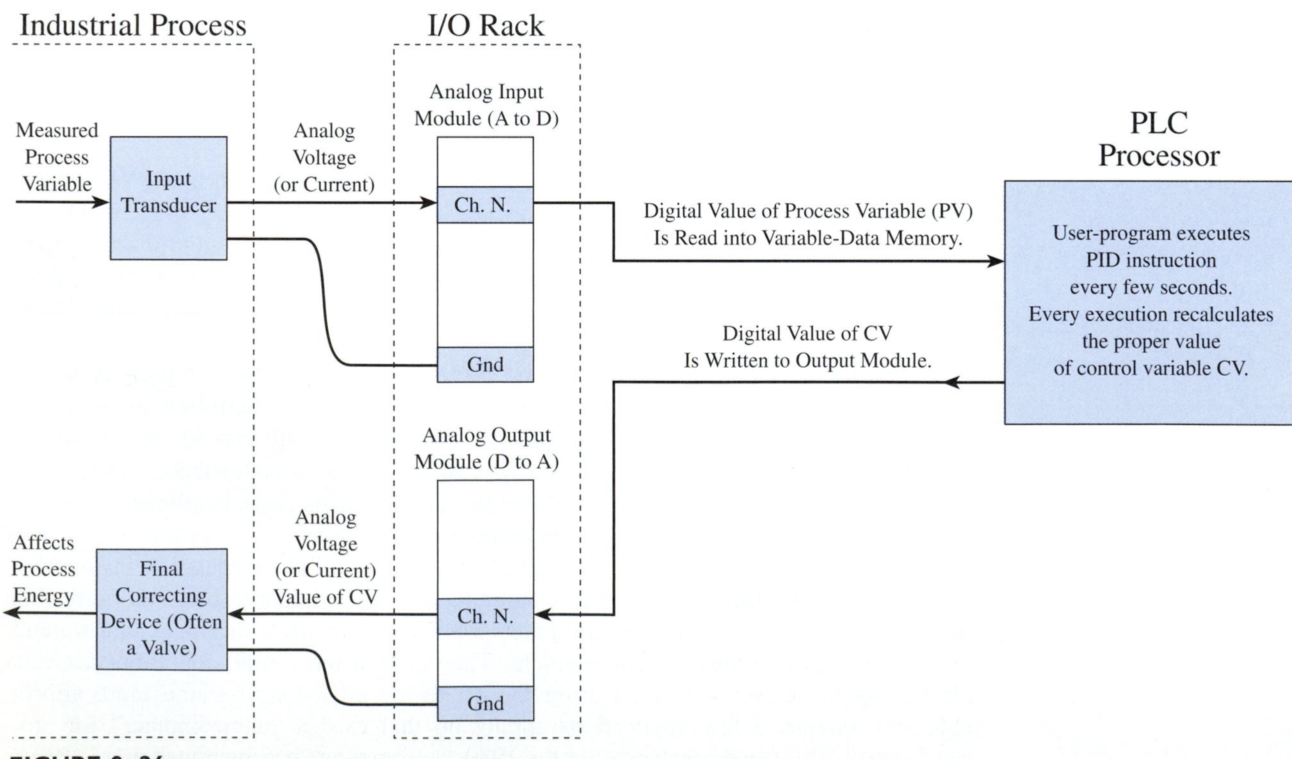

FIGURE 9–26
Performing closed-loop industrial process control with a PLC.

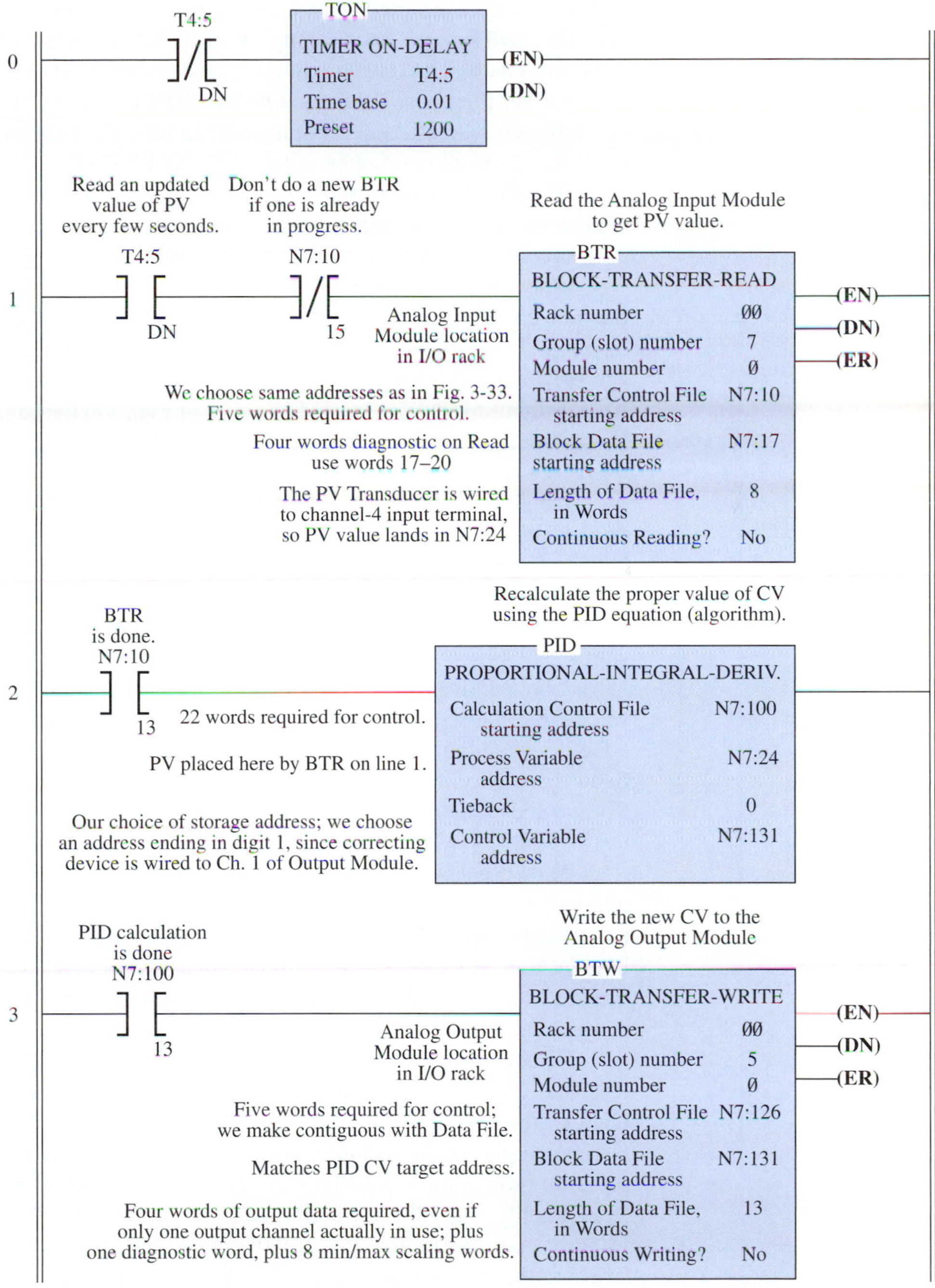

FIGURE 9–27
The PID closed-loop control program has four parts: (1) Update timer; (2) READ the Process Variable PV from the Analog Input Module; (3) Go through the PID digital algorithm to generate an updated digital value of Control Variable CV; (4) WRITE CV to the Analog Output Module, which converts it to analog (DAC) and sends the analog value to the final correcting valve.

The instructions are cascaded so that one can't start until the previous one finishes. Thus in Fig. 9–27 the PID instruction on line 2 can't go TRUE until the examine-On N7:10/13 instruction signals that the BTR's DoNe bit has gone HI. On line 3, the BTW instruction can't go TRUE until the examine-On N7:100/13 instruction signals that the PID algorithm has completed its calculations and the PID's DoNe bit has gone HI.

The entire three-instruction sequence is initiated by the timing out of timer T4:5 on line 0 and the setting of the T4:5/DN bit to HI. The BTR instruction on line 1 can't go

TRUE until the examine-On T4:5/DN instruction on line 1 signals that the timer has timed out. This will happen consistently every 12.00 seconds in Fig. 9–27 because the timer runs continually. The only condition for it continuing to run on every program scan is that it be not yet timed out (examine-Off T4:5/DN on line 0). The T4:5/DN bit is HI for just one program scan after time-out in order to reset the timer to zero. On the very next scan the timer starts all over again with the DoNe bit back at 0.

The Preset value of the timer, which is the PID update time, must be chosen to be compatible with the process-reaction delays and lags in the industrial process. The update time should be greater than the combined value of all process delays and lags.

The usual addressing requirements must be met in order to allow the BTR, PID, and BTW instructions to exchange information properly. We have assumed in Fig. 9–27 that the process variable's input transducer is wired to channel 4 of the Analog Input Module and the final correcting device is wired to channel 1 of the Analog Output Module. The Analog Input Module is located in slot 7 of rack ØØ, the same as it was in Chapter 3. We have chosen to locate the Analog Output Module in slot 5 of rack ØØ, as suggested in the BTW box on line 3.

The BTR Data File begins with an address ending in 7. With four diagnostic words this makes the channel-1 data land in the next higher address that ends in 1, in this case N7:21. The channel-4 data (the PV value) lands in address N7:24, so the PID instruction is told to look for it there (in the PID box on line 2). The BTR Data File must be eight words in length in order to reach channel 4 (four diagnostic words plus four channel-data words).

The PID instruction does quite a lot of calculation. It subtracts the measured PV value from the set point (SP) to find the error signal, then multiplies the error signal by a factor called the proportional gain (which is equivalent to the reciprocal of our proportional band) in order to calculate the Proportional part of its output control value. The SP and the proportional gain (K_P) were entered by us when we programmed the PID instruction.

It then goes on to add this present error signal (first multiplied by the update-time interval) to all its previously accumulated products of error multiplied by time. This accumulated sum is then multiplied by a factor called *integral gain* (the reciprocal of our integral time constant) in order to calculate the Integral part of its output control value. The integral gain (K_I) was likewise entered by us when we programmed the PID instruction.

The instruction then subtracts the preceding value of PV from the present value of PV. That difference is divided by the update time to find the rate of change, then multiplied by the derivative time-constant (called *derivative gain,* K_D) in order to calculate the Derivative part of its output control value. The K_D value was also keyed in by us after software prompting when PID was entered into the user-program.

Finally, the PID instruction adds all three parts, P + I + D, to calculate its new value of control variable, CV. It writes that value to the control variable address that we key in to the PID box on line 2 of Fig. 9–27. In this example we used the address N7:131.

The Calculation Control File in the PID box holds the gain constants K_P, K_I, and K_D, as well as set point SP, PV, error signal, and all temporary results of mathematical calculations. In addition, it handles several other control parameters that we have ignored in our discussion (for example, dead band). This Calculation Control File is 22 words in length, starting at the address that we specify in the PID box. In Fig. 9–27 the Calculation Control File spans from word N7:100 to word N7:121.

The execution time of a PID instruction is about 1 msec. When it has finished, it sets its DoNe bit, N7:100/13 in Fig. 9–27. The BTW instruction then writes a block that is 13 words in length, starting with the address that we specify as the Block Data File starting address in the BTW box. The first word in the block is sent to channel 1 of the Analog Output Module. The second word in the block is sent to channel 2, and so on through channel 4 (an Analog Output Module has only four channels). Therefore, if the final correcting device is wired to the channel-1 terminal, as in our example, the Block Data File starting address of BTW must be the very same address as the CV address of PID, namely N7:131 in Fig. 9–27.

If the final correcting device had been wired to channel 2 of the output module, then we would have told the PID instruction to place its calculated CV result at the *second* word into the Data File, namely, N7:132.

After the four channel-data words, the Analog Output Module needs one configuration word and eight scaling words (a minimum value and a maximum value for each of its four channels). Thus, $4 + 1 + 8 = 13$ words total, as programmed in the BTW box. The Analog Output Module's configuration information and its min and max values are keyed in by us with the help of the Allen-Bradley software, similarly to the Analog Input module, which was described in Chapter 3.

In Section 9-10, we saw that an analog electric PID controller must have its proportional band, integral time constant and derivative time constant properly adjusted to handle the characteristics of the industrial process. The same is true for the PID instruction in a PLC program. It must be given values for K_P, K_I, and K_D that are appropriate for the process reaction delay, transfer lag, dead time, and load-disturbance nature of the controlled process. These values are keyed into words 3, 4, and 5, respectively, of the Calculation Control File that starts at address N7:100 in Fig. 9–27. Thus, K_P is entered into word N7:103, K_I into N7:104, and K_D into N7:105. The act of trying out different values for these gains in order to minimize the process' overshoot and recovery time is called *tuning* the control loop.

TROUBLESHOOTING ON THE JOB

USING A PLC FOR AUTOMATED VARIATION OF PROCESS CONDITIONS AND DATA GATHERING

Look ahead to Troubleshooting on the Job in Chapter 12 on page 556, dealing with the scrubbing process for removing sulfur from the combustion products of a coal-burning electric utility plant. The water-based slurry at the base of the tower in Fig. 12–41 is normally at a temperature of about 60°C (about 140°F) due to the energy that is released by the sulfur-capturing chemical reactions.

A research project is proposed in which electric immersion heaters are installed in the tower's base to heat the slurry to higher temperatures. The purpose of the project is to investigate whether warmer water from the spray nozzles can improve the SO_2-capturing reaction so that a greater percentage of sulfur can be removed from the exhaust stack.

The project design calls for a model PLC 5/12 programmable logic controller to control the temperature of the slurry by varying the electric current in the immersion heaters with a PID instruction. Also, the PLC is assigned the task of varying the temperature set point in 2° Celsius increments, from 60°C up to 80°C. After a particular temperature set-point value is assigned, it will be maintained for a 12-hour period. At the end of the 12-hour period, the PLC program will automatically raise the slurry's set-point temperature by 2°C, then hold that higher value for 12 hours. This will allow SO_2-inflow data and SO_2-outflow data to be gathered and recorded over an extended period.

For example, there will be 12 hours of data with the slurry temperature held steady at 62°C, then 12 hours of data with the slurry temperature held at 64°C, and so on. This approach of a lengthy period of constant temperature will allow the slurry-temperature effect to be averaged for varying grades of coal and of limestone.

An input transducer for measuring SO_2-concentration must be mounted in the duct carrying the exhaust gas from the precipitator; refer to Fig. 12–41. This is called the *SO_2-inflow transducer,* and it will be wired to channel 2 of the Analog Input Module that is located in slot 7 of the I/O rack (refer to Figs. 9–26 and 9–27). Another measurement transducer, the *SO_2-outflow transducer,* will be mounted in the chimney stack itself in Fig. 12–41. It will be wired to channel 3 of the Analog Input Module.

The SO_2 concentrations will be graphed versus time by analog recording instruments. The SO_2-inflow recording instrument will be connected to channel 2 of the Analog Output Module in slot 5 of the I/O rack (Fig. 9–27); the SO_2-outflow recorder will be connected to channel 3 of the output module. Actual measured slurry temperature is also to be graphed versus time; this is accomplished by a temperature recorder connected to channel 4 of the Analog Output Module.

Channel 1 of the output module is used to supply the feedback trigger-control voltage to a triac driver like the one shown in Fig. 6–11(f) on page 223. The triac then controls the ac current through the resistive immersion heater to maintain the process temperature at or near the set point.

YOUR ASSIGNMENT

Anyone who can successfully program and implement a PID control loop on a PLC soon gets a reputation as a PID expert. Since you have done this once before, you are regarded as the resident expert in your plant. Therefore you are assigned the job of writing the user-program for this project.

In your previous experience with the PID instruction of the PLC 5/12, you learned that the set point is stored in word two (the third word) of the Calculation Control File in the PID box. Thus in Fig. 9–27 the temperature set point will be stored in word N7:102. The set-point value, in meaningful units of degrees Celsius, can be directly keyed in to that word through the keyboard (with the Allen-Bradley software displaying the PID Data Monitor Menu), or it can be automatically written to that word by some other instruction in the user-program. In this project the set point will be changed every 12 hours by another instruction in the user-program.

Working with the program instructions shown in Fig. 9–27, add rungs to the user-program to raise the temperature-control set point by 2°C every 12 hours.

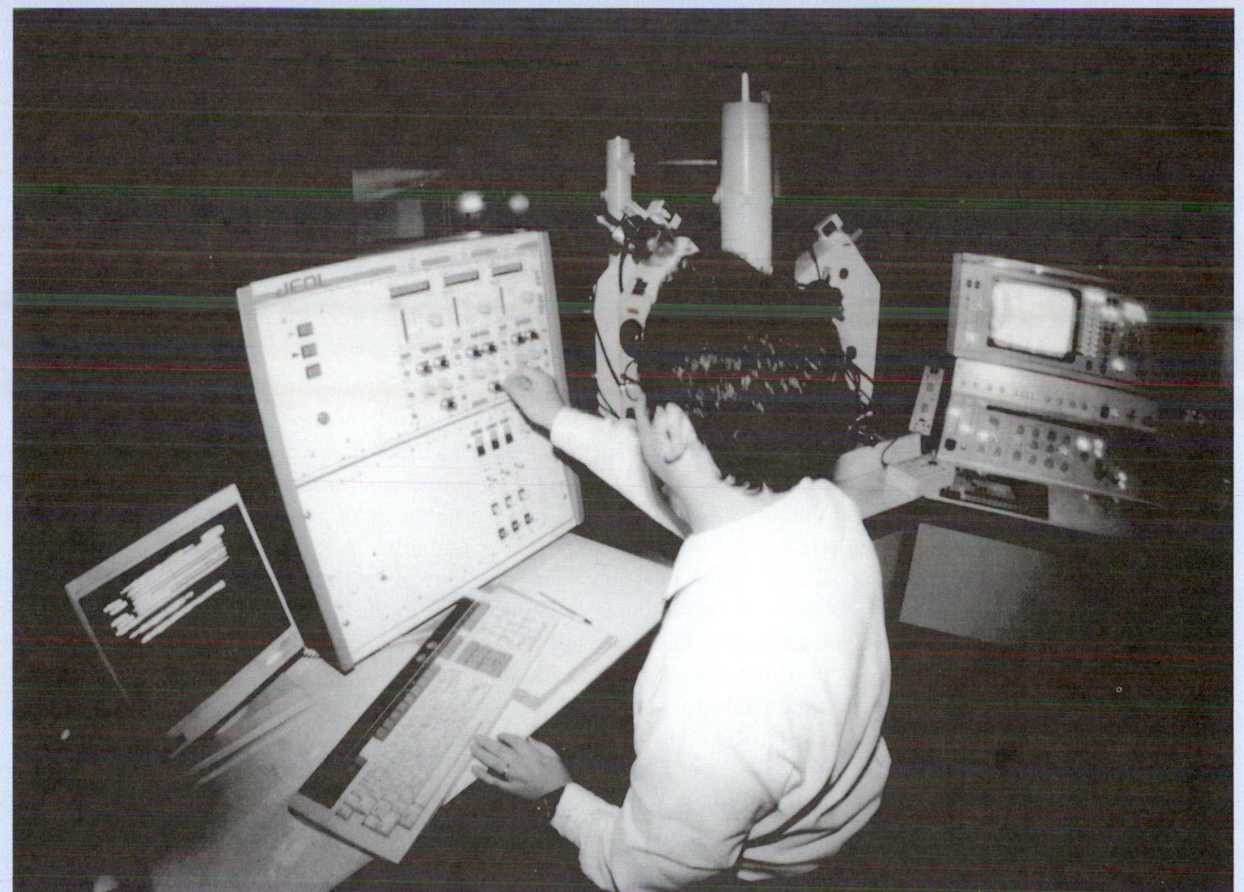

A newly written PLC process-control program can be tested by using an analog signal board to artificially vary the input voltages or currents to the Analog Input Module.
Courtesy of Railway Technical Research Institute of Japan.

Following its completion of the 12-hour shift at 80°C, the program should take the slurry temperature back to 60°C and repeat the 5-day (120-hour) temperature-raising cycle for the batches of coal and limestone then in use.

Keep in mind that the greatest time-base for a T4 timer is 1.0 second, and the largest Preset value that can be entered is 9999. These numbers do not allow a timer to be preset to 12 hours. You will have to work around this problem.

The accumulated-value word of an up-counter is addressed as C5:X.ACC. Do not overlook the need for resetting any counter(s) that you place in your program. Be sure you understand that simply Block-Transfer-Reading the SO_2-inflow value, the SO_2-outflow value, and the measured temperature value will place those values into the Read Block Data File (N7:17 through N7:24), but it does not get them into the Write Block Data File (N:7:131 through N7:143). Your program must take care of the latter task.

SUMMARY

- A closed-loop system is fundamentally different from an open-loop system in that it automatically attempts to maintain the measured value of the process variable close to the set point.
- All closed-loop systems can be understood in terms of a generalized block diagram. This diagram contains a comparer (error detector); a controller; a final correcting device; the process itself, including its disturbances; and a measurement device (transducer).
- Mode of control refers to the manner in which the controller responds to the error signal. There are five basic modes of control: (1) On-Off; (2) Proportional; (3) Proportional plus Integral; (4) Proportional plus Derivative and; (5) Proportional plus Integral plus Derivative (PID).
- In the On-Off control mode, the controller causes the final correcting device to switch abruptly between two extreme positions.
- In the Proportional control mode, the controller causes the final correcting device to change gradually, in proportion to the magnitude of the error signal.
- In the Proportional plus Integral control mode, the controller causes the final correcting device to respond gradually, in proportion to the magnitude of the error signal and in proportion to the amount of time the error has persisted. This control mode is able to reduce offset error to zero, eventually.
- In the Proportional plus Integral plus Derivative (PID) control mode, the controller causes the final correcting device to respond in the same way as the Proportional plus Integral controller, with the additional feature that the controller makes an initial overcorrection if it senses that the measured process variable is changing very rapidly.
- A programmable logic controller equipped with Analog Input and Output Modules can perform PID control of industrial processes.
- The detailed response performance of the industrial process determines which control mode will work best. The process response characteristics also determine the relative amounts of proportional control (proportional band setting), integral control (integral time constant setting), and derivative control (derivative time constant setting) that are required.
- In general, process dead time is more difficult for a PID controller to deal with than process reaction delay or transfer lag.

QUESTIONS AND PROBLEMS

Section 9-1

1. Explain the difference between an open-loop system and a closed-loop system.
2. In Fig. 9–1(b), suppose that the length of the B member is fixed but that the pivot point can be moved to the left or to the right. Which direction would you move it to decrease the proportional band?
3. Explain in step-by-step detail how the closed-loop system of Fig. 9–1(b) would react if the downstream restriction in the exit pipe were decreased. Contrast this to what would happen in Fig. 9–1(a).
4. Explain the general function of the comparer in a closed-loop control system.
5. Explain the general function of the controller in a closed-loop control system.
6. Explain the general function of the measurement device in a closed-loop control system.

7. Explain the general function of the final correcting device in a closed-loop control system.
8. Under what conditions is an amplifier needed with the final correcting device?
9. Name some common final correcting devices used in industrial process control.
10. Define the term *error signal*.
11. When is the error signal considered positive, and when is it considered negative?
12. What does the idea of *control mode* mean?

Section 9-2

13. What characteristics distinguish a good closed-loop control system from one which is not so good?

Section 9-3

14. It was tacitly assumed that all three pulleys had the same diameter in Fig. 9–4. Supposed that the fixed pulley on the left and the movable pulley both have a 3-in diameter and that the fixed pulley on the right has a 6-in diameter. Suppose also that the cord cannot slip on any pulley. If the pointer is now moved 5 in. to the left, how far will the object move?
15. Suppose that the amplifier gain is extremely high in Fig. 9–4, so that even a few millivolts of input causes a large output voltage. Explain why the object's position will never stabilize but will continually oscillate back and forth, "hunting" for the right position.
16. In Fig. 9–5, would you get a more exact copy of the pattern piece if the mounting support were moved slowly or if it were moved quickly? Explain why.
17. In Fig. 9–5, suppose the pattern piece and workpiece are both 12-in. long and $^1/_4$-in. wide (into the page) and have a maximum cut depth of 6 in. The workpiece is wood, and the cutting tool is a circular steel saw. About how fast would you move the mounting support? Express your answer in inches per second or inches per minute. Try to give a justification for your estimate.
18. In Fig. 9–6, if the temperature setting pointer is moved to the right, thereby rotating the rotatable shaft counterclockwise, will this tend to close the mercury switch or open it? Does this increase the temperature set point or decrease it? Is this how a residential thermostat acts? Compare to the one in your house.
19. What control mode is illustrated by Fig. 9–6?
20. In Fig. 9–7, if you desired to increase the set-point pressure, would you tighten the compression spring or loosen it? Explain.
21. In Fig. 9–7, assume that the amplifier's output voltage polarity is the same as its input voltage polarity (the output is positive on top if the input is positive on top). If the output voltage polarity is + on the top and − on the bottom, should the damper drive open or drive closed?

Section 9-4

22. Name the five basic control modes.

Section 9-5

23. Explain the meaning of differential gap in On-Off control.
24. What is the important disadvantage of On-Off control compared to the other control modes?

25. What advantages does On-Off control have over the other four control modes?
26. What is the most widely used control mode in modern American industry?
27. Do you ever see a solenoid-actuated valve as the final correcting element in the proportional control mode? Why?
28. What benefit arises from widening the differential gap in an On-Off controller?
29. What disadvantage arises from widening the differential gap in an On-Off controller?
30. Generally speaking, when is the On-Off control mode acceptable?

Section 9-6

31. Under what general conditions is it necessary to use the proportional mode of control rather than the On-Off mode?
32. In proportional control, if you want the controller to give a "stronger" reaction to a given amount of error, should you widen the proportional band or make it narrower? Explain.
33. If the full control range of a temperature controller is 1000°F and the proportional band has been adjusted to 15%, how much does the measured temperature have to change in order to drive the correcting device from one extreme position to the other?
34. The controller of Question 33 has been adjusted to stroke the final control valve from full closed to full open if the temperature changes by 280°F. What is the proportional band?
35. Suppose that the controller of Question 33 is controlling right on the set point with the final control valve exactly 50% open. The set point is 670°F. Suppose that a process disturbance causes the measured temperature to drop to 630°F, which causes the final control valve to go 100% open (just). How wide is the proportional band?
36. The controller of Question 33 is controlling right on the set point (670°F) with the final control valve 50% open. The proportional band is set at 40%. What measured temperature will cause the control valve to go fully open? What temperature will cause it to go fully closed?
37. The controller of Question 33 is controlling right on the set point of 780°F with the final control valve 75% open. The proportional band is set at 25%. What measured temperature will cause the control valve to go fully open? To go fully closed?
38. A certain temperature controller has a control range of 1500°F to 2200°F. It is controlling right on a set point of 1690°F with the control valve 35% open. The proportional band is 28%. What temperature will cause the control valve to go fully closed? To go fully open?
39. The controller of Question 38 is controlling right on a set point of 1690°F with the control valve 35% open. The proportional band is 45%. What temperature will cause the control valve to go fully closed? To go fully open?
40. Which will produce a larger offset, a wide proportional band or a narrow proportional band?
41. Explain why permanent offset occurs with the proportional control mode.
42. A proportional temperature controller is controlling at 1415°F with the set point at 1425°F. The control valve is 80% open. If the set point is raised to 1430°F, will the offset become larger or smaller? Explain your answer.
43. Define the term *offset* as applied to closed-loop control systems.
44. Explain why the proportional band is made narrower as the resistance of the proportional band adjust pot is increased.

Section 9-7

45. In the proportional plus integral control mode, what two things determine the controller's output signal?
46. What beneficial result arises from using proportional plus integral control compared to straight proportional control?
47. When does a proportional plus integral controller tend to correct the offset faster, when the integral time constant is long or when it is short?
48. How is reset rate related to integral time constant?
49. In Fig. 9–16, how would you increase the reset rate, by increasing the resistance R or by decreasing it?
50. Under what general conditions is it necessary to use the proportional plus integral mode of control rather than the proportional-only mode?
51. In Fig. 9–17, the valve ends up in the same final position no matter what the reset rate. Explain why this is reasonable and is to be expected.
52. In Fig. 9–17, the amount of *time* it takes the control valve to settle into its final position varies depending on the reset rate. Explain why this is to be expected.

Section 9-8

53. In Fig. 9–20, the amount of initial change in valve position varies depending on the rate time. Explain why this is to be expected.
54. In Fig. 9–20, the maximum error after a disturbance depends on the rate time. Explain why this is to be expected.
55. In a proportional plus integral plus derivative controller, if you want the derivative response to be more vigorous, should you increase or decrease the rate time?
56. How would you increase the rate time in the controller of Fig. 9–18, by increasing R or decreasing R?
57. Describe the action of each of the five control modes. That is, tell what kind of orders the controller sends to the final correcting device for every possible signal that it might receive.
58. Under what general conditions is it necessary to use the proportional plus integral plus derivative mode of control rather than the proportional plus integral mode?

Section 9-9

59. Explain the meaning of transfer lag. Why does transfer lag exist in processes?
60. Why is transfer lag considered a serious process control problem, while process reaction delay (time constant delay) is not considered so serious?
61. In the liquid level control system of Fig. 9–1, what provides the system's capacity? What causes the system's resistance?
62. Explain the meaning of transportation lag. Why does transportation lag exist in some processes?
63. Which is the more serious control problem, transportation lag or transfer lag? Why?
64. Define *dead time*. What is the chief cause of dead time in a servo control system?

Section 9-11

65. When a PLC is used to implement PID control of an industrial process, the I/O rack must contain an __________ __________ module and an __________ __________ module.

66. When the processor executes a PID instruction, how does it know the proportional-band setting, the integral time constant setting, and the derivative time constant setting?
67. If the PV input transducer had been wired to channel 12 of the Analog Input Module in Fig. 9–26, instead of channel 4, what program entries would have to be changed in Fig. 9–27? Give their new values.
68. If the CV final correcting device had been wired to channel 3 of the Analog Output Module, rather than channel 1, what program entries would have to be changed in Fig. 9–27? Give their new values.
69. With the PLC in PROGRAM mode, is a PID instruction's Calculation Control File automatically saved to the computer's hard disk during programming? Explain.

CHAPTER

INPUT TRANSDUCERS—MEASURING DEVICES

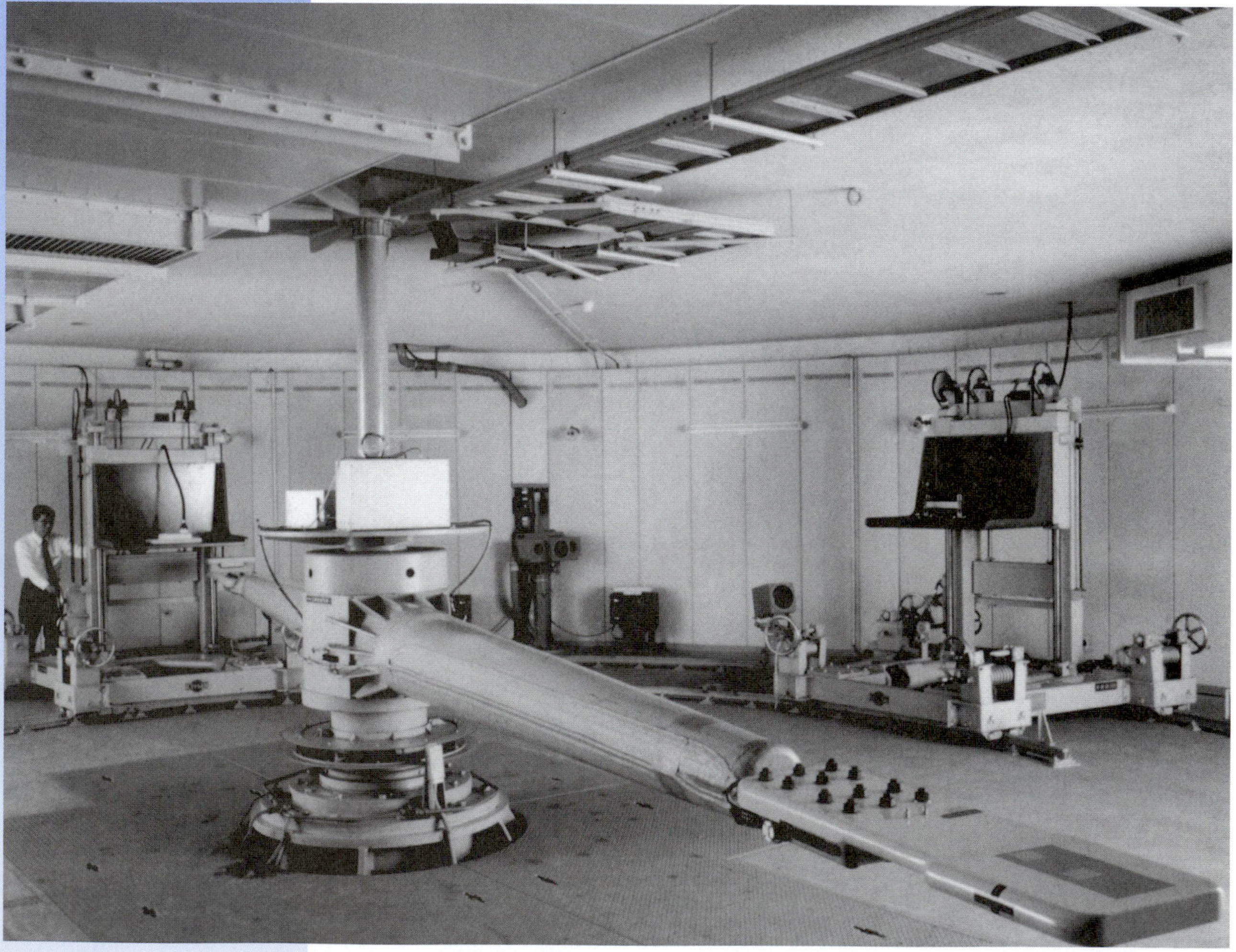

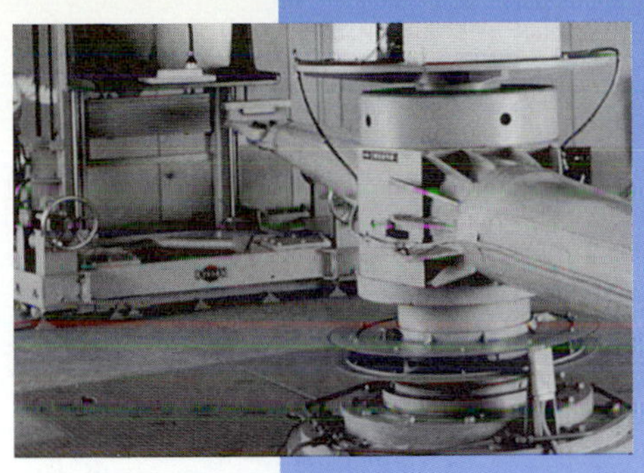

All of industrial control depends on the ability to accurately and swiftly measure the value of the controlled variable. By and large, it has been found that the best way to measure the value of a controlled variable is to convert it into an electrical signal of some sort and to detect the electrical signal with an electrical measuring device. This approach is superior to converting the value of the controlled variable into a mechanical signal because electrical signals have certain advantages over mechanical signals:

1. Electrical signals can be transmitted from place to place much more easily than mechanical signals. (All you need is a pair of wires.)
2. Electrical signals are easier to amplify and filter than mechanical signals.
3. Electrical signals are easy to manipulate to find out such things as the rate of change of the variable, the time integral of the variable, whether the variable has exceeded some limit, etc.

Devices which convert the value of a controlled variable into an electrical signal are called *electrical transducers*. The number of different electrical transducers is very great. Electrical transducers have been invented to measure virtually every physical variable, no matter how obscure. Industrially, the most important physical variables that are encountered are position, speed, acceleration, force, power, pressure, flow rate, temperature, light intensity, and humidity. Accordingly, in this chapter, we will concentrate our attention on electrical transducers which measure these particular variables.

OBJECTIVES

After completing this chapter, you will be able to:

1. Explain the meaning of the terms *linearity* and *resolution* as applied to potentiometers.
2. Explain the operation of a linear variable differential transformer (LVDT).
3. Describe the construction and operation of a Bourdon tube, and list the most popular shapes of Bourdon tubes.
4. Describe the construction and operation of bellows used for pressure measurement.
5. Explain the construction and operation of a thermocouple and how it is compensated against variations in cold-junction temperature; state the temperature ranges in which thermocouples are applied.

6. Describe the operation of a resistive temperature detector (an RTD), and state the temperature ranges in which RTDs are applied.
7. Describe the operation of thermistors, and state the temperature ranges in which they are applied.
8. Describe the operation of solid-state temperature transducers, and cite their limitations.
9. Explain the operation of an optical pyrometer, and state its inherent advantage over other temperature transducers.
10. Describe the behavior of photovoltaic cells and photoconductive cells, and state the relative advantages and disadvantages of these two devices.
11. Explain the operation of an optical shaft-position encoder.
12. Given the design details of an optical position encoder, calculate its resolution and its maximum angle displacement. For any value of measured angle, state the transducer's binary output, and vice versa.
13. Describe how photocell detectors work; describe a chopped photocell detector and discuss its advantages.
14. Describe the following uses of photocells: measuring a material's translucence, automatic bridge balancing, and dc signal chopping.
15. Explain the operation of LEDs, and distinguish between visible LEDs and infrared LEDs.
16. Explain the operation of phototransistors and photodiodes, and discuss their advantages over photoconductive cells.
17. Describe the operation of optical coupler/isolators, and list some of their industrial uses.
18. Describe the construction and operation of an optical fiber; sketch the component arrangement of a fiber-optic signal-transmission system and explain the advantage of such a system.
19. Discuss the industrial applications of ultrasonic waves.
20. Describe the operation of strain gages, and show how they are stabilized against temperature variations.
21. Describe the construction and operation of a strain-gage accelerometer, and state its industrial applications.
22. List the five main types of industrial tachometers; describe the operation of each type and state its relative advantages and disadvantages.
23. Define the Hall effect and explain the construction, operation and desirable features of the following Hall-effect devices: proximity detectors, power transducers, and flowmeters.
24. Describe the operating principles of the following flowmeters: ultrasonic/Doppler effect; turbine; nutating-disk; pressure-drop (venturi).
25. Explain the operation of a four-winding resolver used in single-phase mode for measuring the angular position of a turning shaft.
26. Explain the operation of a four-winding resolver used in two-phase mode for measuring the angular position of a turning shaft.
27. Describe the operation of resistive hygrometers and psychrometers for measuring relative humidity, and take readings correctly from a psychrometer table.

10-1 ■ POTENTIOMETERS

The *potentiometer* is the most common electrical transducer. Potentiometers can be used alone, or they can be attached to a mechanical sensor to convert a mechanical motion into

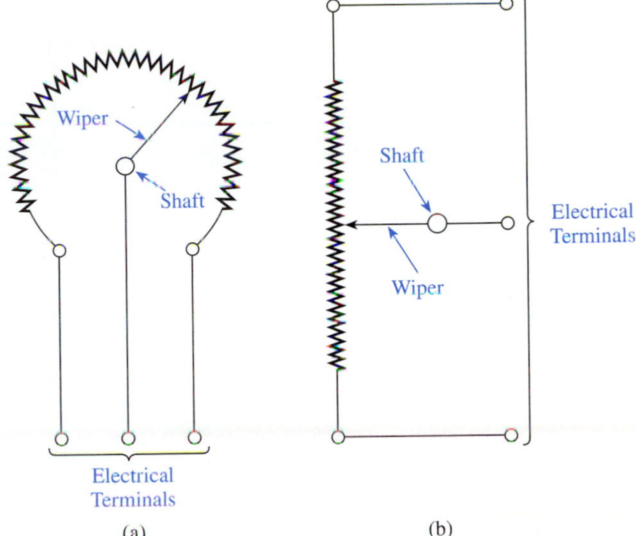

FIGURE 10–1
Schematic symbols of potentiometers. (a) Circular symbol, which suggests the pot's physical appearance. (b) Straight-line symbol.

an electrical variation. A potentiometer is quite simple in conception. It consists of a resistive element and a movable contact that can be positioned anywhere along the length of the element. This movable contact is called by various names, including *tap, wiper,* and *slider*. We will use all three terms interchangeably.

Figure 10–1 shows two schematic representations of a potentiometer. In Fig. 10–1(a), the resistive element is drawn in circular form; this representation suggests the physical construction of most potentiometers, in which the resistive element really is circular and spans an angle of about 300°. The wiper position is then adjusted by turning the shaft to which the wiper is attached. The shaft can be turned by hand or with a screwdriver, depending on whether it has a knob attached or a screwdriver slot in its end.

The more popular electrical schematic representation is shown in Fig. 10–1(b). This is more popular only because it is easier to draw.

10-1-1 Potentiometer Linearity

The great majority of potentiometers are *linear*. The term *linear* means that a given mechanical movement of the wiper produces a given change in resistance, no matter where the wiper happens to be on the element. In other words, the resistance of the element is evenly distributed along the length of the element. The precise degree of linearity of a potentiometer is very important in some applications. Manufacturers therefore specify a *percent linearity* on the potentiometers they make. The meaning of percent linearity, or just linearity for short, can be understood by referring to Fig. 10–2.

Figure 10–2(a) shows a graph of resistance versus shaft angle for a perfectly linear potentiometer. The resistance plotted on the vertical axis can be thought of as the resistance between the wiper terminal and one of the end terminals in Fig. 10–1. The shaft angle plotted on the horizontal axis is the angle through which the shaft has been rotated, with 0° being the position in which the wiper is in direct contact with the end terminal. As can be seen, a perfectly linear potentiometer yields a given amount of resistance change for a given number of degrees of shaft rotation, no matter where the shaft happens to be. That is, a shaft movement from 0° to 60° yields a resistance change of exactly 20% of total resistance; likewise, a shaft movement from 180° to 240°, a 60° rotation, produces a resistance change from 60% to 80% of the total resistance, also an exact 20% change.

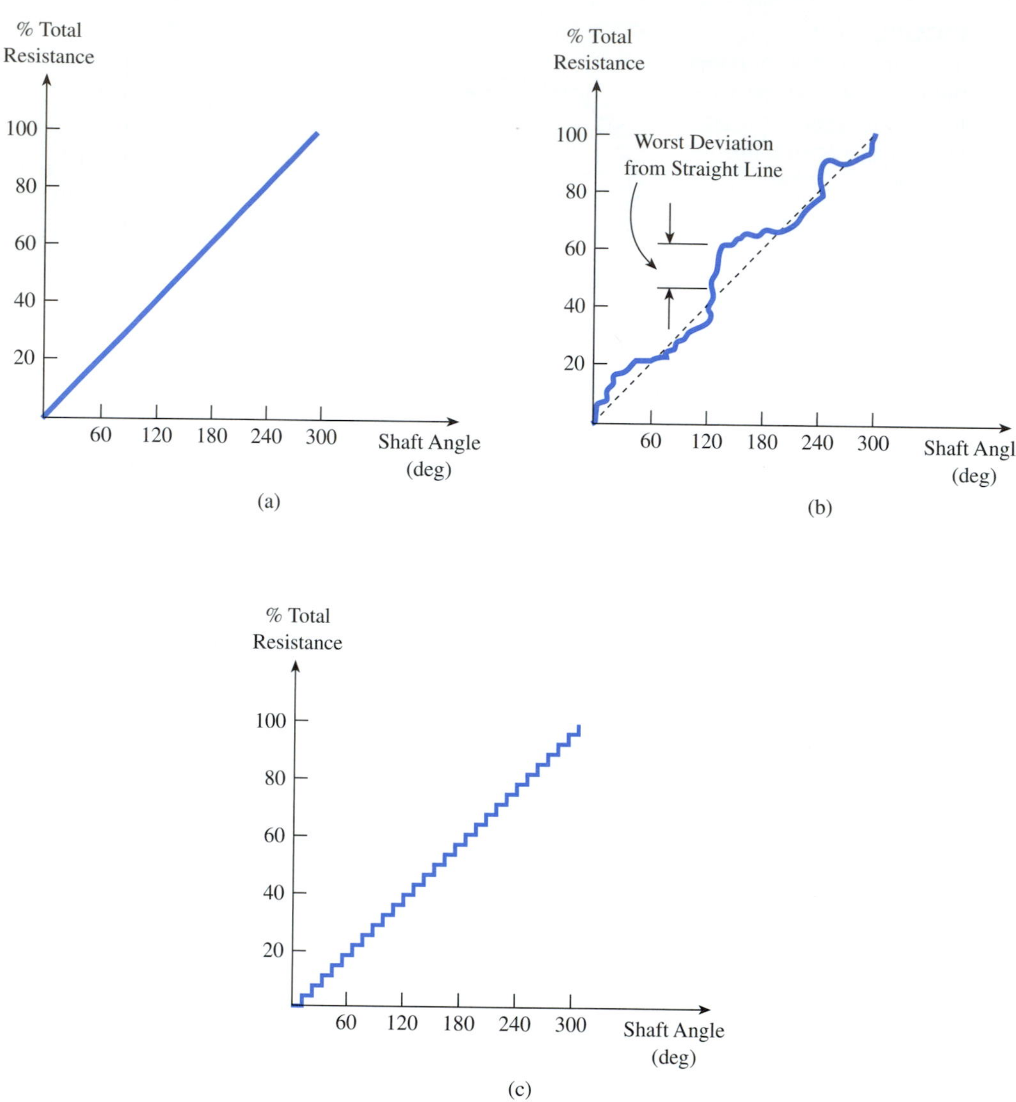

FIGURE 10–2

Graphs of resistance versus shaft angle for a potentiometer. (a) Perfectly linear potentiometer. (b) Real potentiometer, with the resistance deviating from a perfect straight line. (c) Stepwise or noncontinuous variation in resistance.

It is, of course, impossible to manufacture potentiometers that have perfect linearity. The true state of affairs is shown in Fig. 10–2(b), in which the resistance deviates from the ideal straight line. The point of worst deviation from the ideal straight line determines the percent linearity of the potentiometer. For example, in the graph of Fig. 10–2(b), at the worst point, the actual resistance deviates from the ideal straight line by 10%. This means that the actual resistance differs from the expected resistance by an amount which is 10% of the total resistance. The linearity of this potentiometer is therefore 10%.

When a manufacturer specifies a 10% linearity for a potentiometer, it is guaranteed that the resistance will deviate from the straight-line resistance by no more than 10% of total resistance. Thus, a 500-Ω potentiometer with a 10% linearity would have a graph of resistance versus shaft angle in which the actual resistance strayed from an ideal straight line by no more than 50 Ω.

While 10% linearity might be adequate for many industrial potentiometer applications, it almost certainly would not be adequate for a *measuring* application. Usually, potentiometers used as transducers have linearities of less than 1% and sometimes as low as 0.1%. For a 500-Ω pot with a 0.1% linearity, the actual resistance would deviate from the expected straight-line resistance by no more than 0.5 Ω.

10-1-2 Potentiometer Resolution

Many potentiometers are of the *wirewound* variety. In a wirewound pot a thin piece of wire is wound many times around an insulating core. The wiper then moves from one turn of wire to the next as the pot is adjusted. The result is that the wiper resistance does not vary in a perfectly smooth fashion, but in steps. This phenomenon is shown, greatly exaggerated, in Fig. 10–2(c).

The important point here is that there is a limit on the minimum resistance change possible. The smallest possible resistance change is equal to the resistance of one turn of wire. For example, a 500-Ω wirewound pot having 200 turns would have a resistance per turn of 500 Ω/200 = 2.5 Ω. The smallest possible pot adjustment would move the wiper from one turn to the neighboring turn, so the smallest possible resistance change would be 2.5 Ω.This smallest possible change in resistance determines the potentiometer *resolution*.

Resolution of a potentiometer can be considered to be the minimum possible resistance variation, expressed as a percentage of the total resistance. For the pot described in the earlier paragraph, the resolution would be 2.5 Ω/500 Ω = 0.5%.

As a general rule, pots which have inherently good resolution have inherently bad linearity, and vice versa. Of course, by taking great pains in the manufacturing process, it is possible to make potentiometers which have both good resolution and good linearity. Potentiometers used for measuring purposes are usually of this type. They have good resolution, good linearity, and good environmental characteristics (temperature and humidity don't affect them). They are somewhat expensive, costing as much as 20 times what a simple control pot would cost.

Very often, a potentiometer is installed in a circuit with a voltage applied between its end terminals, as illustrated in Fig. 10–3(a). The rotation of the shaft then creates a *voltage* variation between the terminals, instead of just a *resistance* variation between terminals. If the shaft position represents the value of some measured variable, the potentiometer establishes a correspondence between the measured variable and V_{out}.

Another common potentiometer hookup is shown in Fig. 10–3(b). Resistor R_1 and R_2 are equal, and the measuring apparatus is arranged so that the pot wiper is exactly centered for some neutral or reference value of measured variable. This is a bridge circuit. If the bridge is driven by a dc voltage source, the magnitude of V_{out} corresponds to the *amount* by which the measured variable differs from its reference value, and the polarity of V_{out} corresponds to the *direction* of the difference, greater than or less than the reference value. If the bridge is driven by an ac source, the magnitude of V_{out} corresponds to the amount of deviation from the reference value, and the *phase* of V_{out} corresponds to the direction of the deviation. If the measured value is greater than the reference value, the pot wiper moves *up* in Fig 10–3(b). Then V_{out} will be in phase with the ac source. If the measured value is less than the reference value, the pot wiper moves *down* in Fig. 10–3(b), and V_{out} will be 180° out of phase with the ac source.

Another common arrangement of a potentiometer in a bridge circuit is shown in Fig. 10–3(c). Recall that the basic idea of bridge circuits is that the bridge will be

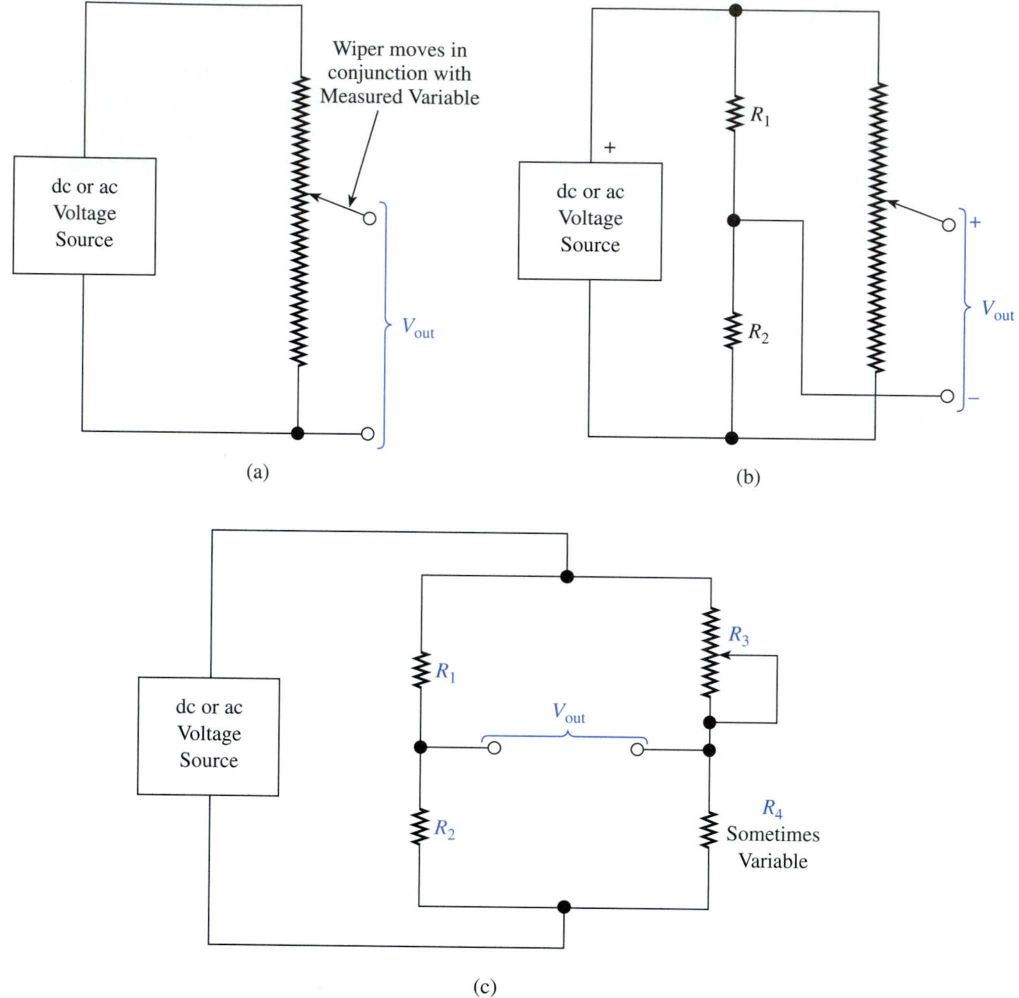

FIGURE 10–3
Potentiometers used in measuring circuits. (a) Potentiometer as a simple voltage divider. (b) Bridge circuit, with the pot comprising one side of the bridge. (c) Bridge circuit with the pot comprising one leg of the bridge.

balanced when the ratio of resistances on the left equals the ratio of resistances on the right. In other words, $V_{out} = 0$ if

$$\frac{R_1}{R_2} = \frac{R_3}{R_4}$$

A bridge of this sort can be used in either of two ways:

1. The measured variable can be used to position the pot shaft, and then the output voltage (V_{out}) from the bridge represents the value of the measured variable.
2. The measured variable can be used to cause one of the resistors, say R_4, to vary. R_4 may be a potentiometer itself, or it may be a resistor that varies in response to some stimulus, such as temperature. R_3 is then adjusted either manually or automatically until V_{out} equals zero, meaning that the bridge is balanced. The position of the R_3

pot shaft then represents the value of the variable. The shaft can be attached to some indicating device to read out the value of the measured variable.

10-2 ■ LINEAR VARIABLE DIFFERENTIAL TRANSFORMERS (LVDTS)

A *linear variable differential transformer* gives an ac output voltage signal which is proportional to a physical displacement. Figure 10–4 shows the construction, schematic symbol, and output waveforms of an LVDT.

Figure 10–4(a) shows that an LVDT has one primary winding and two secondary windings all wrapped on the same form. The form itself is hollow and contains a magnetic core which is free to slide inside the form. As long as the magnetic core is perfectly centered in the form, the magnetic field linkage will be the same for secondary winding 1 and secondary winding 2. Therefore both secondary winding voltages will be equal. If the core moves to the left in Fig. 10–4(a), the magnetic linkage will be greater to secondary winding 1 because more of the core is inside that winding than secondary

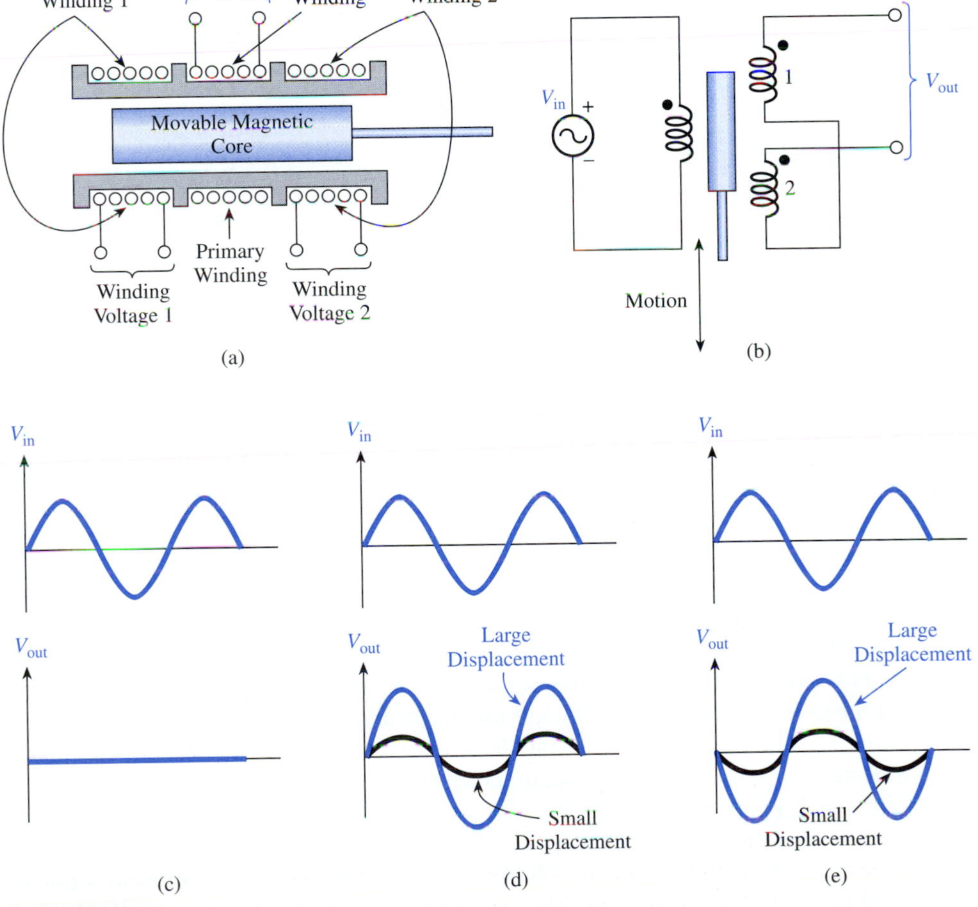

FIGURE 10–4
(a) Physical construction of an LVDT. (b) Schematic drawing of an LVDT. (c) When the LVDT core is perfectly centered, V_{out} equals zero. (d) When the core moves up, V_{out} is in phase with V_{in}. (e) When the core moves down, V_{out} is out of phase with V_{in}.

winding 2. Therefore winding voltage 1 will be greater than winding voltage 2. On the other hand, if the core moves to the right in Fig. 10–4(a), winding voltage 2 will be greater than winding voltage 1, because secondary winding 2 will have more of the core inside it. The LVDT is built so that the *difference* between the two secondary winding voltages is proportional to core displacement.

When the LVDT is in use as a measuring device, the secondary windings are connected together in series opposition, as indicated in Fig. 10–4(b). Therefore, if the core is centered and winding voltage 1 equals winding voltage 2, the net output voltage (V_{out}) is zero. This is shown in Fig. 10–4(c). If the core moves up in Fig. 10–4(b), winding voltage 1 becomes larger than winding voltage 2, so V_{out} becomes nonzero. The farther the core moves, the greater V_{out} becomes. This is shown in Fig. 10–4(d). Also, V_{out} is *in phase* with V_{in} because of the way the output voltage phase is defined in Fig. 10–4(b).

If the core moves down below its center position in Fig. 10–4(b), winding voltage 2 becomes larger than winding voltage 1, and V_{out} again becomes nonzero. This time V_{out} is 180° *out of phase* with V_{in}, as shown in Fig. 10–4(e). Thus the size of V_{out} represents the amount of displacement from center, and the phase of V_{out} represents the direction of the displacement.

Most LVDTs have a displacement range of about plus or minus 1 inch. That is, the core can move up 1 inch from center or down 1 inch from center. If the LVDT is to be used to measure mechanical displacements much greater than 1 inch, an appropriate mechanical ratioing apparatus (gearing) must be used.

As far as voltage values are concerned, most LVDTs are designed to operate on an input voltage of less than 10 V ac. Full-scale output voltages fall in the same general range. That is, full-scale output voltages can range from about 0.5 V ac to about 10 V ac for different LVDT models.

10-3 ■ PRESSURE TRANSDUCERS

The different approaches to industrial pressure measurement are numerous. We will concentrate our attention on just two common classes of pressure-sensing devices, *Bourdon tubes* and *bellows*. These devices detect the measured pressure and convert it into a mechanical movement. The mechanical movement is then transduced into an electrical signal by either a pot or an LVDT.

10-3-1 Bourdon Tubes

A *Bourdon tube* is a deformed metal tube with an oval cross section. It is open at one end and sealed at the other end. The whole tube is elastic because of the elasticity of the metal used in its construction. The fluid whose pressure is being measured is admitted to the inside of the tube at the open end, which is mechanically anchored. The tube then deflects by an amount proportional to the magnitude of the pressure. This deflection is mechanically transmitted to the wiper of a potentiometer or to the core of an LVDT to provide an electrical signal. Figure 10–5(a) through (d) shows the various shapes of Bourdon tubes and the motions which they produce.

Figure 10–5(e) shows how a C-shaped Bourdon tube could be linked to a potentiometer. Figure 10–5(f) shows how a C-shaped tube could be linked to an LVDT. Spiral and helical Bourdon tubes are often preferable to C-shaped Bourdon tubes because they produce greater motion of the scaled tip per amount of pressure. Bourdon tubes are most often used to measure pressures in the range from 10 to 300 psi.

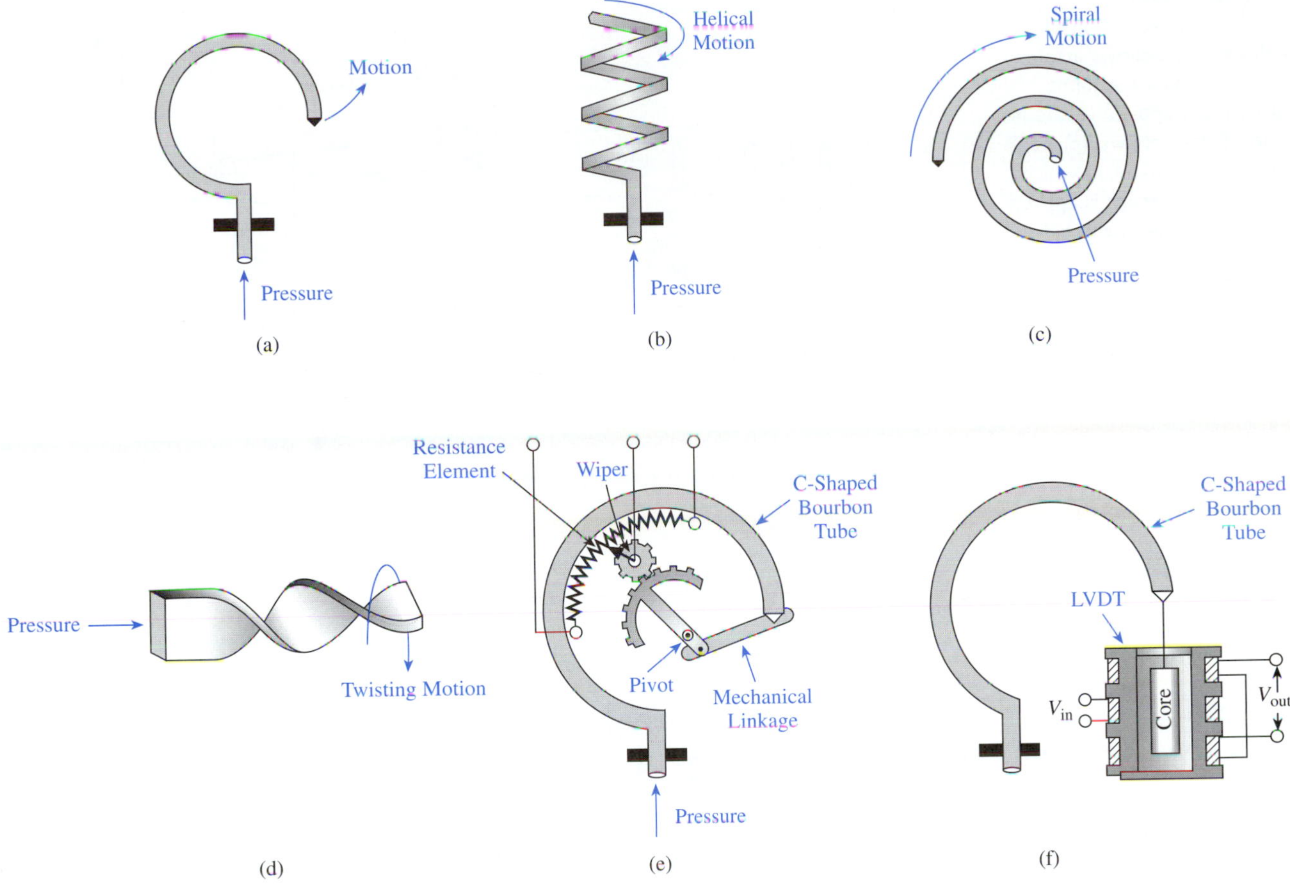

FIGURE 10–5
Bourdon tubes. (a) C-shaped Bourdon tube, the most common type. (b) Helical-shaped Bourdon tube. (c) Spiral Bourdon tube. (d) Twisted Bourdon tube. (e) C-shaped Bourdon tube linked to a potentiometer. (f) C-shaped Bourdon tube linked to an LVDT.

10-3-2 Bellows

A *bellows* is essentially a series of metal diaphragms connected together. When subjected to fluid pressure, a metal diaphragm will distort slightly because of the elasticity of the material used to construct it. When several diaphragms are soldered together in series, the total movement of the end diaphragm can be considerable. Figure 10–6(a) shows a cut-away view of a bellows. With the pressure inlet port anchored, the bellows will expand as the fluid pressure rises, and the output boss will move to the right. As the fluid pressure falls, the bellows contracts, and the output boss moves to the left. The contraction force can be provided by the springiness of the bellows diaphragms themselves or by a combination of diaphragm springiness with an external spring.

Figure 10–6(b) and (c) shows two common bellows arrangements. In Fig. 10–6(b), the pressure is applied to the inside of the bellows and tends to expand the bellows against the pull of the tension spring. As the bellows expands, it actuates a mechanical linkage which moves the wiper of a potentiometer to furnish an electrical output signal.

In Fig. 10–6(c), the measured pressure is applied to the outside of a bellows, forcing it to contract against the push of a compression spring. As it moves, it actuates a mechanical linkage which moves the core of an LVDT to furnish an electrical output signal. These

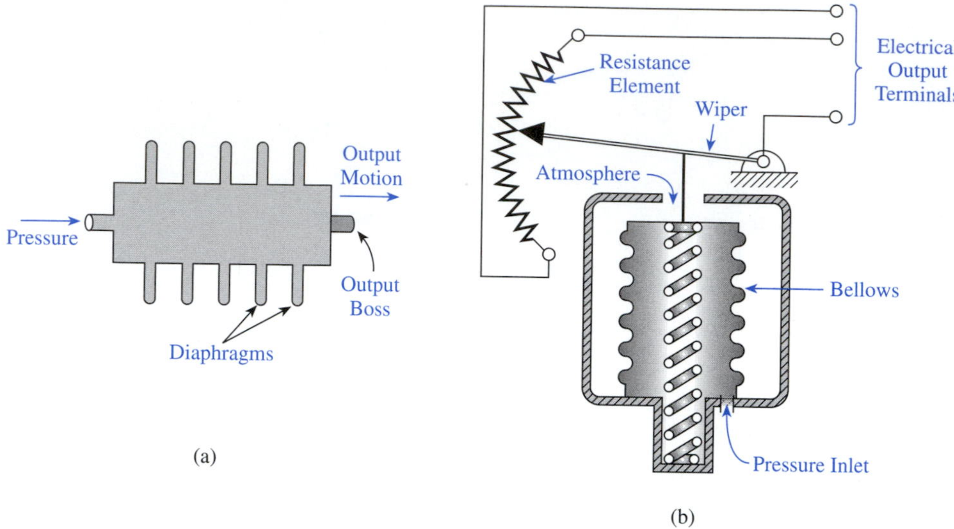

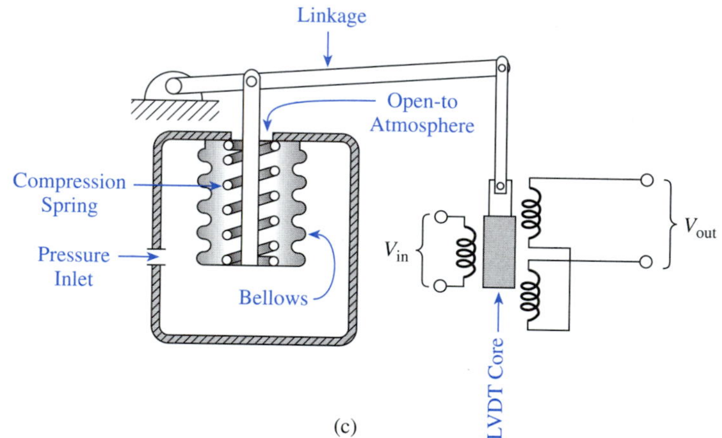

FIGURE 10–6 (a) Basic construction of a bellows. (b) Bellows arrangement in which the input pressure is applied to the inside of the bellows. (c) Bellows arrangement in which the input pressure is applied to the outside of the bellows.

pressure transducers are calibrated by adjusting the initial tension or compression of the return spring. An adjustment nut, not shown in Fig. 10–6, is provided for this purpose.

Bellows-type pressure transducers find their main usefulness in measuring pressures in the range from 0.5 to 20 psi.

10-4 ■ THERMOCOUPLES

The most common device for measuring industrial process temperatures is the *thermocouple*. A thermocouple is a pair of dissimilar metal wires joined together in a complete loop, as shown in Fig. 10–7(a). The dissimilar wires have two junction points, one at each end of the loop. One junction, called the *hot junction*, is subjected to a high temperature; the other junction, called the *cold junction*, is subjected to a low temperature. When this is done a small net voltage is created in the loop; this voltage is proportional to the *difference* between the two junction temperatures.

What happens in a thermocouple loop is that a small voltage is produced at each junction of the dissimilar metals due to an obscure phenomenon called the *Seebeck effect*.

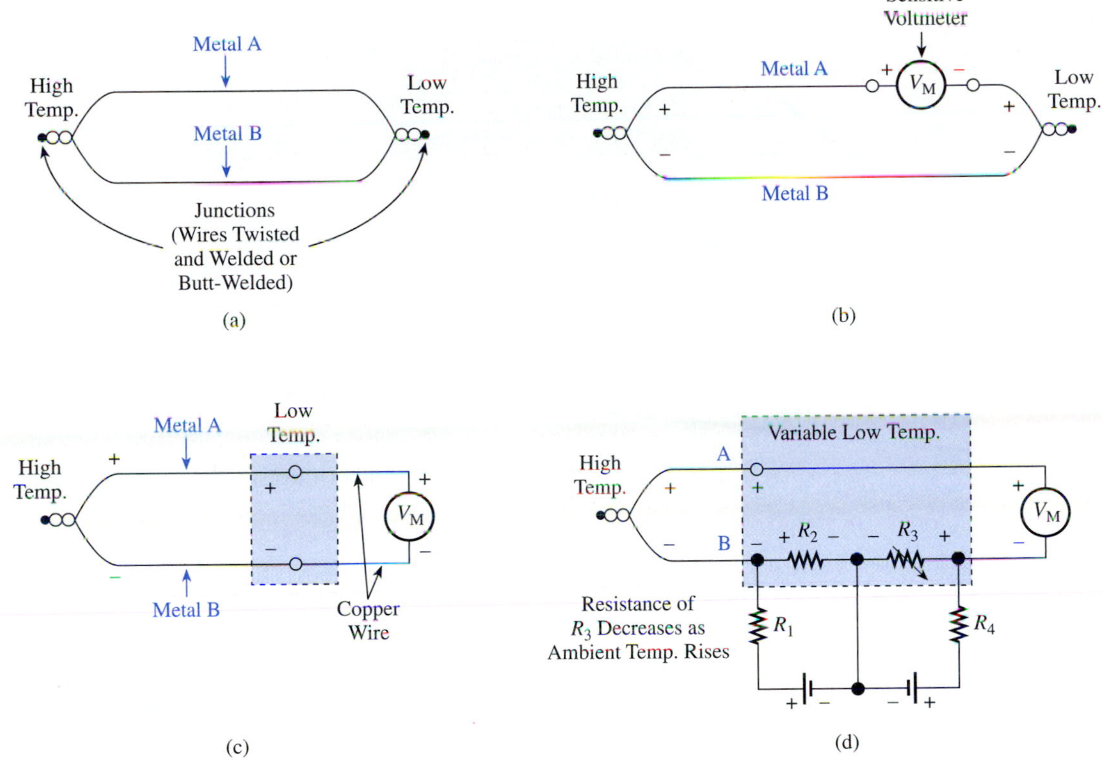

FIGURE 10–7
(a) Basic thermocouple. (b) Thermocouple with a voltmeter inserted in the loop. (c) Thermocouple loop with no cold junction between metal A and metal B. (d) Thermocouple loop which is compensated against the variations in the cold-junction temperature.

The higher the temperature at the junction, the greater the voltage produced by that junction. Furthermore, the relationship between voltage and temperature is approximately linear; that is, a given increase in temperature produces a given increase in voltage. The proportionality constant between voltage and temperature depends on which two metals are being used. Since a complete loop always has two junctions, two voltages are produced. These voltages oppose each other in the loop, as Fig. 10–7(b) shows. The net voltage available to drive current through the resistance of the loop is the difference between the two individual junction voltages, which depends on the difference between the two junction temperatures.

To measure the temperature difference, it is only necessary to break the loop open at a convenient point (at some cool location) and insert a voltmeter. The voltmeter must be rather sensitive, since the voltage produced by a thermocouple loop is in the millivolt range. The voltage reading can then be converted into a temperature measurement by referring to standard tables or graphs which relate these two variables. Graphs of voltage versus temperature difference for several popular industrial thermocouples are given in Fig. 10–8. In each case the first metal or metal alloy mentioned in the thermocouple is the positive lead, and the second metal or metal alloy is the negative lead.

To avoid the problem of identifying thermocouples by proprietary registered trade names, a letter code for thermocouple types has been adopted. Thus, type *J* thermocouples have the response shown in Fig. 10–8 no matter what particular name is used to

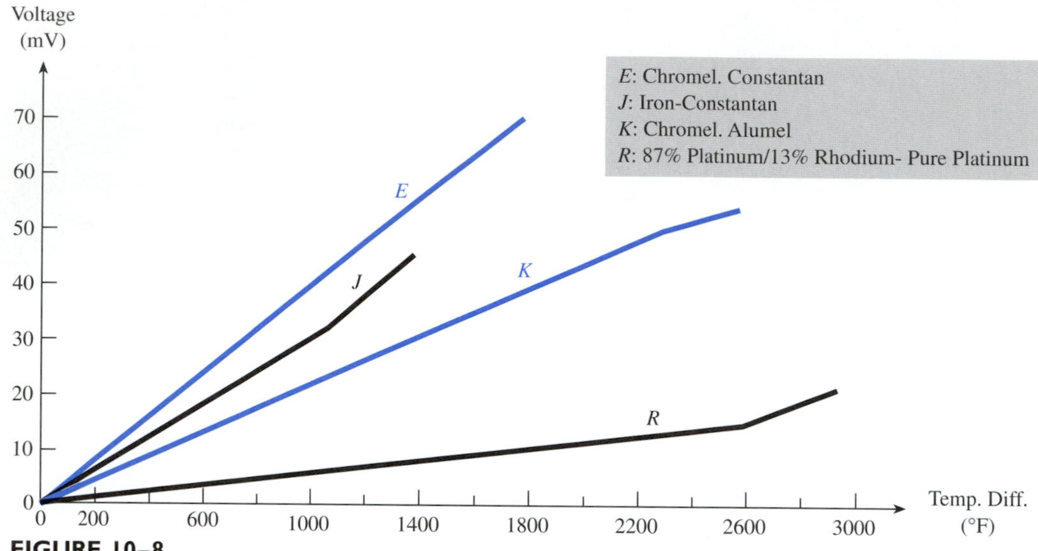

FIGURE 10–8

Voltage versus temperature curves for types *E*, *J*, *K*, and *R* thermocouples. The words *Chromel, Constantan,* and *Alumel* are proprietary trade names of individual manufacturers of thermocouplewire.

identify the metal alloy. The same is true for type *K* and type *R* thermocouples, and for other types not graphed in Fig. 10–8.

When a voltmeter is inserted into the thermocouple loop, it is usually most convenient to insert it as shown in Fig. 10–7(c). In that figure, metal A and metal B don't actually touch each other at the cold junction. Instead, both metals are placed in contact with standard copper conducting wires. The connections are normally made on a terminal strip. The copper wires then lead off and connect to the sensitive voltmeter. It may seem that this would disrupt the total net voltage generated by the thermocouple loop, but that does not happen. The net loop voltage remains the same because there are now *two* cold junctions, one between metal A and copper and the other between metal B and copper. The sum of the two junction voltages produced by these cold junctions equals the voltage that would have been produced by the single cold junction of metal A with metal B. Of course, the two cold junctions must be maintained at the same temperature that the single junction would have felt. This is no problem, since the copper wires and terminals are always inside some enclosure which is thermally insulated from the process being measured and which is subjected to the same temperature to which a single junction would have been subjected, namely, the ambient temperature in the industrial location. Therefore the circuit in Fig. 10–7(c) will yield the same reading as the circuit in Fig. 10–7(b).

One further matter is important in the use of thermocouples in industry. It regards the variation in ambient temperature at cold junctions. Here is the situation: If we knew beforehand the temperature of the cold junctions, then instead of relating the voltmeter reading to the temperature *difference,* we could relate it to the hot-junction temperature itself. This would be possible because we could construct the tables of temperature versus voltage to reflect the fact that the cold junctions are at a certain known *reference temperature,* as it is called.

As an example, consider the type *J* thermocouple in Fig. 10–8. The graph shows that at a temperature difference of 400°F, the thermocouple loop voltage is 12 mV. If we knew that the cold junction was always at 75°F, say, then we could conclude that a 12-mV loop voltage represented a hot-junction temperature of 475°F (475°F − 75°F = 400°F). As long as the cold junction was constantly maintained at the reference temperature of 75°F, we could go right through the thermocouple table and add 75°F to every temperature difference reading. The resulting temperature value would then represent the temperature at the hot junction.

As a matter of fact, this is exactly what is done in industrial thermocouple tables. The figure of 75°F has been chosen because it represents a fairly reasonable guess at the average ambient temperature in an industrial setting. (In thermocouple tables for *laboratory* use, the reference temperature is usually considered to be 32°F, the freezing point of water.)

For this approach to work accurately, the cold junction must be constantly maintained at the 75°F reference temperature. This is usually impractical unless the temperature-measuring instrument happens to be located in an air-conditioned control room. In all probability, though, the measuring instrument is located right out with the industrial equipment and machinery. The ambient temperature may easily vary from about 50°F in the winter to about 100°F in the summer; even wider seasonal swings in ambient temperature are common. Because of this variation in cold-junction temperature, industrial thermocouple loops must be *compensated*.

A simple automatic compensation method is illustrated in Fig. 10–7(d). The two dc voltage supplies and the four resistors are arranged so that the voltages across R_2 and R_3 are in opposition. The voltage polarities across R_1 and R_4 do not matter, since R_1 and R_4 are outside the thermocouple loop. R_3 is a temperature-sensitive resistor, having a negative temperature coefficient. This means that its resistance goes down as its temperature goes up. The circuit is designed so that at 75°F the small voltage across R_3 equals the small voltage across R_2. The voltages across the two resistors exactly cancel each other, and the voltmeter reading is unaffected. Now if the cold-junction temperature should rise above 75°F, the voltmeter reading would tend to decrease because of a smaller difference between the hot and cold junctions. This would tend to yield a measured temperature reading which is *lower* than the actual temperature at the hot junction. However, the resistance of R_3 decreases as the cold-junction temperature rises, resulting in a smaller voltage across its terminals. The R_3 voltage no longer equals the R_2 voltage. Therefore, the R_2-R_3 combination introduces a net voltage into the loop which tends to incease the voltmeter reading. A careful check of the polarities of the voltages in Fig. 10–7(d) will prove that this is so. Because of the design of the compensating circuit, the net voltage introduced by the R_2-R_3 combination exactly cancels the decrease in loop voltage caused by the temperature rise at the cold junction.

If the cold-junction temperature should fall below 75°F, the R_2-R_3 combination introduces a net voltage in the opposite direction. This offsets the increase in loop voltage caused by the larger temperature difference between hot and cold junctions. Verify this to yourself by carefully checking the polarities in Fig. 10–7(d).

Many industrial temperature-measuring/recording instruments use an automatically balancing bridge to indicate temperatures. The thermocouple loop voltage is balanced by moving the wiper of a potentiometer in a Wheatstone bridge circuit. The potentiometer shaft is ganged to another shaft, which operates the temperature-indicating needle. Therefore, for every value of thermocouple loop voltage, there is a corresponding position of the temperature-indicating needle. A temperature scale is then marked off behind the needle.

10-5 ■ THERMISTORS AND RESISTIVE TEMPERATURE DETECTORS (RTDS)

Besides using thermocouple *voltage* to electrically measure temperature, it is also possible to utilize the *resistance* change which occurs in many materials as their temperature changes. Materials used for this purpose fall into two classes, pure metals and metallic oxides.

Pure metals have a fairly constant *positive* temperature coefficient of resistance. The temperature coefficient of resistance, usually just called the *temperature coefficient,* is the ratio of change in resistance to change in temperature. A positive coefficient means that the resistance gets larger as the temperature increases. If the coefficient is constant, it means that the proportionality factor between resistance and temperature is constant and that resistance and temperature will graph as a straight line. Resistance versus temperature graphs for several popular metals are given in Fig. 10–9(a). The *resistance factor* in this graph means the factor by which the actual resistance is greater than the reference resistance at 0°F. For example, a factor of 2 indicates that the resistance is twice as great as it was at 0°F. When a pure metal wire is used for temperature measurement, it is referred to as a *resistive temperature detector,* or RTD.

When metallic oxides are used for temperature measurement, the metallic oxide material is formed into shapes which resemble small bulbs or small capacitors. The formed

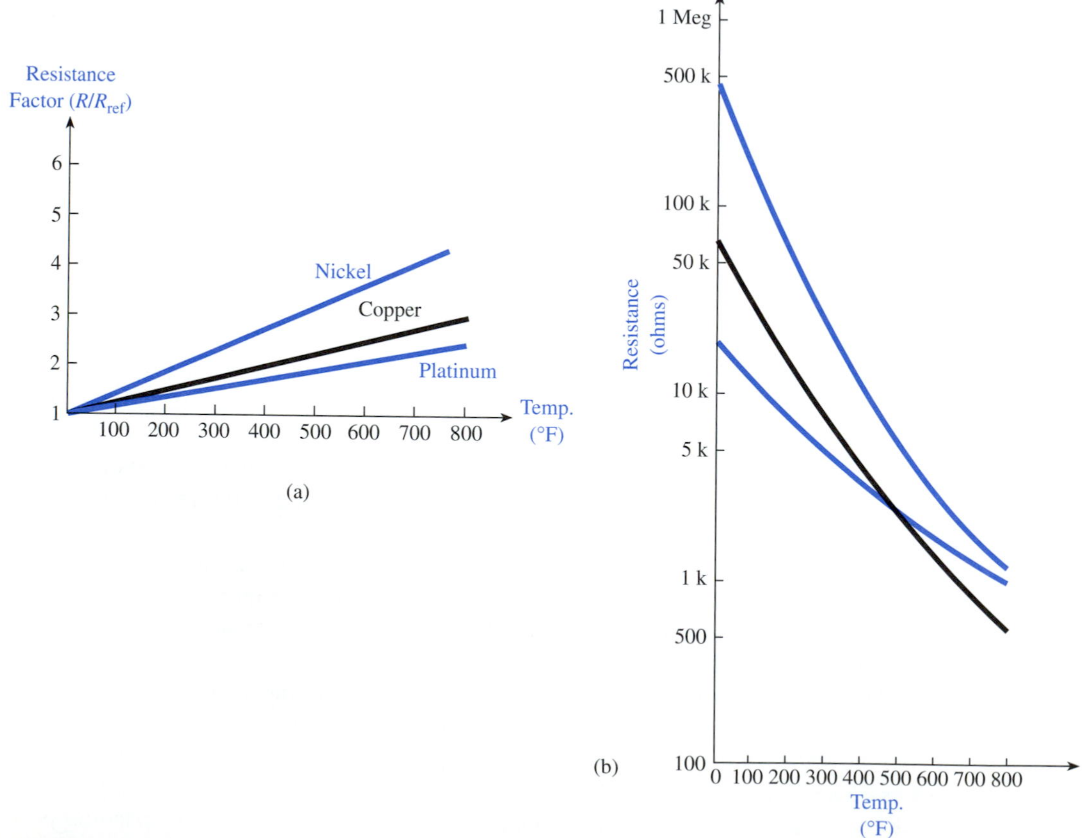

FIGURE 10–9
(a) Resistance versus temperature curves for pure metal RTDs. (b) Resistance versus temperature curves for typical thermistors.

device is then called a *thermistor*. Thermistors have large negative temperature coefficients which are nonconstant. In other words, the resistance change per unit of temperature change is much greater than for a pure metal, but the change is in the other direction—the resistance gets smaller as the temperature increases. The fact that the coefficient is nonconstant means that the change in resistance per unit temperature change is different for different temperatures. Figure 10–9(b) shows graphs of resistance versus temperature for three typical industrial thermistors. Note that the vertical scale is logarithmic to allow the great range of resistances to be shown. The temperature-sensitive resistor which compensated the thermocouple in Sec. 10-4 would be a thermistor.

Figure 10–10 shows three circuits for utilizing thermistors and/or RTDs. In schematic drawings, temperature-sensitive resistors are symbolized by a circled resistor with an arrow drawn through, and the letter T outside the circle. A resistor with a positive temperature coefficient can be indicated by the arrow pointing toward the top of the circle, and a resistor with a negative temperature coefficient can be symbolized by an arrow pointing toward the bottom of the circle. These are not universally accepted rules, but we will use them in this book.

In Fig. 10–10(a), the temperature transducer is shown in series with an ammeter and a stable voltage supply. As the temperature rises, the resistance goes down, and the current increases. If the specific characteristics of the thermistor are known, it is possible to relate the current measurement to the actual temperature. The supply voltage must not change or the current-to-temperature correspondence will become invalid.

In Fig. 10–10(b), the temperature transducer increases its resistance as the temperature rises. This causes a larger portion of the stable supply voltage to appear across its

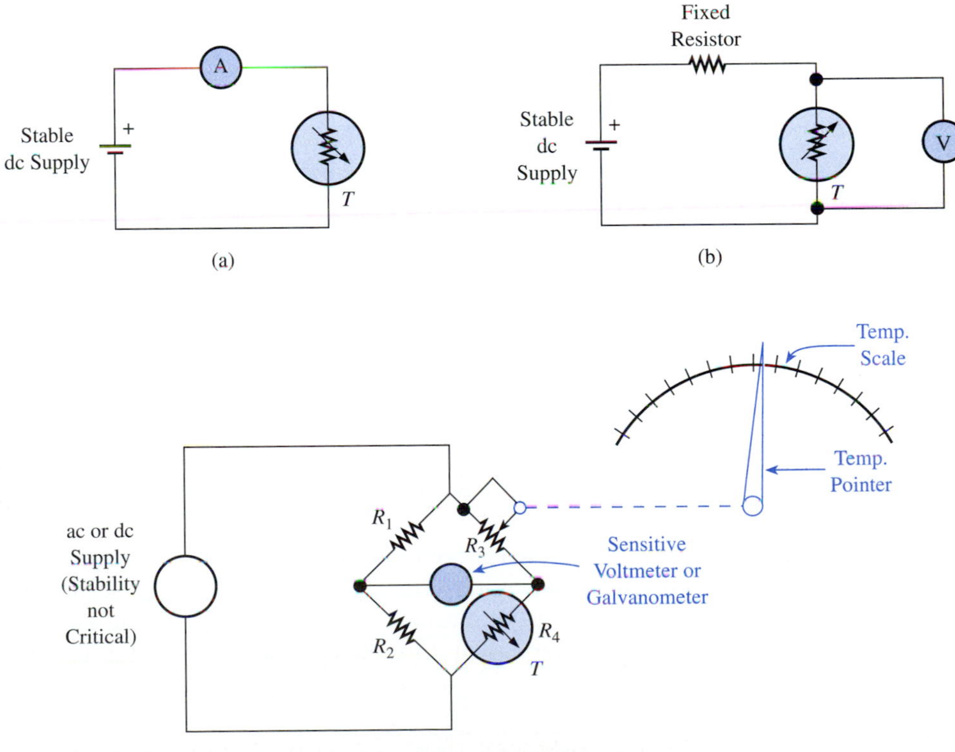

FIGURE 10–10 Circuits utilizing resistive temperature transducers. (a) The ammeter reading corresponds to the measured temperature. (b) The voltmeter reading corresponds to the measured temperature. (c) Bridge arrangement. When the bridge is balanced, the position of the pot wiper corresponds to the measured temperature.

terminals. Thus the voltmeter reading can be related to temperature. If desired, the voltmeter scale can be marked in temperature units instead of volts for a direct temperature readout.

In Fig. 10–10(c), a bridge circuit is used. For precision, bridge circuit measurements are inherently superior to other measurements because the meter which detects bridge imbalance can be made very sensitive. Therefore even a slight imbalance in the bridge can be detected and adjusted out. The bridge detecting meter can be made very sensitive because when the bridge is *close* to being balanced, the voltage across the bridge is near zero; since the detecting meter does not have to measure a large voltage, it can be made to respond vigorously to a small voltage. In other words, it can be made very sensitive. By contrast, the meters in Fig. 10–10(a) and (b) cannot be made very sensitive because they must be able to read (relatively) large values of voltage or current.

The bridge circuit works like this: As the temperature of the thermistor rises, its resistance drops. This changes the ratio of resistances on the right-hand side and throws the bridge out of balance (assuming it was balanced to start with). Either manually or automatically, R_3 is adjusted until the ratio of resistances on the right-hand side is once again equal to the ratio on the left-hand side, bringing the bridge back into balance. The position of the R_3 pot shaft then represents temperature, since for every possible value of R_4 resistance there is only one value of R_3 resistance which will balance the bridge. The R_3 shaft is mechanically linked to another shaft which positions the temperature pointer.

When using the balanced-bridge measurement method, the temperature readout scale will be linear if the transducer is linear. A linear readout scale means that equal *distances* on the scale represent equal *differences* in temperature, or, said another way, that the scale temperature marks are all equally spaced. Since we have seen that a thermistor is very nonlinear, we would expect that the temperature scale in Fig. 10–10(c) would also be nonlinear.* The extreme nonlinearity of thermistors makes them poorly suited for measuring temperatures over wide ranges. However, for measuring temperatures within narrow bands, they are very well suited, because they give such a great response for a small temperature change. This great response is also what recommends thermistors in applications like that described in Fig. 10–7(d), compensating a thermocouple loop over a fairly narrow band of cold-junction temperatures. The hearty response of the thermistor makes it easy to generate sufficient compensation.

The natural nonlinearity of thermistors can be partially corrected by connecting several matched thermistors together in a series-paralled combination. The resulting circuit is called a *thermistor composite network*. These networks are quite linear over a fairly wide temperature range (about 200°F), but they are naturally more expensive than plain thermistors.

As a general rule, thermistors are preferable when the expected temperature band is narrow, and RTDs are preferable when the expected temperature band is wide. Most thermistors are manufactured for use somewhere between −150°F and +800°F, although special thermistors have been developed for use at extremely low temperatures, near absolute zero. RTD thermometers are available for use at temperatures from −400°F to +2000°F.

Besides their uses as measurers of temperature in an external medium, thermistors also have applications which make use of the *internal* heat generated as they carry current. In any external temperature-measuring application, it is important to eliminate the effect of the thermistor's internally generated heat; this is done by making the thermistor current very small. In some applications, however, the ability of a thermistor to change

*For an example of a nonlinear scale, look at the ohms scale on a VOM. The marks representing a given ohms difference are far apart on the right but close together on the left.

its own resistance as it generates I^2R heat energy can be very useful. For example, thermistor self-heating can be used for establishing time delays, protecting delicate components from surge currents, detecting the presence or absence of a thermally conductive material, etc.

10-6 ■ OTHER TEMPERATURE TRANSDUCERS

10-6-1 Solid-State Temperature Sensing

An ordinary silicon diode is temperature-sensitive. For constant current, its forward anode-to-cathode voltage varies inversely with temperature. A typical diode temperature response is shown in Fig. 10–11(a).

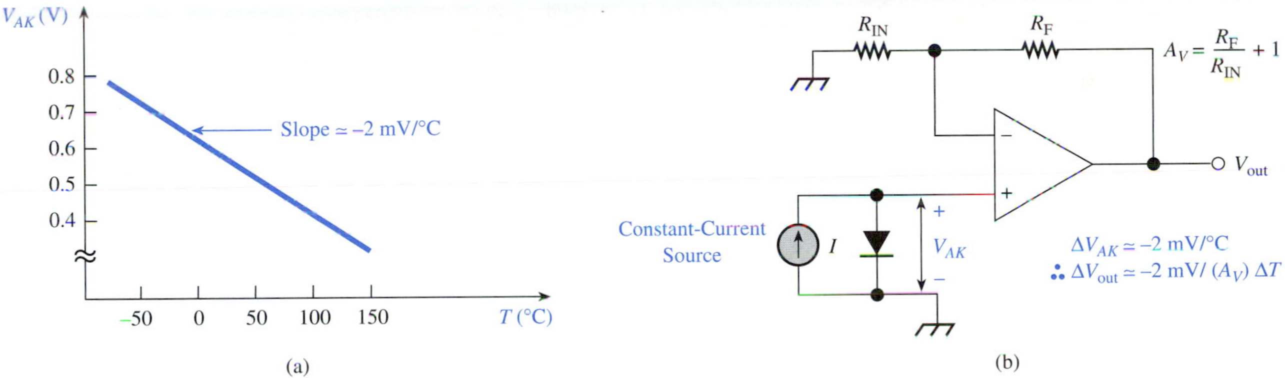

FIGURE 10–11
(a) Typical characteristic graph of junction voltage versus temperature for a silicon diode carrying constant current. (b) Amplifying the junction voltage to detect small temperature changes.

This temperature dependence can be used to measure the change in temperature of a medium containing the diode, or of a device which is in thermal contact with the diode. Figure 10–11(b) shows the general circuit approach. By installing appropriate op amp offset circuitry it is possible to make $T = 0°C$ correspond to $V_{out} = 0$ V. Then the "change" expression in the voltage-temperature relation given in Fig. 10–11(b) becomes the absolute expression

$$V_{out} = -2\text{ mV}(A_V)T$$

Other solid-state devices can also be used as temperature-sensing elements. The universal shortcoming of this method is the inevitable batch instability of any solid-state device. Therefore solid-state temperature sensing is more applicable to temperature-limit detection than to accurate measurement.

10-6-2 Optical Pyrometers

At temperatures higher than about 2200°C (4000°F), contact-type temperature sensors have such a short life expectancy that they are impractical for industrial use. In this high range it is necessary to measure temperature from a distance, by sensing the visible and/or invisible electromagnetic radiation emitted by the hot body.

The visible electromagnetic radiation (light) that is emitted from a hot body is concentrated at a frequency which is indicative of the body's temperature. Therefore, if the

FIGURE 10–12
Optical pyrometer:
(a) Overall appearance; (b) Comparing the slits.

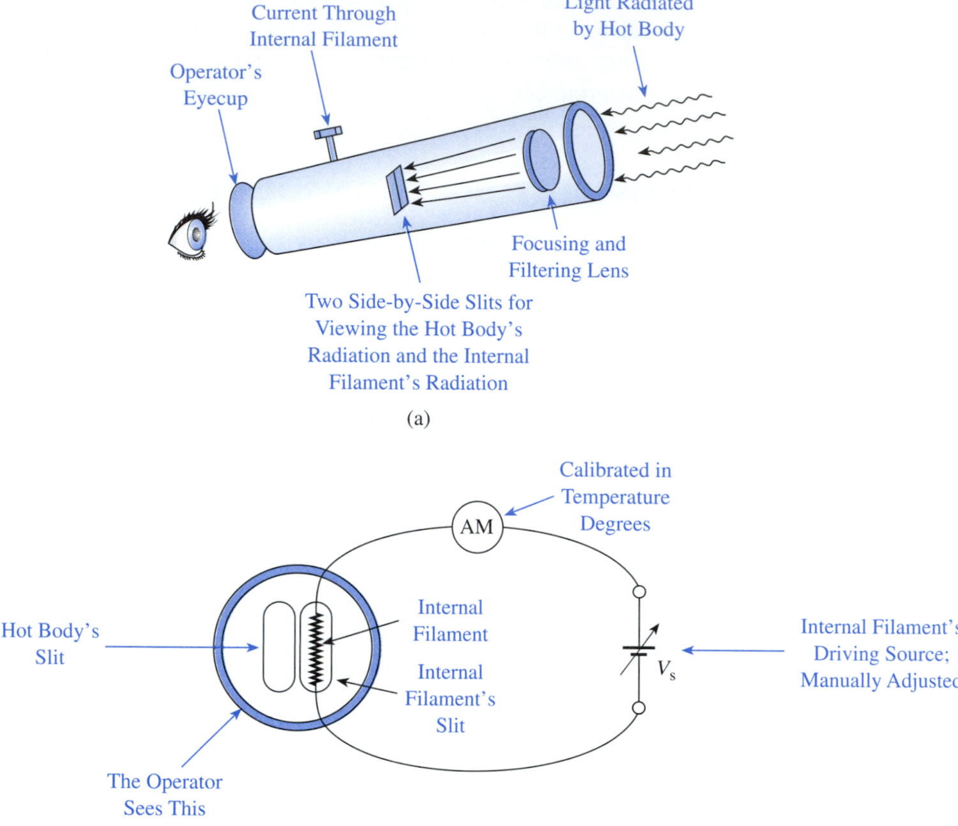

weaker frequency components of its radiated light are optically filtered out, a hot body will take on a temperature-indicative color. We can use this behavior to determine a hot body's temperature by adjusting the temperature of a reference light source until its color matches the color of the hot body. This is the operating principle of an *optical pyrometer,* shown structurally in Fig. 10–12(a).

The pyrometer assembly is held or mounted so that it aims at the body whose temperature is being measured. The body's emitted radiation is filtered and then focused through a slit inside the instrument's structure, where it can be viewed by the human operator. Located alongside the hot-body slit is a second slit which displays the filtered radiation emitted from an internal filament. The operator adjusts the current through the internal filament until the colors of the two slits match each other. At that point the temperatures are equal for both light sources. Because the thermal characteristics of the internal filament are known, we can measure its current and relate that current value to temperature. The current value can be related to temperature by using a look-up table, or, more conveniently, the ammeter can simply be calibrated in units of temperature degrees. These ideas are conveyed by Fig. 10–12(b).

10-7 ■ PHOTOCELLS AND PHOTOELECTRIC DEVICES

Photocells are small devices which produce an electrical variation in response to a change in light intensity. Photocells can be classified as either *photovoltaic* or *photoconductive*.

A photovoltaic cell is an energy source whose output voltage varies in relation to the light intensity at its surface. A photoconductive cell is a passive device, not capable of producing energy; its resistance varies in relation to the light intensity at its surface.

Industrially, applications of photocells fall into two general categories:

1. Sensing the presence of an opaque object:
   a. The sensing may be done on an all-or-nothing basis, in which the photocell circuit has only two output states, representing either the presence of an object or the absence of an object. This is the kind of sensing used to count pieces traveling down a conveyor or to prevent a mechanism from operating if an operator's hands are not safely out of the way.
   b. The sensing may be done on a continuous basis, with the photocell circuit having a continuously variable output, representing the variable position of the object. This is the kind of sensing used to "watch" the edge of a moving strip of material to prevent it from straying too far from its proper position.

The outstanding advantage of photocells over other sensing devices is that no physical contact is necessary with the object being sensed.

2. Sensing the degree of translucence (ability to pass light) or the degree of luminescence (ability to generate light) of a fluid or solid. In these applications the process has always been arranged so that the translucence or luminescence represents some important process variable. Some examples of variables which could be measured this way are density, temperature, and concentration of some specific chemical compound (carbon monoxide, carbon dioxide, water, etc.).

10-7-1 Photovoltaic Cells

The symbols often used for photovoltaic cells are shown in Fig. 10–13(a). The two wavy arrows pointing toward the circled battery suggest that external light energy produces the battery action. Since wavy arrows are inconvenient to draw, the Greek letter λ is often used to suggest light activation.

The open-circuit output voltage versus light intensity is graphed in Fig. 10–13(b) for a typical photovoltaic cell. Notice that the graph is logarithmic on the light intensity axis. This graph indicates that the cell is more sensitive at low light levels, since a small *change* in intensity (say, from 1 to 10 fc) can produce the same boost in output voltage as a larger change in intensity (say, from 100 to 1000 fc) at a higher light intensity level.

The output current characteristics of a photovoltaic cell operating into a load are graphed in Fig. 10–13(c) for various load resistances; as can be seen, a single photovoltaic cell cannot deliver very much current. The output currents are measured in microamps in this example. Photocells can be stacked in parallel, however, for increased current capability.

An example of a photovoltaic cell furnishing all-or-nothing-type information to a logic circuit is illustrated in Fig. 10–14. In Fig. 10–14(a), the light from the light source is gathered and focused at the photovoltaic cell, which is mounted some distance away. Distances of 10 ft or more are not uncommon in industrial situations. When the photovoltaic cell is activated by the light, it picks up sensitive relay *R*, whose contact passes the input signal to the logic circuit. If an object blocks the light path, the photocell deenergizes the relay, and the logic circuit receives no input.

The object blocking the light path could be anything. It could be a moving object whose passage was to be counted by an electronic or mechanical counter; it could be a moving object whose passage alerts machinery farther down the line to prepare for its

FIGURE 10–13
(a) Schematic symbols used for photovoltaic cells. (b) Graph of voltage versus illumination for a typical photovoltaic cell. (c) Graphs of current versus illumination for several different load resistances.

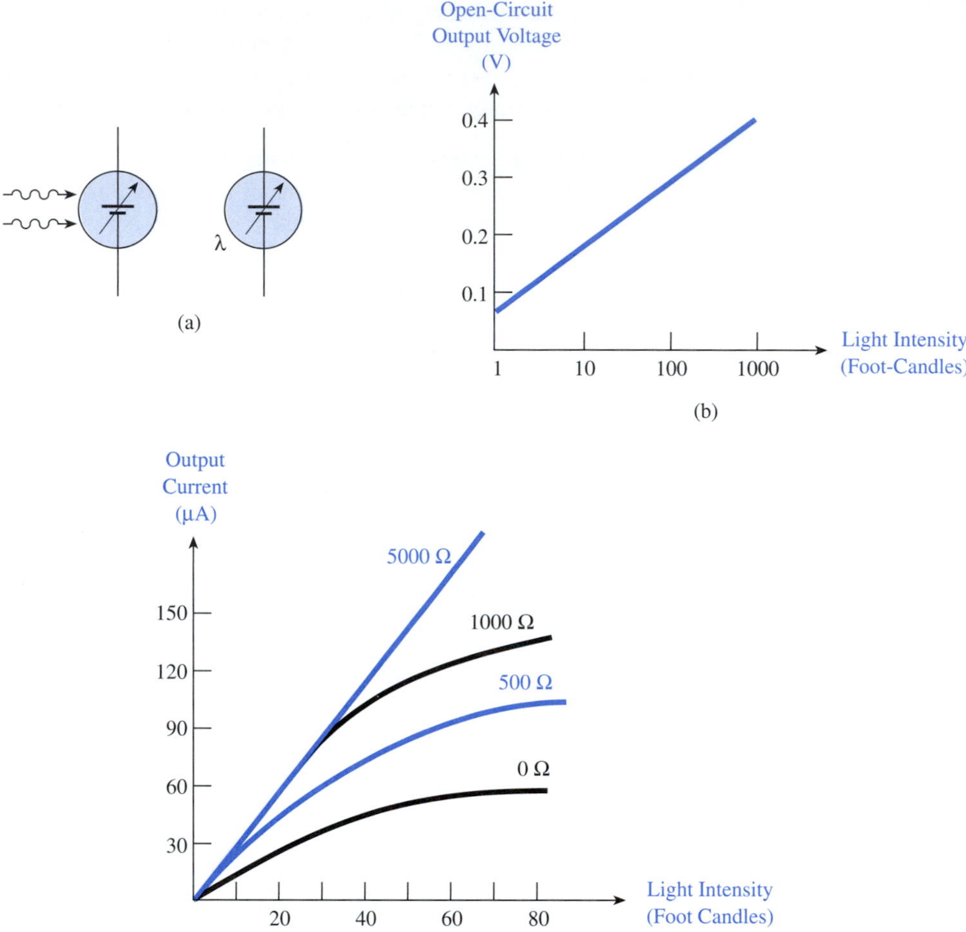

arrival; it could be a workpiece or machine member which is supposed to get out of the way before the logic circuitry will allow some other motion to occur.

If the photovoltaic cell has trouble picking up the relay directly, it can operate through a transistor amplifier, as shown in Fig. 10–14(b). It is a good idea to do this anyway, since photovoltaic cells are subject to *fatigue* when they deliver near their full current for any length of time. The output voltage and current decline when a photovoltaic cell suffers from fatigue.

Sometimes the light source, focusing device, photocell, amplifier, and relay are all included in the same package, as shown in Fig. 10–14(c). Light leaves the package, passes some distance through space, is reflected by a reflecting surface, and reenters through the same aperture. It is then reflected off the one-way mirror and strikes the photocell. The amplifier, relay, and contacts are all contained in the package, so the final output is the switching of the relay contacts to indicate whether or not an object has blocked the light path.

Often the problem arises that the signal light cannot be distinguished from ambient light. The photoelectric system may then be unreliable, because the photovoltaic cell can deliver an output due just to the ambient light. The system then would indicate that there was no object present blocking the path when in fact there *was* an object present. There is a solution to this problem. Instead of just passing the light directly out through the fo-

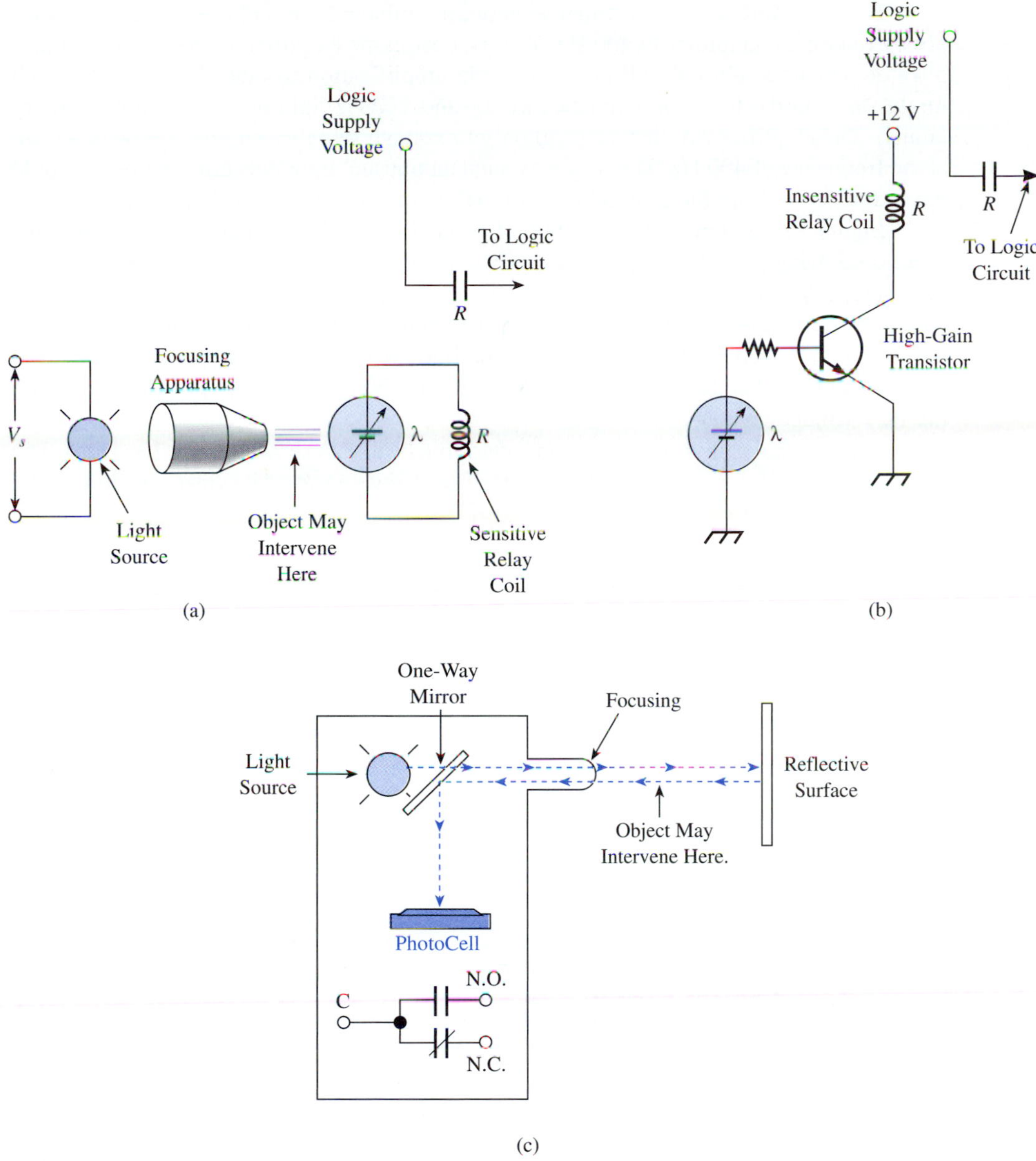

FIGURE 10–14
(a) Photovoltaic cell energizing a relay directly. (b) Photovoltaic cell energizing a relay through a transistor switch. (c) All photocell components contained in a single enclosure.

cusing apparatus, the light beam is "chopped." That is, the beam is periodically interrupted at some specific frequency by a moving object inside the package between the light source and the outlet.

One way to do this is to install a rotating disk between the light source and the one-way mirror in Fig. 10–14(c). Part of the disk is translucent and part of it is opaque, so the light beam is alternately passed and blocked at some constant frequency, usually several hundred hertz. Let us assume for purposes of illustration that the light beam is chopped at a frequency of 400 Hz.

It is now quite easy to distinguish between ambient light and true signal light simply by tuning the amplifier to 400 Hz. That is, design the amplifier so that it will not amplify dc signals at all and will give very little amplification to other frequencies which might sneak in the light aperture (such as 60- and 120-Hz light pulsations from mercury lamps). The amplifier will then respond only to voltage signals from the photovoltaic cell at the frequency of 400 Hz. The only way such an unusual light pulsation frequency could get to the cell is from the true light signal. All extraneous light signals are ignored.

Figure 10–15 shows an application of photovoltaic cells to measure the translucence of a liquid being passed through a sample cell. Suppose that the translucence is known to be a sure indication of the concentration of some impurity in the liquid. The semitransparent mirror passes half of the light from the source to the liquid, and the other half of the light is reflected to PC1. Only part of the light sent to the liquid can pass through it and strike PC2. Therefore the voltages generated by PC1 and PC2 will be different, with the PC1 voltage greater.

Photovoltaic cells 1 and 2 are connected in a bridge as shown in Fig. 10–15(b). The bridge is manually or automatically balanced by adjusting R_2. The final position of the

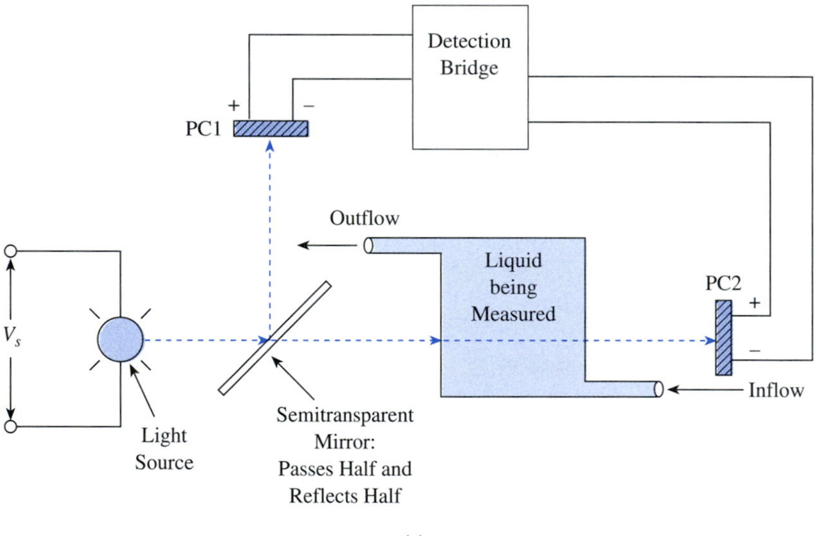

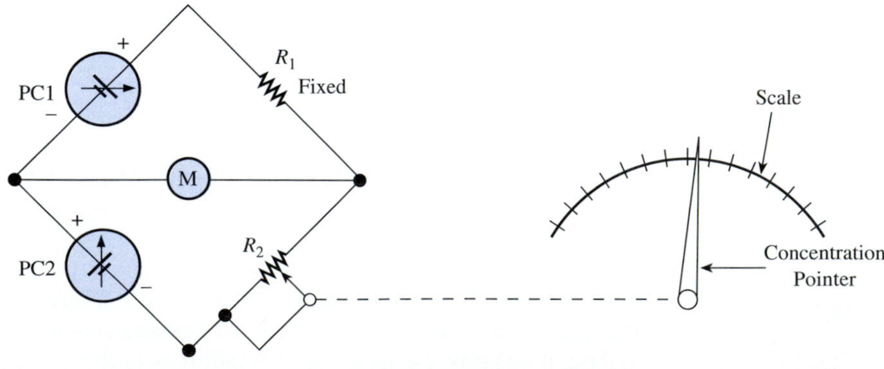

FIGURE 10–15
Photocell bridge circuit for measuring a liquid's translucence. (a) Arrangement of the light source, the photocells, the semitransparent mirror, etc. (b) Electrical schematic of the bridge circuit.

R_2 wiper will depend on the voltage difference between PC1 and PC2, which in turn depends on the concentration of the impurity. Thus, once the bridge has been balanced, every value of R_2 corresponds to some certain value of impurity concentration. The R_2 shaft is mechanically linked to a pointer shaft, which has a scale of concentrations marked beneath it for direct readout.

This measurement arrangement has some stabilizing features that deserve comment. First, both photocells are excited by the same light source. This eliminates the possibility of an error due to one light source changing in intensity more than another. In Fig. 10–15, if the light source changes in intensity due to aging of the bulb or supply voltage variations, both photocells are affected equally. These equal changes are canceled out by the action of the bridge.

Second, photovoltaic cells are somewhat temperature sensitive. That is, their output voltage depends slightly on their temperature. However, if PC1 and PC2 are physically close together, they experience the same temperature changes, so any temperature errors are also canceled out by the bridge.

10-7-2 Optical Position-Encoding

The idea of a rotating disk for alternately passing and blocking a light beam to a photovoltaic cell was suggested for Fig. 10–14(c). This same idea can also be used to measure the amount that a shaft has turned. Usually, the gear linkage between the measured shaft and the rotating disk is designed to produce many rotations of the disk for one turn of the measured shaft, as shown in Fig. 10–16.

Figure 10–16(a) shows a large-diameter gear on the measured shaft linked to a small-diameter gear on the disk shaft. The motion ratio is equal to the gear ratio, inversely. Thus, if the large gear is 10 times as large as the small gear, the disk shaft will turn 10 times as far as the measured shaft. One complete rotation of the measured shaft produces 10 rotations of the disk.

The optical disk has many slit openings. A stationary light source is mounted on one side of the disk, with a photovoltaic cell mounted directly across on the opposite side, as Fig. 10–16(a) makes clear. As the disk turns, it alternately passes light to and blocks light from the photocell. Every slit that moves between the optical devices produces one voltage pulse from the photocell, as indicated in Fig. 10–16(b). For the 24-slit disk shown in that figure, one voltage cycle is produced for each 15° of rotation, since

$$\frac{360^\circ}{24} = 15^\circ$$

The V_{PC} waveform is processed in a signal-conditioning circuit to make it TTL-compatible, the resulting pulses are then passed to a binary counter, as shown in Fig. 10–17. The system control logic will reset the counter to zero before the measured shaft begins to move. When the measured shaft begins moving away from its prior position, the content of the binary counter represents the distance that it has moved. For the construction shown in Fig. 10–16, the counter will increment one bit for each 1.5° movement of the measured shaft. This is so because

$$\frac{1\text{ bit}}{1\ \cancel{\text{pulse of } V_{PC}}} \times \frac{1\ \cancel{\text{pulse of } V_{PC}}}{15\cancel{^\circ \text{ of disk}}} \times \frac{10\cancel{^\circ \text{ of disk}}}{1^\circ \text{ of measured shaft}} = \frac{1\text{ bit}}{1.5^\circ \text{ of measured shaft}}$$

10:1 gear ratio

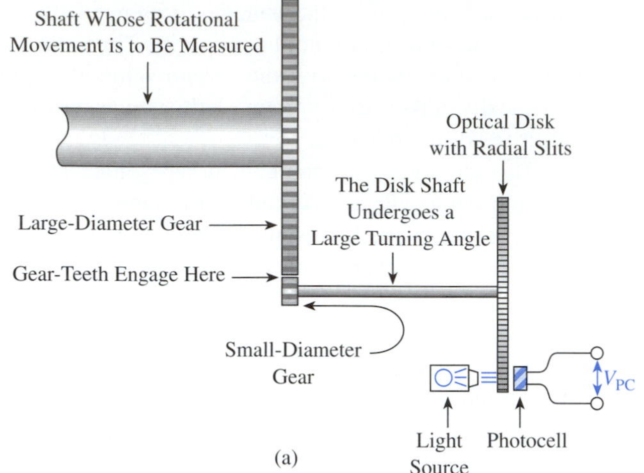

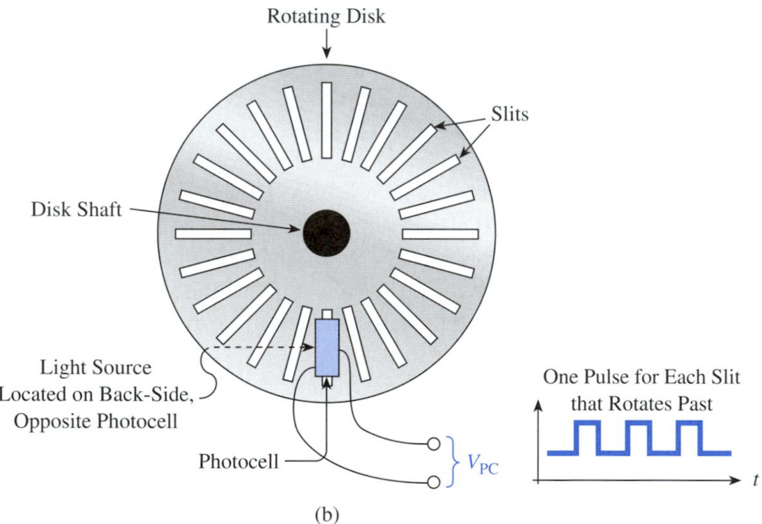

FIGURE 10–16
The layout of an optical position-encoder. (a) Side view. (b) Face view of optical rotating disk. In this example the slits are 15° apart.

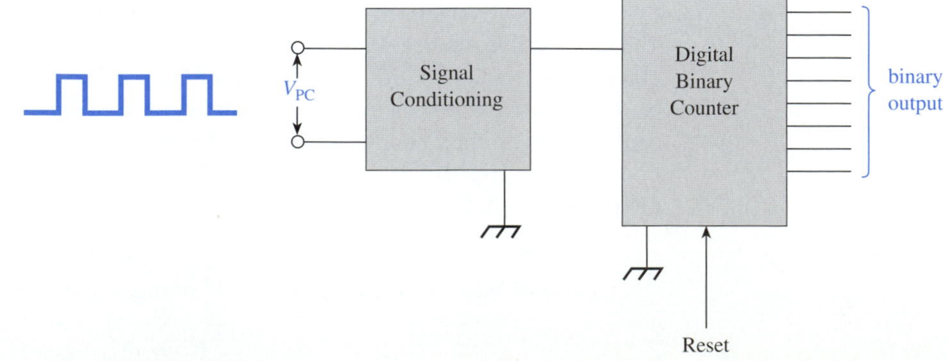

FIGURE 10–17
Counting the photocell pulses to represent the amount of shaft movement. The counter is reset to zero each time before a new movement begins; therefore the counter's binary output represents amount of motion relative to the prior position. It does not represent absolute position relative to a "home" position.

■ EXAMPLE 10-1

Refer to the optical position-encoder of Fig. 10–16.

(a) If the measured shaft moves exactly one-half turn (180°) from its prior position, what is the content of the binary counter?

(b) How far can the measured shaft turn without exceeding the capacity of the 8-bit counter shown in Fig. 10–17?

Solution. (a) As the measured shaft moves 180°, V_{PC} produces a number of pulses given by

$$180° \times \frac{1 \text{ pulse}}{1.5°} = 120 \text{ pulses}$$

Therefore the counter will contain the binary equivalent of decimal 120, or **0111 1000.**

(b) The capacity of the counter is binary 1111 1111, or $255_{(10)}$. The maximum shaft rotation is

$$255 \text{ bits} \times \frac{1.5°}{\text{bit}} = \mathbf{382.5°}$$

or 1 complete turn plus 22.5°.

■ EXAMPLE 10-2

A real optical position-encoder used on a robot axis might have a 10:1 gear ratio, an optical disk with 72 slits, and a 12-bit binary counter. For such a position-encoder,

(a) What is the value of 1 bit? That is, how much movement of the measured shaft is required to cause the counter to increment by 1?

(b) What is the maximum allowable shaft motion to ensure that the counter never over-ranges?

(c) What amount of shaft movement is represented by a binary output of 0101 1011 0010?

Solution. (a) The slits are 5° apart, since

$$\frac{360°}{72 \text{ slits}} = 5° \text{ per slit}$$

Therefore one count will occur for each 5° rotation of the disk. With a 10:1 gear ratio, the equivalent motion of the measured shaft is

$$\frac{5° \text{ disk rotation}}{10} = \mathbf{0.5}° \text{ movement of measured shaft}$$

This is the smallest movement that can be detected by the optical transducer. We say that the transducer has a *resolution* of 0.5°.

(b) Binary twelve 1s (1111 1111 1111) converts to decimal 4095 $[2^{12} - 1 = 4095]$. Therefore the transducer can tolerate 4095 increments of 0.5° before the counter reaches its capacity and is on the verge of over-ranging. This gives an overall turning angle of

$$4095 \times 0.5° = 2047.5°$$

$$2047.5° \times \frac{1 \text{ rotation}}{360°} = 5.6875 \quad \text{rotations of the measured shaft}$$

or five complete rotations plus 247.5°.

(c) Binary 0101 1011 0010 is decimal 1458. Therefore

$$1458 \times 0.5° = \mathbf{729°}$$

or two complete rotations plus 9°. ■

Turning in both directions. The photoelectric position-encoder presented in Figs. 10–16 and 10–17 cannot distinguish between clockwise or counterclockwise turning of the measured shaft. In many position-measurement applications it is necessary to make this distinction. It can be done by placing a second light source/photocell combination across the rotating disk; this second photodetector is not aligned with the slits in the same way as the first. Instead, it is offset by one quarter of the slit distance, as illustrated in Fig. 10–18(a). Therefore the V_A pulses and the V_B pulses are phase-displaced by one-quarter cycle; this is sketched in Fig. 10–18(b) and (c).

In Fig. 10–18(b), photocell A is light at this moment. Photocell B is dark now but will become light after an additional rotation of 11.25°, which is one quarter of the distance between slits. Thus, the voltage pulse from photocell A leads the voltage pulse from photocell B by one-quarter cycle.

In Fig. 10–18(c), the disk is rotating counterclockwise. Photocell B is light at this moment, with photocell A about to become light after another 11.25° of rotation. The V_A pulses lag the V_B pulses by one-quarter cycle.

A *phase-detector* circuit is used to sense the leading versus lagging phase relationship of V_A and V_B. In Fig. 10–19, the phase-detector is designed to produce a logic 0 output if V_A is leading V_B; it produces a logic 1 output if V_A is lagging V_B. The direction logic level output, 0 or 1, is combined with the binary magnitude output from Fig. 10–17. In this way the processing circuitry that interprets the transducer's output can tell both magnitude and direction of the movement of the measured shaft.

When this combination is made, it is common for the binary magnitude counter to be reduced by 1 bit to make room for the direction or "sign" bit. Thus the 8-bit digital counter of Fig. 10–17 would be reduced to 7 bits with a maximum magnitude of decimal 127 (27 − 1 = 127). The direction or sign bit is then placed on the far left of the binary number, where the eighth bit used to be. The combination is then interpreted by the digital processing circuitry as varying from

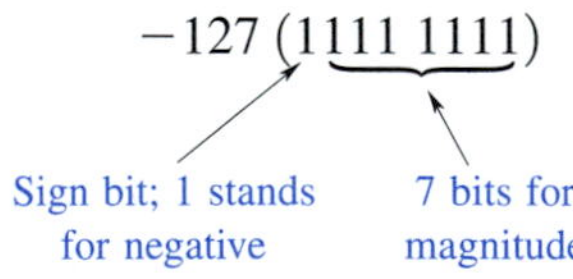

to

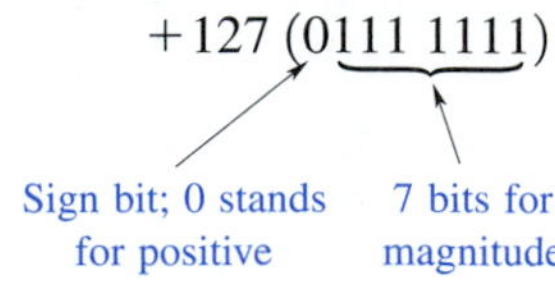

A 12-bit-resolution transducer, like Example 10-2, would then have an indicating range from −2047 to +2047

FIGURE 10–18
The direction of movement is detected by using two photocells, A and B. (a) The photocell displacement is only three quarters of the slit displacement. For clarity, we are showing a 45° slit displacement here. The photocell displacement would then be 33.75°. (b) With the optical disk rotating clockwise (measured shaft is turning counterclockwise), V_A leads V_B by one-quarter cycle. (c) With the optical disk rotating counterclockwise (measured shaft is turning clockwise), V_A *lags* V_B by one-quarter cycle.

Photocell B
45°
33.75°
V_B
11.25°
Photocell A
(a)
V_A

CW Rotation
V_B
11°
V_A
(b)

11°
CCW Rotation
V_B
V_A
(c)

V_A
Cycle Period
T
t
V_B
$\frac{1}{4}T$
$\frac{1}{4}T$
t

V_A
$\frac{1}{4}T$
$\frac{1}{4}T$
t
T
V_B
t

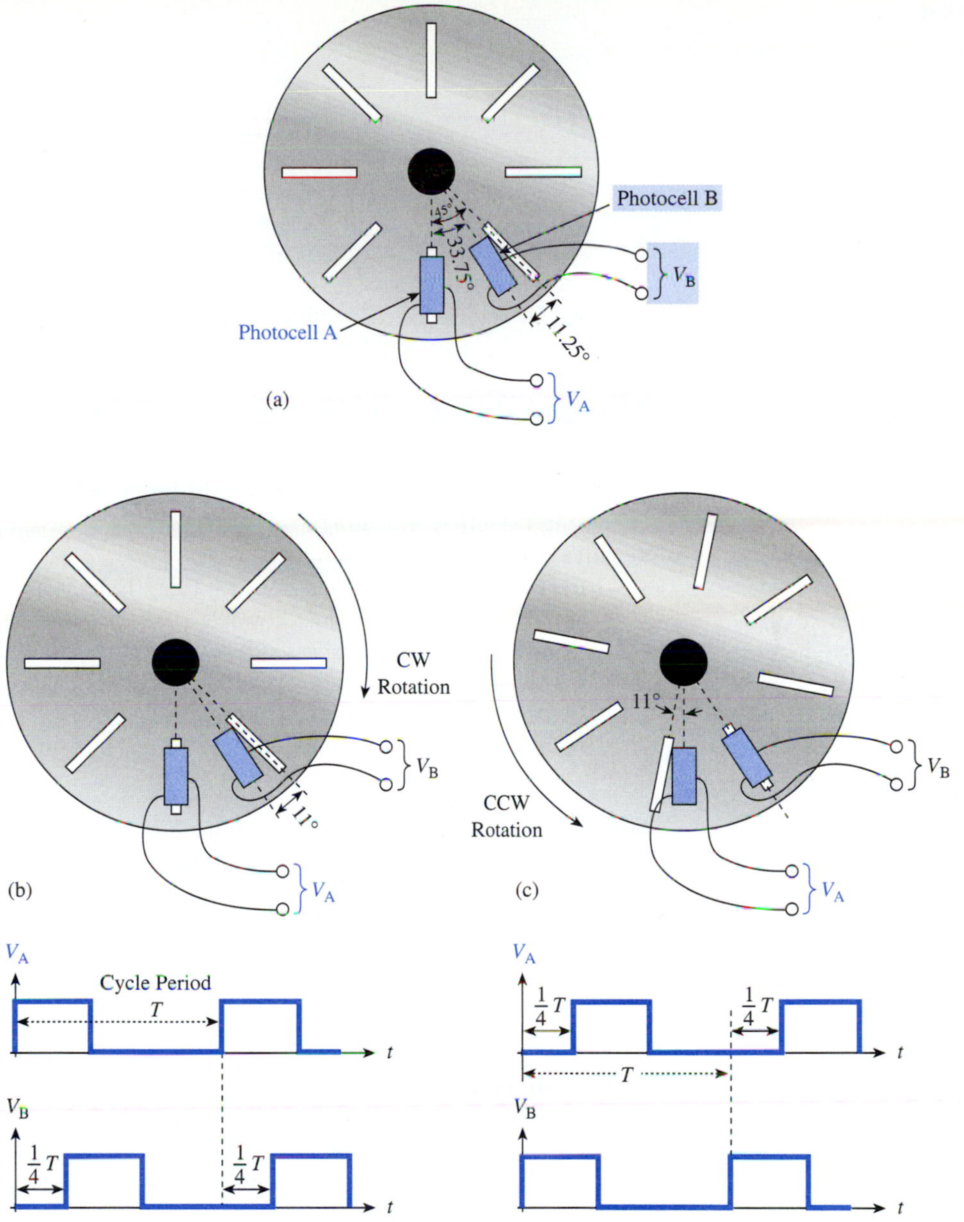

FIGURE 10–19
A phase-detector can tell whether V_A leads or lags V_B. By this means it can determine the direction of movement.

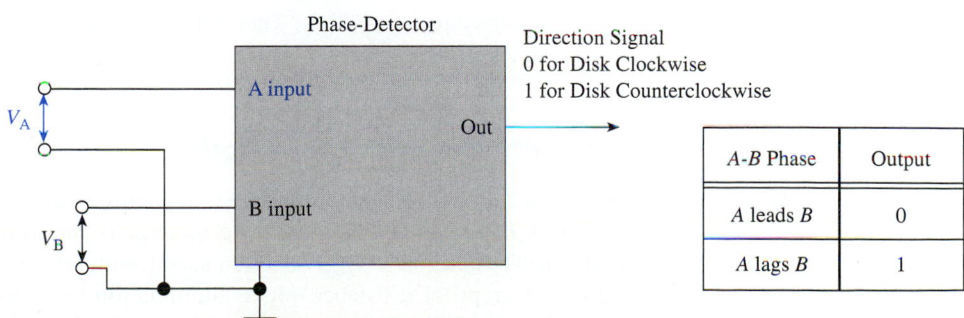

A-B Phase	Output
A leads *B*	0
A lags *B*	1

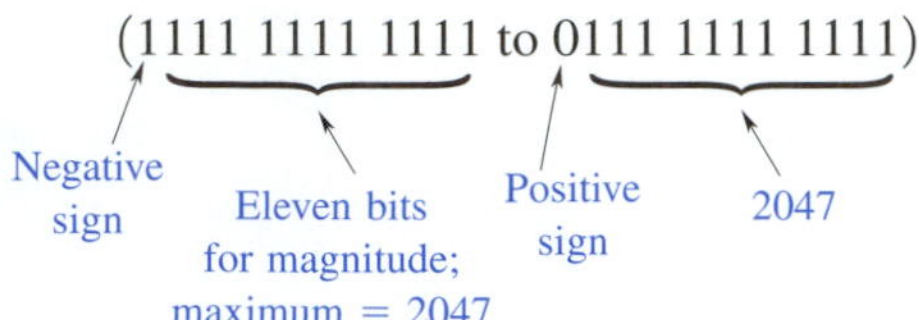

with one bit dedicated to direction and eleven bits dedicated to magnitude of movement.

■ EXAMPLE 10-3

The optical position-encoder of Example 10-2 (10:1 gear ratio, reversing, and 72 slits) has direction-indicating ability. Its output is 12-bit signed-magnitude binary with the twelfth bit on the far left representing sign—0 for positive (disk rotating clockwise), and 1 for negative (disk rotating counterclockwise).

(a) Is the transducer resolution still 0.5°? Explain.

(b) What is the maximum allowable shaft movement from the prior position to ensure that the counter never over-ranges?

(c) What is the magnitude and direction of measured shaft movement for a binary output of 0011 0111 1001?

(d) What is the magnitude and direction of measured shaft movement for a binary output of 1010 1011 1110?

Solution. (a) Yes, the transducer still increments the counter 1 bit for each 0.5° of shaft movement, since the gear ratio and the number of slits on the optical disk are the same.

(b) Maximum binary magnitude = 111 1111 1111 (eleven 1s). This is equal to decimal $2^{11} - 1 = 2047$, which represents a turning angle of

$$2047 \times \frac{0.5°}{1 \text{ count}} = 1023.5°$$

or **two complete rotations plus 303.5°.**

(c) The sign bit is 0, which means a positive number. Our description has defined positive as clockwise rotation of the optical disk, which is produced by **counterclockwise movement of the measured shaft** (mechanical reversal by gears in Fig. 10–16).

The rightmost 11 bits, 011 0111 1001, equate to $889_{(10)}$. Therefore $889 \times 0.5° = 444.5°$ or **one complete rotation plus 84.5° counterclockwise.**

(d) The sign bit is 1, so the measured shaft has moved clockwise. The rightmost 11 bits equate to $702_{(10)}$. Therefore $702 \times 0.5° =$ **351°, clockwise by the measured shaft.** ■

10-7-3 Photoconductive Cells

Photoconductive cells change resistance in response to a change in light intensity (the formal term is *illumination*), as mentioned earlier. As the illumination goes up, resistance goes down. The schematic symbols often used for photoconductive cells are shown in Fig. 10–20(a). A graph of resistance versus illumination for a typical photoconductive cell is given in Fig. 10–20(b). Note that both scales are logarithmic to cover the wide ranges of resistance and illumination that are possible.

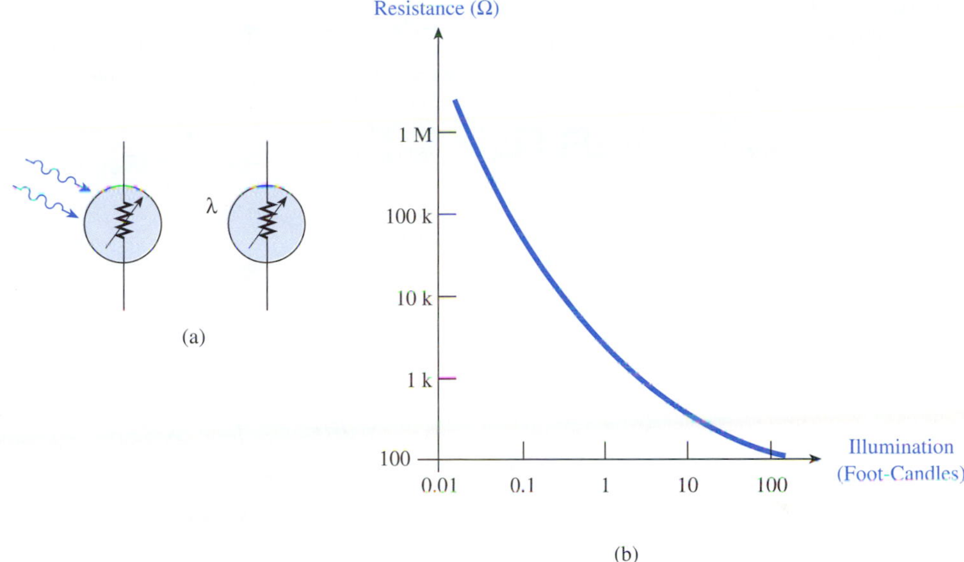

FIGURE 10–20
(a) Schematic symbols for a photoconductive cell. (b) Curve of resistance versus illumination for a typical photoconductive cell.

The chief virtue of the modern photoconductive cell is its sensitivity. As Fig. 10–20(b) illustrates, the cell's resistance can change from over 1 million Ω to less than 1000 Ω as the light intensity changes from darkness (illumination less than 0.01 fc) to average room brightness (10 to 100 fc).

Photoconductive cells can be used for many of the same purposes that photovoltaic cells are used, except, of course, that they cannot act as energy sources. Photoconductive cells are preferred to photovoltaic cells when very sensitive response to changing light conditions is needed.

When fast response is necessary, photovoltaic cells are preferable to photoconductive cells. Likewise, if a photocell is to be rapidly switched on and off, as suggested in Sec. 10-7-1, photovoltaic cells are preferred because they can be switched at higher frequencies than photoconductive cells. As a rough rule of thumb, photoconductive cells cannot be successfully switched at frequencies higher than about 1 kHz, whereas photovoltaic cells can be switched successfully at frequencies up to about 100 kHz and sometimes higher.

Photochoppers. One interesting application of photoconductive cells is in chopping a dc voltage signal for insertion into an ac amplifier. Dc signal chopping will be utilized in Sec. 11-11-2 in conjunction with servo amplifiers. The photoconductive cell is a good alternative to the vibrating mechanical switch chopping method used in that amplifier. This is illustrated in Fig. 10–21.

In Fig. 10–21(a), the square-wave driving voltage is applied to two neon bulb-rectifier diode combinations. When V_{drive} is positive, rectifier diode A is forward biased and rectifier diode B is reverse biased. Therefore neon bulb A is turned on, and neon bulb B is turned off. A neon bulb is capable of quickly turning on and off as voltage is applied and removed. A normal incandescent bulb cannot turn on and off quickly, since it depends on the heating of its filament to emit light.

When V_{drive} goes negative, rectifier diode B is forward biased, and rectifier diode A is reverse biased. Therefore neon bulb B turns on, and neon bulb A turns off. Resistor R is inserted to limit the current through the neon bulbs.

FIGURE 10–21
Photoconductive cells used to chop a dc signal. (a) Circuitry for alternately switching the photocells from light to dark. (b) The photocells wired into the amplifier input. (c) Equivalent circuit when PCA is light and PCB is dark. (d) Equivalent circuit when PCB is light and PCA is dark.

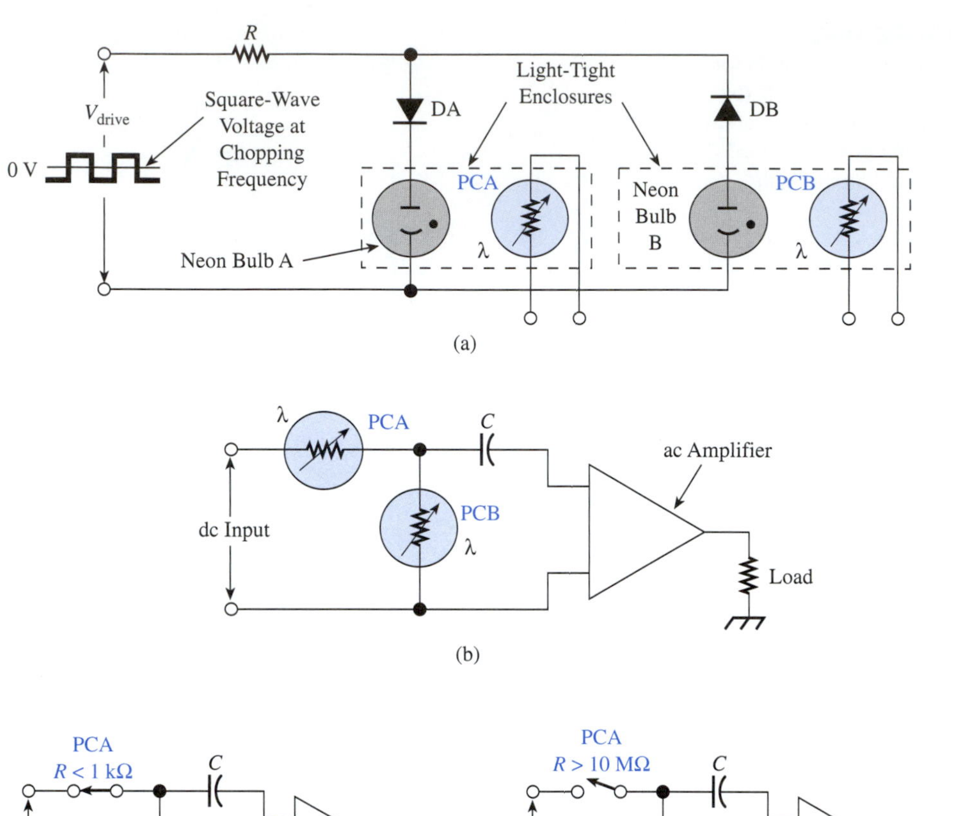

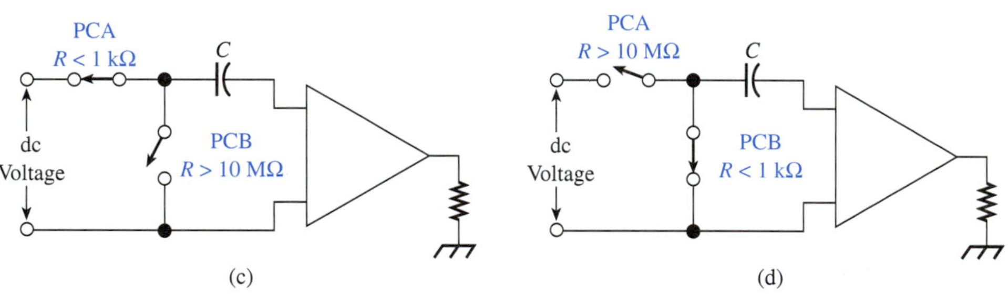

Photoconductive cells PCA and PCB are exposed to neon bulbs A and B, respectively, inside light-tight enclosers. In such an enclosure, no external light can get in to affect the photocells. The photocells are specially chosen to have a large resistance change between light and dark conditions. In this case, let us assume the resistance changes from about 10 MΩ to below 1 kΩ. The ratio of resistances is therefore about 10 000 to 1 (10 MΩ/1 kΩ = 10 000). The cells are said to have a *light-to-dark ratio* of 10 000.

Photoconductive cell A is placed in series with the amplifier input lead, and photoconductive cell B is placed in shunt across the amplifier, as shown in Fig. 10–21(b). Thus when neon bulb A is turned on, the amplifier sees a low resistance in series with its input and very high resistance in shunt. These low and high resistances can be thought of as closed and open switches, as shown in Fig. 10–21(c). Therefore at the instant shown in Fig. 10–21(c), the dc input voltage gets through to the coupling capacitor practically unattenuated (if the amplifier input impedance is much higher than 1 kΩ).

When V_{drive} goes negative, neon bulb B turns on, and the situation at the amplifier is as shown in Fig. 10–21(d). The amplifier sees an open switch in series and a closed switch in shunt. The dc input signal cannot get through to the coupling capacitor at this instant. The dc input voltage is thus being chopped just as shown in Sec. 11-11-2. This

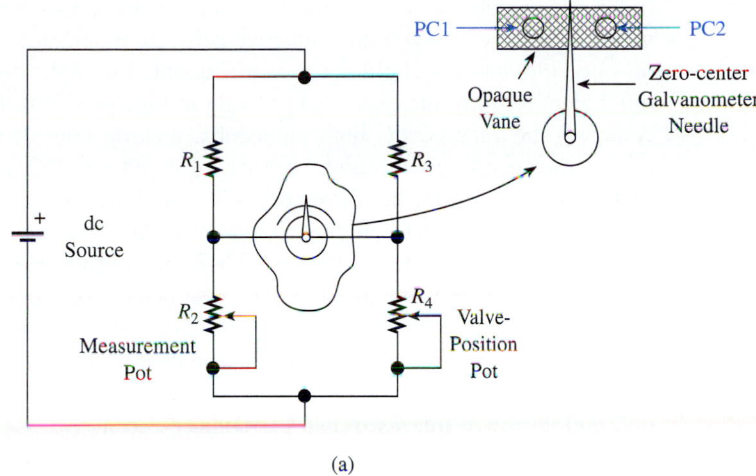

(a)

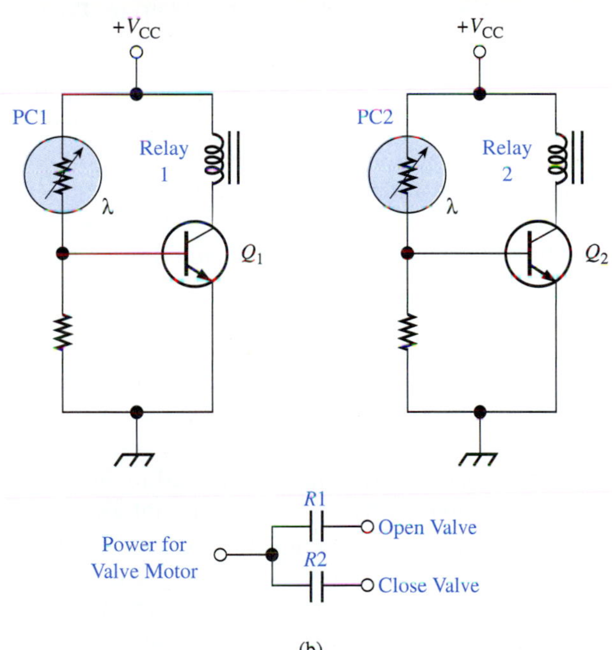

(b)

FIGURE 10–22 Photoconductive cells used to automatically balance a bridge. (a) Bridge circuit, with an opaque vane attached to the galvanometer needle. (b) Photocell circuitry. If both photocells are dark, neither relay picks and the valve motor does not move. If either photocell goes light, the proper relay will pick up to apply power to the valve motor and bring the bridge back into balance.

chopping method has the advantage of electronic reliability (no moving parts), and it would be less expensive than a vibrating mechanical switch.

Photocells for automatic bridge balancing. Figure 10–22 shows another popular use for photoconductive cells. The bridge circuit in Fig. 10–22(a) balances the measurement potentiometer against the valve-position potentiometer to bring about proportional control. The galvanometer and photocell arrangement presented in this figure is a cheap and reliable method for accomplishing automatic balancing of the bridge. Here is how it works.

The galvanometer is a zero-center meter. That is, if there is no current flowing through it, the needle returns to the center of the scale. If current flows through it from left to right, the needle moves to the right of center; if current flows through it from right to left, the needle moves to the left of center. Attached to the needle is a lightweight vane which is opaque. Two photoconductive cells are mounted a slight distance from the vane on one side, and two light sources are mounted a slight distance from the vane on the other side. Refer to the expanded drawing in Fig. 10–22(a). If the galvanometer needle is centered, the vane covers both photocells, making both of their resistances high. If the needle moves off center, either photocell 1 or photocell 2 will be uncovered, depending on the direction of needle movement. When a photocell is uncovered, its resistance drops drastically because of the light striking its surface. The lowered resistance turns ON one of the transistor switches in Fig. 10–22(b), picking up one of the relays. The relay contacts then drive the valve motor either open or closed, moving the valve-position pot until the bridge is back in balance. When the bridge is rebalanced, the galvanometer current drops to zero and the needle returns to the center of the scale. Both photocells again go dark, turning OFF whichever transistor was turned ON. The relay which was picked drops out, and the valve freezes in that particular position.

10-7-4 Optical Coupling and Isolation: Phototransistors, Light-Emitting Diodes

Figure 10–23 shows two ways of building an *optical isolator*. Figure 10–23(a) shows a standard incandescent light source and a photoconductive cell to accomplish the isolation, and Fig. 10–23(b) shows a *light-emitting diode* (LED) and *phototransistor* to accomplish the isolation. We will look at some industrial uses of optical isolators and then explain the operation of both of these designs. The design in Fig. 10–23(b) has certain advantages over that in Fig. 10–23(a), and these advantages will be pointed out.

An optical isolator is basically an interface between two circuits which operate at (usually) different voltage levels. The most common industrial use of the optical isolator is as a signal converter between high-voltage pilot devices (limit switches, etc.) and low-voltage solid-state logic circuits. Optical isolators can be used in any situation where a signal must be passed between two circuits which are electrically isolated from each other. Recall from Chapters 1 and 2 that complete electrical isolation between circuits (meaning that the circuits have no conductors in common) is often necessary to prevent noise generated in one circuit from being passed to the other circuit. This is especially necessary for the coupling between high-voltage information-gathering circuits and low-voltage digital logic circuits. The information circuits are almost always badly exposed to noise sources, and the logic circuits cannot tolerate noise signals.

The optical coupling method eliminates the need for a relay-controlled contact or an isolating transformer, which are the traditional methods of providing electrical isolation between circuits. Refer to Sec. 1-7 and Fig. 1–12 for a review of these methods.

Incandescent bulb-photoconductive cell optical isolator/coupler. The optical isolator in Fig. 10–23(a) has an incandescent bulb wired in series with a protective resistor. This series combination is connected through a pilot device to a 115-V signal. If the pilot device is open, there will be no power applied to the incandescent bulb, so it will be extinguished. The photoconductive cell, being shielded from outside light, will go to a very high resistance, allowing the voltage at the transistor base to rise. The transistor switch turns ON, pulling V_{out} down to ground voltage, a logical 0.

If the pilot device closes, power will be applied to the bulb, causing it to glow. The resistance of the photoconductive cell will decrease, pulling the base voltage down below 0.6 V. The transistor turns OFF and allows the collector to rise to $+V_{CC}$, a logical 1. Therefore if a 115-V input signal is present, the circuit yields a logical HI. If there is no 115-V signal present, the circuit yields a logical LO.

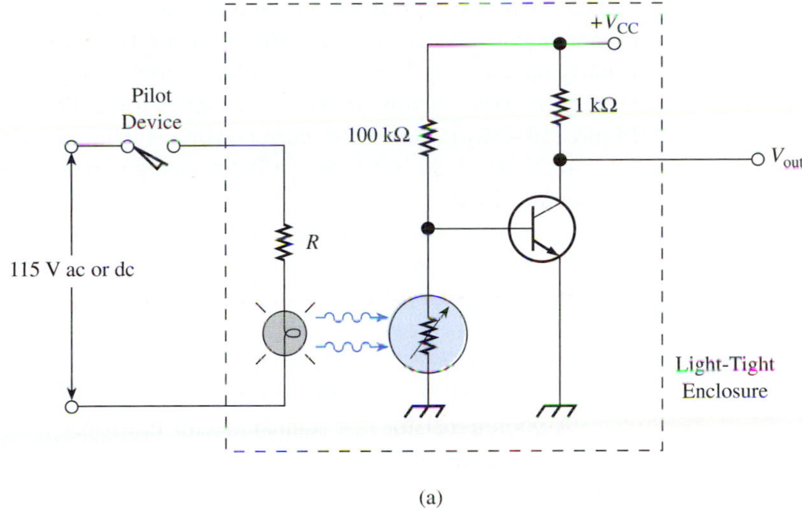

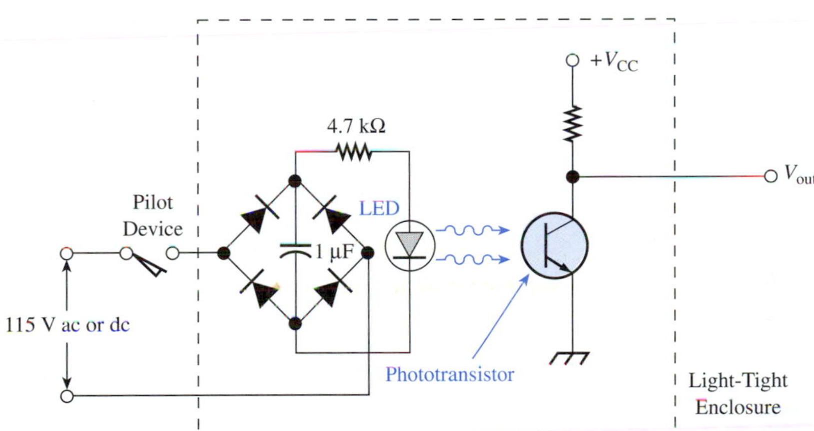

FIGURE 10–23
(a) Optical isolator using an incandescent bulb and a photoconductive cell. (b) Another optical isolator using an LED and a phototransistor.

The optical coupling method is superior in many applications, because it gets rid of some of the less desirable features of relays and reformers. Relays and transformers have certain shortcomings as couplers and isolators, namely,

1. They are fairly expensive.
2. They are bulkier and heavier than optical devices.
3. They create magnetic fields and switching transients which may be a source of troublesome electrical noise.
4. Relay contacts may create sparks, which are very undersirable in certain industrial situations.

The optical coupler works equally well on either ac or dc high-voltage signals. For this reason, signal converters using optical coupling are sometimes referred to as *universal* signal converters.

LED-phototransistor optical isolator/coupler. Figure 10–23(b) shows an optical isolator/coupler using an LED and phototransistor instead of an incandescent bulb and photoconductive cell. An LED is a semiconductor diode which emits light when it carries current in the forward direction. The forward breakdown voltage of an LED is higher than 0.6 V, since LEDs are not made from silicon as rectifier diodes are. They usually have forward breakdown voltages in the range of 1.0 to 2.2 V. Also, LEDs have reverse breakdown voltages which are much lower than those of silicon rectifier diodes. Figure 10–24(a) shows the current-voltage characteristic of a typical LED. Figure 10–24(b) shows the relationship between light power output and forward current for a particular LED.

A visible LED is not very bright compared to a No. 44, 6-V bulb, for example. Some LEDs do not even emit a visible light but emit infrared light invisible to the human eye. Of course, such LEDs must then be used with photodetectors which are sensitive to infrared radiation. In commercially built optical couplers this is standard practice, since no human has to see the light anyway. Also, infrared LEDs are more efficient than visible LEDs, converting more of their electrical energy into light* and less into heat.

A phototransistor is a semiconductor transistor which responds to the intensity of light on its lens instead of to base current. The phototransistors can respond to both incident light and base current. The phototransistor in Fig. 10–23(b) doesn't have a base lead, so it responds to light only. The wavy arrows pointing toward the base location symbolize that the transistor is a phototransistor.

Figure 10–25 shows the characteristic curves of a typical phototransistor. Notice that the family of curves represents different values of light power density (the formal term is *irradiance*), not different values of base current. Phototransistors do not have a response as linear as that of a junction transistor. Notice the inconsistent spacing of the curves, indicating a nonlinear relationship between collector current and light intensity.

The LED-phototransistor combination in Fig. 10–23(b) has some important advantages over the bulb-photoconductive cell combination in Fig. 10–23(a):

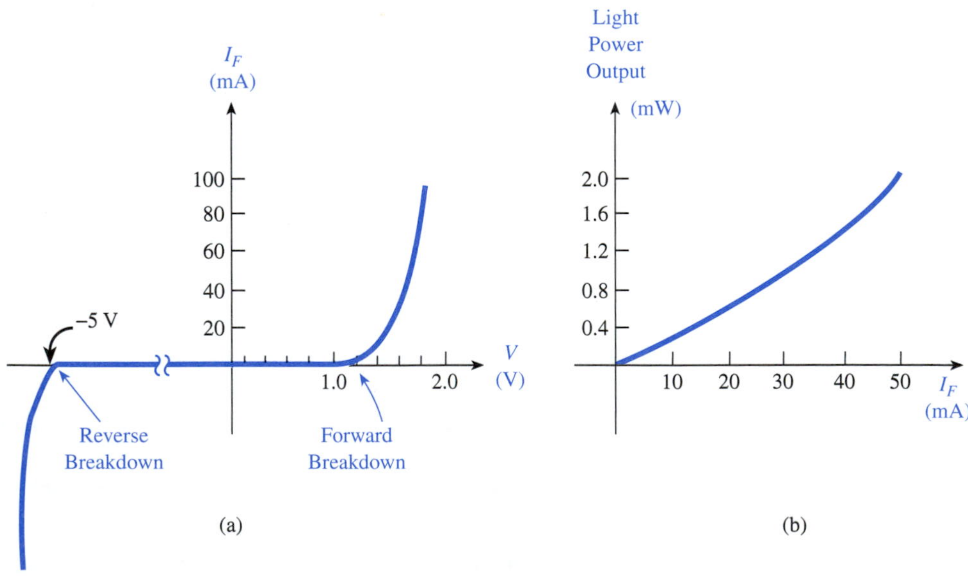

FIGURE 10–24
(a) Current versus voltage characteristic curve of a typical LED. (b) Curve of light output versus forward current for a typical LED.

*Strictly speaking, the term *light* refers to electromagnetic radiation which is visible to the human eye. Infrared radiation, which is at too low a frequency to be visible, is not really light. However, popular use has blurred this distinction, and we hear the phrases "infrared light" and "ultraviolet light." We adopt the less rigorous usage of the term *light,* and we will refer to infrared radiation as light.

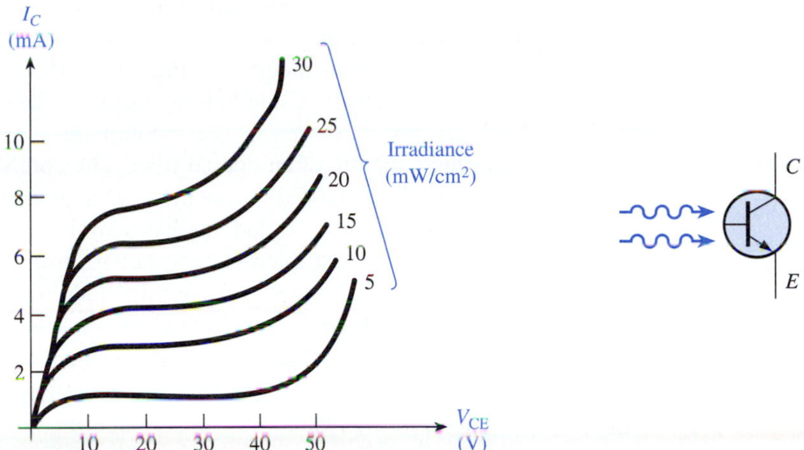

FIGURE 10–25 Characteristic curves of a typical phototransistor.

1. An LED has an extremely long life compared to a bulb of any kind. An LED will emit light forever if operated at the proper current; an incandescent bulb does well if it lasts 10 000 h.
2. An LED can withstand the mechanical vibrations and shocks in industrial environments much better than a filament lamp, thus providing greater reliability.
3. The LED and phototransistor have faster response speed than a lamp and photoconductive cell. This may be an advantage for certain high-frequency switching applications.

Of course, there is no reason an optical coupler/isolator could not combine an LED with a standard photoconductive cell, and this is sometimes done. Usually, though, an LED light source is combined with a phototransistor light detector because of the better match between their operating speeds and between their light emission and detection wavelengths.

10-8 ■ OPTICAL FIBERS

In an industrial environment, when an electrical signal is transmitted by wire, it is vulnerable to interference from a variety of sources, as we know. To protect the integrity of our signals, we have devised various techniques for coping with electrical interference, or noise. For example, we may surround the signal wires with a braided shield in order to block out capacitively coupled noise; we sometimes twist two signal wires together (called a twisted pair) to eliminate magnetically induced noise; for digital signal transmission, we may set up a constant-current loop which is modulated by the signal; we may even route signal wires in a roundabout manner to avoid passing close to a known source of noise.

These techniques solve the problem, more or less, but what we would really appreciate is a signal-transmission method which is not even subject to electrical noise. Such a method is now available. *Optical fibers* are very thin strands of glass or plastic which carry light from the sending location to the receiving location. The crystalline structure of an optical fiber enables the input light to follow the fiber's path with only slight attenuation, even if the fiber bends and turns. Therefore a coated optical fiber can be used like a wire, but without the wire's susceptibility to electric or magnetic interference. The fiber is immune from noise pickup because the signal it is carrying is nonelectrical in nature—it's light.

The basic layout of a fiber-optic transmission system is diagramed in Fig. 10–26(a). As that figure indicates, an alignment fixture must be used at both the sending and receiving ends. Such a fixture is pictured in Fig. 10–26(b).

An optical fiber is able to guide light by virtue of the extremely pure chemical compositions of its *core* and its *cladding*. These components are identified in Fig. 10–27(a), which illustrates the structure of an optical fiber. The core's diameter may be as small as 3 μm (about one ten-thousandth of an inch) or as large as several hundred micrometers, depending on the fiber's type. The cladding has a similar range of dimensions, from a few micrometers to a few hundred micrometers, but in inverse correspondence to the core. That is, fibers with large cores tend to have thin claddings, and vice versa.

The compositions of the core and cladding are selected so that the core is denser* than the cladding. In general, if light is passing through a denser substance and then strikes the boundary of a less dense substance, it will reflect from the boundary almost totally; almost none of the light will enter into the less dense substance. Therefore, in an optical fiber, when light strikes the core/cladding boundary at a fairly shallow angle, it reflects from the boundary rather than entering the cladding. The reflected light travels across the core to strike the boundary on the other side, where it reflects again. In this way, if a light ray is not parallel to the fiber's axis, it tends to follow a zigzag path down the length of the core, as Fig. 10–27(b) illustrates. Of course, the nonparallel situation is bound to occur where the fiber bends, so it is this nearly total-reflection property of optical fibers that gives them their light-bending ability.

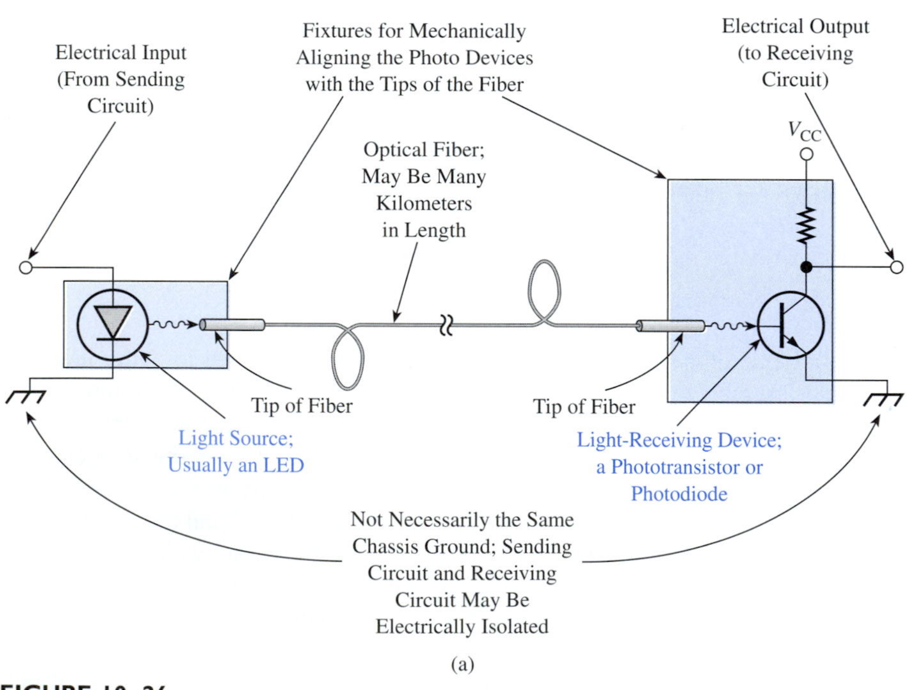

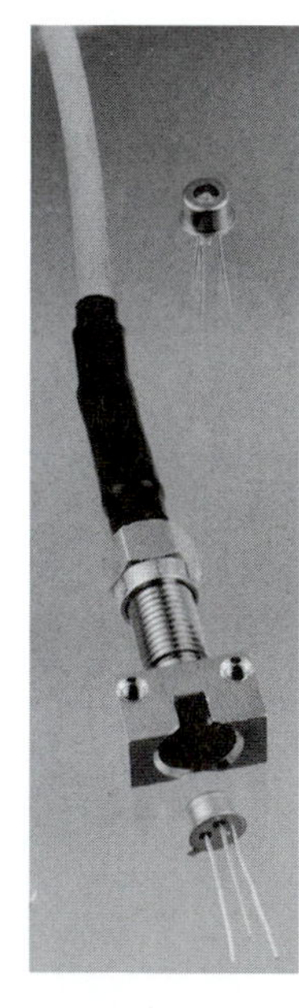

FIGURE 10–26
(a) Fiber-optic signal-transmission system. (b) Fiber-aligment fixture.
Courtesy of Motorola, Inc.

*In more rigorous optical terms, the core has a greater *index of refraction* than the cladding. For chemically similar substances though, index of refraction tends to correlate with density, so we can use the simpler descriptions "denser" and "less dense."

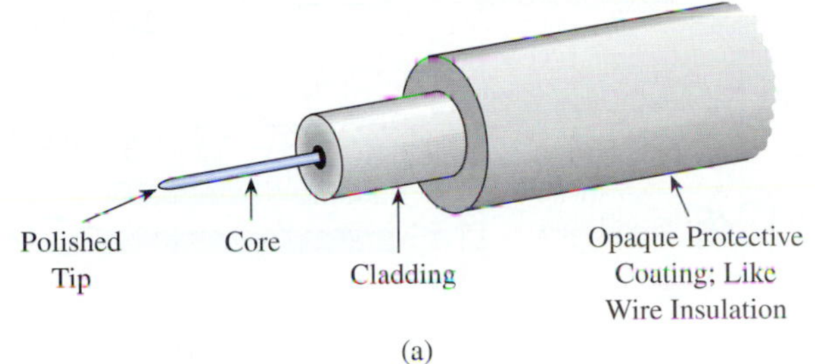

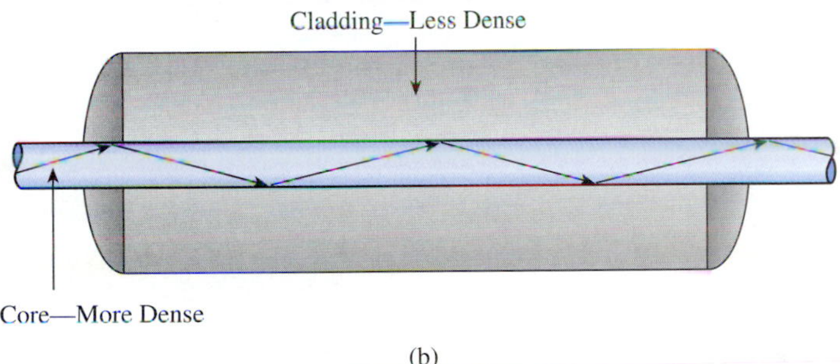

FIGURE 10–27 (a) Structure of an optical fiber. Core and cladding are special plastic or glass. (b) A light ray reflects back and forth down the length of the core.

In Fig. 10–27(a), a phototransistor has been shown as the light-receiving device. Phototransistors perform well at frequencies below a few hundred kilohertz. For higher-frequency applications the *photodiode* makes a better light receiver.

A photodiode is a silicon diode with an opening in its case containing a lens which focuses incident light on the *pn* junction. With the receiving circuit reverse-biasing the diode, its leakage current depends on the light intensity at the junction. The photodiode's leakage current is then detected and amplified to provide a usable output.

Photodiodes are capable of receiving digital optical data at baud rates greater than 50 megabits per second. Analog optical signal reception is restricted to somewhat lower frequencies.

10-9 ■ ULTRASONICS

Some of the industrial measurement and detecting tasks that are commonly handled photoelectrically can also be handled ultrasonically. For example, in Sec. 10-7-1, we used photoelectric devices to determine whether the path between two points was clear or blocked. This task can be accomplished just as well by an ultrasonic system. Instead of a light beam, a high-frequency (ultrasonic) sound wave* is transmitted through the air by the sending unit. If the monitored path is clear, the ultrasonic emission from the sending unit is detected by the receiving unit, which converts the wave into an electrical signal, amplifies the signal, and energizes a relay. If the path is blocked, the receiving unit deenergizes a relay.

An ultrasonic sending unit is structurally simple. It consists of a crystal of *piezoelectric* material sandwiched between two metal plates. One side of the sandwich is mechanically anchored, and the other side is connected to a vibrating diaphragm, as shown in Fig. 10–28. A 20- to 100-kHz ac voltage is applied to the metal plates. The atomic

*A sound wave with a frequency greater than about 20 kHz cannot be heard by humans. Such frequencies are above the range that our sonic senses can detect—thus the description *ultrasonic*.

FIGURE 10–28
Ultrasonic sending unit.

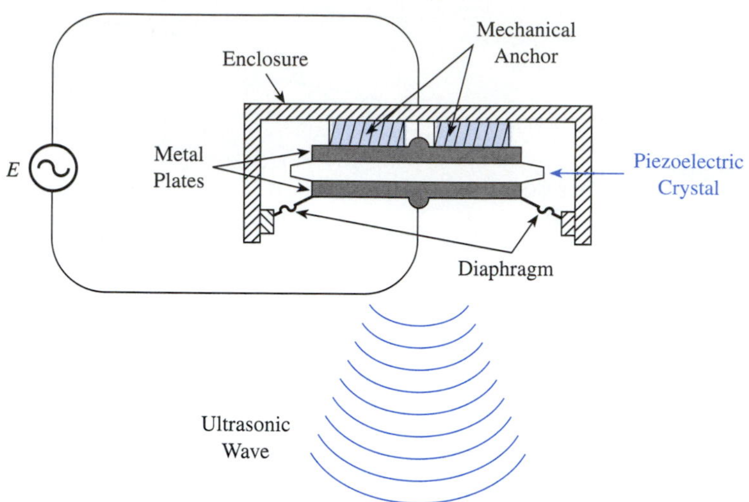

structure of the piezoelectric crystal is such that one polarity of applied voltage causes it to expand and the other polarity causes it to contract; this is called the *piezoelectric effect.* The high-frequency expansions and contractions are imparted to the attached diaphragm, which vibrates against the air in its vicinity, establishing the ultrasonic wave.

With the ac source removed and a voltage amplifier installed in its place, the same assembly can serve as an ultrasonic receiver. The incident sound wave causes the diaphragm to vibrate, which imposes periodic compression/tension on the crystal. Now the piezoelectric effect works in reverse. The high-frequency mechanical vibration of the crystal creates a high-frequency ac voltage between the plates, which can be detected and processed electronically.

A different type of ultrasonic transducer operates on the *magnetostrictive* effect that is exhibited by certain materials. This effect involves mechanical expansion and contraction associated with an alternating magnetic field.

Sound propagation is many orders of magnitude slower than light propagation, as you know. This slowness makes it possible to detect a change in ultrasonic propagation time through a solid or liquid medium, which occurs due to a variation in the thickness or density of the medium. Ultrasonic systems are widely used in industry for monitoring and/or measuring thickness and density. By the same principle, ultrasonic waves can be used to detect cavities and inclusions in castings or abrupt variations in the composition of alloys, and in other similar applications.

10-10 ■ STRAIN GAGES

Strain gages are used in industry to accurately measure large forces, especially large weights. There are also strain gages designed to measure small forces, but they are not as common. A strain gage is basically a resistance wire which is firmly cemented to a surface of a strong object which then receives a force. When force is applied to the object, it distorts very slightly. That is, the object either stretches slightly or compresses slightly, depending on whether it feels a tension or a compression force. The resistance wire, being cemented to the surface of the object, also distorts slightly. The distortion of the wire changes its resistance, which is detected and related to the value of the force.

What a strain gage really measures is *strain,* which is the change in the length of the strong object as a percentage of its original length. The strain of the strong object represents the force applied to the object through Hooke's law, which says that

$$\frac{F}{A} = Y\epsilon = Y\frac{\Delta L}{L_O} \tag{10-1}$$

where F stands for the force applied to the object (in the direction of distortion); A is the cross-sectional area of the object; Y is Young's modulus, which depends on the particular material of which the object is made; and ϵ stands for strain, the change in length per unit of original length ($\epsilon = \Delta L/L_O$). The important point is that the change in the length of the object depends on the force applied to the object and can be related to that force on a one-to-one basis.

The resistance of the wire which comprises the strain gage depends on the length and cross section of the wire, as shown by

$$R = \frac{\rho L}{A} \tag{10-2}$$

In Eq. (10-2), R stands for the resistance of the wire in ohms; ρ is the *resistivity* of the wire material, a property of the particular material used; L is the length of the wire; and A is the cross-sectional area of the wire. It can be seen that if the wire is stretched slightly, the resistance R will increase because the length will increase and the area A will decrease. On the other hand, if the wire is compressed slightly, R will decrease because the length L will decrease and the cross-sectional area will increase. Therefore the resistance of the wire depends on the change in length of the wire and can be related to that change in length on a one-to-one basis.

To sum up, the wire resistance depends on the length and cross-sectional area of the wire, and the wire length depends on the length of the strong object since they are cemented together. The length of the object depends on the applied force, so the end result is that the resistance of the wire depends on the applied force. By precisely measuring the resistance change, we can measure the force.

Figure 10–29(a) shows a top view of a strain gage, looking at right angles to the mounting surface. The resistance wire is usually a copper-nickel alloy with a diameter of about one thousandth of an inch (0.001 in). The wire is placed in a zigzag pattern on a very thin paper backing, called the *base*. The wire is zigzagged in order to increase the effective length which comes under the influence of the strain. The entire zigzag pattern is called the *grid*. Copper lead-wires are attached at the ends of the grid.

Figure 10–29(b) shows a strain gage mounted on the surface which is to undergo the strain. The base is placed flat on the surface (the surface may be curved, as in Fig. 10-29 b), and the entire strain gage is completely covered with special bonding cement. The bonding cement establishes an intimate contact between the wire grid and the strain surface of the strong object. Because of this intimate contact, and because the wire has practically no strength of its own to resist elongation or compression, it elongates or compresses exactly the same distance as the strong object. The strain of the wire grid is therefore exactly the same as the strain of the strong object.

The percent change in resistance for a given percent change in length is called the *gage factor* of the stain gage. In formula form,

$$GF \text{ (gage factor)} = \frac{\Delta R/R}{\Delta L/L} = \frac{\Delta R/R}{\epsilon} \tag{10-3}$$

Most industrial strain gages have a gage factor of about 2. This means that if the length of the object changes by 1% ($\epsilon = 0.01$), the resistance of the strain gage changes by 2%.

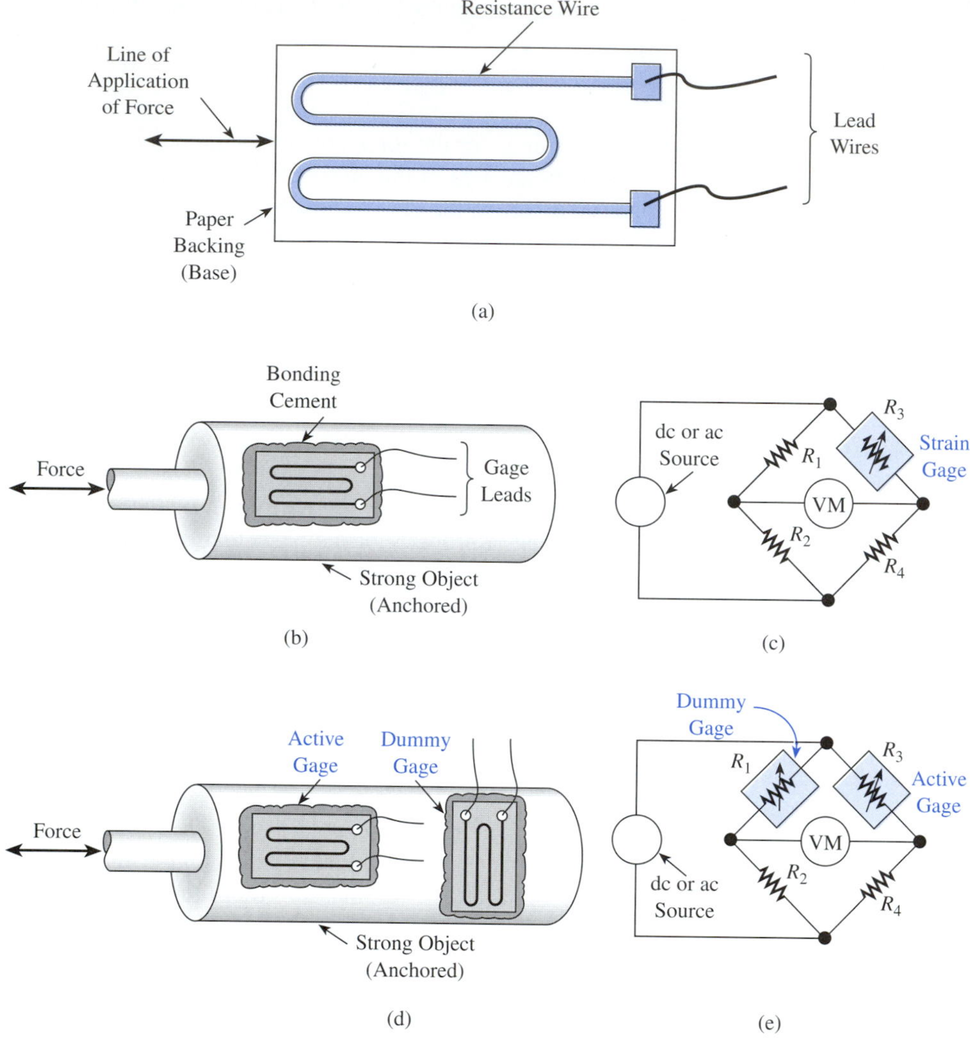

FIGURE 10–29 (a) Physical appearance of a strain gage. (b) Strain gage bonded to a cylindrical object. (c) How the strain gage would be wired into a bridge circuit. (d) Dummy gage bonded to the cylindrical object along with an active gage. (e) How the dummy and active gage would be wired into the bridge to provide temperature stability.

A strain gage wired into a bridge circuit is shown in Fig. 10–29(c). The bridge is usually designed to be in balance when the force exerted on the strong object equals zero. A trimmer pot may be added to one of the legs of the bridge to adjust for exact balance at zero force. As a force is applied, the bridge becomes unbalanced, and the voltage across the bridge can be related to the amount of force. A greater force creates a greater change in gage resistance and a larger voltage output from the bridge.

To compensate for temperature effects, a second strain gage, identical to the first one, can be mounted at right angles to the application line of the force. The force has no effect on this strain gage, since the gage is not aligned with the force. The gage is wired into the bridge as shown in Fig. 10–29(e) and is labeled as a *dummy gage;* the force-sensing gage is then called the *active gage*. The purpose of the dummy gage is to cancel out any temperature-related resistance change in the active gage. With both gages experiencing the same temperature, any resistance change in the active gage due to temperature variations appears in the dummy gage too. Since the error appears on both sides of the bridge, it is canceled out.

For precise weight measurement, a carefully shaped and machined object, containing several strain gages, is used. The gages are strategically placed on the machined surfaces at various angles to yield the utmost temperature stability. The gages themselves are designed to provide a linear relationship between bridge output voltage and force (weight) on the machined object. The object, in combination with its strain gages, is then called a *load cell*. Accurate weighing scales measuring large weights almost always have load cells as transducers.

10-11 ■ ACCELEROMETERS

An *accelerometer* is a device which measures acceleration. Most accelerometers work in an indirect manner. They bring a known amount of mass, called the *seismic mass,* into mechanical junction with the object being measured, so that whatever acceleration the measured object undergoes, the seismic mass must undergo the same acceleration. Then the accelerometer detects the force exerted on the seismic mass. The measured force value is related to the acceleration value by Newton's second law:

$$a = \frac{F}{m_s} \quad \begin{matrix} \longleftarrow \text{Measured by a force transducer} \\ \longleftarrow \text{A fixed known amount of mass} \end{matrix} \tag{10-4}$$

Therefore the force transducer can be calibrated to read out in units of acceleration.

For example, if the known value of the accelerometer's seismic mass is 0.5 kilogram, and if the accelerometer's force transducer detects a force of 2.0 newtons being exerted on that seismic mass, the transducer will be calibrated to read out 4.0 meters per second squared, rather than 2.0 newtons. ($a = F/m_s = 2.0 \text{ N}/0.5 \text{ kg} = 4.0 \text{ m/s}^2$.)

Fig. 10–30 is a diagram of a strain gage-based accelerometer. The accelerometer frame must be firmly attached to the measured object. The seismic mass is attached to the accelerometer frame by a low-deflection elastic link, which can be thought of as a very stiff spring. The seismic mass is constrained in the up/down and in/out directions by guides, but the guides permit free movement in the left/right direction.

When the measured object accelerates to the right, the frame transmits a force via the elastic link to the seismic mass, causing it to accelerate equally. The reaction force stretches the elastic link, which allows the mass to shift very slightly to the left. The strain on the elastic link is manifested as a resistance change in the strain gage, which can be related to

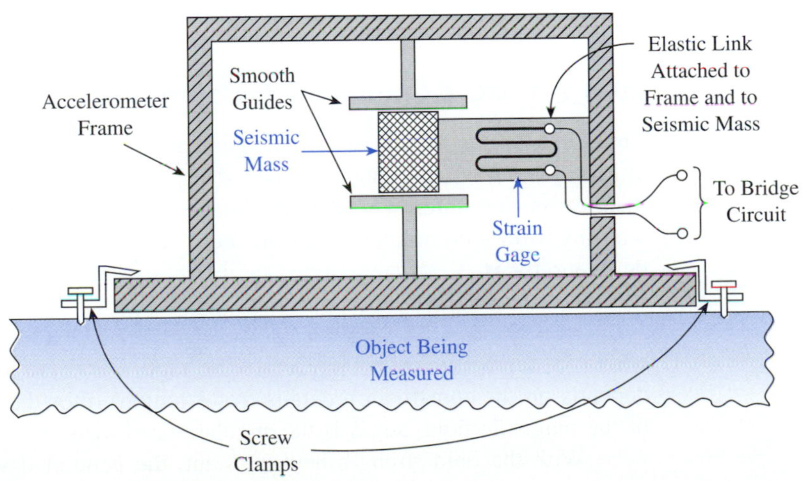

FIGURE 10–30 Structure of a strain-gage accelerometer.

force in the usual strain-gage manner, and then to acceleration by virtue of Newton's second law, as explained previously.

Besides the strain-gage approach, many other force-detection techniques can be employed in the design of accelerometers. Recently, several solid-state devices have been developed that transduce force into an electrical variable. The foremost examples are the piezoresistor and the piezotransistor, both of which are well adapted for use in accelerometers.

Industrially, accelerometers find application in sophisticated servo systems, to furnish an additional feedback signal to the comparer. As the servo system is starting from standstill or slowing to a stop, the comparer takes the acceleration measurement into account in its determination of the augmented error signal, resulting in faster and more stable system response. Advanced motor drive systems make similar use of accelerometers.

Accelerometers also are commonly applied in the area of vibration detection and analysis. Rotating machines and shock-receiving machines are both subject to resonant mechanical vibrations which can be harmful. Such vibrations can be detected and measured by a high-frequency accelerometer, since mechanical vibration is equivalent to rapidly cyclically reversing acceleration. An accelerometer that is designed specifically for vibration analysis is usually referred to as a *vibrometer*.

10-12 ■ TACHOMETERS

A *tachometer* is a device which measures the angular speed of a rotating shaft. The most common units for expressing angular speed are revolutions per minute (r/min) and radians per second. A radian is equal to $1/(2\pi)$ revolutions, or approximately 57 mechanical degrees. We will use the units of r/min exclusively.

Tachometers in industry use one of two basic measuring methods:

1. The angular speed is represented by the *magnitude* of a generated voltage.
2. The angular speed is represented by the *frequency* of a generated voltage.

In the domain of magnitude tachometers there are two principal types: the *dc generator tachometer* and the *drag cup tachometer*.

In the domain of frequency tachometers, there are three principal types: the *rotating field ac tachometer*, the *toothed-rotor tachometer*, and the *photocell pickup tachometer*. These names are not universally accepted, but they describe the action of the various tachometers rather well and we will adopt them in this book.

10-12-1 DC Generator Tachometers

The dc generator tachometer is a dc generator, pure and simple. The field is established either by a permanent magnet mounted on the stator or by a separately excited electromagnet on the stator. The output voltage is generated in a conventional dc armature winding with a commutator and brushes. The equation for true generated voltage in a dc generator is

$$V_G = kBS$$

where V_G represents the true generated voltage, k is some proportionality constant which depends on the construction details (rotor length, rotor diameter, etc.), B is the strength of the magnetic field, and S is the angular speed measured in revolutions per minute.

With the field strength held constant, the generated voltage is proportional to the angular speed of the shaft. It is therefore possible to connect the tachometer shaft to the shaft being measured, apply the generated voltage to a voltmeter, and calibrate the meter in terms of r/min. One nice feature of a dc generator tachometer is that the polarity of the generated voltage reverses if the direction of rotation reverses. Therefore this type of tachometer can indicate rotational direction as well as speed.

10-12-2 Drag Cup Tachometers

A drag cup tachometer has two sets of windings on its stator at right angles to each other, just like an ac servo motor. Refer to Fig. 11–14(a). The rotor is not a squirrel-cage rotor, however. It is a hollow copper cylinder, called a *cup,* with a laminated iron inner core, which does not contact the cup. The cup attaches to the tachometer input shaft and rotates at the measured speed.

One of the stator windings, called the *exciting winding,* is driven by a stable ac voltage source. The other stator winding is the output winding. The exciting winding sets up an alternating magnetic field which induces eddy currents in the copper cup. The eddy currents set up an *armature reaction field* at right angles to the field from the exciting winding. The right-angled field will then induce an ac voltage in the output winding whose magnitude depends on the speed of rotation of the cup. The result is an ac output voltage which varies linearly with shaft speed.

The output voltage frequency is equal to the exciting frequency (usually 60 Hz), and it is 90° out of phase with the exciting voltage. The direction of shaft rotation determines whether the output voltage leads or lags the exciting voltage. Therefore this tachometer also can indicate direction as well as speed of rotation.

All tachometers which rely on a voltage magnitude to represent speed are subject to errors caused by three things:

1. Signal loading.
2. Temperature variation.
3. Shaft vibration.

Regarding problem 1, the voltage delivered by any kind of generator will vary slightly as the current load on the output winding varies. This is because the *IR* voltage drop in an output winding varies as its current varies.

Regarding problem 2, as temperature changes, the magnetic properties of the core change, causing variations in the magnetic field strength. As the magnetic field strength varies, so does the generated voltage.

Regarding problem 3, as the shaft vibrates, the precise spacing between the field and armature windings changes. This change in spacing causes variations in generated voltage.

Modern tachometer designs have minimized these errors and have produced tachometers in which the voltage-speed linearity is better than 0.5%. This is quite adequate for the majority of industrial applications.

10-12-3 Rotating Field AC Tachometers

The rotating field ac tachometer is a rotating field alternator, pure and simple. The field is usually created by permanent magnets mounted on the rotor. The rotor shaft is connected to the measured shaft, and the rotating magnetic field then induces an ac voltage in the stator output windings. The equation for the frequency of the generated voltage in an ac alternator is

$$f = \frac{PS}{120} \tag{10-5}$$

where f is the frequency in hertz, P is the number of magnetic poles on the rotor, and S is the rotational speed in r/min. It can be seen that the output frequency is an exact measure of the angular speed of the shaft.

10-12-4 Toothed-Rotor Tachometers

The toothed-rotor tachometer is the most popular of the frequency tachometers. This tachometer has several ferromagnetic teeth on its rotor. On its stator it has a permanent magnet with a coil of wire wrapped around the magnet. This arrangement is illustrated in Fig. 10–31(a) for a rotor with six teeth.

As the rotor rotates, the teeth come into close proximity with the magnet and then pass by. When a tooth is close to the magnet, the magnetic circuit reluctance is low, so the field strength in the magnet's core increases. When no tooth is close, the magnetic circuit reluctance is high, so the field strength in the magnet's core decreases. Therefore one cycle of field strength is produced for every time a tooth passes by. This variation in magnetic field strength induces a voltage in the coil wrapped on the permanent magnet. One voltage pulse is produced by each tooth. This is shown in Fig. 10–31(b).

The relation between pulse frequency and speed is given by

$$\text{rev/s} = \text{pulses/s} \div 6$$

because it takes six pulses to represent one revolution. The number of revolutions per minute equals the number of revolutions per second multiplied by 60, or

$$S \text{ (in r/min)} = 60 \text{ (rev/s)} = 60 \text{ (pulses/s} \div 6)$$
$$S \text{ (in r/min)} = 10f \qquad \textbf{(10-6)}$$

where f represents the pulse frequency; whatever readout method is used would have to reflect Eq. (10-6), namely, that the rotational speed equals the measured frequency multiplied by the factor 10. For a different number of teeth on the rotor, the factor would be different.

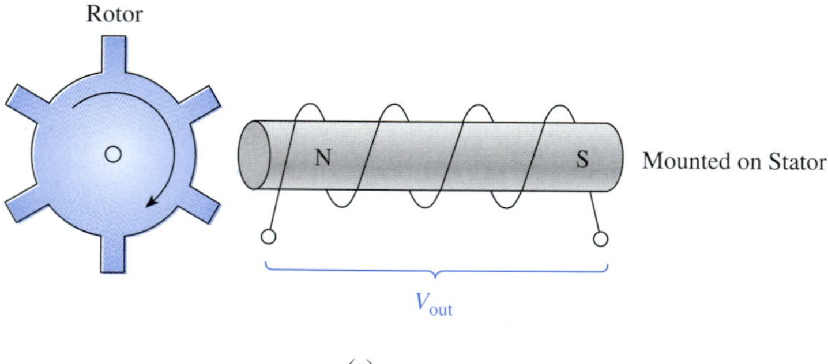

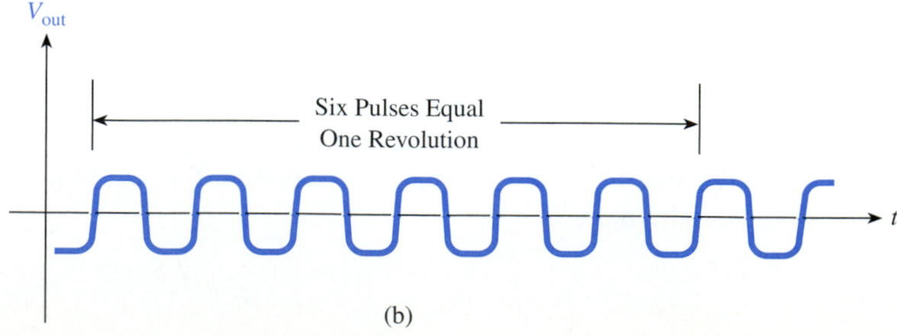

FIGURE 10–31
(a) Arrangement of a toothed-rotor tachometer. (b) Output voltage waveform of a toothed-rotor tachometer.

10-12-5 Photocell Pickup Tachometers

A photocell pickup tachometer is basically the same device that was suggested in Sec. 10-7-1 for chopping a light beam. A rotating disk is placed between a light source and a photovoltaic cell. Part of the disk passes the light beam, and part of the disk blocks the light beam. Therefore the photovoltaic cell is constantly being turned on and off, at a frequency which depends on the angular speed of the disk. By connecting the disk shaft to the measured shaft, a voltage waveform will be generated by the photocell. The frequency of the waveform will then be a measure of the angular speed of the shaft.

For example, if the disk had four light areas and four dark areas, the speed would be given by

$$S\,(\text{in r/min}) = 15f$$

where f is the frequency of the photocell output waveform. You should justify this equation to yourself. Use the same derivation approach that was used in Sec. 10-12-4.

10-12-6 Frequency Tachometers versus Magnitude Tachometers

The main advantage of frequency-measuring tachometers is that they are not subject to errors due to output loading, temperature variation, and shaft vibration. Also, their linearity is perfect. However, frequency tachometers all have the disadvantage of awkwardness in reading out the speed. It takes a lot more effort to convert a frequency into a readable form than it does to convert a voltage magnitude into a readable form. It is true that frequency measurements lend themselves to digital detection and readout, but digital measurement and readout are much more complex than a simple analog meter readout. The digital circuits must repeatedly go through the cycle of count, store, display, reset.

On the other hand, to a human being digital readout is more intelligible than analog readout, because the person taking the reading doesn't have to figure out the value of each meter marking. Thus, as far as readout is concerned, frequency tachometers and magnitude tachometers both have their advantages and disadvantages.

Many times in industrial control the measured speed is to be used as a *feedback* signal. This is certainly the case in a closed-loop speed control system, where the measured speed is compared with the set point to find the speed error signal. Measured speed is also used as feedback in a servo control system, where the speed of approach is used to subtract from the position error signal to prevent overshoot. This process, called *error-rate damping,* is common in servo systems. In cases like these, the speed signal must be expressed as an analog voltage instead of a digital number. Therefore, in feedback systems, magnitude tachometers have the advantage over frequency tachometers because magnitude tachometers automatically provide an analog voltage signal.

Frequency tachometers could provide an analog voltage signal, but only by adding an extra signal-processing circuit (a D-to-A converter or a frequency demodulator). Overall, magnitude tachometers are preferred to frequency tachometers for feedback applications.

10-13 ■ HALL-EFFECT TRANSDUCERS

10-13-1 The Hall Effect Explained

The Hall effect is the phenomenon by which charge carriers moving through a magnetic field are forced to one side of the conducting medium. Refer to Fig. 10–32, which illustrates the Hall effect for a situation in which the motion of the charge carriers is perpendicular to an externally imposed magnetic field. In part (a) a simple electric current is operating in a space which is vacant of magnetic fields. The current is considered to consist of positive charge carriers flowing from top to bottom through a flat piece of metal. Take note of the fact that, in the absence of a magnetic field, the charge carriers are evenly distributed across the width of the flat metal piece.

If the identical circuit is operated in the presence of a magnetic field, however, the charge carriers crowd over to one side of the flat piece. This is illustrated in the drawing of Fig. 10–32(b), which indicates the presence of a magnetic field pointing out of the page. In that drawing the positive charge carriers congregate on the left side of the flat piece because of the forces exerted on them resulting from the interaction of charge motion and magnetic flux. The magnitude of an individual force on a charge carrier is given by the Lorentz relation

$$F = qvB \tag{10-7}$$

in which q is the charge, in coulombs, on an individual charge carrier; v is the charge carrier's velocity in meters per second; and B is the magnetic flux density in webers per square meter, or teslas. The force comes out in basic SI units of newtons. The direction

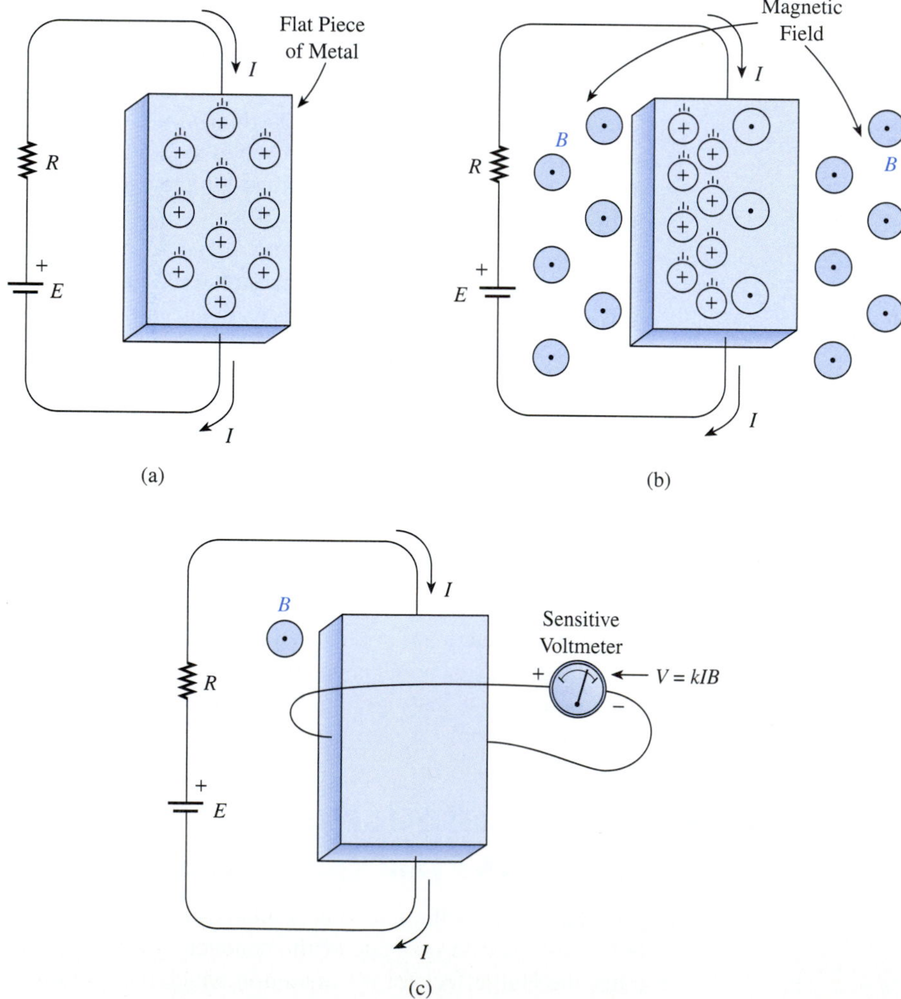

FIGURE 10–32
The Hall effect. (a) In the absence of a magnetic field, there is no Hall effect. (b) With a magnetic field present, the charge carriers are nonuniformly distributed across the flat conductor. (c) The nonuniform charge distribution produces a side-to-side potential difference—the Hall voltage.

of the force is given by the right-hand rule for electric-magnetic interaction, which says, With the fingers of the right hand, rotate the motion (velocity) vector into alignment with the B-field vector; the right thumb then points in the direction of the interactive force. Verify for yourself that the right-hand rule indicates interactive forces to the left in Fig. 10–32(b).

Because of the alteration of the normal charge distribution within the metal, a voltage will appear between the left and right sides of the flat piece. In Fig. 10–32(b), the voltage would be positive on the left side relative to negative on the right side. Working from Eq. (10-7), it can be shown that the magnitude of the Hall-effect voltage is given by

$$V = kIB \tag{10-8}$$

where V is the voltage between the sides, in volts; I is the current in amperes; and B is the magnetic field strength in Wb/m^2, or teslas. The proportionality constant k depends on the thickness of the flat metal piece and the electrical resistivity of that particular metal. The existence of the Hall-effect voltage is pointed out in Fig. 10–32(c).

The current-carrying flat piece need not be metal for the Hall effect to occur. Some Hall-effect transducers use semiconductor material for the flat piece and some use high-resistivity crystalline material. The relation among the three variables V, I, and B never changes, though. As long as the B-field flux lines are perpendicular to the flat face, Eq. (10-8) holds true—voltage is proportional to the product of current and field strength.

Generally, Hall-effect voltage is quite small. Transducers built on the Hall-effect principle usually must contain a high-gain voltage amplifier. For example, if the flat strip were made of copper having the same thickness as AWG No. 18 wire, about 1 mm, and if it were carrying a current of 10 A through a space with magnetic flux density of 1 tesla (rather a strong B-field), the Hall-effect voltage would be only about 0.7 μV.

10-13-2 Hall-Effect Proximity Detector

In industrial automation, it sometimes is desirable to sense the presence of an object without having to touch the object. A magnetic *proximity detector* can accomplish this. A small permanent magnet is attached to the moveable object, and a Hall-effect magnetic detector is mounted so that the magnet's flux impinges on the detector's surface when the object comes near. The circuit construction of a Hall-effect magnetic proximity detector is shown in Fig. 10–33. This entire circuit, including the flat conducting piece, is available as an integrated circuit; it has flat-surface dimensions of about 7 mm × 7 mm and a thickness of about 2 mm.

10-13-3 Hall-Effect Power Transducer

The Hall effect is inherently a multiplying phenomenon, I multiplied by B. Because of this, the Hall effect lends itself to the construction of electrical power transducers, since electrical power is also a multiplying concept, I multiplied by V. To construct such a transducer, we must arrange two things:

1. The current I through the flat conductor must be proportional to the current in the circuit being measured.
2. The magnetic field strength B must be proportional to the voltage in the circuit being measured. In an ac circuit, these relationships can be achieved by using a current transformer (a standard voltage transformer) to drive the coil of the electromagnet that produces the B-field. This arrangement is pictured in Fig. 10–34.

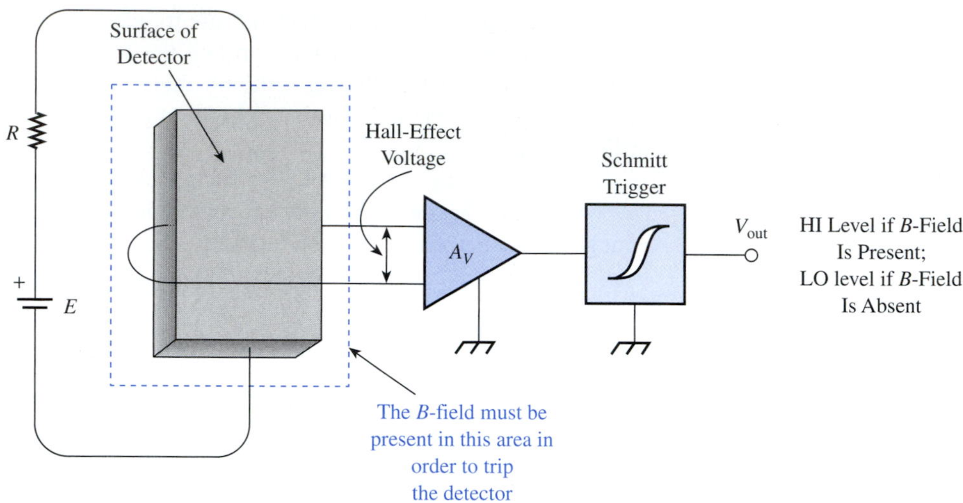

FIGURE 10–33
Layout of a Hall-effect proximity detector.

The Hall-effect voltage waveform obtained from the transducer of Fig. 10–34 consists of a dc component superimposed by an alternating component at twice the line frequency; it is somewhat similar to the waveform produced by a full-wave rectifier. If the variable being measured is the circuit's average power, the alternating component is filtered out and the dc component represents the measured value. If the variable of interest is the circuit's instantaneous power, then the instantaneous output voltage represents the measured value. In either case the current transformer, potential transformer, and electromagnet must all be carefully designed so that any phase displacement between the circuit's current and voltage is matched by an equal phase displacement between the current and magnetic field in the flat conductor.

Some modern microcomputer-based industrial systems use a circuit's maximum instantaneous power as a decision-making parameter. This is usually accomplished by having the microcomputer take very rapid samples of the instantaneous output voltage from a Hall-effect power transducer.

In some transducer designs, the roles of circuit current and voltage are reversed. That is, the circuit current is used to establish a proportional Hall-effect magnetic field and the circuit voltage is used to produce a proportional Hall-effect current in the flat conductor.

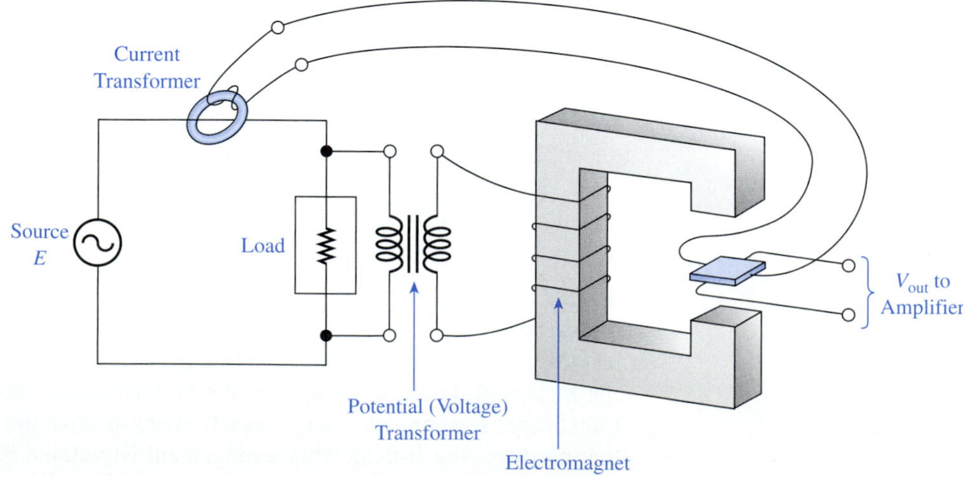

FIGURE 10–34
Hall-effect power transducer. Instantaneously, magnetic field strength is proportional to load voltage and flat-conductor current is proportional to load current.

10-13-4 Hall-Effect Flowmeter

Some liquids and gases are chemically structured in such a way that they contain free charge carriers. The charge carriers are not individual subatomic particles, as is the case for solids; instead the charge carriers are entire atoms or molecular components that have a net imbalance of charge. Such charged pieces of matter are generically called *ions*. If a liquid or gas contains ions, its velocity through a pipe can be measured by using the Hall effect. Furthermore, given the pipe's cross-sectional area, there is a direct correspondence between velocity and volume flow rate. Therefore a Hall-effect apparatus can be calibrated to read in units of either volume flow rate or velocity, whichever is preferred.

A Hall-effect fluid-flow transducer is pictured in Fig. 10–35. A constant-strength magnetic field is established through the space occupied by the pipe. As the ions pass through the *B*-field, they are deflected by the electric-magnetic interactive force. For the magnetic orientation and flow direction shown in Fig. 10–35, any free positive ions will congregate near the top of the pipe, and any free negative ions near the bottom of the pipe. A Hall-effect voltage therefore appears between the top and bottom electrodes that are inserted through the pipe walls. For a given value of *B* and a known constant value of ionic density, the magnitude of this voltage can be proportionally related to the fluid velocity, as suggested by Eqs. (10-7) and (10-8).

Compared to a mechanical-type flowmeter, a Hall-effect flowmeter is superior in the respect that almost no energy is extracted from the moving fluid.

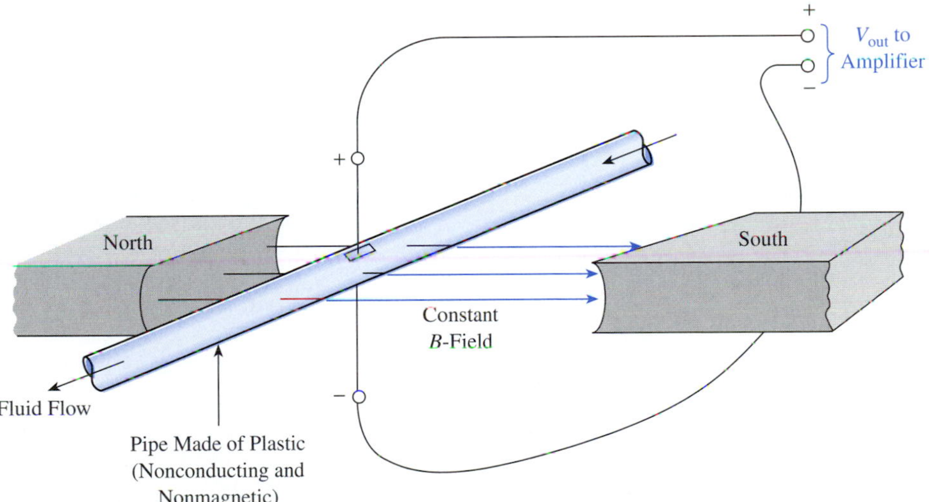

FIGURE 10–35 Arrangement of a Hall-effect flowmeter.

10-14 ■ OTHER FLOWMETERS

The Hall-effect flowmeter of Fig. 10–35 is elegant and efficient. Its efficiency is the result of its being totally electric/magnetic, with no mechanical presence inside the pipe. However, it requires that the fluid be ionic. Pure water is not sufficiently ionic to make practical use of the Hall-effect idea. If a sufficient amount of an ionic compound is dissolved in water, then it can be measured by a Hall-effect flowmeter.

10-14-1 Ultrasonic Flowmeter

Another nonintrusive flow-metering technique, this one usable with nonionized fluid, is the ultrasonic flowmeter illustrated in Fig. 10–36. Two ultrasonic units like the one shown in Fig. 10–28 are mounted adjacent to each other, with physical orientation pointing upstream, on the pipe's outside surface. The sending unit is excited by a sine or square wave of a fixed frequency, say 50 kHz. Its ultrasonic pressure waves propagate through the pipe's surface and impinge on the fluid molecules that are moving toward it. These waves are reflected back to the receiving unit at a frequency greater than 50 kHz due to the Doppler effect.

The Doppler effect can be understood intuitively by referring to Fig. 10–37. Part(a) depicts the situation for the first (positive) half cycle of the emitted pressure wave. Beginning at $t = 0$, the positive half cycle commences its travel through the liquid medium. Assume that it requires 30 μs to travel through the liquid medium and impinge an oncoming molecule. This is reasonable if we assume a propagation velocity of about 330 m/s and an average impingement distance of 1 cm. Then

$$t = \frac{\text{distance}}{\text{velocity}} = \frac{1 \times 10^{-2}\ \text{meter}}{330\ \text{meter/s}} \approx 30\ \mu\text{s}$$

Reflecting from that oncoming molecule, the pressure wave requires another 30 μs to return to the receiving unit, assuming that the sender and receiver are located side by side. Thus, the received pressure wave delivers its positive-going zero crossover at the time instant $t = 60$ μs.

Now study Fig. 10–37(b), which shows the situation for the negative half cycle. The flowing water molecule has moved closer to the ultrasonic units during the 10 μs of elapsed time that was required to produce the positive half cycle. It has moved forward to the right by the distance shown between the vertical dashed lines.

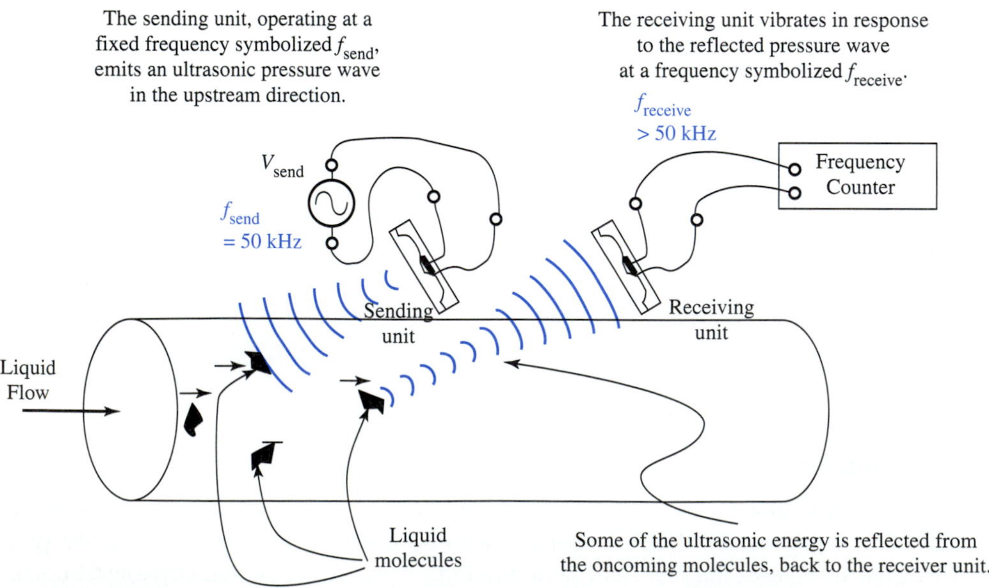
The sending unit, operating at a fixed frequency symbolized f_{send}, emits an ultrasonic pressure wave in the upstream direction.
The receiving unit vibrates in response to the reflected pressure wave at a frequency symbolized $f_{receive}$.
$f_{receive}$ > 50 kHz
Frequency Counter
V_{send}
f_{send} = 50 kHz
Sending unit
Receiving unit
Liquid Flow
Liquid molecules
Some of the ultrasonic energy is reflected from the oncoming molecules, back to the receiver unit.

FIGURE 10–36
Ultrasonic flowmeter.

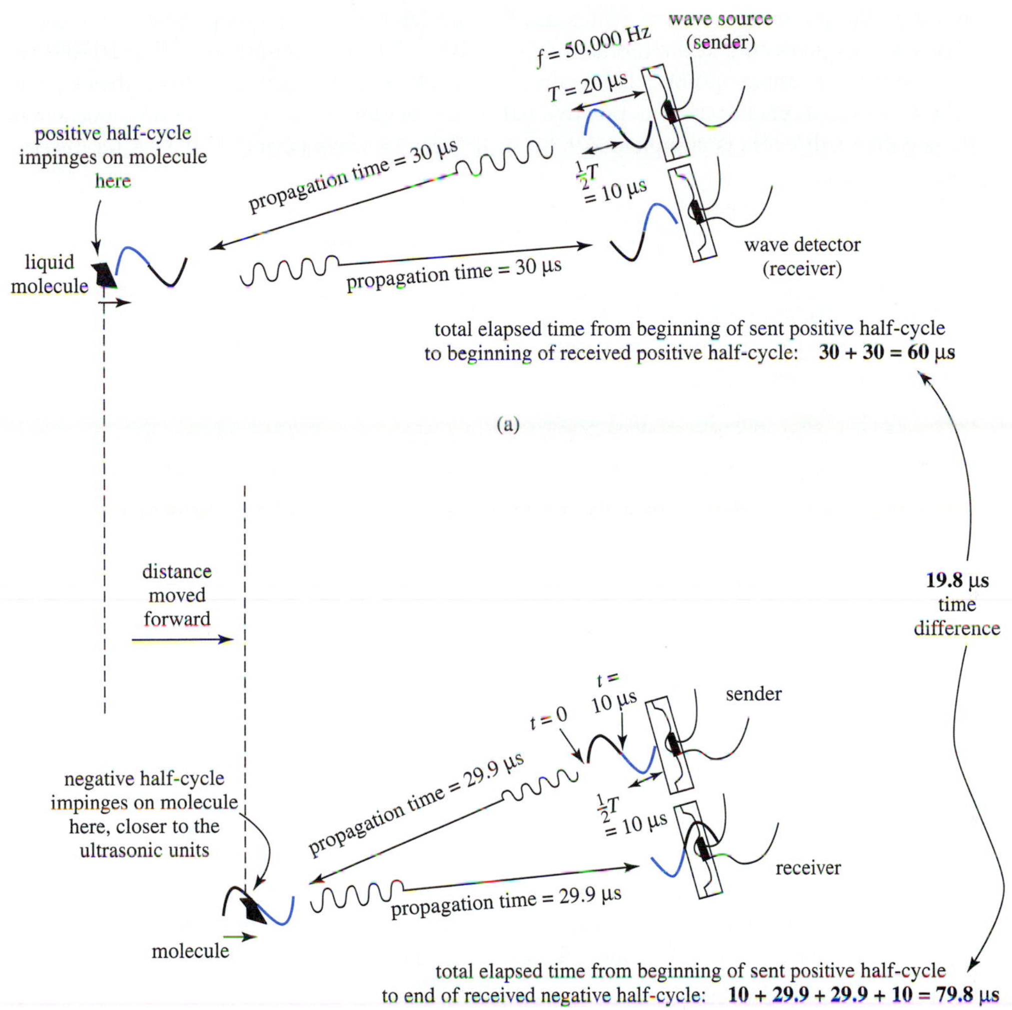

FIGURE 10–37
Understanding the Doppler effect. The reflected wave completes a full cycle at the receiver unit in just 19.8 μs, not the 20 μs that would be expected from a normal (stationary object) reflection. Therefore its perceived frequency is $f = 1/T = 1/19.8\ \mu s = 50\ 505$ Hz. (a) Thinking about the positive half cycle. (b) Thinking about the negative half cycle.

Because the molecule is now closer, the negative half cycle beginning at the $t = 10\ \mu s$ moment will reach it in less time than was required for the positive half cycle. Assume that its travel time is 29.9 μs instead of 30 μs. Therefore the pressure wave delivers its negative-going zero crossover to the molecule at time $t = 10 + 29.9 = 39.9\ \mu s$.

The negative-going zero crossover reflects from the molecule and travels back to the receiving unit in an elapsed time of 29.9 μs, the same as the travel-forward time. The negative-going crossover arrives at the time instant

$$t = 10\ \mu s + 29.9\ \mu s + 29.9\ \mu s = 69.8\ \mu s$$

and then the pressure cycle completes itself 10 μs later. Thus the completion of the pressure wave cycle occurs at the moment 69.8 + 10 = 79.8 μs, as shown in Fig. 10–37(b).

From the vantage point of the receiver, the cycle period is 19.8 μs. This is the elapsed time from the moment when its positive half cycle begins (60 μs) and the moment when its negative half cycle is complete (79.8 μs). It "sees" a wave period $T = 19.8\ \mu s$. Its detected frequency is

$$f = \frac{1}{T} = \frac{1}{19.8\ \mu s} = 50\ 505\ \text{Hz}$$

even though the actual transmitted wave had $f = 50\ 000$ Hz. This is the rationale of the familiar Doppler effect for approaching objects, in which perceived frequency is higher than actual frequency. (For receding objects, the perceived frequency is lower than the actual frequency.)

The greater the velocity of the oncoming fluid molecules in Fig. 10–36 or Fig. 10–37, the greater will be the difference between $f_{receive}$ and f_{send}. Therefore frequency difference can be related to velocity, which in turn is directly related to flow rate, if the fluid is incompressible. So it is possible to calibrate the frequency counter of Fig. 10–36 to read in units of mass- or volume-flow.

10-14-2 Turbine Flowmeters

A turbine flowmeter, sometimes called a paddle-wheel flowmeter, is pictured in Fig. 10–38. A turbine wheel fits tightly against the circular inner surface of the fluid passage, so almost all the fluid must pass across the surfaces of the blades themselves. The blades are shaped to prevent cavitation—all the volume is filled with fluid all the time. Under these circumstances the speed of wheel rotation can be directly related to mass- or volume-flow rate.

Rotational speed is measured by sensing the number of blade passes per second beneath the magnetic pickup coil. When a steel blade is aligned with the coil's vertical axis, its flux Φ increases; between blades, Φ decreases. These flux variations produce one cycle of voltage change across the series resistance. Therefore a measurement of signal frequency can be related to rotational speed, and thus to flow-rate.

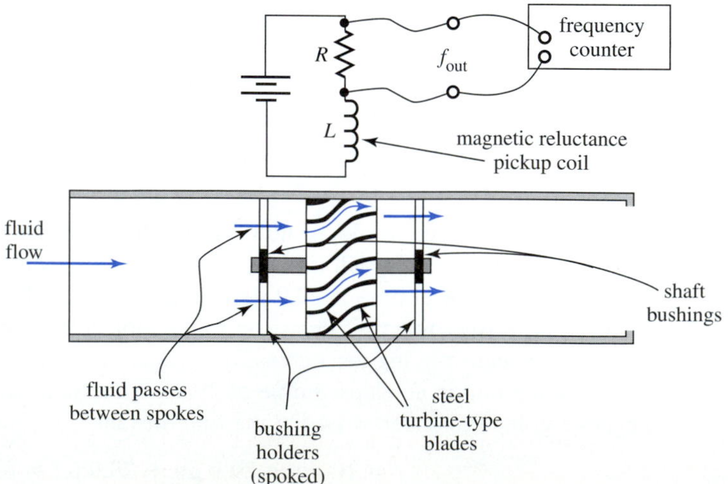

FIGURE 10–38 Turbine flowmeter construction.

Turbine flowmeters, like all intrusive flowmeters, restrict the fluid passage to some extent. Therefore they decrease the flow by the very act of measuring the flow. Because of its surface complexity, a turbine flowmeter cannot be installed in a pipe carrying suspended solid material—a slurry. It is suitable only for totally dissolved material.

10-14-3 Nutating-Disk Flowmeter

Nutation is the act of rotating and wobbling simultaneously, as a spinning top, for example. At one moment the rotational axis points in one direction; at a later moment the axis has changed to a different mechanical orientation. A nutating-disk flowmeter is shown in Fig. 10–39 in side view. The red disk and sphere are bonded together as a unit, with the shaft emerging from the sphere perpendicular to the disk. The disk/sphere unit is contained in a hollow toroidal structure, all surfaces fitting tightly but capable of sliding. That is, in Fig. 10–39(a) the edges of the disk can slide along the curved inside surfaces of the outer wall and the sphere can slide along the inside surface of the hollow spherical center of the structure, enabling the shaft to wobble to the left. Likewise for wobbling in the other three directions—wobble into the page, part (b); wobble to the right, part (c); and wobble out of the page, part (d).

The disk itself does not rotate, so strictly speaking, such a flowmeter is not really nutating, but the tip of the shaft describes a circular locus in space, so it mimics the action of a rotating top. Each shaft nutation corresponds to a fixed volume of liquid passing through the flowmeter's chamber, as explained later. Counting the number of shaft nutations is equivalent to counting the number of increments of liquid volume that have passed. The counting mechanism, not shown in Fig. 10–39, can be coupled to the shaft either mechanically or by magnet force.

One complete nutation is related to a fixed volume of liquid as follows. Refer to Fig. 10–39. The outer wall of the toroidal structure has two openings beside each other, facing forward toward us in Fig. 10–39(a). The opening on the left accesses the upstream side of the pipe; it is for inflow. The opening on the right gives out upon the downstream side of the pipe; it is for outflow. The inlet port opening and outlet port opening are separated from one another by an interior vertical wall, not showable in this figure.

As pressurized water enters the inlet port in Fig. 10–39(a), it is trapped in a chamber formed by the top surface of the disk and the left-side interior space of the toroid, to the point at the exact rear of the structure. Water pressure exerts force on the disk's top surface in the rear, which is free to move in the downward direction. Therefore the disk surface does move downward, causing the shaft to wobble to the rear, as indicated in Fig. 10–39(b). At this moment the fixed volume is trapped in a chamber formed between the top surface of the disk and the rear interior space of the toroid, between the exact leftmost and the exact rightmost points on the disk.

Now in-flowing water is entering the toroid *below* the disk. This below-the-disk water is not counted in our description of the nutation of Fig. 10–39. However, this pressurized water exerts force on the disk's bottom left side, causing the side to continue moving upward. As the left side moves up the shaft wobbles to the right in Fig. 10–39(c). During this motion the previously trapped volume of water starts exiting the toroid via the outlet port on the right.

Pressurized water below the disk is now exerting force tending to raise the disk in the rear. That section of disk is free to move upward and it does so, as shown in Fig. 10–39(d). The shaft therefore wobbles forward, out of the page toward us. That motion continues to eject the previously trapped water from the outlet port, as indicated in part (d).

Now new water is entering above the disk on the left, tending to force it down. (The bottom water at this moment is trapped below the disk in the rear, incapable of propelling

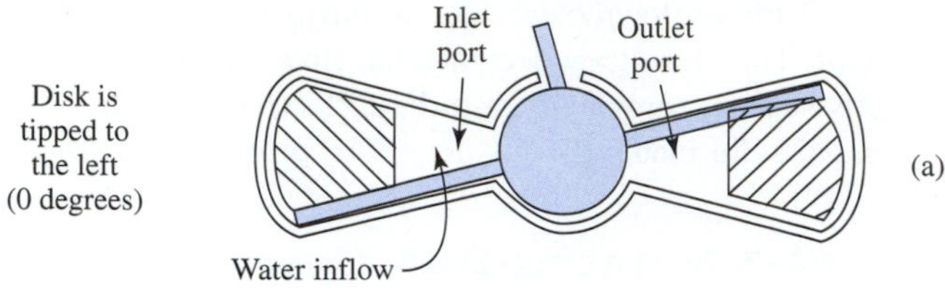

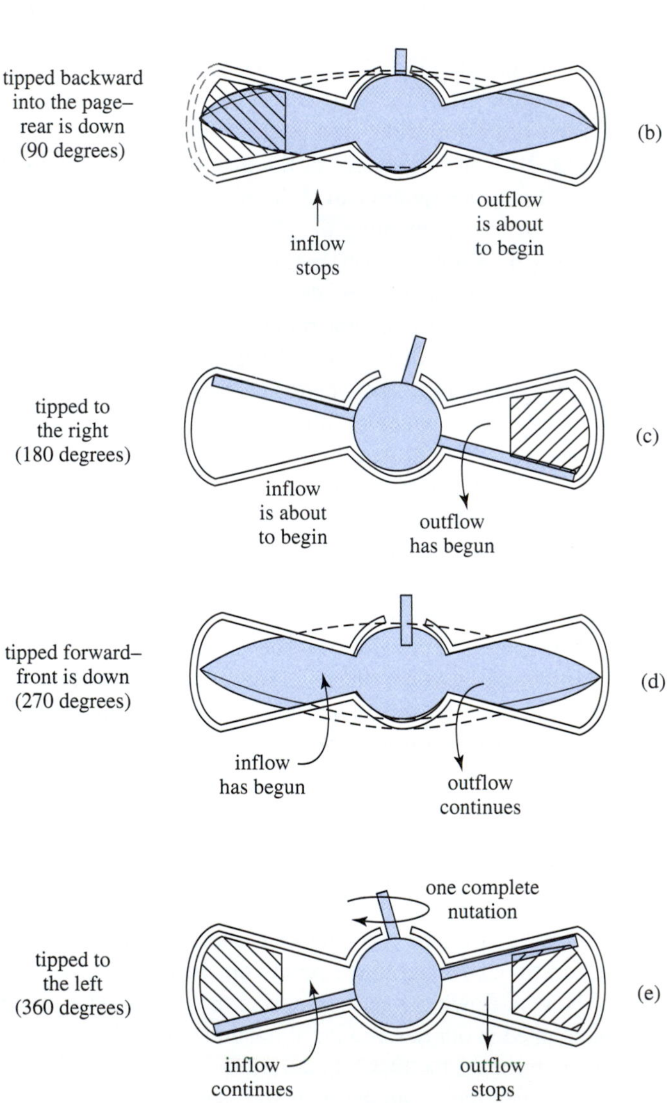

FIGURE 10–39 Nutating-disk flowmeter. These descriptive notes refer to water that passes through the flowmeter *above* the disk. An equal amount passes *below* the disk, out of phase by 180 degrees.

the wobbling action any further.) As the left side of the disk descends, the right side rises, ejecting the last of the above-the-disk trapped volume. This motion causes the shaft to wobble back to the left, its starting position. It has completed one full nutation, which is duly counted by mechanical or magnetic/electrical means.

The water volume that our discussion has focused on is the volume that passes through the flowmeter above the disk; call it *top* water. There is also an equal amount of *bottom* water passing below the disk; this bottom water is out of synchronization with the events occurring to the top water, by one half cycle. Thus, each shaft nutation represents double the amount of top water that we have described. The readout means of the flowmeter are just calibrated to account for this double effect.

10-14-4 Pressure-Drop Flowmeters

In a flowing containment system (pipe or duct) where some locations have large cross-sectional area and others have smaller cross-sectional area, here is the situation regarding pressure and velocity:

1. A large-area location has a low value of velocity and a large value of pressure.
2. A small-area location has a high value of velocity and a low value of pressure.

These facts are illustrated in Fig. 10–40.

On the left of Fig. 10–40 the pipe cross-sectional area is large. Because the area is large, a given-width "slice" of the fluid does not have to move at a rapid speed in order to accomplish a particular volume-flow rate. This is conveyed by the short vector arrow representing velocity v_1. Because relatively little of the fluid's energy is committed to actual motion (kinetic form of energy), a large portion of the fluid's energy is contained in its pressure (static form of energy). This is pointed out in Fig. 10–40 by the large value of pressure P_1 measured on gage 1.

Farther to the right the pipe necks down. In section 2 the cross-sectional area is moderate. Because the area is not as large, a given-width slice of the fluid must now be moving at a faster speed in order to maintain the same volume-flow rate. Observe the longer-length vector arrow for velocity v_2. Because more of the fluid's energy is now appearing in kinetic form, a smaller portion of that energy is available to appear in static form. Therefore gage 2 measures a lower value of pressure P_2.

In section 3 the cross-sectional area is even smaller. Velocity v_3 is greater yet, and pressure P_3 is lower yet, as Fig. 10–40 indicates.

In section 4 the area has returned to the original value. Therefore velocity v_4 and pressure P_4 are the same as v_1 and P_1, ideally.

Working with the equation form of Bernoulli's law, it can be shown that a fluid's flow-rate is proportional to the square root of its pressure difference between two locations. That is,

$$Q = k_p \sqrt{P_1 - P_2} \tag{10-9}$$

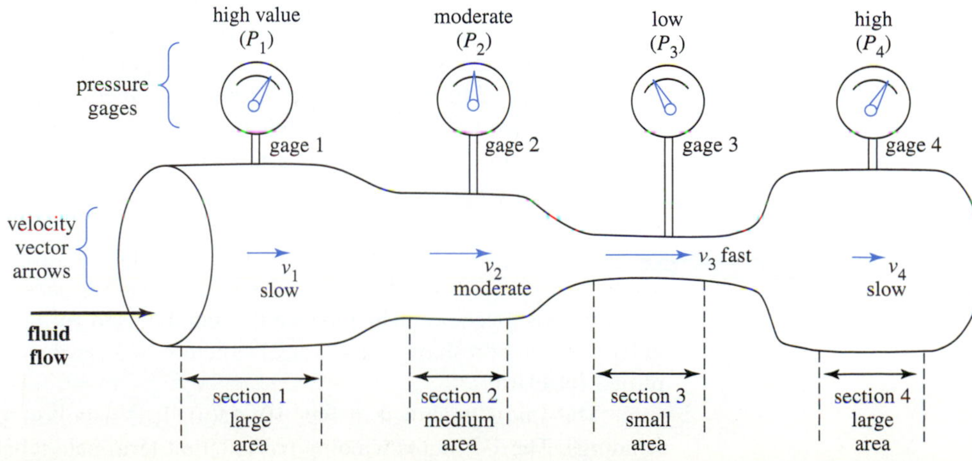

FIGURE 10–40 Illustrating Bernoulli's idea of fluid motion. Sections of large cross section have low velocity and high pressure. Sections of small cross section have high velocity and low pressure. We are assuming that the pipe's elevation is constant.

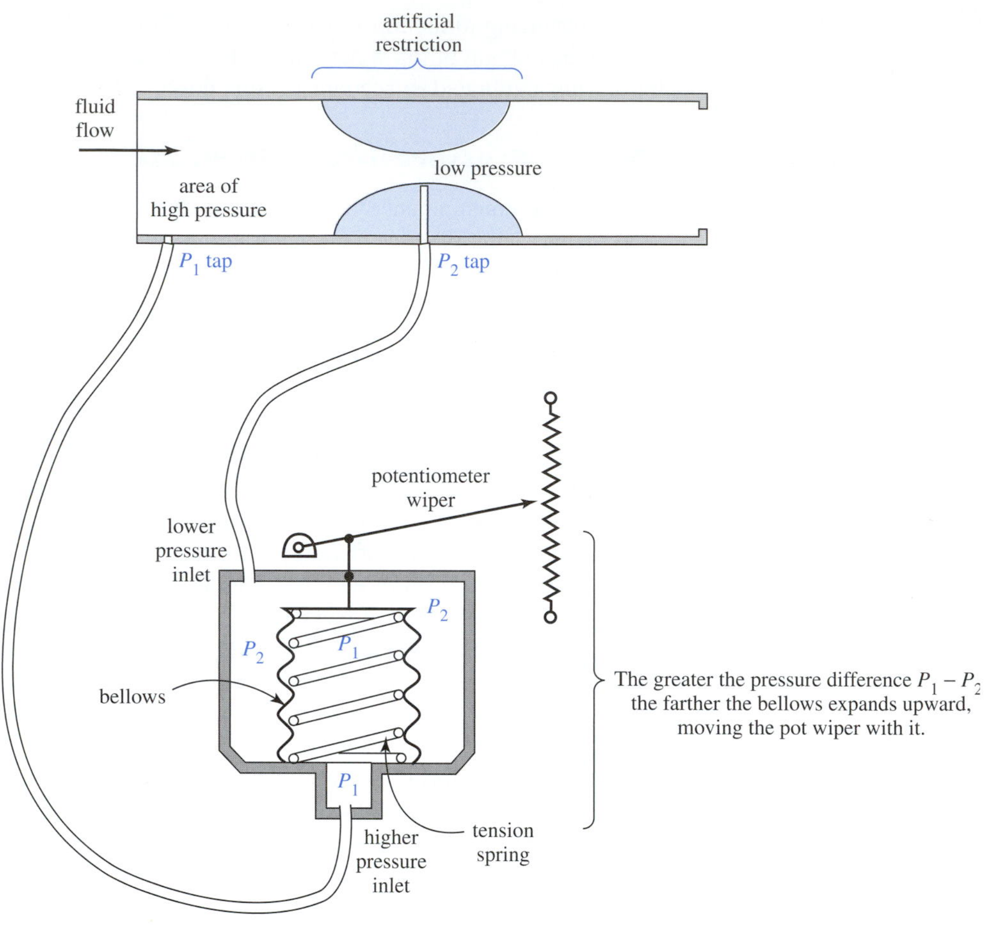

FIGURE 10–41 Measuring the pressure drop $P_1 - P_2$ by taping the pipe at an artificial restriction. This is sometimes called a *venturi*.

in which Q is the symbol for volume-flow rate and k_p is the proportionality factor, which depends on both cross-sectional areas and the fluid's density. The quantity $(P_1 - P_2)$ is called the pressure difference or *pressure drop*.

For a constant-density fluid, it is possible to measure flow rate in a pipe or duct by reducing the cross-sectional area at a certain location and measuring the two pressures P_1 and P_2. Pressure P_1 is measured upstream, away from the restriction; P_2 is measured right at the restriction, or just slightly downstream from it. This arrangement is illustrated in Fig. 10–41. The potentiometer position is proportional to pressure drop, so an instrument can be constructed around the position of the pot wiper. This instrument must be calibrated to respond to the square root of the pot wiper displacement, as Eq. 10-9 calls for.

There are various designs and shapes for the restriction in Fig. 10–41. To the extent that a restriction has smooth surfaces and is not too narrow, it can be used to measure slurries of constant density.

10-15 ■ RESOLVERS

A *resolver* is a generator-like device that can be used for precise measurement of the angular position of a shaft. The physical structure of a resolver is shown in Fig. 10–42, simplified for clarity.

The *stator* is shown in Fig. 10–42(a). It is a hollow cylinder with two multiturn windings. The 1-3 stator winding has external terminals labeled S_1 and S_3. The winding

FIGURE 10–42
Structure of a four-winding resolver. (a) The stator windings are driven by external sine-wave voltages $V_{S1\text{-}3}$ and $V_{S2\text{-}4}$. These two windings acting together create the net stator magnetic flux $\Phi_{S(total)}$. (b) The rotor windings $R_{1\text{-}3}$ and $R_{2\text{-}4}$ are also offset by 90° from one another, like $S_{1\text{-}3}$ and $S_{2\text{-}4}$.

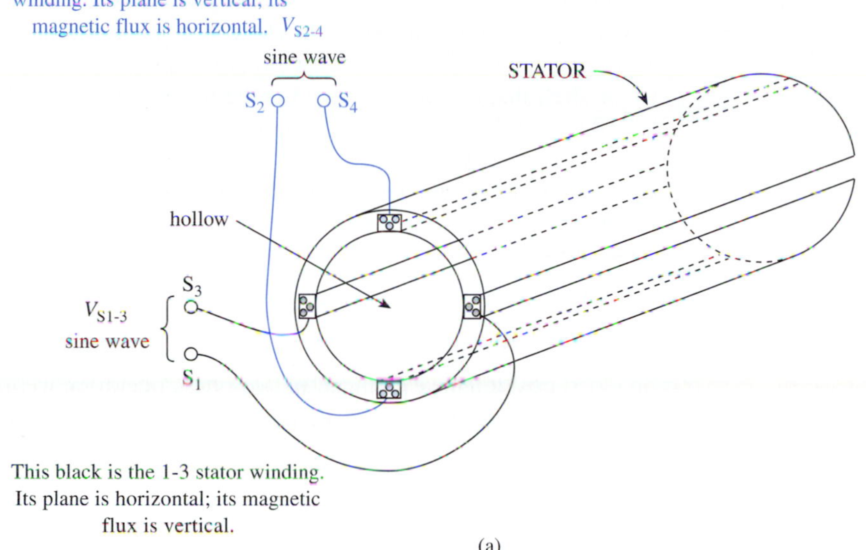
This orange is the 2-4 stator winding. Its plane is vertical; its magnetic flux is horizontal. $V_{S2\text{-}4}$
sine wave
S_2
S_4
STATOR
hollow
S_3
$V_{S1\text{-}3}$
sine wave
S_1
This black is the 1-3 stator winding. Its plane is horizontal; its magnetic flux is vertical.
(a)

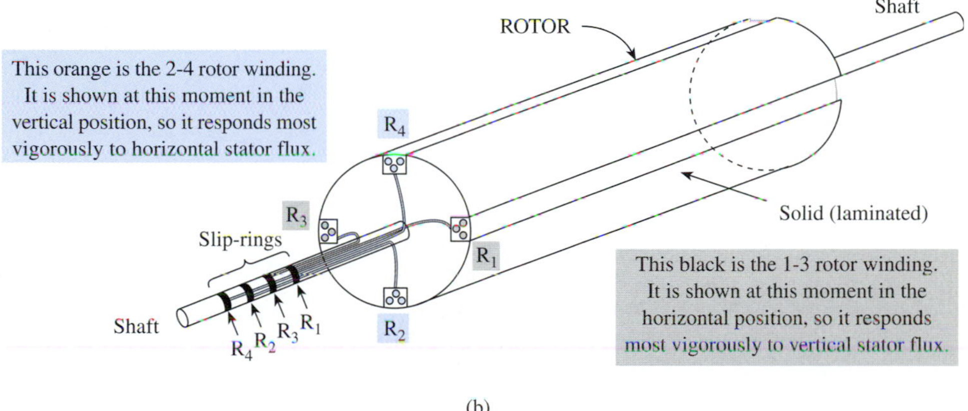
ROTOR
Shaft
This orange is the 2-4 rotor winding. It is shown at this moment in the vertical position, so it responds most vigorously to horizontal stator flux.
R_4
R_3
Slip-rings
R_1
Solid (laminated)
This black is the 1-3 rotor winding. It is shown at this moment in the horizontal position, so it responds most vigorously to vertical stator flux.
Shaft
R_4 R_2 R_3 R_1
R_2
(b)

as a whole is symbolized $S_{1\text{-}3}$. When driven by an ac sine-wave source, the $S_{1\text{-}3}$ winding creates magnetic flux Φ which is vertical in direction (perpendicular to the winding's plane, which is horizontal). The flux $\Phi_{S1\text{-}3}$ is sinusoidal itself, pointing upward toward 12 o'clock during the positive half cycle and downward toward 6 o'clock during the negative half cycle. Its magnitude varies in sine-wave fashion.

The 2-4 stator winding, $S_{2\text{-}4}$, is identical to the 1-3 stator winding but offset by 90 mechanical degrees. $\Phi_{S2\text{-}4}$ points to the right toward 3 o'clock during the positive half cycle of applied voltage $V_{S2\text{-}4}$ (+ on terminal S_2, − on S_4); the flux $\Phi_{S2\text{-}4}$ reverses and points left toward 9 o'clock during the negative half cycle of applied sine-wave voltage $V_{S2\text{-}4}$.

The rotor structure is shown in Fig. 10–42(b). There are two rotor windings, $R_{1\text{-}3}$ and $R_{2\text{-}4}$. They are brought out of the machine via slip-ring to brush contacts, on terminals labeled R_1, R_2, R_3, and R_4. The carbon brushes (blocks) are not shown in Fig. 10–42(b).

The stator windings act like the primary of a transformer, producing magnetizing flux that passes through the coils of the secondary. The rotor windings act as the secondary of the transformer.* The voltages induced in the rotor windings by transformer action are symbolized $V_{R1\text{-}3}$ and $V_{R2\text{-}4}$. These voltages depend in a very precise way on the angular position of the rotor relative to the stationary stator. Therefore, by measuring $V_{R1\text{-}3}$ and/or $V_{R2\text{-}4}$, we can get a precise indication of the shaft's position. With the rotor shaft attached to the shaft that is being measured in the industrial system, the resolver becomes a precise transducer of the angular movement of that shaft.

In some applications, a resolver's stator is driven by just one sine-wave voltage, or by two sine-wave voltages that are in phase with each other. Both of these arrangements can be described as single-phase operation. In other applications the resolver's stator windings' driving voltages, $V_{S1\text{-}3}$ and $V_{S2\text{-}4}$, are electrically phase-shifted by 90°. That situation can be described as two-phase operation.

The output indication of shaft position is quite different for the two different modes of operation. In single-phase operation we look at the rotor voltage *magnitude* to obtain our shaft position measurement. In two-phase operation we look at the rotor voltage *phase* relative to the reference voltage phase (say, $V_{S1\text{-}3}$) in order to obtain our shaft position measurement.

10-15-1 Single-Phase Operation

To understand single-phase operation, we will begin by making the simplifying assumption that only one stator winding is being driven. The other is shorted out or simply left disconnected. Suppose that stator winding $S_{1\text{-}3}$ is driven by a 60-Hz 10-V peak regulated source. This is indicated schematically as shown in Fig. 10–43.

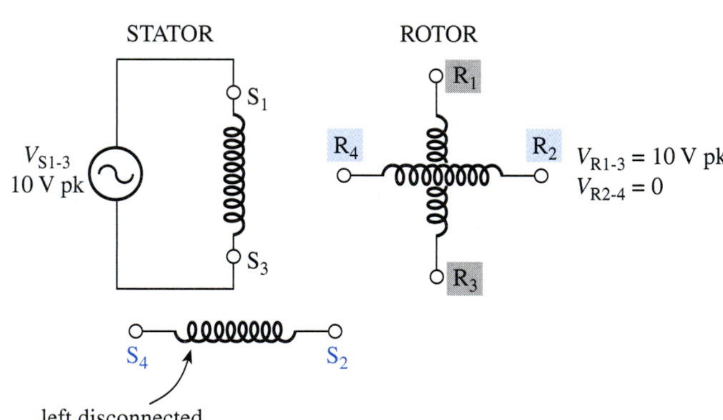

FIGURE 10–43
Schematic representation of single-phase/single-primary-winding operation of four-winding resolver. The rotor position shown and the rotor voltages correspond to the zero position of the rotor shaft.

Compare Fig. 10–43 to Fig. 10–42. In the schematic view of Fig. 10–43, we show the $S_{1\text{-}3}$ winding as oriented vertically. This is consistent with the actual physical view of Fig. 10–42, because the winding's magnetic flux is vertical. Don't be put off by the S_1 and S_3 terminals shown as vertically displaced from each other in the schematic view but horizontally displaced from each other in the actual physical view. A schematic is just a schematic; we can draw it any way that is convenient to us. Here it is convenient to indicate that the winding's flux is oriented vertically.

Likewise with the rotor windings. The schematic appearance of windings $R_{1\text{-}3}$ and $R_{2\text{-}4}$ in Fig. 10–43 corresponds to the rotor position that is shown in Fig. 10–42(b). This

*On a resolver, the roles of stator and rotor can be reversed. That is, on some resolvers the sine-wave external voltage sources are applied to rotor windings $R_{1\text{-}3}$ and $R_{2\text{-}4}$, which become the primary. The output voltages are then induced in the stator windings $S_{1\text{-}3}$ and $S_{2\text{-}4}$, which become the secondary.

is because winding $R_{1\text{-}3}$ will *respond* to vertical flux and winding $R_{2\text{-}4}$ will respond to (induce secondary voltage in response to) horizontal flux.

The 0° position. Assume that the rotor holds steady in the position represented by Figs. 10–42 and 10–43. This is called the *zero position*. The oscillating vertical flux $\Phi_{S1\text{-}3}$ is 100% linked to rotor winding $R_{1\text{-}3}$. That is, *all* the magnetic flux created by $S_{1\text{-}3}$ passes through the plane of $R_{1\text{-}3}$ because $R_{1\text{-}3}$ is in perfect alignment. If the turns ratio is 1.0:1 ($R_{1\text{-}3}$ has exactly the same number of turns as $S_{1\text{-}3}$), the induced voltage $V_{R1\text{-}3}$ will be equal to $V_{S1\text{-}3}$. That is, $V_{R1\text{-}3}$ will have a 10-V peak magnitude.

$V_{R2\text{-}4}$ will be zero, however, because *none* of the flux created by $S_{1\text{-}3}$ passes through its plane. Understand this by referring to Figs. 10–42 and 10–43.

The 30° position. Now suppose the rotor moves to the 30° position, shown in physical cross section in Fig. 10–44(a) and schematically in Fig. 10–44(b). With the $R_{1\text{-}3}$

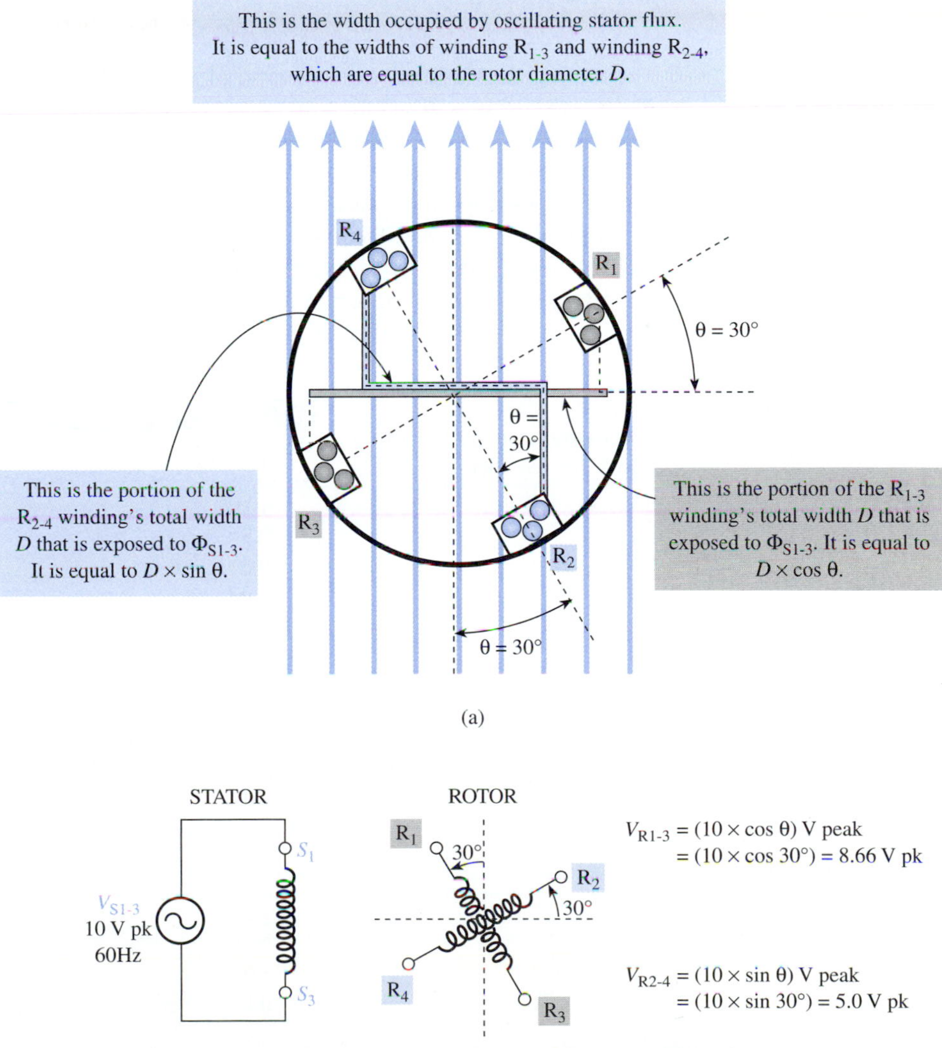

This is the width occupied by oscillating stator flux. It is equal to the widths of winding R1-3 and winding R2-4, which are equal to the rotor diameter D.
R4
R1
θ = 30°
θ = 30°
R3
R2
This is the portion of the R2-4 winding's total width D that is exposed to ΦS1-3. It is equal to D × sin θ.
This is the portion of the R1-3 winding's total width D that is exposed to ΦS1-3. It is equal to D × cos θ.
θ = 30°
(a)
STATOR
ROTOR
S1
S3
VS1-3
10 V pk
60Hz
R1
30°
R2
30°
R4
R3
VR1-3 = (10 × cos θ) V peak
= (10 × cos 30°) = 8.66 V pk
VR2-4 = (10 × sin θ) V peak
= (10 × sin 30°) = 5.0 V pk
(b)

FIGURE 10–44 Conditions for the rotor windings when the rotor moves to $\theta = 30°$. (a) Cross-sectional view. (b) Schematic diagram.

winding tipped 30° off horizontal, it no longer is able to present its entire cross-sectional area to the oscillating magnetic flux. Its length along the axis of the rotor is unaffected, but its effective width is reduced to a factor $\theta = \cos 30° = 0.866$ times its actual width, *D*. This is the portion of the actual width that is perpendicular to the vertical flux. Since the winding's effective width is reduced to 0.866 of its actual value, so is its flux linkage, and so too is its induced voltage. Thus, $V_{R1\text{-}3}$ decreases from 10 V pk to 8.66 V pk. Study Fig. 10–44(a) carefully to understand this.

Meanwhile, the $R_{2\text{-}4}$ winding is tipped 30° away from vertical. This winding now presents an effective width of $\sin\theta$ times its actual width *D*. Assuming identical windings (turns ratio $n = 1.0$), its voltage increases from 0.0 to 5.0 V peak, since $(10 \times \sin\theta)$ V $= 10 \times \sin 30° = 10 \times 0.5 = 5.0$ V peak.

In general, we can say:

$$V_{R1\text{-}3} = V_{S1\text{-}3} \cos\theta \qquad \textbf{(10-10)}$$

and

$$V_{R2\text{-}4} = V_{S1\text{-}3} \sin\theta \qquad \textbf{(10-11)}$$

assuming a rotor-to-stator turns ratio of $n = 1.0$, which is usual.

By using a stored look-up table, a microcomputer can translate every magnitude of $V_{R1\text{-}3}$ into a particular angular position of the shaft. Table 10–1 is a partial look-up table for angles from 0° to +90°.

A look-up table is a collection of paired data values that are permanently stored in a computer's memory. By presenting one variable's value to the computer, it is able to go to the look-up table to retrieve the other variable's value. For a resolver-based transducer, the value of $V_{R1\text{-}3}$ is presented to the computer, and it retrieves the value of the angle of the shaft. This is portrayed in Fig. 10–45.

TABLE 10–1
Look-up table for $V_{R1\text{-}3}$ to θ in 5° increments*

$V_{R1\text{-}3}$ (V)	θ (DEG)	$V_{R2\text{-}4}$ (V)
10.0	0	0.00
9.96	5	0.87
9.85	10	1.74
9.60	15	2.59
9.40	20	3.42
9.06	25	4.23
8.66	30	5.00
8.19	35	5.74
7.66	40	6.43
7.07	45	7.07
6.43	50	7.66
5.74	55	8.19
5.00	60	8.66
4.23	65	9.06
3.42	70	9.40
2.59	75	9.60
1.74	80	9.85
0.87	85	9.96
0.00	90	10.0

*The value of $V_{R2\text{-}4}$ could be used just as well as $V_{R1\text{-}3}$, but the data-value pairings in the table would be different, naturally.

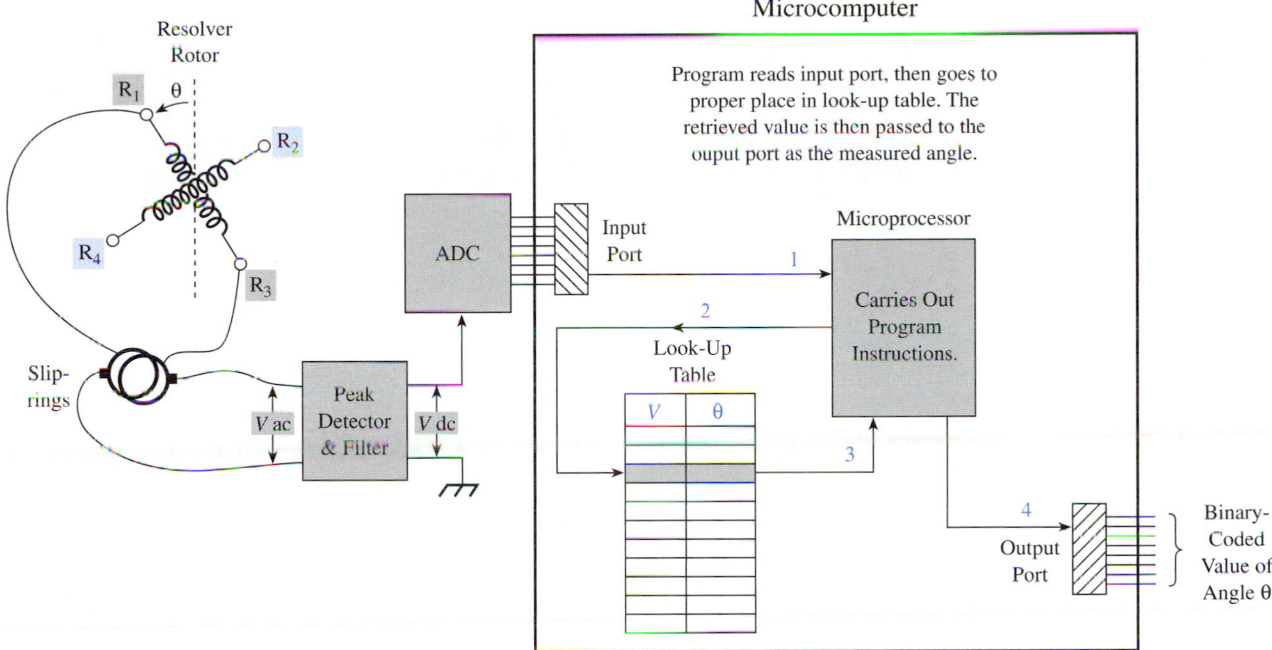

FIGURE 10–45
The idea of a microcomputer using a look-up table for transducer functioning.

In Fig. 10–45, the microprocessor (μP) carries out instructions in its program that accomplish the following:

1. It reads the binary coded value of $V_{R1\text{-}3}$ from the computer's input port.
2. It uses this read value to guide it to the proper location in the look-up table.
3. It retrieves the value of the measured variable from the look-up table.
4. It writes the binary-coded value of the measured variable to the output port.

In some cases a look-up table is stored on one of the computer's disks. It may remain there during use, or the program may bring the table into the computer's RAM for easier and quicker access. Alternatively, if a look-up table is used full-time by the computer, it may be stored in the computer's ROM.

Figure 10–45 conveys the concept of a measurement transducer based on a computer's look-up table action as intermediary, but it is not a complete picture of resolver action. It does not show the mechanism by which the transducer can distinguish 30° from 330° (−30°). And the same for distinguishing 30° from 150° or 210°. These distinctions are made by detecting the phase of $V_{R1\text{-}3}$ relative to $V_{S1\text{-}3}$ and the phase of $V_{R2\text{-}4}$ relative to $V_{S1\text{-}3}$. Figure 10–46 (page 430) shows the phase relations among these three voltages for the four shaft angles 30°, 150°, 210°, and 330°. The microcomputer-based transducer of Fig. 10–45 must take these phase relations into account in order to distinguish among the four angles.

Both stator windings driven in parallel. In some resolver applications the two stator windings are both driven by the ac source, or by two ac sources that may have different magnitudes but are in phase with each other. Let us consider only the former situation, which is shown schematically in Fig. 10–47(a).

With both $S_{1\text{-}3}$ and $S_{2\text{-}4}$ driven by a 10-V peak source, their combined flux $\Phi_{S(\text{total})}$ is located at the 45° midpoint between the windings. This is shown in Fig. 10–47(b). Its

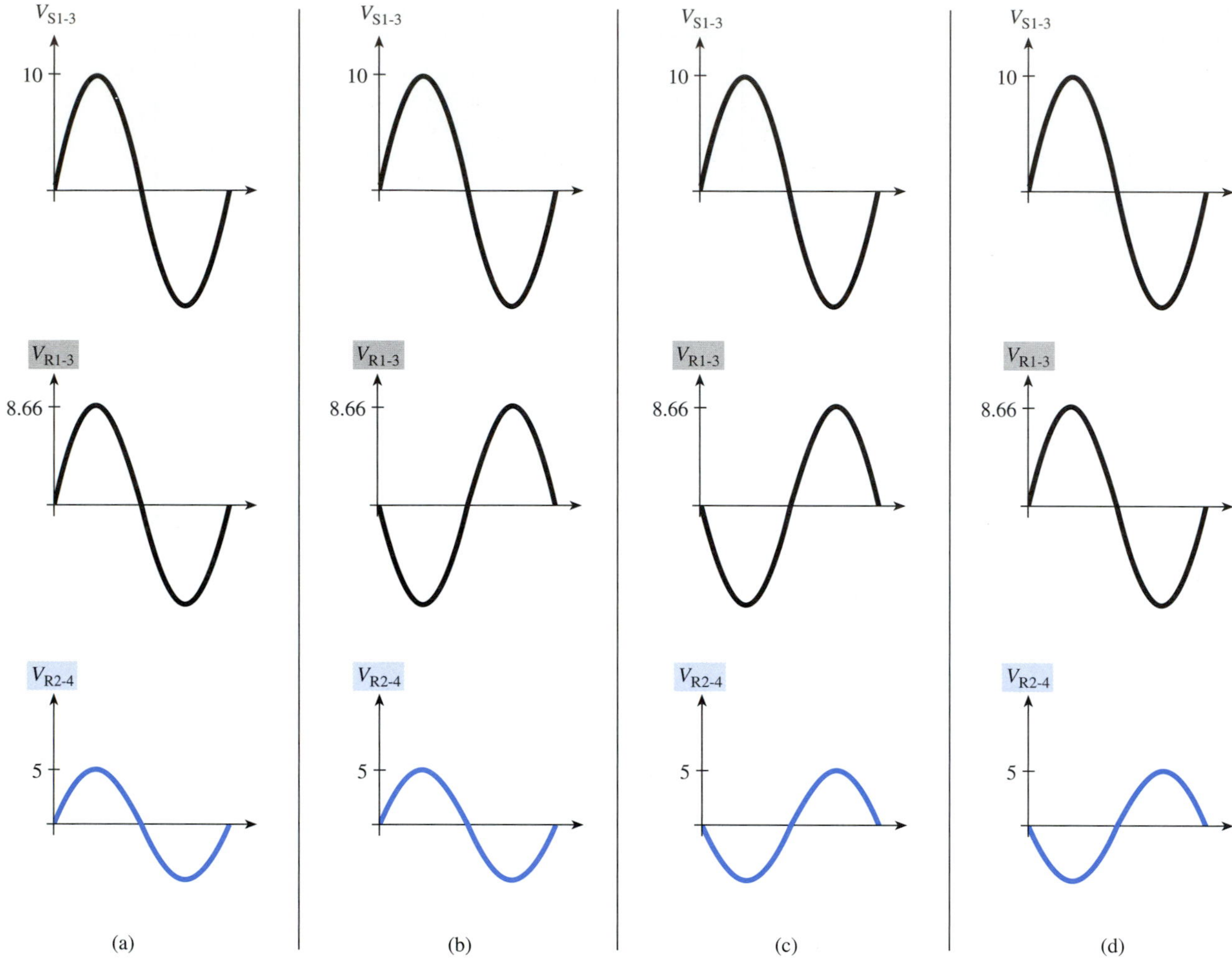

FIGURE 10–46
Phase relations (and relative magnitudes) for the four different angular positions that place the R_{1-3} winding offset by 30° from the horizontal plane (see Fig. 10–44). For resolvers, the ac sine-wave voltages are defined positive when the first subscript number is positive relative to the second subscript number. For example, V_{S1-3} is in its positive half cycle when terminal S_1 is positive relative to terminal S_3; V_{R2-4} is defined positive when terminal R_2 is positive relative to terminal R_4. (a) 30°. (b) 150°. (c) 210°. (d) 330°, also called −30°.

magnitude is the vector sum of the individual winding fluxes Φ_{S1-3} and Φ_{S2-4}. By the Pythagorean theorem,

$$\Phi_{S(total)} = \sqrt{(\Phi_{S1-3})^2 + (\Phi_{S2-4})^2}$$

so with $n = 1.0$ it follows that

$$V_{R1-3(max)} = \sqrt{(10\text{ V})^2 + (10\text{ V})^2} = 14.14\text{ V peak}$$

and

$$V_{R2-4(max)} = 14.14\text{ V peak}$$

These maximum values appear in the voltage equations for V_{R1-3} and V_{R2-4}, namely

FIGURE 10–47 Resolver action with both stator windings driven by the same ac source. (a) Schematic representation. (b) Stator cross section showing the 45° shift in orientation of oscillating flux, and its larger magnitude. (c) For a rotor displacement angle θ, the equations for rotor winding voltage magnitudes. (d) Waveform graphs for shaft displacement angle θ = 30°.

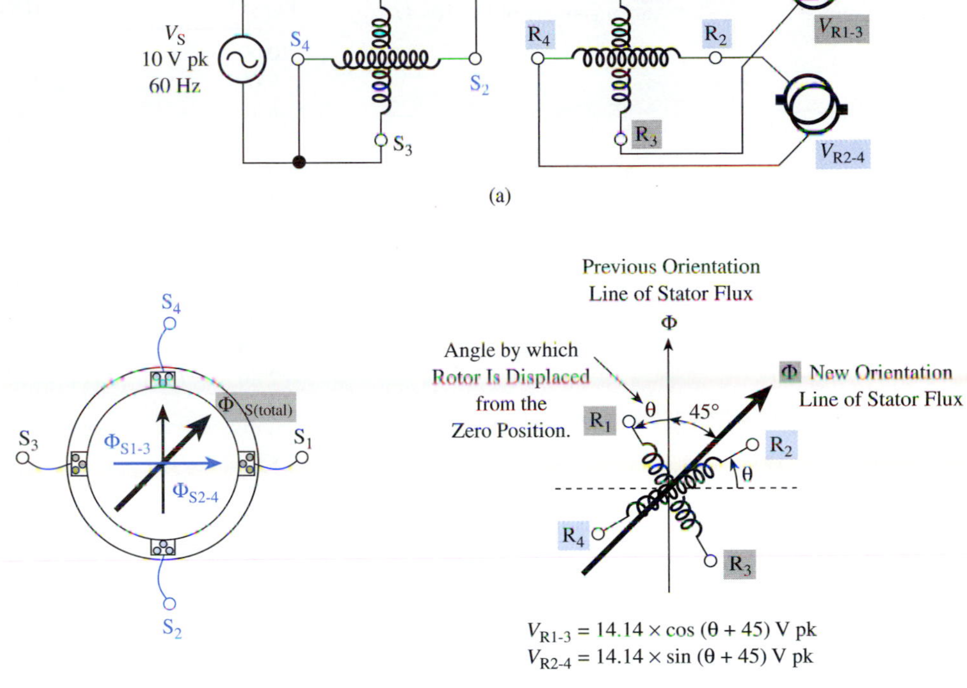

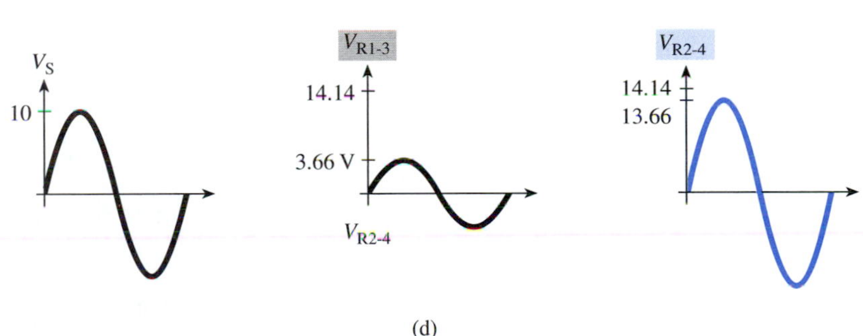

$$V_{R1\text{-}3(pk)} = 14.14 \times \cos(\theta + 45°)\ \text{V} \qquad \textbf{(10-12)}$$

$$V_{R2\text{-}4(pk)} = 14.14 \times \sin(\theta + 45°)\ \text{V} \qquad \textbf{(10-13)}$$

which are shown alongside the displaced rotor schematic diagram of Fig. 10–47(c). The angle 45° must be added to the arguments of both cosine and sine functions because the new stator flux direction (with both stator windings powered) is 45° clockwise from the previous stator flux direction (with just one stator winding powered). Study Fig. 10–47(c) to understand this.

A 30° rotor shaft displacement (θ = 30°) is graphed in Fig. 10–47(d). The peak magnitude of $V_{R1\text{-}3}$ is given by Eq. (10-12).

$$\begin{aligned} V_{R1\text{-}3(pk)} &= 14.14 \times \cos(30° + 45°) \\ &= 14.14 \times \cos(75°) = 14.14 \times 0.2588 = 3.66\ \text{V} \end{aligned}$$

$V_{R2\text{-}4}$ magnitude is given by Eq. (10-12).

TABLE 10–2
Look-up table for the arrangement of Fig. 10–47

$V_{R1\text{-}3}$ (V)	θ (DEG)	$V_{R2\text{-}4}$ (V)
10.00	0	10.00
9.09	5	10.83
8.11	10	11.58
7.07	15	12.25
5.98	20	12.82
4.84	25	13.29
3.66	30	13.66
2.46	35	13.93
1.23	40	14.09
0.00	45	14.14
1.23	50	14.09
2.46	55	13.93
3.66	60	13.66
4.84	65	13.29
5.98	70	12.82
7.07	75	12.25
8.11	80	11.58
9.09	85	10.83
10.00	90	10.00
10.83	95	9.09
11.58	100	8.11
12.25	105	7.07
12.82	110	5.98
13.29	115	4.84
13.66	120	3.66
13.93	125	2.46
14.09	130	1.23
14.14	135	0.00

$$V_{R2\text{-}4(pk)} = 14.14 \times \sin(75°)$$
$$= 14.14 \times 0.9659 = 13.66\text{ V}$$

All three waveforms in Fig. 10–47(d) are in phase with one another because the rotor is displaced less than 90° from the flux orientation line.

For use as a microcomputer-controlled measurement transducer, this configuration of the resolver requires a revised look-up table, of course. It is shown in Table 10–2. Compare the 30° row in that table to the waveform magnitudes in Fig. 10–47(d).

10-15-2 Two-Phase Operation

Figure 10-48(a) is a schematic representation of a resolver being driven by two 90°-phase-shifted equal-magnitude ac sources. We are concentrating on just one of the rotor windings, $R_{1\text{-}3}$. Here is an important and fascinating fact: Driving two 90°-apart stator windings with two 90°-phase-shifted sine-wave sources causes a revolving magnetic field to be produced. That is, the magnetic flux Φ no longer has a fixed orientation line with an instantaneous magnitude that goes through sine-wave oscillations. Instead, Φ has a constant magnitude, but it moves its orientation line at a constant rate. Φ revolves around the rotor, making one complete 360° mechanical revolution during each cycle of the ac sources. To understand how this happens, you may want to look ahead to Fig. 14–1.

FIGURE 10–48
Two-phase operation mode for a resolver. (a) Schematic. (b) Mechanical shaft position alters *phase* of $V_{R1\text{-}3}$, not magnitude.

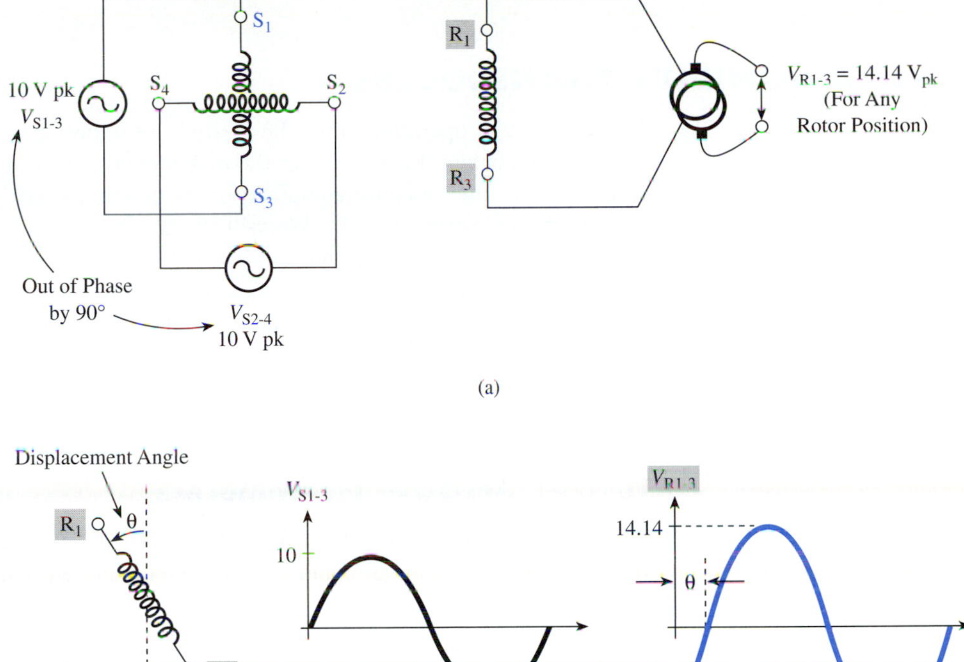

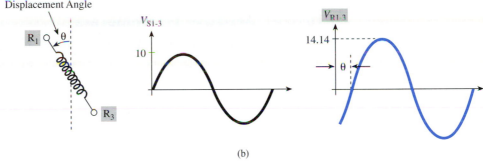

Under this operating condition, the magnitude of the induced voltage in the rotor secondary winding is constant at 14.14 V peak, no matter what the mechanical position of the rotor shaft. The feature that changes is the *phase* of $V_{R1\text{-}3}$. As shown in Fig. 10–48(b), if the shaft is displaced by angle θ from its mechanical zero position, the sine wave $V_{R1\text{-}3}$ is electrically phase-shifted by θ degrees relative to source $V_{S1\text{-}3}$. The transducer must measure this electrical phase shift to represent the angular position of the measured shaft. This is suggested in Fig. 10–49.

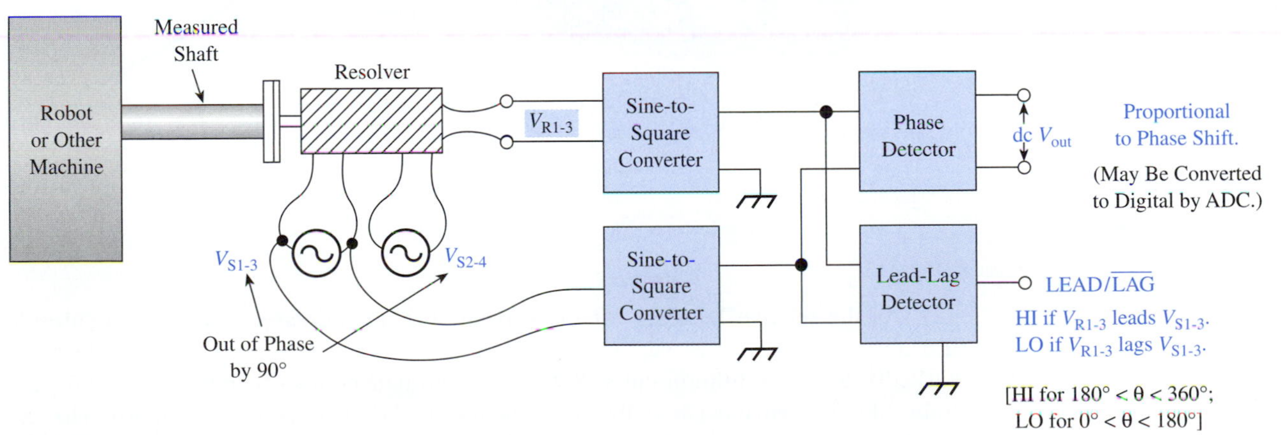

FIGURE 10–49
In a two-phase resolver transducer, the phase shift between $V_{R1\text{-}3}$ and $V_{S1\text{-}3}$ is measured by processing the sine waves into square waves, then running the two square waves into a phase-detector. The average voltage (dc voltage) at the output of the phase-detector is proportional to measured angle θ. An exclusive-OR phase-detector can't distinguish leading from lagging, so a separate digital lead-vs-lag detector must furnish that bit of measurement information.

10-16 ■ HUMIDITY TRANSDUCERS

Many industrial operations must be carried out under specific and controlled conditions of moisture content. In some cases the moisture contained in the ambient air is important; in other cases the moisture contained in the product itself is more important to the success of the industrial process. We will discuss two common methods of measuring the moisture content of ambient air and one method of measuring the moisture content of a sheet product. The most common scale for measuring the moisture content of air is the *relative humidity* (RH) scale. Formally, relative humidity is the ratio of the water vapor (moisture) actually present in the air to the maximum amount of water vapor that the air could possibly hold.

10-16-1 Resistive Hygrometers

A resistive hygrometer is an element whose resistance changes with changes in the relative humidity of the air in contact with the element. Resistive hygrometers usually consist of two electrodes of metal foil on a plastic form. The electrodes do not touch each other, and they are electrically insulated from each other by the plastic form. A solution of lithium-chloride is then used to thoroughly coat the entire device. This construction is illustrated in Fig. 10–50(a).

FIGURE 10–50
(a) Physical appearance of a resistive hygrometer. (b) Curve of resistance versus relative humidity for a resistive hygrometer.

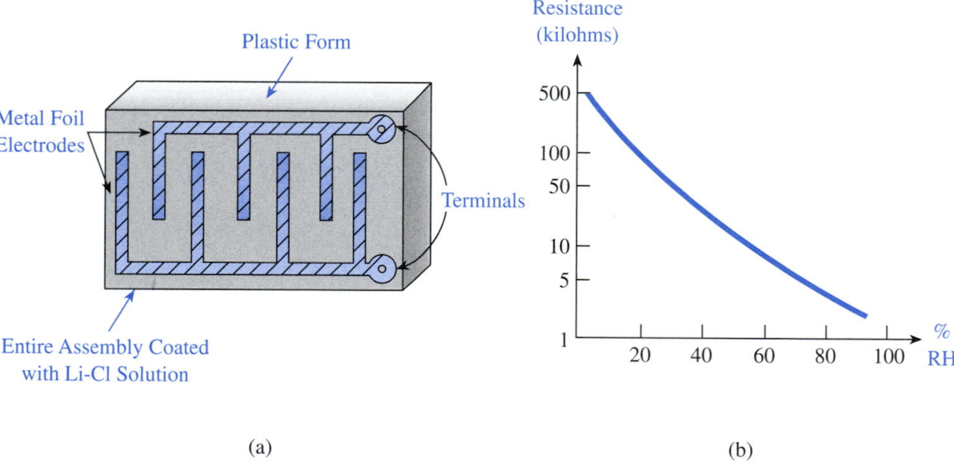

As the relative humidity of the surrounding air increases, the lithium-chloride film absorbs more water vapor from the air. This causes its resistance to decrease markedly. Since the lithium-chloride film is in intimate contact with the two metal electrodes, the resistance between the electrode terminals also decreases markedly. The terminal resistance can then be related to the relative humidity. A typical characteristic curve of resistance versus relative humidity for a resistive hygrometer is shown in Fig. 10–50(b).

Resistive hygrometer transducers cannot be used over the full range of relative humidities, from 0% to 100%. Most of them have an upper safe operating limit of about

90% RH. Exposure to air with a relative humidity higher than 90% can result in excess water absorption by the lithium-chloride film. Once that happens, the resistance characteristics of the hygrometer are permanently altered.

10-16-2 Psychrometers

A psychrometer is a relative humidity-measuring device which has two temperature transducers (thermometers). One of the thermometers measures the temperature of an element which is simply located in the ambient air. This element is called the *dry bulb*. The second thermometer measures the temperature of an element which is surrounded by a fibrous material saturated with pure liquid water. This element is called the *wet bulb*. The ambient air is forced to flow over both the dry bulb and the wet bulb by a blower of some sort. This arrangement is illustrated in Fig. 10–51(a).

The temperature transducers shown in Fig. 10–51(a) are platinum-wire RTDs. The dry bulb stays at the temperature of the flowing ambient air, so the dry bulb temperature is simply equal to the ambient air temperature, regardless of its humidity. The wet bulb, however, is colder than the dry bulb because of the evaporation of the liquid water contained in the fibrous material which surrounds the wet bulb. The greater the evaporation rate of the water, the greater the cooling effect on the wet bulb and the lower the wet bulb temperature reading. The evaporation rate depends on the relative humidity of the moving air. If the air is dry (low relative humidity), the evaporation rate will be great, and the wet bulb will be much colder than the dry bulb. If the air is moist (high relative humidity), the evaporation rate will not be so great, and the wet bulb will be only a little bit colder than the dry bulb. The *difference* in temperatures is therefore an indication of the relative humidity of the air.

To understand why the evaporation rate out of the water-soaked fibrous material depends on relative humidity, think of it this way: If the ambient air were at 100% RH, it would not be able to absorb any more water at all, because it would already be saturated. Therefore the water would not evaporate from the fibrous material at all. It is easy to reason back from this extreme condition to understand that the drier the air is, the better it is at accepting extra water (causing evaporation).

Therefore, the lower the relative humidity, the faster the water evaporates from the fibrous material.

The percent relative humidity can be read from a psychrometer table by knowing two things:

1. The dry bulb temperature.
2. The difference in temperatures between the two bulbs.

An abbreviated psychrometer table of this type is shown in Fig. 10–51(b). More precise psychrometer tables, marked in 0.5°F graduations, are available in psychrometer handbooks.

The psychrometer can be used to take manual readings of percent relative humidity, or it can be used in a control application to automatically maintain a certain desired humidity. We will explore an automatic humidity control system using a psychrometer in Chapter 15.

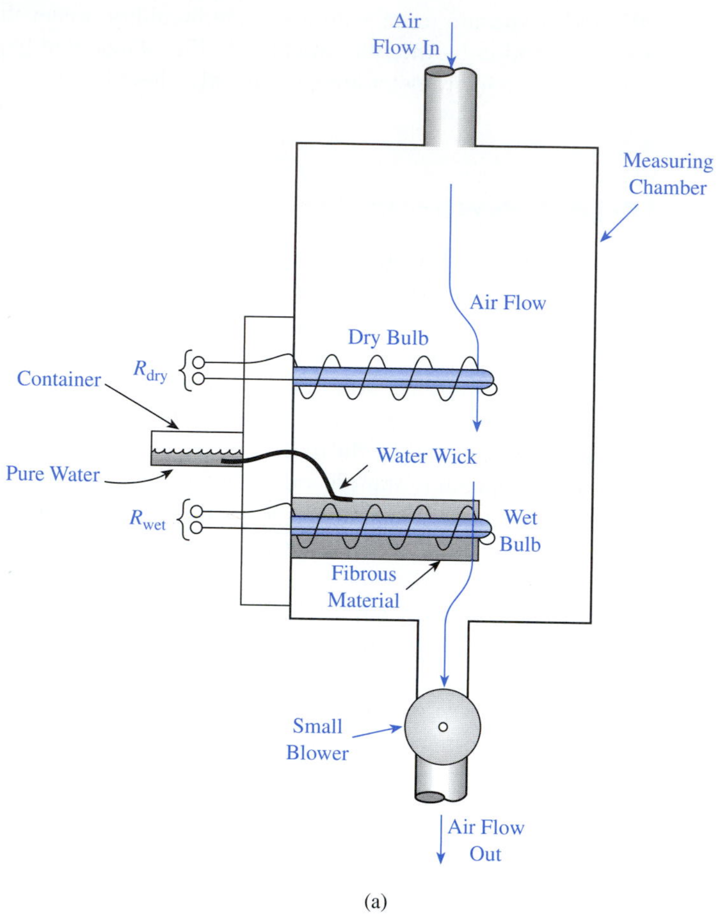

FIGURE 10–51
(a) Arrangement of the dry bulb and the wet bulb in a psychrometer. The example shows RTDs as the temperature detectors. (b) Psychrometer table, which relates the three variables of dry bulb temperature, temperature difference between bulbs, and percent relative humidity.

Difference Between Dry Bulb and Wet Bulb Temperatures (°F)

Dry Bulb Temp. (°F)	1	5	10	15	20	25
40	92	60	—	—	—	—
50	93	68	38	12	—	—
60	94	73	49	26	6	—
70	95	77	55	37	20	3
80	96	79	61	44	29	16
90	96	81	65	50	36	24
100	96	83	68	54	42	31

Relative Humidities (%)

(b)

10-16-3 Detecting Moisture Conditions in a Solid Material

Hygrometers and psychrometers are devices which are capable of measuring the moisture content of air as expressed on a universal well-accepted measurement scale, the relative humidity scale. Given the proper calibration data, they can measure the relative humidity of *any* mixture of gases (not only air) on this well-known scale. Sometimes, though, it is

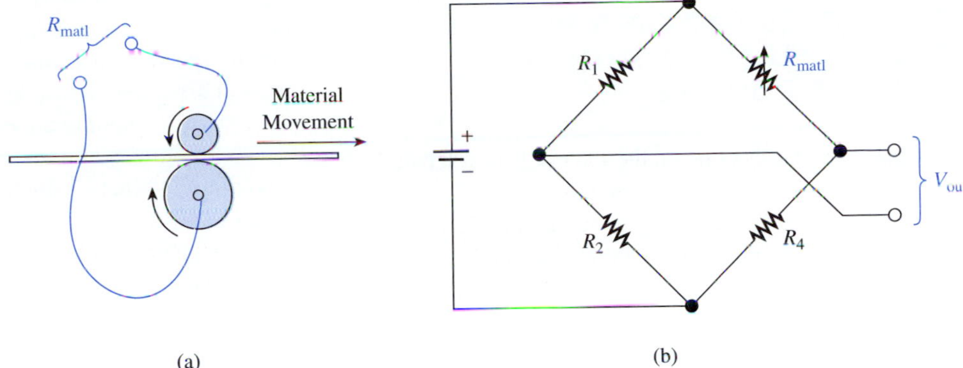

FIGURE 10–52 (a) Method for detecting the moisture content of a sheet material. (b) The bridge circuit establishes a correspondence between the output voltage and the moisture content.

not important to know moisture content as expressed on some universal scientific scale. Sometimes it is only important to know whether the moisture content is above or below a certain desired moisture content, or *reference* content. This is especially true in industrial situations where a continuous strip of material (paper, textiles, etc.) must be handled and processed. When the moisture content of a solid strip of material in relation to some reference content is all that is needed, the usual method of taking the measurement is shown in Fig. 10–52(a). In this figure the strip of material is passed between two rollers which contact opposite sides of the strip. Each roll has a lead wire attached to its stationary bearing support. The resistance measured between the leads is then an indication of the moisture content of the material in question. Virtually all materials will have a low resistance when wet and a higher resistance when dry. The user determines the resistance between the lead wires when the material has the optimum moisture content for his purposes. The system then detects any deviation from that resistance and attempts to correct that deviation.

A circuit for detecting moisture/resistance variations is the familiar Wheatstone bridge drawn in Fig. 10–52(b). Suppose the resistance of the material (R_{matl}) is 3250 Ω when the moisture content is just right. The bridge could then be built with $R_1 = R_2 = R_4 = 3250\ \Omega$. Then, if the material were at the right moisture level, R_{matl} would equal 3250 Ω, and the bridge would balance. V_{out} would equal 0. If the moisture content were to change because of some upset in the system, the bridge would go out of balance. The magnitude of V_{out} would indicate the amount of deviation between actual and desired conditions, and the polarity of V_{out} would indicate the direction of deviation. The V_{out} signal could then be applied to some sort of controller to restore the moisture content to the proper level.

In this type of control application, there is no measurement taken per se. That is, there is no measurement result obtained which could be expressed on a universal measurement scale. There is simply a comparison of an actual moisture condition with a desired moisture condition, and no attempt is made to express these moisture conditions on a numerical scale.

TROUBLESHOOTING WITH A MULTICHANNEL DIGITAL STORAGE SCOPE

In Troubleshooting on the Job in Chapter 8, part **a,** your answer should state that the MagLev system malfunction could be in the Hall-effect sensor mounted on the concrete sidewall beside the track.

All the testing/troubleshooting procedures described in Troubleshooting on the Job in Chapters 4, 5, and 8 would have to be performed repetitively, for one propulsion-coil pair after another, in the section of track containing the problem. This is because there is no way of knowing in advance exactly which particular propulsion-coil pair/control module is causing the train's propulsion problem within that section of track.

Initially, a propulsion problem can be narrowed down to within approximately 25 m (occurring somewhere within a 25-m-long section of track). Then, since the coils are approximately 0.7-m apart, the number of possible propulsion-coils/control modules that could be at fault is given by

$$\text{Number of coil pairs} \approx \frac{25\text{ m}}{0.7\text{ m/coil pair}}$$
$$\approx 36\text{ coil pairs}$$

Thus the testing/troubleshooting procedures of Assignments 4, 5, and 8 would have to be repeated about 36 times. Before beginning such an extensive checkout process, it is sensible to be certain that the system malfunction is *not* in any one of the 36 Hall-effect sensors, or their associated dif amps and shielded wiring (refer to Fig. 8–4). This can be done by using digital storage oscilloscopes with multiple channels. For example, the digital storage oscilloscope in Fig. 10–53 has four channels monitoring four Hall-effect transducer inputs. The scope has read/write digital memory (RAM)* that is partitioned into segments. In this example there are 4096 words of memory (4K) partitioned into four 1024-word segments (1K per segment). After the scope is triggered, each one of its input channels is sampled 1024 times. Every individual sample sees a certain instantaneous analog input voltage, which is converted to digital by an ADC (8-bit resolution, generally) and stored in a particular word within that channel's segment of RAM. The next sample that is taken at that input channel is converted to 8-bit digital code and stored in (written into) the next higher-numbered word within that channel's segment of RAM. This action is carried out 1024 times, so the scope ends up with a digital representation of the input voltage at 1024 separate time instants. The amount of real time that is represented by the digital data is equal to the sample interval (reciprocal of sample rate) multiplied by 1024.

The above sequence of events is carried out for each input channel, in *chopped* fashion. That is, the signal-acquisition controller takes a sample from channel 1, then a sample from channel 2, then from channel 3, then from channel 4. Then it recycles through the four channels, taking a second sample from channel 1, then a second sample from channel 2, and so on. Thus, when the acquisition process is complete, four separate input voltage waveforms have been stored in RAM. In the example of Fig. 10–53 the input waveform that was acquired at channel 1 is stored in words H000 through H3FF; the input waveform acquired at channel 2 is stored in words H400 through H7FF; channel 3 is in words H800 through HBFF; channel 4 is in words HC00 through HFFF. As the user of the oscilloscope, you do not have to concern yourself with the specific addresses that are holding the sampled data. We have specified them here only for explanatory purposes.

Now that the waveform data has been written into read/write memory, we do not retrigger the scope. Doing so would repeat the entire sampling operation, overwriting and destroying the data. Instead, now the scope's control circuitry reads the data back from the RAM, reconverts it to analog (DAC), and displays it on the screen. We say that the scope operates in "real time," because the amount of time that is required for the beam to sweep across the screen and draw the voltage waveform is the same as the actual acquisition time—the amount of time during which the data was originally

*The acronym RAM stands for *R*andomly *A*ccessible *M*emory, which is memory whose words can be accessed in any order, not necessarily sequentially. But it's not the *randomness* of Read/Write memory that is its important defining feature; instead it's the fact that we can perform *both* operations, reading and writing. Thus a better acronym for this kind of memory would be RWM.

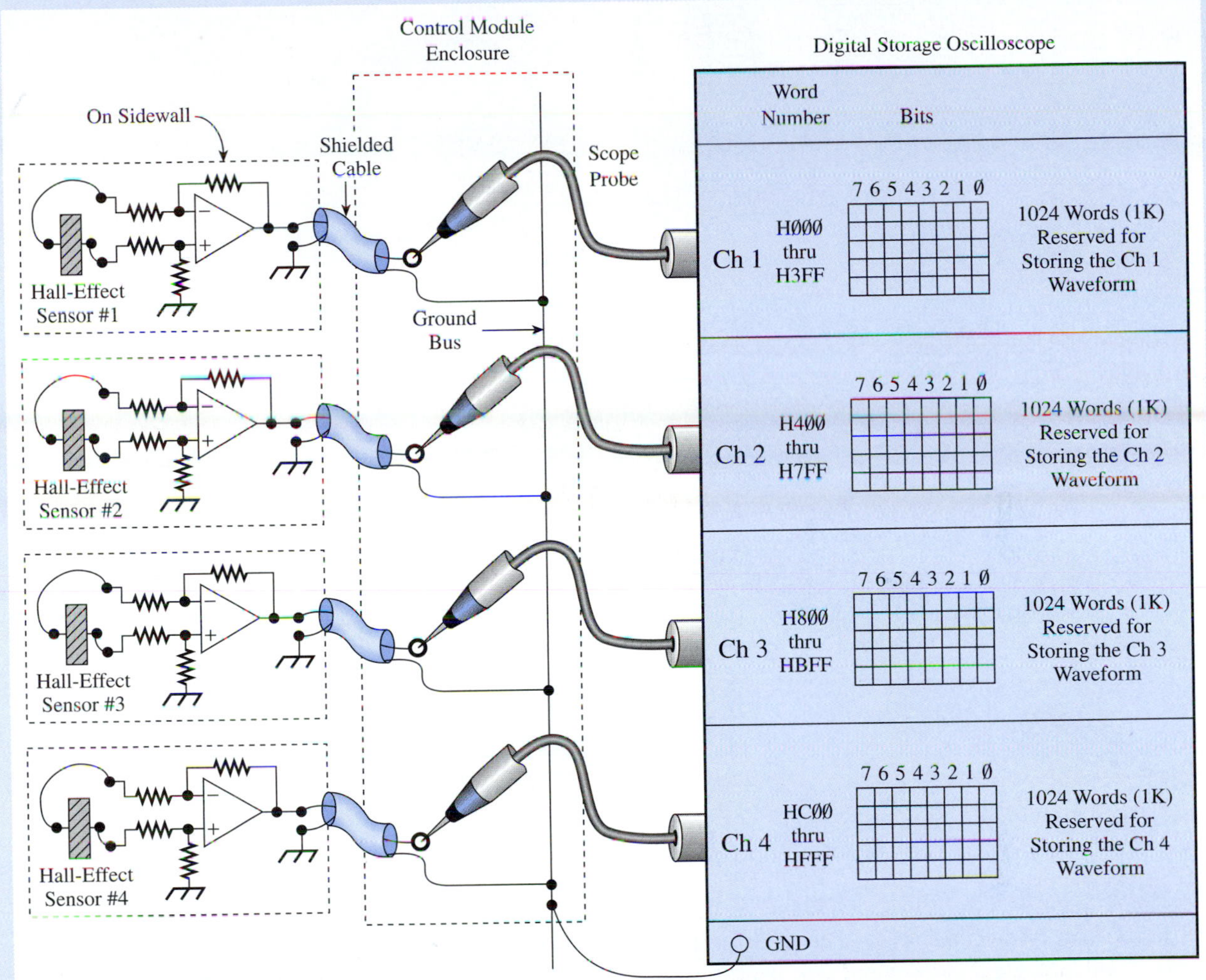

FIGURE 10–53
Four-channel digital storage scope. Its digital memory (RAM) is partitioned into four equal segments. Each RAM segment is dedicated to storing the digital representation of one waveform.

sampled and stored.* By repeating the sweep-and-read-the-memory operation, the scope presents us with a stable waveform display that we can examine carefully, at our leisure. By alternating the display process among the four separate RAM segments, the oscilloscope shows us the timing relations among the four input waveforms. Thus, on its first sweep across the screen the scope reads RAM addresses HØØØ through H3FF sequentially, converts their digital contents to analog, and puts the analog voltage into the vertical deflection amplifier. The waveform of input No. 1 appears on screen. On its second sweep across the screen the cope's control circuit reads RAM addresses H4ØØ through H7FF sequentially, using that information to drive the vertical deflection amplifier—the input No. 2 waveform appears on screen, somewhat lower

*Real-time operation can be contrasted with advanced sampling techniques in which the data is acquired over many cycles of the waveform (a few words from one cycle, a few more words from the next cycle, a few more words from the cycle after that, and so on). For such advanced digital acquisition techniques to work, the waveform must be highly repetitive. The MagLev's V_{sens} waveform consists of only about 2.5 cycles, which is not repetitive enough for such methods to succeed. Therefore V_{sens} must be sampled with a real-time digital storage oscilloscope.

A 16-channel digital storage scope in use.
Courtesy of Tektronix, Inc.

(vertically) than the previous sweep; then again for RAM words H8ØØ through HBFF—input No. 3 on screen, somewhat lower; finally, RAM words HC ØØ through HFFF—input No. 4 on screen, near the bottom. Then repeat the entire display process before the phosphor has stopped glowing from the previous process. This is the principle of the digital storage oscilloscope.

If the maximum number of input channels were four, we would be able to test the performance of only four Hall-effect sensors as a MagLev train zips past, but real-time digital storage scopes are available with 12 input channels. Thus, three such 12-channel scopes should be able to capture all 36 V_{sens} waveforms from the section of MagLev track that is malfunctioning.

YOUR ASSIGNMENT

Show how you would set up three 12-channel digital storage oscilloscopes to inspect the V_{sens} waveforms from all 36 of the Hall-effect input transducers that are in doubt. It is not necessary to see the entire waveform from every Hall-effect transducer; all that is necessary is to see enough of the waveform to verify that its positive (North) peak exceeds $+10$ V and its negative (South) peak goes below -10 V.

Your three scopes have external trigger jacks (TRIG) with standard Level and Slope selectability. The last Hall-effect transducer *before* the problem area of track is guaranteed to produce a proper V_{sens} waveform (see Fig. 8–19[a]). Assume that the next train that is scheduled to come by will be moving to the right in Figs. 4–15, 4–16, and 4–17 (North supermagnet leading).

To make your trigger-timing decisions, you will probably need to look at waveform sketches showing the precise time relationships among the 36 transducers. Therefore, start your assignment by making time-relation sketches of the proper V_{sens} waveforms for the following numbered transducers: number 0 (the last one before the problem area of track, known to be working correctly), number 1, number 2, number 12 (the last input transducer that will be captured by scope #1), number 13 (the first input transducer captured by scope #2), number 24 (last by scope #2), number 25 (first by scope #3), and number 36 (last by scope #3).

To establish trigger timing for the three scopes it will be necessary to construct a circuit using an op amp comparer with a negative clipper (as shown in Fig. 8–18) and two (2) one-shots. You must correctly set the trigger controls of the oscilloscopes (level and slope).

You must set the scopes' controls to select an appropriate sampling-duration width, which is the amount of time during which sampling will continue after the trigger event has been recognized by the scope. The sampling-duration width equals the acquisition time. You must base this time duration on the need to see enough of the *last* waveform so that you can verify its positive peak and its negative peak.

Assume that each digital record is to be 1K words in length, as suggested in Fig. 10–53. After you select that record length for each scope, the scope will automatically set its sampling rate (number of samples per second per input channel) to obtain 1024 samples in its allotted time.

To document your scope setup for future troubleshooting, show the exact sampling duration for each scope (three acquisitions altogether) on the V_{sens} waveform collection that you drew at the beginning of the assignment.

■ SUMMARY

- Potentiometers are the single most common component in the area of industrial measurement transducers.
- An LVDT is a dual-secondary-winding transformer with a movable core. It is commonly used to measure small mechanical displacements of about 1 inch or less.
- Fluid pressure is usually measured with a Bourdon tube or a bellows.
- The thermocouple is the temperature-measurement transducer that gives the best linearity into high temperature ranges. For inexpensive measurement of low to moderate temperatures, RTDs and thermistors are commonly used. RTDs have a positive temperature coefficient and are quite linear. Thermistors have a negative temperature coefficient and are nonlinear.
- A photovoltaic cell produces an output voltage in response to light striking it.
- A photoconductive cell changes its resistance in response to light striking it.
- Optical couplers convert high-voltage input signals to low-voltage logic signals, or vice versa, with complete electrical isolation and noise-rejection between the high- and low-voltage circuits.
- An optical shaft-position encoder counts the number of pulses produced by a rotating optical disk, then relates that counted number to the amount of shaft movement.
- Ultrasonic transducers are useful for measuring or detecting the internal characteristics of solids or semisolids.
- Strain gages are used to measure mechanical force, or acceleration.
- There are two fundamental types of tachometers: (1) those that sense voltage magnitude, and (2) those that respond to waveform frequency.
- Hall-effect transducers are sensitive to variations in magnetic field strength; this makes them useful for detecting when an object is coming near (proximity detection).
- A resolver is a generator-like device that relates the magnitude or phase of measured voltage to the absolute angular position of a measured shaft.
- The most common humidity transducers are the resistive hygrometer and the wet bulb/dry bulb psychrometer.

FORMULAS

$$\frac{F}{A} = Y\frac{\Delta L}{L_O}$$ (Hooke's Law) (Equation 10-1)

$$R = \frac{\rho L}{A}$$ (Equation 10-2)

$$GF = \frac{\Delta R/R}{\Delta L/L}$$ (for a strain gage) (Equation 10-3)

$$f = \frac{PS(\text{in r/min})}{120}$$ (Equation 10-5)

$$V = kIB$$ (Hall-effect voltage) (Equation 10-8)

$$V_{R1\text{-}3} = V_{S1\text{-}3}\cos\theta$$ (for resolvers) (Equation 10-10)

$$V_{R2\text{-}4} = V_{S2\text{-}4}\sin\theta$$ (Equation 10-11)

QUESTIONS AND PROBLEMS

Section 10-1

1. What are the advantages of electrical signal transducers over mechanical signal transducers?
2. If a 10 000-Ω potentiometer has a linearity of 2%, what is the most the actual resistance can differ from the ideal resistance at any point?
3. Roughly speaking, what is considered an acceptable linearity for a potentiometer used as a measuring transducer?
4. If a 1000-Ω wirewound potentiometer has 50 turns, what is its percent resolution?
5. In Fig. 10–3(c), if $R_1 = 5\text{ k}\Omega$, $R_2 = 12\text{ k}\Omega$, and $R_4 = 15\text{ k}\Omega$, what value of R_3 will cause the bridge to be balanced?

Section 10-2

6. The discussion in Sec. 10-2 explained why the words *variable differential* are used in the name of the LVDT. Why do you suppose the word *linear* is used in its name?
7. About how far can the cores of LVDTs move?
8. Roughly, how much signal voltage can you expect from an LVDT?

Section 10-3

9. What are Bourdon tubes made of?
10. Are Bourdon tubes used to measure liquid pressure, or gas pressure, or both?
11. What is the advantage of a spiral or helical Bourdon tube over a C-shaped Bourdon tube?
12. Are Bourdon tubes useful for measuring low pressures? Explain.
13. A Bourdon tube itself is a mechanical transducer. How are Bourdon tubes used to provide an electrical measurement signal?
14. Generally speaking, what range of pressures can a bellows transducer handle?
15. Is a pressure bellows itself a mechanical transducer or an electrical transducer? Explain.
16. What returns a bellows to its original position when its pressure is relieved?

Section 10-4

17. Strictly speaking, does a thermocouple loop measure the hot-junction temperature, or does it measure the difference between hot- and cold-junction temperatures?
18. In industrial thermocouple tables, what cold-junction temperature is usually assumed?
19. What is the purpose of thermocouple compensating circuits?
20. If the cold junction of a thermocouple loop were located in a heated and air-conditioned control room, would a compensating circuit be necessary? Explain.
21. Of the common thermocouple types, which type is the most sensitive?
22. Of the common thermocouple types, which type is best suited to temperatures above 2000°F?
23. Of the common thermocouple types, which type do you suppose is the most expensive?

Section 10-5

24. What is the difference between a positive temperature coefficient of resistance and a negative temperature coefficient of resistance?
25. Does an RTD have a positive or a negative temperature coefficient of resistance? Repeat the question for a thermistor.
26. Which device is more sensitive, a thermistor or an RTD?
27. Which device is more linear, a thermistor or an RTD?
28. Of the common RTD materials, which is the most sensitive?
29. Over what range of temperatures do RTDs find their greatest industrial use? Repeat for thermistors.
30. Draw the schematic symbol for a temperature-sensitive resistor which has a positive temperature coefficient. Repeat for a negative temperature coefficient.
31. What is the difference between a linear measurement scale and a nonlinear measurement scale?
32. What is a logarithmic scale? Why are they used in graphs?
33. When a thermistor is being used to measure the temperature of an external medium, is self-heating good or bad? Explain.
34. Explain how the self-heating of a thermistor could be used to detect whether a liquid was above or below a certain height in a tank.

Section 10-6

35. Explain why solid-state temperature transducers are generally not well suited for accurate temperature measurement.
36. In Fig. 10–11(b), suppose that $R_F = 49\ \text{k}\Omega$ and $R_{IN} = 1\ \text{k}\Omega$. The diode has the temperature characteristic shown in Fig. 10–11(a). How much change in output voltage will occur for a temperature change of 5°C? Repeat for $\Delta T = 100°\text{C}$.
37. What is the fundamental advantage of an optical pyrometer over other temperature transducers?
38. To measure a higher temperature with an optical pyrometer, what change must be made in the filament current, an increase or a decrease?

Section 10-7

39. What is the difference between a photovoltaic cell and a photoconductive cell?
40. What is the main operating advantage of photoconductive over photovoltaic cells?
41. What is the main operating advantage of photovoltaic cells over photoconductive cells?
42. Do photoconductive cells have a positive or negative illumination coefficient of resistance? Explain.
43. What does the term *light-to-dark ratio* mean?
44. Roughly speaking, about how much voltage can a typical photovoltaic cell deliver? Repeat the question for current.
45. What is photovoltaic cell fatigue?
46. Can photovoltaic cells be stacked in parallel? Can they be stacked in series?
47. Explain the operation of a *chopped* photodetector. What is its advantage over a plain photodetector?
48. Which chopping frequencies should not be used for chopped photodetectors? Why?
49. Are photocells at all sensitive to temperature?
50. What are the units of illumination in the English system? (There are two names for the same unit; see if you can track down the other name.)
51. What are the units of illumination in the SI metric system? What is the conversion factor between the English and SI units?
52. What illumination level is considered pitch dark?
53. What is the illumination level outdoors on a bright sunny day?
54. Why can't incandescent bulbs be used for the photochopper of Fig. 10–21?
55. Name some of the industrial uses of optical coupler/isolators.
56. What are the advantages of optical couplers over the couplers discussed in Chapter 1?
57. What are the advantages of LEDs over conventional light sources?
58. What is the difference between visible LEDs and infrared LEDs? State the relative advantages of each.
59. An optical shaft-encoder has a gear ratio of 10:1 and 180 slits on its disk.
    a. How far must the measured shaft move in order to increment the counter by 1 bit?
    b. What is the resolution of this transducer?
    c. If the measured shaft moves positive 87.4°, what is the binary content of the counter. Assume 12-bit counter construction, with eleven magnitude bits and one sign bit.
60. For the transducer in Problem 59, what amount of movement will produce a binary content of 0011 0111 1010? Specify direction and magnitude.
61. Repeat Problem 60 for a binary counter content of 1101 0010 0101.
62. Can the optical shaft transducer of Problems 59, 60, and 61 be built with a single photocell, or are two photocells required? Explain.

Section 10-8

63. Explain the fundamental advantage of fiber-optic signal transmission compared to transmission by wire.
64. At a bend in an optical fiber, approximately how much of the incident light enters the cladding?
65. Which is the faster device, a phototransistor or a photodiode? Which is more convenient to use?

Section 10-9

66. In an ultrasonic thickness-monitoring system, which parameter is used to detect a thickness variation, wave intensity or wave propagation time?

Section 10-10

67. Write Hooke's law, and explain what it means.
68. Upon what three things does the resistance of a piece of wire depend? State the relationship in formula form.
69. Combining your answers to Questions 67 and 68, explain how a strain gage works.
70. Define *gage factor* of a strain gage. About how great is the gage factor of an industrial strain gage?
71. Describe how a dummy gage can eliminate temperature-related errors in the use of strain gages.

Section 10-11

72. In Fig. 10–30, if the measured object starts accelerating to the right, will the strain-gage resistance increase or will it decrease?
73. Repeat Question 72 for the measured object accelerating to the left.
74. If a strain-gage accelerometer is being used to detect vibration in a large synchronous motor spinning at 1800 r/min, what will be the frequency of the output voltage from the strain-gage bridge circuit?

Section 10-12

75. What are the five basic types of industrial tachometers? Classify each type as either a magnitude tachometer or as a frequency tachometer.
76. Describe the principles of operation of each of the five types in Question 75.
77. What are the three chief sources of error in industrial tachometers?
78. Which class of tachometers has the greater linearity, magnitude or frequency?
79. What are the advantages of magnitude tachometers over frequency tachometers?

Section 10-13

80. In Fig. 10–32(c), if the *B*-field direction were reversed, would the polarity of the Hall voltage also reverse? Explain.
81. In the power transducer of Fig. 10–34, with the load wholly resistive, does the Hall voltage reverse polarity when the ac source reverses polarity? Explain.
82. Many fluid flowmeters can measure flow in one direction only. They are unable to provide a measurement if the flow direction reverses in the pipe. Does a Hall-effect flowmeter suffer from this limitation? Explain.

Section 10-14

83. The Doppler effect causes a wave received from an approaching object to have a perceived frequency ____________ than its actual frequency; a wave received from a receding object is perceived to have a frequency ____________ than its actual frequency.
84. True or False: A major advantage of an ultrasonic Doppler-effect flowmeter is that it has no mechanical presence inside the containment system (the pipe).

85. Which word, *intrusive* or *nonintrusive,* properly describes an ultrasonic flowmeter?
86. True or False: A turbine flowmeter is mechanically complex but electrically/magnetically rather simple.
87. Which word, *intrusive* or *nonintrusive,* properly describes a turbine flowmeter?
88. True or False: With an intrusive flowmeter, the measured flow rate is invariably lower than the rate that would have occurred if no measurement had been taken.
89. The mechanical combination of rotation and cyclical change in orientation of the axis of rotation is called ____________.
90. Which word, *intrusive* or *nonintrusive,* properly describes a nutating disk flowmeter?
91. Bernoulli's principle of fluid flow states that a large cross-sectional area implies a ____________ value of fluid velocity and a ____________ value of static pressure; a small cross-sectional area implies a ____________ value of fluid velocity and a ______ value of static pressure.
92. To determine fluid flow rate using Bernoulli's principle, it is necessary to measure the pressure at how many locations?
93. Write the equation that describes a fluid's flow rate in terms of the Bernoulli principle measurement(s).
94. Which word, *intrusive* or *nonintrusive,* properly describes a pressure-drop flowmeter?

Section 10-15

95. A shaft-position resolver resembles what basic electromagnetic machine?
96. How many slip-rings does a four-winding resolver have? Explain why.
97. In the single-phase operating mode, what aspect of the output sine wave, magnitude or frequency, represents the shaft position?
98. In a standard four-winding resolver with both stator primary windings connected in parallel, by what factor is the oscillating flux greater than for a single stator winding being used? Explain why this is so.
99. Comparing the parallel-stator configuration of Problem 98 to the single-winding resolver configuration, what is the new position of the oscillating flux? Explain.
100. We have described resolver operation with the stator serving as the transformer primary and the rotor serving as the secondary, producing the output signal. Is this the only way the resolver can be utilized, or can the roles be reversed? Explain.
101. When a resolver is used in two-phase mode, what aspect of the output sine wave, magnitude or phase, represents the shaft position?
102. For a resolver in the two-phase operating mode, can the net stator flux be described as oscillating? Explain this.
103. A two-phase-configured resolver transducer senses magnitude of shaft displacement by the amount of phase-shift between sine waves. How does it sense *direction* of shaft displacement?

Section 10-16

104. What is a resistive hygrometer? Explain its construction and principle of operation.
105. Are resistive hygrometers linear? Explain.
106. What is the relative humidity limit for a typical resistive hygrometer? What happens if that limit is exceeded?

107. Describe a wet bulb-dry bulb psychrometer's construction and operating principles.
108. Explain why a greater temperature difference between wet and dry bulb temperatures means a lower relative humidity.
109. If the dry bulb temperature is 60°F and the wet bulb temperature is 45°F, what is the % RH?
110. If the dry bulb temperature is 75°F and the wet bulb temperature is 67.5°F, estimate the % RH.

CHAPTER

FINAL CORRECTING DEVICES AND AMPLIFIERS

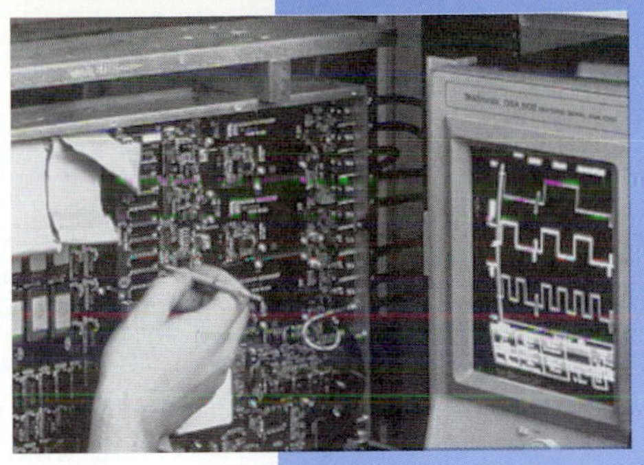

In very many cases, the final correcting device in a closed-loop system is a valve or valve-like device that varies the flow of a fluid into the process. This is usually the case in temperature-control processes, where the heat input into the process is varied by adjusting a valve which controls the flow of combustion air or of liquid or gaseous fuel. Likewise, in pressure-control processes, the pressure is usually corrected by changing a valve opening on either the inlet side or the outlet side of the process. For example, to raise the pressure in a process chamber, the valve regulating the inlet flow may be opened further or the valve regulating the escape flow may be closed further. In general, valves and valve-like devices such as dampers, louvers, sliding gates, etc., are the most common final correcting devices in industrial processes.

Sometimes the final correcting device is a continuously running motor, whose rotational speed determines the amount of process load. Many drying operations operate this way. The amount of heat energy supplied to the drying chamber is held constant, and the dryness of the final product is corrected by changing the rate at which the product moves through the drying chamber. For example, in grain being dried prior to storage, if the moisture content of the output grain is measured as too high, the control system could correct this condition by reducing the rate at which the grain is fed through the drier. In this case, control is accomplished by changing the load rather than by changing the fuel input. In such a system the final correcting device would be a motor, and the rotational speed of the motor would be the *manipulated variable*. The manipulated variable in any control system is that quantity which is varied so as to affect the value of the controlled variable. In any system having a valve as its final correcting device, the manipulated variable is fluid flow rate.

In some industrial control systems the manipulated variable is electric current. The most obvious example of this is an electric heating process. In an electric heating process the current might be controlled proportionally by continuously changing the firing angle of an SCR or triac. Or the current could be controlled in an On-Off manner by using some kind of switch or relay contact. In the former situation the thyristor and associated circuit components would comprise the final correcting device. In the latter situation the relay would be considered the final correcting device.

When the closed-loop system is a servo system, the controlled variable is the position of an object. The final correcting device is then a servo motor combined with a gearing system connecting the motor shaft to the object.

As the foregoing examples indicate, the final correcting devices in use in modern industry are quite varied. In this chapter we will look at a number of the commonly used final correcting devices and study their operation and characteristics.

OBJECTIVES

After completing this chapter, you will be able to:

1. List some of the commonly used final correcting devices in industrial control.
2. Explain the operation of a solenoid valve, and tell the circumstances in which solenoid valves are used.
3. Explain the operation of a two-position motor-driven control valve, and interpret the limit-switch timing diagram of such a valve.
4. Explain the operation of a variable-position motor-driven valve, and tell the circumstances under which such valves are used.
5. Describe the construction and operation of an electropneumatic valve operator using the balance beam principle, and tell the circumstances under which such operators are used.
6. Describe the construction and operation of an electropneumatic converter and a pneumatic power positioner, and explain how they are joined together to accomplish variable-position control.
7. Describe the construction and operation of an electrohydraulic valve positioner using the jet pipe principle, and tell the circumstances under which such positioners are used.
8. Explain how electromagnetic relays and contactors can be used to control electric current in a control system, and discuss the difference between pick-up current and dropout current for relays and contactors.
9. Distinguish between a three-phase delta connection and a three-phase wye connection, and calculate power for each type of connection.
10. Explain the theory of operation of a split-phase ac motor, explain the rotating magnetic field and the creation of torque, and show how the phase shift is accomplished in a split-phase ac motor.
11. Describe the general layout of a servo system, and state some of the benefits arising from servo systems.
12. Distinguish between plain split-phase ac motors and an ac servo motor.
13. Explain the function of a servo amplifier, and state its general characteristics and qualities.
14. Explain the operation of some specific solid-state ac servo amplifiers, demonstrating the following features:
    a. Push-pull power output stage
    b. Chopper-stabilized dc input
    c. Voltage feedback stabilization
    d. Current feedback stabilization
    e. IC op-amp front end followed by discrete driver and power-output stages
15. Describe the operation of a dc servo motor, and tell the circumstances under which dc servo motors are used.
16. Explain the operation of some specific dc servo amplifiers using SCRs as power control elements.

11-1 ■ SOLENOID VALVES

Figure 11–1 shows a cross-sectional view of an electric solenoid-operated valve, or *solenoid valve* for short. In the absence of current through the solenoid coil, there will be no magnetic field to pull the armature up, so the compression spring will push the armature down. The valve stem is attached to the armature, so it also moves down and pushes the *valve plug* tight against the *valve seat*. This blocks the flow of fluid between the inlet and

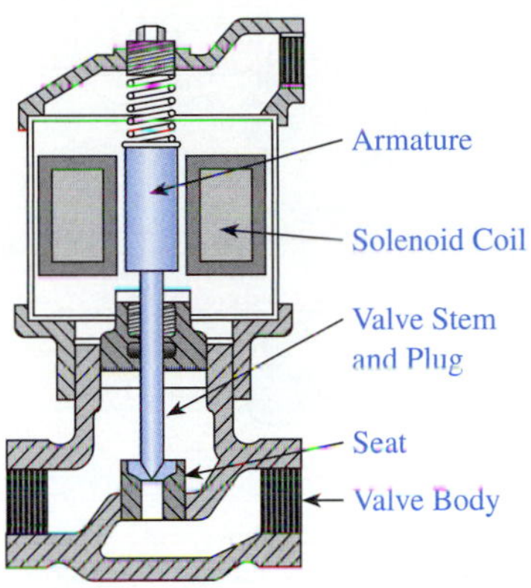

FIGURE 11–1
Cross-sectional view of a solenoid valve.

outlet ports. When the solenoid coil is energized and the coil conductors are carrying current, a magnetic field is established which pulls the armature up. The armature must overcome the spring force tending to push it down in order to move into the middle of the coil. As the armature moves up, it lifts the valve plug off the valve seat and opens the passageway from inlet to outlet. Solenoid valves are inherently two-position devices. That is, they are either all the way on or all the way off. Therefore they lend themselves to being used with the On-Off control mode.

Solenoid coils can be designed to operate on ac voltage or dc voltage, but the ac design is much more common.

Ac solenoid coils have a serious weakness that dc solenoid valves don't have. If an ac solenoid valve sticks in the closed or partially closed position when power is applied to the coil, the coil will probably burn up. This happens because the magnetic armature cannot enter the core of the coil, so the inductance of the coil remains low. (The inductance of an inductor depends strongly on the magnetic permeability of the core material.) With inductance low, the inductive reactance is also low, and a large ac current will flow through the coil indefinitely. This will eventually overheat the windings of the coil.

11-2 ■ TWO-POSITION ELECTRIC MOTOR-DRIVEN VALVES

In situations where the valve is large or where it must operate against a high fluid pressure, it is better to actuate the valve by an electric motor than by a solenoid coil. In this case the valve body and stem would look like the valve shown in Fig. 11–1, but the stem would be attached to some sort of mechanical linkage which is stroked by an electric motor. Most two-position valves of this type are operated by a unidirectional *split-phase induction motor*. The motor is geared down to provide slow output shaft speed and high torque. As the output shaft rotates from 0° to 180°, the attached linkage opens the valve. As the shaft rotates from 180° back to 360°, the home position, the attached linkage closes the valve. Integral limit switches inside the motor housing detect when the valve has reached the 180° position and when it has reached the home position. A diagram showing the windings, limit switches, and controller connections of such a motor is presented in Fig. 11–2.

Here is how the motor-driven two-position system works. If the motor output shaft is in the home position, meaning that the valve is closed, LS1 is mechanically contacted and LS2 is also mechanically contacted. The LS2 N.C. contact is thus open, and the LS1 N.O. contact is held closed. If the controller calls for the valve to open, it does so by closing contact *A*. This applies 115-V ac power to the N.O. terminal of LS1. Since the N.O. contact is closed at this time, power is applied to the motor windings, and the motor

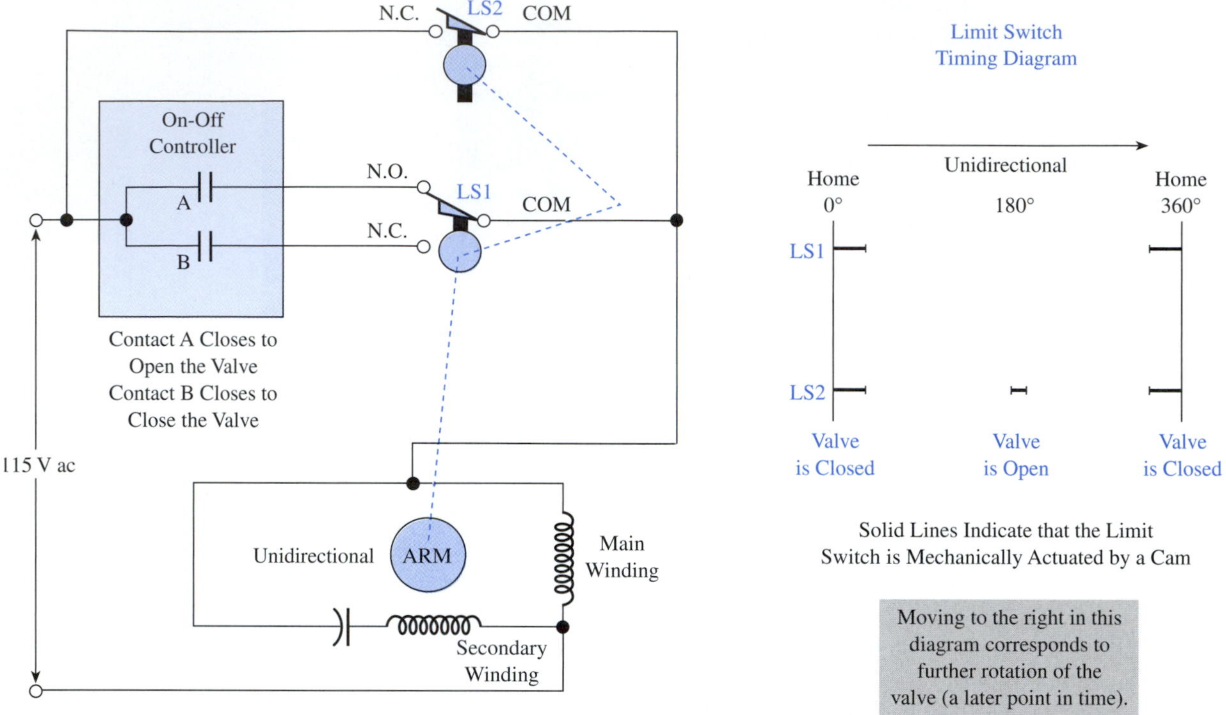

FIGURE 11–2
Circuit for operating a split-phase valve motor. The motor positions the valve (which is not shown) either closed or wide open. The limit switches are shown for the valve in the closed (home) position. The timing diagram shows how the cams operate the two limit switches.

begins to turn. Shortly after it leaves the home position, LS1 is released by its cam, causing the N.O. contact to go open. However, power is maintained to the windings through the LS2 contact, which was also released by its cam. Refer to the timing diagram to see this.

When the motor output shaft reaches the 180° position, meaning that the valve is open, LS2 is once again contacted by its cam, as indicated in the timing diagram. This opens the LS2 contact and disconnects the power from the motor windings. The motor stops in this position, and the valve remains open.

When the controller wants the valve to close, it signals this by closing contact *B*. Power is applied to the motor windings through the N.C. contact of LS2, which is closed at this time. It is closed because LS2 is *not* mechanically actuated by a cam, as shown by the timing diagram. The motor turns in the same direction as before until it reaches the home position. In the home position, both LS1 and LS2 are contacted by their cams, so power is removed from the motor windings and the motor stops. The valve is therefore closed.

Most electric motors used on two-position motor-driven valves have a full travel time of less than 30 seconds. That is, it takes less than 30 seconds for the motor to completely open or to completely close the valve.

Some motors are purposely made to operate very slowly, having a full travel time as long as 4 minutes. When a valve moves this slowly, there is a pretty good chance that the controlled variable will have recovered to set point before the valve completes its motion. If this situation is coupled with a *three-position controller,* the control mode is not really On-Off, and it is not really proportional but a compromise between the two. It is then termed *floating control*. A three-position controller is one which has three output signals instead of only two:

1. The measured value is too low, so drive the valve open.
2. The measured value is too high, so drive the valve closed.
3. The measured value is within the differential gap, so don't drive the valve at all but just let it sit where it is.

Floating control is considered by some people to be a sixth control mode in its own right. However, it is not as important as the five control modes discussed in Chapter 9, so we will not grant it that status.

11-3 ■ PROPORTIONAL-POSITION ELECTRIC MOTOR-DRIVEN VALVES

In proportional control, as we have seen, there must be a method of positioning a control valve at any intermediate position. The usual method is to connect the valve to a reversible slow-speed induction motor.* Figure 11–3(a) illustrates such an arrangement on a variable-position damper.

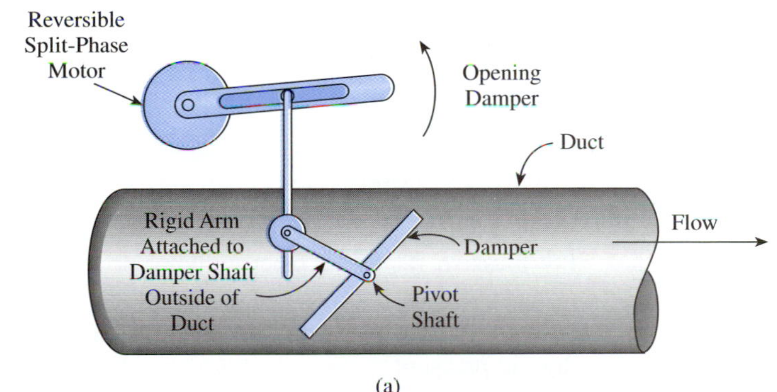

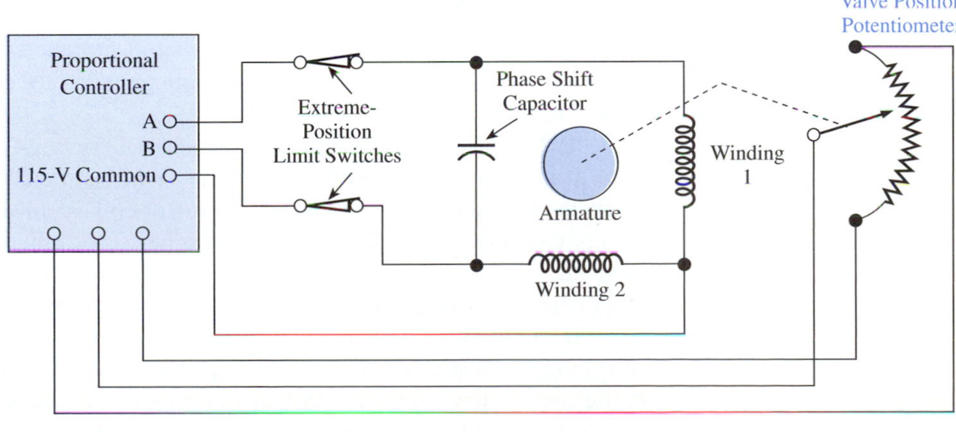

FIGURE 11–3 Proportional-position-control damper driven by a split-phase motor. (a) The mechanical arrangement of the damper, the damper linkage, and the motor. Note that the motor is reversible, in contrast to the motor in Fig. 11–2. (b) The electrical circuit. The controller applies power to either terminal *A* or terminal *B*, depending on whether it wants the damper positioned more closed or more open. The potentiometer provides feedback information to the controller, telling it the present position of the damper.

*Split-phase ac motors are discussed thoroughly in Secs. 14-2 and 14-3.

When the proportional controller sees a *positive* error signal from the comparer, it applies 115-V ac power to terminal *A*. This connects winding 1 across the 115-V line and connects the phase-shift capacitor in series with winding 2; this series combination is also across the ac line. This causes the motor to run clockwise (let us assume) and causes the damper to begin closing off the opening through the duct. When the position potentiometer, which is integrally built into the motor housing, sends the proper position signal back to the controller, the proportional controller is satisfied, and it removes power from terminal *A*. This stops the motor and freezes the damper in that position.

When the proportional controller detects a *negative* error from the comparer, it applies 115-V ac power to terminal *B*. Now winding 2 is directly across the ac line, and winding 1 is placed in series with the phase-shift capacitor. This series combination is also across the ac line. This causes the motor to turn counterclockwise and causes the damper to start opening up. When the motor has turned far enough, the position potentiometer signal matches the error signal, and the proportional controller is satisfied. It removes power from terminal *B*, and the motor stops. The valve therefore stops in a position which is in agreement with the magnitude and polarity of the error signal.

The internal action of a reversible split-phase motor will be described in more detail in Sec. 11-9.

When the motor has traveled to either one of its extreme positions, either clockwise or counterclockwise, one of the limit switches will open up and remove power from the windings. After that, the motor can turn in the opposite direction only. It will do so when the controller orders it to begin turning the other way by applying power to the opposite terminal, *A* or *B*.

11-4 ■ ELECTROPNEUMATIC VALVES

For massive valves, an electric motor drive may not be practical. The inertia and breakaway friction of the valve assembly may preclude the use of an electric motor as the positioning device. In such situations, the valve is moved by pneumatic pressure or by hydraulic pressure.

11-4-1 Electropneumatic Valve Operator

Figure 11–4 is a schematic illustration of an electropneumatic valve operator. The final position of the valve is determined by the magnitude of the electrical input current. Here is how it works.

The balance beam is a small, lightweight, friction-free metal beam, a few inches long. It is pivoted about a fulcrum near its right-hand end. When an input current is fed in via the terminals, the electromagnet actuator coil establishes a magnetic field which interacts with the field of the permanent magnet. The force resulting from this interaction pulls the beam up, which tends to rotate it clockwise. The force tending to rotate the beam clockwise is proportional to the amount of current flowing through the actuator coil.

If the beam rotates slightly clockwise, the left end of the beam will move up and restrict the escape of air from the nozzle. The closer the left end of the beam (called the baffle) gets to the nozzle, the less air can escape from the nozzle. As the air escape is cut off, air pressure increases in the variable-pressure tube leading to the nozzle. This occurs because the movement of air through the fixed restriction is reduced, resulting in a smaller pressure drop across the restriction and consequently higher pressure downstream of the restriction.

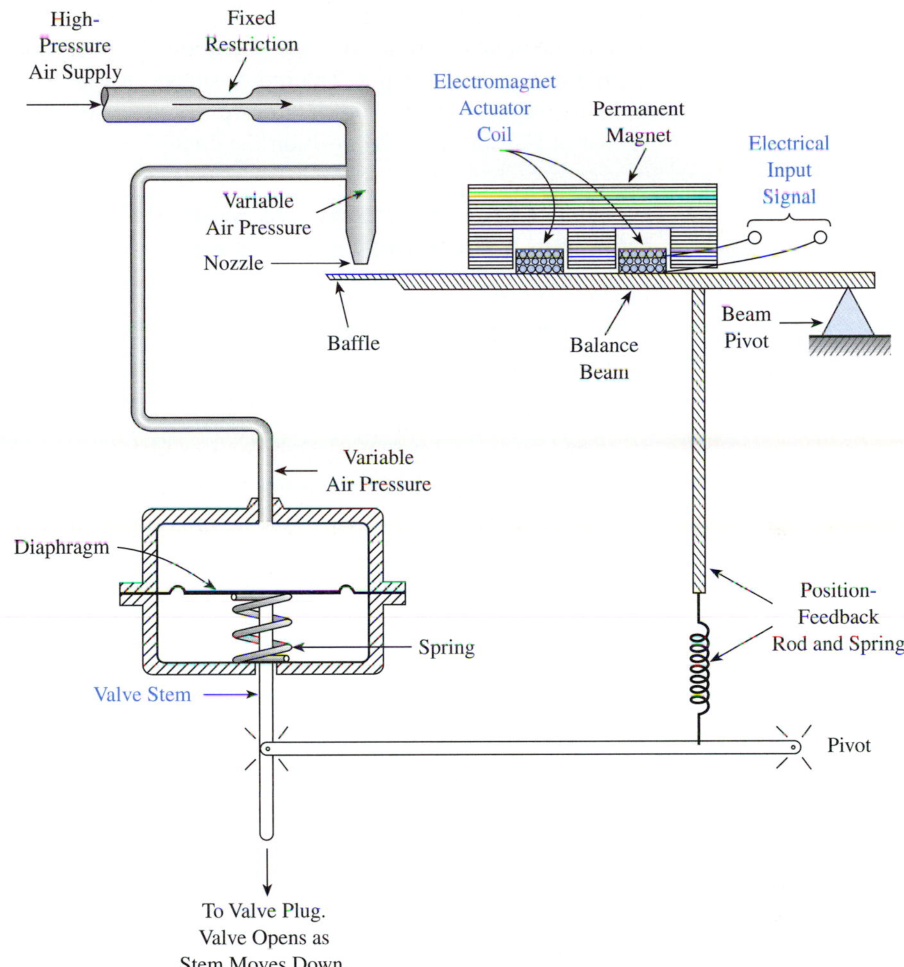

FIGURE 11–4 Electropneumatic valve operator. The position of the valve stem is proportional to the electrical input signal in the upper right of the drawing.

The higher pressure in the variable pressure tube is applied to the diaphragm chamber above the valve diaphragm. This exerts a downward force on the valve stem, thereby opening the valve.* As the valve stem moves down, it causes the feedback spring to exert a counterforce on the balance beam, tending to rotate it counterclockwise. When the countertorque exerted by the feedback spring exactly balances the original torque exerted by the electromagnet, the beam balances in that position. Therefore the final position of the valve stem and thus of the valve opening itself is determined by the electrical input signal (current) to the actuator winding.

If this apparatus were to be used with an electric proportional controller of the type shown in Fig. 9–15, the electrical input signal could be taken from the wiper of the valve position pot on the left. The voltage between that wiper and ground would be applied to

*This is correct as long as the valve is an *air-to-open valve*. Any valve which is *opened* by increasing air pressure applied to its diaphragm is called an air-to-open valve; the spring tends to close the valve. Any valve which is *closed* by increasing air pressure on its diaphragm is called an *air-to-close valve;* the spring tends to open the valve. The choice of which type of valve to use in an industrial system depends on whether the valve should fail open or fail closed in case of pneumatic pressure failure.

the electromagnet coil in Fig. 11–4. The electromagnet coil would have to be designed to draw negligible current from the potentiometer so as not to disturb the voltage division along the length of the pot. The motor shown in Fig. 9–15 would not position the control valve directly but would only serve to rotate the potentiometer shaft. The apparatus shown in Fig. 11–4 would position the control valve.

11-4-2 Electropneumatic Signal Converter Driving a Pneumatic Positioner

Figure 11–5 shows a somewhat different approach to controlling a large valve pneumatically. Again, the original input signal is an electric current through an electromagnet coil. In this design, though, the feedback to the balance beam does not come from the controlled valve itself but from a bellows element. The pneumatic output pressure is then further balanced against the mechanical position of the valve. Here is how it works.

The input current through the electromagnet coil creates an upward force on the balance beam, tending to rotate it clockwise. As it moves slightly clockwise, the nozzle/baffle arrangement causes the air pressure to increase in the tube ahead of the nozzle, as described in Sec. 11-4-1. This pressure signal is applied to the feedback bellows, which exerts a downward force on the beam, tending to rotate it counterclockwise. The beam stabilizes when the clockwise torque from the electromagnet is equal to the counterclockwise torque from the feedback bellows. Therefore the magnitude of the input current exactly determines the pressure signal applied to the bellows. This pressure signal is also taken out via the output tube for use in another location.

To summarize, it takes in an electrical input signal and sends out a proportional pneumatic output signal. The converter is designed so that the relationship between output pressure and input current is very linear.

The output pressure signal is fed to a pneumatic valve positioner, which may be located some distance from the electropneumatic converter. The output signal of the converter becomes the input signal to the valve positioner.*

The input pressure signal to the valve positioner tends to expand the input bellows to the right. The input bellows causes mechanical linkage A to rotate counterclockwise slightly. As this happens, the linkage shifts the spool in the small pilot cylinder. As the spool moves up, it uncovers the outlet ports of the pilot cylinder. When this happens, the blind-end port (top port) is opened to the interior of the pilot cylinder, applying high-pressure air to the blind end of the power cylinder. At the same time, the rod-end port is opened to the vent hole in the bottom of the pilot cylinder, venting the rod end of the power cylinder. The power cylinder thus strokes down, moving the valve stem down.

As the valve stem moves downward, it causes mechanical linkage *B* to rotate clockwise. This compresses the feedback spring, applying a force tending to compress the bellows. When the input pressure force is balanced by the feedback spring force, linkage A returns to its original position. This centers the spool in the pilot cylinder, blocking the outlet ports. The power cylinder stops stroking, and the valve freezes in that position. Therefore we have a condition in which the final valve opening is exactly determined by the input pressure signal. By properly designing the positioner mechanisms, the relationship between valve opening and input pressure can be made very linear.

*The word *positioner* is generally used to mean an apparatus which uses a high-pressure *cylinder* to move a valve or valve-like device. This is in contrast to the example studied earlier in which variable pressure air was applied to a *diaphragm* to stroke the valve. When a diaphragm is used, the apparatus is usually called an *operator*.

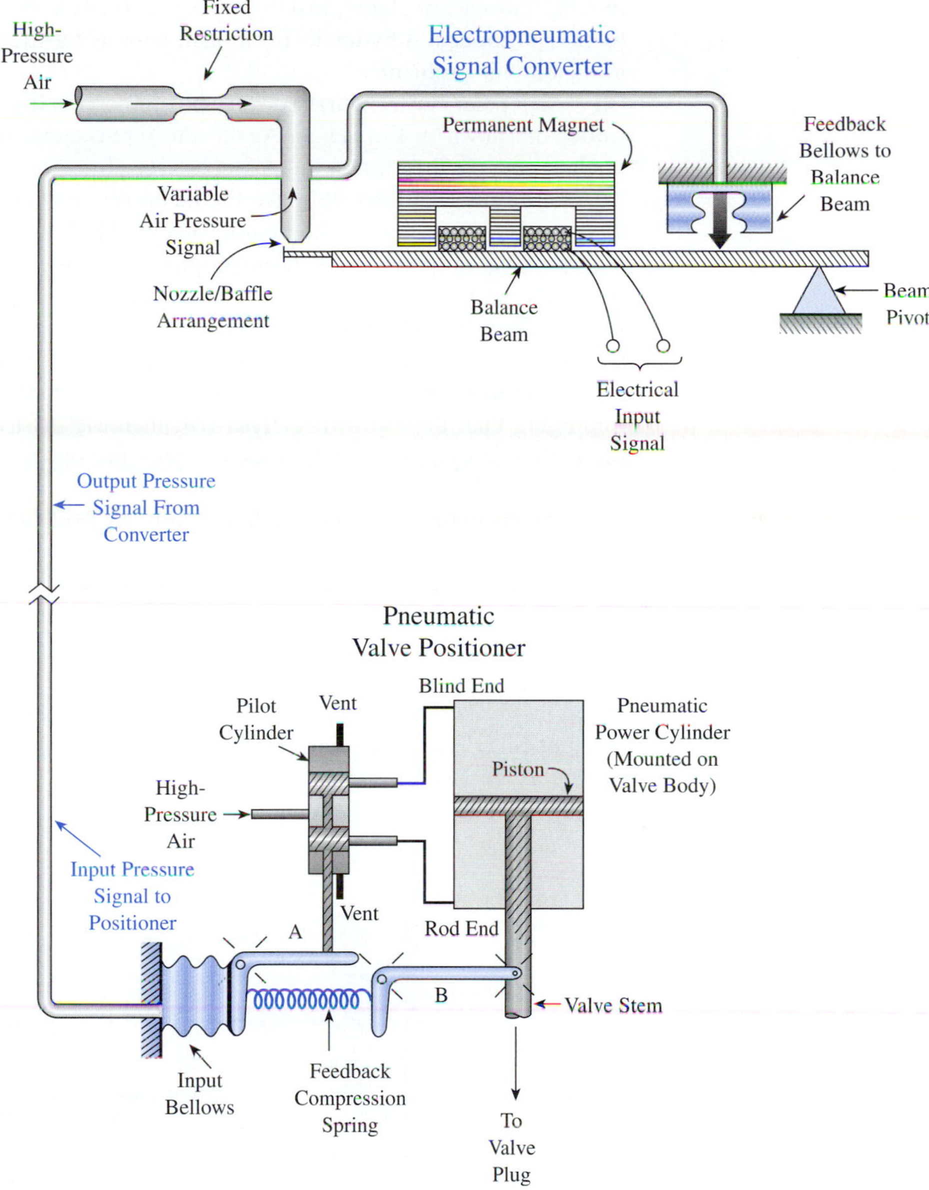

FIGURE 11–5 Electropneumatic signal converter and valve positioner. The apparatus on the top of the drawing converts an electrical signal (current) into a proportional pneumatic signal (air pressure). The apparatus on the bottom positions the valve in proportion to the pneumatic signal.

The overall situation is that the valve opening is linearly related to the input current into the electropneumatic converter. This arrangement is very compatible with an electric or electronic proportional controller.

11-5 ■ ELECTROHYDRAULIC VALVES

In control situations where the valve or damper is very massive, or where it is difficult to hold the valve in a steady position due to large irregular forces exerted by the moving fluid, a *hydraulic positioner* is the best actuator. Also, if a valve seldom moves, it may get stuck in a certain position. This can happen because dirt and debris may build up on moving linkages or shafts, making it very difficult to break them free when the valve is to be repositioned. A hydraulic positioner, with its terrific force capability, may be needed to handle this problem.

A popular electrohydraulic valve positioner, readily adaptable to a proportional controller, is shown in Fig. 11–6. Again, the input signal is a current through an electromagnet coil. As the current increases, a larger force is exerted to the left on the vertical balance beam. This tends to make the beam rotate counterclockwise. On the other side of the pivot point, toward the bottom of Fig. 11–6, is a jet pipe relay. Hydraulic oil at high pressure is forced through the jet pipe, emerging from the jet nozzle at high velocity. If the jet pipe is perfectly vertical, the oil stream impinges equally on the left and right orifices. Therefore there is no pressure imbalance between the two sides of the jet pipe relay, and the hydraulic piston is in force equilibrium. However, if the electromagnet coil moves the jet pipe slightly counterclockwise, the right orifice will feel more oil impinging than the left orifice. This will increase the hydraulic pressure on the top of the hydraulic cylinder and decrease the pressure on the bottom. The hydraulic cylinder will thus stroke downward.

As the cylinder rod moves downward, the feedback lever rotates clockwise, pulled by tension spring A. The linkage on the left of the feedback lever increases the tension in feedback spring B, tending to rotate the balance beam clockwise. Eventually the hydraulic piston will have moved far enough so that the torque exerted by the feedback spring exactly equals the original torque exerted by the electromagnet. At this point the balance beam returns to the vertical position, and the pressure is again equalized between the left and right sides of the jet pipe relay. The piston stops moving, and the valve remains in that position. The final valve position is therefore determined by the magnitude of the input current signal.

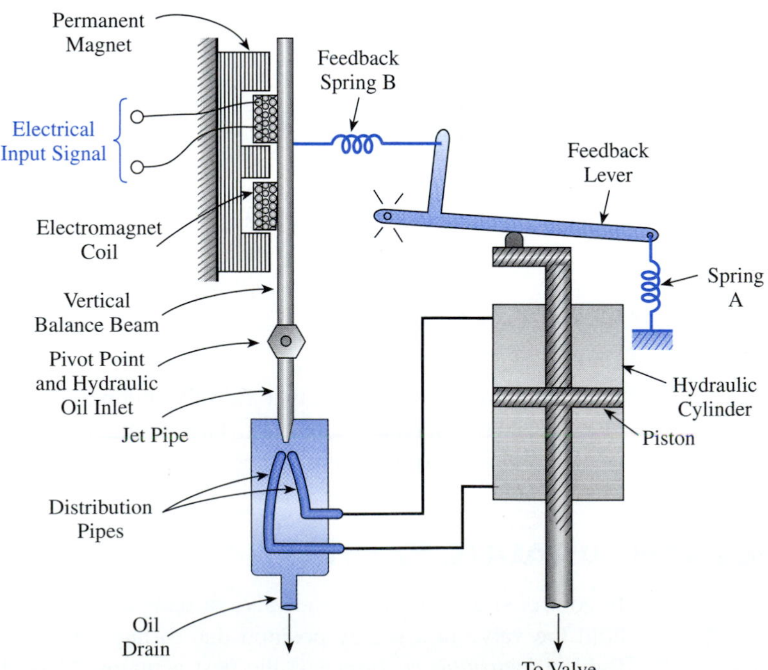

FIGURE 11–6
Electrohydraulic valve positioner, using a jet pipe. The position of the valve is proportional to the electrical input signal.

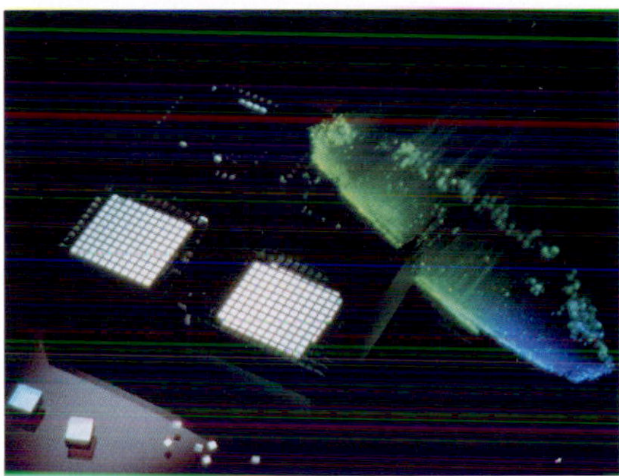

Optical fibers and filters

High-performance filters remove radiation at spurious frequencies.

Magnetic levitation-type bearing

Because there is no mechanical contact between the shaft and the stator, mechanical wear and bearing vibration are avoided.

Variable polarity electric welder

Instantaneous weld polarity is switched electronically, producing a current waveform (similar to that shown in Figure 7–11Cd).

Interchangeable end-effectors for a robot arm

Clockwise, from upper right—an abrasive brush, a two-finger mechanical gripper, a gas-welding torch, and a grinder.

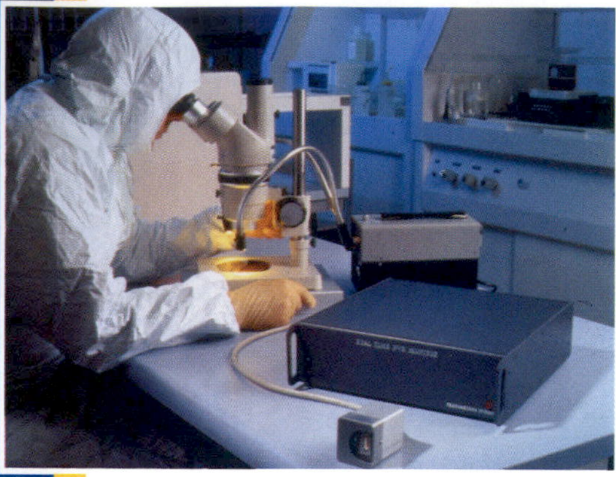

Contaminant Monitor

A chemical vapor detector can detect an air contaminant and identify its specific chemical compound.

FarField-2 Laser

Pulsed laser sources are ideal for making precise cuts in a fast-moving material.

Filmless radiography

Solid-state photoluminescent material can replace traditional film for recording radiographic images, making more precise image analysis and software enhancement possible.

(a)

(b)

High-temperature semiconductor

(a) Sputter deposition is the dominant technology in thin film production. This system heats the substrate by resistance heating before depositing a thin film of conducting material on it.

(b) A substrate heating unit.

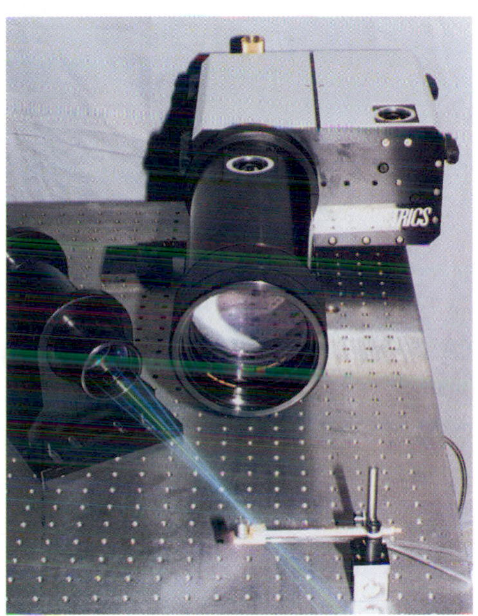

Rainbow

When liquid fuel and air are mixed in an internal combustion engine, the homogeneity of the mixture can be studied by measuring the refraction of single-frequency laser light passing through the mixture.

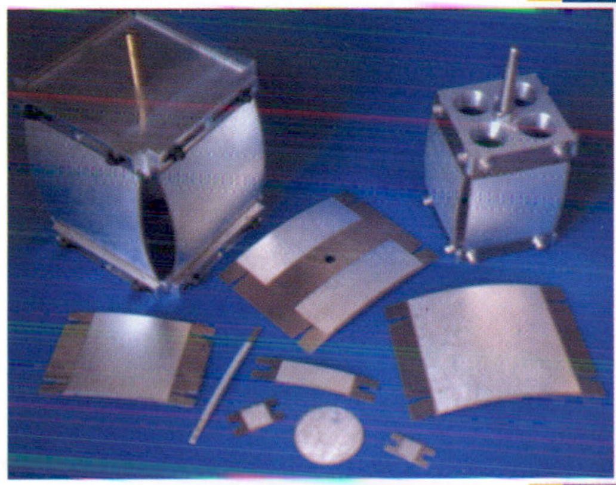

Piezoelectric transducers

The heart of every ultrasonic sender and receiver (see sections 10–9 and 10–14–1).

Illuminating development

With strobed illumination, this video system obtains images of fast-moving particles in a welding or cutting operation.

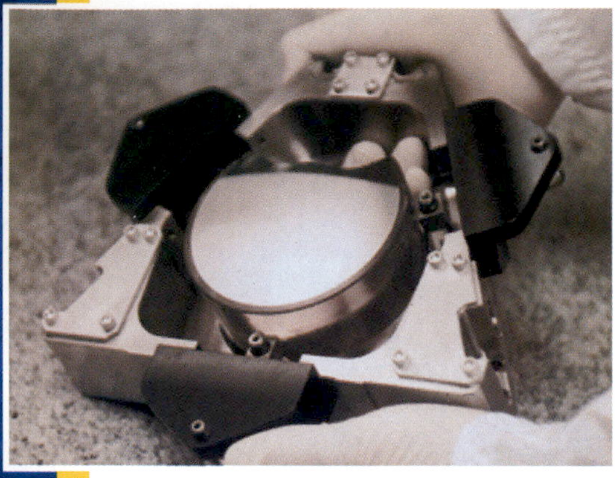

ULTRA-PRECISION OPTICS

This ultraviolet source provides "light" for the optical lithography process of memory chip manufacturing (see section 18-5).

SOFTWARE-ENHANCED STRAIN MAP

A polariscope aims its polarized light source at a stressed object (see section 10-10); the reflected light pattern shows the magnitude and direction of the strain.

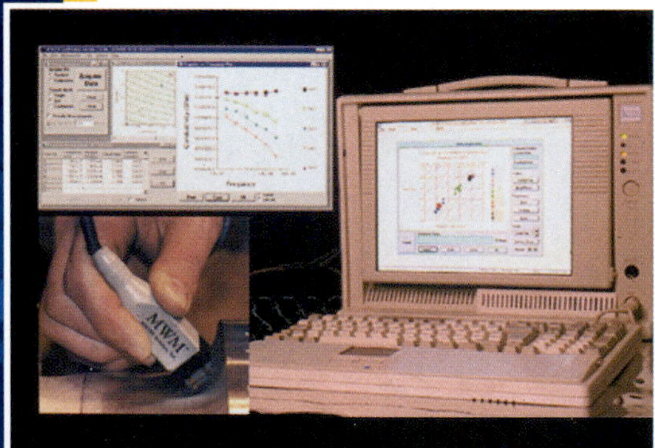

METAL-FATIGUE EARLY WARNING

Imperfections due to aging or stress are quickly located by a hand-guided electrode dielectrometer, operating on the same principle as the capacitive proximity sensor of section 19-10.

OPTICAL FIBER PROTECTION

This fiber-coating system applies a thin film of protective coating to the outer surface of optical fibers (see section 10-8).

11-6 ■ VALVE FLOW CHARACTERISTICS

The ideal flow characteristic in a controlled process is illustrated in Fig. 11–7. As the graph shows, the fluid flow is exactly linear with percent valve opening. That is, with the valve 20% open, the system flow is 20% of maximum; with the valve 40% open, the system flow is 40% of maximum; etc. The actual system flow characteristic depends not only on the valve flow characteristics but also on the flow characteristics of the rest of the piping system.

FIGURE 11–7
Ideal flow characteristic for a controlled process. In real situations, such linear response cannot be achieved.

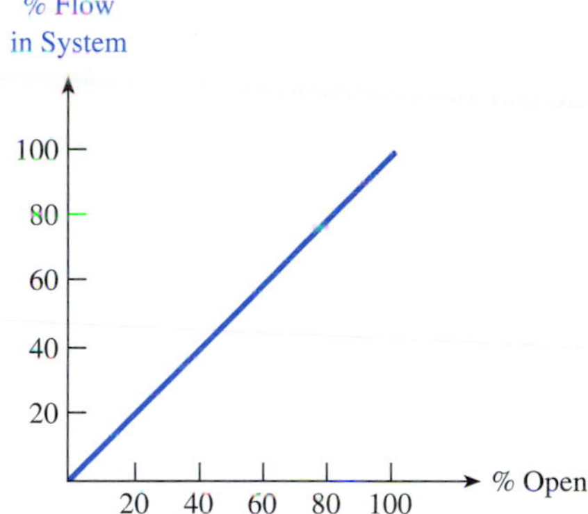

The valve flow characteristic curve shows the percentage of maximum flow versus percent opening, for a *constant pressure drop across the valve*. In a real system it is impossible to maintain a constant pressure drop across the valve as the valve position varies. This is so because as the valve position varies, the flow varies, causing the pressure losses in the rest of the piping system to vary also. Specifically, as flow increases, the pressure drop in the rest of the piping system increases, leaving less pressure drop across the valve. This situation is analogous to a fixed voltage source driving a series combination of a fixed resistor and a variable resistor. As the resistance of the variable resistor is decreased (analogous to opening a valve further) the current flow increases, causing a larger voltage drop across the fixed resistor. Since there is only a given amount of source voltage to begin with, if a larger voltage drop occurs across the fixed resistor, there must be less voltage drop across the variable resistor.

The result of this phenomenon in a piping system is to make the real system flow characteristics quite a bit different from the *valve* flow characteristic. This is illustrated in Fig. 11–8(a) and (b). Figure 11–8(a) shows a perfectly linear valve characteristic. If this perfectly linear valve were installed in a real system, the valve's pressure drop would be reduced at larger percent openings, so the flow response would also be reduced at the larger percent openings. In other words, at the larger openings we get less increase in flow for a given amount of change in valve position. This means that the slope of the system flow curve tapers off, as shown in Fig. 11–8(b).

The flow curve in Fig. 11–8(b) is pretty undesirable. It shows that 80% of the flow change takes place in the first 50% opening of the valve, and that only 20% of the flow change takes place in the second 50% opening of the valve. The disadvantages of this are quite evident.

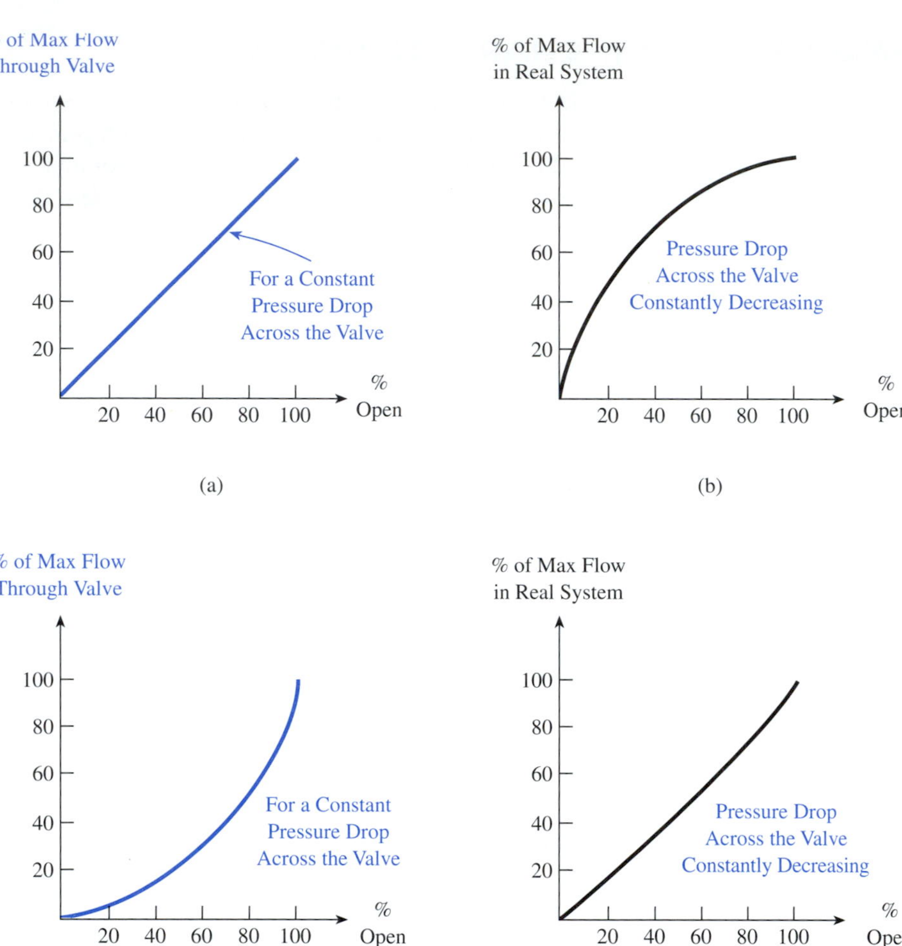

FIGURE 11–8
Flow characteristic curves of real systems. (a) Flow curve of a perfectly linear valve. (b) Overall system flow curve which would result from a perfectly linear valve. (c) Flow curve of a nonlinear valve. This valve gives lively response near the open end of its range and sluggish response near the closed end of its range. (d) Overall system flow characteristic which would result from using the valve in part (c). The valve nonlinearity tends to cancel the system nonlinearity, since the nonlinearities are in opposite directions. The result is an overall system response which is nearly linear.

The general solution to this problem is to design valves to have a flow characteristic like that in Fig. 11–8(c). This shows the valve having a flow characteristic which is *concave up*. When such a valve is installed in a real piping system, which always has a flow characteristic which is *concave down* as in Fig. 11–8(b), the resulting overall flow characteristic is fairly linear. The overall system flow characteristic is graphed in Fig. 11–8(d).

As a general rule of thumb, the greater the pressure losses in the rest of the piping system compared to the pressure drop across the valve, the more nonlinear the valve characteristic itself should be. Valve designers can vary the flow characteristics of a valve almost at will, by varying the shape of the valve plug.

With butterfly valves and louvers (Fig. 11–9a and b) it is of course impossible to vary the shape of the plug, because there is no plug. Instead, the poor flow characteristics of these devices (shown in Fig. 11–9c) are corrected by making adjustments to their operating shaft linkages. The manufacturers of hydraulic and pneumatic valve positioners furnish instructions on how to adjust the linkage which connects the positioner's cylinder rod to the shaft of the butterfly valve. By following these instructions, it is possible for the user to create linear flow characteristics.

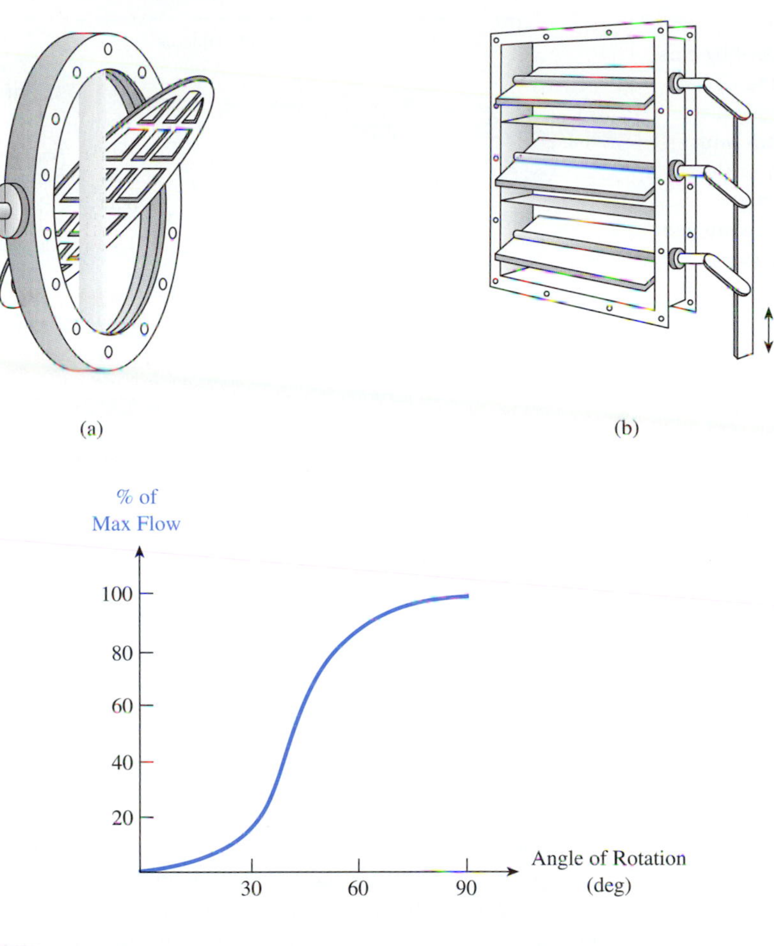

FIGURE 11–9
(a) Butterfly valve or damper. (b) Louver. (c) Typical flow curve of a butterfly valve or louver. Almost all the flow change takes place in the middle 30° of rotation; very little flow change takes place in the first 30° or the last 30°.

11-7 ■ RELAYS AND CONTACTORS

11-7-1 On-Off Control of Current to a Load

When electric current is the manipulated variable in a closed-loop control system, the final correcting device is often a relay or a contactor. For example, in an electric heating process, the temperature might be controlled in the On-Off mode simply by opening and closing a contact leading to the heating element. This is illustrated in Fig. 11–10.

Figure 11–10(a) shows a single-phase heating element driven by a single-phase ac source. When the controller receives a positive error signal (measured temperature greater than set point), it deenergizes contactor coil *CA*. This interrupts the current flow to the heating element and allows the temperature to drop. When the controller receives a negative error signal, it energizes coil *CA*. This closes the N.O. *CA* contact, applying power to the heating element and raising the temperature.

In an application requiring greater heat input, the heating element might be a three-phase element driven by a three-phase supply as in Fig. 11–10(b). In this case the contactor would need three contacts, in order to break each of the three lines.

FIGURE 11–10 Temperature control by closing and opening the circuit to a resistive heating element. (a) Contactor with one contact for a single-phase heating circuit. (b) Three contacts switched by one contactor for a three-phase heating circuit.

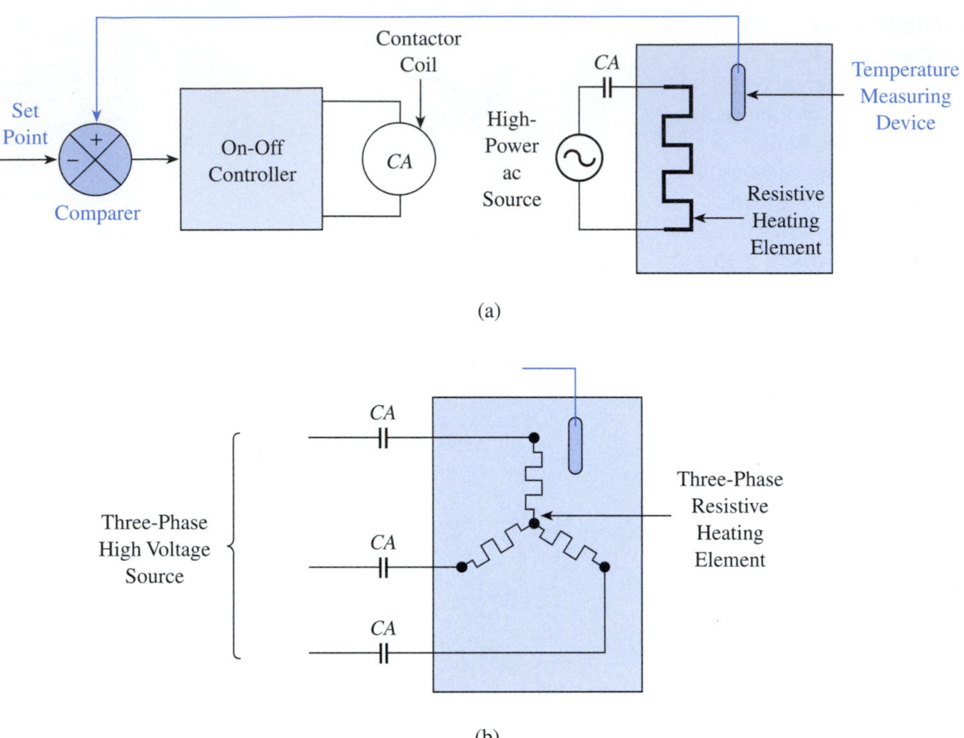

The only difference between a relay and a contactor is in the current-carrying and interrupting capability of the contacts. Contactors are capable of handling large currents, whereas relays can handle only relatively small currents. The contactor in Fig. 11–10 could be replaced by a relay if the load current were small enough.

11-7-2 Relay Hysteresis

An interesting feature of relays and contactors is that they naturally tend to provide a differential gap for On-Off control because of the hysteresis effect inherent in their operation. To cause a magnetic relay to energize, the coil current must rise beyond a certain value, called the *pull-in current* or *pick-up current,* to move the armature and switch the contacts. However, once a relay has been pulled in, the coil current must drop below a certain *lower* value of current to cause the relay armature to return to normal. The lower value of current is called the *hold-in current* or the *drop-out current*. This action is illustrated in Fig. 11–11(a).

The reason for the difference between the pull-in and drop-out currents can easily be understood by referring to the stylized drawing of a relay in Fig. 11–11(b). The spring causes the armature to lift up away from the core when the coil is deenergized. This creates an air gap between the top of the core and metal of the armature. When the coil current starts flowing, it must establish a strong enough magnetic field to pull the armature down against the spring tension. This proves somewhat difficult for two reasons:

1. There is an air gap in the magnetic loop; this causes the magnetic field to be weaker than it would be for a continuous loop of magnetically permeable material.

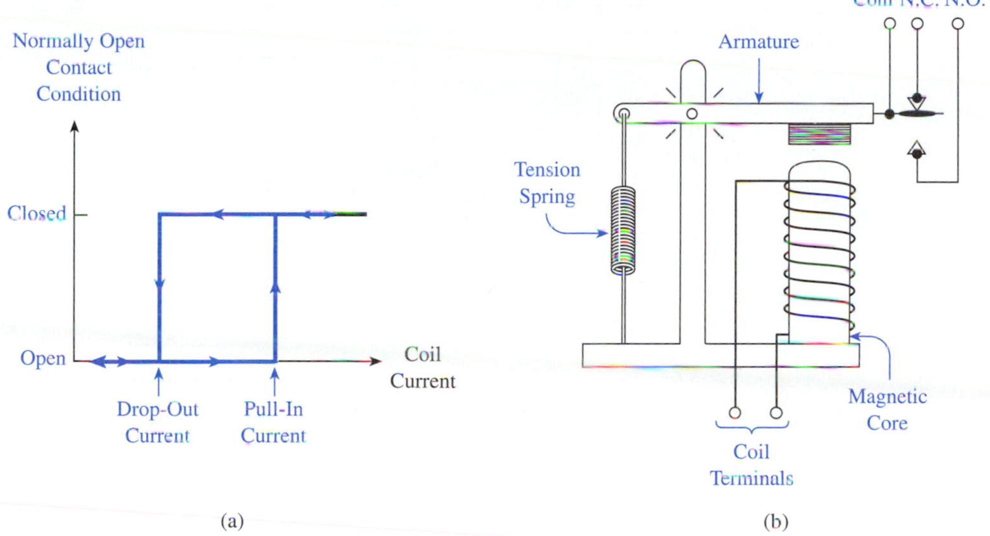

FIGURE 11–11
(a) Graphical illustration of relay hysteresis. As current is increasing, the switching occurs when the pull-in value is reached. As current is decreasing, the switching occurs when the drop-out value is reached. (b) The essential parts of an electromagnet relay. Hysteresis is due to the gap between the armature and the magnet core.

2. The attractive force between the core and armature (opposite magnetic poles) is weak because of the distance between the poles. When magnetic poles are farther apart, the attractive force between them is weaker, all other things being equal.

If the coil current is large enough, it will create a magnetic field which is strong enough to overcome these two handicaps, and the armature will be pulled down.

When the coil current begins to decrease, after the armature has moved, these two handicaps are no longer in effect. Therefore it is easier to *hold* the armature down than it was to pull it down initially. Because of this, the coil current must drop considerably below the pull-in current in order to allow the armature to move back up.

11-7-3 A Three-Phase Contactor Switching between Delta and Wye

An interesting example of the use of a three-phase contactor as the final correcting device in an electric heating process is shown in Fig. 11–12(a). This is basically On-Off control, except that the Off position is not really completely Off. The idea is that when contactor *CA* is dropped, the three-phase heating elements are connected in a wye configuration, and when *CA* is picked, the heating elements are connected in a delta configuration. In the wye configuration, the power delivered to the three-phase heating elements is much less than the power delivered if they are connected in delta. The controller causes *CA* to drop if the measured temperature is above set point. This causes the process temperature to fall because of the reduced power input to the wye connection (with consequent reduction in heating effect). The controller causes *CA* to pick up if the measured temperature is below set point. This causes the process temperature to rise because of the increased power input to the delta connection. Of course, the system must be designed so

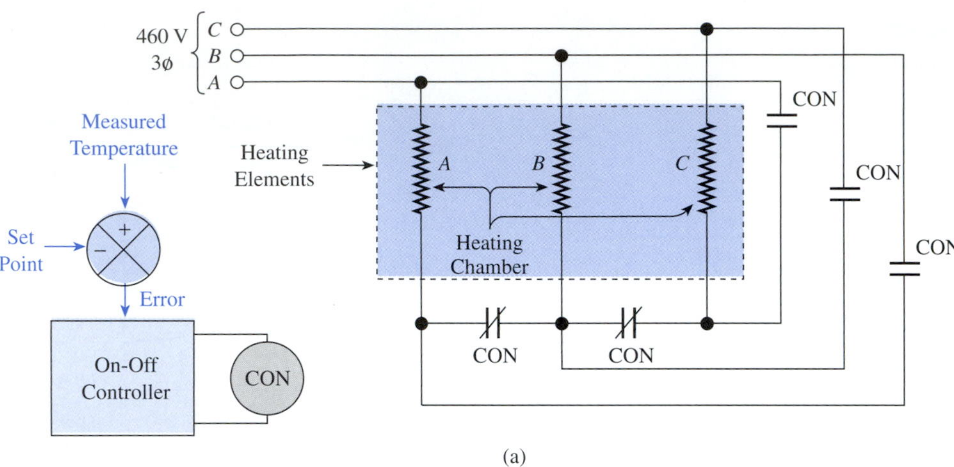

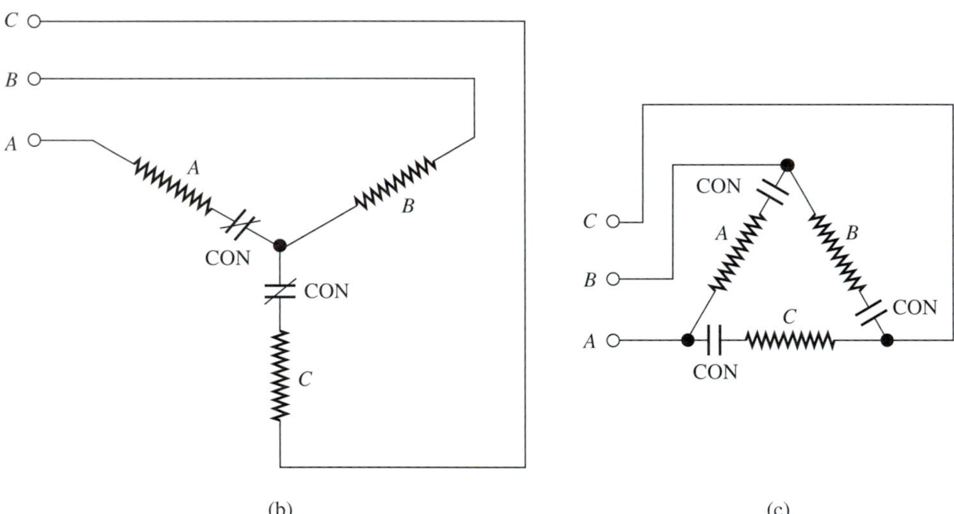

FIGURE 11–12 (a) Delta-wye heating circuit. (b) Simplified view of the heating circuit with the contactor deenergized. The heating elements are connected in wye, and the heat output is lower. (c) Simplified view of the heating circuit with the contactor energized. The heating elements are connected in delta, and the heat output is higher.

that the heat created by a wye connection causes the temperature to decline and the heat created by a delta connection causes the temperature to rise.

To see that the elements are connected in wye with *CA* dropped, study Fig. 11–12(b). To produce Fig. 11–12(b), the normally open *CA* contacts have been removed from the schematic of Fig. 11–12(a), thus uncluttering the drawing. It is apparent that the three elements are connected in a wye configuration.

Figure 11–12(c) shows the situation with the *normally closed* contacts removed from the schematic of Fig. 11–12(a). It is apparent that the elements are connected in delta when *CA* is picked up.

We have stated that the heat created by a delta connection is greater than the heat created by a wye connection because the electrical power delivered to the elements is greater in delta than in wye. This will now be proved. Refer to Sec. 14-5 for a thorough discussion of three-phase ac systems and the distinctions between wye and delta configurations.

In the delta (Δ) configuration, the voltage across any one heating element is equal to the line voltage, or

$$V_{ph} = V_{line} \qquad \text{for } \Delta$$

where V_{ph} is the voltage across one phase of the heating load. If the resistance of one phase of the load (one single element) is symbolized R, then we can say that

$$P_{ph} = \frac{V_{ph}^2}{R} = \frac{(V_{line})^2}{R}$$

where P_{ph} is the average power delivered to one single phase of the load (one element). We are assuming the elements are pure resistances with unity power factor.

The total average power delivered to the entire three-phase load is simply three times the power delivered to any one phase, so total power is given by

$$P_T = \frac{3(V_{line})^2}{R} \qquad \text{for } \Delta$$

Assuming some actual numbers, if the line voltage is 460 V ac and the resistance per phase is 25 Ω, the total power is

$$P_T = \frac{(3)(460)^2}{25} = 25\,400 \text{ W} \qquad \text{for } \Delta$$

Now for wye: The voltage across any one phase is not equal to the line voltage. For a balanced three-phase wye system, we can always say that

$$V_{ph} = \frac{V_{line}}{\sqrt{3}} = \frac{V_{line}}{1.73} \qquad \text{for Y}$$

Therefore, average power delivered to one phase of the load is

$$P_{ph} = \frac{V_{ph}^2}{R} = \frac{(V_{line}/\sqrt{3})^2}{R} = \frac{(V_{line})^2}{3R} \qquad \text{for Y}$$

Again, the total power is just three times the power delivered to any one element,

$$P_T = \frac{3(V_{line})^2}{3R} = \frac{(V_{line})^2}{R} \qquad \text{for Y}$$

Assuming the same numbers as before, namely 460-V line voltage and 25-Ω heating elements, we get

$$P_T = \frac{(460)^2}{25} = 8460 \text{ W} \qquad \text{for Y}$$

The total heat delivered to the process by the wye connection is thus only one third as much as the heat delivered by the delta connection (8460 W/25 400 W = 1/3).

11-8 ■ THYRISTORS

When the manipulated variable is electric current and the delivery must be *continuously* variable, relays and contactors cannot do the job. In modern control systems, power thyristors, namely, SCRs and triacs, are used as final correcting devices for such applications. Thyristors can also be used in On-Off systems, replacing relays and contactors.

Thyristors lend themselves quite well to proportional control of temperature. The conduction angle can be made to vary in proportion to the error between measured temperature and set point. This continuously varies the flow of current into a heating element, providing the benefits of proportional control. The integral and/or derivative control modes can be added to the system by adding the proper electronic circuitry.

Of course, not all applications of thyristors involve closed-loop control of current flow. Thyristors serve as the control device in many open-loop systems. In a thyristor open-loop system, current to the load is continuously variable, but there is no built-in comparison between measured value and set point to bring about *automatic* adjustment of the current.

As an example of an SCR in an open-loop temperature control system, simply substitute a resistive heating element for the load in Fig. 4–8 and imagine that the heating element is supplying heat energy to a process chamber. If R_v is adjusted, the firing delay angle of the SCR varies, varying the current through the heating element and thus the power delivered to the heating process. For a given set of process conditions, a given setting of R_v will produce a specific process temperature. Of course, if any process condition changes (if a disturbance occurs), that particular setting or R_v will cause a different temperature to be established. In an open-loop situation, the firing delay angle of the SCR is *not* automatically altered to correct the temperature. The temperature simply lands wherever the process conditions dictate.

Thyristors have many other industrial applications besides varying current flow to heating elements. They can be used to vary the current through an electromagnet whose magnetic pulling force must be variable; they can be used to vary current to an incandescent light bulb for situations where light intensity must be varied; they can be used to vary welding current to alter the properties of a weld, as we saw in Chapter 7; they can be used as sequencing devices, as shown in Fig. 5–7. But their most important use is in varying the current through a motor winding to adjust the speed of the motor. This is such an important part of industrial electronics that we will devote a separate chapter to the subject. In Chapter 16 we will treat in detail the use of power thyristors as motor speed-control devices.

11-9 ■ SPLIT-PHASE AC MOTORS

Split-phase motors were seen in Secs. 11-2 and 11-3 as motors for opening and closing flow-control valves. Not only do split-phase motors handle positioning of valves, but they also perform many positioning jobs in servo systems. Many servo systems have what is called an *ac servo motor* as their final correcting device. An ac servo motor is basically a split-phase motor with some minor construction differences. In this section we will give a brief explanation of the operation of split-phase motors. In Sec. 11-10 we will examine ac servo motors, their construction, and their operation in mechanical positioning systems. For more thorough coverage of two-phase and split-phase squirrel-cage induction motors, refer to Secs. 14-1 through 14-3.

In any ac induction motor, ac power is applied to the *field winding* on the stationary part of the motor (the stator). Current is then induced in the *armature winding* on the rotating part of the motor (the rotor) by transformer action. The interaction of the magnetic field created by the field winding with the current-carrying conductors of the armature winding creates mechanical forces which cause the rotor to spin.

Let us concentrate first on how the field winding of a split-phase motor establishes a magnetic field, and on the behavior of that magnetic field. Then we will explore how currents are induced in the armature conductors. Finally, we will combine the magnetic field with the armature current to see how torque is produced.

11-9-1 The Rotating Field

Figure 11–13(a) is a schematic representation of the field winding of a split-phase motor. The view shown represents what would be seen looking down the hollow stator from one end of the motor shaft. The rotor has not been drawn in order to keep the drawing clear.

There are two windings shown in Fig. 11–13(a). Each one is usually referred to as a *winding* (singular). We have winding 1, driven by voltage V_1, which tends to establish a magnetic field in the vertical direction; and we have winding 2, driven by voltage V_2, which tends to establish a magnetic field in the horizontal direction. Naturally the ac voltages V_1 and V_2 are continually changing polarity, but it is convenient to assign them a *defined-positive polarity,* as has been done in Fig. 11–13(a). That is, when V_1 is + on the top and − on the bottom, we will consider it as having a positive polarity. When V_1 is − on the top and + on the bottom, we will consider it as having a negative polarity. The same holds for V_2.

Now if V_1 is positive, a current will flow through winding 1 from top to bottom, thereby creating a magnetic field pointing from top to bottom. This field direction is shown in Fig. 11–14(a).

If V_2 is positive, a current will flow through winding 2 from left to right, thereby creating a magnetic field pointing from left to right. This field is shown in Fig. 11–14(c).

If V_1 and V_2 are both positive at the same time, currents will flow in both windings. With each winding creating an individual field, the *net* magnetic field will be a compromise between the individual fields. This means that the net magnetic field will point midway between the individual fields. This is shown in Fig. 11–14(b).

By some means, the voltages applied to the two windings, V_1 and V_2, must be adjusted so that the winding currents are 90° out of phase, as illustrated in Fig. 11–13(b). There are several ways of doing this, which we will investigate later. For now, just remember that by some external method V_1 and V_2 must be adjusted so that I_1 and I_2 are 90° out of phase.

Now study Fig. 11–13(b) and compare it carefully with Fig. 11–14. The arrows in Fig. 11–14 show the direction of the magnetic field created by the two windings at various points along the ac cycle (various points in time). The arrow labeled F_1 indicates the direction of the field due to winding 1; the arrow labeled F_2 indicates the direction of the field due to winding 2. The arrow labeled F indicates the overall net field direction, due to both windings combined.

FIGURE 11–13
(a) Field poles and windings for a split-phase ac motor. This is a view looking down the hollow stator. (b) The phase relationship between the two winding currents. I_1 leads I_2 by 90°.

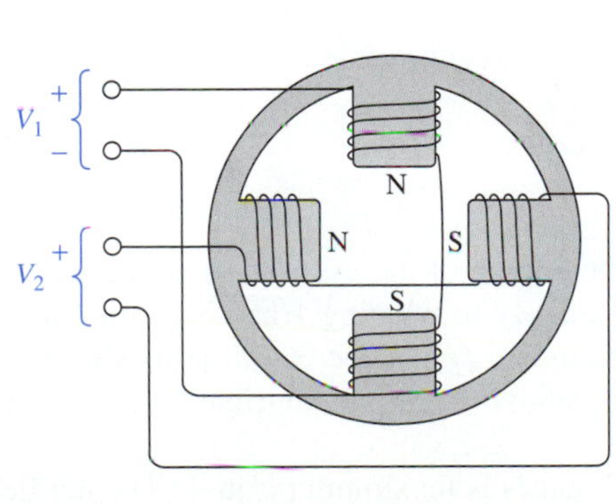

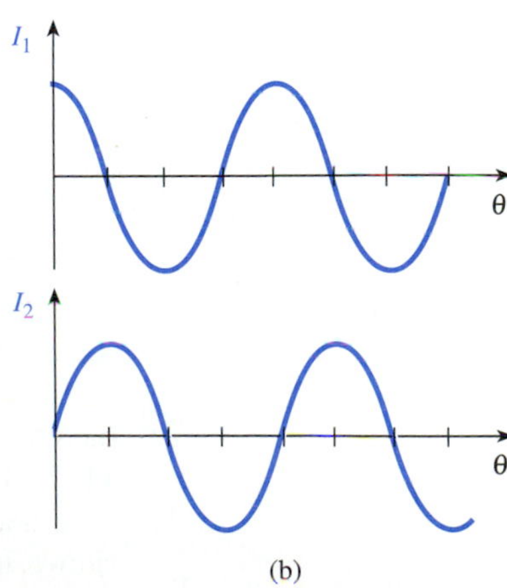

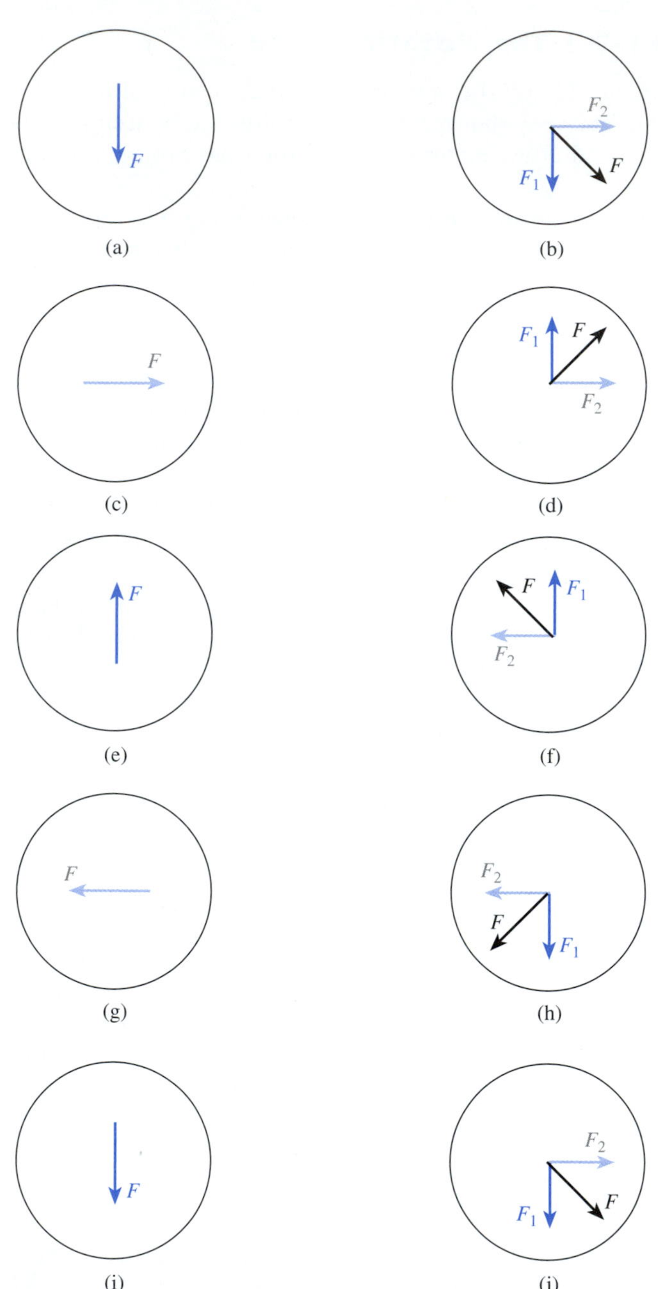

FIGURE 11–14 Individual fields created by the two windings (F_1 and F_2) and the net resultant field (F) at different instants in time. Each drawing represents a 45° progression in the ac cycle, or about a 2.08-ms progression in time for a 60-Hz ac line.

At 0°, at the beginning of the cycle, I_1 is maximum positive, and I_2 is 0. Therefore the net field is due entirely to winding 1 and is in the direction shown in Fig. 11–14(a).

At 45° into the cycle, I_1 has decreased but is still positive, and I_2 has climbed into the positive region. Both windings are contributing to the net field, which is shown in Fig. 11–14(b).

At 90°, I_1 is 0 and I_2 is maximum positive. The net field is due entirely to I_2 and is shown in Fig. 11–14(c).

At 135° into the cycle, I_1 is negative, so the field created by winding 1 is pointing *up,* as shown in Fig. 11–14(d). The field from winding 2 is still pointing to the right because I_2 is still positive. The net field is as shown.

At 180° into the cycle, I_2 has dropped to 0, and I_1 is maximum negative. The field is shown in Fig. 11–14(e).

At 225° into the cycle, I_1 is still negative, and I_2 has also moved into the negative region. Therefore the field due to I_1 points up, and the field due to winding 2 points to the left. The net field points to the upper left. This is shown in Fig. 11–14(f).

At 270° into the cycle, I_2 is maximum negative, and I_1 is 0, so the net field points to the left, as shown in Fig. 11–14(g).

At 315° into the cycle, I_1 is once again positive, and I_2 is still negative, so the net field points to the lower left, as shown in Fig. 11–14(h).

At 360° the winding currents have both returned to starting conditions, so the field is also back in the starting direction. Figure 11–14(i) shows the same net field direction as Fig. 11–14(a).

What is happening here is that the net field is rotating *around the stator,* just as if a field winding were actually physically rotating. The field is making one rotation around the stator for every cycle of the ac supply voltage. Its speed of rotation in revolutions per second is equal to the frequency of the ac voltage in cycles per second (Hz).* The strength of the magnetic field depends on the magnitude of the currents in windings 1 and 2, just as with any electromagnet.

The graphs in Fig. 11–13(b) show I_1 leading I_2 by 90°, and the diagrams in Fig. 11–14 show that the net field is rotating counterclockwise. If the relationship between the winding currents were changed so that I_2 led I_1 by 90°, the net field would rotate in the opposite direction, clockwise. You should prove this to yourself.

This completes our discussion of the action of the *field* in a split-phase motor. To sum up, we have seen that as long as both windings are carrying out-of-phase currents, the net field will rotate. The strength of the net field is determined by the amount of current flowing in the windings. The direction of rotation of the field depends on which current leads the other.

11-9-2 The Armature Conductors

When a rotor is inserted into our split-phase motor, forces will be exerted on it to make it follow the rotating field. If the field rotates clockwise, the rotor will rotate clockwise. If the field rotates counterclockwise, the rotor will rotate counterclockwise. Here is how this happens.

Figure 11–15 shows a three-quarter view of a *squirrel-cage rotor* for a split-phase motor. Most induction motors have rotors that look like this. (The split-phase motor is a specific example of the general class of induction motors.)

The rotor is basically a cylinder with shafts extending out both ends. The shafts are not shown in Fig. 11–15. The shafts are supported in bearings. The rotor is thus free to spin. The rotor material is some ferromagnetic (iron-based) alloy which has good magnetic properties. The rotor has slots, or grooves, cut lengthwise in its body, in which aluminum conductors are inserted. The conductors are joined one to another at each end by an aluminum *end ring*. There is no insulation between the aluminum conductors and the iron core. However, any current which flows from one end of the rotor to the other must

*This is true only for a motor with two poles per phase, as shown in Fig. 11–13(a).

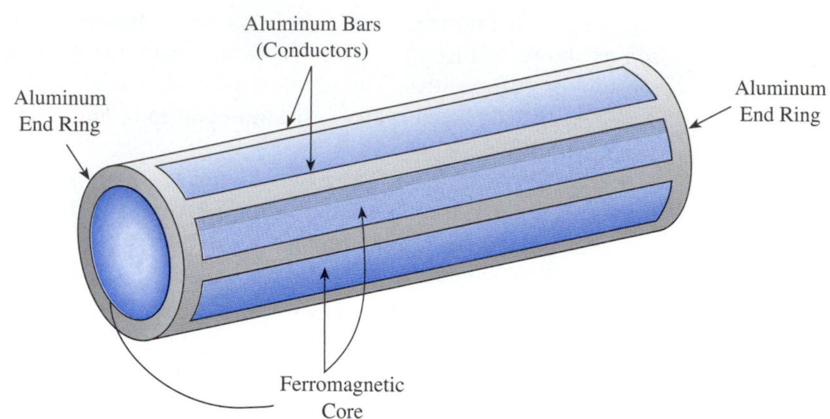

FIGURE 11–15
Isometric view of a squirrel-cage rotor, showing the aluminum end rings connected by aluminum conducting bars. The bars are laid into slots in the magnetic core.

do so by flowing down the aluminum bars because the core is laminated. That is, it is made up of layers of core material separated by layers of insulating material. This makes it impossible for currents to flow from one end to the other through the core material.

As the stator's magnetic field rotates (revolves) around the squirrel-cage rotor, it induces currents to flow in the aluminum conducting bars, by transformer action. Because these currents flow, forces are exerted on the conducting bars which cause the rotor to spin in the same direction as the rotating field. The origin of the forces can be viewed in either of two ways:

1. The currents in the rotor conductors interact with the magnetic field lines to produce mechanical forces according to the familiar *right-hand rule* of electric-magnetic interaction.
2. The current flow in the rotor conductors converts the rotor into a powerful magnet in its own right. The rotor magnetic poles then seek to align with the stator field poles according to the laws of magnetic attraction and repulsion (namely that unlike poles attract each other, and like poles repel each other).

Either way it is viewed, the rotor seeks to follow along behind the rotating field because of the forces. This is how motor torque is created in an ac induction motor.

11-9-3 Creating the Phase Difference between the Two Winding Currents

Split-phase motors will work satisfactorily even if the phase angle between I_1 and I_2 is not exactly 90°. *Any* phase relationship which can create the effect of a rotating field will cause the motor to spin. One easy way to create a phase difference between the winding currents is to insert a capacitor in series with one of the windings and then drive both windings with the same voltage source. Figures 11–2 and 11–3 showed exactly this. The capacitor tends to cause the winding current to lead the applied voltage. It cannot cause the current to lead the voltage by exactly 90° because of the resistance and inductance of the winding, but it does establish *some* phase shift.

If the capacitor is inserted in series with winding 1, it would be indicated schematically as shown in Fig. 11–16(a). This would cause I_1 to lead I_2, as shown in Fig. 11–13(b). The rotor would then rotate counterclockwise if the windings were wrapped exactly as shown in Fig. 11–13(a).

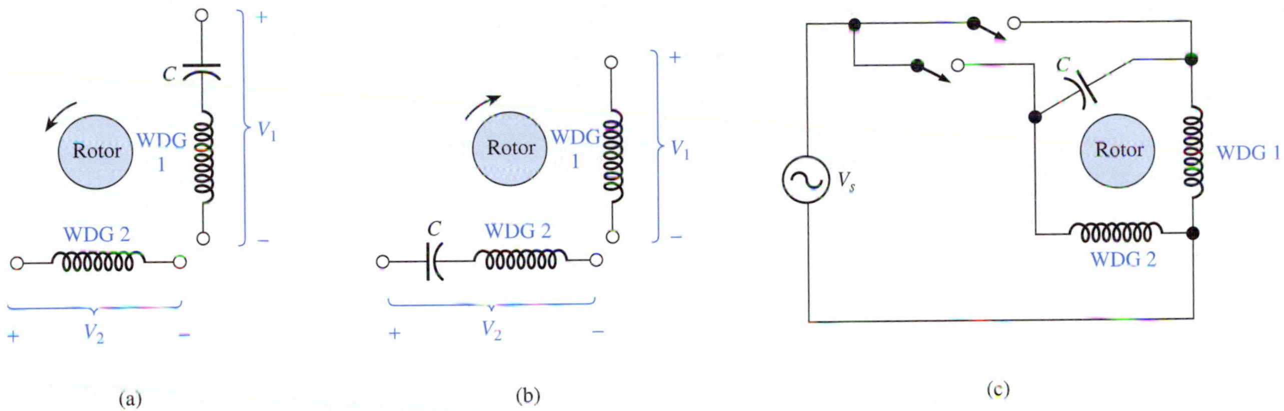

FIGURE 11–16
Creating the phase shift between winding currents. (a) With a capacitor inserted in series with winding 1, I_1 leads I_2; this causes the motor to spin in a certain direction. (b) With a capacitor inserted in series with winding 2, I_2 leads I_1; this causes the rotor to spin in the opposite direction. (c) Closing one of the switches causes one direction of rotation, while closing the other switch causes the opposite rotation.

However, if the capacitor were placed in series with winding 2, then I_2 would lead I_1, and the rotor would rotate clockwise. This situation is illustrated in Fig. 11–16(b).

With these relationships in mind, it is now possible to see why the motor will rotate in different directions depending on which switch is closed in Fig. 11–16(c). This is essentially the same circuit as shown in Figs. 11–2 and 11–3, and it is the most popular way to use split-phase motors in industrial control.

The use of a fixed capacitor is not the only way to create the phase difference needed to make the field rotate. Some split-phase motors *start* with a capacitor in series with one winding, and then switch that winding completely out of the circuit once the motor has accelerated up to operating speed. This can be done with a centrifugally operated switch. The winding that is switched out is called the *starting winding,* and the one that continues to drive the rotor is called the *running winding* or the *main winding*. This technique is possible because split-phase motors can often *run* on a single winding; however, they can not start from a standstill using a single winding.

Several other methods that are used to duplicate the action of a rotating field are presented in Sec. 14-3.

11-10 ■ AC SERVO MOTORS

As we know, when the controlled variable in the closed-loop system is a mechanical position, the system is called a servo system. Two simple servo systems were presented in Sec. 9-3 as examples of closed-loop systems. A more general arrangement of a servo system is shown in Fig. 11–17.

The setting potentiometer on the left is adjusted to express the desired position of the controlled object. There would probably be some sort of scale attached to the setting pot. That scale would relate potentiometer shaft position with mechanical position of the controlled object. For example, if the servo system made it possible to position the controlled object anywhere within a 12-ft range, the setting pot dial might have markings with spacing equal to one twelfth of the total rotation of the pot shaft. Then each mark would "translate" into 1 ft of mechanical movement of the object. The operator could decide

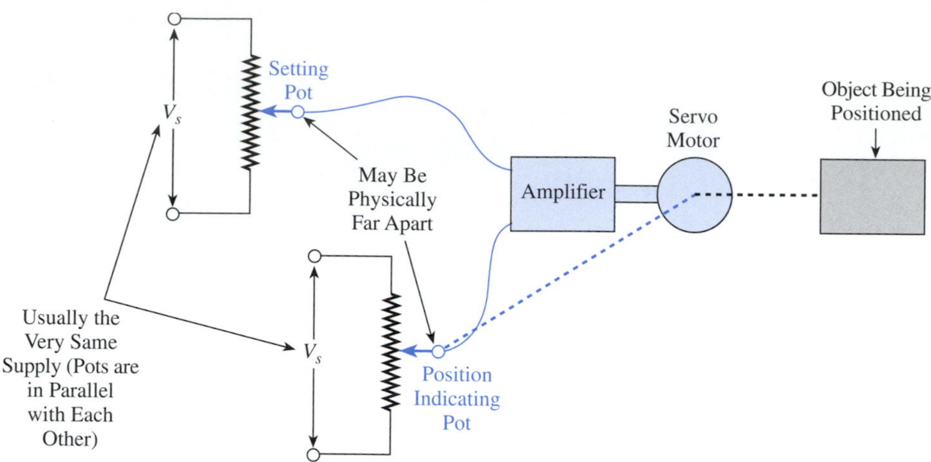

FIGURE 11–17
General layout of a servo system.

what mechanical position of the object he wanted, turn the setting pot dial to the proper number on the scale, and walk away. The servo system would handle the rest of the job of positioning the controlled object where it is supposed to be.

There are many reasons for the use of servo control systems in industry. Among them are the following:

1. The object may be very large and/or heavy, so that a human being cannot handle it directly. It can be handled conveniently only by a specially designed servomechanism dedicated to that task.
2. The object may be inaccessible or inconvenient to get to. The remote setting and adjusting feature of the servo system is then a great benefit.
3. The object may be dangerous to be near. The remote adjustment capability allows the operator to exert control without exposing himself to danger.

For these and other reasons, servo systems are very useful in industrial control. In this section we will study the most common final correcting device of a servo system, the ac servo motor.

An ac servo motor is essentially the same as the split-phase ac motor discussed in Sec. 11-9. There is one important difference between a standard split-phase motor and an ac servo motor. It is that the servo motor has thinner conducting bars in the squirrel-cage rotor, so the conductor resistance is higher. As we proceed with our discussion of ac servo motors, we will point out why this feature is needed.

The two windings of an ac servo motor are referred to as the *main winding* and the *control winding*. Sometimes the main winding is called the *fixed winding*. The word *phase* is frequently substituted for the word winding. Thus we hear the terms *fixed phase* and *control phase* to describe the two separate field windings of a servo motor.

In the great majority of servo control systems, the servo motor is not *switched* on and off as were the split-phase motors in Figs. 11–2, 11–3, and 11–16(c). Instead, a servo motor is driven as indicated in Fig. 11–18(a).

The voltage applied to the control winding, V_C, is taken from the output of an amplifier. The input to the amplifier is the error voltage, V_e, which depends on how far the object is from the desired position.

The fixed winding always has power applied to it by a fixed-voltage ac source, as Fig. 11–18(a) shows. In this case there is a capacitor inserted in series with the fixed winding to shift its current by about 90°.

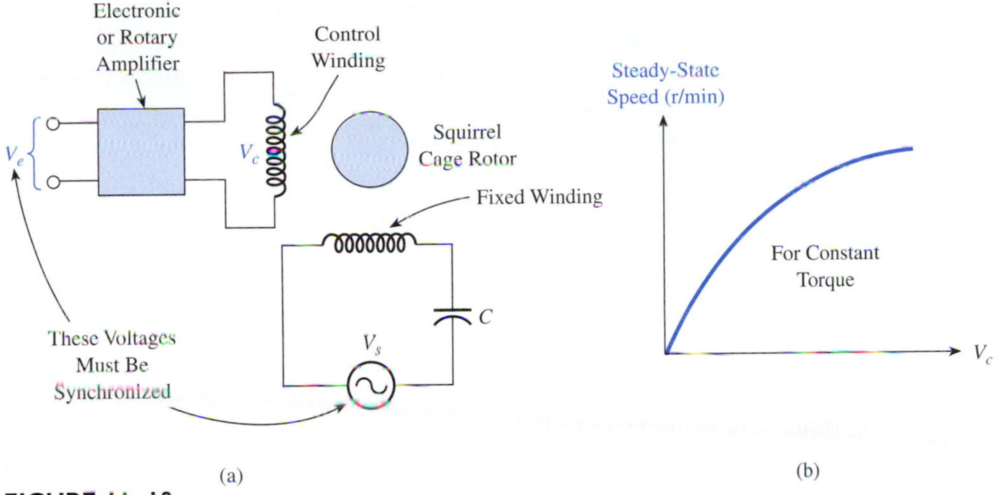

FIGURE 11–18
(a) Circuit of an ac servo motor. The voltage applied to the control winding varies in proportion to the error voltage, V_e. (b) Curve of speed versus control voltage for an ac servo motor.

The method of operation of the servo motor is not hard to understand. If the difference between actual position and desired position is great, V_e will be large (see Fig. 11–17). If V_e is large, V_c will also be large, and the control winding current will be large. This will cause the servo motor to run at high speed. As the difference between the actual position and desired position decreases, the error voltage V_e also decreases, as Fig. 11–17 shows. Therefore V_c decreases, reducing the control winding current and making the motor run slower. The relationship between rotational speed and control voltage is shown in Fig. 11–18(b).*

When the position of the object is precisely correct, the error voltage V_e will shrink to zero. The amplifier will have no input signal, so V_c will also go to zero. The motor curve shows that the motor stops when $V_c = 0$, so the object will stop moving when it gets to the desired position. Of course, the servo motor *must* stop when V_c gets to zero. Recall that some split-phase motors can continue to run with only one winding powered up. This is called *single-phasing*. Servo motors must not be able to single-phase. This is one of the reasons for creating the high-resistance conductor bars in the rotor of an ac servo motor. High-resistance bars prevent single-phasing.

Observe the note in Fig. 11–18(a) which says that V_e and V_s must be synchronized. This means that they must be derived from the same ultimate voltage source, so their phase relationship is either in phase or 180° out of phase. This is necessary so that the final field winding currents are approximately 90° out of phase with each other. In this example, the phase-shift capacitor C causes the fixed winding current to lead the applied voltage by about 90°. The control winding current, on the other hand, is roughly in phase with the control winding voltage.

*The curve in Fig. 11–18(b) shows the *steady* speed which would eventually be established if the control voltage stayed constant long enough for the motor to stabilize, versus the size of the control voltage. Of course, as the positioned object homes in on the desired position in a real servo system, the control voltage doesn't remain constant; it is constantly shrinking toward zero. Also, this curve is for constant shaft torque, which never occurs in a real servo system; the torque delivered by the motor shaft is reduced as the object gets closer to the set position. Therefore the curve in Fig. 11–18(b) is not a valid graph of speed versus control voltage for a servo motor operating in a real servo system. It is only intended to show the characteristics of the motor under ideal and somewhat artificial conditions. Curves of this type are necessary, however, or how could we compare one motor to another?

The control voltage is in phase with error voltage, assuming the amplifier processes the input signal without introducing any phase shift of its own. This last requirement, that the amplifier introduce no phase shift, is easy to accomplish at the low frequencies used for servo systems (60 Hz in most industrial systems).

The end result is one of these two situations:

1. If the error voltage is *in phase* with V_S, the control winding current will *lag* the fixed winding current by about 90°, and the motor will turn in one direction (assume clockwise).
2. If the error voltage is *180° out of phase* with V_S, the control winding current will *lead* the fixed winding current, and the motor will turn in the other direction (counterclockwise).

These two situations are illustrated graphically in Fig. 11–19. The winding currents have been assumed to be in phase with the voltages applied to the windings, which is a convenient assumption to make for purposes of understanding servo motor operation. It is not quite true in reality, however. Also, the fixed winding current is shown leading the source voltage by exactly 90°, this is another assumption which is not quite true in reality but is convenient for explanatory purposes.

Error in Phase with Source

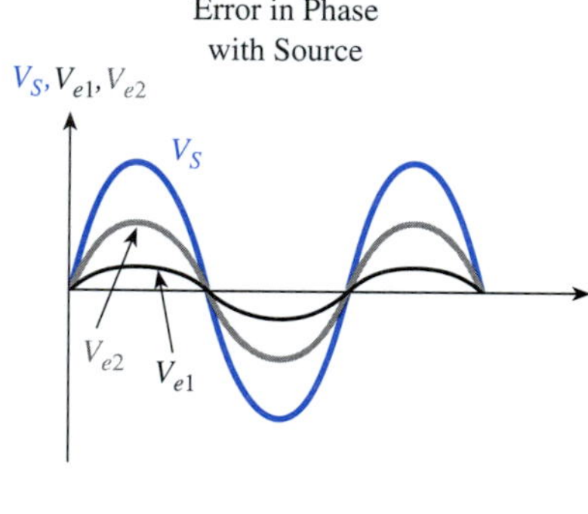

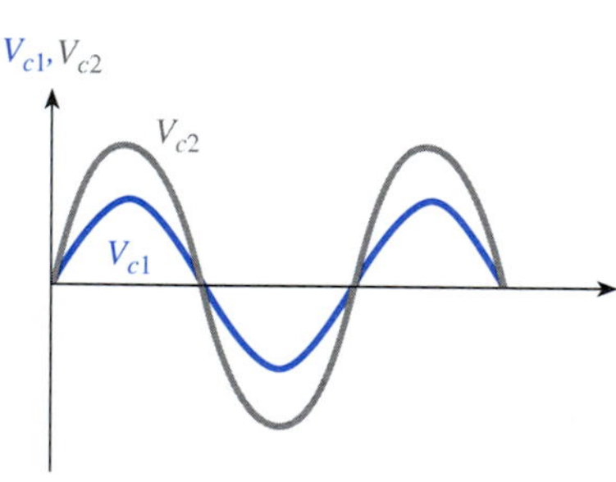

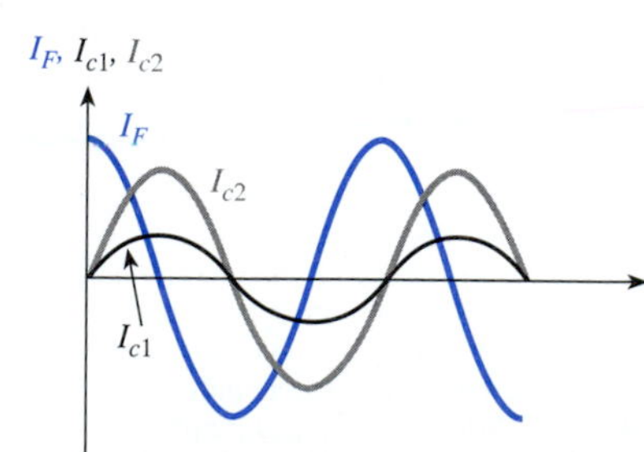

(a)

Error 180° Out of Phase with Source

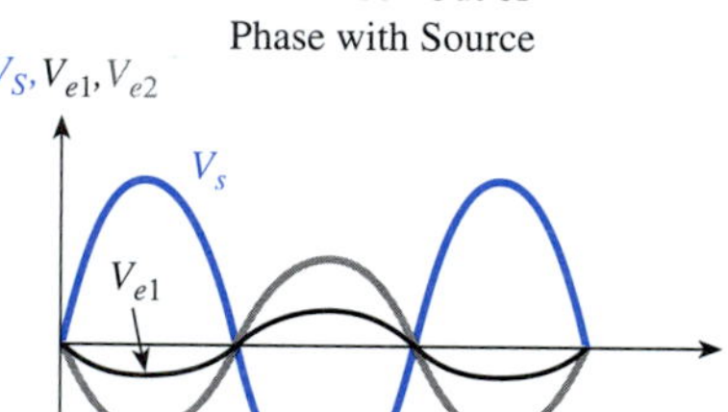

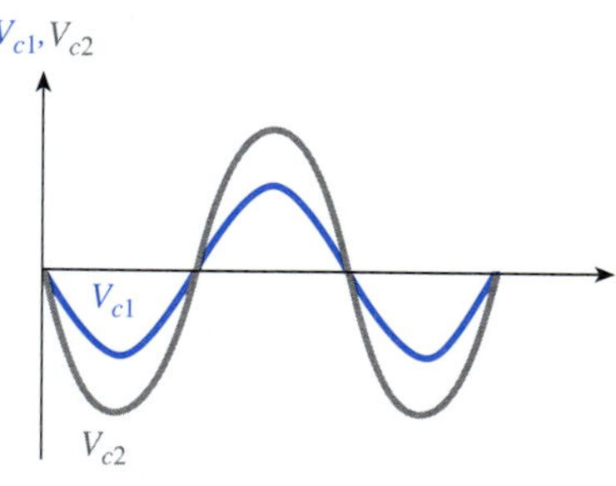

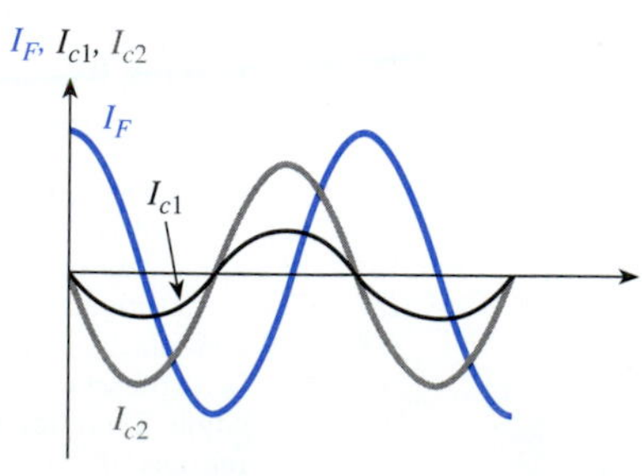

(b)

FIGURE 11–19
(a) Graphs of the various voltages and currents in an ac split-phase servo motor for the error voltage in phase with the source voltage. (b) The same graphs for the case in which the error voltage is out of phase with the source voltage. In the current graphs, the fixed winding current is shown exactly 90° out of phase with the control winding current. This is an idealization.

In Fig. 11–19(a), on the left, V_e is in phase with V_s. Two graphs are plotted for two different error voltages, V_{e1} and V_{e2}, corresponding to two different distances of the controlled object from the desired position. The fact that V_e is in phase with V_s means that the controlled object is in one particular direction away from the desired position (let us say to the east).

In Fig. 11–19(b), on the right, V_e has been taken as 180° out of phase with V_s. This would mean that the controlled object is in the opposite direction away from the desired position (to the west). Again, graphs have been drawn for two different distances from the desired position. For a small distance away, the error voltage is V_{e1}. It results in small control winding current, I_{c1}, which causes slow motor speed. For a large distance away, the error voltage is V_{e2}. It results in large control winding current, I_{c2}, which causes faster motor speed. The symbol I_F stands for current through the fixed winding; note that it is always 90° ahead of V_s.

Occasionally a servo motor will have no phase-shift capacitor in series with the fixed winding. In such cases the required 90° phase shift must be provided by the amplifier. The amplifier would be specially designed so that the V_c output is 90° out of phase with the V_e input. This can be done by making appropriate choices of coupling capacitors and other electronic circuit components. When this is done, the amplifier can be used only for a particular frequency, since the phase-shift angle would change if the frequency changed. This practice is more often seen in aircraft servo systems than in industrial servo systems.

11-10-1 Torque-Speed Characteristics of Ac Servo Motors

As mentioned at the beginning of this section, ac servo motors are essentially the same as standard split-phase ac motors except that the rotor bar resistance is made higher. We have already seen one of the advantages of this, that it prevents single-phasing. Single-phasing would be disastrous in a servo control system because it would mean that the motor would not stop running when the position error was reduced to zero. The other important advantage of high rotor bar resistance is that it makes the torque-speed relationship of the motor better for servo applications. Figure 11–20 shows the torque-speed relationship of a standard split-phase motor compared to that for an ac servo motor. The curve of the servo motor is for a constant value of control voltage.

The curve for the normal split-phase motor is telling us that when the motor is running slow, it has a certain amount of torque-producing ability. As the speed increases, the

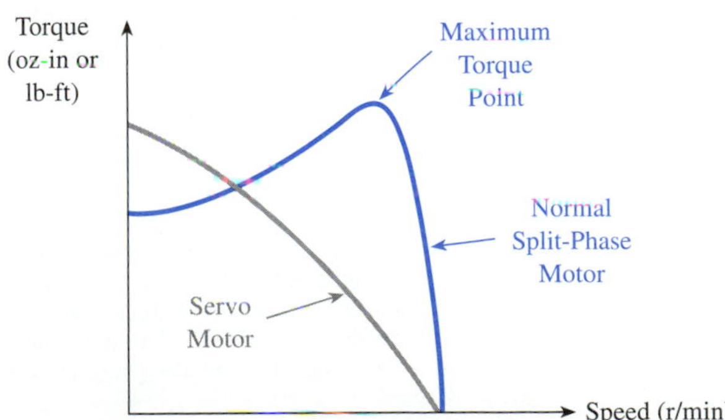

FIGURE 11–20 Torque versus speed curves for a normal split-phase ac motor and for an ac servo motor. This curve assumes a constant control voltage applied to the servo motor.

torque-producing ability of the motor also increases. This relationship holds true up to a certain point, called the *maximum torque point*. After that, any further increase in speed results in a reduction in the torque-producing ability of the motor.

The reason for this behavior is that the torque-producing ability of the motor is determined basically by the amount of current flowing in the rotor conductor bars. This current, though, is determined by two things:

1. The *amount* of voltage induced in the rotor conductor loops
2. The *frequency* of the voltage induced in the rotor conductor loops

Regarding effect 1, the amount of voltage induced in the rotor conductor loops is large when the speed is slow and small when the speed is fast, because the rotating stator field is moving fast *relative to the conductor bars* when the rotor is moving slow. (In other words, the rotor is "slipping behind" drastically.) On the other hand, the rotating stator field is moving fast. (The rotor is not "slipping behind" much.) Therefore the voltage is large when the rotor is moving slowly, and the voltage is small when the rotor is moving fast. This effect tends to *reduce* rotor current at fast speeds.

However, regarding effect 2, the frequency of the induced voltage is high when the rotor speed is slow, and the frequency is low when the rotor speed is fast. This is because the frequency of the induced voltage in a rotor conductor loop is equal to the number of times per second that the rotating field "laps" the rotor loop. That is, one cycle of ac voltage will be induced in a rotor conductor-bar loop every time the rotating stator field gains one revolution on the rotor.*

The higher the frequency, the higher the inductive reactance of the one-turn rotor conductor loops ($X_L = 2\pi fL$). As you know, a higher inductive reactance causes lower ac current flow, and a lower inductive reactance allows higher ac current flow. This effect therefore tends to *increase* the rotor current at fast rotor speeds.

What we have here are two opposite effects. One effect, voltage, tends to *reduce* current and torque at fast speeds, and the other effect, frequency, tends to *increase* current and torque at high speeds. It's just a matter of which effect is stronger. For the normal split-phase motor, the frequency effect is stronger in the low-speed range, so torque-producing ability goes up as speed goes up. At a certain point (maximum torque point) the voltage effect takes over and predominates. Thereafter, the torque-producing ability goes down as speed goes up. This explains why normal split-phase motors exhibit the torque-speed relationship that they do, which is drawn in Fig. 11–20. In fact, this explains why virtually all induction motors have the torque-speed characteristics they do (including the workhorse of industry, the three-phase induction motor).

Unfortunately, though, these torque-speed characteristics are not good for a servomechanism. In servo control, it is desired to produce a lot of torque at slow speeds so that the motor can accelerate the positioned object rapidly. Besides, it is better if the torque-producing ability of the motor is reduced at high speed, because that makes it less likely that the motor will overshoot its mark. That is, it's less likely that the controlled object will shoot past the desired position and have to back up.

---

*You can understand this by picturing a fast race car carrying a magnet on its side and a slow race car carrying a coil of wire on its side. If both cars are going around the race track, the fast car's magnet will create one cycle of induced voltage in the slow car's coil every time the fast car laps the slow car. If the speed of the fast car is held constant (like a rotating stator field), then the speed of the slow car will determine the frequency of the induced voltage. The slower the slow car goes, the more often it will be lapped and the higher the frequency. The faster the slow car goes, the less often it will be lapped and the lower the frequency.

Servo motors are therefore built to have a large rotor conductor-bar resistance. This is easily done by making the bars thinner and shallower. With a large rotor bar resistance, the inductive reactance of the rotor bar loops is swamped out, and the frequency effect is minimized. The voltage effect thus predominates. As seen earlier, the voltage effect tends to *reduce* torque-producing ability at high rotor speeds. The result is the torque-speed characteristic for an ac servo motor shown in Fig. 11–20.

The complete set of characteristic curves for an ac servo motor is shown in Fig. 11–21. The different curves are for different control voltages applied to the control field winding. As Fig. 11–21 shows, the torque-producing ability of a servo motor is greatest at low speed and declines as speed goes up. This is true for either a large or a small control winding voltage.

One interesting fact about servo motors emerges from these curves. At small control voltages (meaning that the controlled object is getting close to the desired position) the servo motor can actually exert a *reverse torque* on its shaft. This is helpful in preventing overshoot because if the object has a large inertia, it may tend to coast past the desired position even if the servomechanism exerts no motive force whatsoever. That is, in a case involving high inertia and low friction, the servo motor could be completely disconnected from the power lines when the controlled object was still some distance from the desired position, and the object would not be able to stop in time.

In a real servo system, the control winding is never disconnected from the amplifier. The amplifier output will be reduced to a very small voltage, though, as the object gets close to the desired position. If the inertia of the controlled object tries to make the rotor spin faster than it wants to, the interaction of the fixed-winding magnetic field with the fast-turning rotor causes current to flow *backward* in the control winding. Under these

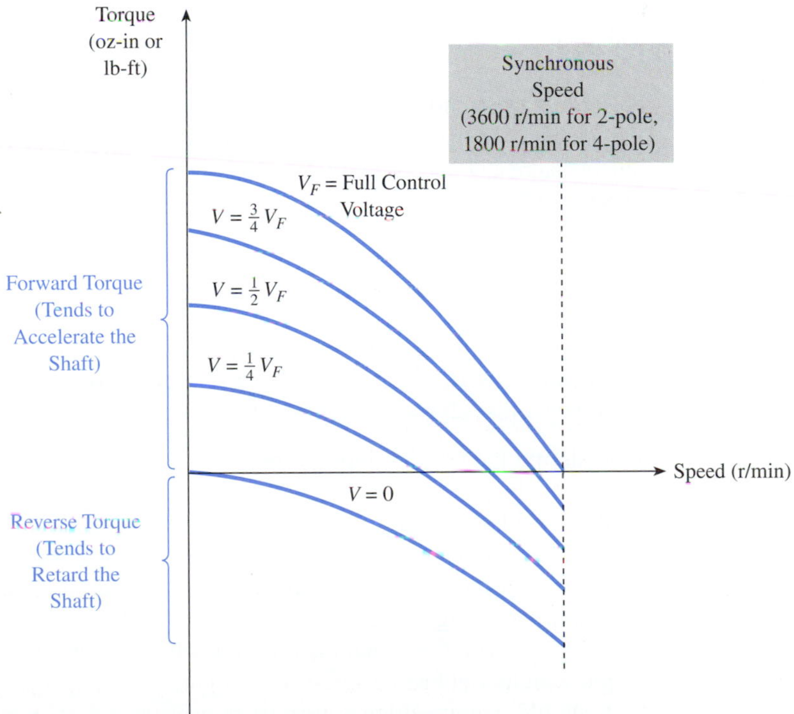

FIGURE 11–21
Torque versus speed curves of an ac servo motor for various values of control voltage.

circumstances, the control winding is no longer being driven by the amplifier but is acting like a generator. Because of the reversed control winding current (180° different from what it normally is), the torque exerted on the rotor is in the opposite direction. This tends to bring the motor shaft to a screeching halt and minimizes overshoot of the desired position.

This is not to say that servo systems never overshoot the mark. Far from it. Usually the reversed torque applied by the servo motor is not sufficient by itself to stop the load in time. An obvious solution to this problem is to allow the motor to run only very slowly, by applying only very small control voltages. This would defeat one of the good features of a closed-loop control system, though, because it would lengthen the correction time. Instead, the usual solution is to provide *damping,* which is the tendency to exert a countertorque when the speed is fast. Many ingenious techniques have been devised to provide damping. Such things as eddy-current drag cups and tachometer feedback loops are examples of damping methods applied to servomechanisms. Any book dealing exclusively with servo systems will give an explanation of such methods.

Ac servo motors are generally preferred to dc servo motors for the same reasons that ac motors are always preferred to dc motors:

1. Their squirrel-cage rotors are simple and rugged compared to the complex armature windings found on dc machines.
2. They have no brush-to-commutator contacts which require frequent inspection and maintenance.
3. There is no insulation around the armature conductors as there is on a dc motor, so the armatures can dissipate heat much better.
4. Because the rotor has no complicated insulated windings on it, its diameter can be made quite small to reduce rotor inertia; this helps prevent overshoot in a servomechanism.

Nevertheless, some servo systems require a dc motor as the final correction device. This usually happens when the positioned object is very massive. When a very massive object is being positioned, the motor must naturally have a high power rating. High-power ac servo motors suffer from the problem of their rotor conductors overheating because of the necessity to make the rotor conductor resistance high. Therefore servo systems which deliver large amounts of power normally use dc motors. Dc servo motors will be discussed in Sec. 11-12.

11-11 ■ SOLID-STATE AC SERVO AMPLIFIERS

An ac servo amplifier amplifies the position error voltage to produce the servo motor control voltage, as shown in Fig. 11–18. An ac servo amplifier has the same general requirements that all good amplifiers have:

1. It should have a high input impedance, so it doesn't load down its signal source (the error voltage source).
2. It should have a hefty voltage gain (A_V) which is fairly independent of temperature changes, component aging, and variations among components.
3. It should have a low output impedance, so it can drive a heavy current load without its output voltage sagging.
4. It should keep the signal distortion down to a reasonable level. An ac sine-wave input should yield an ac sine-wave output.
5. It should operate efficiently so its transistors run cool, especially its final output transistors.

One thing a servo amplifier does *not* need is a wide bandwidth. On the contrary, high-frequency roll-off of its frequency response curve is a blessing because it tends to eliminate high-frequency noise signals which sneak into the amplifier.

We will study three different ac servo amplifiers. Among the three of them, most of the important features and variations among servo amplifiers will be illustrated.

11-11-1 Servo Amplifier 1: A Four-Stage Transistor Amplifier with Push-Pull Output

Refer to Fig. 11–22, which is the schematic drawing of a four-stage servo amplifier. Transistor Q_1 and its associated components comprise the input stage of the servo amplifier. The 60-Hz ac input signal (the error voltage) is fed to the base of Q_1 through coupling capacitor C_1. Q_1 is connected in an emitter-follower configuration to display a high impedance to the error voltage source. Because of the high impedance looking into the input stage, the reactance of C_1 does not need to be very low. Therefore C_1 is quite a bit smaller than the other coupling capacitors in this circuit.

The input resistance (impedance) of the input stage is given by

$$R_{\text{in }1} = R_1 \| R_2 \| r_{b\text{ in }1} \quad \textbf{(11-1)}$$

where $r_{b\text{ in }1}$ stands for the ac resistance looking into the base of Q_1. The ac resistance looking into the base of Q_1 is specified by

$$r_{b\text{ in }1} = \beta_1(r_{ej1} + r_{e\text{ out }1}) \quad \textbf{(11-2)}$$

where β_1 is the current gain of Q_1, r_{ej1} is the ac resistance of the base-emitter junction of Q_1, and $r_{e\text{ out }1}$ is the ac resistance looking out from the emitter of Q_1 to ground. The ac resistance across the base-emitter junction of a forward-biased transistor is rather low, usually less than 50 Ω, so r_{ej1} is negligible compared to $r_{e\text{ out }1}$. Therefore, assuming that β_1 is at least 100 (a conservative estimate), the input resistance of the input stage is given approximately by

$$R_{\text{in}1} = R_1 \| R_2 \| 100(r_{e\text{ out }1})$$

or

$$R_{\text{in }1} = 500\text{ k}\Omega \| 680\text{ k}\Omega \| 100(r_{e\text{ out }1}) = 307\text{ k}\Omega \| 100(r_{e\text{ out }1}) \quad \textbf{(11-3)}$$

Now the ac resistance looking out from the emitter of Q_1 to ground is equal to the parallel resistance of two current flow paths. Path 1 runs directly down through emitter resistor R_3 to ground. Path 2 runs through C_2, through R_4, through the gain-adjust pot, and into the Q_2 circuitry. The resistance of path 1 is simply 47 kΩ, the size of R_3. The resistance of path 2 is equal to the sum of R_4, plus the resistance of the gain-adjust pot, plus the ac resistance looking into the second stage. Considering the worst case, with the gain-adjust pot dialed completely out for maximum amplifier gain, the ac resistance of path 2 is equal to R_4 (10 kΩ) plus the input resistance of the second stage. The input resistance of the second stage is at least 11 kΩ if β_2 is at least 100 (since $R_{\text{in }2} = R_5 \| R_6 \| \beta_2[r_{b\text{ in }2}] = 150\text{ k}\Omega \| 22\text{ k}\Omega \| 100[270\ \Omega] = 11\ k\Omega$). Therefore

$$r_{e\text{ out }1} = r_{\text{path }1} \| r_{\text{path }2} = 47\text{ k}\Omega \| (10\text{ k}\Omega + 11\text{ k}\Omega) \cong 15\text{ k}\Omega$$

Going back to Eq. (11-3), we can now calculate the input resistance of the first stage as

$$R_{\text{in }1} = 307\text{ k}\Omega \| 100(15\text{ k}\Omega) = 307\text{ k}\Omega \| 1500\text{ k}\Omega \cong 250\text{ k}\Omega$$

An input resistance of 250 kΩ is adequate for most applications. That is, most error voltage sources would not be unduly loaded down by an amplifier whose input resistance was this high.

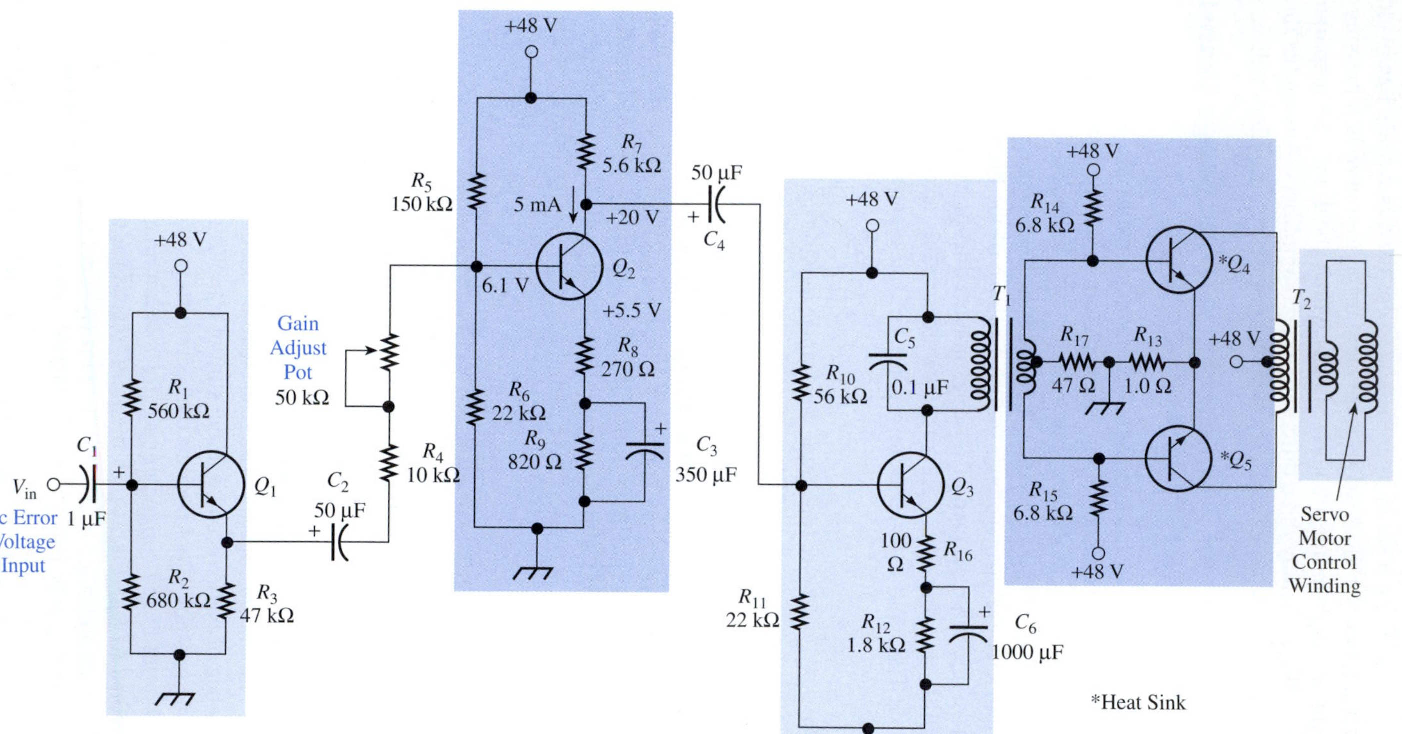

FIGURE 11–22
Straightforward solid-state servo amplifier.

In an ac servo amplifier, as in most ac amplifiers, it is a good practice to keep the reactance of the coupling capacitors less than about 3% of the input resistance of the amplifier stage. In this particular case, for a 60-Hz signal,

$$X_{C1} = \frac{1}{2\pi f C_1} = \frac{1}{2\pi(60)(1 \times 10^{-6})} = 2.6\ \text{k}\Omega$$

so

$$\frac{X_{C1}}{R_{\text{in }1}} = \frac{2.6\ \text{k}\Omega}{250\ \text{k}\Omega} \cong 0.01$$

which means that X_{C1} is only about 1% as large as $R_{\text{in }1}$.

The Q_1 emitter follower can provide no voltage gain. Therefore the voltage appearing at coupling capacitor C_2 is slightly smaller than the input error voltage. The C_2 voltage is then further divided between the R_4/gain-adjust pot combination and the second stage of the servo amplifier. The second stage, comprised of transistor Q_2 and its associated components, is a voltage amplifier with a voltage gain of about 10. The emitter resistors R_8 and R_9 serve to stabilize the dc bias point against temperature changes and transistor batch variations. The emitter bypass capacitor C_3 allows ac current to bypass R_9. This tends to increase the voltage gain. R_8 is left unbypassed to help raise the input impedance of the second stage.

The output voltage from the second stage is taken from the collector of Q_2 and fed through C_4 into the third stage. The third stage amplifies the signal further and applies it across the primary winding of T_1. The primary winding of T_1 is tuned to 60 Hz by C_5. Together, the winding inductance and the capacitor C_5 form a parallel-resonant circuit with a resonant frequency of around 60 Hz. This allows the third stage to provide maximum voltage amplification to the 60-Hz signal frequency and very little amplification to extraneous signals at other frequencies.

The ac output voltage from the third stage is stepped down by T_1 and appears across the center-tapped secondary winding. T_1 is a 10:1 step-down transformer; it reduces the third-state output voltage down to less than 2 V while boosting the current capability by a factor of 10. This provides a proper match with power transistors Q_4 and Q_5, which need very little ac input voltage but do require a fair amount of input current. Power transistors Q_4 and Q_5 are wired in a common-emitter push-pull configuration. Here is how the push-pull power output stage works.

During the half cycle that the T_1 secondary voltage is positive on top, the top half of the winding delivers base current to Q_4. Current flows out the top terminal of the winding, through the base-emitter junction of Q_4, through R_{13} and R_{17}, and back into the center tap. Q_4 then carries collector current via this path: from the +48-V supply, into the center tap of the T_2 primary, through the top half of T_2, through the collector-to-emitter path of Q_4, through R_{13}, and into the ground.

While all this is happening, Q_5 is biased OFF. This is because the bottom terminal of the T_1 secondary is negative relative to ground, robbing Q_5 of the small base current which was supplied through R_{15}. With Q_5 turned OFF, the bottom half of the T_2 primary winding carries no current.

Thus during the positive half cycle of the signal voltage, the T_2 primary winding carries a net current in the *up* direction. This induces a T_2 secondary voltage of a certain polarity, say positive on top. This voltage is applied to the motor control winding. Therefore one half cycle of an ac sine wave is created across the control winding.

During the other half cycle of the signal voltage, the T_1 secondary is positive on the bottom. This allows the bottom half of the T_1 secondary to deliver base current to Q_5, which turns on and carries collector current via this path: from the +48-V supply, into the center tap of the T_2 primary, through the bottom half of T_2, through the collector-emitter path of Q_5, through R_{13}, and into the ground. While all this is happening, Q_4 is biased OFF by the negative voltage applied to its base by the top half of the T_1 secondary. With Q_4 turned OFF, the top half of the T_2 primary carries no current.

Thus during the negative half cycle of the signal voltage, the T_2 primary carries a net current in the *down* direction. This induces a T_2 secondary voltage of the opposite polarity from before, positive on the bottom. In this manner the other half cycle of the sine wave is applied to the motor control winding.

You may wonder why we would use such a complicated method to deliver a complete sine wave of voltage to the control winding. The reason is a very good one. Here it is: This method allows the output transistors to spend half of their time "resting" and cooling down. One transistor in the push-pull stage handles one half of the signal, and the other transistor in the push-pull stage handles the other half of the signal. *Neither transistor has to carry steady dc current at all times.*

This method has a great power advantage over the conventional transistor biasing method, in which a certain dc bias current is established and the ac signal current is superimposed on it. That dc bias current is of no use in driving the load because it is blocked out by the coupling capacitor or transformer. Nevertheless, it creates a power drain on the dc power supply and forces the transistor to dissipate that power. A push-pull output stage as shown in Fig. 11–22 eliminates such power waste. The output transistors still burn a lot of power because of the relatively large motor currents they carry. However, all the power they do burn is the result of ac current, which is at least useful for driving the load. By eliminating the dc power consumption problem, smaller-power transistors can be used, and the size of their heat sinks can be reduced. Virtually all ac servo amplifiers use some variation of the push-pull configuration for their power output stage.

It is not hard to see that if the phase of the error voltage input changed by 180° relative to the fixed winding voltage, the lead-lag relationship would be reversed. That is, if the control winding current had been leading the fixed winding current by 90°, it would now be lagging the fixed winding current by 90°. This would cause the motor to turn in the opposite direction. Therefore the *phase* of the error voltage determines the direction of rotation of the servo motor, and the size of the error voltage determines the speed of rotation.

11-11-2 Servo Amplifier 2: A Four-Stage Chopper-Stabilized Transistor Amplifier with Negative Feedback and Unfiltered Dc Supply to the Control Winding

Figure 11–23 shows an amplifier design which illustrates some other popular features of ac servo amplifiers. First, notice that the error voltage input is a dc signal. Some error detectors are unable, for one reason or another, to provide an ac voltage to the servo amplifier. Instead they deliver a dc voltage. Straightforward amplification of a dc error voltage is difficult because of the drift problems inherent in a dc amplifier. A popular alternative is to *chop* the dc signal to make it resemble an ac signal. The signal chopper is just a vibrating reed switch whose common terminal is alternately connected to the dc input and ground potential. The signal appearing at the common terminal of the chopper switch is a square wave whose height equals the magnitude of the dc input voltage.

The vibration of the mechanical switch is created by the chopper drive coil, an electromagnet which is powered by a 60-Hz supply. The chopper drive supply is synchronized with the fixed motor winding ac supply. On one half cycle of the 60-Hz ac line the

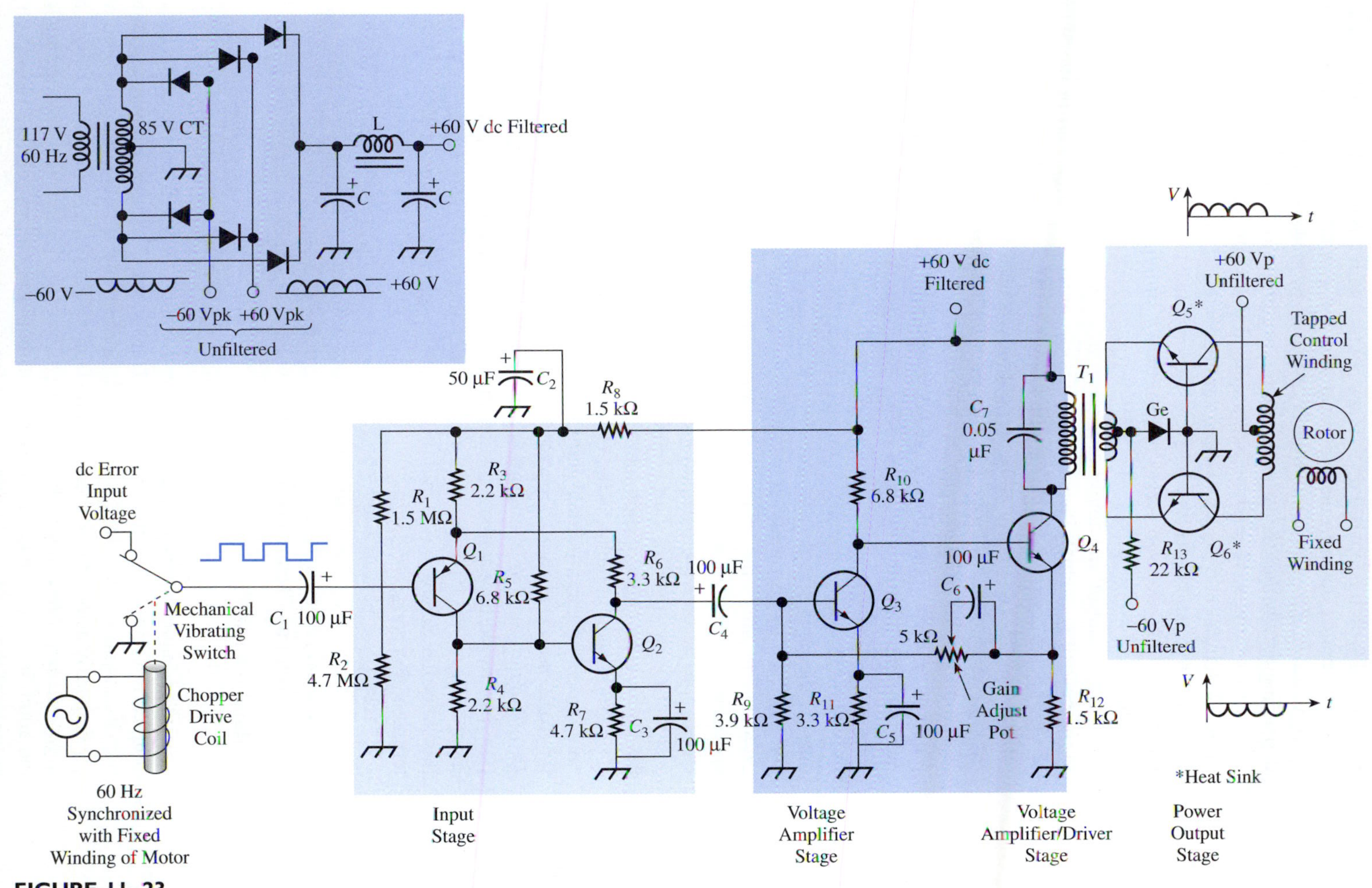

FIGURE 11–23

Another transistor servo amplifier. This design illustrates several popular features of servo amplifiers, including chopped dc input, negative feedback from one stage to another, power supply decoupling between stages, and unfiltered dc supply to the output stage.

mechanical switch is pushed up to make contact with the input terminal; on the other half cycle it is pulled down to make contact with the ground terminal. The result is a square wave which is either in phase with or 180° out of phase with the fixed-winding supply, depending on whether the dc error voltage is positive or negative.

The square-wave input signal is applied to coupling capacitor C_1 and into the base of the Q_1 transistor. Q_1 is a *pnp* transistor connected as a common-emitter amplifier. Emitter resistor R_3 stabilizes the bias currents against temperature changes and transistor batch variations. R_3 also serves to give the Q_1 amplifier stage an inherently high input resistance.

The output of this amplifier is taken from the collector of Q_1 and is direct-coupled to the base of Q_2, which is another common-emitter amplifier. Part of the output voltage of Q_2 appears across R_3, which is the emitter resistor of Q_1. R_3 is used by both transistors Q_1 and Q_2. The portion of the Q_2 output voltage which appears across R_3 *in the emitter of* Q_1 is of such a polarity that it *opposes the original input signal* applied to Q_1. This is called *negative voltage feedback.* It serves to raise the input resistance of the Q_1 amplifier stage even higher than it would be naturally. Negative voltage feedback also further stabilizes the voltage gain of the Q_1-Q_2 combination against temperature change and transistor variations.

To understand why this arrangement provides negative feedback, consider what happens as the input signal at C_1 goes into its negative half cycle. The negative-going voltage tends to turn Q_1 on harder, causing the collector of Q_1 to go more positive relative to ground. This drives the base of Q_2 more positive and tends to turn Q_2 on harder. As Q_2 begins conducting greater collector current down through R_3 and R_6, it causes the emitter of Q_1 to become more negative (closer to ground, further from the $+V_{CC}$ supply). The fact that the ac signal at the Q_1 emitter is forced to become more negative tends to cancel the original effect of the input signal going negative.

To explain in detail, if the Q_1 emitter had merely followed the input signal, a certain amount of current would have flowed through the base-emitter junction of Q_1; the feedback, however, tends to reverse-bias the Q_1 base-emitter junction, which tends to turn Q_1 off somewhat. The end result is that less current flows through the base-emitter junction than would have without the negative feedback. With less input current for the same input voltage, the input resistance of the amplifier is increased. The input resistance of this amplifier approaches 1 million ohms. The input resistance is also stabilized against temperature changes, etc.

The price paid is a reduction in voltage gain of the Q_1-Q_2 input stage, which becomes much lower than it would have been without the feedback. The voltage gain becomes stabilized though, as mentioned earlier.

To sum up, the effects of negative voltage feedback are to

1. Raise the input resistance and stabilize it against circuit variations.
2. Lower the voltage gain and stabilize it against circuit variations.

The output signal of the Q_1-Q_2 input stage is taken from the collector of Q_2 and coupled through C_4 into the base of Q_3. Transistor Q_3 is another common-emitter amplifier. Its output is taken from the Q_3 collector and direct-coupled to the base of Q_4. The Q_4 emitter resistor, R_{12}, serves to help stabilize the Q_4 bias current and also to provide *negative current feedback* to the base of Q_3. Negative current feedback serves to stabilize the amplifier gain against circuit variations but also, unfortunately, to lower the input resistance of the Q_3 stage. At this point in the circuit, however, input resistance has ceased to be so important, because the Q_3 stage is being driven by the Q_1-Q_2 stage, which has good current-delivering ability.

To understand why the R_{12}/gain-adjust pot combination provides negative current feedback, consider what happens as the signal at C_4 enters its positive half cycle. The positive-going voltage tends to turn Q_3 on harder. This causes the voltage at the collector of Q_3 to go more negative (closer to ground). This negative-going voltage tends to turn Q_4 off, causing a reduction in current in the Q_4 emitter lead. The voltage across R_{12} is therefore going more negative at this instant. This negative-going voltage sucks current through C_6, robbing the base of Q_3 of some of the input current coming through C_4. What happens is that the attempt to increase the ac signal current in the Q_3-Q_4 combination causes a reaction which tends to decrease the ac current into Q_3. This is the essence of negative current feedback.

The effectiveness of the negative current feedback depends on the ac resistance between the emitter of Q_4 and the base of Q_3. This ac resistance can be varied by adjusting the gain-adjust pot. This ac resistance between these two points is whatever portion of the pot is not shorted out by C_6. As the wiper is moved to the left, more of the total pot resistance is shorted out by C_6, and the ac resistance is lowered. Lower ac resistance causes the feedback effect to be stronger, thus lowering the voltage gain of the Q_3-Q_4 combination.

If the pot wiper is moved to the right, less of the gain pot would be shorted out by C_6, thereby raising the ac resistance between the Q_4 emitter and the Q_3 base. This would reduce the strength of the negative current feedback and raise the voltage gain.

The output from the Q_4 amplifier appears across the primary of transformer T_1, which is tuned to a resonant frequency of 60 Hz by C_7. This helps reduce the gain of the square-wave harmonic frequencies and causes the output to be more sinusoidal.

Notice that the +60-V dc power supply line to the Q_3 and Q_4 amplifiers is separated from the Q_1-Q_2 supply line by R_8. This is a *decoupling resistor*. It works in conjunction with *decoupling capacitor* C_2 to filter out any power supply line noise signals before they reach the Q_1-Q_2 input stage. That is, any voltage disturbances appearing on the +60-V dc supply line due to the current drawn by Q_3 and Q_4 are prevented from appearing at Q_1 and Q_2. The R_8 and C_2 decoupling components can accomplish this because the reactance of C_2 is much less than the resistance of R_8 at 60 Hz or higher frequencies.

In this example, the reactance of C_2 at 60 Hz is about 50 $\Omega (X_C = 1/2\pi fC)$. This is only about 3% as large as the 1500-Ω resistance of R_8, so only about 3% of any noise disturbance will appear across C_2. Of course, keeping noise signals from appearing across C_2 is the same as keeping them from appearing across Q_1 and Q_2. It is important to keep power supply noise signals away from the input stage because if they get into the input stage, they are eligible for full amplification by the amplifier. Noise signals in the succeeding stages are not as harmful because they are not subject to the full amplification of the amplifier.

In the power output stage, the power transistors again are connected in a push-pull configuration, although this time they are wired as common-base amplifiers. Let us look first at the biasing arrangement. Refer to the waveforms of the power output stage in Fig. 11–24, specifically parts (a), (b), and (c). There is a small dc bias current supplied to Q_5 and Q_6 by the −60-V pk unfiltered supply. The path of bias current flow is this: from ground into the base of the transistor, out the emitter lead and through half the secondary winding, through R_{13}, and into the −60-V pk supply. This bias tendency is pulsating because the −60-V pk supply is full-wave pulsating dc. Biasing the base with unfiltered pulsating dc reduces the transistor's heat-dissipating requirements somewhat.

The T_1 secondary winding delivers emitter current to power transistors Q_5 and Q_6. The emitter current waveforms are drawn in Fig. 11–24(d) and (e). When the T_1

FIGURE 11–24
Various voltage and current waveforms for the amplifier of Fig. 11–23.

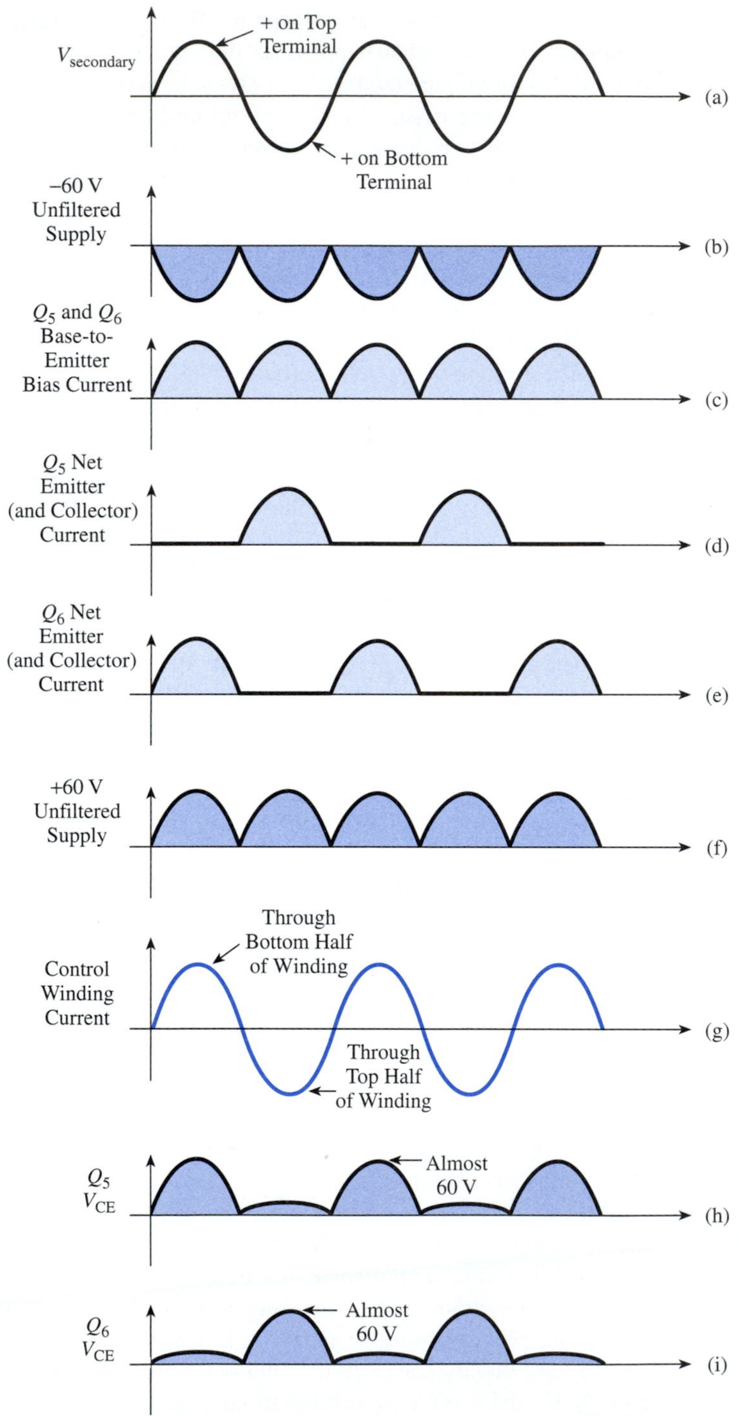

secondary winding goes positive on its top terminal, the bottom half of the secondary winding turns on Q_6. The current flow path is the loop comprised of the bottom half of the secondary winding, the base-emitter junction of Q_6, and the germanium diode. On the alternate half cycle, when the secondary voltage is positive on the bottom, the top of the winding turns on Q_5 in the same way.

The positive supply voltage to the center tap of the motor control winding is also full-wave pulsating dc. This unfiltered pulsating supply voltage helps reduce transistor power consumption considerably, because there is less voltage across the main transistor terminals as the instantaneous collector current climbs toward its peak value. That is, as the collector current rises toward its peak, the power transistor stays just barely out of saturation because the collector supply voltage is following the current up. With the transistor just barely out of saturation, the voltage drop across its main terminals stays small, causing the power consumption to be small ($P_{\text{inst}} = V_{\text{inst}} I_{\text{inst}}$). The waveforms of the positive voltage supply, the control winding current, and the collector-to-emitter voltages across the transistors are drawn in Fig. 11–24(f), (g), (h), and (i).

Besides cutting down on transistor power dissipation, using unfiltered pulsating dc to drive the push-pull power output stage also relieves the load on the *filtered* dc power supply. This reduces the ripple in that supply.

The servo motor control winding is a center-tapped winding, with only one half of the winding carrying current at any instant in time. This does not alter the fact that the magnetic field set up by the control winding changes directions every half cycle. In other words, when the current flows through the control winding from center to bottom (Q_6 conducting), the magnetic field it creates points in a certain direction relative to the stator of the motor. When the current flows through the winding from center to top (Q_5 conducting), it sets up a magnetic field pointing in the *other* direction relative to the stator. This alternation of magnetic field direction is all that is necessary to interact with the fixed-winding field to create a net rotating field.

11-11-3 Servo Amplifier 3: A Hybrid Amplifier Using an IC Op Amp in the Front End with a Discrete Push-Pull Output Stage

A simple servo amplifier using an IC op amp is shown in Fig. 11–25. In this application the op amp is being used as a noninverting ac amplifier. The voltage gain is set by the combination of R_1, R_2, and the gain-adjust pot. The voltage gain of the op amp noninverting amplifier can be adjusted over the range of 5.7 to 55.7 in this circuit since

$$A_{V\text{min}} = \frac{4.7\ \text{k}\Omega + 1\ \text{k}\Omega}{1\ \text{k}\Omega} = 5.7$$

and

$$A_{V\text{max}} = \frac{54.7\ \text{k}\Omega + 1\ \text{k}\Omega}{1\ \text{k}\Omega} = 55.7$$

The input impedance of the noninverting amplifier is naturally very high, so no special techniques are needed to raise the input impedance.

The components C_1, C_2, and R_3 are connected to the proper op amp terminals to determine the frequency response characteristics. The op amp manufacturer's data sheets always give advice on the sizes of these components.

The ac error voltage is applied to the noninverting input terminal, and an amplified signal is taken from the op amp's output terminal. It is passed through R_4 and C_3 into transistor Q_1. This transistor is connected in common-emitter configuration to provide maximum power amplification. R_5, R_6, and R_7 set the bias current of Q_1, and that bias point is stabilized by R_7. C_4 is connected in parallel with the T_1 primary winding to resonate at the signal frequency.

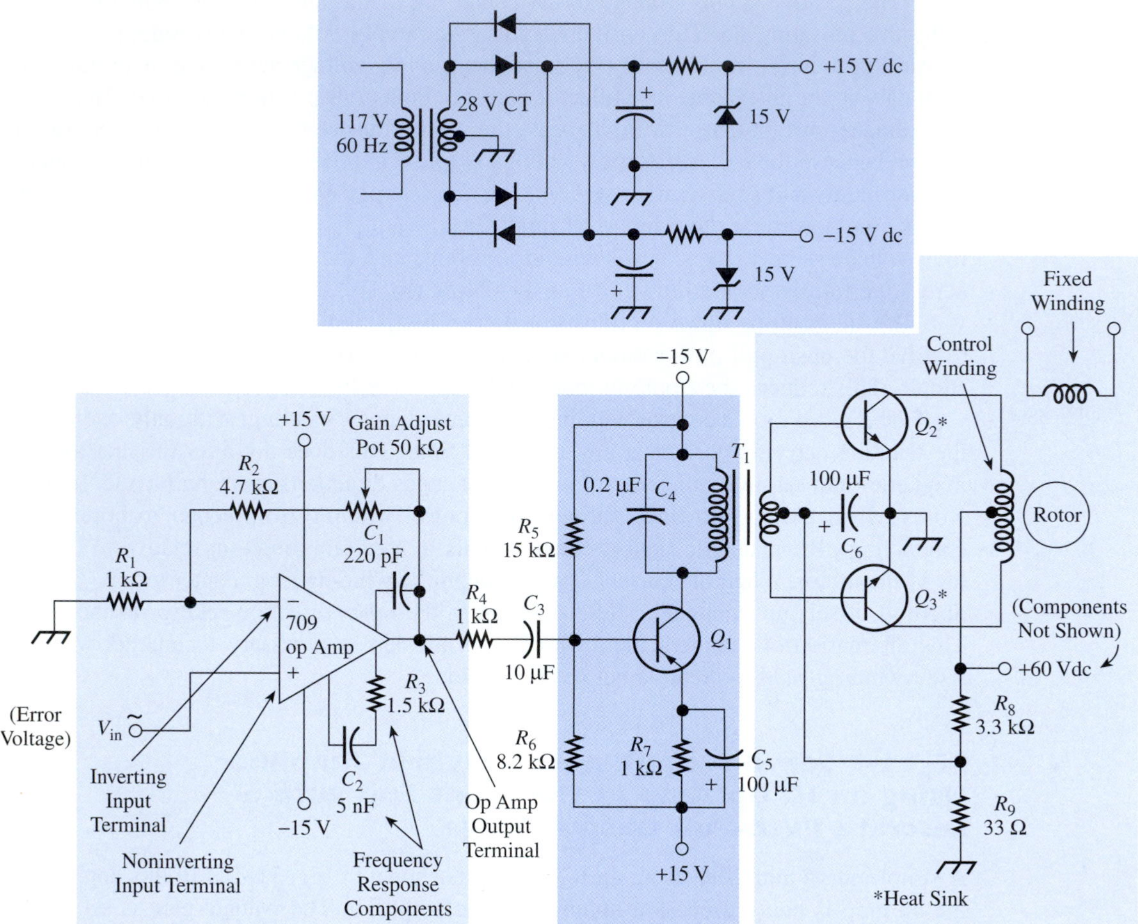

FIGURE 11–25
Hybrid (having both IC and discrete components) servo amplifier.

The T_1 secondary drives the Q_2-Q_3 transistor pair, which is connected in push-pull configuration. The transistors are biased slightly on by the R_8-R_9 voltage divider attached to the +160-V dc supply.

The 100-μF capacitor C_6 provides the path for the ac signal current to flow through the base-emitter junctions of Q_2 and Q_3.

11-12 ■ DC SERVO MOTORS

As mentioned in Sec. 11-10, ac servo motors are generally preferred to dc servo motors, except for use in very high-power systems. For very high-power systems, dc motors are preferred because they run more efficiently than comparable ac servo motors. This enables them to stay cooler. An efficient motor also prevents excessive power waste, although power waste is generally not a primary concern in servomechanisms.

Conventional wound-rotor dc motors are covered thoroughly in Chapter 12. Our present discussion of dc servos presumes that you already have some familiarity with dc motors.

A dc servo motor is no different from any other dc shunt motor for general-purpose use. It has two separate windings: the field winding, placed on the stator of the machine, and the armature winding, placed on the rotor of the machine. Both windings are connected to a dc voltage supply. In many dc shunt motor applications the windings might actually be connected in parallel (shunt) and driven by the same dc supply, but in a servo application, the windings are driven by separate dc supplies. This situation is illustrated schematically in Fig. 11–26(a).

The field winding of a dc motor is usually symbolized schematically as a coil. The field winding is connected to the dc voltage supply labeled V_F in Fig. 11–26(a). The armature winding of a dc motor is symbolized schematically as a circle in contact with two squares. This suggests the physical appearance of a dc armature as a cylinder having brushes bearing against its surface. The armature winding is connected to the dc voltage supply labeled V_A in Fig. 11–26(a).

We will not get into a detailed physical description of dc motor speed-control at this time. That subject is covered in Chapter 16. Suffice it to say that the steady-state speed can be controlled either by varying V_F or by varying V_A in Fig. 11–26(a). In virtually all modern servo systems, the adjustment is made to V_A, the armature supply voltage.

The relation between steady-state speed and applied armature voltage is shown in Fig. 11–26(b) for a constant torque. It is approximately a linear relation. However, in a real servo system, the motor torque is not constant. It varies as the controlled object approaches the desired position and the motor voltage is reduced. The more meaningful torque-speed curves at various armature voltages are shown in Fig. 11–26(c).

These curves tell us that the torque-producing ability of a dc servo motor is greater at low speeds than at high speeds for a given applied armature voltage V_A. This enables

FIGURE 11–26
(a) Schematic of a dc servo motor. (b) Graph of speed versus armature voltage for a constant shaft torque. (c) Torque versus speed curves for various applied armature voltages.

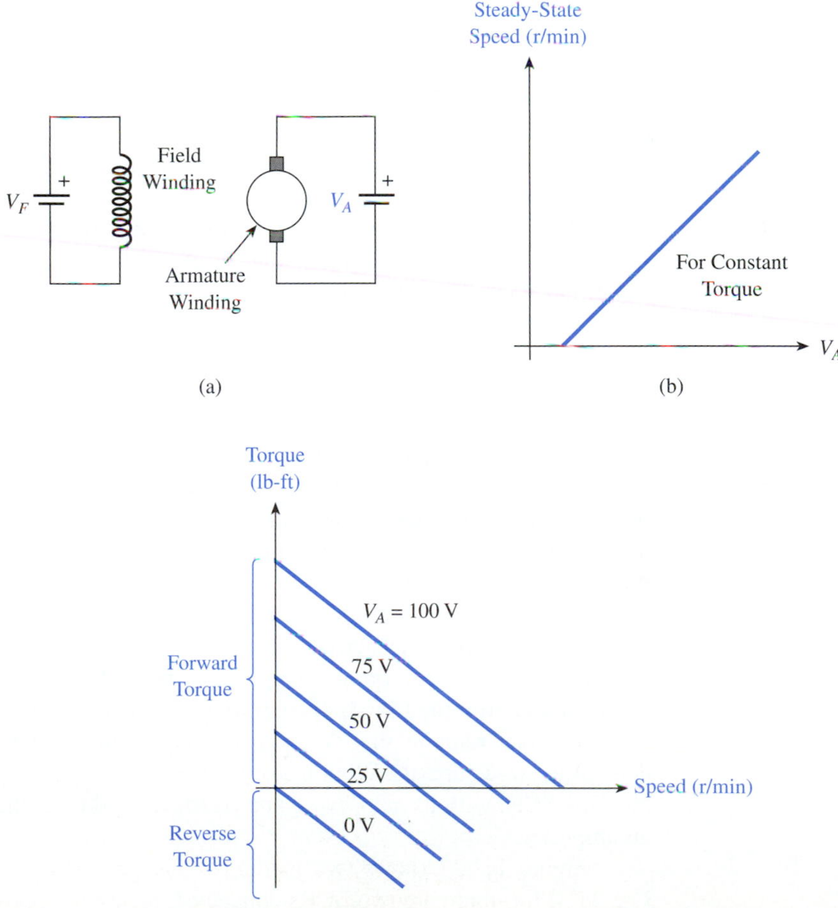

the motor to accelerate the load (the positioned object) rapidly from a standstill. Furthermore, the torque-speed curves show that as the positioned object approaches its desired position and V_A is reduced, the dc motor is capable of delivering a reverse torque to slow the load down if the speed of approach is high. This is possible because the motor armature winding starts to act like a generator under this condition. The current in the armature conductors reverses direction (assuming the armature voltage supply is capable of sinking current), and the reverse current creates reverse torque.

Thus the standard dc shunt motor satisfies the requirements for a good servo motor. It has high torque at low speeds, and it has a built-in *damping effect,* in that a high approach speed results in an automatic slow-down tendency. Of course, as with any dc motor, it cannot be expected to operate with as little trouble as an ac induction motor. A dc motor is inherently a high-maintenance device, because of the complex insulated armature winding and because of the fact that current must be carried to and from the armature through brush-to-commutator contacts.

11-13 ■ AMPLIFIERS FOR DC SERVO MOTORS

Theoretically, there is nothing wrong with the idea of using a dc amplifier to amplify a dc error voltage in a servo system. The amplifier would simply boost the dc error voltage up to a larger dc voltage to drive the armature of the servo motor. There are some practical problems with this idea, though. For instance, all coupling capacitors between the amplifier stages would have to be eliminated, since dc current cannot pass through a capacitor. Likewise, transformer coupling from one stage to another would have to be abandoned, since transformers will not respond to a dc current. In short, all dc amplifiers must be *direct-coupled.* This means that the output of one stage must be connected to the input of the next stage through resistors and/or wires only. Direct coupling is illustrated in the input stage of the amplifier in Fig. 11–23; the Q_1-Q_2 connection contains no capacitors or transformers. The other stages in that amplifier are not direct-coupled, though.

Direct coupling throughout a multistage amplifier causes certain practical problems which are just about insurmountable. In a direct-coupled amplifier, any change in transistor bias voltage due to power supply variation, temperature changes, or component aging is treated exactly like a real input signal. For example, if the dc bias voltage at the collector of a certain transistor were to change slightly due to a temperature rise, the other stages of the amplifier could not distinguish that change from a genuine voltage signal caused by a dc error input voltage.

In other words, direct-coupled amplifiers tend to change their dc output voltage due to internal bias voltage variations which have nothing to do with the externally applied input signal. As can easily be appreciated, this is very bad. This phenomenon is called *drift.* If it occurs in a servo system, the servo motor will run when it isn't supposed to run, and the controlled object will be improperly positioned.

There are various ways to minimize drift, but they all involve circuit complications, and they never work perfectly. Therefore servomechanism builders avoid using dc servo amplifiers. If the error input voltage must be a dc voltage for some reason, it can be chopped and amplified in an ac amplifier as shown in Fig. 11–23.

The fact remains that dc servo motors require rather large dc voltages capable of delivering large currents. What is needed is an ac amplifier capable of delivering a dc output. The SCR, with its *amplifying* and *rectifying* ability, is the ideal output device for such an amplifier.

Figure 11–27 shows two methods of using SCRs to control a dc servo motor. In Fig. 11–27(a) there are two SCRs connected, facing in opposite directions. SCR_1 controls the armature current in one direction, from top to bottom. It thus controls the motor when

FIGURE 11–27 Control circuits for a dc servo motor. (a) A method using just one ac supply winding. (b) A method using two ac supply windings. (c) Waveforms of error voltage and output voltage showing that a larger error voltage causes earlier firing of an SCR.

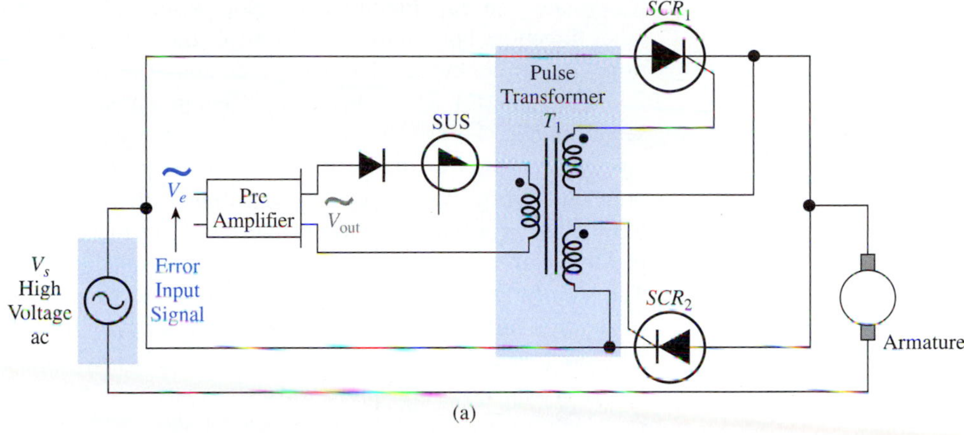

(a)

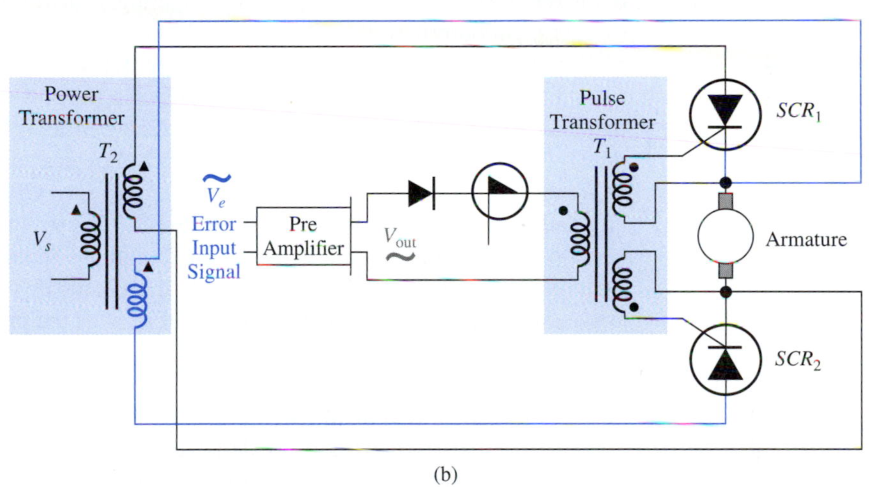

(b)

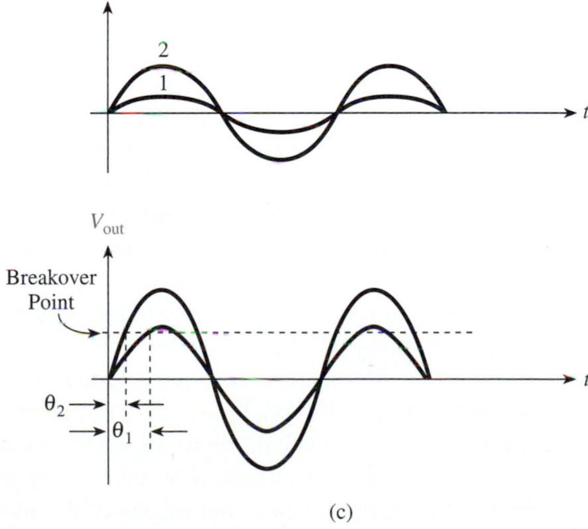

(c)

it is rotating in one direction, say, clockwise. SCR_2 controls the armature current in the other direction, bottom to top. It controls the motor when it is rotating in the other direction, counterclockwise. Here are the details of operation.

The ac error signal (V_e) is amplified in a preamplifier of the same type as those presented in Sec. 11-11. The voltage at the output terminals of the preamplifier (V_{out}) is an amplified version of the error signal and is in phase with the error signal. At a certain point in the positive half cycle, the output voltage will reach the breakover point of the SUS, causing a burst of current into the T_1 primary winding. This occurs during every positive half cycle as long as the input error voltage is above a certain minimum value. The specific point in the half cycle is determined by how large the error voltage is. For a large error voltage, the breakover will occur early in the half cycle; for a small error voltage, the breakover will occur later in the half cycle. This is shown clearly in the waveforms of Fig. 11-27(c). These waveforms apply to either SCR circuit.

For a small V_e, the curves labeled 1 show what happens. V_{out} reaches the breakover point relatively late in the half cycle, resulting in firing delay angle θ_1. For a larger V_e (bigger position error) the curves labeled 2 apply. V_{out} reaches the breakover point of the SUS earlier in the cycle, resulting in a firing delay angle θ_2.

Every time the SUS breaks over to deliver a burst of current into the T_1 primary winding, positive pulses occur at the two T_1 secondary windings in Fig. 11–27(a). Both of these positive pulses tend to fire the SCRs to which they are connected. That is, both SCR_1 and SCR_2 receive a gate current pulse telling them to turn ON. However, only one SCR actually does turn ON, because only one of them is forward biased at this instant. If V_s is positive on the top when the pulses arrive, SCR_1 turns ON and SCR_2 remains OFF because of the reverse bias across its main terminals. If V_s is positive on the bottom when the pulses arrive, SCR_2 turns ON and SCR_1 remains OFF. Thus the phase relationship between V_s and V_e determines which SCR will fire. If V_e is in phase with V_s, SCR_1 fires; if V_e is 180° out of phase with V_s, SCR_2 fires. The phase of the error voltage determines the direction of rotation of the motor and the direction of movement of the controlled object. The size of the error voltage determines the firing delay angle of the SCR and therefore the average voltage (and current) delivered to the armature. It therefore controls the servo motor's speed.

Figure 11–27(b) shows a somewhat different arrangement of SCRs and motor armature, but the action of the gate control circuitry is exactly the same as in Fig. 11–27(a). If V_e is in phase with V_s, then at the instant the gate pulses arrive at SCR_1 and SCR_2 the top T_1 secondary winding will be forward-biasing SCR_1 and the bottom T_1 secondary winding will be reverse-biasing SCR_2. Therefore SCR_1 will fire and SCR_2 will remain OFF, causing current to pass through the armature from top to bottom. To understand this, refer to the phasing marks on the transformer windings. If V_e and V_s are in phase, then when the dotted terminals of the T_2 windings are positive, the triangled terminals of the T_1 windings are also positive. During the negative half cycle, V_{out} will be the wrong polarity to fire the SUS, so no gate pulses are delivered and no SCR turns ON.

If V_e is 180° out of phase with V_s, then at the instant the gate pulses arrive at the SCRs, the bottom T_1 secondary winding will be forward-biasing SCR_2 and the top T_1 secondary winding will be reverse-biasing SCR_1. Therefore SCR_2 will fire and SCR_1 will remain OFF, causing current to pass through the armature from bottom to top. To see this, recognize that if V_e and V_s are 180° out of phase, then when the dotted terminals of the T_2 windings are positive, the triangled terminals of the T_1 windings are negative.

As in the first circuit, if V_e and V_s are in phase, the motor spins in one direction, and if V_e and V_s are 180° out of phase, the motor spins in the other direction. The magnitude of the error input voltage again determines the firing delay angle of the SCR, thereby controlling the average voltage applied to the motor armature.

TROUBLESHOOTING ON THE JOB

TROUBLESHOOTING AN ELECTROPNEUMATIC VALVE OPERATOR

The electropneumatic valve operator of Fig. 11–4 is being used to control the water valve in the textile-moistening process of Fig. 15–16. It is not necessary to make a careful study of the overall process-control system at this point, because you have been assured that the system's input transducer and electronic control components are functioning perfectly, up to the op amp output. Nevertheless, the textile moistening chamber's humidity has declined dramatically in the last 10 minutes. You have been told to make an immediate check of the water-control valve and valve-operator. If the valve-operator has failed in the closed position, you are to open the valve manually to establish water flow to the spray nozzles. Then start investigating the electropneumatic operator to determine the cause of trouble.

Your test instruments consist of a multimeter for measuring the current in the actuator coil, and various resistances, and a manometer for measuring the air pressure at various pressure-tap locations in Fig. 11–4. The electropneumatic valve-operator apparatus is tapped for pressure measurement at three locations: (1) in the supply line, ahead of the fixed restriction; (2) in the variable-pressure chamber just ahead of the nozzle; and (3) in the diaphragm enclosure, above the diaphragm. You also carry a bottle of soapy water with a small paint brush for brushing the water onto tubing joints and fittings. An air leak will cause bubbles to form when soapy water is applied.

Describe your testing/troubleshooting procedure. For each step in the procedure, tell what conclusion(s) can be drawn for each possible outcome and/or what your next logical step will be.

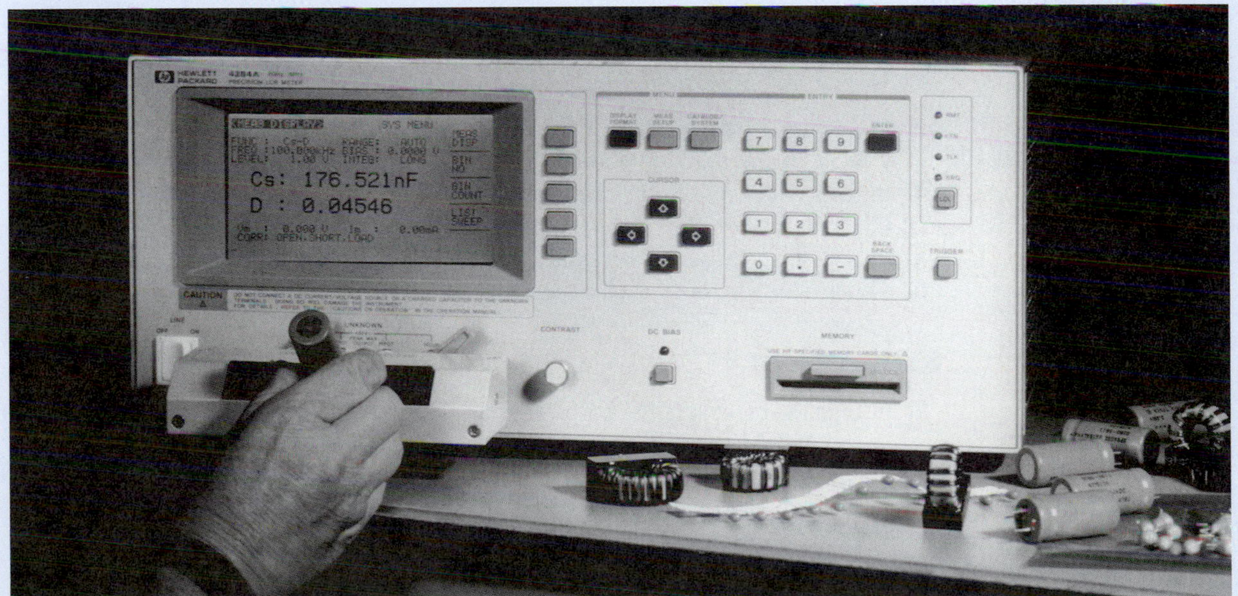

Precise measurement of capacitance or inductance (as for the valve actuator coil of Fig. 11–4) is accomplished by an LCR meter. It can also measure very low resistance values much more precisely and accurately than a standard multimeter.

Courtesy of Hewlett-Packard Company.

SUMMARY

- Valves are one of the most common final correction devices. Two-position valves are either snap-action solenoid-operated, or slow-moving motor-driven.
- Proportional-position pivoting valves must have a pot or other transducer attached for position-feedback.
- Large valves are often proportionally positioned by an electropneumatic or electrohydraulic apparatus.
- Split-phase ac induction motors have two stator windings that carry currents that are phase-shifted by 90°. This produces a revolving magnetic field that pulls the rotor along with it.
- Ac servo motors are conceptually like split-phase motors, but they have thinner rotor bar conductors. This gives them their different speed-vs-torque characteristic.
- Various amplifier designs are utilized to boost the position error-voltage signal so that it can drive an ac servo motor's control winding. The following amplifier design techniques are common: (1) push-pull transformer coupling; (2) dc error signal driving a chopper-stabilized ac amplifier; (3) negative feedback for bias- and gain-stability and noise reduction; and (4) unfiltered (pulsating) dc supply to a center-tapped control winding.

QUESTIONS AND PROBLEMS

Section 11-1

1. For a solenoid-actuated valve, explain the function of each of the following parts: valve seat, valve plug, valve stem, armature, solenoid coil, spring.
2. Why is it that ac solenoid valves can have their coils damaged if the valve stem sticks, but dc solenoid valves don't have that problem?

Section 11-2

3. In an electric motor-driven two-position valve, what does the motor gearing provide: high speed and low torque, or low speed and high torque?
4. In the limit switch timing diagram of Fig. 11–2, LS2 is released slightly before LS1 is released. Why is this necessary?
5. In Fig. 11–2, why is it necessary that LS2 be actuated at the 180° mark?
6. About how long does it take for a two-position control valve to go full travel?

Section 11-3

7. In Fig. 11–3, what is the purpose of the extreme-position limit switches?

Section 11-4

8. Make a distinction between a pneumatic valve *operator* and a pneumatic valve *positioner*.
9. Explain the difference between an air-to-open valve and an air-to-close valve.
10. Generally speaking, under what circumstances are pneumatic and hydraulic valve actuators preferred to electric motor actuators?
11. In Fig. 11–4, what causes the variable air pressure to increase and decrease?
12. Identify the forces which combine to balance the balance beam in Fig. 11–4.

13. In Fig. 11–4, if the feedback spring were to break, what would the valve do if a small electrical input signal were applied to the coil?
14. Repeat Question 13 for the problem of the baffle breaking.

Section 11-5

15. Distinguish between the *blind end* and the *rod end* of a cylinder.
16. In Fig. 11–5, which way must the pilot cylinder move in order to cause the main power cylinder to stroke down?
17. In Fig. 11–5, what causes the power cylinder to stroke up: an increase in the input pressure signal, or a decrease in that signal?
18. Give a detailed step-by-step explanation of what happens in the apparatus of Fig. 11–5 if the electrical input signal is decreased.
19. In Fig. 11–6, identify the two forces that combine to balance the vertical balance beam.
20. Repeat Question 18 for the apparatus of Fig. 11–6.

Section 11-7

21. What is the difference between a relay and a contactor?
22. Define the term *pick-up current* for a relay.
23. Define the term *drop-out current* for a relay.
24. Explain why electromagnetic relays have hysteresis.
25. Is the hysteresis of a relay necessarily a bad thing? Explain.
26. Draw a three-phase wye connection. Repeat for a delta connection.
27. Which load connection, wye or delta, delivers the entire line voltage to each phase of the load?
28. Which load connection, wye or delta, delivers more electrical power to a resistive load? By what factor is the power greater?

Section 11-8

29. Name some of the most important industrial uses of power thyristors. Which is the most important of all?

Section 11-9

Questions 30–46 apply to split-phase ac motors.

30. How far out of phase should the two stator winding currents be, ideally?
31. In Fig. 11–13(a), how many rotations does the magnetic field make for one ac cycle? If the ac line frequency is 60 Hz, what is the rotational speed of the magnetic field expressed in r/min?
32. Describe the construction of a squirrel-cage rotor. What materials are used to construct a modern squirrel-cage rotor?
33. Why does no current flow down the length of the core of a squirrel-cage rotor?
34. What is the most common method of creating a phase shift between the two stator winding currents?
35. Is it ever possible for a split-phase ac motor to run on just a single stator winding?
36. Is it ever possible for a split-phase ac motor to start on just a single stator winding?

Section 11-10

37. What is the difference in construction between a normal split-phase motor and an ac servo motor?
38. Distinguish between the control winding and the fixed winding in an ac servo motor.
39. In an ac servo motor, if the current in the control winding leads the current in the fixed winding, how could the direction of rotation be reversed?
40. What is it about the torque-speed characteristic of a normal split-phase motor that makes it unacceptable for use in a servo system?
41. Name the two effects which interact to produce the rising torque-speed behavior of the normal split-phase motor in Fig. 11–20. Which effect is accentuated to produce the drooping torque-speed behavior of the servo motor? How is this effect accentuated by the motor manufacturer?
42. List some of the reasons ac servo motors are preferred to dc servo motors.
43. Under what circumstances are dc servo motors used instead of ac servo motors?
44. What is *damping* in a servo system? What good does it do? Why is damping better than just running the servo motor at slow speed?

Section 11-11

45. What is the function of the servo amplifier in a servo system?
46. List some of the desirable characteristics of an ac servo amplifier, and explain why each one is desirable.

Questions 47–56 refer to servo amplifier 1, drawn in Fig. 11–22.

47. The input stage does not provide any voltage gain whatsoever, so what good is it?
48. Explain why resistors R_8 and R_9 are placed in the emitter lead of Q_2. What good do they accomplish?
49. What is the purpose of bypass capacitor C_3? What good does it accomplish?
50. What is the purpose of C_5 in the collector circuit of Q_3?
51. What do we call the coupling method used between Q_3 and the power output stage?
52. What is the purpose of resistors R_{14} and R_{15}?
53. About how much base current flows in Q_4 and Q_5 when the error signal is zero?
54. Describe the current flow paths in the push-pull power output stage when an error voltage is present.
55. What happens to the base current in the power transistors when the error voltage becomes nonzero?
56. Discuss the advantage of a push-pull amplifier stage over a conventional ac amplifier stage.

Questions 57–64 refer to servo amplifier 2, drawn in Fig. 11–23.

57. Explain the purpose of chopping a dc signal into an ac signal prior to amplification.
58. What do we call the circuit arrangement in which the Q_1 emitter resistor is part of the Q_2 collector resistance? What benefits are provided by this circuit arrangement?
59. What do we call the circuit arrangement of R_8 and C_2? What benefit is provided by this arrangement?
60. What do we call the circuit arrangement in which the base of Q_3 is connected through a resistance to the emitter of Q_4? What benefit does this provide?
61. Explain in intuitive terms why the gain of the servo amplifier is reduced as the gain-adjust pot wiper is moved to the left.

62. Describe the flow path of the dc base bias current for power transistors Q_5 and Q_6.
63. Does the motor control winding current have to pass through the germanium diode in the push-pull stage? Discuss your answer carefully. Does the germanium diode have to be a power rectifier, or can it be a small-signal diode?
64. Explain why the power transistors tend to run cooler when unfiltered dc voltage is used to drive the push-pull stage.

Questions 65–67 refer to servo amplifier 3, drawn in Fig. 11–25.

65. What is the voltage gain of the op amp noninverting amplifier when the gain-adjust pot is adjusted to its center position?
66. Why is there no provision made to correct the output offset of the op amp? Why isn't it necessary?
67. Do you think the C_1, C_2, and R_3 frequency response components would be sized to provide a low value of upper cutoff frequency or a high value of upper cutoff frequency? Obtain the frequency-response curves of a 709 op amp, and see what these particular component sizes would do.

Section 11-13

68. In Fig. 11–27(a), what is it that determines whether SCR_1 drives the armature or SCR_2 drives the armature?
69. If SCR_1 in Fig. 11–27(a) is driving the armature and correcting the position of the controlled object but the object overshoots its desired position, which SCR will cause it to back up to the proper position?
70. Figure 11–27(c) shows V_{out} to be exactly in phase with V_e. Is this absolutely necessary, or could V_{out} be out of phase with V_e?

CHAPTER

12

WOUND-ROTOR DC MOTORS

Everyone understands the function of an electric motor. When an electric source is connected to it, the motor turns its shaft. We say that a motor converts electrical energy to mechanical energy.

A motor is often used as the final correcting device in an industrial control system. In one such application the motor drives a servomechanism, causing an object to be placed in the desired position. Other applications include opening and closing valves, varying the speed of a pump to adjust liquid flow rate, and varying the speed of a fan to adjust air flow. There are also many other applications besides these.

OBJECTIVES

After completing this chapter, you will be able to:

1. Describe the function of a motor, the function of a generator, and relate the terms *dynamo, motor,* and *generator*.
2. Define the stator and the rotor of a machine.
3. Define the field and the armature of a machine.
4. For an elementary stationary-field generator, explain the moment-by-moment voltage-generation process, both magnitude and polarity, for a complete rotation.
5. Describe the slip-ring/brush arrangement for making electrical connection between the rotor and the external world.
6. Explain the rectifying action of a split slip-ring (a commutator).
7. Describe the moment-by-moment torque-production process of an elementary dc motor through a complete rotation. Use both descriptive methods: magnetic pole attraction/repulsion, and Lorentz relationship.
8. Analyze the electrical performance of a shunt dc motor using Ohm's law and Kirchhoff's voltage and current laws.
9. Explain the origin of motor counter-EMF, E_C. Relate E_C mathematically to flux density B, speed S, and structural proportionality factor k_{EC}.
10. Describe the ideal proportional relationship between magnetic flux density B and field current I_F in a shunt motor. Then compare the actual B-vs.-I_F relationship to the ideal.
11. Explain the automatic response of a motor to a change in the load's torque demand. Describe the interrelationship among torque, speed, motor counter-EMF, and armature current.
12. Relate motor torque production to magnetic field strength, armature current, and the motor's structural proportionality factor.

13. Relate a motor's mechanical shaft power to its speed and its torque. Do this for basic units and also for the common American units—horsepower, rotations per minute, and pound-feet.
14. Contrast the speed-vs.-torque characteristic graphs of a shunt motor and a series motor.
15. Given the specific no-load and full-load speed values for a motor, calculate its speed-regulation percentage.
16. Apply Ohm's law to predict the starting current of a motor, given its armature-winding resistance, R_A, and its field-winding resistance.
17. Contrast the armature current-vs.-torque characteristic graphs of shunt and series motors.
18. Summarize the advantages and disadvantages of shunt motors and series motors.
19. Describe the construction of a compound motor, and compare its operating characteristics to shunt and series motors.
20. Explain the origin of the armature reaction problem; describe how interpole windings are used to deal with this problem.
21. State the two methods that are used to limit the inrush current when a large motor is started.
22. Trace the operation of a relay ladder-logic circuit that accomplishes timed motor-starting using current-limit starting resistors.
23. State the methods that can be used to stop a motor.
24. Contrast the following motor-stopping methods: dynamic braking, regenerative dynamic braking, and plugging.
25. Trace the operation of a relay ladder-logic circuit that accomplishes reversing starting of a compound dc motor.

12-1 ■ MOTOR TYPES

Motors can be broadly divided into two classes—dc and ac. Within the dc class, we can further divide into the important subclasses shown in Fig. 12–1. Within the ac class, we can divide into the subclasses shown in Fig. 12–2.

Of the ac subclasses, induction-type motors are the most important for industrial control. Synchronous motors are certainly used in industry, but they usually are large three-

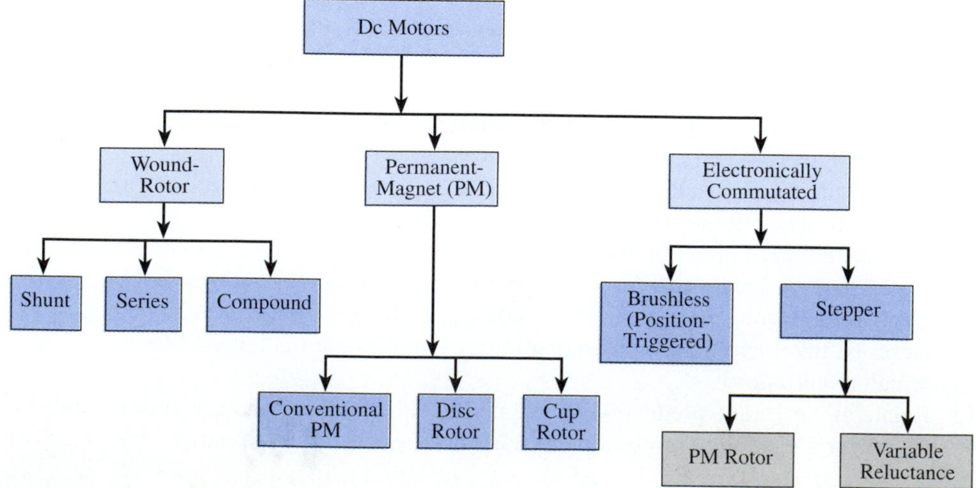

FIGURE 12–1
Subclasses of dc motors.

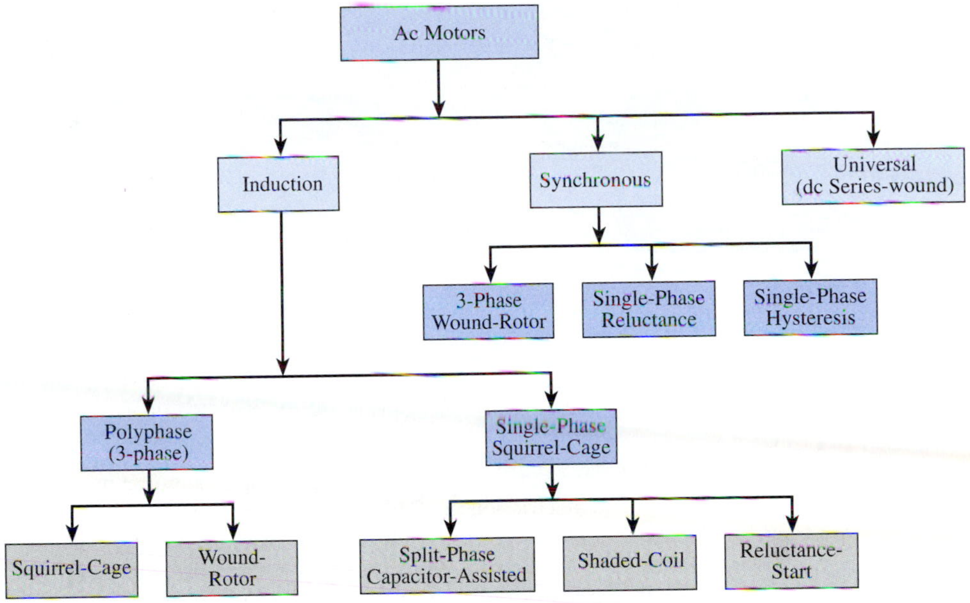

FIGURE 12–2
Subclasses of ac motors.

phase machines used for steady noncontrol purposes, such as running air compressors. The small single-phase synchronous types, namely the reluctance motor and the hysteresis motor, are used mostly in audio equipment and time-keeping clocks. Universal motors are only occasionally seen in control systems, their main use being in handheld variable-speed tools. We will direct our attention to the induction subclass, concentrating on polyphase and single-phase squirrel-cage types.

12-2 ■ WOUND-ROTOR DYNAMO PRINCIPLES

A dc motor receives dc electric power and produces mechanical shaft rotation. The same identical machine is perfectly capable of reversing that energy-conversion process—receiving mechanical rotation from an external engine and producing dc electric power. When the machine works in this direction, mechanical input and electrical output, it is called a *generator*. In other words a dc motor and a dc generator are the same machine but operated in opposite manners. Refer to Fig. 12–3 for clarification of this idea.

Since a disconnected wound-rotor dc machine is not necessarily a motor, nor necessarily a generator, we need a word to refer to the generic machine itself. That word is *dynamo*. When you look at a disconnected wound-rotor dc dynamo, think to yourself "I can make that dynamo perform as a motor, or I can make it perform as a generator, whichever I want."

An exploded view of a wound-rotor dc dynamo is shown in Fig. 12–4. It is called a wound-rotor dynamo because it has an electromagnet coil, a multiturn winding, on its rotating part. It does not have permanent magnets on its rotating part.

An exploded view like Fig. 12–4 is helpful, but there is nothing like seeing an actual dynamo to get a clear understanding of its structure. If possible, disassemble an old dynamo and inspect it.

Our goal in this section is to understand the operating principles of a wound-rotor dc motor. However, it is to our advantage to first understand the dynamo's principles of generator operation. There are two reasons for this. First, most people feel that generator operation is easier to explain than motor operation. Second, even when the dynamo

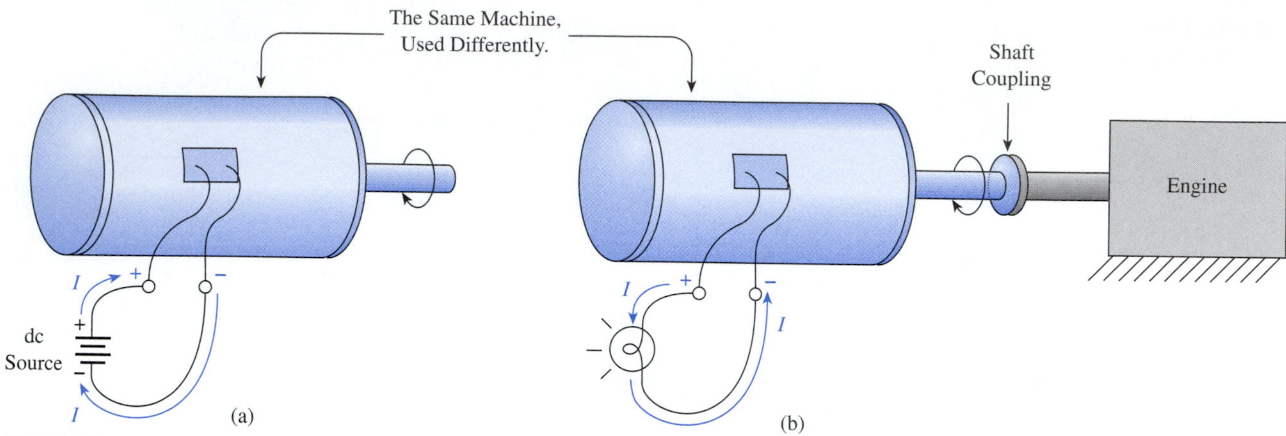

FIGURE 12–3
Two energy-conversion directions for the same dc machine. (a) Operating as a motor the machine receives electric energy from an electric source. Conventional current enters the motor on its positive terminal and exits on its negative terminal. It delivers rotational energy to its mechanical load, which is not shown here. (b) Operating as a generator the machine receives mechanical rotational energy from an engine. It delivers electrical energy to its electric load, in this case a lamp. Conventional current exits the machine on its positive terminal and reenters on its negative terminal.

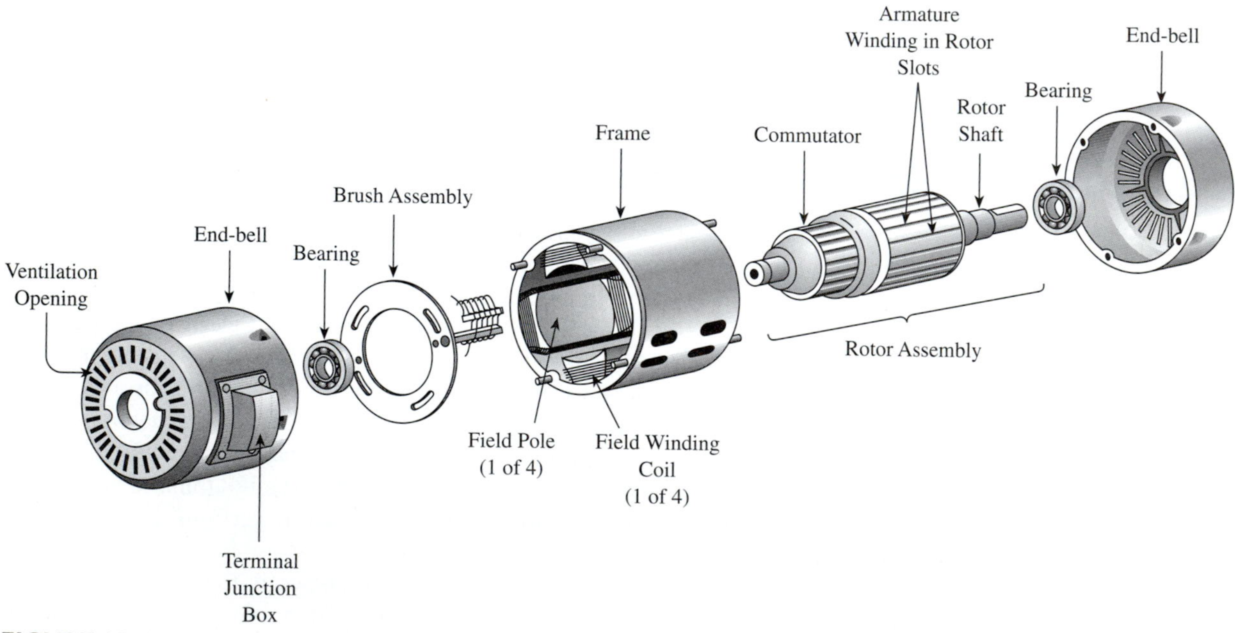

FIGURE 12–4
Essential parts of a wound-rotor dc dynamo.

operates as a motor, it simultaneously generates voltage. We tend not to notice this fact because the voltage that it generates is smaller than the applied source voltage, so it does not determine or coincide with the direction of current flow. However, this generated voltage is the key to understanding the torque-load characteristics of motors. For these two reasons we begin by explaining generator action.

12-2-1 Rotor and Stator

To make a generator work, the inside part, the *rotor,* must be forced to rotate. The outside part, the *stator,* remains stationary. The rotor is a cylinder mounted on a shaft with the shaft supported by bearings at both ends, as Fig. 12–4 shows.

Note carefully that the terms *rotor* and *stator* are mechanical terms. They are not electric/magnetic terms.

12-2-2 Armature and Field

A generator creates voltage because one of its windings, called the *armature winding,* experiences a change in magnetic flux. The magnetic flux, symbolized Φ, is produced by the other generator winding, called the *field winding.* It is called this name because it is an electromagnet that creates a magnetic field of flux density B when current is made to flow in it.

Note carefully that the terms *armature winding* (or just *armature*) and *field winding* (or just *field*) are electric/magnetic terms, not mechanical terms.

Often, but not always, a generator has its armature winding located on the rotor. This has produced an unfortunate piece of confusion in America: Many people have the idea that the rotor assembly and its associated winding can *always* be called the armature, and that the stator structure and its associated winding can *always* be called the field. This is not correct.

The correct way to think of a generator's armature winding is to say that it is the winding that actually produces the output voltage V_{OUT}. It *may* be on the rotor, but not necessarily.

For now, let us assume that the armature winding is indeed on the rotor and the field winding is on the stator, as shown in Fig. 12–4. This arrangement is shown more clearly in Fig. 12–5 for a two-pole machine design; that is, the field winding creates one North magnetic pole and one South magnetic pole, for a total of two poles.

Figure 12–6(a) is a cross-sectional view of a two-pole machine showing detailed field winding wrap and flux (Φ) direction. The armature winding conductors (wires) have been left out—the rotor slots are empty.

Figure 12–6(b) gives a cross-sectional view of a four-pole machine. It shows the flux distribution within the rotor core and through the air gaps between the rotor surface

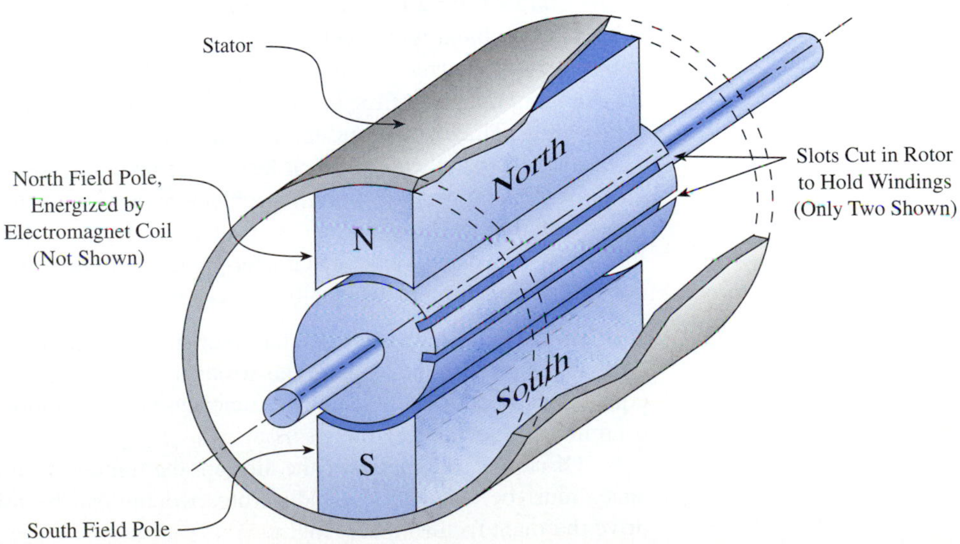

FIGURE 12–5
For a dc generator the field is placed on the stator. The armature is placed on the rotor.

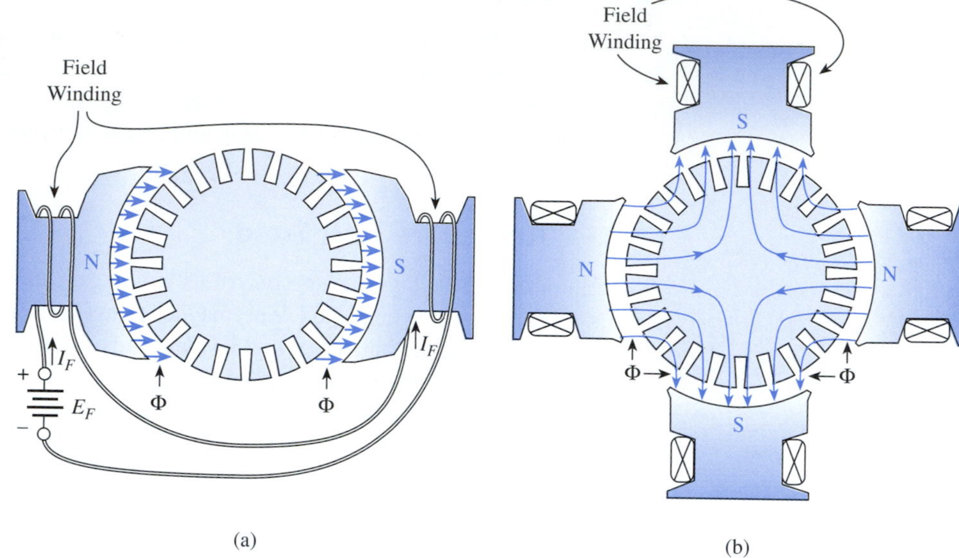

FIGURE 12–6
Magnetic flux distributions in two- and four-pole generators. Notice the shape of the salient poles. (a) Here the details of the electric polarity and the wrapping direction are given; use your right-hand rule (or other method) to verify the magnetic pole polarity—North on the left, South on the right. (b) This is our manner of representing windings when we don't want to be concerned about the wrapping details.

and pole-faces. A salient pole (one that sticks out from the overall structure) almost always widens toward its face, as shown in Fig. 12–6. There are two reasons for this:

1. So the magnetic field flux is distributed more evenly around the surface of the rotor. (Of course, if the field were mounted on the rotor we would say flux is distributed more evenly across the surface of the stator.)
2. So the winding can't slip off the pole. Look back at Fig. 12–5. If an electromagnet winding were wrapped around that nonflared pole core, what's to keep it from falling off?

Laminations. The material used for the rotor cylinder and the field poles is steel. The steel is alloyed and heat-treated to give it high magnetic permeability so that a given amount of magnetizing "effort" exerted by the field windings ("effort" is symbolized H) produces a greater magnetic flux density, symbolized B. Look back at your basic electricity textbook for a review of these magnetic ideas.

The problem with steel is that it's also a good electrical conductor. This makes it subject to *eddy-currents,* which are whirlpool-like currents induced in the body of the steel by changing magnetic flux. Eddy-currents can be a serious problem because they produce I^2R heating. Thus they represent a waste of energy (actually, power), and they contribute to heating of the machine, which leads to temperature-related stress or damage.

To combat the eddy-current problem, manufacturers construct the rotor and the field poles with thin steel laminations, separated by very thin plastic insulating layers. That way the whirlpools cannot have a diameter greater than the thickness of the laminations. This construction technique drastically reduces their associated I^2R power loss.

Air gap. The distance between the rotor surface and the field pole-faces is called the *air gap*. The narrower the air gap, the stronger is the magnetic field, all other things being equal. This boosts the machine's efficiency, giving us more electrical energy output for each unit of mechanical energy input.

Of course, the narrower the air gap, the tighter the machine's manufacturing tolerances must be, and the less radial rotor wobble can be tolerated. These considerations drive the manufacturing cost higher.

Today the highest quality industrial-use machines have an air gap of about 1.5 mm (about 0.06 inch or $\frac{1}{16}$ inch). Lower quality machines may have air gaps in the 5-mm range (the $\frac{1}{4}$-inch range).

12-2-3 How a Generator Produces Voltage—Faraday's Law

To understand generator action you must be familiar with Faraday's law and Lenz's law. Faraday's law can be written

$$V = N\frac{\Delta\Phi}{\Delta t} \tag{12-1}$$

which tells us that when a loop of wire (a winding) is subjected to time-varying magnetic flux, it will induce, or generate, a voltage V. The voltage induced per turn, in volts, is equal to the time rate of change of magnetic flux, measured in webers per second. If the overall winding has many turns, N, then the overall voltage that it produces is N multiplied by the per-turn voltage. Look at Fig. 12–7 to understand how Faraday's law is applied to an elementary single-turn generator.

The single-turn coil is actually mounted in slots cut into the rotor cylinder of Fig. 12–5. It is supported by the rotor assembly; we have removed the rotor assembly from Fig. 12–7 for clarity. The rotor shaft is forced to spin by an external mechanical engine of some kind. This engine is called a *prime mover*. It may be an internal-combustion engine, a steam turbine, or a hydraulic (water) turbine. We don't care exactly how the prime mover gets the rotor to rotate, just so it is able to keep it rotating.

As the single-turn coil rotates in space, the instantaneous voltage appearing at the winding terminals varies in sine-wave manner. To be specific, as the coil rotates from the 0° position in Fig. 12–8 to the 90° position and then onward to the 180° vertical position, the coil's generated voltage v produces a positive half cycle of a sine wave, as shown.

Then, as the coil rotates through the next 180°, namely, from the 180° position back to the 0° position, v produces the negative half cycle. Of course, the negative half cycle is just a polarity reversal. Instead of v_{OUT} being positive on the red-tipped lead in Fig.

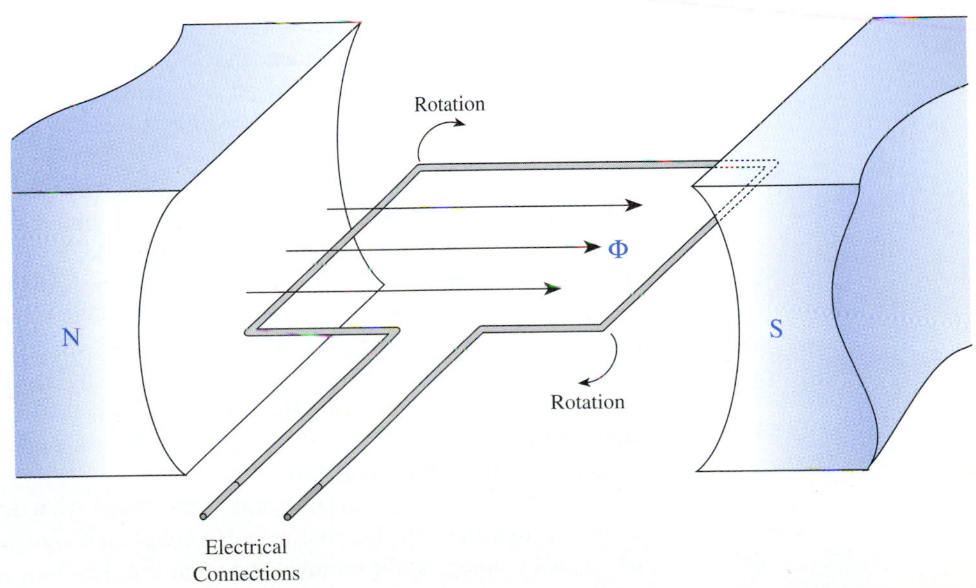

FIGURE 12–7 The spatial relationship between magnetic flux and a single armature coil.

FIGURE 12–8 One complete rotation of the coil produces one complete sine-wave cycle of voltage.

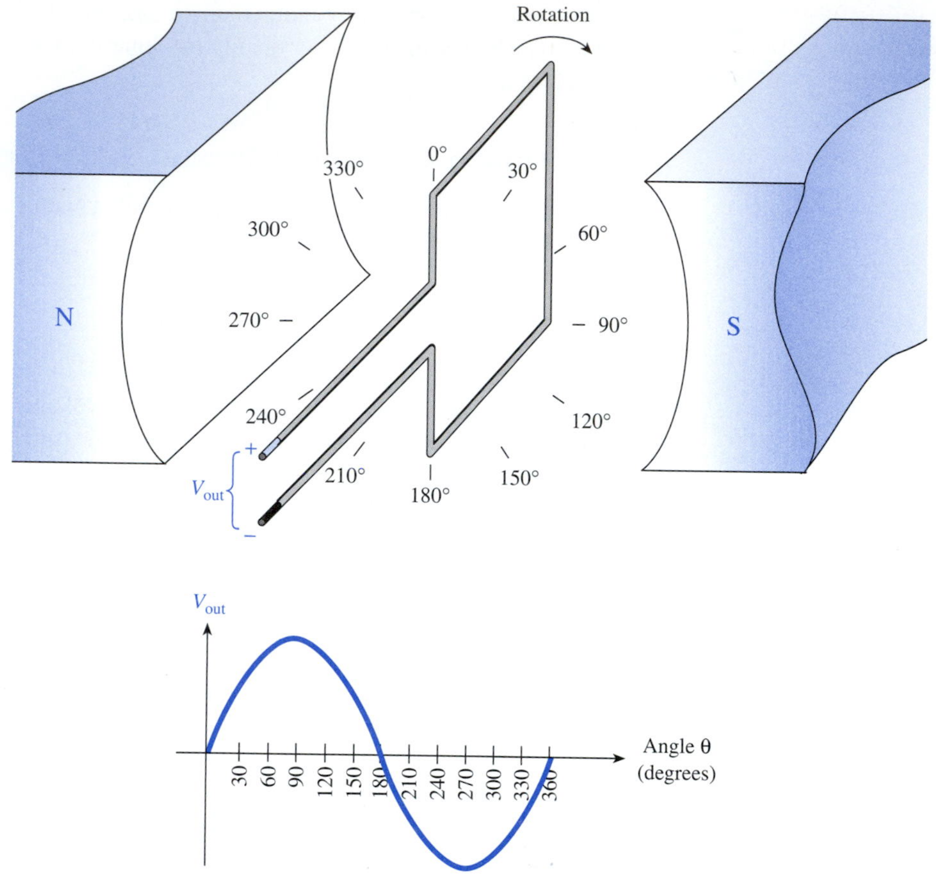

12–8, which is our defined-positive polarity, v_{OUT} becomes negative on the red-tipped lead and positive on the black-tipped, which is our defined-negative polarity.

Let us apply Faraday's law on a moment-by-moment basis through the first 120° of rotation. This will help us to understand why the coil generates a sine-wave voltage.

In Fig. 12–9, we have changed from an isometric view to a straight-on, cross-section-type view. Start in Fig. 12–9(a) with the winding coil in the 0° position. The plane of the coil is vertical, as indicated. The magnetic field shows 13 flux lines, which we can think of as 13 milliwebers of flux (Φ_{total} = 13 mWb). With the coil plane perfectly vertical in part (a), all 13 lines pass through the coil. Imagine this drawing in three dimensions, with the coil and the magnetic field having depth. The flux Φ passing through the coil is at its maximum possible value, 13 mWb. Furthermore, at this instant in the coil's rotation, the through-the-coil flux is not *changing*—it is momentarily constant. You can appreciate this by imagining the coil plane moving slightly away from vertical—say 2° clockwise. Since the top edge of the coil sticks up slightly above the topmost flux line, a tiny 2° movement will not bring the coil edge below the topmost flux line. The same explanation holds on the bottom edge of the coil. Thus a slight rotation of the coil produces no change in through-the-coil flux.

Faraday's law, Eq. (12-1), is quite clear about what action creates voltage. It's not *having* through-the-coil flux that creates voltage; it's *changing* the through-the-coil flux that creates voltage. At the moment shown in Fig. 12–9(a), we have 13 mWb of flux, the

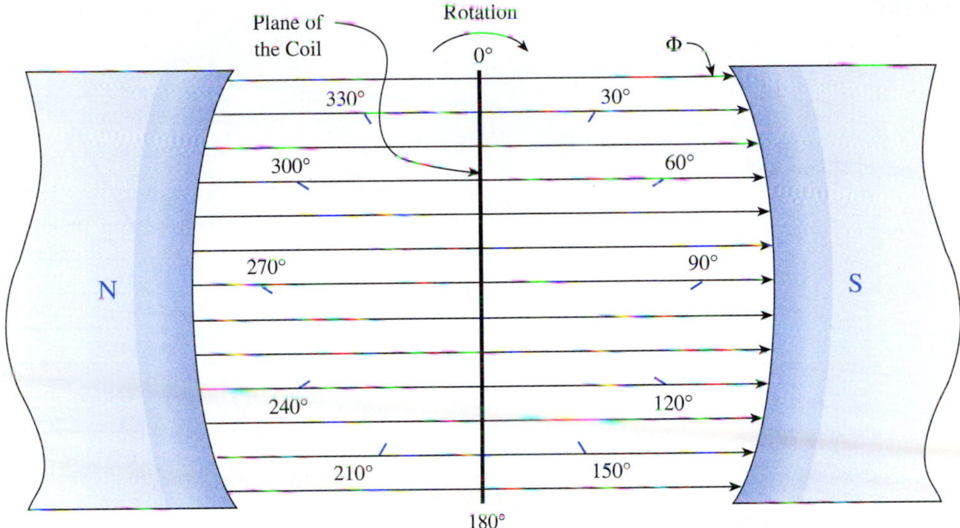

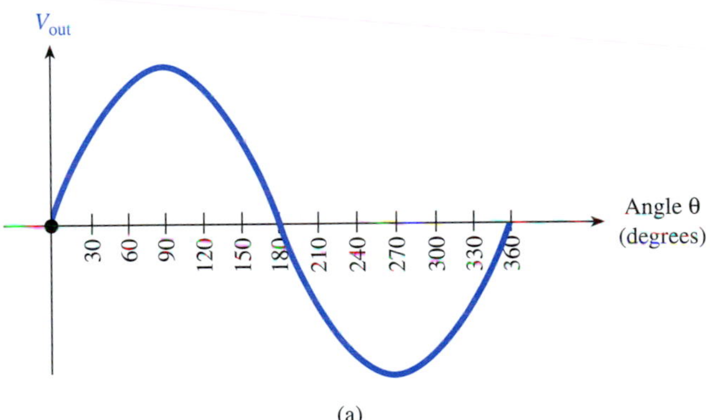

(a)

FIGURE 12–9
(a) Instantaneous coil flux condition at the 0° instant (vertical instant).

maximum, but the time rate of change of flux, $\Delta\Phi/\Delta t$, is zero. Therefore the instantaneous voltage is zero, as indicated on the sine waveform.

Now look carefully at Fig. 12–9(b), showing the coil at an instant when it has rotated by 30°. It is visually clear that the coil no longer has the topmost flux line passing through it—and the same at the bottom. There are now only 11 lines of through-the-coil flux. Thus, the coil has lost two lines (2 mWb) during the time that elapsed as it rotated from 0° to 30°.

Suppose for the sake of discussion that it takes 1 ms for the rotor to move 30° (this would be pretty fast rotation). Then the average rate of change of flux during this interval is given by

$$\frac{\Delta\Phi}{\Delta t} = \frac{13\text{ mWb} - 11\text{ mWb}}{1\text{ ms}} = \frac{2\text{ mWb}}{1\text{ ms}} = \frac{2 \times 10^{-3}\text{ Wb}}{1 \times 10^{-3}\text{ s}} = 2\text{ V}$$

This is a rather small voltage. At the snapshot moment shown in Fig. 12–9(b), the instantaneous voltage is larger than 2 V because our calculation gave the average value. If the voltage is getting larger, and we know it is, then the instantaneous value at the end of the interval must be greater than the average value over the entire interval. Nevertheless, the voltage at 30° is relatively small, so we can show it as only halfway up the hill of the sine waveform in that figure.

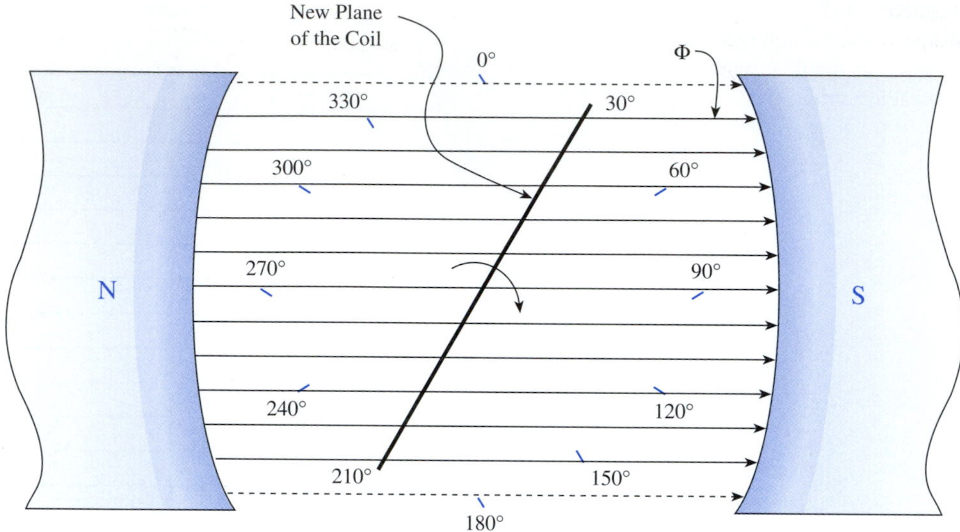

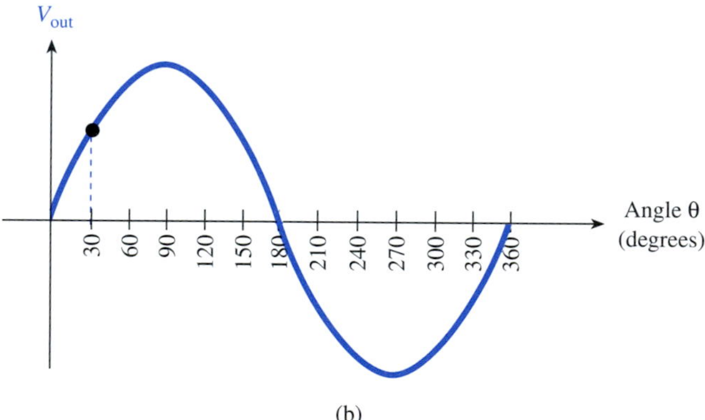

(b)

FIGURE 12–9 *(continued)* (b) Instantaneous coil flux condition at the 30° instant.

Figure 12–9(c) shows the coil plane 1 ms later, having moved an additional 30° to the 60° position. Now the topmost three lines and bottommost three lines are missing the coil. There remain seven lines through the coil. Therefore, during this 1-ms interval, the change in flux per unit of time is

$$\frac{\Delta\Phi}{\Delta t} = \frac{11\text{ mWb} - 7\text{ mWb}}{1\text{ ms}} = \frac{4\text{ mWb}}{1\text{ ms}} = 4\text{ V}$$

by Faraday's law. Again, this is average voltage during the 1-ms interval. The instantaneous voltage at the snapshot moment of part (c) is greater than 4 V. The increase in generated voltage is plotted at the 60° point in the waveform graph of Fig. 12–9(c).

Figure 12–9(d) shows a snapshot of the coil after another 1-ms time interval. The coil plane is perfectly horizontal, so zero flux is able to pass through the coil. All 13 flux lines are shown dashed in Fig. 12–9(d) to indicate that they don't pass through the coil. The rate of change in through-the-coil flux during this 1-ms interval is

$$\frac{\Delta\Phi}{\Delta t} = \frac{7\text{ mWb} - 0\text{ mWb}}{1\text{ ms}} = \frac{7\text{ mWb}}{1\text{ ms}} = 7\text{ V}$$

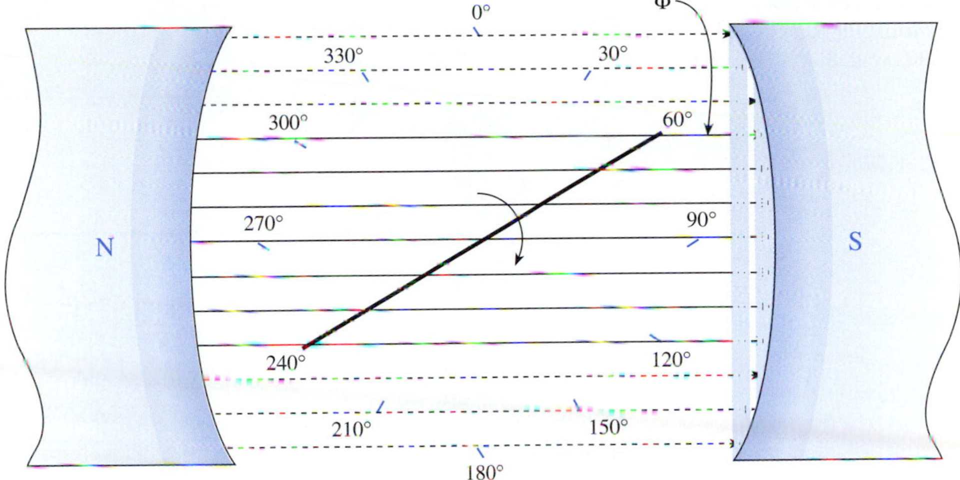

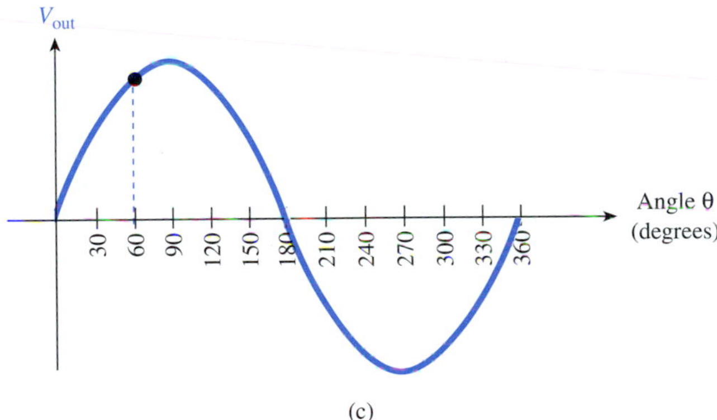

FIGURE 12–9 *(continued)* (c) Instantaneous coil flux condition at the 60° instant.

according to Faraday's law. This is the greatest average generated voltage yet. The instantaneous voltage at the snapshot moment is even greater and is shown at the peak of the waveform in Fig. 12–9(d).

Take a minute to think about the meaning of Fig. 12–9(d), especially compared to Fig. 12–9(a). The instantaneous through-the-coil flux Φ in Fig. 12–9(d) is zero, but the time rate of change of flux, $\Delta\Phi/\Delta t$, is at its maximum possible value. In Fig. 12–9(a), the instantaneous through-the-coil flux Φ was its maximum possible value, but its time rate of change, $\Delta\Phi/\Delta t$, was zero. This is the way sine-shaped variations always behave in all aspects of nature. If you understand this important fact, many technological concepts become easier to grasp.

It would be possible to continue analyzing the coil's voltage-generation performance in 30° increments. However, during the 90° to 180° quarter rotation, we must keep in mind that the instantaneous voltage at the snapshot moment is *less than* the average voltage obtained from Faraday's law. This is because the voltage waveform is now heading downhill rather than uphill. Thus, for example, if you were to analyze the rate of change in through-the-coil flux during the interval from 90° to 120°, you would obtain 7 V, the same *average* value as between 60° and 90°, but now at 120°, the instantaneous value of v is less than 7 V, whereas v was greater than 7 V at the 90°

FIGURE 12–9 *(continued)* (d) Instantaneous coil flux condition at the 90° instant.

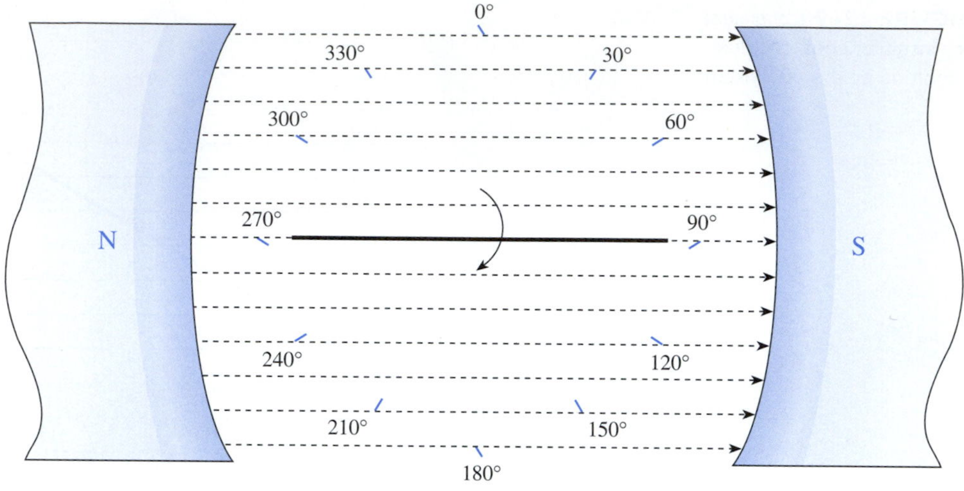

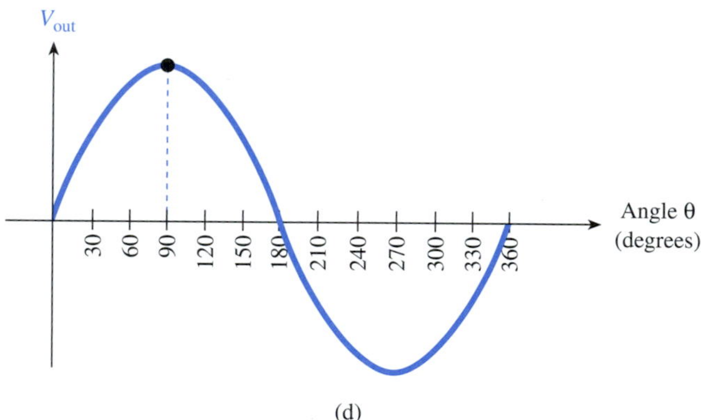

(d)

moment. Therefore the waveform point at 120° would be lower than the peak point occurring at 90°. And likewise for the following 1-ms intervals as the coil goes to 150° and then to 180°.

12-2-4 Polarity of the Generated Voltage—Lenz's Law

We already know from seeing the sine-wave graphs of Figs. 12–8 and 12–9 that the coil's generated voltage must reverse to the negative polarity at the 180° instant. Let us return to Fig. 12–8, where we can tell one coil terminal from the other, and use Lenz's law to understand why the polarity reversal occurs.

There are many ways to state Lenz's law. Here is the way that is most helpful to us:

> *If outside forces cause a winding coil to experience a change in through-the-coil flux, the winding attempts to circulate current in its conductors in the direction that causes its own "internal" flux to* oppose the change *being forced upon it by outside events.*

Thus Lenz's law says that if outside events (forced rotation within a fixed external magnetic field) cause the through-the-coil flux to be decreasing, the coil must induce a voltage polarity that attempts to circulate current in the direction that creates "internal" flux that *helps,* or *aids,* the external flux. By aiding the external flux, the coil is opposing the decrease that is being forced upon it by outside events.

However, if outside events are causing the through-the-coil flux to be increasing, then the coil must induce a voltage polarity that attempts to circulate current in the direction that creates internal flux that *opposes,* or *points opposite to,* the external flux. By pointing opposite to the external flux, the coil is opposing the increase that is being forced upon it by outside events.

In Fig. 12–8, as the coil rotates from 0° to 90°, the external flux passing through the coil from left to right is decreasing. To oppose this decrease, the coil attempts to circulate conventional current flowing *out of the page on the top* conductor, through a load that we imagine connected to the red and black coil-ends, and returning *into the page on the bottom* conductor. By applying the right-hand rule to these current directions, verify that the internal flux passes through the coil from left to right, thereby aiding the external flux. Since conventional current exits from a voltage source (a generator) on its positive terminal and returns to a source at its negative terminal, the top (red) terminal is positive and the bottom (black) terminal is negative during the first 90° of rotation, according to Lenz's law. We imagine an oscilloscope connected with the probe to the top (red) terminal and the ground-clip to the bottom (black) terminal. Therefore this instantaneous polarity deflects the scope up, into the positive half cycle, as the waveform of Fig. 12–8 shows.

During the quarter rotation as the coil moves mechanically from 90° to 180°, magnetic flux increases through the coil in the left-to-right direction. Therefore, by Lenz's law, the coil must circulate a current which flows *out of the page on its right* side (the red-tipped side, which is now below the mechanical horizontal midline). The current passes through the load and returns back *into the page on the left* side (the black-tipped side, which is now above the horizontal midline). With the scope connected probe to red, ground to black, the voltage is still positive. Thus the generator is producing the second quarter cycle of the sine wave as the rotor moves from 90° to 180° mechanically.

At the 180° instant, flux Φ is once again at its maximum possible value, which corresponds to zero rate of change of flux. That is, $\Delta\Phi/\Delta t = 0$ at the instant when Φ is equal to Φ_{max}, occurring when the winding plane is perfectly vertical. Faraday's law predicts 0 V generated at that instant. The sine-wave positive half cycle has just finished at this moment and is crossing through zero.

As the rotor moves from the 180° mechanical position to the 270° position, external flux passing through the coil from left to right is decreasing. To oppose this decrease the winding circulates a current which creates its own internal flux passing through the winding from left to right, aiding the external flux. To do this, current must flow out of the page on the top side (now the black-tipped side), and it must flow back into the page on the bottom side (the red-tipped side). With current leaving the generator voltage source on the black-tipped terminal and reentering the source on the red-tipped terminal, the oscilloscope polarity has changed to negative. Thus the generator begins producing the negative half cycle of a sine wave as the rotor passes the 180° position.

Apply Lenz's law and Faraday's law to the rotation from 270° to 360°. Satisfy yourself that the generator is producing negative voltage, changing from a maximum value (at the 270° horizontal position) to zero (at the 360° vertical position), and that therefore it is making a fourth quarter cycle, completing a sine-wave cycle of voltage generation.

The foregoing explanation, adapted to the specific generator of interest, describes the operation of any electric generating machine.

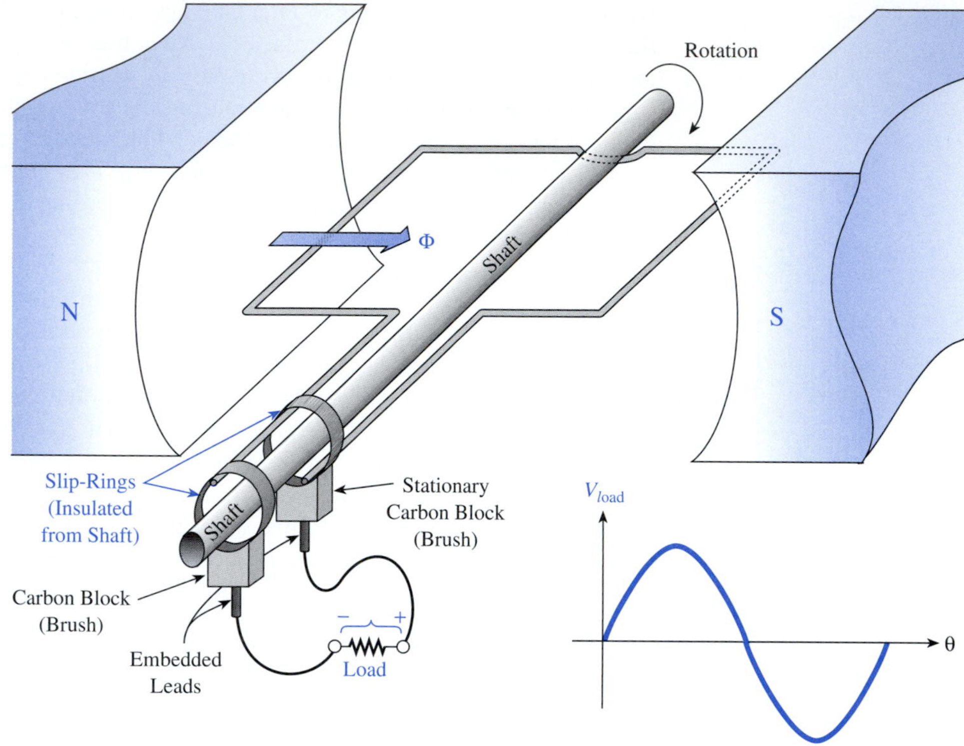

FIGURE 12–10
Slip-rings and brushes for getting the electrical energy out of the generator to the external load.

12-2-5 Adding Slip-Rings and Brushes to Make a Complete Ac Generator

To connect the rotating armature coil to the external world of loads, the coil ends are soldered to copper rings that are mounted on the shaft but insulated from it. This is shown in Fig. 12–10.

Carbon blocks, called *brushes*, are held tight against the slip-rings. Figure 12–10 does not show the compression springs that hold these brushes tight against the slip-rings. Refer back to Fig. 12–4 to see the brush-mounting assembly, or rigging. Embedded in the brushes are copper wires that lead away to the resistive load. By this brush-to-slip-ring arrangement, we are able to connect the moving armature winding to a stationary external load device.

12-2-6 Mechanically Rectifying the Ac to Dc—Commutation

So far, we have an ac generator. We are looking for a dc generator, because once we have it, we will name it a *dc dynamo*. Then we intend to carefully analyze our dynamo to understand dc motor action.

The final feature needed to get us from ac to dc is the mechanical commutator—the split slip-ring.

Figure 12–11 shows the simplest possible commutator. It is a single slip-ring cut into two pieces, or segments. The commutator segments are insulated from the shaft and from each other. The segment that is on top at this instant is soldered to the black-tipped coil side. The segment that is on the bottom at this instant has the red-tipped coil side sol-

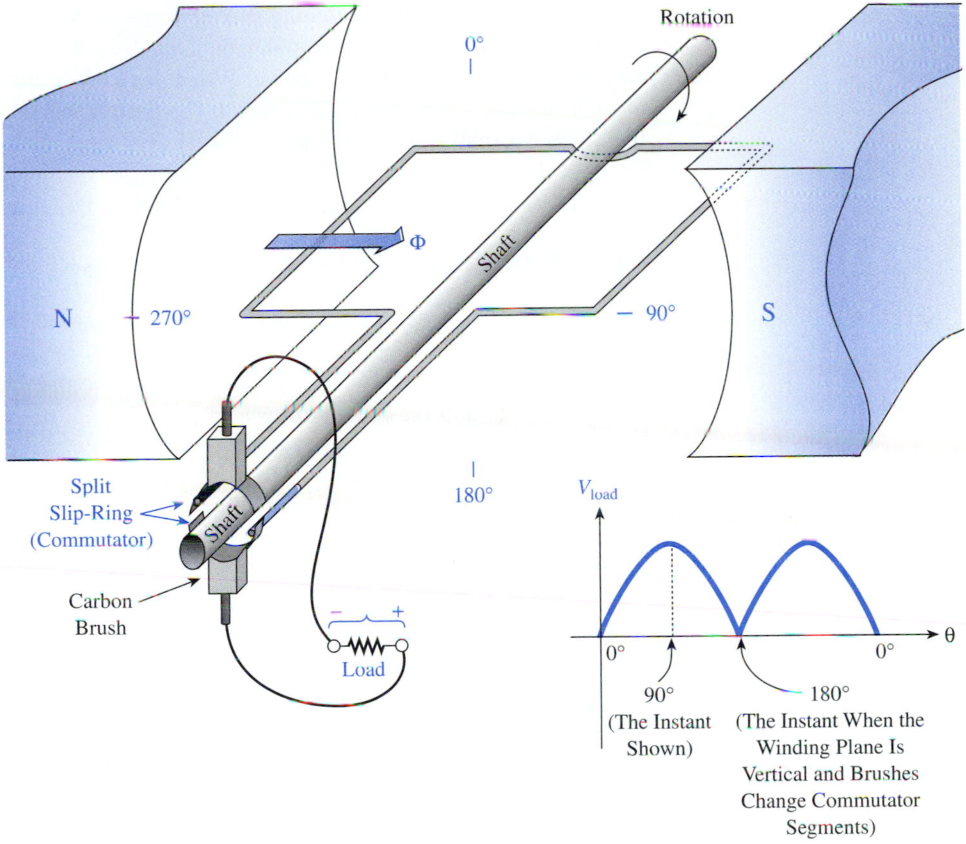

FIGURE 12–11 Splitting the slip-ring into segments for rectifying the ac to dc.

dered to it. The brushes that ride against the commutator segments are stationary—held in place by brush rigging not shown in this figure. The commutator segments rotate with the rotor shaft since they are mechanically attached to it.

At the instant shown in Fig. 12–11, the armature winding is 90° away from its starting position. The magnetic flux passing through the coil is zero, since the coil plane is exactly parallel to the direction of the flux lines—horizontal. With $\Phi = 0$, $\Delta\Phi/\Delta t$ is at its greatest possible value, the peak value of the sine wave. There is nothing new about this description; this was all established in Fig. 12–9.

During the next 90° of rotation, the red-tipped coil side will remain positive; the black-tipped side will remain negative. During that 90° of rotation, the two commutator segments will remain in contact with the brushes that we see them contacting right now. Specifically, the bottom brush will remain in contact with the red-tipped segment; and the top brush will remain in contact with the black-tipped segment.

But at the 180° instant, the brushes switch segments. Shortly after the armature winding passes its 180° position, the red-tipped commutator segment will have moved out of its contact with the bottom brush and into contact with the top brush. Stare at Fig. 12–11 until you are satisfied that you see this.

Likewise with the black-tipped segment. It will leave its contact with the stationary top brush and establish contact with the bottom brush.

Of course, it's exactly the 180° instant at which the winding terminals change electrical polarity. At that instant, the red-tipped terminal changes from electrical positive to electrical negative, and the black changes from negative to positive.

So the mechanical switching of segments with brushes exactly coincides with the electrical changing of segment polarity. Therefore the bottom brush is *always* positive, and the top brush is *always* negative. When we look at the external load, its right side is always positive since it is hard-wired to the bottom brush. The left side is always negative. The internally generated ac sine wave has been full-wave rectified at the external load, as shown in the waveform.

The Fig. 12–11 machine is a full-fledged dc dynamo.

12-2-7 Improving the Dynamo—More Armature Coils

Figure 12–12 shows a second armature coil on the rotor offset from the first by 90°. The commutator has been cut into four segments instead of just two. It is not hard to see that this arrangement will allow one of the coils to drive the load during the peak vicinity of each half cycle and the other coil to take over during the peak vicinity of *its* half cycles.

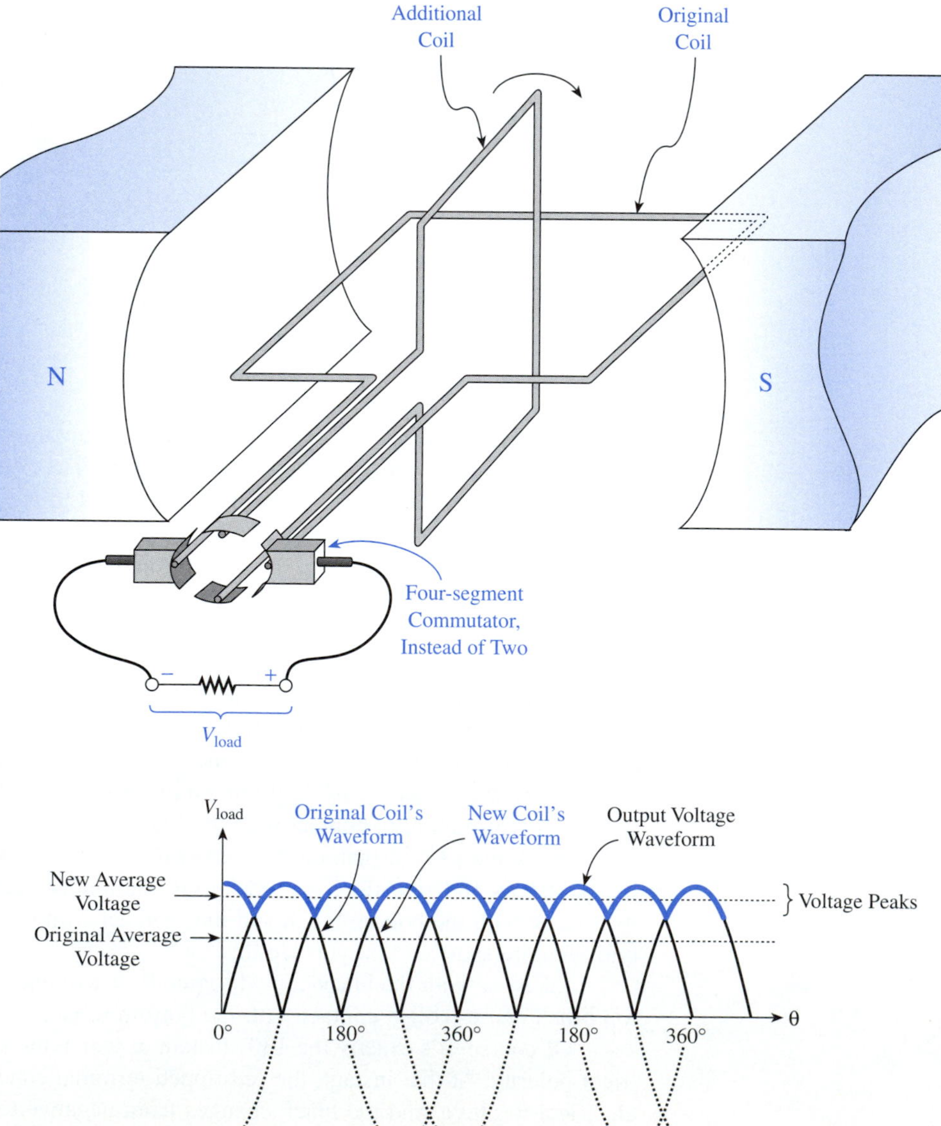

FIGURE 12–12
Placing a second coil on the rotor makes the output waveform smoother, with higher average voltage.

This way neither coil drives the load during the low-voltage portions of its cycle pulsations. The result is sketched in the waveform of Fig. 12–12.

The advantage of additional armature coils should be clear: With only a small amount of additional investment in the machine's manufacture, its output rating is improved dramatically.

If this is a good idea for a second coil, why not a third and a fourth coil, located at the 45° positions between the first two coils? Come to think of it, why not pack as many coils onto the rotor as possible? In fact, this is exactly what the dynamo manufacturers do, as you can see if you look closely at the rotor slots and commutator segments in Fig. 12–4.

Also, in a modern dynamo armature winding, the armature coils do not operate independently of one another. They are all interconnected in series/parallel arrangements so that each individual coil makes a partial contribution to the overall machine output at every instant. If you are interested in the detailed structure and interconnections of modern dynamo armatures, consult a text dedicated to electric machinery.

12-2-8 Using the Dynamo as a Motor

To change from generator to motor operation, we throw away the electrical load and bring in a dc source, as shown in Fig. 12–13. On the shaft we attach the mechanical load device instead of a prime mover.

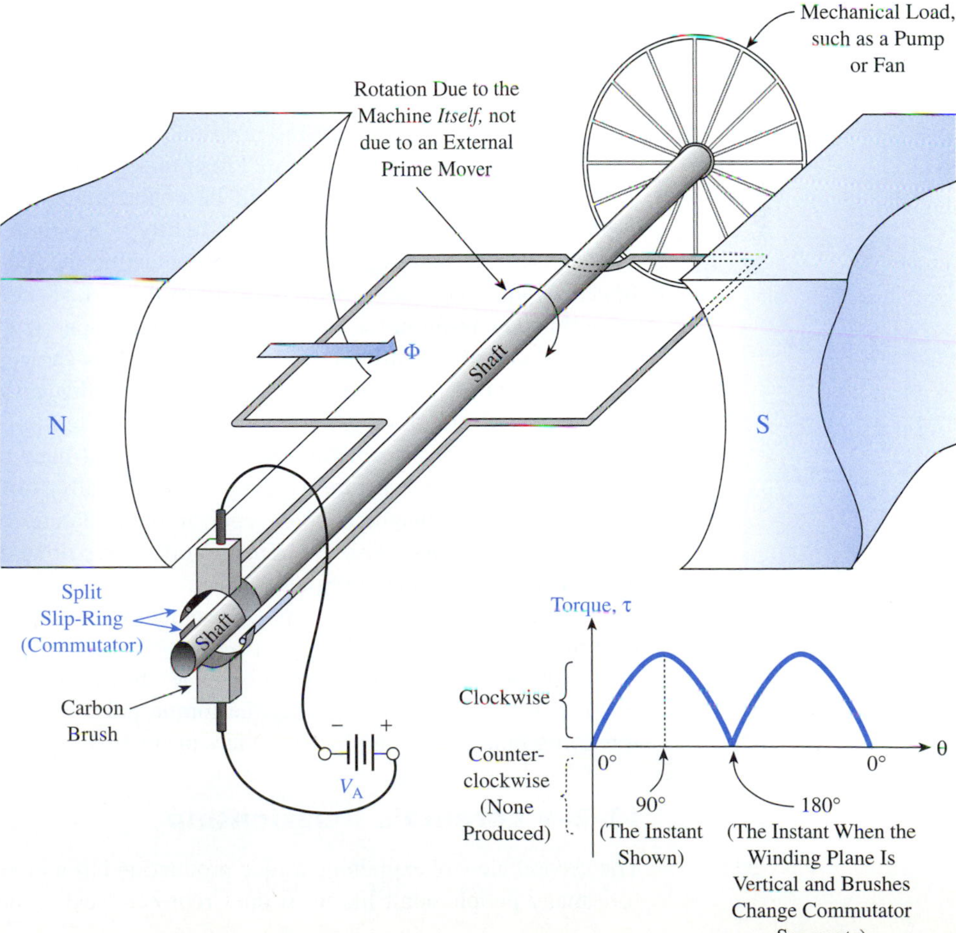

FIGURE 12–13 Converting a dc machine into a motor.

In Fig. 12–13, V_A symbolizes the source voltage applied to the armature. τ is our symbol for torque, which is mechanical twisting action overcoming opposition to motion. In the English measurement system, still widely used in North America, torque is measured in pound-feet (lb-ft). In System International, SI, torque is in newton-meters (N-m).

There are two different views of explaining torque production in motors. One view, which is more intuitive and easy to grasp but less rigorous, can be called the *pole-attraction view*. Here it is: When current is forced through the armature winding coil(s) by armature voltage V_A, the armature coil produces its own magnetic field, independent of the initial magnetic field produced by the field winding. One side of the armature coil becomes magnetic North, and the other side becomes magnetic South. The armature's North side is attracted to the stationary field's South pole; the armature's South side is attracted to the stationary field's North pole. These attraction forces cause the armature/rotor assembly to rotate into magnetic alignment with the armature's North moving as close as possible to the stationary field's South, and vice versa.

Let us apply this magnetic pole attraction view to the dc motor of Fig. 12–13. At the moment shown, conventional current is entering the motor on the bottom brush, flowing into the armature coil on the red-tipped coil-end, flowing down the right side toward the rear of the winding, passing around in back, flowing up the left side toward the front of the winding, coming out of the coil on the black-tipped coil-end, exiting the machine via the top brush, and returning to the negative terminal of V_A. Applying the right-hand rule to the conductor current directions indicates that the top side of the armature coil is North and the bottom side is South. Therefore the top side is attracted to the South main pole on the right side of the stator, and the bottom side is attracted to the North main pole on the left side of the stator. Thus the rotor assembly rotates clockwise, trying to bring unlike magnetic poles together.

After rotating 90° from the position shown in Fig. 12–13, the armature is at the 180° position. There it switches its brush-to-segment connections. A short distance past 180°, the positive bottom brush will be contacting the black-tipped segment, with the black-tipped winding conductor being slightly to the right of the dynamo's vertical midline, and the negative top brush will be contacting the red-tipped segment, with the red-tipped winding conductor being slightly to the left of the midline. You may wish to redraw this new position for yourself, taking an end-on view (the isometric view is hard to draw). In this new position the armature winding is carrying current away from us in the top (black-tipped) conductor, into the page toward the rear of the winding. Current passes around in back and flows toward us in the bottom (red-tipped) conductor, out of the page toward the front of the winding. Applying the right-hand rule to these current directions reveals that now the armature winding is magnetically North on its left side and South on its right side. Like magnetic poles repel, so the armature/rotor assembly is now forced to continue moving clockwise. Stationary North is repelling armature North, and stationary South is repelling armature South.

Thus the armature always produces clockwise torque on the rotor shaft, so the shaft continues to spin.

Torque τ is not smooth. It pulsates as shown in the torque graph of Fig. 12–13. It is not possible to understand *why* the torque pulsates in this manner when we take the foregoing pole-attraction/repulsion view of motor torque production.

12-2-9 Lorentz's Relationship

The second view of explaining torque production has a more serious physics flavor; therefore many people don't like it. It does represent reality, however, and it explains all the details of the moment-by-moment torque production. The Lorentz relationship can be stated:

> *The mechanical force on a current-carrying wire within a magnetic field is given by the equation*
>
> $$F = lIB \tag{12-2}$$
>
> *in which l is the wire length, I is its current, and B is the magnetic field strength (flux density). If basic SI units are used for l, I, and B (meters, amperes, and teslas), the force is in newtons.*

The Lorentz relationship also specifies the direction of the mechanical force, given that we know the directions of current and flux. Here is one way to specify the force direction—you must use your three-dimensional imagination:

> *Assuming that I and B are at right angles to each other, imagine the current directional arrow as pivoted around its tail. With the fingers of your right hand, push the head of the current arrow I into alignment with the flux arrow B. The I arrow is being turned on its tail-pivot; the B (or Φ) arrow is just sitting there. As you perform this pushing, the natural position of your right thumb points in the direction of the mechanical force F.*

The procedure is illustrated in Fig. 12–14(a). If you have studied physics thoroughly, you will recognize this procedure as vector cross multiplication.

An equivalent way of expressing the force direction resulting from known directions of I and B is this: Hold your right hand so that your thumb, index finger, and middle finger are all at right angles to one another. Let your index finger point in the direction of the current I, and let your middle finger point in the direction of magnetic flux B (or Φ). The position of your thumb then indicates the direction of the mechanical force F. This arrangement is shown in Fig. 12–14(b).

Now that we have a grip on the directional aspect of the Lorentz relationship, let us apply it to the dc motor shown in Fig. 12–13.

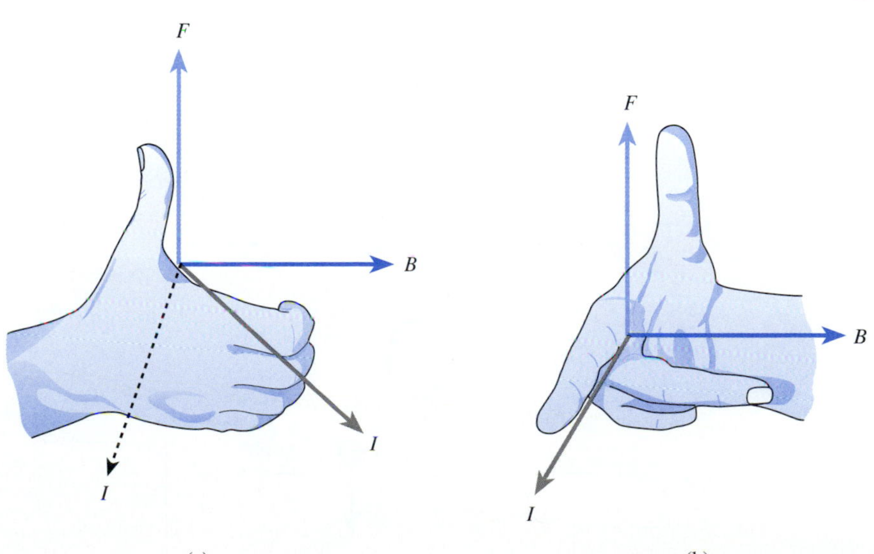

FIGURE 12–14
The direction aspect of the Lorentz relationship. (a) Push on the I arrow with your fingers, pivoting it through the 90° angle toward B (or Φ). Your thumb indicates the direction of mechanical force F. (b) Hold your hand so that index finger represents current I and middle finger represents magnetic field B (or Φ). Your thumb then represents mechanical force F.

We will follow the progress of the rotating armature conductors from the 90° instant, which is the moment shown in Fig. 12–13, through three quarters of a rotation, around to 360°. By looking carefully at the Lorentz relationship between the current-carrying conductors and the magnetic flux, we will satisfy ourselves that the mechanical forces always produce clockwise torque. This is the more rigorous view of wound-rotor motor action.

An end-on (cross-section) view of the motor is shown in Fig. 12–15. In part (a), at the 90° instant, the right-side (red-tipped) conductor is carrying conventional current into

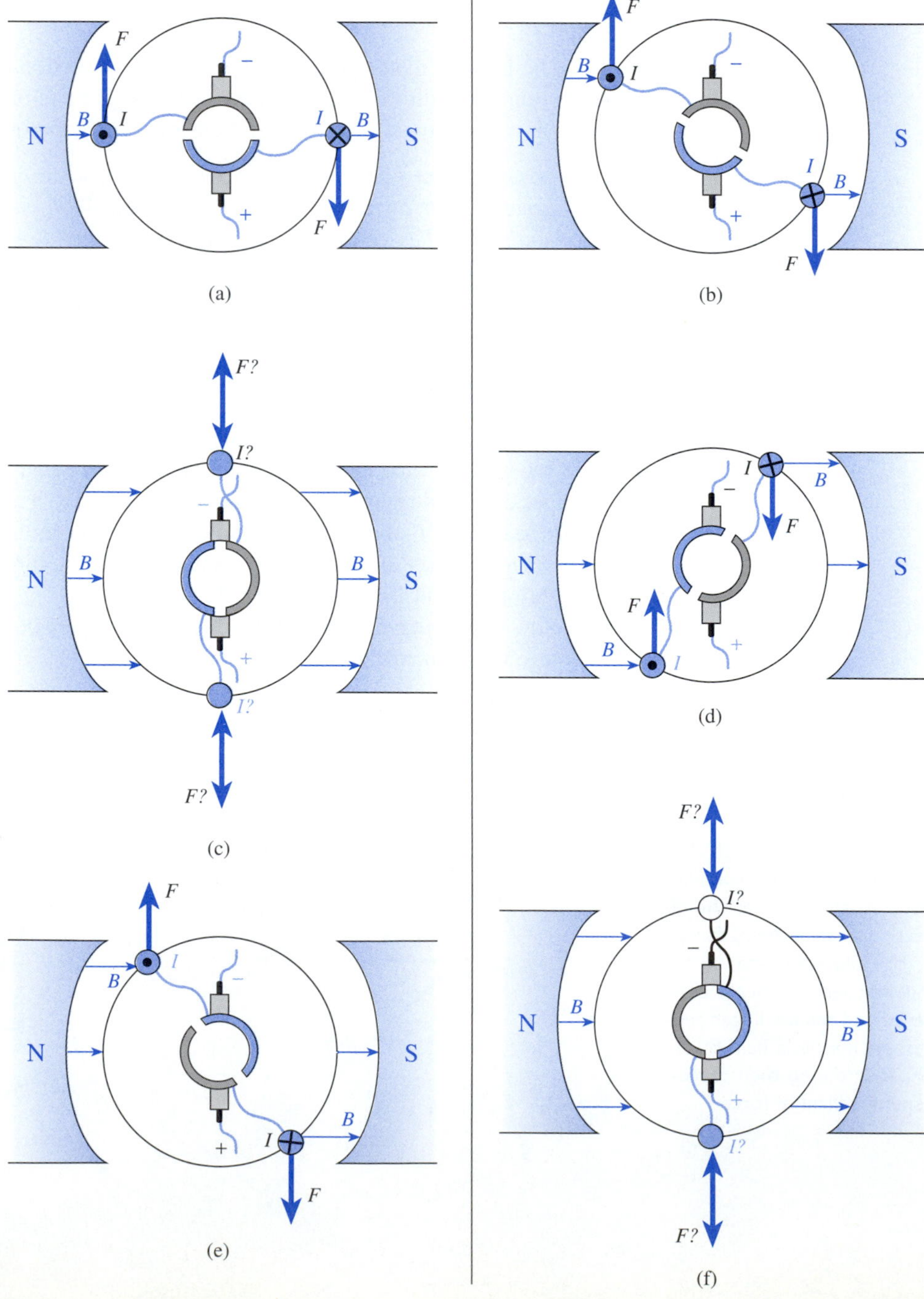

FIGURE 12–15 Torque production at various moments during a rotation.

the page, away from us. We indicate this with a tail-feather. The current flows through the armature coil and emerges out of the page on the left (black-tipped) conductor. This is indicated with an arrowhead. The current reaches the top commutator segment, the top brush, and eventually returns to the negative terminal of the external dc source, V_A.

Applying the Lorentz relation to the right-side conductor, we get a downward force direction. This tends to spin the rotor assembly clockwise. Repeating for the armature coil's left-side conductor, we get an upward force direction. This also tends to spin the rotor clockwise.

In Fig. 12–15(b), the rotor has advanced to the 120° position. There has been no switching of brush-to-commutator segment contacts, so there is no reason for the currents to have reversed directions in the coil sides. Therefore the Lorentz relationship still gives downward force on the right and upward force on the left. The forces themselves are just as strong as they were at the 90° instant because the magnitudes of I and B are no different, but the torque τ created at the 120° instant is less than the peak torque. This is because the line of application of the force gives it a reduced moment arm relative to the shaft's center. Torque is equal to the product of force times moment arm. Therefore the poorer mechanical position of the conductor results in the reduction in torque that is illustrated on the torque-versus-position graph in Fig. 12–13.

Moving ahead to the 180° position, we are unable to state the current directions with certainty in Fig. 12–15(c) because the brushes are in the process of switching commutator segments, but even if we could state the current directions, we would conclude that zero net torque is created. This is because the line of application of the force (either one, top or bottom) passes exactly through the axis of rotation. That is, it has zero moment arm. Thus it produces zero torque. The rotor assembly must *coast* through this position.

Carefully study the situation at the 210° position in Fig. 12–15(d). The brushes have switched segments. The bottom positive brush is now contacting the black segment, not the red segment. The top negative brush is contacting the red segment. Therefore the current directions in the two coil sides have reversed, with current in the red-tipped conductor flowing out of the page (arrowhead) on the left side of the vertical midline. Current in the black-tipped conductor flows into the page (tail-feather) on the right side of the vertical midline. Thus, due to cleverly timed commutation, whichever conductor is to the left of the midline carries current out of the page toward us, and whichever conductor is to the right of the midline carries current into the page away from us. The magnetic field never changes, so this current consistency results in all left-side mechanical forces being upward and all right-side mechanical forces being downward. Torque is always clockwise, or zero. It is never counterclockwise.

Figure 12–15(e) shows the rotor assembly at the 330° position. Trace through the situation for yourself, applying the Lorentz relation to verify that τ is still clockwise.

In Fig. 12–15(f) the rotor is shown back at its 0° starting position. Instantaneous τ is zero, and the motor momentarily coasts until the segment-bridging is interrupted and the brushes return to normal segment contact.

12-3 ■ WOUND-ROTOR DC MOTOR PERFORMANCE

The dynamo structural views of Figs. 12–4 through 12–15 are useful for understanding a machine's fundamental operating principles, but once we're past fundamental operating principles, we prefer to represent our motors in schematic manner. Figure 12–16(a) shows the schematic diagram for a standard-configuration dc motor in which the field winding is connected in parallel with the armature winding. The actual physical wiring arrangement is shown in Fig. 12–16(b) to help you relate the schematic appearance to the motor's structural parts.

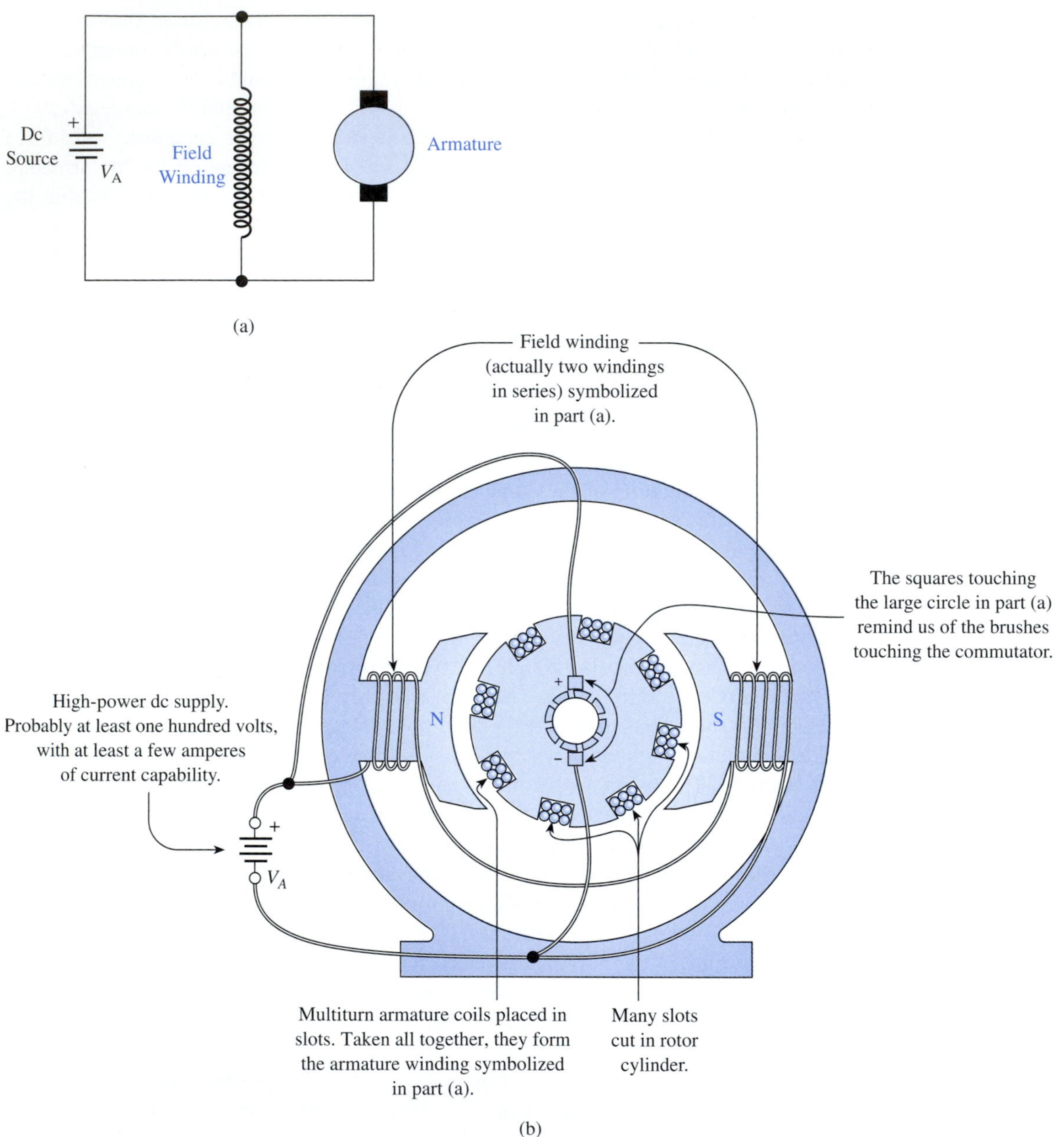

FIGURE 12–16
(a) Schematic drawing of a dc motor with field winding in parallel (in *shunt*) with armature winding. (b) Physical wiring of shunt motor. To avoid crowding, this drawing does not show the wires leading between the armature coil-ends and the commutator segments.

The shunt motor's schematic diagram has been redrawn in Fig. 12–17 to emphasize the relationship among the three currents—total supply current I_T, shunt field current I_F, and armature current I_A. Kirchhoff's current law applies, naturally.

$$I_T = I_F + I_A \tag{12-3}$$

Field current I_F is set by Ohm's law applied to the field winding. Since a shunt motor has its field winding connected directly across the high-voltage source, its resistance

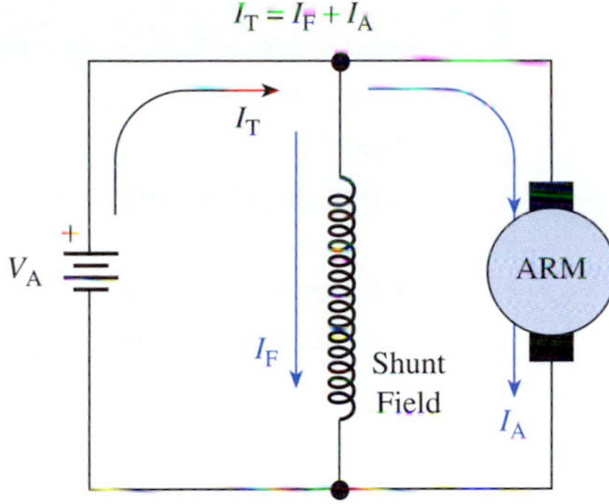

FIGURE 12–17 Identifying the three currents in a shunt-configured dc motor.

R_F is made rather high. The manufacturer makes R_F rather high by using a large number of turns (long overall length of wire) of thin gage. This enables R_F to limit I_F to a reasonably small value according to Ohm's law

$$I_F = \frac{V_A}{R_F} \tag{12-4}$$

If there is an adjustable rheostat placed in series with the field winding (not shown in Fig. 12–17), then I_F becomes adjustable according to

$$I_F = \frac{V_A}{R_F + R_{rheo}} \tag{12-5}$$

Variable (R_{rheo})

Even though its current is limited to a relatively small value, the shunt field winding nevertheless creates a very strong magnetic field B, because the winding has so many turns. Figure 12–16(b) shows only four winding turns per pole; in reality, the number of turns is several tens, or even hundreds.

■ EXAMPLE 12-1

In the motor of Fig. 12–17, suppose we isolate the shunt field winding and measure its resistance with an ohmmeter as $R_F = 153\ \Omega$. Suppose that the applied source voltage V_A is 230 V.

(a) Calculate the field current I_F.

(b) Imagine that a 200-Ω rheostat is inserted into the shunt field circuit of Fig. 12–17. Calculate the range of adjustment of I_F.

Solution. a) From Eq. (12-4)

$$I_F = \frac{V_A}{R_F} = \frac{230\text{ V}}{153\ \Omega} = \mathbf{1.50\ A}$$

(b) With R_{rheo} dialed down to 0 Ω, Eq. (12-5) gives the maximum I_F, still **1.50 A.** With R_{rheo} dialed all in, Eq. (12-5) gives

$$I_{F(min)} = \frac{V_A}{R_F + R_{rheo}} = \frac{230\text{ V}}{153\ \Omega + 200\ \Omega} = \mathbf{0.65\ A}$$

■ EXAMPLE 12-2

Refer to the shunt motor schematic diagram of Fig. 12–17. Suppose that the field current I_F is known as 1.5 A.

(a) If the total supply-line current I_T is 7.8 A, how much current flows in the armature loop?

(b) Suppose that an increase in mechanical "drag" on the motor shaft, caused by the mechanical load device, makes the motor have to work harder. It works harder by automatically "drawing" a greater current from the dc supply. If the new value of total current is $I_T = 12.2$ A, what is the new value of armature current I_A?

Solution. (a) From Kirchhoff's current law, Eq. (12-3),

$$I_A = I_T - I_F$$
$$= 7.8\text{ A} - 1.5\text{ A} = \mathbf{6.3\text{ A}}$$

(b) I_F doesn't change. I_A does all the changing when the mechanical output power has to change. With I_F still 1.5 A, Eq. (12-3) gives

$$I_A = I_T - I_F$$
$$= 12.2\text{ A} - 1.5\text{ A} = \mathbf{10.7A}$$

■

From Example 12-1 and 12-2, come away with this understanding: The field winding current path is as straightforward as can be—Ohm's law, plain and simple. But the armature winding current path is quite a different situation. The armature is continually revising itself so that it "draws" the proper amount of current I_A to create just enough torque to overcome the mechanical drag (opposition torque) of its load device. Ohm's law has only secondary importance with what happens in the armature loop. This is one of the central concepts of motors.

12-3-1 Counter-EMF

To understand how the armature performs the task of automatically revising its current draw, look at the augmented schematic diagram of Fig. 12–18. Inside the armature circle we are showing resistance R_A and variable voltage source E_C. The symbol E_C represents the *counter-EMF*, or *counter-voltage*, that is *generated* by the motor's armature winding. The polarity of counter-EMF is in opposition to the driving source voltage, V_A. Thus, E_C

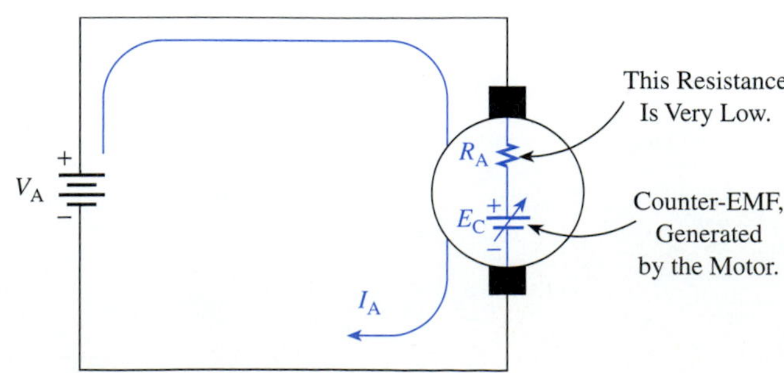

FIGURE 12–18 Visualizing a motor armature winding. The armature's resistance R_A is very low. Most of the voltage drop for satisfying Kirchhoff's voltage law comes from counter-EMF, E_C.

subtracts from V_A in determining the net voltage available to cause current I_A to flow through the winding's resistance R_A. We can write this in Ohm's-law fashion as

$$I_A = \frac{V_A - E_C}{R_A} \tag{12-6}$$

Or we can express the same facts in Kirchhoff's voltage law fashion as

$$V_A = I_A R_A + E_C \tag{12-7}$$

in which $I_A R_A$ is the Ohm's law "voltage drop" in the armature loop and E_C is also regarded as a voltage *drop* since it has the polarity of a drop.

■ EXAMPLE 12-3

Regard the armature loop of Fig. 12–18 as part of the overall shunt motor of Fig. 12–17. Some of that motor's specifications were given in Example 12-1: $V_A = 230$ V, $R_F = 153\ \Omega$. Here is another specification of that motor: The armature winding resistance is $R_A = 2.8\ \Omega$. This is typical of motor armature resistances. They are quite low, usually just a few ohms. In some cases R_A is just a fraction of an ohm.

(a) If the armature current I_A is 6.3 A, how much counter-EMF E_C is being generated by the armature?

(b) If E_C decreases to 200 V, how much current I_A flows in the armature loop?

Solution. (a) From Kirchhoff's voltage law, Eq. (12-7),

$$V_A = I_A R_A + E_C$$
$$E_C = V_A - I_A R_A$$
$$= 230\text{ V} - (6.3\text{ A})(2.8\ \Omega) = 230\text{ V} - 17.6\text{ V} = \mathbf{212.4\ V}$$

(b) From Ohm's law, Eq. (12-6),

$$I_A = \frac{V_A - E_C}{R_A}$$
$$= \frac{230\text{ V} - 200\text{ V}}{2.8\ \Omega}$$
$$= \frac{30\text{ V}}{2.8\ \Omega} = \mathbf{10.7\ A}$$

■

In both operating conditions of Example 12-3 (which are the same conditions as in Example 12-2), the counter-EMF is much greater than the Ohm's-law drop. In part (a), counter-EMF E_C accounts for about 92% of the applied voltage, and Ohm's-law drop accounts for only 8%. That is,

$$\frac{E_C}{V_A} = \frac{212.4\text{ V}}{230\text{ V}} \approx 92\%$$

and

$$\frac{I_A R_A}{V_A} = \frac{17.6\text{ V}}{230\text{ V}} \approx 8\%$$

In part (b), the values are

$$\frac{E_C}{V_A} = \frac{200 \text{ V}}{230 \text{ V}} \approx 87\%$$

and

$$\frac{I_A R_A}{V_A} = \frac{30 \text{ V}}{230 \text{ V}} \approx 13\%$$

Get used to this general condition. Counter-EMF E_C is by far the lion's share of the total armature loop voltage. Ohm's-law drop $I_A R_A$ is small in comparison.

You may be wondering about E_C. Your thoughts may be along these lines: "I thought we were running a motor here. Why are we talking so much about a *generated* voltage?" The response to that question is that yes, we are running a motor, but when a motor runs it also generates. Its generated voltage doesn't represent the output of the machine, as it would in a true generator, because E_C always loses out to V_A, which is larger. Thus the armature loop current I_A flows in the direction that V_A wants it to; I_A does not conform to the direction that E_C pushes.*

The origin of counter-EMF E_C is explained this way: Two things are required for the generation of voltage, according to Faraday's law:

1. The presence of a magnetic field.
2. Change in flux due to movement of the winding through the magnetic field.

In a running motor, both of these requirements are met. The magnetic field is certainly present, and the armature coils are certainly rotating.

Variables affecting counter-EMF. A motor's counter-EMF is determined by three things:

1. The magnetic field strength (flux density), B.
2. The rotational speed of the motor, S.
3. All the structural details of that particular motor, which are rotor length and diameter, number of poles, total number of conductors in the armature winding, and the specific manner (series/parallel) in which the individual coils of the overall winding are interconnected.

We lump together all the structural details of part 3 as a proportionality constant, with a specific numerical value. Let us symbolize this proportionality constant as k_{EC}. Then we write the motor's counter-EMF equation as

$$E_C = k_{EC} B S \qquad \textbf{(12-8a)}$$

This equation properly predicts E_C for that particular motor only. If you change to a new motor, it will have a new value of k_{EC}. Furthermore, you must specify what units you are using for B and S. Magnetic field strength B is everywhere measured in teslas, but rotational speed S has several different units in common use. In North American industrial environments, as distinct from scientific laboratories, rotational speed is usually measured in rotations per minute, or r/min. Most people verbalize this incorrectly as rev-

*As an aside, the equivalent situation applies to all generators. They produce counter-torque. Their job in life is to generate an output voltage. At the same time they also attempt to motor, but in the backward direction. The prime mover's forward torque overcomes the generator's backward counter-torque. Thus the machine-combination spins in a direction that does not conform to the direction that the generator's counter-torque pushes.

olutions per minute, which is a transfer from internal combustion engines, where the word *revolution* does make some sense. In scientific research and development, rotational speed is expressed in its basic units, radians per second, or rad/s. In Asia, rotational speed is usually expressed in rotations per second, or r/s. If you switch from one speed unit to another, you must change your k_{EC} value by the appropriate factor. This is one of the penalties we pay for not getting everyone to agree on world-standardized units.

Let us go along with North American industry and measure S in rotations per minute, r/min. The unit symbol *rpm* was formerly used for r/min, and you will no doubt encounter it.

Some people prefer to talk about the flux per pole, Φ, rather than the flux density, B. Then the counter-EMF equation is written

$$E_C = k_{EC(2)}\Phi S \qquad \textbf{(12-8b)}$$

where $k_{EC(2)}$ is equal to the original k_{EC} value divided by the area of the pole faces, A_{pole}, expressed in square meters ($k_{EC(2)} = k_{EC}/A_{pole}$). As always, flux Φ is in webers.

■ EXAMPLE 12-4

Suppose that our motor from Examples 12-1 through 12-3 has an E_C proportionality constant of $k_{EC} = 0.087$. Its pole faces have area $A_{pole} = 0.006$ square meter, so $k_{EC(2)} = k_{EC}/A_{pole} = 0.087/0.006 = 14.5$. The motor's manufacturer would have to provide all this information on a detailed data sheet. There is no way you could learn it by reading the motor's nameplate.

Suppose further that the I_F value of 1.5 A from Example 12-1 produces a flux density of 0.95 T ($B = 0.95$ T). All these values are typical for an integral-horsepower machine.

(a) If the motor spins at 2480 r/min, what amount of counter-EMF does it generate?

(b) Suppose that the field current I_F is reduced to 1.2 A, using a series rheostat. If B and Φ are proportional to I_F, calculate their new values.

(c) With the motor operating with this new B value, suppose that its counter-EMF decreases to 195 V. Find the motor's new speed, using Eq. (12-8a).

(d) Repeat the calculation of the new speed using Eq. (12-8b).

Solution. (a) From Eq. (12-8a),

$$\begin{aligned} E_C &= k_{EC}BS \\ &= (0.087)(0.95\text{T})(2480 \text{ r/min}) = \mathbf{205\ V} \end{aligned}$$

(b) If the magnetic variables are proportional to the field current, we can write

$$\begin{aligned} \frac{B_{new}}{B_{old}} &= \frac{I_{F(new)}}{I_{F(old)}} \\ &= \frac{1.2\text{ A}}{1.5\text{ A}} = 0.8 \end{aligned}$$

$$B_{new} = 0.8 \times B_{old} = 0.8 \times 0.95\text{ T} = \mathbf{0.76\ T}$$

The original flux, Φ_{old}, was

$$\begin{aligned} \Phi_{old} &= B_{old} \times A_{pole} \\ &= (0.95\text{ T}) \times (0.006\text{ m}^2) \\ &= 5.7 \times 10^{-3}\text{ Wb} \end{aligned}$$

Using the above proportionality method for flux Φ,

$$\Phi_{\text{new}} = 0.8 \times \Phi_{\text{old}}$$
$$= 0.8 \times 5.7 \times 10^{-3}\ \text{Wb} = \mathbf{4.56 \times 10^{-3}\ Wb}$$

(c) Rearranging Eq. (12-8a), we get

$$S_{\text{new}} = \frac{E_{\text{C(new)}}}{k_{\text{EC}} \times B_{\text{new}}}$$
$$= \frac{195\ \text{V}}{0.087 \times 0.76\ \text{T}}$$
$$= \mathbf{2950\ r/min}$$

(d) Rearranging Eq. (12-8b) gives

$$S_{\text{new}} = \frac{E_{\text{C(new)}}}{k_{\text{EC(2)}} \times \Phi_{\text{new}}}$$
$$= \frac{195\ \text{V}}{14.5 \times (4.56 \times 10^{-3}\ \text{Wb})}$$
$$= \mathbf{2950\ r/min}$$

This demonstrates convincingly that the two versions of the E_C equation (12-8a and b) are equivalent to each other. ■

Of the variables in Eq. (12-8), it isn't difficult to measure speed S—a tachometer will do it nicely. And in a realistic motor it isn't difficult to get a pretty good estimate of E_C. We'll see how that is done in a moment. However the magnetic variables B and Φ are very difficult to measure directly. Refer to Fig. 12–19. First of all, you need a magnetometer, which is an expensive instrument. Second, to get the magnetometer's sensing coil into the narrow air gap, it is necessary to remove one of the motor's end-bells (see Fig. 12–4). Then you must insert the sensing coil and bring its leads out through some ventilation opening, then reinstall the end-bell as pictured in Fig. 12–19. After reassembly you must mechanically lock the rotor shaft so that it cannot spin (the locked-rotor

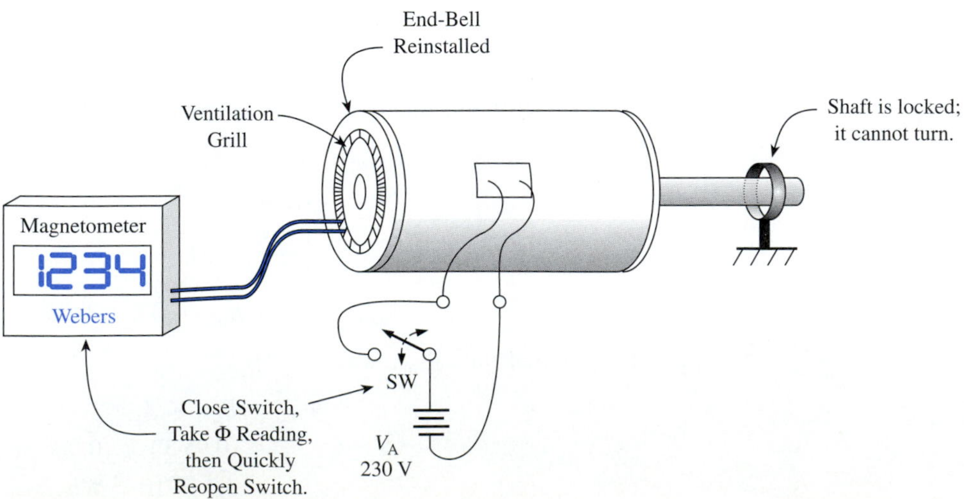

FIGURE 12–19
Using a magnetometer to measure flux Φ directly. This is impractical in an industrial setting.

condition). If the rotor assembly were allowed to spin, any slight shift in the position of the magnetometer sensing coil would be a wreck. Finally, you turn on the power for a few moments to pass current I_F through the field winding; the magnetometer then gives the measurement of magnetic B or Φ. All this is a major chore.

Because direct measurement of B or Φ is so difficult, we often make the simplifying assumption that we used in Example 12-4—that B and Φ are directly proportional to I_F. We can express the assumed B-to-I_F proportionality as

$$B = k_B I_F \tag{12-9}$$

where k_B is their proportionality constant.

Substituting assumed Eq. (12-9) into Eq. (12-8a) gives

$$E_C = k_{EC} BS \tag{12-8a}$$

$$E_C = k_{EC}(k_B I_F)S \tag{12-10}$$

We then define a new proportionality factor $k_{EC(3)}$ as

$$k_{EC(3)} = k_{EC} k_B$$

so that the counter-EMF equation Eq. (12-10) becomes

$$E_C = k_{EC(3)} I_F S \qquad \text{(assumed true)} \tag{12-11}$$

This gets rid of the troublesome magnetic variable B and replaces it with an easy-to-measure variable, field current I_F.

Testing a motor to find its $k_{EC(3)}$. Now refer to Fig. 12–20 to see what we can realistically do.

1. With the motor deenergized, isolate the armature winding and measure its resistance, including brushes, with a low-scale ohmmeter, as shown in Fig. 12–20(a). This requires disconnecting one of the armature leads (embedded brush leads) from the rest of the motor circuit. That usually isn't too difficult. Record the measured value of R_A.

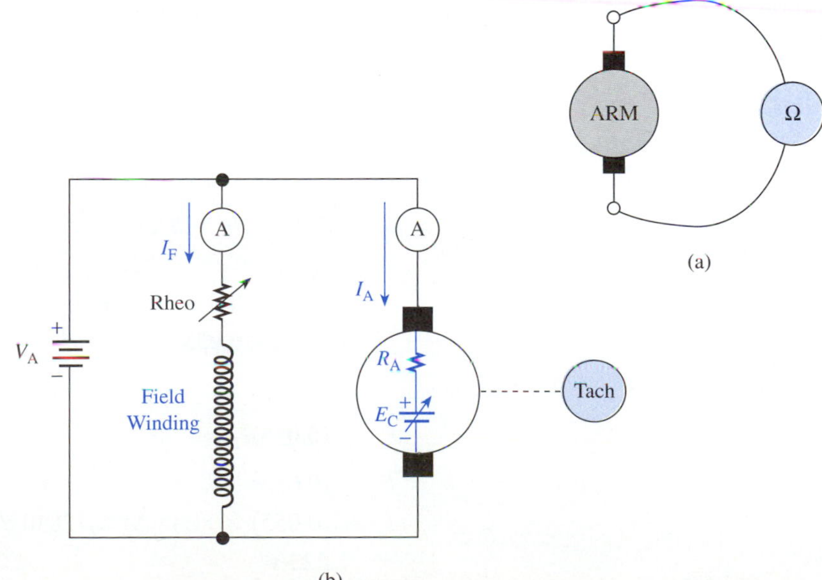

FIGURE 12–20
Experimental method of determining a motor's proportionality factor $k_{EC(3)}$ for counter-EMF E_C versus field current I_F and speed S.
(a) First isolate the armature to measure R_A. (b) With the motor running, measure V_A, I_F, I_A, and S.

2. Place ammeters in series with the field and in series with the armature as shown in Fig. 12–20(b). They will measure I_F and I_A. Make arrangements to measure shaft speed with a tachometer.
3. Turn on the power, let the motor accelerate to full speed, then measure V_A, I_F, I_A, and S. Record all four values. Turn off the motor.
4. Find counter-EMF E_C by rearranging Eq. (12-7).

$$E_C = V_A - I_A R_A$$

Record the E_C value.

5. Rearrange the assumed-true equation Eq. (12-11) to solve for $k_{EC(3)}$.

$$k_{EC(3)} = \frac{E_C}{I_F \times S}$$

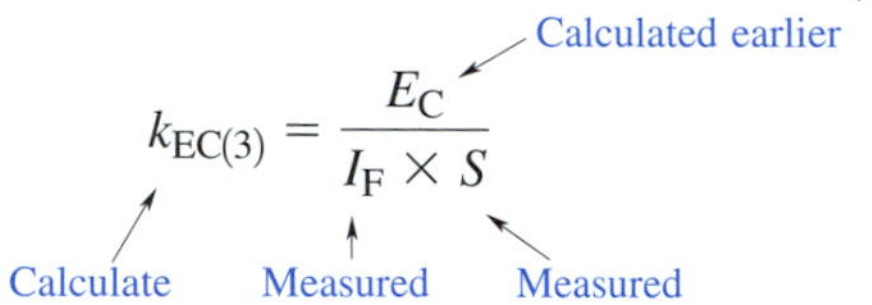

This gives the motor's counter-EMF proportionality constant for field current and shaft speed (I_F and S) assuming that Eqs. (12-9) and (12-11) are true.

■ EXAMPLE 12-5

Suppose that you perform the above testing for the integral-horsepower motor that we've been discussing. Your results are $R_A = 2.8\ \Omega$; $V_A = 230$ V; $I_F = 1.35$ A; $I_A = 3.2$ A; $S = 2970$ r/min.

(a) Find the motor's $k_{EC(3)}$, its proportionality factor for I_F and S. Write the motor's counter-EMF equation for I_F and S.

(b) Suppose that when field current is increased to $I_F = 1.43$ A, speed slows to $S = 2840$ r/min. Find the motor's new value of counter-EMF.

Solution. (a) Referring to step 4, we have

$$\begin{aligned} E_C &= V_A - I_A R_A \\ &= 230\text{ V} - (3.2\text{ A})(2.8\ \Omega) \\ &= 221\text{ V} \end{aligned}$$

Then step 5 gives

$$\begin{aligned} k_{EC(3)} &= \frac{E_C}{I_F \times S} \\ &= \frac{221\text{ V}}{1.35\text{ A} \times 2970\text{ r/min}} \\ &= \mathbf{0.055} \end{aligned}$$

Eq. (12-11) becomes

$$\boldsymbol{E_C = (0.055) I_F S}$$

(b)

$$\begin{aligned} E_C &= (0.055)\, I_F S \\ &= (0.055) \times (1.43\text{ A}) \times (2840\text{ r/min}) \\ &= \mathbf{223\ V} \end{aligned}$$

■

Nonideal relation of B versus I_F. We were careful to say that the counter-EMF equation $E_C = k_{EC(3)}BS$ (Eq. 12-11) was *assumed* to be true. This qualifier is necessary because in reality, magnetic flux density B is not strictly proportional to field current I_F over the entire range of values of I_F. Rather, their proportionality holds true only for moderate values of I_F. At large values of I_F, B loses its proportional response. This effect is graphed in Fig. 12–21.

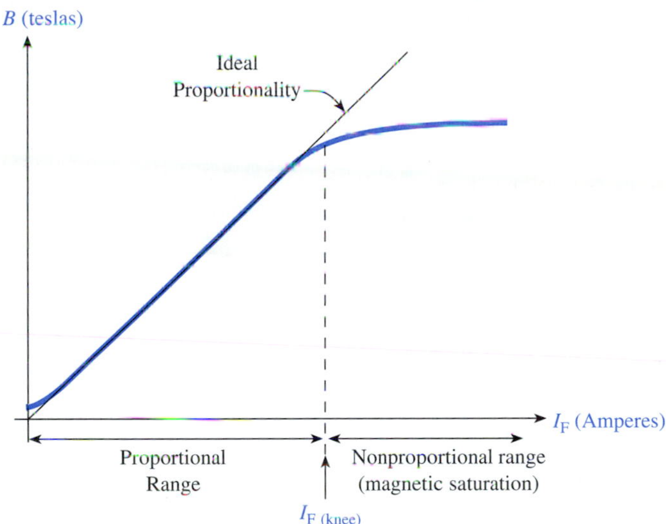

FIGURE 12–21 Magnetic flux density B (or Φ) is fairly proportional to I_F for values less than the knee value. For values greater than the knee value, the magnetic core begins to *saturate*, and proportionality is lost. This graph is often called the *saturation curve* for the dynamo's magnetic core.

An actual magnetic curve like Fig. 12–21 reminds us of a leg that is bent slightly at the knee. Therefore the I_F value where the bend occurs is called the knee current, symbolized $I_{F(knee)}$. For I_F values below $I_{F(knee)}$, there is nearly ideal magnetic proportionality, but for I_F values greater than $I_{F(knee)}$, magnetic proportionality is lost. Thus our assumption in Eq. (12-9), $B = k_B \times I_F$, is true only to the left of $I_{F(knee)}$. Since the counter-EMF equation $E_C = k_{EC(3)}I_FS$ (Eq. 12-11) is based on Eq. (12-9), it also is true only to the left of $I_{F(knee)}$. At larger values of I_F, to the right of $I_{F(knee)}$, the counter-EMF Eq. (12-11) does not work.

The motor manufacturer's detailed data sheet may give the value of $I_{F(knee)}$. If not, you can find it experimentally by running the dynamo as an unloaded generator. Slowly vary I_F, and gather data points of generated voltage V_{gen} versus I_F. Then plot the data points on graph paper to find where the knee occurs. Plotting this data for the dynamo has the added advantage of giving you a clue to what actually does happen for field currents greater than $I_{F(knee)}$.

12-3-2 How the Armature Automatically Adjusts Its Current Draw

Remember why we brought up the counter-EMF concept in the first place. E_C is important because it enables us to understand how a motor armature reacts to variations in the load's torque demand. We can now offer this partial explanation of motor performance.

Suppose that our motor is running at a steady speed S_1, delivering a modest amount of torque τ_1. This situation is represented in Fig. 12–22(a) with the "1" subscripts. It is the mechanical load attached to the motor's shaft that determines the amount of torque

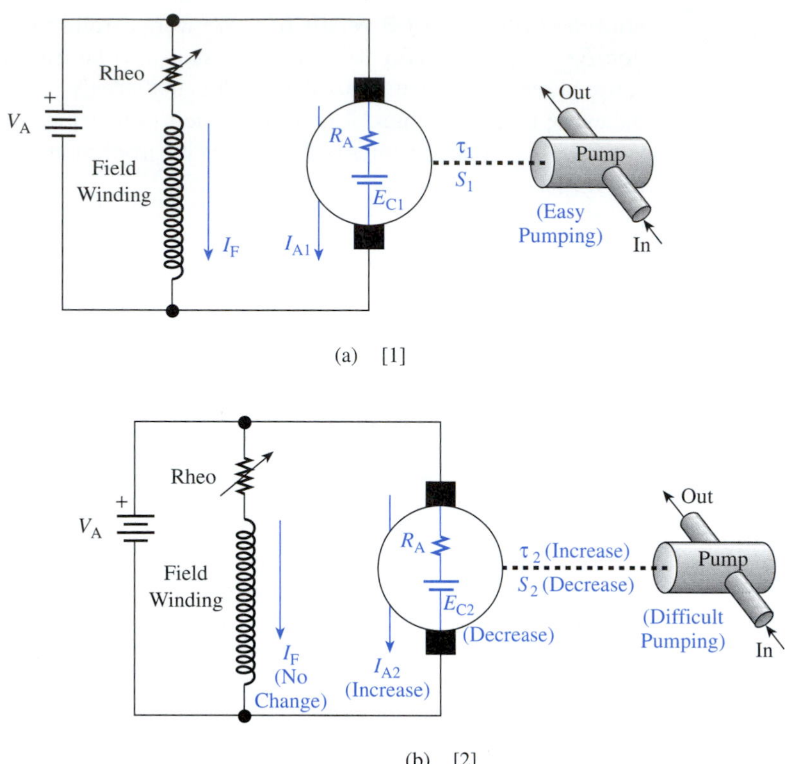

FIGURE 12–22 As the mechanical load changes its torque demand, the motor has to respond. (a) The motor is loafing. (b) Where the load is harder to turn, the motor must make a greater amount of torque. To understant how the motor manages to do this, we keep our eye on counter-EMF.

that the motor must produce. If the mechanical load is easy (a pump that is pumping a light-weight, nonviscous liquid, for example), the motor will not be required to produce much torque.

If the mechanical load is hard to drive (the same pump now working at pumping a dense, viscous liquid, for example), the motor will be required to produce greater torque. This is the situation in Fig. 12–22(b), with all changing variables denoted by "2" subscripts. As the motor-driven pump goes from the **1** situation to the **2** situation, think of the sequence of events as follows:

1. The increase in torque demanded by the load (the pump) causes the motor speed to slow. In Fig. 12–23, look on the left side to see the immediate effect on speed of the load-torque increase. This is the natural mechanical reaction, just like a car engine tending to slow when the car starts up an incline.
2. The decrease in speed results in a decrease in generated counter-EMF, since

$$E_{C2} = k_{EC} B S_2 \qquad \textbf{(12-8a)}$$

or

$$E_{C2} = k_{EC(3)} I_F S_2 \qquad \textbf{(12-10)}$$

Nothing changes in the motor's field loop; I_F maintains a steady value. Therefore the magnetic field strength B also maintains a steady value. As speed S declines, so does E_C in accordance with these counter-EMF equations. This is pictured in the time graphs of Fig. 12–23.

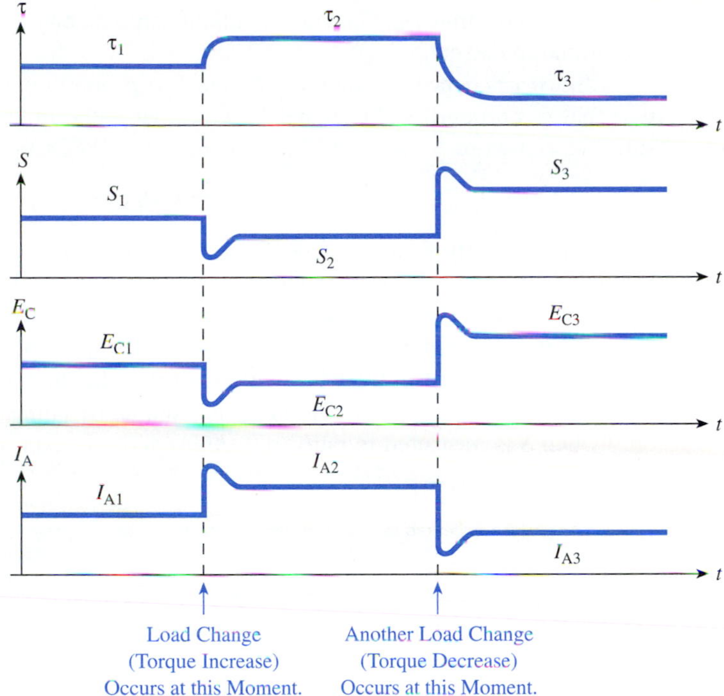

FIGURE 12–23 Graphs of the relationships among torque, speed, counter-EMF, and armature current.

3. The decrease in E_C brings about an increase in I_A. This can be seen by simple inspection of the schematic in Fig. 12–22(b). Or, applying Eq. (12-6),

$$I_{A2} = \frac{V_A - E_{C2}}{R_A} \tag{12-6}$$

   V_A and R_A have constant values, ideally. Therefore, when E_C decreases to E_{C2} in Eq. (12-6), I_A must increase to I_{A2}. The rise in I_A is clearly shown on the left side of Fig. 12–23.
4. The motor may overshoot, as Fig. 12–23 indicates. It will quickly stabilize at a new set of operating conditions, where the new larger value of armature current, I_{A2}, is just sufficient to enable the motor to create the new larger torque, τ_2, to satisfy the load demand.

12-3-3 The Relationship Between τ and I_A

Equation (12-2), $F = lI \times B$, expresses the idea that the mechanical force created by a motor armature conductor is proportional to its current I and also proportional to the strength of the magnetic field, B. The simplified cross-sectional view in Fig. 12–15 helps us see how the motor's rotational torque is related to the mechanical force F. As we know, a modern dc motor's armature winding is much more complex than that picture shows. A real armature has multiple coils interconnected with each other through numerous commutator segments.

All the coils work simultaneously, aiding one another in the overall torque production. This results in smoother, less pulsating torque, with much greater magnitude than the Fig. 12–15 arrangement. These torque improvements are like the generator improvements illustrated in Fig. 12–12. So even though the Fig. 12–15 view is simplified, it does properly demonstrate the fundamental relationship among torque, magnetic field strength, and armature current.

Since τ is dependent on force F, and F is proportional to B and to I, we should expect that τ is proportional to B and I_A. For a particular motor with specific structural design, we can write

$$\tau = k_\tau B I_A \tag{12-12a}$$

The proportionality factor k_τ depends on the same dynamo construction details that k_{EC} depended on in Eq. (12-8a)—namely, rotor diameter and length, number of poles, total number of conducting wires in the armature winding, and the specific series/parallel interconnection method of the armature coils. In fact, if all the variables in Eq. (12-12a) and (12-8a) are measured in their *basic* units, then k_τ for motor torque and k_{EC} for counter-EMF are the same number. That is, k_τ has the same numeric value as k_{EC}. This is true only when S is measured in rad/s (not r/min) and τ is expressed in N-m.

■ EXAMPLE 12-6

Our integral-horsepower motor of Examples 12-1 through 12-5 has $k_{EC} = 0.087$, with S measured in r/min. This fact absolutely determines the value of its k_τ factor, since k_{EC} already takes into account all of the motor's structural details. Our motor has $k_\tau = 0.83$, with speed in r/min and τ in N-m. The manufacturer's detailed data sheet might tell you this fact. Or, it can be calculated by multiplying k_{EC} by the conversion factor from rad/s to r/min, which is 9.551.

$$k_\tau = 9.551 \times k_{EC} = 9.551 \times 0.087 = 0.83$$

From Example 12-4, $B = 0.95$ T when $I_F = 1.5$ A. The motor generated $E_C = 205$ V with $S = 2480$ r/min. Remember that $R_A = 2.8\ \Omega$.

(a) Find the armature current I_A under these running conditions. Use Eq. (12-6).

(b) Calculate the motor's output torque τ. Use Eq. (12-12a).

Solution. (a) From Eq. (12-6),

$$I_A = \frac{V_A - E_C}{R_A} = \frac{230\ \text{V} - 205\ \text{V}}{2.8\Omega} = \mathbf{8.93\ A}$$

(b) Applying Eq. (12-12a) gives

$$\tau = k_\tau B I_A = 0.83(0.95\ \text{T})(8.93\ \text{A}) = \mathbf{7.0\ N\text{-}m}$$ ■

Since the pound-foot unit for torque is still in use in North America, we can use the conversion factor 0.7376 lb-ft/1 N-m to convert, if desired.

$$\tau = 7.0\ \cancel{\text{N-m}} \times \frac{0.7376\ \text{lb-ft}}{1\cancel{\text{N-m}}} = 5.16\ \text{lb-ft}$$

■ EXAMPLE 12-7

Suppose the mechanical load changed from Example 12-6 so that less torque is required. Suppose the new torque is $\tau = 4.5$ N-m. Describe in detail how the motor will react.

Solution. With decreased torque demand, the motor tends to speed up by natural mechanical action. This will increase E_C [see Eq. (12-8a)], which will cause a decrease in I_A. The armature current will decrease to whatever new value is necessary to produce $\tau = 4.5$ N-m.

Magnetic flux density B is unchanged, since no change occurs in the field circuit. All automatic motor responses occur in the armature circuit. From Eq. (12-12a),

$$I_A = \frac{\tau}{k_\tau B} = \frac{4.5\text{ N-m}}{(0.83)(0.95\text{ T})} = \mathbf{5.71\ A}$$

This decrease in armature current is pictured on the right side of Fig. 12–23. No numerical values are shown.

The new value of counter-EMF can be found by rearranging Kirchhoff's voltage law, Eq. (12-7), as

$$\begin{aligned} E_C &= V_A - I_A R_A \\ &= 230\text{ V} - (5.71\text{ A})(2.8\ \Omega) \\ &= 230\text{ V} - 16.0\text{ V} = 214\text{ V} \end{aligned}$$

The rise from $E_C = 205$ V to 214 V also is illustrated on the right side of Fig. 12–23 (no numerical values shown).

The motor will stabilize at a higher speed found from Eq. (12-8a).

$$S = \frac{E_C}{k_{EC}B} = \frac{214\text{ V}}{(0.087)(0.95\text{ T})} = 2590\text{ r/min}$$

The speed increase from 2480 to 2590 r/min is graphed on the right side of Fig. 12–23. ■

12-3-4 Relating a Motor's Torque to the Measurable Currents I_F and I_A

We were able to write the motor's counter-EMF equation in terms of I_F and S by making the assumption that magnetic flux density B is proportional to field current I_F (for $I_F < I_{knee}$). The result was Eq. (12-11).

$$E_C = k_{EC(3)} I_F S$$

Using the same B-vs.-I_F proportionality assumption, we can write the motor torque equation Eq. (12-12a) in terms of measurable variables I_F and I_A. By careful derivation it can be shown that the torque proportionality factor k_τ is equal to $k_{EC(3)}$ multiplied by 9.551, the conversion factor from rad/s to r/min speed units. That is,

$$k_\tau = 9.551 \times k_{EC(3)}$$

so the motor torque equation

$$\tau = k_\tau B I_A \qquad \textbf{(12-12a)}$$

can be rewritten as

$$\tau = (9.551 \times k_{EC(3)}) I_F I_A \qquad \textbf{(12-12b)}$$

The advantage of Eq. (12-12b) is that it deals in easily measured variables I_F and I_A rather than the difficult-to-measure B. Furthermore, $k_{EC(3)}$ can be found by experimentation with the motor; you don't have to rely on the motor manufacturer to furnish the proper proportionality information. This is a great help, since motor manufacturers are notorious for leaving out that information. Of course, Eq. (12-12b) is reliable only for moderate values of field current. It is not applicable for $I_F > I_{knee}$.

■ EXAMPLE 12-8

Our integral-horsepower motor was tested in Example 12-5, giving $k_{EC(3)} = 0.055$.

(a) Write our motor's torque equation in terms of field current I_F and armature current I_A.

(b) Use this equation to calculate the motor's torque for the conditions described in Example 12-6, namely $S = 2480$ r/min, $I_F = 1.5$ A, and $I_A = 8.93$ A. Compare to the result obtained in Example 12-6, part (b).

Solution. (a) From Eq. (12-12b),

$$\begin{aligned} \tau &= (9.551 \times k_{EC(3)})\, I_F I_A \\ &= (9.551 \times 0.055)\, I_F I_A \\ \tau &= \mathbf{0.525}\, \boldsymbol{I_F I_A} \end{aligned}$$

(b)

$$\tau = (0.525)(1.5\text{ A})(8.93\text{ A}) = \mathbf{7.0\ N\text{-}m}$$

This agrees with the torque obtained from Eq. (12-12b) in part **b** of Example 12-6. ■

12-3-5 Mechanical Power

The mechanical power of a rotating shaft is given by the product of torque and speed. That is,

$$P_{mech} = \tau S \qquad \textbf{(12-13)}$$

Rotational mechanical power can be expressed in watts, which is its basic unit. Or, it can be expressed in horsepower, commonly used in North America.

To express mechanical power in watts, first convert speed S to basic units, radians/second. Then apply Eq. (12-13) with τ expressed in its basic units, N-m. In Example 12-8 this would be accomplished as

$$\begin{aligned} S &= 2480\text{ r/min} \\ &= 2480\ \cancel{\text{r/min}} \times \frac{1\text{ rad/s}}{9.551\ \cancel{\text{r/min}}} = 260\text{ rad/s} \end{aligned}$$

Conversion factor

Applying Eq. (12-13) gives

$$\begin{aligned} P_{mech} &= \tau \times S \\ &= (7.0\text{ N-m}) \times (260\text{ rad/s}) \\ &= 1.82 \times 10^3\text{ W or } 1.82\text{ kW} \qquad \text{(kilowatts)} \end{aligned}$$

To express mechanical power in units of horsepower, divide the basic-unit result by the conversion factor

$$745.7\text{ W} = 1\text{ hp}$$

In this example,

$$P_{\text{mech}} = 1.82 \times 10^3 \not{W} \times \frac{1 \text{ hp}}{745.7 \not{W}} = 2.44 \text{ hp}$$

There are other ways to calculate P_{mech} in horsepower units without first converting speed, torque, and P_{mech} to their basic units. If you find yourself working in an environment where such nonbasic units are used, learn the appropriate direct-calculation formulas for P_{mech}. One such formula that is often encountered calculates P_{mech} in horsepower units if τ is measured in lb-ft and S is measured in r/min. Then we have

$$\underset{\text{hp}}{P_{\text{mech}}} = \frac{\overset{\text{lb-ft}}{\tau}\,\overset{\text{r/min}}{S}}{5252} \qquad \text{(12-14)}$$

12-4 ■ CHARACTERISTIC GRAPHS OF SHUNT-CONFIGURED DC MOTORS

The motor equations of Sec. 12-3, Eqs. (12-3) through (12-14), apply to all wound-rotor dc motors in general. If a particular motor's proportionality constants k_{EC} and k_τ are known, then putting those values into the equations makes the equations apply specifically to that particular motor. For instance, the motor that we used for our example problems has the specific equations

$$E_{\text{C}} = 0.087BS \qquad \text{(speed in r/min)}$$

and

$$\tau = 0.83BI_{\text{A}} \qquad \text{(torque in N-m)}$$

The operating characteristics of a particular motor can also be expressed by graphs rather than equations. The most useful graphs are for speed versus torque (S vs. τ) and armature current versus torque (I_{A} vs. τ). These graphs are assumed to be for the condition of constant applied voltage V_{A} and constant magnetic flux density B (due to constant I_{F}).

Figure 12–24 presents the S vs. τ and I_{A} vs. τ *characteristic graphs* for our example motor. These graphs are ideal, and they apply only to the specified operating condition of $V_{\text{A}} = 230$ V, $I_{\text{F}} = 1.5$ A. If either one of those operating variables is changed, it results in a different pair of graphs.

If the motor were actually tested, the real graphs would deviate slightly from these ideal straight lines. Also, the I_{A} graph would intersect the vertical axis slightly above 0 A.

For any motor, the full-load point corresponds to the maximum continuous I_{A} current value that the armature winding can carry without overheating. In our motor that value is $I_{\text{A(FL)}} = 12$ A, identified in Fig. 12–24(b). Continuous prolonged current greater than 12 A produces I^2R power losses that raise the motor's operating temperature to damaging levels.

The torque that is produced by full-load armature current $I_{\text{A(FL)}}$ is called *full-load torque,* τ_{FL}. This assumes that the motor is operating at the rated voltage (230 V), with maximum field current (1.5 A) creating the strongest possible B-field (0.95 T). For our motor, $\tau_{\text{FL}} = 9.5$ N-m.

The rotational shaft speed under these conditions is called *full-load speed.* Here, $S_{\text{FL}} = 2370$ r/min, identified in Fig. 12–24(a). Full-load operation refers to maximum *continuous* running. The motor can exceed its full-load conditions for a short time

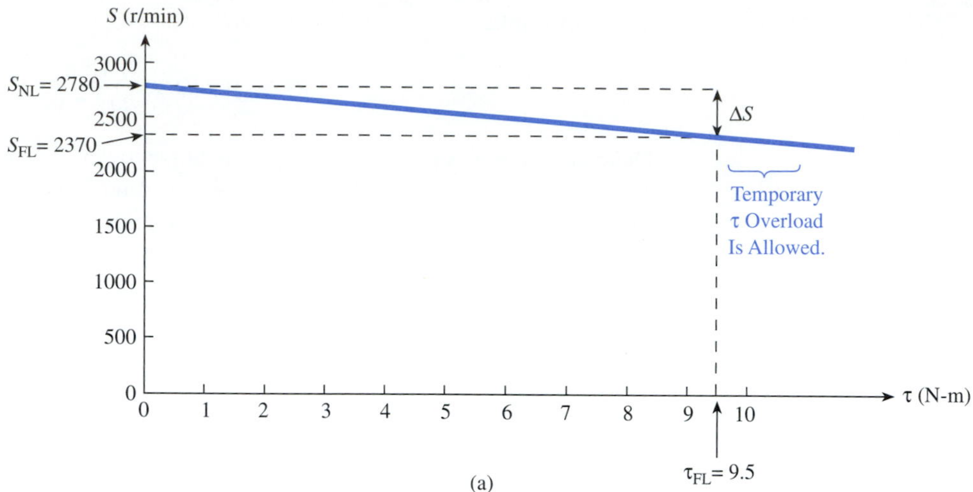

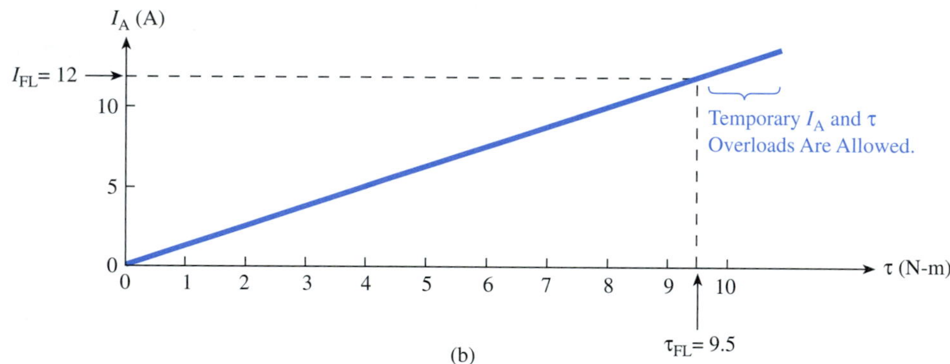

FIGURE 12–24 Ideal characteristic graphs of our particular motor for the operating condition $V_A = 230$ V, $I_F = 1.5$ A.

without harm. In fact this is bound to occur every time the motor is started from a standstill. As it accelerates from $S = 0$, its counter-EMF starts at 0 V. E_C remains lower than normal until the rotor can get up to running speed. Therefore E_C cannot subtract from V_A to help limit the armature current to a reasonable value.

At the instant of starting, I_A surges to a very large value, limited only by Ohm's law for the armature loop, $I_A = V_A/R_A$. For our motor, the starting current surge is

$$I_{A(\text{starting})} = \frac{V_A}{R_A} \tag{12-15}$$

$$= \frac{230 \text{ V}}{2.8 \ \Omega} = 82 \text{ A}$$

Such a large current surge might create noticeable *IR* voltage drop along the supply lines, resulting in momentary dimming of lamps connected to that circuit.

Of course, the advantage of such a great current surge is that it produces a torque surge far greater than the continuous full-load running torque, τ_{FL}. For our motor,

$$\tau_{\text{starting}} = k_\tau \times B \times I_{A(\text{starting})}$$

$$= 0.83 \times (0.95 \text{ T}) \times (82 \text{ A}) = 65 \text{ N-m}$$

which is about 7 times greater than τ_{FL}. This large *starting torque* enables the motor to quickly get its load going. Thus $\tau_{starting}$ is an important motor parameter that is used to judge a motor's usefulness for a particular application. In Fig. 12–24(a), if the downward-sloping line were continued far to the right, it would intersect the horizontal τ axis at the $\tau_{starting}$ value, 65 N-m.

Another motor parameter that affects its usefulness for a particular application is its *speed regulation*. This is a measure of how much relative speed change occurs as the load's torque demand varies from minimum (usually 0) to maximum (τ_{FL}). The defining formula for percentage speed regulation is

$$S_{reg}\,(\%) = \frac{S_{NL} - S_{FL}}{S_{FL}} \times 100\% \qquad \textbf{(12-16)}$$

where $S_{NL}$and S_{FL} refer to the motor's no-load and full-load speeds, respectively.

■ EXAMPLE 12-9

For our motor with its characteristics graphed in Fig. 12–24, find the speed regulation percentage.

Solution. Applying Eq. (12-16) to the speed values marked in Fig. 12–24(a), we get

$$S_{reg}\,(\%) = \frac{S_{NL} - S_{FL}}{S_{NL}} \times 100\%$$

$$= \frac{2780 - 2370}{2370} \times 100\%$$

$$= \frac{410 \text{ r/min}}{2370 \text{ r/min}} \times 100\% = \mathbf{17\%}$$ ■

The configuration of our motor is called *shunt,* because the field winding is electrically in parallel (shunt) with the armature winding. This was pointed out clearly in Fig. 12–16. The general characteristics of shunt-configured dc motors are

1. Shunt motors have fairly good speed regulation. A figure of $S_{reg} = 17\%$ is considered to be fairly good. Other dc motor configurations exhibit speed regulation factors that are much worse than 17%. (However, some ac motors have speed regulations much better than 17%.)
2. Shunt dc motors have fairly good starting torque. A factor-of-7 increase over τ_{FL} is considered fairly good.

Some motors do much better than a factor of 7, and some are not nearly as good.

In Fig. 12–24, notice that we have placed torque τ on the horizontal axis, not the vertical axis. This is the reasonable way to draw motor characteristic graphs because τ is the *independent variable,* while S and I_A are *dependent variables*. The term *independent variable* means the variable that takes on any value that it wants to, not waiting for some other variable to tell it what to do. In the situation where a motor shaft is driving a mechanical load, the torque is the independent variable from the motor's point of view. This is so because the conditions in the *load* decide how much torque the motor has to produce. That is why we call it torque *demand*.

The motor reacts to a change in torque demand by changing its speed and its armature current-draw. In that respect these two motor variables depend on torque. Therefore S and I_A are called dependent variables.

It is customary to place the independent variable on the x axis and the dependent variable on the y axis. That way when we inspect the graph, we imagine the y-axis variable "following" the x-axis variable. Another way of saying this is that the y-axis variable is a "function of" the x-axis variable. Always be careful to follow this philosophy when you draw a graph; otherwise your graph becomes misleading.

However, it can happen in technology that your point of view determines which variable is independent and which is dependent. For example, from a mechanical load's point of view, it usually makes more sense to regard speed as the independent variable and torque as the dependent variable. It's as if the load is pondering the question: "Let's see, if I'm supposed to move this fast, how much torque do I have to demand from my motor?"

The issue of independence/dependence can be tricky. Later we will be taking viewpoints where we want to call speed S the independent variable and torque τ the dependent variable. When that happens, we will explain why we're doing it.

12-5 ■ CHARACTERISTICS OF SERIES-CONFIGURED DC MOTORS

The alternative to shunt configuration is series configuration. As its name implies, the field winding is connected electrically in series with the armature winding, as shown in Fig. 12–25.

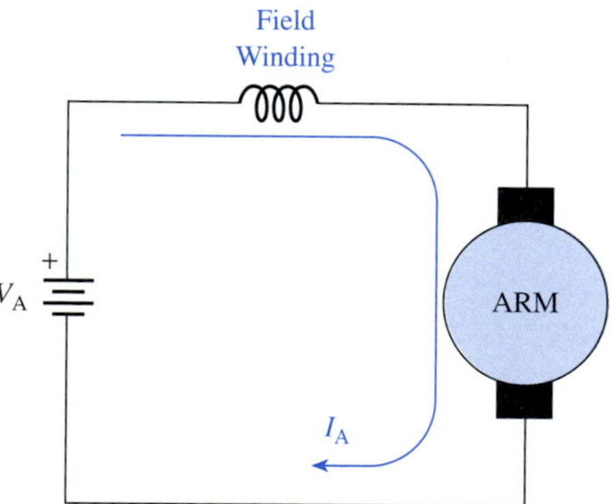

FIGURE 12–25 In a series dc motor the field winding is in series with the armature. There is only one current in the motor, I_A.

The mechanical placement of the field winding is exactly the same as for the shunt-configured motor, shown in Fig. 12–16(b). However, the winding of a series motor must be made of much thicker-gage wire, because it must carry the entire armature current I_A, not just a relatively small shunt field current I_F. This is clear from the Fig. 12–25 schematic diagram.

A series-configured field winding also tends to have fewer turns than a shunt winding. This is reasonable because with its larger current it doesn't need a great number of turns in order to create a strong magnetic field B.

An electromagnet's magnetizing force H is given by

$$H = \frac{NI}{l} \quad \textbf{(12-17)}$$

where N is the number of winding turns, I is the winding current, and l is the length over which the winding's turns are distributed. H is magnetizing force in units of amp-turns per meter. The series-configured motor uses lower N and higher I compared to the shunt-configured motor's higher N and lower I.

The basic counter-EMF and torque equations are the same for a series motor as for a shunt motor. We still have

$$E_C = k_{EC} BS \quad \textbf{(12-8a)}$$

and

$$\tau = k_\tau BI_A \quad \textbf{(12-12a)}$$

What is different about a series motor is that magnetic field strength B is not constant at varying torque demands. Now B is established by the armature current I_A, not by a small constant current I_F. Therefore B increases when the motor is drawing a large amount of armature current to meet heavy torque demand by the load, and B decreases when the motor draws smaller I_A, which occurs when the load's torque demand is lighter.

Because magnetic flux density B tends to change in step with the other torque-producing variable, I_A, the motor's torque production is greatly enhanced when it is carrying a surge of current at start-up. This is the chief advantage of a series motor. It can accelerate very rapidly from a standstill, even with a heavy mechanical load on its shaft. To understand this starting advantage, compare the series motor shown in Fig. 12–26(b) to the shunt motor of Fig. 12–26(a).

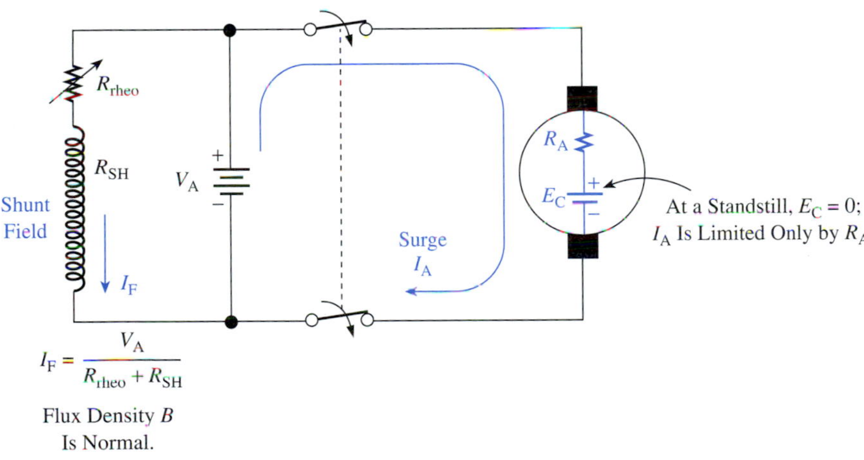

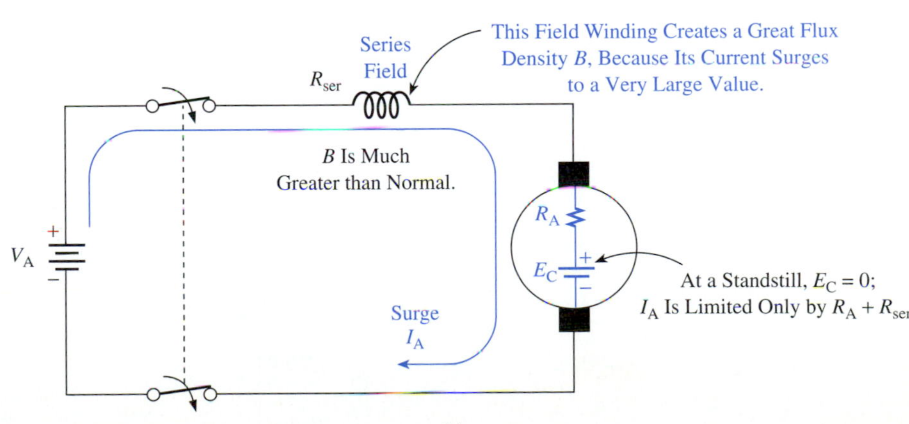

FIGURE 12–26 Contrasting the starting behavior of shunt and series motors. (a) Shunt—no change in flux density. (b) Series—great increase in flux density B.

Instead of a factor-of-7 increase in τ_{starting} over τ_{FL}, as we saw for our shunt motor in the previous section, an equivalent series motor might produce a factor-of-15 or -20 increase in τ_{starting}. Their starting-torque advantage makes series motors especially useful in applications where the motor must frequently stop and start, and acceleration rate is important.

The chief disadvantage of series motors is their poor speed regulation. A few examples will help illustrate these series-motor performance differences compared to shunt motors.

■ EXAMPLE 12-10

If the shunt motor of Sec. 12-4 is disassembled and its field rewound to make it a series motor, it will still have the same proportionality factors as the original shunt motor. This is because all the armature details remain unaltered, and the number of field poles remains unaltered. Thus we can depend on

$$k_{\text{EC}} = 0.087 \quad \text{(speed in r/min)}$$

and

$$k_{\tau} = 0.83 \quad \text{(torque in N-m)}$$

Suppose that the resistance of the field winding, R_{ser} in Fig. 12–27, is measured to be 1.1 Ω. We expect the series field resistance to be very low, since the winding wire is quite thick and much shorter in overall length (fewer turns) than the original shunt winding.

Let us make the assumption that the number of turns in the series field winding is such that this motor's flux density reaches $B = 0.95$ T when the motor is carrying its full-load running current, 12 A. (The armature winding hasn't changed, so it still has the same maximum continuous current limitation that it had previously.) Let us also assume that for all I_{A} values less than 12 A, B is proportional to I_{A}. This is equivalent to saying that the motor is operating in the linear region of its core magnetization curve, to the left of I_{knee}. Or, $I_{\text{knee}} > 12$ A. Refer to Fig. 12–21.

(a) Equation (12-12a) for the series motor is still $\tau = 0.83\, BI_{\text{A}}$. The assumed proportionality between B and I_{A} can be written as

$$\frac{B}{0.95\text{ T}} = \frac{I_{\text{A}}}{12\text{ A}}$$

or

$$B = \frac{I_{\text{A}}}{12\text{ A}} \times (0.95\text{ T})$$

$$B = (0.079)\, I_{\text{A}} \quad \text{(assumed true)} \qquad \textbf{(12-18)}$$

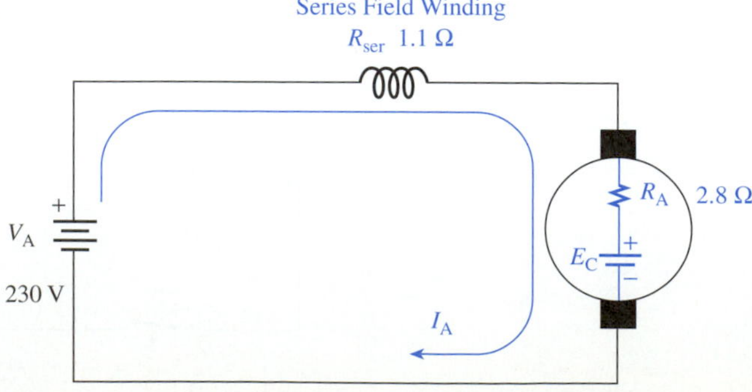

FIGURE 12–27
Our original motor rebuilt as a series motor.

Plug Eq. (12-18) into Eq. (12-19) to get an equation relating torque τ to armature current I_A for our series motor. Give this new equation the number (12-19a).

We have good confidence in Eqs. (12-18) and (12-19a) for $I_A \le 12$ A.

(b) Construct a table with columns labeled τ (N-m), I_A (A), E_C (V), and S (r/min) from left to right. In the torque column, write all the integer values of τ from 1 to 15 N-m. Also include a row for full-load running torque, 9.5 N-m, between the 9 row and the 10 row. For all the integer torque values from $\tau = 1$ N-m up to $\tau = 9$ N-m, rearrange your Eq. (12-19a) to calculate the series motor's armature current. Also calculate I_A for $\tau = 9.5$ N-m, which is τ_{FL}. Enter the calculated I_A values in your table.

(c) We have less confidence in Eqs. (12-18) and (12-19a) for $I_A > 12$ A, because the magnetic saturation effect undoes the proportionality between B and I_A. Nevertheless, to give us something to work with, go ahead and use your rearranged Eq. (12-19a) to estimate the I_A values for all the integer τ values from 10 to 15 N-m.

(d) Plot a graph of points that you have calculated from Eq. (12-19a). Use τ as the independent variable and I_A as the dependent variable.

Solution. (a)

$$\tau = (0.83)\, B I_A$$

and

$$B = (0.079)\, I_A$$

so

$$\tau = (0.83) \times (0.079\, I_A) \times I_A$$

$$\tau = (0.066) \times (I_A)^2 \qquad \textbf{(12-19a)}$$

(b) Rearranging Eq. (12-19a) to solve for I_A gives

$$(I_A)^2 = \frac{\tau}{0.066} = 15.2\,\tau$$

$$I_A = \sqrt{15.2\,\tau}$$

$$I_A = 3.9\sqrt{\tau} \qquad \textbf{(12-19b)}$$

Equation (12-19b) expresses the essential characteristic of a series motor. It tells us that the motor's current doesn't have to increase in direct proportion to its torque, only as the *square root* of the torque. In other words, very large values of starting torque can be produced without necessarily allowing harmful surges of starting current

Beginning with $\tau = 1$ N-m, we get

$$I_A = (3.9)\sqrt{1} = (3.9) \times 1 = 3.9 \text{ A}$$

For $\tau = 2$ N-m, we get

$$I_A = (3.9)\sqrt{2} = (3.9) \times 1.41 = 5.5 \text{ A}$$

and so forth, until for $\tau = 9.5$ N-m we get

$$I_A = (3.9)\sqrt{9.5} = (3.9) \times 3.08 = 12.0 \text{ A}$$

Table 12–1 presents these results.

(c) Continuing on as before, for $\tau = 10$ N-m, we get

$$I_A = (3.9)\sqrt{10} = (3.9) \times 3.16 = 12.3 \text{ A} \qquad \text{(estimate)}$$

and so forth, until for $\tau = 15$ N-m we get

$$I_A = (3.9)\sqrt{15} = (3.9) \times 3.87 = 15.1 \text{ A} \qquad \text{(estimate)}$$

TABLE 12–1
Operating variables for our series motor*

τ(N-M)	I_A(A)	E_C(V)	S(r/min)
1.0	3.9	215	8000
2.0	5.5	209	5480
3.0	6.8	204	4410
4.0	7.8	200	3720
5.0	8.7	196	3250
6.0	9.6	193	2950
7.0	10.3	190	2670
8.0	11.0	187	2450
9.0	11.7	184	2280
9.5	12.0	183	2200
10.0	12.3	182	2140
11.0	12.9	180	2010
12.0	13.5	177	1900
13.0	14.1	175	1810
14.0	14.6	173	1720
15.0	15.1	171	1640

*The shaded part of the table is estimated only, based on assumed proportionality between field flux density B and current I_A. The magnetic saturation effect has been ignored.

We now proceed to fill in the bottom six rows of Table 12–1 with I_A values that are estimates.

(d) The data is plotted in Fig. 12–28. Compare it to shunt-motor Fig. 12–24(b); that straight-line graph has been extended to 15 N-m and overlaid on the series graph. Notice how the series motor can produce overload torque without beating up its armature winding and commutator as badly as the shunt motor does.

■ EXAMPLE 12-11

(a) Referring to the schematic diagram of Fig. 12–27, calculate the starting current, $I_{A(starting)}$.

(b) At very large values of I_A it is not reasonable to assume that B retains its proportionality. Instead, for a change in I_A from $I_{A(FL)} = 12$ A to $I_{A(starting)}$, a more reasonable estimate is that after magnetic saturation begins (the knee) any further increase in B occurs at only half the rate that I_A increases.

Use this method to estimate the increase in B at the starting instant, above and beyond its 0.95-T value that occurs at the full-load condition, $I_A = 12$ A. Then estimate $B_{starting}$ by adding this further increase in B to 0.95 T.

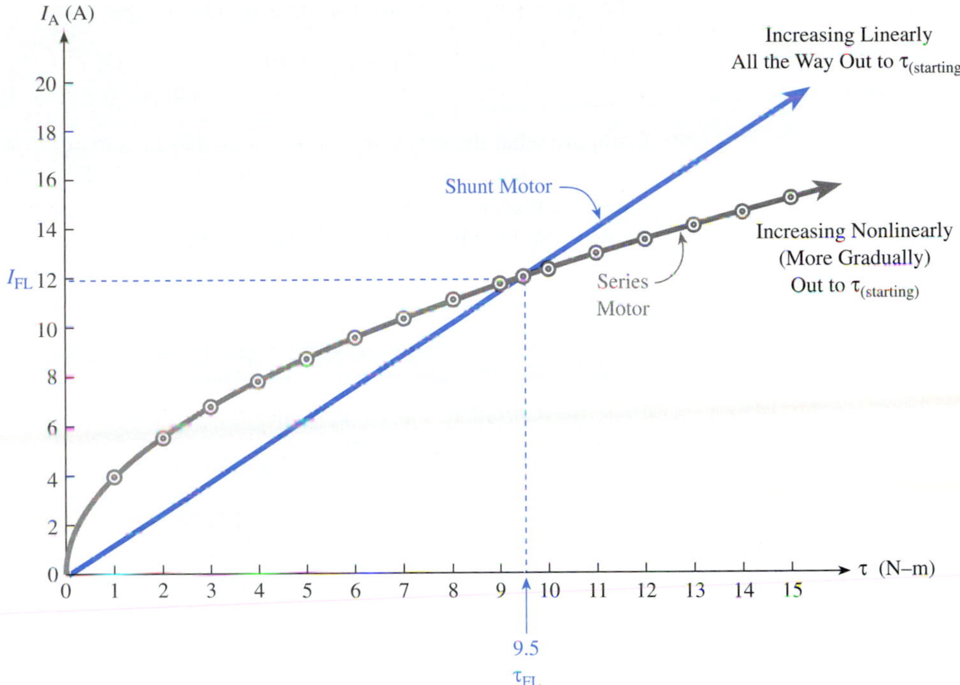

FIGURE 12–28
Characteristic curve of I_A versus τ for a series motor. The graph is realistic up to 9.5 N-m. Beyond that it is idealized.

(c) Estimate the starting torque produced by this series motor.

Solution. (a) At the starting instant $E_C = 0$, so Ohm's law applies. There are now two resistances in the armature loop, R_A and R_{ser}. The total armature loop resistance is given by

$$R_{arm(T)} = R_A + R_{ser}$$
$$= 2.8\ \Omega + 1.1\ \Omega = 3.9\ \Omega$$

Ohm's law gives

$$I_{A(starting)} = \frac{V_A}{R_{arm(T)}}$$
$$= \frac{230\text{ V}}{3.9\ \Omega} = \mathbf{59.0\ A}$$

(b) Comparing the starting condition to the full-load condition, armature current increases by the factor

$$\frac{I_{A(starting)}}{I_{A(FL)}} = \frac{59\text{ A}}{12\text{ A}} = 4.9$$

So if B had retained proportionality, it would have increased to

$$B_{starting} = 4.9 \times B_{FL}$$
$$= 4.9 \times (0.95\text{ T}) = 4.65\text{ T}$$

This would have meant an increase in B from 0.95 T to 4.65 T, which would have been $\Delta B = 3.7$ T; but our estimate is that the actual ΔB is only $1/2 \times 3.7$ T = **1.85 T.**

$$B_{starting} = B_{FL} + \Delta B$$
$$= 0.95\text{ T} + 1.85\text{ T} = \mathbf{2.8\ T} \qquad \text{(estimate)}$$

(c) Using the motor torque equation, we get

$$\tau_{\text{starting}} = k_\tau B I_A \qquad \textbf{(12-12a)}$$
$$= 0.83 \times 2.8 \text{ T} \times 59.0 \text{ A} = \mathbf{137 \text{ N-m}}$$

Compare what the series motor accomplishes at start-up to what the shunt-configured motor accomplished. The series motor produces 137 N-m of torque, which is about double the shunt motor's 65 N-m. And the series motor was able to do that with a surge of only 59 A, considerably less than the shunt's 82 A.

■ EXAMPLE 12-12

(a) Complete the E_C column of Table 12–1 by calculating counter-EMF for each row using Kirchhoff's voltage law.

$$E_C = V_A - I_A \times R_{\text{arm(T)}}$$
$$= 230 \text{ V} - (I_A \times 3.9 \ \Omega)$$

(b) Complete the S column of Table 12–1 by calculating speed as follows:

$$B = \frac{\tau}{k_\tau I_A} \qquad \textbf{(12-12a)}$$

and

$$S = \frac{E_C}{k_{EC} B} \qquad \textbf{(12-8a)}$$

Substituting Eq. (12-12a) into Eq. (12-8a) gives us

$$S = \frac{E_C}{k_{EC} \times (\tau / k_\tau I_A)} = \frac{E_C \times k_\tau I_A}{k_{EC} \times \tau}$$
$$= \frac{0.83}{0.087} \times \frac{E_C \times I_A}{\tau}$$
$$S = 9.54 \times \frac{E_C \times I_A}{\tau} \qquad \textbf{(12-20)}$$

(c) Graph the curve of S versus τ for this series motor.

Solution. (a) In the first row we have

$$E_C = 230 \text{ V} - (3.9 \text{ A} \times 3.9 \ \Omega)$$
$$= 230 \text{ V} - 15 \text{ V} = 215 \text{ V}$$

For the second row,

$$E_C = 230 \text{ V} - (5.5 \text{ A} \times 3.9 \ \Omega) = 230 \text{ V} - 21 \text{ V} = 209 \text{ V}$$

and so on, through the last row,

$$E_C = 230 \text{ V} - (15.1 \text{ A} \times 3.9 \ \Omega) = 230 \text{ V} - 59 \text{ V} = 171 \text{ V}$$

The E_C results in the rows beneath the full-load row, $\tau = 9.5$ N-m, are estimates.

(b) In the first row we have

$$S = 9.54 \times \frac{E_C \times I_A}{\tau}$$

$$= 9.54 \times \frac{215\text{ V} \times 3.9\text{ A}}{1\text{ N-m}} = 8000\text{ r/min}$$

In the second row,

$$S = 9.54 \times \frac{209\text{ V} \times 5.5\text{ A}}{2\text{ N-m}} = 5480\text{ r/min}$$

and so on, through the last row,

$$S = 9.54 \times \frac{171\text{ V} \times 15.1\text{ A}}{15\text{ N-m}} = 1640\text{ r/min}$$

As before, the shaded portion of the table beneath the 9.5 N-m row is estimated and not reliable.

(c) Speed vs. torque is plotted in Fig. 12–29.

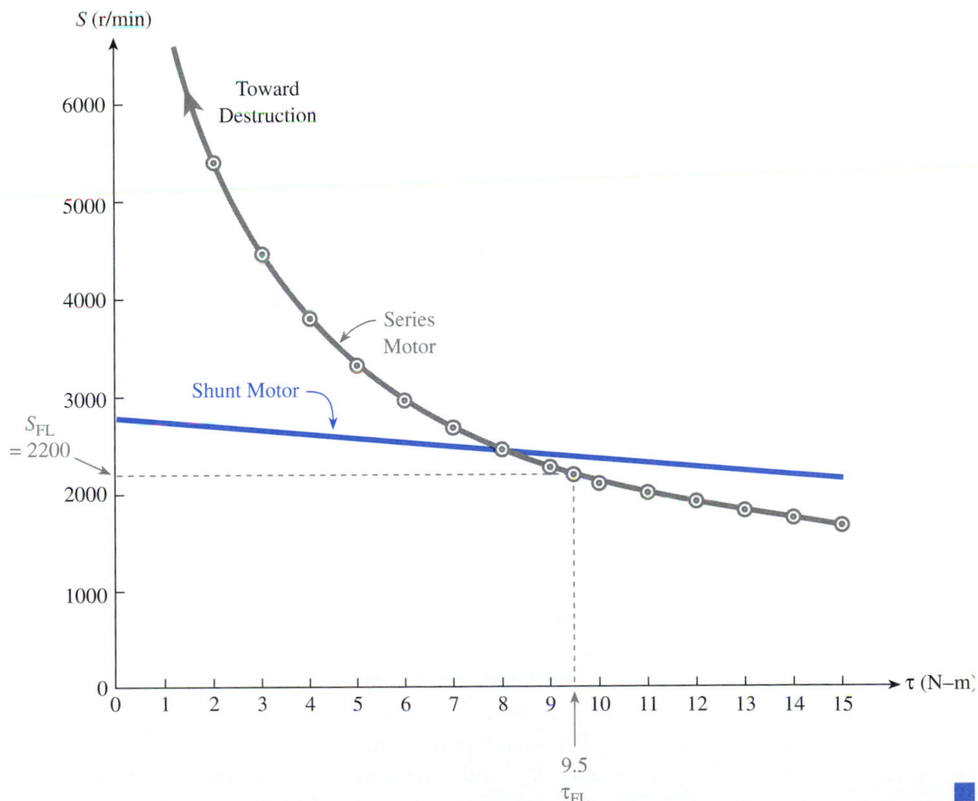

FIGURE 12–29 A series-configured motor has very poor speed regulation. It's impossible to specify a S_{reg} factor based on Eq. (12-16) because a series motor cannot operate at no-load. Even in this graph, the 5500-r/min speed is problematic. It would have to be a very well-made rotor assembly in order to hold together at that speed. For most dc dynamos the maximum safe speed is about 4000 r/min.

Compare the speed regulation of the series motor to that of the shunt motor. The shunt motor's characteristic graph from Fig. 12–24(a) has been extended to 15 N-m and overlaid on the series speed curve. Note the very poor speed regulation of the series motor. It's so bad that we couldn't show the 8000-r/min value corresponding to $\tau = 1$ N-m.

This brings up an important fact about series motors: They cannot be allowed to run with no load (zero load torque), or even with very small values of load torque. A series motor will rotate at an excessive speed, thereby destroying itself, if the load torque is too *low*. It certainly cannot be allowed to run free, with its shaft mechanically disconnected.

12-6 ■ COMPOUND CONFIGURATION

The shunt-configured motor has good speed regulation and fairly good starting torque. The series-configured motor has poor speed regulation and excellent starting torque. A compound-configured motor strikes a compromise between the operating characteristics of the shunt and series motors. It gives fair speed regulation, much better than the series motor, and it produces a good starting torque by taking advantage of the surge inrush current to boost its *B*-field. The schematic diagram of a compound motor is shown in Fig. 12–30.

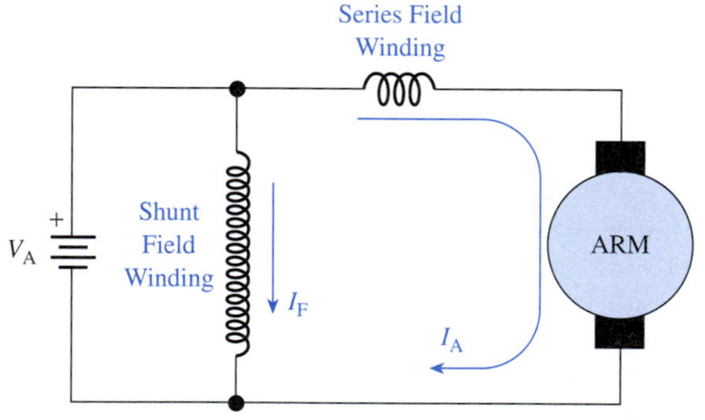

FIGURE 12–30 The compound configuration. The shunt-field winding carries low-value current I_F. The series-field winding carries armature current I_A, which is much greater than I_F under heavy torque load conditions. At the starting instant the series field carries the I_A surge, thereby boosting the motor's flux density *B*. This in turn boosts $\tau_{starting}$.

The shunt-field winding consists of a large number of turns of thin wire. It has a high resistance R_F and carries a relatively small current I_F. The shunt-field winding is in parallel with the series combination of the armature winding and the series-field winding, as Fig. 12–30 makes clear.

The series-field winding consists of relatively few turns of thick wire. It has a low resistance R_{ser} and carries the entire armature current I_A. Figure 12–31 is a cross-sectional view of a compound dc motor, for clarifying its winding details.

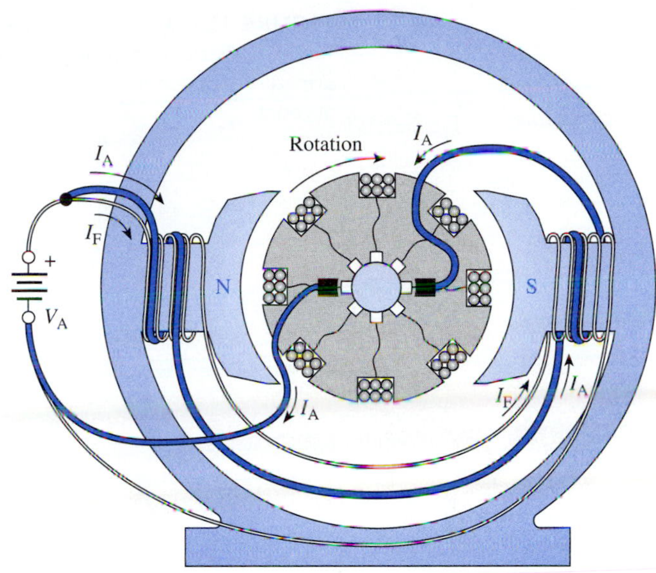
Rotation
I_A
I_F
V_A
N
S

FIGURE 12–31
Cross-section of a compound motor, showing electrical relation among shunt-field winding, series-field winding, and armature winding.

12-7 ■ INTERPOLES

In our discussion of wound-rotor motors, we have assumed that the magnetic flux maintains its spatial distribution shown in Figs. 12–6, 12–7, and 12–9. That is, the Φ lines are perpendicular to the pole-faces, evenly distributed at all points on the pole-faces. In reality the flux doesn't behave this ideally. It becomes distorted when the armature winding draws large current at heavy torque load, because the armature conductors create their own flux, which is at a 90° angle relative to the main field flux (for a two-pole machine). This problem, called *armature reaction,* is pictured in Fig. 12–32.

The net flux resulting from the combination of the horizontal main-field flux and the vertical armature reaction flux becomes bent and distorted, as shown in Fig. 12–33.

FIGURE 12–32
(a) Main-field flux is horizontal. (b) When I_A flows, the armature establishes a flux at a 90° angle to the main flux.

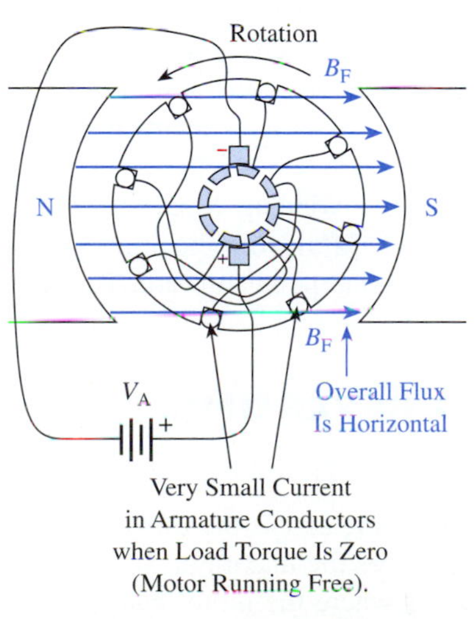
Rotation
B_F
N
S
V_A
Overall Flux
Is Horizontal
Very Small Current
in Armature Conductors
when Load Torque Is Zero
(Motor Running Free).

(a)

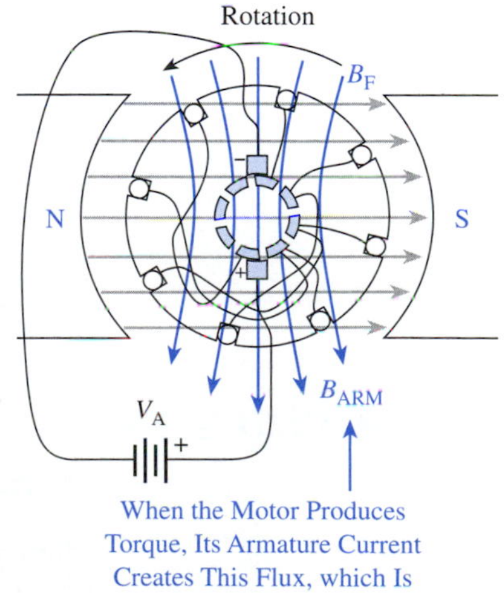
Rotation
B_F
N
S
V_A
B_{ARM}
When the Motor Produces
Torque, Its Armature Current
Creates This Flux, which Is
Basically Vertical.

(b)

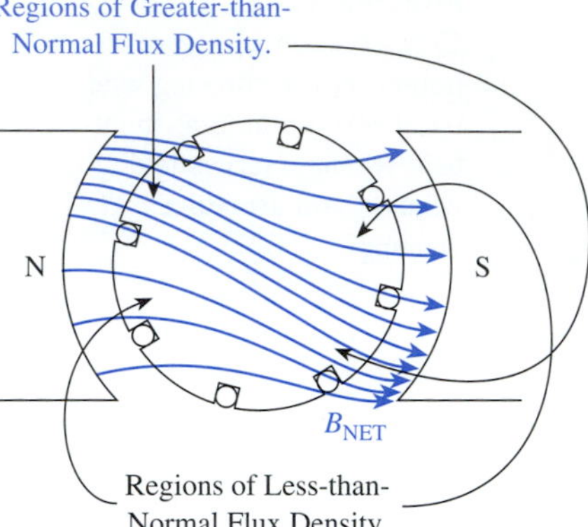

FIGURE 12–33
Net flux distribution, taking armature reaction into account.

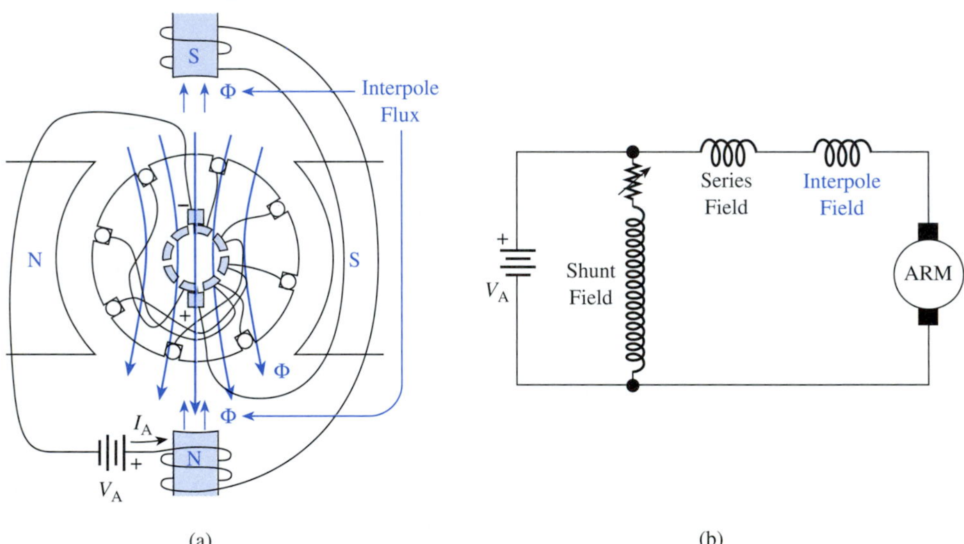

FIGURE 12–34
The interpole idea.
(a) With armature reaction flux pointing vertically downward, the interpole counter-reaction flux must point vertically upward.
(b) Schematic appearance of the interpole winding.

This flux distortion causes two problems. It reduces the motor's torque production, and it worsens the sparking occurring between the brushes and the commutator segments.

To combat the armature reaction problem, some motors are built with *interpoles*. These are physically smaller poles placed between the main poles, as shown in Fig. 12–34(a). The interpole windings are connected in series with the armature, as the schematic diagram of Fig. 12–34(b) makes clear. Their winding wrap direction makes their flux oppose the armature reaction flux. In Fig. 12–34(a), with the armature reaction flux pointing downward, the interpoles' counter-reaction flux points upward. Thus the interpoles cancel the armature reaction effect, eliminating the two problems associated with it.

12-8 ■ STARTING, STOPPING AND REVERSING

12-8-1 Starters

Many dc motors are started in the across-the-line manner, which is shown in Fig. 12–26. This is the method that produces the maximum starting torque and quickest acceleration; but there are some situations where the motor cannot safely be started this way because the inrush current lasts too long. This happens when the load has a great amount of rotational inertia, the motor's rotor itself has great inertia, or both. Large inertia tends to accompany large motors. So, as a general rule, large motors are more likely to need special starting circuits than small motors.

There are two kinds of starting circuits for limiting the armature inrush current to a dc motor:

1. Circuits that gradually increase the applied voltage V_A.
2. Circuits that insert current-limiting resistor(s) in series with the armature, then remove the resistor(s) as the motor speeds up.

The first kind of starting circuit uses one or more SCRs with a gate control circuit that gradually changes the triggering moment, thereby raising the average value of V_A. As shown in Fig. 12–35, the shunt-field winding is driven by a steady dc line voltage. If a rheostat is present in the field loop, it should be set to 0 Ω before applying power to the armature. This makes I_F as large as possible, providing maximum flux density and maximum torque per ampere of armature current.

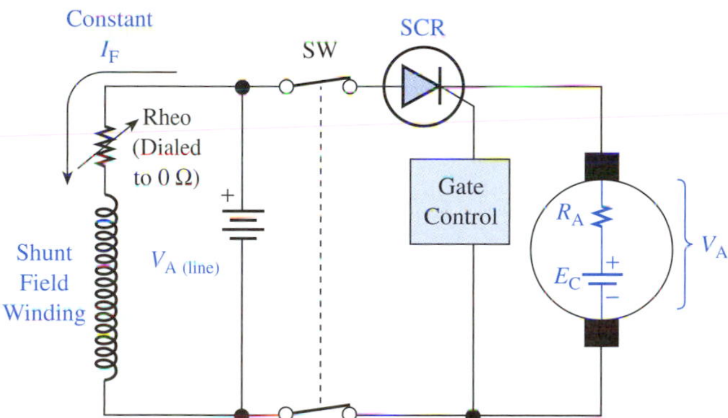

FIGURE 12–35
When SW first closes to power the armature, the gate control circuit triggers the SCR so that it delivers a small applied voltage V_A. Then it gradually raises V_A.

An example of a switched-resistor starting circuit is shown in Fig. 12–36. When the main disconnect switch is closed, $V_{A(line)}$ is applied to the shunt-field winding. Normally closed (N.C.) relay contact *RA*2-2 on line 2 is closed initially, so the rheostat is short-circuited. This guarantees the largest possible values of I_F and magnetic field strength B.

A thermal overload (OL) detector is placed in series with the armature on line 3. As long as it has not overheated, it will not have *tripped* the N.C. OL contact on line 5. Therefore line 5 is powered to the N.C. STOP pushbutton (PB) switch. If STOP is not mechanically actuated it remains closed. When the human operator actuates the normally open (N.O.) START PB switch, motor-starter coil *MS* energizes. It seals itself in the energized state by its light-duty contact *MS*-2 on line 6, which parallels N.O. PB START.

On line 3, when heavy-duty N.O. contact *MS*-1 goes closed, current surges into the armature as shown in Fig. 12–36(b), but it is limited by starting resistors R_1 and R_2, in addition to the low-value winding resistances R_A and R_{inter}. Therefore $I_{A(starting)}$ is limited to a safe value.

The rotor starts spinning and the armature begins to generate counter-EMF E_C, further limiting I_A. This is clearly shown by the graphs in Fig. 12–36(b). A few seconds later, time-delay contact *MS*-3 on line 7 goes closed. That energizes the *RA*1 coil, Relay Acceleration #1. Instantaneous contact *RA*1-1 closes on line 4, bypassing current-limit resistor R_1. Therefore I_A surges again, and the rotor puts on a burst of acceleration. Figure 12–36(b) shows this response.

A few seconds later, time-delay contact *RA*1-2 closes on line 8, energizing coil *RA*2. Heavy-duty contact *RA*2-1 on line 4 goes closed, removing R_2 from the armature current path. I_A increases again, and the rotor accelerates up to full running speed, as indicated in

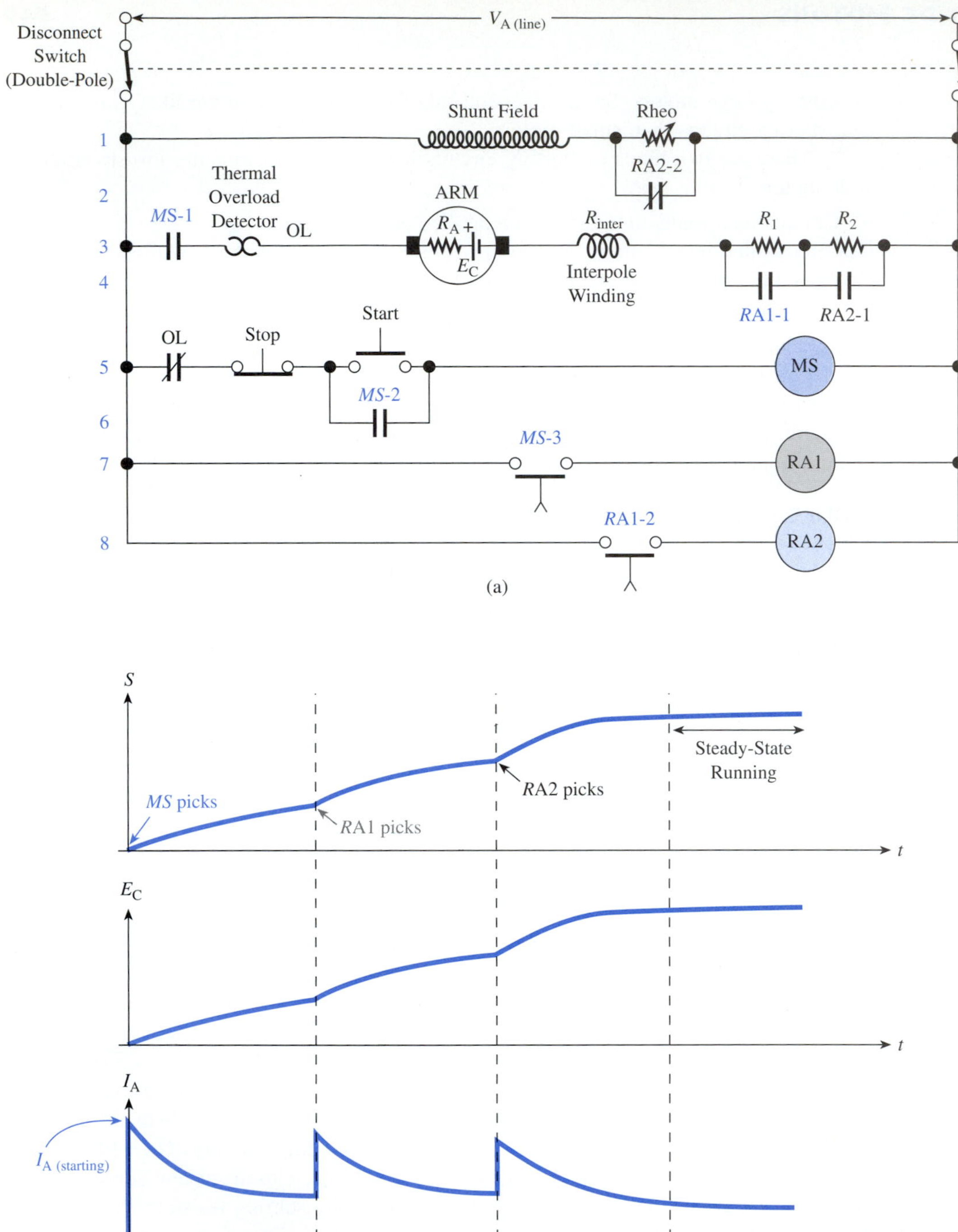

FIGURE 12–36
(a) Time-delay current-limit motor-starting circuit, with overload protection, built with electromagnetic relays. (b) Response of S, E_C, and I_A during acceleration period.

Fig. 12–36(b). If R_{rheo} had been previously adjusted to a nonzero value, the opening of N.C. contact *RA*2-2 places that resistance in series with the shunt-field winding. The setting of R_{rheo} determines the motor's running speed through its effect on flux density B. For a shunt-configured motor, weaker B (reduced field current I_F) results in faster shaft speed.

12-8-2 Stopping

There are four ways of stopping a motor:

1. Turning off the power and letting it coast.
2. Turning off the power and engaging mechanical friction brakes.
3. Dynamic braking.
4. Plugging.

Methods 1 and 2 are self-explanatory. Do keep in mind that when the power is turned off, however, both the armature and the shunt field must be turned off simultaneously, or the armature must be turned off first and the shunt field turned off last. You cannot reverse the order—field off first, then armature off last. It is improper to have the shunt field deenergized when the armature is powered, because with zero magnetic field flux the armature finds it impossible to generate any counter-EMF to limit its I_A, no matter how fast it rotates. Many dc motor control circuits have a *field-failure detector* which automatically disconnects the armature if shunt field current I_F stops flowing for any reason due to machine malfunction or human error.

Method number 2, using friction brakes, is occasionally done for safety purposes.

Dynamic braking. When the applied voltage V_A is removed from the armature brushes, counter-EMF E_C continues to exist there for as long as the motor continues rotating. By switching a resistive electric load onto the armature brushes, we cause the dynamo to go into generator mode temporarily. This is shown in Fig. 12–37.

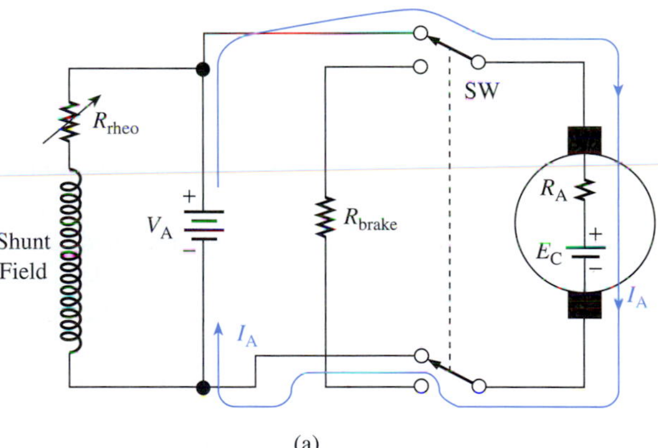

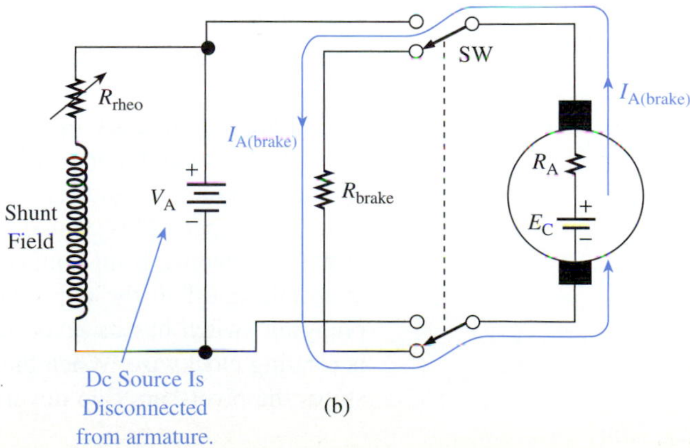

FIGURE 12–37 Dynamic braking. (a) When the armature is driven by V_A, armature current flows in the direction that produces forward torque. (b) When the armature's counter-EMF drives R_{brake}, armature current flows in the direction that produces reverse (backward) torque.

In Fig. 12–37(a) current I_A passes through the armature winding from top to bottom. This current direction produces forward torque (assume it's clockwise), so the dynamo motors in the clockwise direction.

In Fig. 12–37(b) the double-pole switch has been thrown to the down position. This disconnects the dc driving source V_A and connects the low-value resistance R_{brake} across the armature. Now the generated E_C takes over as the driving source in the armature loop. It circulates current I_{brake} through R_{brake}, then through the armature from bottom to top. Since the current in the armature conductors is now flowing in the opposite direction from part (a), it produces torque in the opposite direction, counterclockwise in this case. Thus the counterclockwise torque exerts braking action on the clockwise coasting tendency. The rotor is brought to a stop quickly, as shown in Fig. 12–38. When the rotor does stop rotating, E_C becomes zero, so braking action ends naturally.

FIGURE 12–38
Comparative motor-stopping behavior.

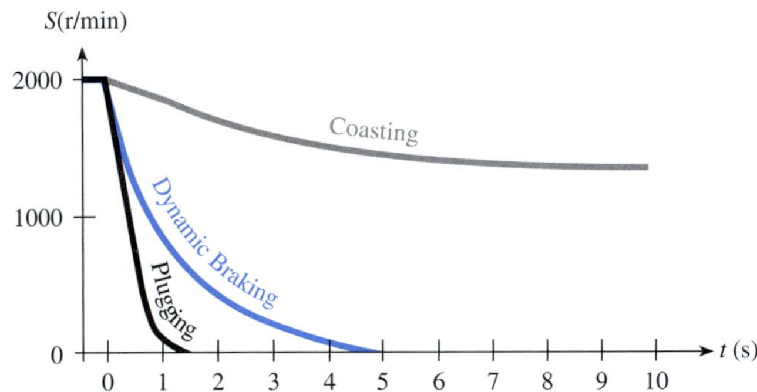

An improvement on plain dynamic braking is *regenerative braking*. In this stopping method the motor is switched into generator mode, but then the dynamo's rotational energy is actually returned to the dc supply instead of being dissipated into the ambient air by resistance R_{brake}. Electric cars use regenerative braking in combination with mechanical brakes. This practice enables them to conserve energy, reduce recharging costs, and extend their operating range in around-town driving.

12-8-3 Plugging

In dynamic braking, the retarding torque is great at first, but it gets weaker as the rotor slows. This is because generated E_C gets smaller. A more vigorous braking procedure is to apply dc source voltage V_A to the armature winding with reversed polarity. This is shown in the schematic diagram of Fig. 12–39.

Plugging actually increases the stopping torque (assume counterclockwise) as the rotor slows. It can be very impressive how rapidly a speeding motor is brought to a halt. At the moment that the rotor stops, a centrifugal switch that is attached to the motor's shaft must open. This *zero-speed switch,* also called a *plugging switch,* disconnects the armature winding from voltage source V_A. It is necessary in order to prevent the motor from taking off in the opposite direction after stopping. In Fig. 12–39 the double-pole plugging switch has its pivot-bars forced upward into the closed position when the shaft is rotating clockwise. When their centrifugal force disappears at the moment that rotation stops, the pivot-bars flop down into their normally open position.

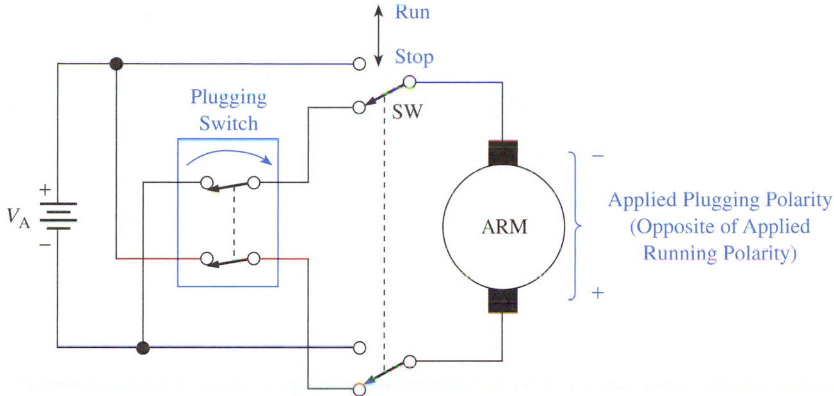

FIGURE 12–39 Plugging is the technique of applying the reverse V_A polarity to the armature winding. It requires the use of an automatic plugging switch to disconnect the armature at the moment the rotor stops.

12-8-4 Reversing

For a shunt-configured motor the direction of rotation can be reversed by reversing the polarity of the armature winding or the field winding, but not both. As a practical matter, it is better to reverse the armature. This helps avoid the problems associated with having the field momentarily deenergized while the armature is powered. Many control circuits will not permit that condition to occur anyway, as mentioned earlier.

Figure 12–40 shows a reversing motor-starting circuit for a compound-configured motor with interpoles. It is important to recognize that the current direction through the series-field winding must not be reversed. If reversed, its flux would oppose the main flux created by the shunt-field winding instead of aiding that main flux, as it is supposed to.

However, the interpole winding current must be reversed when the armature current reverses. Since changing the direction of current in the armature conductors causes the armature reaction flux to reverse its magnetic direction, the interpoles must likewise reverse in order to counteract and cancel the armature reaction flux. Refer back to Figs. 12–32 and 12–34 to understand this requirement.

In Fig. 12–40 the high-power 230-V circuit that actually drives the motor windings is shown in ladder-logic format in lines 1 through 5. The low-voltage (115-V) control circuit is in lines 6 through 11.

With the double-pole disconnect switch closed, current I_F passes through the coil of current-sensing relay *R*FF, then through the shunt-field winding, then bypasses the rheostat via N.C. contact *R*Rheo-1, which is initially in its normal state. The purpose of *R*FF (Relay Field Failure) is to verify at all times that current is actually flowing in the shunt-field circuit. The coil of *R*FF is wrapped with relatively few turns of heavy-gage wire. Its resistance is therefore quite low, so it doesn't affect the current I_F through the high-resistance shunt winding. As long as I_F continues to flow in the shunt field winding, *R*FF will be energized. This closes N.O. contact *R*FF-1 on line 6, which makes motor starting and running possible.

If the field winding current should stop for any reason, the *R*FF coil will deenergize. Contact *R*FF-1 will then return to its open condition, making it impossible to power the armature by either the FOR or the REV contactor.

Pressing the FORWARD Start PB on line 6 energizes the FOR contactor coil. At the same time its N.C. pole on line 8 guarantees deenergization of the REV coil. Such

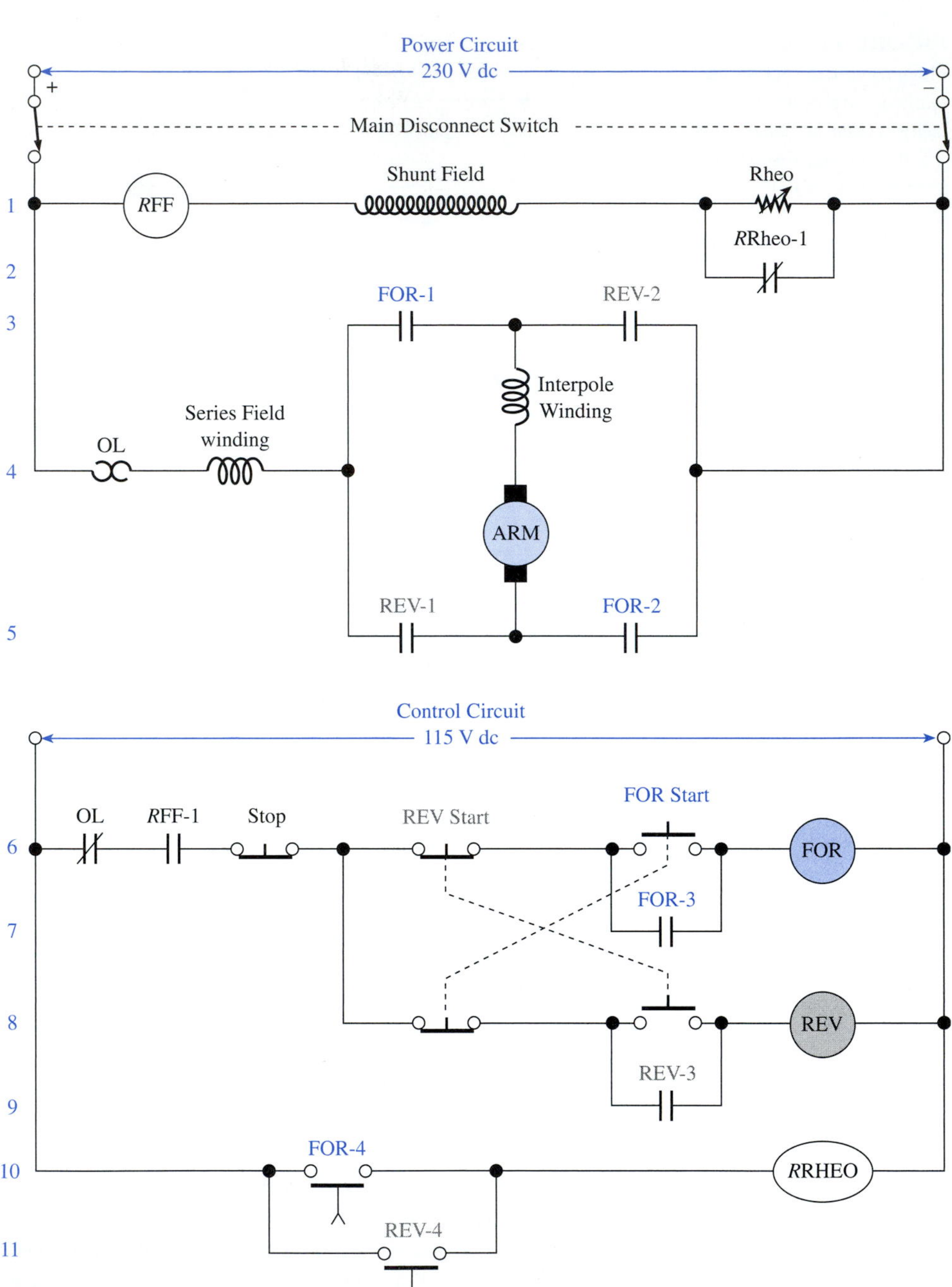

FIGURE 12–40
Reversing motor-control circuit.

interlocking is required in a reversing control circuit because we dare not energize both FOR and REV coils simultaneously.

With FOR energized, it seals itself in via N.O. contact FOR-3 on line 7. Its heavy-duty contacts FOR-1 on line 3 and FOR-2 on line 5 establish a current flow path as follows: Through OL, through the series field winding from left to right, through N.O. contact FOR-1 on line 3, through the interpole/armature winding series combination from top to bottom, and finally through N.O. contact FOR-2 on line 5. The motor starts and runs in the forward direction.

At the starting moment I_F is at its maximum possible value, because the shunt field rheostat is short-circuited. This is necessary in order to provide maximum field strength B contributing to maximum starting torque, $\tau_{starting}$. The motor therefore accelerates rapidly.

A few seconds after the FOR coil has energized, N.O. time-delay contact FOR-4 on line 10 goes closed. This picks up relay coil *R*Rheo, which opens N.O. contact *R*Rheo-1 on line 2. Now the I_F current is controllable by the rheostat, making it possible for the operator to manually adjust the running speed.

Trace the circuit carefully to verify that the REV contactor causes a reversal of current direction through the armature and interpole windings but no reversal of the shunt- and series-field windings.

TROUBLESHOOTING ON THE JOB

REMOVAL OF SULFUR FROM A COAL-FIRED ELECTRIC UTILITY STACK

Most coal contains a considerable amount of sulfur. When the coal is burned, the sulfur reacts with oxygen to form sulfur dioxide, a gas. If this gas is allowed to escape into the atmosphere, it reacts with water vapor to form sulfuric acid. Rain brings the acid back to the earth's surface, with well-publicized harmful effects on the environment.

Removal of sulfur dioxide gas is accomplished in a water-spray tower, also called a *scrubbing tower,* illustrated in Fig. 12–41. The 150-ft high, 50-ft-diameter tower holds liquid slurry about 20 ft deep at its bottom. Powerful pumps lift the water-based slurry to a collection of spray nozzles near the top of the tower.

Having already passed through an intelligent precipitator (see Troubleshooting on the Job in Chapter 6), the SO_2-contaminated exhaust gas is blown into the tower below the nozzles. As the hot gas rises, it must pass through the fine-droplet water spray. As it does so, the initial capturing chemical reaction occurs: The sulfur dioxide, SO_2, reacts with the water, H_2O, to produce liquid hydrogen sulfite, H_2SO_3. Over 90% of the total sulfur in the exhaust is captured by this reaction with water; less than 10% escapes up the chimney.

Second-stage limestone-grinding station. The grinder control motor is on the far left, only partly visible.

Courtesy of General Electric Environmental Services, Inc.

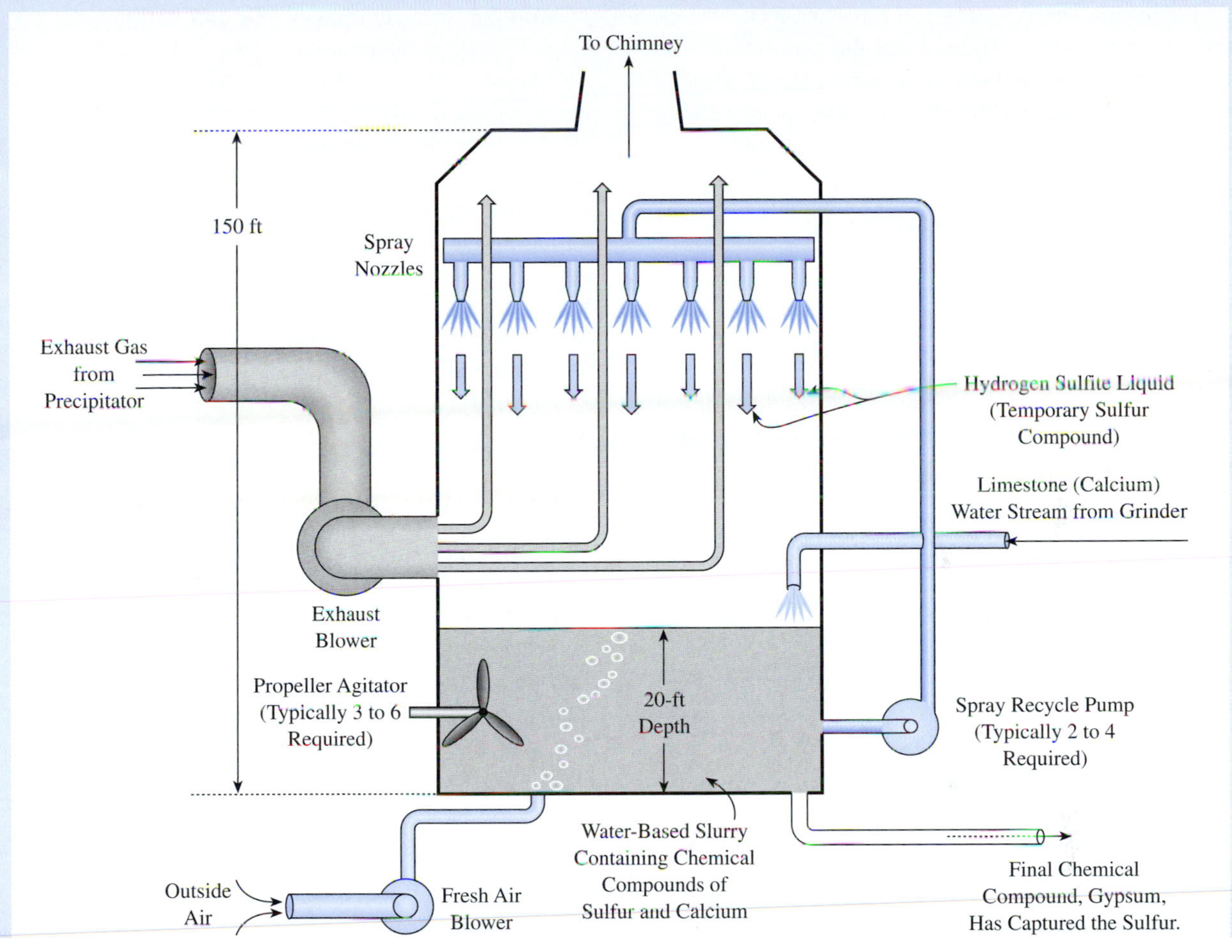

FIGURE 12–41
Sulfur-scrubbing tower.

Hydrogen sulfite is not a stable chemical compound, so the sulfur cannot be permitted to remain in that form, if falls into the liquid slurry at the bottom, where it is mixed with pulverized limestone carried in on a water stream, as Fig. 12–41 shows. Of course, the limestone had to be mined or quarried and hauled to the site by freight train, just like the coal.

Limestone has a large concentration of calcium (Ca) in the form of calcium carbonate, ($CaCO_3$). In the stirred-up slurry, the final capturing chemical reactions occur:

1. The limestone's $CaCO_3$ reacts with the H_2SO_3 to produce $CaSO_3$, calcium sulfite.
2. The $CaSO_3$ picks up one additional oxygen atom from the fresh air that is bubbling up through the slurry. This produces $CaSO_4$, calcium sulfate, also known as *gypsum*. It is a stable compound that forms solid crystals. (There is a viable market for gypsum in the manufacture of residential wallboard.)

The solid gypsum is filtered out of the slurry (pipe leaving Fig. 12–41 at lower right). After the filtering operation, the liquid is returned to the slurry. The return pipe is not shown in Fig. 12–41.

Electronic sensors measure the chemical concentrations in the slurry. To maintain the concentration in an acceptable range, the rate of limestone grinding is automatically controlled by increasing or decreasing the speed of the dc motor that runs the grinder.

YOUR TROUBLESHOOTING ASSIGNMENT

As the technician responsible for maintaining the spray tower, you keep an eye on the H_2SO_3 concentration in the slurry. Today that concentration is too high, with the automatic limestone-grinder control system unable to bring it back down to acceptable range. Upon investigation you find by tachometer measurement that the grinder-control motor is running at only 1100 r/min, even though the applied armature voltage is at its maximum value of 300 V.

1. What is your conclusion regarding the system feedback loop and the armature voltage controller?
2. What further can you say about the origin of the trouble?

Working quickly, it takes you and two helpers 1.5 hours to replace the dc drive motor with a backup unit from repair inventory. During this time the spray tower has remained nearly fully effective at its job of capturing the sulfur dioxide from the exhaust gas because that reaction requires only that there be adequate water from the spray nozzles, but the slurry has been deteriorating during this motor downtime, with the H_2SO_3 concentration rising higher and higher. At some point the H_2SO_3 concentration will become so great that there will be insufficient numbers of H_2O water molecules in the spray to fully react with the SO_2 gas molecules.

Fortunately, after your speedy replacement and reenergization of the motor, the grinder shaft tachometer rises to 3300 r/min, its maximum speed in this system. The grinder starts feeding limestone at the fastest possible rate into the water stream supplying the slurry, and the H_2SO_3 concentration starts to drop. The slurry is recovering.

Taking the faulty motor back to your repair area, you decide to determine the cause of failure. This can be useful information for possible revising of

Suppose that you make a visual inspection of the grinder's hopper and jaws, which reveals that the grinder's jaws are not being bound up by hard-rock contamination from the limestone supply (which sometimes does occur).

3. What is your suspicion at this point?

You make an immediate visual check of the brush-to-commutator electrical condition. It reveals that the carbon brushes are not worn past their useful life—there is still plenty of solid carbon remaining. Also, the springs in the brush rigging appear to be holding the brushes firmly against the copper segments, so the problem can't be solved by a simple brush replacement or spring replacement (see Fig. 12–4).

4. What is your reasonable next step?

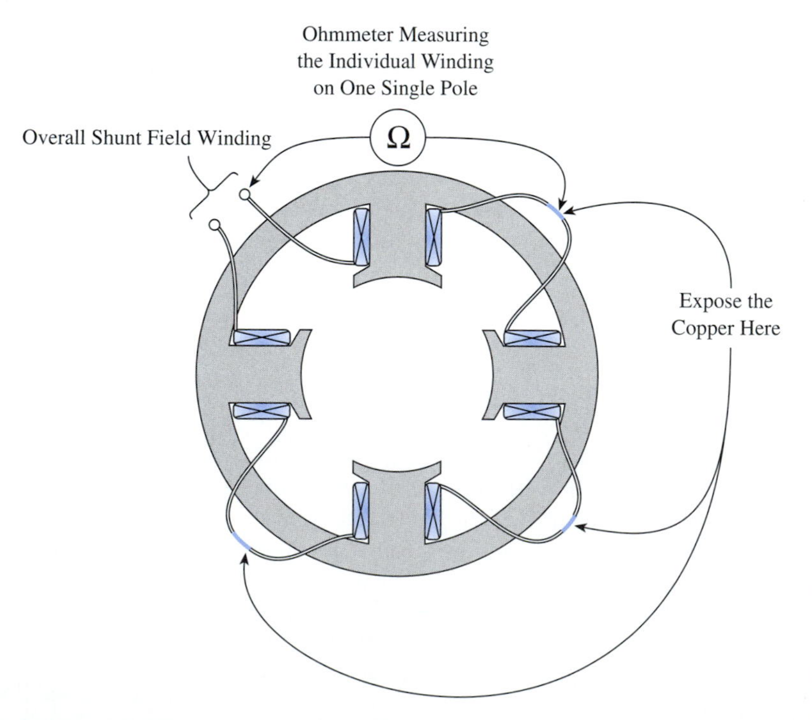

FIGURE 12-42
Measuring all four individual pole-winding resistances.

lubrication schedules or replacement schedules, or perhaps it may reveal that a different model of motor would deliver superior performance in this application.

5. What is the very first thing you would check about the motor?

Suppose that the rotor spins effortlessly, so you can't blame the bearings for the motor failure.

6. What ohmmeter tests can you perform on this wound-rotor dc motor to check its electrical integrity?
7. If you don't have the winding-resistance specifications, which is typical in real life, what other measurements can you make to get an indication whether the windings are intact? Refer to Fig. 12–42.

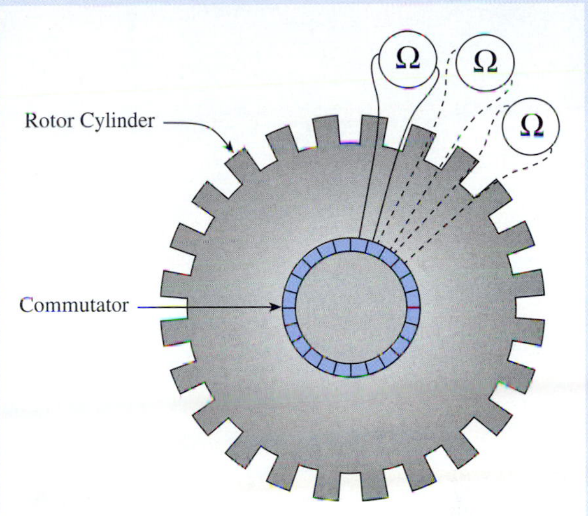

FIGURE 12–43
Using a low-range ohmmeter to measure the resistance between adjacent commutator segments.

If you discover that a winding has suffered damaged insulation, it may have occurred because the motor is improperly matched to the dc operating voltage (unlikely), or because the dc supply has very high-voltage noise spikes that have punched through the insulation. If so, these spikes must be eliminated. Or, the motor may be improperly matched to the ambient air temperature, or to airborne contaminants in the vicinity.

Suppose that your tests reveal that all four pole-windings measure 42 Ω (within 1 Ω). You conclude that your field winding is OK.

8. What further ohmmeter test can you run? Refer to Fig. 12–43.

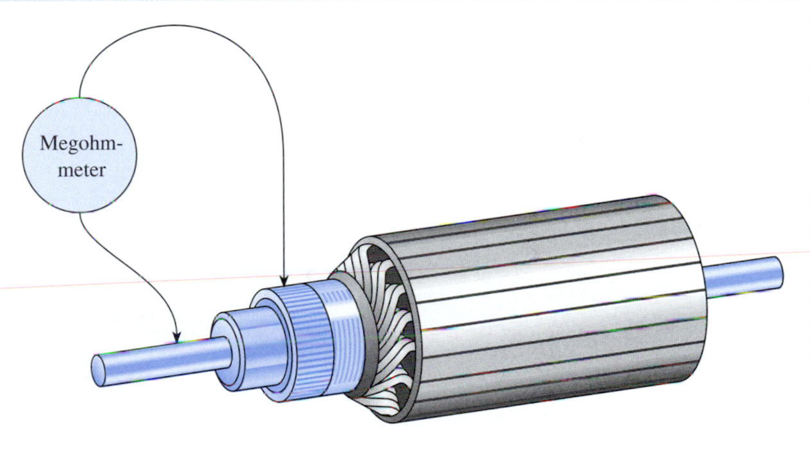

FIGURE 12–44
A megohmmeter applies a high voltage, usually greater than 1000 volts, between the winding and the steel rotor structure that the winding is mounted on. A megohmmeter is often called a *megger,* though that word is a registered trademark of the AVO International Corp.

Suppose that you discover a group of three successive measurements that are much lower than the other twenty-one measurements that you get in Fig. 12–43. Therefore you conclude that the motor's armature winding is damaged. This may have been caused by a spike-ridden dc supply, or it may be due to frequent overheating because the motor spends too much time running in an overloaded condition. Or the motor may be improperly matched to the Starting/Stopping frequency (which is unlikely in this grinding application, where the motor probably seldom stops).

It might be that this failure is part of a pattern of similar failures at your plant location and/or at other stack-scrubbing installations around the country. If enough evidence of motor failures is gathered, it might be the impetus for someone to mount a full-scale investigation into the noise-quality of the systems' dc supplies.

9. With the preceding paragraph in mind, what should you do to finalize your investigation?

Limestone powder separator. Particles that are ground small enough are sent into the limestone feed pipe in Fig. 12–41. Particles too large for quick dissolving in the slurry are sent back for another pass through the grinder.

Courtesy of General Electric Environmental Services, Inc.

If the armature winding had tested OK as you moved the ohmmeter around the commutator, your final test would be to apply a megohmmeter to measure the overall insulation resistance between the armature winding and the rotor core. This is illustrated in Fig. 12–44.

A megohmmeter applies a high voltage, so it indicates the insulation's performance under realistic high voltage conditions, including possible voltage spikes. If the insulation is intact, the megohmmeter will measure many megohms of resistance (ideal insulation would have infinite resistance). If the insulation breaks down or "leaks" under high-voltage conditions, the megohmmeter will measure an improperly low value of resistance between an electrical point on the winding and the steel. Such insulation leakage has the potential to interfere with proper current circulation in the armature winding. It is not necessarily cause for immediate motor replacement, but at least it should be taken as an alert to possible future trouble.

10. What use should you make of a slightly lower-than-expected megohmmeter measurement?

SUMMARY

- A wound-rotor dc machine has a complex multiturn armature winding on its rotor, as opposed to a simple permanent magnet.
- Any wound-rotor dc machine is generically called a dynamo. It can function in either of two ways: (1) as a generator (mechanical to electrical energy conversion); or (2) as a motor (electrical to mechanical energy conversion).
- *Stator* and *rotor* are mechanical structural terms. *Field* and *armature* are electrical/magnetic terms. To be correct, they should not be used interchangeably.
- The principles of generator action are explained by Faraday's law and Lenz's law. Faraday's law deals with voltage magnitude and Lenz's law deals with voltage polarity.
- In a wound-rotor dc generator, the induced individual winding voltage is sine-wave ac, but the output voltage is rectified into pulsating dc by the mechanical commutator.
- In a wound-rotor dc motor armature, an individual winding's current is ac but its resulting torque is unidirectional (like dc) due to the action of the mechanical commutator.
- The principle of motor action can be understood as magnetic pole attraction/repulsion, or it can be understood more thoroughly by applying the Lorentz relationship.
- Counter-EMF, symbolized E_C, is the voltage that is generated by a motor's armature winding in opposition to the applied source voltage. Focusing on E_C is the best way to understand most motor performance.
- Counter-EMF E_C is proportional to magnetic flux density B and to rotational speed S. As an equation, $E_C = k_{EC} B\,S$.
- The magnetic flux density B produced by an electromagnet is not truly proportional to its current I at large values of I, due to magnetic core saturation.
- The torque τ that a motor produces is proportional to magnetic flux density B and to armature current I_A. As an equation, $\tau = k_\tau B I_A$.
- Mechanical power P_{mech} is given by the product of torque and rotational speed. As an equation, $P_{mech} = \tau \times S$ (assuming basic units).
- Shunt-configured dc motors have better speed regulation than series-configured dc motors, but they cannot develop as much starting torque. The compound configuration, containing both a shunt and a series winding, has compromise characteristics.
- Interpoles are used to cancel the effect of armature reaction, which degrades motor performance.
- Starting a large motor may require special circuit arrangements to limit the initial in-rush current.
- A motor can be quickly stopped by dynamic braking; plugging provides even more vigorous stopping action.
- Reversing the rotation of a dc motor is usually accomplished by reversing the applied armature voltage polarity.

FORMULAS

$$V = N\frac{\Delta\Phi}{\Delta t}$$ **(Equation 12-1)**

$$F = l \times I \times B$$ **(Equation 12-2)**

$$I_A = \frac{V_A - E_C}{R_A}$$ **(Equation 12-6)**

$$V_A = I_A R_A + E_C \qquad \text{(Equation 12-7)}$$

$$E_C = k_{EC} B S \qquad \text{(Equation 12-8a)}$$

$$E_C = k_{EC(2)}\, \Phi\, S \qquad \text{(Equation 12-8b)}$$

$$E_C = k_{EC(3)} \times I_F S \quad \text{(assumed true)} \qquad \text{(Equation 12-11)}$$

$$\tau = k_{\tau} B I_A \qquad \text{(Equation 12-12a)}$$

$$\tau = (9.551 \times k_{EC(3)})\, I_F I_A \qquad \text{(Equation 12-12b)}$$

$$P_{mech} = \tau \times S \qquad \text{(Equation 12-13)}$$

$$745.7 \text{ W} = 1 \text{ hp}$$

$$\underset{\text{hp}}{P_{mech}} = \frac{\underset{\text{lb-ft}}{\tau} \times \underset{\text{r/min}}{S}}{5252} \qquad \text{(Equation 12-14)}$$

$$I_{A(starting)} = \frac{V_A}{R_A} \qquad \text{(Equation 12-15)}$$

$$S_{reg}(\%) = \frac{S_{NL} - S_{FL}}{S_{FL}} \times 100\% \qquad \text{(Equation 12-16)}$$

QUESTIONS AND PROBLEMS

Section 12-2

1. True or False: A given wound-rotor dc dynamo can operate as a motor, or it can operate as a generator. The choice is up to the user.
2. A motor receives ___________ input energy; it converts that to ___________ output energy.
3. A generator receives ___________ input energy; it converts that to ___________ output energy.
4. Describe what we mean by the term *stator* on a dynamo.
5. Describe what we mean by the term *rotor* on a dynamo.
6. True or False: It is proper to regard the word *armature* and the word *rotor* as synonyms; the words can simply be interchanged.
7. True or False: In a wound-rotor dc motor, the number of poles is always even (not odd).
8. A magnetic pole-face that flux lines are emerging from is called a ___________ pole.
9. True or False: In a wound-rotor dc motor, the rotor cylinder is made of solid steel alloy.
10. Give a typical air-gap distance for a modern dc motor.
11. True or False: In a dc generator the voltage produced by a single coil is at its maximum instantaneous value when the through-the-coil flux is at its maximum instantaneous value.
12. Faraday's law tells us about the magnitude of generated voltage; Lenz'a law tells us about the ___________ of generated voltage.
13. True or False: Slip-rings must be insulated from the shaft that they are mounted on.
14. The carbon blocks that contact slip-rings are called ___________.
15. When a slip-ring is split into two or more segments for dc functioning, we call it a ___________.

16. In the English measurement system, what are the measurement units of torque? In SI, what are the measurement units of torque?
17. The Lorentz relationship relates three variables; they are magnetic flux density B, current I, and ____________.

Section 12-3

18. A shunt-configured motor has a field winding with ____________ turns, made of ____________ wire.
19. True or False: In a shunt motor, the field winding is connected in parallel with the armature winding.
20. When a shunt motor is working hard (large torque), its armature current I_A is much ____________ than its field current I_F.
21. A certain shunt motor has a supply voltage $V_A = 120$ V. Its field winding resistance R_F is 50 Ω, with a 100-Ω rheostat in series.
    a. Find the maximum field current, $I_{F(max)}$.
    b. Find the minimum field current, $I_{F(min)}$.
22. When a shunt motor is subjected to changes in torque demand, the ____________ current remains constant, but the ____________ current automatically adjusts itself.
23. A certain shunt-configured dc motor has $R_A = 1.7$ Ω, with $V_A = 240$ V.
    a. If armature current $I_A = 12$ A, find counter-EMF E_C.
    b. If $E_C = 207$ V, find I_A.
24. Generally speaking, counter-EMF E_C is what approximate percentage of applied armature voltage V_A?
25. A motor's counter-EMF is proportional to magnetic field strength B and proportional to ____________.
26. For a certain shunt motor, we wish to find the $k_{EC(3)}$ proportionality factor between E_C, I_F, and S. We know the motor's armature resistance $R_A = 2.0$ Ω. With the motor running under part-load, we measure $V_A = 125$ V, $I_A = 6.0$ A, $I_F = 1.2$ A, and $S = 2800$ r/min. Find $k_{EC(3)}$ and write the E_C equation.
27. For the motor of Problem 26, suppose we measure $I_F = 1.3$ A and $S = 2400$ r/min.
    a. Find the motor's counter-EMF under this condition.
    b. Solve for the motor's armature current I_A, using Kirchhoff's voltage law. Assume $V_A = 125$ V.
28. Flux density B is proportional to field current I_A for values of I_F ____________ I_{knee}.
29. A shunt motor is operating at a particular torque. If the load increases its torque demand, tell whether each of these motor variables will increase, decrease, or remain the same.
    a. Speed
    b. Counter-EMF
    c. Field current
    d. Armature current
    e. Total current
30. A certain shunt motor has a k_τ proportionality factor of 0.76. Thus, $\tau = 0.76BI_A$. When $I_A = 7.2$ A, $\tau = 5.2$ N-m. If I_A increases to 9.1 A, find the new value of τ. Assume B remains constant.
31. Suppose you perform the $k_{EC(3)}$ testing procedure on the motor of Question 30. Your results are: $R_A = 2.1$ Ω; $V_A = 250$ V; $I_F = 0.9$ A; $I_A = 6.8$ A; $S =$ 3290 r/min.
    a. Calculate counter-EMF E_C.

b. Find the $k_{EC(3)}$ (proportionality factor for E_C versus field current I_F and speed S). Follow the technique of Example 12-5.
c. Find the motor's torque proportionality factor for I_F and S, and write the torque equation. Follow the technique of Example 12-8.
d. What amount of torque is the motor delivering to the load under these operating conditions?

32. In Problem 31, find the mechanical shaft power as follows:
    a. Convert the speed to basic units of rad/s.
    b. Calculate the mechanical shaft power, in basic units of kW. Use Eq. (12-13).
    c. Convert the shaft power to English units of hp; use the conversion factor 1 hp = 745.7 W.
33. In Problem 32,
    a. Convert the shaft torque to English units of lb-ft; use the conversion factor 1 lb-ft = 1.36 N-m.
    b. Calculate the shaft power in hp units by applying Eq. (12-14). Check against your result from Problem 32, part **c.**

Section 12-4

34. A certain shunt motor has $S_{NL} = 2850$ r/min and $S_{FL} = 2570$ r/min for $\tau_{FL} = 10$ N-m.
    a. Find the motor's speed regulation percentage, S_{reg}.
    b. Assuming a straight-line characteristic graph, find the speed for $\tau = 5$ N-m.
    c. Repeat for $\tau = 7$ N-m.

Section 12-5

35. Draw the schematic diagram of a series-configured motor. Mark it with $V_A = 230$ V, $R_{ser} = 1.5\ \Omega$, $R_A = 2.1\ \Omega$.
36. For the motor of Problem 34, calculate the starting current, $I_{starting}$, with an across-the-line starter.
37. Explain why the series motor of Problem 34 delivers higher starting torque, $\tau_{starting}$, than a comparable shunt-configured motor.
38. Suppose the series motor of Problem 34 has $\tau_{FL} = 12$ N-m and $S_{FL} = 1800$ r/min.
    a. Can the motor operate successfully with its mechanical load disconnected from its shaft? Explain.
    b. Suppose a load is connected to the shaft, demanding $\tau = 2$ N-m. What is the best guess for the shaft speed?

    2000 r/min   2500 r/min   4500 r/min   13000 r/min

Section 12-6

39. Draw the schematic diagram of a compound motor that is designed to operate at $V_A = 240$ V. Estimate reasonable values for R_F, R_{ser}, and R_A.

Section 12-7

40. Severe sparking between a motor's commutator and its carbon brushes is caused by the ____________ ____________ problem.
41. Motor designers can deal with this problem by including ____________ in the motor's design.

Section 12-8

42. In Fig. 12–36(a), suppose $R_A = 0.5\ \Omega$, and $R_{inter} = 0.2\ \Omega$. The limiting resistors are $R_1 = R_2 = 2\ \Omega$. The thermal overload has $0.1\ \Omega$ of resistance. If $V_{A(line)} = 300$ V, calculate $I_{A(starting)}$.
43. At the moment when *R*A1 picks, shorting out starting resistor R_1, is it possible to calculate the value of the I_A surge? Explain.
44. Suppose the motor is in the process of starting (accelerating), with *R*A1 picked but *R*A2 not yet picked, when the STOP PB switch is momentarily actuated, then released, describe precisely how the relay circuit of Fig. 12–26(a) will react.
45. When stopping a shunt dc motor, you must either deenergize both windings simultaneoulsy or else deenergize the ____________ winding first.
46. The motor-stopping method of connecting a resistive load across the armature winding is called ____________ ____________.
47. The motor-stopping method of connecting the V_A source across the armature winding in reverse is called ____________.
48. In the reversing motor-starting circuit of Fig. 12–40, explain why the series-field winding is placed outside the polarity-reversing section but the interpole-winding is placed inside that section.
49. True of False: To reverse the rotational direction of a dc motor, we reverse the current direction in both the armature and field windings.
50. Explain the purpose of the *R*FF current-sensing relay in Fig. 12–40.
51. Explain the purpose of the bottom rung (lines 10 and 11) in the Fig. 12–40 ladder-logic diagram.

CHAPTER

13

NONTRADITIONAL DC MOTORS

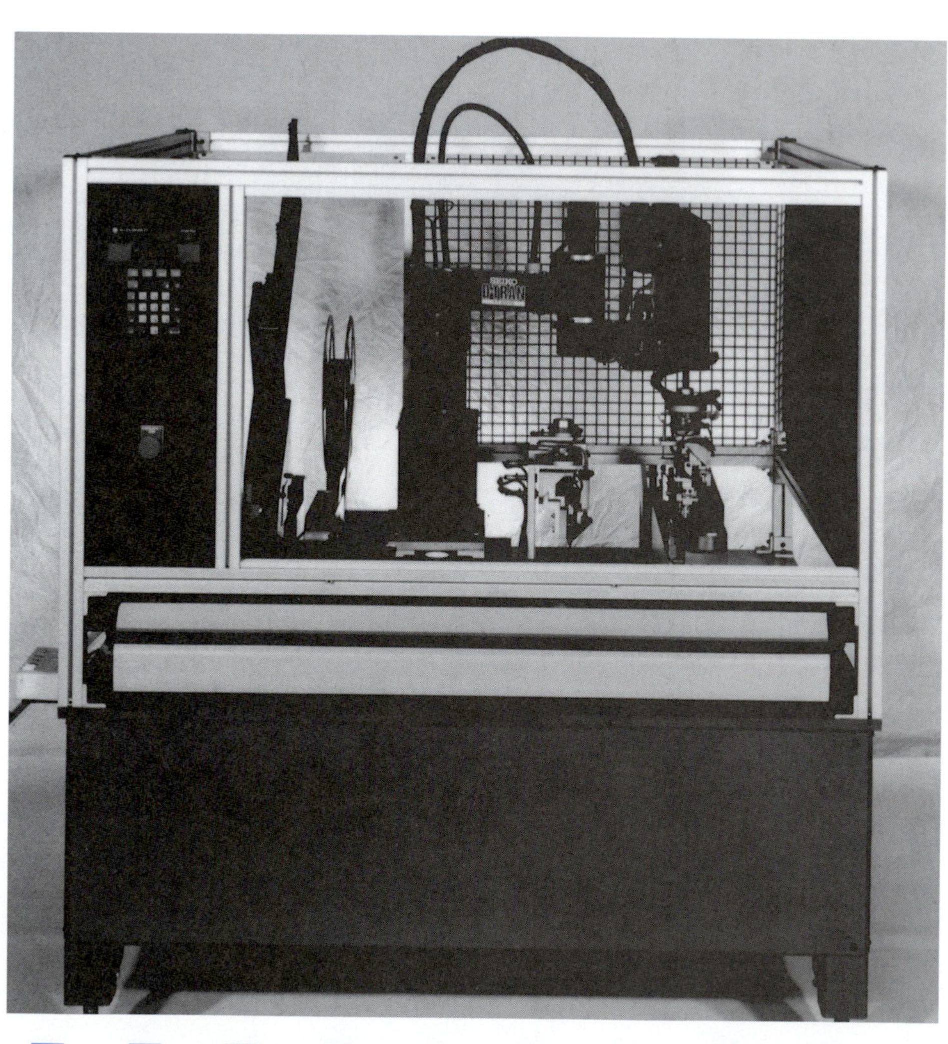

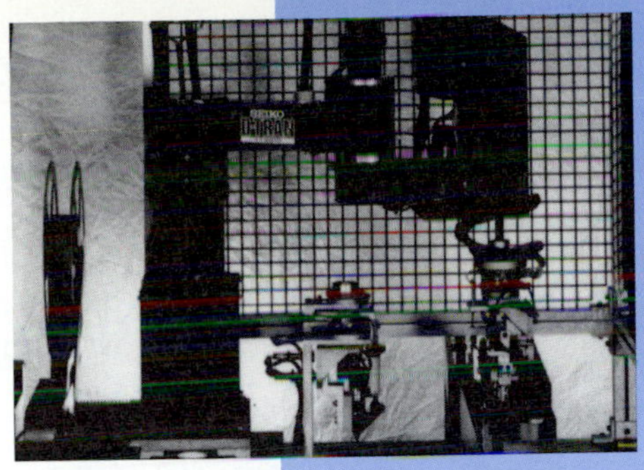

Traditional wound-rotor, wound-field dc motors are most useful in moderate- to high-power applications where the speed must be adjustable (shunt and compound configurations) or where the application calls for frequent starting and stopping with heavy torque loads (series configuration). For low- to moderate-power applications with special performance requirements, other types of dc motors often have advantages over traditional wound-rotor machines.

OBJECTIVES

After completing this chapter, you will be able to:

1. Draw the cross-sections of conventional permanent-magnet motors with:
   a. Radial flux orientation in the magnets.
   b. Concentric circular flux orientation in the magnets.
   c. Linear flux orientation in the magnets.
2. State the fundamental advantage of coreless dc motors.
3. Apply the Lorentz relation to the cup-structure coreless motor to explain its torque development; do the same for the disk-structure coreless motor.
4. Describe the cogging phenomenon of iron-core motors, and explain why coreless dc motors do not exhibit cogging behavior.
5. Distinguish between brush-equipped dc motors and electronically commutated dc motors.
6. Draw the cross section of a permanent-magnet stepper motor. Referring to its transistor switching sequence of the stator poles, explain its principle of operation.
7. Referring to a stepper motor cross section and a transistor switching sequence, explain how a stepper motor can be made to take either full steps or half steps.
8. Explain the use of kickback-suppression diodes with stepper motors.
9. Interpret the characteristic stepping rate-versus-torque curves for a stepper motor.
10. Interpret the ramping-up and ramping-down graphs for a stepper motor.
11. Define the terms *static holding torque* and *detent torque* for a stepper motor.
12. Distinguish between the one-phase and two-phase operating modes for a stepper motor, and state the advantage of each mode.
13. Interpret the truth table of an IC controller for a stepper motor.
14. Define the term *microstepping,* and explain how it is accomplished.
15. Describe overshoot in a stepper motor. Explain how severe overshoot can destroy the motor's step integrity.

16. Show how electrical damping can be used with a stepper motor to minimize overshoot.
17. Draw the cross section of a position-triggered, brushless dc motor, and explain its principle of operation.
18. Draw the schematic diagram of a thyrector used for bidirectional kickback suppression in a brushless dc motor.
19. Discuss the relative advantages, disadvantages, and operating characteristics of the following dc motors: traditional wound-rotor/wound-field; conventional permanent-magnet; coreless; permanent-magnet stepper; variable-reluctance stepper; and position-triggered brushless.

13-1 ■ CONVENTIONAL PERMANENT-MAGNET MOTORS

A conventional permanent-magnet dc motor is conceptually the same as a shunt-configured wound-rotor dc motor. The rotor assembly is identical, with a laminated iron core, armature windings placed in rotor slots, commutator, brush rigging, and everything else the same. The only difference is that the magnetic field is established by permanent magnets rather than electromagnets.

In its simplest form, the motor contains permanent magnets that are magnetized radially, as shown in Fig. 13–1. Magnetic flux emerges from the face of the North pole on the left in Fig. 13–1(a). The flux passes through the air gap, through the conventional rotor core, and reenters the face of the South pole on the right. The outer surface of the right-side magnet is North; the outer surface of the left-side magnet is South. The magnetic flux path is completed through the motor's steel frame, the same as in a wound-field machine.

Four-pole and six-pole designs are also available. When the number of field poles is increased, the armature winding must be redesigned so that individual armature coils

FIGURE 13–1
Cross sections of conventional permanent-magnet motors with radially magnetized poles. (a) Two poles. (b) Four poles.

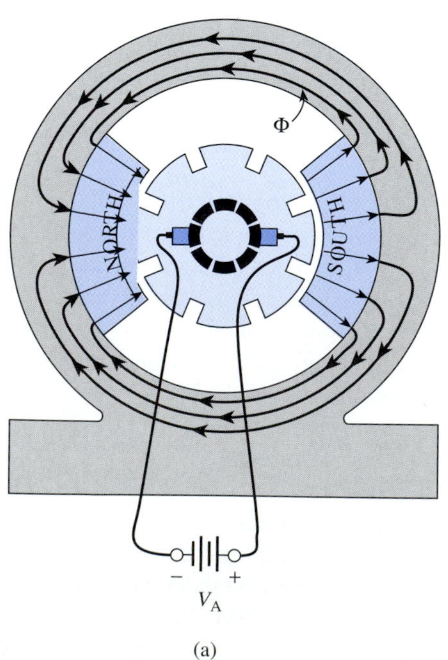

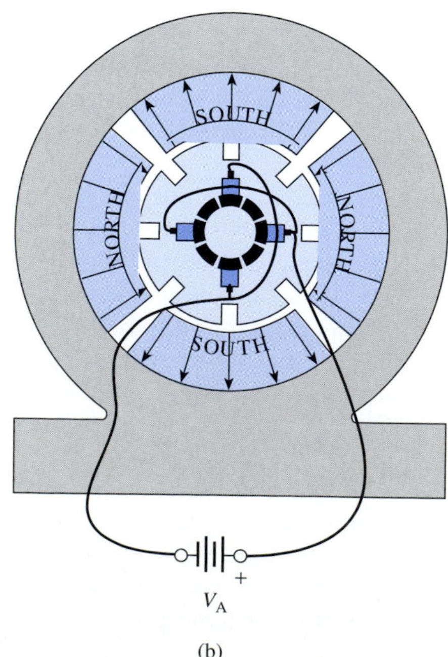

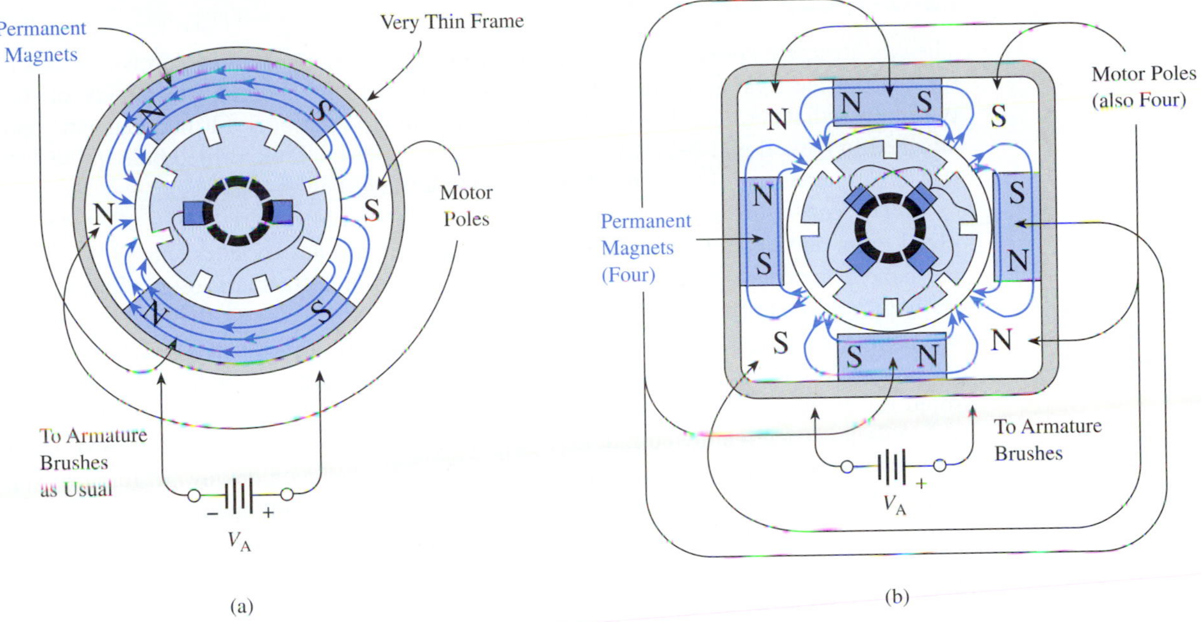

FIGURE 13–2
Nonradial magnetization of the permanent magnets. (a) Magnetizing the two permanent magnets in a concentric (circular) direction. (b) Magnetizing the four magnets in a straight direction, across the width.

have the same mechanical span as adjacent poles. For instance, the four-pole field structure shown in Fig. 13–1(b), with 90° between adjacent poles, would require that the armature coils have their sides placed in rotor slots that also are 90° apart (approximately). Usually there are four commutator brushes as well. These facts regarding multipole motor armatures are just as true for wound-field dc machines as for permanent-magnet machines.

For some exotic permanent-magnet materials, it is easier to establish the magnetic field in the circular direction rather than in the radial direction. In that case, shown in Fig. 13–2(a), the poles are made of highly permeable laminated steel, and they serve to complete the magnetic flux path. The motor's frame is not involved in any magnetic flux path, so it can be very thin. This enables designers to reduce the size and weight of such motors.

The same concept is illustrated in Fig. 13–2(b) for a four-pole motor with rectangular-cross-section permanent magnets.

13–2 ■ CORELESS PERMANENT-MAGNET MOTORS

To reduce rotor inertia and allow very quick acceleration and stopping, designers have developed two clever motor structures that contain no iron (steel) in the rotor. Such motors are called *coreless* motors, also known as *ironless-rotor* motors. (These motors are sometimes identified as *moving-coil* motors, but that name is misleading and should not be used.)

13-2-1 Cup Structure

The *cup structure* coreless motor is pictured in Fig. 13–3. The hollow cup is attached to the shaft at the right side only, just before the commutator. The cup's thin body fits between the radial permanent magnets on the frame and the stationary drilled magnetic core attached to the left end-bell. The shaft's left end slips through the drilled hole in the core and is supported by a bushing or bearing in the left end-bell.

The cup is made of fiberglass or similar lightweight material, with copper wires glued to its outside and inside surfaces. A single-turn coil has its sides spaced 90° apart, the same distance as adjacent permanent-magnet poles. In a standard solid-cylindrical rotor with a normal armature, the coils have crossover wire segments that pass across the surface of the cylinder-end, but in the motor design of Fig. 13–3 there is no cylinder-end to work with. Therefore the far left section of the cylinder surface is dedicated to providing room for the wires to make their 90° offset. Thus the wire closest to us in the Fig. 13–3 view might carry armature current from right to left on the outside of the cup. The Lorentz relationship between left-flowing I and North flux Φ predicts a downward force F. This tends to rotate the cup assembly clockwise, viewed from the left side.

That particular current-carrying conductor turns up toward the top of the cup. At the very top it crosses the cup's wall into the interior. Its current then continues down the inside of the cup in the left-to-right direction (not visible in Fig. 13–3). The top is a region of South flux, so the Lorentz relation calls for a force out of the page toward us. This force tends to rotate the cup assembly clockwise also, so we have a workable motor design.

Because the rotor assembly contains no iron, the rotor has no *detent* tendency—the tendency to prefer to remain in particular rotational locations relative to the field structure. Normal wound-rotor motors have this tendency, which produces the *cogging* effect when they are turning slowly. In cogging, the rotor slips eagerly into one rotational position, then tries to stall there for a moment until its torque makes it advance to its next preferred rotational position.

The absence of iron in the rotor eliminates eddy-current I^2R losses and also magnetic hysteresis power losses. With these losses eliminated in the rotor, coreless motors have high efficiencies.

The cup structure also goes by the names *can structure* and *shell structure.*

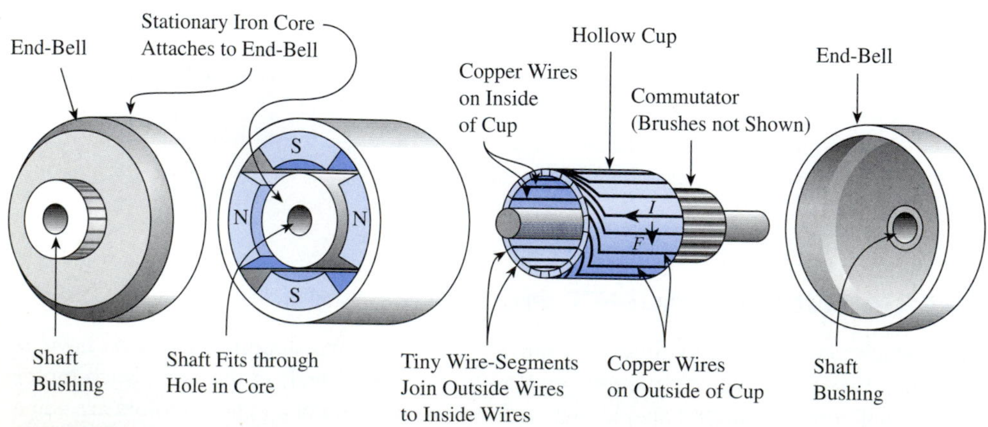

FIGURE 13–3
Exploded view of cup-type coreless four-pole permanent-magnet motor.

13-2-2 Disk Structure

The *disk structure* is quite unlike any other motor structure in that the internal magnetic field flux is not radial. Instead the flux points parallel to the machine's shaft (called the *axial* direction). Flux lines pass from a North pole-face on one side of the thin disk to a South pole-face a short distance away on the other side of the thin disk. For example, at the top in Fig. 13–4 a permanent magnet attached to the right end-plate emits North flux; that flux passes through the thin disk, then enters a South permanent magnet attached to the left end-plate.

Offset by 60°, another pair of North-South permanent magnets sends flux through the disk in the opposite direction, from left to right, and this is repeated in 60° increments all the way around the machine. The magnetic flux paths are completed through the motor's steel frame, not shown in Fig. 13–4.

The light-weight disk has bonded copper tracks, just like an etched printed-circuit board. The tracks are placed to form single-turn armature wraps. The active torque-producing portions of a wrap are the copper track segments in the middle area of the disk, within the space occupied by the magnetic fields. This is pointed out in Fig. 13–4. The two opposite-side middle-area copper tracks are offset from one another by 60°, the same as the permanent magnets. This offset is accomplished by the track segments in the outer area and the inner area of the disk. These arc-shaped track segments lead to a disk-crossing location that is 30° away from the active-conductor segment. Their corresponding arc-shaped segments on the opposite side (invisible in Fig. 13–4) provide an additional 30° offset, so the total offset is 60°.

For example, look at the active copper-track segment that is visible at the top (middle area) of Fig. 13–4. Suppose that the current in that track segment flows in the radially *outward* direction (from bottom to top of the drawing). That wrap's active copper

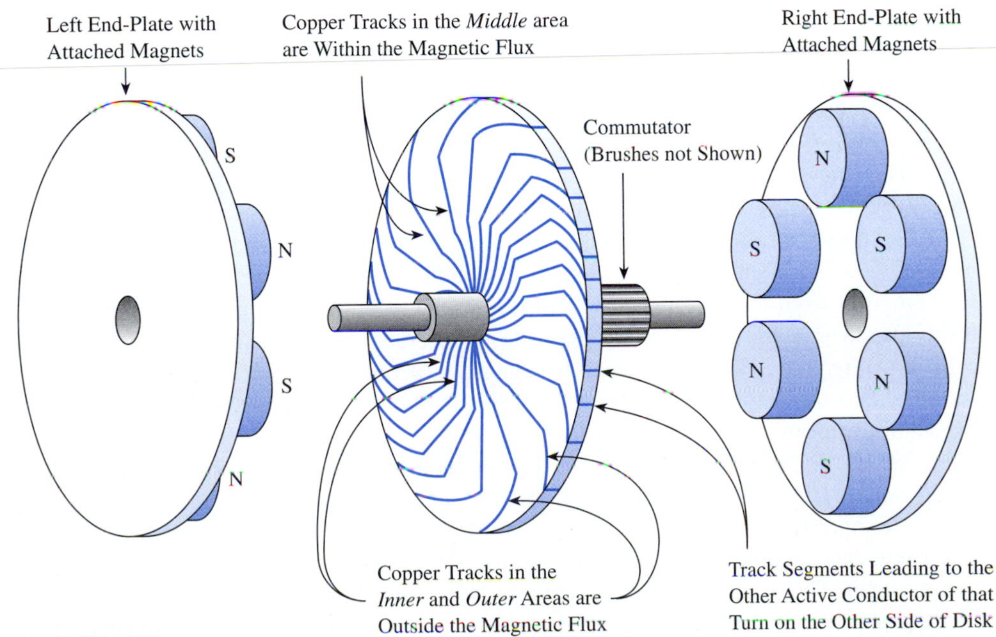

FIGURE 13–4
Exploded view of disk-type coreless six-pole permanent-magnet motor.

track segment on the rear (invisible) side of the disk must flow radially *inward;* of course, the inward current conductor is in the left-to-right flux created by the permanent-magnet pair that is toward us by 60°.

Apply the Lorentz relation to the topmost (front-side) active track segment, carrying current upward within a right-to-left magnetic field. You will conclude that its torque is clockwise, viewing the disk from the front side.

Then apply the Lorentz relation to the turn's rear-side active conductor segment, carrying current inward toward the shaft. That current direction, combined with the left-to-right flux direction, produces a force that also contributes clockwise torque to the rotor disk. Thus the opposite sides of the turn are aiding each other in their torque creation, so we have a workable motor design.

The disk-rotor structure of Fig. 13–4 has even less rotational inertia than the cup structure of Fig. 13–3. Because of its low mass and inertia, the disk-type coreless structure yields the fastest acceleration of any motor design. Some models are capable of astonishing acceleration rates, going from 0 to 5000 r/min in less than 100 ms.

Naturally, coreless motors cannot compete with wound-rotor motors in the high-torque ranges. Their permanent magnets cannot produce magnetic field strengths comparable to multiturn electromagnets, and their single-turn wraps give up the force-multiplying advantage of multiturn armature coils.

13–3 ■ STEPPER MOTORS

Stepper motors and brushless dc motors together make up the class of motors called *electronically commutated motors.* They are conceptually similar to each other. The stepper motor is better suited to moving the shaft an exact rotational amount; the brushless dc motor is better suited to continuous adjustable-speed rotation. We will cover stepper motors now and deal with brushless dc motors in the next section.

Stepper motors are fundamentally different from other dc motors—they have no brushes or mechanical commutator. Instead the switching (commutating) action necessary for the dc motor function is accomplished by external transistors. Furthermore, the rotor has no armature winding. It is simply a collection of salient permanent magnets, as Fig. 13–5 shows.

The four stator-pole windings and their controlling transistors are labeled *A*, *B*, *C*, and *D*. When the control circuit turns ON a particular transistor, current flows from the $+V_S$ dc supply, through that particular winding, through the transistor to ground. When a single winding is energized, it is wrapped so that its pole becomes magnetic North. Its flux emerges from the pole-face, passes through the rotor, then completes its path by entering the face of the pole directly opposite to it. For example, if transistor *A* energizes pole A in Fig. 13–5, the flux created by that pole completes its path through pole C and through the frame of the motor. Thus C automatically becomes a South pole even though its winding carries no current.

In Fig. 13–5 the permanent rotor poles are labeled 1 through 6, with poles 1, 3, and 5 being South. The alternate poles, numbers 2, 4, and 6, are North.

The operating principle of a permanent-magnet stepper motor is this: The energized stator pole that becomes active magnetic North pulls the closest rotor South pole into alignment with it. This torque-producing action is helped by the passive stator South pole (on the opposite side of the stator) pulling the opposite rotor North pole into alignment with *it.*

For example, in Fig. 13–5, if transistor *A* is turned ON, stator pole A at 12 o'clock is active North. At the instant shown it has already pulled South rotor pole 1 into align-

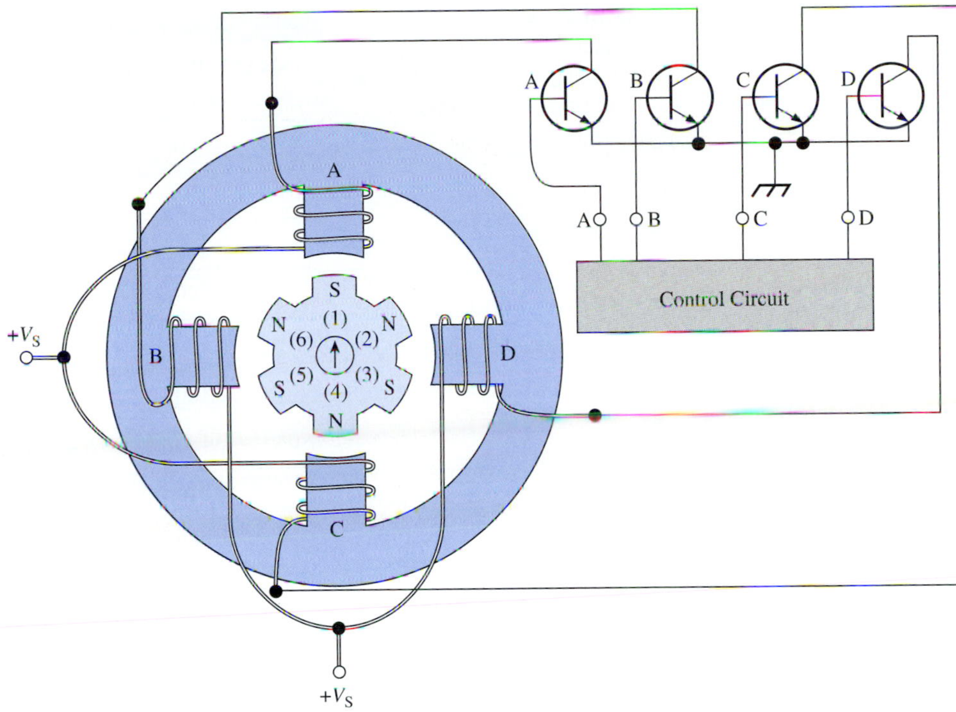

FIGURE 13–5 With four stator poles and six rotor poles, this stepper motor has a natural step angle of 30°. The rotor is shown in the 0° position (imaginary position-arrow pointing up toward 12 o'clock).

ment with it. Also, stator pole C at 6 o'clock is passive South. It has pulled North rotor pole 4 into alignment with it.

(The fact that opposite magnetic poles attract each other is an alternative way of explaining motor action, as mentioned in Sec. 12-2. Actually, however, if we looked at the internal atomic orientation of the permanent-magnet core, we would still find the Lorentz relation at work. It is still "current," in the form of electrons revolving around the atomic nucleus, interacting with the magnetic field, that ultimately creates the attraction force.)

Let us define the rotor position shown in Fig. 13–5 as the 0° position. We have shown an imaginary position-arrow on the shaft pointing up toward 12 o'clock. As the rotor shaft rotates we can describe its new position by giving the pointing direction of the imaginary arrow.

The control circuit now turns OFF transistor *A* and simultaneously turns ON transistor *B*. Stator pole B becomes active North; stator pole D becomes passive South. Poles A and C become neutral—demagnetized. Stator pole B attracts South rotor pole 5; stator pole D attracts North rotor pole 2. The rotor moves clockwise by 30° so that the rotor poles become aligned with the stator poles. We say that the motor takes a 30° step. The imaginary position-arrow now points toward 1 o'clock.

Once the 30° step has been taken, the controller can turn OFF transistor *B* if the mechanical load doesn't cause the rotor shaft to move past the target position, which is 1 o'clock. If the load does tend to cause this problem, the *B* transistor should be left ON to enable the stepper motor to hold its position. The first 30° step is recorded in Table 13–1, going from the table's top row to the second row.

Next the control circuit turns OFF transistor *B* and simultaneously turns ON transistor *C*. This makes stator pole C North and A passive South. Rotor South pole 3 is just 30° away from C at this moment, so it moves into alignment. The motor has taken another 30° step clockwise, as indicated in Table 13–1. The imaginary position-arrow points to 2 o'clock, 60° from its starting position. And so on, with the controller firing the

TABLE 13–1
Sequence of transistor switching for taking full steps in the clockwise direction

SHAFT POSITION (DEG)	TRANSISTOR TURNED ON
0	*A*
30	*B*
60	*C*
90	*D*
120	*A*
150	*B*
180	*C*
210	*D*
240	*A*
270	*B*
300	*C*
330	*D*
360	*A*

transistors in the repeating sequence *ABCD*, shown in Table 13–1. You should follow the stepping of the motor through a complete 360° rotation, all the way to the bottom of Table 13–1. Keep your eye on which stator pole has become North and which rotor South pole is 30° away from it.

Reversing a stepper motor to counterclockwise rotation is simple, conceptually. Just arrange for the control circuit to turn ON the switching transistors in the reverse sequence, *DCBA*. This is shown in Table 13–2. Starting from the 0° Power-On-Reset (POR) position

TABLE 13–2
Sequence of transistor switching to make the Fig. 13–5 motor take full steps in the counterclockwise direction

SHAFT POSITION (DEG)	TRANSISTOR TURNED ON
0	*A*
−30	*D*
−60	*C*
−90	*B*
−120	*A*
−150	*D*
−180	*C*
−210	*B*

which has *A* turned ON, the controller first switches to transistor *D*. In Fig. 13–5 this causes rotor South pole 3 to move 30° counterclockwise to align with stator North pole D, and of course rotor North pole 6 aligns with stator passive South pole B. Follow the counterclockwise rotation of the stepper motor by referring to Table 13–2 and Fig. 13–5.

Half-steps. It is possible to get the motor of Fig. 13–5 to take 15° steps, called *half-steps*. The transistor switching sequence is given in Table 13–3. Compare Table 13–3 to Fig. 13–6(a), which shows the direction of the motor stator's magnetic flux for each row of the table.

Starting at the 0° POR position (12 o'clock) in Fig. 13–5, the control circuit turns OFF transistor *A*, simultaneously turning ON transistors *C* and *D*. Both stator poles C and D become active North; both poles A and B become passive South. Their combined net magnetic flux is in the middle of the space between the poles. This is indicated by the flux arrow labeled C and D in Fig. 13–6(a). Note that the flux arrow points from lower right to upper left (from 135° to 315°). This means that the emerging North flux is entering the rotor at the 135° location. In Fig. 13–5 the nearest rotor South pole is pole 3, which is only 15° away from the C and D flux arrow. Therefore the rotor makes a 15° clockwise step. The imaginary shaft position-arrow in Fig. 13–5 now points 15° clockwise from 12 o'clock, as shown in Fig. 13–6(b).

The next step in Table 13–3 has transistor *B* turned ON alone. In Fig. 13–6(a) the net flux is to the right, indicated by the B arrowhead. In the motor cross section of Fig. 13–5, at this moment rotor South pole 5 is 15° away from the emerging point of the North flux. Therefore pole 5 moves into alignment with that stator flux, stepping the rotor clockwise another 15°. The imaginary shaft position-arrow now points toward 1 o'clock (30° clockwise from 12 o'clock), as shown in Fig. 13–6(b).

TABLE 13–3
Sequence of transistor switching to make the motor of Fig. 13–5 take half-steps (15°) in the clockwise direction

SHAFT POSITION (DEG)	TRANSISTORS TURNED ON
0	*A*
15	*C* and *D*
30	*B*
45	*A* and *D*
60	*C*
75	*A* and *B*
90	*D*
105	*B* and *C*
120	*A*
135	*C* and *D*
150	*B*
165	*A* and *D*
180	*C*

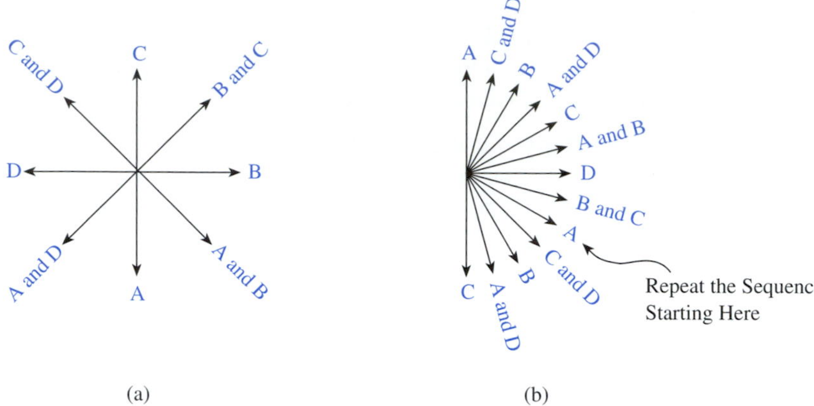

FIGURE 13–6
(a) Showing the direction of the net stator flux for each transistor switching possibility (each row) in Table 13–3. (b) Showing the position of the imaginary shaft arrow for each row in Table 13–3.

Work with Table 13–3, motor cross-section Fig. 13–5, and Fig. 13–6(a) and (b) to follow the stepper motor through its first 180° of movement.

Of course, it is also possible to make the stepper motor take half-steps in the counterclockwise direction. You should be able to create your own table of transistor switching sequence and a net flux diagram (like Fig. 13–6) to show how this is done.

13-3-1 IC Controllers for Steppers

Several manufacturers have developed integrated circuits (ICs) that are designed for controlling particular stepper motors. These ICs typically contain the power switching transistors (*A*, *B*, *C*, and *D* in our example motor) and the digital logic that determines the specific switching sequence. By applying the appropriate digital input levels to the IC logic terminals, the user instructs the IC to run the motor either clockwise or counterclockwise, and either full-step or half-step. The IC shown in Fig. 13–7 also has a logic input terminal labeled $1PH/\overline{2PH}$, which instructs the IC to run the stepper motor in one-phase mode or two-phase mode. We will cover the meaning of this mode selection shortly.

The four high-current switching transistor emitters all share a common internal ground. Their collectors are brought out to current-sinking terminals labeled *A*, *B*, *C*, and *D*. In the motor, the *A-C* winding pair is referred to as a *phase,* and the *B-D* winding pair is called the *other phase.* The *A-C* phase is powered by its own dc supply voltage, symbolized $V_{S(A\text{-}C)}$; the *B-D* phase is powered by a potentially separate dc supply, symbolized $V_{S(B\text{-}D)}$. In Fig. 13–5 and throughout our description of motor operation so far, these two dc supply voltages have been the same, but in some applications the voltage magnitudes are separately modulated to achieve very small stepping angles. We will discuss that practice when we cover the mode-selection issue.

The $V_{S(A\text{-}C)}$ power-supply terminal of the motor is jumpered to the $V_{S(A\text{-}C)}$ terminal of the IC, as Fig. 13–7 shows. This is done solely for the purpose of placing internal kickback-suppression diodes in parallel with the motor windings. The diodes protect the transistors from damage due to large transient voltages generated by the inductive windings when their current is interrupted. The same is done for the *B-D* phase on the right side of Fig. 13–7. Trace these circuits carefully, redrawing them if necessary, to see that the internal diodes are reverse biased when the motor windings are energized but go into forward bias to short-circuit the windings when they induce kickback voltage. Resistors are often placed in the leads to these IC terminals to shorten the inductive time-constant ($\tau = L/R$).

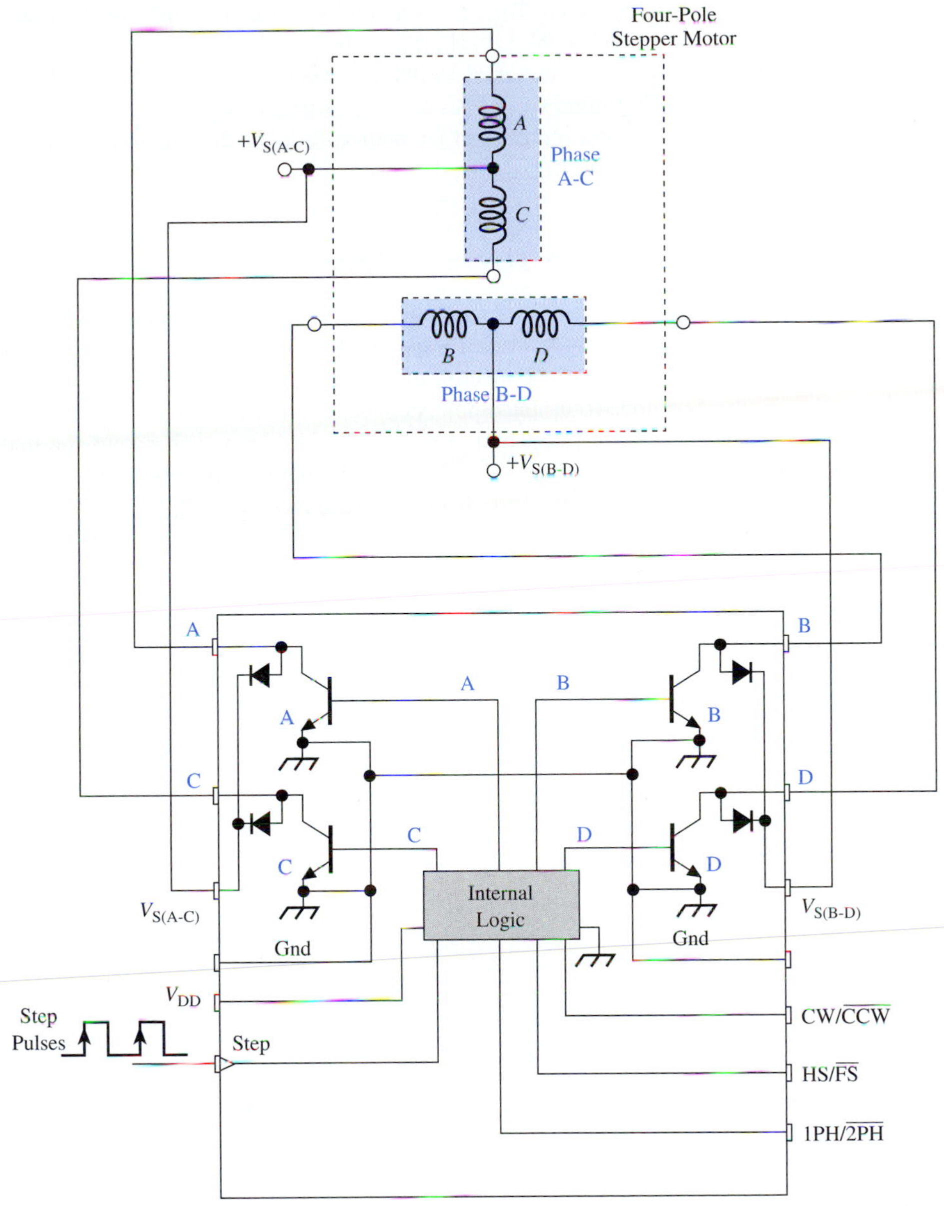

FIGURE 13–7
Typical IC controller for a four-stator-pole stepper motor.

The digital logic's V_{DD} power supply is brought in to the V_{DD} terminal, and digital ground is connected to one of the two common motor-ground GND terminals. The terminal that we have labeled CW/$\overline{CCW}$ instructs the IC control logic to produce an appropriate switching sequence to run the motor either clockwise (if the terminal is HI) or counterclockwise (if the terminal is made LO). This terminal is often labeled DIR. The terminal that we have labeled HS/$\overline{FS}$ causes the control logic to produce a half-step switching sequence, such as Table 13–3, when it is brought HI. It causes the logic to deliver a full-step sequence, such as Table 13–1 or 13–2, when it is made LO. Manufacturers sometimes give this terminal the label HALF.

Thus, for example, if you wanted the motor to turn clockwise in full 30° steps, you would place a logic HI on the CW/$\overline{CCW}$ terminal and a logic LO on the HS/$\overline{FS}$

terminal. To get counterclockwise rotation in half-steps, you would have inputs $CW/\overline{CCW} = 0$, $HS/\overline{FS} = 1$.

The actual switching occurs when a positive-going edge is delivered to the STEP terminal of the IC. Every positive-going edge produces one step. Thus the stepping rate of the motor, and its shaft speed, are determined by the frequency of the external clock circuit that drives the STEP input.

13-3-2 Stepper Characteristic Curves

Naturally there are operating limits on the STEP frequency. As a general rule, the greater the motor's load torque, the lower is its maximum stepping rate. This relationship is quantified for a specific motor by its stepping rate-vs.-torque curves. The pair of curves for our four-stator-pole stepper motor is shown in Fig. 13–8.

Look first at the bidirectional curve. It shows the maximum rate at which the controller is permitted to step the motor, given that the motor's next step may be in the opposite direction from the preceding step. (Clockwise step followed by a counterclockwise step, or vice versa.) For our example motor in Fig. 13–8, if the shaft load torque is zero, the external clock is permitted to run the motor at any rate between 0 and 190 steps per second. Since the motor is taking 30° per step, this is equivalent to a maximum rotational speed of

$$\frac{190 \cancel{\text{steps}}}{\cancel{\text{s}}} \times \frac{30 \cancel{\text{deg}}}{1 \cancel{\text{step}}} \times \frac{1 \text{ rotation}}{360 \cancel{\text{deg}}} \times \frac{60 \cancel{\text{s}}}{1 \text{ min}} = 950 \text{ r/min}$$

If the torque load on the shaft is greater, the clock's stepping frequency is limited to slower rates. For example, for $\tau = 0.04$ N-m, the stepping rate must be no greater than 80 steps/s. This is shown clearly in Fig. 13–8. If the controlling clock exceeds this frequency, the

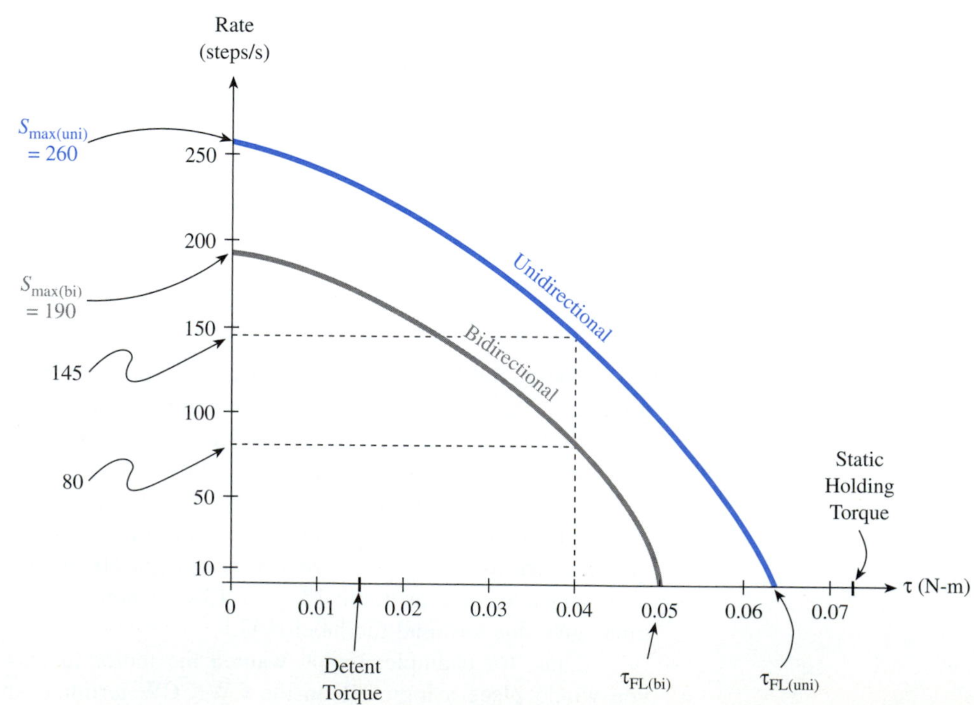

FIGURE 13–8
Characteristic operating curves for our particular stepper motor. Such curves are valid only for a given set of operating conditions, in this case full-step, one-phase operating mode. If the operating conditions change (going into half-step operation, for example), a new set of curves applies to the motor.

motor manufacturer cannot guarantee that the motor will respond properly to a command to make a reverse step—it may miss the step.

In bidirectional operation the maximum allowable load torque is about 0.05 N-m. In America small motors like stepper motors often have their torque expressed in ounce-inches (oz-in). The conversion factor between ounce-inches and Newton-meters is 141 oz-in = 1 N-m. Therefore this value of $\tau_{FL(bi)}$ could be expressed as

$$\tau_{FL(bi)} = 0.05\ \cancel{\text{N-m}} \times \frac{141\ \text{oz-in}}{1\ \cancel{\text{N-m}}} = 7.0\ \text{oz-in}$$

When it must deliver this full-load torque, the motor can step at only a 10 steps/s maximum rate, equivalent to 50 r/min.

If your application is such that the motor never has to reverse directions from one step to the next, but runs in just one direction (at least temporarily), then the unidirectional curve in Fig. 13–8 becomes operative. Now the motor can run faster without danger of missing a step. For example, at no-load the maximum stepping rate becomes 260 steps/s, as shown on the characteristic graph. This stepping rate is equivalent to 1300 r/min. At $\tau = 0.04$ N-m the motor can step unidirectionally as fast as 145 steps/s, and the new value of full-load torque is about 0.063 N-m, with no danger of missing a unidirectional step.

When a stepper motor is to be run at unidirectional speeds, its stepping rate must be gradually increased from the corresponding bidirectional maximum rate. This process is called *ramping up*. The manufacturer provides ramping graphs that show the minimum number of steps that must be taken in accelerating from the bidirectional maximum to the unidirectional maximum. Such a graph is shown in Fig. 13–9 for our example motor.

When our motor is to move unidirectionally, delivering load torque of 0.04 N-m, it must begin stepping at a rate no greater than 80 steps/s. Once that rate is attained it can be accelerated up to a maximum rate of 145 steps/s, but it must take at least 70 steps to reach that faster rate, as the bottom graph indicates in Fig. 13–9.

When it is necessary to stop this fast-running motor, it cannot be stopped by immediately ceasing all STEP pulses to the controller. If that were attempted the motor shaft

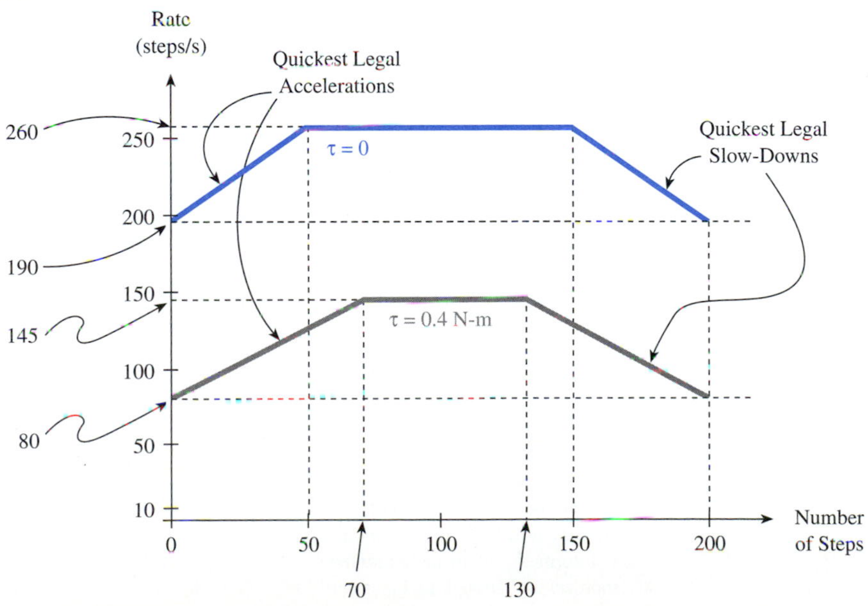

FIGURE 13–9
Stepper motor acceleration (ramping up) and slow-down (ramping down) when bidirectional maximum rates are exceeded.

would move past the desired stopping position. Instead the speed must be *ramped down,* as Fig. 13–9 indicates. The motor must be slowed from 145 steps/s to 80 steps/s in no fewer than 70 steps (200 − 130 = 70 steps). Once it has slowed to 80 steps/s, then the controller's STEP pulses can be abruptly terminated with no danger of the motor shaft overshooting its proper stopping position.

The top graph in Fig. 13–9 shows that the ramping up and ramping down can be accomplished more rapidly when the load torque is lighter. For $\tau = 0$, only 50 steps are needed to change from unidirectional maximum speed to bidirectional maximum speed, and vice versa.

In a simple stepper-motor system, STEP pulses to the IC controller in Fig. 13–7 might be produced by a fixed-frequency clock oscillator, but in a sophisticated stepper-motor control system, STEP pulses and control inputs are provided by a microprocessor. This arrangement is shown in Fig. 13–10. It is the responsibility of the microprocessor's program to ensure that maximum stepping rates are not exceeded and to abide by the motor's acceleration and slow-down requirements.

Holding torque and detent torque. Permanent-magnet stepper motors have the ability to hold a fixed shaft position, as mentioned earlier. If a stator pole winding remains energized, the position-holding ability is quite pronounced. The amount of torque that can

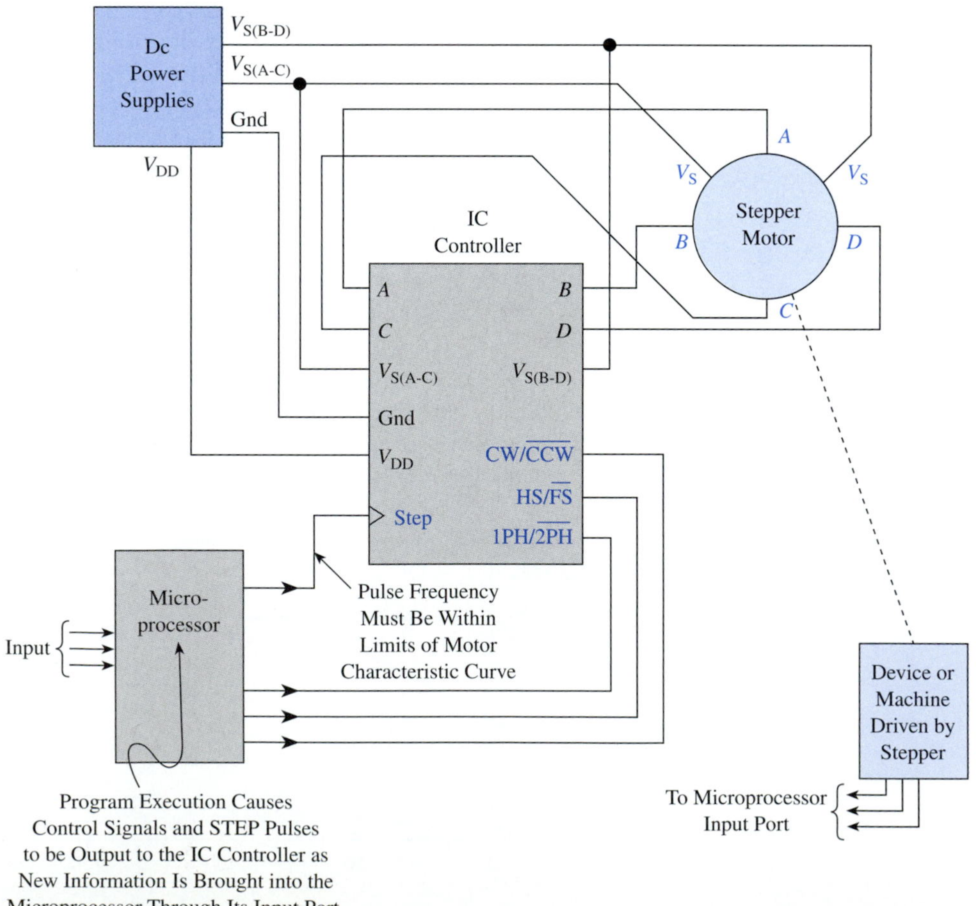

FIGURE 13–10
Microprocessor-based stepper-motor control system.

be applied to the motor shaft by the mechanical load, without the motor losing its grip and slipping out of a steady position, is called its *static holding torque,* or *static stall torque.* Our example motor in Fig. 13–8 has a static holding torque of 0.072 N-m, with a single stator pole energized. If two stator poles are energized (in half-stepping mode), the motor's static holding torque will be greater because the net stator magnetic field is stronger.

Even if the stator pole windings are all deenergized, there is still some attraction force between the rotor's permanent-magnet poles and the stator poles they have aligned with. Therefore the motor can still hold a fixed position, but not as firmly. Then the maximum load-originating torque that the motor can withstand without slipping is called *detent torque.* For our motor detent torque is 0.014 N-m in Fig. 13–8.

Two-phase operating mode. In half-stepping operation, the stator switching sequence has a single pole winding energized, followed by two pole windings simultaneously energized, then back to a single pole energized. Refer to Table 13–3. An important fact about half-stepping is this: The steps that occur with two pole windings energized are stronger than the steps that occur with only one pole winding energized. In other words, we get higher-torque steps alternating with lower-torque steps.

This suggests that it is possible to boost the torque capability of a stepper motor by *always* having two pole windings energized. The transistor switching sequence that accomplishes this is given in Table 13–4. To understand the two-phase operating mode, work with Table 13–4, motor cross-section Fig. 13–5, and Fig. 13–11.

TABLE 13–4
The two-phase operating mode provides full-stepping with increased torque*

	SHAFT POSITION (DEG)	TRANSISTORS TURNED ON
	45	*A* and *D*
CW ↓	75	*A* and *B*
	105	*B* and *C*
	135	*C* and *D*
	165	*A* and *D*
	195	*A* and *B*
	225	*B* and *C*
	255	*C* and *D*
CCW ↑	285	*A* and *D*
	315	*A* and *B*
	345	*B* and *C*
	375 (15)	*C* and *D*
	45	*A* and *D*

*It is called two-phase mode because the motor always has both of its two phases energized (the *A-C* phase and the *B-D* phase).

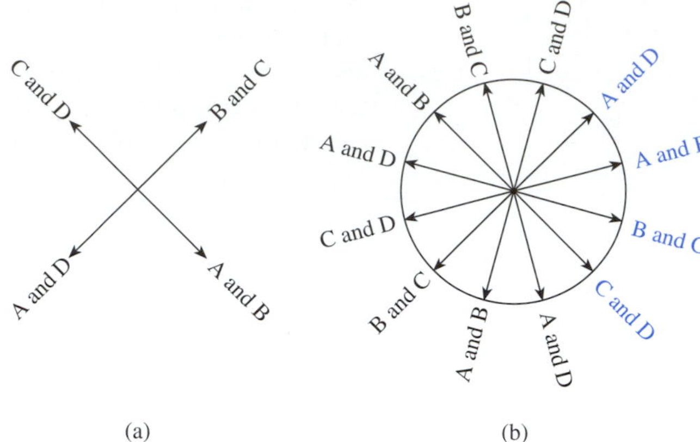

FIGURE 13–11
(a) Stator flux direction for each row of Table 13–4.
(b) Position of the imaginary motor shaft position-arrow for each row of Table 13–4.

Part (a) of Fig. 13–11 shows the net magnetic flux direction resulting from the two poles that are energized. Remember that an energized pole always becomes active North; two adjacent energized poles produce North flux emerging from a point midway between them. Figure 13–11(b) shows the position of the imaginary shaft arrow for each transistor switching combination.

Two-phase operation is achieved with the IC controller of Fig. 13–7 by placing a logic LO on the HS/$\overline{\text{FS}}$ control terminal and a logic LO on the 1PH/$\overline{\text{2PH}}$ terminal. The motor can be stepped in either direction, clockwise or counterclockwise, by having a HI or LO, respectively, on the CW/$\overline{\text{CCW}}$ terminal.

At the moment when the IC controller is initially put into two-phase operating mode (when a pair of LOs arrives at HS/$\overline{\text{FS}}$ and 1PH/$\overline{\text{2PH}}$), the internal logic automatically switches the transistors to the Power-On-Reset (POR) condition. For this particular IC controller, designed for this particular motor, the POR condition has transistors *A* and *D* both ON. Therefore rotor South pole 1 in Fig. 13–5 moves to the 45° position. (Actually, whichever South pole happens to be closest to 45° is the one that moves into alignment with the stator flux. That pole is then considered to be pole number 1.)

The IC controller truth table shown in Table 13–5 summarizes the actions of control terminals HS/$\overline{\text{FS}}$ and 1PH/$\overline{\text{2PH}}$.

TABLE 13–5
IC controller truth table*

HS/$\overline{\text{FS}}$	1PH/$\overline{\text{2PH}}$	MOTOR OPERATION
0	0	Two-phase mode. Full steps, higher torque.
0	1	One-phase mode. Full steps, standard torque.
1	0	Half steps, alternating torque.
1	1	Stop; ignore STEP pulses

*The Stop condition cannot be immediately imposed if the motor is running faster than its bidirectional maximum rate.

13-3-3 Microstepping

To obtain very small stepping angles, the two separate phase supply voltages, $V_{S(A\text{-}C)}$ and $V_{S(B\text{-}D)}$, can be modulated. This practice is called *microstepping*. The modulation can be amplitude-type or pulse-width type. The general topic of pulse-width modulated control is covered in Chapter 16, Secs. 16-8 and 16-9. For now let us consider amplitude-modulated microstepping.

The half-stepping sequence of Table 13–3 tells us that the normal way of producing a 15° half-step from the 30° position to the 45° position is to turn OFF transistor *B* and simultaneously turn ON transistors *A* and *D*. Let us examine a method of microstepping from 30° to 45° in three 5° steps.

Figure 13–12 shows only the *A* and *D* stator poles and only South pole 1 of the rotor. The stator windings are driven by *programmable* dc power supplies. These power supplies have the capability of changing their output voltage in response to a digital command. They are not themselves programmable in the strict sense; that is, they themselves do not store a program. Rather, they receive command data in accordance with a program that is stored and executed by a microprocessor-based device. A better name for these power supplies would be "program-controlled dc power supply."

The program can be designed to vary the supply voltages $V_{S(A\text{-}C)}$ and $V_{S(B\text{-}D)}$ as shown in Fig. 13–13. At $t = 0$, the *A* and *D* transistors are switched ON by a STEP pulse issued by the program. At the same instant the control program sends data to the power supplies calling for $V_{S(A\text{-}C)} = 23.2$ V and $V_{S(B\text{-}D)} = 16.2$ V. These values produce the proper amounts of current in the A winding and the D winding so that their combined net

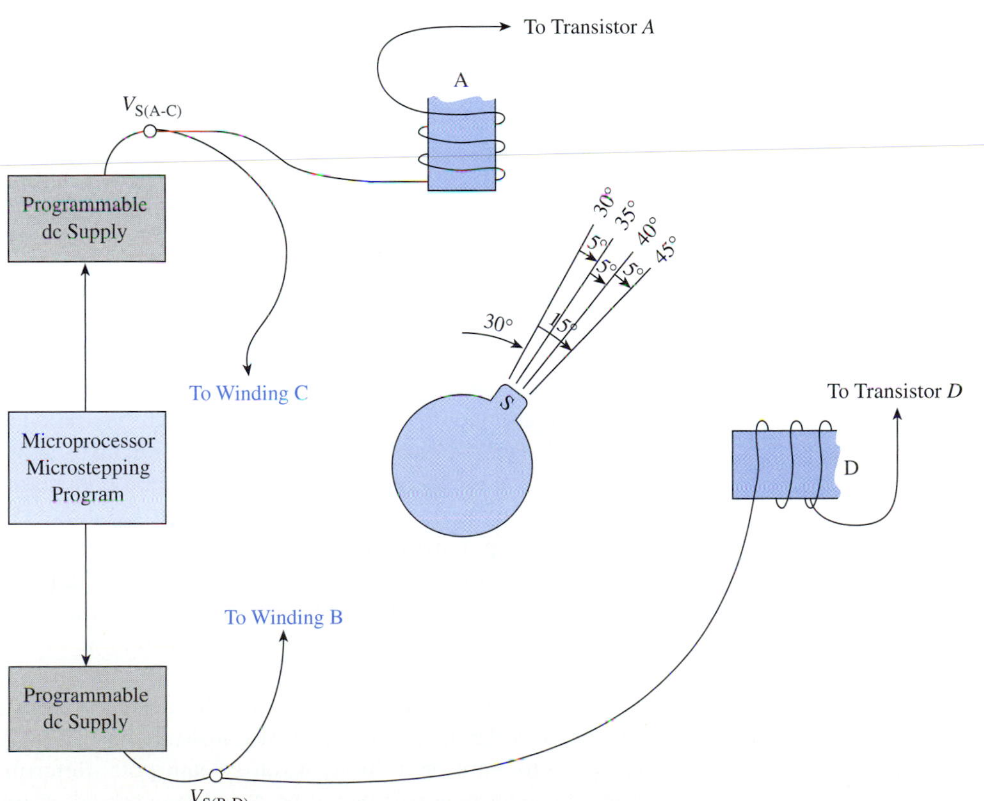

FIGURE 13–12
The microstepping idea. Breaking a normal 15° step into three 5° steps.

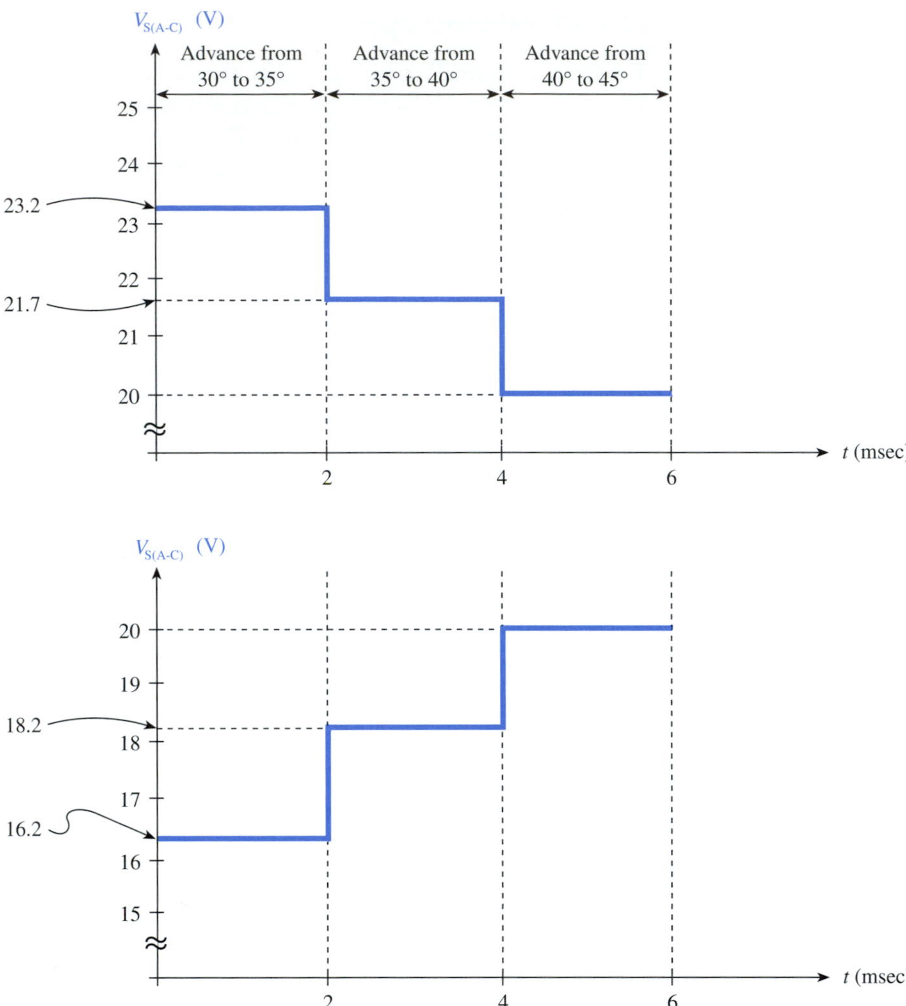

FIGURE 13–13
Waveforms of amplitude-modulated phase-supply voltages to produce 5° steps.

flux is oriented 35° away from vertical. Therefore rotor South pole 1 moves 5° clockwise to align with the net North flux.

At $t = 2$ ms the program sends new data to the phase power supplies, calling for $V_{S(A\text{-}C)} = 21.7$ V and $V_{S(B\text{-}D)} = 18.2$ V, as shown in Fig. 13–13. Performing a rectangular-to-polar conversion of these values shows that they produce a net stator flux that is offset by 40° from vertical. Therefore rotor pole 1 takes another 5° step, from 35° to 40°.

At $t = 4$ ms the power supply voltages are both commanded to become 20 V, producing net flux midway between poles A and D, at the 45° orientation. The rotor takes its third 5° step, moving South pole 1 from the 40° to the 45° position.

At $t = 6$ ms the program could just continue the voltage waveform pattern established in Fig. 13–13. That is, the phase power supplies would be switched to $V_{S(A\text{-}C)} =$ 18.2 V, $V_{S(B\text{-}D)} = 21.7$ V, and so on, to step from 45° to 60° in 5° steps during the following 6 ms.

Alternatively, the program could be designed to issue another STEP pulse that switches ON the appropriate pair of transistors and starts varying the power supply values to go to work on a different rotor South pole. Referring back to Fig. 13–5, the appropriate stator pole-pair would be A and B, working on rotor South pole 5.

Theoretically there is no limit to how small the steps can be using voltage-modulated microstepping.

13-3-4 Disk-Rotor Stepper Motors

The disk rotor structure was devised to reduce the inertia of a coreless dc motor, as explained with Fig. 13–4. The disk structure can also be used for the permanent-magnet rotor of a stepper motor, as shown in Fig. 13–14. Its rotational inertia is not quite as low now, because it is made of exotic elements that are heavier than fiberglass. However, the number of magnetic segments (poles) can be large, since they are nonsalient. Each permanent magnetic segment is North on one side of the disk and South on the other side, as illustrated in the expanded view in Fig. 13–14.

The segment spacing is such that when a disk segment is aligned with stator pole A on top, a neutral space is aligned with stator pole B on the bottom. Therefore, by switching OFF one stator pole winding and simultaneously switching ON the other stator pole winding, the disk rotor steps the distance between a neutral area and a rotor pole.

If the number of rotor segments is 100, which is common, the angular distance between poles is 3.6°. Therefore the step angle is half of 3.6°, or 1.8°, since the rotor moves between a pole and an adjacent neutral area.

Table 13–6 shows the transistor switching sequence for both directions of motion. Verify the operation of these sequences by careful study of Fig. 13–14. To help your understanding, trace the windings' wraps to show that Q_{A1} switched ON causes stator pole A to become South on the front side of the disk and North on the rear side. Q_{A2} reverses that

TABLE 13–6 Switching sequences for clockwise and counterclockwise stepping of disk-rotor stepper

Direction	ON TRANSISTOR	Direction
CW ↓	A1	
	B1	
	A2	
	B2	
	A1	CCW ↑
	B1	
	A2	
	B2	

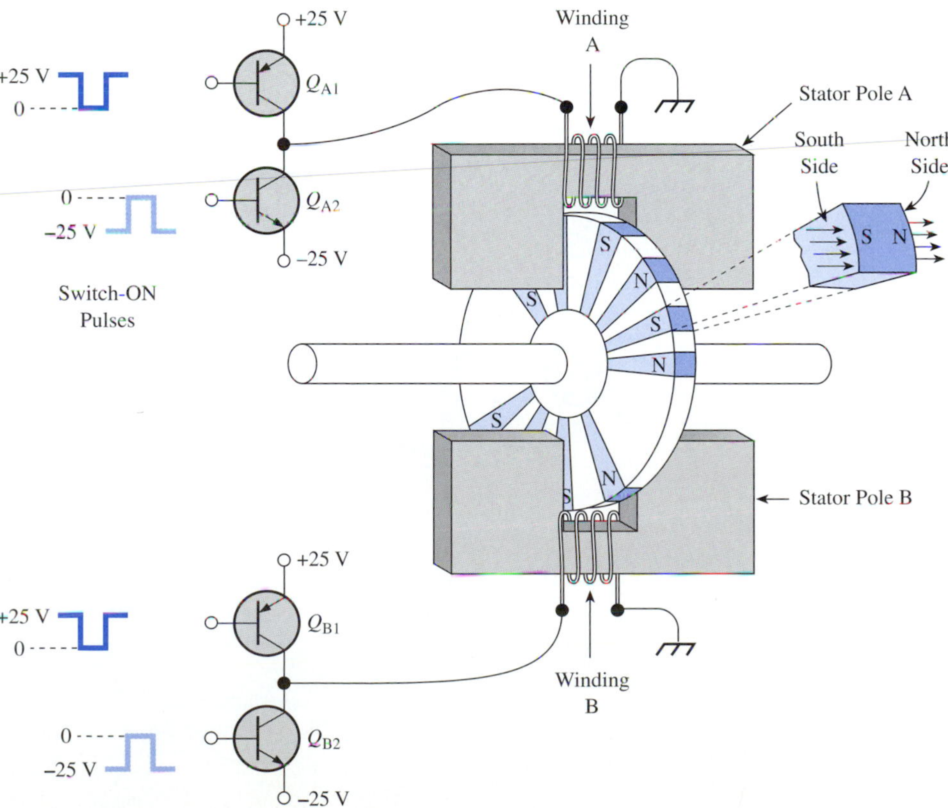

FIGURE 13–14 Permanent-magnet disk-rotor stepper structure and drive circuits.

magnetic polarity. Transistor Q_{B1} pulsed ON causes stator pole B to become North on the front side of the disk and South on the rear side. Q_{B2} reverses that magnetic polarity.

By energizing both windings A and B simultaneously it is possible to obtain half-steps with a permanent-magnet disk stepper motor. For a 100-segment model, described earlier, a half-step is 0.9°.

It is easy to see that stepper motors with a small step-angle can position their load more precisely than large-angle steppers. They also tend to run at higher stepping rates than large-angle motors, but in determining shaft rotation speed, step size generally outweighs stepping rate.

For example, a 1.8° stepper typically could have a maximum unidirectional stepping rate of 2000 steps/s at no load. This is equivalent to a rotational speed of

$$\frac{2000\ \cancel{\text{steps}}}{\cancel{\text{s}}} \times \frac{1.8\ \cancel{\text{deg}}}{1\ \cancel{\text{step}}} \times \frac{1\ \text{rot}}{360\ \cancel{\text{deg}}} \times \frac{60\ \cancel{\text{s}}}{1\ \text{min}} = 600\ \text{r/min}$$

In general, large-angle stepper motors can move a load to its final target position more quickly.

13-3-5 Variable-Reluctance Stepper Motors

Not all stepper motors have permanent-magnet rotors. Stepping action can also be accomplished with a toothed nonmagnetized cylindrical core. A common construction for such a stepper motor is shown in cross section in Fig. 13–15. Such motors are called *vari-*

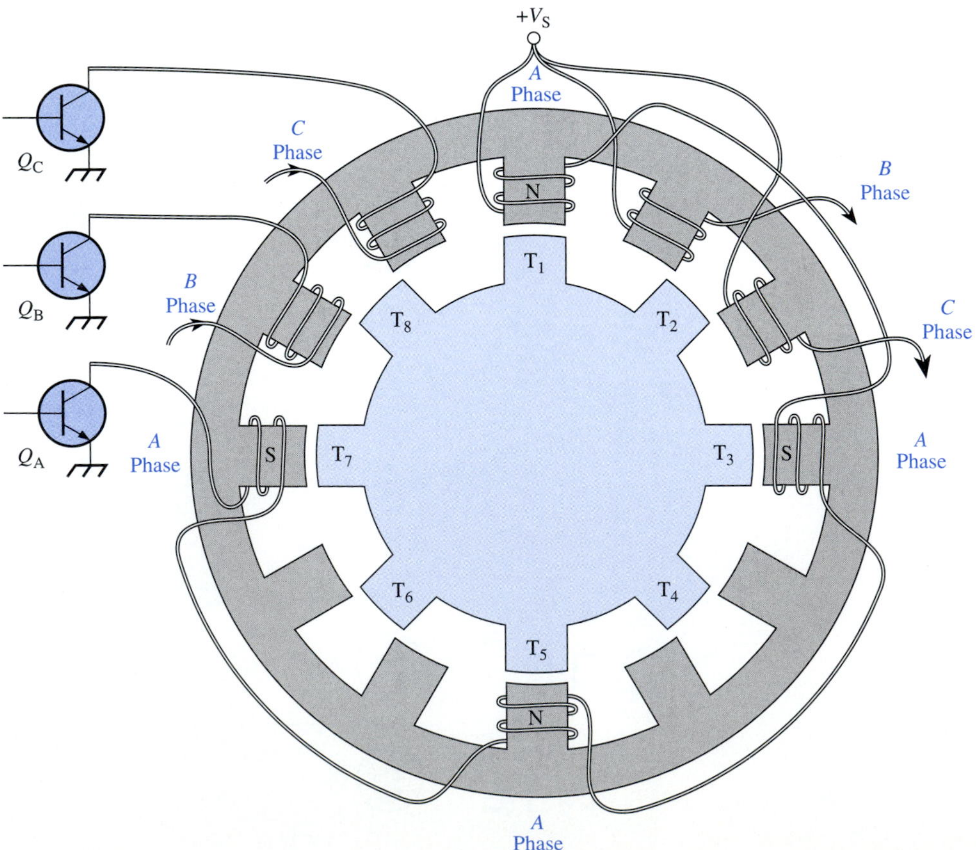

FIGURE 13–15
Cross section of a three phase, four-pole-per-phase, variable-reluctance stepper motor.

able-reluctance steppers because the rotor always moves to the position that minimizes the magnetic reluctance of the overall flux path (minimizes the combined total air gaps in the path).

The transistor switching sequence in Fig. 13–15 is *ABC* for counterclockwise stepping. When the *A* transistor turns ON, the pole windings at 12 o'clock, 3 o'clock, 6 o'clock and 9 o'clock are energized in series with one another. The winding wrap produces North poles at 12 o'clock and 6 o'clock and South poles at 3 o'clock and 9 o'clock. This makes the rotor move to the position shown in Fig. 13–15. It is moved by magnetic attraction forces between the actively magnetized stator poles and the passively magnetized teeth T_1, T_3, T_5, and T_7. Motion stops when the teeth become aligned with the poles. The teeth do not step past this position because they would feel a magnetic restoring force if they did. By aligning themselves in this way, the rotor teeth have minimized the flux's air-gap, or minimized the flux path's reluctance.

The next step occurs when transistor *A* is turned OFF and transistor *B* is simultaneously turned ON. This energizes the four *B*-phase stator poles at 1 o'clock, 4 o'clock, 7 o'clock, and 10 o'clock. The nearest teeth are T_2, T_4, T_6, and T_8. Therefore the teeth step the 15° distance to the poles, moving counterclockwise. When phase *B* is turned OFF and phase *C* is turned ON, the odd-numbered teeth make another 15° step to align and minimize reluctance.

Clockwise rotation is produced by reversing the switching sequence to *ACB*, as you would expect.

Because of the magnetically passive nature of the rotor, a variable-reluctance stepper can operate at much higher stepping rates than a permanent-magnet stepper of the same step angle. Some models using the three-phase design of Fig. 13–15 can be ramped up to 10 000 steps/s, producing rotational speeds of 25 000 r/min.

Because its rotor teeth are magnetically passive, a variable-reluctance stepper has much less static holding torque than a comparable permanent-magnet stepper. And it has almost zero detent torque.

Overshoot. Stepper motors never make perfectly clean steps. Instead, they *overshoot* the step position, then recover and oscillate around that position before settling. This tendency is illustrated in Fig. 13–16. Because they lack a pair of active magnetic poles to cling to each other, variable-reluctance steppers have a much greater overshoot tendency than permanent-magnet steppers.

Under high-inertia, high-speed operating conditions, overshoot can become a serious problem. If oscillation has not died down by the time the next step occurs, it might cause a variable-reluctance stepper to actually make a backward step in the wrong direction. The reason for this is shown in Fig. 13–17.

FIGURE 13–16
Graphing instantaneous position versus time to illustrate overshoot.

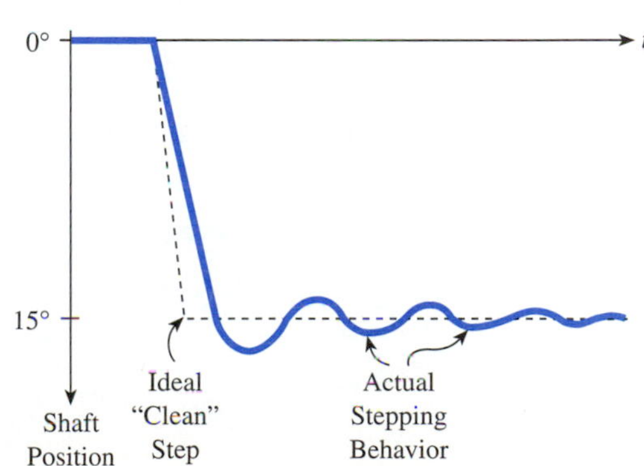

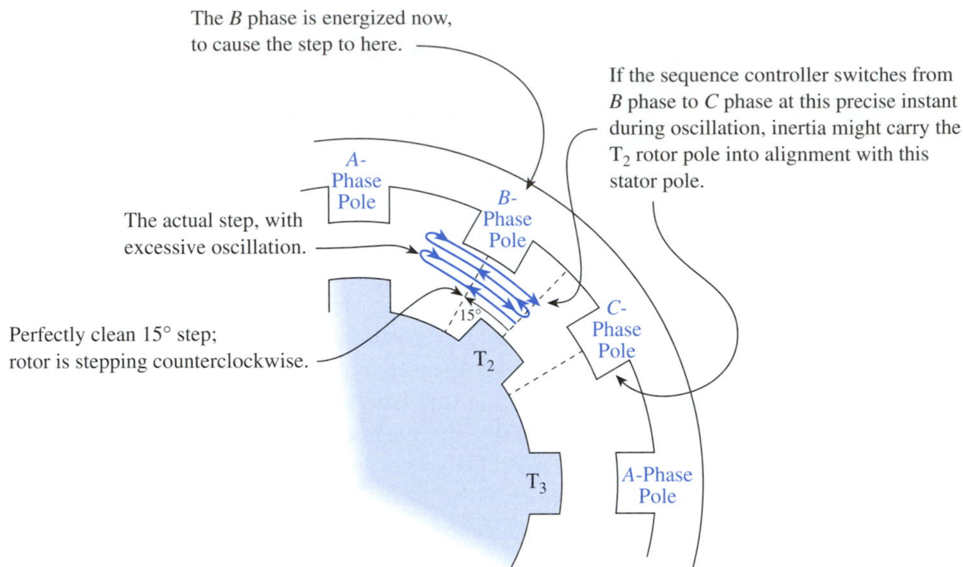
The B phase is energized now, to cause the step to here.
If the sequence controller switches from B phase to C phase at this precise instant during oscillation, inertia might carry the T2 rotor pole into alignment with this stator pole.
A-Phase Pole
B-Phase Pole
The actual step, with excessive oscillation.
15°
C-Phase Pole
Perfectly clean 15° step; rotor is stepping counterclockwise.
T2
T3
A-Phase Pole

FIGURE 13–17
Illustrating the problem of backward stepping in an oscillating variable-reluctance stepper motor.

In Fig. 13–17 the rotor is supposed to be stepping in the counterclockwise direction with the *ABC* switching sequence. When phase *B* is switched ON, rotor tooth T_2 moves counterclockwise, then overshoots the stator pole and goes into oscillation around its proper position. If the oscillations have not yet died out when the sequence controller later switches to the *C* phase, then with bad luck it's possible for the T_2 tooth, rather than the T_3 tooth, to go into alignment with the *C*-phase stator pole. This is a gross malfunction because the rotor has moved clockwise when it was supposed to move counterclockwise.

Damping. To prevent missteps like the one described above, we sometimes make a special effort to minimize the overshoot and suppress the oscillations quickly. This is called *damping*.

Damping can be implemented inexpensively by connecting resistors or capacitors to the motor phases, as shown in Fig. 13–18(a) and (b).

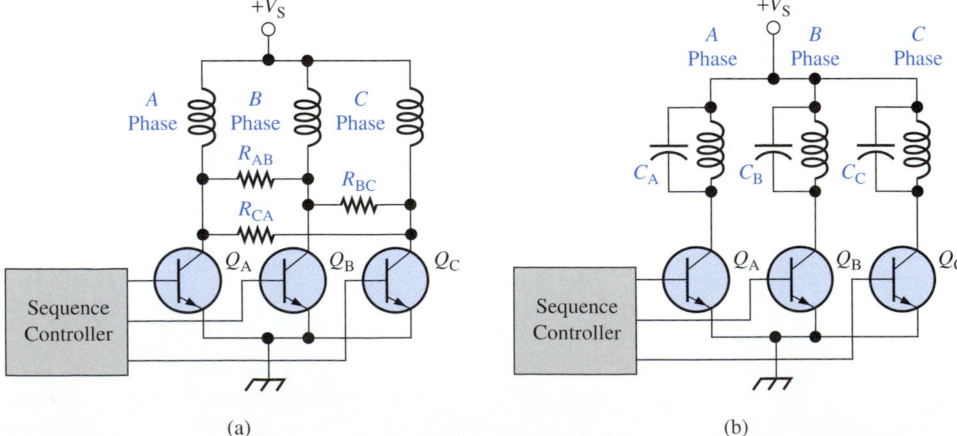
+VS
A Phase
B Phase
C Phase
RAB
RBC
RCA
QA
QB
QC
Sequence Controller
(a)
+VS
A Phase
B Phase
C Phase
CA
CB
CC
QA
QB
QC
Sequence Controller
(b)

FIGURE 13–18
Damping technique.
(a) Resistive. (b) Capacitive.

The resistors in Fig. 13–18(a) allow a small current to flow in the two phases that are deenergized by the sequence controller. For example, when transistor Q_B is switched ON, a small value of current flows through the *A*-phase winding, through resistor R_{AB}, through Q_B to ground, and likewise for the *C*-phase winding and R_{BC}. These currents produce slight reverse torque as the stepping motion begins. Study Fig. 13–15 to understand the reverse (clockwise) attraction of the 12 o'clock stator pole on T_1 and the reverse attraction of the 2-o'clock stator pole on T_2. The OFF-windings' currents produce less reverse torque as the stepping motion nears completion. This initial suppression and delayed release tends to reduce overshoot without greatly reducing the motor's average forward torque.

A similar electrical/mechanical response is obtained from the parallel capacitors in Fig. 13–18(b). The phase winding that is presently being deenergized (the *A* phase at the time represented in Fig. 13–17) has a charged capacitor C_A in parallel with it. The winding that is presently being energized, phase *B*, is momentarily short-circuited by discharged capacitor C_B. This suppresses forward torque at the beginning of the step motion, thereby reducing eventual overshoot.

Damping can also be accomplished by connecting a mechanical slip-clutch assembly to the nonload end of a double-ended motor shaft. Many variable-reluctance steppers are constructed with a double-ended shaft to accommodate such devices.

13-4 ■ BRUSHLESS DC MOTORS

The *brushless dc motor* is the other member of the electronically commutated dc motor family. Actually, a better name for the brushless dc motor would be "position-triggered brushless dc motor" since any stepper motor is also brushless and dc.

The distinguishing feature of the brushless dc motor is its position-sensing mechanism, which is fed back to the sequence controller to tell it when to switch to the next step in the sequence. Thus, as a general rule, the brushless dc motor does not step to the next position, then wait in that position for the sequence controller to switch the stator windings' transistors. Although it *can* operate that way, it is better suited to continuous rotation. As its rotor poles approach alignment with particular stator poles, that approach itself is used to trigger the energization of the next stator poles in the sequence. Therefore a rotor pole just passes by the stator pole that it is approaching, attracted by the next pole or pole-combination on the stator's periphery.

A simplified cross-sectional view of a brushless dc motor is shown in Fig. 13–19 for a two-pole permanent-magnet rotor and nonsalient stator poles. Trace out the wiring between the stator pole windings and their control transistors to see that turning ON a winding's upper *pnp* transistor energizes that winding as magnetic North, while turning ON that winding's lower *npn* transistor energizes that winding as magnetic South (flux passing through the winding from inside to outside). For example, the multiturn winding A that creates stator pole A is at the top of the stator. The winding is placed in a single pair of stator slots in this simplified view. The left lead of that winding connects to the junction of *pnp* transistor Q_{AN} and *npn* transistor Q_{AS}. If the sequence controller circuit turns ON Q_{AN}, current flows out of the $+V_S$ supply, through Q_{AN}, into the page away from us on the left side of winding A, out of the page toward us on the right side, and eventually to ground. Applying the right-hand rule indicates that the resulting flux passes through the plane of the coil from outside to inside. Therefore, from the viewpoint of the rotor, pole A is North.

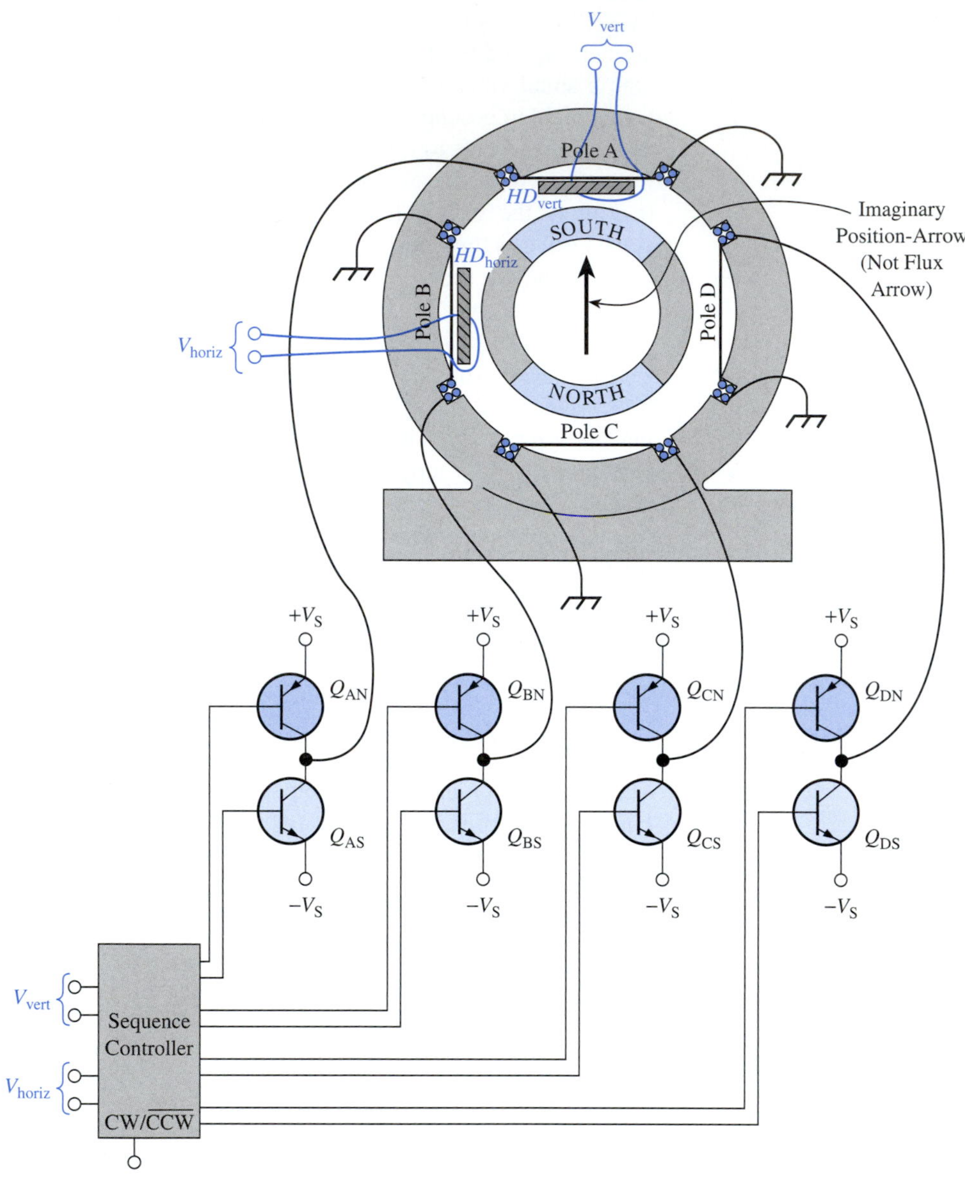

FIGURE 13–19
Brushless dc motor cross section. The Hall-effect detectors are displaced axially (into the page) toward the motor's end-bell.

If Q_{AS} is turned ON instead of Q_{AN}, the current direction reverses in winding A, making it a magnetic South pole. Of course, if both transistors Q_{AN} and Q_{AS} are turned OFF, winding A is deenergized, and pole A disappears. That section of the stator becomes magnetically neutral. The same relation exists among each pole and its two controlling transistors.

The clockwise and counterclockwise switching sequences are shown in Table 13–7. The net magnetic field moves in 45° steps, being offset at a 45° angle from the horizontal whenever all four poles are energized. Work through that table, referring to Fig. 13–19, to understand how the net flux makes a complete 360° revolution, as shown in Fig. 13–20.

TABLE 13–7
Stator-pole magnetic switching sequence for the brushless motor of Fig. 13–19*

	ROW NUMBER	Stator Poles A	B	C	D
↓	1	N	OFF	S	OFF
CW	2	N	S	S	N
	3	OFF	S	OFF	N
	4	S	S	N	N
	5	S	OFF	N	OFF
	6	S	N	N	S
CCW	7	OFF	N	OFF	S
	8	N	N	S	S
↑	1	N	OFF	S	OFF

*N stands for making that pole North by turning on its sub-N transistor; S stands for South.

Two Hall-effect sensors are mounted inside the motor's housing. Since Fig. 13–19 is a cross section, it gives the misleading impression that these sensors are located between the rotor poles and the stator poles. They aren't. They are beyond the rear end of the rotor magnets near the motor's end-bell, actuated by their own separate magnets.

The Hall-effect sensor labeled HD_{vert} detects when the rotor magnets are coming into vertical orientation (look at the imaginary shaft-position arrow in Fig. 13–19). The Hall detector on the left of the figure detects when the shaft arrow is coming into horizontal orientation. Hall-effect detectors produce one polarity of voltage for North flux and

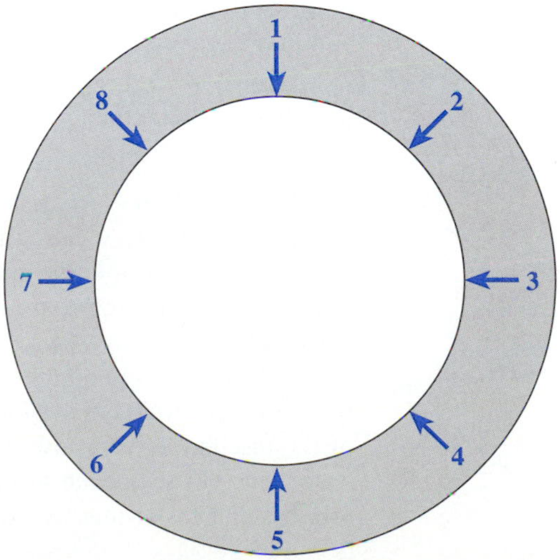

Figure 13–20
The position of the North flux (pointing toward the rotor from the stator) for each row in the Table 13–7 sequence. Remember that whatever the stator location of the North flux, the stator position directly opposite (180° away) receives South flux (pointing out from the rotor back into the stator).

the opposite polarity for South flux, as explained in Sec. 10-13. Thus, when the rotor's South pole is approaching the top 12-o'clock position (imaginary position-arrow pointing upward), HD_{vert} produces one polarity of v_{vert}; when the rotor's North pole approaches the top (imaginary position-arrow pointing downward), v_{vert} becomes the opposite polarity. When the arrow points left toward 9 o'clock, v_{horiz} becomes one polarity; when it points right, v_{horiz} becomes the opposite polarity.

The sequence controller receives v_{vert} and v_{horiz}. Therefore it knows the rotor's position, receiving a specific signal for each 90° of rotation. Its logic circuitry uses this information to trigger the transistors through the switching sequence of Table 13–7. As stated earlier, the sequence controller's electronics triggers the field-creating transistors with slight anticipation. For example, if the motor is rotating clockwise because the controller is going through its clockwise sequence, as the imaginary position-arrow approaches 9 o'clock v_{horiz} informs the controller of that fact. Before the arrow quite arrives at 9 o'clock, the controller steps from the seventh row of Table 13–7 to the eighth row. Thus the stator's net flux becomes North at the upper left and South at the lower right. This is the No. 8 flux position in Fig. 13–20. The flux exerts a new attraction force on the rotor, pulling the imaginary position-arrow toward 10:30 o'clock (toward upper-left position 8 in Fig. 13–20); so the shaft arrow doesn't stop at 9 o'clock but continues right by, heading for 10:30; and so on.

13-5 ■ COMPARING ELECTRONICALLY COMMUTATED TO BRUSH-EQUIPPED DC MOTORS

Reliability and cost. The natural advantage of electronically commutated motors is that they do not need a mechanical split-ring commutator, or brushes. This is a huge advantage, because the commutator and brushes are the leading trouble-causers in dc motors. They also require a great deal of initial manufacturing effort, making the motor more expensive.

Stepper motors and brushless dc motors are commutated at their field windings by electronic semiconductor switches, usually transistors. These switching devices, properly chosen for the job and protected by kickback-suppression diodes, do not need maintenance and seldom cause trouble. They have far greater long-term reliability than a brush and mechanical commutator.

The cost of designing and building the electronic commutation circuit is generally less than the manufacturing expense of a brush-mechanical commutator arrangement.

Heat dissipation. In a wound-rotor brush-equipped motor, the high-current heat-producing armature winding is on the rotor in the center of the machine. Its waste heat must be dissipated into the ambient air from this center location, which is a difficult thermal flow path. Therefore a tendency exists for the motor to retain its waste heat energy and run hotter.

Electronically commutated motors have permanent magnets or simple teeth on the rotor—no electromagnet windings. Therefore the waste heat originating from the center of the machine is very low. The high-current electromagnet windings are located on the stator toward the outside of the machine. The I^2R losses and magnetic heat losses therefore have an easier thermal path to the ambient air. These motors have a natural tendency to run cooler.

Radio-frequency interference (RFI). Brush-equipped motors produce a current arc between the commutator and the brush every time a brush breaks contact with a segment. Interpoles can minimize this effect, but it cannot be entirely eliminated. There is no such arcing in a brushless motor because there are no mechanical contacts. Besides, the inductive kickback voltage arising from their sudden current interruption doesn't rise very high. It is held to a reasonable value by kickback-suppression diodes, as shown in Fig. 13–7. For electronic commutation systems that produce both directions of winding current, a bidirectional suppression device, a *thyrector,* can be installed, as shown in Fig. 13–21.

A thyrector is like two back-to-back zener diodes. It acts like an open circuit until current interruption makes the winding's kickback voltage rise to its rated value, say ±50 V. The thyrector then becomes a short circuit to any further voltage rise, clipping the actual winding transient voltage value at 50 V. By this means the interrupting transistor is protected from high-voltage breakdown of its reverse junction.

This more orderly interruption of current and dissipation of the windings' magnetic energy translates into much less radio frequency interference from brushless dc motors. Noise-sensitive electronic circuitry in the vicinity can be shielded more easily than for a comparable brush-equipped dc motor.

Precision. In object-positioning applications (servo systems), a stepper motor is more precise than a conventional brush-equipped dc motor. With a stepper, the final position is known to within a small percent of one step. With a standard dc servo motor the precision of the final position depends on the servo amplifier gain (see Figs. 9–4, 9–5, and 9–7).

The standard servo is subject to analog noise errors. The stepper motor is fundamentally a digital device, so it is not subject to analog noise unless the noise is so great that it destroys the pulse-to-step integrity of the control system.

Mechanical holding. Stepper motors, especially permanent-magnet types, possess an inherent shaft-holding ability. They have their detent torque with the windings deenergized; they have their static holding torque with the windings energized. Traditional brush-equipped motors do not possess inherent shaft-holding ability. They must first slip out of position in order to produce an error signal before the servo amplifier comes to life; only then can a dc servo motor develop torque to restore an object to its proper position.

FIGURE 13–21
A thyrector, also called a *varistor,* is useful for limiting the winding kickback voltage in either polarity. It would be used in the circuits of Figs. 13–14 and 13–19.

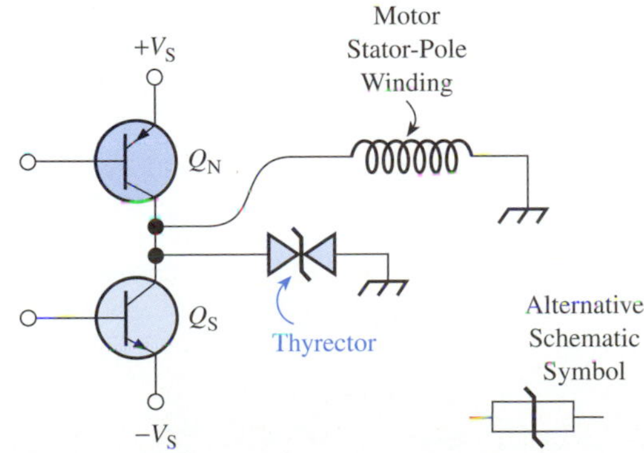

Speed range. Stepper motors have absolutely no trouble in turning the shaft very slowly. We just slow down the STEP frequency. Wound-rotor dc motors tend to stall when they try to run slower than a few hundred r/min. For slow-speed applications such brush-equipped motors need gear-reduction.

Position-triggered brushless dc motors are able to run successfully at fairly slow speeds. Their speed is varied by adjusting the source voltage driving the stator windings in Fig. 13–19.

With their simpler rotor construction, electronically commutated motors have the mechanical potential to spin much faster than wound-rotor machines. While most industrial wound-rotors have an upper speed limit of about 4000 r/min, position-triggered brushless dc machines can reach speeds in excess of 20 000 r/min.

Power. The advantage of traditional wound-rotor brush-equipped dc motors is their greater torque and power, of course. The most powerful stepper motors develop maximum torques on the order of 1 N-m, and that only at slow stepping speeds. Position-triggered brushless motors have comparable torque production. By contrast, an integral-horsepower wound-rotor dc motor develops about 10 times as much torque and does not have to slow down very much in order to do so (shunt or compound configuration). The greater-torque/greater-speed combination gives the wound-rotor motor an enormous power advantage over electronically commutated types. Wound-rotors don't stop at integral horsepower sizes. Full-load shaft ratings of several hundred horsepower are common.

Efficiency. Modern wound-rotor dc motors have full-load efficiencies exceeding 90%. Electronically commutated motors are not as efficient. Stepper motors are poorer than position-triggered brushless motors in this regard.

Table 13–8 summarizes the advantages and disadvantages, performance characteristics, and typical applications of the various kinds of dc motors.

TABLE 13–8
Dc motor comparisons

MOTOR TYPE	ADVANTAGES	DISADVANTAGES	PERFORMANCE CHARACTERISTICS	TYPICAL APPLICATIONS
Wound-rotor, wound-field	High efficiency	Commutator-brush structure limits trouble-free life. Expensive. Produces great deal of electromagnetic interference.	Can be tailored to the application by degree of compounding. Can be built for very high power output.	Medium- to high-power servo or adjustable-speed applications.
Conventional permanent-magnet, brush-equipped (wound-rotor)	High efficiency	Same as wound-rotor, wound-field. Disadvantage compared to wound-field machines in that compounding is not possible.	Good speed regulation ($S_{reg} < 10\%$). Moderate starting torque. Maximum size about 1 hp (about 750 W).	Medium-power servo systems.
Coreless	Very fast acceleration and slowdown. High efficiency. Brushes and commutator don't wear out as quickly as other brush-equipped types because armature winding inductance is lower. Low inductance contributes to fast electrical time constant to match fast mechanical time constant (low inertia).	Low full-load torque production. Rotor material is not thermally rugged; subject to damage.	Maximum size fractional hp (several hundred watts).	Low-power, fast-response servo systems, seen in computer peripherals, office equipment, and instrumentation.
Permanent-magnet stepper	Naturally digital, no D/A conversion needed. Digitally reversible — no high-power polarity switching required. Precise positioning — very small step angles possible. Good shaft-holding ability.	Poor efficiency. Only moderate rotation speeds attainable.	Maximum size fractional hp. Can produce perfect tracking between two motors. Can turn very slowly.	Machine-tool control ("numerical" control) and other servo systems where precise positioning and digital compatibility are important. Standard servo feedback is not required.
Variable-reluctance stepper	Naturally digital. Fairly good positioning precision. Fairly good shaft-holding ability when energized. Digitally reversible.	Poor efficiency. Overshoot problem with high-inertia loads.	Maximum size fractional hp. Can turn very slowly.	Similar to permanent-magnet steppers.
Position-triggered brushless	Moderate power and torque without commutator and brushes. Trouble-free and long-lived. Good efficiency. Digitally reversible.	Hall-effect sensing system susceptible to electromagnetic interference from surroundings.	Maximum size about 1 hp. Capable of extremely high speeds.	Used for low-RFI-emission, high-efficiency servo systems and speed control. Very useful in flammable or explosive environments. With proper shaft bearings, can operate submerged in liquid.

TROUBLESHOOTING ON THE JOB

LEARNING A COMPLEX STEPPER MOTOR TEST CIRCUIT

Your company has purchased a batch of stepper motors that are specified as having the characteristic operating curve of Fig. 13–8.

Your assignment is to test every motor in the entire batch, to verify that each one is able to

1. Step reliably at a rate of at least 190 steps/s in the bidirectional mode (full-step, one-phase operation).
2. Step reliably at a rate of at least 260 steps/s in the unidirectional mode, including appropriate ramping up from 190 steps/s and back down to 190 steps/s.

The testing system is shown in Fig. 13–22. Before you can begin the testing process, you must have a complete understanding of how it works and how it is used. Explain the system in detail, including how you will use it to perform the motor tests.

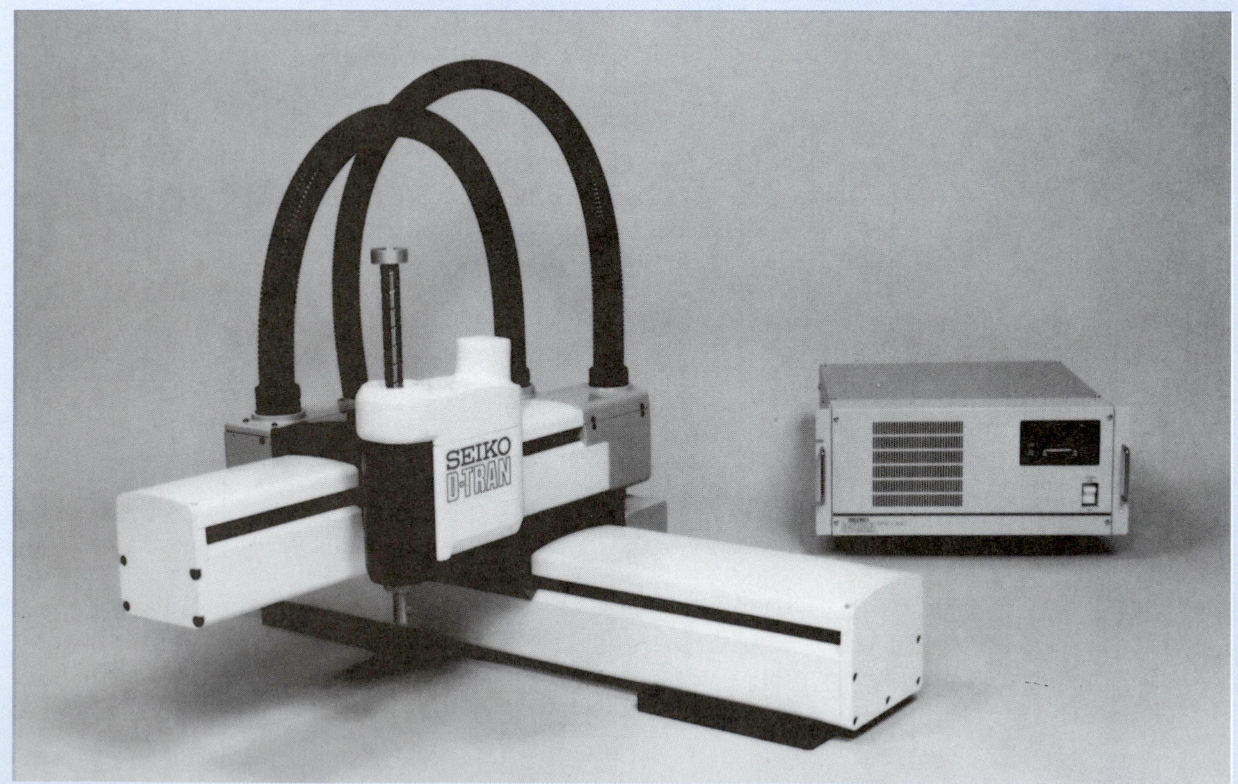

Three stepper motors accomplish three-dimensional motion.
Courtesy of Seiko Instruments USA, Inc.

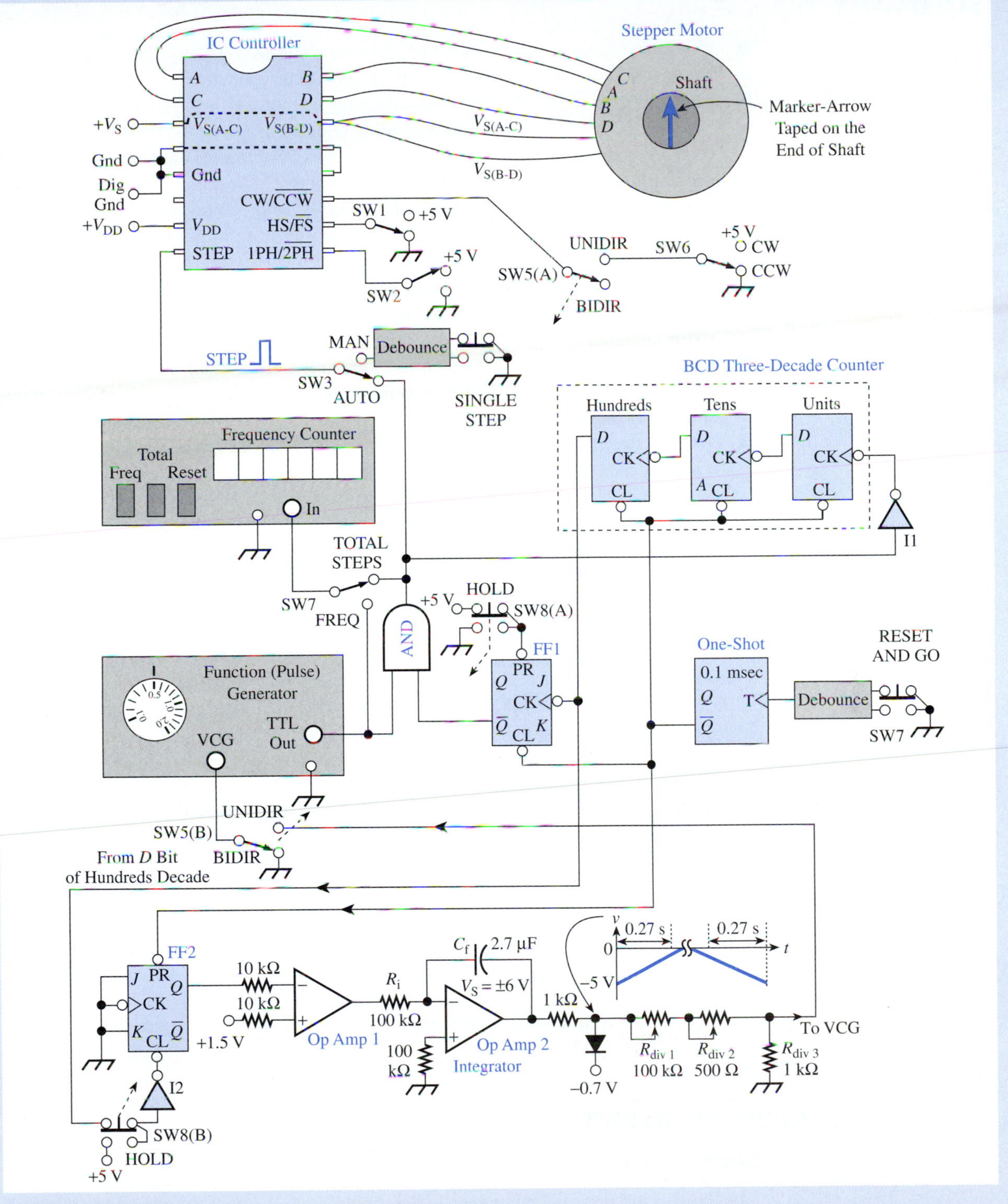

FIGURE 13–22
Stepper-motor test system.

SUMMARY

- A conventional permanent-magnet dc motor simply replaces the wound-field electromagnets with permanent magnets mounted on the stator.
- There are two types of coreless (ironless-rotor) motors, the cup structure and the disk structure. Both have an armature winding consisting of printed copper tracks or wires on a nonconducting surface. Their advantage is very low mechanical inertia, and consequently very quick acceleration and stopping.
- A stepper motor has no armature winding on its rotor, so no brush-to-commutator assembly. The permanent magnets on its rotor step into alignment with the energized stator windings, which are switched On and Off by external transistors.
- Variable-reluctance stepper motors don't actually have permanent-magnet poles on the rotor, just salient "teeth." The teeth become magnetized in response to the transistor energization of nearby stator windings.
- Stepper motors can be made to take either full steps or half steps, and to step either clockwise or counterclockwise, depending on the specific sequence of energizing the stator windings.
- When a stepper motor is operating in bidirectional mode (any step can be followed by a step in the opposite direction), the motor has a certain characteristic curve of maximum allowable speed versus torque. When the same motor is operating in unidirectional mode (never reversing direction), it will have a higher maximum speed-vs.-torque curve, but the higher speed must be reached gradually, by ramping up from the lower bidirectional speed. And likewise when the motor slows down prior to stopping altogether.
- Stepper motors can be made to take small fractions of a full step, by modulating the power supply voltages to the phase windings. This practice is called microstepping.
- The brushless dc motor (position-triggered) contains Hall-effect position-sensors that detect when the rotor is in the proper instantaneous position for the stator electromagnet coils to be switched On and Off. The Hall-effect sensors signal an IC sequence controller, which then switches the control transistors.
- Electronically commutated dc motors have certain advantages over wound-rotor dc motors in the small to medium sizes: They are less expensive, more reliable, run cooler, generate less electronic noise, and can achieve very precise shaft position.

FORMULAS

$$1 \text{ N-m} = 141 \text{ oz-in}$$

$$1 \text{ rad/s} = 9.551 \text{ r/min}$$

$$S \text{ (in r/min)} = \text{stepping rate (in steps/s)} \times \text{step angle (in deg)} \times \frac{1}{60}$$

QUESTIONS AND PROBLEMS

Section 13-1

1. Explain the difference between a traditional wound-rotor/wound-field motor and a conventional permanent-magnet motor.
2. Draw a cross-sectional view of a six-pole conventional permanent-magnet motor with radial magnets. Label the field poles with their magnetic polarity, and show

the flux distribution throughout the machine. Show six armature brushes contacting the commutator.

3. In the six-pole motor of Question 2, how far apart in angular degrees are the sides of an individual armature coil?
4. True or False: A permanent-magnet motor with lengthwise-magnetized arc-shaped permanent magnets on its stator needs an extra-thick frame.

Section 13-2

5. Explain the fundamental advantage of the coreless motor.
6. True or False: A cup-structure coreless motor has the standard radial flux orientation that wound-field motors have.
7. True or False: A disk-structure coreless motor has the standard radial flux orientation that wound-field motors have.
8. True or False: In the cup structure and disk structure, the copper tracks of a single armature "coil" are on opposite sides of the fiberglass thickness.
9. True or False: In the cup structure and disk structure, the copper tracks of a single armature "coil" are directly across from each other through the fiberglass thickness.
10. Do coreless dc motors produce a large amount of torque for their physical size? Explain this.

Section 13-3

11. Draw the cross section of a permanent-magnet stepper motor with a four-pole stator and a two-pole rotor. Label the stator poles A, B, C and D.
12. For the motor you drew in Question 11, if the controller energizes just a single stator winding at a time, how far does the motor step?
13. Repeat Question 12, assuming that the controller alternates between energizing one winding and two windings.
14. In the motor that you drew for Question 11, suppose that we wished to energize *both* windings of a phase whenever the phase was energized at all. In other words whenever winding A was energized winding C was also energized, and vice versa, and the same for the *B-D* phase. Should the windings be wrapped as shown in Fig. 13–5 so that any pole with an energized winding becomes magnetic North? Explain this carefully.
15. An IC controller that is designed for use with a particular model of stepper motor has a digital input terminal labeled CW/CCW. Explain the function of that pin.
16. The same IC controller has a digital input terminal labeled HS/FS. Explain the function of that pin.
17. A stepper IC controller has a digital input terminal labeled DIR. Explain the function of that pin.
18. A certain stepper motor can be operated in one-phase mode or two-phase mode, like the one in Fig. 13–5. If it is taking full steps in the two-phase operating mode, tell what is different about the motor's behavior compared to full steps in the one-phase mode. Be sure to state all differences, not just a single one.
19. Define static holding torque for a stepper motor.
20. Define detent torque for a stepper motor.
21. True or False: When a stepper motor is stepping at a fast stepping rate, it can deliver greater maximum torque to its load than when it is stepping at a slow rate.

22. When a stepper is operating within the __________ region of its characteristic graph, it can instantly change stepping directions from one step to the next if its controller wants it to.
23. To get to the __________ region of a stepper's characteristic graph, the stepping rate must begin slowly, then ramp higher.
24. True or False: To slow down from the condition of Question 23, the stepping rate must be ramped lower instead of abruptly changing in frequency.
25. True or False: When ramping a stepper up or down, you are allowed to write the program so that it takes *more* steps than specified by the manufacturer to get from one rate to another, but you cannot take *fewer* steps.
26. True or False: It would be reasonable for a stepper motor to deliver 5 N-m of full-load torque when operated in two-phase mode.
27. True or False: With programmable dc power supplies, it is possible to write a control program that reduces the step angle to one fifth of the normal half-step angle.
28. True or False: Repeat Question 27 for one fiftieth of the normal half-step angle, speaking theoretically.
29. The process referred to in Questions 27 and 28 is called __________.
30. True or False: A disk-structure permanent-magnet stepper motor tends to have many more rotor poles than a standard cylindrical structure.
31. Based on your answer to Question 30, the disk-type stepper motor takes __________ steps than the cylindrical-rotor type (larger or smaller). Explain your answer.
32. In Fig. 13–14(d), which transistor must be turned ON to make stator pole A North in front (on the front side of the disk) and South on the rear side. Answer by carefully tracing the winding's wrap and applying the right-hand rule.
33. A certain stepper takes 7.5° steps. Its stepping rate is 300 steps/s, and it is delivering 0.4 N-m of average load torque. Find its rotational speed in r/min.
34. For the situation in Question 33, do you have enough information to calculate the mechanical shaft power? Explain.
35. Your answer to Question 34 should have been yes. Go ahead and calculate shaft power.
36. True or False: All other things being equal, variable-reluctance stepper motors produce more torque than permanent-magnet stepper motors. Explain your answer.
37. A three-phase, four-pole-per-phase variable-reluctance stepper has a stator just like the drawing in Fig. 13–15. Its rotor has 10 teeth. What is its full-step angle?
38. Variable-reluctance steppers have a __________ severe overshoot problem than permanent-magnet steppers (answer more or less). Explain why this is so.
39. To suppress overshoot, we use the practice of __________.
40. True or False: If overshoot is severe, it is possible for the motor to step in the wrong direction when its controller receives a STEP pulse.
41. True or False: Stepper motor damping can be accomplished either mechanically or electrically.
42. Explain the difference between salient stator poles and nonsalient poles.

Section 13-4

43. True or False: In a brushless dc motor the switching transistors are switched ON and OFF at a rate determined by the control program.

44. True or False: In Fig. 13–19, as the rotor North pole approaches HD_{horiz} in the clockwise direction, it produces v_{horiz} of one polarity, but if the rotor North pole approaches in the counterclockwise direction, v_{horiz} has the opposite polarity.
45. In the brushless dc motor of Fig. 13–19, how many thyrectors are required to suppress RFI due to inductive kickback?
46. If the dc supply voltages are $\pm V_S = \pm 80$ V, suggest a reasonable voltage rating for the thyrectors of Question 45.
47. The answer to Question 44 is false. What *does* reverse the polarity of v_{horiz}?
48. True or False: In the motor of Fig. 13–19, the switching frequency of an individual transistor is equal to the number of rotations per second.

Section 13-5

49. Which dc motor is best adapted to turning very slowly and guaranteeing very precise final position?
50. Which dc motor is best adapted to accelerating its load extremely rapidly from a standstill?
51. Which dc motor is best adapted to delivering great torques in the 100 N-m range?
52. Which dc motor gives reliable maintenance-free service, is efficient, and can be reversed without switching a high-power dc circuit?
53. True or False: One of the advantages of the wound-rotor/wound-field dc motor is its good thermal conductivity between the heat-wasting part of the machine and the surrounding air.
54. Which types of dc motors have been benefited by the introduction of inexpensive high-voltage, high-current switching transistors?

CHAPTER

14 AC MOTORS

The appeal of dc motors is that their speed is easily adjustable. Most dc motors speed up or slow down as the applied voltage is varied. Wound-rotor/wound-field, conventional permanent-magnet, coreless, and position-triggered brushless types work this way. Stepper motors are different in this regard; they change speed as the step-pulse frequency is varied via digital techniques. However, the common feature among all dc motors is that speed control is *easy*.

The major drawback of dc motors is that they require a switching process. In the traditional wound-rotor/wound-field, conventional permanent-magnet, and coreless types, switching is done mechanically by a commutator-brush arrangement. In brushless and stepper types the switching is done electronically.

On these two issues, ac motors are the converse of dc motors. One of ac motors' major appeals is that they don't need any switching process in order to run (starting can be another matter). Their main drawback is that their speed is not easily adjustable since it is so closely tied to the ac line frequency. Of course, since they connect directly to the ac line, ac motors have the advantage of not requiring a rectified dc power supply.

OBJECTIVES

After completing this chapter, you will be able to:

1. Explain how the rotating magnetic field is developed in a two-phase motor.
2. Draw the stator cross section of a two-pole induction motor with salient poles; do the same for a four-pole motor.
3. Draw the stator cross section of a four-pole induction motor with nonsalient poles.
4. Relate the synchronous speed of an induction motor to its poles (per phase) and the ac line frequency. Calculate any one of these if the other two are known.
5. Describe the construction of a squirrel-cage rotor.
6. Explain the torque-production process in a squirrel-cage rotor.
7. List the advantages of an induction motor over a conventional brush-equipped dc motor.
8. Draw the schematic diagram of a capacitor-assisted split-phase motor, and explain its principle of operation.
9. Draw the cross section of a shaded-pole motor, state its direction of rotation, and explain its principle of operation.
10. Draw the cross section of a reluctance-start motor, state its direction of rotation, and explain its principle of operation.

11. Define the slip of a squirrel-cage induction motor.
12. Draw the fundamental speed-vs.-torque characteristic graph of a squirrel-cage motor, and explain why it has that shape.
13. Identify the no-load speed, the full-load speed and torque, the maximum torque, and the starting torque on the squirrel-cage motor's characteristic curve.
14. Draw the characteristic curve of current versus speed for a squirrel-cage induction motor.
15. Draw the phase voltage waveforms for a three-phase ac system.
16. Draw the schematic diagram, identify polarities, and label the variables in a three-phase wye system.
17. Draw the schematic diagram, identify polarities, and label the variables in a three-phase delta system.
18. List the advantages of three-phase ac systems over single-phase and two-phase systems.
19. Relate line voltage and current to phase voltage and current in both wye and delta systems.
20. Calculate apparent power S and true power P in balanced three-phase systems.
21. Draw the stator cross section of a two-pole, three-phase motor with salient poles, and explain how it develops a rotating field.
22. Draw the stator cross section of a two-pole, three-phase motor with nonsalient poles, and describe the manufacturing process of distributing the windings over the stator surface. Explain why this gives three-phase motors greater power per unit of volume or weight than other ac motors.
23. Contrast the operating characteristics of NEMA type B, C, and D three-phase motors. Describe in a general way how the motor manufacturers achieve these different characteristics.
24. Sketch the general efficiency-vs.-torque and power factor-vs.-torque characteristic curves for three-phase induction motors.
25. Draw the schematic diagram of a three-phase across-the-line starting circuit; do the same for a reversing starting circuit.
26. Properly wire a 9-terminal dual-voltage three-phase motor for higher-voltage or lower-voltage operation.
27. Properly wire a 12-terminal dual-voltage three-phase motor in delta configuration for higher-voltage or lower-voltage operation.

14-1 ■ THE ROTATING FIELD

Refer back to the ac motor organization-classification diagram in Fig. 12–1. It shows three basic classes of ac motors; induction, synchronous, and universal. Of these three the induction and the synchronous classes depend for their operation on the idea of the rotating magnetic field. Since our main interest in this chapter is the induction class, we must first get a clear understanding of the rotating field idea.

Actually, the concept of stator windings creating a moving magnetic field is already familiar to us from our study of electronically commutated dc motors. For example, the brushless dc motor of Fig. 13–19 uses its stator windings to move the stator magnetic field direction from one position to another every time its controller switches ON a different winding or winding-pair in the sequence. All the sequential magnetic flux positions were indicated in Fig. 13–20.

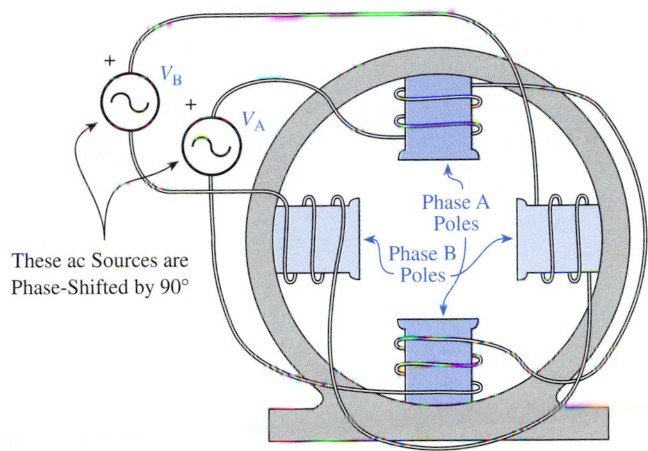

FIGURE 14–1
Stator of two-phase, two-pole-per-phase ac motor. This diagram shows salient poles for clarity, but most ac motors actually have non-salient poles. For them, the windings are placed in stator slots that are perpendicular to the page.

In the realm of ac motors the same general effect is created. However, it is created by gradual sine-wave variations in the stator windings' currents rather than by abrupt ON-OFF switching of stator windings.

The easiest way to understand the creation of a rotating field by ac sine waves is to study a two-phase motor. Look at Fig. 14–1, which is a cross section of the stator of a two-phase motor with salient poles.

Ac sine-wave source A drives the windings that create vertical flux in Fig. 14–1. The ac source and its windings and poles are collectively called phase A. Horizontal flux is created by phase B, consisting of ac source B and its windings and poles.

When V_A is instantaneously positive, its flux is North on the top pole and South on the bottom pole. Therefore its flux contribution through the motor's rotor (through the interior of the stator in Fig. 14–1) is vertical, pointing from top to bottom. Trace the winding wraps carefully in Fig. 14–1, and apply the right-hand rule to satisfy yourself that this is true. When V_A is instantaneously negative, the phase-A flux reverses direction, pointing through the motor interior from bottom to top.

When phase B is instantaneously positive, it contributes right-to-left horizontal flux. When it is negative, its flux contribution reverses to left-to-right. Verify this by tracing the winding wraps.

The two phase waveforms are shown in Fig. 14–2(a). Phase A leads phase B by 90°. The waveforms refer to both voltage and current. In other words, we are assuming for simplicity that a phase winding's ac current I is in phase with its ac source voltage V. This is approximately true for an ac motor driving a heavy torque load. Under that condition the inductive reactive nature of a winding is swamped out by its resistive behavior.

The flux contributions of the A and B phases are shown at 45° sine-wave increments in Fig. 14–2(b) along with the net flux resulting from their vector addition. For example, at the 45° instant, the A flux points downward with 70.7% of its peak magnitude ($i_A = I_{pk} \times \cos 45°$). The B flux points left, also with 70.7% of peak magnitude ($i_B = I_{pk} \times \sin 45°$). The vector sum points to the lower left, rotated 45° from its vertical starting position at the 0° instant.

At the 45° instant, the resultant net flux has a magnitude equal to the peak magnitude of either individual phase's flux. Figure 14–3 demonstrates why this is so.

A similar derivation at any instant on the sine waveform would show that the net flux Φ_{net} is always equal to Φ_{pk} of either individual phase.

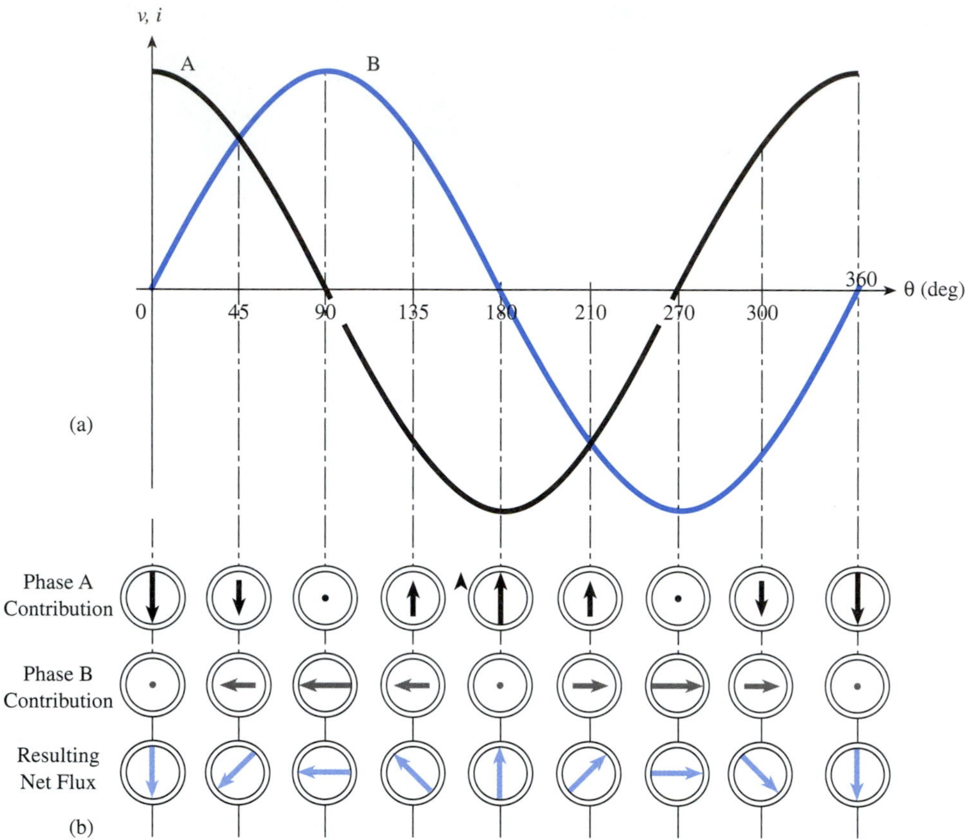

FIGURE 14–2 Creating the rotating field in a two-phase motor. (a) Waveforms of phase voltages and currents. (b) Instantaneous phase fluxes and resulting net flux orientation.

At the 90° instant of Fig. 14–2(a) the current in the A phase has fallen to zero, while the current in the B phase has risen to its peak. Part (b) shows that the flux contribution from A is zero and the flux contribution from B is pointing left, at peak value. Therefore Φ_{net} points left with magnitude equal to Φ_{pk}.

At 135°, the A phase has gone negative. v_A is negative on top and positive on bottom in Fig. 14–1. The phase winding current reverses, making the bottom pole North and the top pole South. In Fig. 14–2(b) the A-phase flux contribution points up. It adds to the

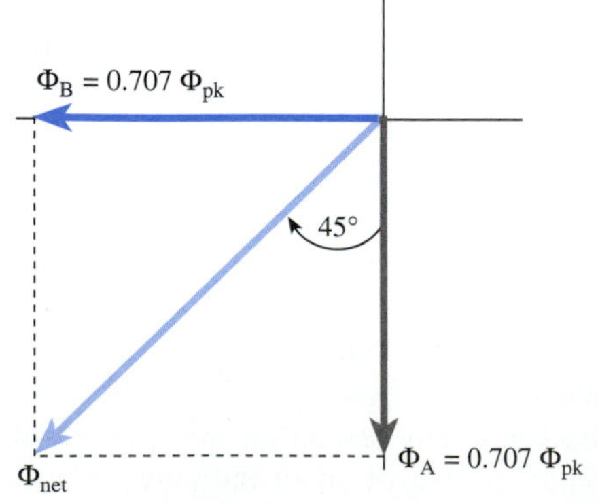

$$\Phi_{net} = \sqrt{\Phi_A^2 + \Phi_B^2}$$
$$= \sqrt{(0.707\ \Phi_{pk})^2 + (0.707\ \Phi_{pk})^2}$$
$$= \sqrt{0.5\ (\Phi_{pk}^2) + 0.5\ (\Phi_{pk}^2)}$$
$$= \sqrt{1.0\ (\Phi_{pk}^2)}$$
$$= \Phi_{pk}$$

FIGURE 14–3 Showing why net flux Φ_{net} produced by both windings is equal to peak flux Φ_{pk} of a single winding.

B-phase left-pointing flux to produce Φ_{net} pointing to the upper left, mechanically displaced 135° from its starting position.

Study the remaining instants in Fig. 14–2 to understand that one cycle of the ac sine wave produces one clockwise revolution (360° mechanical movement) of the stator's net flux. By the way, since the flux is created in the outside stator and moves circularly around the inside rotor, a better name would be *revolving field,* rather than *rotating field*. However, we will abide by common usage and call it rotating field.

More than two poles per phase. The cross-section diagram in Fig. 14–1 shows two poles per phase. At every instant one of them will be magnetically North and the other will be magnetically South, except at the crossover instants when they are deenergized. It is also quite possible to build ac motors with four poles per phase, or any even number of poles per phase. For example, Fig. 14–4(a) shows a stator cross section of a two-phase motor with four poles per phase, salient. The more realistic nonsalient construction is drawn in Fig. 14–4(b).

The winding placement in Fig. 14–4(b) is theoretically OK. The stator slots that hold the opposite sides of a winding are slightly less than 45° apart for this four-pole machine. However, the preferred construction method would have the slots exactly 45° apart, with each slot only half-filled by one pole's winding; the other half of the slot-depth (the half that is radially closer to the machine middle) would be filled to the inside edge of the stator with the neighboring pole's winding. For example, at the very top of the stator there would be just one slot, not two. Half of the total depth (not width) of the slot would be occupied by the winding that spans from 12 o'clock to 3 o'clock; the other half of the slot's depth would be occupied by the winding that spans from 9 o'clock to 12 o'clock. This explanation ignores the issue of distributed winding groups. We will take up the winding-distribution issue later, in Section 14-6, when we study three-phase motors.

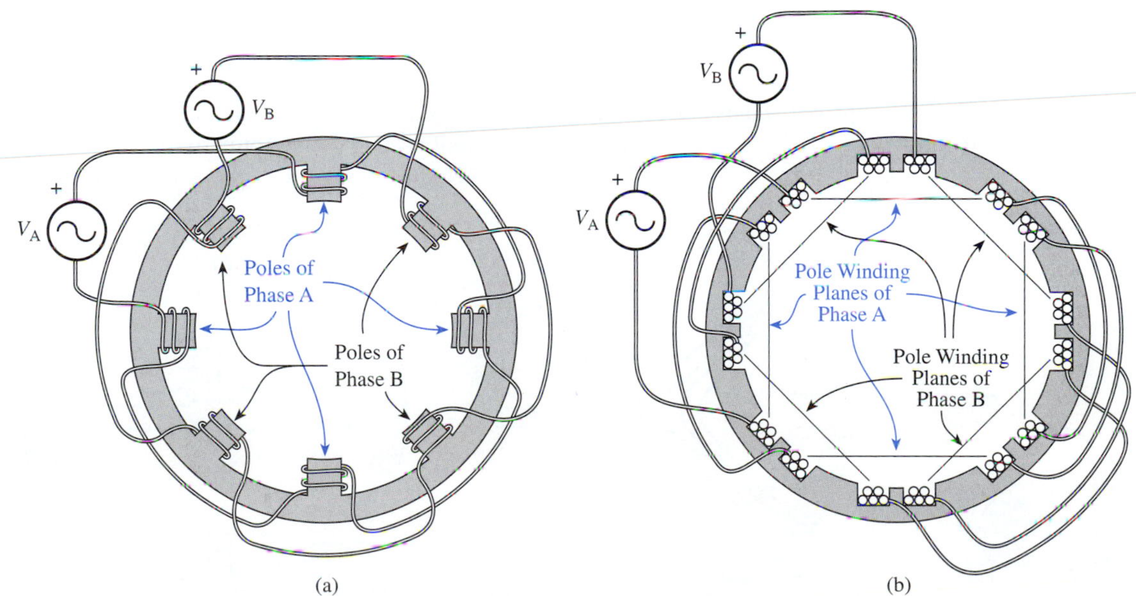

FIGURE 14–4
Four-pole motor. (a) Salient poles (easy to see the poles but not a realistic construction). (b) Nonsalient (embedded) poles (more realistic construction). When we state a motor's number of poles, we mean the number of poles *per phase*, not the total number of poles altogether.

An interesting thing happens when we build our motors with more than two poles per phase: The angular speed of the rotating field becomes *slower*. To understand why this is true, let us study Fig. 14–4(a). At the 0° sine-wave instant the A phase carries peak current and the B phase is at zero—refer to Fig. 14–2(a) but do not look at part (b) (it is no longer valid for a motor that has four poles).

The mechanical flux distribution in the four-pole motor at the 0° instant is drawn in Fig. 14–5(a). The 12-o'clock and 6-o'clock poles are North, and the 3-o'clock and 9-o'clock poles are South. The mechanical field position at the 90° instant is drawn in Fig. 14–5(b). With the phase-B sine wave at its peak, the phase-B poles have their peak flux, but the phase-B poles are offset by only 45° mechanically from the phase A poles. Therefore the field rotates only 45° mechanically as the electrical waveform goes through 90°.

The mechanical field flux orientation at the 180-electrical-degree instant is drawn in Fig. 14–5(c). As the sine wave advances by 180°, or one half cycle, the poles reverse magnetic polarity. This is the natural way that electromagnets always behave. Now we have North at 3 and 9 o'clock and South at 6 and 12 o'clock, so the rotating field has rotated by 90° mechanically as the electrical line advanced by 180°.

We could continue this analysis, but we have already seen enough to realize what is happening: The mechanical rotation is only half as rapid as the electrical oscillation. To make the point even more clearly, the mechanical rotation of the field in a four-pole motor is only half as fast as the rotating field in a two-pole motor.

In North America, where the ac line oscillates at the frequency of 60 cycles per second, the rotating field of a two-pole motor has a speed of 60 rotations per second, but the rotating field of a four-pole motor has a speed of only 30 rotations per second. Putting these speeds into the familiar r/min units, we get

$$\text{for a two-pole motor } S_{\text{syn}} = \frac{60 \text{ rot}}{\text{s}} \times \frac{60 \text{ s}}{1 \text{ min}} = 3600 \text{ r/min}$$

$$\text{for a four-pole motor } S_{\text{syn}} = \frac{30 \text{ rot}}{\text{s}} \times \frac{60 \text{ s}}{1 \text{ min}} = 1800 \text{ r/min}$$

The subscript *syn* stands for synchronous. It is used to make it clear that we're talking about the speed of the rotating field, not the actual speed of the motor shaft.

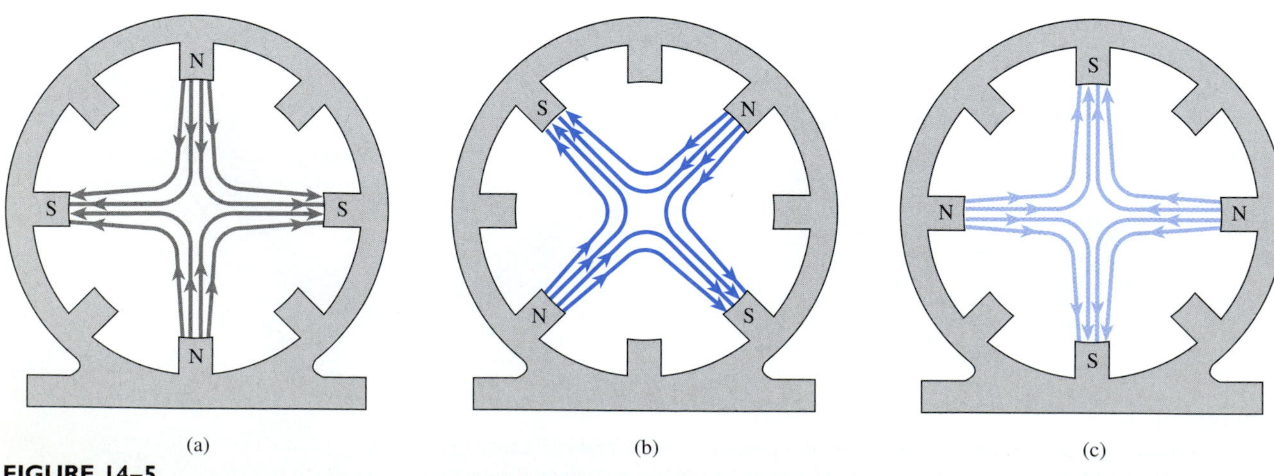

FIGURE 14–5
Rotation of the stator field flux during the first half cycle (180 electrical degrees) of the sine wave. (a) At the 0° instant. (b) The 90° instant. (c) The 180° instant.

In general, for a motor with P number of poles per phase, driven by the 60-Hz ac line, we can say

$$S_{syn} = \frac{(60\text{ Hz}) \times (60\text{ s/min})}{(P/2)} = \frac{(60\text{ Hz}) \times 120}{P} \tag{14-1}$$

where speed S_{syn} is in r/min units. The quantity $P/2$ is called the number of pole-pairs (per phase).

Generalizing to other ac line frequencies (all of Europe and Australia, parts of South America, parts of Africa, and parts of Asia), we have

$$S_{syn} = \frac{f \times 120}{P} \tag{14-2}$$

where f is the ac line frequency in hertz.

■ EXAMPLE 14-1

(a) Find the synchronous speed of the rotating magnetic field for a six-pole induction motor in North America.
(b) At what speed would that same machine's field rotate in Europe, where the ac line has $f = 50$ Hz?

Solution. (a) From Eq. (14-1) or (14-2),

$$S_{syn} = \frac{60\text{ Hz} \times 120}{P} = \frac{60\,(120)}{6} = \mathbf{1200\ r/min}$$

(b) From Eq. (14-2)

$$S_{syn} = \frac{f \times 120}{P} = \frac{50\,(120)}{6} = \mathbf{1000\ r/min}$$

■

If having a greater number of poles reduces the motor's synchronous speed (and consequently reduces its actual shaft speed, as we will learn in a moment), then we ought to gain something to compensate for our speed sacrifice. As you would expect, the thing that we gain is torque. For given physical machine dimensions, and all other things being equal, changing from a two-pole stator design to a four-pole stator design doubles the torque production. This is reasonable, since mechanical rotational power has two constituents, torque and speed, and they can be traded for one another. You know already that torque and speed can be traded for one another with mechanical gearing. Either one can be boosted at the expense of the other by selecting the gear ratio. Now we are saying that the trading can also be accomplished right at the point of original production, in the ac motor. To understand the actual *mechanism* by which torque is traded, we need to understand the actual mechanism by which torque is produced in an ac motor. Of course, in order to do that we must look at the rotor.

14-2 ■ THE SQUIRREL-CAGE ROTOR

For an ac motor the standard rotor structure is the squirrel-cage structure. Isometric and end-on drawings of a squirrel-cage rotor are presented in Fig. 14–6(a) and (b). Refer also to Photo 14–1.

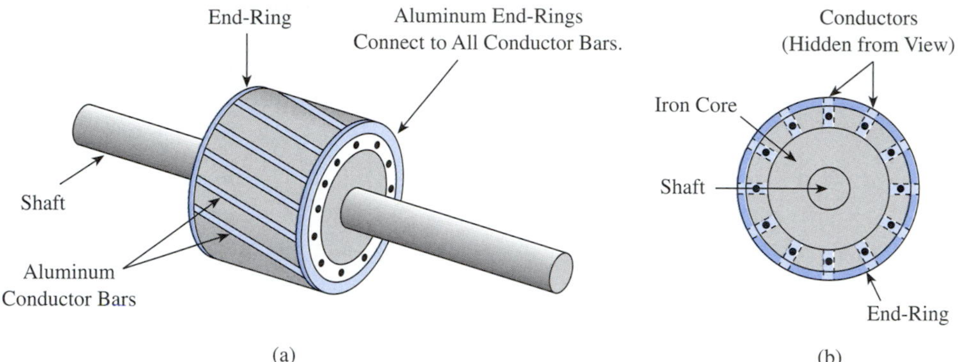

FIGURE 14–6 Squirrel-cage rotor. (a) Isometric view, which shows why this structure is given the name *squirrel-cage*. (b) This rotor has 12 conductor bars. All 12 are connected together at one end by an end-ring, and likewise at the other end.

The conductor bars and end-rings are usually aluminum, formed by a casting process, but in very high-power motors they may be copper. The core is an iron-based alloy with good magnetic permeability. The shaft is high-strength steel, as always. It is supported by two bearings mounted in the motor's end-bells, as usual.

Observe carefully that there are absolutely no electrical connections to the squirrel-cage rotor. That is, no external dc or ac supply is connected by brush to the rotor's conducting paths, neither through split-ring commutator nor continuous slip-rings.

Furthermore, there are no switches, either mechanical or electronic, in the stator circuits of Figs. 14–1 and 14–4. Thus the machine has no switches in the stator field, and not even any external connections—let alone switches—in the rotor armature circuit. This is the absence of switches that we touted in the chapter introduction.

Certainly currents flow in the conductor bars. It always takes electric current interacting with a magnetic field to produce mechanical force. The currents in the conductor

PHOTO 14–1 Photograph of the rotor from a 1/3-hp split-phase motor. The conducting bars are skewed to eliminate torque pulsation (similar to cogging in a dc machine); in this way shaft-to-bearing rattle is minimized.

bars are *induced* in them by the rotating magnetic field. We can understand how this works by making a careful study of three adjacent bars, which are presented in Fig. 14–7.

The three parts of Fig. 14–7 show the same section of the rotor at the same instant in time. We are not advancing in time. This is like three separate snapshots taken by three separate cameras, all triggered at the same instant.

Figure 14–7(a) shows us the overall situation in this vicinity of the squirrel-cage rotor. Figure 14–7(b) shows what is happening from the viewpoint of the "coil" consisting of bar 1, bar 2, and the end-ring segments that join them at both ends of the rotor. Figure 14–7(c) shows what is happening from the viewpoint of the coil consisting of bar 2, bar 3, and the end-ring segments. Notice that bar 2 is shared by, or is common to, both of these two "coils."

Let us assume that the stator flux is moving clockwise, as indicated in Fig. 14–7(a). The North flux is not concentrated at one point, of course. It is spread out over a certain angular range, as parts (b) and (c) make clear. However, it is *denser* in the center of the range; this is represented in parts (b) and (c) by showing the individual flux lines close together in the center, then getting farther apart as we look closer to the edges of the flux's range.

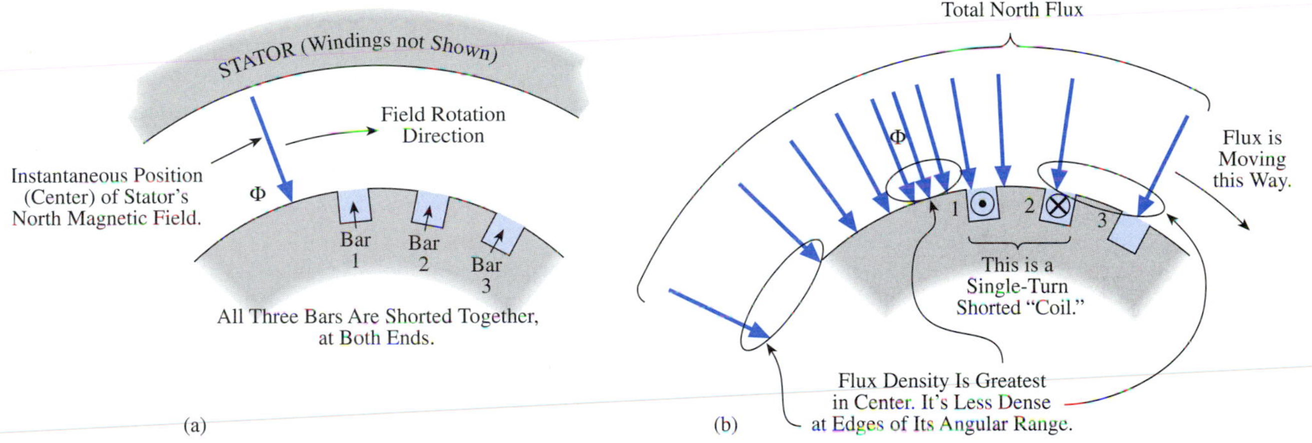

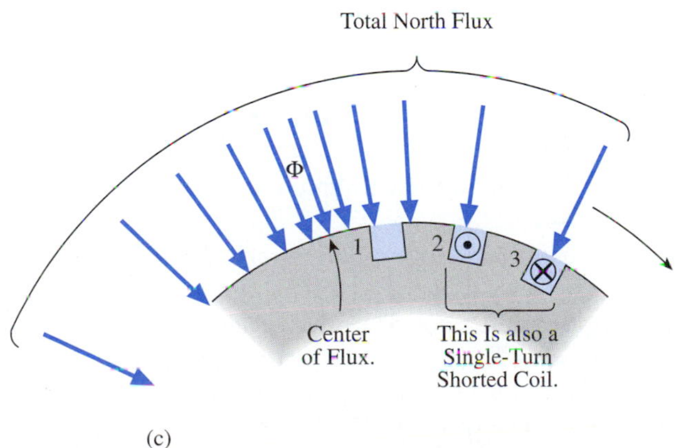

FIGURE 14–7
Making a careful examination of the induced voltage and resulting current direction in a squirrel-cage rotor bar. (a) Focusing on three individual rotor bars that are being approached by North flux from a rotating stator field. (b) What happens in the single-turn coil of bars 1 and 2. (c) What happens in the coil of bars 2 and 3.

The rotor is moving clockwise; we must assert this fact in advance, even before we have shown why. However, it's not moving as fast as the rotating field. Therefore, at the snapshot instant shown in Fig. 14–7, the North flux is sweeping across the three rotor bars from left to right (clockwise).

Figure 14–7(b) focuses attention on the coil made up of rotor bars 1 and 2. As the rotating field gains on the coil, the total North flux passing through the inside of the coil is increasing at some rate $\Delta\Phi_{1\text{-}2}/\Delta t$. This flux change induces a voltage in the coil, by Faraday's law. (The same principle as a transformer.) The induced voltage circulates a large current in the coil because the resistance of the coil is very low. But what is the direction of the current?

According to Lenz's law, the current direction must create its own flux that opposes the increase in North flux from the stator. That is, the coil's own flux must point from inside to outside, opposite to the stator flux. To accomplish this, the coil in Fig. 14–7(b) tries to circulate current into the page on bar 2 (tail-feather) and back out of the page on bar 1 (arrowhead). Apply the right-hand rule to satisfy yourself that this circulation opposes the North flux increase. This current circulation is what the bar 1/bar 2 coil would like to do, based on what it sees happening, strictly from its own viewpoint.

Now look to Fig. 14–7(c). This figure shows what the coil consisting of bars 2 and 3 would like to do. It too is experiencing an increase in North stator flux as the rotating field gains on it. The rate of change is $\Delta\Phi_{2\text{-}3}/\Delta t$. This flux change induces voltage by Faraday's law, which then circulates a large current by Ohm's law. The circulation direction in this coil is into the page on bar 3 (tail-feather), and back *out of the page on bar 2* (arrowhead). Lenz's law makes the 2-3 coil want to do this for the same reason that Lenz's law made the 1-2 coil want to do what is shown in part (b)—namely, so that the coil opposes the increase in North stator flux.

So bar 2 is being pushed and pulled in two opposite ways. The 1-2 coil in part (b) wants it to carry current into the page away from us; the 2-3 coil in part (c) wants it to carry current out of the page toward us. The actual direction of bar 2's current is decided by which induced coil voltage is greater; and that is decided, according to Faraday's law, by which coil is experiencing the more rapid time rate of change of flux, $\Delta\Phi/\Delta t$.

Now there is no doubt that the stator flux itself through the 1-2 coil, $\Phi_{1\text{-}2}$, is greater than the stator flux itself through the 2-3 coil, $\Phi_{2\text{-}3}$. However, does that mean that the rate of change is greater through the 1-2 coil? No. In fact, it means just the opposite. Where the flux is great, there its time rate of change $\Delta\Phi/\Delta t$ is small. Where the flux is small, there its time rate of change $\Delta\Phi/\Delta t$ is great. This is always how nature works, in any sine-wave situation. In our situation $\Delta\Phi_{2\text{-}3}/\Delta t$ is greater than $\Delta\Phi_{1\text{-}2}/\Delta t$, so the 2-3 coil dominates the 1-2 coil. Thus, the actual direction of current in bar 2 is out of the page, as shown in Fig. 14–7(c).

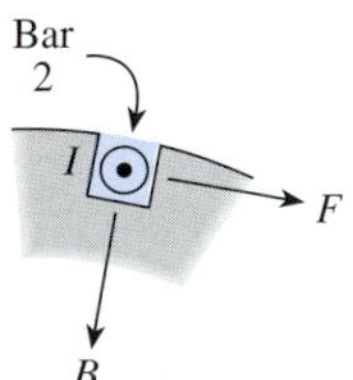

FIGURE 14–8
Applying the Lorentz relation to current flowing out of the page and the *B* field pointing down. Force *F* is to the right, producing clockwise torque.

Applying the Lorentz relation to the current in bar 2 and the downward-pointing magnetic field, we get mechanical force to the right, as Fig. 14–8 illustrates. This produces clockwise torque on the rotor. Thus our earlier assertion that the rotor spins clockwise, the same as the stator's rotating field, is now shown to be correct.

All this description was to gain an understanding of a single rotor bar. We certainly won't repeat it for every one of the rotor bars in Fig. 14–6(b), but if we did, we would find that all the bars that are receiving North flux are carrying current out of the page toward us. Also, all the bars that are experiencing South flux are carrying current into the page away from us. This overall situation is pictured in Fig. 14–9.

The situation for a four-pole motor with a counterclockwise rotating field is shown in Fig. 14–10. The direction of the rotating field could be changed from clockwise (Fig. 14–5)

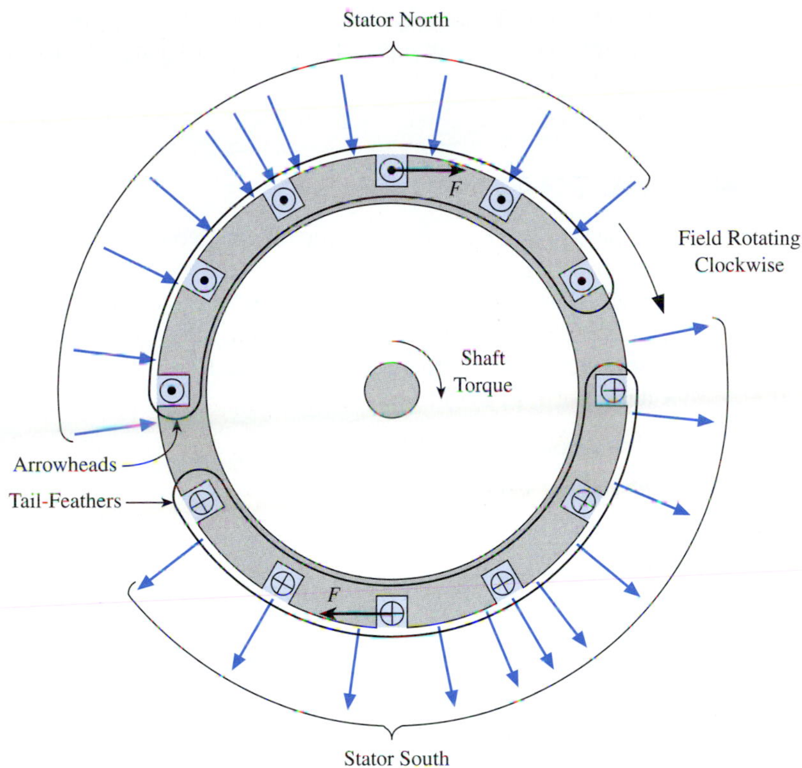

FIGURE 14–9
Instantaneous flux orientation and directions of rotor bar currents for a two-pole motor with a clockwise rotating field. For the arrowhead bars, all six mechanical forces point to the right, producing clockwise torque. Only one of those bar forces is shown, to keep the diagram simple. For the tail-feather group, all six mechanical forces point to the left. Use the Lorentz relation to prove this. Again only one of the six forces is drawn, but they all aid one another, producing clockwise torque.

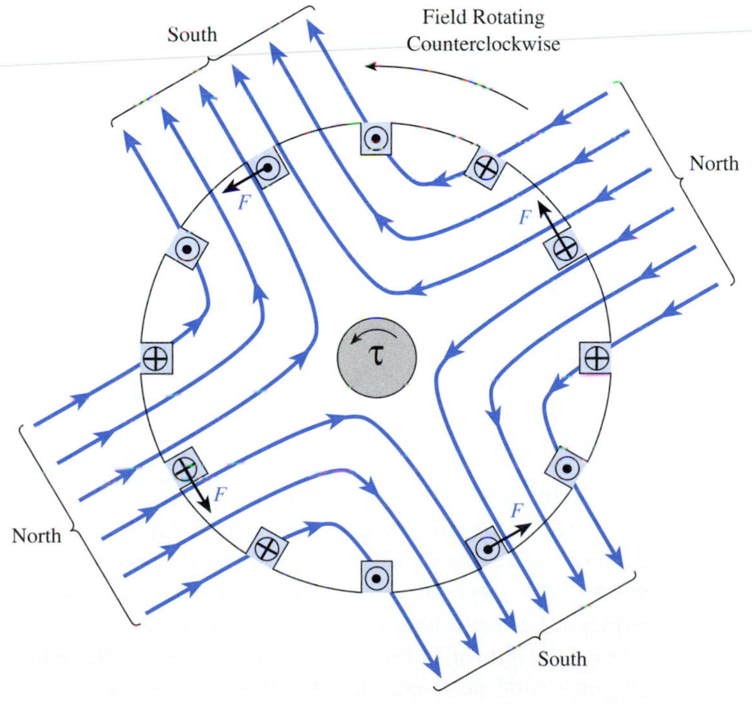

FIGURE 14–10
Instantaneous flux orientation and directions of rotor bar currents for a four-pole induction motor with a counterclockwise rotating field. All 12 rotor bar forces contribute torque in the counterclockwise direction. Only four of the bar forces are shown.

to counterclockwise (Fig. 14–10) by reversing the phase relationship between V_A and V_B—in other words, by arranging for V_B to lead V_A by 90° instead of the relationship shown in Fig. 14–2.

14-3 ■ GETTING THE PHASE-SHIFT

In Sec. 14-1 the creation of a rotating magnetic field was possible only because of the 90° electrical phase displacement between the two ac sources (Fig. 14–2a). It is the 90° electrical phase-shift together with the 90° mechanical offset of the motor windings (assuming a two-pole motor) that gives us the action of the rotating field. So a crucial issue is how we get access to 120-V two-phase ac power, which is represented by the phasor diagram of Fig. 14–11.

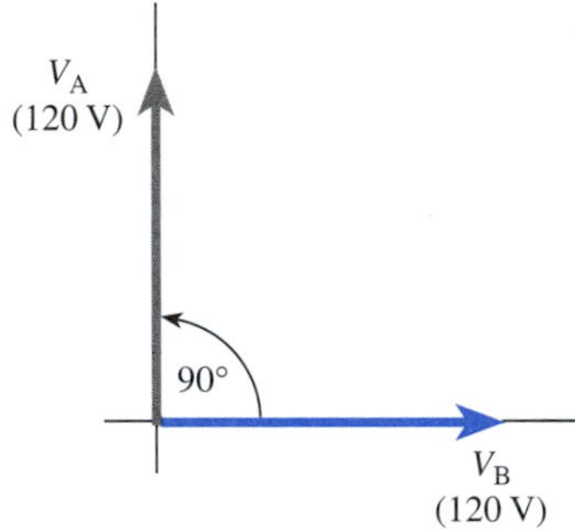

FIGURE 14–11
Phasor diagram of 120-V, two-phase ac power source. In all our phasor diagrams a phasor magnitude is assumed to represent an rms value, not its strict interpretation as a peak value.

Don't think that the 240-V center-tapped ac line in a residence can do the job. The two 120-V ac lines are 180° out of phase, not 90°. This is pictured in Fig. 14–12.

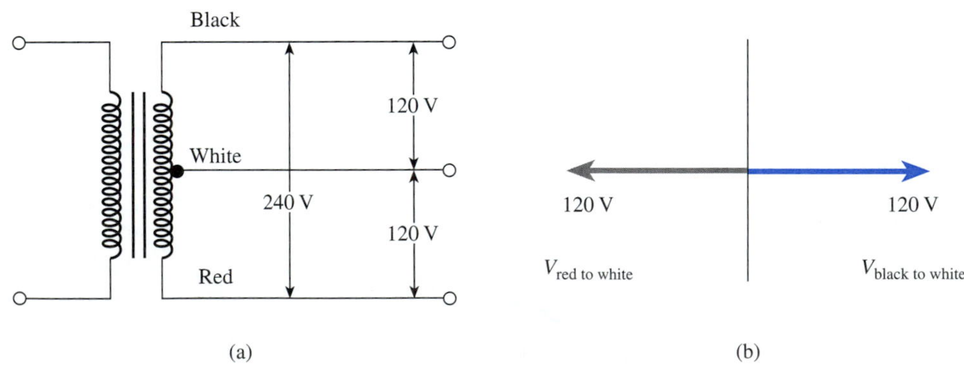

FIGURE 14–12
The electric utility's center-tapped 240-V ac line provides two 120-V sources, but they are out of phase by 180°, not 90°. (a) Schematic diagram of transformer secondary. (b) Phasor diagram.

In the first half of the twentieth century, electric utility companies generated and distributed 90° two-phase ac power, but they have abandoned that for the more economical 120° three-phase power that we will discuss starting in Sec. 14-5. It is possible to convert 120° three-phase power into 90° two-phase power by clever interconnection of two tapped transformers. This technique is called the *Scott tee method,* named after its inventor; it is shown in Fig. 14–13. We will not make a thorough study of how it works, but you should be aware of its existence.

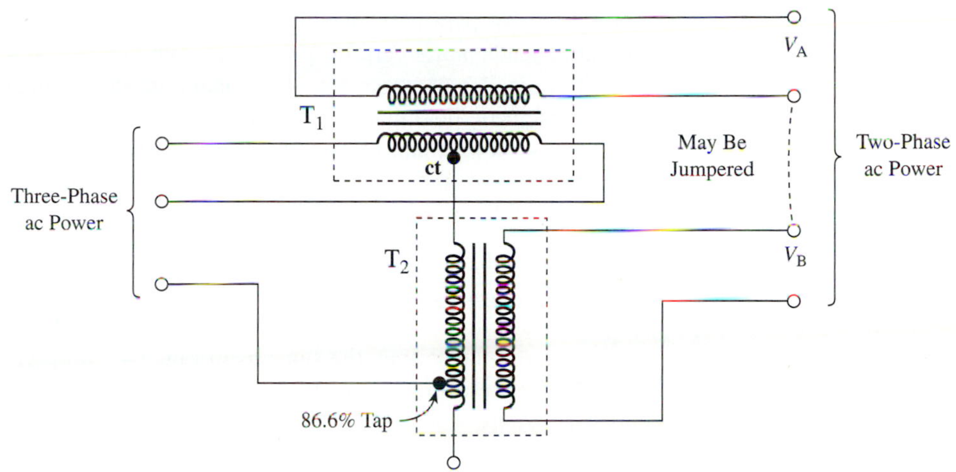

FIGURE 14–13 Scott tee transformer connection, for converting 120° three-phase into 90° two-phase ac power. Transformers T_1 and T_2 are identical, both having identical primary tap-points at 50% (center-tapped) and 86.6%. On T_1 we use the center-tap but not the 86.6% tap; on T_2 we use the 86.6% tap but not the center-tap.

14-3-1 Phase-Shifting Current with Capacitive Reactance—The Split-Phase Motor

Having two ac voltage sources that are phase-displaced by 90° is the ideal way to drive a two-phase induction motor, but it isn't absolutely necessary. If we can just arrange for the *currents* through the two phase windings to be out of phase with one another, that will create a rotating stator field. After all, magnetic fields originate with current, as we know from

$$B = \mu H = \mu \frac{NI}{l} \tag{14-3}$$

Furthermore, while the 90° phase difference of Fig. 4–2(a) is ideal, induction motors can operate with any reasonable amount of phase difference between winding currents I_A and I_B. They don't operate quite as well, but they do operate.

One simple way of getting some phase-shift between the winding currents is by placing a capacitor in series with one of the phase windings. When this is done, the motor is called a *split-phase motor*. It is shown schematically in Fig. 14–14(a).

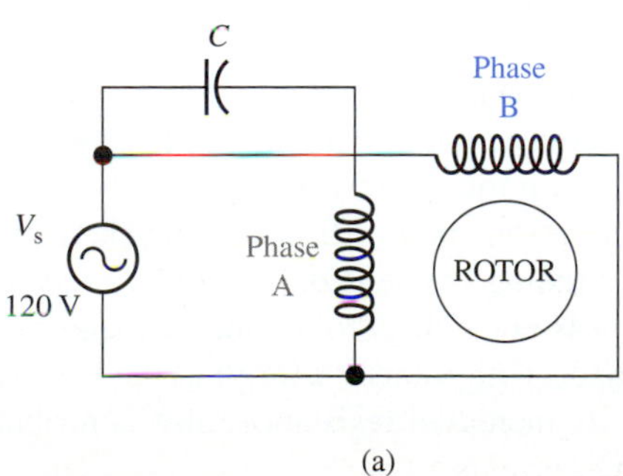

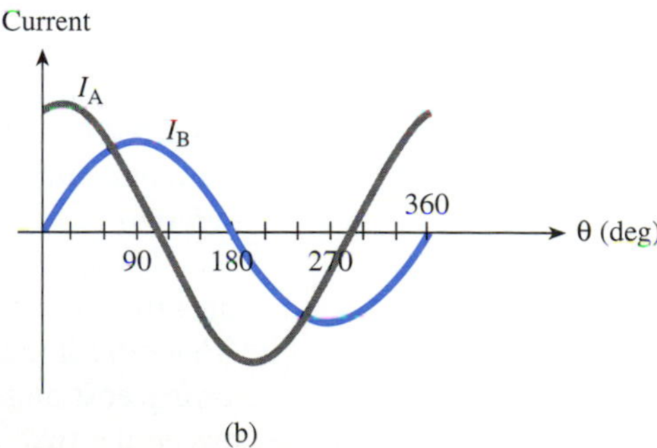

FIGURE 14–14 Placing a capacitor in series with one of the motor's phase windings to create the current phase shift. (a) Schematic showing just a single ac source, V_s, rather than two ac sources, V_A and V_B. (b) Current waveforms, with I_A leading I_B by about 70°.

The capacitive reactance X_C subtracts from the phase-A winding's inductive reactance X_L. If X_C is greater than X_L, which is generally true, this subtraction makes the phase-A circuit path net capacitive. Phase B is net inductive, naturally. Therefore phase current I_A leads phase current I_B by a substantial angle, but less than 90°.

The smaller phase-shift and unequal current magnitudes in Fig. 14–14 result in a rotating magnetic field that doesn't move around the stator at a constant speed and doesn't have a constant resultant magnitude. Therefore the motor's torque is not constant but pulsates. However, the rotating field still obeys the synchronous speed formula, $S_{syn} = f(120)/P$ [Eq. (14-2)].

Reversing. Most 120-V capacitor-assisted motors rotate in one direction only, but if reversibility is needed, it can be achieved with the circuit of Fig. 14–15. With the switch up in the Forward position, phase A is series-connected with capacitor C, and phase B is directly across the ac line, the same as in Fig. 14–14(a). However, with the switch down in the Reverse position, phase B gets the capacitor, and phase A is directly across the ac line. Analyze the stator of Fig. 14–1 with I_B leading I_A instead of the usual I_A leading I_B. You will agree that the rotating field reverses to counterclockwise rotation.

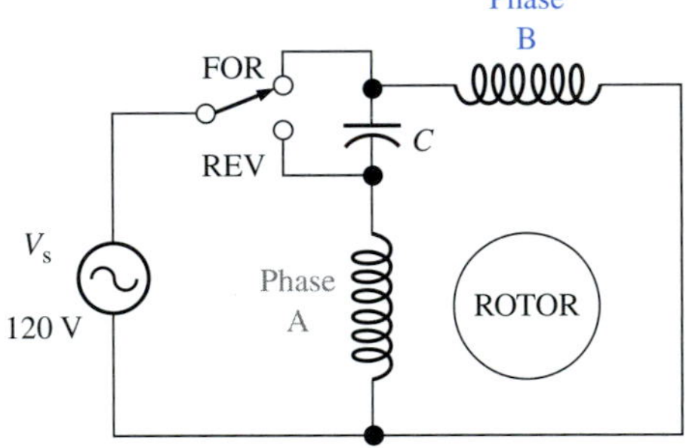

FIGURE 14–15
Reversing circuit for split-phase motor.

14-3-2 Capacitor-Start

Here is an important fact about induction motors that initially may seem strange: The motor doesn't necessarily need a second phase at all in order to keep running once it has reached operating speed.

In order to start from a standstill, the motor absolutely needs two functioning phases, but once the rotor is spinning at close to the synchronous speed of the rotating field, one of the phases can be disconnected. Thereafter, simply reversing the magnetic polarity of a single phase winding makes a fair imitation of a 180° field rotation.

Manufacturers often take advantage of this fact by installing a centrifugal switch on the motor shaft. When the shaft accelerates up to about 50% to 70% of the synchronous speed, centrifugal action opens the normally closed switch. This stops the current in the capacitor-connected phase winding, which is called the *start winding,* but ac sine-wave current continues to flow in the other winding, called the *run winding*. The motor is then referred to as a *capacitor-start motor*. It is shown schematically in Fig. 14–16.

Since the Start winding is energized for only a few seconds as the motor accelerates from a standstill, it is not subject to thermal damage from long-term heat dissipation. Therefore it can be made with thinner wire than the Run winding. This saves manufacturing cost and, with its increased resistance, also contributes to greater phase-shift between the two winding currents.

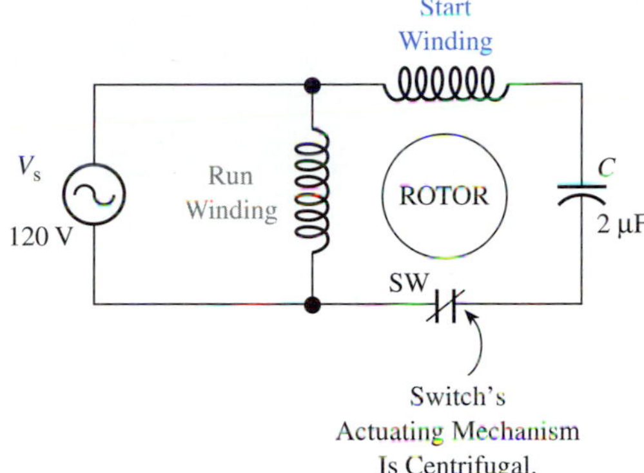

FIGURE 14–16
Capacitor-start motor. The capacitor must be capable of polarity reversal, with 200-V rating or better.

14-3-3 Shaded-Pole Motors

We learned from our brief discussion of capacitor-start motors that just a single phase will keep a squirrel-cage motor running once it has reached its operating speed. We can relax our requirements even further by pointing out that two separate motor phases are not absolutely required even for starting. Just getting the stator flux to sweep across the surface of the squirrel-cage rotor will induce rotor-bar currents and enable the motor to accelerate. This flux-sweeping motion can be achieved with just a single phase, using specially constructed salient poles.

The *shaded-pole motor* is illustrated in Fig. 14–17. Note that it has only a single phase in its stator construction, not two phases. In this example the phase has only two poles for simplicity, but four- and six-pole shaded-pole motors are more common.

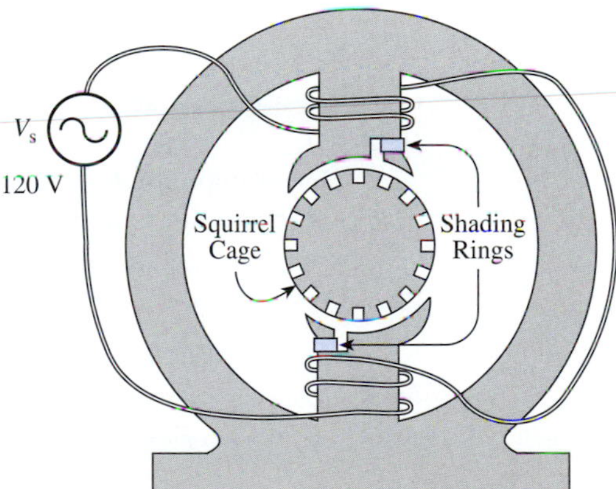

FIGURE 14–17
A shaded-pole motor has notched salient poles, each with a solid copper ring fitted on the notched side, called the *shaded side*.

It works by making the magnetic flux sweep across the pole-face, starting on the unnotched side and moving toward the notched side as the ac half cycle of current proceeds. This sweeping action is produced by the solid copper ring that is wrapped around the notched side.

In the early portion of the winding current's half cycle, the attempt by i_{wdg} to establish flux through the ring is opposed by the ring's inductance. In Fig. 14–18(a) the ring is circulating its own current, i_{ring}, in the opposite direction to i_{wdg}. This produces opposition flux, in accordance with Lenz's law. Therefore the net flux is weak on the right (shaded) side of the pole and strong on the left (unshaded) side of the pole.

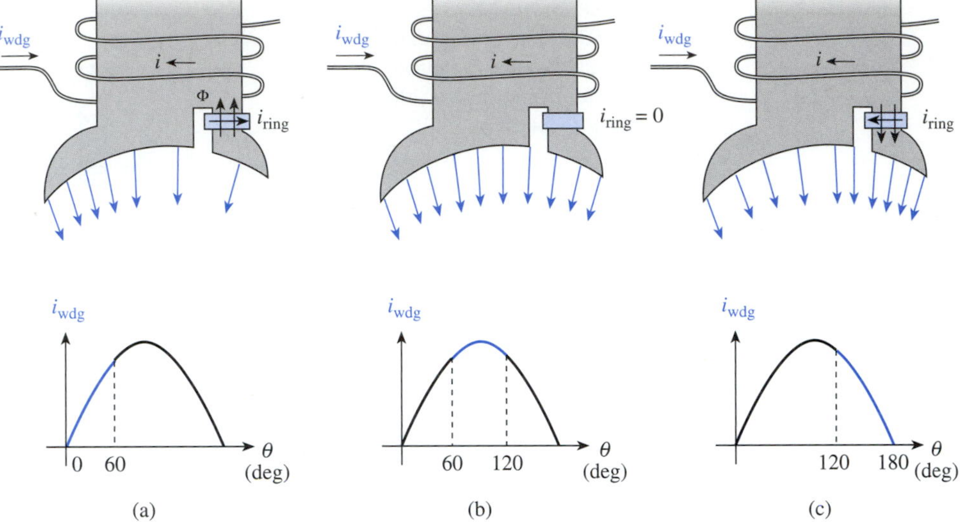

FIGURE 14–18 The sweeping motion of the magnetic flux across the face of the pole. (a) Early in the half cycle flux is concentrated on the unshaded side. (b) In the middle of the half cycle flux is evenly distributed. (c) Toward the end of the half cycle flux is concentrated on the shaded side.

In the middle of the half cycle, when i_{wdg} is not changing rapidly, the copper ring quiets down. At the peak of the half cycle $i_{ring} = 0$, and the net flux is distributed evenly across the pole-face, as shown in Fig. 14–18(b).

In the late portion of the half cycle i_{wdg} is decreasing. Now Lenz's law calls for the copper ring to circulate current i_{ring} in the same direction as i_{wdg}, as shown in Fig. 14–18(c). Therefore the ring-produced flux aids the winding-produced flux, and the resulting net flux is concentrated on the right side of the pole. The overall effect is for the net flux to move from the pole's unshaded side to its shaded side, mimicking a true rotating field.

Thus in Fig. 14–17 the flux-sweeping effect and the periodic pole-reversal combine to mimic a clockwise rotating field. That is sufficient to start the squirrel-cage from a standstill.

Our example motor has only two poles, for simplicity. However, two-pole shaded-pole motors are not common because they have difficulty imitating the fast synchronous speed (3600 r/min) associated with two-pole 60-Hz induction motors. Most shaded-pole motors are four- or six-pole machines.

14-3-4 Reluctance-Start Motors

Another mechanical arrangement that produces the flux-sweeping effect is shown in Fig. 14–19. The salient poles are misshaped, with one side having a wide air-gap to the rotor and the other side a narrow air-gap and greater surface area. The wider section has greater magnetic *reluctance,* analogous to electrical resistance. The narrower pole-section has less reluctance. As the winding current's half cycle begins, magnetic flux is quicker to appear in the high-reluctance wide portion of the pole-face. On the other hand, the more-iron low-reluctance portion of the pole-face has greater inductance L associated with it. This delays the establishment of flux in that part of the pole until later in the half cycle. The time-displacement of the fluxes creates the familiar flux-sweeping effect that imitates a rotating field. In Fig. 14–19, during the defined-positive half cycle of V_S, flux sweeps

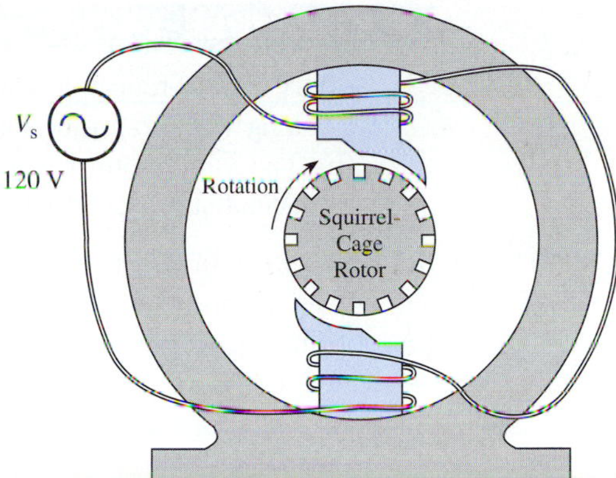

FIGURE 14–19
Cross-section of two-pole *reluctance-start* motor.

across the top pole from left to right, and South flux sweeps across the bottom pole from right to left. Therefore the overall field moves clockwise through the rotor material, so the squirrel-cage starts and runs clockwise.

Most reluctance-start motors have four or more poles. Two poles are shown in Fig. 14–19 for simplicity. Neither the shaded-pole nor the reluctance-start motor is reversible. This is because their field-sweeping actions are created by mechanical contrivances, which are permanent.

Do not confuse the reluctance-**start** motor with the reluctance-**synchronous** motor, which is often called simply the *reluctance motor*. The reluctance-synchronous motor has nonsalient stator poles and a misshaped rotor. It creates a true rotating field, and its rotor locks on to that field, spinning at exact synchronous speed.

14-4 ■ OPERATING CHARACTERISTICS OF SQUIRREL-CAGE MOTORS

For an induction squirrel-cage motor to produce torque and turn its load, the rotor-bar loops must experience a change in flux with time. This was the essential ingredient in our explanation of torque production that accompanied Fig. 14–7. Therefore the rotor and shaft can never spin at the same speed as the rotating field, namely the synchronous speed S_{syn}. If the rotor did spin at the same speed as the field, there would be no relative motion between them, and there would be no change in flux with time.

Thus the rotor's speed S_{rot} is always slower than S_{syn}. We define *slip* as the difference between the two speeds. That is,

$$\text{slip} = S_{syn} - S_{rot} \qquad \textbf{(14-4)}$$

Slip is sometimes called *absolute slip* to distinguish it from percent slip.

Percent slip is the percentage of synchronous speed that slip represents. As a formula,

$$\text{percent slip} = \frac{\text{slip}}{S_{syn}} \times 100\% \qquad \textbf{(14-5)}$$

or

$$\text{percent slip} = \frac{S_{syn} - S_{rot}}{S_{syn}} \times 100\% \qquad \textbf{(14-6)}$$

■ EXAMPLE 14-2

(a) A certain four-pole split-phase induction motor spins at 1760 r/min when running with no torque load. Find its slip and percent slip at no load.
(b) The same motor has a full-load percent slip of 7.5%. Find its absolute slip and actual shaft speed when delivering full-load torque.

Solution. (a) In North America, unless it is explicitly stated otherwise, we always assume that our motors are driven by an ac line with $f = 60$ Hz. The synchronous speed is found from Eq. (14-1) as

$$S_{\text{syn}} = \frac{(60 \text{ Hz})\,120}{P} = \frac{(60)\,120}{4} = 1800 \text{ r/min}$$

From Eq. (14–4), using the subscript (NL) for no-load,

$$\begin{aligned} \text{slip}_{(\text{NL})} &= S_{\text{syn}} - S_{\text{rot(NL)}} \\ &= 1800 \text{ r/min} - 1760 \text{ r/min} = \mathbf{40\ r/min} \end{aligned}$$

From Eq. (14–5),

$$\begin{aligned} \text{percent slip}_{(\text{NL})} &= \frac{\text{slip}_{(\text{NL})}}{S_{\text{syn}}} \times 100\% \\ &= \frac{40 \text{ r/min}}{1800 \text{ r/min}} \times 100\% = \mathbf{2.2\%} \end{aligned}$$

(b) From Eq. (14–5),

$$\begin{aligned} 7.5\% &= \frac{\text{slip}_{(\text{FL})}}{S_{\text{syn}}} \times 100\% \\ 0.075 &= \frac{\text{slip}_{(\text{FL})}}{1800 \text{ r/min}} \\ \text{slip}_{(\text{FL})} &= (0.075)(1800 \text{ r/min}) = \mathbf{135\ r/min} \end{aligned}$$

S_{rot} is found by rearranging Eq. (14–4),

$$\begin{aligned} \text{slip} &= S_{\text{syn}} - S_{\text{rot}} \\ S_{\text{rot(FL)}} &= S_{\text{syn}} - \text{slip}_{(\text{FL})} \\ &= 1800 - 135 = \mathbf{1665\ r/min} \end{aligned}$$

It would also have been possible to rearrange Eq. (14-6) to solve for full-load shaft speed directly, without first finding absolute slip.

■ EXAMPLE 14-3

Calculate the speed regulation of the split-phase motor of Example 14-2.

Solution. Speed regulation for any motor, dc or ac, is given by Eq. (12-16),

$$\begin{aligned} S_{\text{reg}} &= \frac{S_{\text{rot(NL)}} - S_{\text{rot(FL)}}}{S_{\text{rot(FL)}}} \times 100\% \\ &= \frac{1760 \text{ r/min} - 1665 \text{ r/min}}{1665 \text{ r/min}} \times 100\% = \mathbf{5.7\%} \end{aligned}$$

■

Squirrel-cage induction motors tend to have fairly good speed regulation (small regulation percent), as pointed out in this example.

14-4-1 Characteristic Speed-Torque Graph

All squirrel-cage induction motors have a speed-torque graph similar to the one in Fig. 14–20. This particular graph is for a four-pole split-phase motor like the one in Examples 14-2 and 14-3.

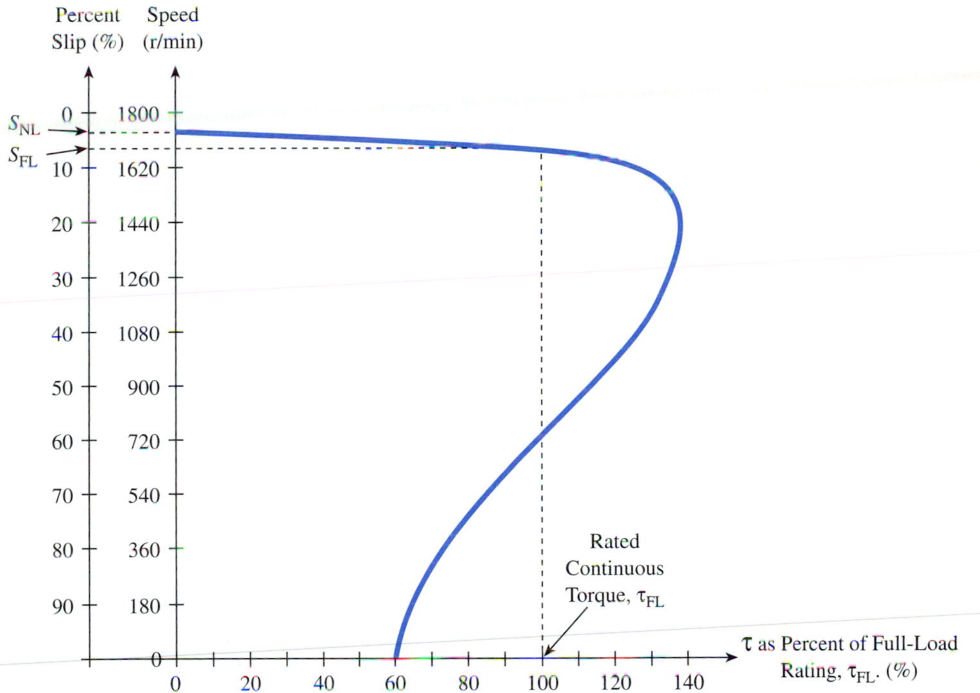

FIGURE 14–20 Typical speed-vs.-torque characteristic curve for a split-phase squirrel-cage induction motor.

To make sense of this speed-torque curve, let us break it into segments. Figure 14–21 shows the graph broken into segments A, B, and C.

Curve segment A is the normal operating region of the motor. Within this region, if the torque demand by the motor's load increases (becoming the heavy-load characteristic line in Fig. 14–21a), the motor slows down slightly. By slowing slightly, the slip between rotor and stator field increases. This creates greater time rate of change of flux, $\Delta\Phi/\Delta t$. By Faraday's law, larger rotor-loop voltages are induced. By Ohm's law larger rotor bar currents flow, so torque production increases by the Lorentz magnitude relation. The motor/load combination will stabilize at the point where the motor's characteristic S-vs.-τ curve intersects the heavy load's S-vs.-τ curve. This becomes the new operating point. It will be to the right of and lower than the former operating point. That former point was the intersection of the light-load characteristic S-vs.-τ graph with the motor's curve.

If the load's torque demand later decreases (becoming again the light-load line in Fig. 14–21a), all variables revert to their former values. This motor, like all motors, is automatically responding to variations in the load conditions; a good visualization example is pumping liquid which sometimes is dense and highly viscous and at other times becomes light and nonviscous. As long as the torque demand does not exceed the motor's

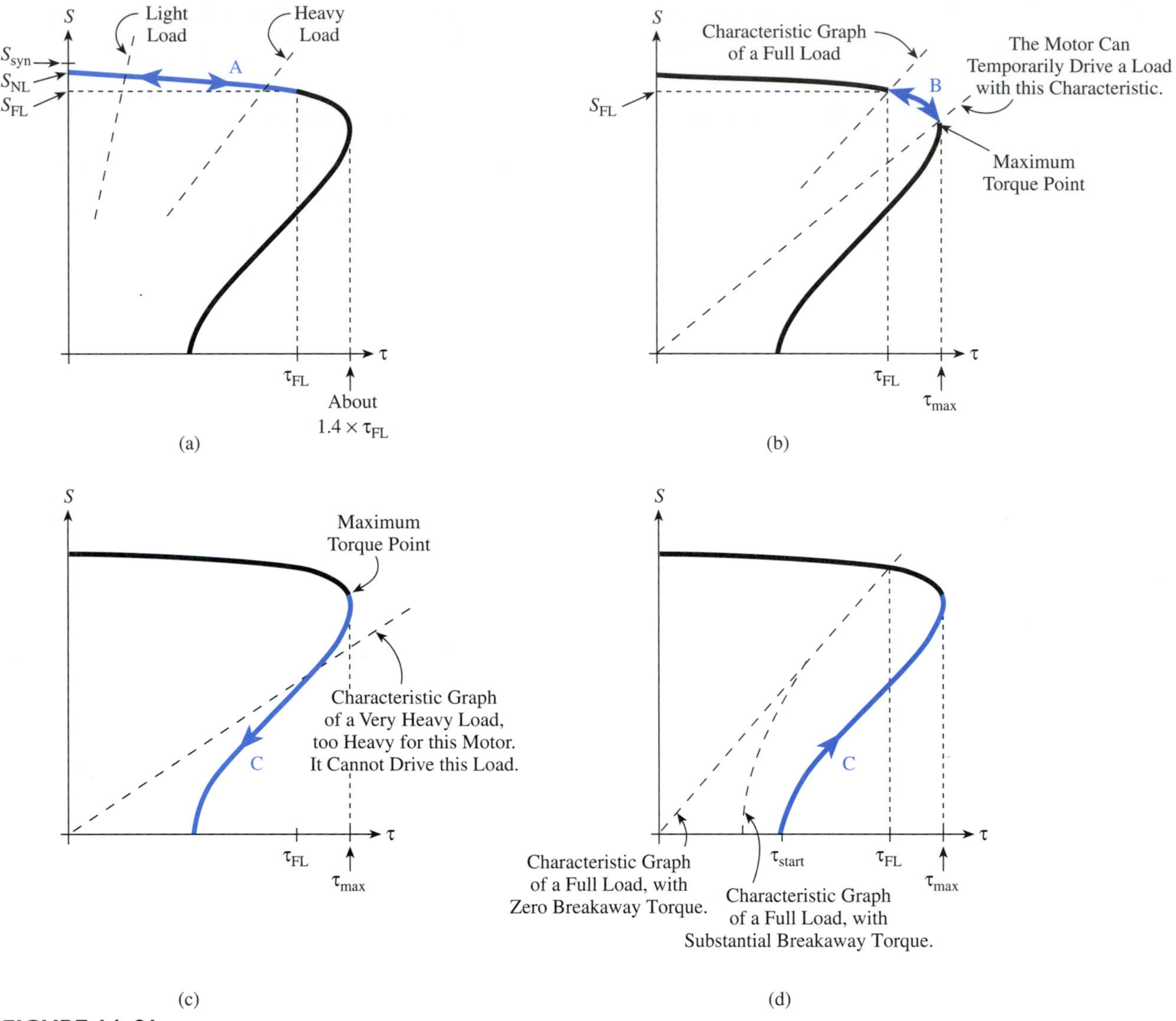

FIGURE 14–21
Understanding the generic speed-vs.-torque characteristic curve for a squirrel-cage induction motor. (a) Normal operating region. (b) Overload region. (c) Stalling. (d) Starting.

full-load rated torque value τ_{FL}, the motor's rotor and stator currents will be low enough that there is no danger of overheating.

Figure 14–21(b) emphasizes segment B of the S-vs.-τ characteristic. In this segment the motor-load interaction is the same as in segment A. Now, however, the torque exceeds the motor's full-load rating; this translates into the motor's currents being too large to be sustained indefinitely. The motor can operate in this region only temporarily. If the load torque demand does not relent and thereby allow the motor to return to the normal operating segment of its curve, overheating and damage will eventually occur.

The extreme right point of the B segment marks the *maximum torque* of the motor. If the motor slows below this point's speed, its torque production begins to *decrease,* as shown on segment C in Fig. 14–21(c). This is the opposite of the behavior that we "expect" from motors. In effect, then, if the load's characteristic curve touches the maximum torque point or lies to the left of it, the motor will be able to drive the load, but if the load's characteristic S-vs.-τ curve lies to the right of the maximum torque point, the motor cannot drive the load. Practically speaking, the motor will either stall or slow tremendously. In either case its rotor and stator currents increase dramatically until the circuit-protecting device blows, disconnecting the ac power.

The maximum torque of a squirrel-cage induction motor is sometimes called by other names. The phrases *pull-out torque, stall torque,* and *breakdown torque* are sometimes used to mean *maximum torque*. Pull-out torque is a phrase borrowed from synchronous motors. It is the maximum amount of torque that the rotor can deliver to the load before it "pulls out" of speed synchronization with the rotating field. It really should not be used to describe squirrel-cage motors. Stall torque has a better claim to legitimacy, because a squirrel-cage motor usually will stall when the load's S-vs.-τ curve is located as shown in Fig. 14–21(c), even if the overload detectors don't blow. *Breakdown* suggests stalling, so this word is not too misleading.

There is another view to be taken of the C segment. It is the starting process, when the motor moves through segment C in the rising-speed direction, as suggested in Fig. 14–21(d). During start-up, interpret the C segment as follows: At the moment when ac power is first applied to the motor's stator field winding, the squirrel-cage rotor produces an amount of torque called *starting torque*, or τ_{start}. As the rotor speed rises, its torque production increases along segment C of the curve. The reason for the torque increase is this: Increased speed means less slip between the rotor and the rotating field, as Fig. 14–22 points out. Therefore the frequency of the induced ac voltage in any rotor-bar loop decreases—the rotating field "laps" an individual loop less frequently. The decrease in frequency results in reduced inductive reactance of a loop, according to the reactance formula $X_L = 2\pi fL$.

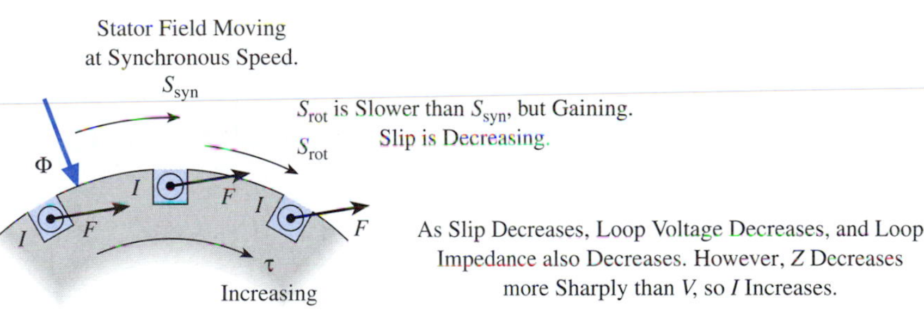

FIGURE 14–22 Picturing the relationships among S_{syn}, S_{rot}, slip, induced loop frequency, induced loop voltage, rotor-bar current I, F, and τ during acceleration from a standstill to the maximum torque point (along segment C in Fig. 14–21d).

At this time induced loop voltage is also decreasing in magnitude because decreasing slip decreases the time rate of flux change, but the decrease in X_L and Z is sharper than the decrease in the voltage magnitude. Therefore, by Ohm's law, $I_{\text{loop}} = V_{\text{loop}}/Z_{\text{loop}}$, the rotor loop current increases. This creates greater mechanical forces on the rotor bars and increases torque production. Study Fig. 14–22 to grasp these relationships.

This description of events holds true until the motor reaches its maximum torque point in Fig. 14–21(d). Beyond that point rotor voltage reduction starts to dominate rotor-loop impedance reduction. Therefore the rotor-bar current I in Fig. 14–22 begins to decrease. This causes rotor torque to begin decreasing, placing the motor into segment B of its characteristic curve.

One further point bears mentioning in Fig. 14–21(d). Some motor loads exhibit the *breakaway torque* phenomenon. This means that the load "sticks" in the stopped position and will not begin to move from that position until a certain amount of torque is exerted on it; that amount is called the load's breakaway torque. It is essential that the motor's starting torque be greater than the load's breakaway torque. Figure 14–21(d) shows this requirement being satisfied, so the motor will be able to successfully start and accelerate. However, if the load's breakaway torque were greater than τ_{start}, the motor could not budge it from a standstill.

Breakaway-type loads include mechanical conveyors, most compressors, some valve-operation mechanisms, and some liquid pumps. Fans are zero-breakaway loads, crossing through the origin of the S-vs.-τ axes in Fig. 14–21(d).

14-4-2 Current Versus Speed

In dc motors the two important characteristics are speed vs. torque and current vs. torque. With ac motors the same two characteristic curves are often seen. However, it is also common to plot the motor's line current (the current flowing from the ac source into the stator winding) versus *speed* as the independent variable. This has been done in Fig. 14–23 for a generic split-phase motor.

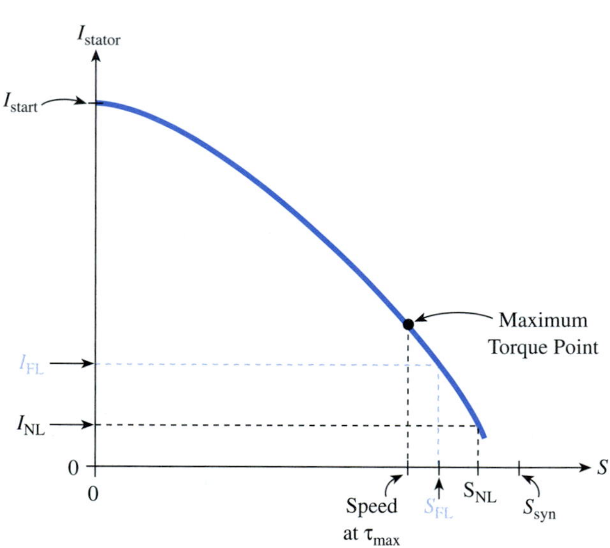

FIGURE 14–23 Current-vs.-speed generic characteristic curve for a split-phase motor. We have always regarded speed S as a dependent variable, plotted on the y axis, but here we take the view that speed is the variable that goes its own independent way and current I_{stator} must function in response to speed.

The main usefulness of the I-vs.-S curve is to point out the tremendous increase in the motor's current-draw at low speeds. This current surge occurs either in starting (Fig. 14–21d) or in stalling (Fig. 14–21c). In the starting case it represents the electrical stress that must be temporarily endured in order to get the motor going; in the stalling case it represents either a malfunction of the load or a misapplication of the motor.

14-5 ■ THREE-PHASE AC POWER SYSTEMS

14-5-1 The Three-Phase Configurations

We mentioned in Sec. 14-3 that two-phase 90°-shifted ac power was generated commercially at one time but has long since been replaced by three-phase 120°-shifted ac power. The three separate phases are produced by locating the generating windings 120° apart mechanically in the *alternator* (ac generator) at the electric utility company. The wave-

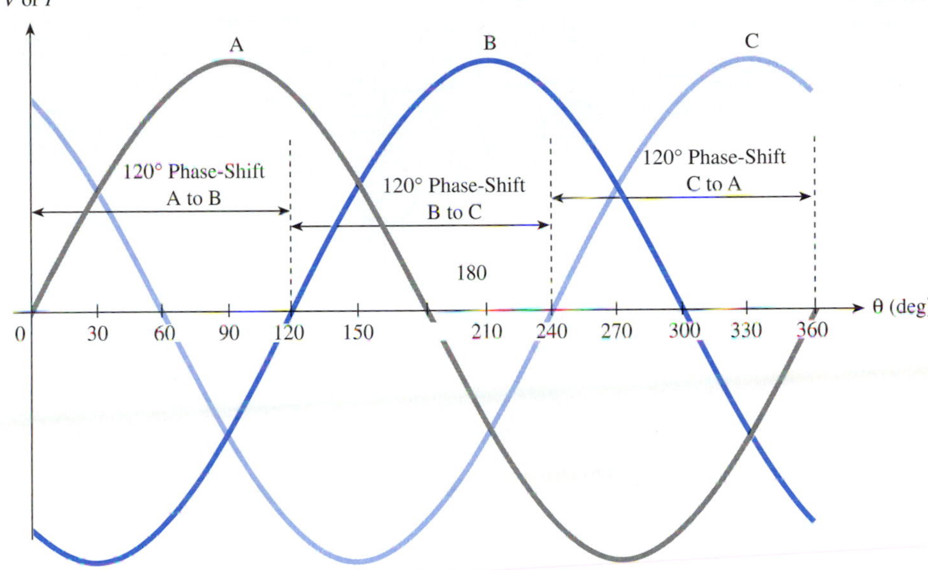

FIGURE 14–24
Waveforms of three-phase ac system showing 120° phase displacement among phases A, B, and C.

forms of the three separate phase voltages or currents are shown in Fig. 14–24; the voltage phasor diagram appears in Fig. 14–25.

In any three-phase system the three individual phase voltages V_A, V_B, and V_C are guaranteed to be equal to one another. The three individual phase currents I_A, I_B, and I_C are not necessarily equal, in general. That depends on the nature of the electrical load. If the electrical load is a three-phase induction motor or motors, then the three individual currents will all be equal to one another. This condition, called the *balanced-load* condition, is the only one that we will consider.

You can visualize the three phases as separate and electrically isolated circuits, as suggested in Fig. 14–26(a). There is no reason why they couldn't be used in this manner. The polarity marks on the three voltage sources represent the *defined-positive polarity*. That is, when the actual instantaneous voltage polarity matches those marks, the voltage is considered positive. When the actual instantaneous polarity is opposite to those marks,

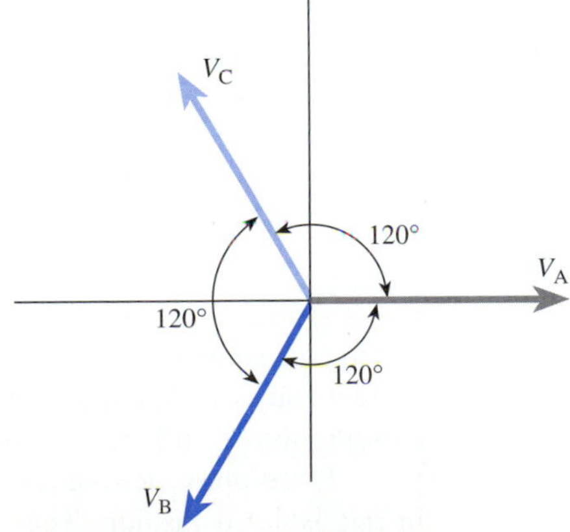

FIGURE 14–25
Phasor diagram of three-phase ac system with voltage as the displayed variable. Phase voltage V_A leads V_B by 120°; V_B leads V_C by 120°.

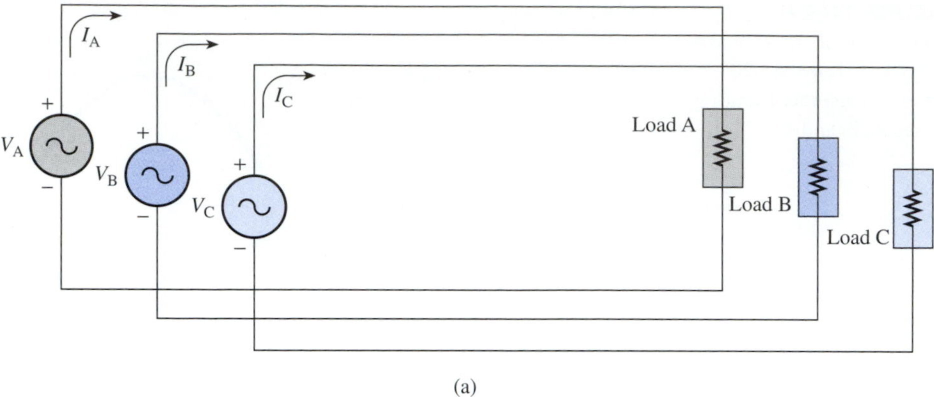

(a)

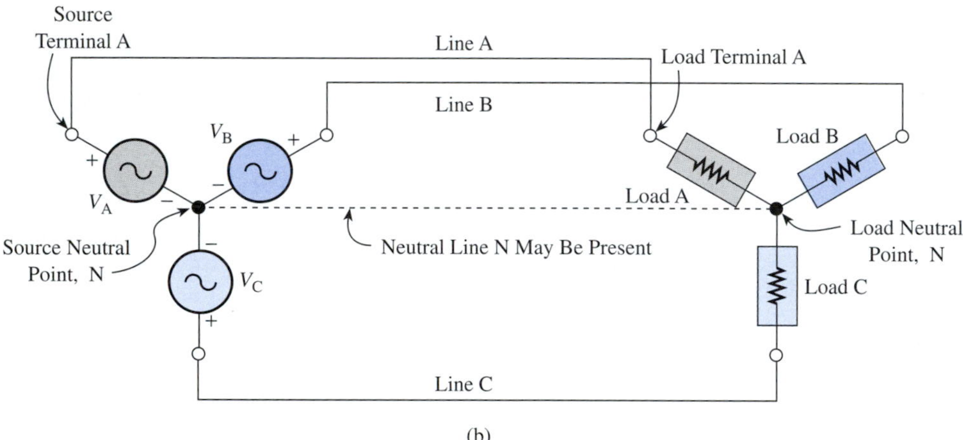

(b)

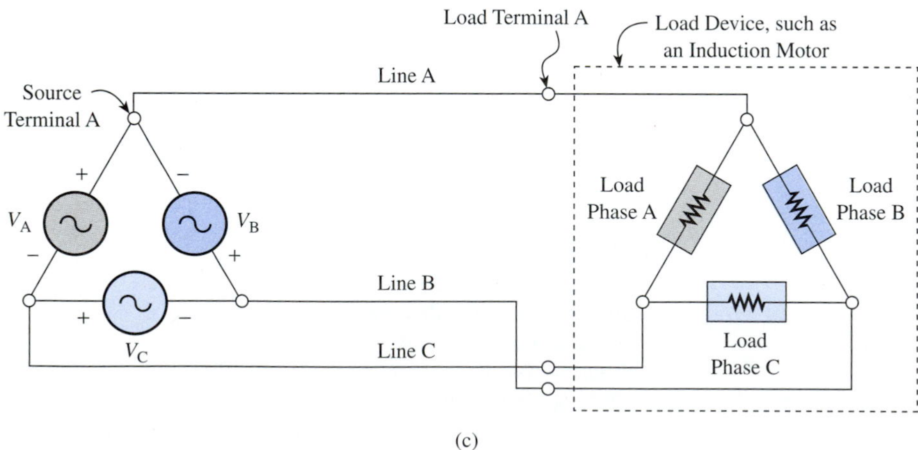

(c)

FIGURE 14–26
Methods of interconnecting the three sine-wave phases. (a) Isolated. (b) Wye, also called *star.* (c) Delta.

it is considered negative. Furthermore, when the instantaneous current (conventional) is leaving the source via the positive terminal, that is considered the positive direction. When current leaves via the negative terminal, the current is considered to be negative. Thus the currents shown in Fig. 14–26(a) are all defined-positive.

Three-phase ac sources, namely alternators and transformers, are not normally used in this isolated fashion. They are normally interconnected in the clever ways shown in

Fig. 14–26(b) and (c), which we will discuss in a moment. But in order to understand the essence of three-phase ac, you should realize that they *could* be used this way. Three-phase ac in its essence is just a collection of three equal-value voltage sources that are 120° phase-shifted in relation to each other.

One of the normal three-phase connection methods is called the *wye* connection, shown schematically in Fig. 14–26(b). All three defined-negative terminals are connected together to form the *neutral* point, labeled N. The defined-positive source terminals are then labeled A, B, and C. The wires leading away from them to the three-phase electrical load (the three-phase induction motor) are called line A, line B, and line C. Sometimes there is a fourth wire, line N, joining the neutral points of the source and load; however, to operate a three-phase motor, line N is not required.

The wye interconnection of part (b) has an important advantage over the separate circuits of part (a). The wye reduces I^2R power losses along the wire-runs. When compared on a fair basis (same amount of power transported, same amount of copper contained in the combined wires), the wye system reduces I^2R losses by 25%. The reason for this savings has to do with the fact that every wire serves a dual purpose: (1) It carries current *to* the load from its associated source phase; and (2) It carries current *back from* the load that was delivered by the other two source phases.

For example, look at the 90° instant in Fig. 14–24. Assume that the waveforms represent currents rather than voltages. At the 90° instant source phase A and line A are carrying their peak current to the load, but the system doesn't have a wire dedicated to the job of carrying I_A back from the load. Instead, the system uses lines B and C (both carrying negative currents at this instant) to serve as the return paths back to the source.

The same description applies at the 210° instant, except that now some roles are reversed. Line B is carrying current to the load, and lines A and C are carrying it back from the load to the source. At any instant there is a balance between current to the load and current back from the load. This must be true in order to satisfy Kirchhoff's current law for the three-phase load as a whole.

A similar description applies at the 270° instant, as a final example. Now two lines are involved in carrying current to the load (lines A and B) and just one line carries it all back (line C).

It is this sharing of duties in the three-phase wye system that is the basic rationale for its reduction in I^2R losses. It has been estimated that the use of three-phase power systems saves America's electric consumers about 50×10^9 kilowatt-hours annually in reduced energy losses during the transmission process alone. This translates into power transmission cost savings of about \$4 billion annually. There are further savings arising from reduced manufacturing costs and greater operating efficiencies for three-phase generators, transformers, and motors. These may account for another \$4 billion to \$8 billion by various estimates. The total annual advantage to the nation of the three-phase idea is on the order of \$10 billion, or roughly 5% of our electric expenditure.

Figure 14–26(c) shows the second method of interconnecting the three phases. It is called the *delta* system because it forms a schematic triangle, reminding us of the Greek letter Δ, the capital delta. The defining feature of the delta system is that the defined-positive terminal of one phase is connected to the defined-negative terminal of the next phase, and then again, and then again, to complete the delta loop. The wire that connects to the positive terminal of phase A is called line A, and so on, as shown in Fig. 14–26(c). The delta system provides the same reduction in I^2R power loss in the wire-runs as the wye system.

Figure 14–27 illustrates some other features about three-phase systems. First, the source phases are usually drawn as windings (coils) rather than circles. This is

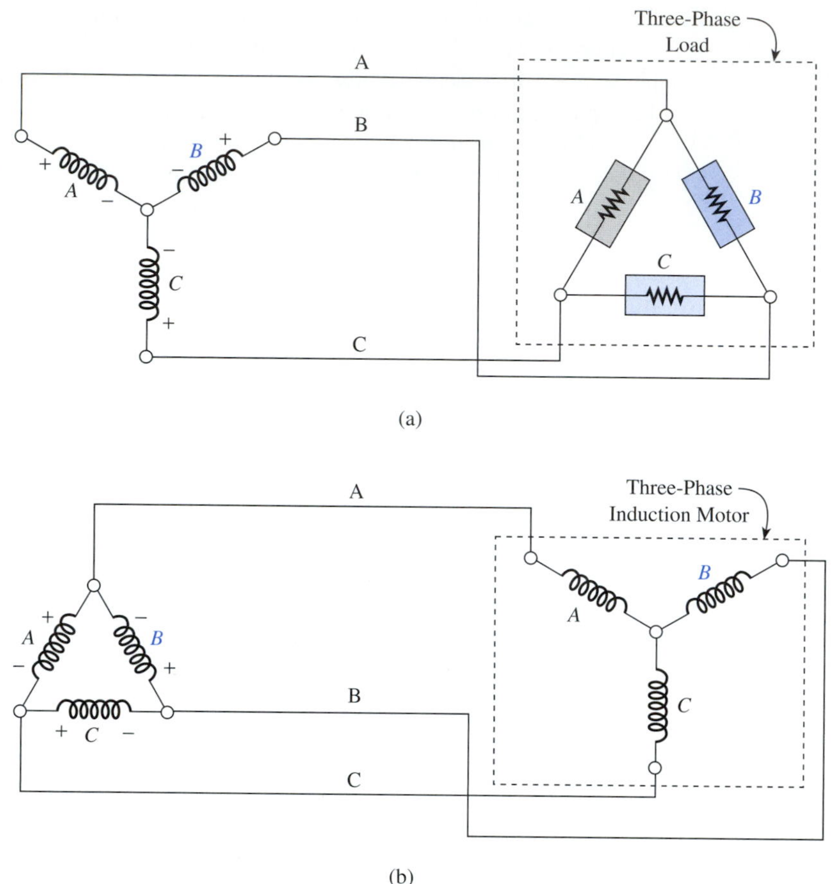

FIGURE 14–27 Showing the ac source as windings, not circles. (a) Wye-connected source driving delta-connected load. (b) Delta-connected source driving wye-connected motor load with the motor phases also shown as windings, not impedance-boxes.

reasonable, since they really are three windings in the actual construction of a three-phase alternator or transformer.

Second, it is not required that the load configuration match the source configuration. It is perfectly all right for a wye source to drive a delta load, as shown in Fig. 14–27(a), and vice versa, as shown in Fig. 14–27(b). Finally, if the electrical load is a three-phase motor, the load phases are themselves often drawn as windings rather than impedance-boxes. This is just as reasonable as the source situation, for the same reason. There really are three coiled-up windings in the motor's construction.

14-5-2 Line Variables and Phase Variables

Wye. In a wye system like Fig. 14–28(a), the voltage between any pair of lines is given by the phasor (vector) sum of the two phase voltages associated with those lines, but one of the phase voltages must be looked at "backward" while the other is looked at "forward."

For example, the voltage from line B to line A, symbolized V_{AB}, is given by the phasor sum of $V_{NB} + V_{AN}$. To get from line B to line A we must look at the B-phase voltage from line B to the N point, then trigonometrically add the A-phase voltage from the N point to line A. In other words we must look at the B-phase voltage "backward" (from defined-positive end to defined-negative end), followed by the A-phase voltage looked at "forward" (from defined-negative end to defined-positive end). This concept is demon-

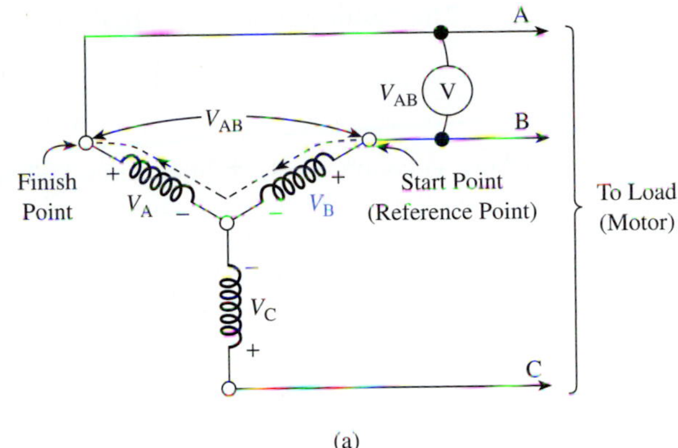

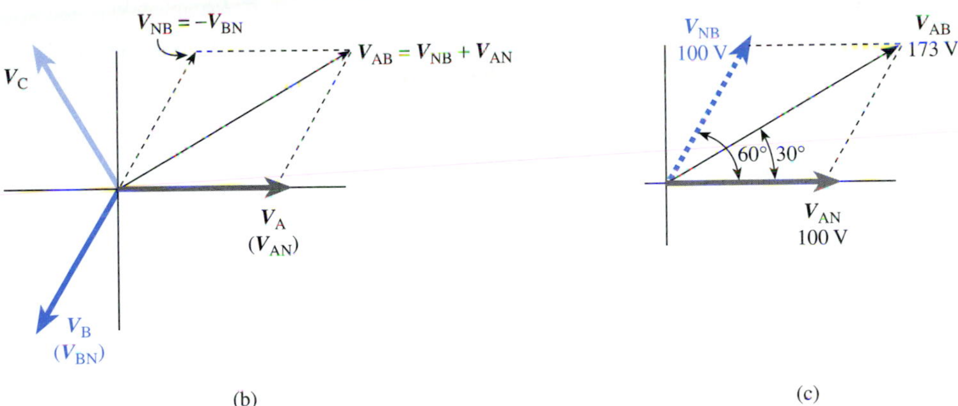

FIGURE 14–28 Relation between line-to-line voltage V_{AB} and individual phase voltages, V_A and V_B (V_{AN} and V_{BN}). (a) Schematic diagram. The second letter in a voltage subscript marks the starting point, or the reference point that we are measuring with respect to. The first letter in a subscript marks the finish point whose voltage is being measured with respect to the start point. (b) Looking at the B-phase voltage backward by inverting V_{BN} to get V_{NB}. (c) Performing the phasor addition $V_{NB} + V_{AN} = V_{AB}$.

strated in the phasor diagram of Fig. 14–28(b). In that diagram the phasor for the B-phase voltage V_{BN} has been flipped around by 180° to produce the V_{NB} phasor. This is what we mean by looking at the B-phase voltage "backward."

■ EXAMPLE 14-4

Assume that the phase-voltage magnitudes are 100 V in Fig. 14–28. Find the magnitude of the line-to-line voltage V_{AB}.

Solution. Look at Fig. 14–28(c). Resolve the V_{NB} phasor into horizontal and vertical components (P → R conversion), giving

$$V_{NB(horiz)} = |V_{NB}| \times \cos 60° = 100\text{ V} \times \cos 60° = 100\text{ V} \times 0.5 = 50\text{ V}$$
$$V_{NB(vert)} = |V_{NB}| \times \sin 60° = 100\text{ V} \times \sin 60° = 100\text{ V} \times 0.866 = 86.6\text{ V}$$

Add the horizontal components by direct algebraic addition, since they have the same phasor position.

$$V_{horiz(total)} = 100\text{ V} + 50\text{ V} = 150\text{ V}$$

Combine the total horizontal value with the total vertical value (R → P conversion), giving

$$V_{AB} = \sqrt{(150\text{ V})^2 + (86.6\text{ V})^2} = \mathbf{173\ V}$$

■

This result can be generalized for any wye system as

$$V_L = 1.73\ V_{ph} \qquad \text{(for wye)} \tag{14-7a}$$

$$V_L = \sqrt{3}\,V_{ph} \qquad \text{(for wye)} \tag{14-7b}$$

where V_L symbolizes the line-to-line voltage between any pair of lines and V_{ph} symbolizes the line-to-neutral (actually neutral-to-line) voltage of any one of its three phases. Line-to-line voltage is verbalized as "line voltage." Line-to-neutral voltage of an individual phase is verbalized as "phase voltage." Many people use the symbol V_ϕ for phase voltage instead of V_{ph}.

Currents in wye. It is immediately clear by inspection of Fig. 14–28(a) that the current in any line is identical to the current in that phase; the two are in series with each other. This is generalized for a wye system as

$$I_L = I_{ph} \qquad \text{(for wye)} \tag{14-8}$$

We are overlooking the fact that all three line currents I_A, I_B, and I_C have equal magnitudes only if the load is balanced (all three load phases have equal impedances).

14-5-3 Delta

Let us generalize the line-versus-phase situation for a delta-configured ac source by referring to Fig. 14–29. It is clear by inspection that the line-to-line voltage is just the same as the voltage across a phase. For example, line voltage V_{BA} equals phase voltage V_B, since the line-to-line voltmeter measuring V_{BA} is directly in parallel with the B-phase winding. In this example $V_{BA} = V_B = 100$ V. As a general statement,

$$V_L = V_{ph} \qquad \text{(for delta)} \tag{14-9}$$

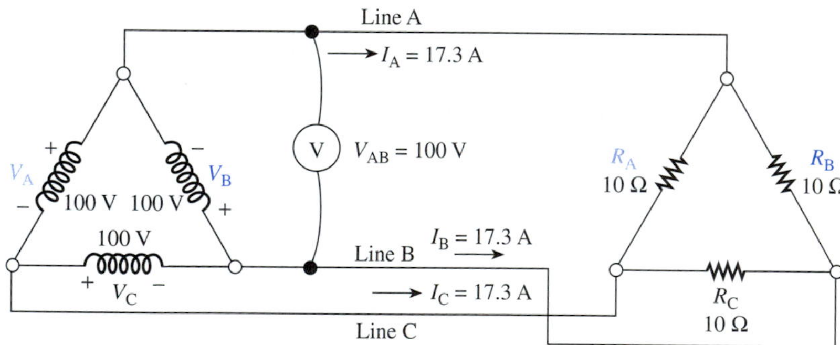

FIGURE 14–29
A delta/delta system for understanding the relation between line variables and phase variables.

As you would suspect, it's the *current* that gets the factor-of-1.73 boost in a delta system. In Fig. 14–29 source phase A is connected directly in parallel with load phase A. Trace the diagram to satisfy yourself that this is true. Applying Ohm's law, we have

$$I_{RA} = I_{\text{source phase A}} = \frac{V_A}{R_A}$$

$$= \frac{100\text{ V}}{10\ \Omega} = 10\text{ A}$$

Thus we expect the current in *line* A to be greater by 1.73, or

$$I_A = 1.73 \times 10\text{ A} = 17.3\text{ A}$$

This is a true statement, and it could be proved mathematically by a phasor addition of individual phase currents at source terminal A. We will forgo doing the mathematical proof since it would just duplicate the procedure in Example 14-4. As a general statement,

$$I_L = 1.73 \times I_{ph} \qquad \text{(for delta)} \tag{14-10a}$$

or

$$I_L = \sqrt{3} \times I_{ph} \qquad \text{(for delta)} \tag{14-10b}$$

14-5-4 Apparent Power and True Power in Three-Phase

Let us review what we know about apparent power and true power in single-phase ac circuits. Apparent power is the simple product of rms voltage multiplied by rms current, without regard to the *V-I* phase relationship. We write

$$S = V_{rms} \times I_{rms} \tag{14-11a}$$

or simply

$$S = V \times I \tag{14-11b}$$

where S symbolizes apparent power, which is expressed in units of voltamperes, or VA.

True power, or power, is given by

$$P = (V \times I) \times \cos \phi \tag{14-12a}$$

$$P = (V \times I) \times \text{PF} \tag{14-12b}$$

or

$$P = S \times \text{PF} \tag{14-12c}$$

with power P expressed in watts. In Eq. (14-12b and c), power factor PF is equal to the cosine of ϕ, the phase angle by which current I is out of phase with voltage V.

The same basic concepts hold in three-phase ac circuits, of course. For example, consider the delta system with balanced resistive load in Fig. 14–29.

■ EXAMPLE 14-5

(a) Find the apparent power S in phase A alone.
(b) Do you expect the apparent powers S_B and S_C to be equal to S_A? Explain.
(c) Based on your answer to part (b), find the apparent power of the entire three-phase system as a whole.
(d) Write a generalized equation for the total apparent power of a balanced delta system in terms of phase voltages V_{ph} and phase currents I_{ph}.
(e) Find the total true power for this particular circuit, then write a generalized equation for the total true power of a balanced delta system.

Solution. (a) In a delta/delta system, load phase A is directly in parallel with source phase A, so the phase voltage appears across the phase impedance. Applying Ohm's law,

$$I_{ph\,A} = \frac{V_{ph\,A}}{Z_{ph\,A}} = \frac{100\text{ V}}{10\ \Omega} = 10\text{ A}$$

From Eq. (14-11b),

$$S_{\text{ph A}} = V_{\text{ph A}} \times I_{\text{ph A}} = 100 \text{ V} \times 10 \text{ A} = \mathbf{1000\ VA}$$

or

$$S_{\text{ph A}} = \mathbf{1\ kVA}$$

which is verbalized as "the apparent power in phase A is one kilovoltampere." Many Americans verbalize kilovoltampere as kVA—"kay–vee–ay."

(b) Yes, we expect the B and C phases to be the same as the A phase because the individual load-phase impedances (simple resistances in this case) are equal. Therefore $S_{\text{ph B}} = \mathbf{1\ kVA}$**,** and $S_{\text{ph C}} = \mathbf{1\ kVA}$**.**

(c) Apparent powers, like true powers, add by simple algebraic addition. Power is not a phasor-type (vector-type) variable, so there is no trigonometry involved in its addition. Thus

$$S_{\text{total}} = S = S_{\text{A}} + S_{\text{B}} + S_{\text{C}}$$
$$= 1 \text{ kVA} + 1 \text{ kVA} + 1 \text{ kVA} = \mathbf{3\ kVA}$$

(d) $$S_{\text{total}} = \mathbf{S = 3 \times V_{ph} \times I_{ph}} \quad \text{(for balanced delta)} \qquad \textbf{(14-13)}$$

(e) With the load being totally resistive, PF = 1.0 because $\phi = 0°$. From Eq. (14-12c),

$$P_{\text{total}} = P = (3 \text{ kVA}) \times \text{PF}$$
$$P = (3 \text{ kVA}) \times 1.0 = \mathbf{3\ kW} \text{ (kilowatts)}$$

In general,

$$\mathbf{P = 3 \times V_{ph} \times I_{ph} \times PF} \quad \text{(for balanced delta)} \qquad \textbf{(14-14)}$$

Equations (14-13) and (14-14) are perfectly valid if you know the phase current I_{ph}. We were able to get the phase current in Example 14-5 because someone told us the phase impedance value, 10 Ω. In real life it would be a rare event when someone could tell you the phase impedance of a particular three-phase load. Besides, if the load is a three-phase motor, the phase impedance changes as the motor's mechanical shaft load changes. So an Ohm's law calculation like the one we performed in part (a) of Example 14-5 is a fine textbook technique for initial learning, but not a likely thing at an industrial site.

Can we measure the phase current I_{ph} with an ammeter? Well, yes, if you are willing to open up the three-phase source or the motor. You would have to open up one of them because all that emerges from their enclosures are the three wires. Draw dashed-line boxes around both source and load in Fig. 14–29 to get a feel for their realistic appearance in the field. Breaking into either of those boxes to install an ammeter is not an easy thing.

So Eqs. (14-13) and (14-14) may not be usable, since I_{ph} may not be knowable. What is needed are formulas for S and P that are based on *line* variables, not phase variables. Line voltage and current are always externally measurable. Substituting Eqs. (14-9) and (14-10b) into Eqs. (14-13) and (14-14) gives us

$$S = 3 \times V_{\text{ph}} \times I_{\text{ph}}$$
$$= 3 \times V_{\text{L}} \times \frac{I_{\text{L}}}{\sqrt{3}}$$
$$S = \sqrt{3} \times V_{\text{L}} \times I_{\text{L}} \quad \text{(for balanced delta)} \qquad \textbf{(14-15a)}$$

and

$$P = \sqrt{3} \times V_L \times I_L \times PF \qquad \text{(for balanced delta)} \qquad \textbf{(14-16a)}$$

■ EXAMPLE 14-6

Verify the total power result from Example 14-5 by applying Eq. (14-16a) to Fig. 14–29.

Solution. For delta,

$$I_L = \sqrt{3} \times I_{ph} = 1.732 \times 10\text{ A} = 17.32\text{ A}$$
$$V_L = V_{ph} = 100\text{ V}$$

From Eq. (14-16a),

$$P = \sqrt{3} \times V_L \times I_L \times PF$$
$$= 1.732 \times 100\text{ V} \times 17.32\text{ A} \times 1.0 = \mathbf{3000\ W}$$ ■

Wye. In a wye system the phase variable that is difficult to measure is voltage, not current. This is true only when no neutral wire is present, as in Fig. 14–30.

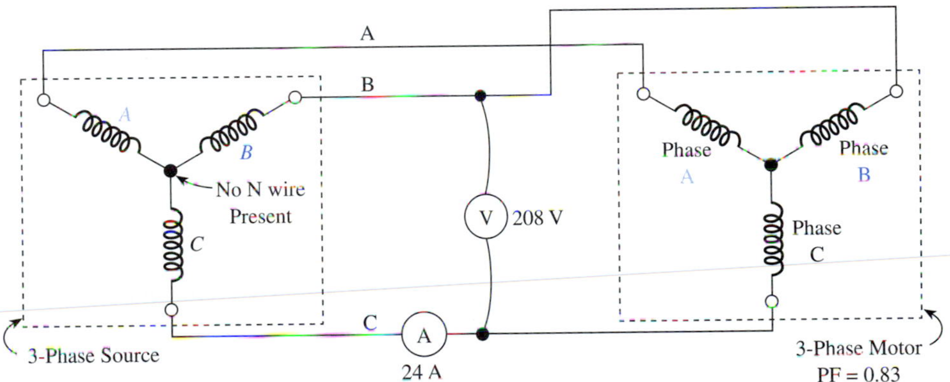

FIGURE 14–30 A wye system for practicing apparent power and true power calculations.

If we derived the apparent power and true power formulas for a wye system, they would be identical to Eqs. (14-13), (14-14), (14-15a), and (14-16a), which we originally derived for delta. Discarding Eqs. (14-13) and (14-14) because of their difficulty of use, we can rewrite Eqs. (14-15a) and (14-16a) as

$$S = \sqrt{3} \times V_L \times I_L \qquad \text{(for balanced wye)} \qquad \textbf{(14-15b)}$$
$$P = \sqrt{3} \times V_L \times I_L \times PF \qquad \text{(for balanced wye)} \qquad \textbf{(14-16b)}$$

■ EXAMPLE 14-7

In the wye system of Fig. 14–30 a voltmeter connected between the B and C lines measures an rms value of 208 V. An ammeter wired into the C line measures 24 A. We know that the motor's power factor is 0.83 under this particular loading condition.

(a) Find *S*.

(b) Find P.
(c) What is the value of phase voltage?
(d) What is the impedance of one phase of the motor under this loading condition?
(e) What is the phase angle ϕ between current and voltage?
(f) Tell specifically what current and what voltage would show the phase relationship of part (e).

Solution. (a) From Eq. (14-15b),

$$S = \sqrt{3} \times V_L \times I_L$$
$$= 1.732(208\text{ V})(24\text{ A}) = \mathbf{8.65\ kVA}$$

(b) From Eq. (14-16b),

$$P = \sqrt{3} \times V_L \times I_L \times \text{PF} = S \times \text{PF}$$
$$= (8.65 \times 10^3) \times (0.83) = \mathbf{7.18\ kW}$$

(c) Rearranging Eq. (14-7b),

$$V_{ph} = \frac{V_L}{\sqrt{3}} = \frac{208\text{ V}}{1.732} = \mathbf{120\ V}$$

This is the familiar 120-V/208-V wye-configured three-phase system often seen in North American industry. If a neutral wire is installed, the 208 V can be used to drive the motors, and the 120 V can be used for the lights. Of course, when this is done, it creates an overall load condition that is unbalanced. None of our formulas work, viewed from the source. This is because the individual line currents will differ from one another depending on the amount of lighting load connected to that particular phase. The formulas still work when applied to an individual motor or to a motorbank, however.

(d) The three-phase source voltages will split across the load phases exactly as they were induced in the load windings, as long as the load is balanced. This is true even without the neutral wire present. Thus, inside the three-phase motor, we would measure

$$V_{ph\ motor} = V_{ph\ A\ motor} = V_{ph\ B\ motor} = V_{ph\ C\ motor} = 120\text{ V}$$

Applying Ohm's law gives

$$Z_{ph\ motor} = \frac{V_{ph\ motor}}{I_{ph\ motor}} = \frac{V_{ph\ motor}}{I_L} = \frac{120\text{ V}}{24\text{ A}} = \mathbf{5.0\ \Omega}$$

in which we relied on the equality of phase current I_{ph} and line current I_L for a wye (Eq. 14-8).

(e)

$$\text{PF} = \cos\phi = 0.83$$
$$\phi = \cos^{-1}(0.83) = \mathbf{34° \ (lagging)}$$

(f) The voltage and the current in **one individual winding** (either source or motor) are out of phase by 34°, with I lagging V. Thus $I_{L(C)} = I_{ph(C)}$ would lag $V_{ph(C)}$ by 34°, and the same for A and B.

A particular line current will not lag either line-to-line voltage associated with that line by 34°. Thus in Fig. 14–30 you would not measure a 34° phase angle between $I_{L(C)}$

and V_{BC} or between $I_{L(C)}$ and V_{AC}. To understand why this is true, look at Fig. 14–28(b) and (c). These diagrams show that any line-to-line voltage is phase-displaced by 30° and 150°, respectively, from the individual phase voltages that it spans. ■

14-6 ■ THREE-PHASE SQUIRREL-CAGE INDUCTION MOTORS

A three-phase induction motor produces a rotating field by the same basic method as the two-phase induction motor described in Sec. 14-1. The phases are mechanically offset from one another by 120° (two poles per phase), and the winding currents are 120° electrically phase-shifted. This structure is shown with salient poles in Fig. 14–31(a); the realistic nonsalient-pole stator layout is shown in Fig. 14–31(b).

The motor of Fig. 14–31 is shown wye configured, so the voltage across any individual phase winding is equal to the line voltage divided by 1.73. For example, if the line voltage is 480 V, then

$$V_{\text{ph(Y motor)}} = \frac{V_L}{1.73} = \frac{480\text{ V}}{1.73} = 277\text{ V}$$

The motor manufacturer, in designing the wye-configured motor to operate from a 480-V three-phase ac line, would construct the motor's windings with the proper number of turns and the proper wire gage and insulation thickness to operate successfully at 277 V.

On the other hand, the manufacturer could choose to interconnect the phase windings in a delta configuration. Then the individual phase winding voltage would be the entire line voltage, 480 V. That is

$$V_{\text{ph}(\Delta\text{ motor})} = V_L = 480\text{ V}$$

In this case the manufacturer would redesign the motor's windings with a greater number of turns, thinner wire gage, and thicker insulation to operate successfully with a greater voltage, 480 V, across them.

14-6-1 Advantage of Three-Phase Motors over Two-Phase Motors

The stator of Fig. 14–31(b) has all of a phase winding's turns concentrated in just one pair of slots. Modern three-phase motors actually have several pairs of slots per phase winding, distributed over an angular span of the stator's surface. For example, Fig. 14–32 shows another three-phase two-pole motor, but in this motor's construction a phase winding actually consists of five series-connected individual windings, distributed over 60 mechanical degrees. To hold five individual windings, the manufacturer must cut five pairs of slots in the stator surface.

As a general description, a collection of windings that are placed in neighboring slots and connected in series with one another is referred to as a *winding group*. In a two-pole motor like Fig. 14–32 there is always just one winding group per phase. In a four-pole motor there tends to be two winding groups per phase. However, by filling the slots only half full with one individual winding, then using a different winding to fill each slot to the stator's inside surface, motor manufacturers create a structure with four winding groups per phase. This construction method was suggested in Fig. 14–4(b) for a two-phase motor.

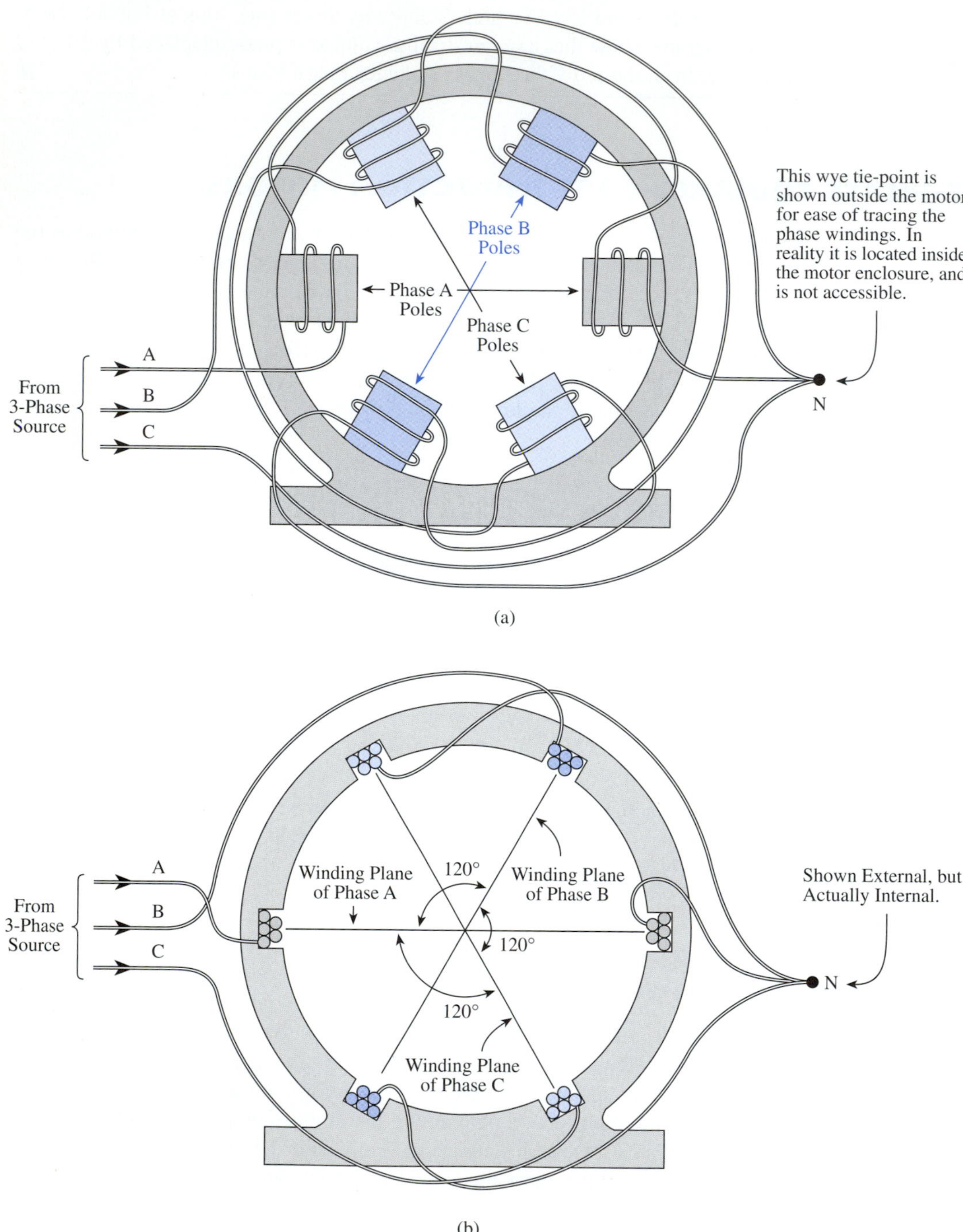

FIGURE 14–31
Stator winding layout of three-phase 2-pole induction motor, using wye configuration. (a) Salient poles; easier to understand but not representative of real three-phase motors. (b) Nonsalient poles; harder to trace and understand, but realistic. There is *not* consistency in flux orientation between part (a) and part (b). Flux is actually perpendicular to the winding plane, so the flux from phase A in part (b) would be vertical; but the flux from phase A in part (a) would be horizontal. Ignore this discrepancy.

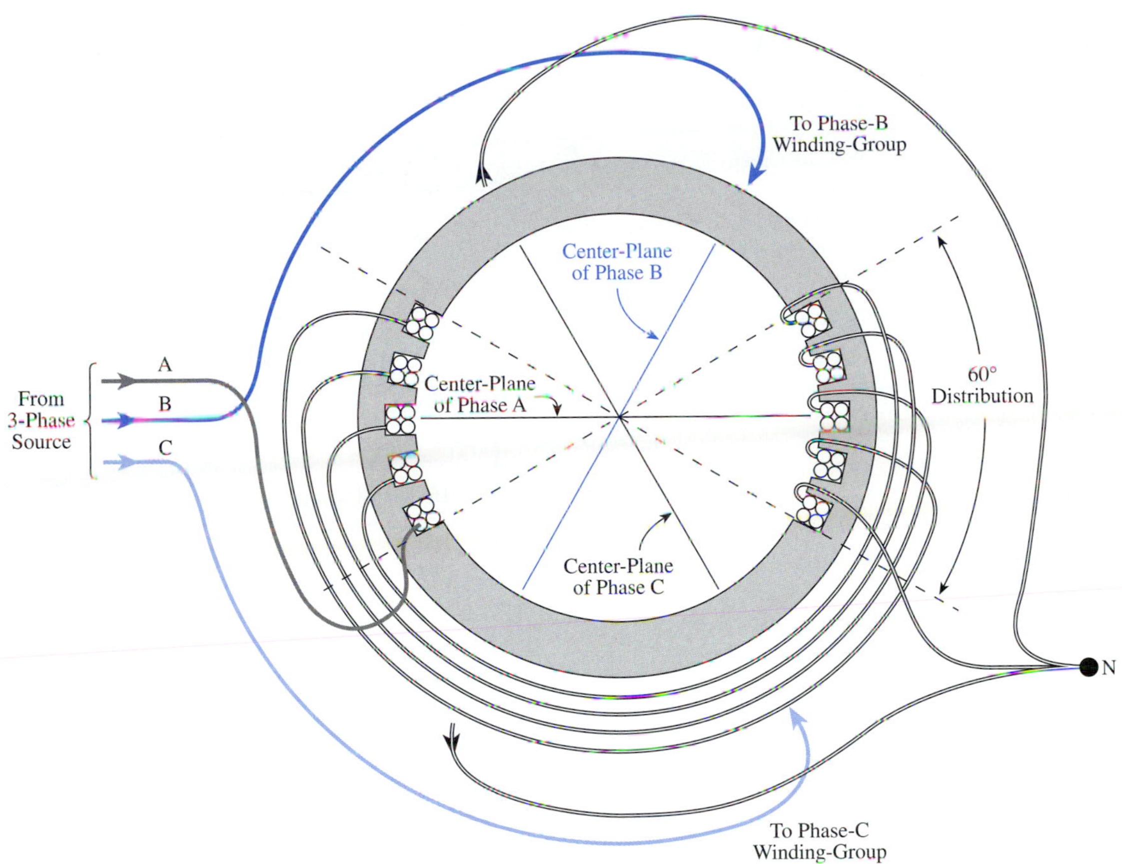

FIGURE 14–32
Structure of distributed winding with five windings per series-connected group. Only phase A is shown in detail to keep the drawing clear. Phase B would also consist of five windings distributed over 60° of the stator surface, on each side; and the same for phase C. At 60° per side, any one phase spans 120° total; this is one third of the entire stator. Thus, with the other two phases in place, the *entire* stator is being used effectively.

In Fig. 14–32 the individual windings that are farthest from the center-plane of the group are 30° away from the center-plane. Therefore their magnetic fluxes are only 30° away from vertical, as shown in Fig. 14–33. Their vertical components of flux are given by

$$\Phi_{\text{vert (worst-position)}} = \Phi_{\text{winding}} \times \cos 30° \approx 0.87 \times \Phi_{\text{winding}}$$

In other words, when a winding group spans a relatively small angular distance, every single winding is able to make a substantial contribution to the net total flux of that group. In a three-phase motor like that shown in Fig. 14–32, where a group spans 60° per side, the worst-positioned windings have only 13% of their created flux become useless (100% − 87% = 13%).

Contrast this three-phase situation to the two-phase situation shown in Fig. 14–4(b). If a manufacturer attempted to use all the space on the stator to hold distributed windings, a winding group would span 90° per side (180° overall). This would place the worst-positioned windings 45° from the group's center-plane. The percent of useful flux from these

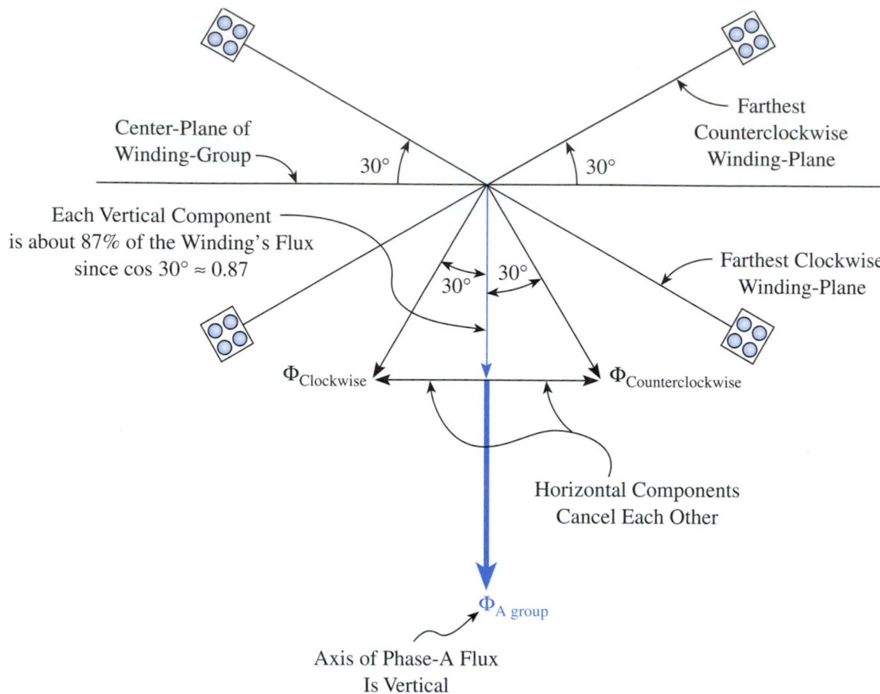

FIGURE 14–33 The flux contribution from the worst-positioned windings in the group still contribute 87% of their flux to the group's net total flux.

windings would be only 70.7%, since cos 45° = 0.707. Taking the other view, 29% of their created flux would become useless.

The ability of a three-phase motor to make more effective use of the stator space is its fundamental advantage over a two-phase motor. Because of this advantage, large, powerful motors can be made physically smaller and less expensive. This is one of the reasons for the move away from early two-phase power generation to the three-phase generation that is standard today.

By the way, three-phase enjoys the same power-density advantage at the generator end of the electric system. That is, a three-phase alternator of a given rating is smaller and less expensive than a comparable two-phase alternator would be. Speaking specifically, most North American utility-grid alternators are three-phase 900-MVA units, made by either General Electric or Westinghouse. They generate about 15 000 V per phase and are wye configured, giving a line voltage of

$$V_L = \sqrt{3} \times V_{ph} = 1.73 \times 15 \text{ kV} \approx 26 \text{ kV}$$

The current models are about 15 ft in diameter, 30 ft in length, and cost about $2 million. To build a comparable two-phase alternator would increase the size and price by perhaps 30%.

14-6-2 Summarizing the Three-Phase Advantages

Three-phase ac has the following advantages over single-phase and two-phase ac:

1. The machines (generators and motors) are more compact, less expensive to manufacture, and more efficient to operate.
2. The transmission of electric power is done more efficiently with reduced I^2R power loss in the transmission lines.
3. The torque produced by a three-phase motor is absolutely constant, with no tendency to pulsate. This is also true of a genuine two-phase motor like Fig. 14–1, but

such motors are virtually extinct. Split-phase motors operating from a single-phase ac line tend to pulsate in torque because we cannot maintain exactly equal phase-winding currents and an exact 90° phase-shift.

The squirrel-cage rotor of a three-phase motor is no different, conceptually, from the squirrel-cage rotor of a split-phase machine; refer to Sec. 14-2, Figs. 14–6 through 14–10. The formulas for synchronous speed [Eqs. (14-1) and (14-2)] and slip [Eqs. (14-4) through (14-6)] are the same for three-phase motors as for single-phase motors. In the synchronous speed formula $S_{syn} = f(120)/P$, keep in mind that P stands for poles per phase, not the total number of poles in all phases combined. Three-phase motors tend to have better speed regulations than single-phase motors.

Three-phase motors are manufactured in a wide range of sizes and power ratings, from about 1 hp (about 0.75 kW) to over 10 000 hp. Motors of several hundred horsepower are not uncommon in heavy manufacturing industries. Small three-phase motors are designed to operate at 208-V or 240-V line voltage, in general. Motors rated above about 5 hp are designed to be used at nominal line voltages of $V_L = 240$ V, $V_L = 480$ V, or $V_L = 600$ V. Large three-phase motors above about 100 hp may be designed for line voltages of 480, 600, or even several thousand volts. In general, the higher a motor's design voltage, the lower its maximum line current demand from the three-phase source.

14-7 ■ CHARACTERISTICS OF THREE-PHASE MOTORS

Three-phase motors have speed-vs.-torque characteristic curves with a basic shape similar to Fig. 14–20 for split-phase motors. However, the difference between maximum torque and starting torque tends to be less dramatic for three-phase motors because a three-phase motor's starting torque, τ_{start}, is almost always greater than its full-load torque, τ_{FL}, as indicated on curve B of Fig. 14–34. All the rationale that was given in Sec. 14-4 for why

FIGURE 14–34
Speed-vs.-torque characteristic curves for different NEMA types of three-phase motors. The class-A curve, not shown, is very close to the class-B curve.

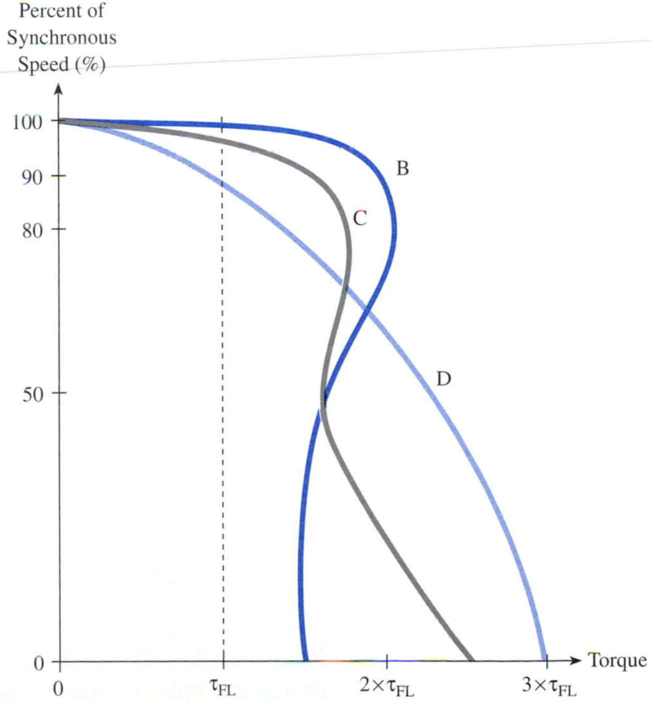

TABLE 14–1
Descriptions of National Electrical Manufacturers Association squirrel-cage motor classes

	Motor Class		
	B*	C	D
Speed regulation	3% to 5%	4% to 5%	5% to 13%
Starting torque (factor of τ_{FL})	1.4 to 1.7	2.0 to 2.5	>2.5
Maximum torque (factor of τ_{FL})	2 to 3	2 to 3	same as starting torque
Starting current (factor of I_{FL})	4 to 6	3 to 6	5 to 7
Efficiency at full-load	>85%	>83%	>80%
Cost factor (class B = 1.0)	1.0	1.1 to 1.2	1.3 to 1.7
Mechanical load characteristics	Needs good speed reg; not frequently started or reversed	Needs fair speed reg; heavy torque demand when starting; not frequently started or reversed	Speed reg not critical; very heavy torque demand when starting; frequently started or reversed
Typical load devices	Fans, centrifugal pumps	Conveyors, compressors	Cranes, stamping machines, and presses

*Class A is very similar to class B.

single-phase motors follow such a curve is still valid for three-phase motors. After all, there is nothing conceptually different between the operation of single-phase and three-phase motors. Their only difference is the use of three rather than two phase-shifted currents to produce the rotating magnetic field.

Motor manufacturers can tailor the exact shape of the speed-torque curve to suit various applications. For example, some applications require fairly good speed regulation and do not require great starting torque. Such an application will be best served by a motor with the type-B characteristics in Fig. 14–34. Other applications don't require good speed regulation but do need to develop great starting torque to accelerate a heavy shaft load. In that case a motor with the type-D curve would be the best choice.

Types of three-phase squirrel cage motors are defined by the National Electrical Manufacturers Association (NEMA). Fig. 14–34 presents the generalized S-vs.-τ characteristics for NEMA types B, C, and D motors. Table 14–1 lists typical characteristic values and appropriate load devices for the three common types.

14-7-1 Squirrel-Cage Cross Sections

A squirrel-cage motor's operating characteristics, and consequently its NEMA type, are determined largely by the cross-section shape of its rotor conductor bars. Actual cross-section examples are shown in Fig. 14–35.

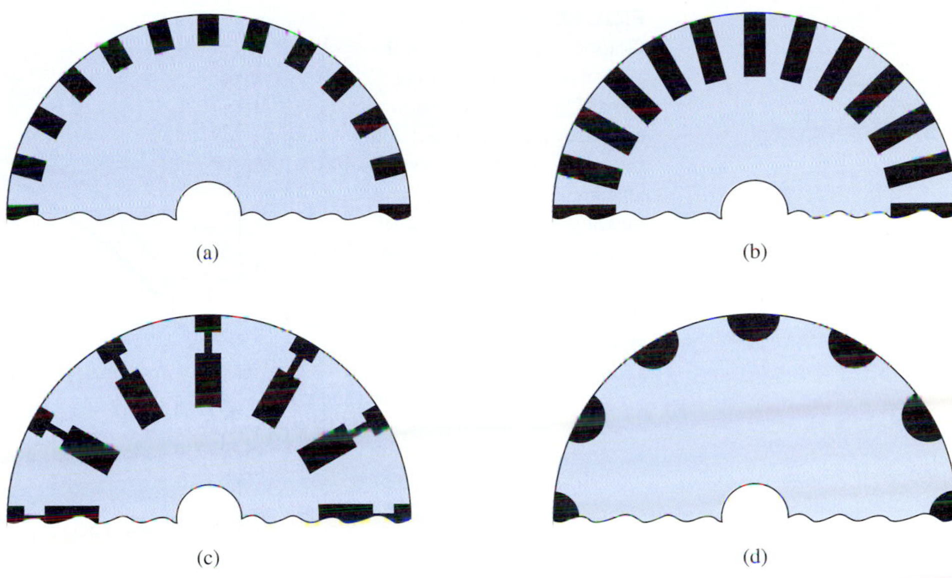

FIGURE 14–35 Cross sections of National Electrical Manufacturers Association squirrel-cage induction motors. (a) Type A. (b) Type B. (c) Type C. The type C rotor is sometimes referred to as a *double-squirrel-cage* design because there is an outer cage connected by a narrow neck to an inner cage. (d) Type D.

The resistance R of a rotor-bar loop is set by the cross-sectional area of the conductor bars. The inductance L of the rotor-bar loops is determined by the depth of the conductor bars into the iron core.

Thus, for example, the NEMA type-A rotor in Fig. 14–35(a) has higher resistance R than the NEMA type-B rotor in Fig. 14–35(b) because the conductor bars have less area. However, type A has lower inductance L than type B because its conductor bars are not buried so deep into the iron.

Upon motor start-up, when slip is high, the induced frequency in the conductor loops is high. Therefore the loop currents are limited mostly by the inductive reactance $X_L = 2\pi fL$, not so much by loop resistance R. Thus the rotor starting current and the motor's starting torque are determined mostly by the rotor-loop inductance, L.

Later, when the rotor has accelerated and slip is low, induced frequency decreases. Then inductance L becomes less significant (the f in $2\pi fL$), and the resistance R becomes the determiner of rotor current and torque. By altering the relative areas and depths of the rotor bars as illustrated in Fig. 14–35, manufacturers are able to achieve the various speed-vs.-torque characteristics in Fig. 14–34.

14-7-2 Efficiency and Power Factor

Like all motors, squirrel-cage induction motors are relatively inefficient at light loads. This is so because with little power going to the mechanical load, a larger portion of their electrical input power goes to wind and bearing friction, I^2R losses, and magnetic core losses. A motor's efficiency improves as its torque load increases; then it may fall off a bit as the motor approaches its full load. This behavior is illustrated in Fig. 14–36 for a NEMA type-B integral-horsepower squirrel-cage induction motor.

An ac induction motor's power factor is also low at light loads. Most of the line current to the motor serves just to magnetize the core. Such magnetizing current is set by the stator winding's inductive reactance X_L, so it lags the source voltage by 90°.

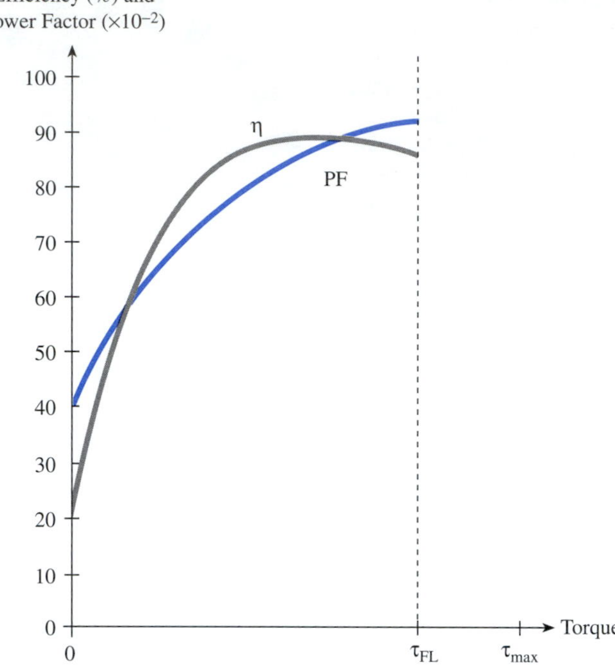

FIGURE 14–36 Graphs of efficiency η and power factor PF for a National Electrical Manufacturers Association type-B 10-hp three-phase induction motor. Efficiency is the ratio of the motor's mechanical output power to its true electrical input power. Power factor is the ratio of the motor's true electrical input power to its apparent input power.

Relatively little current goes into producing shaft power or into internal power losses. Current for these latter two purposes is in phase with the source voltage, lagging by 0°.

As the motor's load increases, the proportion of in-phase power-transferring current increases, relative to 90°-lagging magnetizing current. Therefore the phase angle decreases between the overall motor current and the source voltage. (Overall motor current, the line current flowing to the stator winding, is equal to the phasor sum of the in-phase power-transferring current and the 90°-out-of-phase magnetizing current.) A decreasing phase angle between current and voltage implies a higher power factor. This behavior is shown in Fig. 14–36.

Remember that power factor is important to the electric generation and transmission processes. It determines whether a given amount of true power can be delivered with a reasonably small current (with high PF); or whether the job will require a greater amount of current, and consequently more severe I^2R losses in the generation and transmission (for low PF).

The efficiency and power factor values shown in Fig. 14–36 are typical for a modern integral-horsepower type-B three-phase motor. Estimating full-load η at about 85% and full-load PF at about 0.9, we can develop a rule-of-thumb formula relating voltage, full-load current, and horsepower rating.

$$\begin{aligned} P_{\text{in(FL)}}\,(\text{watts}) &= \sqrt{3}\, V_L I_{L(FL)} \times \text{PF} \qquad \text{(full-load condition)} \\ &= 1.73\, VI_{(FL)} \times (0.9) \\ &\approx 1.56\, VI_{(FL)} \end{aligned}$$

$$\begin{aligned} P_{\text{out(FL)}}\,(\text{watts}) &= \eta \times P_{\text{in}} = 0.85 \times 1.56 \times VI_{(FL)} \\ &\approx 1.3 \times VI_{(FL)} \end{aligned}$$

$$P_{\text{out(FL)}}\,(\text{horsepower units}) = \frac{P_{\text{out(FL)}}\,(\text{watts})}{746\ \text{W/hp}}$$

$$\approx \frac{1.3 \times V \times I_{(FL)}}{746}$$

$$\approx \frac{V \times I_{(FL)}}{574}$$

The approximate factor 574 is usually rounded to 600, giving the rule of thumb

$$\text{Horsepower rating} \approx \frac{V \times I_{(FL)}}{600}$$

or

$$I_{(FL)} \approx \frac{600 \times \text{horsepower rating}}{V} \tag{14-17}$$

Equation (14-17) works fairly well for type-B motors, which account for about 80% of all three-phase induction motors. It works less well for the other NEMA types.

14-8 ■ STARTING, REVERSING, AND TWO-VOLTAGE OPERATION

An across-the-line motor-starting circuit for a three-phase induction motor is shown in Fig. 14–37. A three-phase motor starter is shown in Photo 14–2.

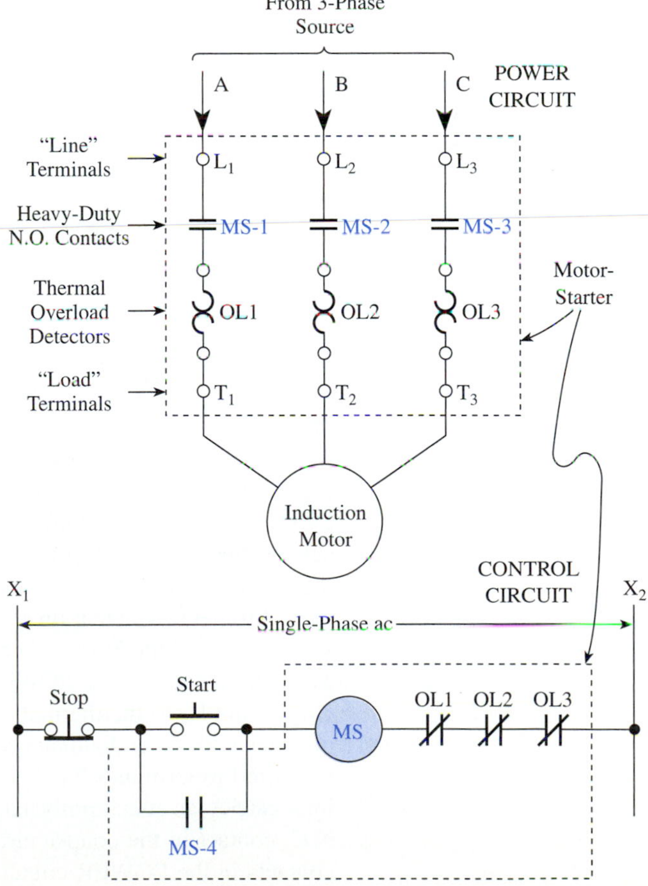

FIGURE 14–37 Schematic diagram of a basic one-station Start-Stop control and overload protection for three-phase motor.

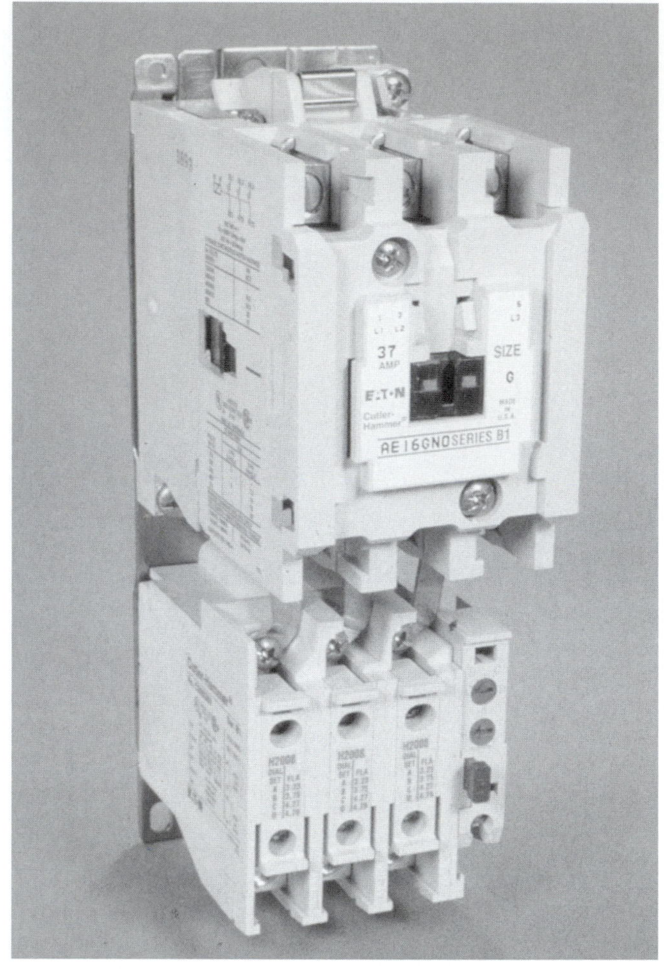

PHOTO 14–2
Physical appearance of National Electrical Manufacturers Association size-1 motor-starter, rated 550 V, 30 A, for use with up to 10-hp motors.
Courtesy of Eaton Corp., Cutler-Hammer Product Division.

The single-phase ac that drives lines X_1 and X_2 in the CONTROL circuit in Fig. 14–37 is usually 120 V, derived from a step-down transformer connected across two lines of the three-phase source. Sometimes the X_1-X_2 control voltage is obtained by connecting directly across two lines of the three-phase source, for example, from line A to line B. This arrangement gives an X_1-X_2 control voltage that is higher than 120 V, perhaps 240 V. Then the control circuit is inherently less safe when the operator reaches into it to actuate the Start or Stop pushbutton switches. Naturally the motor-starter MS coil must then be rated for the higher voltage.

Momentary actuation of the Start PB switch energizes starter coil MS, which then seals in around the Start switch via contact MS-4. Heavy-duty contacts MS-1, MS-2, and MS-3 go closed in the POWER circuit, connecting the three-phase source at terminals L_1, L_2, and L_3 to the thermal detectors OL1, OL2, and OL3. These thermal detectors are in series with the individual phase windings of the motor (imagine a wye connection) that are wired to terminals T_1, T_2, and T_3 of the motor-starter. If any one of the phase windings carries an unnaturally large sustained current, that overload detector will open its N.C. contact in the control circuit. This deenergizes the MS coil, opening all three N.O. contacts in the POWER circuit, which shuts off the motor.

14-8-1 Reversing

To reverse the direction of shaft spin of a three-phase induction motor, we must reverse the direction of the rotating magnetic field. This is done by interchanging any two of the three power leads. To understand why this reverses the rotating field direction, study the motor cross sections in Fig. 14–31(a) or (b). First identify the direction of field rotation if the phase sequence is ABC (A leading B, and B leading C). This means that the A poles in Fig. 14–31(a) are energized in the positive-current direction, followed by the B poles, followed by the C poles, then back to the A poles. You will conclude that the field rotation is clockwise.

Then swap any two motor leads in their connection to the three-phase source. The new phase sequence will become ACB (A leading C, and C leading B). Reanalyzing the magnetic behavior in Fig. 14–31(a) will make you conclude that the field rotation reverses to counterclockwise.

The circuit for accomplishing motor reversal is shown in Fig. 14–38. A reversing motor-starter appears in Photo 14–3.

Figure 14–38 swaps the A and C phases. In Reverse the starter connects incoming line terminal L_1 to motor terminal T_3 and incoming line terminal L_3 to motor terminal T_1.

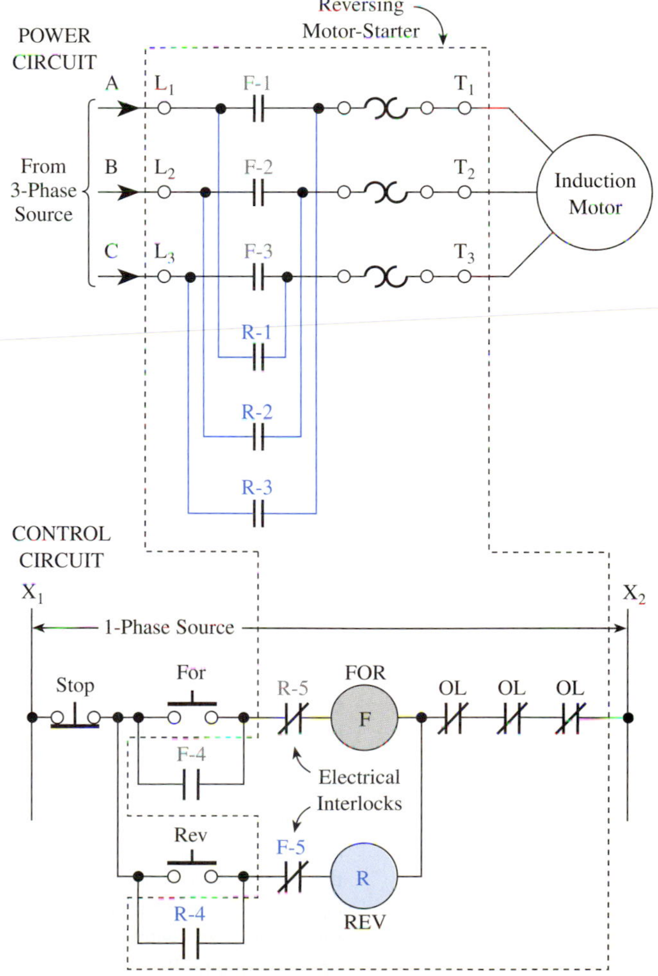

FIGURE 14–38
Wiring schematic for a reversing starter, with interlock arrangement that prevents the Forward and Reverse contacts from closing simultaneously.

The manufacturer brings nine leads out of the motor, numbered 1 through 9. All manufacturers who subscribe to NEMA standards, which is essentially all manufacturers, number the lead-terminals according to the pattern in Fig. 14–40. For high-voltage operation the user makes the following connections: 4 to 7; 5 to 8; 6 to 9; and 1, 2, and 3 to the three-phase source. This is illustrated in Fig. 14–40(a).

For low-voltage operation the user makes the following connections: 1 to 7; 2 to 8; 3 to 9; 4 to 5 to 6; and 1, 2 and 3 to the three-phase source. This is illustrated in Fig. 14–40(b).

Delta option. Sometimes the motor manufacturer gives the user the option of interconnecting the motor phase windings in delta configuration. Of course, the decision to exercise this option must take into account the fact that now an individual phase winding will receive the *entire* source line voltage, not the source line voltage divided by 1.73. The user must ensure that the winding's voltage rating is not exceeded.

Twelve leads, numbered 1 through 12, are brought out of the motor enclosure for this option. For the higher voltage the user must interconnect as shown in Fig. 14–41(a); this gives a series connection between same-location (same-phase) windings. For the lower voltage follow Fig. 14–41(b); this gives a parallel connection between same-location windings.

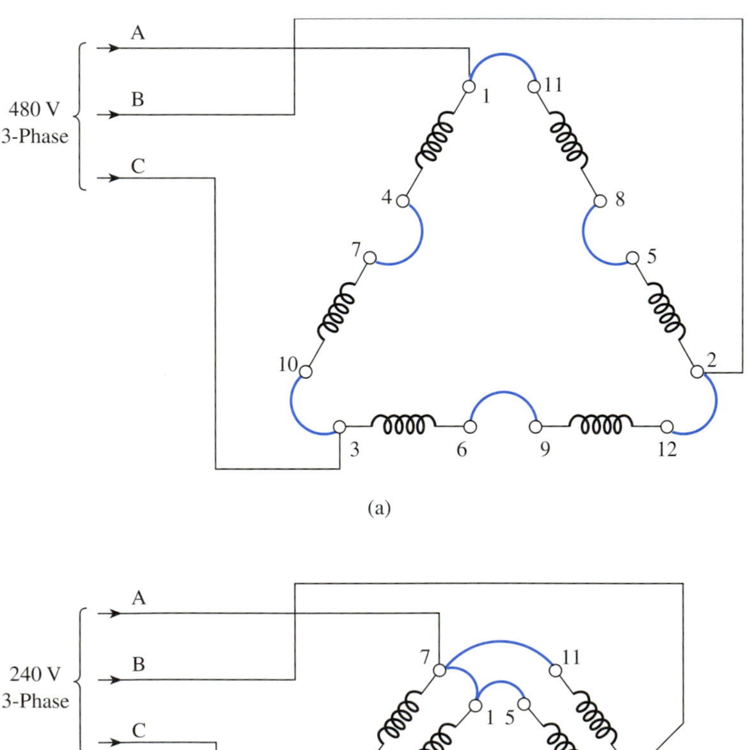

FIGURE 14–41
Dual-voltage delta configuration. (a) High voltage. (b) Low voltage.

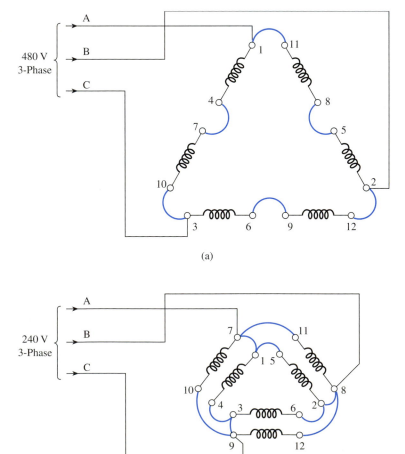

TROUBLESHOOTING ON THE JOB

TROUBLESHOOTING A MOTOR OVERLOAD

The scrubbing tower of Troubleshooting on the Job in Chapter 12 uses two to four Spray Recycle Pumps, as Fig. 12–41 indicates. One of the Spray Recycle Pumping stations is shown in this photo. The three-phase, 600-V, 500-hp squirrel-cage induction motor is on the right, closest to our view. The square unit in the center is a gear box for reducing the shaft speed from about 1750 r/min to about 300 r/min. The pump itself is the large-diameter unit closest to the tower.

The motor is operated by a straightforward Stop-Start circuit like the one shown in Fig. 14–37. The NEMA size-7 motor-starter has three thermal overload detectors rated 600 A. The motor itself has a full-load current rating of 475 A and a full load speed rating of 1735 r/min.

YOUR ASSIGNMENT

One of the Spray Recycle Pumping motors is tripping its overload. You must find out whether the motor is truly being overloaded, there is some fault in the motor itself, or the trip-out is false due to malfunctioning overload detectors.

Your test instrumentation consists of a 1000-V ac voltmeter, a 1000-A clamp-on ac ammeter, and a reflective optical tachometer for measuring shaft speed. Describe the test/troubleshooting procedure that you will follow to determine the nature of the problem.

Courtesy of General Electric Environmental Services, Inc.

SUMMARY

- In a true two-phase motor, the stator's phase windings are driven by two ac sine-wave sources that are 90° phase-shifted. This creates a constant-strength magnetic field that revolves at constant speed. The rotor then attempts to follow the revolving (called rotating) field.
- The angular speed of the rotating field is called synchronous speed, symbolized S_{syn}. It is related to line frequency and number of poles of the motor structure by the formula $S_{syn} = f \times 120/P$.
- An ac induction motor usually has a squirrel-cage rotor. This rotor is very simple, inexpensive, and reliable because it has no conductive path through brushes. Instead, current is *induced* in the rotor bars by transformer action (Faraday's law).
- Since true two-phase ac power is difficult to obtain, most such motors have their stator winding currents phase-shifted by a capacitor and/or by altering the windings' characteristics. Such motors are called split-phase ac motors.
- If the capacitor-assisted winding is automatically switched out of the circuit as a split-phase motor approaches running speed, it is called a *capacitor-start* motor. The winding that is switched out is called the *start winding*.
- The shaded-pole motor and the reluctance-start motor are alternative designs of squirrel-cage ac induction motors that do not require any deliberate phase-shift of current through a second stator winding.
- Squirrel-cage ac induction motors have a recognizable characteristic graph of speed versus torque. By altering the construction features, motor manufacturers can vary the specific shape of this graph to adapt the motor to specific applications.
- Three-phase ac systems have important advantages over single-phase and two-phase systems. The two most important advantages are these: (1) The I^2R losses in the transmission wires are lower for a three-phase system, all other things being equal; (2) three-phase alternators and motors have greater power density than single-phase machines—that is, a three-phase motor or alternator of a given size and cost will produce more power than a comparable single-phase or two-phase machine.
- Since line voltage and current are always accessible for measurement whereas phase voltage and current may not be physically accessible, we prefer to perform all true power and apparent power calculations in terms of line variables, not phase variables.
- In a wye, $V_L = \sqrt{3} \times V_{ph}$ and $I_L = I_{ph}$. In a delta, $I_L = \sqrt{3} \times I_{ph}$ and $V_L = V_{ph}$.
- Apparent power S (voltamperes) is a more meaningful variable than true power P (watts) when rating the maximum capability of an ac source, namely an alternator or transformer. This is because apparent power is independent of the load's power factor, cos ϕ, which cannot be known with confidence when the maximum rating is being assigned to the ac source.
- The National Electrical Manufacturers Association (NEMA) categorizes three-phase squirrel-cage induction motors into classes A, B, C, and D, depending on the motors' rotor-bar cross sections, which affect their characteristic speed-vs.-torque curves and thus their usefulness for particular applications.
- Generally speaking, a motor's efficiency η and its power factor PF rise as its torque load increases. Sometimes efficiency peaks at less than full-load torque, then declines slightly as full-load is approached.
- To reverse the direction of rotation of a three-phase motor, it is necessary to swap any 2 of the 3 power leads.
- A three-phase motor can be stopped very quickly by switching 2 of the 3 motor leads (usually A and C) to create reverse (retarding) torque. This practice is called plugging.

- Many three-phase industrial motors are designed to be operated at either a higher voltage, about 480 V, or at a lower voltage, about 240 V. The winding external leads must be properly interconnected for whichever operating voltage is being used.

FORMULAS

$$S_{syn} = \frac{f \times 120}{P} \qquad \text{(Equation 14-2)}$$

$$\text{slip} = S_{syn} - S_{rot} \qquad \text{(Equation 14-4)}$$

$$\text{percent slip} = \frac{\text{slip}}{S_{syn}} \times 100\% \qquad \text{(Equation 14-5)}$$

$$\text{percent slip} = \frac{S_{syn} - S_{rot}}{S_{syn}} \times 100\% \qquad \text{(Equation 14-6)}$$

$$V_L = \sqrt{3}\, V_{ph} \quad \text{(for wye)} \qquad \text{(Equation 14-7b)}$$

$$I_L = I_{ph} \quad \text{(for wye)} \qquad \text{(Equation 14-8)}$$

$$V_L = V_{ph} \quad \text{(for delta)} \qquad \text{(Equation 14-9)}$$

$$I_L = \sqrt{3} \times I_{ph} \quad \text{(for delta)} \qquad \text{(Equation 14-10b)}$$

$$S = V \times I \qquad \text{(Equation 14-11b)}$$

$$P = (V \times I) \times PF \qquad \text{(Equation 14-12b)}$$

$$P = S \times PF \qquad \text{(Equation 14-12c)}$$

$$P = 3 \times V_{ph} \times I_{ph} \times PF \quad \text{(for balanced delta)} \qquad \text{(Equation 14-14)}$$

$$S = \sqrt{3} \times V_L \times I_L \quad \text{(for balanced delta)} \qquad \text{(Equation 14-15a)}$$

$$P = \sqrt{3} \times V_L \times I_L \times PF \quad \text{(for balanced delta)} \qquad \text{(Equation 14-15b)}$$

$$S = \sqrt{3} \times V_L \times I_L \quad \text{(for balanced wye)} \qquad \text{(Equation 14-16a)}$$

$$P = \sqrt{3} \times V_L \times I_L \times PF \quad \text{(for balanced wye)} \qquad \text{(Equation 14-16b)}$$

$$I_{(FL)} \approx \frac{600 \times \text{horsepower rating}}{V} \qquad \text{(Equation 14-17)}$$

$$1 \text{ N-m} = 0.7376 \text{ lb-ft} \qquad 1 \text{ lb-ft} = 1.356 \text{ N-m}$$

$$1 \text{ r/min} = 0.1047 \text{ rad/s} \qquad 1 \text{ rad/s} = 9.551 \text{ r/min}$$

$$P(\text{in watts}) = \tau(\text{in N-m}) \times S(\text{in rad/s})$$

$$1 \text{ hp} = 745.7 \text{ W} \qquad 1 \text{ kW} = 1.341 \text{ hp}$$

QUESTIONS AND PROBLEMS

Section 14-1

1. True or False: Most ac induction motors have salient poles.
2. In Fig. 14–1, at a moment when V_A has its defined-negative polarity, identify the magnetic polarity of the poles.
3. Repeat Question 2 for V_B having *its* defined-negative polarity.
4. In Fig. 14–4(a), at the moment when V_A is positive and 45° from a zero crossover (at $0.707 \times V_{pk}$) and V_B is also positive and 45° from a zero crossover, identify the instantaneous polarity of every pole.

5. For the moment of Question 4, draw the magnetic flux orientation relative to the stator. Indicate the precise angular position of the flux.
6. A certain ac induction motor has eight poles per phase and is driven by 60 Hz ac sources. Find its synchronous speed S_{syn}.
7. On aircraft the ac line has frequency $f = 400$ Hz. For an ac motor with six poles, find S_{syn}.

Section 14-2

8. In a squirrel-cage, are the conductor bars absolutely parallel to the shaft? Explain.
9. A squirrel-cage rotor has current in its conductor bars due to magnetic ____________, which is how this kind of motor gets its name.
10. In Fig. 14–7 the north flux is rotating clockwise and the conductor bars beneath it carry current out of the page. If that same flux were rotating counterclockwise, then the conductor bars beneath would carry current ____________ the page. Explain.
11. In Fig. 14–10 suppose that the stator flux is rotating clockwise instead of counterclockwise. Show the direction of current in every conductor bar at this instant.

Section 14-3

12. True or False: The 240-V center-tapped residential ac line (black, white, red wires) does a fine job all by itself in driving two-phase induction motors. Explain.
13. In a split-phase ac motor one winding has a ____________ connected in series with it. The current in that winding ____________ the current in the main winding. (Leads or lags?)
14. In a capacitor-start motor, what kind of mechanism actuates the switch that disconnects the starting winding?
15. In a shaded-pole motor the rotor always rotates from the ____________ side to the ____________ side.
16. In a reluctance-start motor the rotor always rotates from the ____________ side to the ____________ side.

Section 14-4

17. A four-pole squirrel-cage motor has $S_{syn} = 1800$ r/min.
    a. If its no-load speed is 1775 r/min, find its percent slip at no load.
    b. If its full-load speed is 1740 r/min, find its percent slip at full load.
    c. Find its speed regulation.
18. Sketch the general speed-vs.-torque characteristic curve of a squirrel-cage induction motor. On that curve identify and label the following: (a) synchronous speed, (b) no-load speed, (c) full-load speed, (d) full-load torque, (e) maximum torque, and (f) starting torque.
19. On the characteristic graph that you drew for Question 18, the top curve segment between $\tau = 0$ and $\tau = \tau_{FL}$ is called the____________ ____________ region.
20. On the graph of Question 18 identify the region where the motor can operate only temporarily, without stalling.
21. True or False: If the motor is made to operate for an extended period in the region of Question 20, it will probably overheat or trip its circuit-protection device.
22. If the motor is operating in the lower part of the characteristic graph of Question 18 (at speeds below the maximum-torque speed), heading lower, it is in process of ____________.

23. If the motor is operating in the lower part of the characteristic graph of Question 18 (at speeds below the maximum-torque speed), heading higher, it is in process of ____________.
24. The situation described in Question 22 arises because the motor's torque production at the maximum-torque speed is ____________ than the torque required to drive the mechanical load at that speed.
25. If a particular pump has a breakaway torque of 3 N-m, the motor that drives that pump must have a ____________ torque greater than 3 N-m.
26. True or False: During the first few seconds of start-up acceleration, the effect of an induction motor's change in loop inductance outweighs the effect of its change in loop voltage.
27. True or False: It is typical for an induction motor's starting current to be at least a factor-of-5 greater than I_{FL}.

Section 14-5

28. A three-phase ac power system originates in an alternator that has windings that are mechanically offset from one another by ____________ degrees.
29. When all three defined-negative phase-winding terminals are tied together, a three-phase system has the ____________ configuration.
30. Draw a schematic diagram of a delta-configured three-phase source. Show defined polarities on the phase windings.
31. When the defined-negative terminal of one phase winding is connected to the defined-positive terminal of its neighboring phase winding, a three-phase system has the ____________ configuration.
32. Draw a schematic diagram of a wye-configured three-phase source. Show defined polarities on the phase windings.
33. In the wye-source to wye-load three-phase system of Fig. 14–26(b), suppose $V_{ph} = 300$ V and $Z_{ph} = 20\ \Omega$ (balanced).
    a. Find V_L, the line voltage.
    b. Find I_{ph}, the phase current.
    c. Find I_L, the line current.
    d. Find S, the total apparent power.
34. In Question 33, if the load is totally resistive with PF = 1.0, find P, the total true power.
35. In Question 33, if the load is partially inductive with PF = 0.82, find P, the total true power.
36. In the delta-to-delta three-phase system of Fig. 14–26(c), suppose the load is balanced with PF = 0.88. An ammeter placed in a line measures $I_L = 25$ A. A voltmeter connected between lines measures $V_L = 600$ V.
    a. Calculate total apparent power S.
    b. Find total true power P.
    c. Find the load's individual phase current, $I_{ph\ load}$.
    d. Find the source's individual phase current, $I_{ph\ source}$.
    e. Find the load's individual phase impedance, Z_{ph}.
37. In the wye-delta system of Fig. 14–27(a), suppose we measure the following: $V_L = 208$ V; $I_L = 12$ A. The load is balanced, and totally resistive. Find
    a. Total S.
    b. Total P.

c. Individual phase power, P_{ph}, in the load.
d. V_{ph} in the load.
e. I_{ph} in the load.
f. R_{ph} in the load.

38. For the system in Question 37, find
    a. V_{ph} in the source.
    b. I_{ph} in the source.
    c. Individual phase power, P_{ph}, in the source.
39. Referring to your results from Questions 37 and 38, compare the individual phase power in the load (Question 37, part **c**) to the individual phase power in the source (Question 38, part **c**). Why is this reasonable?
40. In the delta-wye system of Fig. 14–27(b), suppose we have $V_L = 480$ V and $I_L = 23.5$ A, with the motor operating at full load. Its power factor at full load is $PF_{FL} = 0.84$.
    a. Find the motor's total apparent power, S.
    b. Find its total true power input, P_{in}.
    c. What is the effective impedance Z_{ph} of one individual phase winding, with the motor at full-load condition?
41. For the motor circuit of Question 40, suppose the motor has a full-load efficiency $\eta_{FL} = 91\%$.
    a. Calculate the motor's mechanical output power, in basic units (watts).
    b. Express the motor's shaft output power in the American units of horsepower.
42. For the motor circuit of Questions 40 and 41, suppose the full-load shaft speed is 1735 r/min.
    a. Convert this shaft speed to basic units, radians/second.
    b. Apply the mechanical power formula, $P = \tau \times S$, to calculate the motor's full-load torque, in basic units.
    c. Express this torque in the American units of pound-feet.
43. The motor of Question 42 is operating from the 60-Hz U.S. power grid.
    a. How many poles does the motor have (per phase)?
    b. Find the motor's synchronous speed, S_{syn}.
    c. Calculate the motor's absolute slip at full load.
    d. Find its percent slip at full load.
44. For this motor, the frequency of the induced voltage in the rotor-bar loops is ____________ Hz at full load.

Section 14-6

45. True or False: The major theoretical advantage of three-phase induction motors over genuine two-phase induction motors (now almost extinct) is that the stator flux created by a three-phase motor revolves at a 50% higher speed than the flux created by a two-phase motor.
46. The smallest three-phase squirrel-cage induction motors have a full-load shaft power of about ____________ hp; the largest have a shaft power in excess of ____________ hp.
47. The answer to Question 45 is False. So explain what *is* the theoretical advantage of a three-phase motor over a genuine two-phase motor.

Section 14-7

48. Compared to a NEMA class-B motor, a class-D motor has ____________ starting torque; its disadvantage is ____________ speed regulation.
49. In general an ac motor's power factor ____________ as its shaft load increases.
50. In general a motor's efficiency is ____________ for light shaft loads.
51. Use the rule of thumb, Eq. (14-17), to estimate the full-load line current of a 20-hp, 480-V, three-phase induction motor. How does your result compare to the motor specifications in Question 40 through 43?

Section 14-8

52. From memory, draw the schematic diagram of an across-the-line Start-Stop one-station control circuit for a three-phase motor with overload protection. Explain every item of the circuit, including the origin of the single-phase control voltage.
53. Repeat Question 52 for a reversing three-phase motor control circuit.
54. True or False: To plug a three-phase motor to a stop, the plugging circuit must connect the three-phase source to the motor in the reversing manner.
55. In a two-voltage motor the same-location phase windings must be connected in ____________ for the higher voltage; they must be connected in ____________ for the lower voltage.
56. For all motor manufacturers who subscribe to ____________ standards, their motors' junction lead-wires are numbered in the same identical manner.

CHAPTER

NINE EXAMPLES OF CLOSED-LOOP INDUSTRIAL SYSTEMS

In this chapter, we will look at nine different industrial closed-loop control systems in detail. Among them, these nine systems contain many of the input transducers and final correcting devices studied in Chapters 10 and 11. The control modes represented in these systems include On-Off, proportional, and proportional plus integral.

OBJECTIVES

After completing this chapter, you will be able to:

1. Discuss and explain the process of controlling temperature in a quench oil tank used for quenching heat-treated metal parts.
2. Discuss the operation of soaking pits used to heat steel ingots prior to hot-rolling, and discuss and explain a system for controlling the pressure in a soaking pit recuperator.
3. Explain the operation of an all-solid-state proportional plus reset temperature controller with thermocouple input.
4. Discuss and explain the process of maintaining a constant tension in a strip handling system.
5. Discuss the process of recoiling moving strip, and explain how an edge-sensing system is used to ensure that the strip is wound up straight.
6. Discuss the operation of an automatic powder-weighing system with load cell input, and explain how a servomechanism is used to position an optical shaft encoder to read out the weight.
7. Discuss the steel carburizing process, and explain the operation of a system which controls the carbon case depth by controlling the CO_2 content in the carburizing atmosphere.
8. Discuss and explain the control of relative humidity in a textile moistening process.
9. Discuss and explain the control of relative humidity in granaries and in explosives-storage warehouses.

15-1 ■ THERMISTOR CONTROL OF QUENCH OIL TEMPERATURE

Very often, heat-treated metal parts must be quenched in either oil or water in order to impart the proper metallurgical qualities to the metal. In most such processes, the parts are immersed in a bath of quench oil as soon as they leave the heat-treating chamber. Naturally, the temperature of the quench oil tends to rise due to the continual dunking of the hot metal parts. To accomplish the desired quenching results, the quench oil must be maintained within a certain temperature range; this is done by cooling the oil in a heat exchanger. The situation is illustrated schematically in Fig. 15–1(a). The hot parts slide down a chute into the quench tank, landing on a link belt which catches them on

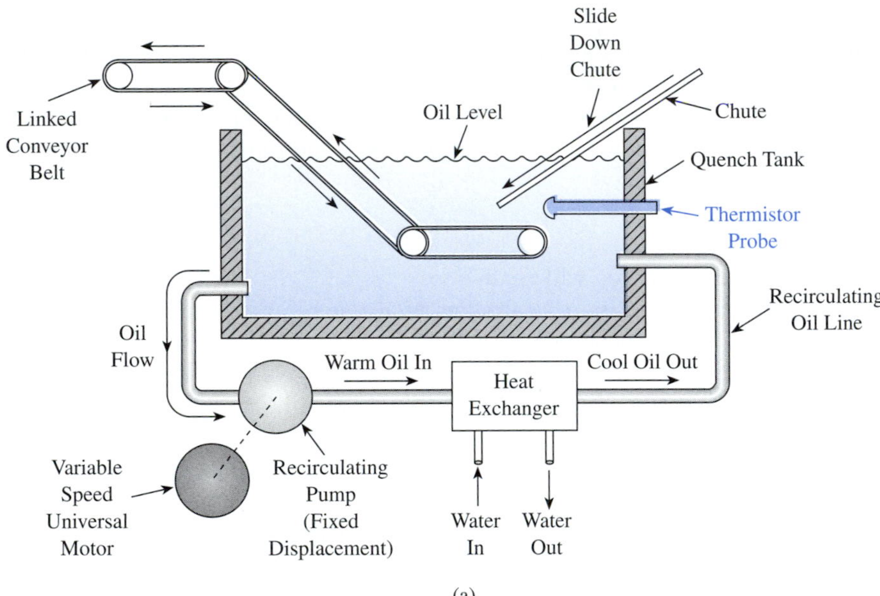

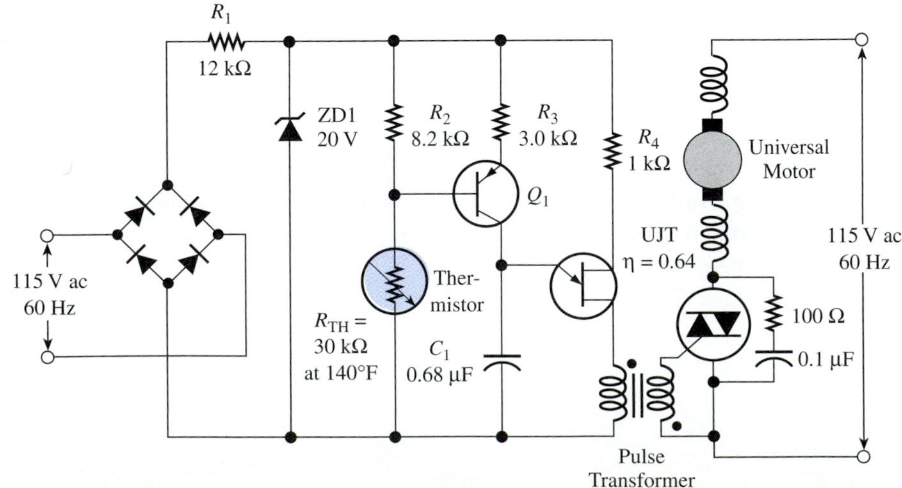

FIGURE 15–1
Quench oil temperature controller. (a) Physical layout of the quench tank and the cooling apparatus. (b) Circuit for controlling the recirculating pump.

its spurs. The moving belt carries them horizontally through the quench oil and then up and out of the tank.

An oil outlet pipe allows oil to flow out of the tank and into the fixed-displacement recirculating pump. A fixed-displacement pump moves a fixed volume of liquid on each revolution, so the speed of rotation of the pump determines how much oil recirculates through the cooling system. The outlet of the pump feeds into a water-cooled heat exchanger. From the heat exchanger, the recirculating oil passes back into the quench tank.

The motor driving the recirculating pump is a series universal motor, capable of operating on either dc or ac. In this system, it is operated on ac voltage. The average voltage applied to the motor terminals determines its speed of rotation. Since the motor drives the pump, the rotational speed of the motor determines how much oil recirculates and thereby determines the amount of cooling that takes place. As the motor speeds up, more oil recirculates, and the oil in the tank tends to cool down. As the motor slows down, less oil recirculates, and the oil in the tank tends to warm up.

The quench oil temperature is sensed by a thermistor mounted inside a *probe*, which is a protective shield. A thermistor is an ideal temperature transducer for this application because it produces a large response for small temperature changes, and because it is suited to the fairly low temperatures encountered in quenching processes (usually less than 200°F). The thermistor is connected into the control circuitry as shown in Fig. 15–1(b). Here is how the control circuit works.

The bridge rectifier, in conjunction with the clipping circuit comprised of R_1 and ZD1, supplies an approximate square wave across the Q_1 circuitry. This square wave has a peak value of 20 V and is synchronized with the ac line pulsations, as we have seen before. At the instant the 20-V supply appears, the R_2-R_{TH} series combination divides it up. The voltage available for driving the base-emitter circuit of Q_1 depends on just how the R_2-R_{TH} voltage divider divides the 20 V. If the thermistor resistance R_{TH} is high, a small voltage will appear across R_2 and the base-emitter drive will be small. If the thermistor resistance is low, a larger voltage will appear across R_2, due to voltage-divider action, and the base-emitter drive will be large.

The voltage available to drive the base-emitter circuit determines the Q_1 emitter current, according to

$$I_E = \frac{V_{R2} - 0.7\ \text{V}}{3.0\ \text{k}\Omega} \tag{15-1}$$

where V_{R2} stands for the voltage appearing across resistor R_2.

Equation (15-1) is just Ohm's law applied to the emitter resistor. It shows that an increase in V_{R2} causes an increase in emitter current.

The Q_1 collector current is virtually the same as the emitter current. As the diagram shows, the Q_1 collector current charges capacitor C_1. When C_1 charges to the peak point of the UJT, the UJT fires. The resulting current pulse is delivered to the triac gate. The triac then turns ON and applies power to the motor terminals.

To summarize the behavior of this circuit, the greater the voltage across R_2, the greater the charging current to C_1. If the C_1 charging current is greater, the UJT will fire sooner in the half cycle and the power delivered to the motor will be greater. This causes the motor and pump to spin faster.

If the voltage across R_2 is small, the Q_1 collector current will charge C_1 slowly. This causes late firing of the UJT and triac, and reduced motor speed.

Now let us see how the measured oil temperature affects the circuit action. An increase in oil temperature results in a lowering of the thermistor resistance R_{TH} due to the thermistor's negative temperature coefficient. Lowering R_{TH} causes an increase in V_{R2} by voltage-divider action. As we have seen, an increase in V_{R2} causes the pump to run faster. This recirculates more oil through the heat exchanger and tends to drive the tank temperature back down.

This particular circuit is designed to start recirculation when the tank oil temperature reaches 140°F. Below 140°F the motor does not run at all. Above 140°F the UJT and triac start firing, and the motor begins running. Therefore, for a temperature of exactly 140°F, the triac should be on the verge of firing. This is equivalent to saying that the firing delay angle should be 180° when the temperature is 140°F. Then any slight increase in temperature beyond that point will reduce the firing delay angle to less than 180° and cause the motor and pump to start running.

The thermistor characteristic is such that at 140°F, $R_{TH} = 30\ \text{k}\Omega$, so

$$\frac{V_{R2}}{20\ \text{V}} = \frac{R_2}{R_2 + R_{TH}} = \frac{8.2\ \text{k}\Omega}{8.2\ \text{k}\Omega + 30\ \text{k}\Omega}$$

$$V_{R2} = 4.3\ \text{V}$$

The emitter current is given by Eq. (15-1):

$$I_E = \frac{4.3\ \text{V} - 0.7\ \text{V}}{3.0\ \text{k}\Omega} = 1.2\ \text{mA}$$

Therefore I_C, the capacitor charging current, equals 1.2 mA also. Assuming that the UJT has a standoff ratio η equal to 0.64, the UJT peak voltage is given by

$$V_P = (0.64)(20\ \text{V}) + 0.6\ \text{V} = 13.4\ \text{V}$$

Therefore the capacitor must charge to 13.4 V to fire the UJT and triac. The time required to do this can be found from

$$\frac{\Delta V}{\Delta t} = \frac{I}{C}$$

which expresses the voltage buildup rate for a capacitor. Rearranging, we obtain

$$\Delta t = \frac{C}{I}(\Delta V) = \frac{(0.68\ \mu\text{F})(13.4\ \text{V})}{1.2\ \text{mA}} = 7.6\ \text{ms}$$

Therefore the UJT should fire at about 7.6 ms after the start of the cycle. This time can be expressed as an angle by saying that

$$\frac{\theta}{360°} = \frac{7.6\ \text{ms}}{16.67\ \text{ms}}$$

where 16.67 ms is the period of the 60-Hz ac line. Thus the firing delay angle is calculated as 164° when the oil temperature is 140°F.

This means that the triac is just barely firing and is supplying a very small average voltage to the motor. Any further temperature rise from this point will cause the firing delay angle to be reduced and, consequently, the motor and pump to begin running faster. The pump is then able to hold the oil temperature close to 140°F.

If it were desired for some reason to have a variable temperature set point, this could easily be accomplished. The R_2 resistor could be replaced by a potentiometer. Then as the pot resistance was increased, the temperature set point would be lowered. As the pot resistance was decreased, the temperature set point would be raised.

15-2 ■ PROPORTIONAL MODE PRESSURE CONTROL SYSTEM

15-2-1 Soaking Pits for Steel Ingots

In the steel industry, a *soaking pit* is an underground pit used to heat steel ingots to about 2400°F prior to rolling. The ingots are placed in the pit by a crane. The pit cover is placed on top, also by a crane, and the gas burners are turned on to bring the pit temperature up to 2400°F. The combustion of natural gas with air creates waste gases, which leave the pit through an exhaust duct. Some of the heat energy contained in the hot waste gases is recovered and used to preheat the fresh combustion air coming into the burners. The preheating takes place in a heat exchanger called a *recuperator.* This process is illustrated schematically in Fig. 15–2(a).

The recuperator in Fig. 15–2 is just a large-diameter duct. It has hot waste gases entering on the left at a temperature of about 2400°F and leaving on the right at about 1800°F. The reduction in waste gas temperature represents the fact that some of the heat energy has been recovered and transferred to the cold combustion air, thus making the soaking process more energy-efficient. Incoming combustion air for the burners is drawn into the cold air intake by the combustion air blower, a large powerful fan. The incoming air passes through a duct to a butterfly valve, which opens to the proper position to maintain the correct upstream air pressure ahead of the recuperator. The cold air makes two or three passes through the recuperator, picking up more heat energy on each pass. It finally emerges from the recuperator at a temperature of about 900°F. From there it travels to the burner air control valve, which passes the air to the burners as it is called for by the temperature controller. The temperature controller is not shown, since we are concentrating on the pressure control system. When air arrives at the burners it is mixed with natural

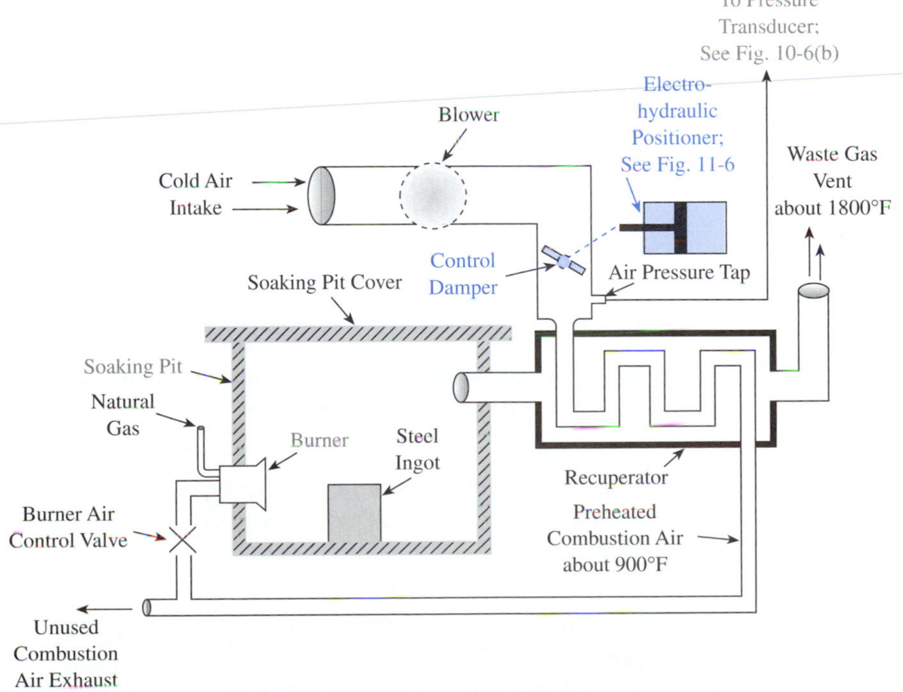

FIGURE 15–2 Soaking pit with recuperator: (a) physical layout; (b) electronic circuit for positioning the control damper.

gas, the process fuel. Unused preheated air can be exhausted through a restriction in the supply header, as shown in Fig. 15–2(a). Sometimes part of the unused preheated air is routed back to the blower input to mix with the new air entering the cold air intake.

It is important to maintain the proper value of pressure in the cold air duct just ahead of the recuperator to prevent the recuperator tubes from overheating. The desired pressure may vary from one set of operating conditions to the next. Therefore the pressure controller must have provision for adjusting the pressure set point. The controller opens or closes the air butterfly valve to correct any deviation of the measured pressure from set point. If the controller sees that the measured pressure is below set point, it opens the butterfly valve further to raise the air pressure upstream of the recuperator. If the measured pressure is above the set-point pressure, the controller closes the butterfly valve further.

The input pressure transducer is a bellows-potentiometer transducer of the type shown in Fig. 10–6(b). The pressure signal for the transducer is taken from a pressure tap in the air duct upstream of the recuperator, as shown in Fig. 15–2(a). The transducer potentiometer has −15 V dc and ground voltage applied to its two end terminals, so the transducer output is a dc voltage which varies between 0 and −15 V. As the measured pressure increases, the pot wiper moves closer to the −15-V terminal. Thus higher pressures are represented by more negative voltages. This is indicated on the left of Fig. 15–2(b).

The final correcting device is an electrohydraulic positioner driving the shaft of the butterfly valve. The positioner is the same kind as that shown in Fig. 11–6. The position of the cylinder rod is controlled by the amount of current flowing through the sensing coil. In this positioner, the coil has a resistance of 2000 Ω. A coil current of 0 mA causes the cylinder to fully retract, completely closing the butterfly valve. A current of 5 mA causes the cylinder to fully extend, driving the butterfly valve wide open. A hydraulically operated positioner is needed for this application because of the great imbalancing forces exerted on the butterfly valve by the large volume of combustion air.

The electronic pressure control circuit is shown in Fig. 15–2(b). The set point and the measured pressure are the two electrical inputs to this circuit. These inputs are compared, and the difference between them, the error, causes control action. The output of the controller is the dc current delivered to the 2000-Ω sensing coil on the far right of the drawing. The control mode is strictly proportional. That is, the sensing coil current is varied in proportion to the error between set point and measured pressure.

15-2-2 Electronic Comparer/Controller

To begin understanding the operation of op amp 1 and its input circuits, let us temporarily make two simplifying assumptions:

1. Assume the span-adjust pot is turned all the way to the top. This will allow the entire wiper voltage from the set-point pot to be applied to R_4.
2. Assume the set-point zero pot wiper is adjusted to exactly 0 V. This effectively eliminates this pot and R_2 and R_3 from consideration, since they supply only a 0-V signal to op amp 1.

In a little while, we will come back and see why these two potentiometers are necessary.

With these assumptions made, we can simplify the op amp 1 circuitry as shown in Fig. 15–3. It is not difficult to see that this is an unweighted summing circuit. The set-

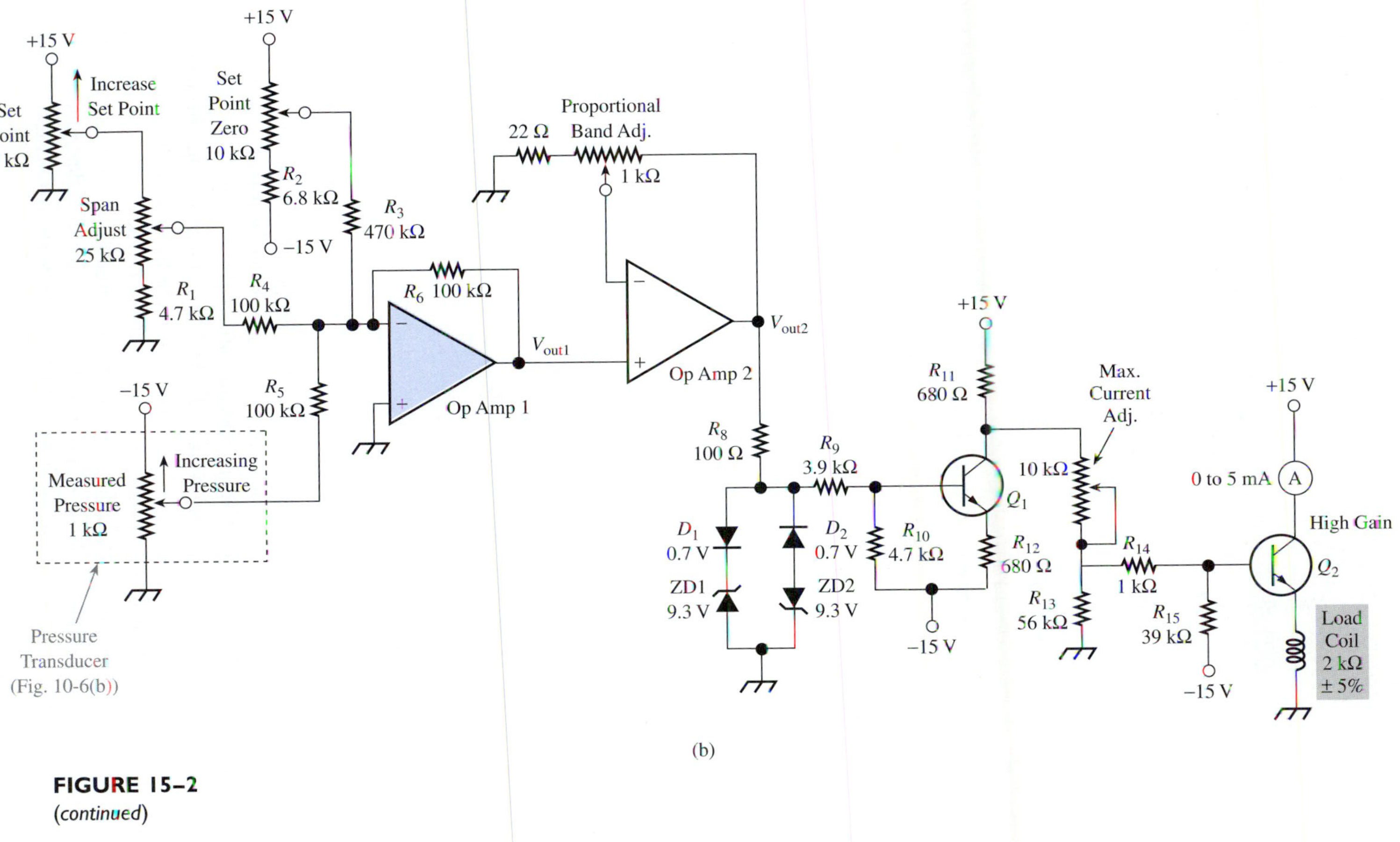

(b)

FIGURE 15–2
(continued)

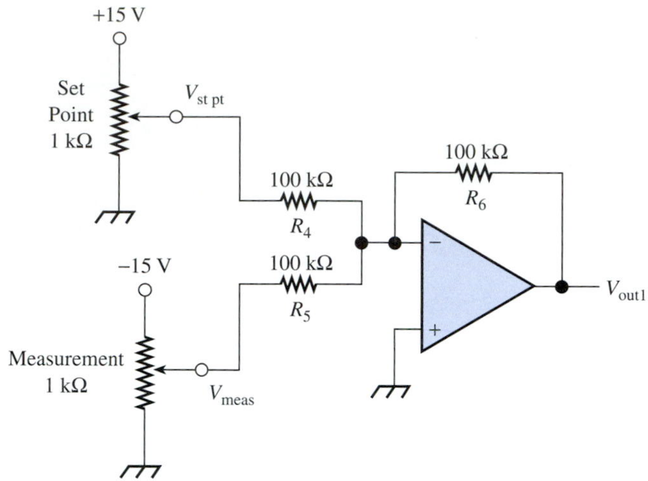

FIGURE 15–3
Simplified schematic diagram of the op amp 1 circuitry of Fig. 15–2(b).

point voltage, which is positive, is added to the measurement voltage, which is negative. The sum of these voltages appears in inverted form at V_{out1}. In equation form,

$$-V_{out1} = \frac{R_6}{R_4}(V_{st\ pt}) + \frac{R_6}{R_5}(V_{meas}) = \frac{100\ k\Omega}{100\ k\Omega}(V_{st\ pt}) + \frac{100\ k\Omega}{100\ k\Omega}(V_{meas}) \quad (15\text{-}2)$$

$$-V_{out} = V_{st\ pt} + V_{meas}$$

Keep in mind that V_{meas} is a negative voltage.

Equation (15-2) tells us that if the measured pressure agrees exactly with the set-point pressure, the output of op amp 1 equals 0 V. If the measured pressure is less than the set point, the output of op amp 1 is negative. If the measured pressure exceeds the set point, the output of op amp 1 is positive. The magnitude of V_{out1} represents the amount of deviation of the measured pressure from the set-point pressure, and the polarity of V_{out1} tells us the direction of the error.

Now let us go back and look at the set-point zero circuitry: as can be seen from Fig. 15–2(b), the voltage from the zero pot wiper is a third input to the op amp 1 summing circuit, but it carries very little weight, since R_3 is so large compared to R_4 and R_5. This third input is necessary to compensate for the fact that the set-point pot and the measurement pot cannot go all the way to 0 Ω. Even when these pots are turned down as far as they go, there will be some *end resistance* left. Therefore the wiper voltages will not be absolutely zero. This would be no problem if we could be sure that the two voltages were the same. However, we cannot be sure of that. Since we definitely want V_{out1} to be 0 V when both input pots are turned all the way down, we arrange for that to be so by injecting the lightly weighted signal from the zero pot. This cancels out any imbalance between the low ends of the two main input potentiometers.

Now let us consider the span-adjust pot. One reason the span pot is necessary is that the range of the pressure input transducer might exceed the desired range of the set point. For example, suppose the pressure input transducer is designed to move its pot wiper from bottom to top as the bellows inlet pressure varies from 0 to 3 psig.* If the recuperator

*The "g" following the psi units stands for gage pressure, or pressure above atmospheric pressure.

never under any circumstances needs an inlet pressure greater than 2 psig, then we would like the top of the set-point pot to represent a set-point pressure of 2 psig; that is, when the set-point pot is all the way to the top, the measurement pot should be able to balance it by moving to the 2-psig point, which is only two thirds of its total distance. We can cause this to happen by delivering less than the full set-point wiper voltage to R_4. Instead of full wiper voltage, we arrange for only two thirds of the set-point wiper voltage to be delivered to R_4. This is done by turning the span pot down until the span pot wiper voltage is only two thirds of the set-point wiper voltage.

To summarize, the span pot reduces the set-point pot wiper voltage so that it can be balanced by the measurement pot wiper moving *less than its full range*. It is called a span pot because it determines the "span" of pressure values which can be set on the set-point pot. It would be adjusted by the system operators to provide whatever set-point span they desired.

Now consider op amp 2. It is wired as a noninverting amplifier with a variable gain. The output of this amplifier, V_{out2}, drives the Q_1-Q_2 discrete circuitry, which supplies current to the sensing coil of the butterfly valve positioner. Therefore the value of V_{out2} determines the final position of the butterfly control valve. Since this is so, the voltage gain of the noninverting amplifier determines the controller's proportional band. If the voltage gain is high, it takes only a small error (small V_{out1}) to cause a large change in V_{out2}, and consequently a large change in valve position. This means the proportional band is narrow. If the voltage gain is low, it takes a larger error to cause a given change in V_{out2}. Therefore it takes a large error (large V_{out1}) to cause a given change in valve position, which makes the proportional band wider.

The maximum voltage gain occurs when the pot is adjusted to the far left. At that point, $R_F = 1\ \text{k}\Omega$ and $R_{IN} = 22\ \Omega$, so

$$A_{V\text{max}} = \frac{R_F}{R_{IN}} + 1 = \frac{1000}{22} + 1 = 46.5$$

The minimum voltage gain occurs when the pot is moved all the way to the right. At that point, $R_F = 0$, so

$$A_{V\text{min}} = \frac{0}{R_{IN}} + 1 = 1$$

The voltage gain of the noninverting amplifier can thus be varied from 1.0 to 46.5.

Although V_{out2} can vary from approximately +12.0 to −12.0 V, the voltage at the junction point of R_8 and R_9 is limited to a range of +10 to −10 V. The diode network below the junction point ensures this. Here is how it works.

The D_1-ZD1 pair prevents the R_8-R_9 junction from rising above +10V. If the junction tries to rise above +10 V, the breakdown voltage of the D_1-ZD1 combination will be exceeded, causing that diode path to short out any excess over 10 V. This occurs because D_1 is forward biased by the positive output voltage, and zener diode ZD1 will have reached its breakdown point (10 V − 0.6 V = 9.4 V). Any difference between V_{out2} and +10 V is then dropped across 100-Ω resistor R_8.

The D_2-ZD2 diode pair prevents the R_8-R_9 junction voltage from going more negative than −10 V. If V_{out2} goes below −10 V, D_2 will be forward biased by the negative output, and zener diode ZD2 will have reached its breakdown point. Any difference between V_{out2} and −10 V will again be dropped across R_8.

The voltage appearing at the left end of R_9 can thus take on any value between +10 and −10 V, but it cannot exceed that range. Positive voltages mean that the measured pressure is above set point, and negative voltages mean that the measured pressure is below set point. Zero voltage means that the measured pressure agrees with the set point.

Let us consider the action of the discrete circuitry when zero volts appear at the left of R_9. The R_9-R_{10} voltage divider determines the voltage at the base of Q_1. The voltage dropped across R_9 can be found by the voltage-division formula

$$\frac{V_{R9}}{V_T} = \frac{R_9}{R_T} = \frac{R_9}{R_9 + R_{10}}$$

where V_{R9} symbolizes the voltage drop across R_9 and V_T refers to the total voltage drop from the left end of R_9 to the -15-V supply. With 0 V on the left end of R_9, the total voltage drop is simply -15 V. Therefore

$$V_{R9} = (15\text{ V})\frac{3.9\text{ k}\Omega}{3.9\text{ k}\Omega + 4.7\text{ k}\Omega} = 6.8\text{ V}$$

With 6.8 V dropped across R_9, the base voltage relative to ground is simply -6.8 V. This forward-biases the base-emitter junction of Q_1, causing Q_1 to conduct. The Q_1 emitter voltage will be 0.7 V below its base voltage, so $V_{E1} = -6.8\text{ V} - 0.7\text{ V} = -7.5\text{ V}$ relative to ground. With the Q_1 emitter voltage at -7.5 V relative to ground, the Q_1 collector voltage must be $+7.5$ V relative to ground, since the Q_1 circuit is perfectly symmetrical. (You should prove this to yourself.)

Now let us assume for a moment that the maximum current-adjust pot is dialed completely out. This will simplify the explanation. With this pot shorted out, the full $+7.5$ V appears at the top of R_{13}. It is then further divided by the R_{14}-R_{15} voltage divider to determine the voltage at the base of Q_2. By the voltage-division formula, we can say that

$$\frac{V_{R14}}{V_T} = \frac{R_{14}}{R_{14} + R_{15}}$$

where V_{R14} stands for the voltage drop across R_{14} and V_T symbolizes the total voltage drop between the left side of R_{14} and the -15-V supply. The total voltage drop is given by

$$V_T = +7.5\text{ V} - (-15\text{ V}) = 22.5\text{ V}$$

so

$$\frac{V_{R14}}{22.5\text{ V}} = \frac{1\text{ k}\Omega}{1\text{ k}\Omega + 39\text{ k}\Omega}$$

$$V_{R14} = 0.6\text{ V}$$

Therefore the voltage at the Q_1 base is given by

$$V_{B1} = +7.5\text{ V} - 0.6\text{ V} = +6.9\text{ V}$$

This voltage forward-biases transistor Q_2, causing it to turn on and conduct. Q_2 is wired as an emitter follower. The base-emitter voltage drop is 0.7 V, so $V_{E2} = 6.9\text{ V} - 0.7\text{ V} = 6.2\text{ V}$. Therefore transistor Q_2 will force enough current through the load coil to cause a voltage drop of 6.2 V across it. The current required to do this is given by

$$I_{\text{coil}} = \frac{V_{\text{coil}}}{R_{\text{coil}}} = \frac{6.2\text{ V}}{2\text{ k}\Omega} = 3.1\text{ mA}$$

The final conclusion of this derivation is that a 0-V input signal coming from op amp 2 causes a load coil current of 3.1 mA. Keep in mind that the entire range of currents necessary to stroke the positioner from fully closed to fully open is only 0 to 5 mA. A coil current of 3.1 mA would cause the positioner to drive the butterfly valve about 62% open, since 3.1 mA/5.0 mA = 0.62.

As the output voltage of op amp 2 takes on values other than 0 V, it causes the positioner to stroke the valve further open or closed.

Consider what happens if V_{out2} goes positive. A positive V_{out2} causes the voltage at the base of Q_1 to go more positive, thereby turning Q_1 on harder. This causes collector voltage V_{C1} to become smaller, which tends to turn off Q_2 and reduce the load coil current. Thus positive values of V_{out2} cause the butterfly valve to close.

A negative V_{out2} causes the base voltage V_{B1} to go further negative, thereby reducing the Q_1 collector current. Collector voltage V_{C1} therefore rises, which tends to turn on Q_2 to increase the load coil current. Therefore negative values of V_{out2} cause the butterfly valve to open.

The Q_1-Q_2 circuitry is designed so that -10 V at the left end of R_9 will cause the load current to equal 5 mA and $+10$ V at the left end of R_9 will cause the load current to equal nearly 0 mA. The exact adjustment of this response is done with the 10-kΩ maximum current-adjust pot. Variations in load coil resistance and in circuit component values may cause the actual load current to differ from the proper 5-mA value when $V_{out2} = -10$ V. Such discrepancies are adjusted out with the maximum current-adjust pot.

The purpose and operating principles of the pressure control system should now be clear. Any tendency of the measured pressure to drop below the set-point pressure causes the controller to deliver more current to the load coil. This opens the butterfly valve and admits more combustion air to bring the measured pressure back up toward the set point. Conversely, any tendency of the measured pressure to rise above set point causes the controller to reduce the current to the load coil. This closes the butterfly valve to bring the measured pressure back down toward set point.

In some soaking pit systems the control of combustion air is done on the basis of flow rate instead of recuperator upstream pressure. The control method is exactly the same as explained here, except that the pressure transducer becomes a *differential* pressure transducer, responding not to a single gage pressure but to the pressure *drop* across an orifice in the air duct. This pressure drop across an orifice is proportional to air flow rate, so the controlled variable becomes flow rate instead of recuperator pressure.

15-3 ■ PROPORTIONAL PLUS RESET TEMPERATURE CONTROLLER WITH THERMOCOUPLE INPUT

In industrial processes the most frequently controlled variable is temperature. In drying processes, melting processes, heat-treating processes, chemical reaction processes, etc., temperature is of prime importance. When the process temperature is above a few hundred degrees Fahrenheit, the preferred transducer is usually a thermocouple. One of the most common temperature control schemes is a thermocouple input into a proportional plus reset electronic temperature controller, with the final correcting device being a variable-position fuel valve. In this section we will investigate such a control scheme in detail.

15-3-1 Thermocouple-Set Point Bridge Circuit

On the left in Fig. 15–4(a) is the thermocouple bridge measurement circuit. This circuit combines the thermocouple millivolt signal with the temperature set-point signal to generate an error signal. The magnitude of the error signal represents the deviation between

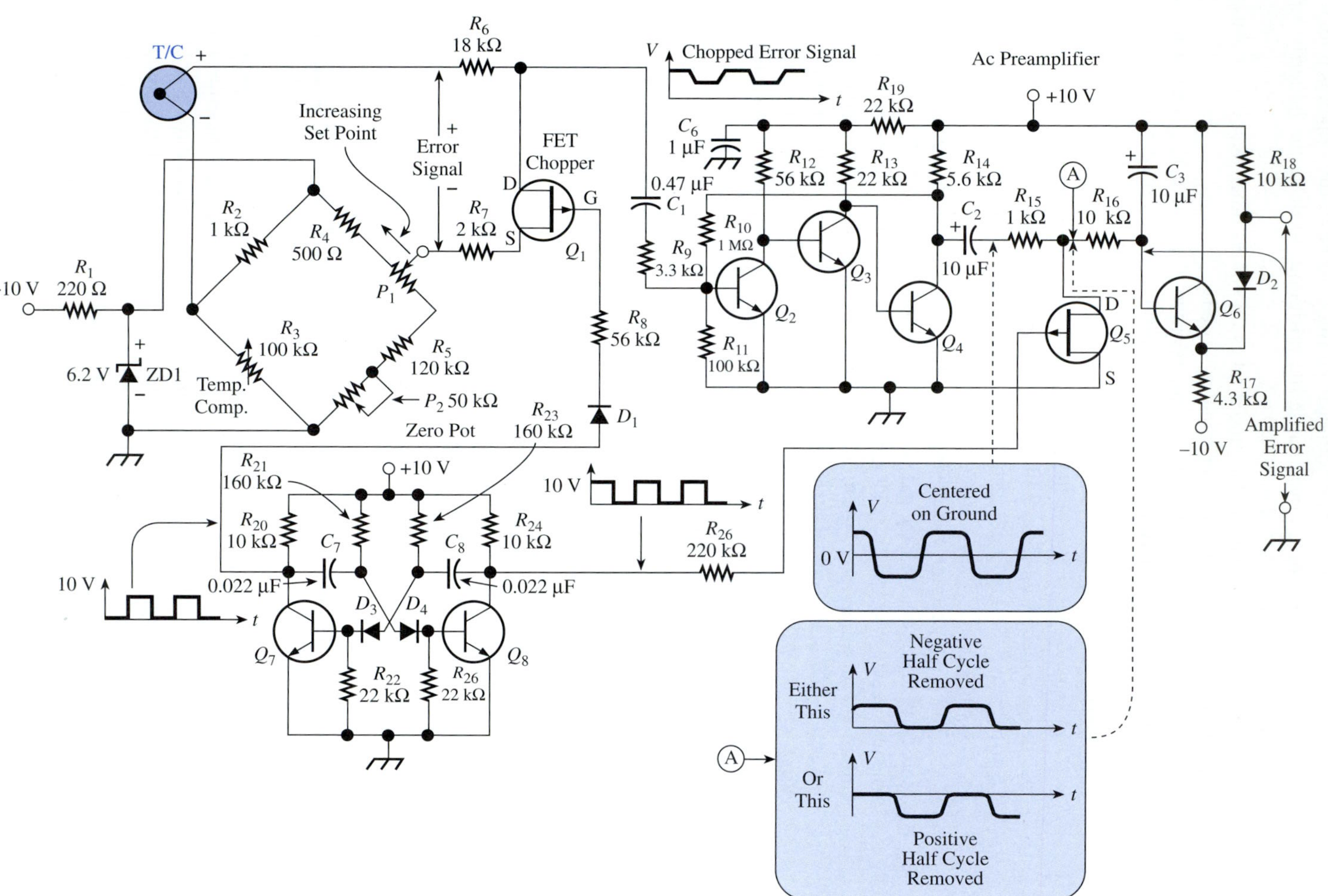

FIGURE 15–4
Thermocouple temperature-control circuit: (a) themocouple bridge-input, chopper, preamplifier, and demodulator; (b) proportional plus integral control circuit which positions the valve.

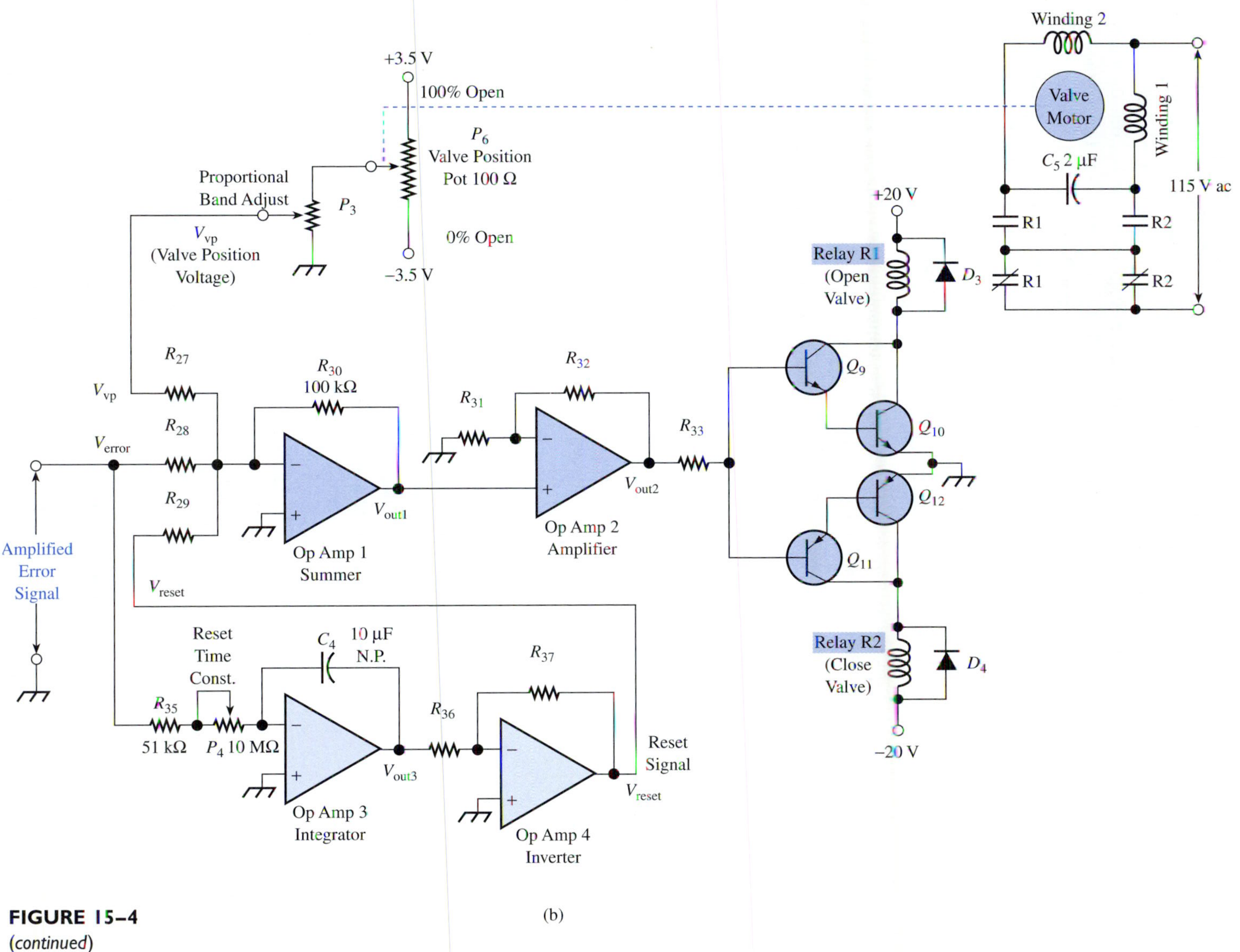

FIGURE 15–4
(continued)

measured temperature and desired temperature. The desired temperature is represented by the position of the set-point potentiometer. Here is how the circuit works.

The 6.2-V zener diode, ZD1, provides the stable dc supply voltage to the bridge. The bridge is designed so the voltage at the top of R_3 equals the voltage at the top of R_5; both of these voltages are measured relative to ground, the bottom of the bridge. The 50-kΩ zero pot, P_2, is adjusted to accomplish this. Causing these two voltages to be equal ensures that an extreme setting on the set-point pot P_1 exactly balances a 0-V signal from the thermocouple. That is, when P_1 is dialed all the way down with its wiper touching the junction of P_1 and R_5, the error signal will be zero when the thermocouple puts out a zero signal. This will happen only if the voltage across R_3 equals the voltage across the series combination of P_2 and R_5.

The system operator selects the desired set-point temperature by adjusting the position of P_1, the set-point pot. This pot has a pointer attached to its shaft, which points to a marked temperature scale. The marked temperature scale is not shown in an electronic schematic diagram.

Once the set-point pot has been adjusted, the system seeks to bring the measured temperature into agreement with the set point. When the two are in agreement, the thermocouple signal voltage exactly equals the voltage between the wiper and the bottom terminal of P_1. If the measured temperature should rise above the set point, the thermocouple (T/C) signal will be greater than the set-point signal and the error voltage will be positive, as marked in Fig. 15–4(a). If the measured temperature should drop below the set point, the T/C signal will be smaller than the set-point signal and the error voltage will be negative. Its polarity would then be the opposite of the polarity shown in Fig. 15–4(a). The greater the deviation between measured temperature and set-point temperature, the greater is the difference between these two voltages and the greater is the magnitude of the error signal.

15-3-2 Preamplifier, Chopper, and Demodulator

The rest of the electronic circuitry in Fig. 15–4(a) serves the purpose of amplifying the tiny dc error signal. Remember from Sec. 10-4 that thermocouples generate a very small signal voltage, no more than a few tens of millivolts. The error signal, being the *difference* between a thermocouple signal and another signal in the millivolt range, is much smaller yet. The error signal is only a fraction of a millivolt when the measured temperature is close to set point. It is virtually impossible to build a dc amplifier that is so drift-free that it can reliably handle a dc signal this small.

For this reason, the preamplifier in this temperature control system is chopper-stabilized. This is, the dc error signal is converted into an ac signal by chopping, is then amplified in an ac amplifier where drift is of no consequence, and is then converted back into a dc signal at the output of the amplifier. Before beginning the discussion of the preamplifier itself, let us deal with the chopping circuitry.

The chopping circuitry consists of an FET (field-effect-transistor) chopper and the chopper drive generator. The chopper drive generator is an astable clock, which was mentioned before in Sec. 2-9. The astable clock develops a square-wave signal at the collector of transistor Q_7, which is then applied to the *gate* lead (marked *G*) of the FET chopper. The frequency of this square-wave signal is given by the approximate equation

$$f \cong \frac{0.7}{R_B C} = \frac{0.7}{(160\ \text{k}\Omega)(0.022\ \mu\text{F})} = 200\ \text{Hz}$$

As this 200-Hz square wave is applied to the gate terminal of the FET chopper the FET behaves as follows:

1. When the square wave goes positive, the gate goes positive relative to the *source* terminal (marked S). This turns the FET OFF and causes it to exhibit a high resistance between the *drain* terminal (marked D) and source terminal. It can be thought of as an open switch.
2. When the square wave goes down, it removes the positive bias from the gate terminal, allowing the FET to turn ON. The FET then exhibits a low resistance from source to drain terminal and can be considered a closed switch.

The FET therefore alternates between an open switch and a closed switch.* When it is an open switch, the top line of the error signal (the T/C voltage) is connected to the preamp coupling capacitor, C_1. When the FET is a closed switch, it effectively applies the bottom of the error signal (the set-point voltage) to the preamp input. This is true because the bottom line is coupled to the preamp input through a Thevenin resistance of about 2 kΩ, whereas the top line (T/C signal) is coupled through a Thevenin resistance of about 18 kΩ. Therefore the bottom line signal overwhelms the top line signal when the FET is ON (when it is a closed switch from source to drain).

The signal delivered to C_1 is therefore a square waveform whose peak-to-peak value equals the magnitude of the dc error signal.

The chopped error signal is coupled through C_1 and R_9 into the base of transistor Q_2. Transistors Q_2, Q_3, and Q_4 constitute a high-gain ac amplifier. There are no stabilizing resistors in the emitter leads of Q_2, Q_3, and Q_4. Instead, bias stability is provided by negative dc feedback from the collector of Q_4 into the base divider (R_{10} and R_{11}) of Q_2. The absence of degeneration resistors in the emitter leads provides a high voltage gain in this three-stage amplifier.

Notice that the Q_2 and Q_3 stages are decoupled from the Q_4 stage by R_{19} and C_6. This technique minimizes the appearance of dc supply line noise in the initial stages of the preamplifier, which is where it could do the most harm. This technique was discussed in Sec. 11-11-2.

The ac signal appearing at the collector of Q_4 rides on a dc level of about 6.5 V, which is the Q_4 collector bias voltage. The dc component is removed by output coupling capacitor C_2. The ac output signal is therefore centered on ground when it appears at R_{15}. The ac signal is converted to a dc signal by the action of Q_5, an FET *demodulator*. This FET is also being used in a switching mode. Its gate is driven by the square wave at the collector of Q_8, which is 180° out of phase with the square waveform at Q_7 which drives the gate of the FET chopper (Q_1). As Q_5 alternately switches open and closed, it removes one half of the ac signal appearing at the left end of R_{15}. When it turns ON, Q_5 shorts the $R_{15}R_{16}$ junction to ground, causing the instantaneous voltage on the left of R_{15} to be removed. (It is dropped across R_{15}.) When Q_5 turns OFF, it disconnects the R_{15}-R_{16} junction from ground and allows the instantaneous voltage on the left of R_{15} to come through R_{15} only a little bit diminished. (R_{16} is quite a bit larger than R_{15}.)

The initial polarity of the dc error signal from the bridge determines whether the FET demodulator removes the negative half cycle or the positive half cycle of the ac wave-

*An FET is unlike a bipolar transistor in that it is a normally ON device; you must deliver an external signal to the gate in order to turn it OFF (junction-type FET). This is the opposite of a bipolar transistor, which is normally OFF and requires an external base signal to turn it ON. An FET is superior to a bipolar transistor in a chopping application of this sort. Its superiority is due to the fact that there are no *pn* junctions between the drain and source of an FET, as there are between the collector and emitter of a bipolar transistor.

form. If the dc error signal is positive as shown, the FET demodulator removes the *negative* half cycle of the ac output. If the dc error signal is negative (measured temperature is below set point), then the FET demodulator removes the *positive* half cycle of the ac output. Try to reason these last statements out for yourself.

The voltage waveform which appears at point A is filtered by R_{16} and C_3. This low-pass filter converts the square waveform at point A into a dc voltage with just a slight ripple component. This dc voltage is called the *amplified error signal*. The amplified error signal may be positive or it may be negative, depending on the polarity of the original dc error signal. It will have the same polarity as the original dc error.

The amplified error signal is applied to Q_6, an emitter follower, which furnishes a high input impedance. The voltage appearing at the emitter of Q_6 is 0.7 V more negative than the amplified error signal because of the voltage drop across the base-emitter junction. This 0.7 V is restored by the R_{18}-D2 combination; there is a 0.7 rise across silicon diode D2. The purpose of passing the amplified error signal through Q_6 and D2 is to buffer it from the demodulator. This results in an amplified error signal which can deliver a fairly large current into the circuit that it drives without disturbing the demodulator.

15-3-3 Proportional Plus Reset Control

The amplified dc error signal is brought into the op amp summing circuit on the left in Fig. 15–4(b). Let us first concentrate on the proportional aspect of the control. We will then investigate the circuitry which furnishes the reset control action.

Proportional action. The amplified error signal is applied to the op amp 1 summer through R_{28}. Assume for a moment that this signal is positive and has just now suddenly increased in magnitude because of a load disturbance. Here is what will happen.

The positive voltage at the left of R_{28} will tend to drive the inverting input of op amp 1 positive. This will cause the output to go negative. When V_{out1} goes negative, it applies a negative voltage to the noninverting input of op amp 2, which is a high-gain noninverting amplifier. The noninverting amplifier output, V_{out2}, becomes a large negative voltage and will forward-bias Q_{11} and Q_{12}, which are connected as a *Darlington pair*. Because of the very large current gain of a Darlington pair (the total current gain is the product of the two individual transistor current gains), a small trickle of electron current through R_{33} into the base of Q_{11} will cause Q_{12} to turn ON and saturate. When Q_{12} switches ON, it completes a circuit to relay R_2, causing that relay to pick up. The R_2 contacts change state in the 115-V motor control circuit, causing the valve motor to run. The motor runs in the proper direction to close the fuel valve, since a positive error signal from the preamp means that measured temperature is too high (above the set point). As the valve closes, the valve position pot P_6 moves downward into its negative-potential region. The negative voltage appearing at the wiper of P_6 is applied to P_3, the proportional band-adjust pot. A portion of the P_6 negative voltage is picked off by P_3 and fed back to R_{27} and into the summing circuit. This negative voltage tends to cancel the positive error voltage applied to R_{28}. Eventually, if the valve position pot moves far enough, the negative signal applied to R_{27} will cause the summing circuit output to return to zero. At this point, V_{out2} also goes to zero, so it can no longer keep Q_{11} and Q_{12} turned ON, and relay R2 will drop out. This stops the valve motor and causes the fuel valve to freeze in that position. The reduction in fuel flow should drive the measured temperature back down toward set point.

Considering only the R_{27} and R_{28} inputs to the summing circuit, the general equation which describes the summing circuit is

$$-V_{\text{out1}} = \frac{10\ \text{k}\Omega}{10\ \text{k}\Omega}(V_{\text{error}}) + \frac{10\ \text{k}\Omega}{68\ \text{k}\Omega}(V_{\text{vp}})$$

Where V_{vp} stands for *valve position voltage,* which is the voltage fed back from P_6 and P_3 into R_{27}. Any nonzero V_{out1} will cause one of the Darlington pairs to turn ON. If V_{out1} is negative, V_{out2} is also negative and Q_{11} and Q_{12} turn ON, as we have seen. If V_{out1} had been positive, Q_9 and Q_{10} would have turned ON.

Whichever Darlington pair turns ON picks its associated relay, either R1 or R2. The relay contacts then cause the motor to run the fuel valve and the valve-position pot in whatever direction tends to reduce V_{out1} to zero. When V_{out1} reaches zero, the energized relay drops out and the valve motor stops.

Knowing that the circuit always acts to bring V_{out1} to zero, we can rewrite the preceding equation as

$$0 = \frac{10\ \text{k}\Omega}{10\ \text{k}\Omega}(V_{\text{error}}) + \frac{10\ \text{k}\Omega}{68\ \text{k}\Omega}(V_{\text{vp}})$$

$$-V_{\text{vp}} = 6.8(V_{\text{error}})$$

This equation expresses the proportional nature of the control. It tells us that the greater the magnitude of V_{error}, the greater the magnitude of V_{vp}. Thus the valve correction is *proportional* to the amount of deviation from set point. This is the essence of proportional control.

If P_3 is adjusted toward the top, so that a large portion of the P_6 wiper voltage is fed back to R_{27}, the proportional band is wide. If P_3 is adjusted toward the bottom, so that only a small portion of the P_6 wiper voltage is fed back, the proportional band is narrow. This can be understood as follows.

If a large portion of the P_6 signal is fed back, it will be relatively easy for the valve-position pot to cancel the effect of V_{error} and thereby return V_{out1} to zero. Therefore the valve will not have to move very far. This being the case, it would take a large V_{error} to force the valve to go to an extreme position (fully open or fully closed). When it requires a large error to drive the final correcting device between its limits, the proportional band is wide.

On the other hand, if P_3 feeds only a small portion of the P_6 signal back to R_{27}, it will be difficult for the valve-position pot to cancel V_{error}. In other words, a small V_{error} will cause a large change in valve position. Thus the error required to drive the valve from one limit to the others is not as great as before, and the proportional band is narrower.

Reset action. As with any purely proportional controller, the correction imparted to the valve will never quite return the temperature to set point. All it will do is drive the temperature back toward set point. To get the actual measured temperature back to set point, the reset mode of control must be added. In Fig. 15–4(b) the reset control action is supplied by op amp 3 and op amp 4 and their associated components.

To understand how this circuitry works, consider the positive V_{error} which appeared before. The appearance of the positive V_{error} resulted in a readjustment of the fuel valve in the closed direction; the corresponding reduction in fuel flow caused the measured temperature to return to the neighborhood of the set point. However, a small positive V_{error} will persist.

This small V_{error} is applied to R_{35}, which is part of the input resistance of the op amp 3 integrator. Recall from Sec. 8-4 that the output of an integrator is proportional to *how long* the input has been present. In this particular case.

$$-V_{out3} = \frac{1}{R_{IN}C_4} V_{error}\, t$$

R_{IN} is the sum of the R_{35} and P_4 resistances, and t is the time in seconds that V_{error} has been present. Op amp 4 is simply an inverter with a gain of 1, so its output, V_{reset}, is the opposite polarity of V_{out3}, or

$$V_{reset} = \frac{1}{R_{IN}C_4} V_{error}\, t \qquad \textbf{(15-3)}$$

V_{reset} is applied to the summing circuit through R_{29}, as Fig. 15–4(b) shows. It causes the inverting input of op amp 1 to go positive, even though V_{error} and V_{vp} have nearly canceled each other. V_{out1} then goes negative, causing V_{out2} to go negative and turn ON the Q_{11}-Q_{12} pair *again.* Thus the motor runs the fuel valve a little further closed, tending to reduce the measured temperature some more, bringing it into better agreement with the set point. If V_{error} continues to persist, V_{reset} will continue to increase as time passes. As it increases, it causes more and more correction in fuel valve position. Eventually, V_{error} will reach zero. At that point, the input to the integrator will equal zero, and the integrator will stop building up (stop integrating). V_{reset} will stop growing, and no further fuel valve correction will be made. Therefore the reset corrective action depends on how much time an error has been present, which is the essence of reset control.

The reset time constant is adjusted by 10-MΩ potentiometer P_4. When P_4 resistance is dialed out, the time constant is short, and the integrator builds up quickly. The reset circuitry is then quick to make its effects felt. When P_4 resistance is dialed in (higher resistance), the reset time constant is lengthened, and the integrator builds up slowly, as Eq. (15-3) shows. The reset circuitry is then slow to make its effects felt.

As mentioned in Chapter 9, on commercial temperature control instruments the reset adjustment is usually called *reset rate.* The high numbers on the *reset rate* scale mean quick reset action (small P_4 resistance), and the low numbers mean slow reset action (large P_4 resistance). The amount of reset action used depends on the nature of the specific temperature process, as explained in Chapter 9. The same holds true for the proportional band adjustment.

15-4 ■ STRIP TENSION CONTROLLER

Many industrial processes involve the handling of moving sheets or strips of material. One example of this is the textile process, which will be described in Sec. 15-8. Other examples are the heat-treating, galvanizing, or pickling of steel strip; the finishing of plastic strip; and the drying of paper strip. In all of these applications, it is important to maintain the proper amount of tension in the strip. Too great a tension will cause the strip to stretch and deform, and possibly break. Too low a tension will cause the strip material to sag. This may cause it to tangle up in the handling machinery.

The tension of a strip can be controlled by making adjustments to the relative speeds of the *leading roll* and *following roll* in the strip-handling apparatus. This is shown in Fig. 15–5(a). With the following roll spinning at a given speed, the strip tension can be increased by increasing the speed of the leading roll. Strip tension can be decreased by decreasing the speed of the leading roll.

A popular way of making these slight speed adjustments is to change the position of a drive belt on two conical pulleys. This is illustrated in Fig. 15–5(b).

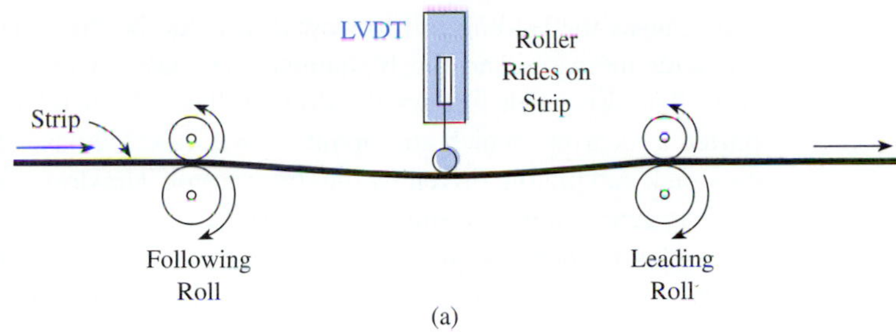

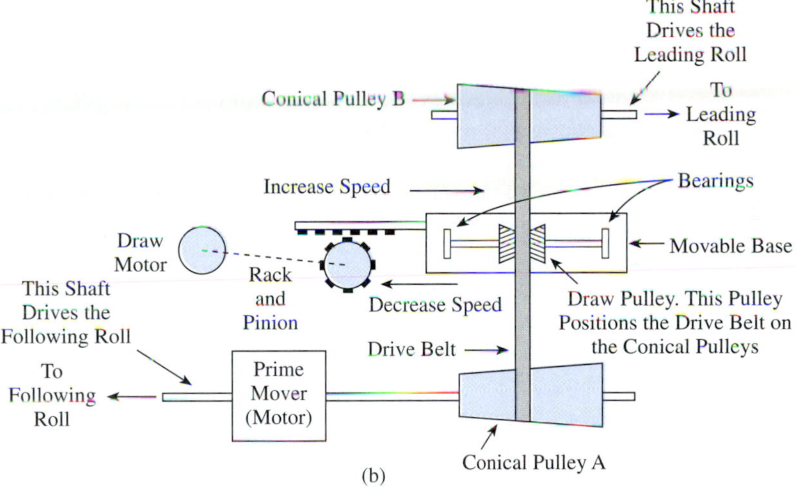

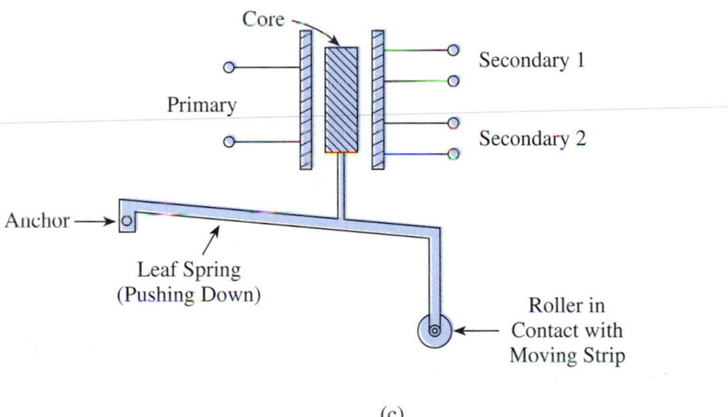

FIGURE 15–5
Strip tension controller. (a) The LVDT rides on the moving strip between the leading roll and the following roll. (b) Mechanism for controlling the speed of the leading roll relative to the following roll. (c) Close-up view of the LVDT sensor.

The prime mover, either a constant-speed or an adjustable-speed electric motor, has a double-ended shaft. One end attaches to the following roll, causing that roll to spin at some reference speed. The other end of the shaft goes to conical pulley A, causing it to spin at the same reference speed.

The drive belt transmits power from conical pulley A to conical pulley B, which then drives the leading roll. If the drive belt is centered on pulleys A and B, then pulley B spins at the same speed as pulley A. However, by moving the *draw pulley* to the left or right the drive belt can be moved to the left or the right on the conical pulleys. As the drive belt is moved to the left, the A diameter decreases while the B diameter increases.

This causes the leading roll to slow down. As the drive belt is moved to the right, the A diameter increases and the B diameter decreases, causing the leading roll to speed up.

The drive belt follows the draw pulley, which is an idling pulley. Its shaft is supported in bearings which are mounted on a movable base. The movable base is controlled by a rack and pinion, driven by the *draw motor.* The draw motor is a dc shunt motor whose shaft is geared down to run at a slow speed.

Control is accomplished by running the draw motor in short bursts. In this way, the movable base can be shifted to the left or right a little bit at a time. The base carries the draw pulley, which then positions the drive belt on the conical pulleys. This imparts slight speed adjustments to the leading roll to adjust the strip tension.

The transducer, which supplies tension information to the control circuitry, is an LVDT, shown in Fig. 15–5(a) and (c). The core of the LVDT is attached to a springy metal arm, which is moved up and down by a roller which rides on the strip. If the strip tension increases, the strip of material rises a little bit, causing the core of the LVDT to rise. If the strip tension decreases, the strip drops down a little. The roller also moves down due to the spring action of the arm. This causes the core of the LVDT to move down. Therefore the output voltages from the secondary windings of the LVDT are an indication of the strip tension.

Refer to Fig. 15–6, which is a schematic diagram of the electronic control circuitry. The frame of the LVDT is situated so that when the strip tension is in the middle of the acceptable range, the LVDT core is centered. In this condition the two secondary voltages are both equal ($V_{S1} = V_{S2}$). These secondary voltages are rectified and filtered and applied to the inputs of differential amplifier op amp 1. D_1 and D_2 are germanium small-signal diodes with a low forward-bias voltage. Thus the dc voltages appearing at the top of C_1 and C_2 are very nearly equal to the peak values of V_{S1} and V_{S2}.

If the strip tension is somewhat tighter than the midpoint of the acceptable range, the C_1 voltage will be larger than the C_2 voltage. If the tension is somewhat looser than the midpoint of the tension range, the C_2 voltage will be larger than the C_1 voltage. This can be seen by looking at the direction markings by the LVDT core. Resistors R_1 and R_2 are bleeder resistors to allow C_1 and C_2 to discharge to continually reflect the peak values of V_{S1} and V_{S2}.

The voltages across C_1 and C_2 are applied to R_3 and R_5, which are the input resistors of a differential amplifier having a gain of 4. In equation form,

$$V_{\text{out1}} = \frac{20\text{ k}\Omega}{5\text{ k}\Omega}(V_{C2} - V_{C1}) = 4(V_{C2} - V_{C1})$$

If strip tension is tighter than the midrange value, V_{out1} is a negative dc voltage. If tension is looser than the midrange value, V_{out1} is a positive voltage. V_{out1} is applied to two voltage comparers, op amp 2 and op amp 3. These comparers have the function of determining if the measured tension is *too* tight or *too* loose. In other words, a certain deviation from the tension midrange will be tolerated, but beyond a certain point, corrective action will be taken. The op amp 2 comparer checks for tension exceeding the limit for looseness, while the op amp 3 comparer checks for tension exceeding the limit for tightness. The limits themselves are adjustable and are set by potentiometers P_1 (looseness) and P_2 (tightness).

For purposes of discussion, suppose that P_1 and P_2 are set at +8 and −8 V, respectively. Then if V_{out1} goes more positive than +8 V, it means the measured tension has exceeded the looseness limit. When this happens, the positive input of op amp 2 goes more positive than the negative input, so V_{out2} switches from negative saturation to positive saturation (from about −13 to +13 V). This causes a +5-V signal to appear at the top of zener diode ZD1. Therefore the appearance of +5 V at ZD1 indicates that tension is too loose and that corrective action is necessary.

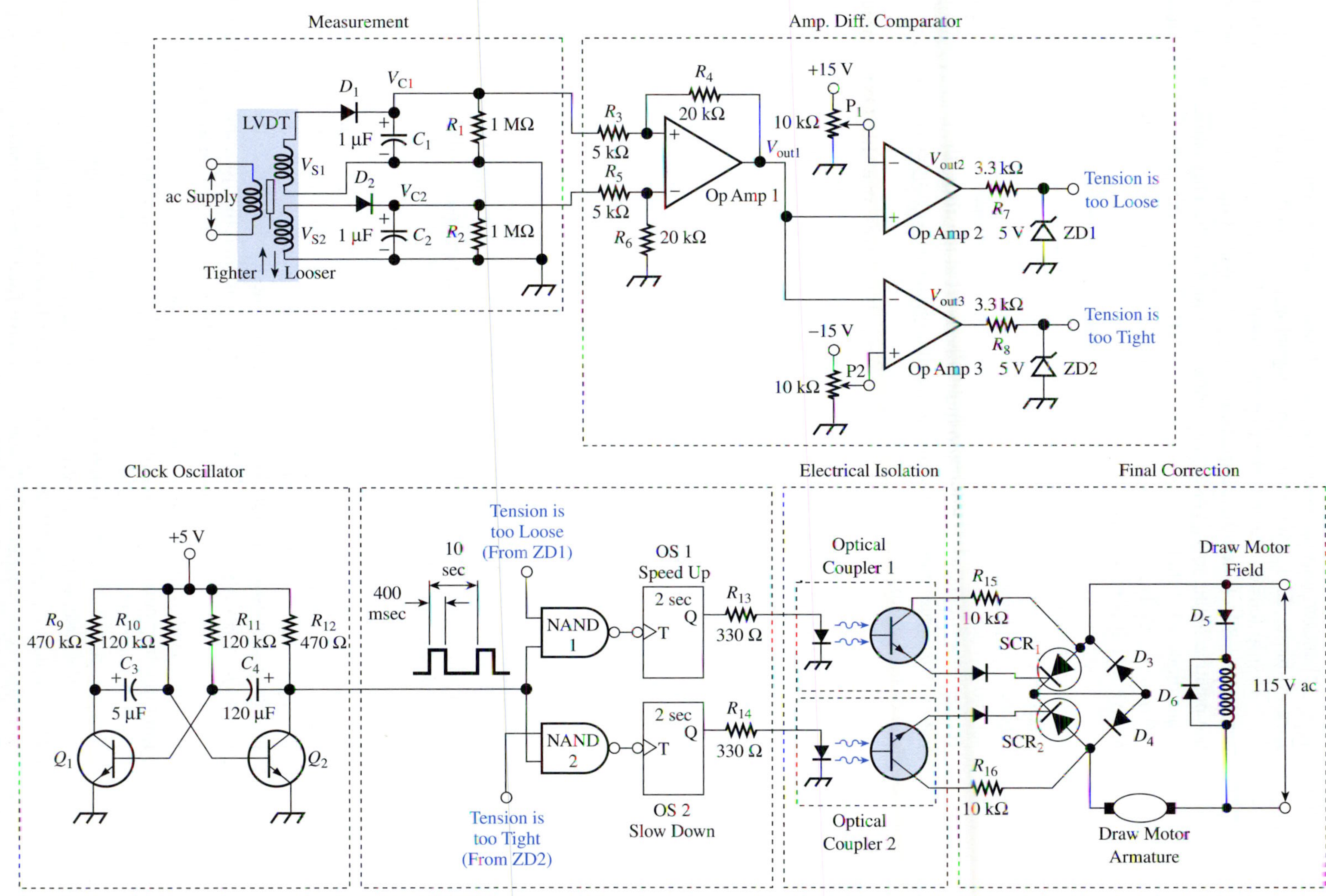

FIGURE 15–6
Schematic diagram of the strip tension control circuitry.

Whenever tension does not exceed the looseness limit, V_{out2} is -13 V, which forward-biases ZD1, causing a -0.6 V signal at the cathode terminal of ZD1.

If strip tension exceeds the tightness limit, V_{out1} will go more negative than the P_2 setting of -8 V. When this happens, the negative input of op amp 3 is more negative than the positive input, so V_{out3} switches from -13 to $+13$ V. This causes the same result at zener diode ZD2 that was seen above for ZD1. That is, the cathode level changes from -0.6 to $+5$ V. This $+5$-V signal represents the fact that tension is too tight and that corrective action is necessary.

The "too loose" and "too tight" signal lines are applied to logic gates NAND1 and NAND2. Assume that the logic family used here operates on a $+5$-V supply level. These NANDs also receive the pulse train output of the astable pulse generator on their bottom inputs. The pulses have a duration of about 400 ms and occur approximately every 10 seconds. When the positive edge of the pulse arrives, it partially enables both NAND gates. If either the "too loose" or the "too tight" line is HI at this time, the corresponding NAND gate will change states, delivering a negative-going triggering edge to the *T* terminal of one of the one-shots.

If the "too loose" line is HI when the pulse arrives, one-shot OS1 is fired, signaling the leading roll to speed up. If the "too tight" line is HI, one-shot OS2 is fired, signaling the leading roll to slow down. Here is how the firing of the one-shots brings about the speed adjustment of the leading roll.

Assume that OS1 fires. Its *Q* output will go HI for a pulse duration of 2 seconds, during which time it will turn ON the LED in optical coupler 1. The detecting device in the optical coupler is a phototransistor, which switches ON during the presence of the one-shot's output pulse. This effectively connects the gate of SCR_1 to its anode through a 10-kΩ resistor. When the anode to cathode voltage enters the positive half cycle, the SCR will be supplied with enough gate current to fire. The anode voltage itself supplies the gate current, through the 10-kΩ resistor. Therefore SCR_1 will fire just after the zero crossover and will continue doing so for the full 2-second duration of the one-shot pulse. With SCR_1 firing on every positive-going zero crossover, the draw motor armature will be supplied with dc current. The flow path is as follows: from the top 115-V supply line, through SCR_1, through the bridge cross-connection wire, through D4, and through the motor armature from left to right. During this 2-second period, the draw motor will run and will shift the movable base to the *right* in Fig. 15–5(b). At the end of 2 seconds, the draw motor will stop running because the one-shot output pulse will terminate. This will disable the optical coupler and break the path of SCR gate current.

Having moved the drive belt on the conical pulleys, the system now waits for the next pulse from the astable pulse generator, which will arrive about 10 seconds later. During this interval, the speed adjustment to the leading roll has the opportunity to effect an increase in strip tension. This should bring the tension above the loose limit and remove the $+5$-V signal from ZD1. If it does accomplish that job, the next pulse from the pulse generator will have no effect on NAND1, since the top input will then be LO. On the other hand, if the speed increase was not enough to bring strip tension above the loose limit, another correction will be made when the next pulse arrives. OS1 will fire again, allowing SCR_1 to turn ON and drive the draw motor for another 2-second burst. This action is repeated until the tension is brought back into the acceptable range.

Of course, if the original tension error had been a "too tight" condition, zener diode ZD2 would have been HI and NAND2 would have been enabled when the pulse arrived from the pulse generator. OS2 would then have fired for 2 seconds instead of OS1. This would energize optical coupler 2, thereby turning on SCR_2. The draw motor armature current would then be reversed, passing through the armature from right to left, through SCR_2, through the bridge cross-connection wire, and through D3. This would shift the movable base and draw pulley to *left* in Fig. 15–5(b). The leading roll would slow down and loosen the strip tension. As before, as many corrections as are needed would be made by the system.

The draw motor field winding is driven from the 115-V ac line through half-wave rectifier D_5. D_6 allows current to continue circulating through the field winding during the negative half cycle. As the field current declines when the ac line passes its positive peak, the induced voltage in the field winding is the proper polarity to forward-bias D_6. Therefore there is no abrupt halt to the field current as the ac line reverses. Diode D_6 allows current to continue flowing in the field winding throughout the negative half cycle.

15-5 ■ EDGE GUIDE CONTROL FOR A STRIP RECOILER

When a strip material has completed its processing, it is wound up into a coil for subsequent handling and shipping. This operation is called *recoiling* and is illustrated in Fig. 15–7. The compass directions indicated in Fig. 15–7(a) will be used to specify directions of movement in the discussion of the recoiling system. This will help avoid confusion.

FIGURE 15–7 Strip recoiler: (a) physical layout of the recoiling mechanism; (b) close-up view of the edge-sensing apparatus.

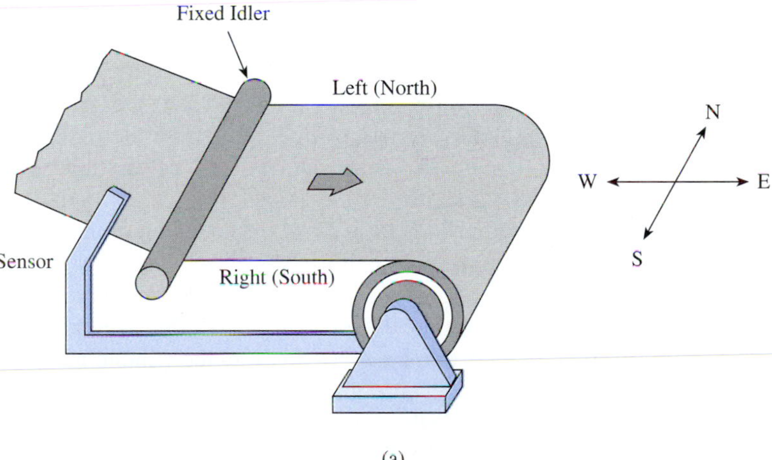

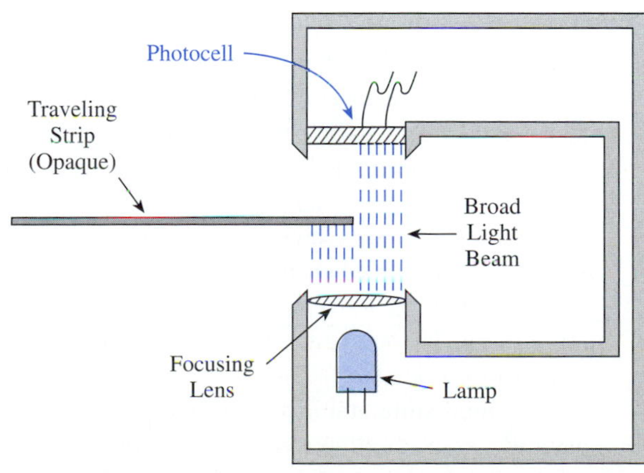

The moving strip passes under a fixed idler roll and is recoiled on the *windup* reel. If the recoiling operation consisted merely of rotating the windup reel and feeding the strip onto it, the coils produced would almost certainly be crooked. That is, the individual coil laps would not be aligned with each other: The sides would either be wavy or "telescoped." This would make handling and shipping the coil more difficult and would increase the likelihood of edge damage.

To produce straight coils, some type of control is necessary to ensure that every coil lap is aligned with every other lap. There are two ways to do this:

1. Guide the traveling strip to correct any tendency it has to move laterally (back and forth) relative to a fixed windup reel.
2. Shift the windup reel back and forth to follow any lateral movement by the traveling strip.

Of these two methods, the second is preferred for more strip materials, especially metal strip.

The position of the windup reel is usually controlled by a hydraulic cylinder. The cylinder rod is attached to the mounting base of the reel, as shown in Fig. 15–7(a). The position-sensing device is a large photoconductive cell, with a diameter of about 1 in. A view of the photoelectric edge-sensing assembly is presented in Fig. 15–7(b); this is the view seen looking toward the east along the top of the traveling strip. If the strip is properly positioned, it will block exactly half of the broad light beam radiated by the lamp. The other half of the light beam will strike the photocell.

If the traveling strip should wander to the south, deeper into the photoelectric assembly, a smaller amount of light will strike the photocell. This will be detected electrically and will initiate control action to move the windup reel to the south, to keep it aligned with the traveling strip.

If the strip should wander to the north, out of the photoelectric assembly, the increasing illumination of the photocell will cause the windup reel to move to the north. Figure 15–7(a) shows that the edge sensor is mounted on an arm which is attached to the windup reel base. The edge sensor thus moves with the windup reel, always maintaining a fixed strip edge position relative to the reel. In this way the windup reel is kept constantly aligned with the strip. Every coil lap edge aligns with every other coil lap edge, and the coil winds straight.

The electrohydraulic circuit which accomplishes this control is shown in Fig. 15–8. Here is how it works. The photoconductive cell in the edge sensor has an approximate resistance of 5 kΩ when the traveling strip blocks half of the light beam. The bias adjustment pot, P_1, is adjusted to turn on transistor Q_1 enough to bring its collector voltage to about 2 V. Q_1 collector voltage is applied to the base of power transistor Q_2. This causes Q_2 to conduct, establishing a current flow in the 320-Ω actuator coil that swings the hydraulic jet pipe.

The jet pipe assembly is designed so that a 10-mA current through the actuator coil causes the jet pipe to be perfectly centered. Adjustment of the center position is done with the balancing spring adjustment. The mechanical force exerted by the balancing spring is equal and opposite to the force created by the permanent magnet and actuator coil.

In Fig. 15–8(a), bias pot P_1 is manually adjusted to provide exactly 10 mA of actuator coil current when the light beam is exactly half blocked. Lowering the P_1 resistance increases the current in the actuator coil, and raising the P_1 resistance decreases the coil current. The adjustment of P_1 would be done by the system users prior to putting the system in service.

Temperature stability is provided to the electronic circuit by negative feedback resistor R_2, connected between the emitter of Q_2 and the base of Q_1. R_2 provides negative current feedback, which lowers the overall circuit gain but at the same time stabilizes the gain and bias point. A stable bias is very important in this circuit, since drift in the bias point will change the actuator coil current and move the jet pipe.

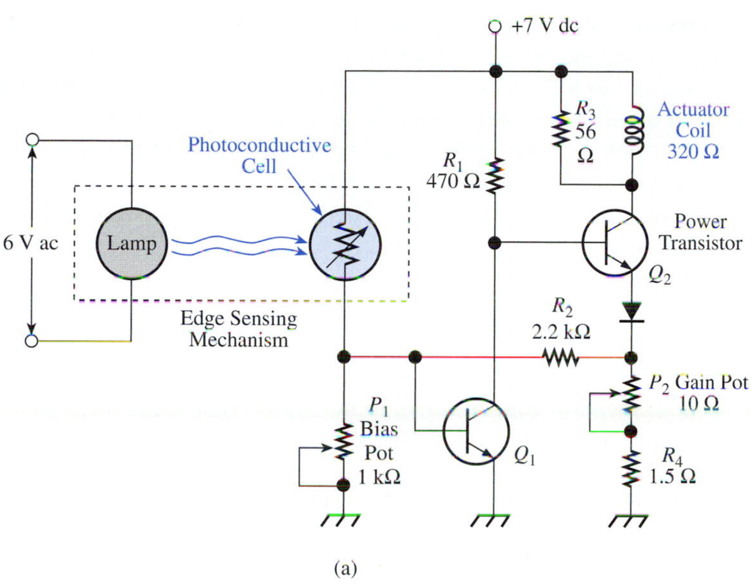

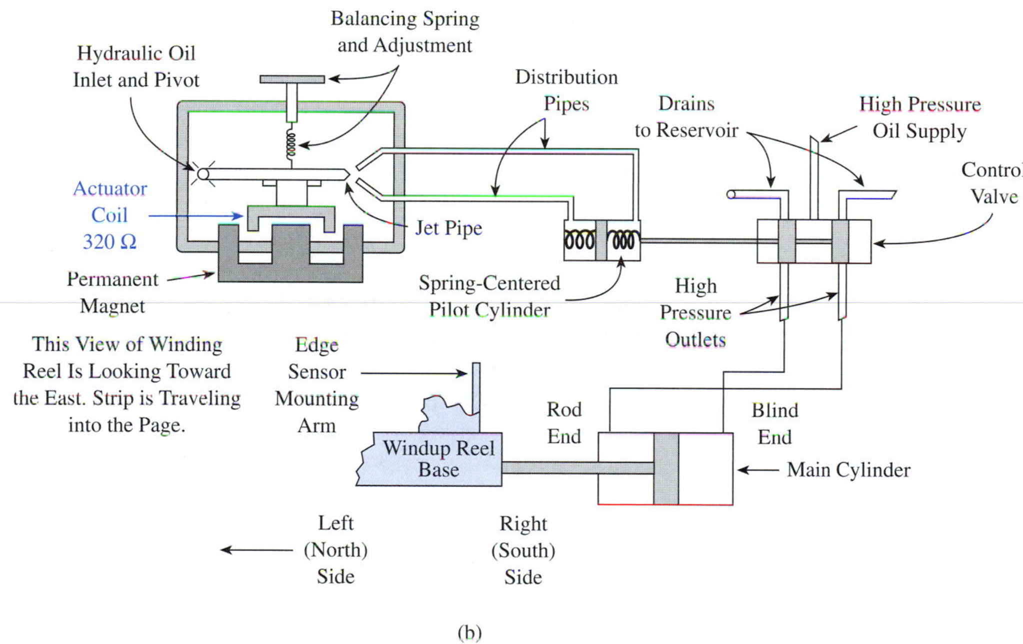

FIGURE 15–8
(a) Edge-sensing and controlling circuit. (b) Electrohydraulic actuator for moving the base of the windup reel.

By adjusting the resistance of potentiometer P_2, the amount of feedback can be varied and the gain of the circuit is varied. If the P_2 resistance is increased, the feedback increases, and the gain decreases. That is, a given change in photocell resistance will result in only a small change in actuator coil current. If the P_2 resistance is decreased, the feedback is decreased, and the gain is increased. This makes the circuit more sensitive to changes in photocell resistance.

Let us imagine that all the electronic adjustments have been made and that the recoiling system is in operation. We will trace out the sequence of actions as the system causes the windup reel to follow the lateral movements of the traveling strip. Refer to Fig. 15–8 and keep in mind the physical layout of the system as shown in Fig. 15–7(a).

If the edge of the traveling strip is passing directly through the middle of the edge sensor, 10 mA of current will flow through the actuator coil and the jet pipe will be exactly centered. Neither distribution pipe will have a higher hydraulic pressure than the other, so the pilot cylinder in Fig. 15–8(b) will be centered by its spring. The pilot cylinder rod is attached to the spool of the main hydraulic control valve. With the pilot cylinder centered, the main control valve does not pass oil to either end of the main cylinder; therefore the windup reel base remains stationary. As long as the edge of the traveling strip remains centered in the light beam, the system does not move the windup reel, and the coil winds straight.

Now suppose the strip edge wanders to the north. This will tend to move the edge *out* of the center, thus exposing more of the light beam and lowering the photocell's resistance. As the photocell resistance drops, Q_1 turns on harder and the Q_1 collector voltage decreases. This reduces the Q_2 conduction and causes the actuator coil current to fall below 10 mA. The jet pipe cannot remain centered but moves up and creates a pressure imbalance in the distribution pipes. In this case the top pipe has a higher pressure than the bottom pipe, so in Fig. 15–8(b) the pilot cylinder moves to the left. The pilot cylinder rod shifts the control valve spool to the left, thereby connecting the high-pressure hydraulic supply port to the blind end of the main cylinder while allowing the rod end of the main cylinder to drain to the reservoir. This causes the main cylinder to extend, moving the windup reel to the north. Thus the windup reel follows the wandering strip edge. As soon as the edge is back in the center of the sensor, the coil current will return to 10 mA, and all corrective action will stop. The windup reel will continue rewinding the coil in this new position as long as the strip edge remains centered in the sensor.

If the traveling strip edge wanders to the south, it will move *into* the sensor and block more of the light beam. The photocell resistance will increase, causing Q_1 to conduct less current. This will raise the Q_1 collector voltage and cause Q_2 to conduct more heavily, thus increasing the actuator coil current above 10 mA.

The magnetic force on the jet pipe now exceeds the mechanical spring force, so the jet pipe moves down. This time the bottom distribution pipe receives the higher pressure, so the pilot cylinder moves to the right; this shifts the main control valve spool to the right and applies high-pressure oil to the rod end of the main hydraulic cylinder. This cylinder strokes to the south, driving the windup reel base to the south. The windup reel thus follows the wandering strip until the sensor is once again centered on the traveling edge. At that point the actuator coil current returns to 10 mA, and the hydraulic control devices return to center. The main cylinder freezes in that particular position, and the coil continues winding straight.

15-6 ■ AUTOMATIC WEIGHING SYSTEM

Automatic weighing systems are frequently used in industry. These systems automatically transfer a preset weight of material into a hopper or container of some type. They are used for the manufacture of products which require a number of carefully weighed ingredients.

15-6-1 Mechanical Layout

A system for weighing one single ingredient is shown in Fig. 15–9(a). The ingredient being handled and weighed is a powder. In an industrial process, the best way to handle powders is with a *screw conveyor*. A screw conveyor is a large pipe, perhaps 1 ft in

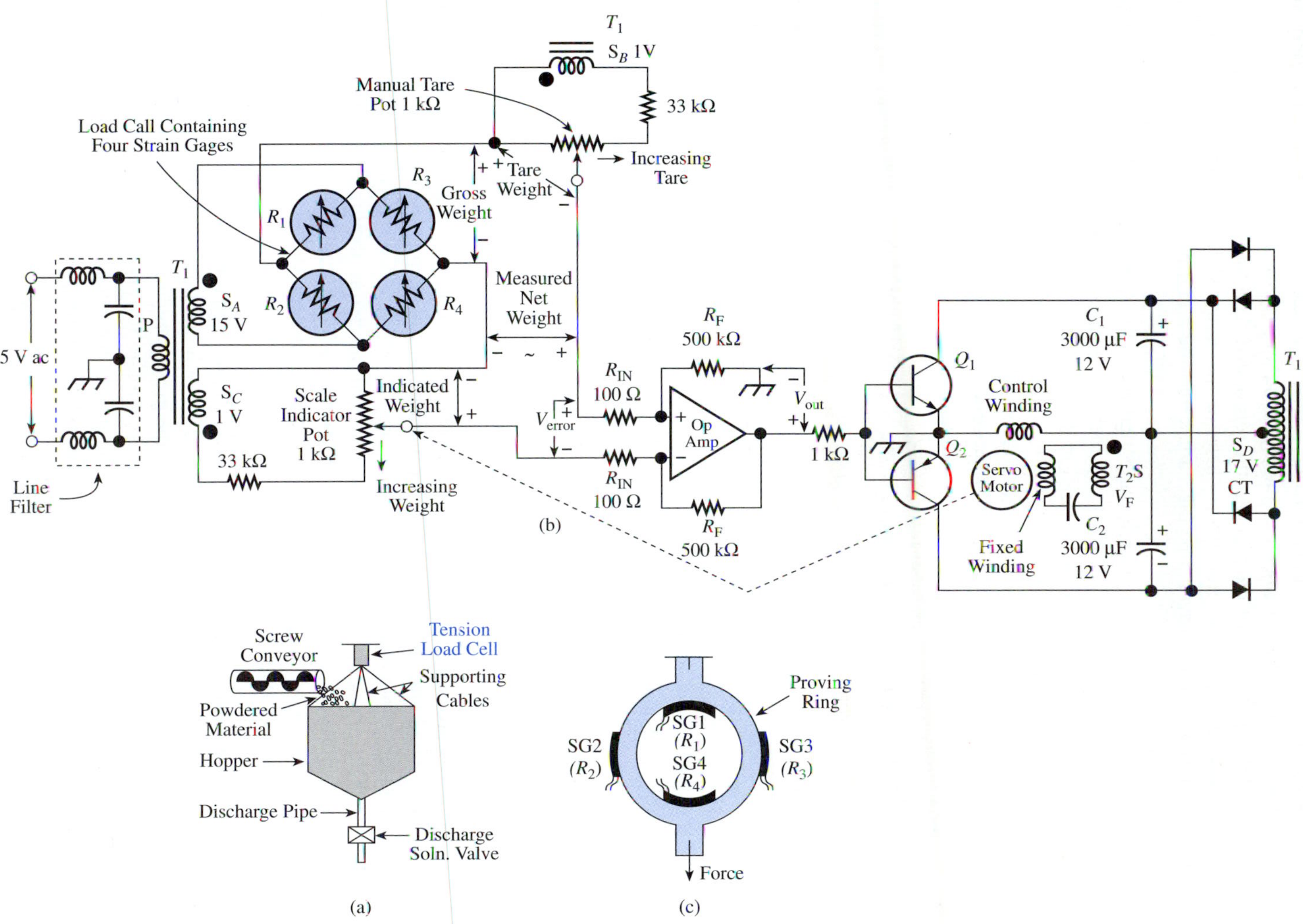

FIGURE 15–9
(a) Mechanical layout of the screw conveyor, the hopper, and the load cell. (b) Schematic diagram of the weighing circuitry. The servo system causes the indicated weight to equal the gross weight minus the tare weight. (c) Close-up view of the load cell. It is comprised of four strain gages carefully bonded to a proving ring.

diameter, with a wide-pitch internal screw. The screw thread diameter is only slightly less than the inner diameter of the pipe, so there is close clearance between the threads and the inner wall of the pipe. As the screw shaft is rotated, powdered material is forced down the pipe. The greater the speed of the screw shaft, the greater the flow rate of material. The conveyor pipe discharges into the weighing hopper, as illustrated in Fig. 15–9(a).

The weighing hopper is supported from above by steel cables which attach to a strain gage load cell. The load cell therefore senses the weight of the hopper itself, its supporting cables, and the material in the hopper.

At the bottom of the hopper is a discharge pipe, containing a solenoid-operated dump valve. This is used to remove the weighed material from the hopper and transfer it to the next stage of the production process.

15-6-2 Electronic Weighing Circuitry

The electronic weighing circuit schematic is drawn in Fig. 15–9(b).

Load cell bridge. The weight signal is taken from the load cell, which is a strain gage assembly. The four strain gages are mounted on a *proving ring,* as shown in Fig. 15–9(c). The proving ring and strain gages, taken together, comprise the load cell. As the proving ring is subjected to a tension load, strain gages 2 and 3 are stretched, causing their resistances to increase. Strain gages 1 and 4, being mounted on the top and bottom of the inside of the ring, are compressed as the ring is loaded. Therefore their resistances decrease. The gages are wired into the Wheatstone bridge circuit so that the two sides of the bridge tend to produce opposite changes in resistance ratio. That is, as the ratio of R_1 to R_2 gets smaller, the ratio of R_3 to R_4 gets larger. By making all four arms of the bridge respond to the load on the load cell, the available output voltage of the bridge is increased. A typical industrial load cell produces an output voltage of 30 mV at full load when excited by a 15-V ac supply.

Since load cell output voltages are so puny, it is very important to keep the weighing signal circuits free from electrical noise. This is the purpose of the *ac line filter* wired into the 115-V supply lines to transformer T_1 in Fig. 15–9(b). Any high-frequency noise signals appearing on the ac lines are filtered out before they reach the T_1 primary winding. In addition to this precaution, all the signal wires leading up to the op amp input terminals would be shielded. The shielding has not been shown in Fig. 15–9(b) in order to avoid cluttering the diagram.

Tare circuitry. Since part of the gross weight signal is due to the weight of the hopper and its support cables, provision is made to subtract this weight from the load cell signal. The weight which is subtracted from the gross weight indicated by the load cell is called the *tare weight.* The pot which produces the tare weight subtraction is called the *tare pot.* The final signal obtained after the tare weight has been subtracted from the gross weight is called the *measured net weight.* To understand the tare weight subtraction process, refer to the circuitry in Fig. 15–9(b).

The load cell bridge is excited by a stable 15-V ac supply from secondary winding A of transformer T_1. This winding is identified as S_A in the schematic diagram. The phase relationships between the various voltages in this circuit are important, so the phases are clearly marked by phasing dots. The ac signal taken from the Wheatstone bridge is a small voltage, only a few millivolts, and it represents the gross weight supported by the load cell. This signal is labeled Gross Weight in Fig. 15–9(b). At some instant in time it will be positive on the top and negative on the bottom, as indicated. The Tare Weight signal is taken from the manually adjusted tare pot, which is excited by the S_B winding. This signal is positive on the left and negative on the right at that same instant in time. This polarity relationship is established by the phase relationship between the two secondary windings S_A and S_B.

Since the Gross Weight signal and the Tare Weight signal are opposite in phase, the resultant signal is the *difference* between these two voltages. In other words, the Tare Weight signal has been subtracted from the Gross Weight signal. This voltage difference is labeled the Measured Net Weight signal in Fig. 15–9(b).

Of course, someone had to adjust the manual tare pot to the proper position before any material was loaded into the hopper. This is done by simply turning the tare pot until the scale weight indicator reads zero when the hopper is empty.

Scale weight indicator (a servo system). The scale weight indicator is a servo system with the positioned object being the scale-indicating pointer. The scale-indicating pointer is a needle moving over a calibrated weight dial, just like a penny scale in a shopping arcade. As shown in Fig. 15–9(b), the position of the scale-indicating pointer is represented electronically by the wiper position of the scale indicator potentiometer. The scale indicator pot wiper is attached to the same shaft as the scale-indicating pointer. This is brought out clearly in Fig. 15–10(a). Therefore, both the scale indicator pot wiper and the scale-indicating pointer indicate the measured net weight. The scale-indicating pointer does it mechanically/visually, and the scale-indicating potentiometer does it electronically.

The Indicated Weight signal in Fig. 15–9(b) is positive on the bottom at the reference time instant. This is due to the S_C secondary winding driving the scale indicator pot. The Indicated Weight signal and the Measured Net Weight signal are thus in phase opposition. The difference between these two signals is the error signal, labeled V_{error} in Fig. 15–9(b).

V_{error} is applied to an op amp, which furnishes the input stage of the servo amplifier. The op amp is connected as a differential amplifier, with a voltage gain of 5000 ($500\ k\Omega/100\ \Omega = 5000$). Therefore V_{out} from the differential amplifier is 5000 times as large as V_{error}. This very high voltage gain is necessary because the signals which are being handled are so small.

V_{out} from the op amp differential amplifier is used to drive the control winding amplifier, which is a *complementary symmetry amplifier*. The complementary symmetry amplifier is distinguished by the use of one *npn* and one *pnp* transistor. The complementary symmetry transistors are specially matched to have identical current-voltage characteristics, except, of course, that the polarities are opposite from each other. Q_1 handles and amplifies the positive half cycles of V_{out}, and Q_2 handles the negative half cycles of V_{out}. This amplification scheme is an alternative to the push-pull amplifiers discussed in Chapter 11. It provides the same advantage, namely, cooler operation of the output transistors because they dissipate no dc power.

Notice that the Q_1-Q_2 emitter tie point is grounded; the +12-V and −12-V supplies are not grounded. These supplies deliver current to the servo motor control winding only, and they are completely separate from the ±15-V supplies to the op amp. There is no ground reference between these two pairs of power supplies.

V_{out}, the op amp output signal, swings positive and negative relative to ground. As it goes positive it forward-biases the base-emitter junction of Q_1, causing that transistor to conduct. Q_1 then passes current through the servo motor control winding from left to right. The source of this current is the +12-V supply.

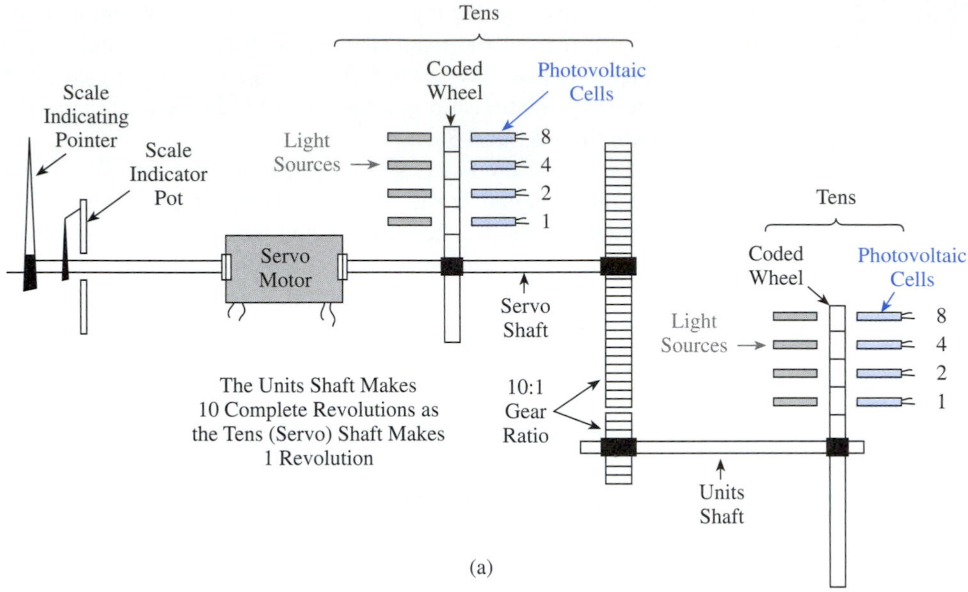

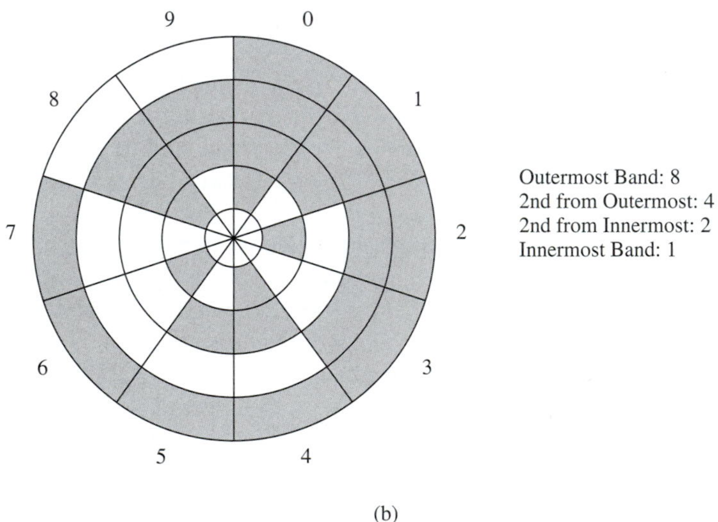

FIGURE 15–10 (a) Side view of the servomechanism. The servo motor shaft on the left positions the weight-indicating pointer and the scale indicator-pot wiper. The servo motor shaft on the right positions the two binary-coded wheels. These wheels optically convert the shaft position into a digital weight signal. (b) Face view of one of the binary-coded wheels.

As V_{out} from the op amp goes negative relative to ground, it forward-biases the base-emitter junction of *pnp* transistor Q_2, causing it to conduct. Q_2 then passes current through the control winding from right to left. The source of this current is the -12-V supply. The control winding current is therefore an ac current synchronized with the voltages in the measuring circuits, which are themselves synchronized with the ac power line. The servo motor fixed-winding current is 90° out of phase with the ac power line because the fixed winding is excited by the T_2 secondary, through a phase-shift capacitor. Transformer T_2 is powered by the same ac power lines that drive T_1. The T_2 primary winding, although not shown in Fig. 15–9(b), would be connected to the ac lines to the left of the line filter.

The servo motor will run whenever an error voltage (V_{error}) exists. As illustrated in Fig. 15–10(a), the servo motor shaft is linked to the wiper arm of the scale indicator pot, and always drives it in the proper direction to reduce V_{error} to zero. Thus the servo system continually equalizes the Indicated Weight signal and the Measured Net Weight signal. In this way the scale-indicating pointer constantly points to the correct net weight on the calibrated weight dial.

15-6-3 Optical Weight Readout

A side view of the servomechanism is shown in Fig. 15–10(a). The servo motor is geared down so that its output shaft spins slowly. The output shaft is referred to as the *servo shaft,* and it is double-ended. One end of the servo shaft is used to position the scale-indicating pointer and the scale indicator pot wiper. The other end is used to position two binary-coded wheels. The binary-coded wheels enable the optoelectronic circuits to read the indicated weight in a digital manner.

Here is how the coded wheels read the weight. The face view of one of the binary-coded wheels is shown in Fig. 15–10(b). Both wheels are identical, but let us temporarily concentrate on the tens wheel only. As can be seen, the wheel is divided into 10 equal sectors, each sector representing one of the decimal digits 0 to 9. The distance that the servo shaft rotates determines which one of these sectors will come into the topmost position, between the light sources and the photocells. If the number 5 sector, for example, comes into the top position, between the light sources and photocells, the information given out by the photocells should represent the decimal number 5. If the number 6 sector comes into the top position, the photocells should represent the decimal number 6, and so on.

The decimal numbers associated with the wheel sectors are coded in the familiar 8421 binary code by virtue of the light and dark areas in the sectors. Notice that the wheel (and each sector) is composed of four concentric bands, or rings. The outermost band indicates the presence or absence of the binary 8 bit. The next to the outermost band indicates the 4 bit. The next to the innermost band represents the 2 bit, and the innermost band represents the 1 bit. With this in mind let us from now on refer to the bands as the 8 *band,* the 4 *band,* and so on.

When a given sector moves into position between the lights and photocells, it will pass the light beams in those bands where it is light, and it will block the light beams in those bands where it is dark. The presence of light at the photocell receptor stands for a binary 1, and the absence of light stands for a binary 0. In this way, a four-bit binary number can be represented.

As an example, consider the number 5 sector in Fig. 15–10(b). Its 8 band and its 2 band are both dark, but its 4 band and its 1 band are light. If the number 5 sector moves into the top position, photocell 4 and photocell 1 will be illuminated. Photocell 8 and photocell 2 will remain dark. The output of the photocells is thus 0101, reading from outermost to innermost (from 8 to 1). This is the binary code for the decimal number 5. It is a fairly easy task to verify that the coded wheel satisfies the binary code for each of the 10 decimal digits.

The units wheel does exactly the same thing as the tens wheel, except that it rotates 10 times as far. This is accomplished by mechanical gearing, as shown in Figs. 10–16(a) and 15–10(a). The gear on the servo shaft has a diameter 10 times as great as the gear on the units shaft. Therefore, for one revolution of the servo shaft, the units shaft goes through 10 revolutions.

To get the idea of how the entire readout mechanism works, assume that the full-scale weight is 100 lb. That is, the servo shaft goes through a complete revolution when 100 lb of powder is loaded into the hopper. This being so, the servo shaft will go through one tenth of a revolution for each 10 lb of powder loaded into the hopper. This translates into one complete revolution of the units shaft.

Therefore, as 10 lb of material slowly pour into the hopper, the units wheel goes through and indicates each one of its 10 digits in turn. When it returns to 0 after a complete revolution, the tens wheel has just completed one tenth of a revolution, and it is changing from the 0 sector to the 1 sector. This action is repeated as the weight of material goes from 10 to 20 lb, and again for every 10 lb thereafter, all the way up to 99 lb (the mechanism cannot read 100 lb with only two wheels).

As an example, suppose that 72 lb of material are loaded into the hopper. The servo shaft will turn a little more than seven tenths of a revolution, so the number 7 sector of the tens wheel will be in the top position. The units shaft will have turned seven complete revolutions plus two tenths of another revolution. Therefore the number 2 sector of the units wheel will be in the top position. The number 7 sector is dark-light-light-light, reading from the outermost ring to the innermost ring, and the number 2 sector is dark-dark-light-dark, again reading from outermost to innermost. Therefore the output of the two groups of photocells will be

0111 0010

which represents 72 in BCD.

15-6-4 Automatic Cycle Logic

Figure 15–11 shows the logic circuitry for controlling the system. Notice first that each readout photocell is amplified by a transistor switch and then put into a logic inverter. The inverters then feed into BCD to 1-of-10 decoders of the same type that we saw in Chapter 2 (Fig. 2–10). The net weight of material in the hopper thus appears at the output of the two decoders. For example, if there are 72 lb of powder in the hopper, the 7 output terminal of the Tens Decoder in Fig. 15–11 will go HI and the 2 output terminal of the Units Decoder will go HI. All the other 18 output terminals will stay LO.

The decoders feed two pairs of selector switches. The first pair of switches, which are called the Desired Weight Switches, set the *desired* weight of material. The second pair of switches, which are called the Slow Feed Switches, set the weight at which the loading of the hopper shifts from fast speed to slow speed. As the material in the hopper approaches the desired weight, the hopper loading speed is shifted in order to prevent overshoot of the final desired weight.

In Fig. 15–11, AND2 is the gate that detects when the slow feed weight has been reached. Its inputs come from the common terminals of the Slow Feed selector switches. AND3 is the gate that detects when the full desired weight has been loaded into the hopper. Its inputs are from the common terminals of the Desired Weight selector switches. In addition, there is a Hopper Empty detection gate, AND1. Its inputs are hard-wired to the 0 output terminals of the Tens and Units decoders. The output of AND1 goes HI when the hopper is empty, and the net weight equals 0 lb.

The cycle operation is not complicated. After the scale has been manually tared by adjusting the tare pot, the operator presses the Start Feed pushbutton. The output of the switch filter applies a HI to NOR1, which delivers a negative clock edge to the Fast Feed Flip-Flop. Both J and K are HI, so the flip-flop toggles into the ON state. This enables output amplifier OA1, which energizes the Fast Feed Motor Starter, MSF. The screw conveyor starts running at high speed, feeding powder into the hopper rapidly.

This continues until the optical weight detectors reach the number set on the Slow Feed selector switches. At this instant, both inputs of AND2 go HI, and its output goes

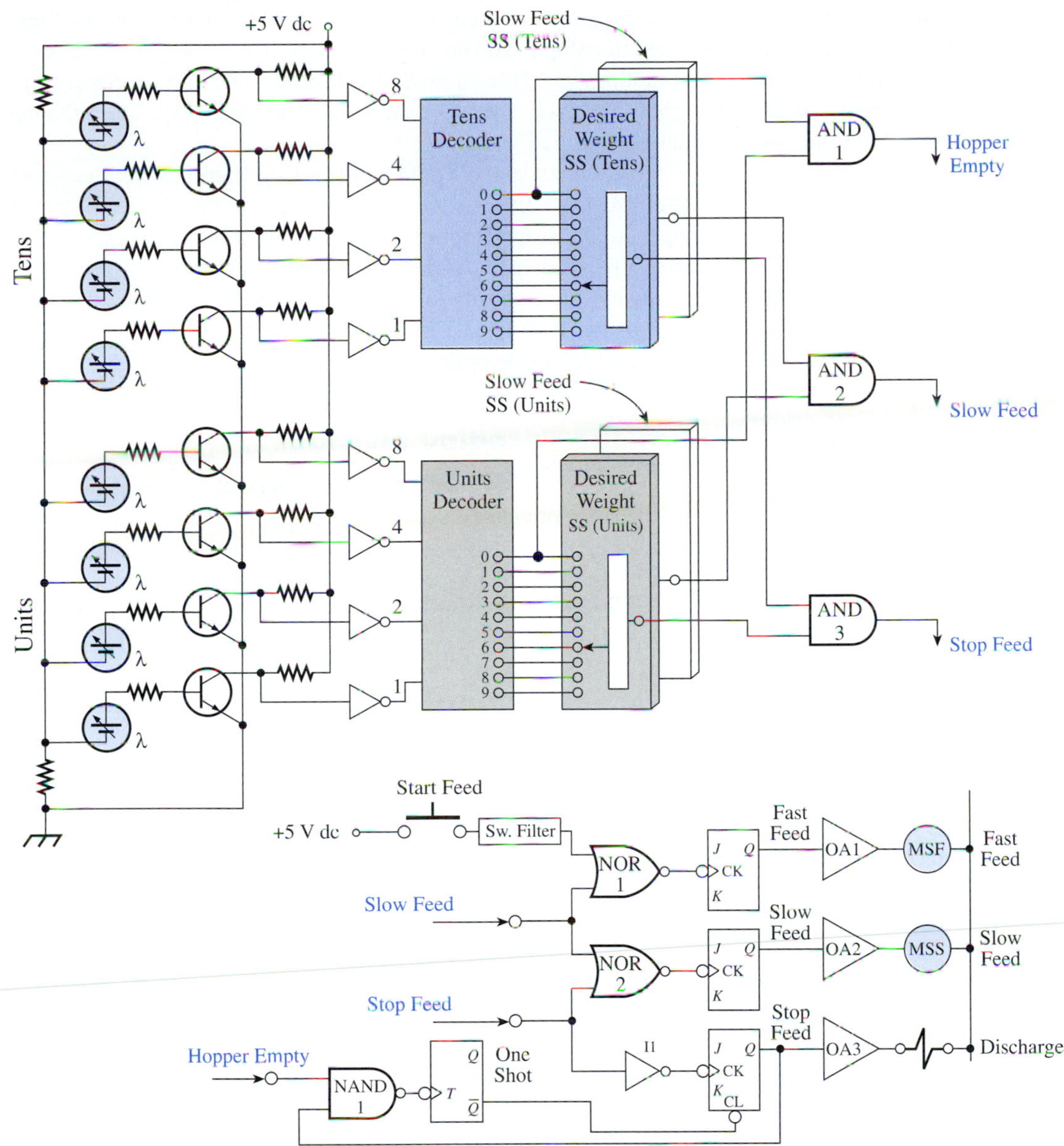

FIGURE 15–11
Weighing cycle control circuitry. The photocell detectors on the far left are the ones shown physically in Fig. 15–10(a). When the outputs of the two decoders match the settings on the two Slow Feed selector switches, the powder feed rate slows down. When the decoder outputs match the settings on the two Desired Weight selector switches, the feed conveyor stops altogether.

HI. This HI appears at the inputs of NOR1 and NOR2, causing their outputs to go LO. Both the Fast Feed and the Slow Feed flip-flops receive a negative clock edge, and they both toggle. The Fast Feed flip-flop turns OFF, deenergizing MSF, and the Slow Speed flip-flop turns ON, energizing MSS, the Slow Feed Motor Starter. The screw conveyor slows down and continues feeding powder into the hopper at a slower rate.

This continues until the weight detectors reach the number set on the Desired Weight selector switches. At this instant, both inputs of AND3 go HI, and the output delivers a

HI to NOR 2 and I1. As the negative clock edges arrive at their CK terminals, the Slow Feed flip-flop turns OFF, and the Stop Feed flip-flop turns ON. The *Q* output of the Slow Feed flip-flop goes LO, disabling OA2. The Slow Speed Motor Starter deenergizes, stopping the rotation of the screw conveyor. Therefore the flow of powder stops when the net weight in the hopper equals the desired weight.

At the same instant, the *Q* output of the Stop Feed flip-flop energizes the discharge solenoid through OA3. The discharge valve opens, and the weighed powder flows out of the hopper. When the hopper is empty, the Hopper Empty line from AND1 goes HI. This causes both NAND1 inputs to be HI, which delivers a negative edge to the trigger terminal of the one-shot. The one-shot fires, causing a LO to appear at its *Q* output. This LO is applied to the clear input of the Stop Feed flip-flop, causing it to turn OFF. The discharge solenoid deenergizes, and the cycle is finished.

15-6-5 Other Codes and Encoding Methods

The optical encoder presented in Sec. 15-6-3 uses true binary code to represent the number of each sector. This is the most obvious way to represent a number on an encoding wheel, and it has the desirable feature of being easily decoded. That is, it is simple to decode the wheel information by inserting it into a BCD to 1-of-10 decoder, as shown in Fig. 15–11. However, the true binary code has a very serious drawback when used on an encoding wheel. The problem arises when the wheel stops right on the dividing line between two sectors.

For example, consider what might happen if the wheel of Fig. 15–10(b) stopped exactly on the line between the number 2 sector and the number 3 sector. Since the 1 band changes from dark to light at the sector dividing line, photocell 1 will be half On and half Off and will not know what to do. Photocells 2, 4, and 8 have no ambiguity, since there is no color change at the sector dividing line in those bands. Therefore, the overall photocell output might be a decimal 2 (0010) or it might be a decimal 3 (0011); there is no way to know what it will do. In this particular example the problem is not too serious because if the position is halfway between 2 and 3 it doesn't really matter what the encoder calls it; either result is very close to the truth.

However, suppose that the wheel lands on the dividing line between the number 7 sector and the number 8 sector. In this case, *all* the bands change color at the dividing line. With all four photocells in an ambiguous condition, it is conceivable that the photodetectors might all read 0. In this case, the error is serious. We wouldn't have minded if the encoder had called the number a 7 or if the encoder had called the number an 8, because either of those results is quite close to the truth. But we definitely *do* mind when the encoder calls the number a 0.

The essential difference between the 2-3 situation and the 7-8 situation is that only a single band changed color at the 2-3 dividing line, while several bands (four) changed color at the 7-8 dividing line. Whenever more than one band changes color at a sector dividing line, serious encoding errors are possible. To solve this problem, other codes have been invented in which *only one bit changes at a time.* That is, as you move from one number to the next number, only one of the constituent binary bits changes state. These other codes are still binary codes because they involve only 0s and 1s, but we make a distinction between them and the familiar 8421 binary code, which is referred to as *true binary.*

The most popular of these other binary codes is the *Gray code.* Table 15–1 shows the Gray code equivalent for each of the decimal numbers 0 to 15. For purposes of comparison, the true binary representation is shown alongside the Gray code.

TABLE 15–1
Gray code and binary code equivalents of the decimal numbers 0 through 15

DECIMAL	GRAY CODE	TRUE BINARY
0	0000	0000
1	0001	0001
2	0011	0010
3	0010	0011
4	0110	0100
5	0111	0101
6	0101	0110
7	0100	0111
8	1100	1000
9	1101	1001
10	1111	1010
11	1110	1011
12	1010	1100
13	1011	1101
14	1001	1110
15	1000	1111

In Table 15–1, notice that one and only one bit changes state in the Gray code as the coded number changes. This eliminates the possibility of serious encoding errors of the type described earlier. It is not worthwhile to memorize the Gray code, since it is not encountered nearly as often as true binary. Its chief application is in optical (and mechanical) position encoders. The Gray code is sufficiently popular for use with position encoders that the IC manufacturers package and sell Gray code to 1-of-10 decoders.

Another item worth mentioning is that two encoding wheels geared together as shown in Fig. 15–10(a) is not a popular arrangement. The preferred method is simply to divide the encoding wheel into more sectors and encode each sector uniquely. After all, the true binary code and the Gray code do not stop at 9; they go on forever. For this example, the wheel could have been divided into 100 sectors, and each sector could have had its own unique Gray code representation. The number of bands would have to be greater, of course. It would require seven bits to encode up to the decimal number 100. Therefore it would take a seven-banded encoding wheel to encode a 100-lb scale in 1-lb graduations. The decoding circuity would also have to be different, but such decoding circuitry is available.

15-7 ■ CARBON DIOXIDE CONTROLLER FOR A CARBURIZING FURNACE

15-7-1 Carburizing Process

A steel part can be given a very hard exterior layer by diffusing free carbon into its surface. The metallurgical process of diffusing carbon into steel is called *carburizing*. Carburizing is normally accomplished by subjecting the steel to a fairly high temperature, about 1700°F, for several hours, in the presence of a *carburizing atmosphere*. A carburizing atmosphere is a mixture of normal combustion products with a specially manufactured gas. The manufactured gas contains heavy concentrations of carbon monoxide (CO) and carbon dioxide (CO_2). By adjusting the composition of the gases in the carburizing atmosphere, the carbon content and carbon depth of the steel can be varied to meet different requirements. Usually the depth of carbon penetration is about 0.050 in. into the surface.

The physical arrangement for controlling the carburizing atmosphere is illustrated in Fig. 15–12. The combustion of fuel and air takes place at the burner mounted on the sidewall of the furnace. The combustion products expand and occupy the interior space of the carburizing furnace. At the same time, the special carburizing gas flows through a variable valve and into the furnace chamber. If the valve is opened further to admit a greater flow of carburizing gas, the concentration of carbon compounds in contact with the steel increases. If the valve is closed somewhat, the flow of carburizing gas decreases, and the concentration of carbon compounds in the furnace decreases.

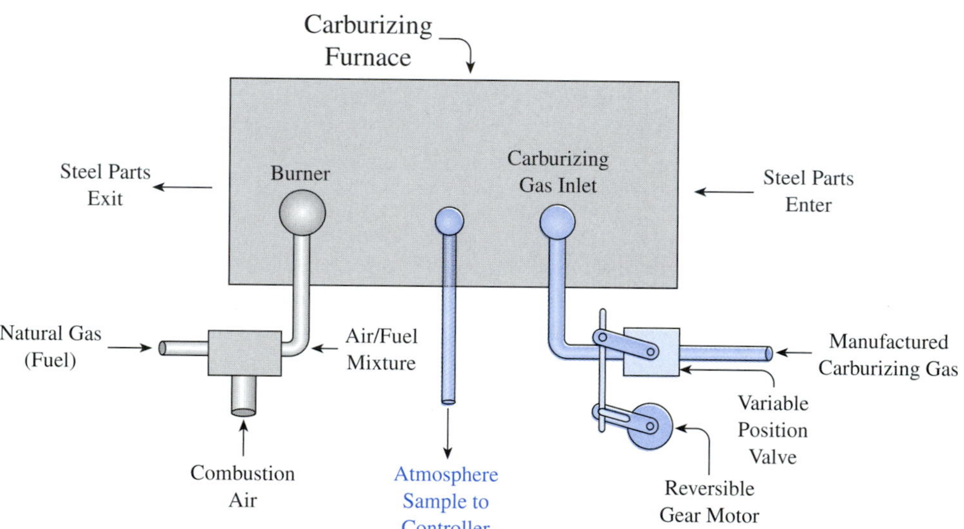

FIGURE 15–12
Physical layout of a carburizing furnace. The fuel and combustion air enter on the left, the carburizing gas enters on the right, and an atmosphere sample is drawn from the center.

The valve position is controlled by a system which compares the actual composition of the atmosphere to the desired composition. If the control circuitry finds any discrepancy between actual and desired composition, it adjusts the valve opening accordingly.

The control system obtains a sample of the atmosphere via a small tube leading out of the furnace chamber. A blower at the other end of the tube continually sucks a fresh sample of atmosphere gas through the tube to the system's measurement device.

It has been found that the best all-around method of controlling a carburizing atmosphere is to control its carbon dioxide concentration. Therefore modern atmosphere control systems measure the concentration of CO_2 in the atmosphere sample and control that concentration.

The CO_2 control system circuitry is fairly extensive, and is illustrated in Figs. 15–13, 15–14, and 15–15. Figure 15–13 shows the measurement circuitry for determining the CO_2 concentration. Figure 15–14 shows the mechanical construction of the measurement device itself. Figure 15–15 shows the error detector, controller, and final correcting device circuitry, which make up the rest of the closed-loop control system.

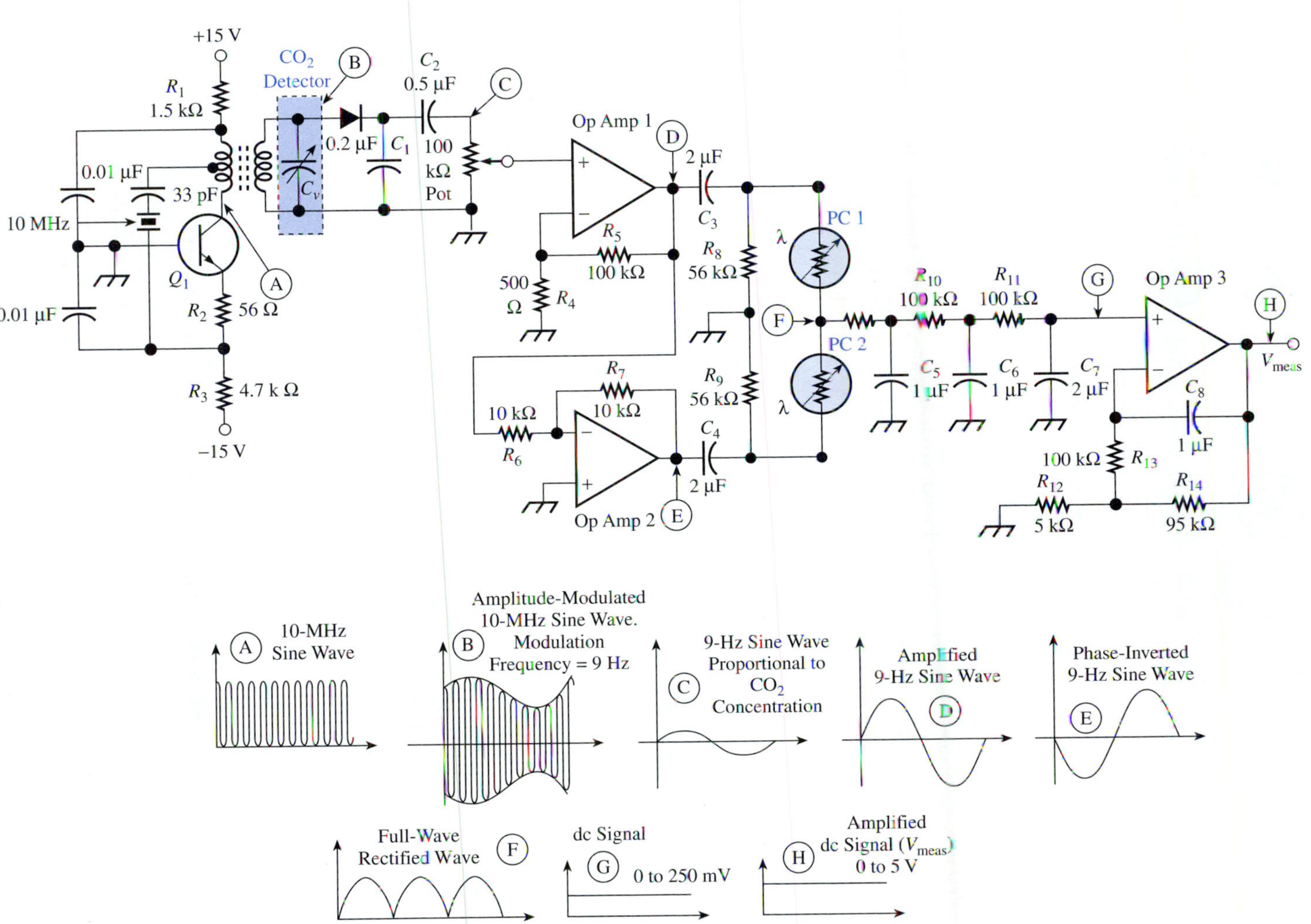

FIGURE 15–13

Circuity for measuring carbon dioxide (CO_2) concentration. The waveforms at various points in the circuit are drawn in A through H.

Courtesy of Beckman Instruments, Inc.

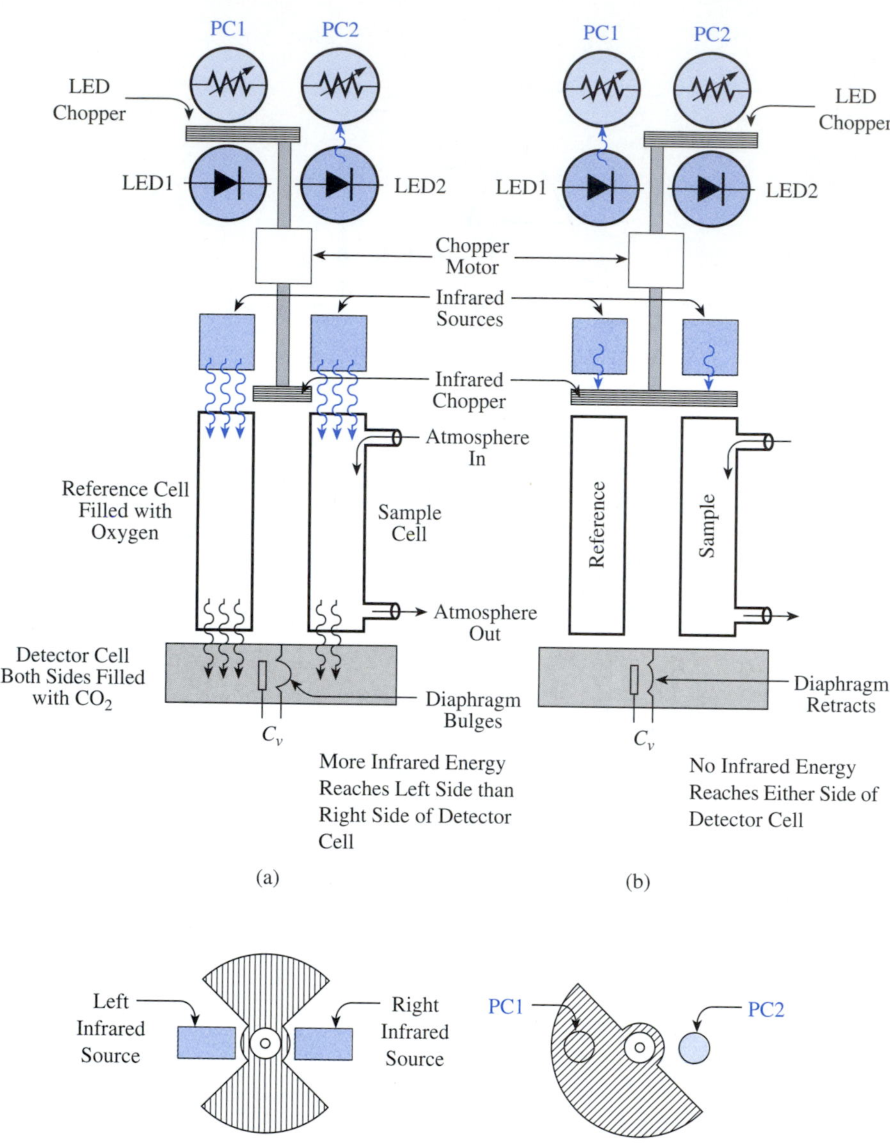

FIGURE 15–14
Physical appearance of the carbon dioxide (CO_2) measurement transducer. (a) At an instant when the infrared chopper is *not* blocking the infrared sources, infrared radiation is being admitted to the reference and sample cells. (b) At an instant when the infrared chopper *is* blocking the infrared sources, no infrared radiation enters the cells. (c) Face view of the infrared chopper. (d) Face view of the LED chopper. The angular position shown corresponds to the angular position shown in part (c) for the infrared chopper.
Courtesy of Beckman Instruments, Inc.

15-7-2 Measuring the CO_2 Concentration

Let us look first at Fig. 15–13. On the far left is an oscillator. It is crystal controlled at an operating frequency of 10 MHz. Transistor Q_1 is biased in the conducting state by the +15- and −15-V supplies in conjunction with R_1 and R_2. The oscillator feedback path is from the center of the transformer primary winding to the R_2-R_3 junction in the emitter lead. The transistor is wired in common-base configuration, because oscillators built around a common-base amplifier are inherently more temperature-stable than oscillators built around a common-emitter amplifier.

This oscillator has a frequency stability of better than 0.01% under normal operating conditions. That is, its output frequency will not stray more than 0.01% from the nominal operating frequency of 10 MHz. The 10-MHz ac signal at point A creates a 10-MHz sine-wave signal across the secondary winding of the coupling transformer. This winding

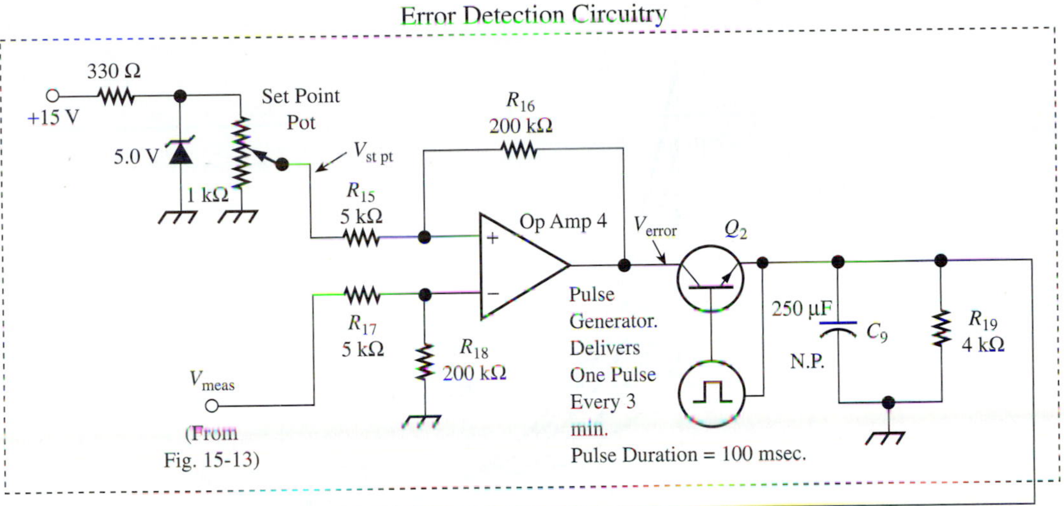

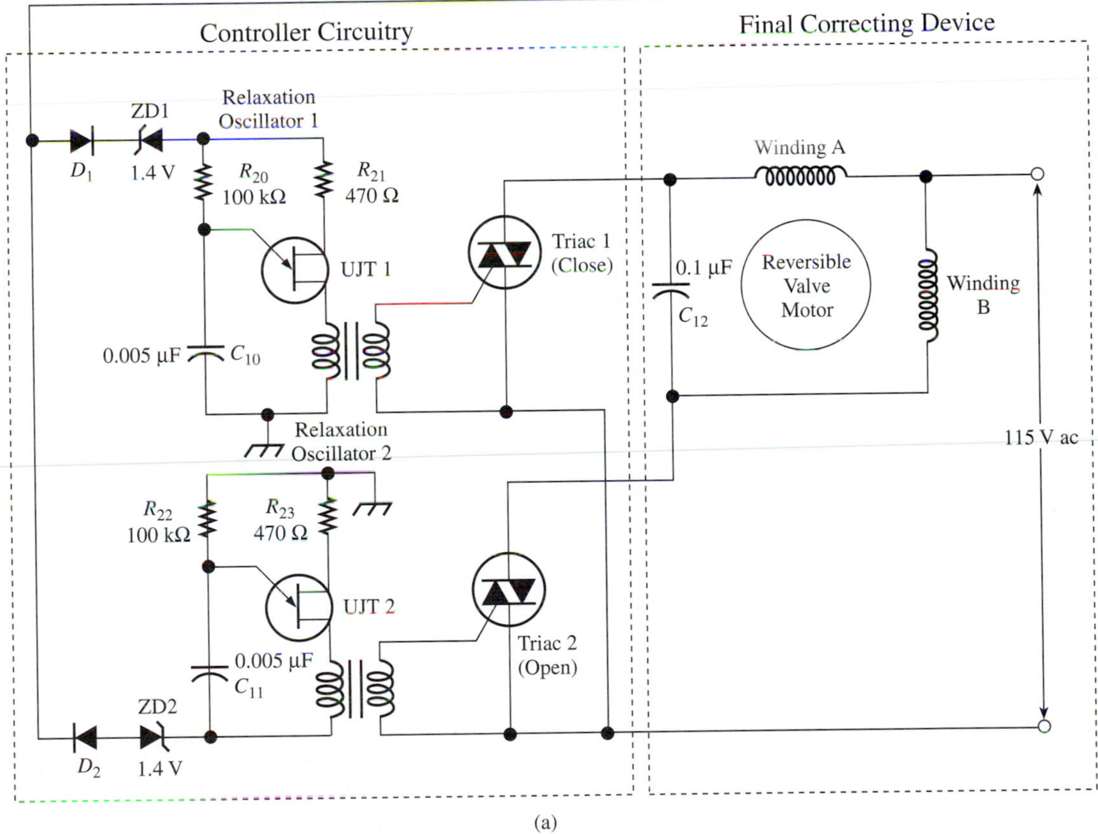

(a)

FIGURE 15–15
(a) Schematic diagram of the error-detection circuitry, the controller circuitry, and the final correcting device.

is connected directly to capacitor C_v, which is a variable capacitance. Its exact capacitance varies in relation to the CO_2 concentration in the atmosphere sample being measured. We will see shortly how this is accomplished.

The secondary winding inductance in combination with the C_v capacitance comprise an *LC* tank circuit. The values of the winding inductance *L* and the capacitance C_v

FIGURE 15–15
(*continued*)
(b) Family of discharge curves for C_9. The greater the error signal (V_{error}), the longer it takes for C_9 to discharge to 3 V. (c) Waveforms of ac supply voltage, triac main terminal voltage, and gate voltage. As long as the gate pulses occur, the triac continues to turn ON immediately after a zero crossover. When the gate pulses cease, the triac remains OFF.

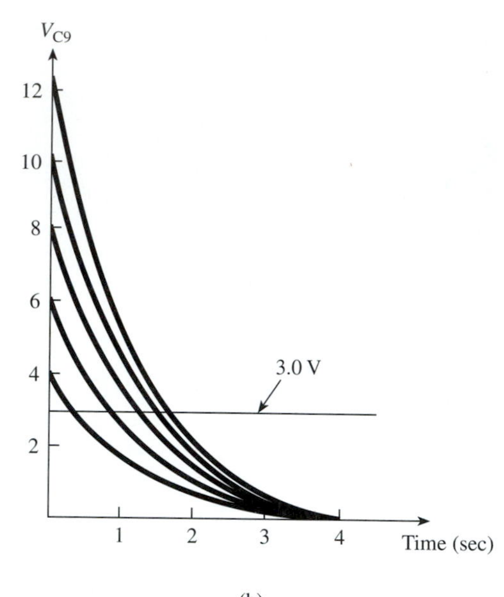

(b)

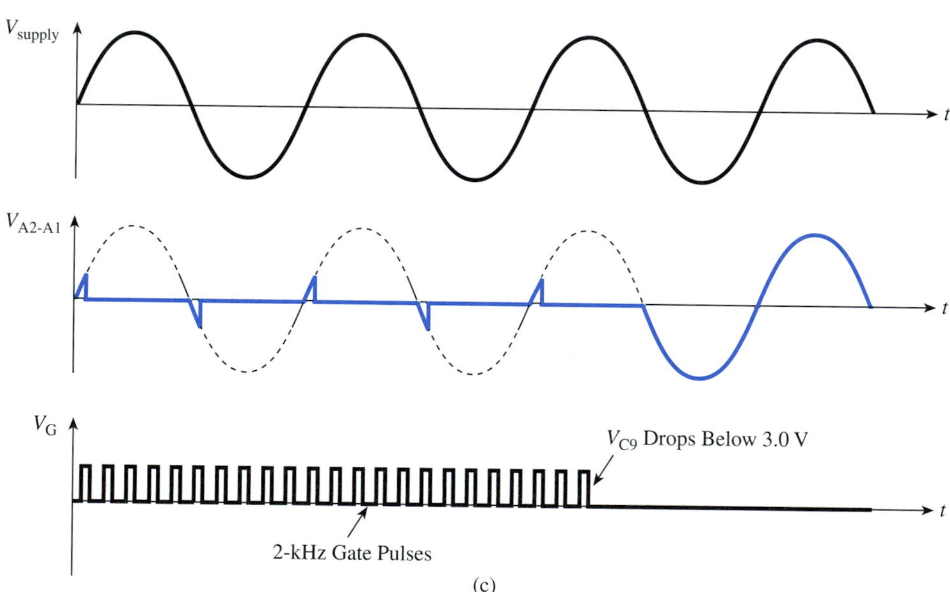

(c)

are such that this tank circuit has a resonant frequency of about 10 MHz. However, the *LC* circuit is thrown slightly out of resonance as the size of C_v varies. If the value of C_v is varied at a frequency of 9 Hz (which is the case, as we will see), the 10-MHz carrier signal appearing across the secondary winding will be *amplitude modulated* at a frequency of 9 Hz. This happens because the 10-MHz current circulating in the C_v tank circuit changes in magnitude as the tank circuit comes in and out of resonance. The amplitude-modulated 10-MHz signal is shown at point B in the circuit.

Let us stop here in our investigation of Fig. 15–13 and transfer our attention to Fig. 15–14. This figure illustrates the construction of the apparatus that causes C_v to vary in relation to the CO_2 concentration. The apparatus has two infrared radiation sources, which radiate energy into two gas-filled cylinders, or cells. These two cells are called the *reference cell* and the *sample cell,* as shown in Fig. 15–14(a). The reference cell is sealed tight and contains pure oxygen. The sample cell experiences a constant flow of fresh atmosphere gas. It "samples" the carburizing atmosphere of the furnace.

Gaseous oxygen and gaseous CO_2 show an interesting and useful contrast, on which this whole measuring method is based. This contrast is that oxygen gas will not absorb infrared energy, but CO_2 gas will absorb infrared energy. Furthermore, the amount of infrared energy that a gas *mixture* absorbs depends on the concentration of CO_2 in the gas. If the CO_2 concentration is light, not very much infrared radiant energy will be absorbed; if the CO_2 concentration is heavy, a lot of infrared radiant energy will be absorbed.

The two infrared sources radiate an equal amount of energy into each cell. All the energy radiated into the reference (oxygen-filled) cell passes through the cell without being absorbed. It emanates from the other end of the reference cell and passes into a *detector cell.* The detector cell is divided into two separate compartments, sealed off from each other by a thin metal diaphragm. Both compartments are filled with pure CO_2. The CO_2 gas in the left compartment, therefore, receives and absorbs *all* the infrared energy that was radiated from the infrared source on the left of Fig. 15–14(a). This causes a tiny increase in the temperature of the CO_2 in the left side, resulting in a tiny increase in pressure on the left side of the thin metal diaphragm.

Now consider what happens on the right-hand side. The energy radiated from this infrared source does not all pass through the sample cell. A certain amount is absorbed by the CO_2 in the atmosphere sample, and a reduced amount of energy emanates from the other end. This is suggested by the fact that there are only two wavy lines leaving the sample cell, although three wavy lines are shown entering the sample cell in Fig. 15–14(a). The CO_2 in the right side of the detector cell absorbs this diminished infrared energy, and it too experiences an increase in temperature. The right side does not experience as much change as the left side, though, because it absorbs less energy than the left side. Therefore the pressure on the right side does not increase as much as the pressure on the left side of the metal diaphragm. The pressure imbalance causes the flexible metal diaphragm to bulge to the right, as shown. There is a lead wire connected to the metal diaphragm, and there is a stationary metal plate next to the diaphragm, also having a wire leading out of the cell. These two metal objects (the stationary plate and the metal diaphragm) form a capacitor, whose capacitance naturally depends on the spacing between the objects. If they are close together, the capacitance C_v is high; if they are farther apart, C_v is low.

It should be clear that the capacitance of C_v depends on the CO_2 concentration in the sample gas. The greater the CO_2 concentration, the more energy is absorbed in the sample cell. The more energy absorbed in the sample cell, the less energy is delivered to the right side of the detector cell. The less energy delivered to the right side, the lower is the pressure on the right side. This lower pressure results in a greater bulging of the metal diaphragm and consequently a reduction in capacitance. The overall result is that a high concentration of CO_2 causes a low value of C_v.

The infrared energy sources are not permitted to continually radiate into the reference and sample cells. A slow-speed gear motor (the *chopper motor*) slowly rotates the *infrared chopper,* which is a blade shaped as shown in Fig. 15–14(c). As the chopper motor shaft rotates, the blades alternately expose and block the infrared sources. Each rotation of the shaft causes two cycles of exposure and blocking. When the infrared sources are blocked, no energy enters the reference cell, the sample cell, or the detector cell. The temperatures and pressures on the left and right sides of the detector cell have time to equalize, and the diaphragm returns to normal shape. This situation is illustrated in Fig. 15–14(b), in which the infrared chopper blade has rotated 90° and is blocking the sources.

The chopper blade rotates at 4.5 revolutions/s (270 r/min). Each revolution produces two cycles of bulging and retraction of the metal diaphragm and therefore two cycles of capacitance variation. Thus the capacitance C_v varies at a frequency of 9 cycles/s, and the magnitude of the capacitance variation is proportional to the concentration of CO_2 in the sample.

The 9-Hz variation of C_v produces the AM signal appearing at point B in Fig. 15–13. We will now continue with our discussion of the Fig. 15–13 circuit.

The AM signal is applied to a *demodulator* consisting of a diode and a 0.2-μF capacitor. These components demodulate the signal, removing the 10-MHz carrier while retaining the 9-Hz information signal. This 9-Hz signal appears at point C in the circuit.

The 9-Hz sine wave is applied to a 100-kΩ pot, where a portion of it is picked off and fed into op amp 1. Op amp 1 is wired as a noninverting amplifier, having a voltage gain of about 200. The 9-Hz sine wave therefore appears in an amplified condition at point D. The amplified signal is routed down to op amp 2, which is wired as an inverting amplifier with a gain of 1 (a phase inverter). Therefore the only action of op amp 2 is to invert the 9-Hz signal. The signal appearing at E is thus 180° out of phase with the signal at D.

Both of the signals are delivered to the R_8, R_9, PC1, PC2 network. This network creates a full-wave rectified signal at F. Here is how it works. The LED *chopper* in Fig. 15–14 is mounted on the other end of the chopper motor shaft. It is shaped as shown in Fig. 15–14(d), and it is mounted so that it blocks the beam to photocell 1 (PC1) and passes the beam to photocell 2 (PC2) at the time the infrared chopper is passing the infrared radiation. This is illustrated in Fig. 15–14(a). When the shaft rotates 180°, the LED chopper blocks the beam to PC2 and passes the beam to PC1; at this instant the infrared chopper is blocking the infrared radiation. This is illustrated in Fig. 15–14(b). To visualize the synchronization between the infrared chopper and the LED chopper, look at Fig. 15–14(c) and (d) together. The positions of the infrared sources and photocells are clearly shown. Imagine these two choppers rotating in unison, and you will see the relationship between the photocell chopping and the infrared chopping.

PC1 becomes nearly a short circuit (it is illuminated) during the positive half cycle of the D sine wave, while PC2 is nearly an open circuit (it is dark). Therefore the positive half cycle of D appears at F. PC2 becomes nearly a short circuit during the negative half cycle of D, which is the positive half cycle of E. Therefore the positive half cycle of E appears at F. The resultant waveform at F is the full-wave rectified signal which is illustrated.

This full-wave rectified signal is passed into a dc filter consisting of R_{10}, R_{11}, and the three associated capacitors. The output of the filter is a small dc voltage, which falls between 0 and 250 mV. This is shown at point G in Fig. 15–13. The voltage is made to equal 250 mV when the CO_2 concentration equals some arbitrarily selected maximum value. This adjustment is made by the 100-kΩ pot feeding op amp 1.

Op amp 3 is a noninverting amplifier with a voltage gain of 20. Therefore the output of op amp 3, labeled V_{meas}, falls between 0 and 5 V, depending on the CO_2 concentration. The purpose of the 1-μF feedback capacitor is to slow down the transient response of the amplifier. A sudden temporary variation in CO_2 concentration cannot cause a sudden variation in V_{meas}; a concentration change must persist for several seconds before V_{meas} will reflect it. This eliminates the effect of short-lived fluke variations in the sample concentration.

15-7-3 Error Detector, Controller, and Final Correcting Device

The error detector is an op amp differential amplifier, shown at the left of Fig. 15–15(a). The inputs to the differential amplifier are two voltages, V_{meas} and $V_{st\,pt}$. V_{meas} represents the measured CO_2 concentration, and $V_{st\,pt}$ represents the desired CO_2 concentration. $V_{st\,pt}$ is derived from the set-point pot, which is excited by a stable 5-V supply. The voltage gain of the differential amplifier is 200 kΩ/5 kΩ = 40. The output, V_{error}, is given by

$$V_{error} = 40(V_{meas} - V_{st\,pt})$$

Of course, V_{error} is subject to the saturation restriction of the op amp. For ±15-V supplies, V_{error} cannot exceed about ±13 V.

V_{error} is applied to the parallel combination of C_9 and R_{19}, but it is applied only once every 3 minutes, when the Q_2 transistor switch is pulsed ON. The turn-ON pulse has a duration of 100 ms, which is enough time to charge the 250-μF cap to the full value of V_{error}. The C_9-R_{19} combination has a discharge time constant given by

$$\tau = (4\ \text{k}\Omega)(250\ \mu\text{F}) = 1\ \text{s}$$

Therefore, when Q_2 turns back OFF after charging C_9 to V_{error}, the capacitor voltage (V_{C9}) discharges back to zero with a 1-second time constant (5 seconds for full discharge).

Consider the graph in Fig. 15–15(b). These curves represent the discharge behavior of V_{C9} for various values of V_{error}. The curves show a discharge from a *positive* voltage to zero, but keep in mind that the actual situation could be a discharge from a *negative* voltage to zero. The polarity of V_{error} depends on the direction of the deviation from set point.

The 3-V line is important in the discharge curves, and it is specially marked in Fig. 15–15(b). As long as V_{C9} is greater than 3 V, it can operate one of the relaxation oscillators shown in Fig. 15–15(a). First let us consider what happens if V_{C9} is more positive than +3 V. Then we will consider what happens when V_{C9} is more negative than −3 V.

If V_{C9} is more positive than +3 V, it will forward-bias diode D1 and still be able to break down the 1.4-V zener diode ZD1. It will require at least 2.1 V to accomplish this, since the voltage necessary to turn on this diode pair is given by

$$V_{diodes} = 0.7\ \text{V} + V_z = 0.7\ \text{V} + 1.4\ \text{V} = 2.1\ \text{V}$$

The remaining 0.9 V is then available to drive the relaxation oscillator, which requires a supply voltage of about 0.9 V to operate properly. Therefore, as long as V_{C9} is greater than 3 V, the pair of diodes will be broken down and sufficient voltage will appear at the top of relaxation oscillator 1. Relaxation oscillator 1 has an operating frequency given by Eq. (5-5):

$$f = \frac{1}{R_E C_E} = \frac{1}{(100 \times 10^3)(0.005 \times 10^{-6})} = 2\ \text{kHz}$$

Gate pulses are thus delivered to triac 1 at a frequency of 2 kHz. This barrage of gate pulses keeps triac 1 turned completely ON during the time that the pulses continue. With gate pulses arriving at such a high frequency, the triac is bound to be fired very shortly after the supply voltage passes through 0 V. The circuit waveforms are shown in Fig. 15–15(c). The gate pulses which appear during the remainder of the half cycle have no effect on the triac, since it has already fired. However, the continuation of the rapid string of pulses ensures that the triac will fire again just as soon as the *next* half cycle begins.

The upshot is that as long as V_{C9} exceeds 3 V, triac 1 will remain ON, and the valve motor will continue to run. The direction of rotation is such that it *closes* the control valve. This is the proper direction of rotation, since a positive V_{error} means that the measured CO_2 concentration is greater than desired. The closing of the atmosphere control valve will restrict the flow of carburizing gas and decrease the CO_2 concentration.

The length of time that the motor runs is determined by the length of time that V_{C9} stays above 3 V. As Fig. 15–15(b) clearly shows, that length of time depends on the initial charge on C_9, that is, on V_{error}. If V_{error} is large, V_{C9} remains above 3 V for a longer time, resulting in a greater correction to the flow of carburizing gas. This is the proper action because a large positive V_{error} means that the measured CO_2 concentration is too high by a large amount.

Now consider what happens if V_{error} is more negative than −3 V. In this case, D_2 and ZD2 are turned on, and a negative voltage can appear at the bottom of relaxation oscillator 2. This relaxation oscillator delivers 2-kHz gate pulses to triac 2, which drives the valve motor in the opposite direction. With triac 2 firing, capacitor C_{12} is effectively in series with motor winding A, whereas when triac 1 was firing, C_{12} was in series with motor winding B. This reverses the direction of rotation, as explained in Chapter 11.

Again, the length of time that V_{C9} stays more negative than −3 V depends on the initial charge on C_9. The curves of Fig. 15–15(b) still apply, even though the initial voltage is negative instead of positive. If V_{error} is a large negative voltage, the motor will run for a longer time, opening the control valve a greater amount. This is the proper action, since a large negative V_{error} means that measured CO_2 concentration is too low by a large amount.

This atmosphere control system has a very long transfer lag and transportation lag. The transportation lag is due to the necessity to suck a new atmosphere sample through the tube leading to the CO_2 detector. This tube is usually quite long for a carburizer, perhaps 50 ft or more. The transfer lag depends on how quickly a change in carburizing gas flow can make its effects felt throughout the carburizing chamber. As can be imagined, this does not happen quickly. Because of these long lags, *continuous control* of carburizing atmospheres is never attempted. Instead, a control action is followed by 3 min of no control. At the end of 3 min, another control action is initiated if needed.

15-8 ■ CONTROL OF RELATIVE HUMIDITY IN A TEXTILE MOISTENING PROCESS

In certain textile finishing processes, the moving textile strip is passed through a moistening chamber to be moistened and softened. The relative humidity in the chamber is maintained at a high level by spraying water into a duct through which the chamber air recirculates. This system layout is illustrated in Fig. 15–16(a).

The moving textile strip enters the moistening chamber through the wall on the left. It passes over several moving rolls as it moves through the chamber to where it exits through the wall on the right. The strip is dry when it enters the chamber and wet when it leaves the chamber, so it is constantly removing water vapor from the air. This loss of water vapor must be continually replenished to maintain the relative humidity at the proper value (about 80%).

The water vapor is replenished by sucking the chamber air into the recirculating duct on the right-hand side of Fig. 15–16(a). As the air passes through the recirculating

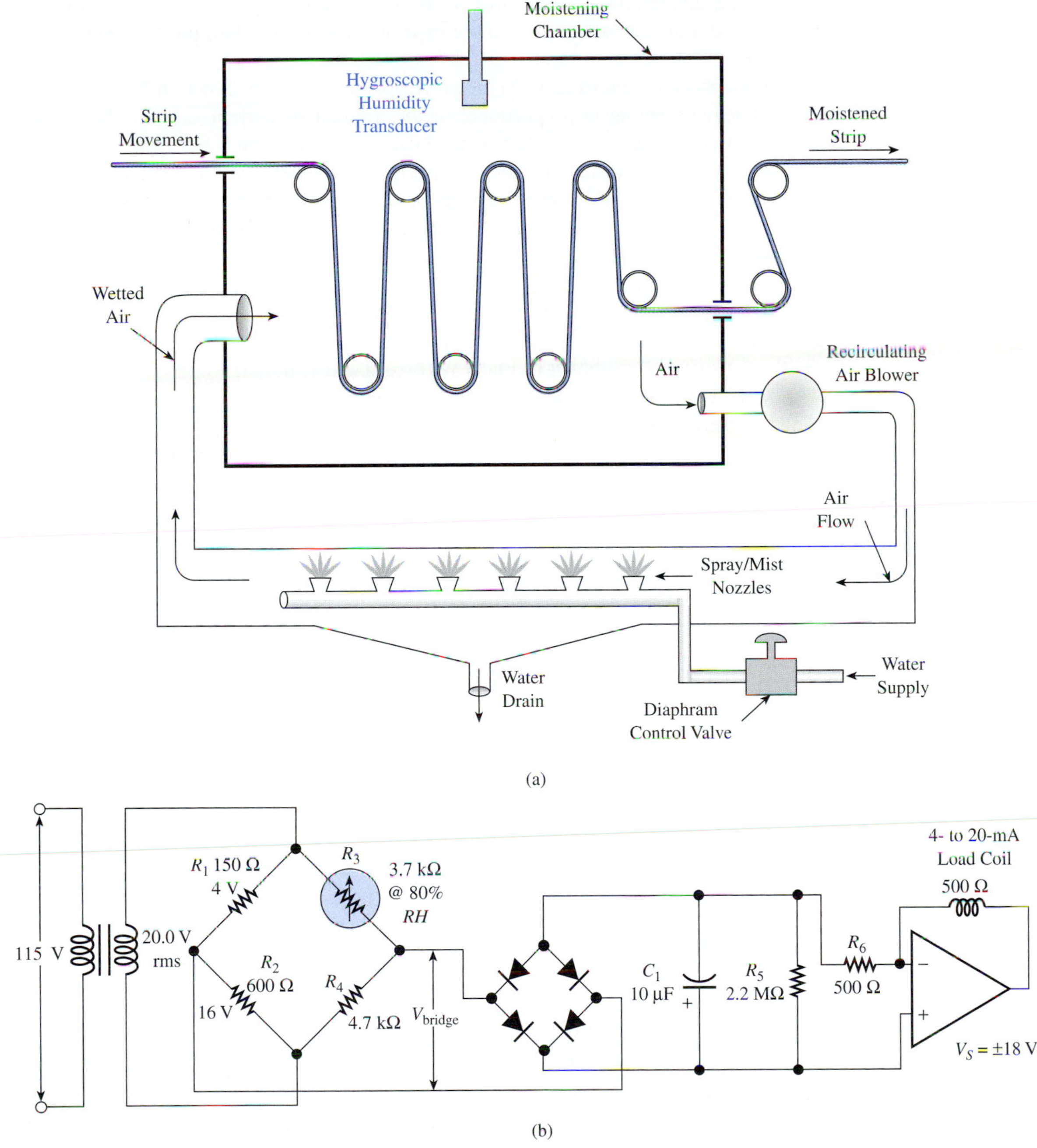

FIGURE 15–16
Strip moistening system: (a) physical layout of the moistening chamber and water feed pipes; (b) circuit for detecting and controlling the humidity.

duct, it encounters a battery of mist nozzles. It picks some of the water vapor from the mist and reenters the moistening chamber on the left. The air reenters much wetter than when it left the chamber.

The amount of water vapor that the air absorbs as it flows through the recirculating duct depends on the water flow admitted to the mist nozzles. This flow is controlled by the pneumatic diaphragm control valve in the water supply line. If the valve is further open, it allows a greater water flow to the mist nozzles, which then provide more vigorous misting.

The diaphragm control valve is the electropneumatic type shown in Fig. 11–4. The amount of valve opening is proportional to the amount of current through the electromagnet input coil which moves the balance beam. Therefore the water flow rate and the amount of mist are determined by the amount of current delivered to the coil. The electropneumatic apparatus is designed to respond to currents in the range of 4 to 20 mA. That is, if the coil current drops to 4 mA or less, the valve shuts completely off. If the current rises to 20 mA, the valve opens wide. For current values between 4 and 20 mA, the valve is somewhere in the throttling range between wide open and completely closed.

The sensing coil is driven by an op amp voltage-to-current converter. The operation of a voltage-to-current converter was explained in Sec. 8-3. The electronic circuit which drives the sensing coil is shown in Fig. 15–16(b). Here is how it works.

The Wheatstone bridge is driven by a stable 20-V rms ac supply. A hygroscopic humidity transducer (hygrometer) is used for resistor R_3. A resistive hygrometer must be excited by ac voltage only. If dc current passes through it for any length of time, it will become chemically polarized, and its characteristics will change. Therefore, it cannot be used in a dc Wheatstone bridge. This transducer has the resistance characteristics shown in the graph of Fig. 10–44(b). The important data point from that graph that concerns us is

Relative Humidity (RH)	***Resistance***
80%	3.7 kΩ

Let us assume that the humidity in the textile moistening chamber is at 80%, which is in the center of the acceptable range. The value of R_3 is then 3.7 kΩ. The Wheatstone bridge ac output voltage, V_{bridge}, is equal to the difference between the voltage across R_2 and the voltage across R_4, or

$$V_{\text{bridge}} = V_{R2} - V_{R4}$$

V_{R2} can be calculated as

$$\frac{V_{R2}}{20\text{ V rms}} = \frac{R_2}{R_2 + R_1} = \frac{600\ \Omega}{600\ \Omega + 150\ \Omega}$$

$$V_{R2} = 16.0\text{ V rms}$$

When the relative humidity equals 80%, V_{R4} is given by

$$\frac{V_{R4}}{20\text{ V rms}} = \frac{R_4}{R_4 + R_3} = \frac{4.7\text{ k}\Omega}{4.7\text{ k}\Omega + 3.7\text{ k}\Omega}$$

$$V_{R4} = 11.2\text{ V rms}$$

Therefore, at 80% *RH*,

$$V_{\text{bridge}} = 16.0\text{ V rms} - 11.2\text{ V rms} = 4.8\text{ V rms}$$

The peak voltage supplied to the ac terminals of the bridge rectifier is given by

$$V_{\text{pk}} = (1.41)(V_{\text{bridge}}) = 1.41(4.8\text{ V}) = 6.8\text{ V peak}$$

The bridge rectifier introduces a total voltage drop of about 0.8 V, since the applied voltage must overcome two germanium diodes at about 0.4 V each, carrying the short-lived current surge that recharges C_1. Therefore,

$$V_{\text{in}} = 6.8\ V_{\text{pk}} - 0.8\text{ V} = 6.0\text{ V dc}$$

which is the dc input voltage to the op amp voltage-to-current converter. The current through the load coil can be calculated from Eq. (8.15):

$$I = \frac{V_{\text{in}}}{R_6} = \frac{6.0\ \text{V}}{500\ \Omega} = 12.0\ \text{mA}$$

Therefore the sensing coil is carrying a current of 12.0 mA when the relative humidity in the moistening chamber is 80%. The value 12.0 mA is the middle of the proportioning current range of the diaphragm-actuated control valve, so the control valve should be about halfway open.

Now if the humidity in the moistening chamber should drop below 80% for some reason, here is what would happen. The resistance of the hygrometer will rise, causing a reduced voltage drop across R_4. This will throw the bridge further out of balance, causing V_{bridge} to become larger. The greater V_{bridge} will result in a greater dc input to the op amp circuit, causing a larger current flow through the sensing coil. This opens the water supply valve further and tends to drive the humidity back up.

Let us try to calculate how far the relative humidity would have to drop in order to open the water supply valve wide open.

To drive the valve wide open, a sensing coil current of 20 mA is required. Therefore, V_{in} is given by

$$V_{\text{in}} = (20\ \text{mA})(500\ \Omega) = 10.0\ \text{V}$$

To have 10.0 V delivered to the 10-μF filter capacitor, the peak input voltage to the rectifier bridge must be about 10.8 V because about 0.8 V will be dropped across the diodes. Therefore the rms output voltage from the Wheatstone bridge must be

$$V_{\text{bridge}} = \frac{V_{\text{pk}}}{1.41} = \frac{10.8\ \text{V}}{1.41} = 7.7\ \text{V rms}$$

For the output voltage to be 7.7 V rms, the voltage across R_4 must be

$$V_{R4} = V_{R2} - V_{\text{bridge}} = 16.0\ \text{V} - 7.7\ \text{V} = 8.3\ \text{V rms}$$

If the voltage across R_4 equals 8.3 V rms, the voltage across R_3 is given by

$$V_{R3} = 20.0\ \text{V} - 8.3\ \text{V} = 11.7\ \text{V rms}$$

Therefore we can find the resistance of R_3 from

$$\frac{R_3}{R_4} = \frac{V_{R3}}{V_{R4}}$$

$$\frac{R_3}{4.7\ \text{k}\Omega} = \frac{11.7\ \text{V}}{8.3\ \text{k}\Omega}$$

$$R_3 = 6.6\ \text{k}\Omega$$

This means that if the resistance of R_3 rises to 6.6 kΩ, the sensing coil will cause the water valve to open wide. From the graph of Fig. 10–44(b) we can find the relative humidity which would cause the transducer resistance to be 6.6 kΩ. It appears to be about 70% *RH*. Therefore a drop in relative humidity to 70% is necessary to call for and produce full water flow to the mist nozzles.

15-9 ■ WAREHOUSE HUMIDITY CONTROLLER

In some storage warehouses, it is essential that the relative humidity be maintained above a certain level. Two good examples are warehouses that store explosives and warehouses that store grain. For both of these commodities, it is dangerous to allow the relative humidity to fall below about 50%. If grain dust gets too dry, spontaneous combustion may occur. Explosives also are dangerous to handle and store under very dry conditions. Figure 15–17(a) shows a method of maintaining a warehouse atmosphere's humidity above a safe value.

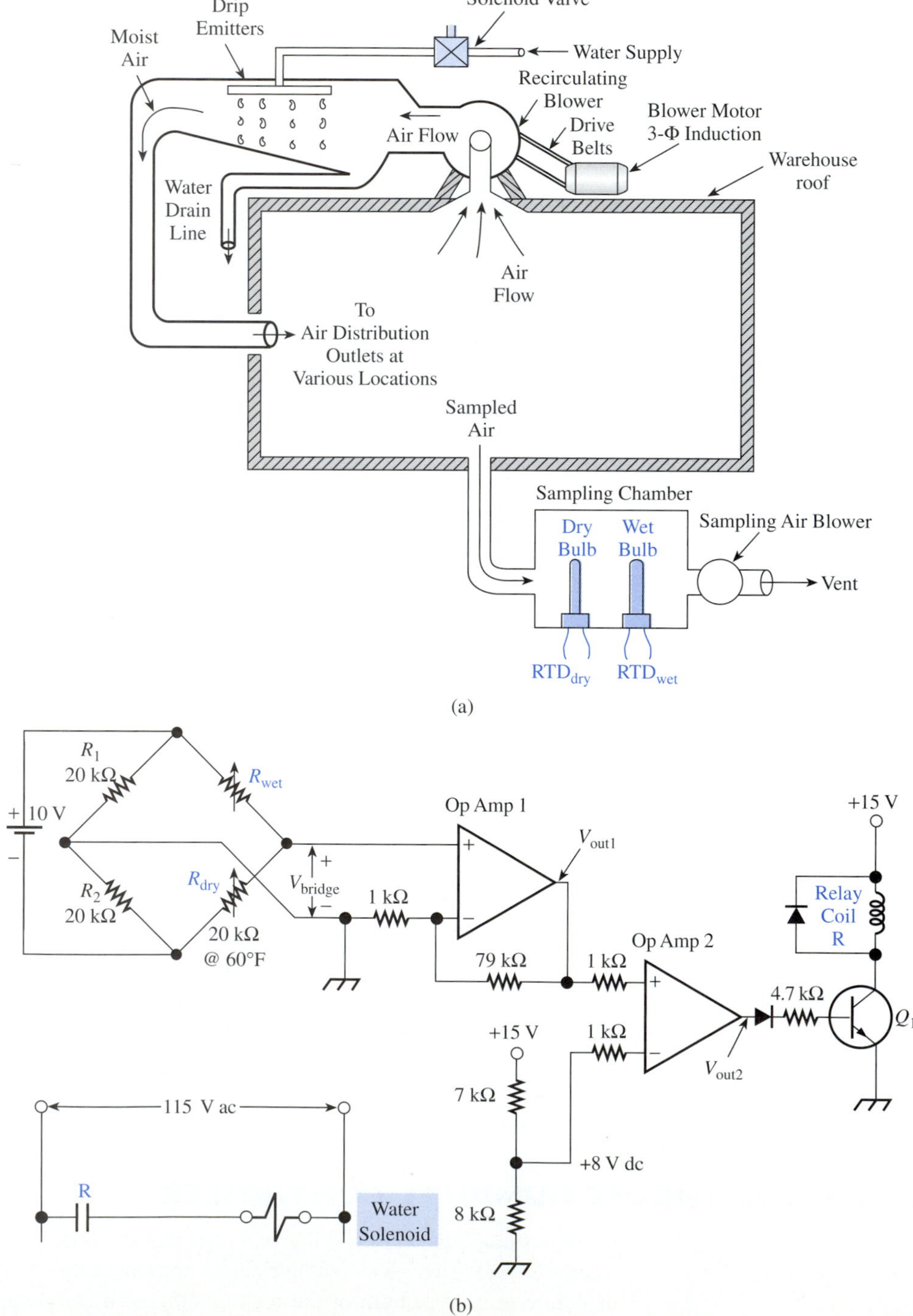

FIGURE 15–17 (a) Physical system for maintaining warehouse humidity. (b) The psychrometer detector and water flow-control circuit.

This system is an On-Off control system. The recirculating air blower runs at all times, distributing air evenly to all parts of the storage area. This ensures a uniform atmosphere throughout the warehouse, preventing any dry pockets from forming. When the water solenoid valve opens, it allows water to reach the series of drip emitters located in the air duct downstream from the blower. The recirculating air blower sucks air out through the roof of the warehouse and forces it past the drip emitters. The moving air absorbs some of this dripping water, and the wetted air is then distributed via ductwork to various locations throughout the warehouse.

The signal to turn on the water originates in the wet bulb-dry bulb sampling chamber. A continuous stream of air from the storage area is sucked into the sampling chamber by a small sampling blower. The sampled air passes over both temperature bulbs, as described in Sec. 10-15-2. The air is always *pulled* into the sampling chamber rather than *pushed* in, so that no heat energy is imparted to the sampled air by the blades of the sampling blower. This might cause erroneous temperature readings. Also, the dry bulb is always located nearer the entry to the sampling chamber. This is so the dry bulb cannot be affected by moisture picked up by the air as it passes over the wet bulb.

In this system, the temperature detectors are nickel wire RTDs.

The nickel RTDs have a resistance of 20 kΩ at 60°F. The dry bulb RTD is placed in a Wheatstone bridge in the R_4 position, and the wet bulb RTD is placed in the R_3 position, as shown in Fig. 15–17(b). The resistances of the RTDs are labeled R_{dry} and R_{wet}, respectively. The left side of the Wheatstone bridge divides the 10-V dc supply evenly, since R_1 equals R_2. The right side will not divide the supply voltage evenly because R_{wet} will be less than R_{dry} due to its lower temperature. Therefore the bridge will be unbalanced, with the bridge output voltage furnishing the input signal to op amp 1. This input voltage is identified as V_{bridge} in Fig. 15–17(b).

The relationship between the relative humidity and V_{bridge} is this: As the relative humidity goes down, the temperature difference between the bulbs becomes greater (refer to Fig. 10–45b). As the temperature difference becomes greater, the difference between R_{wet} and R_{dry} also increases, throwing the bridge further out of balance. Therefore, a decrease in relative humidity causes an increase in V_{bridge}.

In this system, the temperature of the storage area is maintained near 60°F by an independent temperature control system. If the relative humidity drops below the accepted level, 50%, the temperature difference between bulbs will reach 10°F. This is shown in Fig. 10–45(b). Find the row which indicates a dry bulb temperature of 60°F. Move along that row to the column which indicates a temperature difference of 10°F; the relative humidity value is given as 49%. Therefore, if the bulb temperature difference becomes as great as 10°F, it means the humidity is too low and must be brought back up.

We will now calculate the value of V_{bridge} for a temperature difference of 10°F. Nickel wire has a temperature coefficient of resistance of about 0.42%/°F(0.0042/°F). This value can be obtained from Fig. 10–9(a).

Therefore, if the temperature difference is 10°F, R_{dry} and R_{wet} will differ by 4.2%, since (0.42%/°F)(10°F) = 4.2%. This means that R_{wet} will be 19.19 kΩ for a temperature difference of 10°F, since

$$\frac{20\text{ k}\Omega - 19.19\text{ k}\Omega}{19.19\text{ k}\Omega} = 0.042 = 4.2\%$$

Under these conditions, V_{bridge} is given by

$$V_{bridge} = V_{Rdry} - V_{R2} = V_{Rdry} - 5.00\text{ V} \qquad (15\text{-}4)$$

The voltage across R_{dry} is given by

$$\frac{V_{Rdry}}{10\text{ V}} = \frac{R_{dry}}{R_{dry} + R_{wet}} = \frac{20\text{ k}\Omega}{20\text{ k}\Omega + 19.19\text{ k}\Omega}$$

$$V_{Rdry} = 5.10\text{ V}$$

From Eq. (15-4), V_{bridge} is given by

$$V_{bridge} = 5.10\text{ V} - 5.00\text{ V} = 0.10\text{ V}$$

This means that if V_{bridge} reaches a value of 0.10 V, the relative humidity has dropped too low and must be corrected.

V_{bridge} is amplified in the op amp 1 noninverting amplifier, which has a voltage gain of 80. Therefore when V_{bridge} reaches 0.10 V, V_{out1} reaches

$$V_{out1} = 80(V_{bridge}) = 80(0.10\text{ V}) = 8.0\text{ V}$$

Op amp 2 is a voltage comparer. It compares the value of V_{out1} to +8.0 V. The 8-V reference is supplied by the 7-kΩ/8-kΩ voltage divider in Fig. 15–17(b). If V_{out1} is less than +8.0 V, the output of the voltage comparer is −13 V. If V_{out1} is greater than +8.0 V, the output of the voltage comparer is +13 V. Therefore V_{out2} switches from −13 V dc to +13 V dc when V_{out1} reaches +8.0 V. When this happens, transistor Q_1 switches ON and energizes relay *R*. This relay closes a contact in the 115-V ac circuit, which energizes the water solenoid valve. This adds moisture to the recirculating air, thereby driving the relative humidity back up above 50%.

In this control system, the desired relative humidity can be changed easily, just by changing the value of the 8-kΩ resistor in the voltage divider which feeds op amp 2.

TROUBLESHOOTING ON THE JOB

TROUBLESHOOTING AN OP-AMP PROPORTIONAL CONTROL CIRCUIT

The textile-moistening process of Fig. 15–16 demanded attention in Troubleshooting on the Job in Chapter 11. On that occasion the problem was the system's final correcting device, the electropneumatic operator for the water valve. On this day the moistening system is again unable to maintain its proper 80% humidity level. Suspecting another electropneumatic operator problem, you place your ammeter in series with the actuator coil (see Fig. 11–4). The ammeter gives a measurement of zero, so the valve-operator cannot be blamed for failing to open the water-control valve. Describe your testing/troubleshooting procedure from this point.

SUMMARY

- A recuperator is a heat exchanger for capturing and reusing some of the heat energy contained in an exhaust gas. Recuperators are used in steel soaking pits.
- An op-amp summing circuit can be used as the comparer in the block diagram for a closed-loop control process.
- An op-amp inverting or noninverting amplifier circuit can be used to adjust the proportional band of an electronic proportional controller.
- An op-amp integrator circuit, in combination with an op-amp summing circuit, is a straightforward way to accomplish proportional plus integral control.
- Strip tension control is accomplished by adjusting the relative speeds of the leading roll and the following roll.
- An LVDT is an effective device for monitoring strip tension.
- When strip material is being recoiled on a windup reel, its edges must be automatically aligned.
- A screw conveyor is often used to feed powder material into an automatic weighing hopper.
- A load-cell is a strain-gage-based weighing component.
- A complementary symmetry amplifier is a push-pull type of amplifier that contains no transformer. It uses one *npn* and one *pnp* transistor.
- The Gray code is superior to the standard binary code of mechanical/optical absolute encoding because it eliminates gross encoding errors.
- Carburizing is the process of diffusing free carbon into steel to increase its hardness. It is accomplished by passing the steel parts through a high-temperature carburizing atmosphere, rich in carbon monoxide.
- Some strip materials must be moistened by passing through a moistening chamber before further processing.

QUESTIONS AND PROBLEMS

Section 15-1

1. In Fig. 15–1(b), what is the purpose of the *RC* circuit in parallel with the triac?
2. In Fig. 15–1(a), what would be the best physical location for the thermistor probe?
3. In Fig. 15–1(b), if R_2 were larger, would the motor run faster or slower? Explain.

Section 15-2

Questions 4 to 8 refer to Fig. 15–2.

4. Explain the purpose of the span-adjust pot.
5. Explain the purpose of the set-point zero pot.
6. If V_{out2} goes to +12.5 V, what would the voltage drop be across R_8?
7. Explain just why it is that for transistor Q_1, the collector voltage relative to ground is the same as the emitter voltage relative to ground (but of opposite polarity).
8. If the maximum current through the load coil were 6 mA and you wished to correct it to 5 mA, which way would you turn the maximum current-adjust pot to increase or decrease its resistance? Explain.

Section 15-3

Questions 9 and 10 refer to Fig. 15–4.

9. Why is the dc signal from the Wheatstone bridge chopped?
10. Explain why V_{out3} has to be inverted before being applied to R_{29}.

Section 15-4

Questions 11 and 12 refer to Fig. 15–6.

11. If it was found that the amount of time that the draw motor was being run was not long enough, that it was taking too many corrections to get the tension back in the acceptable range, what could you do about it? What would you change in the circuit?
12. Explain carefully the operation of D_5 and D_6.

Section 15-5

Questions 13 and 14 refer to Fig. 15–8.

13. If the lamp got dimmer, explain the consequences.
14. What is the purpose of R_2?

Section 15-6

Questions 15 to 18 refer to Fig. 15–9.

15. What is the purpose of the tare pot?
16. What would happen if the tare winding (S_B) were hooked up backward (if its phase were reversed)?
17. Why is it so important that the weighing circuitry itself be well shielded from noise?
18. What is the purpose of slow feed in an automatic weighing system?

Section 15-7

Questions 19 to 22 refer to Figs. 15–13, 15–14, and 15–15.

19. What does the demodulator do (the diode and C_1)?
20. What does the C_5, C_6, C_7, R_{10}, R_{11} combination do?
21. Why is it important to have the reference cell filled with oxygen and tightly sealed against leaks?
22. Why does C_9 have to be a nonpolarized capacitor?

Section 15-8

Questions 23 to 25 refer to Fig. 15–16.

23. Why must the Wheatstone bridge be an ac bridge instead of a dc bridge?
24. What is the purpose of the R_5-C_1 combination ahead of the voltage-to-current converter?
25. If R_2 were larger, would that tend to raise the humidity set point or lower it? Explain.

Section 15-9

Questions 26 to 28 refer to Fig. 15–17.

26. What control mode is being used in Fig. 15–17?
27. What is the purpose of the diode in parallel with relay coil R?
28. If it were desired to raise the RH set point, would you increase the 8-kΩ resistor or decrease it? Explain.

CHAPTER

16 MOTOR SPEED-CONTROL SYSTEMS

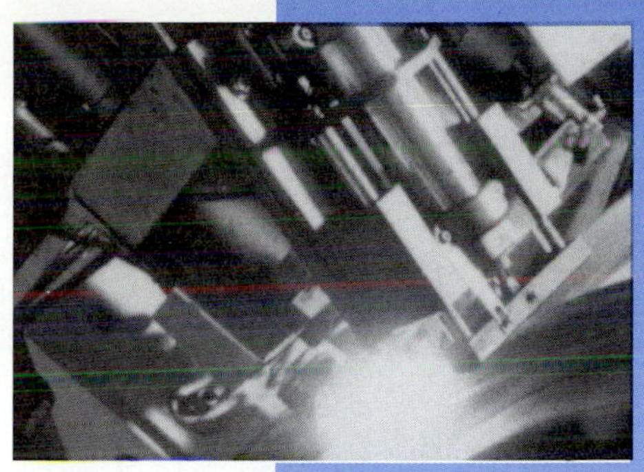

In most industrial situations, motors are operated directly from the ac or dc supply lines. That is, the motor winding terminals are connected directly to the lines which supply the electric current. In these situations, the operating behavior of the motor is determined by the nature of the mechanical load connected to the motor's shaft. In simple terms, if the load is easy to drive, the motor will tend to deliver relatively little torque, and it will run at a high speed. If the load is difficult to drive, the motor will tend to deliver a lot of torque, and it will run at a lower speed. The point is that the operating behavior of the motor is set by its *load* (for a fixed supply line voltage), and the operator has no control over motor behavior.

In modern industrial situations, there are many applications which require that the operator be able to intervene to control the motor speed. Such control is usually accomplished with thyristors. The combination of the motor, the controlling thyristor(s), and the associated electronic components is referred to as a *speed-control system* or a *drive system*.

OBJECTIVES

After completing this chapter, you will be able to:

1. Explain the two basic methods of adjusting the speed of a dc shunt motor.
2. Discuss the relative advantages and disadvantages of speed adjustment of a dc motor from the field and from the armature.
3. Explain why armature control with thyristors is superior to all other dc motor speed-control methods.
4. Discuss how counter-EMF feedback can be used to improve a motor's load regulation.
5. Calculate a motor's load regulation, given a graph of shaft speed versus load torque (or shaft power).
6. Explain the operation of the single-phase and three-phase thyristor drive systems which are presented.
7. Explain the operation of a switchgear-controlled reversible drive system.
8. Explain the process of power control by pulse-width modulation of a switching transistor.
9. Analyze the operation of a dual-555 pulse-width-modulated motor-drive system.
10. For a three-phase inverter circuit, analyze the SCR switching schedule and relate it to the three-phase ac output waveforms.

11. Recognize the necessity for forced commutation of the SCRs in a dc-to-ac inverter, and describe the operation of the method that is presented.
12. For an ac drive system, explain why the voltage magnitude must be varied in proportion to the frequency; show how this can be accomplished by combining a variable rectifier with an inverter.
13. For a 6- or 12-SCR single-phase cycloconverter, describe the SCR switching sequence and relate it to the ac output waveform.
14. Describe the techniques for improving the output waveshape of a cycloconverter.
15. Show how three single-phase cycloconverters are connected to drive a three-phase motor.

16-1 ■ DC MOTORS—OPERATION AND CHARACTERISTICS

Dc motors are important in industrial control because they are more adaptable than rotating-field ac motors to adjustable speed systems.

Figure 16–1 shows the schematic symbol of a dc *shunt motor*. The motor's field winding is drawn as a coil. Physically, the field winding is composed of many turns of thin (high-resistance) wire wrapped around the *field poles*. The field poles are ferromagnetic metal cores, which are attached to the stator of the machine. The high resistance of the field winding limits the field current to a fairly small value, allowing the field winding to be connected directly across the dc supply lines. However, the relatively small field current (I_F) is compensated for by the field winding's large number of turns, enabling the winding to create a strong magnetic field.

The field winding is unaffected by changing conditions in the armature. That is, as the armature current varies to meet varying load conditions, the field winding current stays essentially constant, and the strength of the resulting magnetic field stays constant. The field current can be found easily from Ohm's law as

$$I_F = \frac{V_S}{R_F} \tag{16-1}$$

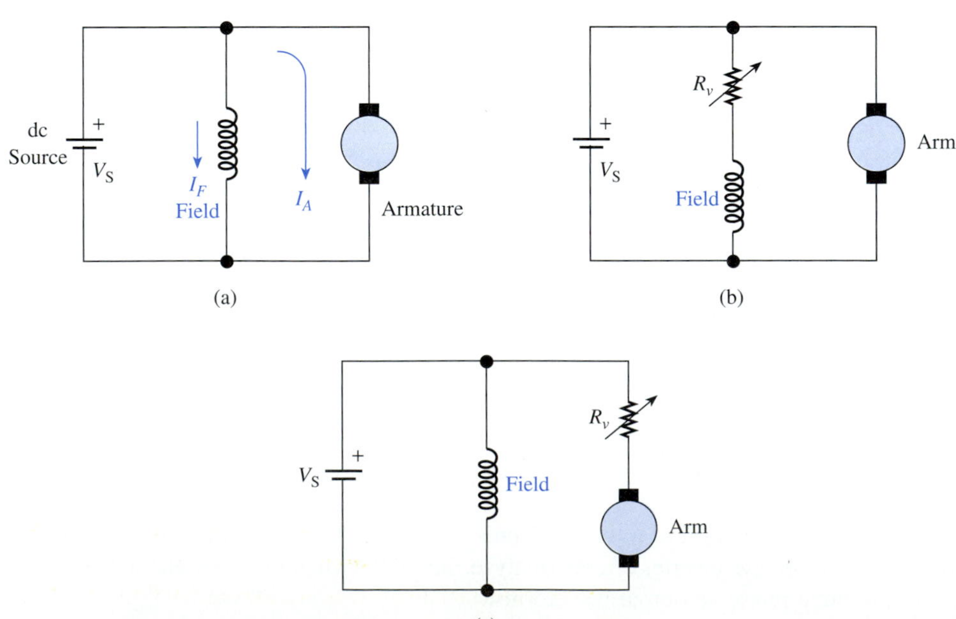

FIGURE 16–1
(a) Schematic representation of a dc shunt motor. (b) A rheostat in series with the field winding for controlling the motor's speed. (c) A rheostat in series with the armature for controlling the motor's speed.

where V_S is the source voltage applied to the field winding (Fig. 16–1a) and R_F is the winding's dc resistance.

The armature winding is shown in Fig. 16–1(a) as a circle contacted by two small squares. It is drawn like this because the armature winding is constructed on the cylindrical rotor of the machine and current is carried to and from the armature winding via carbon brushes contacting commutator segments.

The armature winding of a dc motor is constructed of relatively few turns of thicker wire, so it has a low dc resistance. The armature winding resistance of a medium- or large-sized dc motor is usually less than 1 Ω.

When power is first applied to the armature winding, only the dc ohmic resistance of the winding is available to limit the current, so the inrush current surge is quite great. However, as the motor begins to accelerate, it begins to induce (generate) a *counter-EMF* by the usual generator action. This counter-EMF opposes the applied source voltage and limits the armature current to a reasonable value.

When a dc motor has reached normal operating speed, its counter-EMF is about 90% as great as the applied armature voltage (V_S in Fig. 16–1a). The *IR* voltage drop across the armature winding resistance accounts for the other 10% of the applied voltage, neglecting any voltage drop across the carbon brushes.

The exact size of the counter-EMF generated by the armature winding depends on two things:

1. The strength of the magnetic field. The stronger the magnetic field, the greater the counter-EMF tends to be.
2. The speed of rotation. The greater the speed, the greater the counter-EMF tends to be.

Equation (16-2) expresses the dependence of counter-EMF on field strength and rotational speed:

$$E_c = k_g BS \tag{16-2}$$

In Eq. (16-2), E_c stands for the counter-EMF created by the spinning armature winding, B stands for the strength of the magnetic field created by the field winding, and S is the rotational speed in revolutions per minute. The proportionality constant k depends on the construction details of the armature (the number of winding turns, the length of the conductors, etc.).

Kirchhoff's voltage law for the armature loop is expressed in Eq. (16-3), which simply states that the applied armature voltage is equal to the sum of the voltage drops in the armature. The sum of the voltage drops in the armature winding equals the counter-EMF added to the *IR* resistive voltage drop, again neglecting the minor effect of the brush drop.

$$V_S = E_c + I_A R_A \tag{16-3}$$

In Eq. (16-3), R_A stands for the dc resistance of the armature winding, and, of course, I_A is the armature current.

16-1-1 Varying the Speed of a Dc Shunt Motor

Basically, there are two ways of varying the running speed of a dc shunt motor:

1. Adjusting the voltage (and current) applied to the field winding. As the field voltage is increased, the motor *slows down*. This method is suggested by Fig. 16–1(b).
2. Adjusting the voltage (and current) applied to the armature. As the armature voltage is increased, the motor speeds up. This method is suggested by Fig. 16–1(c).

Field control. Here is how method 1, adjusting the field voltage, works. As the field voltage is increased, by reducing R_v in Fig. 16–1(b), for example, the field current is increased. This results in a stronger magnetic field, which induces a greater counter-EMF in the armature winding. The greater counter-EMF tends to oppose the applied dc voltage and thus reduces the armature current, I_A. Therefore, an increased field current causes the motor to slow down until the induced counter-EMF has returned to its normal value (approximately).

Going in the other direction, if the field current is reduced, the magnetic field gets weaker. This causes a reduction in counter-EMF created by the rotating armature winding. The armature current increases, forcing the motor to spin faster, until the counter-EMF is once again approximately equal to what it was before. The reduction in magnetic field strength is "compensated" for by an increase in armature speed.

This method of speed control has certain good features. It can be accomplished by a small, inexpensive rheostat, since the current in the field winding is fairly low due to the large R_F. Also, because of the low value of I_F, the rheostat R_v does not dissipate very much energy. Therefore, this method is energy-efficient.

However, there is one major drawback to speed control from the field winding: To increase the speed, you must reduce I_F and weaken the magnetic field, thereby lessening the motor's torque-producing ability. The ability of a motor to create torque depends on two things: the current in the armature conductors and the strength of the magnetic field. As I_F is reduced, the magnetic field is weakened and the motor's torque-producing ability declines. Unfortunately, it is just now that the motor needs all the torque-producing ability it can get, since it probably requires greater torque to drive the load at a faster speed.

Thus there is a fundamental conflict involved with field control. To make the motor spin faster, which requires it to deliver more torque, you must do something which tends to rob the motor of its ability to produce torque.

Armature control. From the torque-producing point of view, method 2, armature control, is much better. As the armature voltage and current are increased (by reducing R_v in Fig. 16–1c), the motor starts running faster, which normally requires more torque. The reason for the rise in speed is that the increased armature voltage demands an increased counter-EMF to limit the increase in armature current to a reasonable amount. The only way the counter-EMF can increase is for the armature winding to spin faster, since the magnetic field strength is fixed. In this instance, the ingredients are all present for increased torque production, since the magnetic field strength is maintained constant and I_A is increased.

The problem with the armature control method of Fig. 16–1(c) is that R_v, the rheostat, must handle the armature current, which is relatively large. Therefore the rheostat must be physically large and expensive, and it will waste a considerable amount of energy.

Of the two methods illustrated in Figs. 16–1(b) and (c), the field control method is usually preferred.

16-2 ■ THYRISTOR CONTROL OF ARMATURE VOLTAGE AND CURRENT

As we saw in Chapter 4, an SCR can perform most of the duties of a rheostat in controlling the average current to a load. Furthermore, an SCR, or any power thyristor, does not have the shortcomings of high-power rheostats. SCRs are small, inexpensive, and energy-efficient. It is therefore natural to match the dc shunt motor and the SCR to provide armature control of motor speed. The general layout of an SCR speed-control system is illustrated in Fig. 16–2.

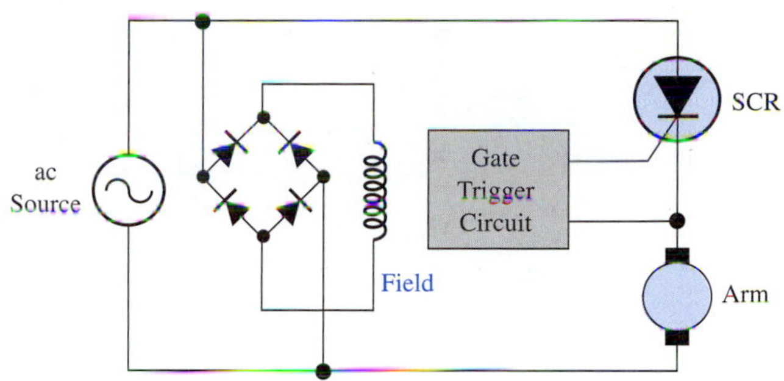

FIGURE 16–2
SCR in series with the armature to control motor speed.

In Fig. 16–2, the ac power is rectified to produce dc power for the field winding. The SCR then provides *half-wave* rectification and control to the armature winding. By firing the SCR early, the average armature voltage and current are increased, and the motor can run faster. By firing the SCR later (increasing the firing delay angle), the average armature voltage and current are reduced, and the motor runs slower. The gate trigger control circuit can be either an open-loop circuit or a closed-loop, automatically correcting circuit.

Of course, Fig 16–2 is not the only arrangement of an armature with an SCR that is acceptable. Either of the circuits shown in Figs. 4–11 or 4–13(b) will also work, with the motor armature being the circuit's load. The circuits of Figs. 4–11 and 4–13(b) may even be preferable to the circuit in Fig. 16–2 because they provide full-wave power control instead of half-wave control.

16-3 ■ SINGLE-PHASE HALF-WAVE SPEED-CONTROL SYSTEM FOR A DC SHUNT MOTOR

Figure 16–3 shows a simple half-wave speed-control circuit for a dc motor. The motor speed is adjusted by the 25-kΩ speed-adjust pot. As that pot is turned up (wiper moved away from ground) the motor speed increases. This happens because the gate voltage relative to ground becomes a greater portion of the ac line voltage, thus allowing the gate-to-cathode voltage to reach the firing voltage of the SCR earlier in the cycle.

As the speed-adjust pot is moved downward, the gate-to-ground voltage becomes a smaller portion of the line voltage, so it takes longer for V_{GK} to reach the value necessary to fire the SCR.

The relationship between speed and firing delay angle for this system is graphed in Fig. 16–3(b). As can be seen, the speed-control action is accomplished in a fairly narrow range of firing-delay adjustment, from about 70° to 110°.

This system has a desirable feature which tends to stabilize the motor speed even in the face of load changes. The feature is called *counter-EMF feedback*. Here is how it works.

Suppose the speed-adjust pot is positioned to provide a shaft speed of 1500 r/min. If the torque load on the motor now increases, there is a natural tendency for the motor to slow down. It does this so that the counter-EMF can decrease slightly, thus allowing an increased armature current to flow. The increased armature current provides the boost in torque needed to drive the heavier load. This is the natural reaction of all motors.

In the system of Fig. 16–3, though, when the counter-EMF decreases, the cathode-to-ground voltage (V_K) decreases, since V_K depends in large part on the counter-EMF generated by the armature winding. If V_K decreases, the firing of the SCR takes place earlier

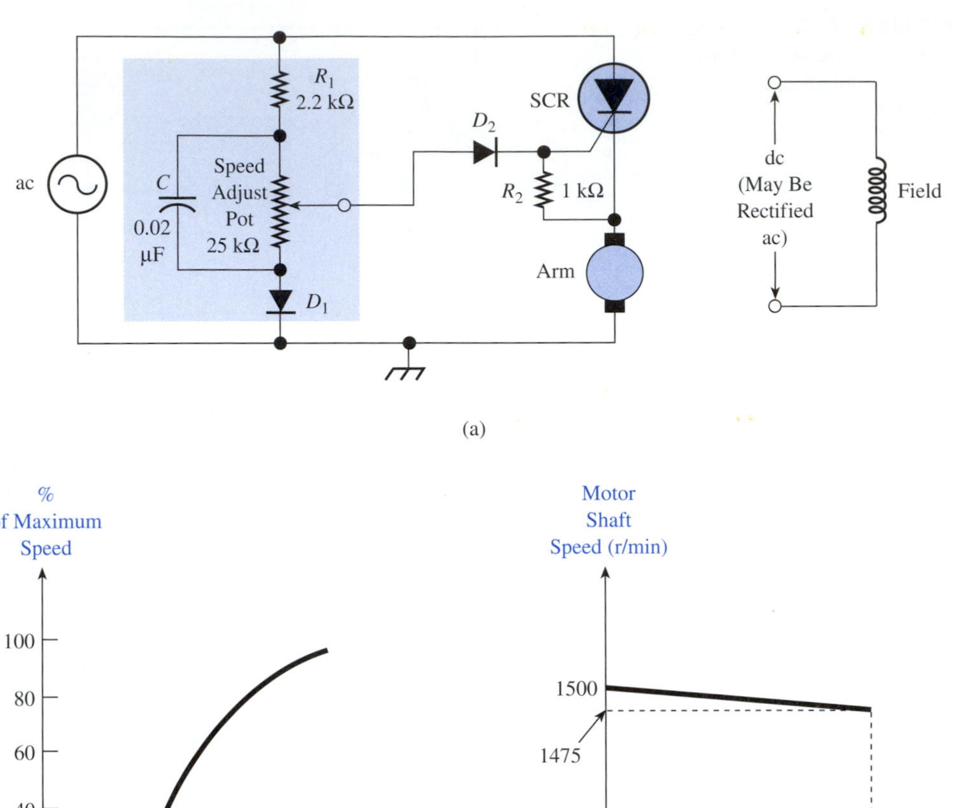

FIGURE 16–3

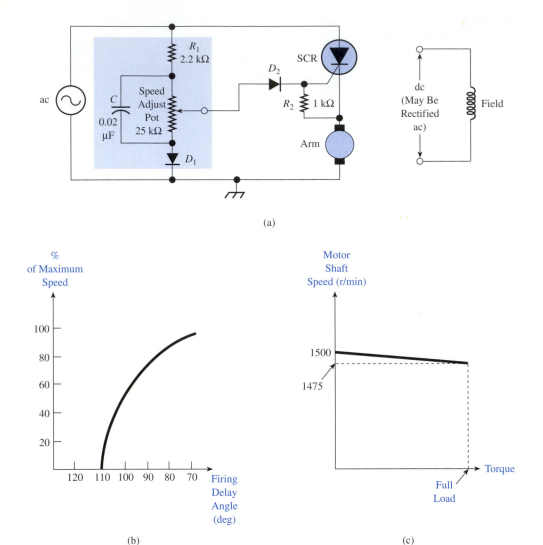

(a) Schematic diagram of a half-wave SCR drive circuit. (b) Graph of shaft speed versus firing delay angle for the circuit of part (a). (c) Graph of shaft speed versus torque for a fixed setting of the speed-adjust pot in part (a).

because V_G does not have to climb as high as before to make V_{GK} large enough to fire the SCR. Therefore, an increase in torque load *automatically* produces a reduction in firing delay angle and a consequent increase in average armature voltage and current. This action holds the motor speed almost constant, even in the face of varying torque load. The graph of motor speed versus load torque is presented in Fig. 16–3(c), assuming an initial no-load shaft speed of 1500 r/min.

The ability of a speed control system to maintain fairly constant motor speed in the face of varying load is called its *load regulation*. In formula form, the load regulation is given as

$$\text{load reg.} = \frac{S_{NL} - S_{FL}}{S_{FL}} \tag{16-4}$$

where S_{NL} stands for the rotational speed at *no load.* The phrase no load means that the load countertorque tending to slow down the motor shaft equals zero. S_{FL} stands for the rotational speed at *full load,* meaning that the load countertorque tending to slow down the motor shaft is at maximum. It can be seen from Eq. (16-4) that the smaller the change in speed from the no-load condition to the full-load condition, the smaller the load regulation. Therefore, the smaller the load regulation figure, the better the control system.

The drive system of Fig. 16–3 provides good load regulation. This is another advantage over the speed control methods described in Sec. 16-1.

As a specific example of calculating load regulation, refer to Fig. 16–3(c). We can see that the no-load speed is 1500 r/min and that the full-load speed is 1475 r/min. Therefore, the load regulation is given by

$$\text{load reg.} = \frac{1500\ \text{r/min} - 1475\ \text{r/min}}{1475\ \text{r/min}} = 0.017 \text{ or } 1.7\%$$

For many industrial applications, a load regulation of 1.7% is quite adequate.

16-4 ■ ANOTHER SINGLE-PHASE SPEED-CONTROL SYSTEM

Figure 16–4 shows another speed-control circuit. Here is how it operates.

The incoming ac power is rectified in a full-wave bridge, whose pulsating dc output voltage is applied to the field winding and to the armature control circuit. Capacitor C is charged by current flowing down through the low-resistance armature winding, through D_2 and the speed-adjust pot, and on to the top plate of the capacitor. The capacitor charges until it reaches the breakover voltage of the SUS. At that instant, the SUS allows part of the capacitor charge to be dumped into the gate of the SCR, firing the SCR. The firing delay angle is determined by the resistance of the speed-adjust pot, which sets the charging rate of C.

Diode D_3 suppresses any inductive kickback which is produced by the inductive armature winding at the completion of each half cycle. When the SCR turns OFF at the end of a half cycle, current continues circulating in the armature-D_3 loop for a short while. This dissipates the energy stored in the armature inductance.

The purpose of the R_1-D_1 combination is to provide a discharge path for capacitor C. Recall that an SUS does not break back all the way to 0 V when it fires. Therefore the

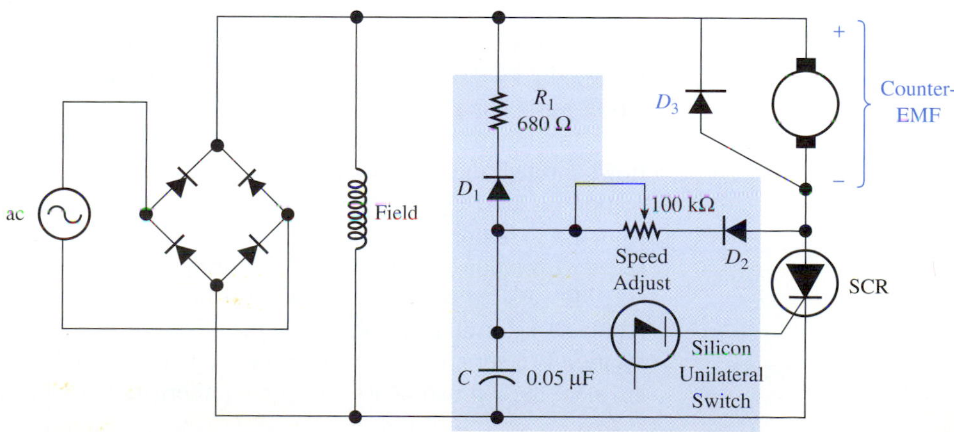

FIGURE 16–4 Another SCR drive circuit. The load regulation of this circuit will be superior to the load regulation in Fig. 16–3.

capacitor is not able to dump *all* its charge through the gate-cathode circuit of the SCR. Some of the charge remains on the top plate of C, even after the SCR has fired. As the dc supply pulsations approach 0 V, the remaining charge on C is discharged through R_1 and D_1 into the field winding. Therefore the capacitor starts with a clean slate on the next pulsation from the bridge.

This system also provides counter-EMF feedback, and it therefore has good load regulation. Here is how the counter-EMF feedback works.

Suppose the speed-adjust pot is set to give a shaft speed of 2000 r/min at a certain torque load. If the load should increase for some reason, the first thing the motor wants to do is slow down a little to admit more armature current. When this happens, the armature's counter-EMF decreases a bit.

As the counter-EMF decreases, the voltage available for charging capacitor C increases. This happens because the voltage available for charging C is the difference between the bridge pulsation voltage and the counter-EMF created by the armature. This can be understood by referring to the counter-EMF polarity markings in Fig. 16–4.

With more voltage available to charge C, it is natural that C will charge to the firing voltage sooner, thus increasing the average voltage supplied to the armature. This corrects the tendency of the motor to slow down, and brings it back up to virtually the same speed as before.

16-5 ■ REVERSIBLE SPEED CONTROL

Some industrial speed-control applications require that the rotation of a motor be *reversible*. That is, the motor must be able to spin either clockwise or counterclockwise, besides having adjustable speed. Reversal of the direction of rotation can be accomplished in either of two ways:

1. Reversing the direction of the armature current, leaving the field current the same.
2. Reversing the direction of the field current, leaving the armature current the same.

The circuits of Fig. 11–27 show how armature current direction can be reversed in a half-wave control system. Figure 16–5 shows how armature current can be reversed in a full-wave speed-control system. The most straightforward method of reversing armature or field current is by the use of two separate motor starter contactors. The *forward* contactor causes current to go through the armature in one direction, while the other contactor, the *reverse* contactor, causes current to flow in the opposite direction.

In Fig. 16–5(a), the FOR contactor is energized by pressing the FORWARD START pushbutton. As long as the REV contactor is dropped out at that time, the FOR contactor will energize and seal itself in around the N.O. pushbutton switch. The operator can then release the FORWARD START button, and the contactor will remain energized until the STOP pushbutton is pressed.

In Fig. 16–5(b), it can be seen that when the FOR contacts are closed, the current flows through the armature from bottom to top, thereby causing rotation in a certain direction (assume clockwise). When the REV contacts are closed, armature current flows from top to bottom, thus causing rotation in the counterclockwise direction. As usual, the speed of rotation is controlled by the firing delay of the SCRs.

Reversible full-wave control can be accomplished without the use of switchgear (contactors, pushbuttons, etc.) with the circuit of Fig. 16–6. In Fig. 16–6, the direction of rotation is determined by which trigger circuit is enabled. If the forward trigger circuit is enabled, the top two SCRs will fire on alternate half cycles of the ac line, and they will

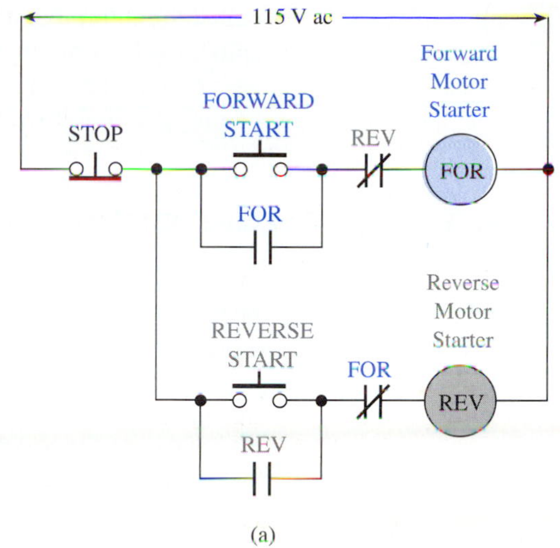

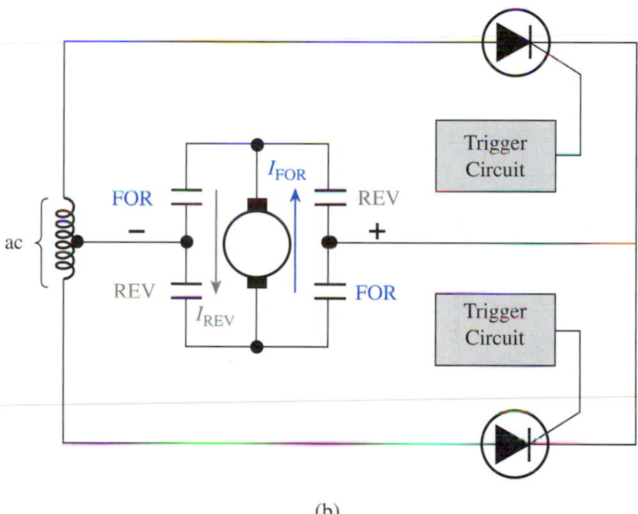

FIGURE 16–5 Reversible full-wave SCR drive system: (a) motor starter control circuit; (b) armature circuit. The SCRs fire on alternate half cycles, causing the armature voltage to have the polarity shown. The armature current direction depends on whether the FOR or REV contacts are closed.

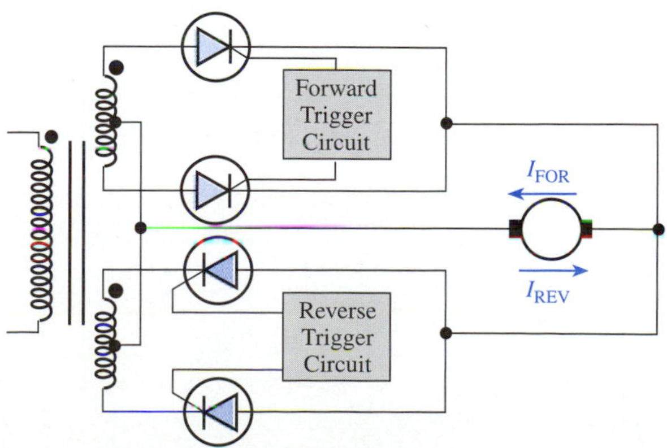

FIGURE 16–6 All-solid-state reversible full-wave drive system.

pass current through the armature from right to left. If the reverse triggering circuit is enabled, the bottom two SCRs will fire on alternate half cycles of the ac line, and they will pass current through the armature from left to right, as indicated. The method of enabling one trigger circuit while disabling the other has not been shown in Fig. 16–6.

16-6 ■ THREE-PHASE DRIVE SYSTEMS FOR DC MOTORS

For dc motors larger than about 10 hp, a three-phase drive system is superior to a single-phase system. This is because a three-phase system provides more pulsations of armature voltage per cycle of the ac line and thus gives greater average armature current flow.

The simplest possible three-phase system is illustrated in Fig. 16–7(a). Although this system gives only half-wave control, it is capable of keeping current flowing through the armature continually. It can do this because when any one phase goes negative, at least one of the other phases is bound to be positive. If a certain phase is driving the armature, at the instant it reverses polarity, one of the other two phases is ready to take over. Thus it is possible to keep armature current flowing continually.

FIGURE 16–7
(a) Three-phase drive system containing a neutral fourth wire. (b) Three-phase drive system with no neutral wire. The rectifier diodes complete the circuit of the armature loop.

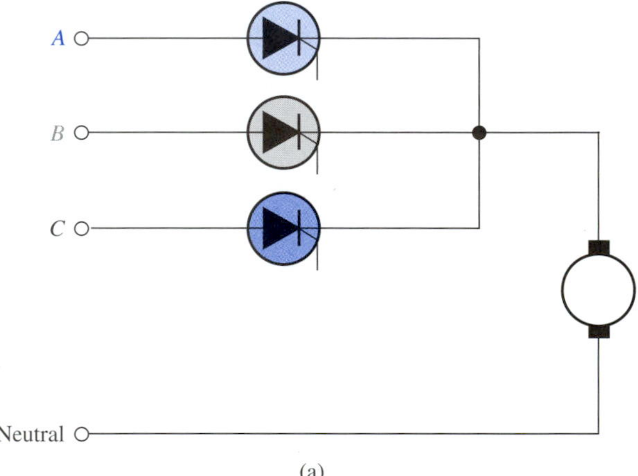

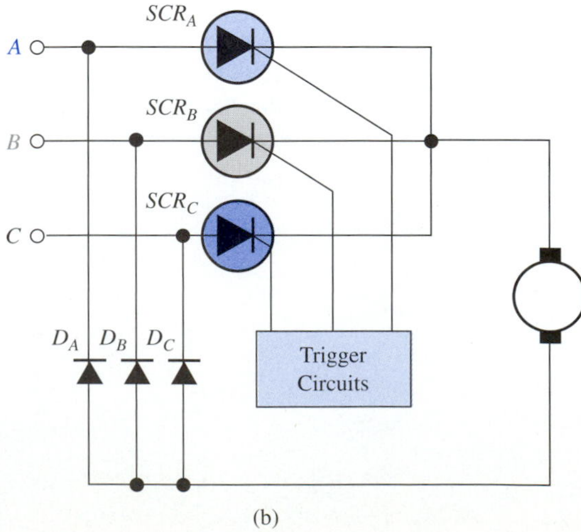

If no neutral fourth wire is available, three-phase half-wave control can be accomplished by the addition of three rectifier diodes, as shown in Fig. 16–7(b). During the time that line voltage *AB* is driving the armature, the flow path is down line *A*, through SCR_A, through the armature, and through D_B into line *B*. When line voltage *BC* is driving the armature, the armature current passes through SCR_B and D_C. At the instant that line voltage *CA* is driving the armature, current passes through SCR_C, through the armature, and back into line *A* through diode D_A.

16-7 ■ AN EXAMPLE OF A THREE-PHASE DRIVE SYSTEM

Figure 16–8 shows the complete schematic diagram of a three-phase drive system. The 230-V, three-phase power is brought in at the upper left in that figure. *Thyrectors* are installed across each of the three phases to protect the solid-state drive circuits from high-

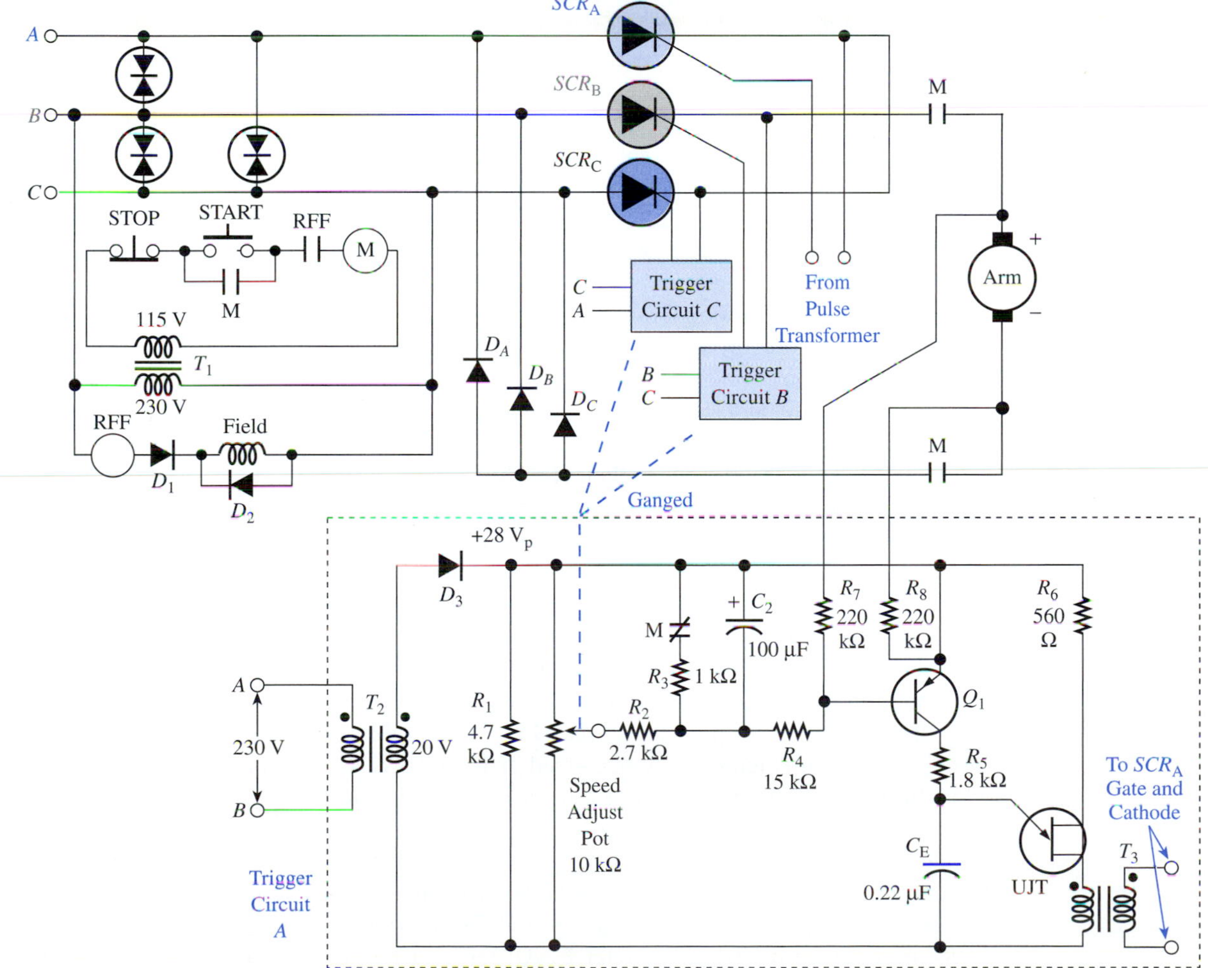

FIGURE 16–8
Complete schematic diagram of a three-phase drive system. Trigger circuits *A*, *B*, and *C* are all identical, so only trigger circuit *A* is drawn. Pulse transformer T_3 in the trigger circuit is wired to the gate and cathode of the corresponding SCR. This drive system would have very good load regulation due to the counter-EMF feedback to Q_1.

voltage transient surges which may appear on the power lines. A thyrector acts like two zener diodes connected back to back. If any momentary voltage surge appears which exceeds the breakdown voltage of the thyrector, the thyrector shorts out the excess. That is, the thyrector acts like a zener diode in that it will allow only a certain amount of voltage to appear across its terminals. If a greater voltage tries to appear, the thyrector becomes a short circuit to any amount of voltage in excess of its rating. In this example, with 230 V rms between the ac power lines, the peak voltage is about 324 V. The thyrectors would be chosen to have a breakdown voltage rating somewhat larger than the peak line voltage. In this case they would probably have a rating of about 400 V.

The 230-V ac appearing between power lines *B* and *C* (V_{BC}) is half-wave rectified by diode D_1 and applied to the motor field winding. Diode D_2, called a *free-wheeling diode,* allows the field winding current to continue flowing during the negative half cycle of the V_{BC} voltage. When V_{BC} enters its negative half cycle, the field winding induces a voltage which is positive on the right and negative on the left. This induced voltage forward-biases D_2, causing current to flow through the loop comprised of the field winding and D_2 in the same direction as before.

Relay RFF is a *field-failure relay*. Its job is to constantly monitor the current in the field winding. Since it must sense current, rather than voltage, its coil is comprised of just a few turns of relatively heavy wire. As long as the field current is flowing, RFF will remain energized. If the field current is interrupted for any reason, RFF drops out. If it does drop out, it causes all power to be removed from the armature winding, as described later. This is necessary because a dc motor can be destroyed if its armature remains powered up when there is no magnetic field present. With no magnetic field present, the armature winding is incapable of generating enough counter-EMF to limit the armature current to a safe value. Under this condition, the armature current would quickly surge up to a destructive level, thereby overheating and ruining the armature conductors and/or insulation. Even if the armature winding could withstand the electrical strain, the rotor speed would increase drastically, in a vain attempt to induce sufficient counter-EMF. This drastic increase in rotational speed can cause mechanical destruction due to overheated bearings or armature windings thrown out of their slots by centrifugal force.

Transformer T_1 steps the 230-V line voltage down to 115 V for use in the START-STOP circuit. This is done to provide safety to the operators who use the START and STOP pushbuttons. When the START button is depressed, motor starter M will be energized, as long as RFF is energized. The auxiliary M contact then seals in around the START button, allowing the START pushbutton to be released. The main M contacts are in series with the motor armature itself; when those contacts close, armature current can flow, and the motor can begin running.

The motor will stop whenever the M coil deenergizes. This will happen if the N.C. STOP pushbutton is temporarily depressed or if the RFF contact opens.

In the speed control circuit proper, SCR_A controls armature current when V_{AB} is in its positive half cycle (past the 60° point). Diode D_B then carries the armature current back to line *B*. During the interval when V_{BC} drives the armature, the armature current flows through SCR_B and D_C. When V_{CA} drives the armature, current flows through SCR_C and D_A. SCR_A is triggered by Trigger Circuit *A*, SCR_B is triggered by Trigger Circuit *B*, and SCR_C is triggered by Trigger Circuit *C*. All three SCRs are fired at approximately the same firing delay angle within their respective conduction cycles. This is accomplished by ganging the three control potentiometers together.

Trigger Circuits *A*, *B*, and *C* are all identical. The schematic diagram of Trigger Circuit *A* is the only one of the three which is drawn schematically in Fig. 16–8.

Trigger circuit. In the upper left of the trigger circuit, transformer T_2 steps V_{AB} down to 20 V ac. The T_2 output voltage is rectified by diode D_3 to provide a 28-V peak supply for the triggering circuitry. To understand the operation of the rest of the trigger circuit, assume for a moment that the N.C. M contact has been open for a while and that C_2 is fully charged. We will return later to see what these components do.

The base current of *pnp* transistor Q_1 is set by the speed-adjust pot. The base current flow path is down from the +28-V supply, through the emitter-base junction of Q_1, through R_4 and R_2, and into the 10-kΩ pot. As the pot wiper is adjusted downward, the voltage available for driving the Q_1 base increases, and the base current increases. As the pot wiper is moved up, Q_1 base current decreases. The Q_1 base current determines the Q_1 collector current, which charges capacitor C_E through R_5. As we saw in Chapter 5, the faster C_E charges, the earlier V_P of the UJT is reached, and the earlier the SCR fires.

This trigger circuit has built-in speed regulation (load regulation) also, just like the circuits of Figs. 16–3 and 16–4. The regulation is provided by 220-kΩ resistors R_7 and R_8. Here is how it works.

Suppose that the speed-adjust pot has been adjusted to provide 2000 r/min at a certain load torque. Recognize that the counter-EMF voltage developed by the armature tends to *reverse-bias* the emitter-base junction of Q_1, through R_7 and R_8. The reverse-bias tendency due to this circuit is overcome by the R_4-R_2-speed-adjust-pot combination, which keeps Q_1 turned on and conducting. If the torque load on the motor shaft increases, the motor slows down a little bit and the counter-EMF decreases. This reduces the reverse-bias tendency through R_7 and R_8. Therefore the Q_1 base current increases a little bit, and the transistor is able to charge C_E faster. Because of this, the UJT fires earlier, and the SCR fires earlier. This raises the average voltage that the SCR applies to the armature, tending to correct the motor speed.

Let us return now to consider the circuit comprised of C_2, R_3, and the N.C. contact of M. The purpose of this circuit is to accelerate the motor up to speed slowly when it is first started. It does this by limiting the armature current for a certain amount of time after the starter is energized. Prior to starter M being energized, the N.C. M contact is closed and is holding capacitor C_2 discharged. There is thus a short circuit across the path consisting of R_4 and the Q_1 base-emitter junction. Q_1 is held cut-OFF at this time.

When the N.C. M contact opens as the motor starter energizes, C_2 begins charging through resistor R_2 and the speed-adjust pot. As C_2 charges, it begins acting more and more like an open circuit. As this takes place, the Q_1 base current slowly builds up to the steady-state value determined by the pot setting. Until that steady-state value is reached, the firing of the UJT and SCR is delayed past the normal firing instant. In this manner the armature current is temporarily retarded for a while after the starter energizes. Therefore the motor accelerates slowly to its set speed, and the great initial inrush of armature current is avoided.

16-8 ■ PULSE-WIDTH-MODULATED CONTROL

Large dc motors are best controlled by high-power thyristors, as described in Secs. 16-3 through 16-7, but small to medium permanent-magnet dc motors and some brushless dc motors are more successfully controlled by series-connected switching transistors operated in a *pulse-width-modulated* manner. Let us first examine the general idea of power control by pulse-width modulation. Then we can discuss specific pulse-width-modulation techniques for dc motor control.

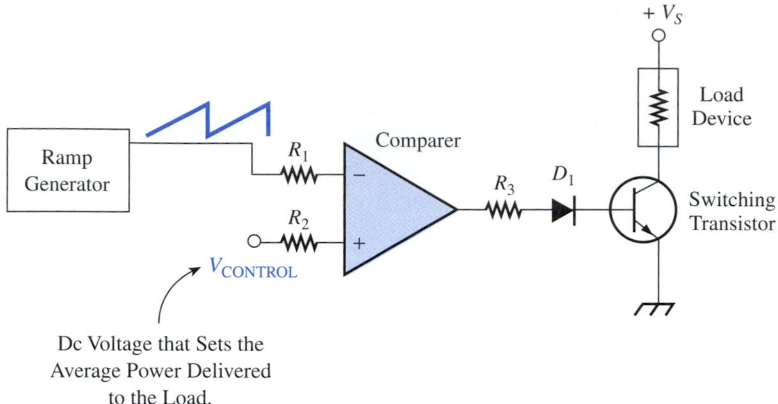

FIGURE 16–9
Basic organization of a pulse-width modulation power-control system.

There are three essential parts in any pulse-width modulation system, as illustrated in Fig. 16–9:

1. A ramp waveform generator, usually operating at a constant frequency.
2. A comparer, for detecting when the ramp voltage has exceeded the control signal voltage.
3. An electronic device that switches the load power at the moment when the comparer detects the critical point on the ramp waveform.

In Fig. 16–9 the comparer is implemented by an op amp, and the electronic switch is a bipolar transistor operating in common-emitter configuration. Here is how it works. Refer to the waveform in Fig. 16–10.

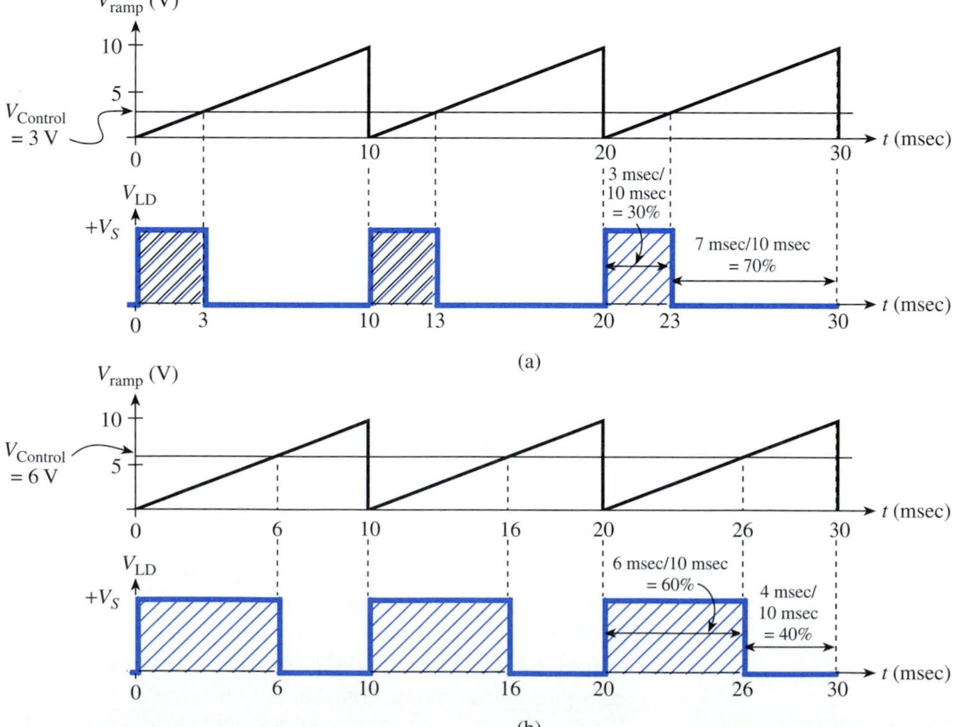

FIGURE 16–10
(a) With a small value of control voltage, the load's duty cycle is low (pulse width is narrow). (b) With a larger value for $v_{Control}$, the load's duty cycle is higher (pulse width is wider). This delivers greater average electrical power, so the load produces a greater amount of its output product.

At the moment that the ramp begins its run-up, the positive control voltage exceeds v_{ramp}. Therefore the op amp comparer produces positive saturation. This saturates the transistor, turning the power-switch ON. Full supply voltage V_S appears across the load. This situation is shown clearly in Fig. 16–10(a).

v_{ramp} increases toward its peak value of 10 V. At the instant when v_{ramp} exceeds v_{control}, the op amp comparer switches to negative saturation. This occurs at 3 V in Fig. 16–10(a). When the comparer output goes negative, it reverse-biases diode D_1 and the B-E transistor junction. Therefore the transistor immediately switches to cutoff, and the load becomes deenergized. It remains deenergized for the remainder of the ramp cycle, from 3 to 10 ms in Fig. 16–10(a).

This process repeats for each cycle of the ramp oscillator, producing an overall load voltage waveform with a 30% duty cycle. Therefore the average load voltage in Fig. 16–10(a) is given by

$$V_{\text{LD(avg)}} = 0.30 \times V_S$$

A larger control voltage is shown in Fig. 16–10(b). Now the ramp must climb to 6 V in order to trigger the comparer and switch OFF the power transistor. Therefore the load's duty cycle increases to 60% and its average voltage increases to

$$V_{\text{LD(avg)}} = 0.60 \times V_S$$

By varying the control voltage, we have varied, or modulated, the pulse width to the load. This varies the load's average voltage and power, which is the idea of pulse-width modulation.

■ EXAMPLE 16-1

In Figure 16–10, suppose $V_S = 50$ V and $R_{\text{LD}} = 5\ \Omega$.

(a) Find the average load power when $v_{\text{Control}} = 3$ V.

(b) Repeat for $V_{\text{Control}} = 6$ V.

Solution. (a) During the time interval that the load is pulsed ON, its power is

$$p_{\text{LD}} = \frac{V_S^2}{R_{\text{LD}}} = \frac{(50\text{ V})^2}{5\ \Omega} = 500\text{ W}$$

However, since the load is pulsed ON only 30% of the total cycle time, the average power is only 30% as large as it would be if the load were ON 100% of the cycle time. Therefore

$$P_{\text{avg}} = 30\% \text{ of } p_{\text{LD(ON)}} = 0.30 \times 500\text{ W} = \mathbf{150\ W}$$

(b) With the ON-pulse time equal to 60% of the cycle period, we have

$$P_{\text{avg}} = 0.60 \times 500\text{ W} = \mathbf{300\ W}$$ ■

If the load were a heating element, it would produce twice as much heat per unit of time under condition (b), compared to condition (a). If it were a lamp load, it would produce twice as much light energy. If it were a motor, it would deliver twice as much mechanical power (product of torque times speed) when pulse-width is modulated at 60% compared to 30%.

Advantage of pulse-width modulation. Transistor-switched pulse-width modulation has a fundamental advantage over a linear transistor amplifier driving the load. It is the same advantage that we get from thyristor switching—better efficiency. Like an SCR, a pulse-width-modulated transistor is either all the way ON, saturated, or all the way OFF, cutoff. When it's ON, its current is large, but its terminal voltage is nearly zero so its internal power consumption is nearly zero. When it's OFF, its terminal voltage drop is large, but its current is essentially zero. So its power consumption is still zero. With the control device consuming nearly zero power, all the power extracted from the dc power supply is delivered to the load. None is wasted by the control system itself. This is quite unlike a linear amplifier, where the load power is controlled by shifting the transistor away from a dc bias condition. A linear transistor consumes from the power supply, and then wastes, an amount of power given by

$$P_{(\text{dc bias})} = V_{\text{CE(bias)}} \times I_{\text{C(bias)}}$$

Filtering the pulse-width modulated waveform. In some applications the rectangular waveforms of Fig. 16–10 are acceptable. In other applications we must smooth out the waveform. Ideally, we can smooth it to a dc voltage equal to the waveform's average value. This idea is pictured in Fig. 16–11.

Figure 16–11(a) shows a low-pass *LC* filter. The inductor in series with the load tends to maintain a steady current through the load. The capacitor in parallel tends to maintain a steady voltage across the load. The *freewheeling* or *kickback* diode *D* provides a complete current path for the inductor-load combination when the transistor turns OFF, and the inductor creates voltage v_L, positive on its bottom, as shown in Fig. 16–11(b). Diode *D* is reverse-biased when the transistor switches ON.

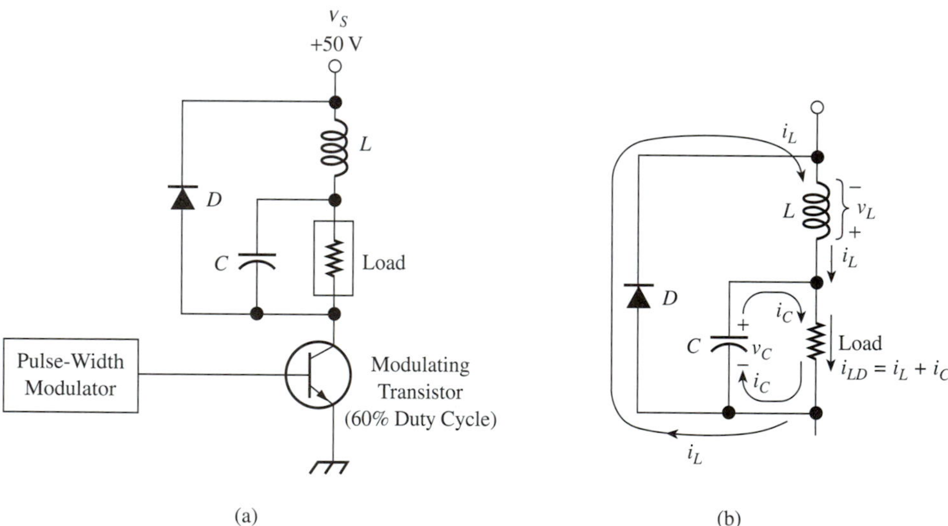

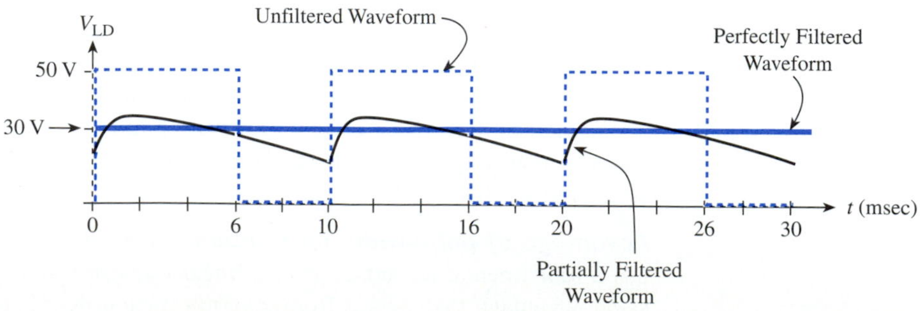

FIGURE 16–11 Filtering the load waveform. (a) Inductance *L* goes in series, and capacitance *C* goes in parallel with the load. (b) Diode *D* provides a path for the inductor component of load current when the transistor is OFF. (c) For pulse width equals 60% of the period, the average value of V_{LD} is $V_{LD} = 0.6 \times 50\text{ V} = 30\text{ V}$.

Generally, a filtering circuit can approach the ideal dc waveform. The more we invest in the filter circuit's construction, the closer we can get to the ideal smooth dc.

16-9 ■ THE TYPE 555 TIMER-OSCILLATOR

One practical means of accomplishing pulse-width modulation is to use a particular integrated circuit called the *five fifty-five*. This name is taken from the type number NE555, that was allocated to it by its inventor, the Signetics Corporation.

The 555 is available in an 8-pin dual-in-line package. A dual-555 unit (two 555s sharing the same dc power supply) is available as a 14-pin DIP, called a 556.

The 555 can perform various functions. To begin understanding its operation, let us first look at its internal model, then put it to work as an *RC* oscillator. Figure 16–12(a)

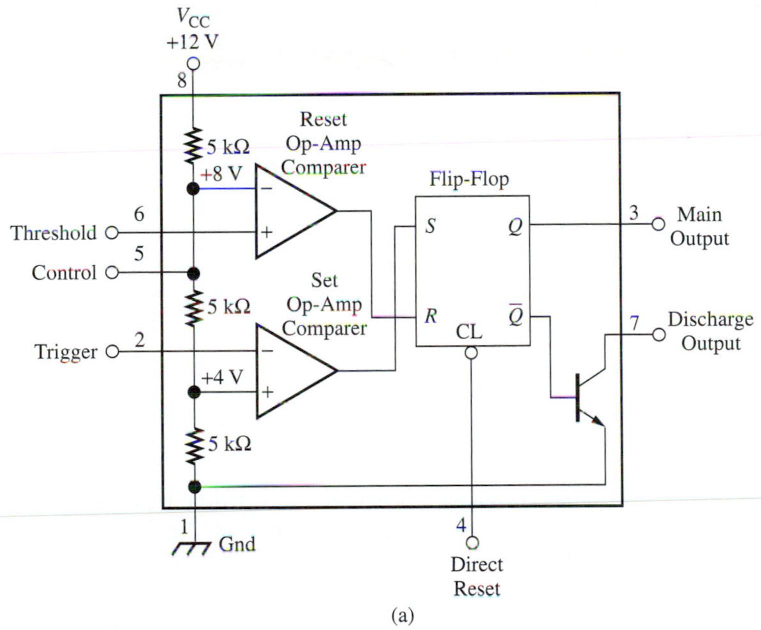

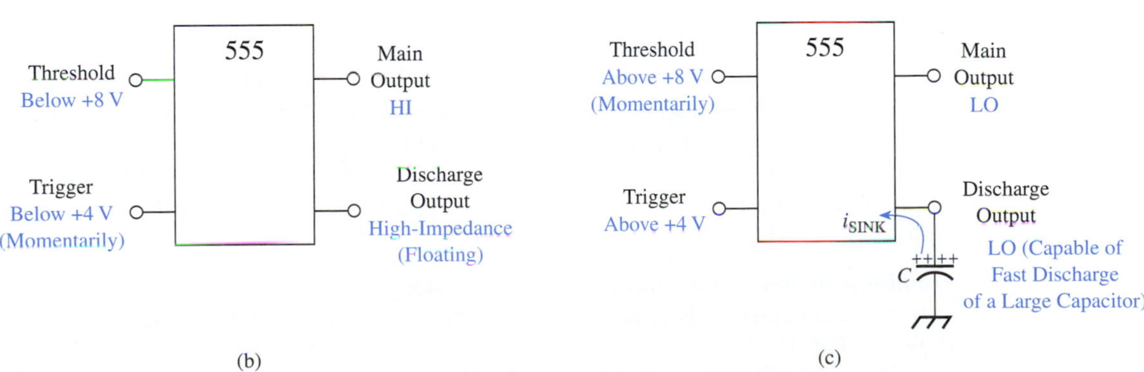

FIGURE 16–12
The 555 Timer/Oscillator. (a) Internal layout. Always refer to this diagram to keep your thoughts straight about 555 operation. (b) Setting the flip-flop—output goes HI. (c) Resetting the flip-flop—output goes back LO.

shows the 555's internal layout, consisting of two op amp comparers, a plain *RS* flip-flop, and a current-sinking transistor. Here is how it functions.

The three-resistor voltage-divider along the left side divides the V_{CC} supply voltage into thirds. Thus, for $V_{CC} = 12$ V, there will be +4 V at the noninverting (positive) input of the Set op amp comparer and +8 V at the inverting input (negative) of the Reset op amp comparer.

16-9-1 Oscillator Mode

As we start our explanation, assume that the flip-flop is reset (Q is LO; Q is HI). Therefore the Main Output at pin 3 is a digital LO, ground potential. Furthermore, assume that the active-LO Direct Reset terminal is disabled (pin 4 wired to inactive HI), and assume that Control is disabled (pin 5 is capacitor-bypassed to ground by 0.01-μF). Thus we have only four terminals to worry about: Trigger (2), Main Output (3), Threshold (6), and Discharge Output (7).

Action begins when the voltage at the Trigger terminal is pulled below +4 V by external circuit action. When this occurs, the Set comparer switches its output from 0 to $+V_{SAT}$, nearly +12 V. This Set signal is applied to the flip-flop's *S* input, as shown in Fig. 16–12(a). Therefore the FF becomes set, sending its Q Main Output to HI (nearly 12 V).

At the same time, the FF brings Q to LO, which cuts off the pull-down transistor. Discharge Output, pin 7, goes to high-impedance—not capable of sinking any current to ground. All the while, the Threshold terminal, pin 6, must be less positive than +8 V so that the Reset comparer is LO. External circuit conditions must be arranged so that this condition is met. Figure 16–12(b) shows the status of the 555 at this point. After the flip-flop has set, it will remain in that state until external circuit action brings about two events:

1. Trigger terminal pin 2 must go back above (more positive than) +4 V, removing the HI from the *S* terminal of FF.
2. Threshold terminal pin 6 must go above +8 V.

When Threshold goes above +8 V, the differential input of the Reset comparer will match the op amp's polarity marks in Fig. 16–12(a). This sends the comparer output to $+V_{SAT}$, placing a HI on the *R* terminal of the flip-flop. The FF resets, bringing Main Output back to ground. Also, the HI on Q drives the sinking transistor into saturation. Therefore the Discharge Output goes to LO, and it is quite capable of sinking a large current to ground if there are any external-circuit capacitors needing a fast discharge. Figure 16–12(c) shows the status of the 555 at this point.

The V_{CC} dc supply voltage in Fig. 16–12 is +12 V. Actually, the 555 can be operated with any dc supply voltage from 5 to 16 V. The Trigger reference voltage will always be $(1/3) \times V_{CC}$, and the Threshold reference will be $(2/3) \times V_{CC}$. For example, for $V_{CC} = 15\text{ V}, V_{\text{Trigger}} = (1/3) \times 15\text{ V} = 5\text{ V}$, and $V_{\text{Threshold}} = (2/3) \times 15\text{ V} = 10\text{ V}$. The HI output voltage from the Main Output is nearly equal to V_{CC}, or +15 V in this example. When operated with a +5-V value for V_{CC}, the 555's Main Output is compatible with TTL and NMOS digital devices.

Oscillator action. It is simple to produce rectangular-wave oscillations from a 555. The only external components required are two timing resistors and one timing capacitor, as shown in Fig. 16–13.

To understand the oscillator's operation, imagine that the voltage across the capacitor, v_{CT}, has just now dropped below 4 V, thereby triggering the FF into the HI state. This is the $t = 0$ instant in Fig. 16–14(a).

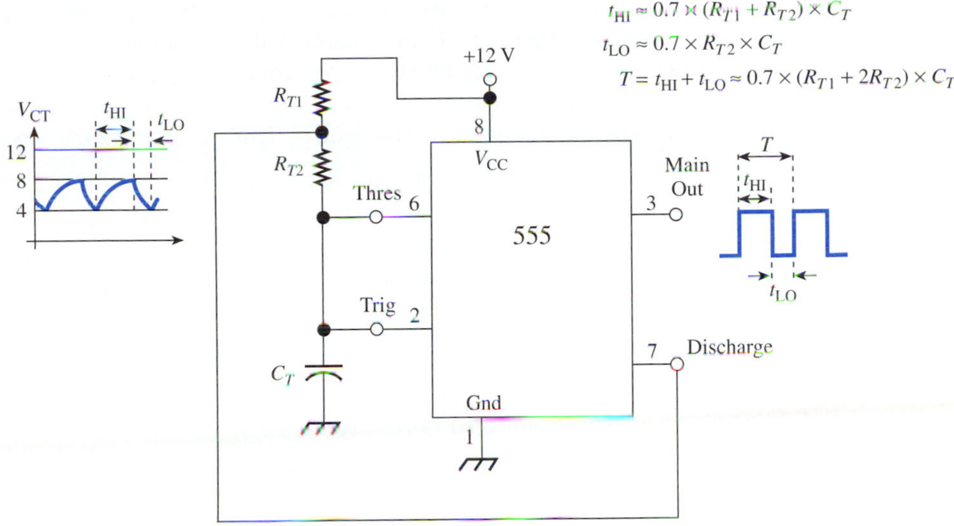

FIGURE 16–13 Schematic arrangement for 555 oscillator. Trigger and Threshold are jumpered together, both inputs receiving capacitor voltage v_{CT}. The Direct Reset and Control terminals are not used. It is best to deactivate Direct Reset (pin 4) by connecting it to the V_{CC} bus, and to deactivate Control by connecting a 0.01-μF cap between pin 5 and ground. These connections are not shown in order to keep the schematic as simple as possible.

With Main Output HI and Discharge Output floating (disconnected from ground), no current can flow into the Discharge terminal. This makes R_{T1}, R_{T2}, and C_T a series combination, with a charging time constant given by

$$\tau_{chg} = (R_{T1} + R_{T2}) \times C_T \tag{16-5}$$

At this moment the capacitor begins charging more positive on top, heading from its instantaneous voltage $v_{CT} = 4$ V toward 12 V, since C_T is connected to the +12-V supply through series resistance $R_{T1} + R_{T2}$. This intention for the capacitor to charge to 12 V is shown by the dashed line in the v_{CT} graph of Fig. 16–14.

Halfway through the charging process, v_{CT} reaches +8 V. This voltage is applied to *both* Trigger and Threshold, since they are jumpered together in Fig. 16–13. When $v_{Threshold}$ reaches +8 V, it switches its internal op amp, resetting the internal flip-flop. Therefore Main Output plunges to LO (ground) as Fig. 16–14 shows. At that same instant the Discharge Output in Fig. 16–13 is also driven LO, causing the R_{T1}-R_{T2} tie-point

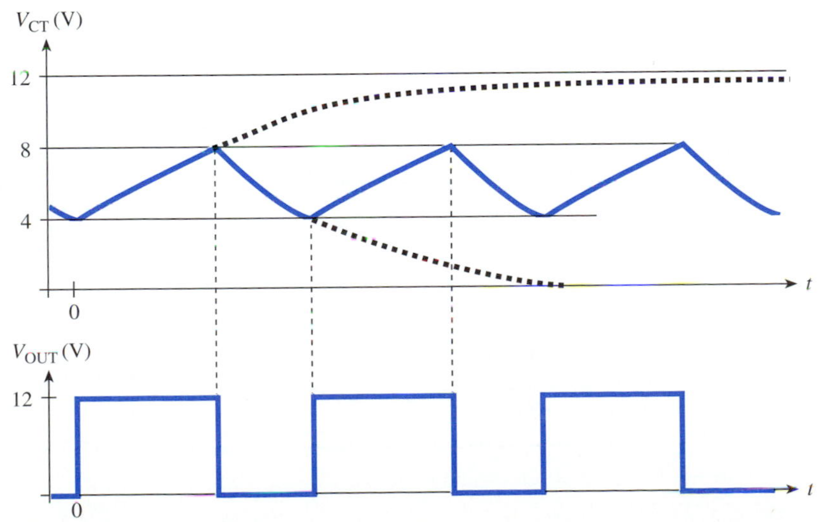

FIGURE 16–14 In oscillator mode, the 555 produces a rectangular waveform at Main Output. The waveform across C_T consists of short pieces of time-constant curves, which are almost linear, like ramps.

to be shorted to ground. This immediately stops the charging of C_T and starts the discharging process. The discharge path is up through R_{T2}, around the outside connection in Fig. 16–13, into the Discharge terminal to ground.

The original intention of C_T to charge all the way to +12 V has been interrupted. The interruption occurred exactly halfway through the charging process, since

$$\frac{\text{Actual change in } v_{\text{CT}}}{\text{Originally intended change in } v_{\text{CT}}} = \frac{8\text{ V} - 4\text{ V}}{12\text{ V} - 4\text{ V}} = \frac{4\text{ V}}{8\text{ V}} = \frac{1}{2}$$

Remember this rule for time-constant transient curves: It requires an elapsed time of 0.7 τ to go halfway through a charging process (verify this in Appendix A). Therefore the amount of time that the 555's Main Output remains HI is given by

$$t_{\text{HI}} = 0.7 \times \tau_{\text{chg}} = 0.7 \times (R_{T1} + R_{T2}) \times C_T \qquad \textbf{(16-6)}$$

which is indicated in Fig. 16–13.

The C_T discharge path is distinguished from the charge path in Fig. 16–15. Part (b) of that figure makes it clear that R_{T2} is the only resistance in the discharge path. Therefore

$$\tau_{\text{dischg}} = R_{T2} \times C_T \qquad \textbf{(16-7)}$$

The capacitor begins discharging with the intention of going all the way from +8 V down to 0 V, as indicated by the falling dashed line in Fig. 16–14. But at the instant it reaches 4 V, the Trigger input switches its op amp and sets the internal flip-flop. The Discharge Output immediately returns to open-circuit condition, interrupting the discharge process. So here too the capacitive transient completes only one half of its original intention, since

$$\frac{\text{Actual } \Delta v_{\text{CT}}}{\text{Originally intended } \Delta v_{\text{CT}}} = \frac{8\text{ V} - 4\text{ V}}{8\text{ V} - 0\text{ V}} = \frac{4\text{ V}}{8\text{ V}} = \frac{1}{2}$$

Therefore the elapsed discharging time is $0.7 \times \tau_{\text{dischg}}$, giving a LO-output time of

$$t_{\text{LO}} = 0.7 \times \tau_{\text{dischg}} = 0.7 \times R_{T2} \times C_T \qquad \textbf{(16-8)}$$

which is indicated beside the output in Fig. 16–13.

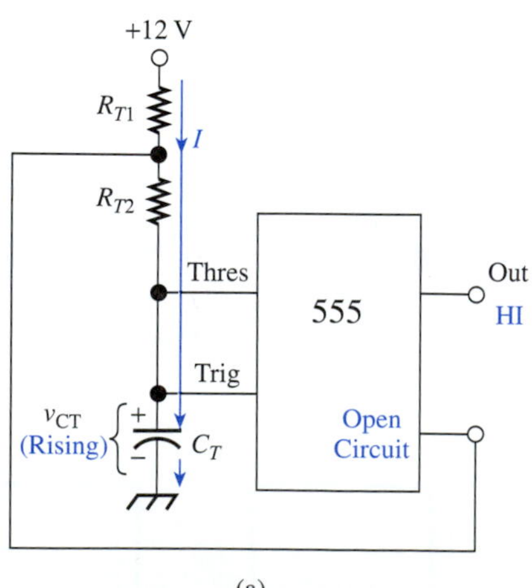

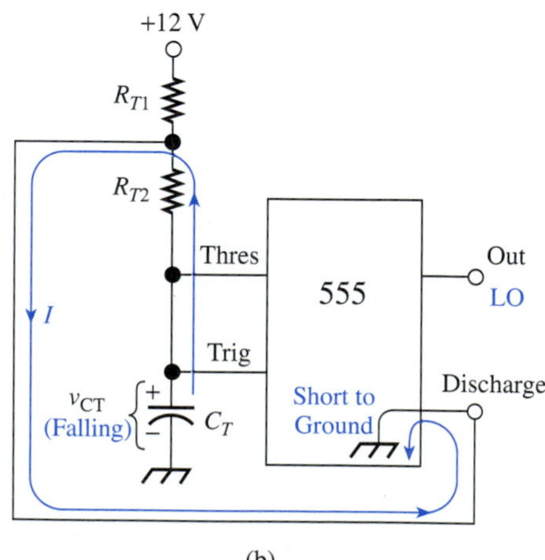

FIGURE 16–15
555 oscillator charge and discharge paths. (a) Charging. v_{CT} is rising; Main Output is HI. (b) Discharging. v_{CT} is falling; Main Output is LO.

■ EXAMPLE 16-2

In Fig. 16–13, suppose $R_{T1} = 2.7\ \text{k}\Omega$, $R_{T2} = 1.8\ \text{k}\Omega$, and $C_T = 0.5\ \mu\text{F}$.

(a) Find the Output-HI time, t_{HI}.
(b) Find the Output-LO time, t_{LO}.
(c) Find the cycle period T and frequency f.
(d) Find the output waveform's duty cycle, in percent.

Solution. (a) From Eq. (16-6)

$$\begin{aligned} t_{\text{HI}} &= 0.7 \times (R_{T1} + R_{T2}) \times C_T \\ &= 0.7 \times (2.7\ \text{k}\Omega + 1.8\ \text{k}\Omega) \times 0.5\ \mu\text{F} = \mathbf{1.58\ ms} \end{aligned}$$

(b) From Eq. (16-8)

$$\begin{aligned} t_{\text{LO}} &= 0.7 \times (R_{T2}) \times C_T \\ &= 0.7 \times (1.8\ \text{k}\Omega) \times 0.5\ \mu\text{F} = \mathbf{0.63\ ms} \end{aligned}$$

(c)

$$\begin{aligned} T &= t_{\text{HI}} + t_{\text{LO}} \\ &= 1.58\ \text{ms} + 0.63\ \text{ms} = \mathbf{2.21\ ms} \\ f &= 1/T = 1/2.21\ \text{ms} = \mathbf{452\ Hz} \end{aligned}$$

(d) For a rectangular wave, duty cycle is given by

$$\begin{aligned} D &= \frac{t_{\text{HI}}}{T} \\ &= \frac{1.58\ \text{ms}}{2.21\ \text{ms}} = 71\% \end{aligned}$$

■ EXAMPLE 16-3

Derive an equation for the frequency of oscillation of the standard 555 astable oscillator shown in Fig. 16–15.

Solution. Using the results from Example 16-2, we can say

$$\begin{aligned} t_{\text{HI}} &= 0.7(R_{T1} + R_{T2})C_T \\ t_{\text{LO}} &= 0.7(R_{T2})C_T \\ T &= t_{\text{HI}} + t_{\text{LO}} = 0.7(R_{T1} + R_{T2})C_T + 0.7(R_{T2})C_T \\ &= 0.7[(R_{T1} + R_{T2}) + R_{T2}]C_T \\ &= 0.7(R_{T1} + 2R_{T2})C_T \end{aligned}$$

Reciprocating T gives

$$f = \frac{1}{T} = \frac{1}{0.7(R_{T1} + 2R_{T2})C_T}$$

$$f = \frac{1.4}{(R_{T1} + 2R_{T2})C_T} \qquad \textbf{(16-9)}$$

■ EXAMPLE 16-4

(a) Explain the operation of the 555 oscillator in Fig. 16–16.
(b) Draw waveforms of v_{OUT} and v_{CT}.

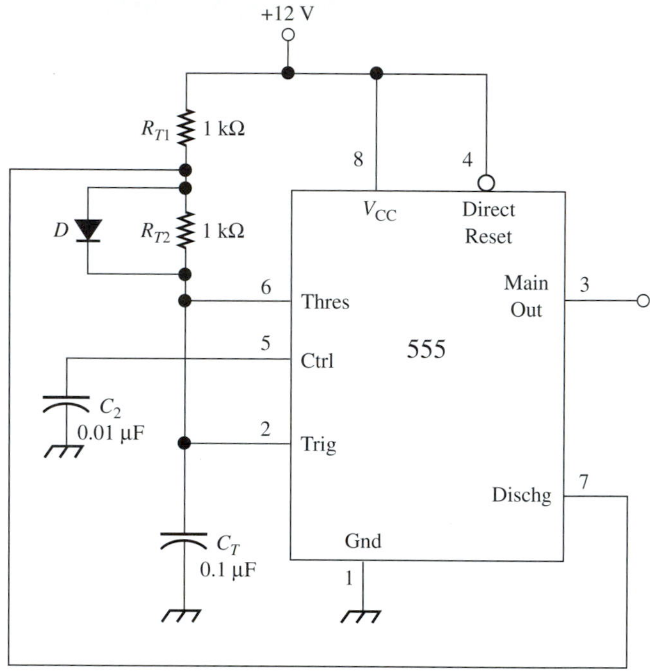

FIGURE 16–16
Variation on the 555 oscillator.

Solution. (a) During capacitor charging, R_{T2} is short-circuited by diode D, which is forward biased. Neglecting the diode's forward voltage, we can write

$$\tau_{chg} = R_{T1} \times C_T$$
$$= 1\ \text{k}\Omega \times 0.1\ \mu\text{F} = 0.1\ \text{ms}$$

Therefore,

$$t_{HI} = 0.7\tau_{chg} = 0.7 \times (0.1\ \text{ms}) = 0.07\ \text{ms}$$

During capacitor discharge, diode D_1 is reverse biased. C_T must discharge through R_{T2} so

$$\tau_{dischg} = R_{T2} \times C_T$$
$$= 1\ \text{k}\Omega \times 0.1\ \mu\text{F} = 0.1\ \text{ms}$$
$$t_{LO} = 0.7\tau_{dischg} = 0.7 \times (0.1\ \text{ms}) = 0.07\ \text{ms}$$

The diode has enabled us to make the HI and LO times equal. This will give an output duty cycle of 50%, which is impossible with the straightforward design of Fig. 16–13.

(b) With $V_{CC} = 10$ V, $V_{Trig} = 3.33$ V and $V_{Thrs} = 6.66$ V. Therefore v_{CT} oscillates between those two values.

The Main Output switches between 0 and 10 V, with period T given by

$$T = t_{HI} + t_{LO}$$
$$= 0.07 \text{ ms} + 0.07 \text{ ms} = 0.14 \text{ ms}$$

The waveforms are drawn in Fig. 16–17. ■

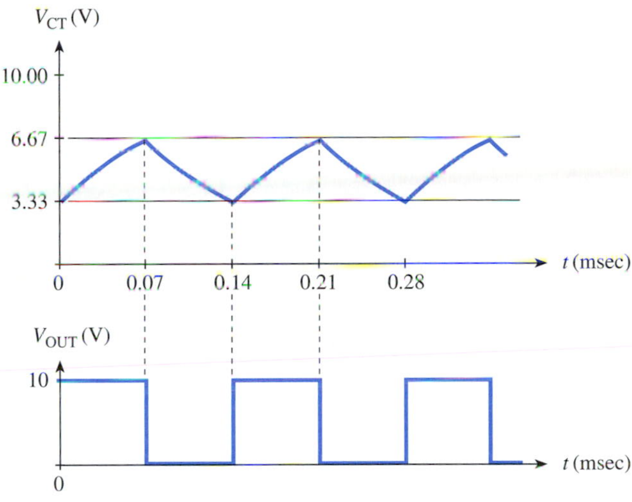

FIGURE 16–17 Waveforms for Fig. 16–16. v_{out} has a 50% duty cycle.

16-9-3 One-Shot Operating Mode

The 555 can also function as a one-shot. To get one-shot operation, we separate the Trigger (pin 2) and Threshold (pin 6) terminals. An external trigger source must deliver a negative-going pulse or spike to the Trigger input, as shown in Fig. 16–18(a).

Here is how the 555 one-shot works. In the resting state the internal flip-flop is reset, making Main Output LO and Discharge shorted to ground. With the top of timing capacitor C_T connected to Discharge (as well as Threshold), the capacitor is held in a discharged state, $v_{CT} = 0$. Any current flowing down through timing resistor R_T goes right to ground via the Discharge transistor. The 555 will remain in this condition indefinitely, waiting for a trigger signal.

When the trigger pulse hits, the Trigger input switches its Set op amp, which forces the flip-flop into set condition. Main Output rises to HI, and Discharge disconnects from ground. Therefore C_T starts charging from 0 V, with charging time constant $\tau_{chg} = R_{T1} \times C_T$. It originally intends to charge all the way to V_{CC}, as shown by the dashed curve in Fig. 16–18(b).

As v_{CT} rises toward V_{CC}, it reaches $^2/_3\ V_{CC}$. This causes $v_{Threshold}$ to switch the Reset op amp, resetting the flip-flop's output to LO, as shown by the v_{OUT} waveform of Fig. 16–18(b). At that same moment Discharge shorts directly to ground, discharging C_T almost instantly.

The duration of the one-shot's output pulse, t_p, is about 1.1 τ. This is because it requires 1.1 time constants to go through 67% of a charging event. We can write

$$t_p \approx 1.1 \times \tau_{chg} = 1.1\ R_T C_T \quad \textbf{(16-10)}$$

FIGURE 16–18
555 used as a one-shot.
(a) Schematic setup, ignoring Control and Direct Reset.
(b) Waveforms.

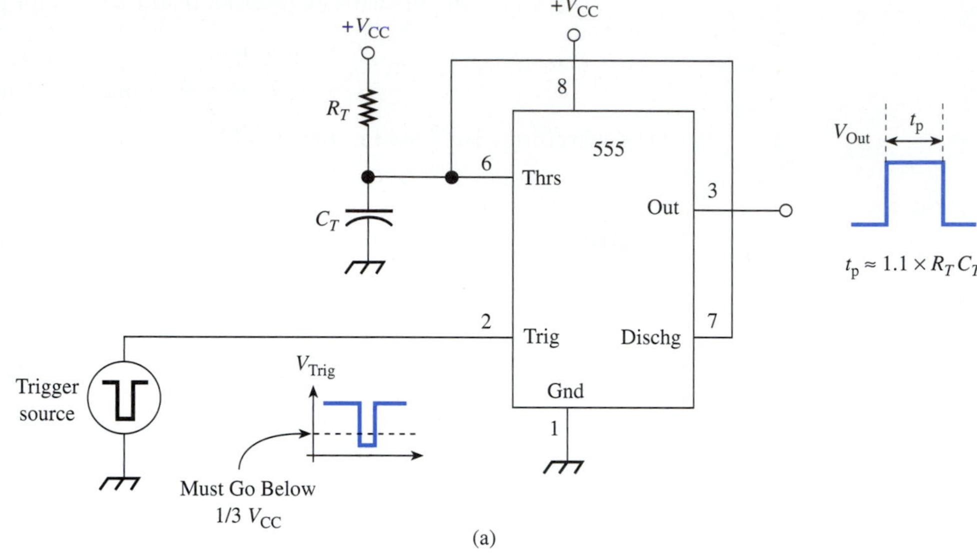

(a)

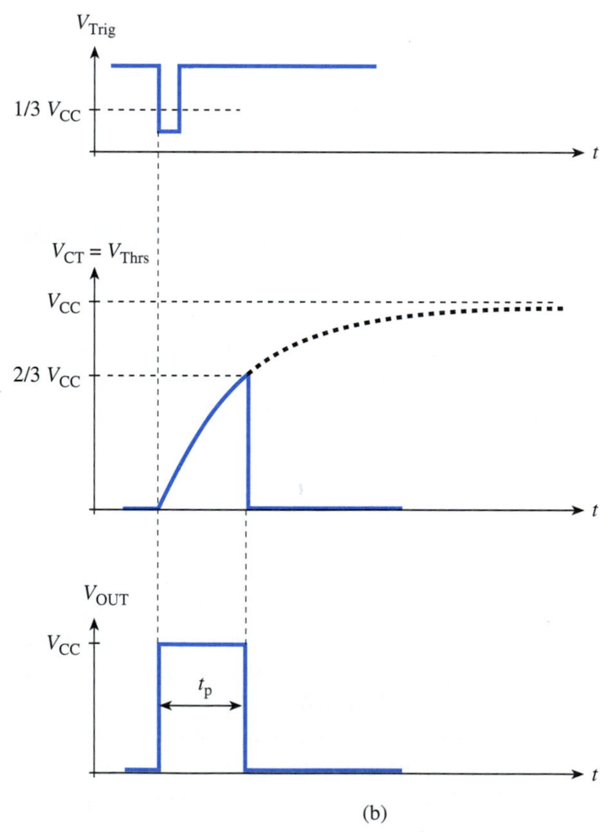

(b)

■ EXAMPLE 16-5

In Fig. 16–18, suppose $V_{CC} = 5$ V, $R_T = 10$ kΩ, and $C_T = 2$ μF.

(a) Describe the nature of the external trigger pulse.

(b) Describe the one-shot's output pulse.

Solution. (a) In order to trigger the 555, Trigger must pass through $\frac{1}{3}V_{CC}$ heading downward. For $V_{CC} = 5$ V, $V_{Trig} = \frac{1}{3}$ (5 V) = 1.67 V. The trigger pulse waveform must rest at a value more positive than +1.67 V, then fall to a value less positive than +1.67 V.

(b) The output pulse rises from 0 V to **5 V.** Its duration is given by Eq. (16-10) as

$$t_p = 1.1\,\tau_{chg} = 1.1\,R_TC_T$$
$$= 1.1\,(10 \times 10^3\Omega)(2 \times 10^{-6}\text{F}) = 1.1(20\text{ ms}) = \mathbf{22\ ms}$$ ■

16-9-4 Pulse-Width Modulation with Two 555s

The control input The normal situation in a 555 is for the three-resistor voltage divider of Fig. 16–12(a) to set the critical trigger voltage at $\frac{1}{3}V_{CC}$ and the critical threshhold voltage at $\frac{2}{3}V_{CC}$. However, this situation can be altered by applying an external voltage to the Control input terminal (pin 5). Until now, we have not used the Control terminal; we have just suppressed noise pickup at its pin by connecting a bypass capacitor to ground.

Figure 16–19 shows an external voltage source applying a variable control voltage, $v_{Control}$. This overrides the normal split-into-thirds operation. Instead it forces the critical threshold voltage to equal $v_{Control}$, and it makes the critical trigger voltage equal to $\frac{1}{2} \times v_{Control}$.

For example, suppose $v_{Control} = 4$ V. Figure 16–19 makes it clear that this value is applied directly to the negative input of the Reset op amp, so it becomes the critical threshold voltage. The bottom two 5-kΩ resistors now voltage-divide $v_{Control}$ into halves. Therefore $\frac{1}{2} \times (4\text{ V}) = 2$ V appears at the positive input of the Set op amp. That 2-V value becomes the new trigger voltage level.

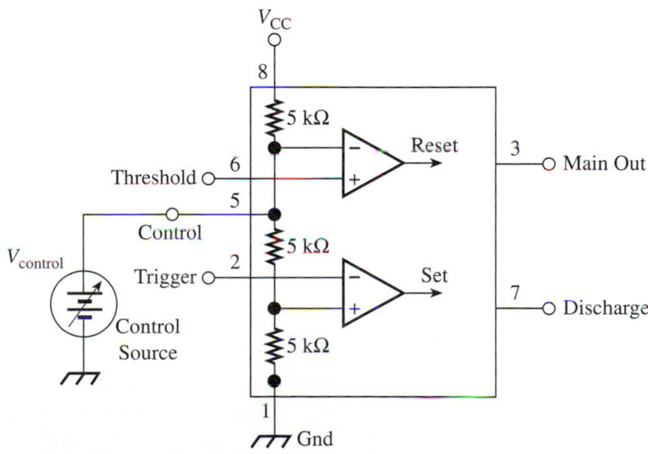

FIGURE 16–19 Using the Control input. $V_{Control}$ must never be larger than $\frac{2}{3}V_{CC}$. Other than that restriction, V_{CC} has nothing to say about the critical Trigger and Threshold voltages.

Variable one-shot. Figure 16–20 shows a variable pulse-width one-shot. The *RC* timing circuit is connected to the Threshold input and Discharge, just as in Fig. 16–18. The Trigger input also is driven in the same manner as in that figure. The only difference is that now v_{Control}, applied to the Control input, sets the critical value $V_{\text{Threshold}}$. Instead of C_T having to charge to $\frac{2}{3}\,V_{CC}$ to terminate the Output pulse, now C_T has to charge to the same value as v_{Control}. Since v_{Control} is not allowed to be greater than $\frac{2}{3}\,V_{CC}$, pulse duration t_{p} will be equal to or less than its normal value of $t_{\text{p}} = 1.1\,R_T C_T$.

■ EXAMPLE 16-6

In Fig. 16–20, suppose $V_{CC} = 15$ V, $R_T = 10$ kΩ, and $C_T = 2$ μF. Referring to the universal-time-constant curve in Appendix A,

(a) Find t_{p} for $V_{\text{Control}} = 4$ V.
(b) Find t_{p} for $V_{\text{Control}} = 8$ V.
(c) Find t_{p} for $V_{\text{Control}} = 10$ V.

Solution. (a) For $V_{\text{Control}} = 4$ V, the timing capacitor C_T must charge to 4 V as it heads toward 15 V; reaching 4 V will terminate the pulse. This is illustrated in Fig. 16–20(b).

The percentage of its "final" value that C_T must reach is given by

$$\frac{4\text{ V}}{15\text{ V}} = 0.27 = 27\%$$

A careful check of the universal-time-constant curve in Appendix A shows that it takes 0.3 time constant to reach 27% of final value. In this circuit, τ_{chg} is the same as in Example 16-5, 20 ms, because we have the same R_T and C_T values. Therefore,

$$t_{\text{p}} = 0.3(\tau_{\text{chg}}) = 0.3 \times 20\text{ ms} = \mathbf{6\text{ ms}}$$

as Fig. 16–20(b) makes clear.

(b) For $V_{\text{Control}} = 8$ V, C_T must climb to 8 V out of 15 V in order to terminate the pulse.

$$\frac{8\text{ V}}{15\text{ V}} = 53\%$$

From the universal-time-constant curve, it requires 0.75 τ to reach 53% of final voltage. Therefore

$$t_{\text{p}} = 0.75 \times 20\text{ ms} = \mathbf{15\text{ ms}}$$

as shown in Fig. 16–20(b).

(c) For $V_{\text{Control}} = 10$ V, the Threshold critical value is its normal value, $\frac{2}{3}$ (15 V). Therefore

$$t_{\text{p}} = 1.1 \times 20\text{ ms} = \mathbf{22\text{ ms}}$$ ■

Example 16-6 is a demonstration of practical pulse-width modulation. Compare the waveforms of the 555 variable one-shot in Fig. 16–20(b) to the theoretical pulse-width-modulation waveforms of Fig. 16–10.

For a complete practical pulse-width-modulated motor-control system, we can construct the circuit of Fig. 16–21. In that figure one 555 runs as an oscillator, using its Main

FIGURE 16–20
(a) Using a 555 as a one-shot with controllable pulse duration, t_p. (b) Waveforms.

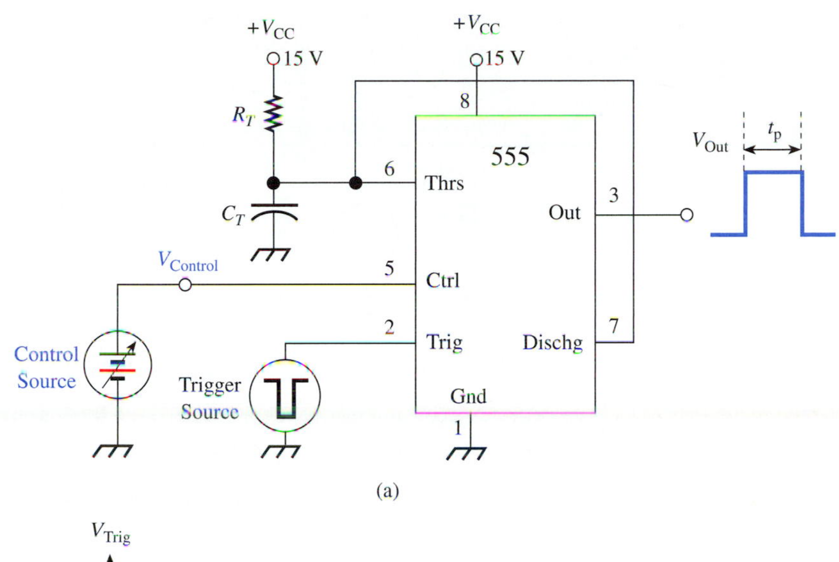

(a)

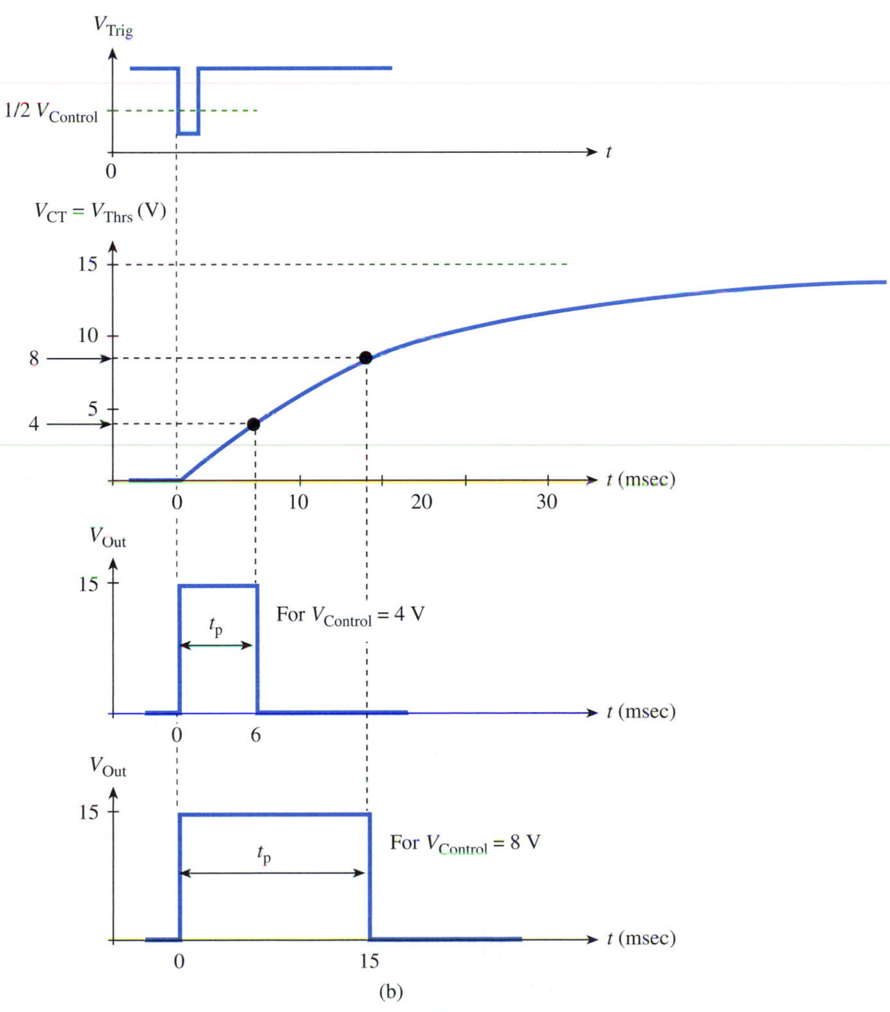

(b)

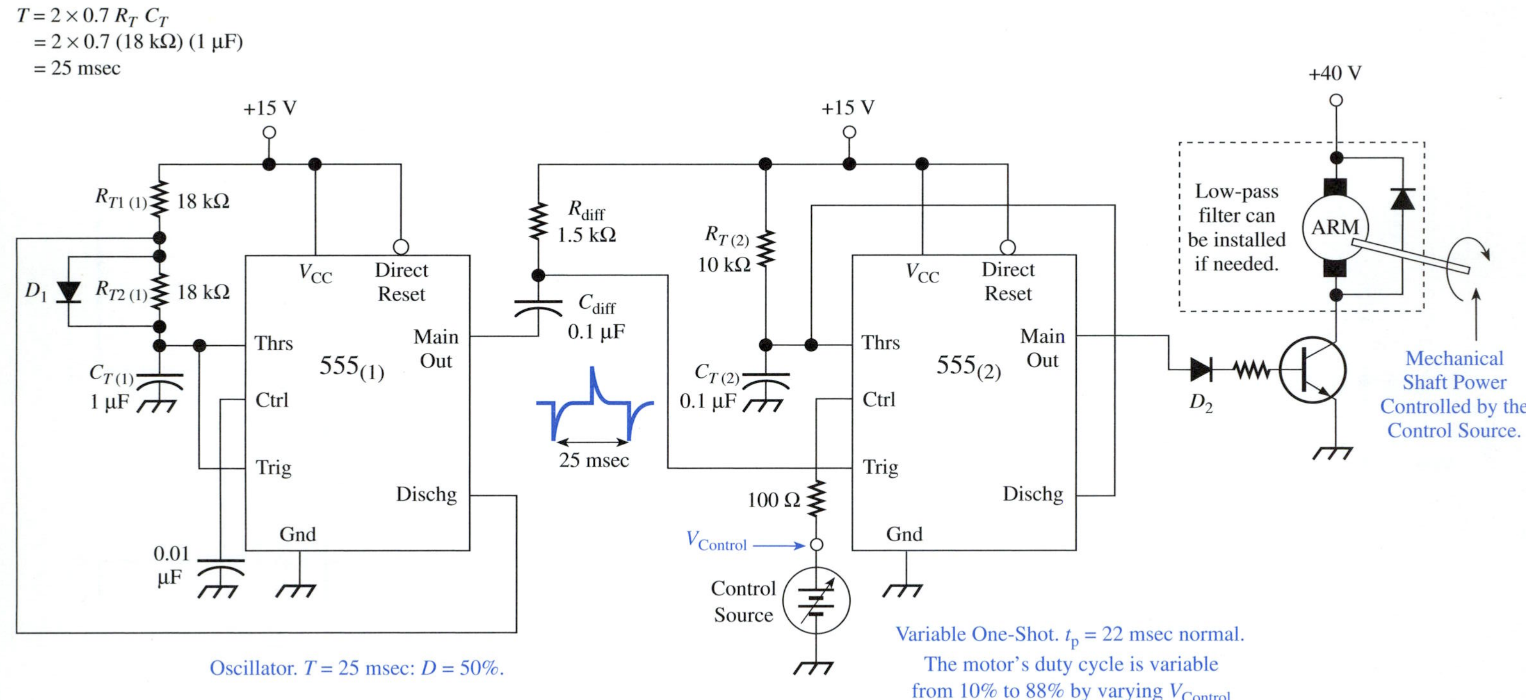

FIGURE 16–21
Pulse-width modulation control system for a small permanent-magnet motor. R_{diff} and C_{diff} form an *RC* differentiator circuit. Every time Main $Out_{(1)}$ goes LO, the differentiator delivers a fast negative-going spike to the Trig input of $555_{(2)}$, with the spike bottoming at 0 V. This initiates another pulse from the one-shot.

Output negative-going edge to trigger a second 555 operating as a variable one-shot. Then we amplify the one-shot's Main Output waveform to drive the armature winding of a permanent-magnet dc motor.

Figure 16–22 shows the waveforms for the pulse-width-modulated control circuit of Fig. 16–21. For a low value of V_{Control}, the load's duty cycle is low. Figure 16–22 shows t_{HI} = 6 ms out of T = 25 ms when V_{Control} = 4 V. The one-shot's (and load's) duty cycle is low, only 6 ms/25 ms = 24%. Low average power is delivered to the motor armature, so the motor creates low torque and runs its mechanical load at a low speed.

For a larger value of V_{Control}, the load duty cycle and average load power increase. In Fig. 16–22, with V_{Control} = 8 V, the one-shot's t_{p} increases to 15 ms, giving a load duty cycle of 15 ms/25 ms = 60%. This drives the armature winding much harder, creating more torque and spinning the shaft at a higher speed.

A maximum value of V_{Control}, 10 V, results in the maximum average load power. The duty cycle = 22 ms/25 ms = 88% in Fig. 16–22.

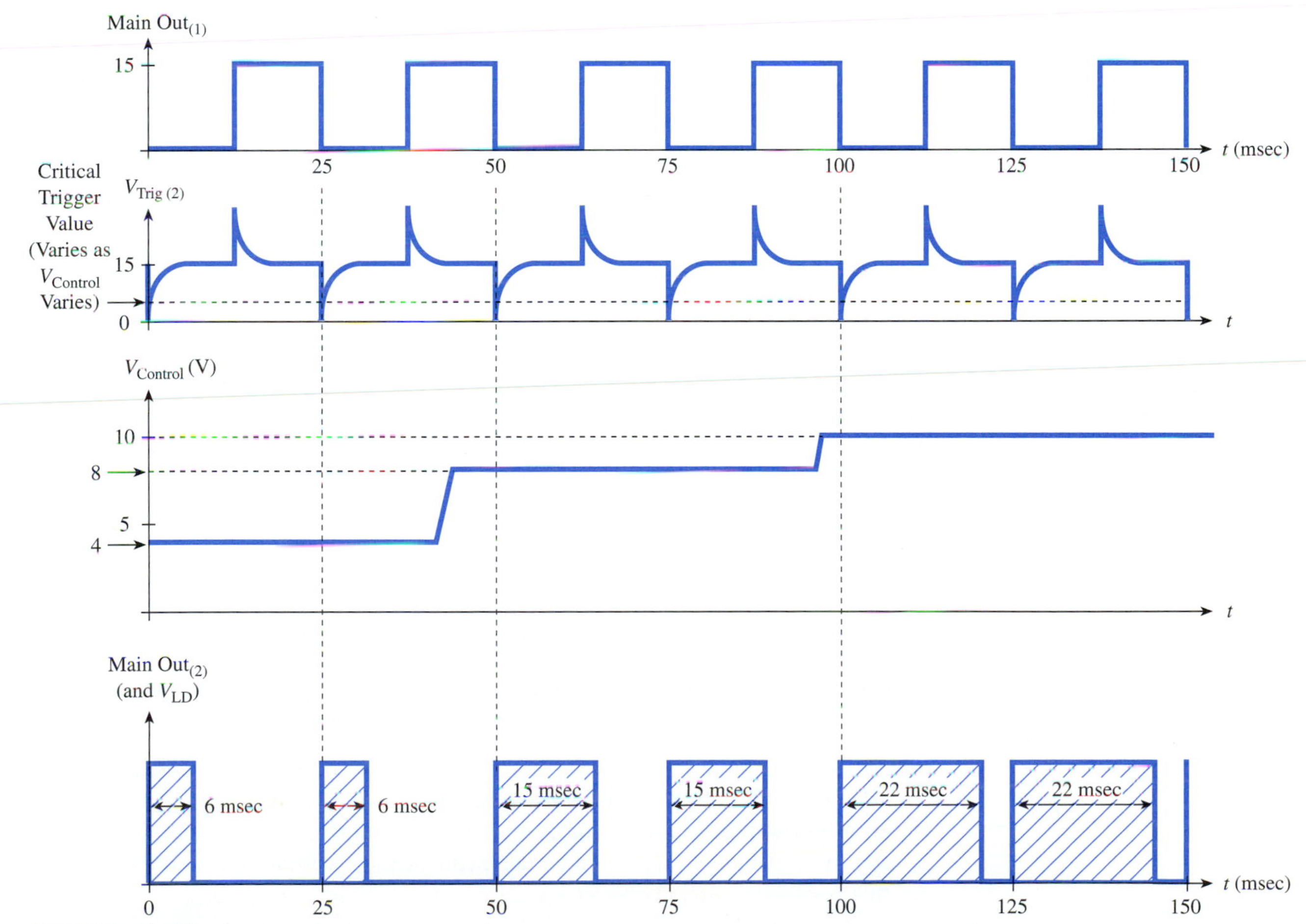

FIGURE 16–22
Waveforms for the pulse-width-modulated motor-control system shown in Fig. 16–21.

■ EXAMPLE 16-7

In the pulse-width-modulation controller of Fig. 16–21, what value of control voltage would be required to produce a load duty cycle of 50%?

Solution. Fifty percent of the oscillator's period is $0.50 \times 25 \text{ ms} = 12.5 \text{ ms}$. This equates to a number of one-shot charging time constants given by

$$\frac{12.5 \text{ ms}}{20 \text{ ms per } \tau} = 0.625\ \tau_{\text{chg}}$$

From the universal-time-constant curve, the percentage of final voltage reached after $0.62\ \tau$ is 46%. Therefore V_{Control} must be 46% of the capacitor's 15-V target voltage, for

$$V_{\text{Control}} = 0.46 \times 15 \text{ V} = \mathbf{6.9\ V}$$ ■

16-10 ■ VARIABLE-FREQUENCY INVERTERS

Inherently, ac motors are not as well suited to variable-speed applications as are dc motors, because their speed cannot be satisfactorily controlled by simple supply-voltage variation. Reducing the supply voltage to a 60-Hz three-phase induction motor will reduce its speed all right, but it also drastically worsens the motor's speed-regulating ability. That is, an ac induction motor operating at reduced voltage is unable to maintain a reasonably steady shaft speed in the face of slight changes in torque demand imposed by the mechanical load.

Satisfactory speed control of an ac induction motor can be accomplished only by varying the supply frequency while varying the supply voltage. If the source of power is the 60-Hz ac utility line, frequency variation is a much more difficult task than voltage variation. Nevertheless, we are sometimes willing to take the trouble to build a variable-frequency drive circuit for controlling the speed of an ac motor so that we can take advantage of certain intrinsic superior characteristics of ac machines. These are the intrinsic superiorities of ac induction motors over dc motors:

1. An ac induction motor has no commutator or friction-slip type electrical connections of any kind. It is therefore easier to manufacture and less expensive than a dc machine. With no brushes to wear down, its maintenance costs are lower.
2. Because it has no mechanical commutator, an ac motor produces no sparks; it is therefore safer.
3. With no electrical connections exposed to the atmosphere, an ac motor holds up better in the presence of corrosive gases.
4. An ac motor tends to be smaller and lighter than a comparable-power dc motor.

There are two basic methods of producing a variable-frequency, high-power, three-phase source for adjusting the speed of an industrial ac induction motor:

1. Convert a dc source into three-phase ac by firing a bank of SCRs in a certain sequence at a certain rate. A circuit which does this is called an *inverter*.
2. Convert a 60-Hz three-phase ac source into a lower-frequency three-phase ac source, again by firing banks of SCRs in a certain sequence at a certain rate. A circuit which does this is called a *cycloconverter*.

We'll deal with inverters now and take on cycloconverters in Sec. 16-12.

Figure 16–23(a) is a schematic diagram of a three-phase inverter driving a wye-connected induction motor. The SCR gate-triggering circuits are not shown, since we are con-

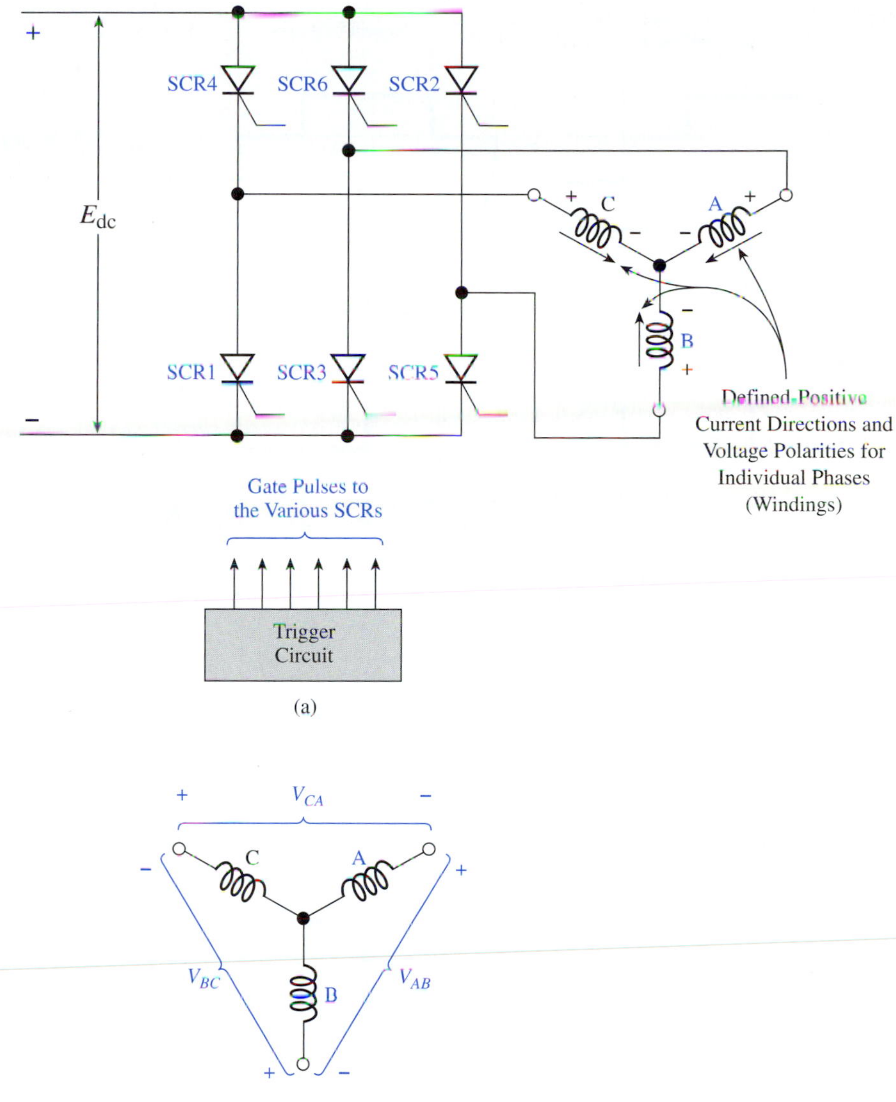

FIGURE 16–23
(a) A three-phase inverter driving a wye-connected motor. The SCR numbering convention has been chosen so that the triggering sequence is in ascending order (or descending order, if the motor is reversible). (b) The defined-positive line voltage polarities.

centrating on the action of the SCRs' main terminals at this time. We want to understand how turning the SCRs ON and OFF in the proper sequence causes the dc supply to be switched across stator windings* A, B, and C of the motor in such a way that a rotating magnetic field is created, thus duplicating the action of a three-phase ac source.

To understand how this happens, it is necessary to refer to the schematic diagram of Fig. 16–23(a), the stator winding current and voltage waveforms (the top three waveforms) of Fig. 16–24(a), the SCR switching sequence of Fig. 16–24(b), and the magnetic field vector diagrams of Fig. 16–24(c).

Because of the physical placement of stator winding A, it produces a magnetic field component oriented from the 60° mechanical position, when current is flowing through

*Stator windings A, B, and C of a three-phase machine are often referred to as *phases A, B,* and *C.* In this context, we will use the terms *stator winding* and *phase* interchangeably.

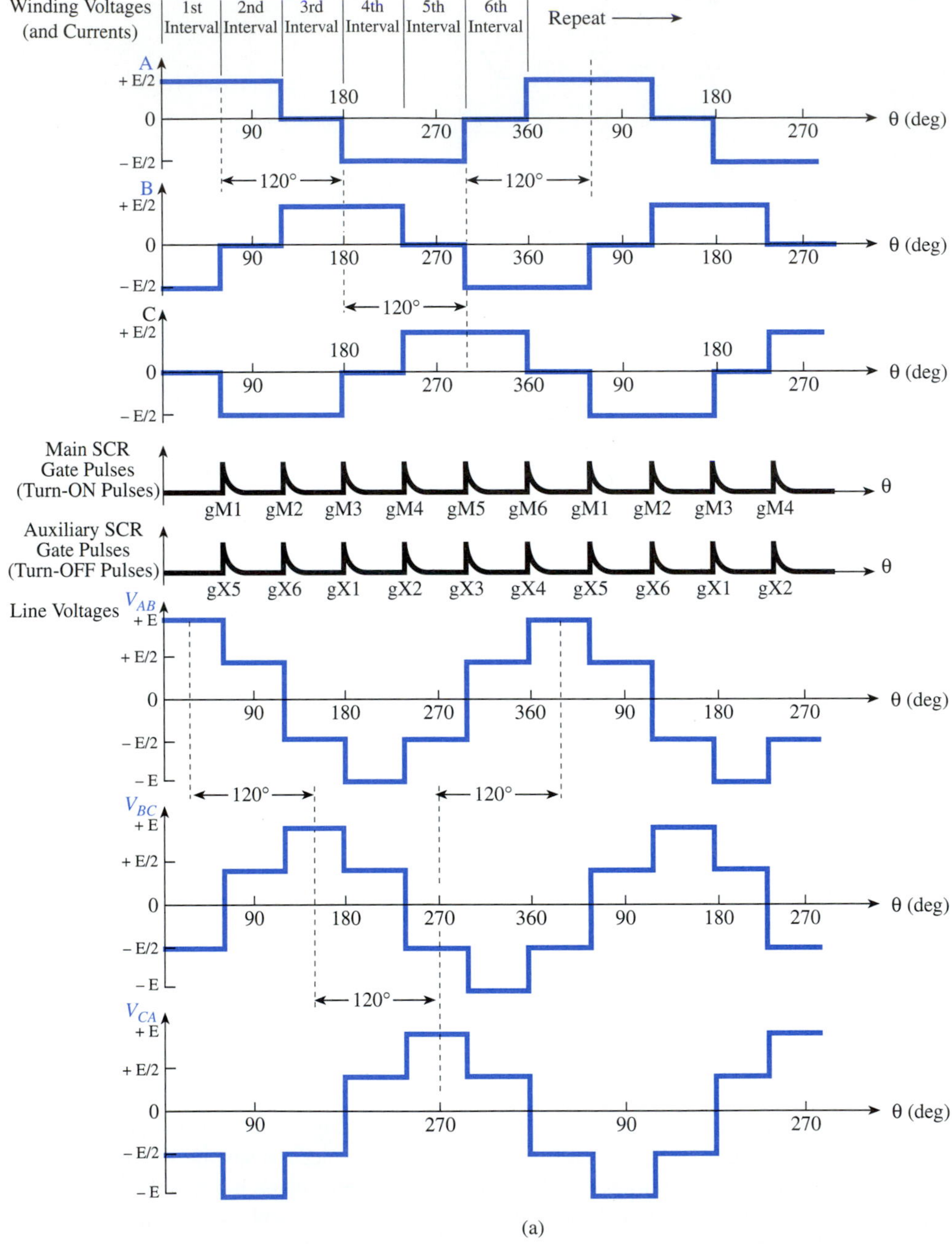

FIGURE 16–24
(a) Idealized waveforms obtained from the three-phase inverter when its SCRs are fired in ascending order. For a real motor load the current waveforms would be smoothed out due to winding inductance. (b) Conditions during each of the six time intervals that comprise a full cycle of inverter output. (c) Magnetic field components and net magnetic field during each of the six time intervals. The net field is seen to make one 360° rotation for one cycle of inverter output (two-pole motor).

FIGURE 16–24
(continued)

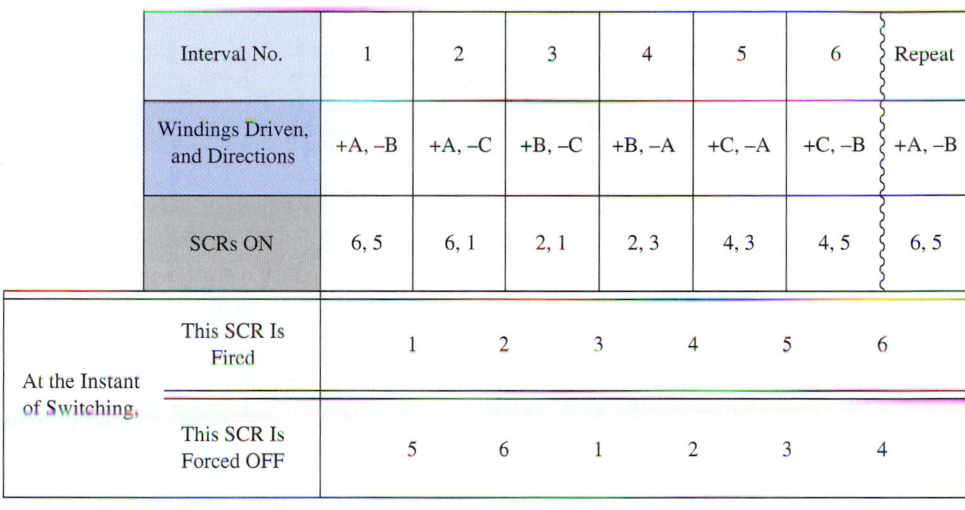

Interval No.	1	2	3	4	5	6	Repeat
Windings Driven, and Directions	+A, –B	+A, –C	+B, –C	+B, –A	+C, –A	+C, –B	+A, –B
SCRs ON	6, 5	6, 1	2, 1	2, 3	4, 3	4, 5	6, 5

At the Instant of Switching,							
This SCR Is Fired	1	2	3	4	5	6	
This SCR Is Forced OFF	5	6	1	2	3	4	

(b)

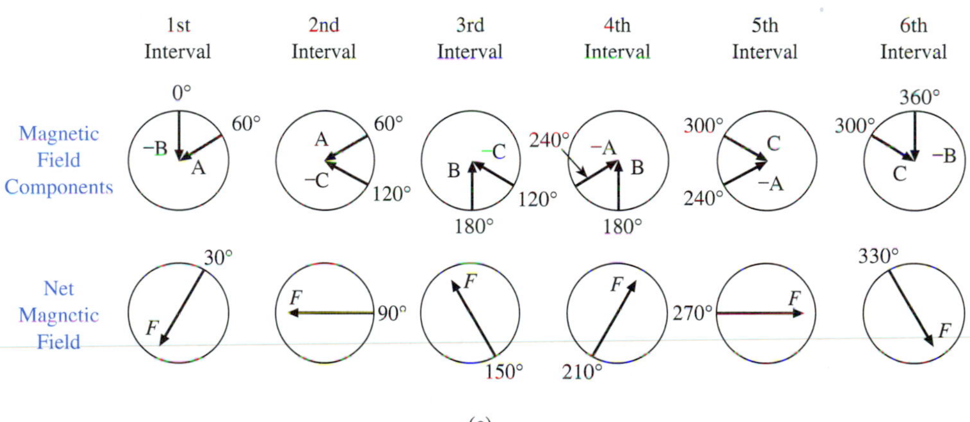

(c)

it in the defined-positive direction.* If current flows through winding A in the negative direction, its magnetic field component reorients by 180 mechanical degrees, coming from the 240° position. (Visualize a compass rose.)

Similarly for stator windings B and C, positive current through B produces a field component from 180°, and negative current through B gives a field component from 0°. For winding C, positive current gives a field component from 300°, and negative current gives a field component from 120° since $300° - 180° = 120°$.

The cyclical operation of the SCR bridge circuit in Fig. 16–23(a) is divided into six equal-duration time intervals. Let us define our first time interval to be that time when the electronic triggering circuitry causes SCRs 6 and 5 to be turned ON and all other SCRs to be turned OFF. This state of affairs is indicated by the first column of Fig. 16–24(b). Never mind how the triggering circuitry causes this; we'll consider that topic later.

*All motor phases have defined-positive current directions as indicated in Fig. 16–23(a), with current flowing from the outside supply line toward the wye tie-point. Since a motor winding is an electrical load, each one has its defined-positive voltage polarity as positive on the current-entry end and negative on the current-exit end, as that figure shows.

With SCRs 6 and 5 turned ON, there is a current flow path as follows: from the positive dc supply terminal, through SCR6, through stator winding A in the positive direction, through stator winding B in the negative direction, through SCR5, and down the dc supply line to the negative dc supply terminal. No current exists in stator winding C at this time because SCRs 4 and 1 are both OFF.

The top three waveforms of Fig. 16–24(a) show these stator winding currents graphically. During the first time interval, the A winding current (and voltage) is positive, the B winding current is negative, and the C winding current is zero. These winding currents produce magnetic field components oriented as shown in the first time interval of Fig. 16–24(c). The net magnetic field F resulting from these components is oriented from the 30° mechanical position on the stator, as the figure indicates.

At the end of the first time interval, the triggering circuitry fires SCR_1 and forces SCR_5 to turn OFF. These events are tabulated at the bottom of Fig. 16–24(b). The new combination of turned-ON SCRs is the 6-1 combination. Therefore throughout the second time interval the motor winding currents are positive through phase A, negative through phase C, and zero in phase B. Trace out the current flow path on Fig. 16–23(a) to verify this for yourself. The waveform graphs of Fig. 16–24(a) illustrate these current conditions during the second time interval.

In Fig. 16–24(c) the magnetic field components are seen to come from 60° due to the positive A current, and from 120° due to the negative C current; these components combine to produce a net field from the 90° mechanical position on the stator. Thus the switching of the SCRs as we proceeded from the first to the second time interval has produced a 60° rotation of the net stator field (from the 30° position to the 90° position).

At the end of the second time interval the triggering circuit fires SCR_2 and forces OFF SCR_6, leaving SCRs 2 and 1 ON. The winding currents are positive through B and negative through C during the third time interval. This is documented in Fig. 16–24(b) and graphed in Fig. 16–24(a). You can trace out the current path in Fig. 16–23(a). The new magnetic field components resulting from these winding currents cause another 60° rotation in the net stator field, to the 150° position, as shown in Fig. 16–24(c).

As this process continues through the fourth, fifth, and sixth time intervals, the SCRs are triggered ON and forced OFF in accordance with the schedule listed in Fig. 16–24(b). That sequence of SCR combinations produces the stator winding current waveforms drawn in Fig. 16–24(a), which causes the net stator magnetic field to keep advancing in 60° jumps. In this way the rotating field effect of a three-phase ac line is reproduced.

The motor torque is not instantaneously constant as it would be if the motor were driven by a three-phase sine-wave, but neither is it as abrupt as the Fig. 16–24(a) waveforms might suggest, since the inductance of the motor windings tends to smooth out the steep edges of the current waveforms.

Stand back and get an overall view of the A, B, and C waveforms in Fig. 16–24(a). If each time interval is regarded as a 60° portion of the complete operating cycle, then these three waveforms are all out of phase from one another by 120°, like three-phase sine-wave ac. This is a consequence of the manner in which the triggering/turn-OFF circuit has handled the SCRs—allowing an individual SCR to remain conducting for 120° (two time intervals) but changing the combining SCR halfway through the 120° conduction angle.

The line-to-line voltages V_{AB}, V_{BC}, and V_{CA} are also 120° phase-displaced from one another, as the three waveforms at the bottom of Fig. 16–24(a) make clear. Also note that the line voltages lead the phase voltages by 30°, just the same as for sine-wave three-phase ac.

The line voltage waveforms can be derived by subtracting one phase waveform from another phase waveform. For example, V_{AB} is derived by subtracting the V_B waveform from the V_A waveform; as always, subtraction is equivalent to sign inversion and addition.

Thus, during the first time interval, the instantaneous value of V_{AB} is obtained by changing the sign of the instantaneous value of V_B (which is $-E/2$), yielding $+E/2$, then adding that to the instantaneous value of V_A $(+E/2)$. The result is $+E/2 + E/2 = +E$, which agrees with the first interval of the V_{AB} waveform. Every other interval of every other line voltage can be derived in the same manner.

Of course, the whole appeal of the inverter is that its waveform frequency is variable. The frequency is determined by the rate at which the trigger/turn-OFF circuit delivers gate pulses to the six SCRs. The time between gate pulses always corresponds to 60° of the waveform cycle. If the gate pulses are spaced closer together in time, then 60° becomes a shorter time, a full cycle takes a shorter time, the frequency rises, and the motor speeds up. Conversely, if the gate pulses are spread further apart in time, the motor slows down.

The motor's direction of rotation can be reversed by altering the gate pulse sequence, again through the medium of the trigger/turn-OFF circuit. If we changed the trigger/turn-OFF circuit so that the SCR bridge worked its way through the schedule of Fig. 16–24(b) from right to left, instead of from left to right, the motor would reverse direction.

In a dc-supplied circuit, getting an SCR to turn ON is no problem; it's getting it to turn OFF that's the problem. Naturally, an inverter must contend with this problem. Many different circuit arrangements have been devised for forcing an SCR to turn OFF (*forced* commutation, as opposed to *natural* commutation). One popular method, mentioned in Sec. 4-7, is to use a solid-state switching device to connect a charged capacitor across an ON SCR so that the capacitor voltage temporarily reverse-biases the main terminals of the SCR. Such a scheme is illustrated in Fig. 16–25, in which the solid-state switching devices are

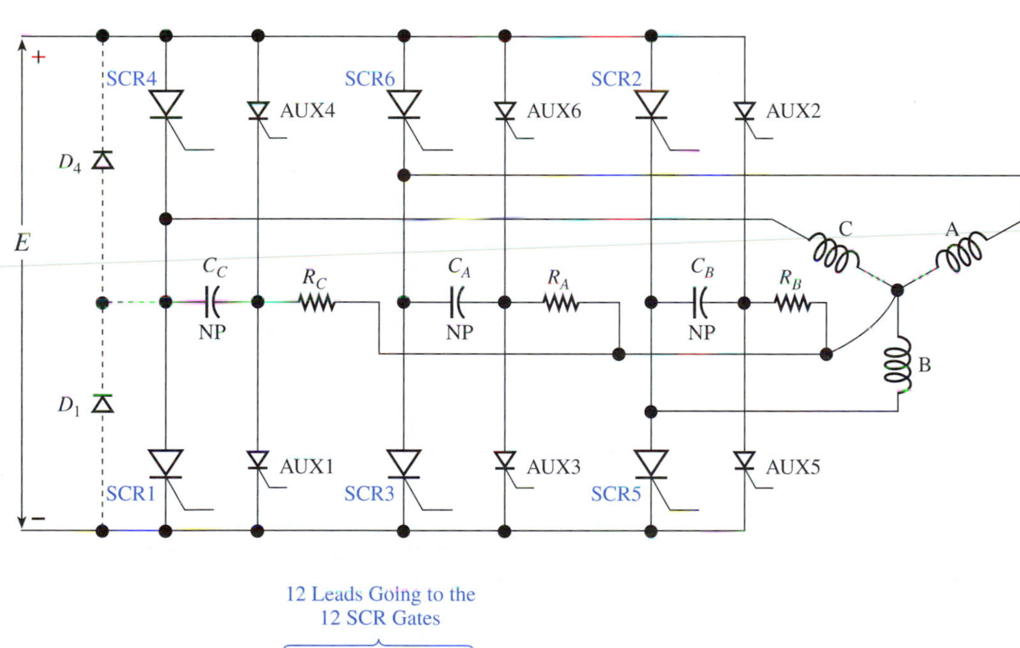

FIGURE 16–25
Three-phase inverter with commutating (turn-OFF) components. Firing an auxiliary SCR causes the like-numbered main SCR to be forced OFF.

smaller "auxiliary" SCRs. They are labeled with the letters AUX to clearly distinguish them from the main SCRs that actually carry the load current. Here is how they work.

If any main SCR is ON, the stator winding that it controls has a voltage of magnitude $E/2$ developed across it. This voltage charges the associated commutating capacitor through associated resistor R, always with the proper polarity for reverse-biasing the conducting SCR. When the associated auxiliary SCR is gated ON, the charged commutating capacitor forces the main SCR to turn OFF.

For example, suppose we are presently in the first time interval of Fig. 16–24(b). Negative current is flowing through phase B of the motor, developing a voltage across that phase which is positive at the inside wye tie-point and negative at the outside line. This voltage appears across the R_B-C_B series combination, causing C_B to charge positive on the right and negative on the left. At the end of the first time interval, the trigger/turn-OFF circuit sends a gate pulse to AUX5 at virtually the same instant that it sends a gate pulse to SCR_1. These gate pulses are shown just below the three winding current waveform graphs in Fig. 16–24(a). When AUX5 turns ON, it acts like a short circuit connecting the right (positive) side of C_B to the cathode of SCR_5. The left (negative) side of C_B is wired directly to the anode of SCR_5. The resulting reverse bias across SCR_5 persists until C_B can discharge through R_B and winding B, which is long enough to force SCR_5 OFF.

Now consider that we are in the 4th time interval of Fig. 16–24(b). SCR_2 is ON, carrying positive current through winding B. The voltage across winding B charges C_B positive on the left and negative on the right. At the end of the fourth time interval a gate pulse is sent to AUX2; see the gate waveforms in Fig. 16–24(a). This causes C_B to be connected in parallel with SCR_2, with the positive (left) side of C_B on the cathode and the negative side on the anode. Here again C_B succeeds in commutating OFF the main SCR that was passing current through winding B.

In the same manner, commutating capacitor C_A forces OFF SCRs 6 and 3 at the proper moments, through the action of AUX6 and AUX3, respectively; and C_C forces OFF SCRs 4 and 1 at the proper moments through AUX4 and AUX1.

Usually in capacitor-based forced-commutation circuits, it is desirable to maintain the reverse bias across the SCR's main terminals for only a short time, just long enough to remove the charge carriers that have accumulated near the internal junctions of the SCR, thereby reducing the current to less than the hold-ON value I_{HO}. For medium- and high-current SCRs, the time required to do this is generally less than 100 μs.

The arrangement in Fig. 16–25 allows a commutating capacitor to maintain reverse bias until it discharges through its associated series resistor and motor winding. Because the series resistance value must be fairly high in order to isolate the three commutating circuits from one another, the capacitor discharge time constant tends to be fairly long, considerably longer than 100 μs. So there is a conflict between the necessity of isolating the commutating circuits and the desirability of removing the commutation voltage from the SCR quickly. This conflict is often resolved by the installation of discharge diodes in parallel with the main SCRs, to provide a low-resistance discharge path. The placements of discharge diodes D_4 and D_1, associated with SCR_4 and SCR_1, are indicated by the dashed-line connections at the far left in Fig. 16–25. As that circuit schematic makes clear, the diode installation establishes a low-resistance path back around to the auxiliary SCR, by which the commutating capacitor can discharge quickly. Discharge diodes 6, 3, 2 and 5 would be placed across their associated SCRs in the same manner.

Concerning the design of the trigger/turn-OFF circuit in Fig. 16–25, there are several methods that can be employed to generate the sequence of gate pulses graphed in Fig. 16–24(a). One approach is to use the sequential switching circuit of Fig. 5–7, extended to six stages. Load 1 would be a pulse transformer with a dual secondary, one winding driving the gate of SCR_1 in Fig. 16–25 (delivering pulse gM1 in Fig. 16–24a), and the other winding driving the gate of AUX5 (delivering pulse gX5 in Fig. 16–24a). Load 2 in Fig. 5–7 would be a pulse transformer supplying simultaneous gate pulses to SCR_2 and AUX6 in Fig. 16–25 (pulses gM2 and gX6 in Fig. 16–24a), and so on.

By ganging together all of the 1-MΩ adjustment pots in Fig. 5–7, the pulse intervals would be made uniform and adjustable, providing the inverter with its variable frequency capability.

An alternative method of generating the gate pulse sequence is by microprocessor. To terminate the first time interval, the μP program writes a 1 to a particular bit in an output port, from which the HI signal is appropriately processed to produce the gM1 and gX5 gate pulses. The program then jumps to a delay subroutine to wait until the end of the second time interval. At the proper moment in real time, the μP returns from its delay subroutine to the main program, where it encounters an instruction to write a 1 to a different bit of the output port. The HI signal on this bit would be hard-wired to SCR_2 and AUX6 (after appropriate processing). In this way the gate pulses gM2 and gX6 are delivered to the inverter, causing it to switch out of the conditions for the second time interval and into the conditions for the third time interval. The program then jumps back into the delay subroutine to spend the same amount of time as before, waiting for the end of the third time interval, and so on.

Variable-frequency control is attained by altering the amount of time the program spends in the delay subroutine. This can be done by writing the subroutine so that its duration depends on the contents of a particular RAM location, and then using the μP's monitor program to vary the RAM value.

To achieve closed-loop motor speed control, the user-program itself would be designed to make the necessary alterations to the delay subroutine in order to automatically adjust the inverter's output frequency. The user-program would make any time-delay alteration in response to the error signal it calculates by subtracting the set-point speed from the actual measured motor speed.

16-11 ■ VARYING THE VOLTAGE ALONG WITH FREQUENCY

Whenever variable-frequency speed control is employed, the motor supply voltage cannot be allowed to remain at a steady value. The magnitude of the motor voltage must be increased or decreased in proportion to the frequency. That is, the voltage-to-frequency ratio, V/f, must remain constant (approximately).

For instance, if the motor has a nameplate rating of 240 V at 60 Hz, the voltage-to-frequency ratio is 4 ($240 \div 60 = 4$). If the motor is speeded up by adjusting its variable-frequency inverter to, say, 90 Hz, the voltage magnitude must be increased to 360 V, since $4 \times 90 = 360$. If the motor is slowed down by adjusting the inverter frequency to 45 Hz, the voltage magnitude must be decreased to 180 V, since $4 \times 45 = 180$.

Here is the reason it is necessary to maintain a constant V/f ratio: The stator's magnetic field strength must remain constant under all operating conditions. If the stator field strength should happen to rise much above the design value, the motor's core material would go into magnetic saturation. This would effectively lower the core's permeability, thereby inhibiting proper induction of voltage and current in the rotor loops (or bars), thus detracting from the motor's torque-producing capability. On the other hand, if the stator field strength should happen to fall much below the design value, the weakened magnetic field would simply induce lower values of voltage and current in the rotor loops, in accordance with Faraday's law. This likewise would detract from the motor's torque-producing ability.

So the sinusoidal magnetic field produced by the stator windings must hold a constant rms value, regardless of frequency, but what determines the value of the stator's magnetic field? The stator's magnetizing current, that's what. The magnetizing current of an induction motor is the current that flows through the stator winding when the rotor is spinning at steady-state speed with no torque load; this is in the same way that the magnetizing current of a stationary transformer is the current that flows through the primary winding when the secondary winding is operating under no electrical load. Just as for a stationary transformer, the magnetizing current for an induction motor is given by Ohm's law,

$$I_{\text{mag}} = \frac{V}{X_L} \tag{16-11}$$

where V is the rms value of the applied stator voltage and X_L is the inductive reactance of the stator winding.*

In Eq. (16-11), X_L does not remain constant as the supply frequency is adjusted; it varies in proportion to the frequency ($X_L = 2\pi fL$). Therefore V must also be varied in proportion to the frequency, so that the Ohm's law division operation yields an unvarying value of magnetizing current.

Alternatively, using $X_L = 2\pi fL$, we can rewrite Eq. (16-11) as

$$I_{\text{mag}} = \frac{V}{X_L} = \frac{V}{2\pi fL} = \frac{1}{2\pi L}\,\frac{V}{f} \tag{16-12}$$

In Eq. (16-12), the factor $1/2\,\pi L$ is a constant, determined by the motor's construction details that have a bearing on the inductance of the stator winding. Equation (16-12) makes it plain that a constant I_{mag} can be achieved only by maintaining a constant V/f ratio.

The most convenient circuit for producing a high-power variable dc voltage to drive the inverter of Fig. 16–23(a) is a three-phase six-pulse variable rectifier built with six SCRs. Figure 16–26 is a schematic diagram of such a rectifier.

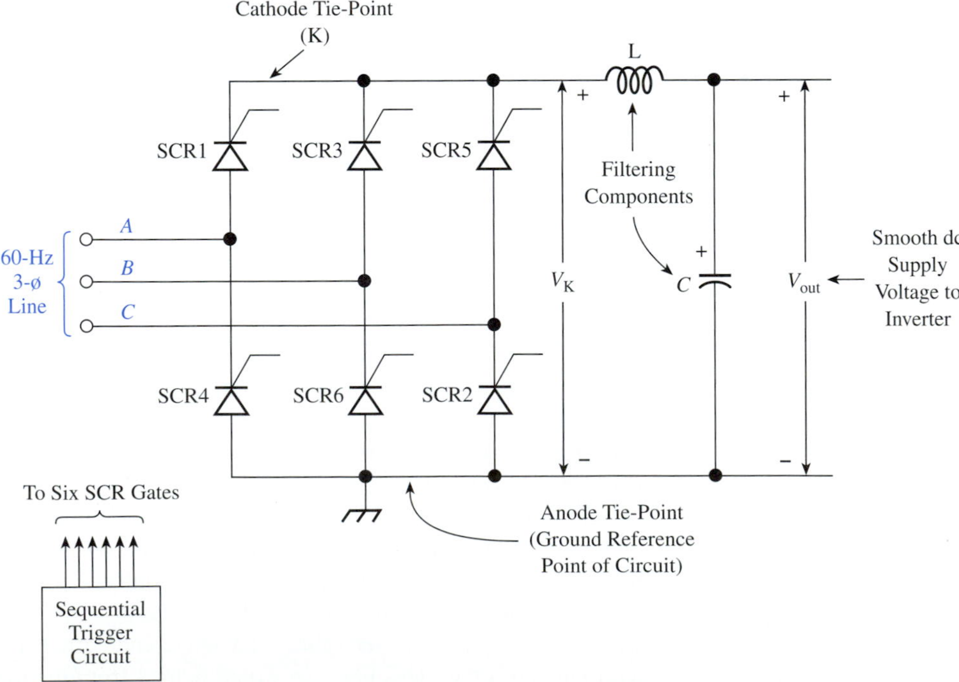

FIGURE 16–26 Three-phase variable-voltage rectifier.

*We are neglecting the resistance of the motor winding, assuming that its impedance is wholly reactive. This is a reasonable assumption for motors above the fractional-horsepower range.

The trigger circuit in Fig. 16–26 delivers line-synchronized gate pulses to the six SCRs sequentially. Because the gate pulses are line-synchronized, the pulse rate is not variable; six equal-interval pulses occur during every cycle of the 60-Hz ac line, one pulse every 2.78 ms. However, the trigger circuit can vary the delay angle of the pulses, with the delay angle referenced to the E_{AB} line voltage of Fig. 16–27(a). That is, the trigger circuit can deliver gate pulses according to either schedule 1 or schedule 2, as follows:

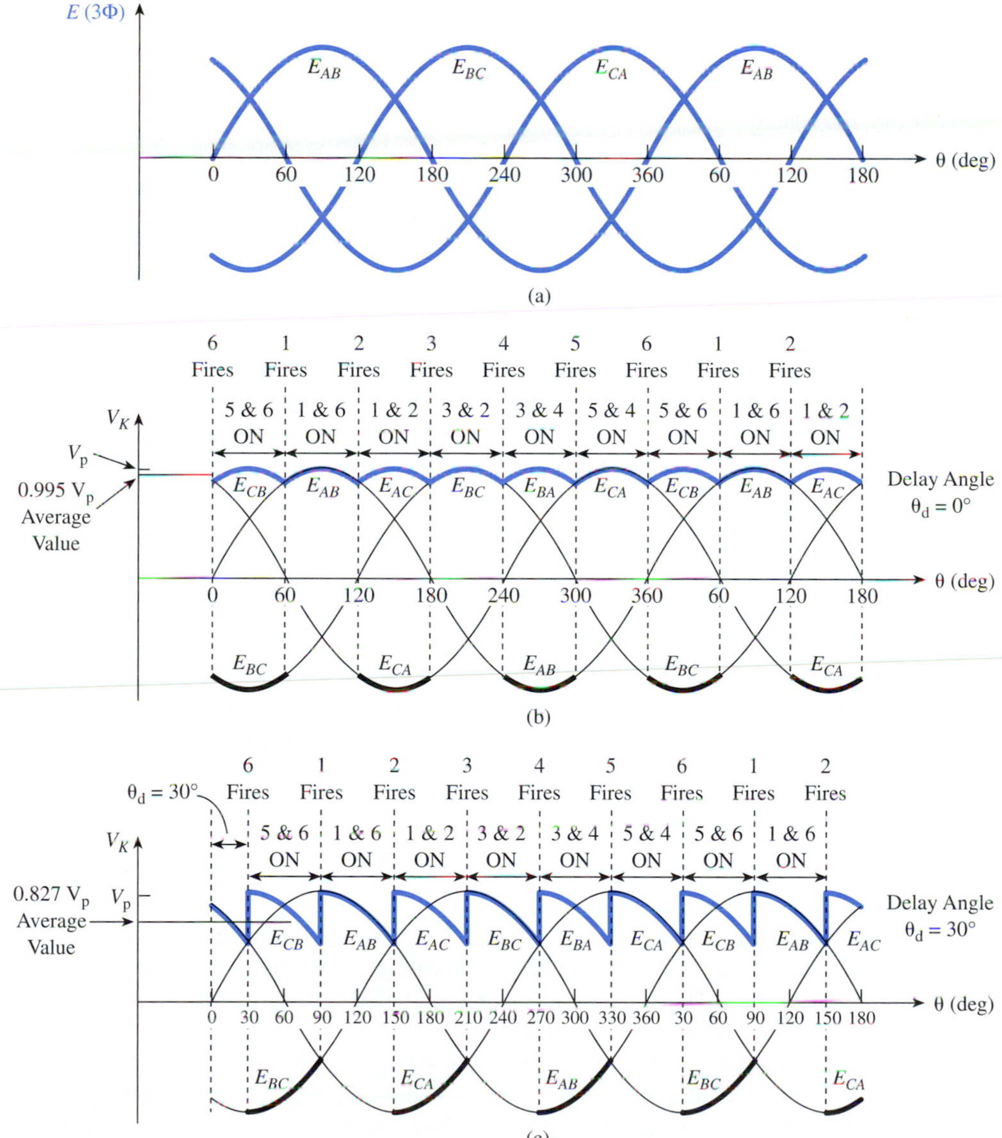

FIGURE 16–27

Waveforms associated with the three-phase variable-voltage rectifier of Fig. 16–26. (a) The three line voltages. (b) The rectifier's unfiltered output voltage for a firing delay angle of zero. (c) The rectifier's unfiltered output voltage for $\theta_d = 30°$. The average value (dc component) has been reduced by the increase in firing delay angle. (d) Conditions during each of the six time intervals that comprise a full cycle of the ac line.

FIGURE 16–27
(continued)

Interval No.	1	2	3	4	5	6	Repeat
Supply Lines Accessed	C+, B–	A+, B–	A+, C–	B+, C–	B+, A–	C+, A–	C+, B–
SCRs ON	5, 6	1, 6	1, 2	3, 2	3, 4	5, 4	5, 6

At the Gate-Pulse Instant							
This SCR Is Fired	1	2	3	4	5	6
And This SCR Turns OFF Naturally	5	6	1	2	3	4

(d)

1. Deliver one gate pulse at the instant E_{AB} makes a positive-going zero crossover; deliver the next gate pulse when E_{AB} is 60° into its cycle, deliver the next gate pulse when E_{AB} is 120° into its cycle, and so on. This schedule is depicted in Fig. 16–27(b).
2. Deliver one gate pulse 30° after E_{AB} makes a positive-going zero crossover; deliver the next gate pulse when E_{AB} is 90° into its cycle (60° after the previous gate pulse), deliver the next gate pulse when E_{AB} is 150° into its cycle (again, 60° after the previous gate pulse), and so on. This schedule is depicted in Fig. 16–27(c).

Schedule 1 corresponds to a delay angle of zero, and schedule 2 corresponds to a delay angle of 30°. Actually, the delay angle is continuously variable over the range from 0° to 90° by the trigger circuit in Fig. 16–26. As the delay angle is varied, the average value of the V_K waveform varies along with it. In that figure voltage V_K appears between the cathode tie-point and the anode tie-point. The *LC* low-pass filter removes the ac content of the V_K waveform and delivers a smooth dc voltage to the output terminals of the rectifier. The magnitude of V_{out} is equal to the average value of the six-pulse-per-cycle V_K waveform.

16-11-1 Delay Angle = 0°

To understand why the V_K waveform has so many pulsations per ac line cycle, refer to the schematic diagram of Fig. 16–26 and the waveforms of Fig. 16–27(b), which are for a 0° firing delay angle. During the first 60° of the ac line cycle (considering line voltage E_{AB} as the reference), the greatest magnitude of voltage is on the E_{BC} waveform, which is near its negative peak. This negative instantaneous value corresponds to a polarity which is positive on line *C* and negative on line *B*. To apply the positive side of that voltage to the cathode tie-point and the negative side of that voltage to the anode tie-point, it is necessary to turn ON SCR_5 and SCR_6. This 5 and 6 combination is indicated in the first 60° interval of Fig. 16–27(b), which also illustrates how the negative peak of the E_{BC} waveform is being accessed in the reverse direction (as E_{CB}, note the reversed subscript order) to produce one positive pulsation of the V_K waveform.

At the 60° instant in the ac line cycle, the magnitude of the E_{AB} waveform becomes equal to the magnitude of the E_{BC} waveform—note the intersection of the E_{CB} pulsation with the E_{AB} waveform. At this point it behooves us to stop accessing the E_{BC} waveform backward and start accessing the E_{AB} waveform forward. This can be accomplished by delivering a gate pulse SCR_1. The turning ON of SCR_1 automatically turns OFF SCR_5 by natural commutation. Cathode tie-point K is thus disconnected from line C and reconnected to line A, while the anode tie-point remains connected to line B through SCR_6. Therefore the positive peak region of the E_{AB} waveform constitutes the second pulsation of the V_K waveform.

The natural commutation of SCR_5 can be understood by studying the circuit conditions just after SCR_1 has been fired, say, at the 61° point. The conditions at this instant are presented in Fig. 16–28, for an assumed line voltage value of 240V rms. As that figure reveals, the firing of SCR_1 causes a reverse bias to be applied to SCR_5 shortly after the firing instant. There is no need for additional components in the rectifier circuit to force the SCR commutation—it happens naturally.

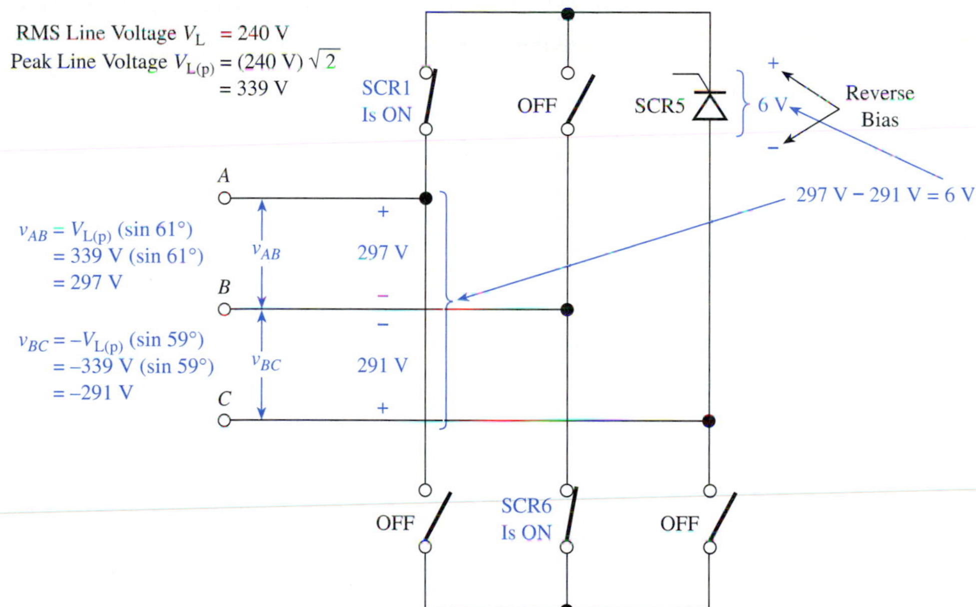

FIGURE 16–28 Instantaneous conditions in the rectifier shortly after SCR_1 has been fired (assuming $\theta_d = 0°$). SCR_5 is seen to be reverse-biased by the natural action of the three-phase ac line; it therefore does not require forced commutation.

The remainder of the V_K waveform is pieced together in like manner. It is instructive to make a pulsation-by-pulsation check of the waveform, verifying that the ascending-number gate pulse sequence always keeps the V_K waveform as positive as possible, and also causes SCR commutation to take place as required. The schedule of SCR conditions for a complete ac line cycle (six pulsations of V_K) is given in Fig. 16–27(d).

16-11-2 Delay Angle = 30°

If the trigger circuit is made to delay the delivery of gate pulses, the SCR switching schedule remains the same, but the SCRs no longer access the ac supply lines during their segments of greatest magnitude. In other words, instead of switching SCRs at the instant when a voltage becomes available with greater magnitude than the voltage that is currently being accessed, we allow the currently accessed voltage to run downhill a while before we switch over. This action is portrayed in Fig. 16–27(c) for a 30° delay of the gate pulses. By inspection, it is apparent that this V_K waveform has an average value less than the average in Fig. 16–27(b). By varying the gate pulse delay angle from 0° to 90°, we can vary V_K's average value all the way down to 0 V.* As explained earlier, the *LC* filter in Fig. 16–26 extracts the average value of V_K and applies it to the inverter of Fig. 16–23, thereby affecting the magnitude of voltage supplied to the motor's stator windings. In this way we are able to vary the motor voltage to achieve the constant V/f ratio that is necessary for maintaining a constant magnetic field strength.

We will not discuss the specific methods of correlating the delay angle of the rectifier trigger circuit in Fig. 16–26 to the pulse rate of the inverter trigger circuit in Fig. 16–23. Suffice it to say that this correlation can be accomplished either by analog electronic techniques or by microprocessor.

16-12 ■ CYCLOCONVERTERS

A cycloconverter has as its input the three-phase ac line and produces as an output a nonsinusoidal ac voltage at a lower frequency. A single cycloconverter produces a single-phase output voltage. To control the speed of a three-phase ac motor, induction or synchronous, we use three cycloconverters. We just arrange for their gate firing pulses to be offset in time, so that their three output voltages are phase-displaced by 120°. The three individual cycloconverters then drive the three individual stator windings of the three-phase ac motor.

16-12-1 Six-SCR Cycloconverters

A cycloconverter can be built with either six SCRs or 12 SCRs. Figure 16–29(a) shows the circuit schematic for the six-SCR design. By convention, the SCRs are labeled with odd numbers only. They are labeled this way so that the gate pulsing sequence is always in ascending order, whether the cycloconverter is built with six SCRs or 12 SCRs. This idea will become clear when we make a close examination of cycloconverter waveforms.

In the cycloconverter of Fig. 16–29(a), three particular SCRs are responsible for producing the positive half cycle of the output waveform; they are SCRs 1, 3, and 5. The remaining three SCRs, 7, 9, and 11, are responsible for producing the negative half cycle. We find it convenient to mentally group the SCRs together on this basis. Fig. 16–29(b) is the same circuit as Fig. 16–29(a) but with the SCRs grouped as described. Let us adopt the phrase *positive triplet* to refer to the group that produces the positive half cycle (numbers 1, 3, and 5) and *negative triplet* for the negative half cycle group (numbers 7, 9, and 11). This nomenclature is used in Fig. 16–29(b).

The cycloconverter's trigger circuit delivers gate pulses to the SCRs, basically at the rate of one gate pulse for each pulsation of the three-phase ac line (basically 180 gate pulses per second for the six-SCR design used with a 60-Hz ac line). The output frequency is determined by the number of gate pulses per half cycle of the output waveform. In plain terms, if the trigger circuit delivers only a small number of sequential gate pulses to one triplet before it changes over to deliver the same number to the other triplet, then each triplet will remain in conduction for only a short time. This corresponds to a short time

*In fact, by delaying the pulses more than 90°, we can actually make the average value of V_K *negative*. This corresponds to a reversed dc output voltage in Fig. 16–26, with a polarity positive on the anode tie-point (the bottom) and negative on the cathode tie-point (the top), but the SCRs cannot reverse their current direction. They insist that current always flows away from the cathode tie-point and toward the anode tie-point. Therefore, if the dc output voltage undergoes a polarity reversal (because $\theta_d > 90°$) the entire SCR bridge circuit undergoes a change in its basic nature, from electrical source to electrical load, since current must now *enter* the circuit via its positive terminal. This phenomenon is very useful for *dynamic braking* of dc motors. A book devoted exclusively to motor drives will give a detailed explanation of how it works.

FIGURE 16–29
(a) Six-SCR cycloconverter.
(b) The same cycloconverter redrawn to show the positive and negative triplets separated.

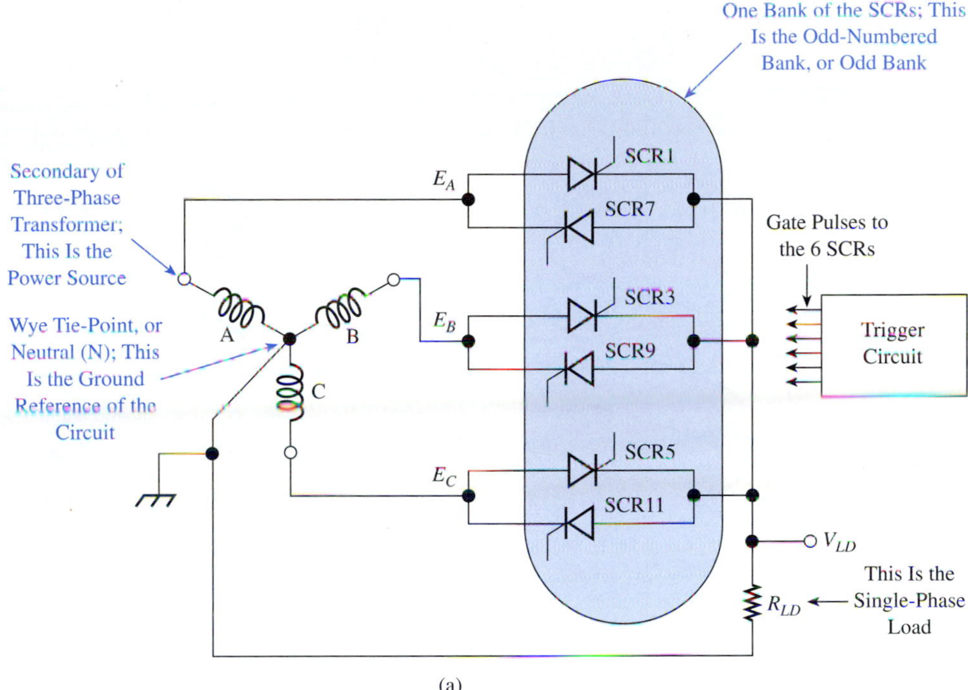

(a)

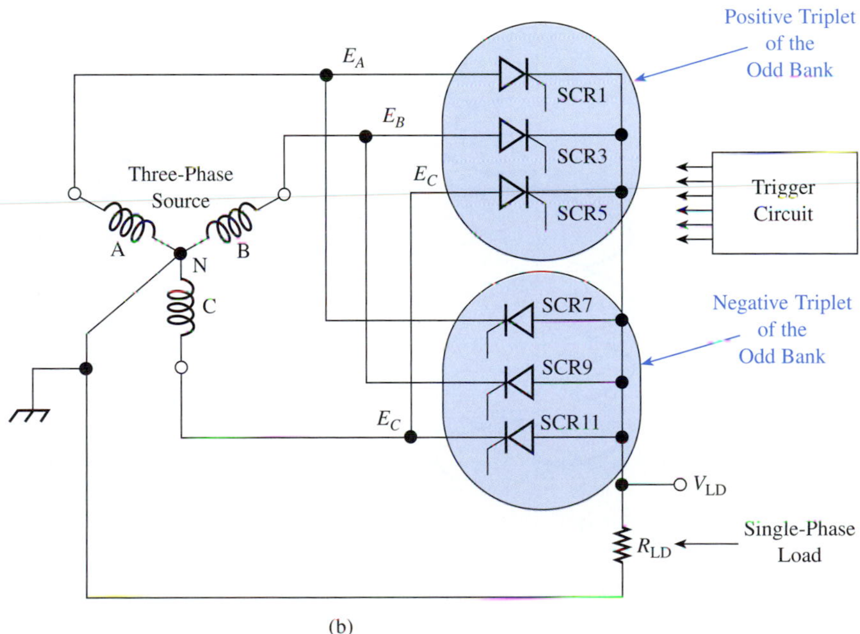

(b)

duration for each half cycle of the output waveform, causing the output frequency to be high. On the other hand, if the trigger circuit delivers a large number of sequential gate pulses to each triplet before changing over, then each triplet will remain in conduction for a long time, causing the output frequency to be low. To clarify this concept, refer to the waveforms of Fig. 16–30.

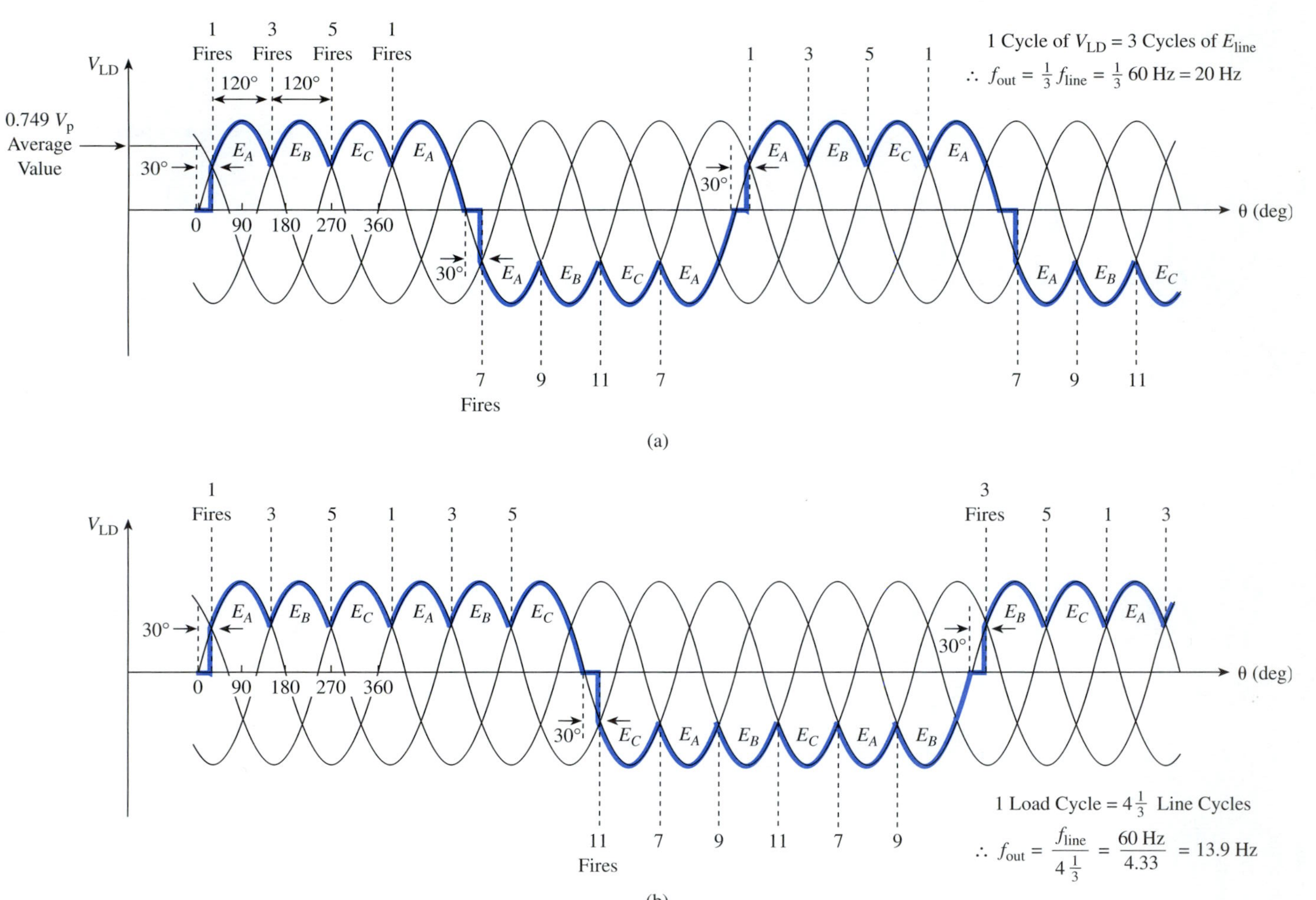

FIGURE 16–30
(a) Output waveform of a six-SCR cycloconverter delivering four pulsations per half cycle.
(b) Six pulsations per half cycle.

In Fig. 16–30(a), the trigger circuit is delivering four sequential gate pulses to each triplet. The pulses are delivered in ascending order, as they were for inverter operation in Secs. 16-10 and 16-11. In this case the pulses are timed to produce a steady firing delay angle of 30°, that is, every SCR is gated ON 30° after its associated ac line phase crosses through zero. To keep things simple for now, we'll assume that the cycloconverter's trigger circuit operates this way, giving a steady firing delay angle.

At the left of Fig. 16–30(a), the four sequential gate pulses turn ON one SCR at a time, in the order 1, 3, 5, 1. The load voltage waveform thus consists of segments of the phase voltages, with the segments 120° wide and centered on their positive peaks. Commutation of the SCRs is natural, because firing at or later than 30° enables the new SCR to apply a reverse bias to the previously ON SCR.

When the trigger circuit is finished with the positive triplet, it delivers a matching pulse sequence to the negative triplet, thereby forming the negative half cycle of V_{LD}. It then returns to the positive triplet to begin the next cycle of load voltage.

In Fig. 16–30(a), one cycle of V_{LD} corresponds to three cycles of the ac line voltage, so the output (load) frequency is one third of the ac line frequency. Verify this for yourself by examining the waveform.

In Fig. 16–30(b) the trigger circuit has been adjusted to give six sequential gate pulses per triplet, again at a steady 30° delay angle. The greater number of gate pulses causes the output frequency to decrease. In this case one output cycle takes four line cycles plus 120°, or $4\frac{1}{3}$ line cycles. For a 60-Hz line,

$$f_{out} = \frac{60 \text{ Hz}}{4\frac{1}{3}} = 13.9 \text{ Hz}$$

Verify this for yourself.

For this frequency setting, the trigger circuit does not begin each cycle of load voltage with the same SCR. Note that the first cycle in Fig. 16–30(b) begins with SCR_1 but the next cycle begins with SCR_3. The third cycle, not shown in Fig. 16–30(b), would begin with SCR_5. It is the job of the trigger circuit to keep track of this. The trigger circuit usually has help from a microprocessor.

16-12-2 Twelve-SCR Cycloconverters

A 12-SCR cycloconverter is shown in Fig. 16–31(a). The additional SCRs enable the cycloconverter to produce six pulsations of load voltage for each cycle of the ac line, rather than just the three pulsations per ac-line cycle that we obtained in the waveforms of Fig. 16–30. This increase in the pulsation rate from 180 Hz to 360 Hz causes the harmonic content of the load waveform to be concentrated at higher frequencies, further from the fundamental output frequency. It therefore becomes easier to filter out the harmonic content to obtain a sinusoidal output, if that is desired.

The six additional SCRs are labeled with even numbers, by convention. Let us adopt the nomenclature *odd bank* and *even bank* to distinguish between the original group of six SCRs and the newly added group of six.

This cycloconverter design can be identified by any of several names. It can be called a *12-SCR* cycloconverter, a *dual-bank* cycloconverter, or a *six-pulsation* cycloconverter (six load pulsations per ac-line cycle). We will feel free to use any of these names.

Notice that in a dual-bank cycloconverter the load is connected between the SCR banks, not to the wye tie-point of the three-phase ac source. Therefore no neutral wire is required, as was the case for the single-bank cycloconverter of Fig. 16–29. In fact, the three-phase source can just as well be delta-connected as wye-connected, even though we show a wye connection in Fig. 16–31(a).

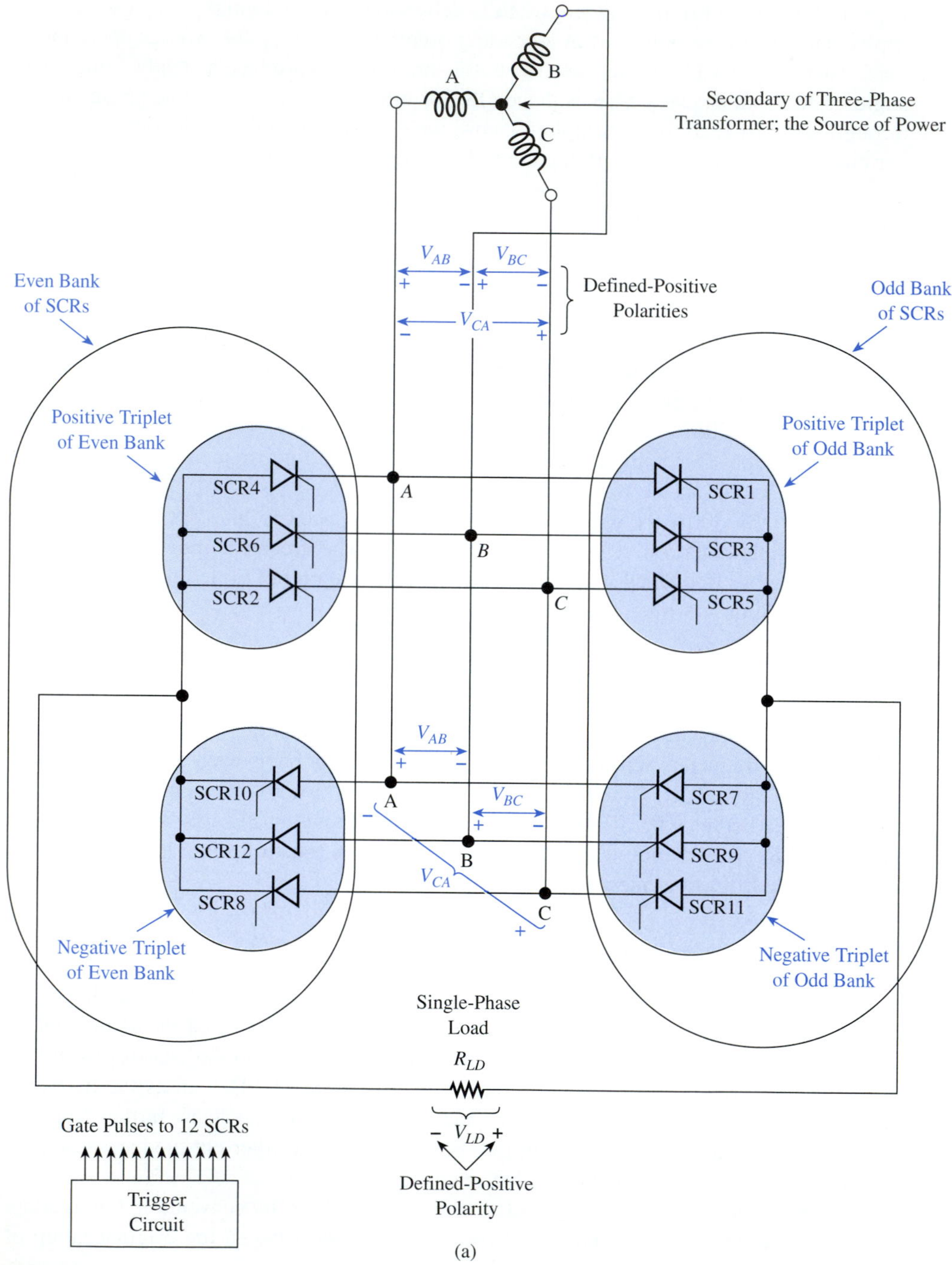

FIGURE 16-31
(a) A 12-SCR or dual-bank cycloconverter. The two positive triplets jointly produce the positive half cycle of the output waveform and the two negative triplets jointly produce the negative half cycle. (b) Output voltage waveform of the dual-bank cycloconverter delivering five pulsations per half cycle. The ripple frequency is 360 Hz for a 60-Hz ac line.

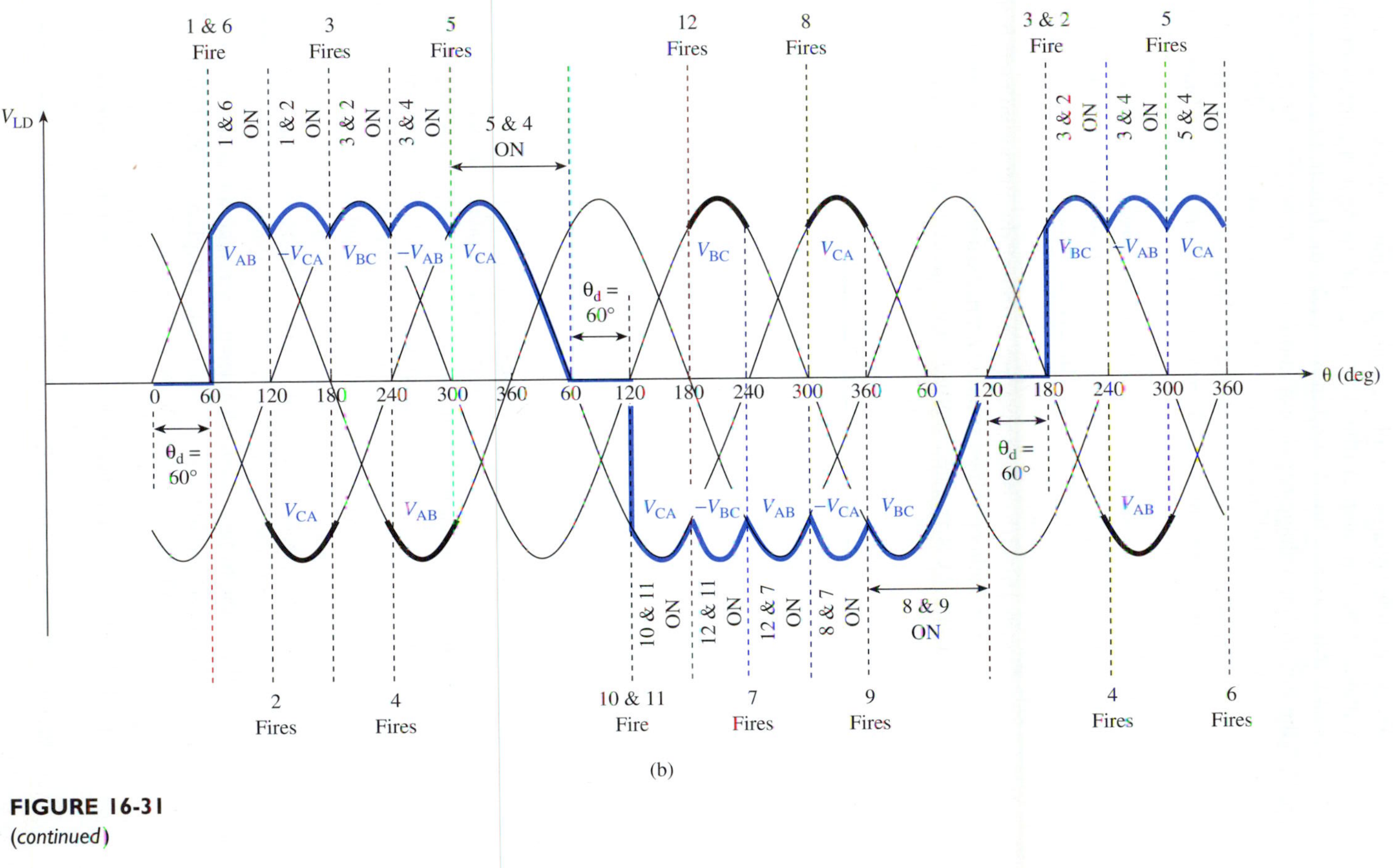

FIGURE 16-31
(continued)

If the three-phase source is wye-connected, the voltages that the SCRs successively access are line voltages, not phase voltages. That is, the cycloconverter never accesses the voltage from A to neutral (the wye tie-point) to form a portion of the load voltage; instead, if it accesses line A at all, it must take the line-to-line voltage between A and B or the line-to-line voltage between A and C.

Consistent with the fact that we are now accessing voltages between two lines of the three-phase source is the fact that a dual-bank cycloconverter always has two SCRs ON at a time, one from the odd bank and one from the even bank. It is the responsibility of the trigger circuit to fire the proper two SCRs simultaneously to initiate a half cycle of load voltage. Once a half cycle is under way though, the trigger circuit fires the SCRs singly, as it always has done. Firing a new SCR always results in natural commutation of one of the previously ON SCRs. The cycloconverter therefore maintains itself in a state of two SCRs ON at all times.

Figure 16–31(b) is a waveform graph of the output voltage from a dual-bank cycloconverter. To keep things simple, we're assuming a steady firing delay angle of 60°. That is, every SCR is gated ON 60° after its associated ac line crosses through zero.* Thus, on the left of the waveform, the positive half cycle of load voltage is initiated by turning ON SCR_1 at the instant that is 60° into the E_{AB} cycle, since SCR_1 is associated with line A. Here at the beginning of the half cycle, the trigger circuit must also turn ON SCR_6, associated with line B, in order to complete the circuit through the load to line B. Therefore the first pulsation of the V_{LD} waveform consists of the 60° segment of E_{AB} that is centered on its positive peak. Sixty degrees later, the trigger circuit sends a gate pulse to SCR_2, whose associated line, line C, made a negative-going zero crossover 60° ago; look at the E_{CA} waveform. It is at this instant that the E_{CA} magnitude becomes greater than the E_{AB} magnitude. With E_{CA} instantaneously negative, the instantaneous polarity between lines C and A positive on A and negative on C, which means that line C becomes more negative than line B at this instant. Therefore SCR_6 commutates OFF naturally, since its cathode is hard-wired to line B and its anode is held at the line C potential by SCR_2.

With SCRs 1 and 2 ON, the next pulsation of the V_{LD} waveform consists of the 60° segment of E_{CA} that is centered on its negative peak, but because SCRs 1 and 2 connect the load to E_{CA}"backward,"** the V_{LD} polarity remains positive.

Sixty degrees later the trigger circuit sends a pulse to SCR_3, commutating SCR_1 OFF, so the load voltage tracks the E_{BC} waveform through the region of its positive peak for the next 60°. These events are indicated by the waveform of Fig. 16–31(b).

In this manner, the positive half cycle of the V_{LD} waveform is formed by piecing together 60° segments of the various line voltages. Note that only SCRs 1 through 6 are used to create the positive half cycle; SCRs 7 through 12 remain OFF.

When the trigger circuit decides to terminate the positive half cycle of V_{LD}, it simply stops sending gate pulses to the positive triplets, SCRs 1 through 6. The trigger circuit waits for all the positive SCRs to commutate OFF naturally, then begins sending gate pulses to the negative triplets, SCRs 7 through 12. The negative half cycle of V_{LD} is pieced together in the same manner as before, using a different collection of SCRs.

Each half cycle is formed by firing the SCRs in ascending order, as usual. The numbering convention of the SCRs has been chosen to establish this condition. However, sub-

*An SCR's associated ac line is the line that the SCR's noncommon electrode is connected to. Thus SCR_4 has line A associated with it; verify this by inspecting Fig. 16–31(a). SCR_3 has line B associated with it, as do SCRs 6, 9, and 12.

**A "forward" connection to E_{CA} would be with SCRs 5 and 4 turned ON.

sequent half cycles do not necessarily begin with the same SCRs as previous half cycles, as is apparent from Fig. 16–31(b).

In the waveform of Fig. 16–31(b), one load cycle takes two line cycles plus 120°, or $2\frac{1}{3}$ line cycles. For a 60-Hz ac line frequency, the cycloconverter's output frequency is

$$f_{out} = \frac{f_{line}}{2\frac{1}{3}} = \frac{60\text{ Hz}}{2\frac{1}{3}} = 25.7\text{ Hz}$$

16-12-3 Reducing the Average Voltage

The average voltage that a cycloconverter delivers to its load can be reduced by increasing the firing delay angle. For a single-bank 6 SCR cycloconverter, the delay angle must be increased beyond 30°. For a dual-bank 12-SCR cycloconverter, firing must be delayed beyond 60°.

Figure 16–32 shows the effects of altering the delay angle to a steady 45° (part a) and to a steady 90° (part b), for a single-bank cycloconverter. Compare those two waveforms to the load voltage waveform for $\theta_d = 30°$ in Fig. 16–30(a), which likewise has four pulsations per half cycle. It is clear by inspection that the average voltage value in Fig. 16–32(a) is less than in Fig. 16–30(a), and that the average value in Fig. 16–32(b) is lower yet. This progressive reduction in average voltage is a consequence of the progressively increasing firing delay angle. It can be shown that the average voltage values for $\theta_d = 30°$, 45°, and 90° are, respectively, 0.749 V_p, 0.714 V_p, and 0.424 V_p. These average values are indicated on the waveform graphs.

Like an inverter motor-drive system, a cycloconverter motor drive must vary its average voltage in proportion to the output frequency (constant V/f ratio) in order to maintain a constant magnetic field strength. Again, the responsibility for accomplishing this falls to the trigger circuit and its support system, usually microprocessor-based.

16-12-4 Nonsteady Firing Delay

As long as such tough demands are being made on the trigger circuit, let's go even further and ask it to change the firing delay from one pulsation to the next. If this is done properly it yields a reduction in the harmonic content of the cycloconverter's output voltage. Such reduction is worthwhile because then the output can be more easily filtered to obtain a sinusoidal final load voltage, if desired. Visually, the cycloconverter's output voltage waveform can be seen to take on an overall sine-wave shape. The waveform of Fig. 16–33(a) illustrates this idea for a single-bank cycloconverter producing a fundamental output frequency of 10 Hz. The firing delay angle changes from one pulsation to the next, as that drawing indicates. The overall average shape of this waveform is visibly less squarish and more sinusoidal than the overall average shapes of the waveforms in Figs. 16–30 and 16–32.

Dual-bank cycloconverters are superior to single-bank cycloconverters for this practice, especially at frequencies greater than 10 Hz. They have twice as many pulsations to work with, so it is reasonable that they can produce a better overall sine shape. Figure 16–33(b) shows a fundamental 20-Hz output waveform produced by a dual-bank cycloconverter operating with nonsteady firing delay. Compare the overall average shape of this waveform to the waveform of Fig. 16–31(b), which was produced by the same type of cycloconverter operating with steady firing delay.

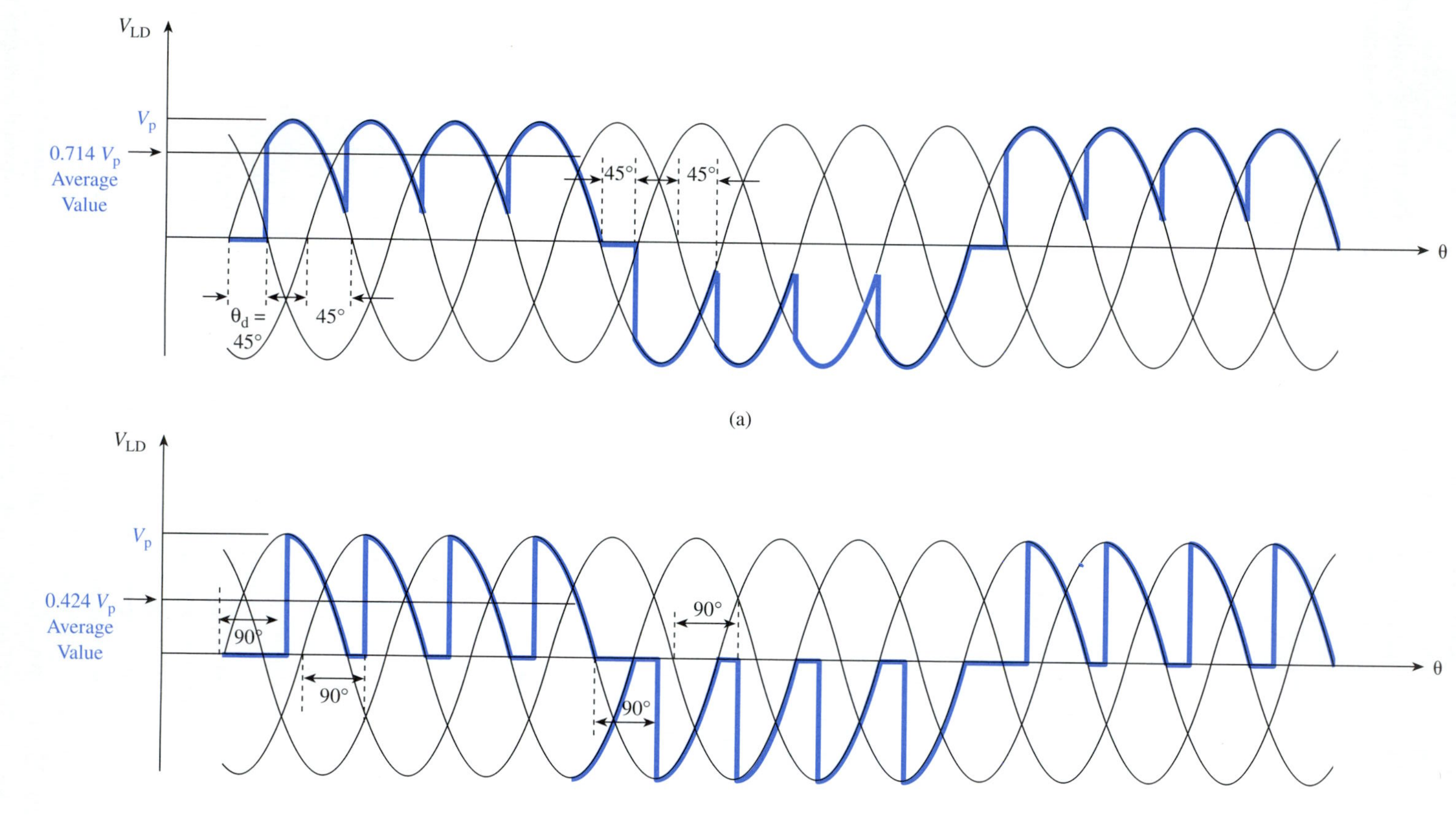

FIGURE 16–32
Output waveforms from a single-bank cycloconverter showing the effect of increasing the firing delay angle. (a) With four pulsations per half cycle, increasing the firing delay angle from 30° [in Fig. 16–30 (a)] to 45° causes the average voltage to decline from 0.749 V_p to 0.714 V_p. (b) Increasing θ_d to 90° reduces the average voltage to 0.424 V_p.

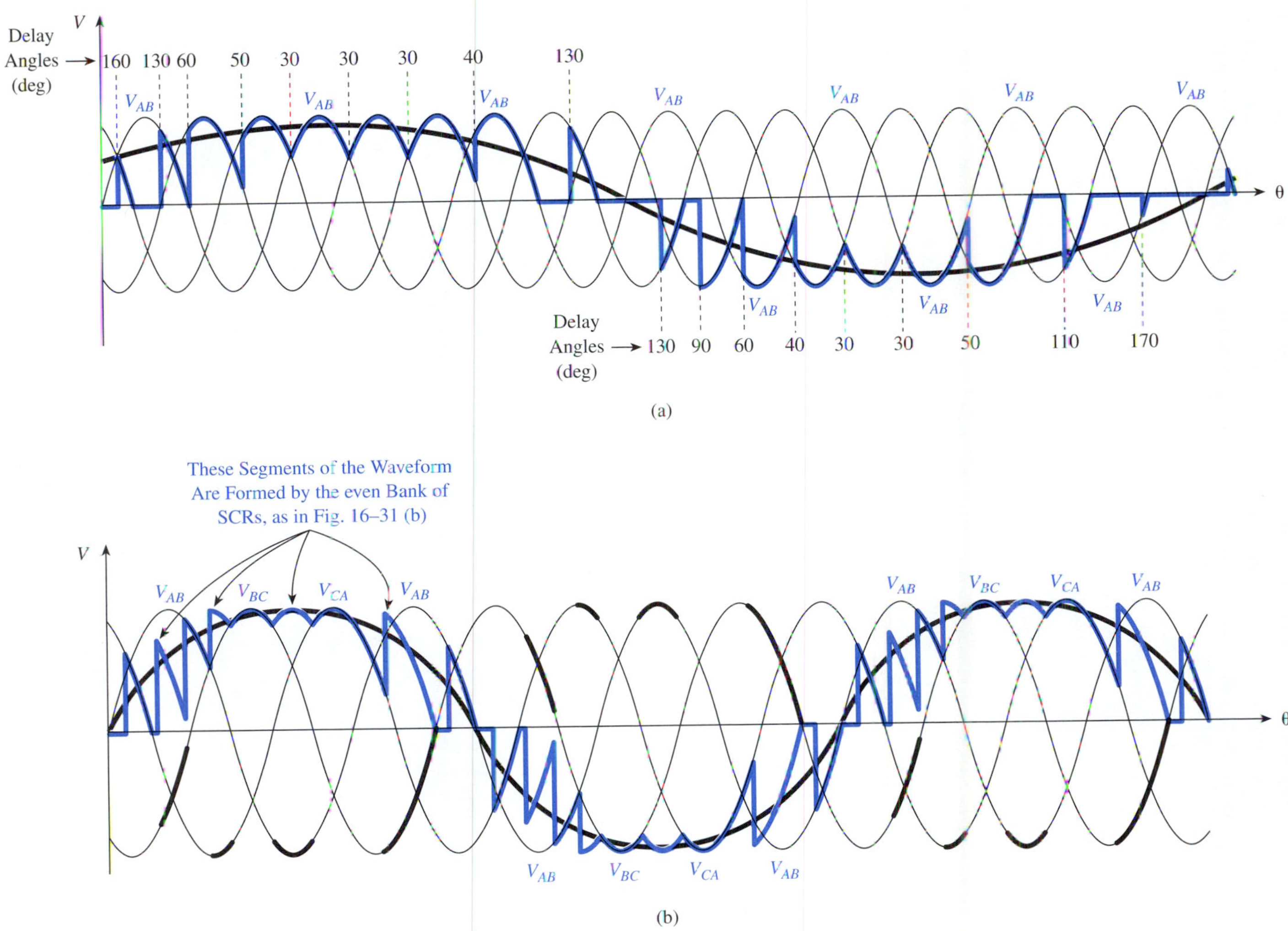

FIGURE 16–33
Improving the overall shape (reducing the harmonic content) of the output waveform by nonsteady firing delay: (a) For a single-bank cycloconverter with nine pulsations per half cycle, fundamental frequency of 10 Hz; (b) For a dual-bank cycloconverter with nine pulsations per half cycle, fundamental frequency of 20 Hz.

16-12-5 Three-Phase Cycloconverters

A three-phase cycloconverter is just three single-phase cycloconverters arranged so that their output waveforms are phase-displaced by 120°. The individual single-phase cycloconverters can be either single-bank units containing 6 SCRs or dual-bank units containing 12 SCRs. The single-bank approach uses a total of 18 SCRs; they are usually labeled with all the odd numbers from 1 to 35. The dual-bank approach uses a total of 36 SCRs, usually labeled with all the integers from 1 to 36.

Figure 16–34 shows a schematic diagram of a single-bank three-phase cycloconverter driving a three-phase induction or synchronous motor. The presence of the neutral wire N will help you keep straight just which phase of the three-phase voltage source is being applied to which phase (stator winding) of the motor at any instant in time. How-

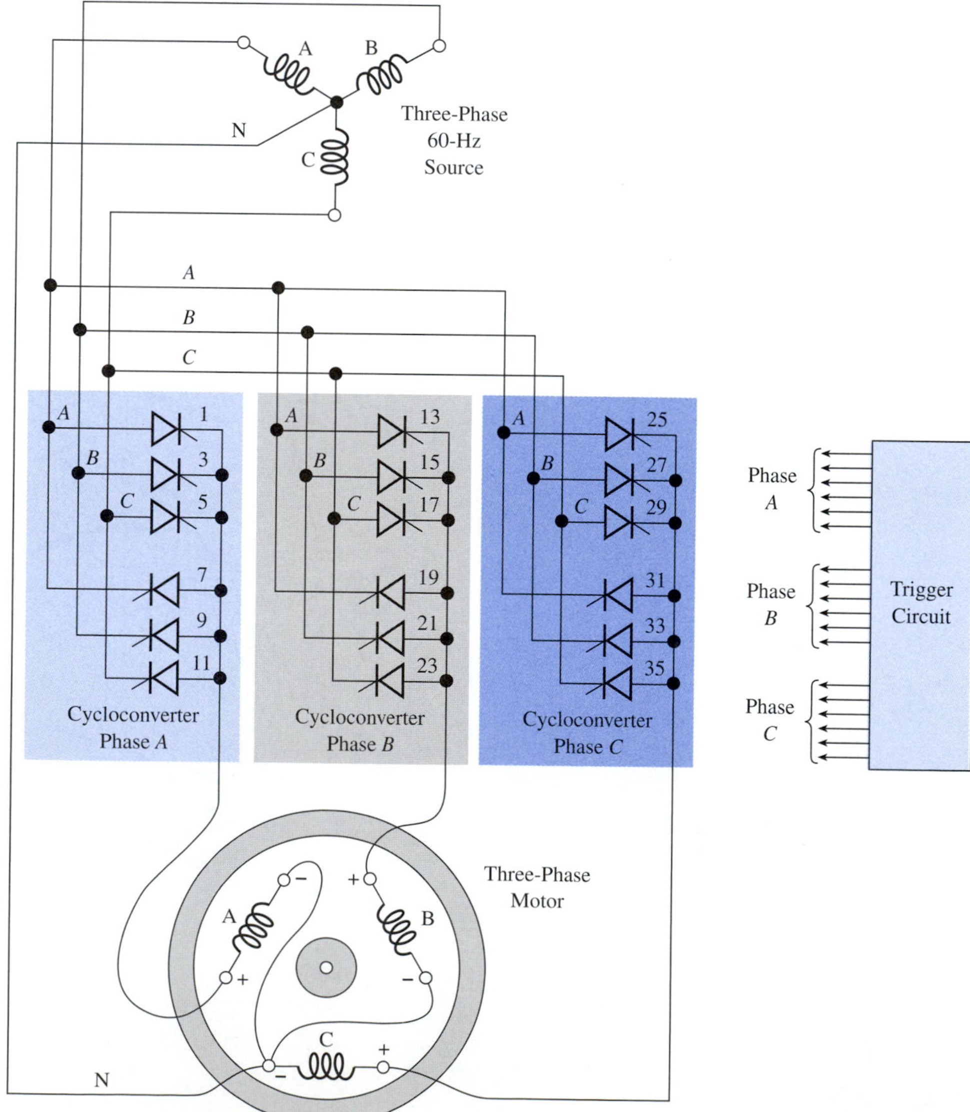

FIGURE 16–34
A three-phase cycloconverter is a combination of three single-phase cycloconverters. It can drive a wye- or delta-connected three-phase load from a wye- or delta-connected three-phase source. This diagram shows a single bank of SCRs per phase, 18 SCRs total. A dual-bank unit would contain 36 SCRs.

ever, if the motor phases are balanced (identical impedances), which they certainly will be unless the motor is a wreck, and if the cycloconverter phases are balanced, each one delivering the same voltage magnitude and frequency as the other two, then the neutral wire is not needed. This is true because whenever one cycloconverter phase is sourcing current into one phase of the load, the other two cycloconverter phases will be sinking current through the other two phases of the load, with their combined amounts exactly equal to the sourced current. The same is true at those times when two cycloconverter phases are sourcing current and just one is sinking current. Therefore no current flows in the neutral wire and it can be removed.

By similar reasoning, the load phases (motor stator windings) could be connected in a delta configuration just as well as a wye, and the same is true of the three-phase source.

Example waveforms for the three load phases are shown in Fig. 16–35. These waveforms portray basic operation at 20 Hz. There is no reduction in voltage magnitude by virtue of firing delay angle extension. Neither is there any waveshape improvement by virtue of firing pulses arriving on a nonsteady schedule. The gate firing-pulse synchronization is indicated right on the load-phase waveforms, for the individual SCRs. All 18 gate pulses are collected on their own time graph at the bottom of Fig. 16–35, displaying the complete firing sequence.

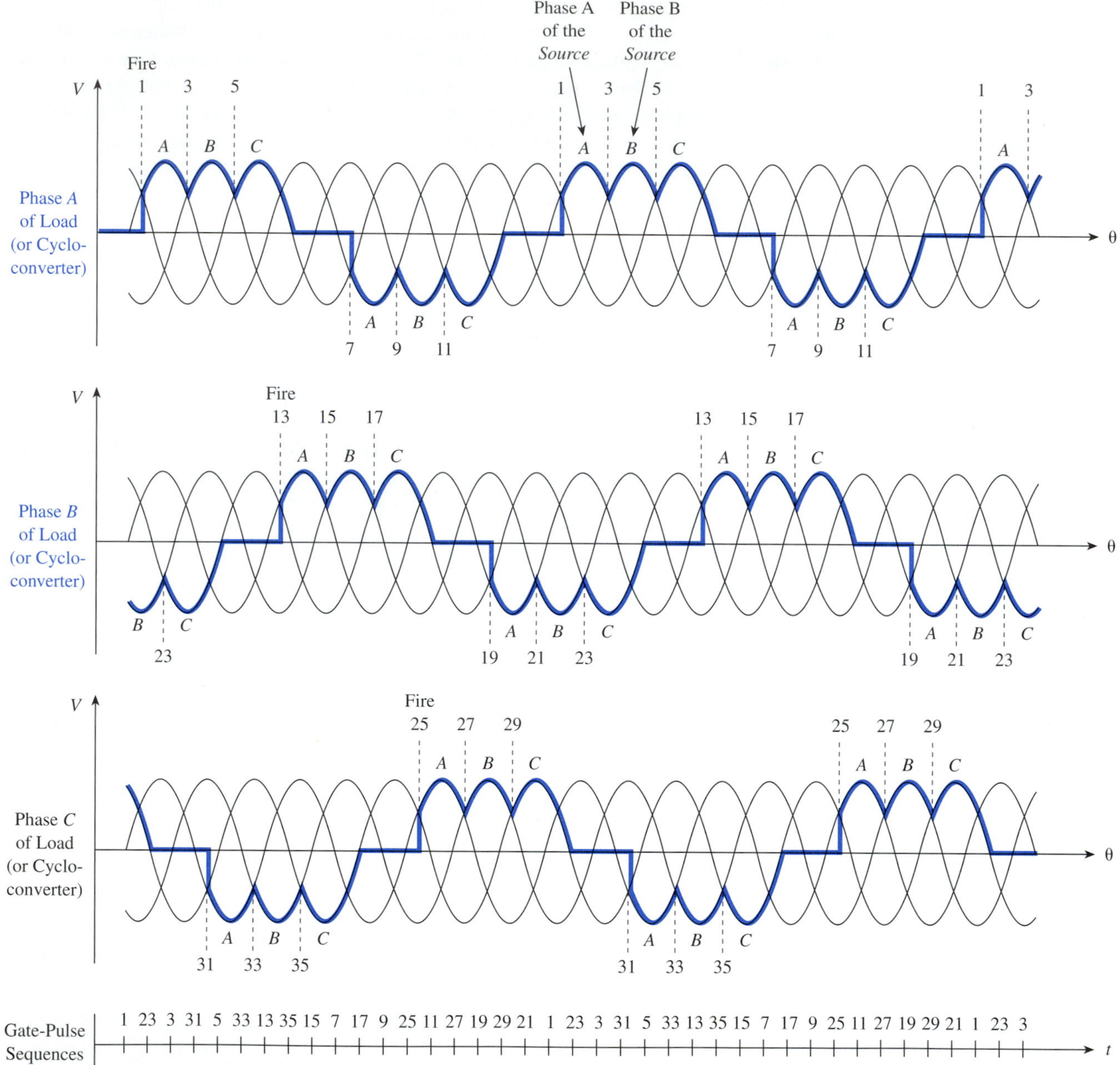

FIGURE 16–35

Phase-voltage output waveforms from the three-phase cycloconverter of Fig. 16–34. The letters *A*, *B*, and *C* near the pulsation peaks identify which phase of the three-phase *source* is being accessed to produce that pulsation.

TROUBLESHOOTING A HIGH-POWER SCR-BASED DC MOTOR DRIVE

In Troubleshooting on the Job in Chapter 12 you dealt with a problem in the limestone-feed apparatus for a sulfur-scrubbing tower. In that job the problem was the dc grinder motor itself.

Today the hydrogen sulfite detector again indicates that the slurry's H_2SO_3 concentration is beyond its normal range. Upon checking the limestone grinder, you discover that the grinder motor is running at only 1850 r/min. The grinder jaws are moving freely, with no hard-rock interference.

The grinder motor is a 20-hp unit driven by the dc drive system shown in Fig. 16–8. In that circuit the three speed-adjust pots are ganged to each other and to the sulfite-concentration pot, which is R_4 in Fig. 17–35(b) of Troubleshooting on the Job in Chapter 17. (Figure 17–35 is essentially the same as Fig. 10–15.) That Wheatstone bridge is automatically nulled by an instrument servo motor that is run by the amplified detector voltage. That is, the bridge-imbalance detector in Fig. 17–35(b) runs a tiny servo motor that turns R_4 and the three speed-adjust pots of Fig. 16–8.

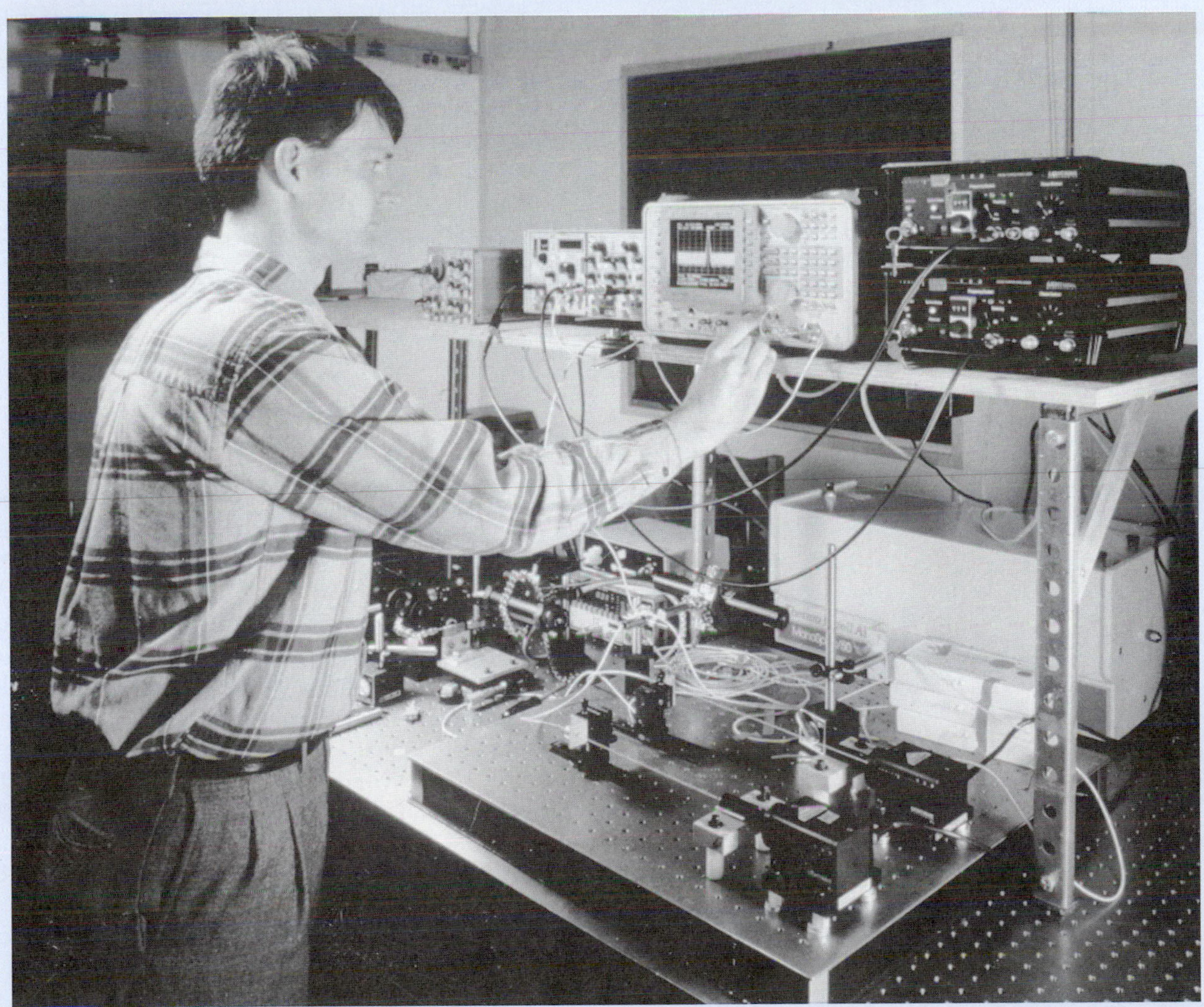

The oscilloscope is the technician's or engineer's most useful all-around instrument.
Courtesy of Tektronix, Inc.

The translucence of the sampled slurry in Fig. 17–35(a) represents the concentration of H_2SO_3, as will be explained in Troubleshooting on the Job in Chapter 17. Higher sulfite concentration causes the instrument servo to move the speed-adjust pot wipers toward the bottom in the Trigger Circuit schematic of Fig. 16–8. This increases the drive on current-source transistor(s) Q_1, firing the UJTs earlier. Therefore the SCRs fire earlier, raising the applied armature voltage to the grinder motor. Faster grinding speed increases the feed rate of pulverized limestone into the slurry stream, which speeds up the final capturing reaction in Fig. 12–41. Thus the H_2SO_3 concentration should be held down to a proper level. This is how the closed-loop control system is supposed to work.

Placing a voltmeter across the motor armature, you measure only 155 V dc. This is far below the 233 V maximum that the drive system is supposed to apply to the motor's armature if the H_2SO_3 concentration rises to the maximum value in its acceptable range.

1. What is your immediate conclusion about the functioning of the overall closed-loop motor speed-control system? Is the problem in the motor itself, or is it elsewhere in the control system?
2. Suppose that you measure conditions in the photocell bridge with a DVM, finding that $V_{\text{detector}} = 0.00$ V and $V_{R2} = 0.5$ V. Is the bridge responding properly to a high sulfite concentration in the slurry sample? Explain.
3. Using a dual-trace scope to investigate the performance of Trigger Circuit A in Fig. 6–8, you measure the voltage waveform between the wiper of the speed-adjust pot and the emitter of Q_1. The scope shows 60-Hz half-wave pulsations, 27 V in magnitude, synchronized with the V_{AB} positive half cycles. Has the speed-adjust pot been properly positioned by the bridge-detector servo motor? Explain your answer.
4. What will you do from here? Describe the testing/troubleshooting procedure that you will use.

SUMMARY

- There are two methods of controlling the speed of a wound-rotor, wound-field dc motor: (1) Varying the voltage V_A applied to the armature winding. Increasing V_A increases speed S. (2) Varying the current I_F through the field winding. Decreasing I_F increases the speed S.
- The speed-control method of varying I_F (called field control) has this great disadvantage: Reducing I_F reduces the motor's magnetic flux density B, which is one of its torque-producing constituents. Thus, the motor's ability to create torque is degraded at the same time that the mechanical load is demanding greater torque in order to turn faster.
- The speed-control method of varying V_A (called armature control) does not have the torque reduction disadvantage. However, it does require an efficient power-control device to handle the large armature current I_A. Generally, variable resistance is not acceptable due to its inefficiency; an SCR or other switching semiconductor must be used.
- When an SCR motor-drive circuit is constructed so that increased armature counter-EMF tends to retard triggering (increase firing delay angle) the system tends to have good speed regulation.
- Reversible dc motor drives often use electromagnetic contactors to reverse the polarity of applied armature voltage V_A. It is also possible to construct totally electronic (no electromagnetic contacts) reversible dc drives.

- High-power dc drive systems are often supplied by a three-phase ac source. Then there must be three separate trigger-control circuits, one for each of the three line SCRs.
- Low-to medium-power dc drives may use switching transistors that vary the average applied armature voltage by pulse-width-modulation techniques.
- The type 555 timer-oscillator is an integrated circuit that is useful for accomplishing pulse-width modulation. One 555 serves as a constant-frequency oscillator and a second 555 serves as a variable one-shot to supply the turn-On pulse to the load-switching transistor.
- It is more difficult to control the speed of an ac induction motor than to control a dc motor because standard ac motors are more dependent on line frequency than on voltage magnitude.
- The ac motor speed-control circuit that converts a dc supply voltage into a variable-frequency ac waveform is called an inverter.
- The ac motor speed-control circuit that converts a 60-Hz three-phase ac source into a variable-frequency three-phase ac waveform is called a cycloconverter.
- Most ac inverter and cycloconverter drive systems maintain a fairly constant voltage-to-frequency ratio, V/f. This is necessary in order to keep the motor's stator magnetizing current, I_{mag}, nearly constant; this in turn holds the stator flux density B nearly constant.
- In a cycloconverter, the positive triplet(s) produces the positive half-cycle of the ac load waveform and the negative triplet(s) produces the negative half-cycle. Each triplet contains 3 SCRs, one for each phase of the 3-phase ac line.

FORMULAS

$$\text{load regulation} = \frac{S_{NL} - S_{FL}}{S_{FL}} \qquad \textbf{(Equation 16-4)}$$

For a 555 oscillator with two timing resistors:

$$t_{\text{HI}} = 0.7 \times \tau_{\text{chg}} = 0.7 \times (R_{T1} + R_{T2}) \times C_T \qquad \textbf{(Equation 16-6)}$$

$$t_{\text{LO}} = 0.7 \times \tau_{\text{dischg}} = 0.7 \times R_{T2} \times C_T \qquad \textbf{(Equation 16-8)}$$

$$T = t_{\text{HI}} + t_{\text{LO}}$$

$$f = \frac{1}{T} = \frac{1}{0.7(R_{T1} + 2R_{T2})C_T}$$

$$f = \frac{1.4}{(R_{T1} + 2R_{T2})C_T} \qquad \textbf{(Equation 16-9)}$$

For a 555 one-shot with no Control input:

$$t_{\text{p}} \approx 1.1 \times \tau_{\text{chg}} = 1.1\, R_T C_T \qquad \textbf{(Equation 16-10)}$$

QUESTIONS AND PROBLEMS

Section 16-1

1. What is the advantage of dc motors over ac motors in industrial variable-speed systems?
2. Which is greater in a dc shunt motor, armature current or field current? Would this be true of a dc series motor?

3. Explain why *decreasing* the field current in a dc shunt motor *increases* its rotational speed.
4. What is the main drawback of field control of a dc shunt motor?
5. Explain why increasing the average armature voltage to a dc shunt motor causes it to speed up.
6. What is the main drawback of armature control using a series rheostat?
7. Describe the sequence of events as a shunt motor is started with an "across-the-line" starter. Describe how the following three variables change: armature current, counter-EMF, and shaft speed.

Section 16-2

8. Why is armature control by thyristor better than armature control by series rheostat?

Section 16-3

9. In Fig. 16–3(b), why can't the motor reach 100% of its full rated speed?
10. Does Fig. 16–3(c) represent speed variation for different pot settings or for a fixed pot setting? Explain.
11. If the drive system of Fig. 16–3 could provide load regulation of 0%, what would the graph of Fig. 16–3(c) look like?

Section 16-4

Questions 12 to 15 refer to Fig. 16–4.

12. What is the purpose of R_1 and D_1?
13. In which direction should the speed-adjust pot wiper be moved to speed up the motor? Should it be moved to the right or to the left?
14. What is the purpose of D_3?
15. Do the bridge rectifier diodes have to be heavy-current diodes, or can they be relatively light-current diodes? Why?

Section 16-5

16. In a dc shunt motor, will the rotation be reversed if *both* the field current and armature current are reversed? Explain.
17. In Fig. 16-5, what is the purpose of the N.C. REV and FOR contacts?
18. Explain the distinction between *switchgear* (relay) control and *electronic* control.

Section 16-6

19. Generally speaking, when is a three-phase drive system used instead of a single-phase drive system?

Section 16-7

Questions 20 to 27 refer to the three-phase drive system of Fig. 16–8.

20. Why do the manufacturers of drive systems install thyrectors across the incoming power lines?

21. Explain the purpose and operation of the field-failure relay, RFF.
22. What is the maximum number of degrees per half cycle for which any SCR is allowed to conduct? Why can't the SCRs be allowed to conduct for 180°?
23. Why is step-down transformer T_1 used? Why don't we simply design the motor starter coil to operate on 230 V ac?
24. Give a step-by-step explanation of why the motor slows down as the speed-adjust pot wiper is moved up.
25. If the firing delay angle of one SCR is changed, do the other two SCRs also change, or are they all independent? Explain.
26. Explain how R_7 provides counter-EMF feedback to the trigger control circuit to yield improved load regulation.
27. What is the purpose of the circuit consisting of C_2, R_2, R_3, and the N.C.M contact? Explain how it works.

Section 16-8

28. In Fig. 16–10, suppose $V_S = 80$ V and $R_{LD} = 7.5\ \Omega$.
    a. Find the load power P_{LD} when $V_{Control} = 4$ V.
    b. To double the power from part **a,** we would have to ____________ the value of $V_{Control}$. Explain why this is so.
29. For the circuit of Question 28, what value of control voltage would produce $P_{LD} = 500$ W?

Section 16-9

30. In a 555 IC, what input, Trigger or Threshold, signals the Main Output to become HI? Which input signals the Main Output to Reset to LO?
31. The Discharge Output (pin 7) goes LO when the Main Output (pin 3) goes LO. Discharge Output goes to the high-impedance state when the Main Output goes HI. Then why do we need a separate Discharge Output?
32. The 555 oscillator of Fig. 16–13 has $R_{T1} = 2.2\ \text{k}\Omega$, $R_{T2} = 4.7\ \text{k}\Omega$, and $C_T = 0.05\ \mu\text{F}$.
    a. Find the oscillator's period T and frequency f.
    b. Find its duty cycle D.
33. Leaving the other components the same, what value of R_{T2} in Question 32 will produce $f = 3$ kHz?
34. Suppose that the one-shot of Fig. 16–8 has $V_{CC} = 15$ V, $R_T = 10\ \text{k}\Omega$, and $C_T = 0.2\ \mu\text{F}$. Calculate the normal pulse duration t_p.
35. In the one-shot of Question 34, suppose the Control input is not bypassed to ground but receives an externally applied control voltage $V_{Control}$.
    a. For $V_{Control} = 6$ V, find t_p.
    b. Repeat for $V_{Control} = 9$ V.
36. In the pulse-width motor speed-control system of Fig. 16–21, it is desired to apply an average voltage of 32 V to the motor armature.
    a. What duty cycle D is required?
    b. What value of control voltage $V_{Control}$ will accomplish this?

Section 16-10

37. State some of the advantages of ac induction motors over dc motors.
38. What term is used to refer to a circuit that accomplishes dc-to-ac conversion?

39. What term is used to refer to a circuit that accomplishes ac-to-lower frequency ac conversion?
40. In Fig. 16–23(a), which SCR must be turned ON in order to carry current through winding B in the positive direction?
41. In Fig. 16–23(a), which SCR must be turned ON in order to carry current through winding B in the negative direction?
42. The topmost waveform of Fig. 16–24(a) shows that winding A carries no current during the third and sixth intervals. Which two SCRs must be turned OFF in order to produce this deenergization of winding A?
43. The top three waveforms of Fig. 16–24(a) show that the magnitude of voltage across a single motor winding is half the dc supply voltage. Explain why this is reasonable.
44. Three commutating capacitors, C_A, C_B, and C_C, are shown in Fig. 16–25. Describe the charge existing on each one of those capacitors during the second time interval of Fig. 16–24(a).

Section 16-11

45. For a variable-frequency ac motor drive system, explain why the magnitude of voltage applied to the motor must be varied in proportion to the frequency.
46. For the circuit of Fig. 16–26, it is necessary to have ON SCRs 1 and 6 in order to access the positive-peak region of E_{AB} (line *A* instantaneously positive relative to line *B*). Which SCRs should be ON in order to access the negative peak region of E_{AB}? Explain.
47. Sketch the corresponding waveform of Fig. 16–27 for a delay angle of 60°. By inspection of your waveform, is its average value of V_K less than it was in Fig. 16–27(c)? If you are mathematically inclined, calculate the new average value by integrating the sine function over the range 120° to 180° ($\frac{2}{3}\pi$ to π radians).
48. Sketch the corresponding waveform of Fig. 16–27 for a delay angle of 90°. By inspection, what is the average value of V_K under this condition? If you are inclined, prove this result by integrating the sine function over the range 150° to 210° ($\frac{5}{6}\pi$ to $\frac{7}{6}\pi$ rad).

Section 16-12

49. For the six-SCR cycloconverter shown in Fig. 16–29, sketch the waveform for the condition of the trigger circuit delivering five sequential gate pulses to a single triplet (five pulsations per half cycle of output voltage). What is the output frequency?

Questions 50 to 54 refer to the dual-bank cycloconverter of Fig. 16–31(a).

50. If SCRs 3 and 4 are ON, which line voltage is being accessed? Is it being accessed "forward" (near its defined-positive peak) or "backward" (near its defined-negative peak)? What is the instantaneous polarity of V_{LD}?
51. Repeat Question 50 for SCRs 7 and 8 ON.
52. Repeat Question 50 for SCRs 8 and 9 ON.
53. If we wish to access E_{CA} near its defined-positive peak to produce a pulsation in the positive half cycle of V_{LD}, which two SCRs must be ON?
54. If we wish to access E_{CA} near its defined-negative peak to produce a pulsation in the positive half cycle of V_{LD}, which two SCRs are needed?

Questions 55 and 56 refer to the three-phase single-bank cycloconverter driving a three-phase motor, as indicated in Figs. 16–34 and 16–35.

55. During the time interval between the firing of SCR_{31} and the firing of SCR_5, which three SCRs are ON? Tell the direction of current through each motor winding. Tell the direction of current through each phase of the three-phase source.
56. Repeat Question 55 for the time interval between the firing of SCR_{13} and the firing of SCR_{35}.

CHAPTER

17

TELEMETRY

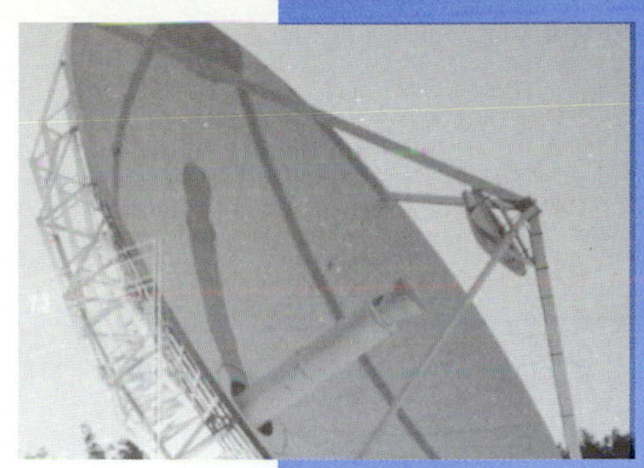

In most industrial control situations, the measurement transducer and the final correcting device are in the same vicinity. There are occasional applications in which the measured value must be transmitted a rather long distance, perhaps several hundred feet or more. In still other instances, electric power distribution for example, the measurement must be transmitted many miles back to the controller. Whenever a measurement must be sent a long distance by wire or optical fiber, it cannot be kept in its original analog form. Noise and signal degradation over long distances undermine the integrity of analog voltages.

Instead, the original analog voltage must be converted to some form of pulse modulation, or else it must be converted to a digital encoded value and transmitted bit by bit. *Telemetry* is the technology of changing an analog measurement to one of these two forms, transmitting the altered form over a long distance, and then reconverting the received information back to analog.

OBJECTIVES

After studying this chapter, you should be able to:

1. State the advantages of transmitting measured values by pulse modulation rather than in direct analog form.
2. Show how a pair of 555 ICs can accomplish pulse-width modulation.
3. Show how a pulse-width-modulated signal is demodulated by a low-pass filter.
4. Explain how pulse-position modulation is derived from pulse-width modulation.
5. Explain why pulse-frequency modulation is more noise-immune than pulse-width or pulse-position modulation.
6. Show how pulse-frequency modulation is accomplished with a 555.
7. Describe the operation of a phase-locked loop.
8. Use a type 565 phase-locked loop for demodulation of a pulse-frequency modulated signal.
9. State the advantage of multiplexing in a telemetry system.
10. Draw the block diagram of a multiplexed telemetry system, and explain its operation.
11. Discuss the use of radio transmission rather than wire or fiber transmission in telemetry.
12. Distinguish between analog methods of pulse modulation and digital pulse encoding, and state the digital advantage.

13. Trace the detailed circuit operation of a 4-bit digitally encoded telemetry system.
14. Give the restrictions on the signal-sampling rate of a telemetry system, and explain why they are necessary.

17-1 ■ TELEMETRY BY PULSE-WIDTH MODULATION

The pulse-width-modulation technique, discussed in Secs. 16-8 and 16-9 in the power-control context, can also be applied to long-distance measurement transmission. Figure 17–1 shows the hardware arrangement.

The input transducer converts the value of the physical variable to an analog voltage, v_{MEAS}. After appropriate conditioning, often involving dc-shifting (superimposing a dc bias) and rescaling, $v_{CONTROL}$ is derived from v_{MEAS}.

Two 555s are used in the usual way to produce a string of pulses with variable width but constant overall waveform period. As the measured variable changes, so will the width of the pulses applied to the long transmission line. Figure 17–2(a) shows the relation between measurement value and pulse width for a situation where v_{MEAS} has only positive values.

The advantage is that the *width* of the pulses is unlikely to be affected by electrical noise injected along the transmission path, and if the pulses are reduced in amplitude by the transmission run, it doesn't matter. The receiving circuitry isn't watching their amplitude. It's watching their time duration and their width.

At the receiving end there must be a circuit that can reconvert the pulse width into an analog voltage that represents v_{MEAS}. It is not necessary that the reconverted voltage be exactly *equal* to v_{MEAS}. It is only necessary that the reconverted voltage be related to the value of the measured variable in a well-defined, known way.

The circuit that performs reconversion back to an analog voltage is called a *pulse-width demodulator*. Its output voltage can be symbolized v_{DEMOD}. Figure 17–3 illustrates this idea.

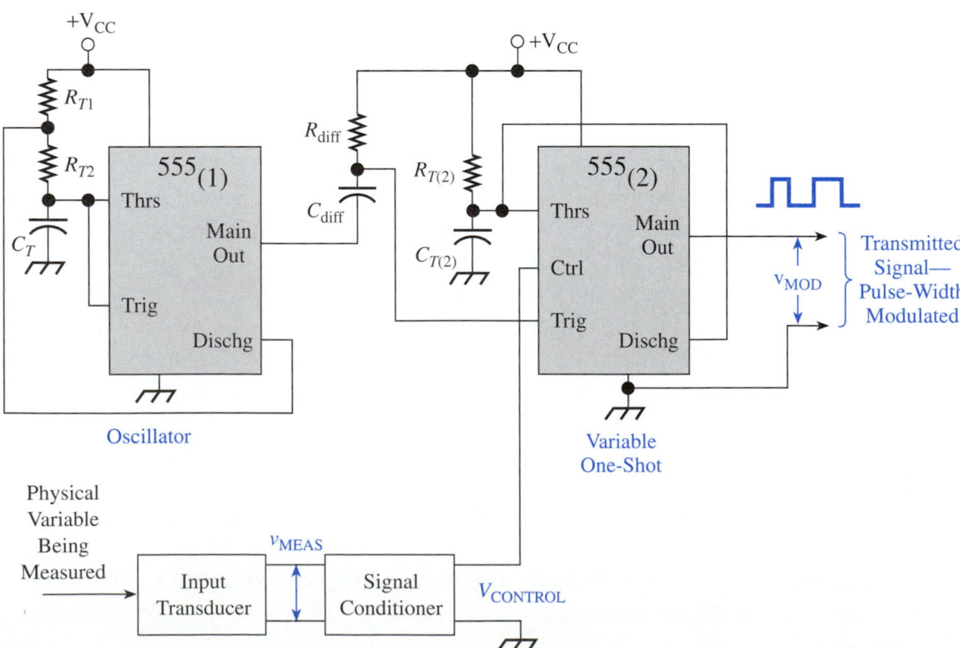

FIGURE 17–1
Telemetry by pulse-width modulation.

FIGURE 17–2 How the measured value of the variable modulates the pulse width. Here the measured value is changing quite quickly in comparison to the overall cycle-period of the transmitted waveform v_{MOD}. If v_{MEAS} were changing in its usual slow manner, there wouldn't be such dramatic differences in width from one pulse to its neighbor. (a) Only positive values of v_{MEAS}. The pulses can only get wider than normal. (b) Both positive and negative values of v_{MEAS}. The pulses can be wider or narrower than normal.

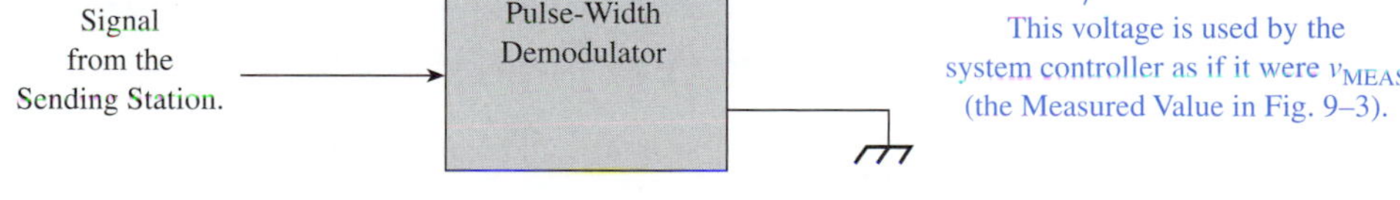

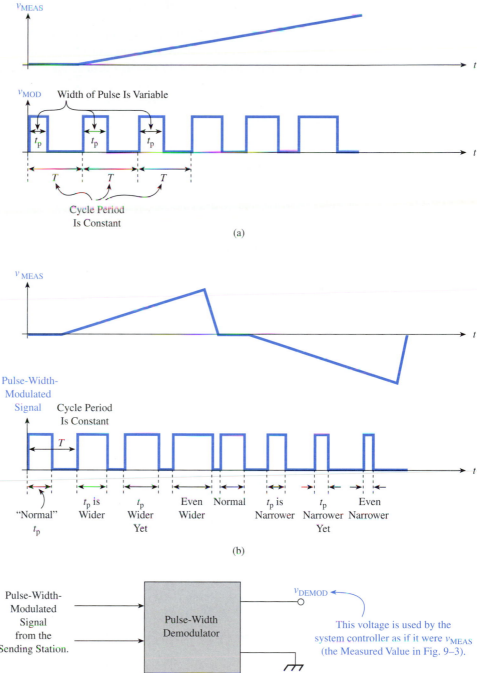

This voltage is used by the
system controller as if it were v_{MEAS}

FIGURE 17–3 Receiving station in a telemetry system. The voltage v_{DEMOD} represents the measured variable; it is used to make the comparison to set point.

17-1-1 Demodulation

The act of demodulating a pulse-width-modulated signal involves two requirements:

1. Make the actual received pulse amplitudes irrelevant.
2. Filter the pulse waveform resulting from step 1 to recover the waveform's average (dc) value.

One straightforward method of accomplishing such demodulation is shown in Fig 17–4.

Pulses, which may be ragged, arrive at input terminal A. They are amplified by a factor of about 13 by noninverting amplifier 1. This guarantees positive saturation at point B if the pulse amplitude is just greater than about 1 V. Waveforms of v_A (input pulse) and v_B (op amp 1 output) are shown in Fig. 17–4(b). At this point the duty-cycle intelligence that was placed on the transmission lines at the sending station has been recovered.

Dropping resistor R_{drop}, acting with the 7-V zener diode, clips the saturated pulse waveform at 7 volts. Duty cycle D is maintained at point C. Then the 7-V pulses are filtered by the low-pass active filter, op amp 2. This filter is designed to provide a gain of 1 (it's a voltage-follower) to the dc component of the v_C waveform, but it greatly attenuates

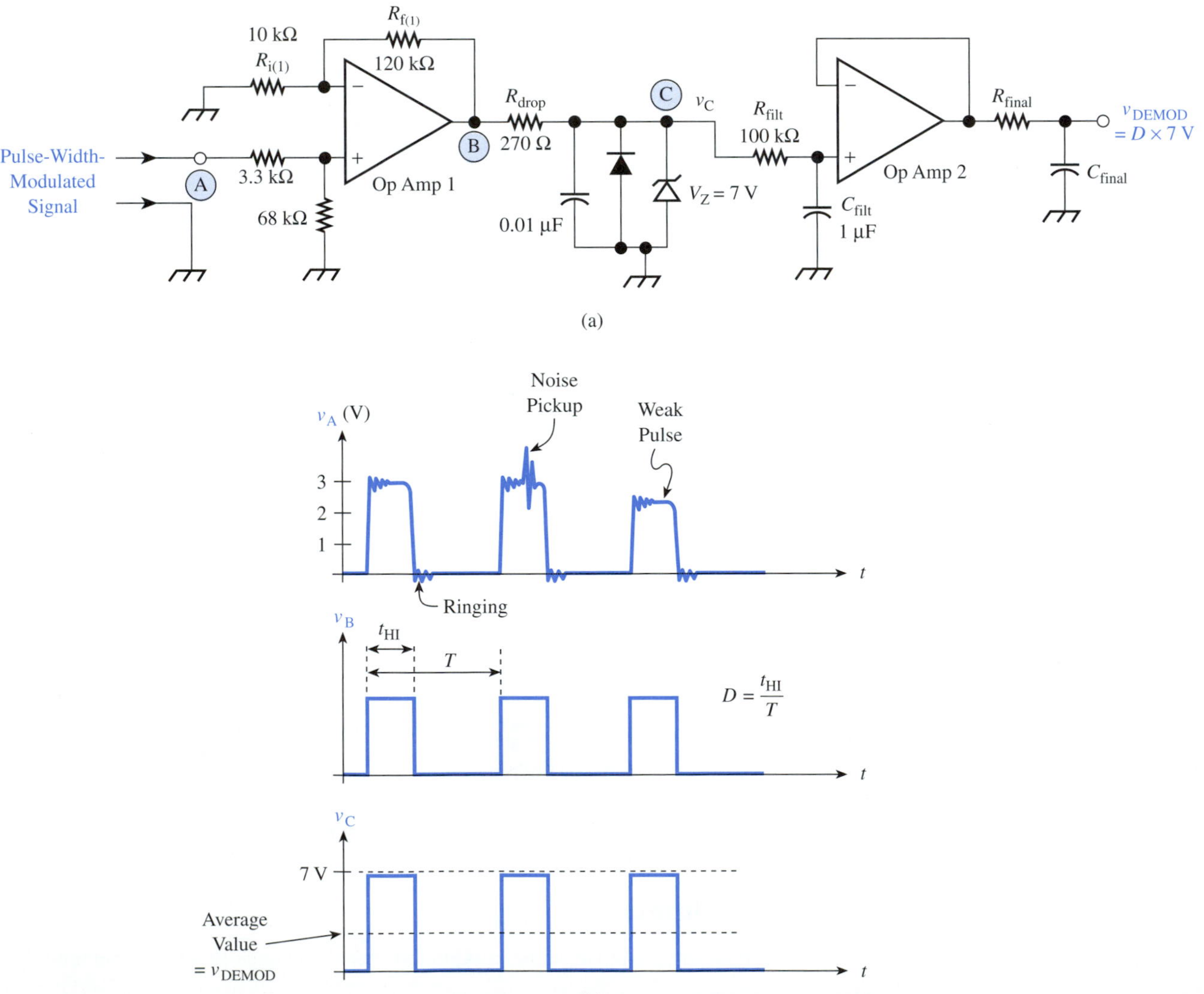

FIGURE 17–4
Pulse-width-demodulating circuit. (a) Schematic. (b) Waveforms.

the fundamental frequency (the oscillator frequency in Fig. 17–1). All harmonic frequencies are attenuated even more than the fundamental. The final result is a dc voltage that theoretically varies between 0 and 7 V, as the pulse-width modulation from the sender varies from 0% to 100%. As a practical matter, pulse-width modulation is never allowed to vary over the entire 100% range. It is restricted to the range from about 10% to 90%, or narrower.

17-1-2 Pulse-Position Modulation

A variation on pulse-width modulation is *pulse-position modulation.* In this encoding method a very short pulse of fixed duration is produced at the moment the width-modulated pulse terminates. That is, every falling edge of the pulse-width-modulated signal produces a fast pulse. This idea is illustrated in Fig. 17–5 for a situation where v_{MEAS} can be either positive or negative.

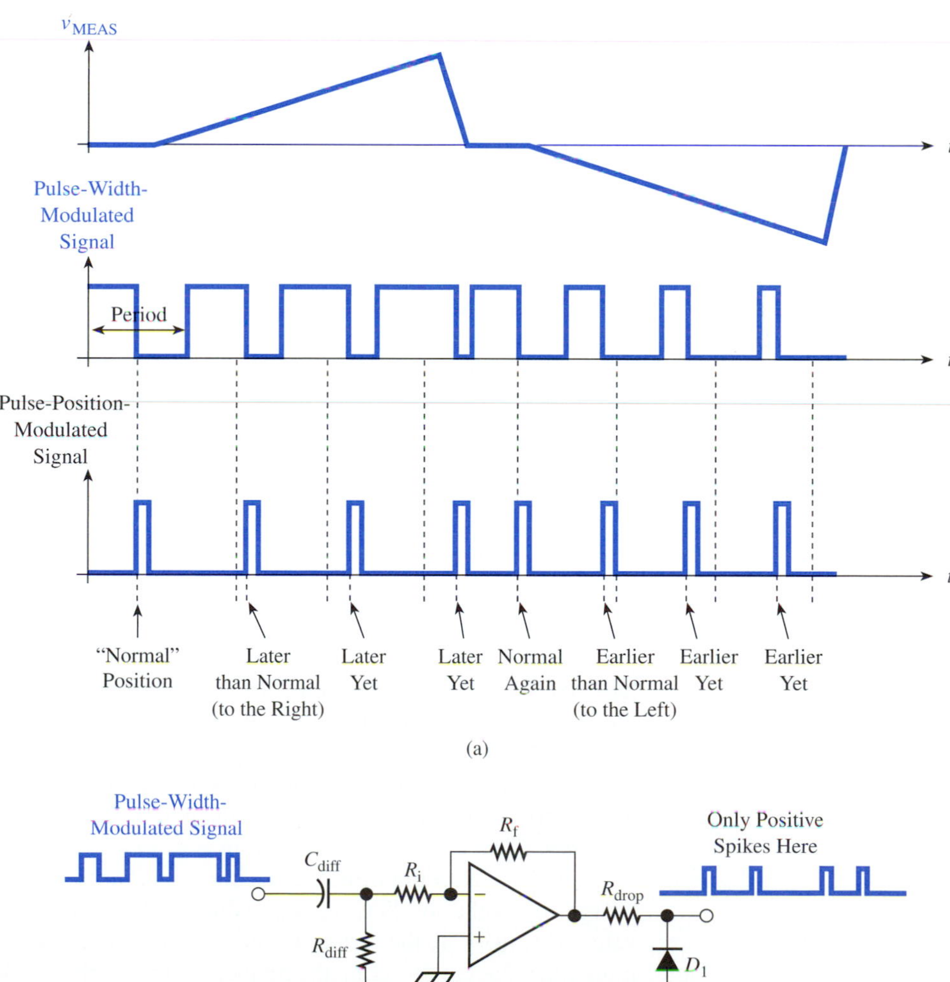

FIGURE 17–5
Pulse-position modulation. (a) Waveforms. (b) Circuit schematic.

In the circuit of Fig. 17–5(b), the short-τ RC differentiator produces a fast pulse (a spike) on every transition, rising or falling, of the width-modulated signal. The spikes are inverted and buffered by the op amp. Then the resistor-diode clipping circuit on the output eliminates the negative spikes. It allows the positive spikes to remain, representing the falling edges of the width-modulated input.

17-2 ■ TELEMETRY BY PULSE-FREQUENCY MODULATION

Pulse-width modulation is quite an effective means for transmitting analog intelligence over long distances, but it's somewhat susceptible to noise injection, especially when the basic oscillation frequency is high. This will be the case if v_{MEAS} is not slow-changing but is itself capable of high-frequency variation.

Under that condition the rise and fall times of the modulated pulse may not be negligible compared to the oscillator period. The situation is illustrated in Fig. 17–6.

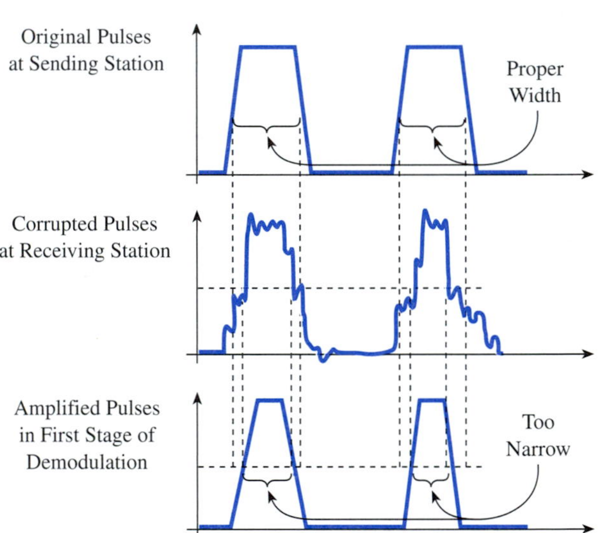

FIGURE 17–6 When the signal has high-frequency content, the modulator oscillation frequency must also be high. Then the pulse rise-time may be significant. When noise is superimposed on rise-time-sensitive transmitted pulses, the demodulator's amplifier stage can reconstruct the pulses incorrectly.

To overcome this problem, we often turn to *pulse-frequency modulation* as an alternative method of telemetry. As its name implies, pulse-frequency modulation changes the actual frequency (and period too, of course) of the transmitted pulses. This is unlike pulse-width modulation, where the cycle period is constant.

To frequency-modulate our pulses, we operate a 555 in astable oscillator mode, with the Control input used to vary the frequency of oscillation. This is the *voltage-controlled oscillator* idea, or VCO. It is shown in Fig. 17–7.

In that setup the timing capacitor will charge to the value of $v_{CONTROL}$, then discharge to a value of $\frac{1}{2}\,v_{CONTROL}$, which is the critical trigger voltage. The greater the value of $v_{CONTROL}$, the further C_T must charge. When C_T must charge further, it requires more time, thereby lowering the frequency. This is shown clearly in Fig. 17–8. Conversely, lowering $v_{CONTROL}$ raises the pulse frequency.

FIGURE 17–7
Pulse-frequency modulation.

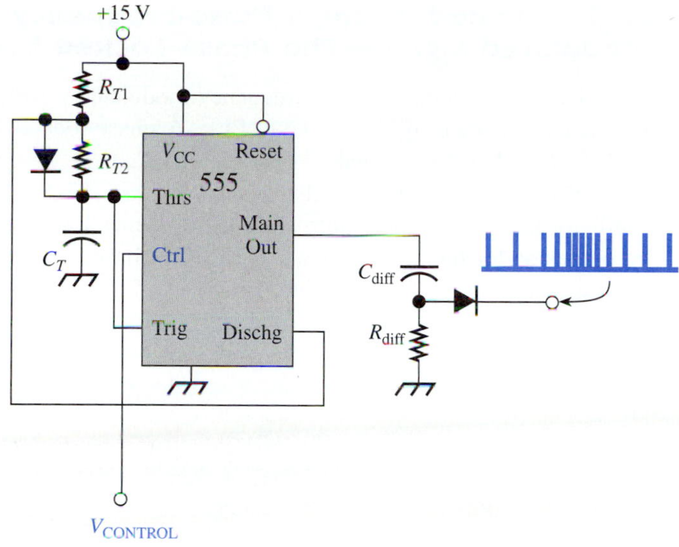

FIGURE 17–8
Waveforms for a 555 frequency-modulated oscillator. HI time does all the changing. LO time stays nearly constant.

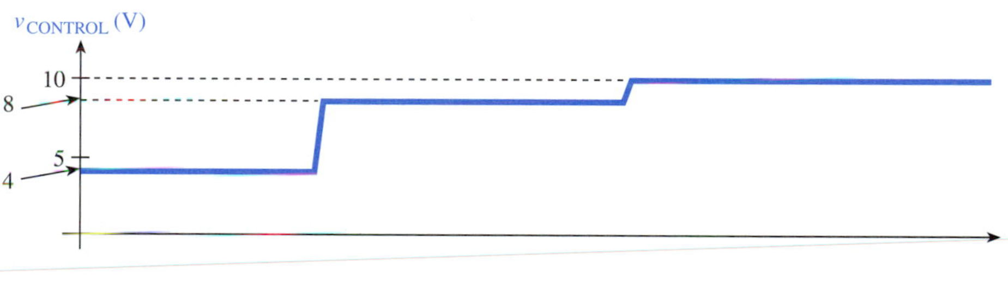

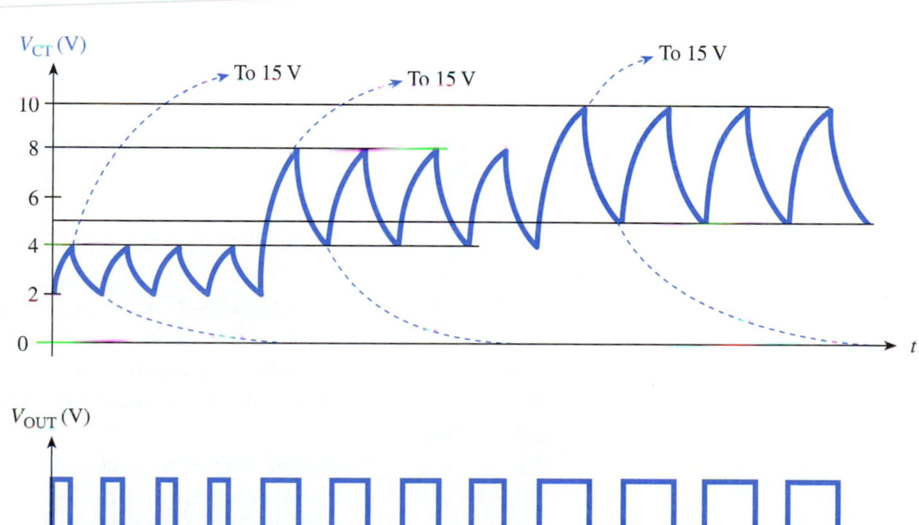

17-2-1 Demodulating a Pulse-Frequency Modulated Signal—The Phase-Locked Loop

To understand demodulation of a frequency-modulated waveform, we must study the *phase-locked loop,* known by the acronym PLL. A phase-locked loop is available on a 14-pin DIP IC, type No. 565, called a "five sixty-five".

The block diagram of a PLL is shown in Fig. 17–9. It contains a voltage-controlled oscillator in which higher control voltage produces lower frequency. This is the same relationship as for the VCO that we just studied based on the 555 IC. Therefore there is a natural compatibility between the 565 and the 555.

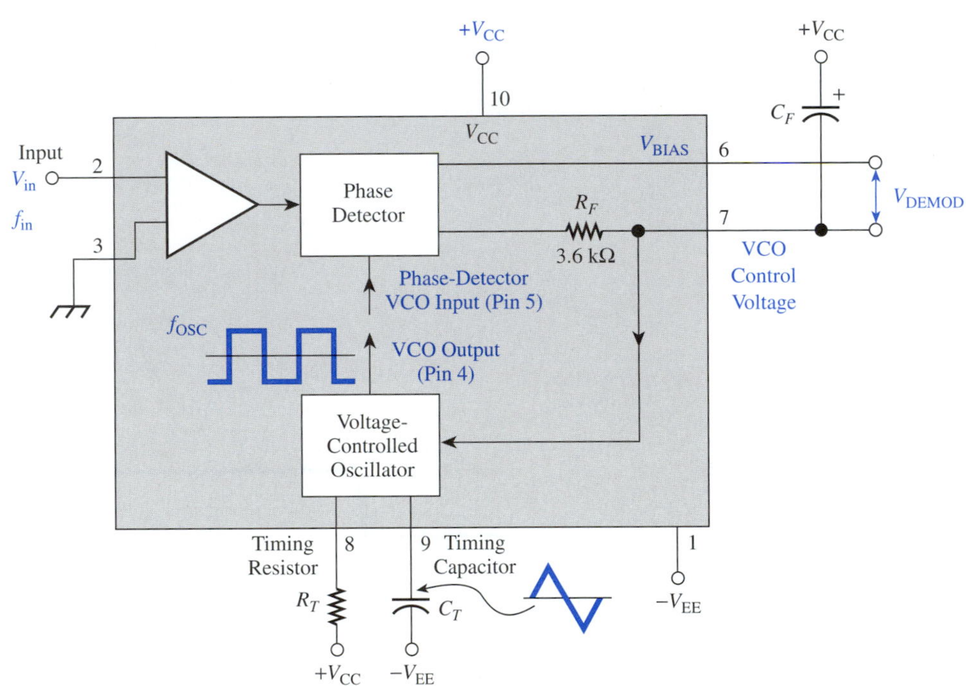

FIGURE 17–9
Phase-locked loop block diagram. With the VCO output connected to the phase-detector VCO input, f_{out} will become exactly equal to f_{in}, even if the V_{in} waveform is very ragged. The 565 PLL comes in a 14-pin DIP, but only pins 1 through 10 are used.

The phase-locked loop's VCO has a natural center frequency, f_{ctr}. That frequency is set by the values of timing resistor R_T and timing capacitor C_T, according to the equation

$$f_{ctr} = \frac{0.3}{R_T C_T} \tag{17-1}$$

The VCO delivers a triangle waveform at pin 9 and a square waveform at pin 4 of the 565. In normal operation of a PLL, the square-wave VCO output is connected directly to the Phase-Detector VCO input. That is, pin 4 is jumpered directly to pin 5 on the 565 IC. In the block diagram of Fig. 17–9, imagine the connection to be made between the VCO and the Phase Detector.

The job of a phase-locked loop is to *change* the frequency of its VCO away from f_{ctr} so that the actual operating frequency of the oscillator, f_{osc}, exactly equals the frequency of the input signal, f_{in}. That is, when the PLL is functioning,

$$f_{osc} = f_{in}$$

The PLL accomplishes this frequency-matching by automatically generating a VCO control voltage, which appears at pin 7 of the 565. The control voltage is internally connected to the VCO; that connection is not made by the user.

The VCO control voltage is produced by the phase detector—the top block inside the PLL. Here is how a phase detector works. The input signal V_{in} drives its amplifier into saturation, delivering a square-wave signal to one of the phase detector's two input points (the one on the left in Fig. 17–9). If, by luck, the VCO center frequency f_{ctr} happens to be exactly the same as the external input frequency f_{in}, then the phase detector produces zero output. The VCO control voltage is then equal to its natural dc bias value—it is shifted away from that natural bias value by 0 V.

In reality, of course, it is extremely unlikely that the VCO's natural center frequency will be exactly equal to f_{in}. If f_{ctr} is lower than f_{in}, the phase detector produces a negative output voltage, causing the VCO control voltage to slip below its natural dc bias value. When this voltage is applied to the VCO, the actual oscillator frequency, f_{OSC}, increases.

This feedback-and-adjust operation continues until f_{OSC} is exactly equal to f_{in}, but *lagging in phase*. The phase-detector circuitry uses the phase difference between the two square waves to produce its output voltage and thereby shift the VCO control voltage. If the voltage-controlled oscillator has to change from f_{ctr} by only a small amount (because there wasn't a large difference between f_{in} and f_{ctr}), the VCO output (at frequency $f_{OSC} = f_{in}$) will lag V_{in} by 90° plus only a few degrees. The phase detector needs only a small amount of phase difference between its two square waves (beyond its natural difference of 90°) to produce a small-magnitude output voltage.

However, if the voltage-controlled oscillator has to change from f_{ctr} by a large amount, the VCO output square wave will lag V_{in} by a large amount beyond 90°. This must happen, because the phase detector needs a larger amount of phase difference between its two square waves in order to produce a larger-magnitude output voltage.

Figure 17–10 shows the relationship among f_{in}, final phase difference, and VCO control voltage for two operating conditions: f_{in} being only slightly greater than f_{ctr}, and f_{in} being substantially greater than f_{ctr}. The value of f_{ctr} is 1.0 kHz.

Figure 17–10 and the foregoing discussion dealt with the situation where f_{in} was higher than f_{ctr}. The entire explanation would be the same if f_{in} were lower than f_{ctr}, but all changes would be reversed. Instead of the VCO square wave moving to the right of the 90° point, it would move to the left. This would cause the VCO control voltage to go more positive than V_{BIAS} instead of more negative. However, the overall PLL outcome would be the same: f_{OSC} would lock on to f_{in} exactly.

565 Capture range and lock range. The voltage-controlled oscillator in the 565 cannot lock on to *any* input frequency; f_{in} must come reasonably close to f_{ctr} in order for successful lock-on to occur.

A phase-locked loop's *capture range* is the range of input frequencies that the PLL can lock on to if it is not already locked on. In other words, if f_{in} is far away from f_{ctr}, with the VCO oscillating at frequency $= f_{ctr}$, the question is "How closely does f_{in} have to approach f_{ctr} to produce lock-on?" The capture-range idea is illustrated in Fig. 17–11(a).

A phase-locked loop's *lock range* is the range of input frequencies that the PLL can remain locked on to if it has already been captured. In other words, if f_{in} has already approached close enough to f_{ctr} to capture the PLL, the question is "How far away from f_{ctr} can f_{in} stray without losing the lock-on?" The lock-range idea is illustrated in Fig. 17–11(b).

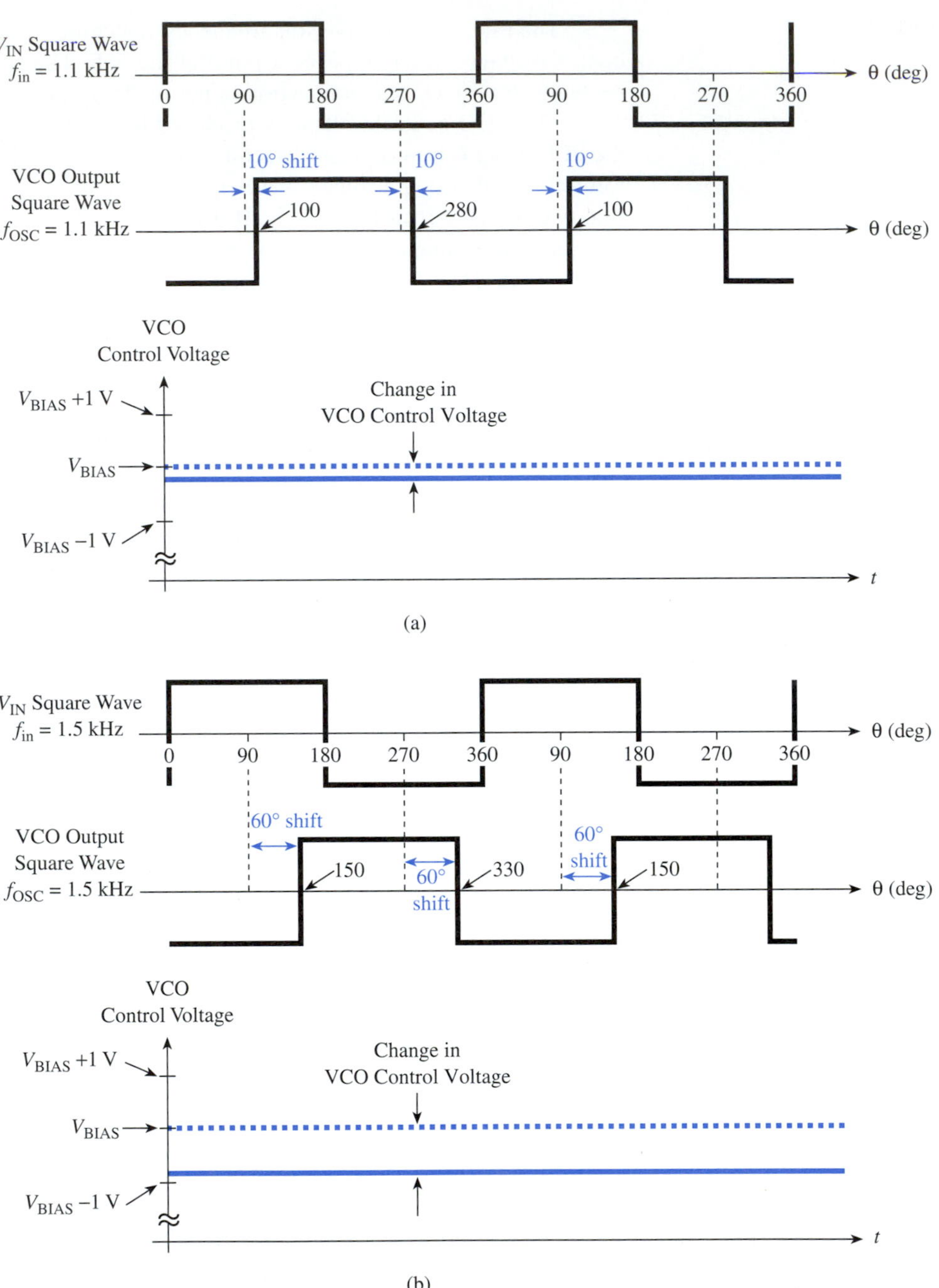

FIGURE 17–10 Phase-detector performance. (a) If f_{osc} must increase by only 10% (from f_{ctr} = 1.0 kHz to f_{in} = 1.1 kHz), the phase difference between the two square waves is not much greater than 90° (only 10 degrees greater). The phase detector changes the VCO control voltage by only a small amount. (b) if f_{OSC} must increase by 50% (from f_{ctr} = 1.0 kHz to f_{in} = 1.5 kHz), the phase difference between the two square waves is considerably greater than 90° (60 degrees greater). With this large a phase shift, the phase detector changes the VCO control voltage by a substantial amount. Of course, this substantial change in control voltage was necessary in order to force the VCO to make such a large change in its frequency.

The capture range is always narrower than the lock range for any phase-locked loop. For the 565 in particular, we have these approximate equations for capture and lock ranges:

$$\text{Capture range} \approx \sqrt{\frac{10 f_{ctr}}{R_F C_F V_{S(T)}}} \qquad \textbf{(17-2)}$$

where C_F is the low-pass filter capacitance connected between pin 7 (VCO control voltage) and the $+V_{CC}$ power supply in Fig. 17–9. $V_{S(T)}$ is the total power-supply voltage

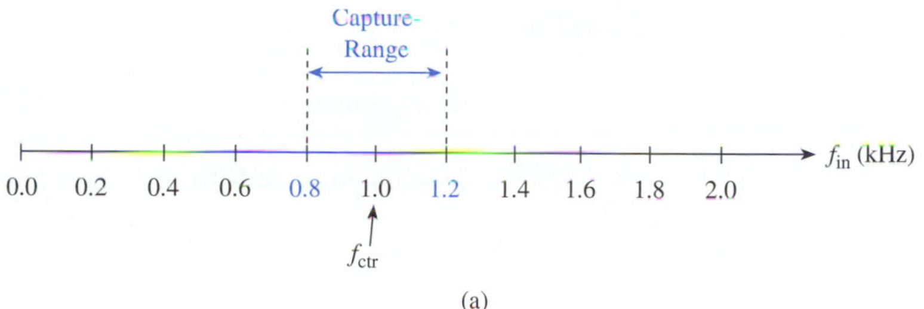

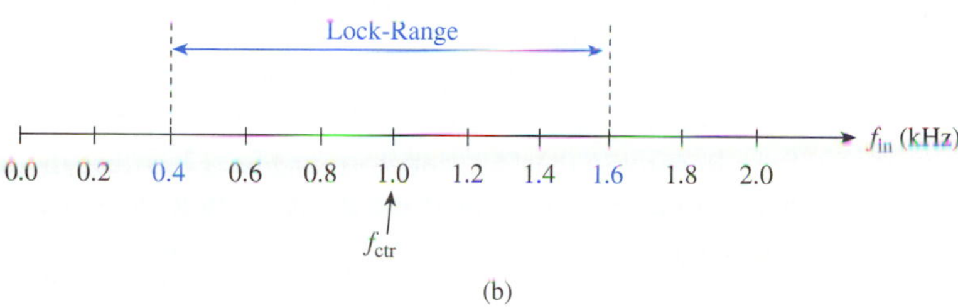

FIGURE 17–11
(a) Capture range. If f_{in} is far away from f_{ctr}, it must approach to within ±200 Hz in order for the PLL to lock on to it (the input source "captures" the PLL). Here capture range = 400 Hz.
(b) Lock-range = 1200 Hz. If f_{in} has already captured the PLL, it can move away from f_{ctr} by ±600 Hz and still maintain the lock. If it strays beyond ±600 Hz, the lock is lost.

difference between V_{CC} (pin 10) and V_{EE} (pin 1) in Fig. 17–9. R_F is internally fixed at 3.6 kΩ.

For lock-range, we have

$$\text{Lock range} \approx \frac{16 f_{ctr}}{V_{S(T)}} \tag{17-3}$$

■ EXAMPLE 17-1

Suppose the 565 PLL of Fig. 17–9 is powered by $V_{CC} = +6$ V, $V_{EE} = -6$ V. The external timing components are $R_T = 3.3$ kΩ and $C_T = 0.1$ μF. The demodulator filter capacitance is $C_F = 2.7$ μF.

(a) Find the center frequency, f_{ctr}. Use Eq. (17-1).
(b) Find the capture range, and state the exact frequencies that are the limits of this range. Express this range as a percentage of f_{ctr}.
(c) Find the lock range, and state the exact frequencies that are the limits of this range. Express this range as a percentage of f_{ctr}.

Solution.

(a) $f_{ctr} = \dfrac{0.3}{R_T C_T} = \dfrac{0.3}{(3.3 \times 10^3\ \Omega)(0.1 \times 10^{-6}\,\text{F})} = \mathbf{909\ Hz}$

(b) $V_{S(T)} = V_{CC} - V_{EE} = +6\text{ V} - (-6\text{ V}) = 12\text{ V}$

From Eq. 17-2,

$$\text{Capture range} \approx \sqrt{\frac{10 \times 909\text{ Hz}}{(3.6 \times 10^3\ \Omega)(2.7 \times 10^{-6}\text{ F})(12\text{ V})}}$$

$$= \sqrt{77.9 \times 10^3} \approx \mathbf{280\ Hz}$$

Assuming the capture range is centered on f_{ctr}, we have $\pm\frac{1}{2}(280) = \pm 140$ Hz. So 909 Hz ± 140 Hz gives a capture range extending from **769 Hz to 1049 Hz.**

140 Hz/909 Hz = 0.154; about **±15%** around f_{ctr}.

(c) From Eq. (17-3),

$$\text{Lock range} \approx \frac{16 \times 909 \text{ Hz}}{12 \text{ V}} = \mathbf{1212\ Hz}$$

which is ±606 Hz. The lock range extends from 303 Hz to 1515 Hz. This is about **±67%** around f_{ctr}. ■

All capture-range and lock-range calculations are approximate. Actual ranges depend on temperature and on the magnitude of the input voltage V_{in}, appearing between pins 2 and 3 in Fig. 17–9. Also, the lock range can be deliberately narrowed by connecting external resistance between pins 6 and 7.

Regarding the 565, notice from Fig. 17–9 that the input amplifier has a terminal that is grounded (pin 3), but the IC as a whole does not have a GND power-supply terminal. The 565 is normally operated from a dual-polarity supply, with V_{CC} being positive and V_{EE} being negative. Then the triangle (pin 9) and square (pin 4) VCO oscillations are centered on 0 V ground potential. However, the 565 can also be operated from a single-polarity dc power supply, with the V_{EE} terminal (pin 1) connected to power-supply ground. Of course, then the VCO output waveforms are not centered on ground.

Another thing. We have shown the input amplifier as single-ended in Fig. 17–9. It can operate this way, but the input is really an op amp that can operate in differential mode to help reject common-mode noise on the input lines. This is shown in Fig. 17–12.

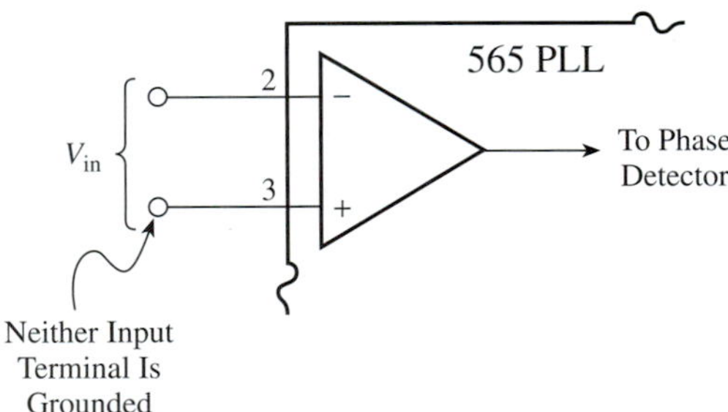

FIGURE 17–12
Differential input to the 565.

17-2-2 The Pulse FM Demodulator

To demodulate the pulse intelligence shown in the bottom waveform of Fig. 17–8, we must focus our attention on the 565's change in VCO control voltage, as shown in Fig. 17–10. It is this change that represents the incoming pulse frequency from the sending station. This is true because the incoming pulse frequency sets the amount by which the VCO must change from f_{ctr}.

The normal dc bias value of the VCO control voltage appears at pin 6 on the 565. This is suggested by the terminal function label V_{BIAS} in Fig. 17–9. Therefore the change in VCO control voltage is found by looking at the difference between pins 7 and 6. Figure 17–13 shows the pulse FM demodulating circuit.

In that figure v_{DEMOD} between pins 6 and 7 is amplified by $A_v = 10$ in the op-amp differential amplifier. Since v_{DEMOD} has a maximum magnitude of only about 1 volt, the final value V_{out} is a more reasonably scaled representation of the analog input voltage at the sender.

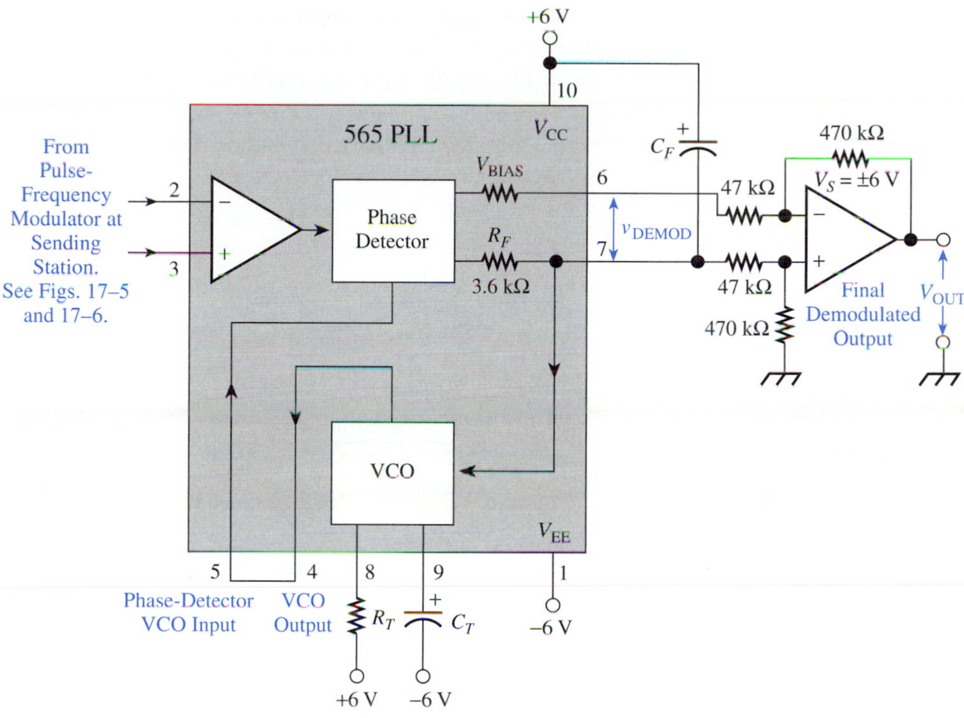

FIGURE 17–13 Demodulator for a pulse-frequency modulated signal. v_{OUT} from this circuit represents the $v_{CONTROL}$ value (the measured variable) at the sending station. The entire range of modulated frequencies must be within the lock-range of the PLL.

It is not possible to say whether increasing values of $v_{CONTROL}$ at the sender produce increasing value of v_{OUT} from the demodulator of Fig. 17–13. That depends on whether the normal astable oscillation frequency of the modulator is lower than, equal to, or higher than the center frequency f_{ctr} of the 565's VCO. This can be an important issue when dc or slowly changing values of the measured variable are being telemetered. We will address this concern in the next paragraph. Before we do, let us consolidate our FM telemetry hardware in a single diagram. Figure 17–14 shows the essential components of a pulse-frequency-modulated telemetry system implemented with 555 and 565 ICs.

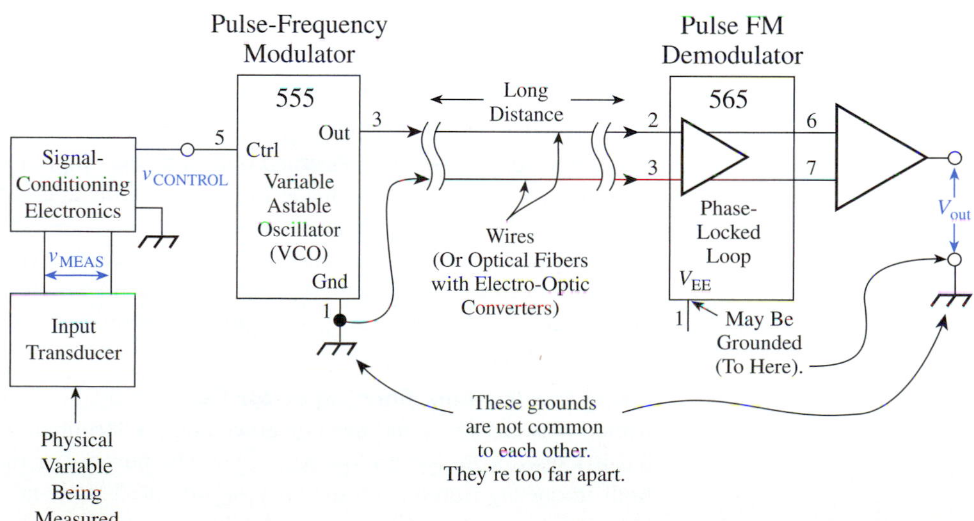

FIGURE 17–14 Overview of pulse FM telemetry system.

The way to eliminate our measurement-to-output transfer function worries is to convert the transducer's dc signal to an ac signal. This can be done by chopping the dc signal. Or, if the transducer is essentially resistive in nature, we can place it in an ac bridge, then use the bridge imbalance voltage, superimposed on some dc bias, as the input control voltage to the modulator. This is demonstrated in Fig. 17–15.

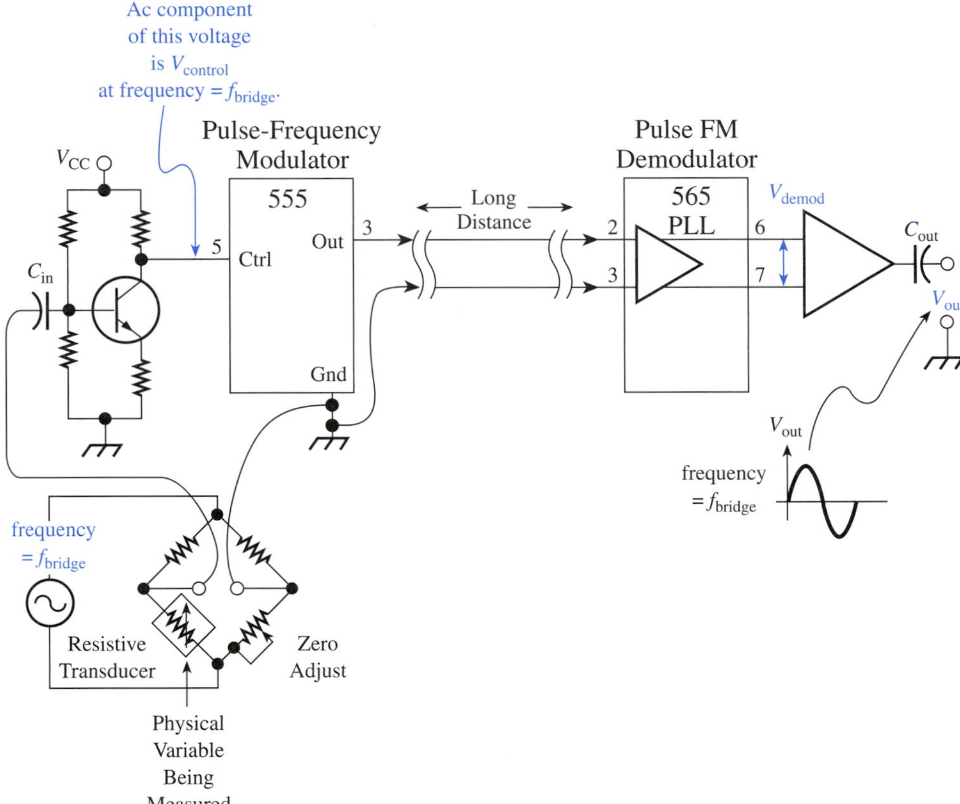

FIGURE 17–15 Pulse-frequency modulated telemetry system in which the transmitted signal is made into sine-wave ac rather than dc. The V_{out} signal amplitude represents the measured value, plain and simple. This is usually preferable to having to figure out the conversion function between dc v_{OUT} and dc $v_{CONTROL}$ in a straightforward dc transducer scheme.

The demodulated signal V_{out} will then be an ac sine wave at the same frequency as the bridge supply. We are concerned only with the amplitude of the V_{out} ac signal. Polarity of dc voltage variation is made irrelevant by this approach.

There are some applications where you can't use an ac sine-wave V_{out} signal; for example, process-control systems usually require a dc-expressed measured value. In the ac approach of Fig. 17–15, the V_{out} sine wave can be perfect-rectified for reconversion to dc. Figure 17–16 illustrates this method of obtaining a final dc output signal at the receiving station.

Sizing the low-pass filter capacitor C_F. Figure 17–9 shows filter capacitor C_F connected between pin 7 and the V_{CC} power-supply terminal, which is an ac ground. C_F and 3.6-kΩ resistor R_F form a low-pass filter. The purpose of the low-pass filter is to remove high-frequency ripple from the v_{DEMOD} output. This is necessary because a real phase-detector does not produce an absolutely smooth dc output. Its output actually contains

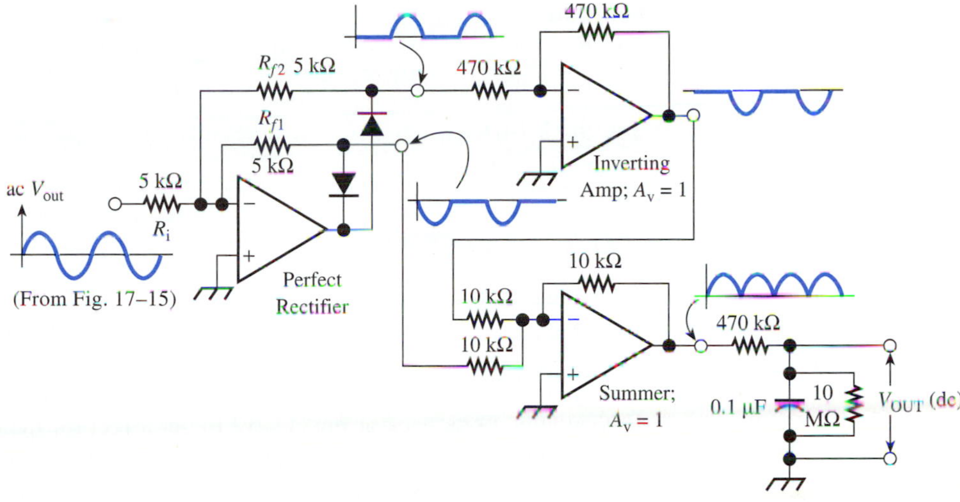

FIGURE 17–16 Reconverting a sine-wave ac voltage to a dc value precisely equal to the ac peaks.

some sine-wave ripple at frequency f_{OSC}; there is also some ripple at a frequency close to $2 \times f_{OSC}$ (if the real phase-locked loop allows a very slight difference between f_{in} and f_{OSC}). And there is other ripple content at harmonics of f_{OSC}. The low-pass filter has the job of removing these ripple components, as described in Fig. 17–17.

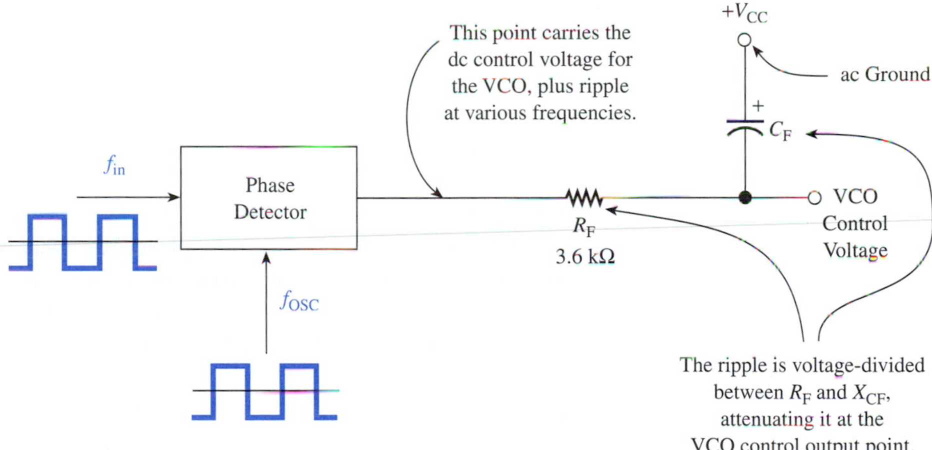

FIGURE 17–17 The low-pass filter at the output of the Phase Detector. R_F is internal in the 565. C_F is external, selected by the user.

A workable rule for sizing C_F for most applications is

$$C_F \geq \frac{2 \times 10^{-3}}{f_{ctr}} \tag{17-4}$$

with f_{ctr} in hertz and C_F in farads. This rule produces a −40 dB (factor-of-100) attenuation of the ripple voltage at the frequency $2 \times f_{OSC}$.

■ EXAMPLE 17-2

In Example 17-1 we had a 565 PLL with f_{ctr} = 909 Hz and C_F = 2.7 μF. Does this satisfy the capacitor sizing rule of Eq. (17-4)?

Solution.

$$C_F \geq \frac{2 \times 10^{-3}}{f_{ctr}} = \frac{2 \times 10^{-3}}{909 \text{ Hz}} = 2.2 \times 10^{-6} \text{ F}$$

Yes, the 2.7-μF capacitor is a bit larger than the 2.2-μF value obtained by the sizing rule. ■

17-3 ■ MULTIPLEXED TELEMETRY

In a telemetry system it may not be necessary to have at the receiving station an absolutely up-to-the-moment representation of the measured value. If there is some method of maintaining a *recent* value of demodulator v_{OUT}, we may not have to be continually watching for any change in the measured value. This is especially true if the measured value changes quite slowly. Look back at Fig. 17–12 to get a feel for the situation.

In that figure imagine that the measurement voltage changes slowly, when it changes at all. Perhaps it changes at the rate of 0.1 V per minute. If that's true, it seems a shame to be devoting the pulse-frequency modulator, the long transmission wires, and the pulse-frequency demodulator to the task of keeping up-to-the-moment track of the measured variable. It would be acceptable if we could look at the measured variable once every 4 seconds, say.

Let us suppose that it requires an elapsed time of 1 second to take a good look at the measured variable. Then we have 3 seconds of time during which we could use our system for other measurements. If the other variables that we need to telemeter all have changing rates like the first one, then we can squeeze four measured values onto our one system. In other words we can get four times as much use out of it. This is the essential idea of multiplexed telemetry.

A *multiplexer* is a device that has several inputs, but only one output. At any point in time the multiplexer is passing one of its inputs to the output; all the other inputs are blocked. At a later point in time the multiplexer can switch to a different input. This concept is pictured in Fig. 17–18.

As a help in picturing multiplexer action, you can think of a rotary-tap switch. This is shown in Fig. 17–19. A real electronic multiplexer has a feature that a rotary switch does not have: An electronic MUX can change from one input to another in any order. Thus, in Fig. 17–18, if the MUX is passing input D at this time, the control signals can instruct it to switch to any other input, not necessarily a "neighboring" input. On the other hand, the switch in

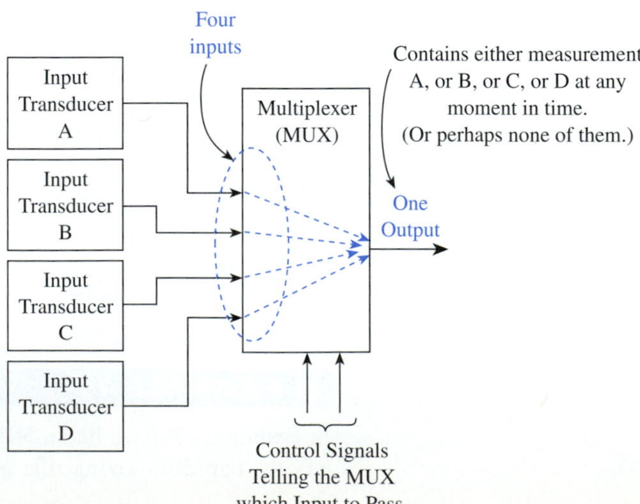

FIGURE 17–18
The multiplexer idea.

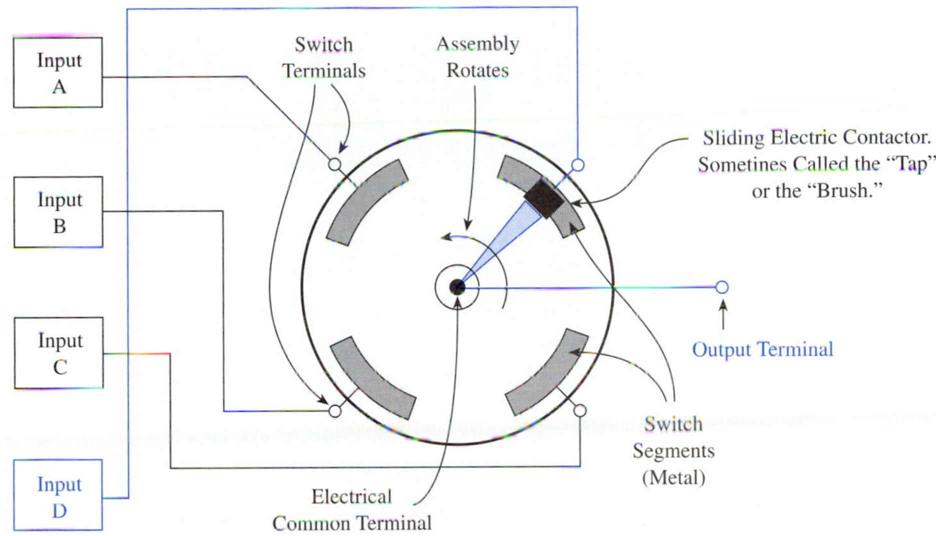

FIGURE 17–19
A rotary-tap switch is a mechanical multiplexer. It can be used as a mental model for the multiplexing idea. Modern MUXs are electronic, not mechanical.

Fig. 17–19 cannot change to *any* other input. It is presently passing input D to the output. If we turn the shaft clockwise, it will change to input C. If we turn counterclockwise, it will change to input A. But there's no way to go immediately from input D to input B.

Whenever we have a device like Fig. 17–19, which can go from one location to only a neighboring location, we describe it by the general term *sequential*. The term *commutating* is sometimes used to mean the same thing. Whenever we have a device that can go immediately from its present location to *any* other location, we describe it by the general term *random*, or *random-access*. An electronic multiplexer is random.

Even though an electronic MUX is random, we normally use it sequentially when performing multiplexed telemetry. In other words the sequence of input selection is A-B-C-D-A-B-C-D (repeating). It is almost never anything like A-C-D-A-B-D-B-C.

If there is a multiplexer, there must be a demultiplexer. This is illustrated in Fig. 17–20. To help in forming a mental picture of a demultiplexer, just reverse the function of the rotary-tap switch of Fig. 17–18.

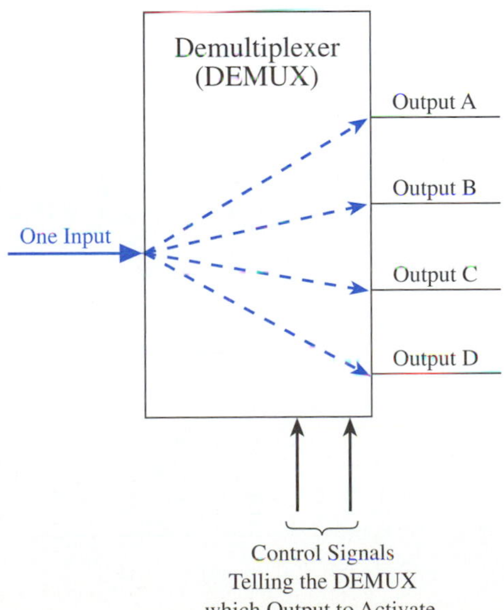

FIGURE 17–20
A demultiplexer has one input but several outputs. At any point in time it passes the input signal to only one of the outputs.

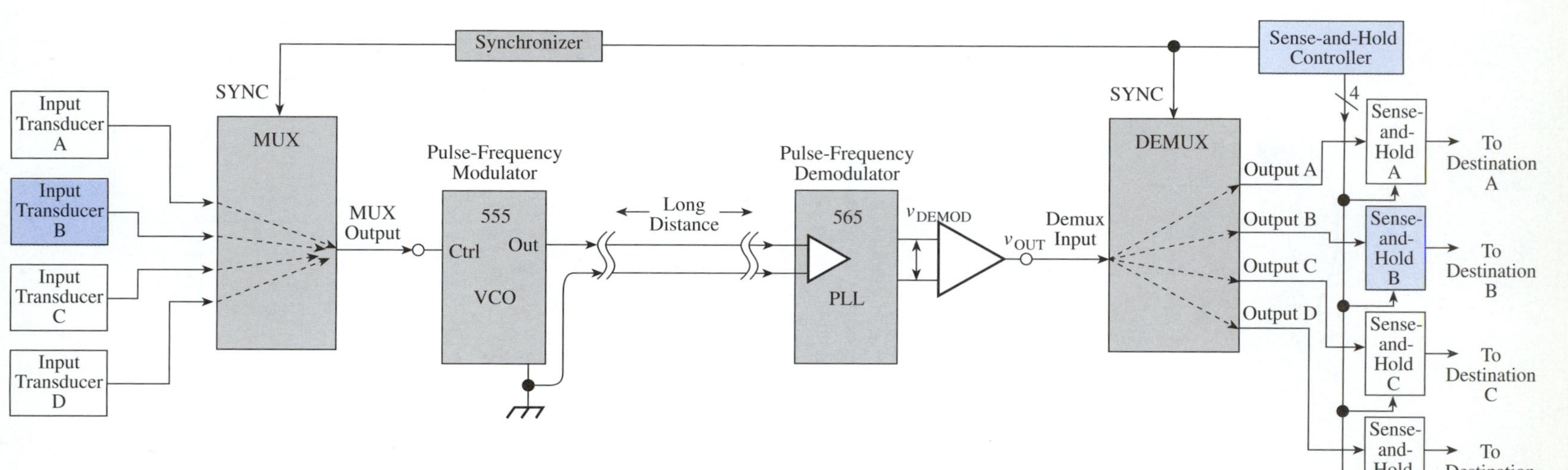

FIGURE 17–21
Conceptual layout of multiplexed telemetry. The advantage of multiplexing is its ability to handle many measurement signals with just one set of telemetry hardware. Multiplexing as we are defining it is sometimes called *Time-Division Multiplexing*, or *TDM*, as distinct from *Frequency-Division Multiplexing*, or *FDM*, which is not at all the same.

A multiplexed telemetry system is shown in Fig. 17–21. For the whole concept to work, the multiplexer at the sending station must be synchronized with the demultiplexer at the receiving station. That is, during a time interval when the MUX is passing Input signal B, the DEMUX must be passing v_{OUT} to the B destination, not some other destination. An electronic synchronizer has the job of guaranteeing that the MUX and the DEMUX stay in step with one another. This is suggested in Fig. 17–21 by the SYNC signals being sent to both MUX and DEMUX. To handle this synchronization signal, it seems that there has to be an additional pair of wires running between the sending and receiving stations. Sometimes that is so. However, sometimes we can arrange to send the synchronizing signal over the same wires that carry the measurement information, saving ourselves the expense of the extra wires.

The synchronizing circuit can be physically located at either end of the telemetry system, sender or receiver.

On the far right of Fig. 17–21 are "Sense-and-Hold" blocks. These electronic circuits are necessary so that the measurement information is *always* present at its destination, even when the multiplexing is working on any of the other signals. Thus, for example, the B Sense-and-Hold circuit must be told the precise moment when it must forget the last v_{OUT} value that it was holding and replace that with a new value coming from the B output of the DEMUX. It must perform this action when the multiplexing hardware is passing the B signal. Every one of the Sense-and-Hold circuits is told to act by the Synchronizer, working through the Sense-and-Hold Controller block in Fig. 17–21.

A Sense-and-Hold circuit is electronically the same as a Sample-and-Hold circuit. We are giving it a different name here because it is performing a somewhat different system function than a standard Sample-and-Hold. Figure 17–22 shows the schematic of a Sense-and-Hold circuit.

In that figure, when the Sense pulse goes LO, it cuts off bipolar transistor Q_1, causing a +10-V pulse to be sent to the gate of MOS transistor Q_2. With the G terminal much more positive than the substrate terminal, the enhancement-type MOSFET is driven into saturation. This causes the v_{OUT} voltage from the DEMUX output terminal to be applied to high-quality capacitor C, through the MOSFET's small internal saturated resistance. Capacitor C then presents that updated v_{OUT} value to the op amp voltage-follower. The voltage-follower's input resistance is huge, so there is virtually no current drain from the capacitor. Therefore C holds its voltage perfectly steady until the next Sense pulse arrives. The op amp output terminal applies the v_{OUT} value to its destination.

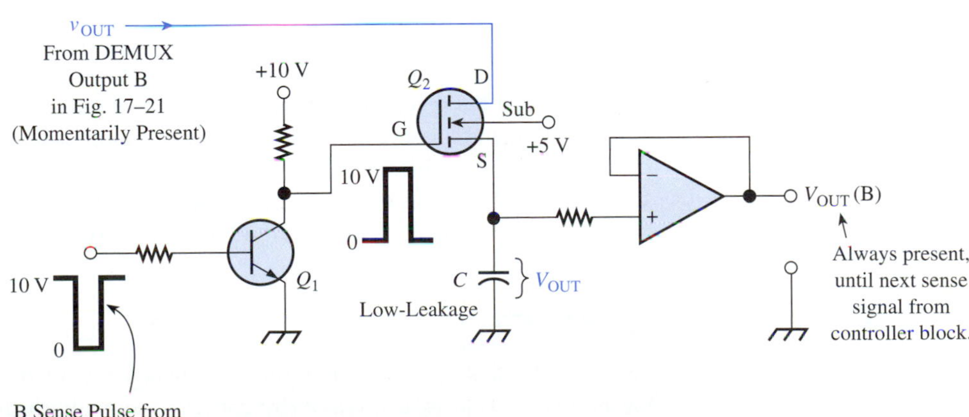

FIGURE 17–22
Sense-and-Hold circuit. Also called Sample-and-Hold.

Multiplexing considerations. Multiplexing is not so much a technique for *control* of processes as it is a technique for *recording data* about processes. For example, when a process is under temperature control, the feedback temperature value is measured at only one location, usually. However, we may wish to have a permanent data record of the time-varying temperatures at many locations within the process. Such a record could be very useful for judging the overall design-effectiveness of a particular process structure.

Another example: In wind-tunnel testing we need to know the pressures at many points on the surface of the object being tested. There must be numerous pressure transducers, each one giving time-varying pressure measurement. In situations like this, as illustrated in Fig. 17–23, multiplexing is an economic necessity. The Sense-and-Hold function may not be necessary if the tape recording is made in analog fashion. This is unlike a control application, where the transmitted measurement value must be present at the control comparer at all times.

In general, the more rapidly a measurement signal changes, the more often it must be looked at in a multiplexing system. If a measured variable is not looked at and telemetered back to the data-logger often enough, its high-frequency components of variation will be missed. Then the collection of recorded values may not be meaningful. This consideration places an upper limit on how many different measurements can be multiplexed over one telemetry channel (one pair of wires). In Sec. 17-5, dealing with the digital-coding of analog voltages for telemetry, we will have more to say about this question of how often we must look at a time-varying signal.

17-4 ■ RADIO TELEMETRY

So far in our explanation of telemetry, we have assumed that the transmission medium is wire, or optical fibers. For very long-distance telemetry, radio transmission of the signal is often preferable.

Accomplishing radio transmission of frequency-modulated pulses may be just a matter of filtering out the harmonic content, amplifying the resulting fundamental sine wave, and applying it to a transmitting antenna. This direct approach is illustrated in Fig. 17–24.

Since the pulse frequency itself is being radiated into the electromagnetic environment, the VCO modulator must be set up to run in a permitted telemetering band, such as 890 to 960 MHz.

For FM radio telemetering at high frequencies, we can skip the pulse-generation process altogether. The input measurement voltage, after appropriate signal-conditioning, can be used directly as the modulation input signal to a true FM modulator, as shown in Fig. 17–25. At the receiving station, the sine-wave radio frequency signal is processed by a true FM demodulator, usually a phase-locked loop design.

Either one of the systems in Figs. 17–24 or 17–25 can be multiplexed. Of course, since there are no wires or fibers connecting the sender to the receiver, the multiplex synchronization must be radio-transmitted too. This arrangement is pictured in Fig. 17–26.

17-5 ■ DIGITAL TELEMETRY

Pulse-width, pulse-position, and pulse-frequency modulation are all fundamentally analog in nature. That is, a *single pulse* conveys the analog value of the measured variable. The waveforms of Figs. 17–2 and 17–5 make this clear.

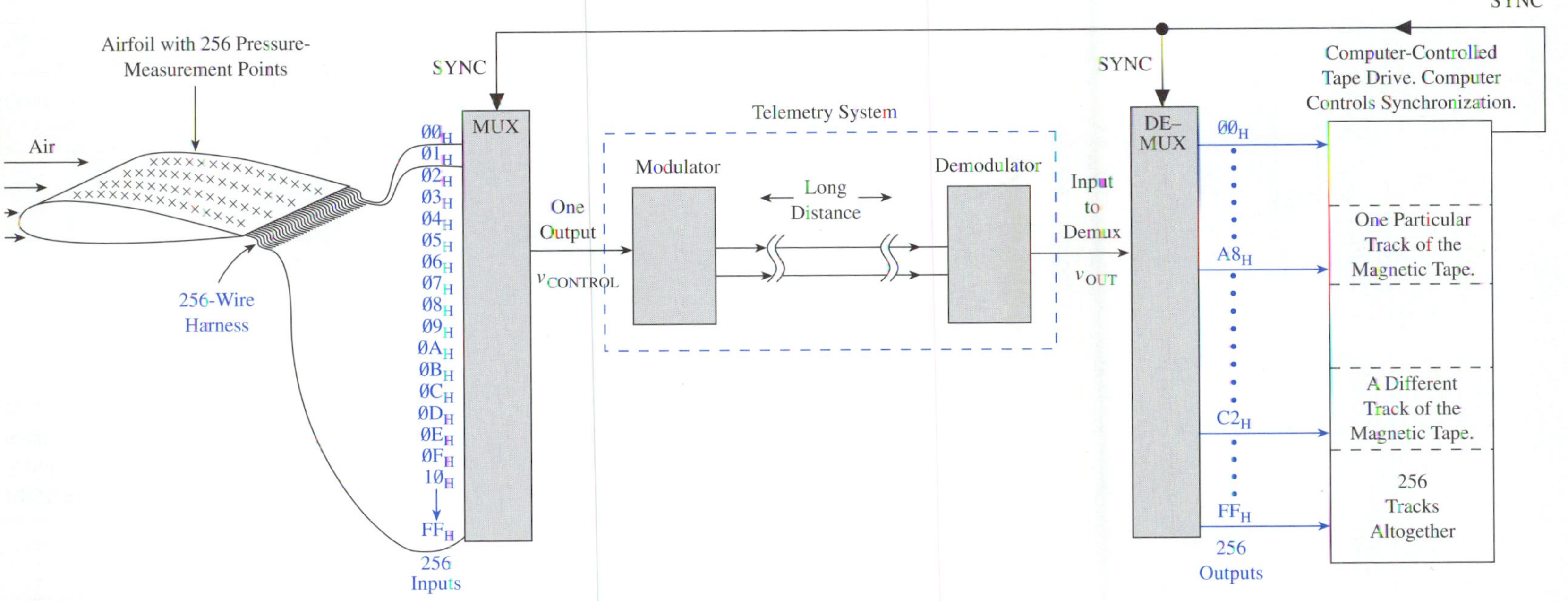

FIGURE 17–23
In a data-logging application with a large number of inputs, multiplexing is essential. There is no Sense-and-Hold circuit shown in this example, but if that function were needed, it would be a single circuit on the left side (input side) of the demultiplexer, not 256 circuits on the right (output) side.

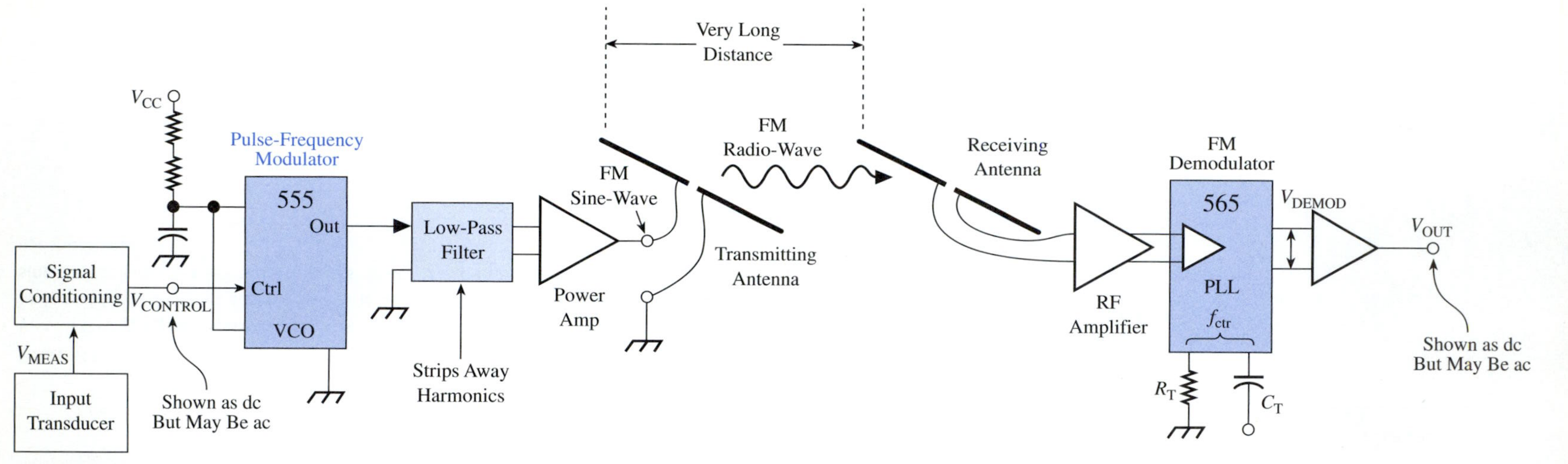

FIGURE 17–24
Simplest radio telemetry method. The radio carrier frequency is the same as the pulse frequency itself.

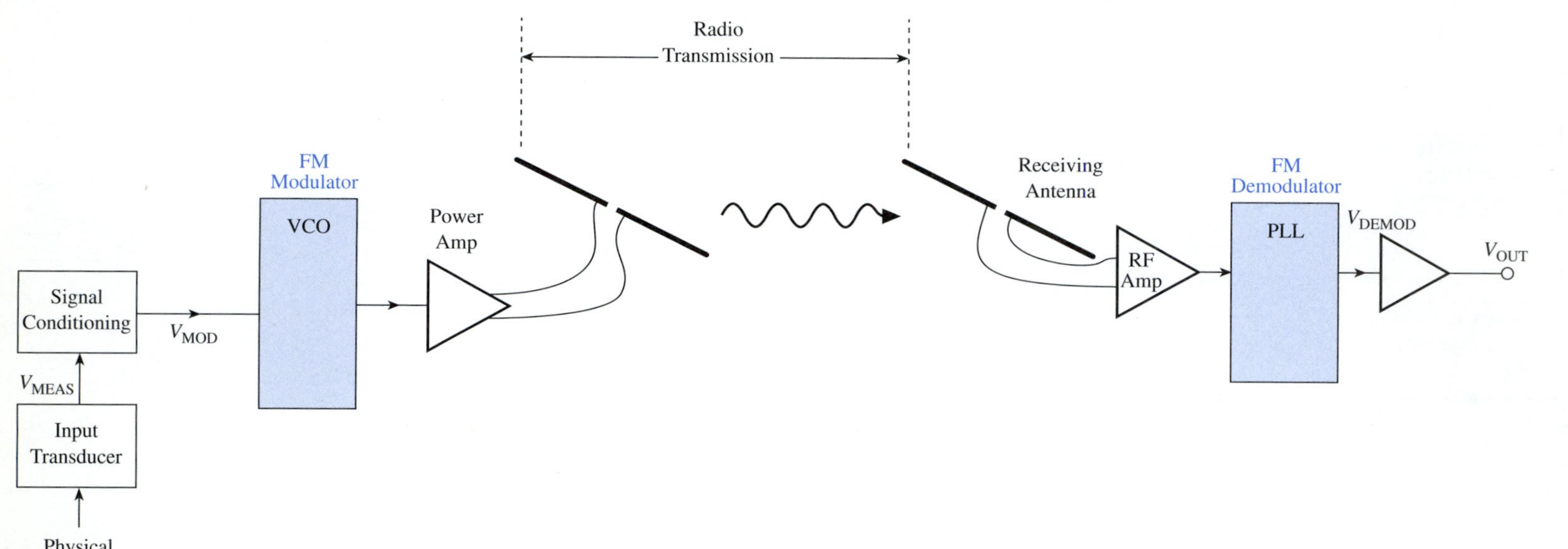

FIGURE 17–25
With radio telemetry there is no worry about analog signal degradation over long wire runs, so we can skip the pulse-modulation process and transmit the analog voltage directly.

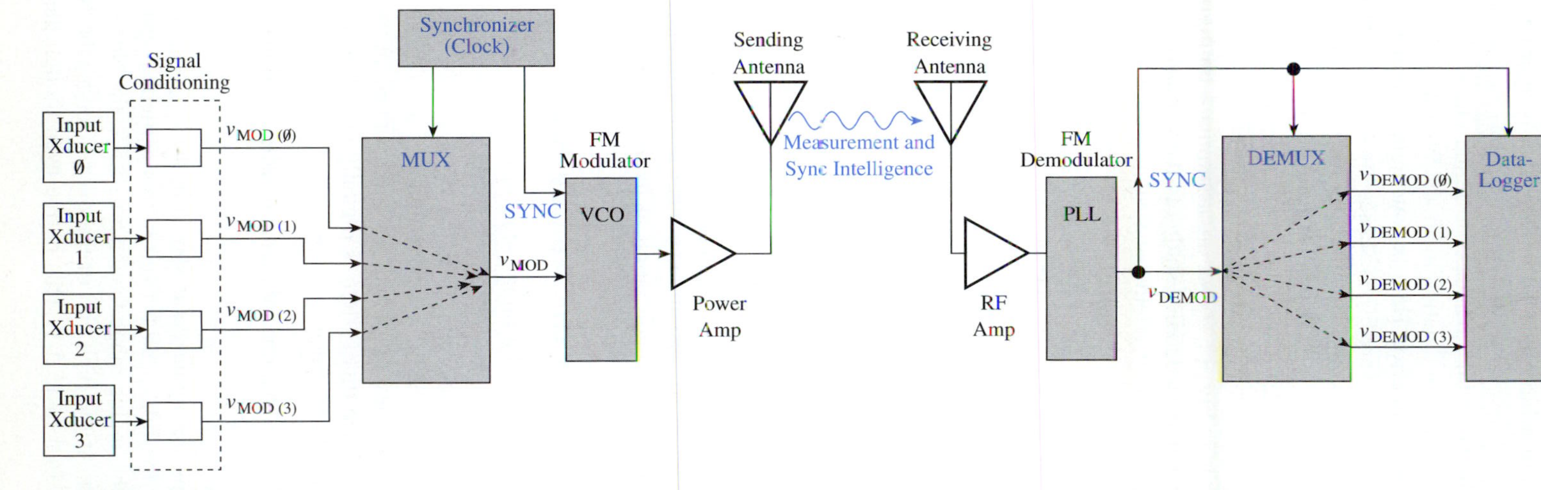

FIGURE 17–26

By radio-transmitting the synchronizing signal we can get the multiplex advantage in a radio telemetry system. The sync information can be transmitted at the same carrier frequency as the measurements, in its assigned sequence just like an extra measurement.

Telemetry is not restricted to analog methods of pulse encoding. A digital approach can also be used. The fundamental difference with the digital method is that several pulses are required to convey one value of the measured variable. The individual pulses represent the bits of a binary-coded number that is approximately equal to the analog measured value.

The fundamental advantage of digital encoding is better noise immunity than any of the analog pulse-modulation methods. With digital, you have to lose an *entire* pulse in order to produce an error. Here is how digital telemetry works. Refer to Fig. 17–27.

The measured variable is sensed by the input transducer to produce v_{MEAS}, then processed through a signal-conditioner, as usual. A *Sample-and-Hold* circuit, just like the one shown in Fig. 17–22, samples the analog output from the signal conditioner at the moment that it is told to do so by a synchronizer circuit pulse. It then holds that constant analog value, presenting it to the ADC for at least as long as the ADC needs to perform its conversion. In Fig. 17–27 we are assuming 4-bit digital encoding. This has been done so that we can make a detailed study of bit combinations and waveforms as we go through our explanation. A real telemetry system would not use such a small number of bits, because the digital resolution would be so poor. Real telemetry systems commonly use 8, 10, or 12 bits.

After the required conversion time has elapsed and the ADC is guaranteed to have reliable digital data on its output bus, the synchronizer starts sending pulses to the four tri-state line-drivers. The first pulse enables line driver 0, allowing it to assert bit 0 (the least significant bit) on the transmission line. It remains there for a short while, until the synchronizer sends the next line-driver-enable pulse. This pulse disables line-driver 0 and enables line-driver 1. It asserts digital bit 1 on the transmission line. And so on through all the remaining bits produced by the ADC.

After all the bits have been sent down the transmission line, the synchronizer could pulse the Sample-and-Hold circuit again for an updated look at the input variable. Whether the synchronizer does this immediately depends on two design considerations:

1. How rapidly does the input change—what is its high-frequency content?
2. Does the telemetry system have other responsibilities that are being multiplexed through it, or can it dedicate all of its time to this one input signal?

17-5-1 Sampling the Analog Input

Let us describe in detail a sample-convert-send sequence, with specific circuit schematics and waveforms. We must begin by considering the range of possible values for v_{ANALOG}, the analog voltage emerging from the signal-conditioner in Fig. 17–27.

By adjusting the offset and the gain of the signal-conditioning block, we can make v_{ANALOG} take on any range that we find convenient. Assume that it has the range from 0 to +15 V. Since this system is digitally encoding with four bits, the number of divisions of that voltage range is given by

$$\text{Number of divisions with } n \text{ coding bits} = 2^n$$
$$= 2^4 = 16$$

The divisions are organized as shown in the v_{ANALOG} waveform of Fig. 17–28.

Each division spans 1.0 volt in this particular system. For example, the No. 7 division spans the analog range from 6.5 to 7.5 V, the No. 8 division spans from 7.5 to 8.5 V, and so forth. At the high end of the overall range, the No. 15 division spans the analog range from 14.5 to 15.5 V. This is true even though the analog voltage itself cannot be

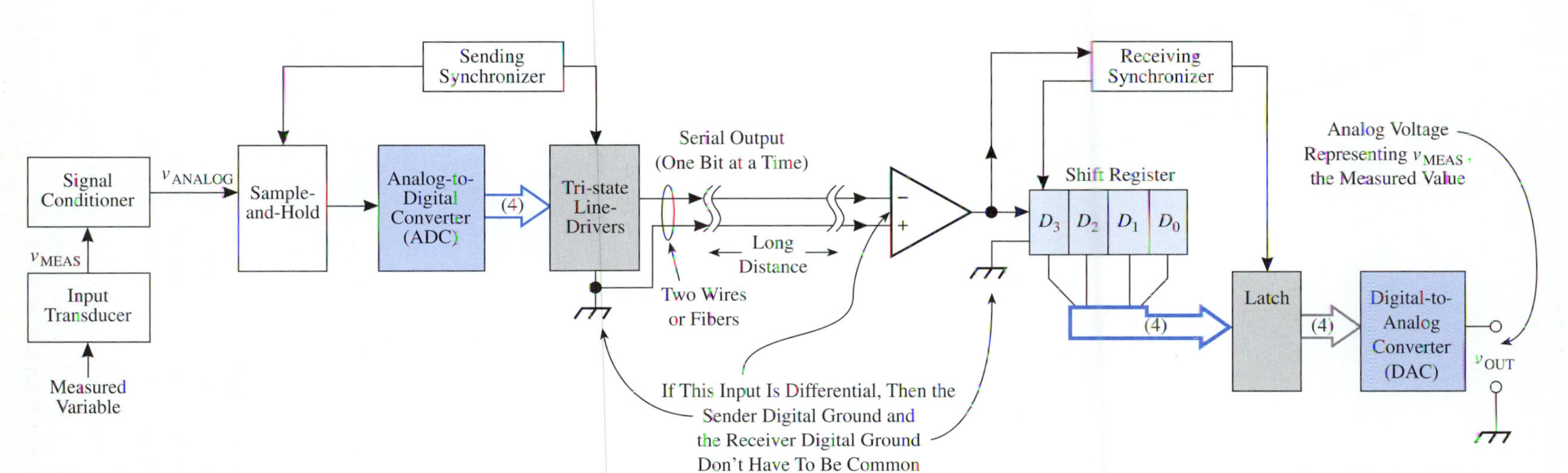

FIGURE 17–27
Block diagram of digital telemetry system.

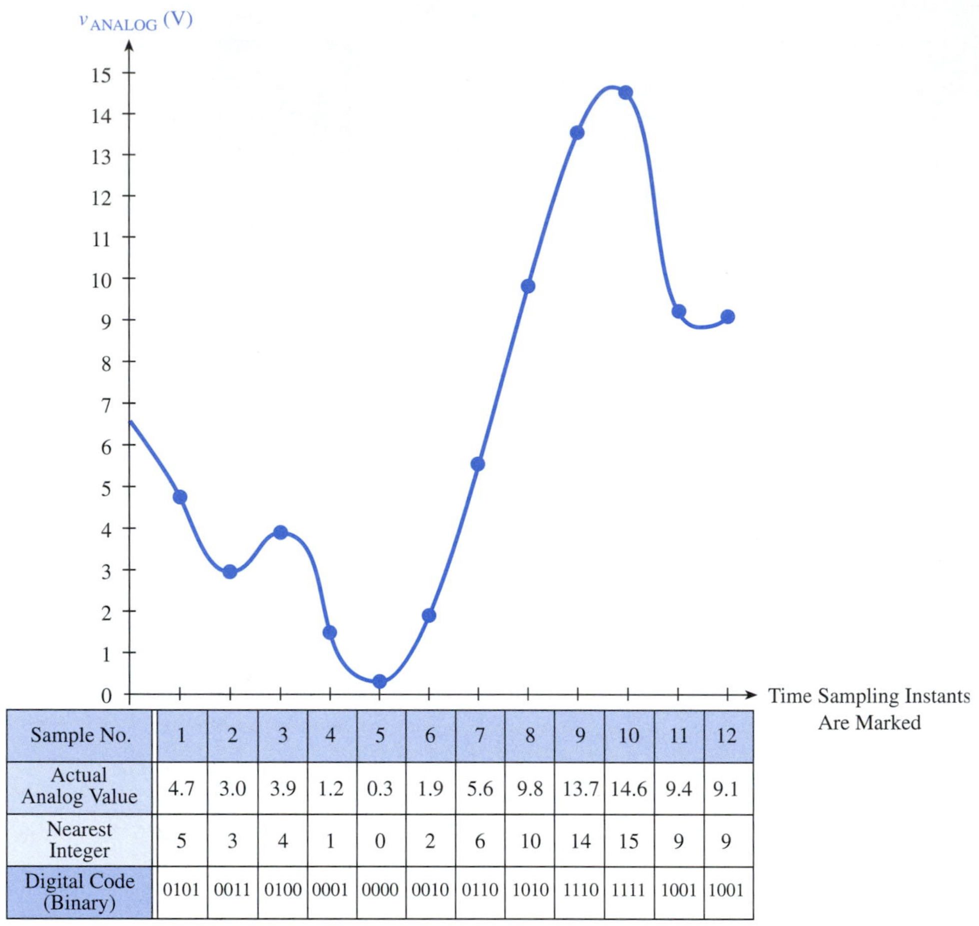

Sample No.	1	2	3	4	5	6	7	8	9	10	11	12
Actual Analog Value	4.7	3.0	3.9	1.2	0.3	1.9	5.6	9.8	13.7	14.6	9.4	9.1
Nearest Integer	5	3	4	1	0	2	6	10	14	15	9	9
Digital Code (Binary)	0101	0011	0100	0001	0000	0010	0110	1010	1110	1111	1001	1001

FIGURE 17–28 Sampling an analog waveform.

greater than 15.0 V due to the way that it has been scaled by the signal-conditioner in Fig. 17–27. Even so, we define the No. 15 division as extending up to 15.5 V to have consistent size with all the other divisions. The same situation occurs at the bottom of the overall range, with the No. 0 division spanning from −0.5 to +0.5 V even though v_{ANALOG} can't really go below 0 V, into the negative region.

When the Sample-and-Hold circuit presents a sampled analog value to the A/D converter in Fig. 17–27, the ADC converts it to the 4-bit binary code that is equivalent to the division number. For example, the first sampling instant (No. 1) in Fig. 17–27 gets an analog value of 4.7 V. The DAC produces the binary code 0101, equivalent to decimal 5.

It is immediately apparent that sampling an analog waveform and converting the value to digital code produces some error. In the preceding example the error is 0.3 V. The worst possible error for the organization shown in Fig. 17–28 would be 0.5 V. This kind of error is called *quantizing error.* It can be severe when the number of encoding bits is small, as in our 4-bit example system. When the number of encoding bits is larger, quantizing error shrinks dramatically. In general the worst-case quantizing error is approximately half the span of a single encoding division. As an equation,

$$\text{Quantizing error (worst case)} \approx \left(\frac{1}{2}\right) \times \frac{\text{Analog voltage range}}{2^n} \qquad \textbf{(17-5)}$$

where n is the number of encoding bits.

■ EXAMPLE 17-3

(a) What is the span of an encoding division, and what is the worst possible quantizing error for a 0- to 15-V analog range encoded by eight bits?
(b) What would the worst-case error be for 12-bit encoding?

Solution. (a) Referring to Eq. (17-5),

$$\text{Span of one encoding division} = \frac{15\text{ V}}{2^n} = \frac{15\text{ V}}{2^8} = \frac{15\text{ V}}{256} = \mathbf{0.0586\ V}$$

$$\text{Worst quantizing error} \approx \left(\frac{1}{2}\right) \times 0.0586\text{ V} = \mathbf{0.0293\ V}$$

which is about a factor-of-10 reduction in the error, compared to 4-bit encoding.

(b)

$$\text{Quantizing error} \approx \left(\frac{1}{2}\right) \times \frac{15\text{ V}}{2^{12}} = \left(\frac{1}{2}\right) \times \frac{15\text{ V}}{4096} = \mathbf{0.001\ 83\ V}$$

which is very small. ■

17-5-2 Digital Encoding and Sending

Figure 17–29 shows the circuit schematic and synchronizer timing waveforms for the digital encoder at the sending station. The input transducer produces an output that varies from 1 to 5 V. In the signal-conditioner block, this v_{MEAS} is first amplified by a factor of 3.75, making the range −3.75 to −18.75 V. Zero reference is established by combining that signal with +3.75 V in the summing circuit. Thus v_{ANALOG} to the Sample-and-Hold block ranges between 0 and +15 V, as diagrammed in Fig. 17–28.

The Sending Synchronizer contains four 555s, which are labeled 555_1, 555_2, 555_3, and 555_4. The basic overall sample-convert-send cycle is set by astable oscillator 555_1. Its timing components are selected to give a 75-ms period, of which 25 ms is HI and 50 ms is LO. This oscillation waveform is drawn at the bottom of Fig. 17–29. When the waveform goes HI, the output terminal of 555_1 enables the Sync Analog Switch, thereby placing a 25-ms-long 5-V pulse on the transmission line. All of the other possible line-driving terminals (the four outputs from the Quad Analog Switch) are floating at this time. This is true because all four enable terminals, $E_\emptyset$ through E_3, are held LO by the output terminals of the Mod-5 counter/decoder. LOs on decoder outputs 0 through 3 are guaranteed when the counter/decoder is forced into the 4 state, by the active-LO on its Reset to 4 terminal, from inverter I_1.

The 25-ms pulse sent down the transmission line allows the receiving circuitry to synchronize its actions with the sending circuitry. It is called a *SYNC pulse*. Look ahead to the transmission line waveform of Fig. 17–30 to see the SYNC pulses clearly.

At the moment the 555_1 output rises, inverter I_2 triggers 555_3, which is wired as an 18-ms one-shot. With the 555_3 Output HI, the Trigger input of 555_4 is also HI. Therefore the 555_4 10-μs one-shot cannot fire. Its Output remains LO, I_3 remains HI, so the Sample-and-Hold circuit is not sampling the analog signal at this moment.

At the time instant 18 ms into the SYNC pulse, the 555_3 Output returns to LO. This triggers 555_4, producing a 10-μs-duration LO from I_3. This LO pulse tells the Sample-and-Hold circuit to sample v_{ANALOG}. Sampling occurs from the 18-ms mark till the 18.01-ms mark within the SYNC pulse, as indicated in the Synchronizer waveforms at the bottom

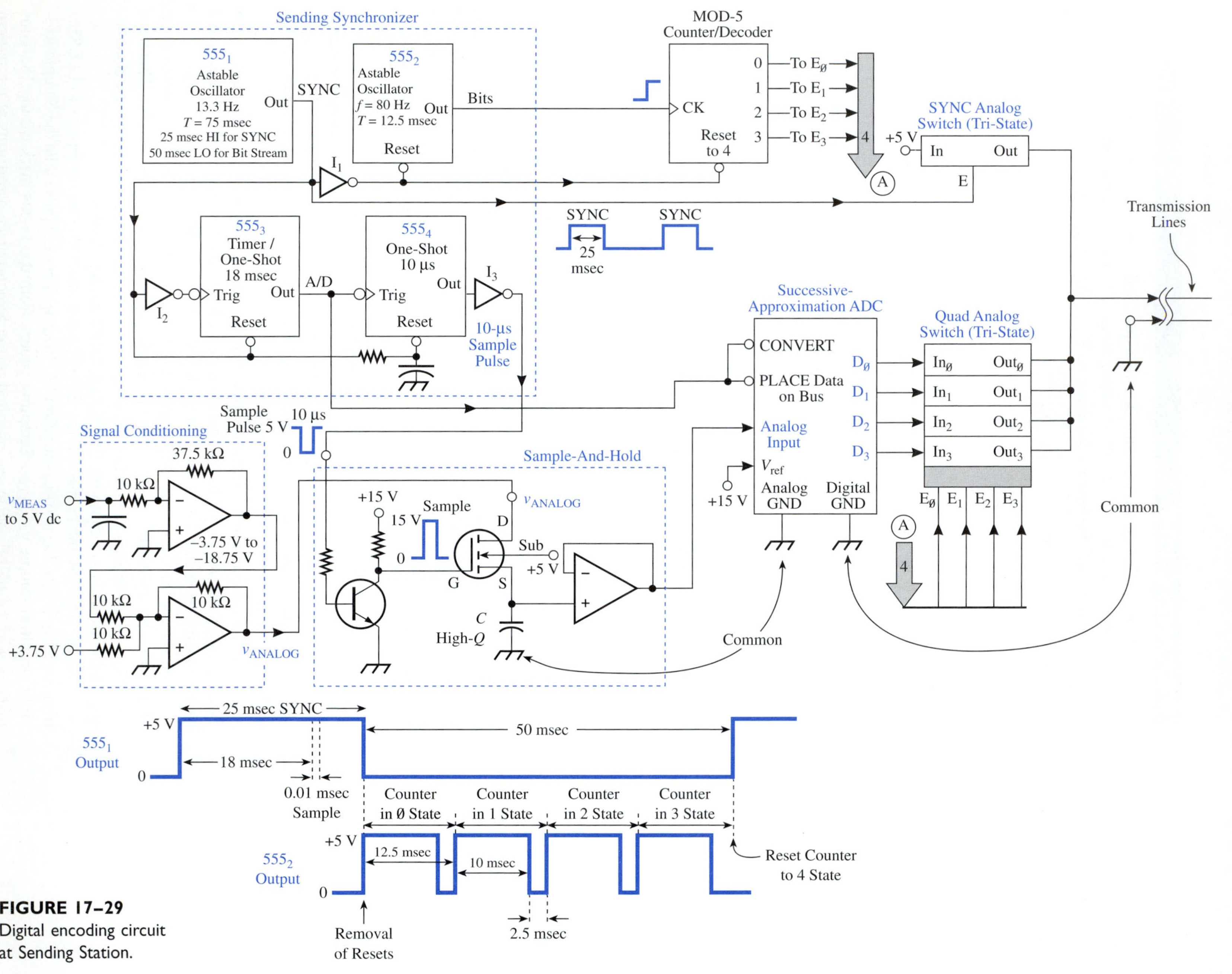

FIGURE 17–29
Digital encoding circuit at Sending Station.

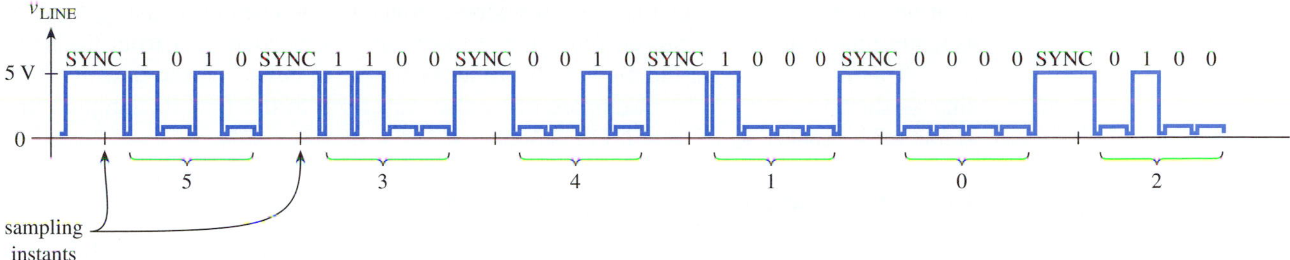

FIGURE 17–30
Transmission line waveform for the first six samples of the analog signal of Fig. 17–28.

of Fig. 17–29. Thereafter the result of this sample is maintained at the ADC's Analog Input terminal until the next sampling action within the next SYNC pulse.

At the same time, the LO Output from 555_3 is routed to the two active-LO inputs of the ADC: CONVERT and PLACE Data on Bus (READ and WRITE on the type 0804). It is not necessary to wait for the successive-approximation conversion to be completed before the digital bits are placed on the output data bus. Random switching of data lines $D_\emptyset$ through D_3 during conversion is no problem because the Quad Analog Switch is totally disabled by the Mod-5 counter/decoder, as described earlier. At some point in time, well before the end of the SYNC pulse, the ADC will stabilize with its best binary approximation of the Analog In voltage.

When the SYNC pulse ends, inverter I_1 removes the active LO from the Reset terminal of 555_2 and from the Reset to 4 terminal of the Mod-5 counter. The 555's timing capacitor is waiting in a completely discharged state at this moment, so 555_2 triggers immediately. This act delivers a positive-going edge to the CK clock terminal of the counter, stepping it from the 4 state to the 0 state. This is indicated on the 555_2 output waveform at the bottom of Fig. 17–29.

The 0 output of the counter/decoder goes HI, enabling the 0 switch in the Quad Analog Switch IC. At the same time, the LO from the output of 555_1 has disabled the SYNC analog switch. Therefore, the $D_\emptyset$ bit of the ADC's digital code is asserted on the transmission line. It remains there for the duration of the oscillation cycle of 555_2, namely 12.5 ms.

At the end of the oscillation cycle, the next positive-going edge from 555_2 advances the counter/decoder to the 1 state. E_1 of the Quad Analog Switch becomes the active control input, so data bit D_1 is asserted on the transmission line for 12.5 ms.

As the 555_2 oscillator continues, it causes D_3 and then D_4 to be sequentially placed on the transmission line, as the Synchronizer waveforms suggest. For the No. 1 sample in Fig. 17–28, the data bit sequence is $D_\emptyset = 1$, $D_1 = 0$, $D_2 = 1$, $D_3 = 0$, starting with the least significant bit and ending with the most significant bit. Look at the v_{LINE} waveform of Fig. 17–30 to see this bit sequence. It is the digital code for 5.

A sequence of binary bits that represents a sampled analog value is called a *data word,* or just a *word.* In this example, the bit sequence

1	0	1	0
LSB			MSB

is one word.

At the end of the 12.5-ms duration of MSB D_3, the 555_1 75-ms oscillation period is also complete. Its Output rises again, placing a new +5-V SYNC signal on the transmission line at the same time that I_1 resets the Mod-5 counter to 4. The Quad Analog Switch is again totally disabled, with all four of its tri-state drivers disconnected from the line.

This process repeats indefinitely. It produces a train of pulses on the transmission line as shown in Fig. 17–30. Compare this pulse waveform to the first six sampled instants of the analog waveform in Fig. 17–28.

17-5-3 Receiving and Digital Decoding

The v_{LINE} waveform in Fig. 17–30 shows short "gaps" between individual data bits and between SYNC pulses and data bits. These gaps occur because of turn-ON delay by the analog switches that assert the data and SYNC pulses in Fig. 17–29. A gap occurs every 12.5 ms, *except* when a SYNC pulse is occurring. This fact is used by the Receiver to recognize a SYNC pulse. The Receiver's circuitry must know when a SYNC pulse is occurring so that the Receiving Synchronizer in Fig. 17–27 can reliably synchronize the events in the Receiver with events in the Sender.

In Fig. 17–31, the components that recognize the presence of a long SYNC pulse are op amp 4, 555_5, op amp 5, and op amp 6. Here is how they work.

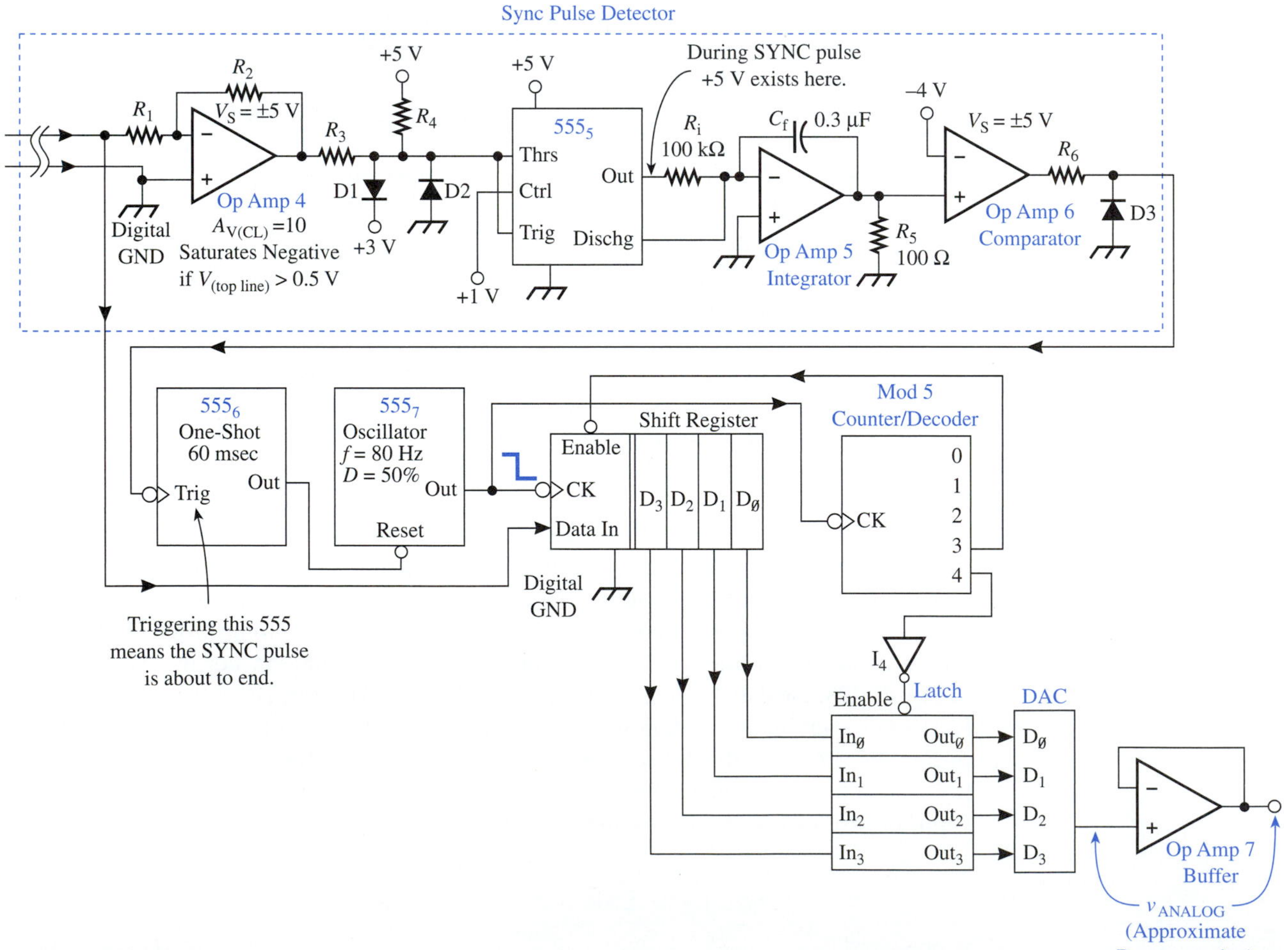

FIGURE 17–31
Digital decoding circuit at Receiving Station.

The op amp 4 inverting amplifier watches v_{LINE} continually. It has gain of -10 and $\pm$5-V dc supplies, so it will drive to negative saturation whenever v_{LINE} is more positive than about +0.5 V. This causes the D1-D2 junction to be clipped at -0.7 V, keeping 555_5 in the HI state since its -0.7-V Threshold voltage is below its Control voltage of +1 V. In Fig. 17–32 the waveforms of v_{LINE}, $v_{D1\text{-}D2\,junction}$, and 555_5 v_{OUT} illustrate this relationship.

If 555_5 maintains its HI output, the op amp 5 integrator continues accumulating charge on its feedback capacitor, C_f. Therefore the output terminal of op amp 5 continues ramping in the negative direction, heading toward -4 V. Figure 17–32 shows the integrator waveform. It takes the integrator precisely 24 msec to ramp down to -4 V. You can prove this to yourself by applying the integrator equation

$$-V_{OUT} = \frac{1}{R_i C_f} V_{IN}(t)$$

with the R_i and C_f values specified in Fig. 17–31 and V_{IN} = +5 V from the Output of 555_5.

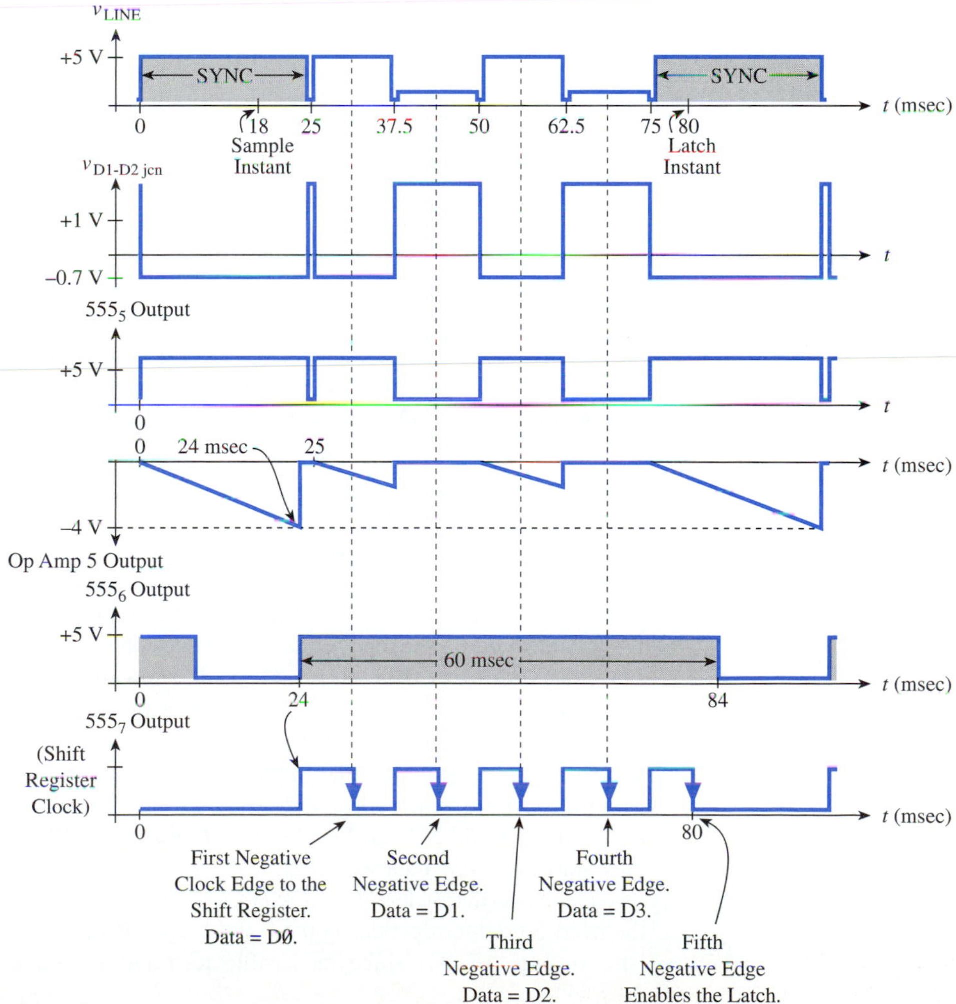

FIGURE 17–32
Waveforms in the Receiving circuit of Fig. 17–31. The particular data bits shown here correspond to sample No. 1 in Fig. 17–28, which is also the first data stream appearing in Fig. 17–30.

If the integrator reaches −4 V, it switches the op amp 6 comparator from positive saturation into negative saturation. The R_6-D3 output circuit then pulls down the Trigger input of 555_6, firing the 60-ms one-shot. Its Output goes HI, removing the active-LO Reset at oscillator 555_7. As Fig. 17–32 makes clear, all this activity occurs *just prior* to the end of the 25-ms-long SYNC pulse. Therefore the 555_7 oscillator in the Fig. 17–31 Receiving circuit starts oscillating with $f = 80$ Hz at virtually the same instant that the 555_2 oscillator starts oscillating with $f = 80$ Hz in the Sending circuit. This is what synchronizes the clocking of the shift-register in Fig. 17–31 with the appearance of data bits coming down the transmission line. In Fig. 17–32, compare the appearance of the negative-going clock edges from the 555_7 oscillator with the appearance of the individual data bits in the v_{LINE} waveform.

For the Receiver to function as described earlier, v_{LINE} from the Sending circuit must be an uninterrupted HI for a duration of 24 ms. Any momentary gap in v_{LINE}, which means that the Sender is *not* sending a SYNC pulse, sends the integrator back to its beginning with 0 V across C_f. This is true because a low-voltage gap in v_{LINE} causes op amp 4 in Fig. 17–31 to lose its negative saturation. Then the R_3-R_4-D1 circuit at the output causes the diode tie-point to go positive, higher than 1 V. This raises the Threshold input of 555_5 above its Control voltage, putting it in the LO-output state. The Discharge output terminal of 555_5 allows C_f to discharge to 0 V virtually instantaneously. This performance is shown in Fig. 17–32 at the 25-ms instant (the end of the SYNC pulse). It occurs again at 37.5 ms and at 62.5 ms. (These latter two instants would be short gaps in v_{LINE}, rather than 12.5-ms-long LO periods, if the next data bits were HI rather than LO.)

Later, when v_{LINE} returns to HI, the integrator must begin ramping downward from its starting condition, namely 0 V. The integrator is prevented from integrating in pieces down to −4 V. It can reach −4 V and initiate the 80-Hz shift pulses only in the ending moments of a 25-ms SYNC pulse. This is how the telemetry system guarantees synchronization between Sender and Receiver.

When the 555_6 one-shot rises at the 24-ms mark in Fig. 17–32, the 80-Hz oscillator rises to digital HI simultaneously. This is a consequence of removing the active-LO Reset from 555_7. The oscillator Output falls back to digital LO about 6.25 ms later, as sketched in Fig. 17–32. This HI-to-LO transition is an active edge to the 4-bit Shift Register in Fig. 17–31. The digital data bit appearing on the transmission line is present at the DATA IN terminal of the Shift Register at this time. Check the waveform of v_{LINE} against the waveform of 555_7 Output in Fig. 17–32. Data bit $D_\emptyset$, which happens to be a 1 in Fig. 17–32, is shifted into the left-most flip-flop of the shift register.

At this same instant, the Mod-5 counter/decoder advances from its 4 state to its 0 state, since its CK terminal also receives the same negative edge. Refer to Fig. 17–31.

The second, third, and fourth clock edges on the 555_7 Output waveform in Fig. 17–32 cause the shifting of data bits D_1, D_2, and D_3, in succession. The receiver depends on the fact that its negative-going clock edges occur in the approximate center of the data duration. Therefore, even if either 80-Hz oscillator has a slight frequency drift, they cannot slip so far out of sync that the fourth negative edge in the Receiver "misses" D_3, the last data bit from the Sender. The system works only because the two 80-Hz oscillators, 555_2 in Fig. 17–29 and 555_7 in Fig. 17–31, are guaranteed to be properly synchronized when they start on a new data word.

The fourth clock edge in Fig. 17–32 fills the shift register with four bits and places the Mod-5 counter/decoder in the 3 state. The HI on the 3 output disables the Shift Register in Fig. 17–31, since its Enable terminal is active LO. At the 80-ms mark in

Fig. 17–32, 12.5 ms later, the fifth negative edge arrives from the 555_7 oscillator. It has no effect on the Shift Register, but it advances the Mod-5 counter to the 4 state. When the 4 output goes HI, inverter I_4 applies an active LO to the Latch's Enable terminal. The Latch receives the shift register data in parallel and presents it to the DAC converter.

When the new 4-bit word combination appears at the input terminals of the DAC in Fig. 17–31, the DAC quickly changes its analog output voltage to the corresponding integer value. After buffering by voltage-follower op amp 7, the reconstructed approximate v_{ANALOG} is ready to be applied to the Receiver's load. There is a 62-ms time lag between sampling the original analog signal and latching that value at the Receiver's load. This is clear from the v_{LINE} waveform of Fig. 17–32, showing the sample instant at the 18-ms mark and the latch instant at the 80-ms mark.

The new value of reconstructed v_{ANALOG} is maintained for the next 75 ms, until another sampling cycle is completed. Then it will be updated to reflect the most recent sampled voltage value taken by the Sender.

A waveform of reconstructed v_{ANALOG} is shown in Fig. 17–33 for comparison to the original v_{ANALOG}. Its quantizing error and time delay are clearly demonstrated. Time delay can be identified as phase-shift, if the analog input is a repeating waveform.

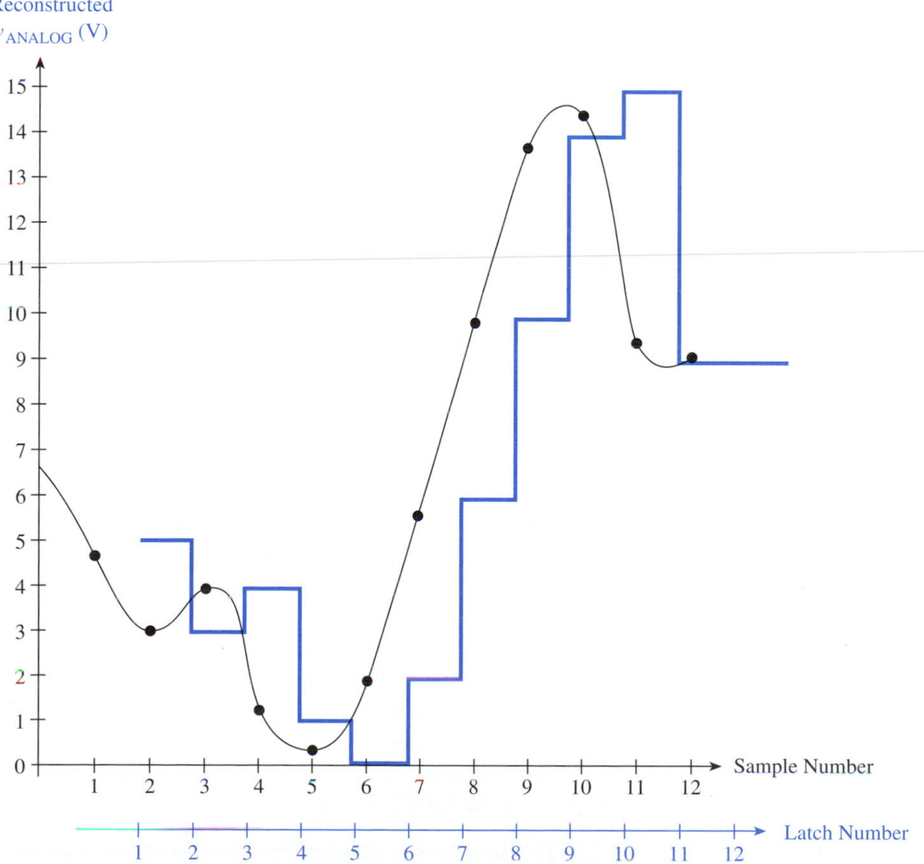

FIGURE 17–33 The digitally encoded telemetry system's original analog signal and its reconstructed analog output waveform.

17-5-4 Improving the Reconstruction

Equation (17-5) makes it clear that a digital telemetry system's quantizing error can be dramatically reduced by encoding the sampled analog value with a greater number of bits. If we had encoded with eight bits instead of four bits in Fig. 17–29, the quantizing error in Fig. 17–33 would be reduced by a factor of 16, a great improvement. Of course, increasing the number of coding bits requires that more bits be transmitted during the data stream. In Figs. 17–30 and 17–32, for example, we would have to squeeze eight bits into the 50-ms duration that was previously occupied by four bits. This could be accomplished with only a slight increase in circuit complexity. The amount of time allotted per bit would be halved, from 12.5 ms to 6.25 ms. The start-synchronized oscillators, 555_2 in the Sender and 555_7 in the Receiver, would have to be increased from 80 Hz to 160 Hz. In other words, the frequency response requirement of the system would increase; we say that the *bandwidth* requirement increases.

Notice also that the easy time-synchronization of the Shift Register clock edges with the sending of the data bits might not be so easy any more. In Fig. 17–32, imagine eight bits on v_{LINE} and eight clock cycles at the 555_7 Output during the 50-ms data stream. Under this condition a 6% combined frequency drift of the two oscillators would result in the most significant data bit, D_7, being missed. Understand that the oscillators are not truly synchronized with each other; only their starting moments are truly synchronized by the SYNC pulse. During the actual data word transmission they operate independently of one another.

If we were to increase the digital resolution to 12 bits, the oscillator sync issue becomes even more critical. Then a 4% combined drift would result in a missed bit at the end of the word.

Quantizing error is not the only problem with the reconstructed v_{ANALOG} waveform of Fig. 17–33. It is visually obvious that this system spends too much time "stuck" at a fixed voltage value. This destroys the smoothness of the wave and produces the severe time-lag mentioned earlier. Both of these defects could be improved by sampling more often. To get the sampling frequency higher, it is necessary to allocate less *frame time* for the data word's bit stream (less than 50 ms) and shorten the SYNC pulse. The 555_1 oscillator in Fig. 17–29 would have to run faster, and the other 555 oscillators and one-shots would have to be speeded up to be compatible. These design changes would place more stringent bandwidth requirements on the circuitry and the transmission line. In almost all real systems, the bandwidth limiter, which is the slowest device in the system, is the Sender's ADC. However, in some very long-distance systems, the limiting element is the transmission line itself, due to its *LC* characteristic.

Multiplexing puts greater bandwidth requirements on the telemetry hardware, naturally. All other things being equal, a two-input multiplexed operation must have two times as great a frequency-response as a nonmultiplexed (single-signal) operation. A three-input multiplexer demands three times the bandwidth, and so on, as you would expect.

Nyquist sampling rate. There is another consideration regarding sampling operations: You must not sample at too low a frequency, or you will overlook essential information in the original analog signal.

As an absolute theoretical limit on how low the sampling frequency can be, we have the Nyquist Sampling Theorem:

> *The sampling frequency must be no lower than twice the highest-frequency variation that you wish to capture.*

For example, in Fig. 17–28 or 17–33, the highest frequency (most rapid) variation occurs in the vicinity of Sample No. 2 and Sample No. 3. Look carefully to see that Sample instant No. 2 coincides exactly with the negative peak of the oscillatory variation, and Sample No. 3 occurs just prior to the positive peak of that oscillatory variation. Therefore the sampling interval, or period, is slightly less than half the "period" of that high-frequency oscillatory variation. This is the same as the sampling frequency being slightly greater than twice the "frequency" of the rapid variation in the analog signal. Therefore the Nyquist sampling limit is satisfied, barely. We will succeed in capturing the variation.

In summary, sampling frequency must be between two limiting values. The low limit is the Nyquist sampling rate, two times the highest frequency harmonic in the signal. Thus the low limit is determined by the nature of the analog signal—what high-frequency content it has.

The sampling frequency high limit is set by the system hardware's response time. The faster its response time, the higher its bandwidth frequency. The sampling frequency, in combination with the specific details of data transmission (how many encoding bits are used, how long the SYNC pulse is) must not result in any system component being required to react more quickly than it is able. This is equivalent to saying that switching events do not occur at a frequency exceeding the system's bandwidth frequency.

Generalizing the Nyquist sampling theorem. We have presented the Nyquist sampling rate idea in the context of digitally encoded telemetry. But it applies to all telemetry methods—pulse-width modulated, pulse-position modulated, and pulse-frequency modulated.

In pulse-width modulation, the pulse period is fixed. The reciprocal of pulse period is pulse frequency, which is equivalent to the sampling rate. Look at the waveforms of Fig. 17–34(a) to see the equivalence of sampling rate and pulse frequency.

For pulse-position modulation in Fig. 17–34(b), the pulse period is not constant. However, we can identify the "normal" pulse period as the pulse cycle time for an unchanging analog signal, shown on the far right of Fig. 17–34. Then the normal pulse frequency is equivalent to the sampling rate.

The situation for pulse-frequency modulation is similar, except that the "normal" pulse frequency, corresponding to some intermediate analog value, is a multiple of the sampling rate. The waveform of Fig. 17–34(c) conveys this idea.

In all three methods of modulation, the sampling rate must not be lower than the critical Nyquist rate—two times the highest-frequency variation in the original analog signal. If this rule is violated, the modulated waveform will not be able to capture all of the original analog intelligence.

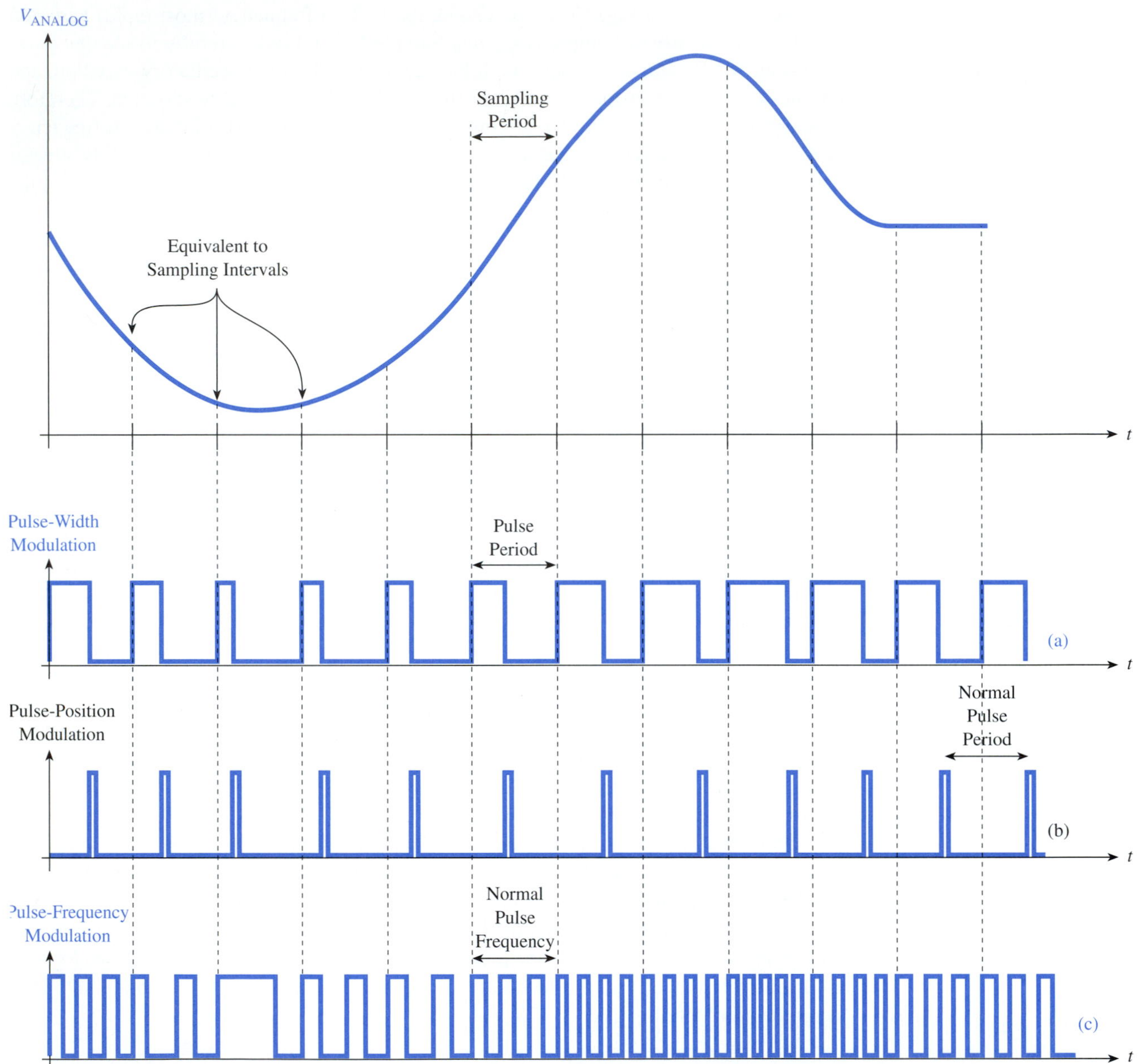

FIGURE 17–34

Relating sampling rate to pulse frequency (sampling period to pulse period). (a) When a pulse train is pulse-width modulated by a signal, the pulse period is equivalent to the sampling period even though there may be no *samples* taken in the strict sense. (b) A similar comparison can be drawn for pulse-position modulation if we consider "normal" pulse period as the period for an unchanging analog value. (c) Similarly again for pulse-frequency modulation.

TROUBLESHOOTING ON THE JOB

TESTING A PULSE-FREQUENCY-MODULATED TELEMETRY SYSTEM

Troubleshooting on the Job in Chapter 12 dealt with the limestone-grinder motor for the sulfur scrubbing tower pictured in Fig. 12–41. In the tower's slurry, the hydrogen sulfite (H_2SO_3) concentration is constantly monitored, as you know from Job Troubleshooting on the Job in Chapter 16. The sulfite concentration is the controlled variable in the closed-loop limestone-feed system for the tower. The sulfite-measurement apparatus is shown in Fig. 17–35.

A flow of slurry is pumped through the small sample chamber in Fig. 17–35(a) (very similar to Fig. 11–15a). The translucence of the sampled slurry represents the concentration of H_2SO_3. This is because hydrogen sulfite molecules, H_2SO_3, are especially able to absorb infrared "light" energy at frequency $f = 3.5 \times 10^{13}$ Hz, which is the frequency of the infrared radiation emitted by the light source in Fig. 17–35(a). The ability of a specific chemical molecule to absorb, and thereby block, a specific frequency of light radiation is the *absorption spectroscopy* idea.

Thus, at greater H_2SO_3 concentrations, a greater portion of the electromagnetic radiation is absorbed while trying to pass through the sample chamber. This causes photoconductive cell PC2 to become darker, thereby raising its resistance. In the automatically balanced bridge of Fig. 17-35(b), higher resistance in photocell PC2 causes R_4 to be adjusted to a higher resistance. This is accomplished by a smal instrument-servo motor, which is not explicitly shown in that figure. As R_4 is raised, its voltage increases. This voltage, v_{R4}, becomes v_{MEAS}, representing the sulfite concentration. Higher concentration produces a larger value of v_{MEAS}.

As v_{MEAS} goes higher, the limestone-grinder motor of Troubleshooting on the Job in Chapters 12 and 16 runs faster. This is so because the shaft of R_4 is mechanically ganged to the motor speed-control pots of Fig. 16–8.

It is necessary to keep a continuous record of sulfite concentration. This is accomplished by telemetering v_{MEAS} to a data-logging computer which is located here at the electric power plant site. In Fig. 17–35(b), v_{MEAS} is processed by an op amp differential amplifier. The Dif Amp output becomes $v_{CONTROL}$ to a 555 pulse-frequency modulator, as shown in Fig. 17–14. In the computer room, a 565 pulse FM demodulator applies v_{OUT} to a computer port. Every few seconds the value of v_{OUT} is sampled by the program and recorded on magnetic disk.

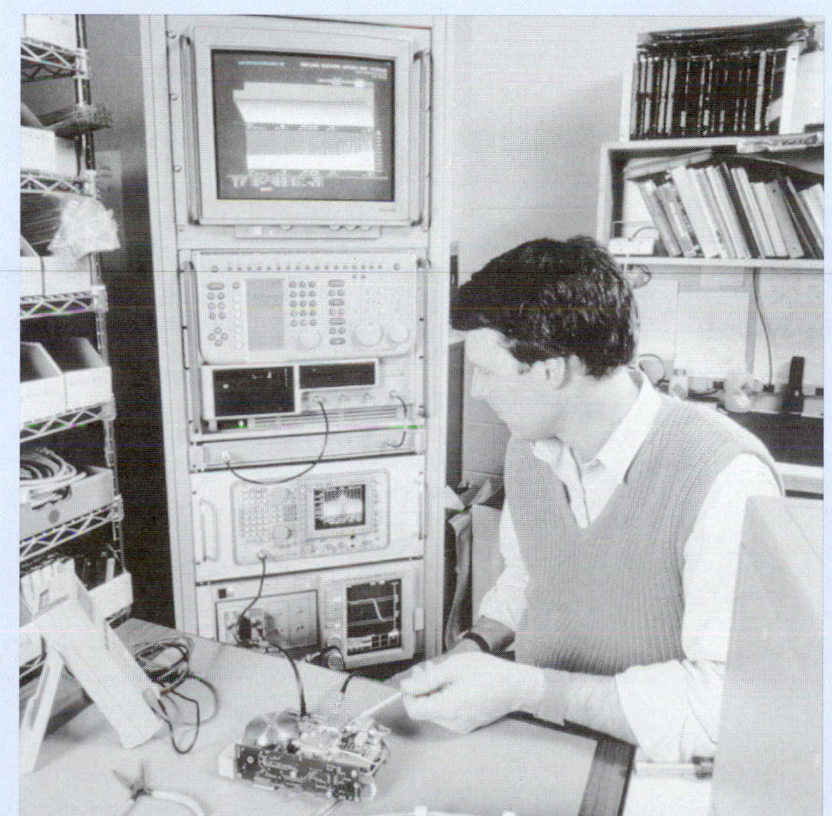

Rack-mounted (as distinct from bench-top) test equipment.
Courtesy of Tektronix, Inc.

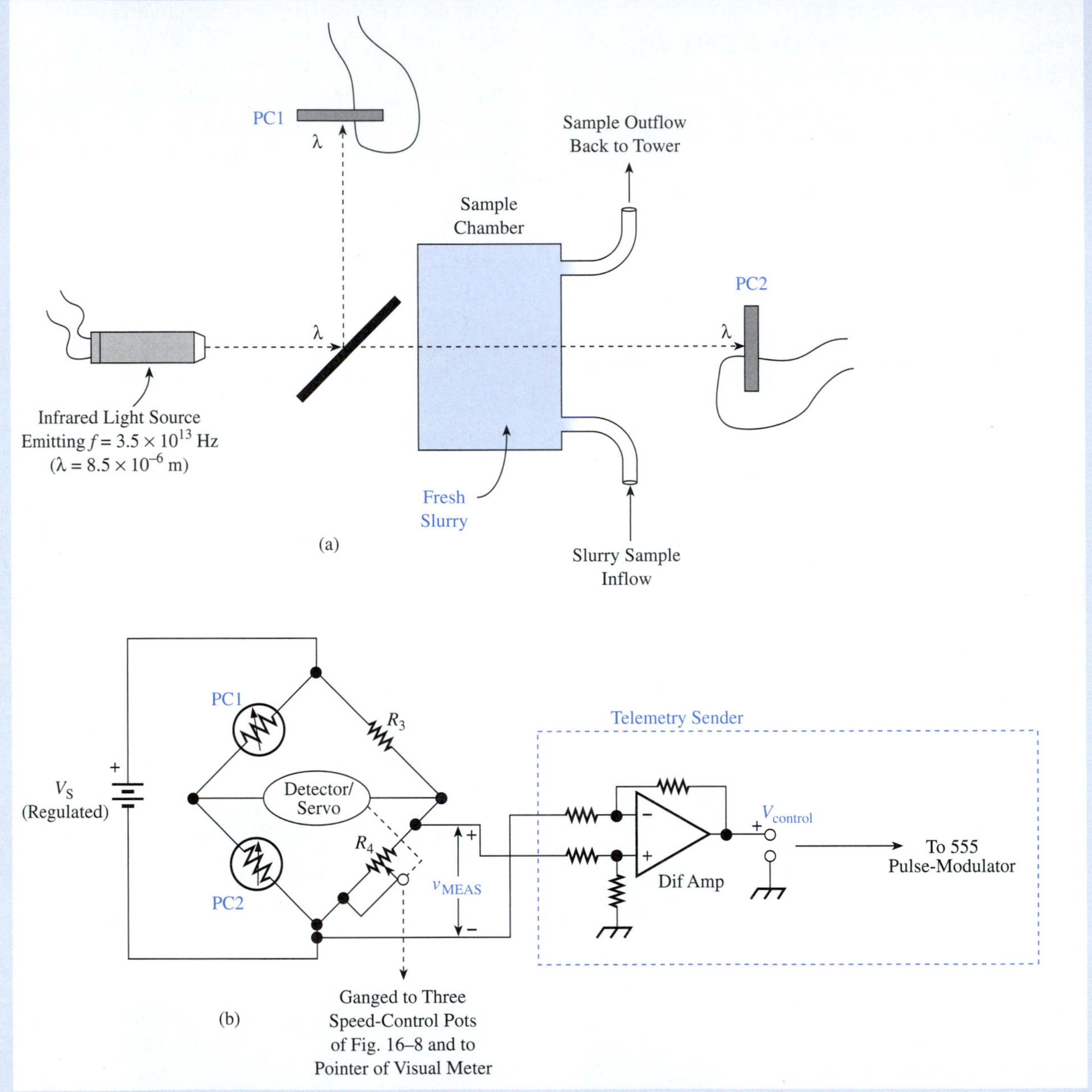

FIGURE 17–35
Measurement system for hydrogen sulfite concentration in slurry. (a) Apparatus for shining critical-frequency electromagnetic radiation (light) on the sample. (b) With the bridge balanced, $v_{PC2} = v_{MEAS}$; therefore v_{MEAS} represents the chemial concentration of hydrogen sulfite in the tower's slurry.

YOUR ASSIGNMENT

Once every 8-hour shift, the sulfite concentration data is printed out on paper, checked by a human being, and backed up on magnetic tape. You have just received word that a computer printout 15 minutes ago gave a data listing showing sulfite concentrations that are out of acceptable range, too low. According to the printout the sulfite has been below the normal range for the past 4 hours and varying erratically. You must find out what is wrong with the slurry-control system, or, if nothing is wrong with it, discover the problem with the telemetry system or computer data port. In addition to voltmeters and oscilloscopes, you have a supply of distilled water and a supply of water containing heavily concentrated H_2SO_3.

1. What is the first thing that you will do? Explain.
2. If the slurry concentration is all right at this time, what arrangements must you make to test the functioning of the telemetry hardware?
3. Describe the specific test/troubleshooting procedure that you will use to locate the telemetry problem.

SUMMARY

- Telemetry is the process of reliably transmitting measured data over long distances by converting the original analog value to pulse modulation or digital code.
- A pair of 555 ICs can be used to convert an analog value into pulse-width-modulated intelligence.
- In a telemetry receiving station, demodulation of a pulse-width-modulated signal is accomplished by amplification, clipping, and filtering to extract the dc average value.
- A 555 used as a voltage-controlled oscillator can be used to accomplish pulse-frequency modulation for telemetry.
- A phase-locked loop, IC type 565, can accomplish pulse-frequency demodulation.
- A phase-locked loop works by automatically changing the control voltage that is applied to its voltage-controlled oscillator to make the oscillation frequency exactly equal to the received input signal frequency. The control voltage sets the value of v_{DEMOD}, representing the measured value of the transmitted data.
- A PLL's control voltage must have its harmonics removed by a properly designed *RC* low-pass filter.
- Multiplexing is the process of time-sharing a telemetry channel among two or more measurement transducers. It requires synchronization between the multiplexer at the sending station and the demultiplexer at the receiving station.
- Telemetry can be accomplished by radio transmission as well as wire- or fiber-transmission.
- Pulse-width, pulse-position, and pulse-frequency modulation are all analog telemetry methods. There is also digital telemetry, in which we digitally encode the measured variable and transmit the bits serially.
- In digital telemetry, encoding the sampled analog value with a greater number of bits improves the reconstruction of the original measurement at the receiving station, but it increases the bandwidth (frequency-response) requirement of the telemetry hardware. The same remarks apply to sampling the measured value more rapidly.
- Telemetry systems must not sample the measured value too slowly, or some of the high-frequency variation in that value will be missed. The low-limit sampling frequency is specified by the Nyquist Sampling Theorem.

FORMULAS

For a 565 PLL:

$$f_{ctr} = \frac{0.3}{R_T C_T} \qquad \text{(Equation 17-1)}$$

$$\text{Capture range} \approx \sqrt{\frac{10 f_{ctr}}{R_F C_F V_{S(T)}}} \qquad \text{(Equation 17-2)}$$

$$\text{Lock range} \approx \frac{16 f_{ctr}}{V_{S(T)}} \qquad \text{(Equation 17-3)}$$

$$C_F \geq \frac{2 \times 10^{-3}}{f_{ctr}} \qquad \text{(Equation 17-4)}$$

$$\text{Quantizing error (worst case)} \approx \left(\frac{1}{2}\right) \times \frac{\text{Analog voltage range}}{2^n} \qquad \text{(Equation 17-5)}$$

QUESTIONS AND PROBLEMS

Section 17-1

1. Explain the fundamental advantage of telemetering by pulses compared to direct transmission of the original v_{MEAS} analog value.
2. There are three popular methods of conveying analog values by pulses (not considering digital encoding). These three methods are called pulse-____________ modulation, pulse-____________ modulation, and pulse-____________ modulation.
3. True or False: To demodulate a pulse-modulated signal, it is necessary to amplify and clip the received pulses so that actual received amplitude is irrelevant.
4. True or False: A high-pass filter is required for demodulation of a pulse-width-modulated signal.
5. The kind of pulse modulation that is derived directly from pulse-width modulation is ____________-____________ modulation.
6. True or False: It is common to see pulse-width modulation in which the pulse width varies from extremely narrow up to virtually the entire width of the pulse period.
7. True or False: To obtain pulse-position modulation from pulse-width modulation, it is necessary to integrate the pulse-width-modulated signal.
8. The answer to Question 6 is False. Revise that statement so that it is correct (true).
9. The answer to Question 7 is False. Change one word in that statement to make it true.

Prepare for Questions 10 through 13 by following these instructions: Using a piece of $8\frac{1}{2} \times 11$-in. paper turned sideways, draw a triangle wave that ramps upward from 0 V to its positive peak, then ramps back down to 0 V. Make the duration of this cycle almost as wide as the 11-in. dimension of the paper; make the peak value about 2 in. above the time axis. Divide the triangle cycle into 10 equal time increments, 5 on the left side of the peak and 5 on the right side. Imagine the left edge of each increment to be a sample instant.

10. Below the triangle wave, draw another horizontal time axis; leave about 1 in. vertically between the two time axes. Sketch a pulse-width-modulated signal on this time axis. Assume that the pulse duty cycle (width) is 20% for $v_{TRIANGLE} = 0$ V,

and assume that it reaches 80% at the peak voltage of the triangle. Don't expect the pulse waveform to look perfectly symmetrical to the left and right of the peak, because the center sample interval is not centered on the peak. Expect the pulse waveform to be denser on the right. However, make use of whatever symmetry you can find, resulting from equal values of sampled voltage.

11. Draw another horizontal time axis about 1 in. lower. Sketch a pulse-position-modulated signal on this time axis. Make the pulses very narrow, coinciding with the falling edges of the pulse-width-modulated signal in Question 10.

Section 17-2

12. Draw another horizontal time axis about 1 in. lower. Sketch a pulse-frequency-modulated signal on this time axis. Assume that there is just one pulse cycle for the first sample interval, with sampled value of $v_{TRIANGLE} = 0$ V. Assume that the second sample interval produces exactly two pulse cycles. (Frequency increases as $v_{TRIANGLE}$ increases, the opposite of $v_{CONTROL}$ for a 555.)
13. Comparing the three different pulse-modulation methods that you have drawn, which one appears to be the least susceptible to demodulation error if the pulse edges are noisy and ragged?
14. A 555 IC is easily used as a fixed-frequency square-wave oscillator. To convert a 555 into a voltage-controlled oscillator, which terminal is used?
15. Draw the schematic diagram of a 555-based VCO.
16. In your 555-based VCO, raising the control voltage higher produces a ________________ frequency of oscillation.
17. True or False: When using a 555-based VCO for pulse-frequency modulation, the lowest possible frequency is obtained when $v_{CONTROL}$ is set at $\frac{2}{3} V_{CC}$.
18. In your 555-based VCO, suppose $R_{T1} = 5.6$ kΩ, $R_{T2} = 2.7$ kΩ, and $C_T = 0.05$ μF. Find the natural (lowest) frequency of oscillation.

In all questions regarding 565 phase-locked loops, assume that the VCO is connected to the phase detector in the usual way (pin 4 is jumpered to pin 5).

19. In a 565 phase-locked loop, what determines the VCO's center oscillating frequency, f_{ctr}?
20. Suppose that the input signal has the same frequency f_{in} as the f_{ctr} of the VCO. Describe the phase relationship between the VCO and the input signal.
21. A 565 has $R_{T1} = 4.3$ kΩ. Calculate the proper value of C_T to give a center frequency of 50 kHz.
22. When f_{in} is higher than f_{ctr}, a 565's VCO lags the input signal by ________________ than 90°; this causes the phase detector to make the VCO's control voltage more ________________ than its bias value, V_{BIAS}.
23. When f_{in} is lower than f_{ctr}, a 565's VCO lags the input signal by ________________ than 90°; this causes the phase detector to make the VCO's control voltage more ________________ than its bias value, V_{BIAS}.
24. A 565's FM demodulation signal is obtained between which two pins? Give their names and pin numbers.
25. In order for a 565's VCO to lock on to an input signal, how close must the input frequency come to f_{ctr}? Give a word answer only, not a numerical answer.
26. After a 565 has locked on to an input signal, how far can the input frequency deviate from f_{ctr} and still maintain the lock? Give a word answer only, not a numerical answer.

27. The PLL of Question 21 is operated with ±6-V dc power supply voltages. Its low-pass filter capacitance is $C_F = 0.05\ \mu F$. Calculate its approximate capture range.
28. For the same PLL, calculate its approximate lock range.
29. Assuming a symmetrical lock range, find the minimum frequency and the maximum frequency that can be tracked by this PLL.
30. In FM telemetry, explain what steps can be taken to avoid the hassle of having to relate a dc v_{MEAS} variation in the Sender to a dc v_{OUT} variation in the demodulator.

Section 17-3

31. In telemetry, the practice of using the same hardware to handle two or more signals on a time-sharing basis is called ______________.
32. Explain the difference between sequential and random-access electronic devices.
33. In multiplexed telemetry, which actual operation is more common, sequential or random-access?
34. Explain the necessity for a synchronizing signal in multiplexed telemetry.
35. Explain the purpose of the Sense-and-Hold circuits at the outputs of a telemetry demultiplexer.
36. True or False: It is rare to see more than four signals being multiplexed through a telemetry system.

Section 17-5

Questions 37 through 42 refer to digitally encoded telemetry.

37. In digital telemetry, explain why it is necessary to have a Sample-and-Hold circuit in the Sender.
38. Suppose v_{ANALOG} can range from 0 to 5 V and the system uses 8-bit encoding. For equally spaced divisions, what amount of voltage does 1 bit represent (the LSB)?
39. For the system of Problem 38, what is the worst-case quantizing error?
40. On any individual sample, is the quantizing error more likely to be negative (digitally coded value less than actual analog value), more likely to be positive (digitally coded value greater than actual analog value), or equally likely to be either?
41. Based on your answer to Question 40, would it be correct to say that quantizing errors are absolutely random in polarity and magnitude?
42. Based on your answer to Question 41, would it be fair to use the term *noise* to describe quantizing errors? Get used to the phrase *quantizing noise*.

Questions 43 through 48 refer to Fig. 17–29.

43. In the Sending circuit of Fig. 17–29, which 555 IC determines the overall system speed (the sample rate)?
44. Which 555 IC determines the precise moment within the system cycle that the sample is actually taken?
45. Which 555 IC determines the time duration of the sample?
46. Why can't the sample duration be extremely short, say, 1 nanosecond?
47. Which 555 IC determines the amount of time that one bit remains on the line?
48. Give the reason why there will never be contention between the SYNC Analog Switch and the Quad Analog Switch, with both trying to drive the transmission line simultaneously.
49. Figure 17–30 shows the v_{LINE} waveform for Samples No. 1 through No. 6 from Fig. 17–28. Complete the waveform for Samples No. 7 through No. 12.

50. In Fig. 17–31, does the SYNC pulse detector react to the amplitude of the SYNC pulse or to the time duration of the SYNC pulse?
51. In Fig. 17–31, the 555_6 IC one-shot has a pulse duration of 60 ms, which is long enough to permit about $4\frac{1}{2}$ cycles of the 555_7 oscillator. With only four bits being sent in the data word, why does the one-shot's pulse time need to be long enough for $4\frac{1}{2}$ cycles?
52. What prevents the Shift Register in Fig. 17–31 from falsely shifting a fifth time when the fifth negative edge is delivered by the 555_7 oscillator?
53. If the sampled analog signal is human speech with a highest frequency component of 3500 Hz, what is the minimum sampling frequency that will guarantee successful high-fidelity capture?
54. True or False: The slowest-responding element in a telemetry system determines the system's bandwidth.

CHAPTER

18 CLOSED-LOOP CONTROL WITH AN ON-LINE MICROCOMPUTER

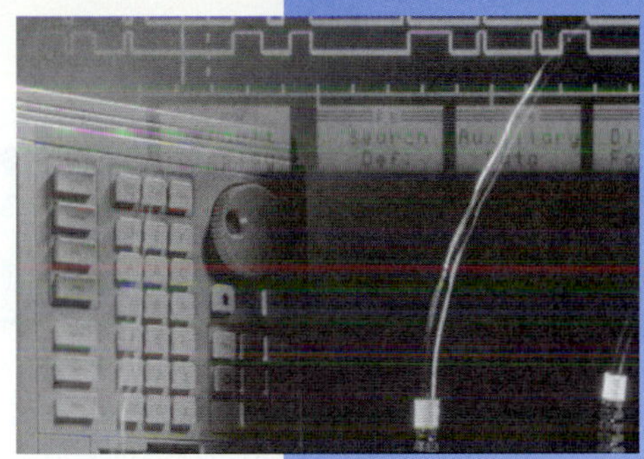

The advent of the microcomputer has greatly expanded the range of industrial operations for which closed-loop computer-based control is economically feasible. In this chapter we will become acquainted with the architecture and program structure of microcomputer-based control systems.

OBJECTIVES

After completing this chapter, you will be able to:

1. Describe the characteristic differences between a dedicated microcomputer and a programmable logic controller.
2. Interpret the flowchart for an on-line control program.
3. Describe the functions of the data bus, the address bus, and the control lines within a microcomputer.
4. Explain the use of tri-state buffers in the various chips of a bus-organized computer.
5. Discuss the functions of the following IC chips within an on-line microcomputer: the microprocessor, the ROM, the RAM, the address decoder, the input buffer, and the output latch.
6. Describe the purpose and function of the following registers within a 6800-series microprocessor: the program counter, the A and B registers, the condition-code register, and the index register.

18-1 ■ A COAL-SLURRY TRANSPORT SYSTEM CONTROLLED BY A MICROCOMPUTER

Sometimes it is necessary to perform mathematical calculations rapidly and repetitively in order to properly control an industrial system. As we learned in Chapter 3, programmable logic controllers often lend themselves to such control tasks, since they can perform arithmetic operations. However, a PLC's arithmetic functions are somewhat cumbersome by comparison to the calculating proficiency of a dedicated microcomputer.

In truth, a PLC actually *is* a microcomputer, but provided with accessories that make it easy to use. No specialized knowledge of a computer assembly language or compiler language is required in order to program a PLC; neither must the user have any specialized knowledge of microprocessor hardware or memory hardware. Naturally, because a PLC has these ease-of-use advantages, it costs much more than a raw microcomputer with comparable specifications.

So, considering hardware costs and computational power, it may be preferable to employ a microcomputer rather than a PLC in certain industrial situations. If that is to be done, someone on the team must have software skill (the ability to write programs in a computer language) and special knowledge of the computer's hardware.

A hypothetical example of such a situation is the coal-slurry pumping system illustrated in Fig. 18–1. Coal slurry is a mixture of pulverized coal and water. It can be pumped to a destination through underground pipes or pipes laid over rugged terrain. After it arrives at the destination, the coal is recovered from the slurry by holding it in large tanks and evaporating the water. This is the preferred method of overland coal transport where wheeled vehicles or trains are impractical.

18-2 ■ THE SYSTEM CONTROL SCHEME

Pulverized coal is fed into the transport pipe by a screw conveyor, where it mixes with the incoming water and is pumped away as slurry, as pictured in Fig. 18–1. The rate at which coal dust is fed into the transport pipe depends upon the speed of the screw conveyor, which is driven by a wound-rotor dc motor. It is necessary to control the coal concentration in the slurry in order to achieve efficient transport operation without overburdening the pumping gear. Control is accomplished by varying the armature voltage applied to the motor to adjust the speed of the screw conveyor.

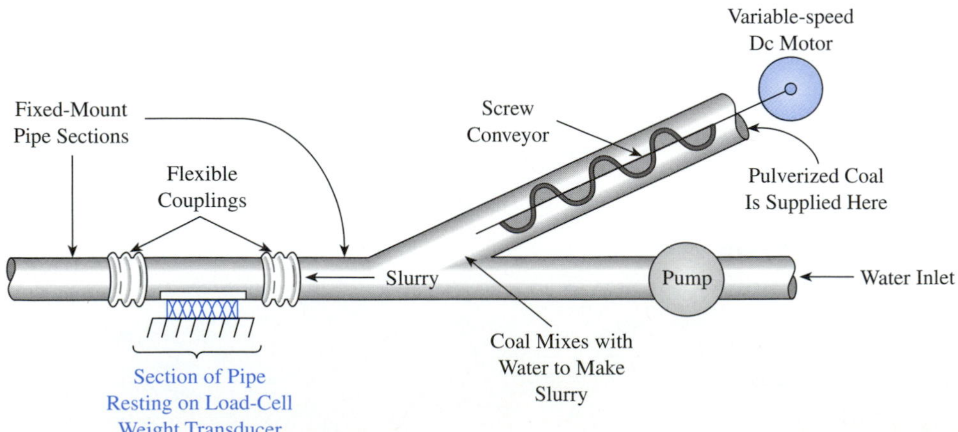

FIGURE 18–1
Mechanical layout of the coal-slurry system.

Just downstream of the first pump* in Fig. 18–1 is an independently suspended section of pipe. This section rests on a load cell, which furnishes an analog signal representing the weight of slurry in the section. By monitoring the average weight of slurry in the weighing section, we can determine if the coal concentration is within the acceptable range, or if it should be corrected. To smooth out variations caused by the pulsating delivery tendency of the screw conveyor, instantaneous weight must be repeatedly sampled at quick intervals throughout an extended time duration. The average weight is then calculated by adding the instantaneous weights and dividing the sum by the number of samples.

Let's suppose a sampling rate of 2 per second, or a sampling interval of 0.5 seconds; let the sampling duration be 10 seconds. Therefore 20 instantaneous weights will be measured, with the average weight given by

$$\text{AVGWT} = \frac{\text{INSTWT}_1 + \text{INSTWT}_2 + \cdots + \text{INSTWT}_{20}}{20} = \frac{\sum_{i=1}^{20} \text{INSTWT}_i}{20} \qquad \textbf{(18-1)}$$

in which the variable name AVGWT stands for "average weight" and INSTWT stands for "instantaneous weight."

Suppose further that the ideal weight is 1750 lb, but an average weight between 1500 and 2000 lb is considered acceptable. If the microcomputer calculates the average slurry weight to be in that range, it makes no correction to the conveyor feed rate; the conveyor is allowed to continue turning at its present speed.

If the average weight is found to exceed 2000 lb, the screw conveyor is slowed down. The amount of speed reduction is proportional to the difference between the actual weight and 2000 lb.

If the average weight is calculated between 1000 and 1500 lb, the screw conveyor is speeded up. The amount of speed increase is proportional to the difference between 1500 lb and the actual weight.

If the average weight is found to be less than 1000 lb, the microcomputer causes the screw conveyor to run at maximum speed by raising the motor's armature voltage to its maximum value of 255 V. This is accomplished by passing the largest possible 8-bit binary number, namely 1111 1111 corresponding to decimal 255, to the motor-drive system's digital input via the microcomputer's output port.

Let us adopt the symbol NEWSP for the new speed (the speed *after* this speed adjustment is made) and OLDSP for the old speed (*before* this speed adjustment is made). Then we can summarize the four possible outcomes from one sampling duration as follows:

1. Average weight is greater than 2000 lb, so

$$\text{NEWSP} = \text{OLDSP} - \text{OLDSP}\left(\frac{\text{AVGWT} - 2000}{\text{AVGWT}}\right) \qquad \textbf{(18-2)}$$

2. Average weight is in the acceptable range, so

$$\text{NEWSP} = \text{OLDSP}$$

*There are probably additional booster pumps farther down the line.

3. Average weight is below 1500 lb but above 1000 lb, so

$$\text{NEWSP} = \text{OLDSP} + \text{OLDSP}\left(\frac{1500 - \text{AVGWT}}{1500}\right) \tag{18-3}$$

4. Average weight is below 1000 lb, so

$$\text{NEWSP} = \text{maximum} = 255 \text{ V armature voltage}$$

After the microcomputer has performed its calculations and made any required adjustment to the speed, it waits 15 seconds for the new speed to take effect. Then it goes back and starts over, sampling the instantaneous slurry weight at 0.5-second intervals. Figure 18–2 shows a timing diagram of the process.

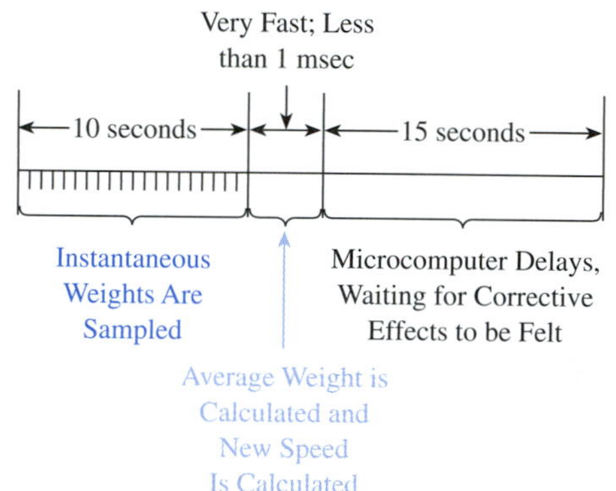

FIGURE 18–2 Timing diagram for the control program. The actual computing all occurs in a tiny fraction of the overall 25-second execution time.

18-3 ■ PROGRAMMING A MICROCOMPUTER

The microcomputer accomplishes its control function by executing a stored program repetitively. The program consists of a series of coded instructions, each one very elementary in itself. This is the same concept that we encountered in our discussion of PLCs in Chapter 3; the only difference is that microcomputer instructions are much more basic than PLC instructions. You, as the user/programmer, must bring together a series of very elementary operations in order to get the microcomputer to accomplish a more complex overall operation.

As an example of this difference, consider the job of comparing two numbers, *A* and *B*, and energizing an output device if *A* is less than *B*. With a PLC, this is done rather easily by the following ladder-logic program:

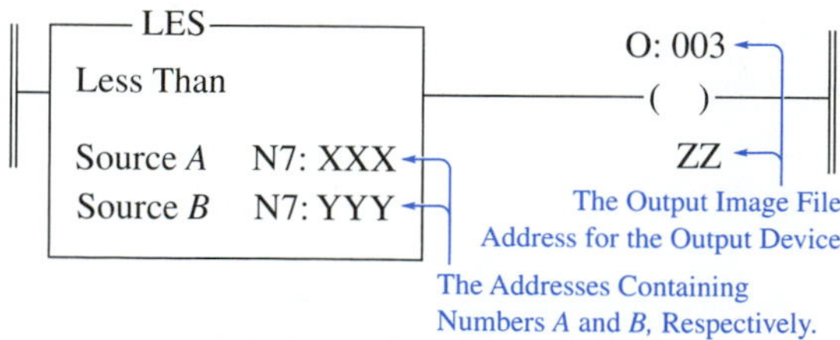

By contrast, performing this comparison in a raw microcomputer would require a lot more programming effort on your part. It would go something like this:

1. Tell the CPU (the microprocessor)* that you want it to load one of its internal storage registers with numeric value *A*, which is stored at a certain location in variable-data memory (RAM).
2. Tell the microprocessor the address of the memory location that is holding *A*.
3. Tell the microprocessor that you want it to load another of its internal storage registers with the numeric value *B* stored at another address in RAM.
4. Tell the microprocessor the address of the number *B*.
5. Tell the microprocessor to subtract the number *A* from the number *B*. If *A* is less than *B*, then when the microprocessor performs the subtraction, it will not have to "borrow"; if *A* is greater than *B*, the microprocessor will have to "borrow" in order to perform the subtraction. If it has to borrow, the microprocessor signals that fact by setting a *borrow bit* to 1; if it does not have to borrow, the microprocessor clears the borrow bit to Ø.
6. Tell the microprocessor to inspect the borrow bit. If it is cleared to Ø (meaning $A < B$), have the microprocessor continue on to instructions 7, 8, and 9 which follow immediately. But if the borrow bit is set to 1 (meaning $A > B$), have the microprocessor branch around instructions 7, 8, and 9, and go directly to instruction 10.
7. Tell the microprocessor to place a 1 at a particular prearranged bit location in one of its *output ports*. (That 1 must then be sensed by an output amplifier or output module, which converts it to a high voltage to energize the output device, just like a PLC.)
8. Tell the microprocessor the address of the output port.
9. Tell the microprocessor to skip (branch around) instructions 10 and 11 and go directly to instruction 12.
10. Tell the microprocessor to place a Ø at the prearranged bit location in the output port. (That Ø must be sensed by an output module, which will then remove power from the output device.)
11. Tell the microprocessor the address of the output port (the same address that was specified in instruction 8).
12. Continue with the rest of the program.

Do you get the general idea? Programming a raw microcomputer requires very exacting reasoning on your part and very close attention to every little detail, more so than for programming a PLC.

18-4 ■ THE PROGRAM FLOWCHART

To help us organize our thoughts when writing microcomputer programs, we begin by drawing a *flowchart*. A flowchart graphically depicts the sequence of events that occur during program execution, especially the various branches that the program may take as it works toward the end of the execution. A flow chart for the coal-slurry control program is presented in Fig. 18–3.

As the flowchart indicates, the conveyor speed is initially set at a moderate value when the program starts operating. Then the 20 weight samples are taken over a 10-second

*Note carefully that the microprocessor is not synonymous with the microcomputer. The microprocessor chip is only a component part of the entire microcomputer, albeit the principal component part. The microcomputer chip must be combined with memory chip(s), input and output chips, bus hardware, and control switch(es) in order to make a complete microcomputer.

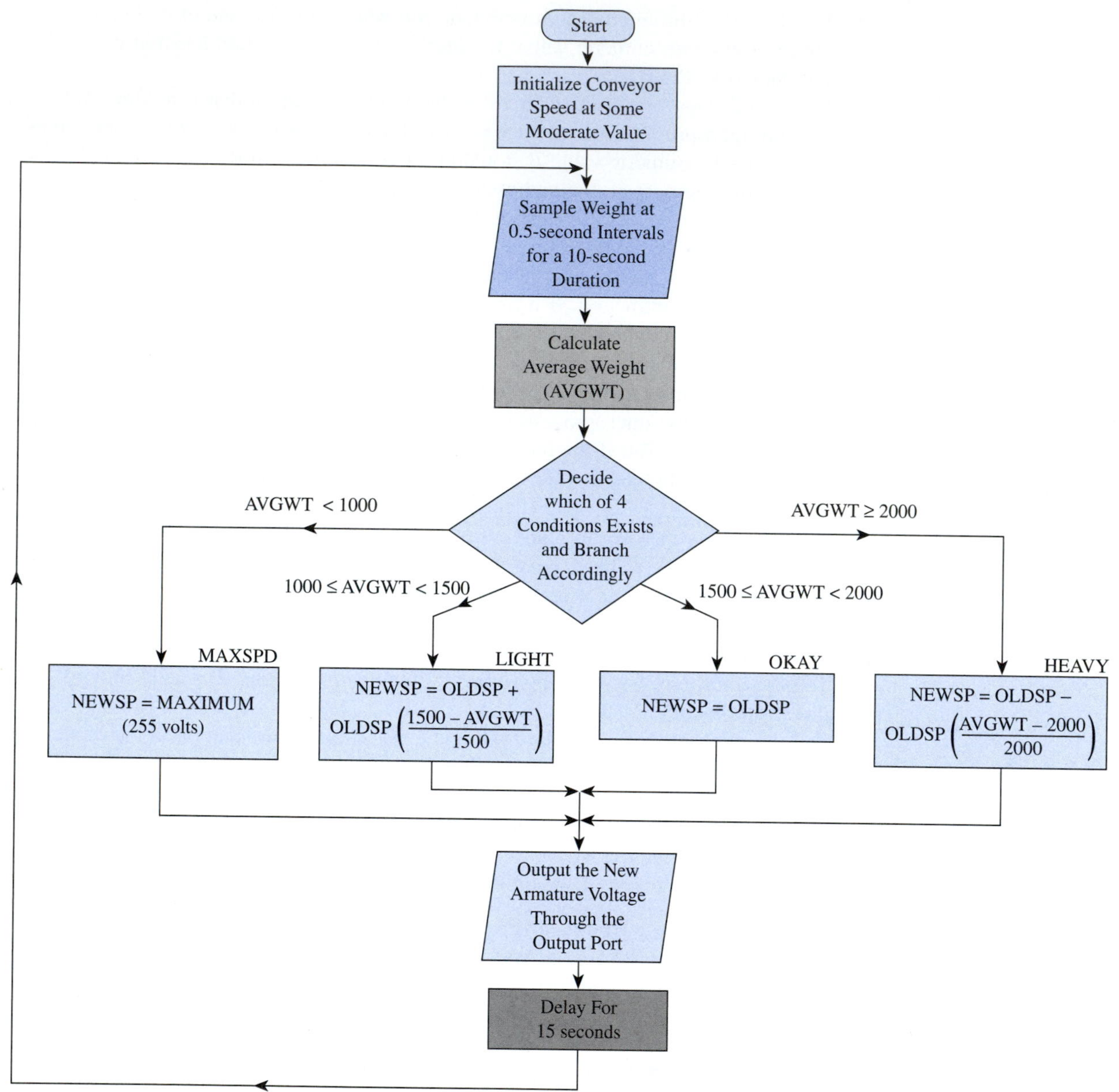

FIGURE 18–3
Flowchart of the control program. The weight-sampling input operation is repeated 20 times for each execution of the program.

period. The actual sampling process that the computer performs is extremely fast, taking perhaps 15 μs or so. Since the elapsed time between samples is 500 000 μs, it is clear that the computer spends most of the available time just waiting for another sample instruction. The terrific speed of the microcomputer is not being effectively utilized during this 10-second period.

When the twentieth (last) weight sample has been gathered, the computer goes to work in earnest. It adds all 20 weight values and divides by 20 to find AVGWT, as expressed in Eq. (18-1). In a typical industrial microcomputer this calculation might take a few hundred microseconds.

Once the average weight is known, the program follows one of the four branches shown in Fig. 18–3. Each branch leads to a different method, or formula, for calculating the new armature voltage and speed NEWSP. The time required for the computer to decide which branch to take and then to carry out the NEWSP calculation, depends on which branch is actually taken, but it can be conservatively estimated at several hundred microseconds.

The new armature voltage (speed) value is passed through an output port, which is a hardware chip coupling the microcomputer to the outside world. In this case the outside-world device that is receiving the speed data is a motor-driven system that adjusts armature voltage V_A to equal the calculated value NEWSP. Therefore the conveyor motor commences its new speed and the corrective action is complete in Fig. 18–3.

Because of the transportation lag associated with changing the screw conveyor's speed and noticing the results of that change in the weighing section, the computer now goes into a 15-second delay period; during this time it does nothing for the system's control. After the 15-second delay the program returns to a new sampling period, as the feedback path in Fig. 18–3 suggests. This constitutes one execution of the program.

When a computer is used in an *on-line application* like this one to control a continuous process, its program just keeps reexecuting over and over again, unless the user deliberately stops it.

18-5 ■ THE MICROCOMPUTER'S ARCHITECTURE

The hardware layout of the microcomputer is sketched in Fig. 18–4, including the input and output devices associated with the overall slurry-control system. The load cell and analog-to-digital converter on the far left of that diagram furnish an 8-bit digital representation of the slurry weight. For the weight values that are encountered in this system, it is convenient to calibrate the load cell so that 1 bit represents 10 pounds. Then the minimum weight is zero (eight Øs from the ADC) and the maximum weight that can be expressed is 2550 lb (eight 1s from the ADC, equivalent to decimal 255).

The digital data from the ADC is always present on the eight lines leading into the *input buffer*. However, the input buffer passes the data to the eight wires that make up the microcomputer's *data bus* only at specific time instants—the time instants when a weight sample is being taken. There are two *control lines* connecting to the input buffer: CE, standing for *Chip Enable,* and R, for *Read*. When both of these control lines go HI, the input buffer passes the weight data to the microcomputer's data bus. Whenever either control signal is LO, the input buffer is disconnected from the data bus by the action of its internal circuitry.

The foregoing idea sounds rather trifling, but it is the fundamental organizing concept of all modern computers. Because we can use control signals to disconnect any device (in this case the input buffer) from the data bus, we can wire many devices onto the data bus. Some of these devices serve only to put data *onto* the bus, some of them serve only to take data *from* the bus, and some of them put data onto the bus at certain times and take data off of the bus at other times. We just have to be careful to design the chip-enable logic and write the program so that control signals never allow more than one data sender to be connected to (putting data onto) the bus at any instant in time.

In Fig. 18–4 there are four devices that can put data onto the bus: the input buffer, the microprocessor chip, the ROM chip, and the RAM chip. Each of these devices has

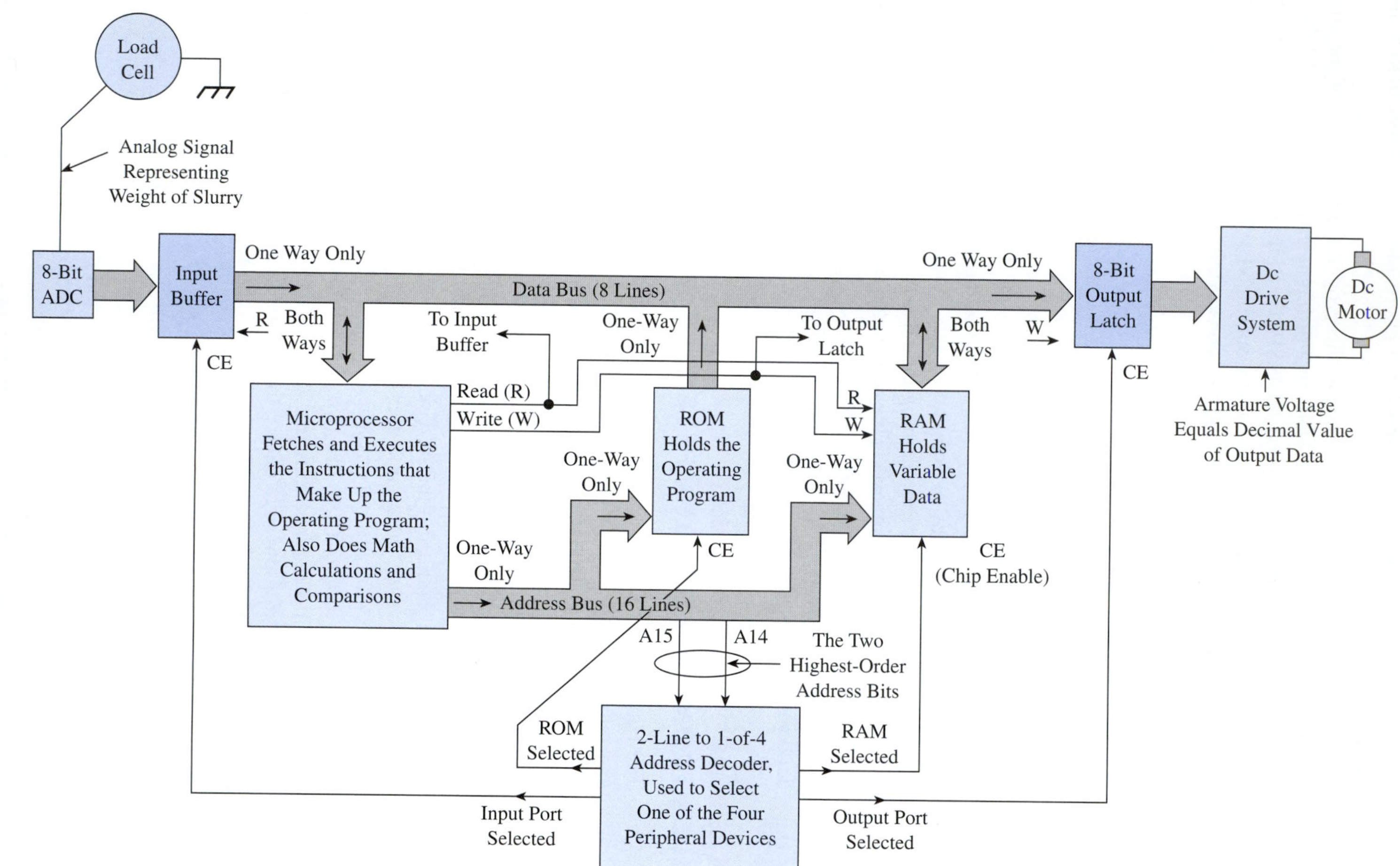

FIGURE 18–4
Block diagram of the microcomputer. Each block in this figure represents one integrated-circuit package, or one chip.

eight data-output lines wired to the data bus. These lines are denoted as D7, D6, D5, D4, D3, D2, D1, and D∅, in order from most significant bit (MSB) to least significant bit (LSB).

Now, for instance, if the internal data in the input buffer called for D7 to be ∅ at the same time that the internal data in the ROM called for D7 to be 1, then we dare not let both devices have access to the bus at the same time, because we would have a *fight for the bus* on our hands. Even if the D7 bits of the input buffer and the ROM agreed, it's a virtual certainty that at least one of the other seven pairs of bits would disagree. So we must make certain that no more than one data-sending device is *enabled* at any one time. This is ensured by our careful design and programming of the control signals that originate in the microprocessor chip. The responsibility is solely ours to prevent fights for the bus, or *bus contention*.

18-5-1 Tri-State Output

In order for this approach to work, the output circuits of the data-sending devices must have *tri-state* capability. That is, each output line, or bit, must have three capabilities: It must be capable of going to ∅ (ground potential), going to 1 (+5 V potential), or disconnecting from the bus—going to the so-called *high impedance* or *high-Z state,* also called the *floating state*. A method of accomplishing tri-state output capability using bipolar electronics is shown in the schematic diagram of Fig. 18–5. It is not essential that you make

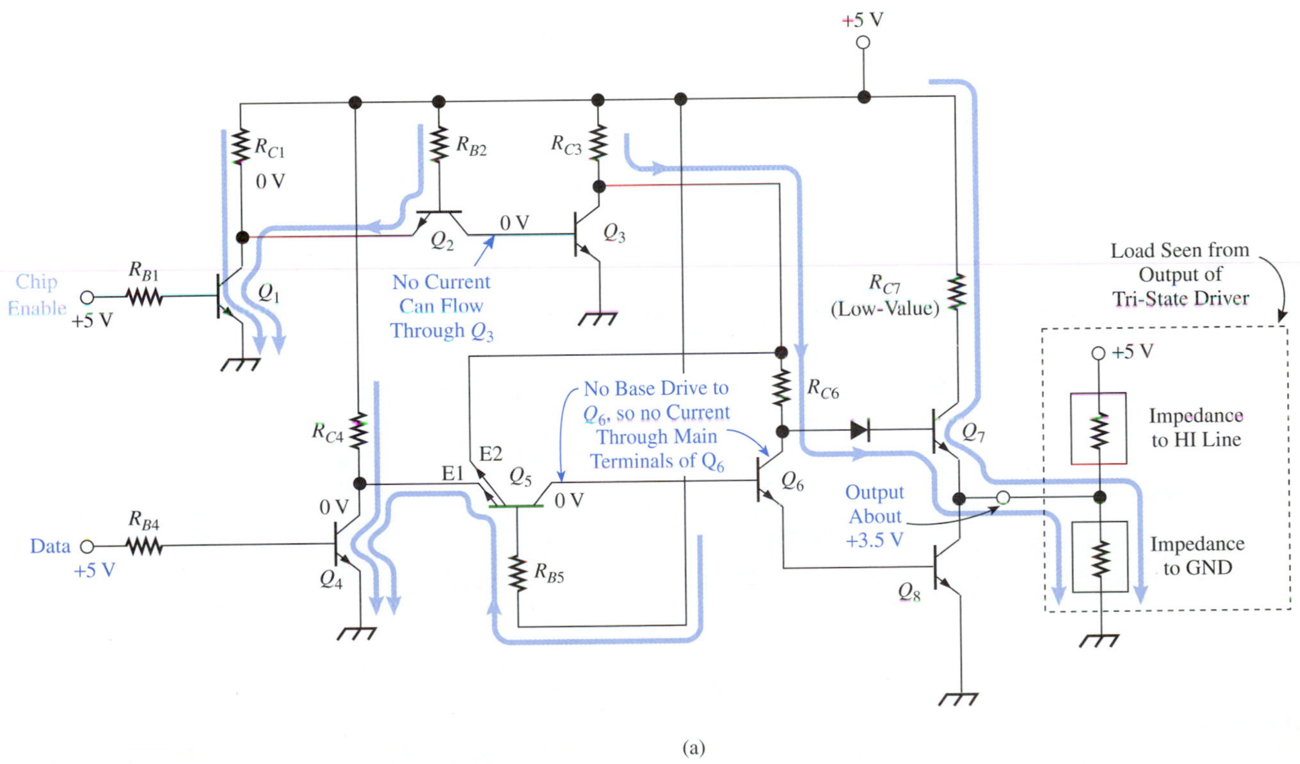

FIGURE 18–5
Tri-state output circuit for a single bit. A device handling 8 bits would have eight identical circuits like this, all with a common CE point. Circuit action with (a) CE HI and Data HI, (b) CE HI and Data LO, (c) CE LO, Data is don't-care.

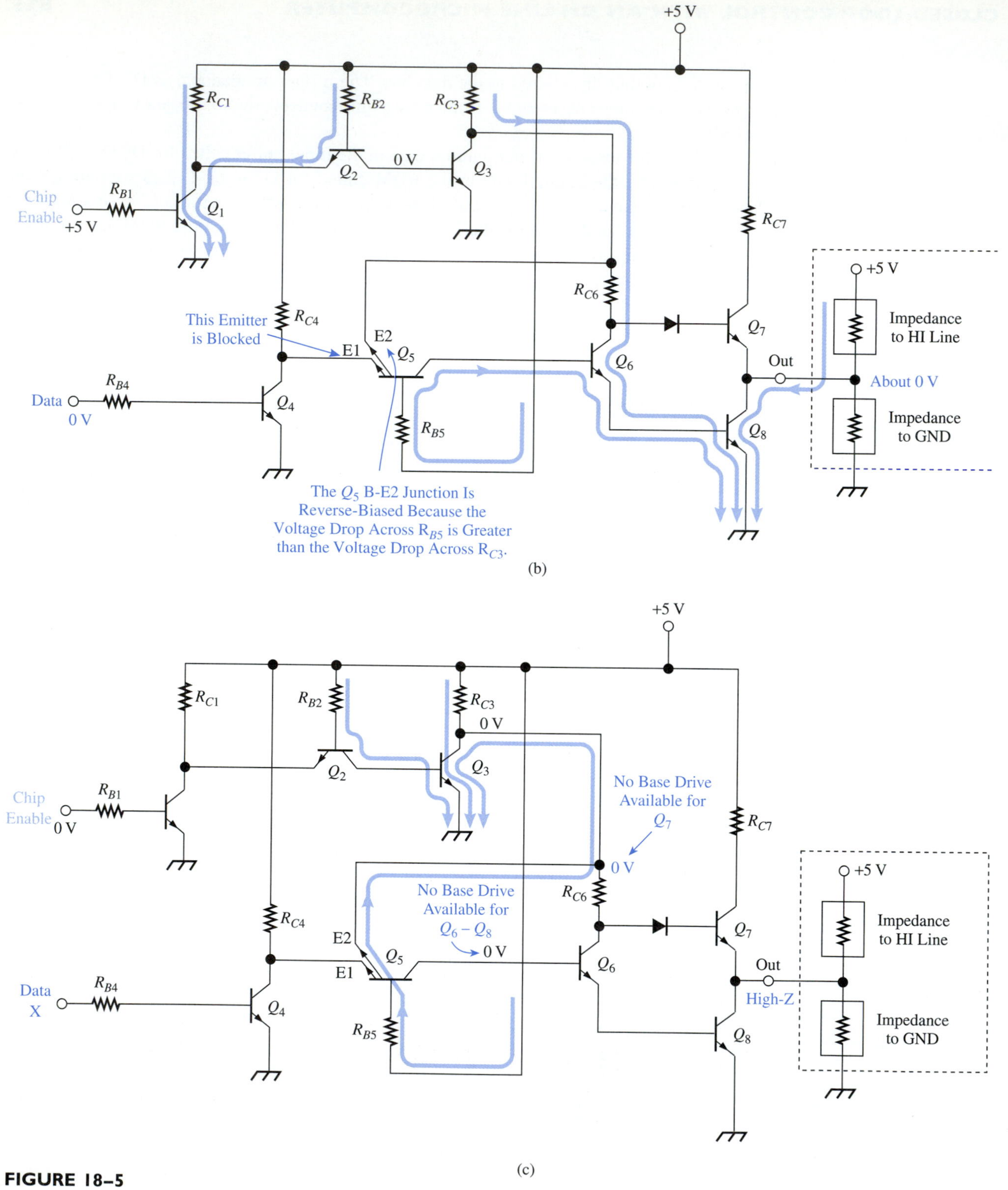

+5 V

R_{C1} R_{B2} R_{C3}

0 V

Q_2 Q_3

Chip Enable +5 V

R_{B1} Q_1

R_{C7}

+5 V

Impedance to HI Line

R_{C6}

This Emitter is Blocked

R_{C4}

E1 E2 Q_5

Q_6 Q_7

Out

About 0 V

Data 0 V

R_{B4} Q_4

R_{B5}

Q_8

Impedance to GND

The Q_5 B-E2 Junction Is Reverse-Biased Because the Voltage Drop Across R_{B5} is Greater than the Voltage Drop Across R_{C3}.

(b)

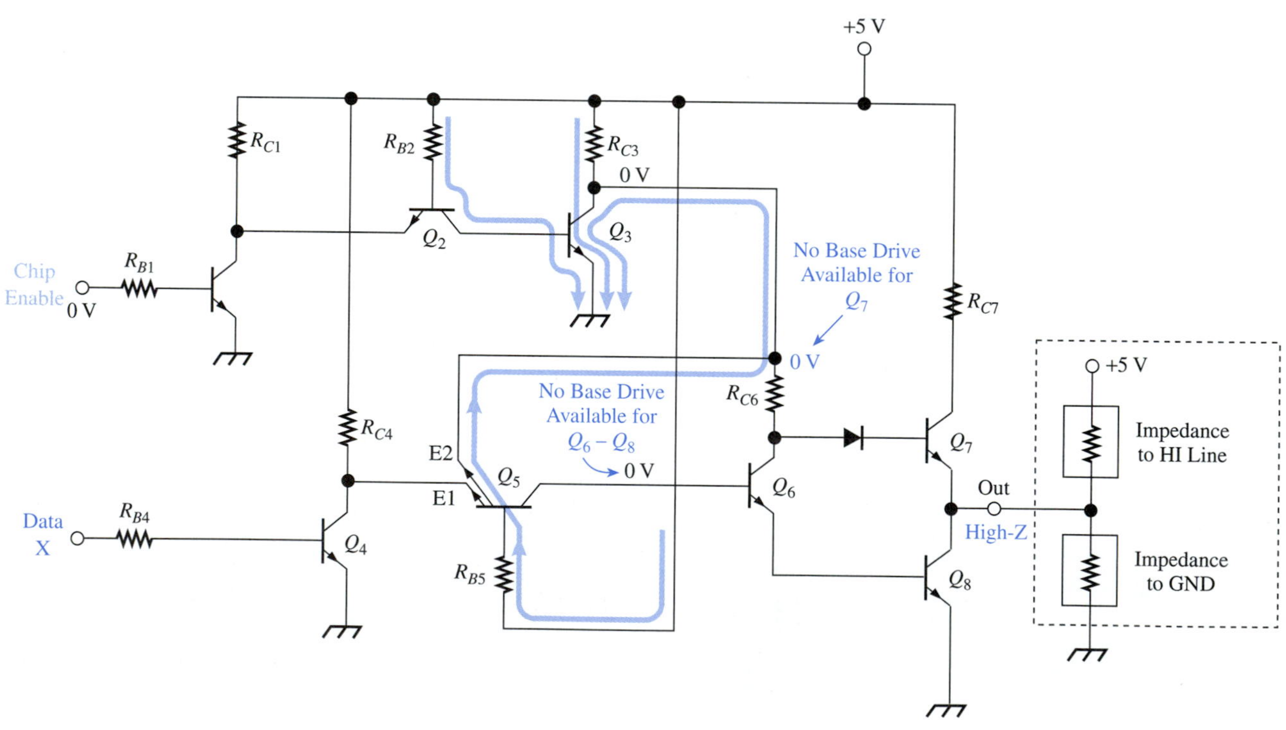

(c)

FIGURE 18–5
(continued)

a careful study of the tri-state circuit right now. You can skip it and come back some other time if you want to understand the electronic details of its operation. Only a conceptual grasp of tri-state output is essential to understanding microcomputer organization.

The tri-state circuit is constructed so that if Chip Enable is HI, then Output is the same as Data. That is, Output goes HI if Data is HI; and Output goes LO if Data is LO. However, if Chip Enable (CE) is LO, then Output goes to its high-Z state, effectively disconnected from the rest of the circuit, no matter what the binary level of Data.

Let's take the three possibilities one at a time:

1. CE is HI and Data is HI.
2. CE is HI and Data is LO.
3. CE is LO and Data is X (don't care).

Refer to Fig. 18–5(a) for condition 1. The HI on CE causes Q_1 to turn ON, thereby turning ON Q_2 and turning OFF Q_3.

With Data at a HI level, Q_4 turns ON, establishing a current flow-path as follows: from the +5 V supply, through R_{B5}, out emitter E1 of the dual-emitter transistor Q_5, and through Q_4 to ground. With Q_5 saturated, its collector drops almost to 0 V, removing any base drive from Q_6. Therefore Q_6 turns OFF and so does Q_8. With Q_3 and Q_6 both cut off, a current flow-path is established through R_{C3} and R_{C6} into the base of Q_7. Therefore Q_7 turns ON and pulls the Output up to a HI level through the low-value resistor R_{C7}.

Condition 2 is shown in Fig. 18–5(b). With CE HI, Q_3 is turned OFF as in condition 1. The LO at Data causes Q_4 to turn OFF, blocking the path through the B-E1 junction of Q_5. Any current that managed to sneak through the B-E2 junction of Q_5 could not pass around through R_{C6} to drive the base of Q_7, because if it did, the emitter potential at Q_7 would rise toward +5 V and shut down the flow path of its own base current. Therefore the B-E2 junction of Q_5 is also blocked. With both emitter junctions blocked, current will flow through the base-collector junction of Q_5, thereby providing base drive to Q_6. Collector current reaches Q_6 through R_{C3} and R_{C6}, enabling it to turn ON Q_8. With Q_8 saturated, the Output is pulled down to a LO level.

Condition 3 is illustrated in Fig. 18–5(c). With CE LO, Q_1 turns OFF and blocks the B-E junction of Q_2. Current therefore passes through the B-C junction of Q_2, thereby driving the base of Q_3. Transistor Q_3 saturates, pulling its collector down to virtually 0 V. This 0-V potential appears at the top of R_{C6}, precluding any possibility of base drive to Q_7. Transistor Q_7 shuts OFF, disconnecting the Output from the +5-V line.

The B-E2 current through Q_5 makes that transistor saturate, causing its collector to fall to nearly 0 V. Such a low voltage cannot deliver any drive current to the Q_6-Q_8 pair, so Q_8 remains cut OFF. With Q_8 OFF, the Output terminal is disconnected from ground.

Thus, no matter which way the Output terminal looks, it sees an open-circuited transistor—it is electrically isolated from the rest of the circuit. The Data level is irrelevant because the conditions at the B-E1 junction of Q_5 cannot have any effect on that transistor—it's already saturated anyway.

The large-scale integrated circuits that are used in a microcomputer usually are not built with bipolar transistors; they're built with field-effect transistors. Even so, they work in a manner similar to the working of the bipolar circuit of Fig. 18–5.

We need a simple schematic symbol to represent a tri-state circuit. The symbol that is most often used is shown in Fig. 18–6(a), along with its truth table. Some tri-state circuits have an active-LO enable function; that is, the Enable terminal must go to 0 in order to transfer data to the output. The standard bubble notation is used to symbolize an active-LO control line, as shown in Fig. 18–6(b).

FIGURE 18–6
Schematic symbol and truth table for a tri-state output circuit: (a) with active-HI enable; (b) with active-LO enable. Active-LO enable could be obtained by removing transistor Q_1 from Fig. 18–5 and feeding the CE signal directly to the emitter of Q_2.

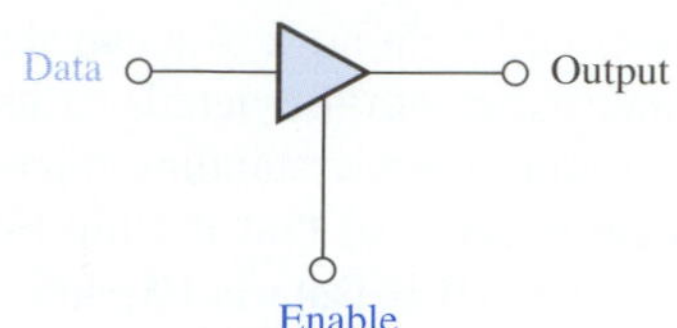

Enable	Data	Output
1	Ø	Ø
1	1	1
Ø	X	High-Z

(a)

Data
Output
Enable

Enable	Data	Output
1	X	High-Z
Ø	Ø	Ø
Ø	1	1

(b)

18-5-2 Bidirectional Devices

Some microcomputer devices handle data in both directions, sending and receiving. The microprocessor chip and the RAM chip in Fig. 18–4 are examples of such *bidirectional* devices. Note the double-ended arrows labeled Both Ways that are shown at the data bus junctions of these two devices. Inside these devices, the data bus splits into two paths, creating a "sending" data bus and a "receiving" data bus. Like the sending data bus, the receiving data bus also has a set of eight tri-state circuits, so that the receiving circuits in the device are not forced to look at the data bus all the time. This internal structure is shown schematically in Fig. 18–7.

The receiving tri-state circuits, collectively called a *buffer,* are enabled only when the bidirectional device is supposed to be receiving data from somewhere else in the microcomputer. This synchronization is accomplished by control lines, but not the same combination of control lines that enables the sending tri-state buffer, naturally. Here are specific descriptions of the control conditions for the two bidirectional microcomputer devices in Fig. 18–4, the RAM and the microprocessor.

RAM buffer control. For the RAM, the sending tri-state buffer is enabled when CE goes to its active level (HI) and R (READ) also goes to its active level (HI). (We will consistently assume that the control lines' active levels are HI, but that isn't always the case in real life.)

In microcomputer parlance, the words *read* and *write* refer to what the *microprocessor* is doing. Thus the word *read* implies that the microprocessor is bringing data from some other device into one of its own registers. Therefore, when the microprocessor chip *reads* the RAM, the RAM is expected to *send* data down the data bus. This is why the control line combination of CE and R both HI is required to enable the *sending* tri-state buffer inside the RAM.

The RAM's receiving tri-state buffer is enabled by the control-line combination of CE and W (write) both being HI. In microcomputer parlance the word *write* implies that the microprocessor is transferring data from one of its own registers into some other device within the microcomputer. Therefore, when the microprocessor chip *writes to* the RAM, the RAM is expected to *receive* data sent down the data bus. This is why the control line combination of CE and W both HI is required to enable the receiving tri-state buffer inside the RAM.

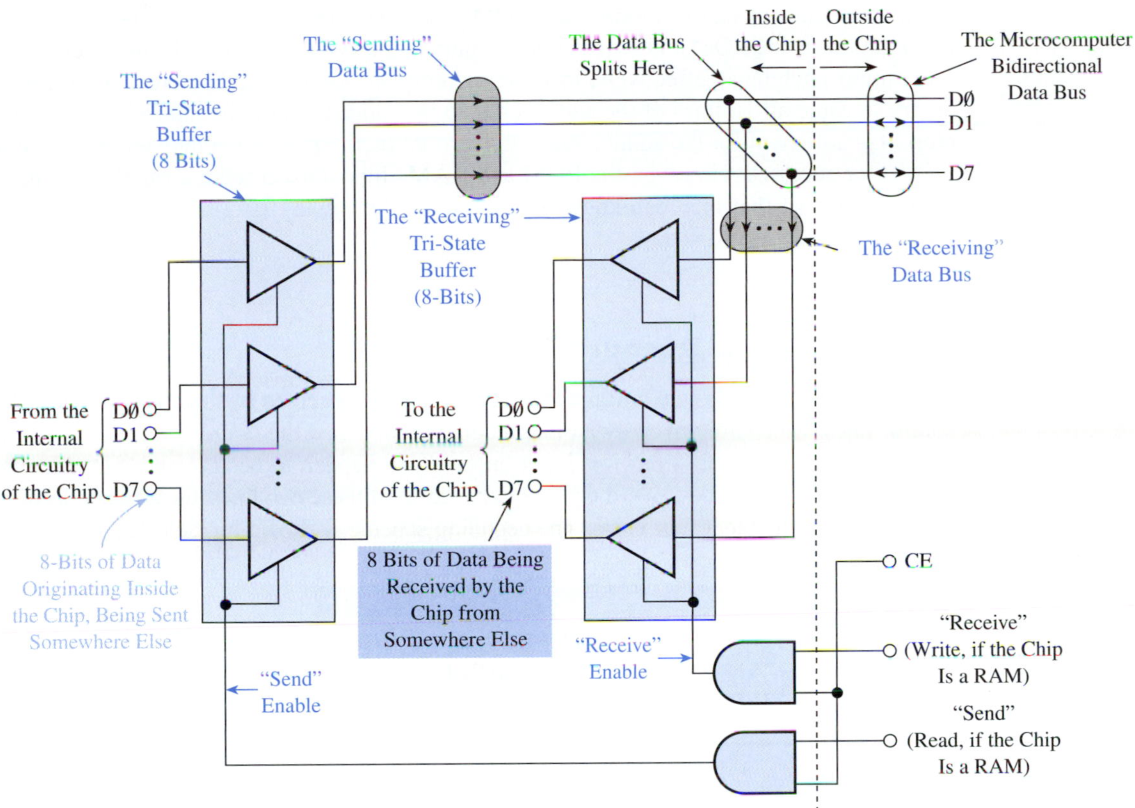

FIGURE 18–7
The internal chip circuitry that permits bidirectional data handling.

Microprocessor buffer control. Now stop considering the RAM and start considering the microprocessor chip itself in Fig. 18–4. Inside the microprocessor chip, the relationship of read and write to the sending and receiving buffers is reversed. The microprocessor's receiving buffer is enabled during a read operation, and the sending tri-state buffer is enabled during a write operation. Be sure you understand why this is reasonable.

Well, that's the data bus. If you understand the data bus, the *address bus* is relatively simple, even though it has more lines.*

18-5-3 The Address Bus

In the system of Fig. 18–4, the address bus is strictly a one-way affair. The microprocessor chip places 16-bit binary combinations on the address bus and sends those combinations to the peripheral chips. The 16-bit binary combination, called the *address,* tells the other chips which one of them is being summoned by the microprocessor, and, if the chip contains many memory locations, it tells the chip being summoned exactly *which* location the microprocessor is interested in. Our system contains four peripheral chips that

*Our example microprocessor has an 8-bit data bus and a 16-bit address bus, for simplicity. Modern industrial on-line control computers have larger buses. Specifically, the standard data bus is now 32 bits wide.

the microprocessor can summon—the ROM, the RAM, the input buffer, and the output latch. Of these, the ROM and RAM contain many memory locations, while the input buffer and output latch have, effectively, just one memory location each. Therefore it isn't necessary to route all 16 lines of the address bus to the input-buffer chip and the output-latch chip; it is sufficient for the address bus simply to let them know when they are being summoned, or accessed. However, the ROM and RAM chips must receive virtually the entire address bus* so that they can know which one of their many internal locations the microprocessor wishes to access.

18-5-4 Address Decoding

The four peripheral chips are informed that they are being accessed by their chip enable (CE) control lines, which are driven by an intermediary chip called the *address decoder*. The address decoder in Fig. 18–4 has two input lines, which happen to be A15 and A14, the two highest-order address lines. Based on the binary status of these two lines, the address decoder enables one of the other four chips. The decoding scheme is shown in Table 18–1.

TABLE 18–1
Truth table for the two-line to 1-of-4 address decoder

A15	A14	PERIPHERAL DEVICE ENABLED
∅	∅	Output latch
∅	1	Input buffer
1	∅	RAM
1	1	ROM

The addressing arrangement can be summed up like this: The microprocessor chip is the brains of the outfit. When it wants to summon the output latch it makes the two highest-order address bits ∅s—the other 14 bits don't matter. That is, A15 = ∅, A14 = ∅, and A13 through A∅ don't matter. In response to A15 = ∅ and A14 = ∅, the address decoder puts a HI on the output latch's CE control line, and it puts LOs on the other three peripheral chips' CE control lines. Thus the output latch is enabled to receive from the data bus, and everybody else is disabled.

When the microprocessor wants to summon the input buffer, it makes A15 = ∅ and A14 = 1; the other 14 bits don't matter. The address decoder puts a HI on the input buffer's CE line** and LOs on the other three CE lines. Thus the input buffer is enabled to place data on the data bus, and everybody else is disconnected from the data bus.

When the microprocessor wants to summon the RAM, it makes A15 = 1 and A14 = ∅. The address decoder then makes the RAM's CE HI, and the other three CE

*These chips may not need to receive the entire address bus, but they need most of it. Just how much of the address bus a memory chip needs to receive depends on how big its memory is (how many address locations it contains).

**This method of accessing the input and output devices is known as *memory-mapped I/O*. Some microcomputers use the so-called *standard I/O* method, which does not require that a memory address be reserved for each I/O port.

lines LO. The other 14 address bits are sent to the RAM to tell the RAM exactly which one of its memory locations the microprocessor wants to read from or write to.

When the microprocessor wants to summon the ROM, it makes A15 = 1 and A14 = 1. The address decoder enables the ROM with a HI on its CE line and disables the other three chips. The other 14 address bits are sent to the ROM to tell the ROM which one of its memory locations the microprocessor wants to read from.

18-6 ■ EXECUTING A PROGRAM

Let us suppose that the user has entered a workable program into the ROM chip. The details of how this is done depend on the type of ROM chip that is being used (regular ROM, PROM, EPROM, or EEPROM), but those procedural details do not concern us right now. All we care about is that there is a collection of instructions, a program, stored in the ROM, and this program is ready to execute.

Someone must press the RESET button (not shown in Fig. 18–4), which is wired to the microprocessor. This causes the microprocessor to access the ROM (A15 = 1, A14 = 1) and to place on lines A13 through AØ the address of the ROM location that contains the first instruction of the program. The microprocessor also brings the R line to its active-HI level, thereby fulfilling all the conditions necessary to enable the ROM's tri-state output (sending) buffer. The ROM now has the data bus; it places the first instruction on the bus, in binary-coded form. The microprocessor's input (receiving) tri-state buffer is also enabled at this time because the computer is in a READ condition. Therefore the microprocessor takes in the 8-bit binary-coded instruction, and it stashes those 8 bits in its *instruction-decoding register*. Some internal logic circuitry looks that register over, figures out what the instruction means, and decides whether the microprocessor can execute the instruction right now, using only information that is currently present in the microprocessor chip, or whether the microprocessor must gather more information from the peripheral chips before it can execute the instruction.

If the instruction can be executed right now, the internal logic circuitry and/or internal arithmetic circuitry proceeds to do it. For example, in a Motorola 6800-series microprocessor, the binary code Ø1ØØ 11ØØ means "increment register A"; register A is one of the microprocessor's internal registers that is used for temporary storage and handling of data. This is an instruction that can be executed right away, because no further information is needed. The arithmetic circuitry in the microprocessor simply adds 1 to the binary number residing in register A and the job is done. The microprocessor can then move to the next sequential address in the ROM, fetch the second instruction, and continue going about its business.

Figure 18–8 shows pictorially how the microprocessor handles an INCA (increment register A) instruction, which is a representative example of that class of instructions that can be executed right away. The class of instructions is referred to as *inherent*.

Many instructions can't be executed right away; they require that more information be brought into the microprocessor. When such an instruction arrives in the microprocessor's instruction-decoding register, the internal logic circuitry tells the arithmetic circuit to wait awhile until the additional information can be brought in. The internal logic circuit then advances the microprocessor's *program counter;* the program counter is a 16-bit register within the microprocessor that stores the current ROM address, thereby keeping track of our current location in the program. When the program counter sends the next sequential address down the address bus to the ROM, the ROM replies with the additional information that was required to carry out the instruction.

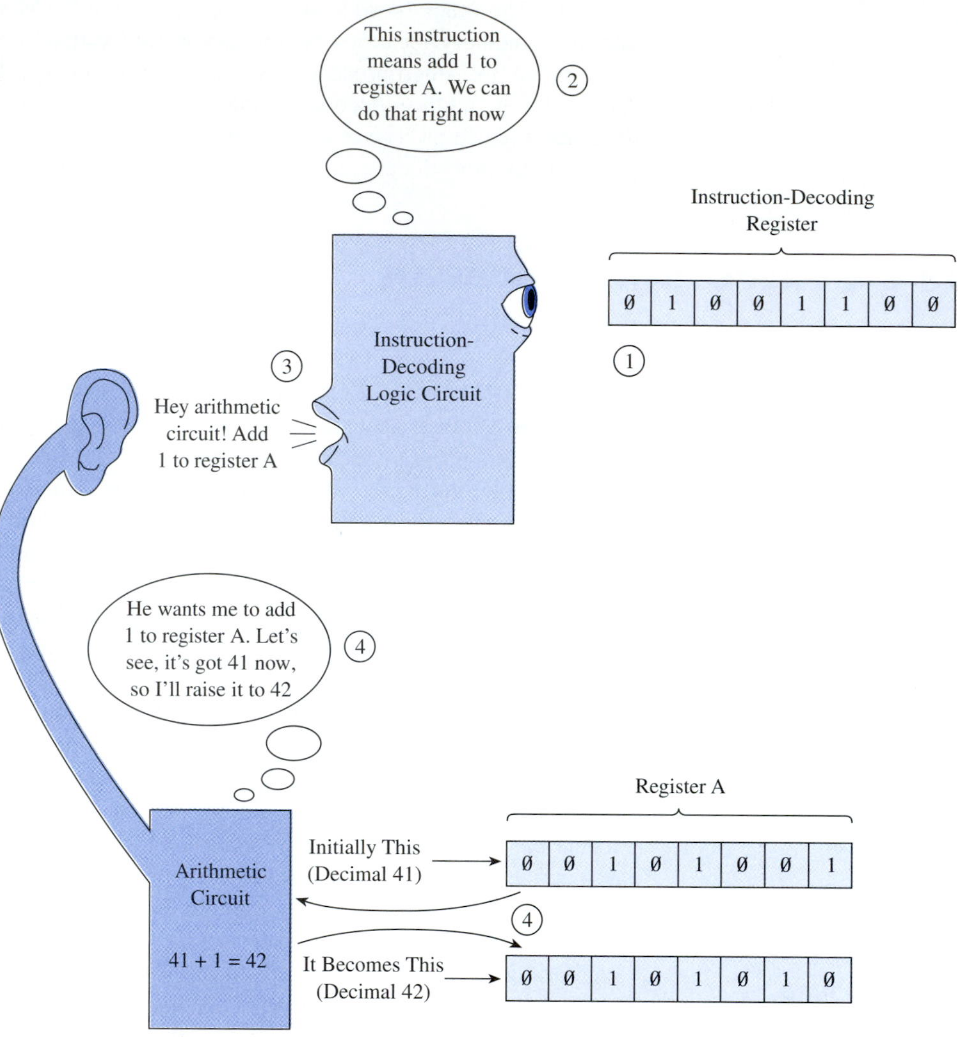

FIGURE 18–8
The sequence of events that occurs inside the microprocessor for the instruction "Increment Register A," which is the type of instruction that is executed right away. ① The coded instruction is brought into the instruction-decoding register from the ROM, via the data bus. ② The decoding logic circuit figures out what the instruction means. ③ The logic circuit issues a command to the arithmetic circuit. ④ The arithmetic circuit hears the command and carries it out.

For example, in the 6802 microprocessor, the binary code 1ØØØ 1Ø11 means "Add the number in the *next* ROM location* to the number in register A." The internal logic circuit must arrange to get that next number out of the ROM and into the microprocessor's memory data register (MDR). Once the number has arrived, the arithmetic circuit performs the addition, and the job is done. Then the microprocessor can move to the next sequential address in the ROM, fetch the second *instruction,* and go about its business.

Figure 18–9 shows pictorially how the microprocessor handles an ADDA (add the next number in ROM to register A) instruction, which is a representative example of that class of instructions that cannot be executed right away.

So this gives us a glimpse of what's happening as a microprocessor chip in an on-line microcomputer executes the program which is stored in the ROM chip.

---

*Note that ROM locations can contain nonvariable numeric values as well as instruction codes. ROM locations can contain other kinds of information also (half of an address, a user-defined code, an American Standard Code for Information Interchange, etc.).

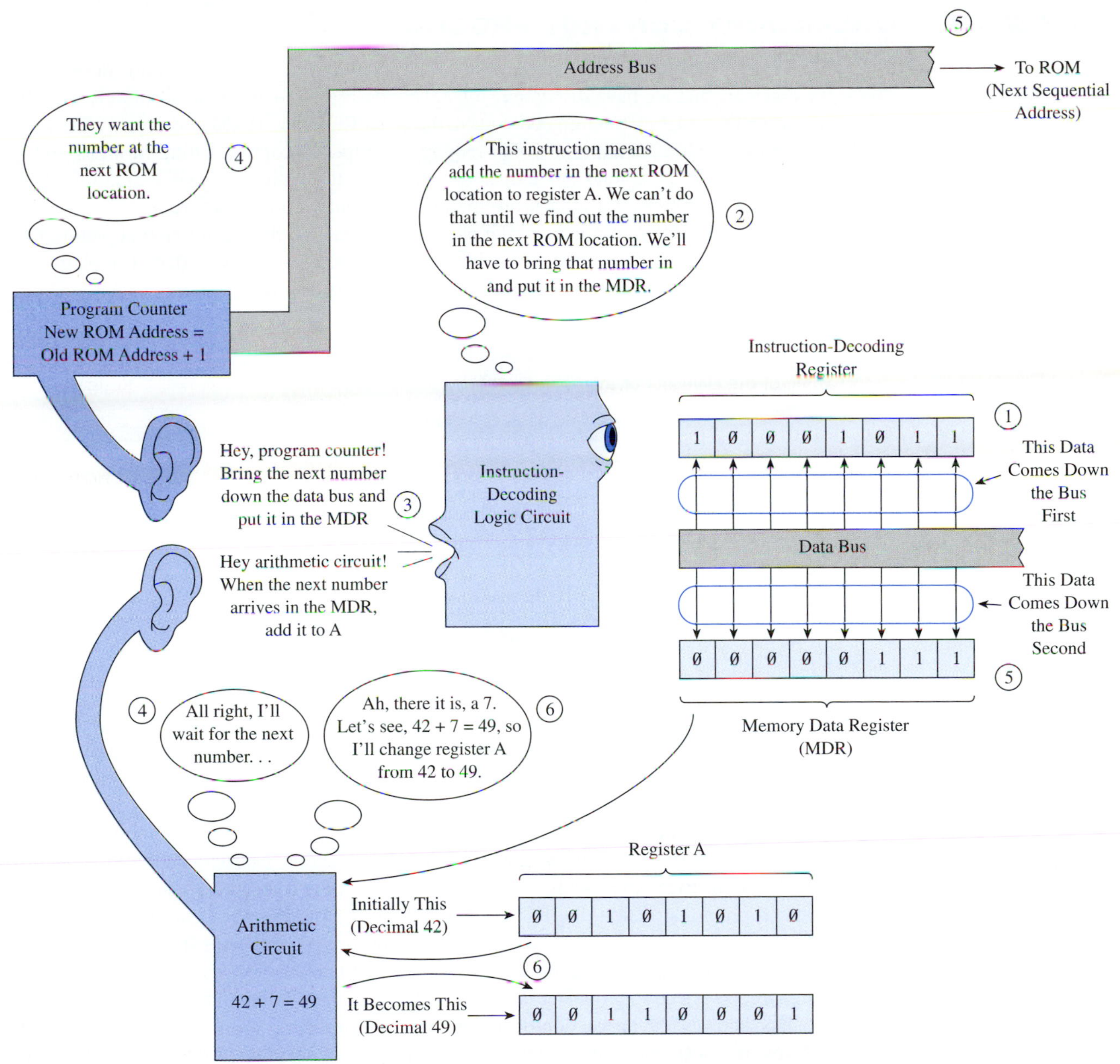

FIGURE 18–9
The sequence of events that occurs inside the microprocessor for the instruction "Add to Register A," which is the type of instruction that must wait for more data to arrive in the microprocessor. ① The coded instruction is brought into the instruction-decoding register from the ROM, via the data bus. ② The decoding logic circuit figures out what the instruction means and realizes that the instruction can't be executed until more information is brought into the microcomputer. ③ The decoding logic circuit issues one command to the program counter to increment the ROM address and a second command to the arithmetic circuit to wait for a new number, then add it to A. ④ The program counter increments the address bus while the arithmetic circuit waits for the new number. ⑤ The next sequential address is sent to the ROM, which sends the number at that address down the data bus to the MDR. ⑥ The arithmetic circuit detects the arrival of the new number in the MDR, and it carries out the addition process with register A.

18-7 ■ THE COAL-SLURRY CONTROL PROGRAM

Let us describe the overall performance of the coal-slurry control program, in terms of the concepts that we have discussed so far. This description cannot be absolutely specific regarding the exact instruction codes used, the exact ROM and RAM addresses referenced, and so on. Such a description would require a complete program listing in an assembly language, which is beyond our scope. We are trying to obtain just a general comprehension of the operation of a typical on-line microcomputer control program.

To successfully comprehend the program's operation, the partial mental picture that we now have of the microprocessor's internal structure must be expanded. The picture as it now stands looks something like the diagram of Fig. 18–10.

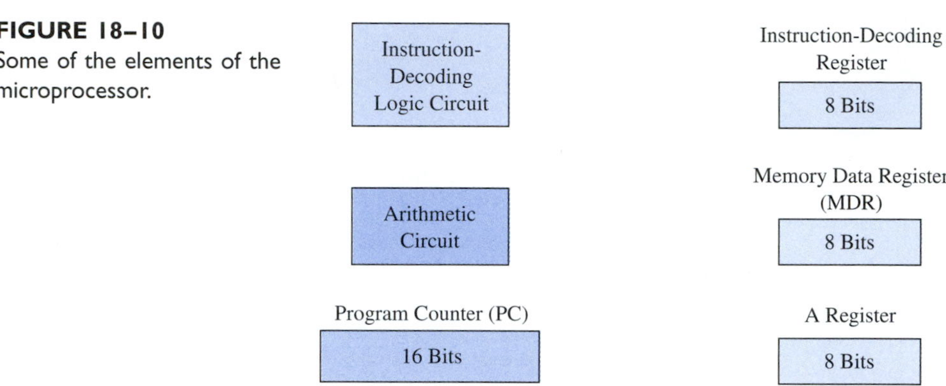

FIGURE 18–10 Some of the elements of the microprocessor.

In that figure there are only two elements that we must keep constantly in mind—register A and the program counter (PC).* As we know from our earlier discussion, register A is an 8-bit register used to handle data. As such, it serves a variety of purposes, as we will see when we trace through the coal-slurry program. The PC always contains the ROM address which identifies our current location in the program. Taking a simplified view, when the microprocessor is working on one program instruction, the PC contains the address in ROM that holds that instruction; when the microprocessor goes on to another program instruction, the PC will contain the ROM address of that new instruction.

Don't worry about the remaining four microprocessor elements in Fig. 18–10. They aren't necessary for grasping the overall picture of the control program's operation. Yes, the instruction-decoding register is in there, receiving instruction codes from ROM via the data bus; and yes, the instruction-decoding logic is figuring out what the instructions mean; and yes, the arithmetic circuit actually performs the mathematical operations. But, concentrating on all this stuff is taking too close a view of things. To get the overall picture, we need to stand back and take a broader view.

Well, that's encouraging. We have eliminated two thirds of the clutter that was weighing on our minds regarding the internal structure of the microprocessor. Don't feel too relieved though, because now we have to introduce some new elements that we must keep in mind as we grapple with the control program. There are four new elements, namely: the B register, the index (X) register, the memory address register (MAR), and the condition-code register (CCR). These are illustrated in Fig. 18–11, which will serve from now on as our mental image of the microprocessor's internal structure. This mental im-

*Note that now the acronym PC means "program counter," not "personal computer" or "printed circuit."

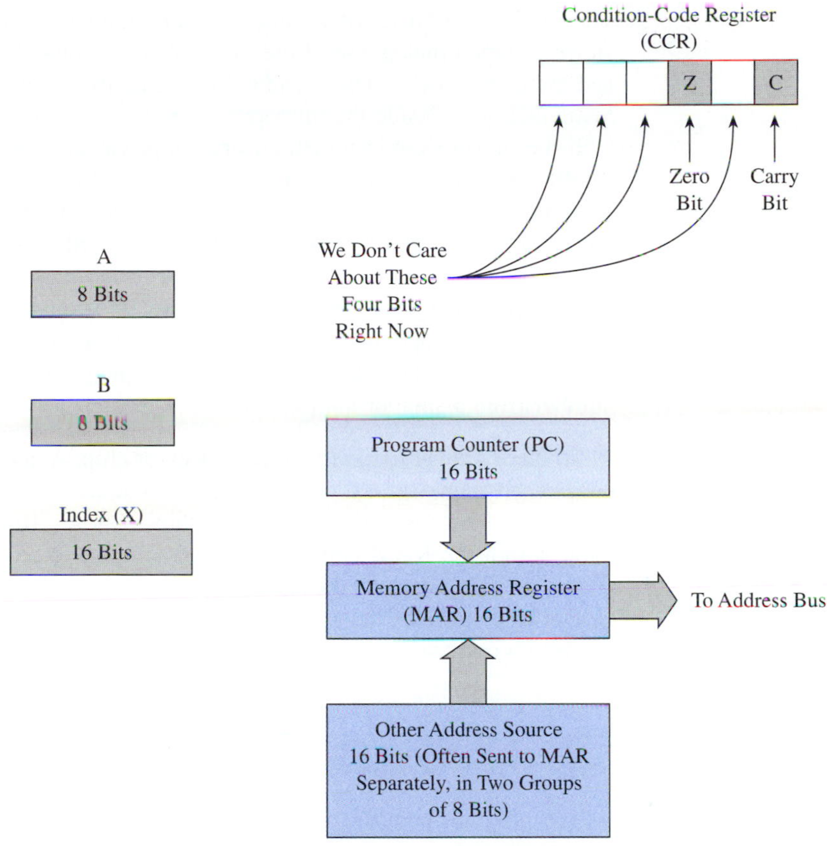

FIGURE 18–11
Partial programming model of the microprocessor. We keep an image of this model in our minds as we think about the operation of the microprocessor.

age of the microprocessor is not thoroughly complete in all respects, but it is sufficient for our present purposes.

These four new elements really aren't too hard to fathom. One at a time, here they are:

1. The B register is just like the A register. It's the same length; it's used for the same type of data-handling jobs; it's alike in all ways. We need B as well as A because sometimes the action gets so fast and furious that a single data-handling register just isn't enough.
2. The X register is a different breed from A and B. It is 16 bits long because it must hold *addresses* rather than data. In the coal-slurry program, X is used to hold the address in RAM that we intend to use next, when we're writing the instantaneous values of slurry weights throughout the sampling duration. X is used for the same purpose during the calculation period, when we're reading those weights back to find the average weight. Refer to the timing diagram of Fig. 18–2 to refresh yourself concerning these matters.
3. The MAR is the 16-bit register that actually places addresses on the address bus. The PC does not drive the address bus directly. Instead the PC transfers its current address to the MAR, and the MAR then places that address on the bus.

It is necessary to have the MAR standing between the PC and the address bus because addresses can originate elsewhere, other than in the PC. In such cases the MAR blocks the PC from the bus and passes the other address instead. This selecting function of the MAR is suggested in Fig. 18–11.

4. The CCR is a 6-bit register whose bits are independent of one another.* The coal-slurry program makes use of only two of the six bits. These two are the zero bit (Z) and the carry bit (C). The Z and C bits signal the results of the most recent arithmetic operation inside the microprocessor.

   If the most recent arithmetic operation produced a numerical result equal to zero (0000 0000), the Z bit goes HI; otherwise the Z bit stays LO.

   If the most recent arithmetic operation produced a numerical result greater than 255 (1111 1111), that number cannot be contained in an 8-bit register, so the C bit goes HI to signal to the program that part of the numerical result has to be *carried* into a separate register. The C bit also goes HI if the most recent arithmetic subtraction-type operation required a *borrow* from a register other than the register that was holding the subtrahend. In plain language, if you tried to subtract a larger number from a smaller number, the C bit goes HI to warn you.

So there's the story on the microprocessor chip. A and B handle data, which always comes in 8-bit chunks.** PC contains the 16-bit addresses that specify where we presently are in the program residing in the ROM chip. X holds 16-bit addresses that specify the location within the RAM chip where we next want to read or write variable data. The MAR passes addresses onto the address bus. Z tells us when an arithmetic operation yields an answer of zero, and C tells us when an arithmetic operation can't be handled with standard 8-bit registers.

18-7-1 Starting the Program—Getting the Conveyor Moving

When the program begins executing, the first thing it must do is store a medium-value number in the output latch of Fig. 18–4. This will cause the dc drive system to develop a moderate armature voltage and the motor will start turning the screw conveyor at a medium speed.

The data numbers in the coal-slurry program will range from 0 to 255, since this is the range of nonsigned decimal numbers that can be expressed with eight bits, as stated earlier. The motor's speed is related to the program's data numbers on the basis of 1 volt per bit. For example, if the number stored in the output latch is 0101 1010, or decimal 90, the drive will apply an average armature voltage of 90 V. If the number passed from the output latch is 1001 1000, or decimal 152, the dc motor will receive $V_A = 152$ V, and so on.

A reasonable starting speed for the motor/conveyor is about half the maximum speed. Therefore the program initializes the motor armature voltage at 128 V, which is about half of 255 V. This is accomplished by the first two instructions in the program. The first instruction *loads* the number 1000 0000 (decimal 128) into the B register, and the second instruction *stores* the contents of the B register in the output latch. (The word *load* implies movement of data *into* a microprocessor register, while the word *store* implies movement of data *out of* a microprocessor register.)

---

*By independent, we mean that the bit to the left is not twice as significant as the bit to its right. Rather, each bit stands alone, signifying a particular fact, or condition, that has nothing to do with the facts that the other bits are signifying.

**An 8-bit chunk is called a *byte*.

When the RESET button is pressed, the microprocessor automatically sets its program counter to the starting address of the program, connects the PC through the MAR to the address bus, and goes into READ mode (R = 1, W = Ø). For concreteness, suppose the program's starting address is 11ØØ ØØØØ ØØØØ ØØØØ, as specified in the flowchart of Fig. 18–12. When the address decoder in Fig. 18–4 gets a look at the two 1s at A15 and A14, it enables the ROM, allowing the ROM to drive the data bus (see Table 18–1). The fact that R is 1 and W is Ø causes the microprocessor's tri-state receiving buffer to be enabled and its sending buffer to be disabled. The microprocessor is thus in a receiving posture, which is appropriate for receiving the first instruction code from ROM.

The program's first instruction is the instruction that means "Load the B register with the numeric value that is now stored in the next sequential location in the ROM." The Motorola code for this instruction happens to be 11ØØ Ø11Ø. When this first instruction code arrives in the microprocessor from the ROM via the data bus, the microprocessor realizes by inspecting the code that the next data byte in the ROM is a numeric value, not another instruction code. Therefore the microprocessor has sense enough to:

1. Increment the PC.
2. Put the PC contents onto the address bus, thereby enabling the output buffer of the ROM.
3. When the contents of the next ROM address show up at the microprocessor via the data bus, load those contents into B.

As long as the user has programmed the ROM with 1ØØØ ØØØØ in the second address location, the numeric value decimal 128 will be successfully loaded into B. So far, so good.

The *third* address location in ROM contains the *second* instruction code. The microprocessor is wise enough to realize this crucial fact, because the second ROM byte provided it with the information it needed to successfully execute the first instruction. So the microprocessor increments the PC, puts the PC on the address bus (which keeps the ROM enabled), grabs the instruction code coming down the data bus from the ROM, and inspects that code. In our program the Motorola code is 1111 Ø111, which means "Store the contents of register B at a memory address which will be given to you in the next two bytes of the program (in ROM)." Therefore the microprocessor realizes that it must get two more bytes out of the ROM so that it knows at what address it should store the contents of B. It increments the PC, reads the fourth ROM address for the *high-order byte* of the destination address, and tucks that byte away in a temporary holding register (the "other address source" depicted in Fig. 18–11). Again, the microprocessor increments the PC, to the fifth byte of the program (at the fifth sequential ROM address), and treats that fifth byte as the *low-order byte* of the destination address. It puts the low-order byte into the lower half of the MAR and transfers the high-order byte from the temporary holding register into the upper half of the MAR. Let's say that our program has ØØ11 ØØØØ as the high-order byte and ØØØØ ØØØØ as the low-order byte. At this point, the microprocessor has acquired all the information it needs to execute the Store instruction. It knows which of its internal registers (the B register) is supposed to have its contents placed on the data bus, and it knows which address (the 16-bit address presently in the MAR) those contents are supposed to be sent to. The microprocessor now does its duty. It disconnects the PC from the address bus (that's the first time that has happened so far in the program) and places the new contents of the MAR on the address bus.

With the MAR contents ØØ11 ØØØØ ØØØØ ØØØØ on the address bus, the ROM is disabled for the first time during the program, and the output latch is enabled. Refer to Fig. 18–4 and Table 18–1. Furthermore, the microprocessor pulls the READ line to the

FIGURE 18–12 Detailed flowchart of microprocessor actions for the first three program instructions. These three instructions cause decimal 128 to be stored in two places—the output latch and memory location SPEED. The machine code that corresponds to this chart is given in Table 18–2.

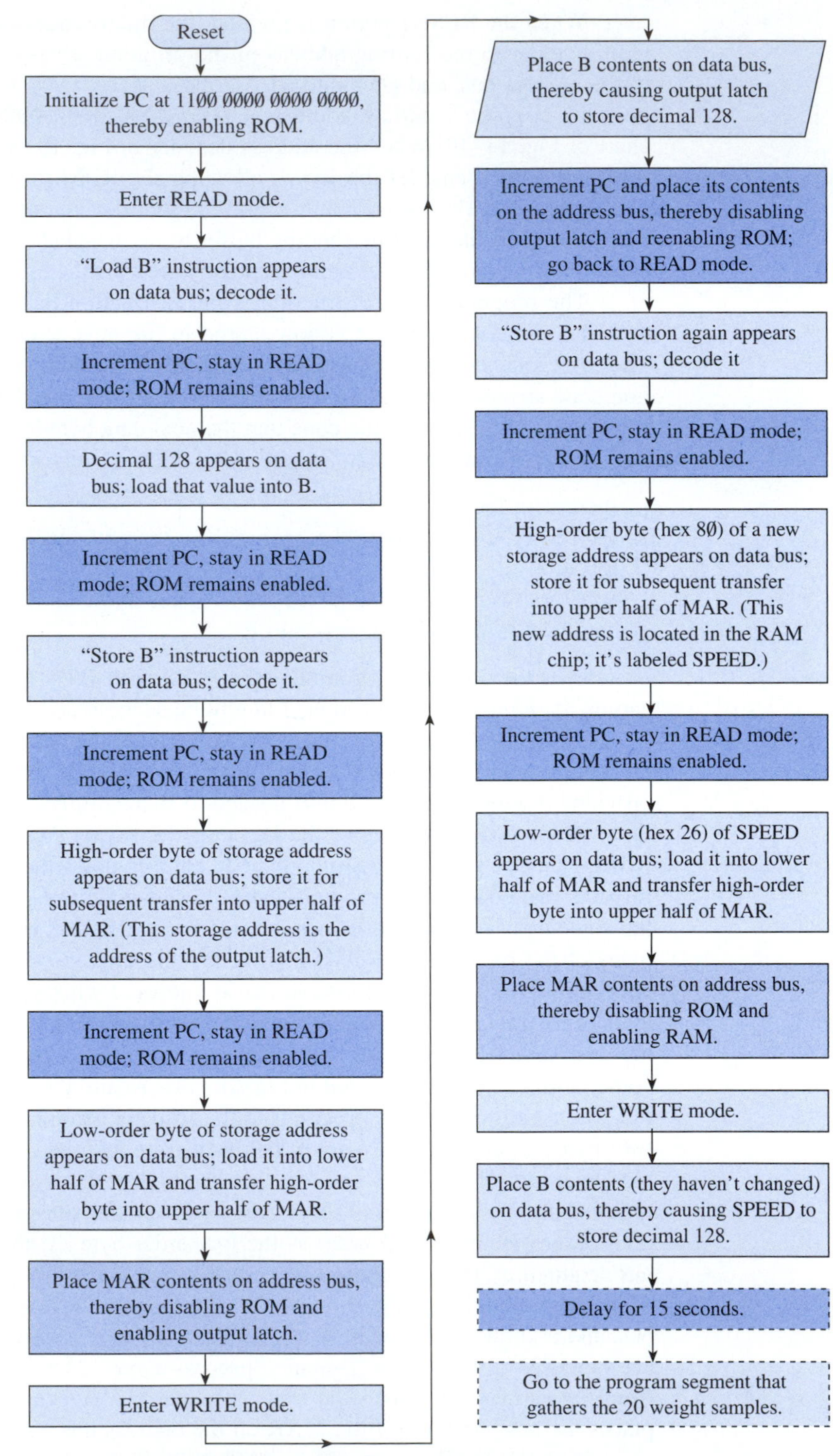

Reset
Initialize PC at 1100 0000 0000 0000, thereby enabling ROM.
Enter READ mode.
"Load B" instruction appears on data bus; decode it.
Increment PC, stay in READ mode; ROM remains enabled.
Decimal 128 appears on data bus; load that value into B.
Increment PC, stay in READ mode; ROM remains enabled.
"Store B" instruction appears on data bus; decode it.
Increment PC, stay in READ mode; ROM remains enabled.
High-order byte of storage address appears on data bus; store it for subsequent transfer into upper half of MAR. (This storage address is the address of the output latch.)
Increment PC, stay in READ mode; ROM remains enabled.
Low-order byte of storage address appears on data bus; load it into lower half of MAR and transfer high-order byte into upper half of MAR.
Place MAR contents on address bus, thereby disabling ROM and enabling output latch.
Enter WRITE mode.
Place B contents on data bus, thereby causing output latch to store decimal 128.
Increment PC and place its contents on the address bus, thereby disabling output latch and reenabling ROM; go back to READ mode.
"Store B" instruction again appears on data bus; decode it
Increment PC, stay in READ mode; ROM remains enabled.
High-order byte (hex 80) of a new storage address appears on data bus; store it for subsequent transfer into upper half of MAR. (This new address is located in the RAM chip; it's labeled SPEED.)
Increment PC, stay in READ mode; ROM remains enabled.
Low-order byte (hex 26) of SPEED appears on data bus; load it into lower half of MAR and transfer high-order byte into upper half of MAR.
Place MAR contents on address bus, thereby disabling ROM and enabling RAM.
Enter WRITE mode.
Place B contents (they haven't changed) on data bus, thereby causing SPEED to store decimal 128.
Delay for 15 seconds.
Go to the program segment that gathers the 20 weight samples.

inactive-LO state and puts the WRITE line in the active-HI state, thereby disabling its (the microprocessor's) own receiving tri-state buffer and enabling its own sending buffer. The microprocessor internally connects the B register to its sending buffer, and, another first for this program, the microprocessor is now driving the data bus.

The data bus has 1∅∅∅ ∅∅∅∅ on its lines and the output latch is enabled by the address decoder (any address would have worked in the fourth and fifth program bytes, just as long as A15 and A14 were both ∅), so the output latch takes in the decimal value 128. This condition lasts for only one *machine cycle* (1 μs, let's say), but that's all the time that is needed by the output latch. At the end of this machine cycle the output latch finds itself disabled once again when the address bus changes, but by then it has the number, and will hold it forever, or until the program comes along and gives it a new number.

The preceding is a blow-by-blow account of the execution of the first two program instructions. This control program has over 100 instructions, occupying over 300 bytes of ROM. It doesn't take a great deal of foresight to realize that we can't afford such a detailed description of the whole program. For the rest of the way, we will content ourselves with a more cursory description.

Another thing: Have you noticed how cumbersome it is to write binary numbers? The 8-bit numbers aren't too bad, but those 16-bit numbers are a drag. To make this task easier, our common practice is to use the *hexadecimal* number system to express the status of the data bus or address bus. The hexadecimal number system (hex, for short) lends itself to such expressions because one hex digit always corresponds to a unique combination of four binary digits, and the other way around. Review your digital electronics textbook for an explanation of hex numbers. We will frequently use hex notation from now on.

Now that the coal-feeding conveyor is running, the program stores the armature voltage (speed) value in a RAM location. This is necessary so that the speed value can be retrieved later in the program. Recall from the flowchart of Fig. 18–3 that the microprocessor must have the old speed in order to calculate the new speed after the slurry-weight samples have been averaged. The speed value being stored now will serve as the old speed when the first round of calculations begins.

Let us choose the RAM address 1∅∅∅ ∅∅∅∅ ∅∅1∅ ∅11∅ (hex 8∅26) as the location which keeps track of the speed. Since keeping numeric addresses straight is such a mental strain even with hex notation, we will assign that address a label by which we can refer to it. This is standard practice. Let the label be SPEED.

Thus the third program instruction stores the contents of register B in SPEED. In the ROM the instruction code byte is followed by two address bytes, the high-order and low-order bytes of SPEED. During the execution of this instruction, the RAM chip will be enabled by the address decoder, since our choice of address has A15 = 1 and A14 = ∅. The W control line will be pulled HI by the microprocessor, so the RAM's receiving buffer will be enabled. The RAM's sending buffer will be disabled by the R line being ∅—refer to Fig. 18–7.

The foregoing actions are accomplished by the first eight-bytes (C∅∅∅ through C∅∅7) of the eleven-byte program segment shown in Table 18–2. This program segment is represented in flowchart format in Fig. 18–12.

The program now goes into a delaying process. It allows 15 seconds to pass so that the coal has a chance to reach the weighing section in Fig. 18–1. Such time delays are common in microcomputer control schemes. They can be accomplished by making an unneeded register count from empty to full many times over. For example, since we are not

TABLE 18–2
Machine code for storing decimal 128 in the output latch and in RAM location SPEED, then jumping to the 15-second delay subroutine—corresponding to the chart in Fig. 18–12*

ROM ADDRESS	HEX CONTENTS OF ADDRESS	DESCRIPTION OF CONTENTS OF ADDRESS	COMMENTS (OUTPUT)
C000	C6	Instruction to load B with the value contained in the next ROM byte	
C001	80	The initial speed	Hex 80 = decimal 128 steps/s
C002	F7	Instruction to store the contents of B at the address specified by the next two bytes	Write the initial value to the output latch
C003	30	High byte of address of output latch (hex 30)	Address of output port (the output latch chip)
C004	00	Low byte (hex 00)	
C005	F7	Instruction to store the contents of B at the address specified by the next two bytes	Write the present value of stepping rate to RAM location SPEED
C006	80	High byte of SPEED (hex 80)	The SPEED address
C007	26	Low byte of SPEED (hex 26)	
C008	7E	Instruction to jump the PC to the address specified by the next two bytes	Jump away to the 15-second delay subroutine
C009	C7	High byte of the ROM address that contains the first instruction of the 15-second delay subroutine (hex C7)	The 15-second delay subroutine begins at ROM address C700
C00A	00	Low byte (hex 00)	

*This code sequence is executed immediately when the slurry system is started up and the microcomputer is brought on line, but the program never returns to it as long as the system is kept in operation.

using the X register for anything else right now, we could make it count up by one on each pass through a *delay loop* that we would insert in the program. (A delay loop is a series of instructions that increments the register, inspects to see if it is full, and if not full increments it again.) X is a 16-bit register and 2^{16} = 65 536, so it would require 65 536 passes through the delay loop to fill X just once. You can see how it is possible to burn up a lot of time just making the program spin its wheels in this fashion.

18-7-2 Sampling the Weight

After the 15-second delay, the sampling process begins. We must set aside 20 RAM locations to hold the 20 instantaneous slurry weights. Let's use RAM addresses hex 8000 through hex 8Ø13. Verify for yourself that these hex numbers span 20 sequential locations. All the addresses start with A15 A14 = 1 Ø, so they will all cause the address decoder to enable the RAM.

Now here comes an important idea, the idea of *indexed-mode addressing*. We load the X register with the first sequential RAM address (that is, the program does so). Then we load register A with the number at the input buffer. For accessing the input buffer, we can use any address that has A15 = Ø, A14 = 1 (hex 4ØØØ for instance)—see Table 18–1. This incoming number from the load cell/ADC/input buffer combination relates to the instantaneous slurry weight on the basis of 10 lb per bit. For example, if enabling the input buffer causes 1ØØ1 Ø111 (hex 97) to appear on the data bus, the instantaneous slurry weight is 1510 lb. Verify this for yourself.

All right. We have got the weight into register A. We can't leave it there, because we will need register A again in half a second for the next weight sample. We must move the weight data out of A into the first RAM address. Saints be praised, there exists an instruction (coded 1Ø1Ø Ø111, hex A7) that means "Store the contents of the A register in the memory location whose address is presently in the X register." Slip one of these into the program, and the weight data is securely stored into the RAM. After this, we increment the X register (instruction code ØØØØ 1ØØØ, hex Ø8), and we're ready for the second sample. We delay for half a second, then instruct the program counter to *jump back* to the ROM address that held the instruction that loads A from the input buffer. So we reexecute the load A instruction, and by the normal advancing of the PC we reexecute the indexed-mode store instruction. Of course, the second time around, the X register contains the next sequential RAM address, because we incremented X during the first pass. Therefore the second weight sample gets sent to the second RAM address.

We continue recycling through the program in this manner until all 20 RAM locations are full of weight data. Do you see the great usefulness of the indexed-mode store instruction? It saves us the trouble of writing the same group of instructions into the program 20 times, each time with a different RAM address. The way we have done it, we write the group of instructions just once, and then reuse that section of the program 20 times.

To help yourself appreciate the advantage of this approach, contemplate a control program that takes not just 20 samples, but 2000 samples (which is quite realistic).

You're probably wondering how the program knows when it's time to quit sampling, that is, when 20 samples have been taken. It uses the B register to count passes through the sampling loop (the repetitive group of instructions). Following each store instruction, we increment the B register. After that we subtract 20 from the B register. Following the subtraction instruction, we include an instruction that means "If the Z bit is HI, do not continue in the normal manner of incrementing the program counter, but *jump the PC ahead* by as many bytes as are specified in the next ROM location." On the first 19 passes, the answer to the subtraction instruction comes out negative and *nonzero,* so the Z bit of the CCR remains LO. Therefore we do not jump the PC ahead, but rather continue recycling through the program as usual. However, after the twentieth incrementation of B, following the twentieth storage into RAM, the subtraction operation yields a result of zero. The Z bit of CCR goes HI, and thus the condition is satisfied to carry out the jump-ahead instruction.

Table 18–3 and Fig. 18–13 show the detailed structure of the program section that we have been discussing. The table begins with the first program byte following the 15-second delay after the screw conveyor starts turning. It ends with the program byte that begins the calculation of the average slurry weight.

TABLE 18–3
Machine code and explanatory remarks for the program segment that gathers the 20 weight samples—corresponding to the chart in Fig. 18–13*

ROM ADDRESS	HEX CONTENTS OF ADDRESS	DESCRIPTION OF CONTENTS OF ADDRESS	COMMENTS
CØ10	5F	Instruction to clear B to zero	Since B is keeping track of how many passes we've made, it must start counting from zero.
CØ11	CE	Instructions to load X with the address specified in next 2 bytes	Load X with the first RAM address.
CØ12	8Ø	High-order byte of address (hex 8Ø)	The first RAM address
CØ13	ØØ	Low-order byte of address (hex ØØ)	
CØ14	B6	Instruction to load A with the data that resides at the address specified by the next two bytes	Load A with the weight value from the input buffer.
CØ15	4Ø	High byte of address of input buffer (hex 4Ø)	Address of input port (the input buffer chip)
CØ16	ØØ	Low byte (hex ØØ)	
CØ17	A7	Instruction to store the contents of A at the memory location whose address is presently in the X register	Indexed-mode store instruction
CØ18	ØØ	Ignore this zero byte	
CØ19	5C	Instruction to increment B	B counts the storage operation
CØ1A	C1	Instruction to subtract the number in the next ROM byte from contents of B	
CØ1B	14	The number to be subtracted from B	Hex 14 equals decimal 20
CØ1C	27	Instruction to jump the PC ahead by as many bytes as the number in the next ROM byte, if Z = 1	Jump ahead if the subtraction result is Ø, indicating that sampling is complete.
CØ1D	16	The number of bytes to jump ahead (hex 16, or decimal 22)	22 is the number of bytes to jump ahead from the normal next location, which is CØ1E.
CØ1E	Ø8	Instruction to increment X	Change X so that it points to the next sequential RAM location.
CØ1F ↓		The delay loop that delays ½ second before the next sample; this loop uses 17 bytes of ROM	
CØ3Ø			
CØ31	7E	Instruction to jump the PC back to the address specified in the next two bytes	The jump-back instruction that enables us to recycle through a section of the program

Jump the PC back to CØ14

A 22-byte jump ahead

TABLE 18–3
(continued)

ROM ADDRESS	HEX CONTENTS OF ADDRESS	DESCRIPTION OF CONTENTS OF ADDRESS	COMMENTS
C032	C0	High byte of address we are jumping back to (hex C0)	We are jumping back to the load A instruction, which is at ROM address hex C014.
C033	14	Low byte of address we are jumping back to (hex 14)	
C034		This is the ROM address that we will jump the PC ahead to, if Z = 1 for the conditional instruction at address hex C01C. The ROM address from which we are jumping, however, is hex C01E	

*Note especially that some ROM address locations contain instruction codes and some contain numerical data. The numerical data can represent several things, including but not limited to (1) half of an address, as hex 80 in ROM location C012, (2) numerical information that the microprocessor needs in order to execute the previous instruction, as hex 14 in ROM location C01B, or hex 16 in ROM location C01D, and (3) the value of some system-relevant variable, as hex 80 (stepping rate value decimal 128) in ROM location C001, back in Table 18–2.

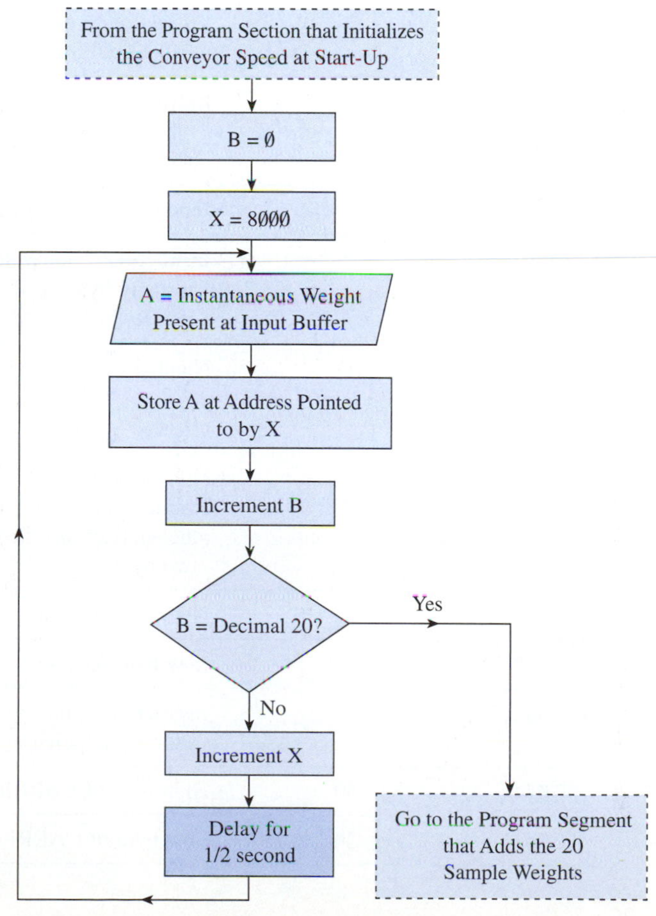

FIGURE 18–13
Flowchart of the program segment that samples the weight 20 times. Table 18–3 gives the machine code that corresponds to this chart.

TABLE 18–4
Machine code and explanatory remarks for the program segment that adds the 20 sample weights—corresponding to the chart in Fig. 18–14

ROM ADDRESS	HEX CONTENTS OF ADDRESS	DESCRIPTION OF CONTENTS OF ADDRESS	COMMENTS
C034	7F	Instruction to clear the RAM address specified in next two bytes	
C035	80	High-order byte of address (hex 80)	8020 is the RAM address that will accumulate addition carrys—label CARREG.
C036	20	Low byte (hex 20)	
C037	FE	Instruction to load X with address specified in next two bytes	Get X pointing to the first RAM address.
C038	80	High byte (hex 80)	The first RAM address, presently holding the first weight sample
C039	00	Low byte (hex 200)	
C03A	4F	Instruction to clear register A	A accumulates the sum of the weight values.
C03B	5F	Instruction to clear register B	B keeps track of how many sample weights have been retrieved and added.
C03C	AB	Instruction to add to A the weight value contained in the RAM address pointed to by X	Indexed-mode add instruction
C03D	00	Ignore this zero byte	
C03E	25	Instruction to jump the PC ahead by as many bytes as the number in the next ROM location, if C = 1	If C bit is 1, it means that the latest addition caused register A to overflow.
C03F	03	Number of bytes to jump ahead	If the conditional jump takes place, we'll jump from C040 to C043, which will land us on the carry-accumulate instruction
C040	7E	Instruction to jump the PC ahead to the address specified in the next two bytes	Unconditional jump that bypasses the carry-accumulate instruction at address C043
C041	C0	High byte of address we are jumping to	Jumping to ROM address C046
C042	46	Low byte (hex 46)	
C043	7C	Instruction to increment the RAM address specified in the next two bytes	The carry-accumulate instruction
C044	80	High byte of CARREG	RAM location 8020 is CARREG
C045	20	Low byte of CARREG	

TABLE 18–4
(continued)

ROM ADDRESS	HEX CONTENTS OF ADDRESS	DESCRIPTION OF CONTENTS OF ADDRESS	COMMENTS
C046	5C	Instruction to increment B	Counting the number of sample weights retrieved from RAM
C047	C1	Instruction to pseudosubtract the number in the next ROM location from B	
C048	14	The number to be subtracted from B	Hex 14 = decimal 20
C049	27	Instruction to jump the PC ahead by as many bytes as the number in the next ROM byte, if Z = 1	If the subtraction result is 0, it means that all twenty weights have been retrieved and added.
C04A	04	Number of bytes to jump ahead	From C04B to C04F
C04B	08	Instruction to increment X	X now points to the RAM location containing the next sample weight.
C04C	7E	Instruction to jump the PC back to the address specified in the next two bytes	Go back for the next sample weight.
C04D	C0	High byte	Jumping back to ROM address C03C
C04E	3C	Low byte	
C04F	B7	Instruction to store register A at RAM address specified in next two bytes	
C050	80	High byte (hex 80)	The lowest eight bits of the sum are stored at address 8021, where they can be concatenated with the contents of 8020(CARREG)
C051	21	Low byte (hex 21)	
C052		First instruction of the program's divide-by-20 routine	

18-7-3 Calculating the Average Weight

To calculate the average weight, we make the program perform a similar ritual. Refer to the Fig. 18–14 flowchart and the corresponding machine-language program-code in Table 18–4.

First we reset the X register to the address of the first RAM location. Then we cycle 20 times through a loop that brings a sampled weight in from RAM and adds it to the total weight that has been accumulated so far in the A register (A is cleared to zero prior to the first pass). Of course, it is easy to see that if we continue accumulating 8-bit num-

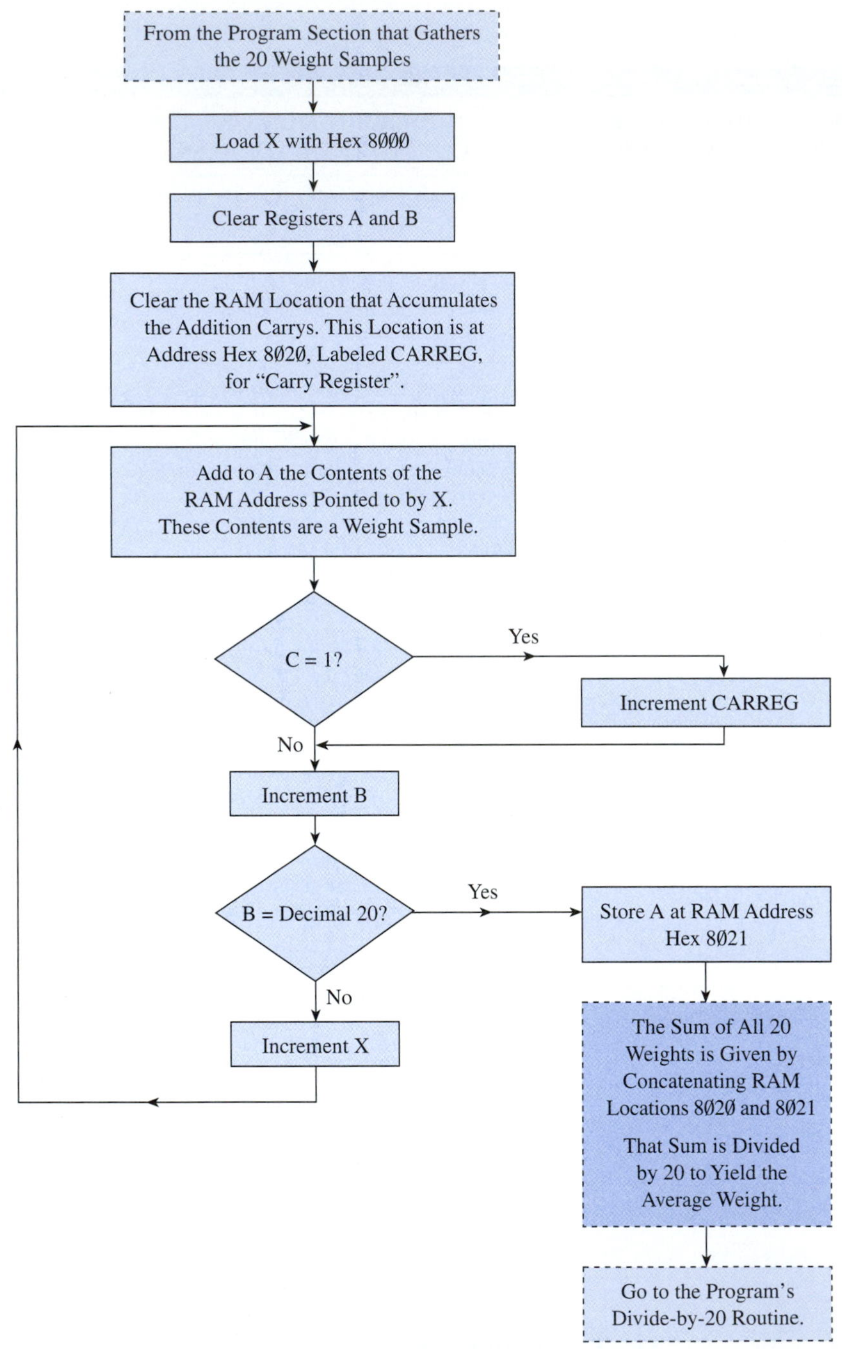

FIGURE 18–14 Flowchart of the program segment that adds together the 20 sample weights. Table 18–4 gives the machine code that corresponds to this chart.

bers in an 8-bit register, sooner or later we are going to overflow the register. To handle this problem, we reserve another location in RAM to accumulate the overflow bits. Each time an addition of a weight value to the A register causes A to overflow,* the C bit of

*Be careful with the word *overflow*. In the context of standard arithmetic with unsigned numbers, it has the straightforward meaning which we are ascribing to it here. But in 2's-complement arithmetic (binary signed-number arithmetic) it has a more abstruse meaning. Check a microprocessor book for a full discussion of 2's-complement overflow.

the CCR will be set. We insert an instruction which means "If the C bit is HI, do not continue in the normal manner of incrementing the program counter, but jump the PC ahead by as many bytes as are specified in the next ROM location." The jump-ahead destination contains an instruction which increments the RAM location that is accumulating the carrys.

Therefore, if the C bit is LO following an addition to A, the PC is advanced in the normal manner, thereby avoiding the instruction which increments the carry-accumulating location in RAM. However, if the C bit is HI following a weight addition to A, the program encounters the carry-accumulate instruction. At the completion of the twentieth weight-addition, the carry-accumulating location is concatenated* with the A register to express the weight total.

Once a weight total has been calculated, the program enters a divide-by-20 routine. We'll forgo a discussion of this routine. At the conclusion of the divide-by-20 routine, the quotient represents the average weight of slurry in the transfer pipe's weighing section throughout the 10-second sampling period. This result, which will never be more than eight bits in length, is written in a single RAM location labeled AVGWT.

18-7-4 Calculating the New Speed

The program then enters its *categorizing section,* which determines which of the four possible categories the average weight falls into; refer to the flowchart of Fig. 18–3. The program's categorizing section operates by a process of elimination. First an instruction determines whether AVGWT is greater than or equal to 2000 lb. This is accomplished by bringing AVGWT back down the data bus into register A and then subtracting the binary number that represents 2000 lb (1100 1000); this numeric value is stored in the ROM byte immediately following the byte containing the code for the subtraction instruction. If the C bit (behaving now like a "borrow" bit) is clear, the average weight must be greater than or equal to 2000 lb. We insert an instruction which inspects the C bit of the CCR and, if $C = 0$, causes the program to branch around the other categorization instructions and go directly to the HEAVY calculation routine. Conversely, if $C = 1$, the average weight must be less than 2000 lb, so the program proceeds to the next categorization instruction, as shown in Fig. 18–15 and Table 18–5.

The next instruction subtracts 1500 lb (1001 0110) from the average weight. If C is still clear, then $1500 \leq \text{AVGWT} < 2000$, and the program branches to the OKAY calculation, but if $C = 1$, the program moves to the next categorization (subtraction) instruction, and so on.

If the current average weight falls into any one of the top three categories (HEAVY, OKAY, or LIGHT), calculation of the new speed requires that the old speed be brought down the data bus from the SPEED location in RAM; the calculation formulas shown in Fig. 18–3 make this plain. The program enters whichever calculation routine is called for (OKAY and MAXSPD don't actually require much in the way of calculation) and the NEWSP value winds up in register A at the end of the routine. This new speed value is written from the microprocessor to two destinations within the microcomputer system of Fig. 18–4. One instruction causes the speed value to be written to the address of the output latch (hex 3000), and a second instruction causes it to be written to the RAM address labeled SPEED. Writing to an output latch or a RAM location causes that location's previous number to be overwritten and lost forever.

**Concatenated* means joined together to form a continuous string of bits.

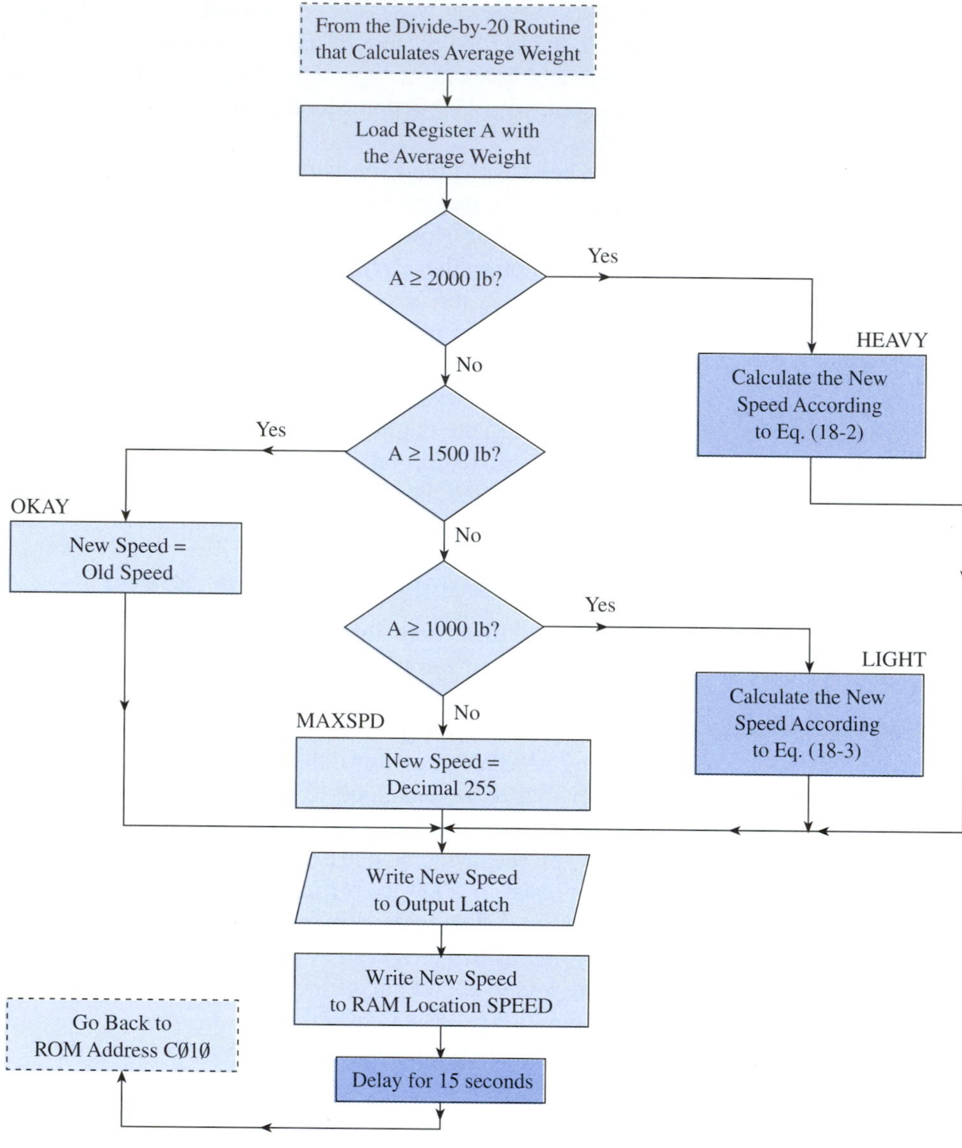

FIGURE 18–15 Simplified flowchart of the program segment that categorizes the average weight and calculates the new speed (armature voltage). Table 18–5 gives the machine code through the second categorization test (AVGWT ≥ 1500 lb).

The output latch then continues to hold the NEWSP number until it is changed again, a little more than 25 seconds from now. Therefore the conveyor motor maintains this new speed until that time. RAM location SPEED holds the number until it is called back into the microprocessor, also a little more than 25 seconds from now, to serve as OLDSP in one of the speed-calculation routines.

Following storage of the new speed value in these two locations, the program enters a 15-second delay loop to allow the effects of the new conveyor speed to show up in the pipe's weighing section. After exiting the delay loop, the last instruction of the program tells the program counter to jump back to ROM address hex CØ1Ø—the first instruction of the weight-sampling section of the program, at the top of Table 18–3.

TABLE 18–5
Machine code and explanatory remarks for the beginning portion of the program's categorizing section

ROM ADDRESS	HEX CONTENTS OF ADDRESS	DESCRIPTION OF CONTENTS OF ADDRESS	COMMENTS
CØ9Ø	B6	Instruction to load A with contents of address specified by next two bytes	
CØ91	8Ø	High byte (hex 8Ø)	Following the divide-by-20 routine, RAM location 8Ø24 contains AVGWT.
CØ92	24	Low byte (hex 24)	
CØ93	81	Instruction to pseudosubtract the number in the next ROM byte from A	First categorization test
CØ94	C8	The number to be subtracted	Hex C8 = decimal 200; 2000 lbs
CØ95	24	Instruction to jump the PC ahead by as many bytes as the number in the next ROM byte, if C = Ø	If C = Ø, it means that no borrow was required, so AVGWT ≥ 2000 lb
CØ96	ØD	The number of bytes to jump ahead (hex ØD)	The HEAVY routine starts at ROM address CØA4, and it uses Eq. (18-2) to calculate NEWSP. (CØ97 + ØD = CØA4).
CØ97	81	Instruction to pseudosubtract the number in the next ROM byte from A	Second categorization test
CØ98	96	The number to be subtracted	Hex 96 = decimal 150; 1500 lb.
CØ99	24	Instruction to jump the PC ahead by as many bytes as specified in the next ROM byte, if C = Ø	If C = Ø, it means that no borrow was required, so AVGWT ≥ 1500 lb
CØ9A	1F	The number of bytes to jump ahead	The OKAY routine starts at ROM address CØBA. (CØ9B + 1F = CØBA)
CØ9B		The next instruction in the program's categorizing section	

To CØA4

To CØBA

TROUBLESHOOTING ON THE JOB

USING A LOGIC ANALYZER FOR DEBUGGING

The microcomputer-controlled coal-slurry system is malfunctioning in that the coal slurry is too dense—pulverized coal is being fed too rapidly into the water stream. This is overloading the pump motors, causing them to blow their overloads.

Your assignment is to determine whether this is being caused by a software problem or a hardware problem. A software problem is an improper system outcome occurring from proper execution of the program by the microcomputer; in other words there is a flaw in the program logic that was not foreseen by the programmer and has not been noticed until now.

A hardware problem, on the other hand, is either

1. Improper execution of the program due to an electronic fault in the microcomputer's architecture (Fig. 18–4).
2. Malfunction of one of the peripheral devices that interfaces with the microcomputer. If such a malfunction occurs at the input port (the load-cell, ADC, or Input Buffer wiring, in our case), it causes the microcomputer to be presented with wrong input data from the controlled system. If a malfunction occurs at the output port (the Output Latch, dc drive system, dc motor, or their inter-

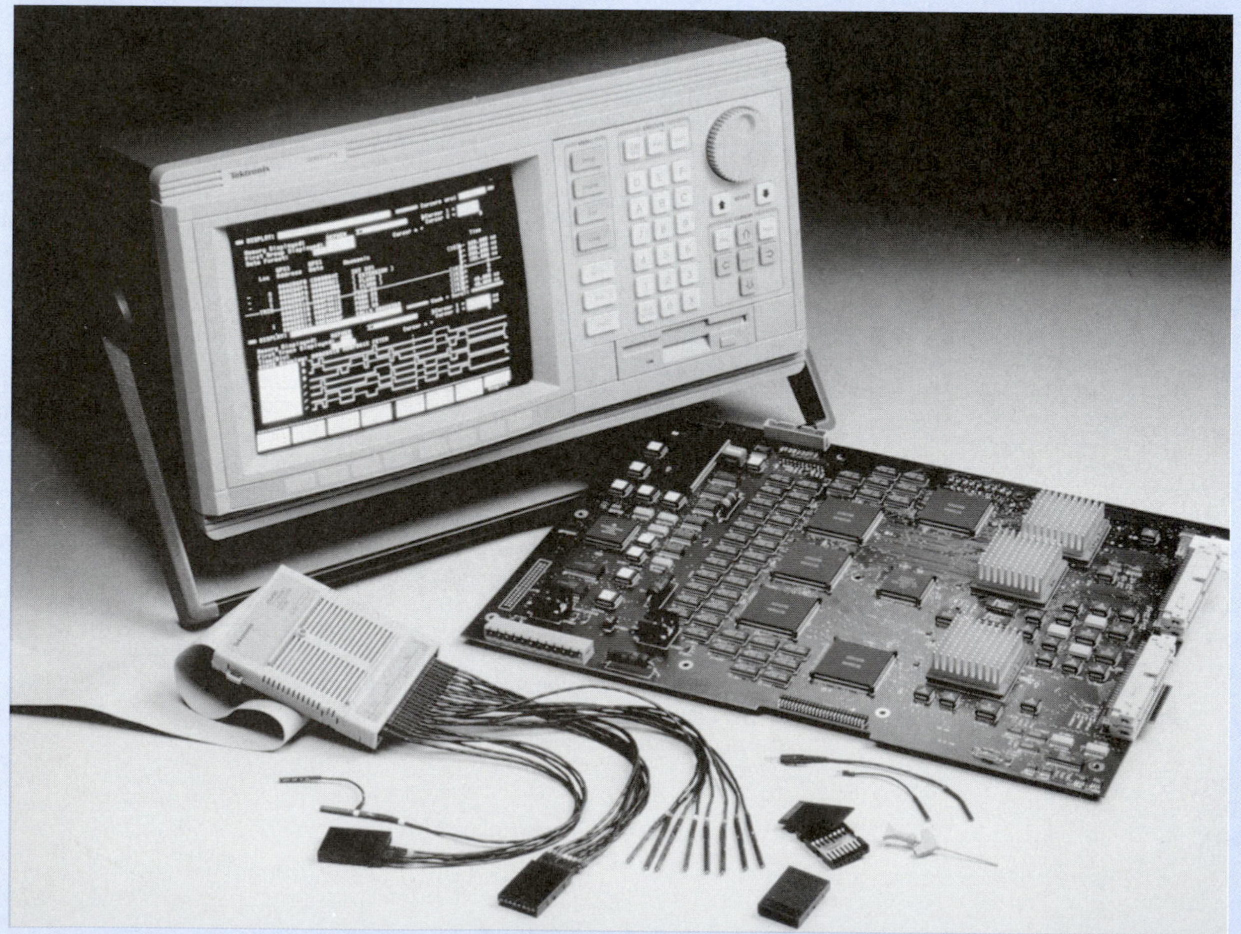

A logic analyzer is used to carefully examine the program execution of a microprocessor.
Courtesy of Tektronix, Inc.

connecting wiring, in our case), it causes the microcomputer's correctly generated output data to incorrectly affect the final control device (the screw conveyor, in our case).

To uncover software problems, engineers and technicians use a *logic analyzer*. It is an instrument that constantly monitors the entire address bus and the entire data bus through its multiconnector cable that physically grips the microprocessor chip from the top.

The photo to the left shows a logic analyzer with its cable and gripping probes. They attach to the spike-like adapters sitting atop the three microprocessor chips on the right side of the printed circuit board.

The logic analyzer stores in its digital memory the binary content of the address bus and the binary content of the data bus for every change in the status of the address bus (or for every machine cycle, if a more exhaustive examination of the software execution is desired). By your trigger selection of the logic analyzer, you cause it to cease its continuous writing of the buses' contents into its memory. The cessation of writing freezes the most recently written data in that memory. You can then search through the analyzer's memory slowly and carefully, to learn the exact point in the program at which the software deviated from the plan of the human programmer. Your searching is physically accomplished by looking at the CRT screen display on the front panel of the logic analyzer. Such a display is shown in the photo below.

Effective use of a logic analyzer requires judgment by you in your choice of its trigger conditions. These are the bus conditions that the instrument watches for in order to know when to cease writing. Usually, you

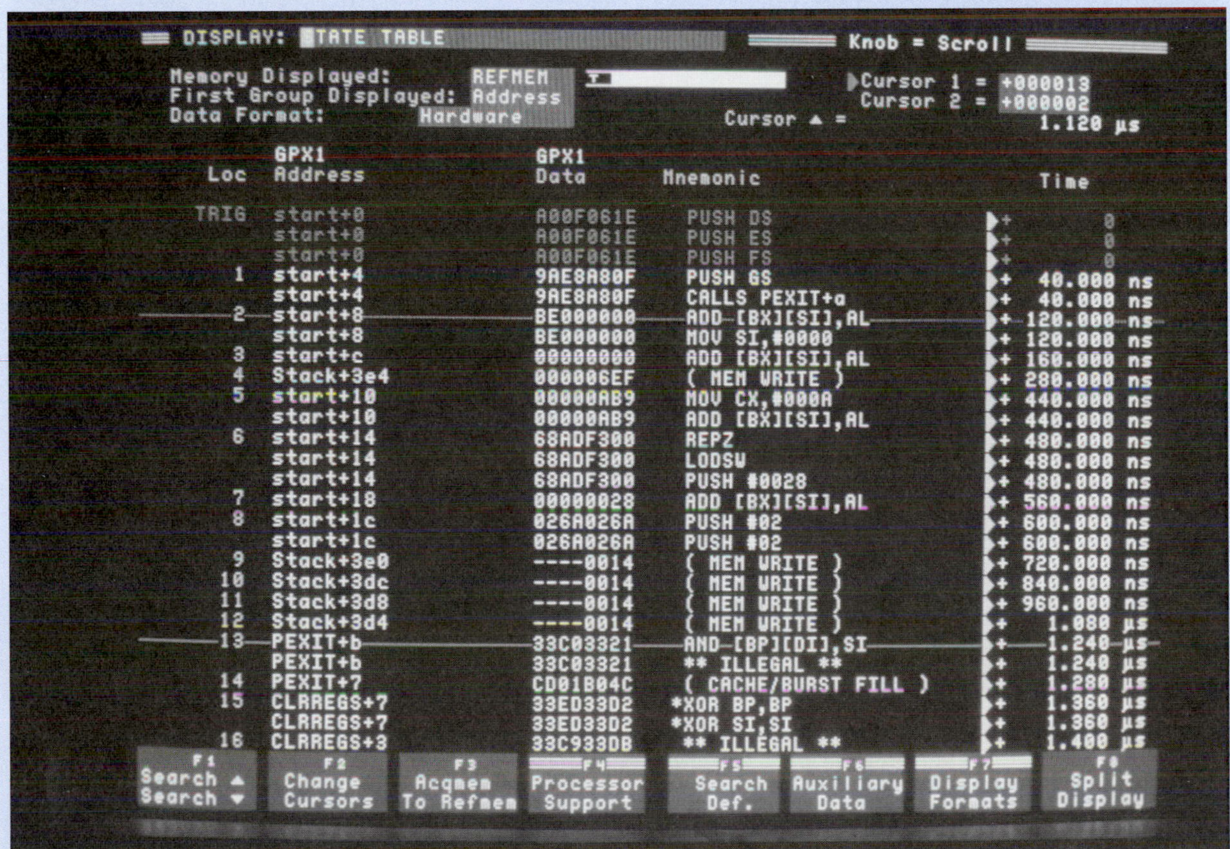

Logic analyzer program display for a microcomputer with a 32-bit data bus. In this particular display setup, assembly language code is shown in the column titled Mnemonic.

Courtesy of Tetronix, Inc.

don't want the writing of the program's execution steps to cease immediately when the trigger conditions are satisfied. Instead, a selectable number of instructions are written into memory *after* the trigger conditions are satisfied. That way, when you come to examine the analyzer's memory contents on its screen, you have the opportunity to observe the program steps preceding the trigger condition (called *pretrigger data*), as well as the program execution steps following the trigger condition (called *posttrigger data*). The ability to examine the program's actual performance leading up to and immediately subsequent to your selected trigger condition is the key to discovering software bugs.

Your supervisor gives you two suggestions on what to look for in the logic analysis, things that could account for the too rapid conveyor feed:

1. Perhaps there is a software problem that can sometimes prevent the program from recognizing that the average weight AVGWT is greater than 2000 lb.
2. Perhaps there is a hardware problem at the input port, causing the load-cell, ADC, and input wiring to sometimes supply bad data. Specifically, perhaps some or all of the input data bytes are lower than the true actual weight that exists in the weighing section of pipe.

YOUR ASSIGNMENT

1. From your understanding of the program's operation, choose a good logic-analyzer trigger condition for investigating suggestion 1. State whether you want to examine the program execution in assembly language format or in actual machine code—on an individual machine cycle-by-machine cycle basis. Explain why you believe that your choice of trigger conditions and display format will be effective at discovering whether this problem (suggestion 1) really exists.
2. Choose a good logic-analyzer trigger condition and display format for investigating suggestion 2. Explain why you believe your choice is a good one. Also explain precisely what you will be watching for as you examine the data bus status at various moments in the posttrigger memory display.

SUMMARY

- Sometimes a dedicated microcomputer is preferable to a PLC for calculation-intensive industrial control applications.
- A microcomputer control program executes repetitively like a PLC user-program. It may contain a deliberate delay during which it waits for its most recent corrective action to take effect before it resamples the input data.
- A flowchart graphically represents the comparisons, decisions, and branches that a microcomputer program must make, as well as its input from and output to the industrial process.
- Input data from the industrial process is brought in through the microcomputer's input buffer. Output to the industrial process is sent through the microcomputer's output latch.
- The program instructions are held in read-only memory, ROM. When the microprocessor IC chip finishes executing one program instruction, it fetches the next instruction from the ROM IC chip.
- The program's temporary numerical results are stored in the read/write memory, or RAM chip. Data can be written to a certain address in the RAM via the data bus and later read back from that RAM address via the data bus. The data bus moves binary information in both directions.
- Several IC chips in the microcomputer can assert data on the data bus, namely the microprocessor, the ROM, the RAM, and the input buffer. All these devices must have tri-state outputs, enabling them to "let go" of the data bus when they are sent to their

high-impedance (floating) state. Only one data bus-driving device can be enabled at any moment in time during the program's execution. This prevents bus contention, which is a fight between two devices over what data should be placed on the data bus.

- The microcomputer's control lines, called Chip Enable, Read, and Write, determine which device is enabled to drive the data bus.
- The address bus is unlike the data bus in the following respects: (1) It is one-way, not two-way. It carries address information *from* the microprocessor chip *to* all the other chips in the microcomputer. (2) There is no problem of bus contention for the address bus because there is only one device that ever drives it—the microprocessor chip.
- The data on the data bus can represent one of four things: (1) actual numeric data—a number being used in calculation or comparison; (2) a binary code that represents an instruction. The microprocessor manufacturer defines all the codes for all the legal instructions in that microprocessor's instruction set. (3) Part (or all) of an address. In our example, with the data bus being 8 bits wide and the address containing 16 bits, the data bus can hold one half of an address. (4) Some binary code that has been invented and defined by the programmer for a particular purpose which is relevant to that program only.
- Execution of a single program instruction usually takes several machine cycles, where one machine cycle is the time required to make one fundamental change in the status of the microprocessor. The time duration of one machine cycle is the period of the microcomputer's internal clock oscillator. A modern microprocessor has a clock frequency greater than 10 MHz, so a machine cycle lasts less than 100 ns, or about 0.1 μs.
- Subroutines are program segments that are reused several times during the execution of the main program, or that are set apart from the main program for some other reason. When the main program needs to perform the operation that is handled by the subroutine, it jumps out of its normal address sequence to the starting address of the subroutine. When the microprocessor reaches the ending address of the subroutine (the subroutine has finished), it usually jumps back into the main program at the point where it left off.

QUESTIONS AND PROBLEMS

Section 18-1

1. What considerations might cause us to choose a dedicated microcomputer in preference to a programmable controller?

Section 18-3

2. True or False: In an on-line application, after a microcomputer has finished executing its operating program, it waits for a signal from the human user before reexecuting the program.

Section 18-5

3. A modern bus-organized computer uses a single group of wires to transfer data among several different circuits. In general terms, explain how this is accomplished.

Questions 4 through 11 refer to the microcomputer architecture diagram of Fig. 18–4.

4. What control signal(s) are necessary to enable the input buffer to drive the data bus?
5. What control signal(s) are necessary to enable the ROM to drive the data bus?
6. What control signal(s) are necessary to enable the RAM to drive the data bus?
7. What control signal(s) are necessary to enable the data bus to transfer its data to the output latch?
8. What control signal(s) are necessary to enable the data bus to transfer its data to the RAM?
9. What control signal(s) are necessary to enable the data bus to transfer its data to the microprocessor?
10. Explain why only two lines of the address bus are needed for controlling four peripheral chips.
11. Of the four peripheral chips (ROM, RAM, input buffer, and output latch), which ones can have their contents change during the program execution?
12. Describe the three possible states of a tri-state output device.
13. Explain the difference between a tri-state device that has active-HI enable and one that has active-LO enable.
14. True or False: A RAM's tri-state receiving buffer is enabled when the microprocessor is in write mode.
15. True or False: A microprocessor's tri-state receiving buffer is enabled when the microprocessor is in read mode.

Section 18-6

16. For the ROM's tri-state sending buffer to be enabled, which mode must the microprocessor be in, read or write?

Section 18-7

17. In our programming model of a 6800-series microprocessor in Fig. 18–11, there are six specific registers, A, B, X, CCR, PC, and MAR. Explain the function of each of these six registers.
18. Define the terms *load* and *store* as used in the microcomputer context.

Questions 19 to 21 refer to the program segment listed in Table 18–2.

19. ROM address C000 contains the instruction code C6. Is this an inherent instruction, or is it the type of instruction that requires additional information for its execution?
20. ROM address C001 contains hex 80. This is not an instruction code. What type of information is it?
21. ROM addresses C006 and C007 contain hex 80 and hex 26, respectively. These are not instruction codes and they are not the same type of information as hex 80 in ROM address C001. What type of information are they?
22. Based on your answers to questions 19 to 21, discuss the various meanings that can be represented by a ROM byte.

Questions 23 to 26 refer to the program segment that samples the slurry weight 20 times, shown in Table 18–3 and flowcharted in Fig. 18–13.

23. We chose to use RAM addresses 8000 through 8013 to store the 20 sample weights. If we had chosen to use addresses 81C0 through 81D3 instead, what

change(s) would be necessary in the operating program. Specify which ROM address(es) would need to be changed and what the change(s) would be.

24. We chose to gather 20 samples over 10 seconds. If we had chosen to gather only 10 samples over 5 seconds, what change(s) would be necessary in the program?
25. We chose to address the input buffer as hex 4000. Based on the address-decoding scheme being used in Fig. 18–4, would the program still control the system properly if the contents of ROM byte C016 were hex A4, rather than hex 00? Explain.
26. Repeat Question 25 for the contents of ROM byte C015 being hex 7F rather than hex 40.

Questions 27 to 29 refer to the program segment that adds the 20 weight samples shown in Table 18–4 and Fig. 18–14.

27. We chose to accumulate the total weight in RAM locations 8020 and 8021. If we had chosen instead to use RAM locations 9A40 and 9A41, what change(s) would be necessary in the program?

To answer Questions 28 and 29, suppose that after the tenth weight value has been added to register A, the contents of A are hex 2C. Suppose that the eleventh weight value is hex B3, stored in RAM address 800A, and the twelfth weight value is hex AE, stored in RAM address 800B.

28. a. When the eleventh weight value is added to A by the instruction at ROM location C03C, what will be the new state of the C bit, 1 or 0? Explain.
    b. Will the program obey the instruction at ROM location C03E? Explain.
    c. Will the program obey the instruction at ROM location C040? Explain.
    d. Will the program obey the instruction at ROM location C043? Explain.
    e. Will CARREG be incremented? Explain.
    f. Will the program obey the instruction at ROM location C046? Explain.
29. a. When the twelfth weight value is added to A by the instruction at ROM location C03C, what will be the new state of the C bit, 1 or 0? Explain.
    b. Will the program obey the instruction at ROM location C03E? Explain.
    c. Will the program obey the instruction at ROM location C040? Explain.
    d. Will the program obey the instruction at ROM location C043? Explain.
    e. Will CARREG be incremented? Explain.
    f. Will the program obey the instruction at ROM location C046? Explain.
30. The flowchart of Fig. 18–15 shows the new calculated speed value being written to RAM location SPEED (hex 8026) after it is written to the output latch. Why must it be written to a RAM location?
31. In Table 18–5, the instruction at ROM location C093 calls for a pseudo subtraction from register A rather than an outright subtraction. Why do we prefer a pseudo subtraction at this point in the program? What disadvantage would attend an outright subtraction?

CHAPTER

INDUSTRIAL ROBOTS

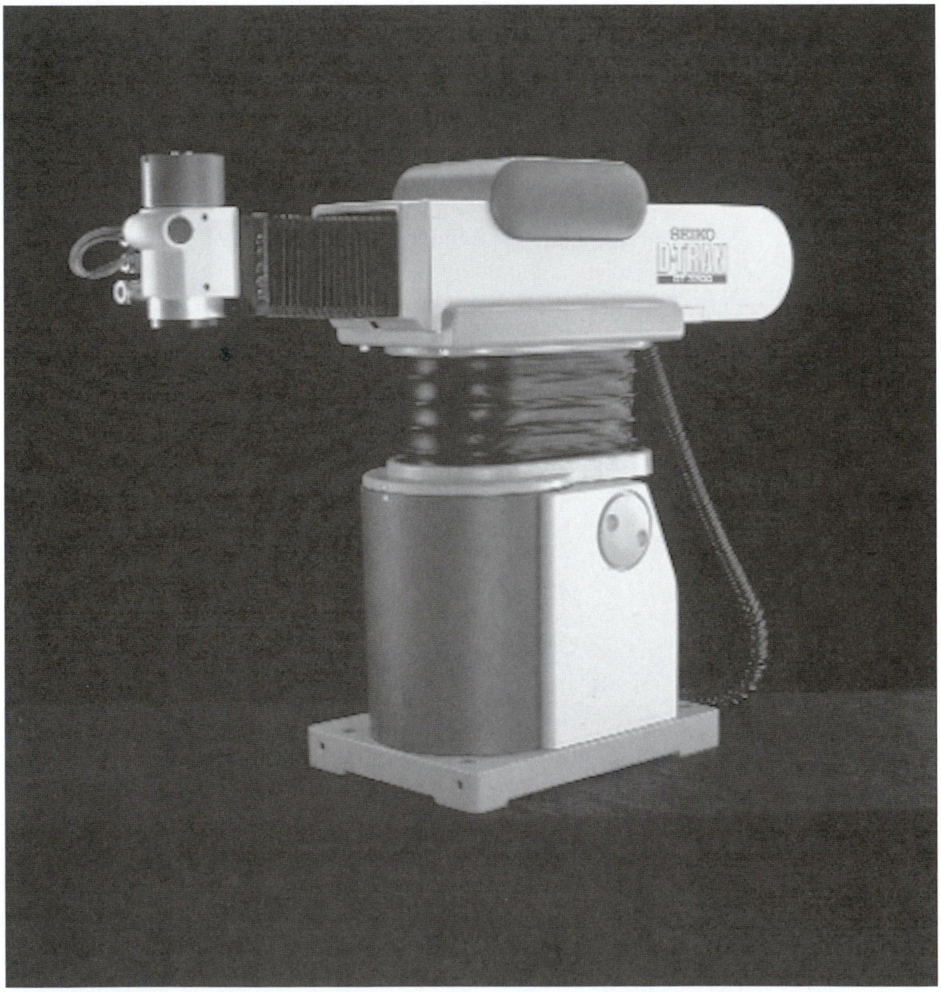

The arrival of the low-cost microcomputer has tremendously enhanced the adaptability of industrial manipulative machinery. When a machine's manipulative cycle is controlled by a microcomputer software program rather than by hard wiring, that manipulative cycle can be completely rearranged simply by changing the program. That's the idea that is propelling industrial robots into the limelight. In this chapter we will discuss the basic concepts of robot hardware and software.

OBJECTIVES

After completing this chapter, you will be able to:

1. Describe the three most common mechanical configurations for industrial robots.
2. Identify by name the various robot axis motions.
3. Describe the characteristic features of each of the three categories of robot software: positive-stop, point-to-point, and continuous-path.
4. Explain the use of a hardware interrupt to stop a robot motion.
5. Explain the operation of position encoders on point-to-point and continuous-path robots.
6. Discuss the operation of the position-comparison routine used in point-to-point and continuous-path programs.
7. Explain how the equal-displacement organization of the various axes' destination values in RAM makes possible the indexed addressing of those values.
8. Describe the process of creating an operating program by using a teach pendant to guide a robot through its manipulative sequence.
9. Discuss the operation of the sampling routine in the creation of a continuous-path operating program.
10. Name and describe the various end-of-arm gripping devices.
11. Explain the operation of inductive and capacitive proximity detectors for a robot.

19-1 ■ THE ROBOT CONCEPT

Some industrial operations require a tool-type device to be manipulated through space repetitively, with the added requirement that the repetitive manipulation actions must be easily changeable by the user. For example, in spray-painting tractor body panels, the tool-type device is a spraying nozzle. An automated painting operation involves moving a body panel into position, opening a valve to turn on the spray, then causing the spraying nozzle to move through the appropriate points in space to apply paint to all areas of the panel. After the motion of the spray nozzle is complete, the manipulating mechanism returns it to the home position, where it waits for another panel.

A typical manufacturing situation might call for the production line to be running left fender panels on Monday, right door panels on Tuesday, engine hoods on Wednesday, and so on. Naturally, the spray nozzle must follow different paths through space depending on which panel is being run. An appropriate path for painting a fender panel might be like the one sketched in Fig. 19–1(a); for a door panel, Fig. 19–1(b) might be appropriate.

PHOTO 19–1
Like all computer applications, programming a robot is made simpler if icon-based menus are used.
Courtesy of Seiko Instruments USA, Inc.

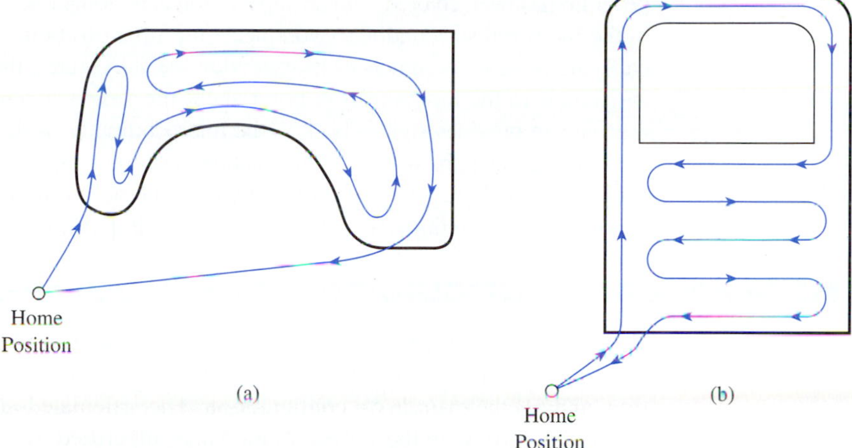

FIGURE 19–1
Paths taken by a painting robot as it sprays tractor body panels.

Whenever these two requirements must be met, namely, manipulation through space of a tool-like device and an easily changeable manipulation path, an industrial robot can probably do the job most effectively.

A precise and thorough definition for a robot is still being debated in robotics circles. However, everybody agrees that, at least, a robot must be able to perform mechanical manipulations, and its manipulations must be controlled from a reprogrammable source—that is, a computer.

An industrial robot installation can be visualized as suggested in Fig. 19–2. A human user arranges for a control program to be entered into a microcomputer. Once a correct

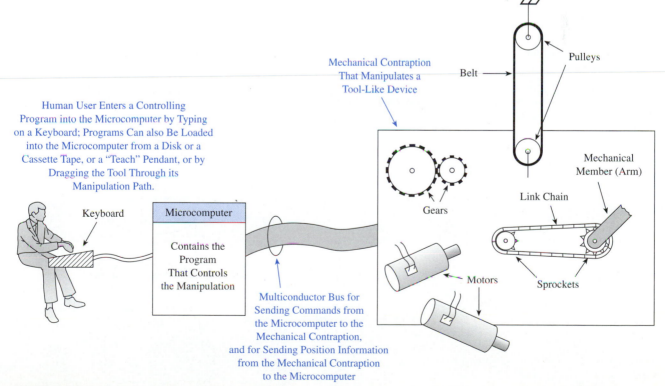

FIGURE 19–2
Matter-of-fact picture of an industrial robot system.

program has been entered, human supervision is no longer needed. When conditions are right in the industrial surroundings (workpiece in proper position, workpath clear of interference, etc.), the program commences its execution. As it executes, the program sends the appropriate signals to the mechanism to bring about the desired manipulations. In many robots the mechanism provides signals back to the microcomputer, which enables the program to keep track of the positions of the various members of the mechanism. After the manipulation has been completed, the mechanism returns to its home position and the program stops. When conditions in the industrial surroundings are again right, the entire process is repeated.

19-2 ■ MECHANICAL CONFIGURATIONS OF INDUSTRIAL ROBOTS

The most common mechanical configurations for industrial robots are (1) the *articulated-arm* configuration, also called the *jointed-arm* configuration, (2) the *spherical* configuration, and (3) the *cylindrical* configuration. The articulated-arm configuration is illustrated in Fig. 19–3(a), with the various motion axes identified by name. The spherical and cylindrical configurations are shown in Figs. 19–3(b) and (c).

All of the configurations in Fig. 19–3 permit rotational movement around six different axes (not counting the gripper, if it exists). We say that these robots have six *degrees of freedom*. The phrase *degrees of freedom* is derived from the fact that movement about one axis is hardware-independent of movement about any other axis. In other words the actuating device (either a motor or a cylinder) that produces shoulder movement, for example, is entirely separate and independent from the actuating device that produces

FIGURE 19–3
Common mechanical configurations for industrial robots: (a) articulated-arm; (b) spherical; (c) cylindrical. If the robot must grasp objects, it will have a mechanical gripper, or *hand*.

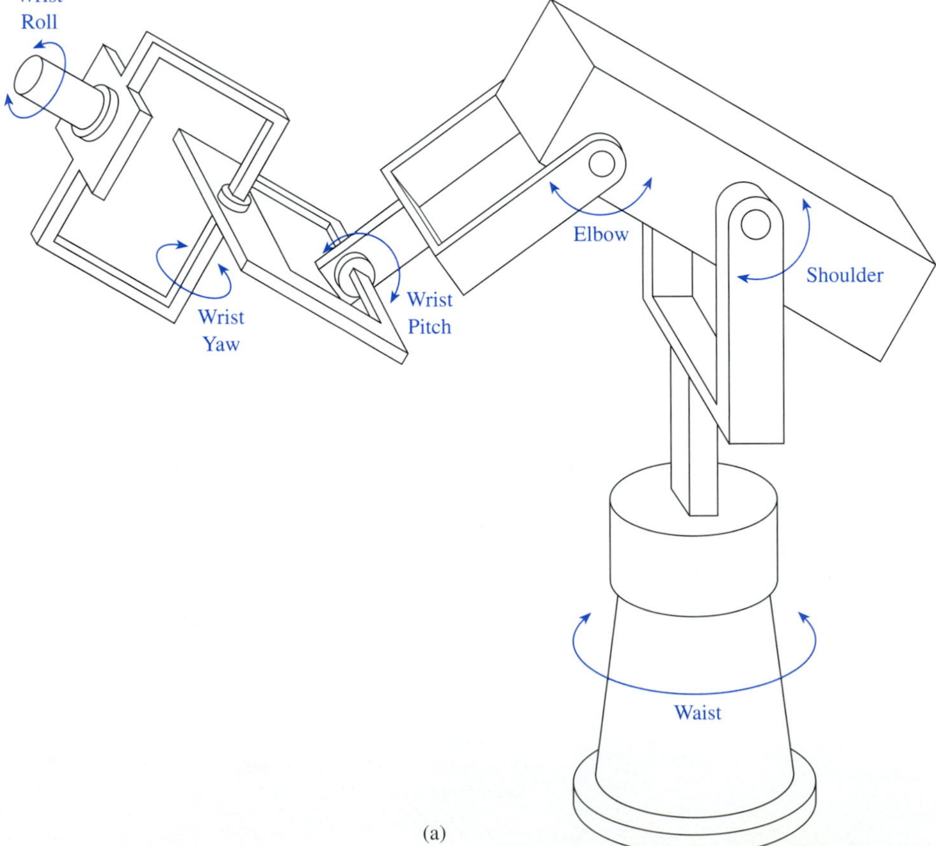

FIGURE 19–3
(continued)

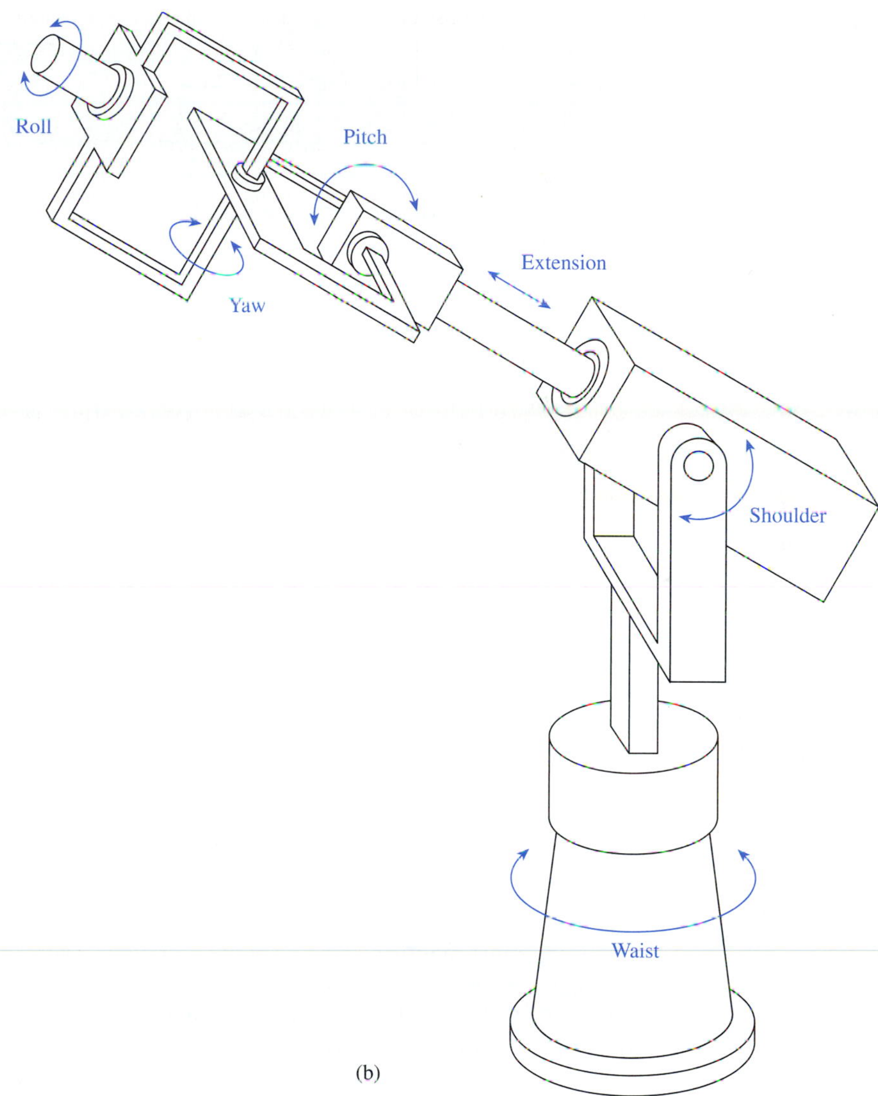

(b)

waist movement. As far as the robot's *hardware* is concerned, the shoulder is free to go wherever it wants to go, regardless of what the waist is doing,* and vice versa.

Many robots are used in applications where fewer than six degrees of freedom are needed. For example, if a painting robot is fitted with a standard nozzle that emits paint in a conical spray pattern, wrist roll is not needed. Such a robot would not contain the apparatus for accomplishing wrist roll motion; it would therefore possess five degrees of freedom. Some simple material-handling robot applications don't require any wrist movement at all. A robot used in such an application would essentially have no wrist; it would

*Of course, when the robot is actually working at performing its manipulations, it is operating in accordance with the user-program instructions stored in the microcomputer. For some robots, these instructions produce specific relationships among the various motions, thus taking away their independence from one another. Such elimination of the freedom of the shoulder to act independently of the waist, for example, is *software imposed*. It is not a consequence of the robot's mechanical hardware.

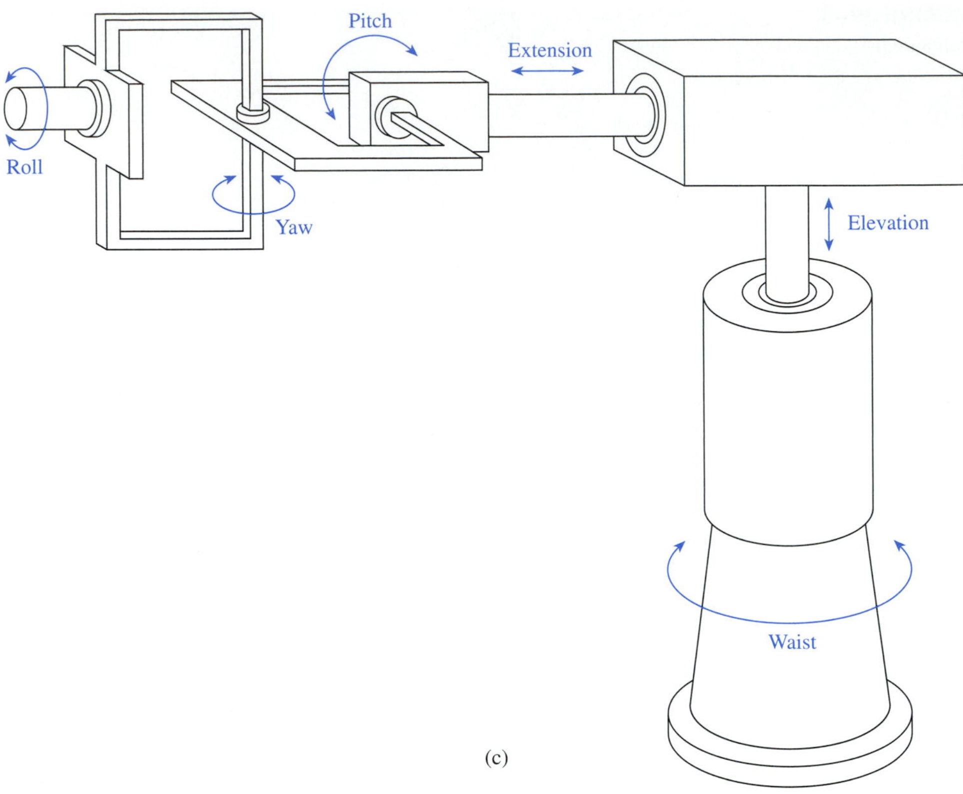

FIGURE 19–3 *(continued)*

just have a gripper attached to the end of a rigid *forearm* (the member extending from the elbow to the wrist). A robot of this design would possess only three degrees of freedom: waist, shoulder, and elbow/extension (the gripper doesn't count).

Notice that industrial robots are not ambulatory, unlike science-fiction robots. They are usually mounted stationary to the floor, with their work brought to them by other equipment. Occasionally an application requires a robot to be mounted on a platform which moves on rails. *Traverse* motion on rails may or may not be considered to be an extra degree of freedom, depending on whether such motion is controlled by the robot's microcomputer or by auxiliary automated equipment.

An inspection of Fig. 19–3 makes it clear that there are mechanical constraints on the range of motion for shoulder, elbow/extension, wrist pitch, and wrist yaw. There are variations from one robot to another, but usually these four motions are limited to less than 180° of rotation about the axis. In the spherical and cylindrical configurations, extension is a linear motion rather than rotational, so the constraint is a linear distance rather than an angular rotation. The maximum extension for any present-day spherical or cylindrical robot is about 5 ft.

Notice from Fig. 19–3, however, that there are no inherent mechanical constraints on waist or wrist roll motion. In principle, these two motions could just keep on going indefinitely—*spin*. It so happens that the potential spinning capability of these axes is seldom used in industry. Waist rotation is usually limited to 330° or less. Maximum wrist roll varies considerably among robots, with some models limited to about 90° and other models capable of two or more complete revolutions (720° or more).

The actuating devices for the various motions are either electric motors, fluid cylinders, or fluid motors. Some robots use all electric motors, some robots use all fluid actu-

TABLE 19–1
Advantages and disadvantages of the three means of actuation

ACTUATOR TYPE	ADVANTAGES	DISADVANTAGES
Electric	Lower initial cost than a fluid system Much lower operating costs than a hydraulic system Clean—no oil leaks to wipe up Accurate servo-type positioning and velocity control can be achieved	Not such great force capability as a hydraulic system Very little holding strength when stopped—will allow a heavy load to sag; mechanical brakes are required
Hydraulic	Great force capability—can handle heavy loads Great holding strength when stopped—hydraulic cylinder will not allow a heavy load to sag Accurate servo-type positioning and velocity control can be achieved Intrinsically safe in flammable environments such as painting	High initial cost High operating costs Messy—tends to leak oil
Pneumatic	Lower initial cost than a hydraulic system Lower operating costs than a hydraulic system Clean—no oil leaks to wipe up Quick response	Programming of accurate positioning and velocity control are impossible; mechanical stops are required Weak force capability Not so much holding strength when stopped as a hydraulic system—will allow a heavy load to sag somewhat

ators, and some robots use a combination of electric motors for some motions and fluid devices for other motions. The usual trade-offs prevail. The advantages and disadvantages of each actuator type are summarized in Table 19–1.

As pointed out in Table 19–1, the major disadvantage of using an electric motor to drive the robot movement is that the motor itself has virtually no ability to hold a member in a steady fixed position. When handling heavy loads, or when precise positioning is required, many robot manufacturers install mechanical brakes to cope with this problem. The brakes engage the electric motor's shaft as soon as it stops running; this prevents the load-derived countertorque from turning the motor shaft and thereby prevents sag. The sag problem is more severe for the elbow and shoulder motions than for any of the wrist motions due to the longer moment arms associated with the robot's forearm and *bicep* (the member extending from the shoulder to the elbow). Sag is nonexistent with regard to the waist.

Hydraulic cylinders are the champions when it comes to strength, either while in motion or when holding a steady fixed position after the motion stops. The unfortunate features of hydraulic systems are that they are very expensive to buy and to operate, and they are forever leaking oil on the floor. Their utility costs are so high because the hydraulic pump must be kept running at all times to maintain adequate oil pressure, even during periods when the robot is not in motion. By contrast, an electric motor consumes energy only when it is actually producing motion.

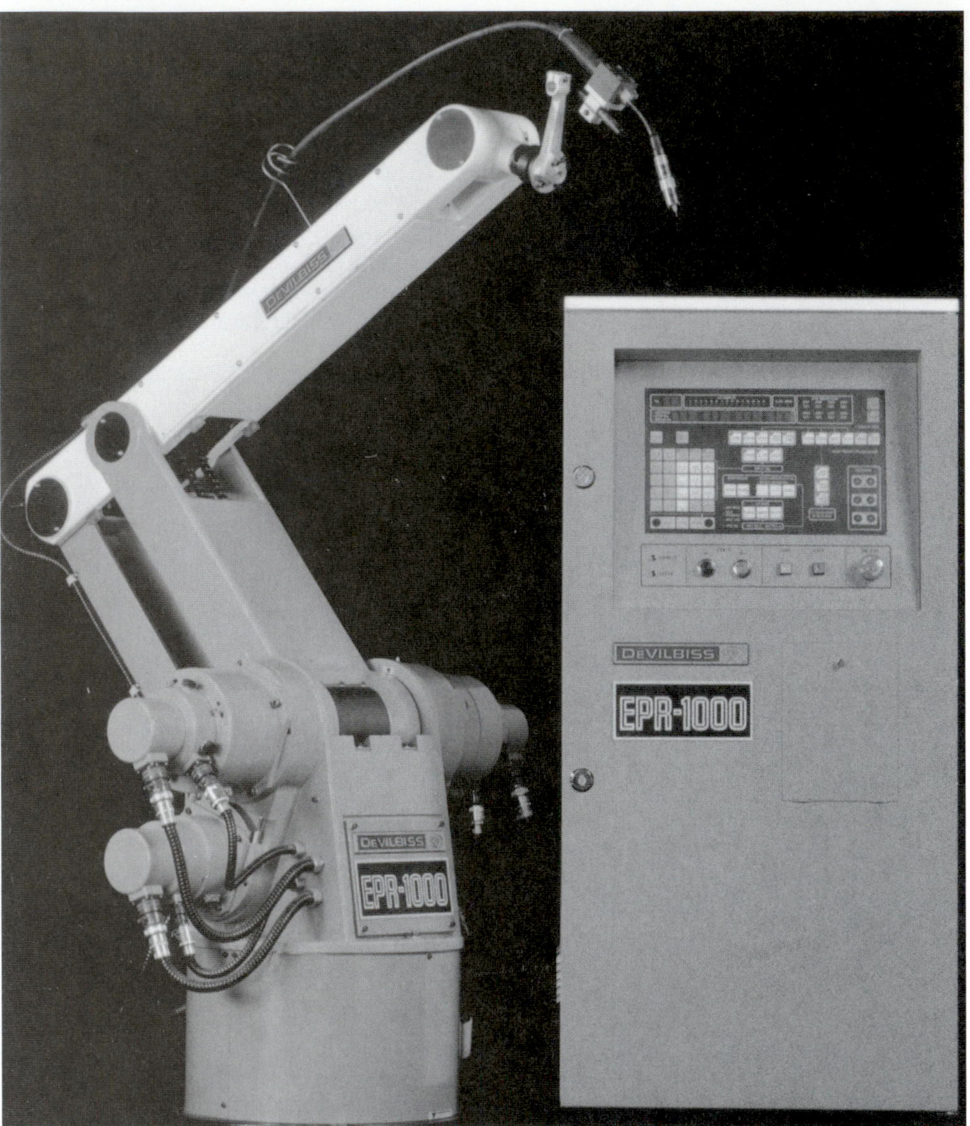

PHOTO 19–2
Several industrial robots. (a) DeVilbiss model EPR 1000 welding robot and its microcomputer control unit. This articulated-arm configuration has five degrees of freedom—waist, shoulder, elbow, wrist pitch, and wrist roll. All five axes are actuated by dc motors. The ranges of motion are waist—300°; shoulder—90°; elbow—75°; wrist pitch—180°; wrist roll—380°. From one manipulation sequence to the next, its position repeatability is ±0.008 in. With its forearm level, the robot is 54 in high. Its width is 34 in. and its depth at the base is 24 in. The volume of its work envelope (the locus of points in space where it can position its tool) is about 90 ft^3. It weighs 650 pounds.
Courtesy of the DeVilbiss Company.

Pneumatic robot actuators have lesser capabilities than electric or hydraulic actuators because of the inherent disadvantages associated with air's compressibility. A pneumatic cylinder will do for actuating an all-or-nothing motion, but not a controlled motion.

Photographs of several industrial robots are shown in Photo 19–2, with brief descriptions of their features.

19-3 ■ CATEGORIES OF SOFTWARE FOR INDUSTRIAL ROBOTS

Robots are able to perform crude and simple manipulations or elaborate and sophisticated manipulations according to whether the microcomputer's user-program is simple or sophisticated. We find it convenient to classify the user-programs that control robots into three categories. We will name these categories, in order of increasing sophistication, as

PHOTO 19–2
(b) A Prab model 5800 material-handling robot, loading and unloading steel sheet stock for a stamping press. This spherical-configured robot is actuated by hydraulic cylinders for shoulder and extension motions, and a hydraulic motor for waist rotation. The ranges of motion are waist—300°; shoulder—20°; extension—58 in. Between manipulation cycles, position repeatability is ±0.008 in. When its extension rod is level, the robot is 55 in. high. It has width of 35 in., and a depth at the base of 74 in. The volume of its work envelope is about 250 ft^3. It weighs 2600 pounds and has a 50-lb payload (maximum weight of the material being handled plus the gripping tool).
Courtesy of Prab Robots, Inc.

1. Positive-stop programs.
2. Point-to-point programs.
3. Continuous-path programs.

When a robot is operating under the control of a particular category of program, it is common practice to refer to the robot itself as that particular category of robot. That is, if a robot is being controlled by a positive-stop program, the common practice is to call the robot a positive-stop robot. This can be misleading, because it implies that the robot itself has hardware limitations that render it incapable of performing more elaborate and sophisticated manipulations, and that implication *may* not be true. Just because our robot happens to be operating right now under the control of a positive-stop program doesn't *necessarily* mean that it couldn't also operate under the control of a point-to-point program, or even a continuous-path program.

So that is the idea you should get straight—the popular categorizing of robots is done on the basis of the *program's* complexity, not the robot's mechanical complexity. Said another way, the common practice is to categorize a robot according to the job that it *is* doing, rather than what it *could* do.

Of course, it is silly and wasteful to have a high-falutin' robot performing simple manipulations. Therefore when a user selects a robot, he usually tries to match the hardware capabilities of the robot to the type of program (the type of manipulation) that he has in mind. This reasonable and widespread practice tends to establish some correlation

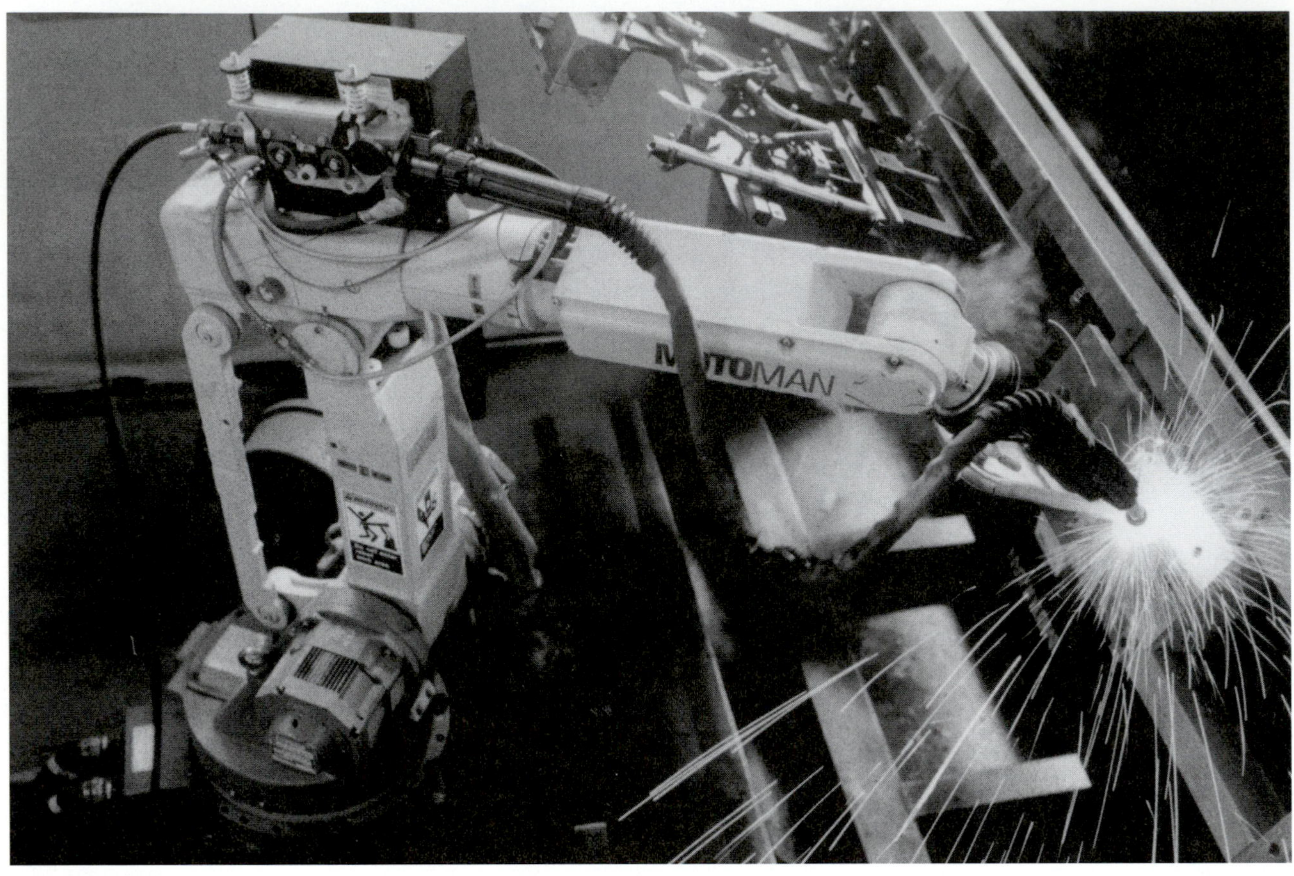

PHOTO 19–2
(c) A MotoMan model K10S welding robot with cylindrical configuration. It has 5 degrees of freedom. Waist and shoulder are electric-motor actuated; extension is hydraulic. Wrist pitch and roll are electric.
Courtesy of MotoMan Incorporated.

between the program category and the robot's mechanical capabilities, but it's not an absolutely reliable correlation.

19-4 ■ POSITIVE-STOP PROGRAMS

19-4-1 Their Two-Position Nature

A positive-stop program produces only two-position motion about any individual robot axis. By this we mean that there are only two possible positions that the waist can stop in, there are only two possible positions that the shoulder can stop in, and so on, for each axis of motion.

This does not mean that a positive stop program is able to move the tool-like device itself between only two positions in space. Since the motion axes are mechanically independent (free) from one another, a positive-stop robot with n degrees of freedom can have its tool device sent to 2^n different locations in space.

For example, imagine that our robot has three degrees of freedom, namely, motions about the waist, shoulder, and elbow axes. We can identify the individual axis positions

TABLE 19–2
Three-degrees-of-freedom positive-stop robot

(a) AXIS POSITIONS

MOTION AXIS	FIRST POSITION	SECOND POSITION
Waist	At counterclockwise limit	At clockwise limit
Shoulder	At down limit	At up limit
Elbow	At in limit (toward base)	At out limit (away from base)

(b) TOOL POSITIONS

POSITION	DESCRIPTION OF POSITION
1	waist = counterclockwise, shoulder = down, elbow = in
2	waist = counterclockwise, shoulder = down, elbow = out
3	waist = counterclockwise, shoulder = up, elbow = in
4	waist = counterclockwise, shoulder = up, elbow = out
5	waist = clockwise, shoulder = down, elbow = in
6	waist = clockwise, shoulder = down, elbow = out
7	waist = clockwise, shoulder = up, elbow = in
8	waist = clockwise, shoulder = up, elbow = out

as in Table 19–2(a). Accordingly, the tool device on this robot can be placed in any one of eight positions, since $2^n = 2^3 = 8$. These eight positions are listed in Table 19–2(b). It may make it easier to see the organization of Table 19–2(b) if you notice its similarity to the truth table for a three-input logic gate.

Of course, the number of positions that the robot's tool actually reaches will depend on the details of the program. Table 19–2(b) just indicates that for a three-axis robot, a positive-stop program can position the tool in as many as eight places.

There is no rule that says a robot can move only one axis at a time. In fact, most programs produce motion on several axes at once. Such motion is called *compound* motion. Whether a control program is written to produce single-axis motion or to produce compound motion is determined by the robot's industrial surroundings. If a certain portion of the program has the job of moving the tool from spatial position A to spatial position B, compound motion would probably be preferred unless there is some object located between positions A and B that the tool might collide with. In that situation it might be necessary to write the program so that one single-axis motion is produced, and when that motion is completed another single-axis motion is produced, thereby moving the tool on a roundabout path that avoids the intervening object.

In Table 19–2, notice the use of the word *limit*. In this usage, limit does not mean the absolute farthest position that the robot's axis is inherently capable of moving to. Rather, it means the position of an adjustable limit switch, which selects the position that the positive-stop robot will cause that axis to move to. Positive-stop programs must be married to the robot's mechanics in this way. If a robot is to be controlled with a positive-stop program, two limit switches must be provided for each axis. One switch signals to the microcomputer that the axis has reached its limit in one direction, and the second switch signals that the axis has moved to its limiting position in the other direction.

For example, if our robot has a construction that allows inherent waist rotation from 0° to 180°, but we adjust the limit switches so that they are actuated at 30° and 140°, then the 30° position becomes the waist's counterclockwise limit position in Table 19–2(a), and the 140° position becomes the waist's clockwise limit position. When the positive-stop program makes the waist move, only two possibilities exist: rotation from 30° to 140°, or rotation from 140° to 30°. The waist cannot stop at an in-between position because the instructions in a positive-stop program always call for the motion to continue until a stop signal is received from one of the limit switches. This inability to stop a robot axis at any position other than one of the two limit positions is the defining characteristic of the positive-stop program category.

Output from the microcomputer. To substantiate our understanding of positive-stop programs, imagine that we are controlling a three-axis robot having a gripper as its tool device, using an 8-bit microcomputer like the one shown schematically in Fig. 18–4. Suppose that the output latch is used to produce the various robot motions according to the schedule shown in Fig. 19–4. For instance, if data line 2 goes HI and is latched by the output latch chip, output amplifier 2 applies power to the elbow-actuating device (an electric motor, say) so that the elbow moves in the inward direction, toward the robot's base. As long as OA2 continues to receive a HI from the latch, the motor will continue moving the elbow inward.

To deenergize the elbow motor, the microcomputer must place a LO on D2 and write that LO to the output latch at address hex 3ØØØ (same address as before, in Sec. 18-7). The output latch will pass the LO to OA2, which will shut off the elbow motor.

To move the elbow in the outward direction, the program must arrange for a 1 to be written to the output latch on line D6. To stop outward motion of the elbow, D6 = Ø must be written to the output latch.

Similar descriptions apply to the waist (lines DØ and D4), the shoulder (lines D1 and D5), and the gripper (lines D3 and D7).

Input to the microcomputer. On the left side of Fig. 19–4, the input buffer is set up to receive information from the robot's limit switches. For instance, when the elbow reaches its inward limit position, it actuates the "Elbow Is In" limit switch, placing a HI on data line 2 on the left side of the input buffer. When the program causes the microprocessor to read the input buffer at address hex 4ØØØ (the same address that was used before), this HI is transferred onto the D2 line of the data bus itself, making the microcomputer aware that the elbow is at its inward limit position. The next instruction in the program would cause a Ø to be written to the output latch on line D2, thereby stopping the elbow motor.

19-4-2 An Example Program

Suppose that it were necessary to accomplish the following sequence of motions with a positive-stop program.

1. Starting from the robot's home position (defined as waist counterclockwise, shoulder down, elbow in, gripper open), turn the waist to the clockwise position.
2. Move the shoulder up while moving the elbow out (a compound motion).
3. Close the gripper.

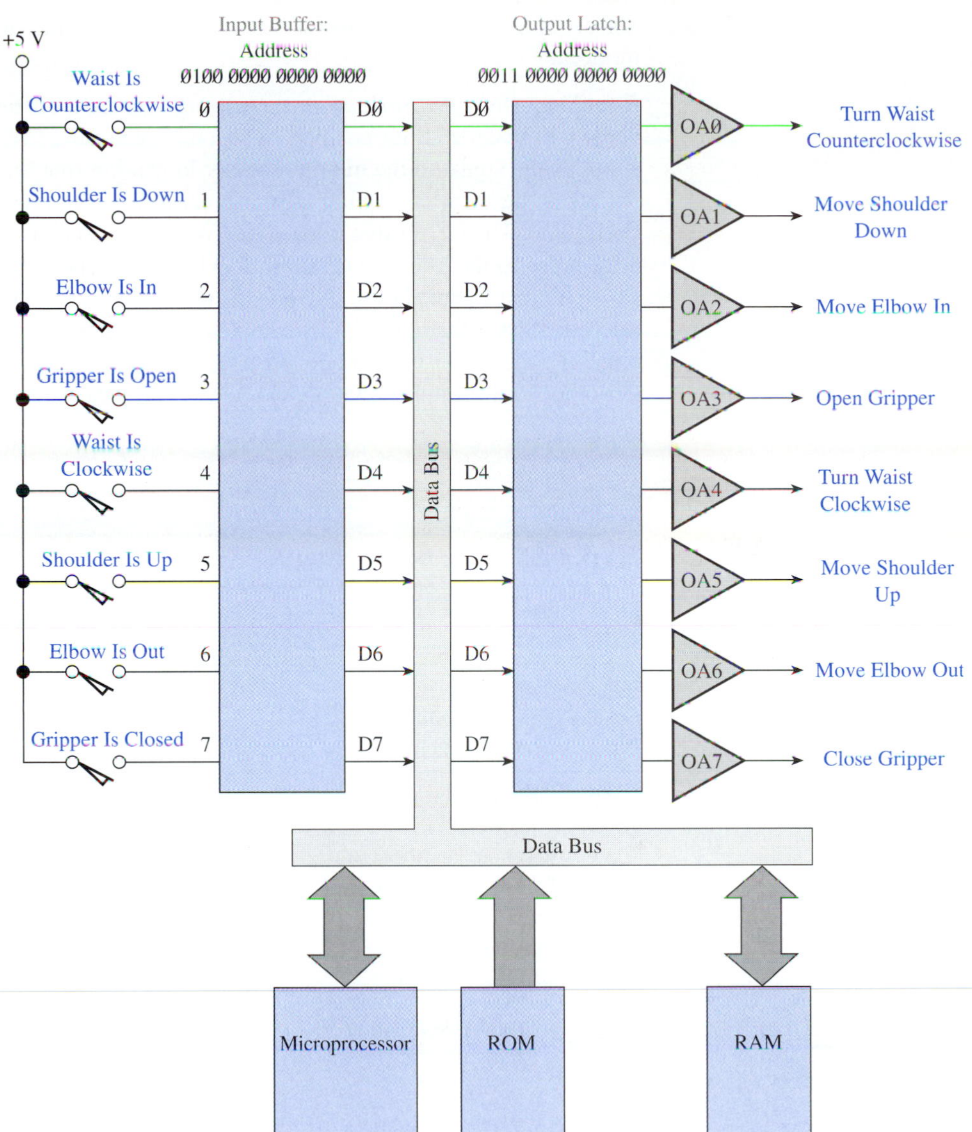

FIGURE 19–4 Input to and output from the microcomputer for a three-axis (plus gripper) positive-stop program.

This three-motion sequence could be accomplished by a program segment which performs as follows. Refer to the list of Table 19–3.

The microprocessor receives an instruction code from the current ROM address telling it to load 0001 0000 into register A. Then the next instruction tells the microprocessor to write the contents of A to address hex 3000, the output latch. This causes output amplifier 4 to receive a HI while all other output amplifiers receive LOs from the output latch. Therefore the waist motor is the only actuating device that is energized; it turns clockwise according to the legend in Fig. 19–4. The first motion has begun.

DELAY subroutine. The next instruction that the microprocessor fetches from ROM tells it to jump its program counter to a separate section of the ROM, where it finds a group of instructions set apart from the main program. Such a group of instructions is called a *subroutine*. In this case the subroutine that the microprocessor jumps to does not accomplish anything useful—it just burns up time until the robot can complete its motion.

Reaching the destination—interrupting the microprocessor. Eventually the robot waist will reach its limit position. A second pole of the "Waist Is Clockwise" limit switch supplies the signal to the microprocessor indicating that the motion is complete. This signal takes the form of a quick HI pulse applied to the microprocessor's *interrupt request* (IRQ) line. The circuit that allows the "Waist Is Clockwise" limit-switch closure to apply a HI pulse to the microprocessor's IRQ line might be built as shown in Fig. 19–5.

As that figure shows, any one of the eight limit switches is capable of signaling the microprocessor that a motion is complete. Of course, each limit switch speaks for its axis only. If the motion is a compound one, one limit switch is bound to be actuated before the other limit switches that are associated with the compound motion. In that case it is

TABLE 19–3
Outline of a positive-stop control program*

MAIN PROGRAM INSTRUCTION	COMMENT
Load register A with 0001 0000 Write the data in A to the output latch	These instructions cause the robot's waist to start turning.
Jump to the DELAY subroutine Load A with 0110 0000 Write A to output latch	These instructions cause the compound motion of the shoulder and elbow to begin.
Jump to DELAY subroutine Load A with 1000 0000 Write A to output latch Jump to DELAY subroutine • • •	These instructions cause the gripper to start closing.
DELAY Subroutine ↓ Return from subroutine to main program	Group of instructions which just loops indefinitely in order to keep the microprocessor occupied while it is waiting for the robot to complete its motion
Interrupt Service Routine Instructions ↓ Read the input buffer and load its contents into B	Group of instructions that compares the B register to the A register to make sure that every HI bit in A is matched by a HI bit in B
If B matches A, branch to the end of the DELAY subroutine	Robot's motion has been completed, so return to the main program.
If B does not match A, branch to the beginning of the DELAY subroutine	Robot's motion has not yet been completed, so wait for another interrupt signal.

*The branching pattern is always from the main program to the DELAY subroutine to the interrupt service routine, then touching down in the DELAY subroutine on the way back to the main program.

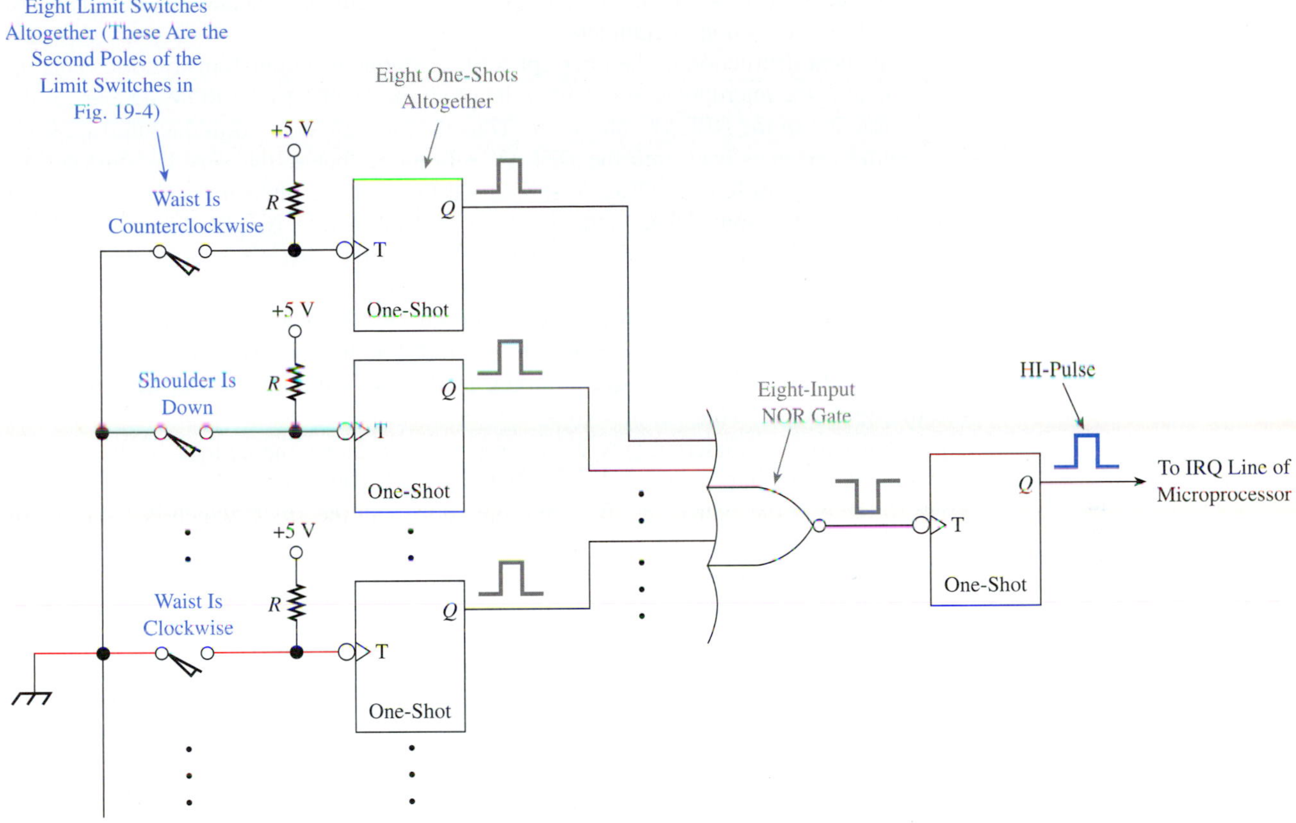

FIGURE 19–5
Circuit for signalling the microprocessor that an axis motion is complete in a positive-stop program. (This circuit ignores the problem of how to handle two nearly simultaneous limit-switch closures from a compound motion.)

the program's responsibility to realize that the overall compound motion is still in progress. In other words, just because one of the robot's axes has reached its limit position does not entitle the program to move on to the next sequential motion; rather, the program must wait for the other axes to reach their limit position before it can move on to the next sequential motion.

Servicing the interrupt. When the HI pulse hits the IRQ line of the microprocessor, the microprocessor drops whatever it was doing (in this case just passing time) and jumps the program counter to yet another section of the ROM, where it encounters another group of instructions separate from the main program. This group of instructions is called the *interrupt service routine*. Our interrupt service routine (see Table 19–3) first reads the input buffer into the B register. At this point in our example program, the only motion away from the robot's home position that has occurred is the clockwise turning of the waist. Therefore the input buffer will supply the data ØØØ1 111Ø to register B. Refer to Fig. 19–4 to see this.

The next activity of our interrupt service routine is to compare the B register to the A register. The comparison instructions do not seek to find B *identical* to A, only to make sure that wherever there is a 1 in A there is also a 1 in B, which proves that all the individual axis motions have been completed. In this case A has a 1 in bit 4 only, as Table 19–3 specifies. Since B also has a 1 in bit 4, the comparison operation indicates that B matches A, and therefore the motion is complete.

The next instruction in the interrupt service routine is a branch instruction. This instruction tells the microprocessor to jump its program counter to the ROM address of the last instruction of the DELAY subroutine. Thus the program exits from the interrupt service routine and goes back into the DELAY subroutine, but at the very last instruction. (When the program left the DELAY subroutine to go to the interrupt service routine, it was churning away somewhere in the middle of that subroutine, caught in an endless loop. The only way that the program can reach the end of the DELAY subroutine is by the method just described.)

It so happens that the last instruction in the DELAY subroutine* tells the microprocessor to change is program counter to the ROM address of the next instruction that is waiting for us back in the main program. In this example, the next instruction is the fourth one down in Table 19–3.

You may be wondering how the microprocessor knows the address of the next instruction in the main program. It knows that address because before it left the main program to jump to the subroutine, it stored that address in the *stack,* which is a section of RAM that is specifically set aside for just this purpose. This stacking action** is automatic on the part of the microprocessor.

We have now completed one motion in the robot's sequence.

Proceeding to the next motion. Here comes motion number 2, the compound motion of lifting the shoulder up while moving the elbow out. On the fourth and fifth lines of Table 19–3, the program writes the data Ø11Ø ØØØØ to the output latch. An inspection of Fig. 19–4 shows that this will energize both OA6 and OA5, causing both the elbow and shoulder motors to start running. The microprocessor then jumps away to the DELAY subroutine to wait until a switch closes.

Let's say the elbow motion finishes first, before the shoulder motion. The "Elbow Is Up" limit switch in the circuit of Fig. 19–5 (it's not actually shown) interrupts the microprocessor to let it know that something has happened. The microprocessor jumps out of the DELAY subroutine into the interrupt service routine, as usual. The first thing it does in that routine is read the input buffer, which yields Ø1Ø1 1ØØØ. Bits 1 and 5 are both Ø because the shoulder is neither down nor up—it's still in motion. When this data is loaded into B and compared to A, the number 5 bits do not match. Bit 5 of A is 1 but bit 5 of B is Ø. Therefore, at the bottom of the interrupt service routine in Table 19–3, the microprocessor jumps back to the DELAY subroutine all right, but at the beginning, not the end. The microprocessor finds itself right back where it came from, passing time waiting for something to happen.

When the shoulder does reach its limit position, the "Shoulder Is Up" switch in Fig. 19–5 (not actually shown) pulses the IRQ line and the microprocessor goes to the interrupt service routine for the second time during this motion. It reads the input buffer into register B, getting Ø111 1ØØØ. Register A hasn't changed in all this time; it still contains the original data that was written to the output latch at the commencement of the motion, namely, Ø11Ø ØØØØ. The comparison instructions verify that every 1 in A is

*For a Motorola 6800-series microprocessor, the last instruction is "Return from Subroutine to main program," assembler mnemonic RTS, OP code hex 39.

**We didn't mention it above because we've got enough other things to worry about, but the stack is also involved in jumping out of a subroutine into an interrupt service routine. A book devoted exclusively to microcomputers will explain all the details about the stack.

matched by a 1 in B, causing the program to branch to the end of the DELAY subroutine. From there it makes an immediate jump to the next main program instruction, which has been kept warm in the stack.

The next motion. Presumably, the robot is now properly positioned to grab a part; for motion number 3 it is closing of the gripper. This is accomplished by the program in the usual way: Write to the output latch, wait for a signal, read the input buffer, compare the byte that was written to the byte that has just been read back, and, if they match, move on to the next motion. The gripper action is probably much faster than any other robot motion, so the program won't have to wait as long for an interrupt signal—that's the only difference.

At the end of the program there will be instructions to return the robot to its home position. It will then wait there until a hardware signal is delivered to the microprocessor, telling it to begin reexecuting the main program from the beginning.

Necessity of mechanical stops. Positive-stop programs are called by that name because a robot controlled by such a program is usually equipped with adjustable mechanical *stops* which prevent it from overshooting its limit switches. This precaution is necessary because if an axis did coast through its limit position, there is no provision in the program for enabling it to recover. Inspect the Table 19–3 instruction list and the schematic diagrams in this section to verify for yourself that this is so. The other two categories of robot programs, point-to-point and continuous-path, do allow an axis to back up and recover if it overshoots its destination.

Remarks about nomenclature. Positive-stop programs are best adapted to applications in which the robot must pick up a part and put it down in a different location. These are called *pick-and-place* applications. Sometimes the phrase "pick-and-place" is used as a synonym for positive-stop.

Other common names for positive-stop programs (or robots) are *limited-sequence, bang-bang,* and *non-servo-controlled*. The phrase *non-servo-controlled* means that the position of an axis is not under automatic closed-loop control, which implies, as mentioned earlier, that the machine is unable to recover from an overshoot.

19-5 ■ POINT-TO-POINT PROGRAMS

19-5-1 Their Multiposition Nature

The essential feature that distinguishes a point-to-point program is its ability to move a robot axis to any position within its range, rather than only the two limit positions. Thus, if the mechanics of our robot give it an inherent range of movement of 0° to 128° on its shoulder, a point-to-point program with 8-bit resolution (1 part in 256) could position the shoulder at 0.0°, at 0.5°, at 1.0°, at 1.5°, at 2.0°, and so on, up to 127.5°.

Destination positions. When the tool device is to be moved from one spatial location to a new spatial location, the point-to-point program must digitally specify a destination position for each axis, such that when each axis reaches its destination position, the tool will be at the desired new spatial location.

For example, suppose we have a three-axis robot (waist, shoulder, and elbow) and it is desired to position the tool device (a gripper, say) at the spatial location corresponding to

waist = 1001 0111 (decimal 151 out of 255)

shoulder = 0011 0001 (decimal 49 out of 255)

elbow = 1101 1010 (decimal 218 out of 255)

The point-to-point program reads the waist's present position from an input device and compares it to the waist's destination position of 151. If the present position is less than 151, the program sends a signal to an output latch which causes the waist motor to run forward; if the present position is greater than 151, the program sends a signal to the waist's output latch causing the waist motor to run in the backward, or reverse direction.

Once it has gotten the waist moving in the right direction, the program moves on to consider the shoulder. If the present shoulder position is less than 49, the program causes the shoulder motor to run forward; if the present position is greater than 49, the program causes the shoulder to reverse.

Then it's on to the elbow. If the elbow's present position <218, the elbow motor runs forward; if the elbow's present position >218, the elbow motor runs reverse.

In this manner the program initiates a three-axis compound motion.

The program then loops back and repeats the comparison sequence, namely

1. Whatever the actual waist position relative to its destination position, the program makes the waist motor run in the proper direction to bring it closer to the destination.
2. Whatever the actual shoulder position relative to its destination, the program latches the shoulder motor running in the proper direction to bring it closer to that destination, in other words, to correct the position error.
3. Whatever the actual elbow position, the program latches the elbow motor running in the proper direction to correct the elbow's position error.

Reaching the destinations. The program continues cycling through steps 1, 2, and 3 repetitively. Eventually, one of the axes will reach its destination. On the next pass through the program loop, the comparison instruction for that axis will show agreement between the actual position and the destination position—the position error equals zero. The program therefore sends a stop signal to that axis' motor, freezing that axis in its proper position. The program's looping continues until all three axes are in their destination positions. When that condition is achieved, the program breaks out of its loop and proceeds on to the next instructions, which produce the next action(s) in the robot's manipulative sequence.

19-5-2 The Microcomputer Architecture and Program

Let's get more concrete about the structure of a point-to-point program by referring to the schematic diagram of Fig. 19–6 and the instruction list in Table 19–4.

On the left side of Fig. 19–6 are three 8-bit position encoders, one each for the waist, shoulder, and elbow. A position encoder is a device that converts a mechanical position into a digital signal that represents that position, like the optical position encoder presented in Section 10-7-2. Let us suppose that a robot position encoder's digital output represents the absolute position of that axis, relative to its home position.* The resolution is usually quite good, 1 part in 4096 being typical (12 bits). Some very accurate robots have 15-bit encoders, for a resolution of 1 part in 32 768. For the sake of simplicity, let us suppose that our robot has encoders with only 8-bit resolution, as specified in Fig. 19–6.

At any instant in time the robot's microcomputer can find out the position of any axis, just by performing a read of the input buffer connected to that axis' position encoder. Thus, to learn the actual position of the waist, the microcomputer would read the WAIN

---

*This is the absolute position-encoding idea, which is conceptually simpler than the relative position-encoding idea that many robots use. In relative position-encoding, the encoder furnishes information about how far and in what direction the robot axis has moved *from its most recent position*.

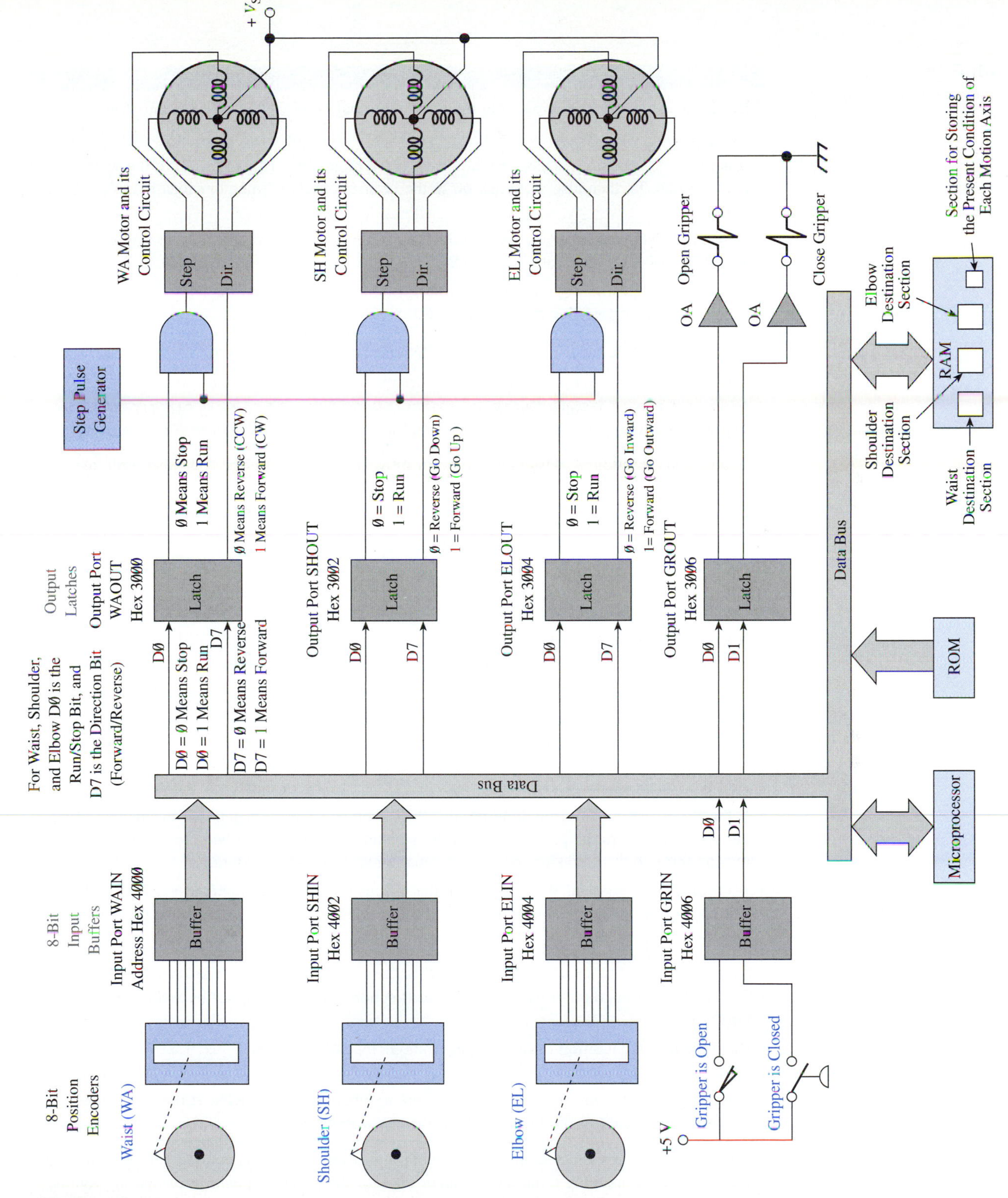

FIGURE 19–6
Input to and output from the microcomputer for a three-axis (plus gripper) point-to-point program. The destination addresses for the three axes must be stored in RAM prior to the start of the program.

TABLE 19–4
Outline of a point-to-point control program*

- Read the actual waist position from the waist input buffer WAIN and load it into register A.
- Compare the actual waist position to the waist destination for this particular move.
- If actual waist position equals destination, stop the motor by writing hex ØØ to the waist output latch (WAOUT); then also store hex ØØ in the "waist condition" memory location (WACON) in RAM.
- If actual waist position is less than destination, run motor forward (For) by writing hex 81 to WAOUT; also store that data in WACON.
- If actual waist position is greater than destination, run motor in reverse (Rev) by writing hex Ø1 to WAOUT; also store that data in WACON.

- Read the actual shoulder position from SHIN and load it into register A.
- Compare the actual SH position to the SH destination for this particular move.
- If actual SH position equals destination, stop motor by writing hex ØØ to SHOUT; also write to SHCON in RAM.
- If actual SH position < destination, run For by writing hex 81 to SHOUT; also write to SHCON.
- If actual SH position > destination, run Rev by writing hex Ø1 to SHOUT; also write to SHCON.

- Read the actual elbow position from ELIN and load it into A.
- Compare actual EL position to the EL destination for this particular move.
- If actual EL position equals destination, stop motor by writing hex ØØ to ELOUT; also write to ELCON in RAM.
- If actual EL position < destination, run For by writing hex 81 to ELOUT; also write to ELCON.
- If actual EL position > destination, run Rev by writing hex Ø1 to ELOUT; also write to ELCON.

- Clear the B register.
- Add the contents of WACON, SHCON, and ELCON to the B register.
- If the B register contains anything besides Ø (if Z bit of CCR is clear), that means that at least one motor is still running, so the overall tool destination has not yet been reached. Therefore recycle through the preceding position-comparing section of the program.
- If the B register contains Ø (Z bit is set), that means that all the motors are stopped, so the overall tool destination has been reached. Therefore do not recycle through the preceding position-comparing section of the program, but increment the X register and jump the program counter to the instruction for the next robot action.

- This is the first instruction for the robot's next action: close the gripper by writing hex Ø2 to GROUT.
- Read the actual condition of the gripper from GRIN and load into A.
- Compare A to the desired gripper condition (hex Ø2). If they are not equal, the gripper has not closed tightly yet, so recycle the program back to where it reads GRIN. If they are equal, jump the program counter to the instruction for the next robot motion.

- This begins the next robot motion: Start another position-comparing routine, but this time with the X register containing a number 1 higher than for the previous position-comparing routine. Therefore the WA, SH, and EL destination addresses in RAM are all 1 higher than they were for the previous position-comparing routine (for the previous robot motion).

*The waist, shoulder, and elbow position-comparing routines are reused for every compound motion in the robot's manipulative path.

input port at address hex 4ØØØ. To learn the actual position of the shoulder it would read the SHIN port, hex 4ØØ2. The elbow's position is available at the ELIN port, hex 4ØØ4.

The gripper condition can be read from the GRIN port at address hex 4ØØ6. The gripper's condition is not expressed as a numerical position value, since all the microcomputer really needs to know is whether the gripper is opened or closed. The open condition can be detected by a mechanical limit switch. But the closed condition is better detected by a pressure switch connected to the blind end of the gripper's actuating cylinder, because the closed position of the gripper jaw will vary depending on the physical size of the part being gripped.

The GRIN input buffer has only two active lines, DØ and D1. It places Øs on lines D2 through D7 when it is read. Therefore the gripper condition is related to the data byte by the following schedule:

Data byte	*Gripper condition*
ØØØØ ØØØ1 (hex Ø1)	Gripper is open.
ØØØØ ØØ1Ø (hex Ø2)	Gripper is closed.
ØØØØ ØØØØ (hex ØØ)	Gripper is in process of stroking—either grabbing or releasing, we can't tell which.

Motor control signals. The motors that move the waist, shoulder, and elbow axes are constant-speed stepper motors, let's suppose. Their control circuits receive only two bits from the output latches: bit Ø is the signal to run or stop, and bit 7 is the direction signal. The bit values have the following meanings:

Bit Ø		*Bit 7*	
Level	*Meaning*	*Level*	*Meaning*
Ø	Stop	Ø	Reverse (toward home position)
1	Run	1	Forward (away from home position)

To signal a stepper motor to run forward, the microprocessor must send a data byte 1XXX XXX1 to the motor's output latch. To signal a motor to run in reverse, the microprocessor must send ØXXX XXX1. To signal a motor to stop, the microprocessor must send XXXX XXXØ. Let us adopt the convention of sending Øs for all the don't-care bits. Then the motor control bytes are

Run forward	hex 81
Run in reverse	hex Ø1
Stop	hex ØØ

As Fig. 19–6 makes clear, the waist output latch is at address hex 3ØØØ, with symbolic label WAOUT. The shoulder output is at hex 3ØØ2, labeled SHOUT; and the elbow output is at hex 3ØØ4, labeled ELOUT.

The gripper is operated by a pair of solenoids controlled from output latch GROUT at address hex 3ØØ6. We will adopt the convention of sending data byte hex Ø1 to cause the gripper to open, and data byte hex Ø2 to cause the gripper to close. Refer to Fig. 19–6 to verify that this convention is appropriate.

Destinations stored in RAM. Before the point-to-point program can begin executing, the user must store the destinations of each robot motion in the microcomputer's RAM. (We will talk later about how this is actually done.) That is, the first motion in the robot's manipulative sequence must have its waist destination position stored at one RAM address, its shoulder destination position stored at another RAM address, and its elbow destination position stored at a third RAM address. These three addresses must be related to each other in a certain way that makes it easy for the program to access them during the robot's first motion. Specifically, all three addresses must be displaced from one another by a given fixed amount.* Let us use a displacement amount of hex 40 for our example.

The second motion in the robot's manipulative sequence must have its waist, shoulder, and elbow destinations stored at the three RAM addresses that are 1 higher than the three RAM addresses that stored the first motion's destination values. The third motion in the sequence must have its WA, SH, and EL destinations stored at the three RAM addresses that are 1 higher than the three RAM addresses that stored the second motion's destinations, and so on.

The reason the axis destinations must be stored in this pattern is because such a pattern allows all three of the current destination-storage addresses to be accessed repetitively as the program recycles through the position-comparison sequence for a given motion. On the first robot motion, we load the X-index register in the microprocessor with hex 8001. Then, whenever an instruction in the program's position-comparing sequence needs to know the waist destination, it is told to read the RAM address contained in the X register; whenever some other instruction needs to know the shoulder destination, it is told to read the RAM address obtained by *adding hex 40 to the contents of the X register;* whenever some other instruction needs to know the elbow destination, it is told to read the RAM address obtained by *adding hex 80* to the contents of the X register. This is another example of the great usefulness of the indexed-addressing technique, which we encountered earlier in Sec. 18-7.

Before it begins a robot motion, the program must arrange for the proper RAM address to be present in the microprocessor's X register. If it is about to start the robot's first motion, X must contain hex 8001; if it is starting the second motion, X must contain hex 8002; and so on, as listed in Fig. 19–7.

Position-comparison routine. So here is the whole story; refer to Table 19–4 and Fig. 19–6, as well as the flowchart of Fig. 19–8. The position-comparing sequence in Table 19–4 begins with an instruction to read the actual waist position from input port WAIN (hex 4000) and load it into register A of the microprocessor. The next instruction fetches this motion's waist destination from the appropriate RAM address and compares it to the actual waist position in A. The comparison is accomplished by a *pseudo-subtract* operation, in which the microprocessor goes through the procedure of subtracting the destination value from the actual value, but it never records the difference. The only things it does record are whether or not the difference is zero and whether or not the subtraction required a borrow. These facts are recorded in the Z bit and the C bit of the CCR, as we know.

At first, as the robot's motion is just beginning, the comparison will not yield a zero result because the actual waist position and the destination waist position will be different (probably). So the program instructs the microprocessor to inspect the C bit to decide which way to move the waist. If C is 1, a borrow was required, so the actual position must be less than the destination value (the subtraction is actual minus destination). Therefore the program writes a hex 81 to the WAOUT port in order to get the waist's stepper motor running forward. If C is 0, no borrow was required, so the actual position must be greater than the destination. Therefore the program writes hex 01 to WAOUT to make the waist motor run in reverse.

When the microcomputer writes an output signal to WAOUT, it immediately writes the same output byte to a RAM location called WACON (waist condition). This output

*This is one method of organizing the destinations in memory. Other methods could be used.

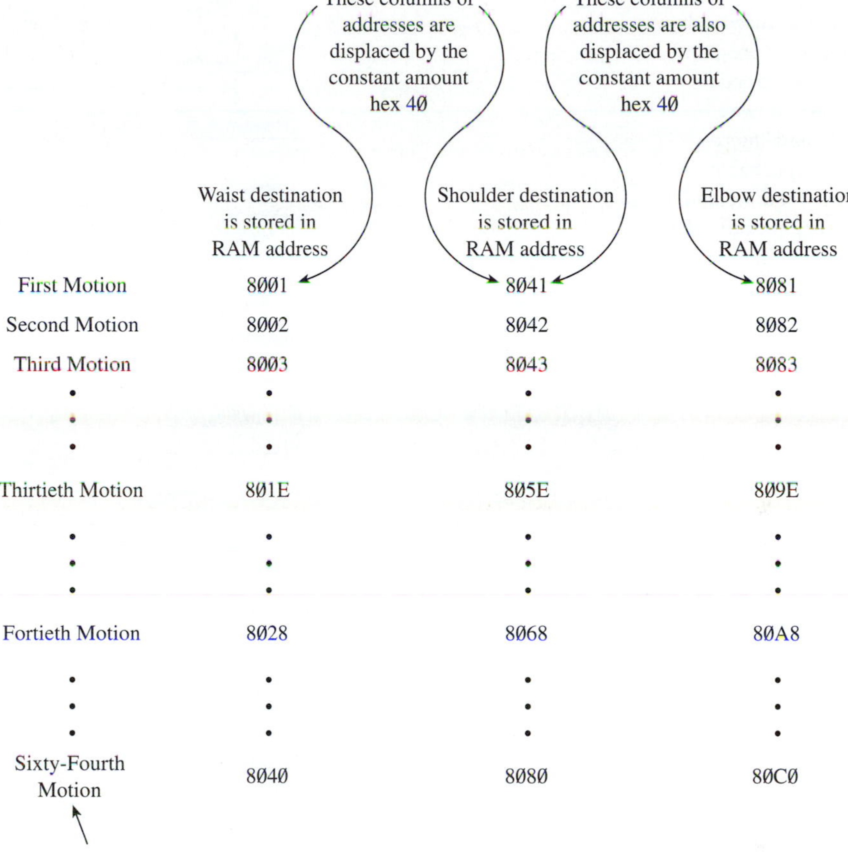

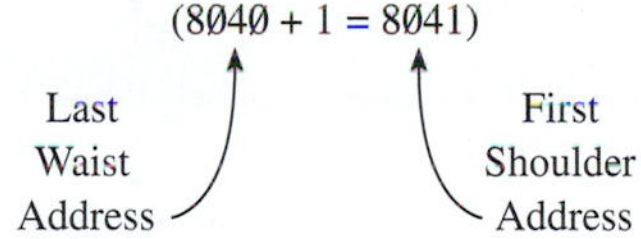

FIGURE 19–7 Destination address storage pattern in RAM. Addresses are given in hex.

byte is used later in the position-comparing routine to detect when the overall motion is finished. For concreteness, suppose WACON to be at hex 8FFØ.

The program then repeats an identical instruction sequence for the shoulder. It uses different addresses of course, but its principle of operation is exactly the same as the waist's. The addresses are hex 4ØØ2 for actual shoulder position, X + hex 4Ø for shoulder destination position, hex 3ØØ2 for signaling the shoulder stepper motor, and SHCON (hex 8FF1) for keeping track of the signal sent to the shoulder output port.

The elbow routine is another carbon copy of the first two routines. Actual position is brought into the microcomputer from ELIN, destination is fetched from X + hex 8Ø, the output signal is passed to the elbow motor through port ELOUT, and the output signal is copied into RAM location ELCON (hex 8FF2).

After the program has sent the proper signal to the elbow, it immediately checks to see if the overall motion is complete. The motion is complete only when all three motors have stopped. If all three motors have stopped, the three RAM locations WACON, SHCON,

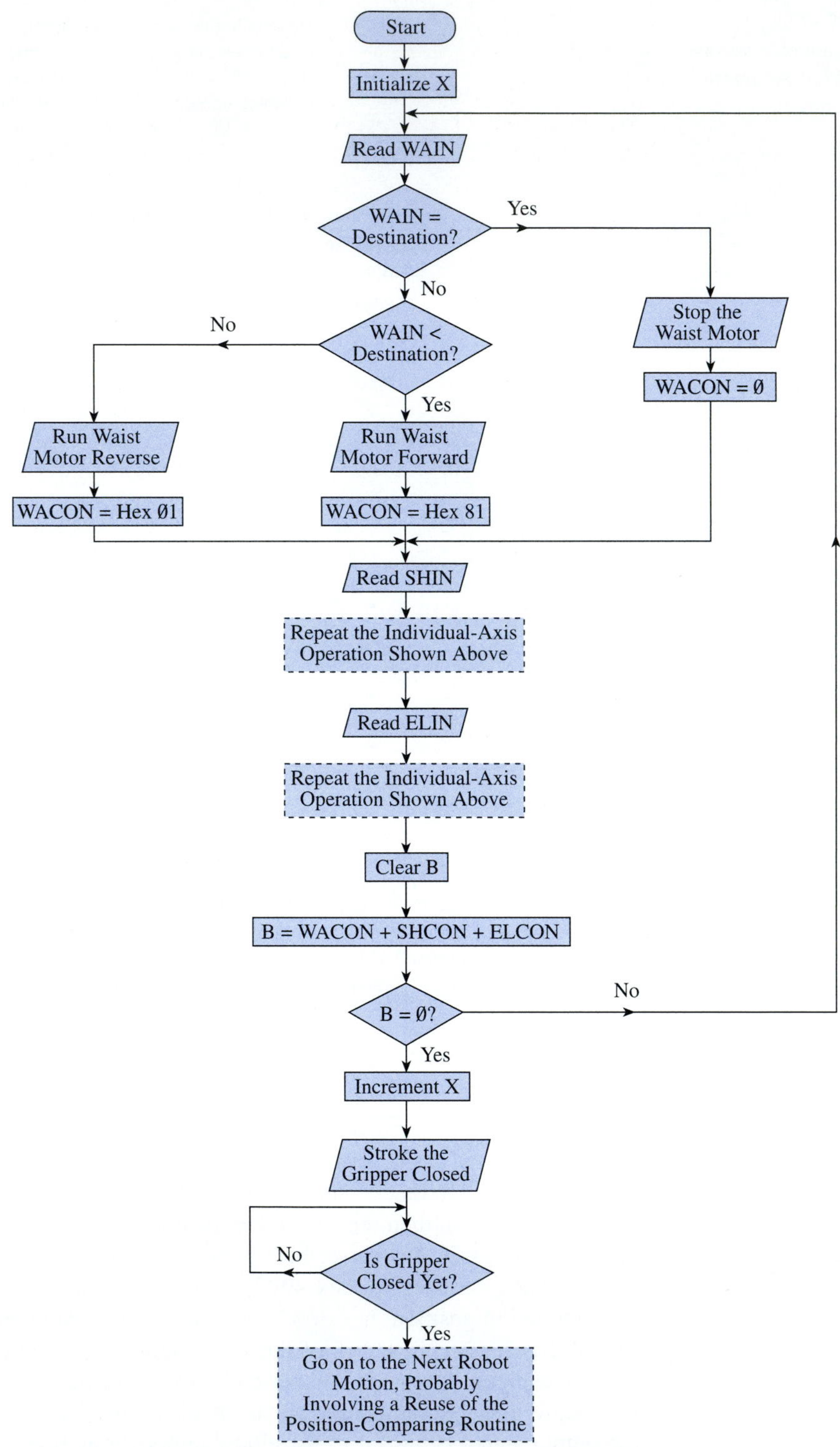

FIGURE 19–8 Flowchart for the three-axis position-comparing program of Table 19–4. The portion of the flowchart between "Read WAIN" and "Increment X" would probably be written as a subroutine and reused for every motion of the robot.

and ELCON will each contain hex ØØ. The program fetches the contents of these locations and adds them together. If the sum is not zero (Z bit of CCR is clear), the overall motion is not complete, so the program jumps back to the beginning of the position-comparison routine and does it all over again. If the sum is zero (Z bit is set), the overall motion is complete, so the program increments the X register and jumps the microprocessor's program counter to the ROM addresses containing the first instruction for the next robot action. Incrementing of the X register is necessary so that during the next robot motion (after the closing of the gripper in Table 19–4) the position-comparison routine will fetch its three destination values from the next-higher RAM addresses in the list of Fig. 19–7.

Note that even if a particular axis reaches its destination, it continues to be tested on subsequent passes through the position-comparison routine. Thus, if it should slip out of its destination position due to the jostling it is getting from the axes that are still in motion, the program puts it right back where it belongs. There's your closed-loop position control.

An axis in motion cannot overshoot its destination due to the software failing to catch it while it is there. Because the program executes so fast, an axis cannot possibly move far enough to advance its position encoder by more than one bit during the time that elapses between passes through the comparison routine. Said another way, as the program proceeds from one pass through the comparison routine to the very next pass through the routine, the greatest possible change that can occur in the position encoder is one bit.

Of course, if the load momentum is great, the robot axis may overshoot its destination due to *hardware* effects.

Proportional servo control. We have described a robot whose axes move at constant speed. Some point-to-point and continuous-path robots are controlled in this manner. They go full speed ahead until they reach their destination, and then they stop on a dime—they do not slow down as they get close. On the other hand, many point-to-point and continuous-path robots move at a fixed speed when they're far from their destination, then slow down as they get close to their destination. Such control can be termed *proportional servo* control, in keeping with its similarity to strict proportional closed-loop control (Sec. 9-6).

To implement proportional servo control, it would be necessary to replace the pseudo-subtraction instructions in the position-comparison routine with genuine subtraction instructions. A genuine subtraction instruction would leave the magnitude of the position error in the microprocessor's A register. That number could then be passed to a full-byte output latch, from which it could be applied to a *programmable* clock. A programmable clock generates pulses at a frequency proportional to its digital input signal. If the clock's output pulses were then used to drive a stepping motor, the motor would step rapidly when the position error was great, but it would step slowly when the position error got smaller. This would tend to reduce the possibility of hardware-caused overshoot, and would make the system more stable in general. The clock frequency would be limited to a certain maximum value in order to fix the maximum speed.

Of course, proportional servo control could also be achieved using regular dc motors to turn the robot axes. We would write the magnitude of the position error to an 8-bit latch and perform a D-to-A conversion. The resulting analog voltage would be applied to an SCR drive containing a gate-control circuit similar to that shown in Fig. 6–11(b).

Ac servo motors are also gaining favor among robot manufacturers.

Point-to-point distinguished from continuous-path. In a program of the type just discussed, only the starting point and finishing point are defined. The path by which the tool device moves between those points is not specifically defined. We can predict roughly what path the tool will follow between the two points, based on hardware considerations—the distance between those points for each axis and the speed of movement of each axis. But that's only an after-the-fact prediction; there is nothing in the software itself that causes the robot to adhere to that path.

Besides, the relative speeds of movement of the various axes might vary depending on the load (imagine that the robot is a material handler rather than a welder or painter).

A heavy load might cause the shoulder axis to lift more slowly than it would under light load, changing the relative speed of the shoulder axis to the waist axis, since the waist-axis speed is unaffected by load. Thus the exact path through space that the tool follows might be affected by the amount of load that it is carrying, even though the starting and finishing points are fixed.

A nonspecific path can be regarded as the defining criterion for a point-to-point robot, as opposed to a continuous-path robot. We will see in the next section that a continuous-path robot is constrained by the program itself to a predefined path as it moves through space.

19-5-3 Program Entry by Teach Pendant

We have imagined that our robot's operating program is stored in a ROM chip. This is a convenient initial supposition, since it gives us a clear and simple way to make the distinction between a program's instructions and its data—instructions are stored in ROM and data are stored in RAM, if they are stored in memory at all. And actually, robots can have their operating programs stored in a ROM device, but usually an *erasable programmable* ROM, or EPROM. Remember, one of the characteristics of a robot that renders it so useful is its reprogrammability—it can paint doors on Monday, hoods on Tuesday, etc. A true ROM does not allow its contents to be altered; once it's programmed, it's programmed for life. An EPROM allows its contents to be erased by exposure to ultraviolet light through an opening in the body of the IC chip. A new program can then be entered by the slow process (comparatively) of serially applying the appropriate signal to every single bit of every address in the program. So an EPROM is preferred to a true ROM for storing a robot's operating program, because it possesses the virtue of reprogrammability.

A robot's operating program can be changed even more easily by storing it in RAM, right along with the data. Many robots are set up this way. One section of RAM is reserved for the operating program, and another section handles the data. Of course, when an operating program is stored in RAM, the RAM's dc power supply must be uninterruptible. Loss of dc power for even a few milliseconds will result in loss of the program.

To accomplish the storing of the operating program in RAM, most point-to-point robots use a *teach pendant* in conjunction with a manufacturer-supplied *monitor program,* which is stored in ROM. The teach pendant is a hand-held switch enclosure linked by cable to the robot's microcomputer. Typically, it has buttons that jog the various robot axes for moving the tool, a button for identifying the current position of the tool as a destination position, buttons to indicate at what positions the tool device should be turned on and off (open and close a paint valve, energize and deenergize welding electrodes, engage and disengage a gripper, etc.), and a button to indicate when the end of the manipulative sequence has been reached, and others. An industrial teach pendant is shown in Photo 19–3.

With the robot switched into TEACH mode, the human user guides the robot through the desired manipulative sequence by pressing buttons on the teach pendant. The monitor program in ROM continually scans the teach pendant's switches. Contingent on these switch closures, and on the data supplied from the robot's various position encoders, the monitor program writes the appropriate instructions and data into RAM, thereby creating the operating program. When the human-guided manipulative sequence is complete and the robot has been returned to home position, the operating program is also complete—the robot has "learned" its program. Then when it is switched from TEACH mode into RUN mode, the robot will perform that manipulative sequence repetitively, as we know.

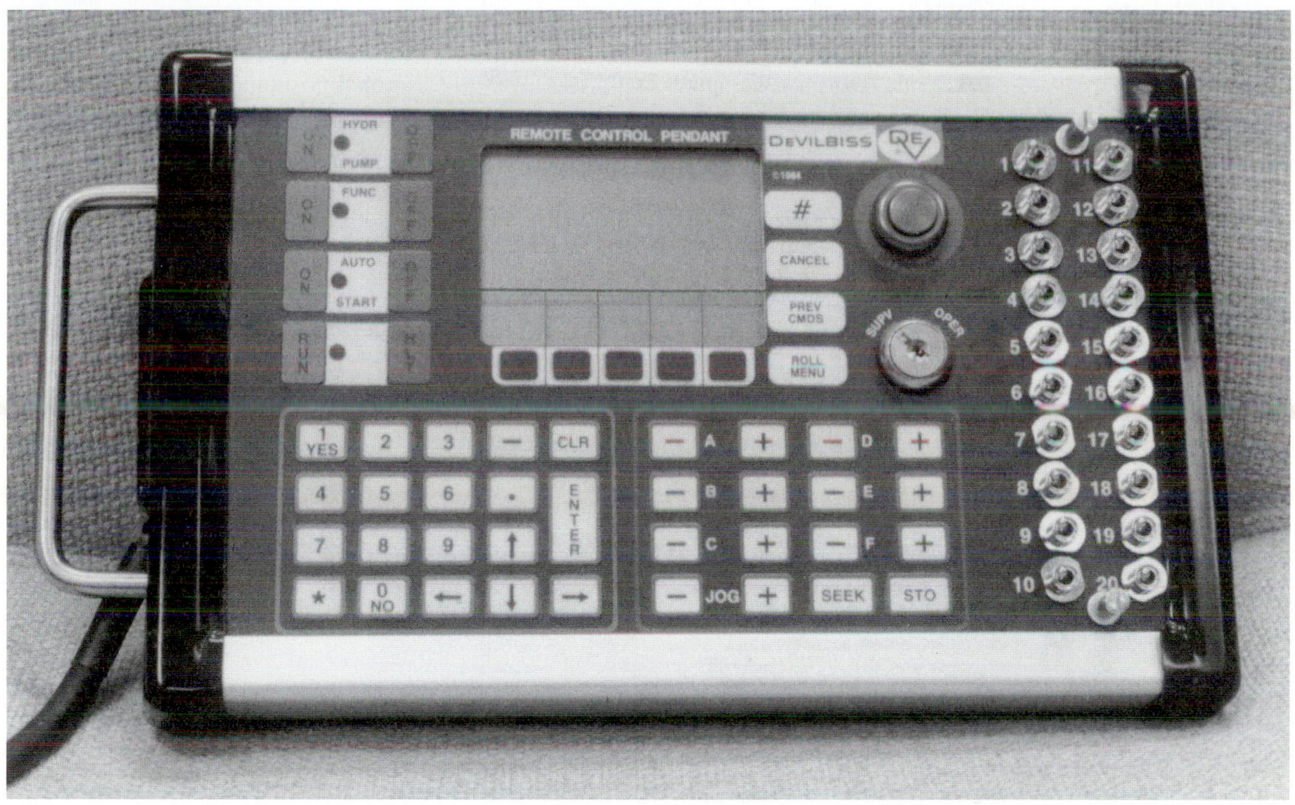

PHOTO 19–3
Teach pendant for a DeVilbiss/Trallfa spray-painting robot.
Courtesy of the DeVilbiss Company.

19-6 ■ CONTINUOUS-PATH PROGRAMS

A continuous-path program is like a point-to-point program, but with the destination positions very close together. That is, a continuous-path program has a position-comparison loop similar to the one listed in Table 19–4. The continuous-path program is able to move the tool device to a destination position very quickly though, and to move it via a virtually invariable path, for the simple reason that the initial actual position of each axis is very close to the destination position.

The usual teaching process for a continuous-path robot is this: The manipulative sequence, or path, is broken down into a great number of tiny compound motions by having the monitor program rapidly sample the various position encoders as the tool device is maneuvered through its desired path by a human. Understand that the execution of the microcomputer's monitor program's sampling routine is spectacularly fast compared to the human-propelled motion that is being sampled. The microcomputer has a clock speed of at least 1 MHz (6800-series), so one machine cycle takes less than 1 μs. It requires probably four microcomputer machine cycles, elapsed time of 4 μs, to read an 8-bit position encoder into a microprocessor register from an explicit address (the address of the input buffer). To write that position data into the RAM using indexed addressing requires probably six machine cycles, using another 6 μs, for an elapsed time of 10 μs per axis position. Even if the machine has six degrees of freedom and the tool device is continually sampled as well, the total elapsed time for a complete robot sample is just 70 μs. At that rate, the microcomputer could take over 14 000 robot-position samples per second. You see why the destination positions can't be too far away from the starting positions.

In actual practice, the axis positions are usually resolved to 12 or more bits by the encoders, so the microcomputer would need more time than outlined above. However, adequate continuous-path control can be achieved by sampling the entire robot 10 to 60 times per second,* so there's still time to spare, which the monitor program must burn up in a delay routine. A simplified flowchart of a continuous-path monitor program is shown in Fig. 19–9.

When the robot is switched into RUN mode and begins its path-reproducing action, its speed of motion is controlled by a timing routine in the operating program. Even though the robot may get to a destination position more quickly than did the human teacher, the operating program can make it wait there until the software timer times out. By adjusting

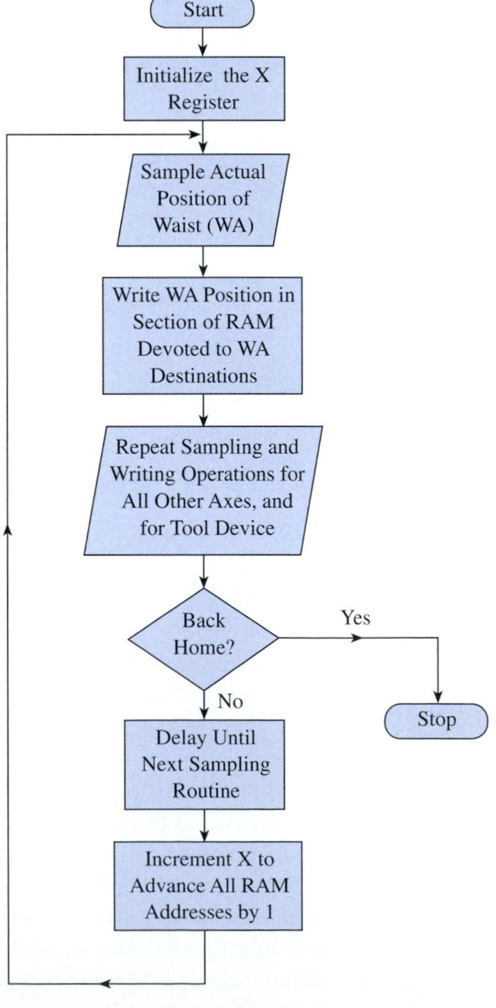

FIGURE 19–9
Simplified flowchart of the monitor program that executes when a continuous-path robot is in its TEACH mode. The robot manufacturer supplies the monitor program, usually in a true ROM. Because such a monitor program cannot be altered, it is not really software; neither is it hardware. It is described as *firmware*.

*Some monitor programs enable the user to select the desired sampling rate.

the duration of the program's software timer, it is possible to adjust the speed at which the robot traces the path. For instance, painting robots are usually programmed to operate at the same speed as the human teacher, but seam-welding robots often can operate faster than the human teacher, depending on the weld parameters.

A simplified flowchart of a continuous-path operating program is shown in Fig. 19–10.

FIGURE 19–10 Simplified flowchart of the (software) operating program for a continuous-path robot. This program executes when the robot is in its RUN mode. The magnitudes of position errors are not calculated, so this robot does not have proportional servo control.

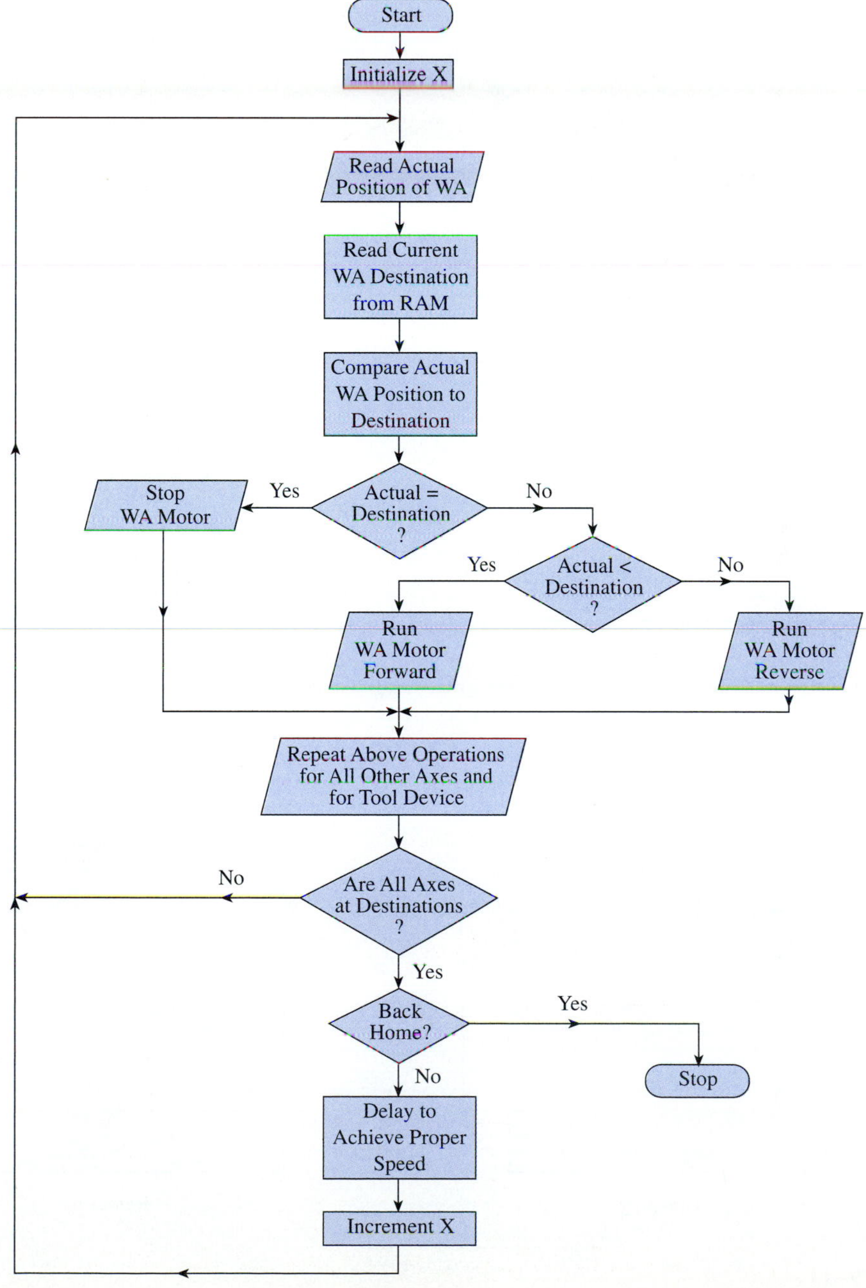

19-7 ■ MECHANICAL GRIPPERS

In material-handling and assembly applications a robot must grasp a piece of work or a part and move it to another location. The robot's end-of-arm gripper must be able to (1) grasp the object, (2) sense the fact that it has been successfully grasped, (3) move the object and perhaps hold it steady while it is operated on by some other tool, and (4) release the object cleanly.

The most straightforward way to accomplish these tasks is by the use of a two-finger gripper. Figure 19–11(a) shows the structure of a parallel-acting two-finger gripper; an

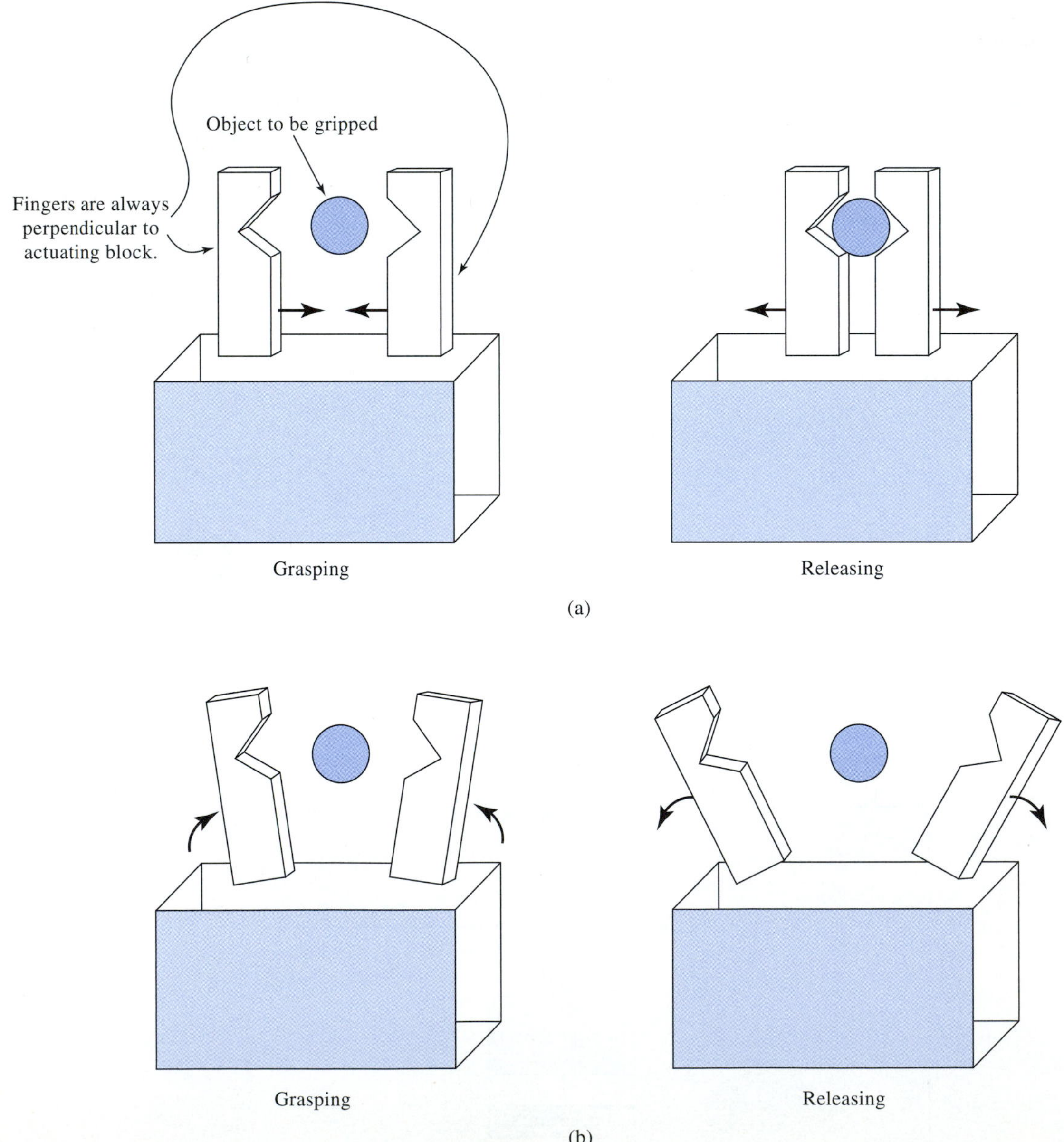

FIGURE 19–11
Finger grippers can have a parallel construction (a), or an angular construction (b).

angular-action version is illustrated in Fig. 19–11(b). The actuating energy source can be an electric motor, a pneumatic cylinder, or a hydraulic cylinder, like a robot axis. For grippers, pneumatic is the most common of the three.

Some objects are better grasped with three fingers or four fingers. Such gripper constructions are shown in Fig. 19–12(a) and (b). The mechanical linkage between actuating cylinder rod and gripper fingers is shown in Fig. 19–13. It is to our advantage to maximize the distance between the cylinder rod and the fingers' pivot points. Such a design maximizes the gripping torque, thus also maximizing the gripper force for holding the

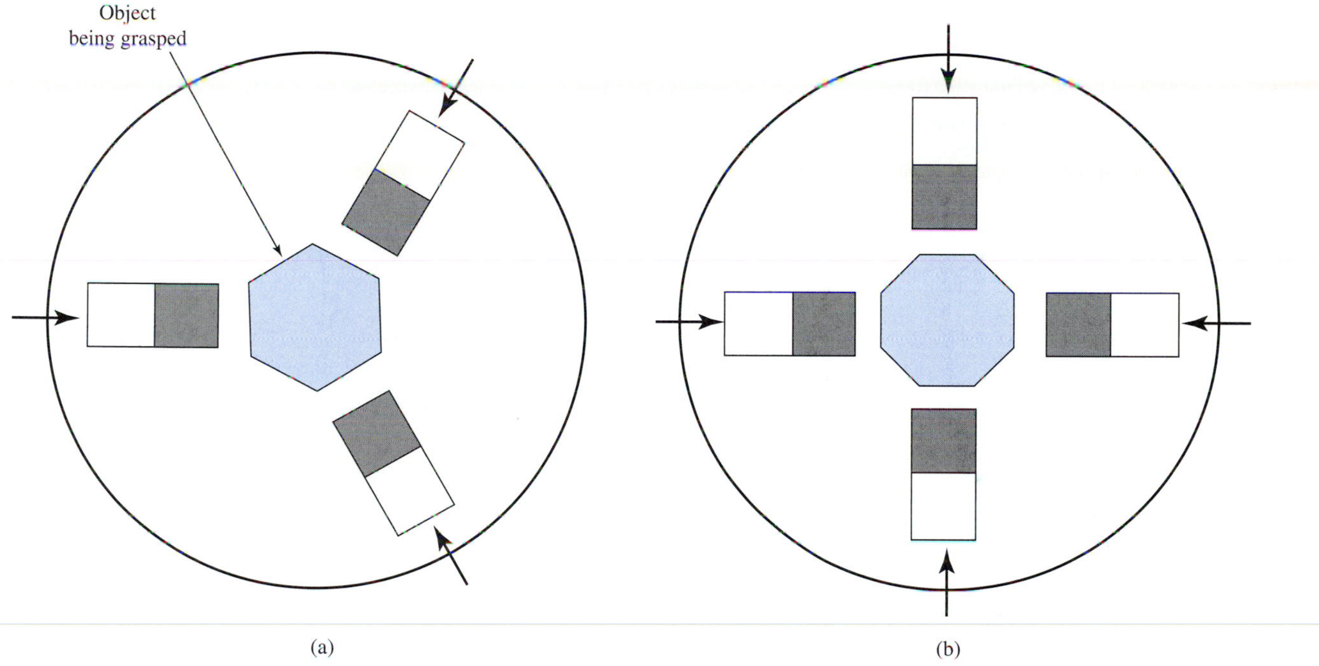

FIGURE 19–12
Shape of the object dictates the optimum number of fingers in a robot gripper. A hexagon cross section (a) is best handled by three fingers; an octagon (b) by four fingers.

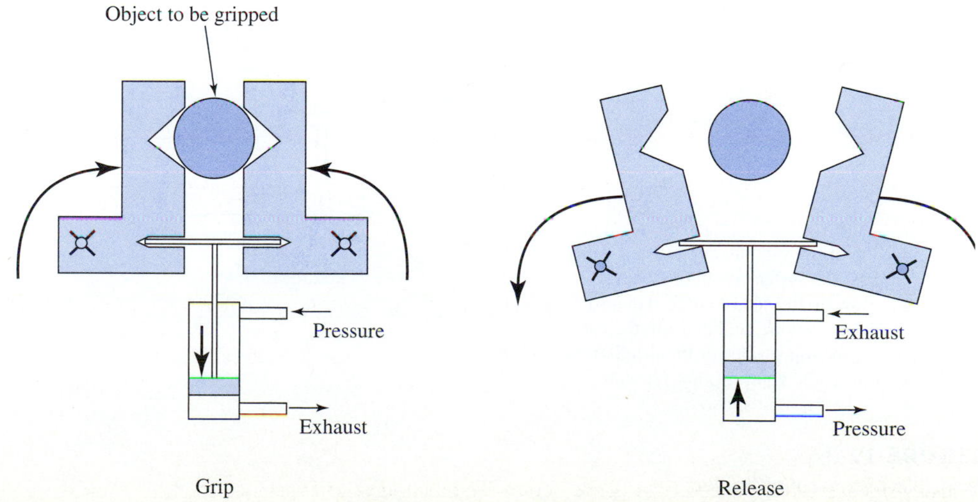

FIGURE 19–13
Detail of actuation mechanism for an angular-acting gripper.

object. In applications where large disruptive forces are exerted on the object while it is being grasped, gripping force is of course the prime consideration.

19-8 ■ VACUUM HOLDERS

It is a simple matter to create a partial vacuum (an absolute pressure lower than the standard atmospheric pressure at the earth's surface) by forcing compressed air to flow through a venturi-type restriction. This is illustrated in Fig. 19–14. Figure 19-14(a) shows the situation before the vacuum cup has pressed against the surface of the object to be lifted. Bernoulli's principle (refer to Sec. 10-14-4 dealing with pressure-drop flowmeters, especially Figs. 10–40 and 10–41) asserts that the static air pressure at the venturi neck will be lower than the pressure of the compressed air at the entry port on the left side of the

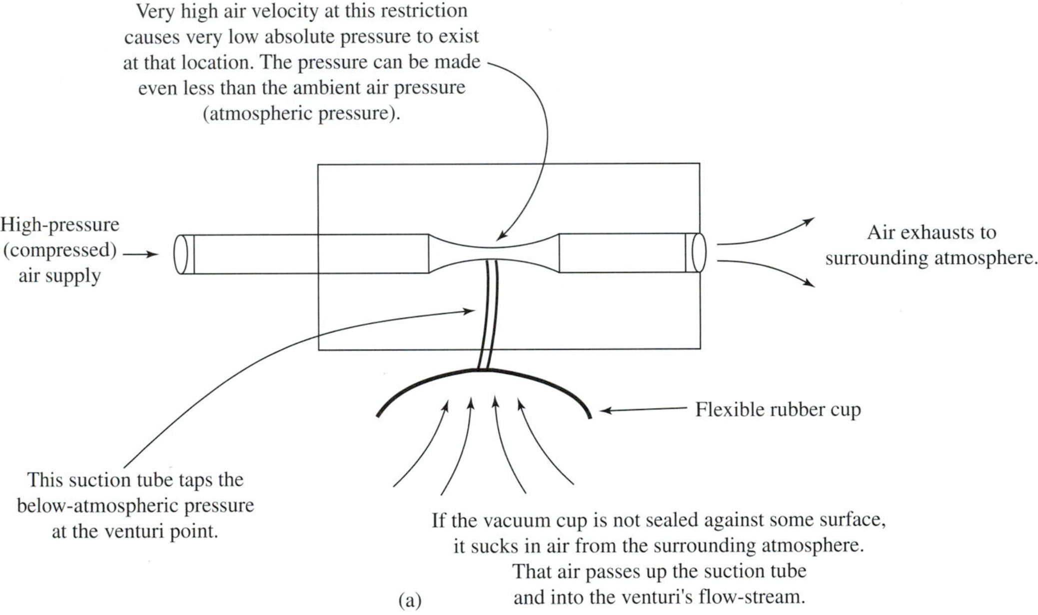
Very high air velocity at this restriction causes very low absolute pressure to exist at that location. The pressure can be made even less than the ambient air pressure (atmospheric pressure).
High-pressure (compressed) air supply
Air exhausts to surrounding atmosphere.
Flexible rubber cup
This suction tube taps the below-atmospheric pressure at the venturi point.
If the vacuum cup is not sealed against some surface, it sucks in air from the surrounding atmosphere. That air passes up the suction tube and into the venturi's flow-stream.
(a)

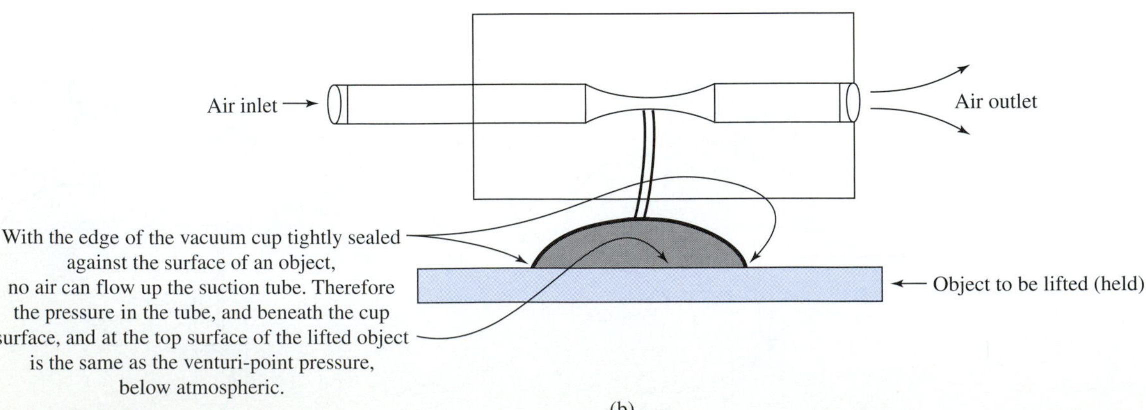
Air inlet
Air outlet
With the edge of the vacuum cup tightly sealed against the surface of an object, no air can flow up the suction tube. Therefore the pressure in the tube, and beneath the cup surface, and at the top surface of the lifted object is the same as the venturi-point pressure, below atmospheric.
Object to be lifted (held)
(b)

FIGURE 19–14
Principle of a vacuum holder.

vacuum device. It remains for us now to make the further assertion that if the venturi effect is drastic enough (if the neck diameter is very small) the static pressure at that point can be even lower than the standard atmospheric pressure of the surrounding air. Therefore it is intuitively clear that the cup will suck air, as part (a) indicates.

When the outer rim of the vacuum cup is pressed against a smooth surface on the object, air will no longer be sucked. Instead the pressure inside the cup, including on the object's surface, becomes equal to the venturi's static pressure. This is pictured in Fig. 19–14(b). But the venturi pressure is below atmospheric. Therefore there is a net upward force on the object. This force is equal to the difference between the upward ambient pressure-related force and the downward vacuum pressure-related force inside the cup, as shown in Fig. 19–15.

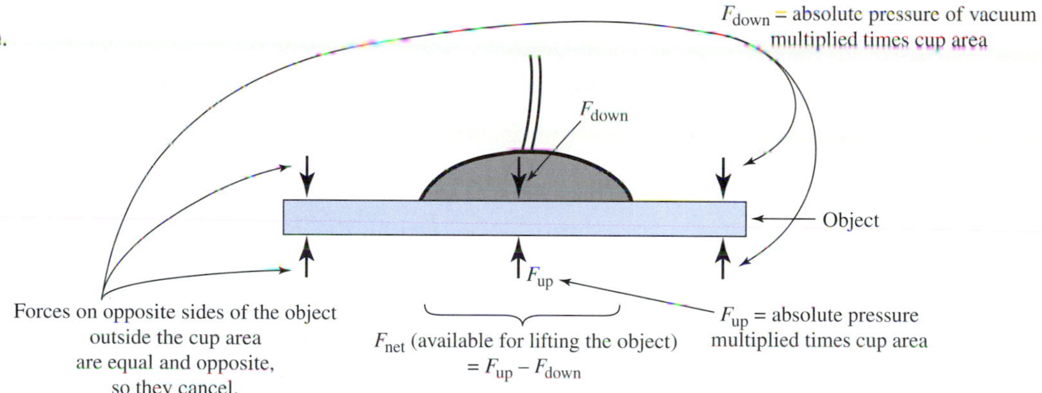

F_{down} = absolute pressure of vacuum multiplied times cup area
F_{down}
Object
F_{up}
Forces on opposite sides of the object outside the cup area are equal and opposite, so they cancel.
F_{net} (available for lifting the object) $= F_{up} - F_{down}$
F_{up} = absolute pressure multiplied times cup area

FIGURE 19–15
Origin of the lifting force.

■ EXAMPLE 19-1

In Fig. 19–15, suppose the vacuum suction cup has a diameter of 3.5 in. Suppose also that the venturi pressure is 0.7 atmosphere; that is, the absolute static pressure is less than atmospheric, equal to 0.7 times the atmospheric pressure.

(a) Find the lifting force available from this single vacuum cup.
Earth's atmospheric pressure is approximately 101×10^3 N/m^2 (newtons per square meter), called Pascals, symbol Pa. That is,

$$1 \text{ atm} \approx 101 \times 10^3 \text{ Pa}$$

or

$$1 \text{ atm} \approx 101 \text{ kiloPascals} \quad \text{(101 kPa, pronounced "kips")*}$$

The conversion factor between inches and meters is 1 m = 39.4 in.

(b) Can the vacuum cup reasonably be expected to hold on to and maneuver an object with a weight equal to that value of lifting force? Explain.

Solution. (a)

$$\begin{aligned} F_{net} &= F_{up} - F_{down} \\ &= (P_{atmosphere})(\text{Area of cup}) - (P_{vacuum})(\text{Area of cup}) \\ &= (P_{atmosphere} - P_{vacuum}) \times \text{Area of cup} \end{aligned}$$

*In the outdated English measurement system, the equivalent value is 14.7 pounds/in^2, or 14.7 psi.

Since $P_{\text{vacuum}} = 0.7(P_{\text{atmosphere}})$, we can say

$$F_{\text{net}} = (P_{\text{atmosphere}} - 0.7\,P_{\text{atmosphere}}) \times \text{Area of cup}$$
$$= P_{\text{atmosphere}} \times (1.0 - 0.7) \times \text{Area of cup}$$
$$F_{\text{net}} = (0.3) \times P_{\text{atmosphere}} \times \text{Area of cup} \qquad \textbf{(19-1)}$$

The area of the cup is given by $A = \frac{\pi}{4} d^2$ in which d is the diameter of the cup. In SI units,

$$d = 3.5\text{ in.} \times \frac{1\text{ meter}}{39.4\text{ in.}} = 0.088\,8\text{ m}$$

so

$$A = \frac{\pi}{4} d^2 = \frac{\pi}{4}(0.088\,8\text{ m})^2 = \frac{\pi}{4}(7.89 \times 10^{-3})\text{ m}^2$$
$$= 6.20 \times 10^{-3}\text{ m}^2 \quad \text{(square meters)}$$

Substituting this value for area into Eq. 19-1 produces

$$F_{\text{net}} = (0.3) \times P_{\text{atmosphere}} \times 6.20 \times 10^{-3}\text{ m}^2$$
$$= (0.3) \times 101 \times \cancel{10^3}\frac{\text{N}}{\cancel{\text{m}^2}} \times 6.20 \times \cancel{10^{-3}}\,\cancel{\text{m}^2}$$
$$= \mathbf{188\ N}$$

In English pounds,

$$188\text{ N} \times \frac{1\text{ lb}}{4.45\text{ N}} = \mathbf{42.2\ pounds}$$

Conversion factor 4.45 N/1 lb

(b) The vacuum cup could grip and slowly lift an object weighing 188 newtons, but it could not accelerate the object (could not change its speed or direction). Acceleration requires application of a net force equal to the object's mass multiplied by the intended acceleration ($F = mA$). Therefore this vacuum force is not sufficient to maneuver a 188-N object through a real industrial activity. For realistic industrial handling, this vacuum device with 188 N of gripping force would be limited to objects weighing less than 188 N. ■

Vacuum surface. If the venturi tap tube is terminated in a manifold with many small holes, the manifold's multihole surface is called a *vacuum surface*. It is drawn in Fig. 19–16. A vacuum surface is well adapted to pickup of nonrigid objects, such as pieces of cloth or paper.

FIGURE 19–16
Vacuum surface.

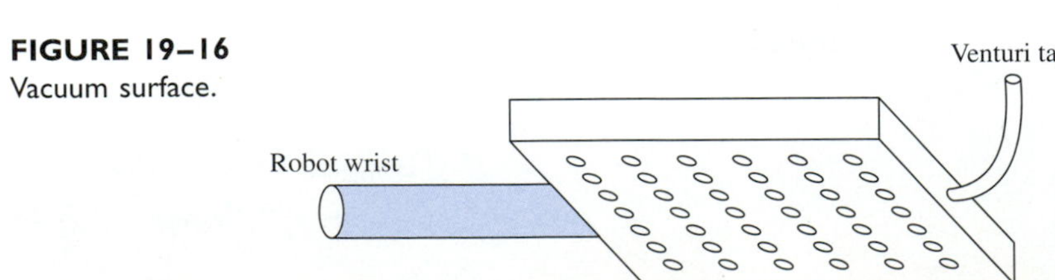

19-9 ■ PNEUMATIC GRIPPERS

Bellows grippers. Air pressurized above atmospheric pressure also can be applied to the task of gripping parts. Figure 19–17 shows the general structure of such a gripper. In Fig. 19–17(a), with the inlet port not pressured up, the springiness of the bellows material returns it to its natural shape, moving the gripping surfaces toward the outside. When air pressure is applied to the inlet port, shown in Fig. 19–17(b), the bellows expand and cause the friction surfaces to move to the inside, gripping the part.

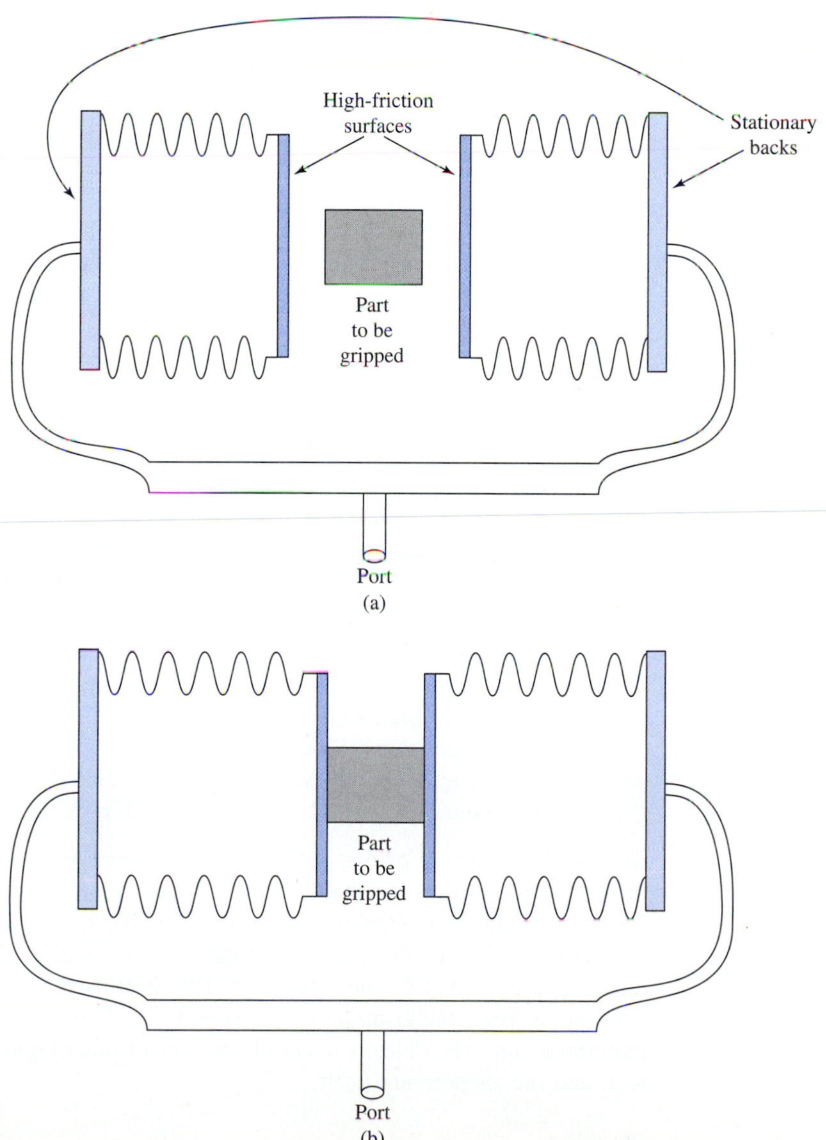

FIGURE 19–17
(a) Pressure released, bellows retract to normal, releasing the part.
(b) Pressure applied, bellows extend, moving friction surfaces into contact with part.

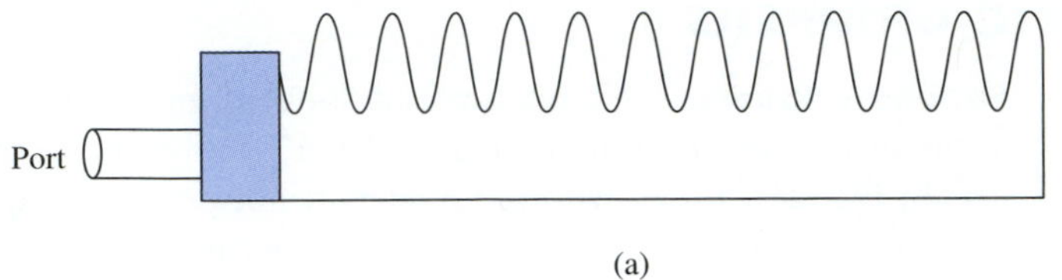

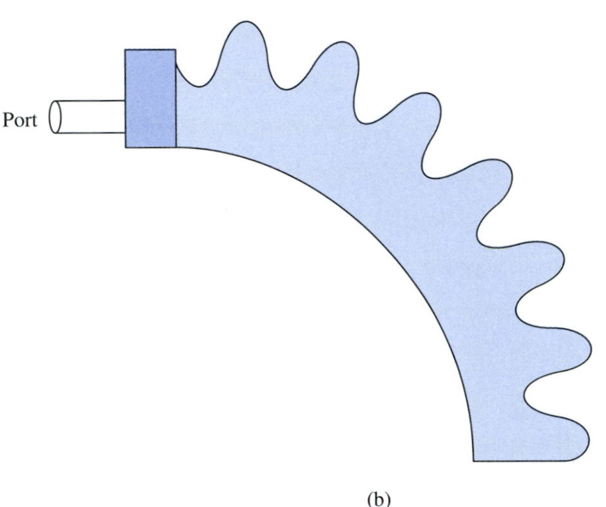

FIGURE 19–18
Bellows finger. (a) Air pressure released, bellows side of cylindrical finger retracts, straightening the finger. (b) Pressure applied, bellows side expands (lengthens), causing finger to bend.

Bellows finger. A variation on the standard bellows idea is the *bellows finger,* shown in Fig. 19–18. A bellows finger is an air-tight cylindrical tube with one side of the tube smooth and the other side made of bellows material. The natural shape of the bellows is such that the cylindrical tube is straight when the air pressure inside is simple atmospheric. This is shown in Fig. 19–18(a). When air pressure is applied in Fig. 19–18(b), the ribbed side can expand and lengthen while the length of the smooth side is fixed. Therefore the tube must bend like a finger.

Bellows fingers are seldom used singly; they tend to be in pairs. Thus, a pair of fingers with high-friction surfaces can grip a part directly, as suggested in Fig. 19–19(a). But they are better adapted to holding a flanged part, as shown in Fig. 19–19(b).

Mandrel and pin grippers. A *mandrel gripper* has an air-tight expandable diaphragm on the outside surface of a mandrel that is inserted into a cavity on the object to be gripped. This is illustrated in cutaway view in Fig. 19–20. A *pin gripper* is the reverse of a mandrel gripper. The object itself must have a protruding "pin," over which the concave diaphragm is placed by the robot arm. This is illustrated in Fig. 19–21. As for all such friction grippers, the gripping force depends on the friction coefficient of the diaphragm material against the object's material, the area of diaphragm that is in contact with the object, and the air pressure itself.

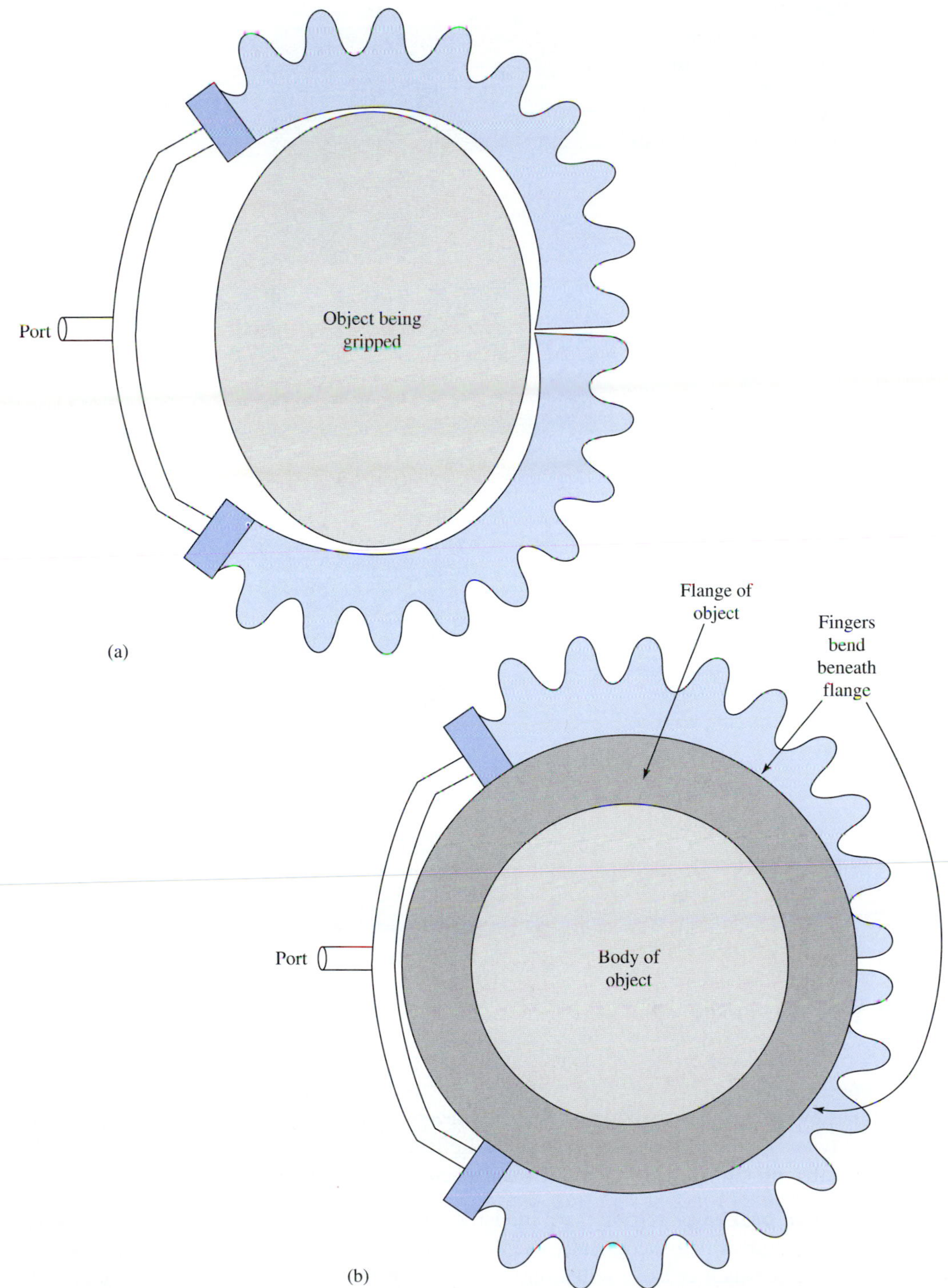

FIGURE 19–19
(a) Fingers gripping a nonflanged object. Grip depends on friction between surfaces. (b) Gripping a flanged object.

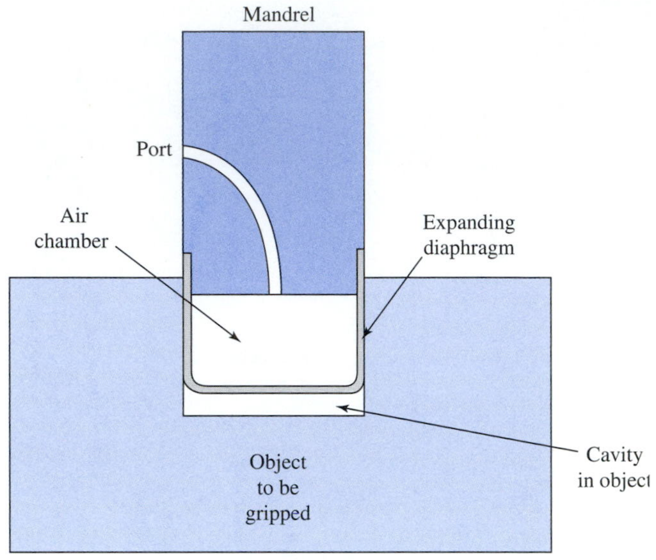

FIGURE 19–20
The robot arm inserts the mandrel gripper into the object's cavity. When the inlet port is pressured up, the diaphragm expands so that its friction surface touches the walls of the cavity.

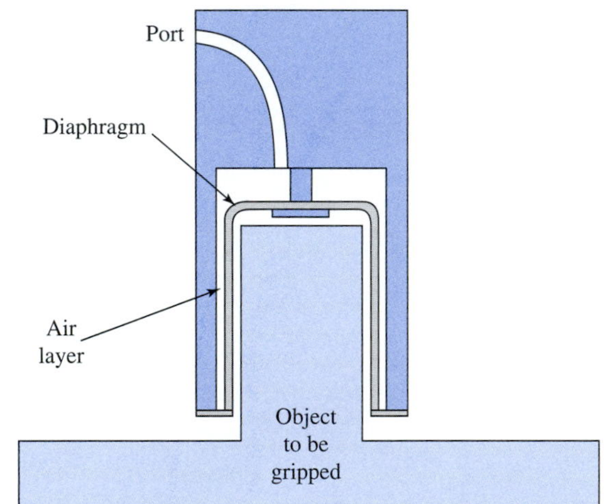

FIGURE 19–21
The robot arm places the pin gripper over the object's pin. When the inlet port is pressured up, the diaphragm expands so that its friction surface touches the walls of the pin.

19-10 ■ PROXIMITY SENSORS

A robot usually must detect the presence of an object before actuating its gripper. Sometimes the detection is done by contact-type devices (the limit switch and its relations), sometimes by photoelectric methods, and sometimes by a proximity sensor. There are three general types of proximity sensor: (1) The Hall-effect sensor described in Sec. 10-13-2, (2) the inductive proximity sensor, and (3) the capacitive proximity sensor.

Inductive proximity sensor. An inductive proximity sensor is an *LC* (inductor/capacitor) oscillator, which successfully oscillates when there is no metal or conductive object nearby, but which ceases oscillation when a conductive object enters the magnetic field produced by the inductor.

For example, Fig. 19–22(a) shows an Armstrong oscillator design (transformer used to couple the output signal back to the input). The spatial layout of the primary winding (inductor) and secondary winding (inductor) is portrayed in Fig. 19–22(b). If a metal conductive object enters the space between the primary inductor L_P and the secondary

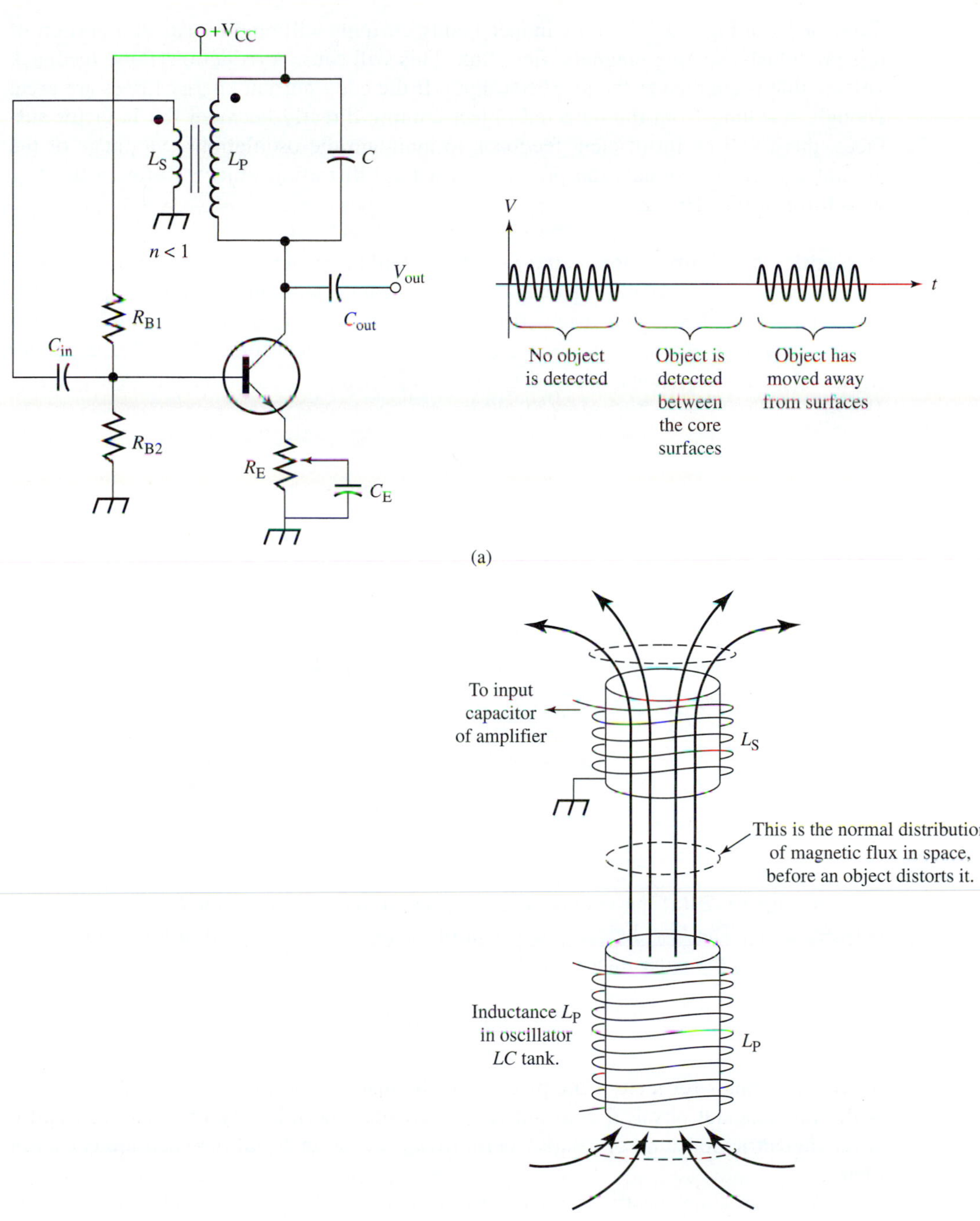

FIGURE 19–22
Inductive proximity sensor.
(a) *LC* tank in output of an amplifier, with a portion of the *LC* oscillation voltage fed back in positive synchronization with the input (positive feedback). (b) Normal, flower-like magnetic field distribution.

inductor L_S in Fig. 19–22(b), its induced eddy currents will tend to cancel a portion of this previously existing magnetic field flux. This will cause a reduction in the feedback voltage that is coupled to the amplifier input. If the eddy-current energy losses are great enough, resulting from the detected object coming directly between the inductor surfaces, there will be insufficient feedback to maintain the oscillations. Stoppage of the natural oscillations signals the presence of a field-distorting object. Refer to the V_{out} waveform in Fig. 19–22.

Capacitive proximity sensor. It is possible to build a capacitance-based proximity sensor by combining a variable-capacitance *LC* tank circuit with an op amp bandpass filter. The value of variable capacitance is determined by the closeness of a detected object between two widely spaced plates of a capacitor, called the detection plates. The capacitor structural situation is shown in Fig. 19–23. A simplified circuit schematic is shown in Fig. 19–24.

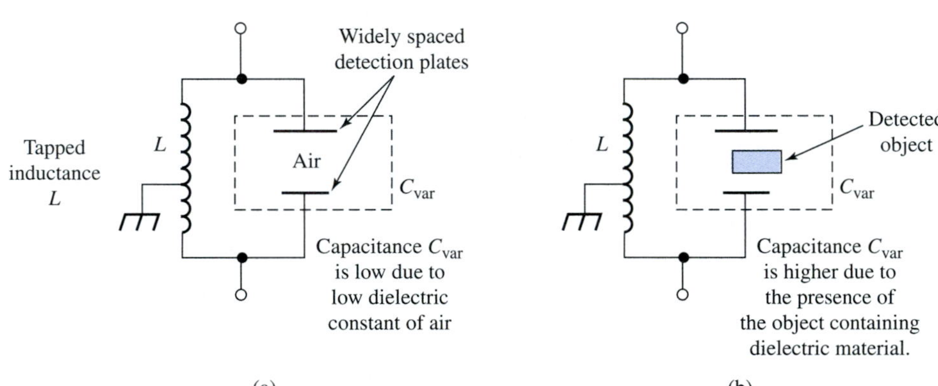

FIGURE 19–23 Variable capacitance *LC* tank oscillator in a capacitive proximity detector.
(a) With no object present, *C* is low so oscillation frequency f_r is high.
(b) With an object present, *C* is high so oscillation frequency f_r is low.

In Fig. 19–23(a) the widely spaced plates of variable capacitor C_{var} have only air between them. The low dielectric constant of air ($\kappa \approx 1.0$) results in a low value of capacitance, from the capacitance-defining formula

$$C = (8.85 \times 10^{-12})\kappa \frac{A}{d} \qquad \textbf{(19-2)}$$

in which *A* stands for area of the plates, *d* is the distance between plates, 8.85×10^{-12} is the fundamental physical constant for the absolute permittivity of a vacuum, and κ is the dielectric constant, or relative permittivity, of the material between the capacitor plates.

Air, consisting mostly of nonpolarized diatomic nitrogen molecules (N_2) has essentially the same permittivity as free space (a vacuum), so it multiplies that absolute constant by about 1.0. That is, air has a dielectric constant of about 1.0. If an object with polarized molecular structure ($\kappa > 1$) enters the space between the capacitor plates, as shown in Fig. 19–23(b), capacitance C_{var} rises per Eq. (19-2).

The natural (resonant) oscillation frequency of the *LC* tank combination is given by the equation

$$f_r = \frac{1}{2\pi\sqrt{LC}} \qquad \textbf{(19-3)}$$

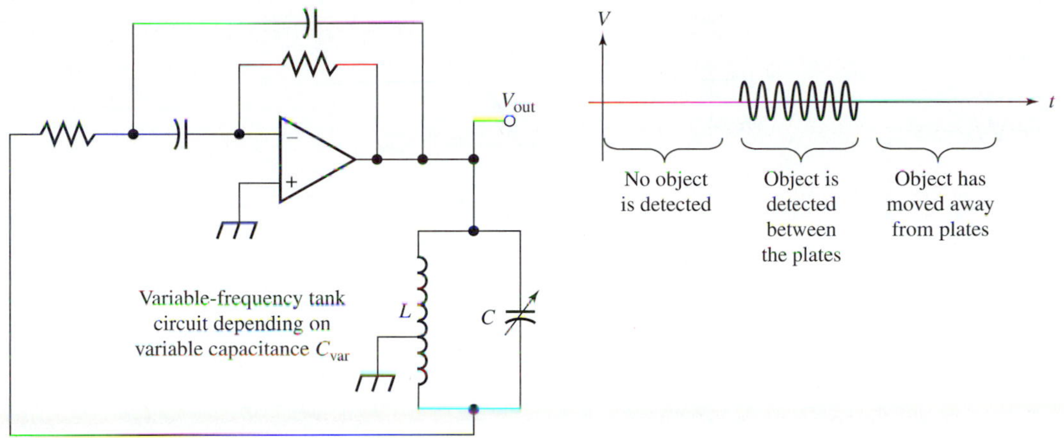

FIGURE 19–24
Complete oscillator circuit consisting of variable-frequency *LC* tank and an op amp active band-pass filter to close the loop.

so it is mathematically clear that the presence of an object between the plates causes oscillation frequency to decrease. Figure 19–24 shows how the oscillating *LC* tank circuit can be connected to a band-pass op amp active filter to form a sustained oscillation loop. The *R* and *C* values of the op amp circuit are chosen so that the pass band of the filter includes the lower oscillation frequency (Fig. 19–23b), but does not include the higher oscillation frequency (Fig. 19–23a). This frequency relationship is indicated in Fig. 19–24. Thus, there is sufficient voltage gain in the op amp filter to satisfy the oscillation criterion of loop gain >1 for the lower frequency (with the object present), but the op amp does not provide sufficient voltage gain when the object is absent. Therefore, monitoring the V_{out} terminal in Fig. 19–24, the presence of sine-wave oscillations indicates that an object is being detected. The absence of oscillations indicates that no object is present.

Op amp active filtering. An op amp inverting amplifier can be made into a low-pass filter simply by placing a capacitor in the feedback position. This is shown in Fig. 19–25(a), with the corresponding frequency-response curve shown in Fig. 19–25(b). Figure 19–25 can be understood intuitively as follows: At low frequencies the capacitive

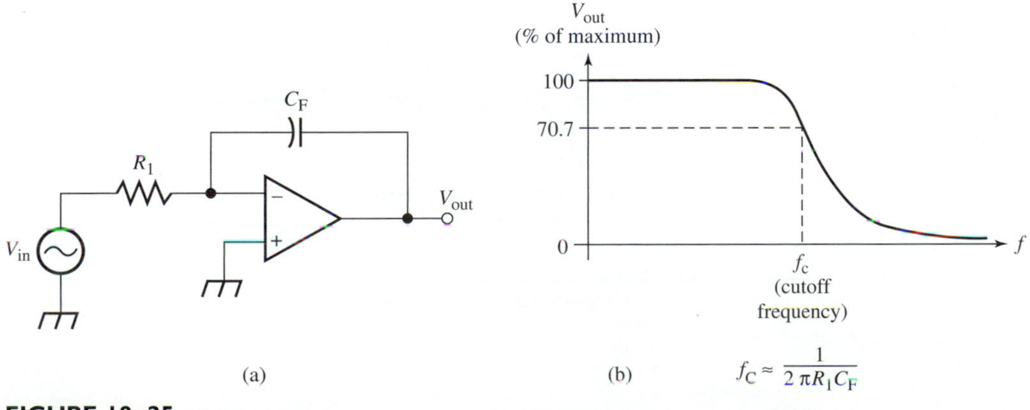

FIGURE 19–25
Low-pass active filter. (a) Schematic arrangement. (b) Response curve.

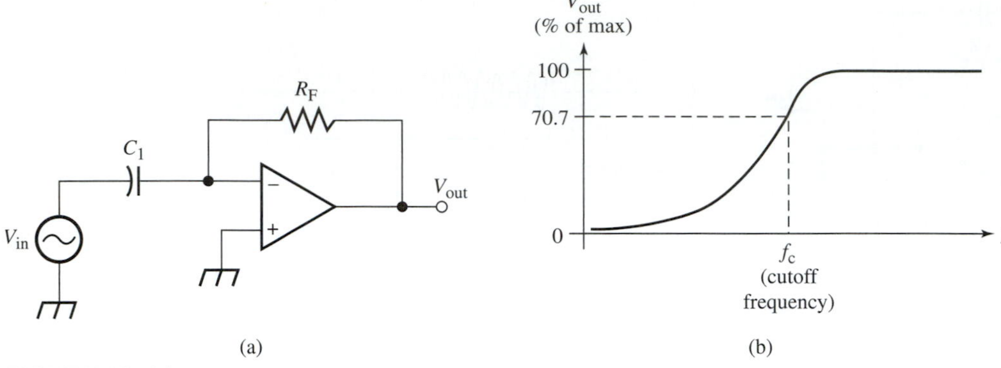

FIGURE 19–26
High-pass active filter. (a) Schematic. (b) Response curve.

reactance X_{CF} is large; therefore, the ohmic value of the feedback impedance is large and the voltage gain of the circuit tends to be large, from

$$A_v = \frac{R_F}{R_1} \tag{8-1}$$

At high frequencies the reactance decreases, so the op amp amplifier circuit has a low ohmic value in its feedback position. By Eq. (8-1), the voltage gain declines and with it declines the value of V_{out}. These response facts are illustrated in Fig. 19–25(b).

A high-pass active filter is obtained by swapping the positions of the capacitor and resistor. This is illustrated in Fig. 19–26. By combining both *RC* combinations on the same op amp as shown in Fig. 19–27(a), a band-pass filter can be constructed. It is necessary to choose R_1 and C_F to produce a low-pass cutoff frequency that is higher than the high-pass cutoff frequency set by C_1 and R_F. Thus there is an active high-gain band between the cutoff frequencies, as shown in Fig. 19–27(b).

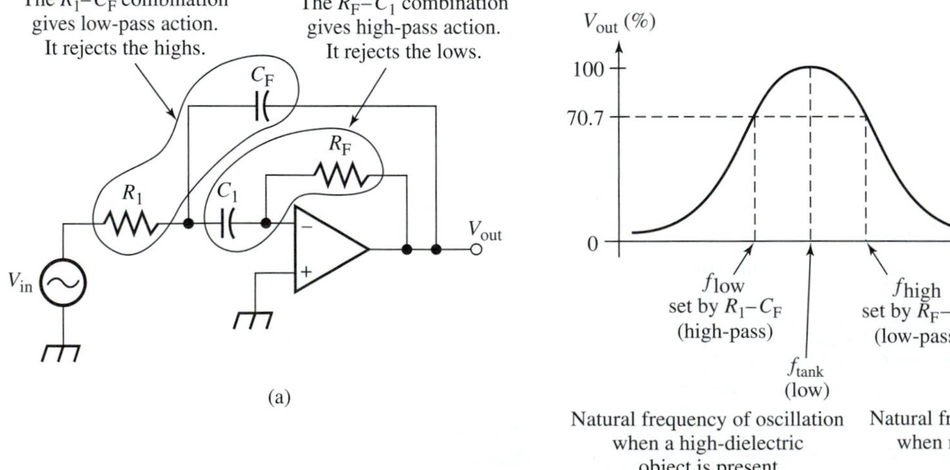

FIGURE 19–27
Bandpass filtering. (a) Schematic. (b) Freqeuncy-response curve.

TROUBLESHOOTING ON THE JOB

UNDERSTANDING ROBOT MALFUNCTIONS

You have worked successfully several times on the point-to-point robot described in Sec. 19-5, and you are developing a reputation as a knowledgeable fellow. In anticipation of your leaving on a 2-week vacation, your supervisor has asked you to write out an explanation of the causes of certain robot malfunctions that occasionally happen. It ought to be possible for less knowledgeable employees to deal with possible trouble while you're gone, your supervisor believes, if they can work with a straightforward explanation of such malfunctions.

It becomes apparent to you during your conversation with him that your supervisor isn't quite the unobservant dolt that you thought. He has noticed, over time, that the robot's rare malfunctions fall into one of the following four categories:

1. A particular axis positions itself too far from the home position on every motion of the program. This distance between the actually achieved axis position and the desired axis position is constant. For instance, the waist positions itself 15° beyond the desired programmed position, always the same amount, on each and every motion.
2. The opposite of Problem 1. A particular axis, say the waist, positions itself too close to the home position, always by the same constant amount, on each and every motion.
3. After the robot has been reprogrammed to perform a new task, it consistently skips one of its programmed motions. For instance, after it makes its seventh compound motion and reaches the correct seventh position (all axes OK), it then proceeds to make the *ninth* motion of the program. It never makes the expected eighth motion.
4. After initially functioning correctly, the robot begins to go to an incorrect spatial position on a particular motion, consistently. It then gets back on track with the following motion. For instance, after the seventh compound motion is made correctly, the robot goes to an unintended position on its eighth motion, always the same position. Then, on its ninth motion, it goes to the proper ninth position.

YOUR ASSIGNMENT

From your experience and your understanding of the robot's controls, provide a written explanation of the cause of each of these four problems. Your fellow employees, who might have to read these explanations, will have access to the I/O architecture diagram, Fig. 19–6, and to the destination position memory map, Fig. 19–7, and also to the position-comparison routine flowchart, Fig. 19–8. Therefore, refer to these figures as needed to make your explanations clear.

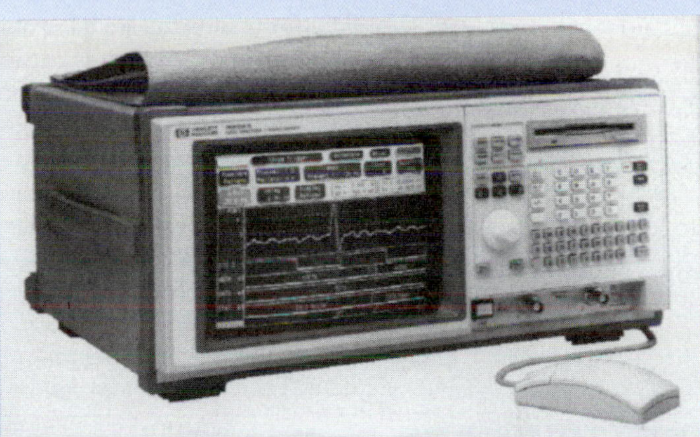

The logic analyzer is again the instrument of choice for troubleshooting computer-related problems.

Courtesy of Hewlett-Packard Company.

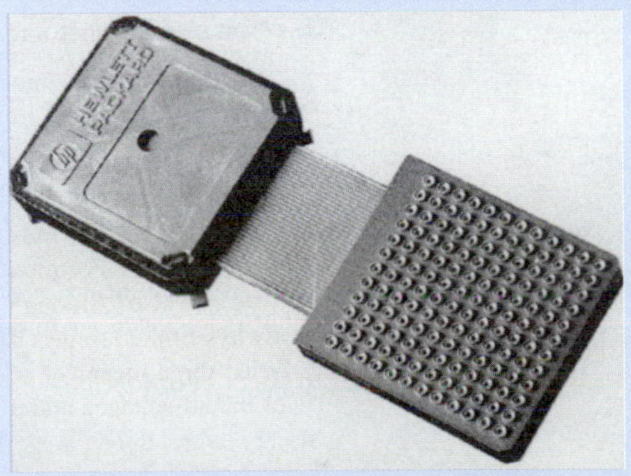

HP's low-profile adapter for probing surface-mount microprocessors.

Courtesy of Hewlett-Packard Company.

■ SUMMARY

- Robots are useful where a tool must be manipulated and the manipulation path must be easily adjustable.
- There are three basic mechanical configurations of industrial robots: articulated arm, spherical, and cylindrical.
- The number of independent motions (axes) that a robot can accomplish is called its number of degrees of freedom.
- There are three means of actuation for a robot: electric motor, hydraulic cylinder, and pneumatic (air) cylinder.
- Robots can be categorized by the sophistication of their computer software, which determines the sophistication of their mechanical motion. In ascending order, the categories used by us are (1) positive-stop, (2) point-to-point, and (3) continuous-path.
- Positive-stop software is capable of moving any single degree of freedom between two extreme positions only. It cannot stop at a point in between.
- Point-to-point software is capable of moving any single degree of freedom to any intermediate point between the two extreme positions. However, point-to-point software cannot guarantee such close motion synchronization among the various degrees of freedom that the tool follows a precisely known path as it moves from one spatial point to the next spatial point.
- Continuous-path software is like point-to-point software with the additional feature that the successive destination points for the degrees of freedom are so close together that there is virtual synchronization among the motions of the various degrees of freedom. Therefore the tool follows a precisely known path as it moves through space.

■ QUESTIONS AND PROBLEMS

Section 19-1

1. What distinguishes a robot from standard automated manipulative machinery?

Section 19-2

2. What are the three most common mechanical configurations for industrial robots?
3. What is meant by a robot's number of degrees of freedom?
4. True or False: For most robots, waist rotation is limited to less than 360°.
5. True or False: For most robots, the maximum shoulder rotation is about 270°.
6. True or False: Among robots that possess wrist roll capability, the maximum roll is always limited to less than 90°.
7. What three means of actuation are commonly used by industrial robots? State some of the advantages and disadvantages of each.

Section 19-3

8. Name the three categories of robot software.

Section 19-4

9. Consider a positive-stop robot with six degrees of freedom. In how many spatial positions can its tool be placed?

10. Define a compound motion.
11. What is the defining characteristic of a positive-stop control program (positive-stop robot)?

Questions 12 through 17 refer to the data transfer arrangement for a positive-stop program that is sketched in Fig. 19–4.

12. Suppose that the input buffer is read when the positive-stop robot is in the following position:

Waist is clockwise	Elbow is up
Shoulder is down	Gripper is open

    Describe the byte that appears on the data bus. That is, tell the digital level of each data line.
13. Repeat Question 12 for the following robot conditions:

Waist is counterclockwise	Elbow is in motion
Shoulder is up	Gripper is open

14. What byte would be written to the output latch to initiate the compound motion of moving the waist counterclockwise while moving the elbow inward?
15. Would it be legal to write the byte 0000 1010 to the output latch? Explain.
16. Would it be legal to write the byte 1101 0000 to the output latch? Explain.
17. Would it be legal to write the byte 0101 0001 to the output latch? Explain.
18. In Table 19–3, the interrupt service routine determines whether B matches A. As explained in Sec. 19-4-2, we understand "match" to mean that every 1 in A is matched by a 1 in B, but it is not necessary for every 1 in B to be matched by a 1 in A. This matching test is not a straightforward subtraction-type instruction, because subtraction-type instructions are able to indicate only whether A and B are *identical* (every bit the same).

    The 6800-series assembly code for the matching test is given below. Explain how it works.

Mnemonic	*Operand*	*Comment*
LDAB	$4000	Load B from the input buffer.
STAA	$8FFF	$8FFF is an address in RAM. Register A still contains the byte that was last written to the output latch.
ANDB	$8FFF	
CBA		
BEQ	SUBEND	SUBEND is the label of the ROM address that contains the last instruction in the DELAY subroutine (the RTS instruction).
JMP	DELAY	DELAY is the label of the ROM address that contains the first instruction of the DELAY subroutine.

19. If a compound motion is initiated that moves all three axes—waist, shoulder, and elbow—how many times will the program encounter DELAY (the beginning of the DELAY subroutine) before the motion is complete? Explain.
20. For the same motion as in Question 19, how many times will the interrupt service routine be executed? Explain.
21. For the same motion as in Question 19, how many times will the program encounter SUBEND (the end of the DELAY subroutine)? Explain.

Section 19-5

22. What characteristic distinguishes a point-to-point program (robot) from a positive-stop program?
23. In Table 19–4, the actual position of each motion axis is compared to its destination position by subtracting the destination from the actual position. Explain why C = 1 indicates that the axis motor should run forward, and C = Ø indicates that the axis motor should run in reverse.
24. Explain why the program of Table 19–4 and Fig. 19–8 is capable of repositioning an axis that reaches its destination position, stops, and is later knocked out of its destination position.
25. The destination storage pattern shown in Fig. 19–7 is conceptually the simplest pattern to use. However, it has two practical disadvantages: (1) It limits the maximum number of motions in the robot's manipulative sequence, and (2) it ties up a section of RAM which may be much larger than necessary, if the manipulative sequence contains only a small number of motions.
    a. Suggest a different destination storage pattern that would eliminate these disadvantages.
    b. If you are familiar with the 6800-series instruction set (or any other microprocessor instruction set), write the appropriate program instructions for accessing the three axis destinations in RAM during the position-comparing sequence.
26. In Fig. 19–6, we depict the robot's operating program as stored in ROM. Discuss the merits of storing the operating program in RAM instead.
27. What characteristic distinguishes a continuous-path program (robot) from a point-to-point program?

Section 19-6

Questions 28 and 29 refer to Fig. 19–9, which is a flowchart of the monitor program for a continuous-path robot.

28. If the time delay (second box from the bottom) is shortened, does that provide more accurate or less accurate duplication of the human-propelled manipulative path? Explain.
29. What effect does shortening the time delay have on the amount of computer memory that is devoted to storing destination positions?
30. From your answers to Questions 28 and 29, describe the relationship among the following robot parameters:
    a. The available memory in the microcomputer.
    b. The overall distance that the tool must travel in the manipulative path.
    c. The degree of accuracy with which the program can duplicate the human-propelled manipulative path.

CHAPTER

20 SAFETY

It is not enough to have knowledge of industrial electronics systems if you do not also have knowledge about the physical dangers that such systems can present. In this chapter, we study the hazards associated with industrial electricity and electronics systems, and we explain the safety practices that are used to cope with those hazards.

OBJECTIVES

After studying this chapter, you will be able to:

1. State the voltage level that is generally considered to be dangerous to a healthy person.
2. List the personal protection precautions that should be taken when working around industrial feeder and motor voltages (<600 V).
3. State the additional precautions that are necessary when working around very high-voltage distribution circuits (>600 V).
4. Describe the safety hazard posed by large capacitors, and explain the proper procedure for rendering them harmless.
5. Explain the function of a grounding wire in a three-wire single-phase system or a four-wire three-phase system.
6. Describe the action of a ground-fault interrupter (GFI), and explain why GFIs provide greater safety than a simple grounded three-wire system.
7. Describe the preparation that is required for dealing coolly with an electric-shock emergency.
8. Explain the steps that must be taken in an electric-shock emergency to render aid to the victim.
9. Describe the danger associated with highly inductive loads carrying large currents. State the precautions for safely deenergizing such a load.
10. List the common-sense precautions that are necessary for protecting your head and eyes in an industrial setting and for protecting yourself against long-term hearing loss.
11. List and describe the four types of fires that are defined by the National Fire Prevention Association. State which types can be fought with water and which cannot.
12. State the meanings for the following colors under Occupational Health and Safety Administration color-coding regulations: red, yellow, green, orange, and purple.

20-1 ■ ELECTRIC SHOCK

A shock hazard exists whenever the voltage level exceeds approximately 50 V. The rationale for this value can be demonstrated mathematically by combining two facts:

1. Current on the order of 10 mA through chest organs can interrupt the neurologic activity that sustains heartbeat and breathing.
2. Under worst-case conditions, the resistance of human skin (two separate skin-contact points taken in series) can be as low as about 5 kΩ.

Applying Ohm's law to this worst-case scenario,

$$\begin{aligned} V_{\text{danger}} &= I_{\text{lethal}} \times R_{\text{skin}} \\ &= (10 \times 10^{-3}\ \text{A}) \times (5 \times 10^{3}\ \Omega) = 50\ \text{V} \end{aligned}$$

Thus, for example, the 120-V ac line actually is quite dangerous, even though people usually regard it casually. The 240- and 480-V line voltages that commonly are encountered in industrial settings are even more dangerous.

Obviously, a person can be electrically shocked if he touches two supply lines simultaneously, as illustrated in Fig. 20–1(a). Because people tend to be alert to this obvious danger, it seldom happens. The more frequent problem occurs when one side of the supply is connected to the earth (the "grounded" or "common" side) and the person touches the other side (the "hot" side) while some other part of his body is in direct or partial electrical contact with the earth. This situation is shown in Fig. 20–1(b).

We deal with the problem in Fig. 20–1(b) by following these safety practices.

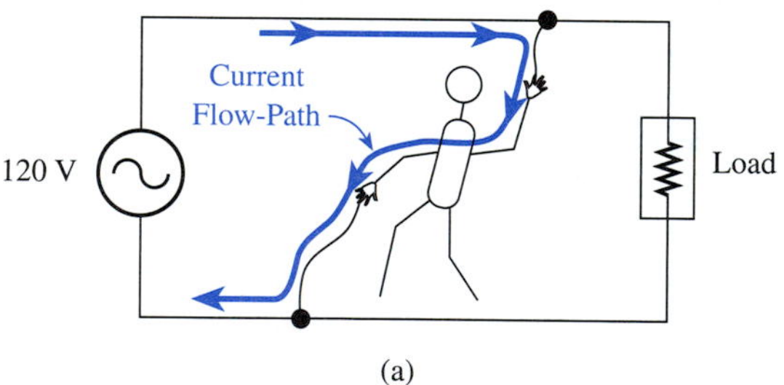

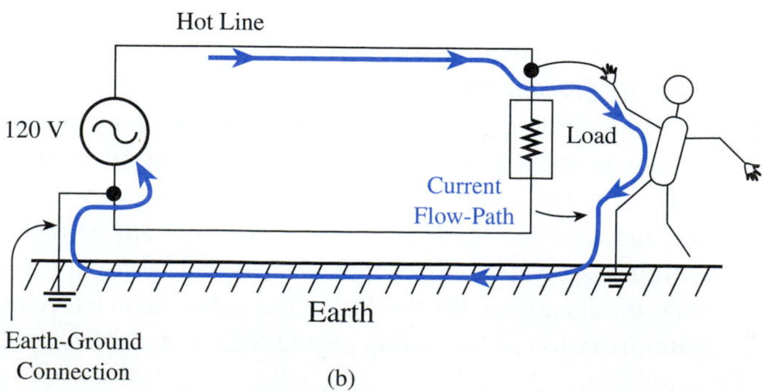

FIGURE 20–1 The danger of electric shock. (a) Touching both wires simultaneously seldom happens. (b) In an earth-grounded system, touching just the hot wire while some other part of the body makes partial contact with the earth also produces a current path through the body.

1. If possible, turn off the power before working on the circuit. Even after the disconnect switch is thrown to Off, check with a voltmeter or voltage-tester that the circuit is truly deenergized.
2. Where large power factor-correction capacitors are present, there may be a dc voltage maintained between the branch load wires even after the disconnect switch has been opened. Such a dc voltage can be of either polarity, and it may or may not be detected by an ac voltmeter, depending on that voltmeter's design. This problem is illustrated in Fig. 20–2. To protect yourself from this situation, connect a dc voltmeter first with one polarity, then with the other. In a three-phase system, six voltmeter measurements are necessary: A to B, both polarities; B to C, both polarities; and C to A, both polarities. If a residual-charge voltage is present, it must be eliminated by discharging the capacitor(s) with an approved discharge clip, as shown in Fig. 20–2(b).

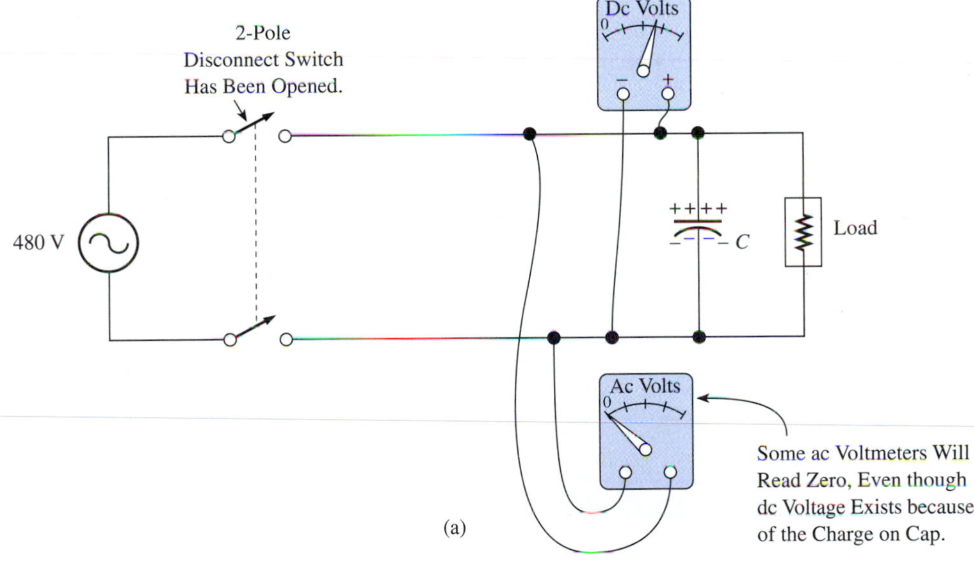

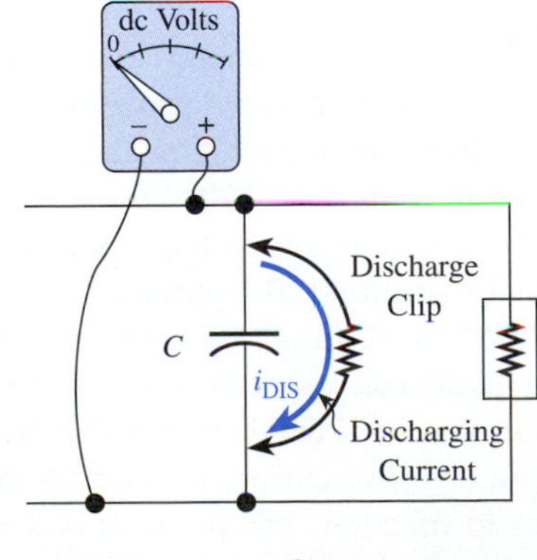

FIGURE 20–2
(a) When the power is turned Off by opening the disconnect switch, capacitors may maintain dc voltage because of residual charge. Residual capacitor charge is a common problem in electronic bench gear also. (b) Use a discharge clip, also called a *bleeder clip,* to allow the capacitor(s) to discharge.

3. In the industrial environment, where another person can throw the disconnect switch back to On, the switch should be padlocked in the Off position. Place a lockout tag on the switch handle giving your name, time and date of the disconnect, and a description of what equipment has been disabled by locking Off this switch. Well-organized industrial operations provide their qualified personnel with lockout hardware and authorization tags.
4. In critical situations, the load wires are shorted to each other and/or to a firm earth ground by mechanical shorting bars designed for that purpose. This makes it absolutely impossible for the load wires to become energized for any reason.

If test measurements are required on an energized circuit, certain precautions must always be taken:

1. Use proper footwear so that your feet are well insulated from the floor surface. Concrete factory floors often contain steel reinforcing rods, placing them in good electrical contact with the earth. For working near very high-voltage circuits (>600 V), special insulating boots are required.
2. Use only wooden or plastic ladders, never aluminum.
3. Never work in a wet area. Floor and cabinets must be dry before beginning work.
4. Wear proper gloves for the electrical environment that you are working in. For "low-voltage" environments (<600 V), special gloves are not absolutely required. However, in high-voltage environments (>600 V) rubberized gloves with a leather-like outer covering are required.
5. Remove metal rings and watches from your fingers and wrists. The skin underneath the surface of a watch-band, for instance, becomes moist. The metal-to-moist skin contact has much lower resistance than dry skin itself, which places you in greater danger.
6. Use only insulated-handle tools. Check the handle insulation often, and replace it at the first sign of damage or cracking. Likewise for the probes of voltmeters and voltage-testers.

20-2 ■ GROUNDING WIRES

In most industrial wiring systems, a separate grounding wire is run along with the current-carrying power leads. The purpose of the separate grounding wire is to connect the load enclosure(s) directly to the earth. To understand why this is necessary for personnel safety, look first at a nongrounded system, as illustrated in Fig. 20–3 for a single-phase circuit. The bottom terminal of the source is deliberately connected to earth ground at the power-distribution location. This is necessary to prevent any part of the ac circuit from becoming more than 120 V away from earth potential, which otherwise could happen if a neighboring higher-voltage circuit developed a short-circuit fault that brought it into contact with this 120-V circuit.

When the circuit functions normally, current flows down the hot supply wire, through the load device, and back to the source by the common supply wire. If an accidental short circuit should occur between the hot wire and the metal enclosure, as shown in Fig. 20–3(a), no overcurrent occurs. So the circuit-breaker does not open, and the load continues to function. The problem is that the metal enclosure now has the same 120-V potential as the hot supply wire. If a person touches the enclosure at the same time that she

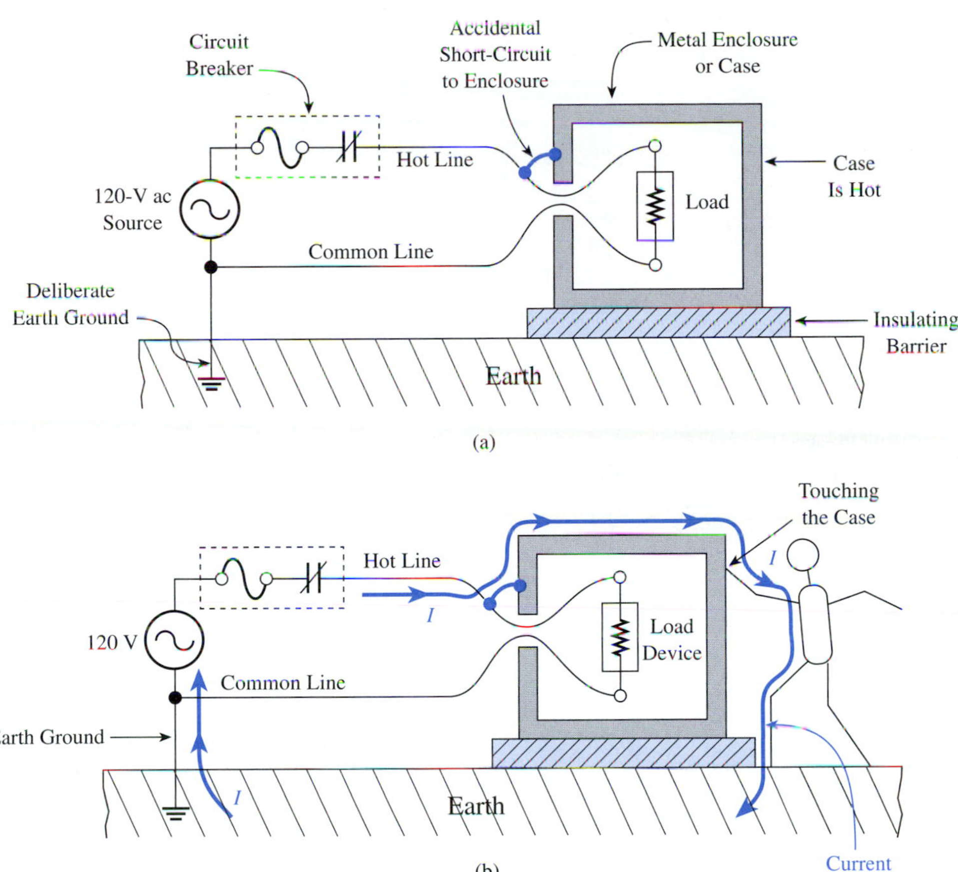

FIGURE 20–3
Accidental short-circuit from hot wire to load enclosure. (a) Circuit continues to function, but enclosure is at 120-V potential relative to earth. It is hot. (b) A person touching the enclosure can be shocked.

contacts the earth, current will flow through her body, as indicated in Fig. 20–3(b). The circuit breaker will not protect her because the current through her body is quite low by the circuit's standards, even though it is harmful to the person.

A grounded system is shown in Fig. 20–4. The third wire is used to connect the enclosure directly to the earth at the branch distribution point. When the circuit is functioning normally, no current flows in this grounding wire. This is shown in Fig. 20–4(a).

However, if a short-circuit fault should occur, as shown in Fig. 20–4(b), an overcurrent flows immediately. Its path is through the metal enclosure, then through the grounding wire back to the source. The circuit breaker opens, shutting off the circuit and eliminating the dangerous condition. It is impossible to reset the circuit breaker until the short-circuit fault has been repaired.

Figure 20–4(c) shows the grounding-wire idea for a three-phase industrial system.

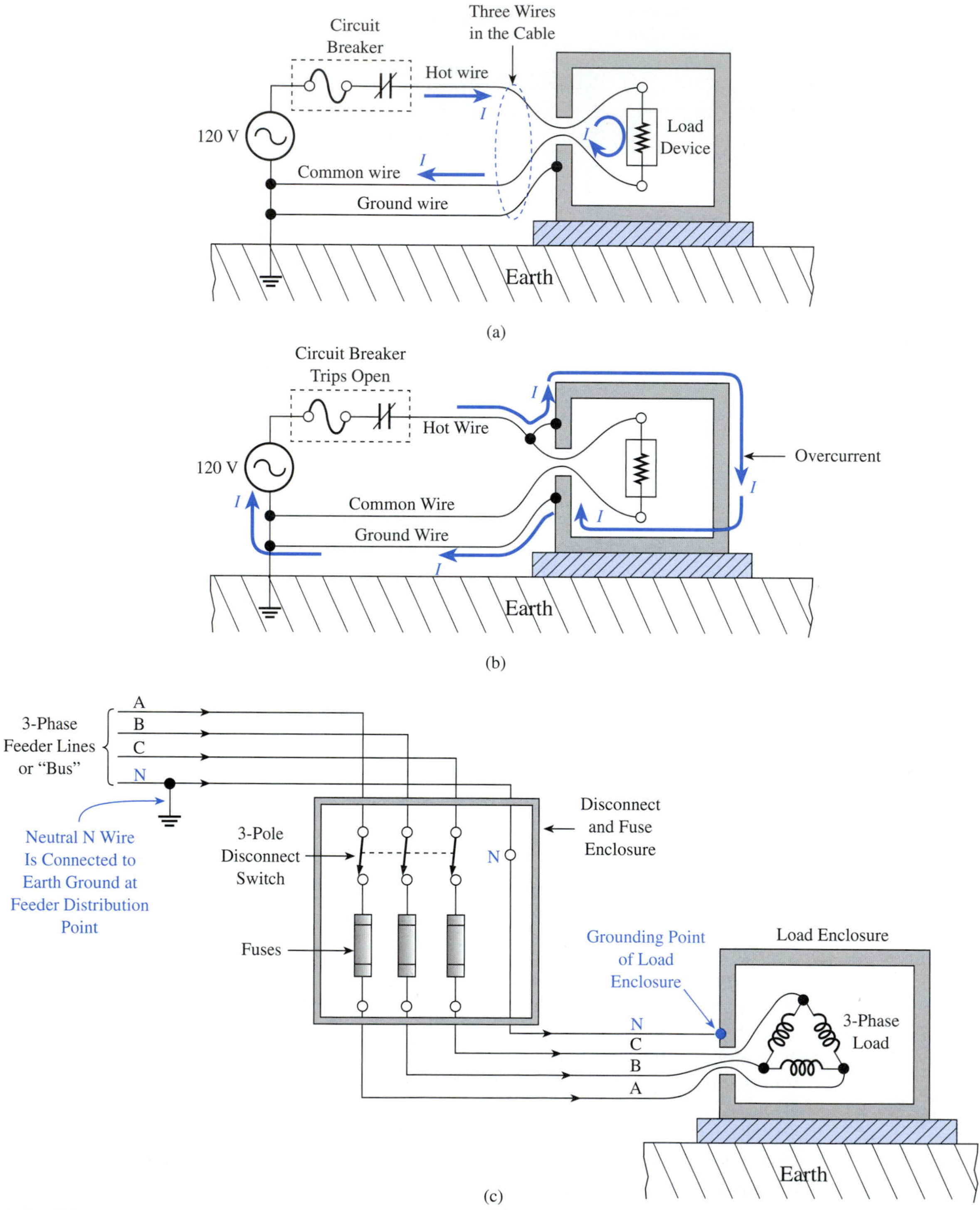

FIGURE 20–4
Grounded systems. (a) Normal operation. (b) Overcurrent blows the circuit breaker if the hot wire shorts to the frame. (c) Three-phase industrial situation.

20-3 ■ GROUND-FAULT INTERRUPTERS

Ground-fault interrupters are used only in single-phase systems. A GFI responds not to an overcurrent but to an *imbalance* between the currents flowing in the hot wire and the common wire. A GFI is a four-terminal device, as shown in the schematic symbol of Fig. 20–5(a). Refer to Fig. 20–5(b), (c), and (d) to understand the functioning of a ground-fault interrupter.

Figure 20–5(b) shows a three-wire system that is working properly. There are no short circuits, either partial or dead, between the hot supply wire and the frames of the load devices; neither are there any short circuits between the hot supply wire and the earth. Under these conditions, the current that flows out via the hot supply wire must be exactly equal to the current that flows back via the grounded supply wire. In Fig. 20–5(b), $I_{\text{black}} = I_{\text{white}}$.

The GFI senses both I_{black} and I_{white}, and it responds to the *difference* between them. A GFI can be considered as measuring the *difference current,* sometimes called the *leakage current* I_{leak}. Symbolically, in Fig. 20–5(b),

$$I_{\text{leak}} = I_{\text{black}} - I_{\text{white}}$$

If the leakage current is zero or very nearly zero, the GFI allows its contact to remain closed. This is the situation in the properly working circuit of Fig. 20–5(b).

However, if I_{leak} should somehow exceed the GFI's critical value, usually about 5 mA, the GFI trips. This opens the GFI contact and disconnects the circuit, just like a regular circuit breaker. The leakage current I_{leak} can be nonzero for either of two reasons:

1. A partial short circuit (or a dead short) might develop between the hot supply wire and the metal enclosure frame, as shown in Fig. 20–5(c). A certain amount of current I_{leak} would flow through the resistance of the partial short, through the metal frame, and back down the third wire to the grounded terminal of the source. A difference then exists between I_{black} and I_{white}, with I_{white} being less than I_{black}. ($I_{\text{white}} = I_{\text{black}} - I_{\text{leak}}$). The GFI senses this difference between the two supply-line currents and opens its contact.
2. A partial short circuit (or a dead short) might develop between the hot supply line and the earth, as shown in Fig. 20–5(d). The most likely reason for this is the partial breakdown of wire insulation, with that insulation touching a conduit wall that is mounted on a structural member of the building. Or, more important, a partial short circuit might occur because a person has accidentally touched the hot supply wire while another part of his body is in contact with the earth (through the floor, probably).

In either case, the partial short will cause some current, I_{leak}, to flow through the resistance of the short circuit, through the earth, up through the special grounding wire into the distribution box and thus back to the grounded terminal of the source, as indicated in Fig. 20–5(d). The white supply-line current I_{white} therefore is reduced by the amount of I_{leak}. The GFI senses the resulting difference between I_{black} and I_{white} and opens its contact to shut down the circuit.

Ground-fault interrupters provide more safety than circuit breakers or fuses, because they react to the existence of a small leakage current rather than a large overcurrent. In Fig. 20–5(d), for example, a standard circuit breaker would not open to save a person who is touching the hot wire, because I_{leak} would be far below its current rating.

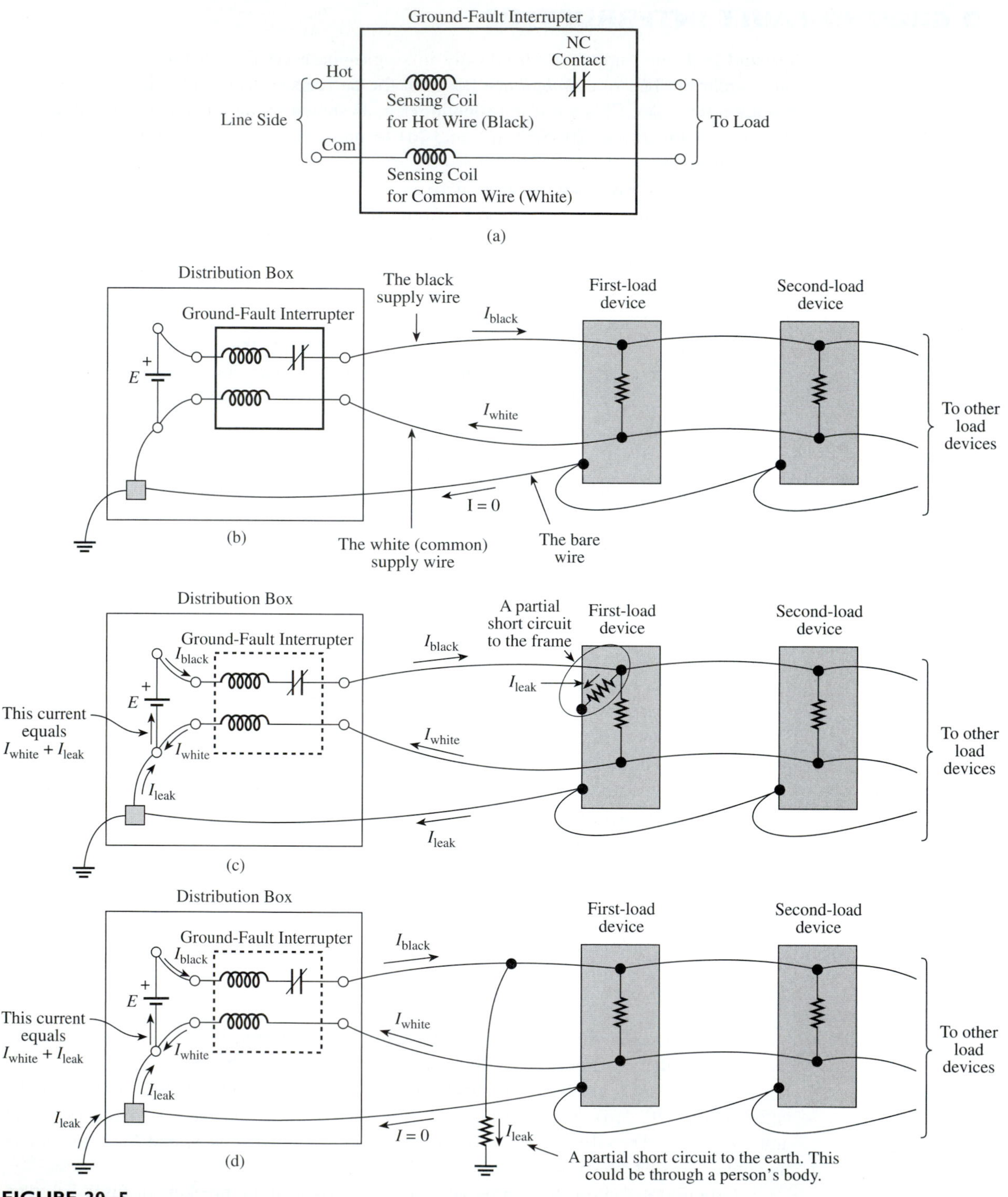

FIGURE 20–5

(a) Schematic symbol for a ground-fault interrupter. (b) Ground-fault interrupter installed in a grounded wiring system. (c) The current flow-paths for a short-circuit to a metal frame, causing I_{black} to not equal I_{white}. (d) The current flow-paths for a short-circuit to the earth.

20-4 ■ ADMINISTERING AID TO A VICTIM OF ELECTRIC SHOCK

To help a person who has suffered an electrical shock, you must be certain that he or she is no longer in contact with an energized circuit. In some cases, the shock throws the victim clear of the electric line; in other cases, the current through the victim's body causes muscle contraction that prevents him or her from pulling free. In the latter case, the rescuer must first deenergize the circuit before rushing to the victim; otherwise, on touching the victim, the rescuer may himself receive a severe shock.

For this reason, it is essential that you know in advance where the disconnect switches are located for all areas where you work. Learn every switch's location, then rehearse going up to it by the most direct route and throwing it to the Off position. Know that for all vertically oriented disconnect switches, the *down* position is Off (open). For vertically oriented rocker-type disconnect actuators, pushing the *bottom* rocker lobe turns the power Off.

For old horizontally oriented disconnect actuators, to the *left* usually is Off, but there are some exceptions to this. You must inform yourself in advance regarding every disconnect switch and rehearse your action, because most of us cannot think clearly in an emergency.

Whenever you open a disconnect switch to deenergize a circuit, stand with your body off to the side of the enclosure and your head turned slightly away. Thus, if the switch actuator is on the right side of the disconnect enclosure, stand off to the right with your head turned further to the right, and use your left hand to actuate the switch. This way, if the disconnecting action causes an explosion inside the enclosure, the force of the blast will be directed out the front of the enclosure and miss you on your left side. Such explosions are very rare and never occur in an installation which has scrupulously followed the National Electrical Code.

If it is impossible to turn off the power to remove a shock victim from an energized conductor, removal can be attempted using an insulating object or tool. A wooden or plastic pole, or a dry rope looped around the victim's arm, leg, or abdomen, may be able to break him free. A fuse-puller may be used to grab the victim's wrist and yank his or her hand free, or insulated lineman's pliers may be used to grab clothing and pull the body free. These actions must be calmly rehearsed with a work colleague, because it is very difficult to accomplish them for the first time in a real emergency.

After the victim is free from the energized source, check for respiration and heartbeat. If they have ceased, administer cardiopulmonary resuscitation (CPR) immediately. Everyone who works around industrial electricity should be formally trained in CPR procedures. Local Red Cross chapters offer such training, as do many hospitals and continuing education departments of school districts.

20-5 ■ BURNS

All working electrical devices rise in temperature to dissipate the heat energy that is produced by current passage or magnetic core reversals. In some situations, their surfaces become so hot that touching or bumping into them can produce a severe burn.

A more serious danger is the burn that can be sustained by using a hand tool to break open (interrupt) the current path in a highly inductive circuit. In Fig. 20–6, the disconnect enclosure contains specially designed gear for successfully interrupting the circuit current and extinguishing the resulting arc. This arc is a product of the very large transient voltage that is induced by the inductive load device when its current is suddenly changed, in accordance with the formula

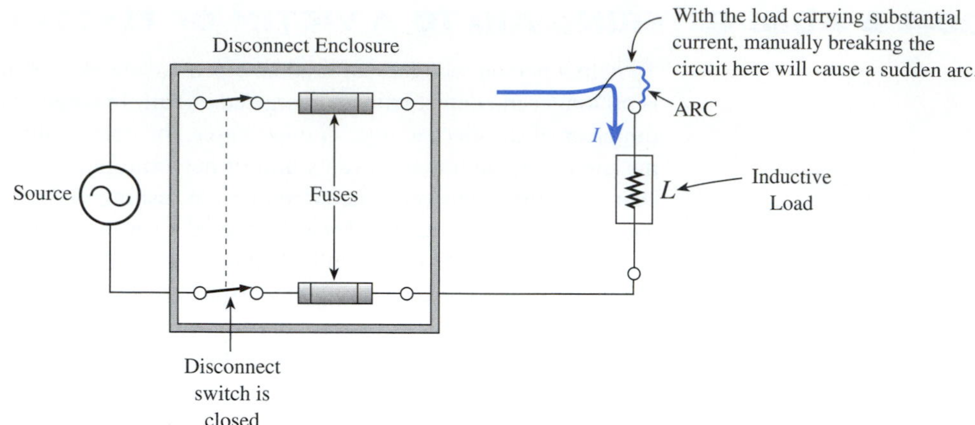

FIGURE 20–6 Highly inductive loads like motors and welding transformers can produce dangerous arcing between separating surfaces if they are manually disconnected. This is sometimes called the *inductive kickback* phenomenon.

$$V = L\frac{\Delta i}{\Delta t}$$

or

$$V = L\frac{di}{dt}$$

where the expressions $(\Delta i/\Delta t)$ or (di/dt) refer to the rate of change of current with respect to time, which is extraordinarily rapid when the conductive path is suddenly opened.

If a person with hand tools attempts to interrupt the inductive circuit path by disconnecting a conductor anywhere outside the enclosure, the large induced voltage may cause such a great arc between the parting metal surfaces that it burns the hand, arm, or face. Never attempt to manually disconnect a conductor supplying current to a highly inductive load like a large motor.

20-6 ■ EYE AND HEAD PROTECTION

In the manufacturing environment, high-speed drilling, sawing, shearing, and other mechanical processes occasionally can send material flying through the air at high speed. Such missiles are a great danger to the eyes and also the head in general. Never go into an industrial environment without shatter-proof safety glasses with side-guards and a hard hat.

Some companies collect and place on display the safety goggles and hard hats that have stopped flying debris from penetrating the eyes or skulls of their employees. Such exhibits are very sobering and useful.

Dangerous Light Sources

Besides the dangers of flying missiles, our eyes are subject to retinal damage by the following:

1. Flashes, both visible light and ultraviolet radiation, produced by arc-welders.
2. Errant x-rays from inspection and scanning equipment.
3. Laser light from object-detection devices and bar-code scanners.

To protect yourself from arc-welding radiation, your eyes must be shielded by properly darkened lenses. X-ray sources should themselves be marked by warning notices and their emissions contained by shielding metal.

Ear Protection

Damage to the ears is seldom sudden and traumatic. Instead, it tends to be slowly cumulative as a result of extended exposure to loud sound. Ear-canal plugs or muffs that cover the whole ear must be used in some industrial settings.

20-7 ■ FIRE

Many fires are electrical in their origin. The proximate cause of an electrical fire is overheating of a conductor, conductor-junction, or component, usually resulting from an overcurrent condition.

Fire Classification

The National Fire Protection Agency (NFPA) defines four classifications of fires, called classes A, B, C, and D. Class-A fires consist of burning paper, wood, rags, plastics, and general trash. Such fires can be extinguished with water, but they also can be fought successfully with the commonly installed modern fire extinguishers that use carbon dioxide foam or a dry smothering powder. To combat a fire most effectively, stand about 8 feet from it, aim the nozzle at the base of the fire, and sweep the spray slowly across the fire's base.

Generally speaking, class-A fires are preventable by diligent housekeeping. Clear away debris of all kinds, recycling when possible or disposing in outside trash bins. Place oily rages and other very combustible materials in specially designated receptacles.

Class-B fires contain burning liquids or gases, such as petroleum fuels and lubricants, hydraulic oil, manufactured hydrocarbons, and like materials. Such fires should not be fought with water, because the water stream tends to spread the burning fluid farther and expand the fire's range. Instead, they are fought by fire extinguishers that emit smothering foam or powder.

Class-C fires contain burning electrical equipment, including plastic insulation and insulating varnish on electromagnet windings. They must be fought with a foam or powder extinguisher. It is not safe to use water on them, because the person wielding the hose can be shocked through the water stream.

Class-D fires contain burning metals, often aluminum or magnesium alloys. They are hotter than other fires and more difficult to put out. They cannot be fought with water, but need an extinguisher specifically marked for class-D fires.

Every fire extinguisher has printed on it the classes of fires for which it is appropriate. Make it your business to familiarize yourself with the exact locations and class ratings of all the fire extinguishers in your work areas.

20-8 ■ OSHA COLOR CODES

All the griping and silly anecdotes notwithstanding, passage of the federal Occupational Health and Safety Act of 1970 has greatly improved the safety and livability of American industrial workplaces. During the 1960s, it was only enlightened employers who made serious organized efforts to ensure personal safety. The federal agency created to enforce the act, OSHA, has instituted many reforms and standards, among them the color-coding of certain safety-related equipment and controls. The following color code is now in place:

1. Red
   a. Fire extinguishers and other fire-fighting apparatus.
   b. Portable containers of flammable liquids.
   c. Emergency STOP pushbuttons for machines.
2. Yellow
   a. Marking of boundaries that should not be crossed by unauthorized personnel.
   b. General caution alerts to unusual conditions.
   c. Waste receptacles for disposing of combustible items.
   d. Electrical power source and disconnect switch for a machine.
3. Green
   a. First-aid material or emergency equipment.
4. Orange
   a. Exposed edges of cutting mechanisms.
   b. Parts of machines that can inflict injury by virtue of their high-speed rotation or by sudden movement.
5. Purple
   a. Radiation sources, including laser, ultraviolet, microwave, and x-ray emitting equipment.

■ SUMMARY

- As a general rule, all voltages above 50 V should be regarded as dangerous because of their ability to source harmful amounts of current through the human body.
- Working around standard industrial line voltages ($<$600 V) demands careful attention to footwear, tools, and measurement probes.
- Working around very high-voltage circuits ($>$600 V) requires specialized gloves, boots, and other equipment.
- Grounding wires are used to reduce the possibility of electric shock resulting from a short-circuit fault to an enclosure. A grounding system must never be defeated deliberately.
- Ground-fault interrupters provide a greater measure of safety than straightforward grounding, because they trip with only a slight partial short circuit to an enclosure or with only a small current to earth ground through the body of a person. However, GFIs cannot protect against a direct line-to-line shock situation in which the victim is simultaneously in contact with both lines of the voltage source.
- After a disconnect switch has been opened, careful checking of the lines with a voltmeter is required to guard against residual capacitor charge and other danger possibilities.
- Highly inductive loads can be turned Off only by a dedicated disconnect switch.
- Every industrial electrical worker should rehearse the location of disconnect switches in his or her areas of work.
- Every industrial electrical worker should take training to become certified in CPR.
- Every industrial electrical worker should rehearse the locations of fire extinguishers and learn clearly the distinction between class-A fires, which can be fought with water, and class-B, -C, and -D fires, which cannot be fought with water.

QUESTIONS AND PROBLEMS

Section 20-1

1. Generally speaking, what voltage value marks the point of danger to a human being?
2. True or False: No harm can come to a human unless the current through the chest organs exceeds several amperes.
3. Explain why rings, metal watchbands, and bracelets should be removed from your fingers and wrists before working on an industrial circuit.
4. Why is it dangerous to wear shoes with badly worn heel-soles when working with industrial electricity?
5. Why is it important that your pliers, screwdrivers, and other hand-tools have plastic-insulated handles?
6. The device that can store charge and present a shock hazard even after the power has been turned Off is the ________.
7. Describe what steps you must take to deal with the danger represented in Question 6.

Section 20-2

8. With a diagram, explain how a grounding wire protects you from electric shock in a three-wire single-phase ac circuit.

Section 20-3

9. Referring to the diagram of Question 8, explain the functioning of a GFI.
10. Referring to the diagram of Question 8, explain one reason why a GFI provides better protection from shock than a straightforward grounded system.

Section 20-4

11. Why is it so important that you rehearse the location of every disconnect switch in your work areas?
12. True or False: A victim of electric shock might remain in contact with the hot line because his or her hand has closed around it because of involuntary muscle contraction and paralysis.
13. True or False: Only medical professionals, doctors, and nurses should risk giving CPR, because anyone else might do more harm than good.
14. For the situation described in Question 12, what is the danger to you as you attempt to give aid?

Section 20-5

15. True or False: A running three-phase motor can be safely brought to a stop by simply grabbing one of its fuses with a fuse-puller and pulling it out of the fuse-holder. Explain.
16. An inductive device, when its current is suddenly interrupted, induces a transient voltage that is large enough to ________ across the separating metal surfaces.
17. True or False: For the reason referred to by Question 16, dc circuits containing inductive devices are more difficult to interrupt than ac circuits containing comparable inductance. Explain your answer.

Section 20-6

18. For head protection from falling objects and high-speed missiles, a ________ is required.
19. For eye protection from flying missiles, cuttings, and molten liquid such as solder, ________ ________ are required.
20. List some of the light and light-like sources that you must be alert to in order to avoid retinal injury.

Section 20-7

21. List the NFPA classifications of fires, describing the nature of each class.
22. Of the classifications you listed in Question 21, which one(s) must not be fought with water?
23. Why is it so important to rehearse the locations and types of all the fire-extinguishers in your areas of work?

Section 20-8

24. True or False: It is generally agreed by all informed people that OSHA has been a drain on federal tax monies that has produced very little benefit to the average citizen.

APPENDIX A UNIVERSAL TIME-CONSTANT CURVES

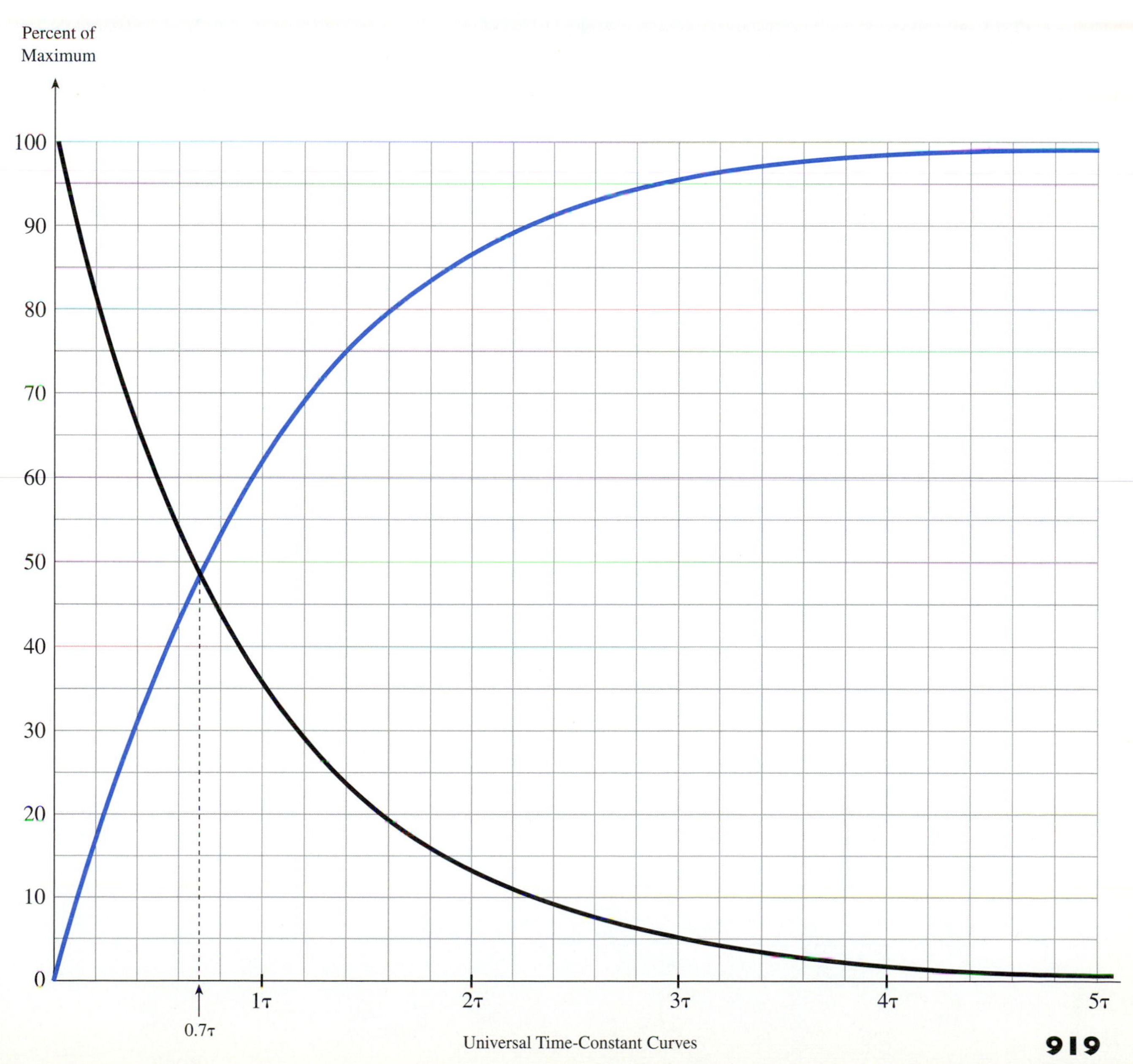

Universal Time-Constant Curves

GLOSSARY

555 One-shot A circuit for delivering a fixed-duration pulse, utilizing an integrated circuit known as the 555 (from a manufacturer's code number). Sometimes used as a variable one-shot to supply the turn-on pulse to the load-switching transistor in a pulse-width modulation system.

555 Oscillator An oscillator utilizing an integrated circuit known as the 555. Often used in pulse-width modulation to provide constant-frequency trigger signals.

565 Phase-locked loop Used for pulse-frequency demodulation. See *Phase-locked loop (PLL).*

Accelerometer A device that measures acceleration by detecting the strain on a known amount of mass (the seismic mass) which is attached to the object being measured.

Actuating device section See *Output section.*

Address bus A one-way bus that carries address information from the microprocessor chip to all the other chips in the microcomputer; there is no bus contention because only one device drives it (the μP chip).

Address decoder An intermediary chip between the two highest-order address lines and the CE control lines of the four peripheral chips.

Alternator The proper term for an ac generator.

Apparent power (*S*) The simple product of rms voltage multiplied by rms current without regard to the *V-I* phase relationship.

Armature control (of Dc Motor speed) Varying the voltage applied to the armature winding to control the speed of a dc shunt motor. See *Field control (of Dc Motor speed).*

Armature reaction A problem that arises when the magnetic flux of a wound-rotor motor becomes distorted due to the flux of the armature at heavy torque load. See also *Interpoles.*

Armature resistance (R_A) The resistance of a motor's armature winding.

Armature winding An electromagnet winding within a generator or motor that produces the machine's output product (voltage in a generator, torque in a motor). See also *Field winding.*

Balanced load (for a 3-Phase system) A condition in which the three individual phase impedances are equal to each other.

Bank, odd or even (of a 12-SCR Cycloconverter) When six additional SCRs are added to a cycloconverter to produce six pulsations of load voltage for each cycle of the ac line, they are arranged in groups of six that are known as the odd bank and the even bank.

Bellows A series of connected metal diaphragms that distort when subjected to fluid pressure; used to sense pressure and convert it into mechanical movement.

Bidirectional mode The way of operating a stepper motor in which any step may be followed by the next step in the opposite direction. This mode limits speed but allows the motor to stop quickly.

Bounce eliminator A device that smoothes the output when the contact surfaces of a mechanical switch bounce before making permanent closure, thus preventing unwarranted on/off fluctuations that can cause malfunctions in logic circuitry.

Bourdon tube An oval tube of metal having elasticity that, through its distortion, is used as a sensing device to measure fluid pressure and convert it into a mechanical movement.

Breakaway torque The amount of torque required by a load device to move it off the stopped position; the motor's starting torque must be greater than the load's breakaway torque.

Breakover device A device that serves to "even out" the electrical variations and temperature sensitivities of SCRs and other thyristors by reacting vigorously to a specific voltage (the "breakover voltage").

Brushes Carbon blocks held tight against the slip-rings or commutator of a rotating energy-conversion machine (generator or motor); designed to connect the moving armature winding to the stationary external world.

Brushless Dc motor (Position-triggered) A motor that contains Hall-effect position sensors that detect when the rotor is in the proper instantaneous position for the stator electromagnet coils to be switched on and off. The sensor signals an IC sequence controller, which then switches the control transistors. See *Hall effect.*

Buffer An output amplifier. Also called a *Driver.*

Bus contention A situation in which two or more data sources are attempting to assert themselves on the data bus simultaneously, thereby conflicting with one another. This must be prevented by programming of the control signals that originate in the microprocessor chip.

Capacitor-start motor A motor in which a centrifugal switch on the motor shaft opens when the shaft has reached some critical speed, usually about half the synchronous speed. The switch disconnects the start winding and its series phase-shifting capacitor.

Capture range The range of input frequencies that a PLL can lock onto. See *Lock range.*

Carburizing The metallurgic process of diffusing carbon into steel.

Center frequency (of a PLL) The natural oscillation frequency set by the values of the timing resistor and the timing capacitor.

Characteristic curve (of a Motor) A graphic representation of the relationship of one motor variable to another, such as speed vs. torque.

Chip-enable (CE) The terminal on an IC chip that instructs the chip to do its assigned job.

Chopper stabilization The process of "chopping" a dc signal to make it resemble an ac signal, so that it can be amplified by an ac amplifier.

Clock A circuit that provides a continuous stream of square-edged pulses, used to synchronize various digital devices with each other.

Clocked (Edge-triggered) flip-flop A single-bit memory device (flip-flop) that responds to its synchronous inputs (*S* and *R* or *J* and *K*) only at the instant when its clock terminal makes a transition (from LO to HI—a positive edge-triggered flip-flop—or from HI to LO—a negative edge-triggered flip-flop).

Closed-loop system A system that is self-correcting, utilizing an error-detector and a controller to force the measured variable toward its set point value.

Cold junction The junction in a thermocouple that is subjected to the lower temperature.

Commutation (of an SCR) The interruption of main terminal current within the SCR by connecting a temporary short circuit from the anode to the cathode; used for turning OFF an SCR in a dc circuit.

Comparator (comparer) An op amp circuit that drives into one saturation polarity if the input voltage is more positive than a certain reference value, and drives into the opposite saturation polarity if the input voltage is more negative than that reference value.

Comparer A mechanical, electrical, or pneumatic device that compares the measured value of a process variable to the set point (the desired value); also called error detector or difference detector.

Complementary symmetry amplifier An amplifier of the push-pull type that contains no transformer, using one *npn* and one *pnp* transistor.

Compound motion A robot motion involving simultaneous movements of two or more axes.

Compound-configured Dc motor A motor that combines shunt and series windings to achieve a compromise between the operating characteristics of each configuration.

Conduction angle The number of degrees of an ac cycle during which an SCR is On. See also *Firing delay angle.*

Continuous-path robot A robot that can produce a precisely specified tool path through space, no matter what the tool's load condition.

Control lines Lines connected to a microcomputer's support chips (CE for chip enable, R for read, and W for write), instructing them when to become active. See *Input buffer, Data bus.*

Control mode One of five manners in which a closed-loop system controller reacts to an error.

Control winding The winding of a servomotor that receives the amplified error voltage.

Controller (of Closed-loop system) The device that receives the error signal from the comparer and generates an output signal to the final correcting device, thus correcting deviations from the set point. See also *Comparer, Final correcting device, Set point.*

Cool subinterval The portion of the weld interval during which there is no current in the welding electrodes. See *Weld interval.*

Coreless motor Developed to allow very quick acceleration and stopping, coreless motors have no steel in the rotor, thus achieving very low mechanical inertia; also called Ironless-Rotor Motors.

Counter-EMF Also called the counter-voltage, it is the voltage generated by the motor's armature winding in opposition to the applied voltage source.

Counter-EMF feedback In a motor speed-control system, the technique of sensing the motor's counter-EMF and applying it to the thyristor trigger-control circuit so that variations in the motor's speed tend to be automatically corrected.

Cup motor A type of coreless motor having a cup-shaped assembly made of fiberglass or other material with printed copper tracks or individual wires serving the function of armature conductors. See *Coreless motor.*

Cycloconverter A sequentially triggered SCR circuit that is capable of converting 3-phase ac, usually 60 Hz, into single-phase ac of a lower frequency.

Damping (of a Stepper motor) The practice of minimizing mechanical overshoot (exceeding of the desired step position) and suppressing the oscillations quickly when overshoot occurs.

Data bus The collection of wires that carry binary codes between the microprocessor chip and the support chips in a microcomputer.

Dead time (in Process control) The actual amount of time that a correction action remains undetected by the measurement device due to transfer lag and/or transportation lag.

Decade counter A circuit that counts, in binary-coded-decimal, resetting to zero from nine when a tenth count pulse is received.

Decision-making section See *Logic section.*

Decoder A device that takes binary-coded information and converts it to decimal information that human beings can understand.

Deenergized The state of having no current through the active electrical element, usually an electromagnet. For a relay, the state of having no current in the control coil, thereby allowing the contacts to return to their normal states.

Degrees of freedom In robotics, the number of axes around which movement can be programmed.

Delta configuration A three-phase connection method in which the individual windings are connected in a way that appears schematically like the shape of the Greek letter delta.

Demodulation (Pulse-width) The process of recovering the original measured value (the modulating value) from a pulse-width modulated signal.

Derivative time constant (in PID control) The common measure of the effectiveness or vigorousness of the derivative action in a proportional plus integral plus derivative controller.

Detent torque The maximum amount of load-originating torque a stepper motor can withstand without slipping when the stator pole windings are all deenergized.

Diac A breakover device used in the gate lead of a triac's gate-control circuit; it guarantees reliable triggering because it delivers a pulse of gate current rather than a sinusoidal gate current.

Differential amplifier (Op amp) An op-amp circuit that amplifies an input voltage derived from two terminals, neither of which is grounded.

Differential gap (in On-Off controller) The smallest range of values the measured variable must pass through to cause the correcting device to cycle from Off to On to Off.

Differentiator (Op amp) A circuit whose output is proportional to how quickly the input is changing.

Digital telemetry Telemetry in which the measured variable is digitally encoded and the bits transmitted serially.

Disk motor A type of coreless motor utilizing a lightweight disk with bonded copper tracks placed to form single-turn armature wraps; it is unique in that the magnetic field flux is not radial but points parallel to the shaft (axial). See *Coreless motor.*

Distributed winding group A motor structure in which individual windings are placed in neighboring slots and connected in series to form a group, thereby distributing the windings around the stator.

Driver An output amplifier. Also called a *Buffer.*

Dropped out Also simply "dropped" refers to a relay whose coil is deenergized.

Duty cycle Percentage of the full cycle time spent in the up or On state.

Dynamic braking A technique for stopping a motor quickly by disconnecting the voltage source and connecting a resistive load across the armature to reverse the current in the armature conductors. See also *Regenerative braking.*

Dynamo The generic term for a wound-rotor dc machine that can act either as a generator (mechanical to electrical energy conversion) or as a motor (electrical to mechanical energy conversion).

Eddy current A whirlpool-like (eddy) current that tends to circulate in the core material of a motor as the magnetic flux direction goes through its reversals. If not suppressed by the lamination technique, eddy currents cause considerable power loss.

Efficiency (of a motor) The ratio of the motor's mechanical shaft output power to its electrical input power.

Electrohydraulic valve A variable-position valve that is moved by hydraulic oil pressure in response to the electric current through its actuating coil.

Electronically commutated motors A class of motors having no mechanical commutator, including both stepper and brushless dc motors; they have certain advantages over conventionally commutated motors. See *Stepper motor, Brushless Dc motor.*

Electropneumatic valve A variable-position valve that is moved by pneumatic (air) pressure in response to the electric current through its actuating coil.

Encoder A device that receives a decimal number and converts it to a binary number; it is the reverse of a decoder.

Energized The state of having current through the active electrical element, usually an electromagnet. For a relay, the state of having current in the control coil, thereby forcing the contacts to assume their nonnormal states.

EPROM (Erasable programmable ROM) A ROM that allows its programming to be erased by ultraviolet light through an opening in the body of the IC chip, thus making it useful for reprogramming robots.

Error detector See *Comparer.*

Error signal (in a Closed-loop system) A signal produced by the comparer when the measured value varies from the set point (the desired value).

Examine-OFF instruction An instruction within a PLC program that produces logic continuity (like a closed electrical contact) when power is not present at the I/O terminal referred to by its address, but which produces logic discontinuity (like an open electrical contact) when power is present at the I/O terminal referred to by its address. Sometimes called a normally closed instruction.

Examine-ON instruction An instruction within a PLC program that produces logic continuity when power is present at the I/O terminal referred to by its address, but which produces logic discontinuity when power is not present.

Faraday's Law This fact of nature: When a loop of wire is subjected to time-varying magnetic flux, the voltage induced per turn (in volts) is equal to the time rate of change of magnetic flux (in webers/second).

FET chopper An electronic circuit that utilizes a field-effect transistor (FET) for chopping a dc signal prior to amplification in an ac-coupled amplifier.

Fiber-optic cable Collection of optical fibers used in telecommunications because of their immunity to electrical interference.

Field control (of Dc motor speed) The technique of controlling the speed of a dc shunt motor by adjusting the field current. It has the serious disadvantage of lessening the motor's torque-producing ability. See *Armature control (of Dc motor speed).*

Field winding An electromagnet winding within a motor or generator that produces the machine's main magnetic flux when current passes through it.

Field-effect transistor chopper See *FET chopper.*

Field-failure detector A current-sensing device in a dc motor control circuit that automatically disconnects the armature winding if shunt field current stops flowing for any reason.

Final correcting device (in a Closed-loop system) The device (often a valve) that varies the energy input to the system.

Firing delay angle The number of degrees of an ac cycle that elapse before an SCR is triggered On. See also *Conduction angle.*

Fixing winding The winding of an ac servo motor that receives a fixed, nonvarying voltage.

Flash-on effect The tendency for thyristor power-control circuits to change from zero load power to substantial load power as the control pot is slowly adjusted, requiring the reversal of the control pot to establish a minimal load power.

Floating state See *High-impedance state.*

Flow meter A device for measuring the flow rate or total accumulated flow of a fluid.

Flowchart A graphic depiction of the sequence of events that occur during program execution.

Flux density The variable that describes the strength of a magnetic field. Measured in teslas, equivalent to webers per square meter.

Flux, Magnetic See *Magnetic flux.*

Frame time (in Digital telemetry) The amount of time allocated for applying the serial bit stream to the transmission wires, exclusive of SYNC time.

Free-wheeling (Kickback diode) A diode connected in parallel with an electromagnet that conducts when the electromagnet generates a kickback voltage as the external driving source changes polarity or switches Off.

Full step In Stepper Motor operation, the larger of the two possible angular stepping amounts, which occurs when just a single transistor is switched On at any instant in time.

Gate Trigger Current (I_{GT}) The amount of current into the gate lead that is necessary to fire an SCR or triac.

Generator An electromagnetic rotating machine that converts mechanical input to electrical output. See also *Dynamo.*

Gray code A popular binary code in which only one bit changes at a time; chiefly used in optical and mechanical position-encoders.

Gripper The device on the end of a robot's arm that grabs the object that is to be manipulated by the robot.

Hall effect The phenomenon by which charge carriers moving through a magnetic field are forced to one side of the conducting medium.

Half step In stepper motor operation, the smaller of the two possible angular stepping amounts, which occurs when a single transistor alternates with a pair of transistors being switched On.

Heat subinterval The portion of the weld interval during which there is current in the welding electrodes. See *Weld interval.*

High-impedance state The state of a tristate digital device that disconnects the output terminal from both the +5-V bus and the ground bus, thus allowing it to remain connected to a data bus line without interfering with some other device asserting itself on the data bus line.

Hold interval In automatic welding, the time after the current is turned off during which the electrode cylinders maintain pressure while the weld cools. It lasts about 1 second and helps avoid distortion of the molten metal.

Holding current (I_{HO}) The amount of main-terminal current necessary to maintain an SCR or triac in the On conducting state once it has been fired.

Holding torque In a permanent-magnet stepper motor, the maximum amount of load-originating torque the motor can withstand without losing its grip and allowing the shaft to slip out of position, assuming that the stator windings remain energized.

Horsepower The common North American unit of measurement for the mechanical power of a rotating shaft; equivalent to 746 watts.

Hot junction The junction in a thermocouple that is subjected to the higher temperature.

Humidity The amount of water vapor in the air, measured as the ratio of water vapor actually present in the air to the maximum the air could possibly hold (percent relative humidity).

Hygrometer A variable-resistance device for measuring relative humidity.

Hysteresis In general, the situation in which a given value of the independent (x) variable can produce two different values of the dependent (y) variable, determined by whether the independent variable is either increasing or decreasing. Hysteresis can be used as a description of the triac flash-on effect, among other common phenomena. See *Flash-on effect.*

I/O (Input/Output) rack A mechanical enclosure with slots to hold printed circuit boards (modules) that contain either 16 input signal converts or 16 output amplifiers.

Ignitron A large, three-electrode mercury-arc rectifying tube that behaves similarly to an SCR but has the capacity to handle tremendous surges of current (as large as 10 000 amperes).

Industrial robot A tool-controlling device that can be programmed through a computer to perform mechanical manipulations.

Information-gathering section See *Input section.*

Input buffer An integrated circuit that admits external data into a microcomputer.

Input image file The portion of a PLC's processor memory that stores input conditions from the machine or process being controlled.

Input scan The portion of a PLC's overall scan cycle during which the current updated status of every terminal is stored in the input image file.

Input section All the devices that supply system information and human operator settings to the logic section of an industrial control system. Also known as *Information-gathering section.*

Inrush current (of a Motor) The surge of current that enters a motor as it begins to accelerate from a standstill.

Integral gain factor (in PID control) The common measure of the effectiveness or vigorousness of the integral action in a proportional plus integral plus derivative controller, when the PID controller is implemented digitally by a PLC.

Integral time-constant (in PID control) The common measure of the effectiveness or vigorousness of the integral action in a proportional plus integral plus derivative controller, when the PID controller is regarded in an analog fashion.

Integrator (Op Amp) An op-amp circuit whose output is proportional to the input voltage and to the amount of time the input has been present.

Interpoles Smaller poles placed between the main poles of a dc dynamo, with electromagnet windings connected in series with the armature to prevent flux distortion.

Inverter drive (for an Ac induction motor) A variable-frequency motor-drive circuit that converts dc into ac motor voltage, as opposed to performing an ac-to-ac conversion, like a cycloconverter.

Inverting amplifier (Op Amp) An op-amp circuit that amplifies single-ended voltage and current, with polarity inversion. See *Op Amp.*

Ironless-rotor motor See *Coreless motor.*

Ladder-logic format The method of drawing a relay control circuit schematic or a PLC program that makes it easier to trace the logic of the control circuit. So called because it resembles a ladder, with siderails and rungs. See also *Rung.*

Lamination The construction technique of using very thin plastic insulating layers to separate thin steel sheets. The purpose is to limit the effect of eddy-currents in the core of a rotating energy-conversion machine. See *Eddy-current.*

LED (Light-emitting diode) A semiconductor that emits light when it carries current in the forward direction.

Lenz's law This fact of nature: If outside forces cause a winding coil to experience a change in through-the-coil flux, the winding attempts to circulate current in its conductors in the direction that causes its own flux to oppose the change.

Light-emitting diode See *LED.*

Line variables (in a 3-Phase system) Voltage and current associated with access to the transmission lines only; not dependent on access to the internal phase windings of the 3-phase source or the 3-phase load (motor).

Linear variable differential transformer See *LVDT.*

Linearity (of a Pot) A measure of the degree to which the resistance of the potentiometer element is evenly distributed along the length of the element.

Load operation (by a μP) Movement of data from a peripheral location (another chip, usually) into the microprocessor's internal registers.

Load cell A strain-gage assembly used for measuring weight or any force.

Lock range The range of input frequencies that a PLL can remain locked onto if it has already captured the input signal. See *Capture range.*

Logic gates Also simply "gates"; individual logic circuits that are the building blocks of more complex circuits. The five basic logic gates are AND, OR, NOT, NAND, and NOR.

Logic input interfacer See *Signal converter.*

Logic section The part of an industrial control system that acts upon the information provided by the input section in a prescribed, logical way. Also known as decision-making section. See *Input section.*

Lorentz's relationship This fact of nature: The mechanical force on a current-carrying wire within a magnetic field is equal to the wire length multiplied by the current multiplied by the magnetic field strength. If the current direction and the magnetic flux direction are perpendicular to each other, the mechanical force is perpendicular to both of them, by the cross-multiplication technique.

LVDT A dual-secondary winding transformer with a moveable core that gives an ac output voltage signal proportional to its physical displacement (generally small, an inch or less).

Magnetic flux The lines that describe the spatial orientation of the magnetic effect.

Magnetic pole A surface from which magnetic flux lines emerge (North pole) or reenter (South pole).

Magnetic saturation The effect within a magnetic core that causes its magnetic flux density to lose its proportionality with winding current.

Main terminals (of a Triac) The two triac terminals between which the main current passes (the load current passes).

Maximum torque point (for an Induction motor) The maximum amount of torque the rotor can deliver to the load before it loses its ability to follow the rotating field. Alternatively, the maximum amount of torque the motor can produce; if any more torque is required, the motor will stall.

Measured value (of Closed-loop system) The value determined continuously by a measurement device so that it can be compared to the set point (desired value).

Measurement device A device that continuously monitors the value of the process variable so that it can be compared with the desired set point; also called a detecting device, detector, or transducer. See *Transducer.*

Mechanical power The product of torque and rotational speed; it can be expressed in units of watts or horsepower.

Microprocessor chip (μP) The integrated circuit that fetches program instructions, decodes them, and executes them.

Microstepping A stepper motor technique for producing very small stepping angles, achieved by modulating the power supply voltages to the phase windings.

Modulation (Pulse-width) The technique of continuously varying the average power to an electrical load by varying the duty cycle (width) of pulses applied to the load.

Monostable multivibrator See *One-shot.*

Motor A rotating machine that converts electrical to mechanical energy.

Motor-starter A relay with heavy-duty contacts, capable of carrying inrush currents to a motor and capable of safely interrupting the motor's running current.

Multiplexer A device that receives several inputs, sending just one of them to the output at any moment; typically, it spends a short time passing one input, then switches to another input.

Negative feedback (for an Amplifier) The practice of applying a portion of the output signal back to the input in such a way that it opposes the original input signal. A stabilization technique.

NEMA classification (of 3-Phase motors) The National Electrical Manufacturers Association categorizes squirrel-cage induction motors into classes A, B, C, and D, depending on the motors' rotor-bar cross-sections, which affect their characteristic speed-vs.-torque curves and thus their usefulness for particular applications.

Neutral (of a Wye) The central point where the three individual phase windings are connected. See *Wye configuration.*

Noninverting amplifier (Op Amp) An op-amp circuit that amplifies single-ended voltage and current with no polarity inversion. See *Op Amp.*

Normally closed contact A contact that is closed when its relay coil is not energized, or when its actuating mechanism is deactuated.

Normally open contact A contact that is open when its relay coil is not energized, or when its actuating mechanism is deactuated.

North pole The surface of a magnet from which flux lines emerge into the surrounding space.

Nyquist sampling theorem This theorem states that the sampling frequency must be no lower than twice the highest-frequency variation to be captured.

Off-delay timer A timer that starts accumulating time when its control coil is de-energized, or in solid-state when its digital input is LO, or in PLC realm when its rung conditions are logically false. The opposite of an On-delay timer.

Offset (in Proportional control) The permanent difference between the set point and the actual measured value.

On-Off control The simplest mode of control, in which the only conditions for the final correcting device are fully On and fully Off.

One-shot A digital circuit whose output temporarily goes HI for a fixed time following a trigger event. Formally, a *monostable multivibrator.*

Op Amp Operational amplifier, an integrated circuit amplifier with a differential input, having very large voltage gain and very high input resistance.

Open-loop system A system that is not self-correcting when variations in input and process conditions occur spontaneously; this system requires human monitoring.

Optical coupling The technique of converting high-voltage input signals to low-voltage logic signals, or vice versa, by a photoelectric-linked isolated interface between the two circuits.

Optical fibers Very thin strands of glass or plastic that carry light from a sending to a receiving location without susceptibility to electric or magnetic interference.

Optical position-encoder An optical position-encoder counts the number of pulses produced by a rotating optical disk and relates that number to the amount of shaft movement; also called an optical shaft-encoder.

Original input converter See *Signal converter.*

Output amplifier A device inserted between a logic circuit and its actuating device to increase the low-voltage/low-current logical power to higher-voltage/higher-current output power.

Output image file The portion of a PLC's processor memory that stores output conditions resulting from the most recent program execution.

Output scan The portion of a PLC's overall scan cycle during which the output conditions in the output image file are passed to the output module in the I/O rack.

Output section The part of an industrial control system that takes output signals from the logical section and converts or amplifies the signals into useful form (examples: motor-starters, solenoids, lamps). See *Logic section.*

Output-energize instruction An instruction within a PLC program that refers to a specific output terminal address and which causes that output terminal to become powered up if the instruction rung is TRUE (logic continuity), or causes that output terminal to become deenergized if the instruction rung is FALSE (logic discontinuity).

Overload detector A thermal or magnetic device in a motor circuit that is designed to detect an overcurrent condition and shut down the motor-starter.

Overshoot (of a Stepper motor) The phenomenon in which a stepper motor passes by the desired step position, then recovers and oscillates around the position before settling.

Peak voltage (V_p) In a UJT, the value of emitter-to-base 1 voltage below which the UJT blocks and above which the UJT fires. Upon firing, the UJT creates a short-circuit-like current surge from emitter to base 1.

Permanent-magnet Dc motor Conceptually the same as a shunt-configured wound-rotor motor, except that the magnetic field is established by permanent magnets mounted on the stator.

Phase detector In a phase-locked loop, the circuit that responds to a phase difference between two square waves to produce its output voltage, which then becomes the control voltage to a voltage-controlled oscillator.

Phase variables (in a 3-Phase system) The voltages and currents associated with the individual phase windings within the source or load. These windings are not accessible for measurement of both voltage and current from outside the enclosure of the 3-phase source or load.

Phase-locked loop (PLL) Used for pulse-frequency modulation, the loop works by automatically changing the control voltage applied to its voltage-controlled oscillator to make the oscillation frequency exactly equal to the received input-signal frequency.

Photocell A device that produces an electrical variation in response to a change in light intensity; used to sense the presence of an opaque object or the degree of translucence or amount of luminescence of a fluid or solid.

Photoconductive cell A passive photocell in which resistance varies in relation to the light intensity at its surface; see *Photocell.*

Phototransistor A transistor that responds to the intensity of light on its lens instead of to base current; often used with an LED in optical coupling.

Photovoltaic cell A photocell that produces an output voltage that varies in relation to the light intensity at its surface; see *Photocell.*

Pick-and-place robot A robot that picks up an object and puts it down in another place, without regard to the path from the pickup place to the put-down place; sometimes used as a synonym for positive-stop robot.

Picked up Also simply "picked"; refers to an energized relay coil.

PID control Proportional plus integral plus derivative control.

PLC See *Programmable logic controller.*

Plugging A technique of stopping a running motor very quickly by applying the reverse polarity to the armature winding of a dc motor through an automatic plugging switch, or applying the reversing phase sequence to an ac induction motor.

Point-to-point robot A robot that can move any degree of freedom to any intermediate point between its two extreme position. In other words, a robot that can put its tool anywhere it is instructed.

Pole, Magnetic See *Magnetic pole.*

Position-encoder (Optical) See *Optical position-encoder.*

Position-triggered brushless Dc motor See *Brushless Dc motor (Position-triggered).*

Positive-stop robot A robot in which each axis (degree of freedom) can position itself only at one of its two extreme positions.

Pot See *Potentiometer.*

Potentiometer A three-terminal variable resistor, capable of variable voltage division of the voltage applied across the end terminals. The most common component in industrial measurement transducers.

Power factor (of an Ac Motor) The ratio of the motor's in-phase, power-producing component of current to its total current (the in-phase component combined with the out-of-phase component). Alternatively, the cosine of the phase angle between the motor's total current and its supply voltage. Power factor is a measure of how effectively an ac system is using the current-carrying capacity of its transmission wires.

Power, Mechanical See *Mechanical power.*

Prime mover An external mechanical engine that forces a generator's rotor shaft to spin.

Process reaction delay The time delay between the application of a corrective action and the final result of that action; also called Time Constant Delay.

Process variable (of Closed-loop system) The variable whose value is measured for comparison to the set point, and which the system attempts to correct if it deviates from the set point. See *Measured value (of Closed-loop system).*

Processor (of a PLC) The portion of a programmable logic controller that stores and executes the user program. As distinct from the other two portions of a PLC, namely the I/O rack and the programming device.

Program counter A register (16 bits long in our example) within the microprocessor that stores the current ROM address, thus keeping track of current location in the program.

Programmable logic controller A logic system that combines hardware and software, so that its coded instructions may be modified (reprogrammed) by use of a keyboard; it provides "flexible," rather than "dedicated," automation.

Programmable unijunction transistor (PUT) A semiconductor device similar to a standard UJT, except that its peak voltage is determined by external circuitry (rather than an intrinsic standoff ratio), which makes the device programmable.

Programming device The portion of a PLC used for entering the user-program into the processor's memory and editing it. It usually has a monitor screen that allows the user to observe the rung-by-rung execution of the program.

Proportional band (in a Closed-loop system) The percent of full controller range by which the measured value must change in order to cause the correcting device to change by 100%.

Proportional control A mode of control in which the final correcting device has a continuous range of possible positions, with the exact position taken being proportional to the error signal; that is, the output from the controller is proportional to its input.

Proportional gain factor (in PID Control) The common measure of the effectiveness or vigorousness of the proportional action in proportional plus integral plus derivative controller, when the PID controller is implemented digitally by a PLC.

Proportional plus integral control A mode of control in which the position of the control valve is determined by (1) the magnitude of the error signal, and (2) the time integral of the error signal—or the magnitude of the error multiplied by the time it has persisted.

Proximity sensor A device that gives a binary indication that an object is near (in the proximity of) a certain surface, which is often a robotic gripper.

Psychrometer A device for measuring relative humidity by comparing the evaporation rates of a wet bulb and a dry bulb.

Pulse transformer A specially designed transformer used for coupling the quick voltage pulses that are seen in thyristor gate trigger circuits.

Pulse-frequency demodulation The process of recovering the original modulating voltage (the measured voltage that was transmitted) from a pulse-frequency modulated signal.

Pulse-frequency modulation A technique of encoding the value of a measured voltage by arranging for the voltage to vary the frequency of a string of pulses.

Pulse-position modulation A technique of encoding the value of a measured voltage by arranging for the voltage to vary the position at which a very short pulse occurs relative to the edge of a reference pulse.

Pulse-width demodulation for telemetry See *Demodulation (Pulse-width).*

Pulse-width modulated Dc drive A speed-control system for a dc motor in which transistor-switched, pulse-width modulation is used. It has the same fundamental efficiency advantage as an SCR—the transistor is either all the way On or all the way Off, so internal power consumption is near zero.

Pulse-width modulation for telemetry See *Modulation (Pulse-width).*

Push-pull amplifier An amplifier in which one half-cycle of the output is produced by one transistor and the other half-cycle is produced by a separate transistor. The push-pull idea has a fundamental efficiency advantage over mid-range-biased transistor amplifiers, and is popular in servo applications.

Quenching (of Heat-treated parts) Immersing heat-treated metal parts in either oil or water to impart the proper metallurgical qualities to the metal.

Rate time (in PID control) The common measure of effectiveness or vigorousness of the derivative action in a proportional plus integral plus derivative controller, when the PID controller is implemented in an analog fashion.

Read operation (R) To access digital data without modifying it.

Read-only memory chip (ROM) A microprocessor chip in which data can be stored but not modified.

Read/write memory chip (RAM) A microprocessor chip that carries stored data that can be both read and modified.

Recoil After strip material has been processed it is wound into a coil (recoiled) for handling and shipping.

Reference cell Used for comparison in carbon dioxide concentration measurement, it is a tightly sealed cylinder that contains pure oxygen. See *Sample cell.*

Regenerative braking An improvement on dynamic braking that switches the motor into generator mode, returning the dynamo's rotational energy to the dc supply instead of allowing it to dissipate into the ambient air by resistance. See *Dynamic braking.*

Relaxation oscillator A simple-to-build oscillator that uses an *RC* timing circuit and a UJT to discharge the capacitor periodically.

Relay A magnetic coil and its controlled contact(s) with the coil being energized or deenergized by the operation of conditional switches or contacts. A relay can perform logic operations. See *Energize, Deenergize.*

Release interval In welding, the time it takes to retract the electrodes from the welded metal.

Reluctance-start motor A motor having a mechanical arrangement that produces a flux-sweeping effect by utilizing salient poles that are misshapen, with one side having a wide air-gap to the rotor and the other side a narrow air-gap. The wider section has greater magnetic reluctance and thus creates a time-displacement that imitates a rotating field.

Reset rate (in PID control) The common measure of effectiveness or vigorousness of the integral action in a proportional plus integral plus derivative controller, when the PID controller is implemented in analog fashion.

Resistive temperature detector See *RTD.*

Resolution (of a Pot) The smallest possible resistance change—that is, the resistance of one turn of wire in a wire-wound potentiometer.

Resolver A generator-like device that relates the magnitude or phase of measured voltage to the absolute angular position of a measured shaft; also called a Shaft-Encoder.

Retentive timer A timer that retains a partially accumulated value of time when its accumulation action stops. When the timer starts timing again in the future, it begins accumulating from that retained value. Such a timer must be explicitly reset by a separate action (separate rung in the PLC realm).

Reversing motor-starter A dual motor-starter (two coils, mutually exclusive) that is capable of reversing the direction of a motor by reversing the polarity of the applied armature voltage (for dc) or of reversing the phase sequence and thus the direction of the rotating magnetic field (for 3-phase ac).

Revolving field A constant-strength magnetic field produced by a motor's stator that revolves at constant speed around the rotor. It can be derived from two ac sine-wave sources that are 90 degrees phase-shifted or by three ac sine waves that are 120 degrees phase-shifted.

Robot See *Industrial robot.*

Rotating field The more common but less correct phrase for a revolving field. See *Revolving field.*

Rotor The cylinder-shaped inner part of a generator or motor that is attached to the shaft and that rotates while the outside (the stator) remains stationary.

RTD (Resistive temperature detector) A temperature-measuring transducer using pure metal wire, having a linear resistance-vs.-temperature characteristic.

Rung A group of instructions within a PLC program, controlling a single output-type instruction. Or, a group of conditional contacts controlling a relay coil within a relay ladder-logic schematic.

Sample cell Used in detecting carbon dioxide concentration, it is a cylinder that experiences a constant flow of fresh atmospheric gas, which has its infrared energy absorption compared to a reference cell that contains pure oxygen. See *Reference cell.*

Sample-and-hold circuit See *Sense-and-hold circuit.*

Saturation, Magnetic See *Magnetic saturation.*

Scan cycle (of a PLC) The overall repetitive sequence of events of a running PLC, consisting of input scan, program scan (execution), and output scan.

SCR See *Silicon-controlled rectifier.*

Screw conveyor A large pipe with a wide-pitched internal screw only slightly smaller in diameter than the pipe itself. When rotated, it moves powdered material down the pipe.

Sense-and-hold circuit Electronic circuit that maintains a recently demodulated measurement value at its destination even when multiplexing is working on other signals; electronically the same as a Sample-and-Hold Circuit.

Series-configured Dc motor Motor whose heavy-gage field winding is electrically in series with the armature. Such a motor has good starting torque and poor speed regulation. See also *Shunt-configured Dc motor.*

Servo Ac motor Conceptually like a split-phase motor, but with thinner rotor-bar conductors that give the motor its unique speed-vs.-torque characteristic.

Servo amplifier An amplifying circuit that boosts the position error voltage to produce the servo motor control voltage for driving its control winding.

Servomechanism A closed-loop system that acts to maintain an object at its mechanical set point.

Set point The desired value of the measured variable in a closed-loop system.

Shaded-pole motor An ac motor having notched salient poles with a copper ring wrapped around the notched side of each pole. This creates the time-displacement-of-flux appearance that imitates a rotating field.

Shaft-encoder See *Resolver.*

Shaft-encoder (Optical) See *Optical position-encoder.*

Shift register A string of flip-flops that transfer their contents from one to another.

Shunt-configured Dc motor A motor whose high-resistance field winding is electrically in parallel (shunt) with the armature winding; such a motor has good speed regulation but low starting torque. See also *Series-configured Dc motor.*

Signal converter A device that converts a high-voltage input signal to a low-voltage logic signal. Also called Original Input Converter, Logic Input Interfacer, etc.

Silicon-controlled rectifier (SCR) A three-terminal solid-state device that acts like a very fast switch and is used to control large currents delivered to a load; when On, it acts like a closed switch; when Off, it acts like an open switch.

Silicon bilateral switch (SBS) A bidirectional breakover device for triggering triacs that is used instead of a diac in low-voltage trigger control circuits. Compared to a diac, an SBS features a lower breakover voltage, a more vigorous switching characteristic, better temperature stability, better symmetry, and less batch spread.

Silicon unilateral switch (SUS) A popular low-voltage unidirectional breakover device that can have its breakover voltage varied through the use of the gate terminal.

Slip (absolute) The difference between the synchronous speed and the rotor speed of an induction motor. See *Slip (percent).*

Slip (percent) The percentage of synchronous speed represented by absolute slip. See *Slip (absolute).*

Slip-rings Copper rings mounted on a motor's or generator's shaft, but insulated from it, to which the coil ends are soldered. Carbon brushes riding on the slip-rings serve to connect the electromagnet coil to the external world.

Soaking pit (for Steel ingots) An underground pit used to heat steel ingots prior to rolling.

Solenoid A specialized electromagnet that functions as an actuating device (for valves, contactors, etc.).

Solenoid valve A solenoid-operated valve that is useful for the on-off control mode.

South pole The surface of a magnet into which flux lines reenter from the surrounding space.

Speed regulation (of a motor) A measure of how much relative speed change occurs as the load's torque demand varies from minimum to maximum.

Split-phase Ac motor A motor with two stator windings that carry currents phase-shifted by about 90 degrees, which produces a revolving magnetic field that pulls the rotor along with it.

Squeeze interval In an automatic welding system, the time allotted for welding electrodes to conform to the curvature of the surfaces being welded and make perfect electrical contact (usually about 1 second).

Squirrel-cage rotor A rotor consisting of end rings connected by aluminum conducting bars laid into slots in the magnetic core; used in an induction motor.

Standby interval In automated welding, the time between removal of a welded part and the positioning of the next pieces to be welded.

Standoff ratio (η) In a UJT, the ratio of the resistance between the emitter and base 1 to the overall internal resistance between base 2 and base 1. This ratio, multiplied by the source voltage, sets the UJT's peak voltage (V_P).

Start winding The winding in a capacitor-start motor that has the capacitor in series with it, and which is disconnected after the motor shaft achieves a certain speed. See *Capacitor-start motor.*

Starting current (of a motor) The current produced by a starting circuit that either gradually increases the applied voltage or that automatically inserts and then removes current-limiting resistors in series with the armature to avoid damage from prolonged inrush current. See *Inrush current.*

Starting torque (of a motor) The amount of torque produced by a motor at the moment it begins accelerating from a standstill. The torque produced at speed = 0.

Stator The outside part of an energy-conversion machine (generator or motor) that remains stationary while the inner part rotates.

Stepper motors Stepper motors have no brushes or armature winding; instead, their rotors step into alignment with energized stator windings that are switched on and off by external transistors.

Store operation (by a μP) Movement of data out of one of the microprocessor's internal registers into a peripheral chip or device.

Strain gage A device designed to measure mechanical force, consisting of a resistance wire cemented to the surface of a strong object that receives the force.

SYNC Pulse (in digital telemetry) A pulse sent down the transmission line that allows the receiving circuitry to synchronize its actions with the sending circuitry.

Synchronizer (for multiplexing) A circuit that maintains consistency between which input is being instantaneously passed by the multiplexer at the sending end of a telemetry system, and which output is instantaneously receiving the output of the demultiplexer at the receiving end of the system.

Synchronous speed The angular speed of the rotating field in an induction motor; it is derived from the line frequency and the number of poles of the motor structure.

Tachometer A device that measures the angular speed of a rotating shaft either through sensing voltage magnitude or through responding to waveform frequency.

Tare weight The weight of the container, which must be subtracted from the total measured weight to find the actual weight of the material in the container.

Teach pendant A handheld switching device linked by cable to a robot microprocessor so that when it is in TEACH mode the human user can guide the robot through a desired manipulative sequence.

Telemetry The process of reliably transmitting measured data over long distances by converting the original analog value to pulse modulation or digital code.

Tesla The unit used to measure magnetic flux density.

Thermistor A temperature transducer with a large negative temperature coefficient of resistance, thereby enabling it to provide very vigorous, but nonlinear, response to temperature change.

Thermocouple A temperature transducer consisting of a pair of wires of dissimilar metals joined to form a loop. By producing a low-value loop voltage, it yields measurement of very high temperatures with good linearity.

Thyrector Protection device against high-voltage transient surges; acting like two zener diodes connected back-to-back, the thyrector shorts out the portion of the voltage transient that exceeds its rated voltage value.

Thyristor A generic term for gated power-switching semiconductors, including both SCRs and triacs. See *SCR, triac.*

Time-constraint delay (in Process control) See *Process reaction delay.*

Timer A device that creates a time delay between two events, one called the trigger event and the other called the output event.

Torque Mechanical twisting action overcoming opposition in motion; measured in pound-feet (lb-ft) or newton-meters (N-m).

Transducer A measurement device that gives an electrical output signal, usually voltage or resistance.

Transfer lag (in Process control) The kind of lag between application of a corrective action and the appearance to the result of that action that occurs in a two-capacity process where the first change must be well advanced before the second cascaded change can begin in earnest.

Transportation lag (in Process control) The kind of lag between application of a corrective action and the appearance of the result of that action that occurs while material is being transported from the energy-input location to the measurement device.

Tri-State output A digital electronic design in which each output line of a data-sending device has three capabilities: going to 0 (ground potential), going to 1 (+5 V potential), and disconnecting from both power-supply buses (floating state). See *High-Impedance state.*

Triac A bidirectional thyristor used to control the average current to an ac load; it differs from an SCR in that it can conduct current in either direction.

Triplet (of a Cycloconverter) A group of three SCRs responsible for producing one of the half-cycles of the output waveform. The positive triplet produces the positive half-cycle and the negative triplet produces the negative half-cycle.

True Power (*P*) Actual real electric power, taking into account the rms voltage and the rms current, and also the *V-I* phase relationship—as distinct from apparent power (*S*), which does not take into account the *V-I* phase relationship.

Two-phase induction motor A motor in which the stator's phase windings are driven by two ac sine-wave sources that are 90 degrees phase-shifted, thus creating a constant-strength magnetic field that revolves at constant speed as the rotor attempts to follow it.

Ultrasonic detection The use of a high-frequency sound wave to detect whether a path is open or blocked; ultrasonic transducers are useful for measuring or detecting the internal characteristics of solids or semi-solids.

Unidirectional mode The mode of operating a stepper motor in one direction only. It makes possible a higher maximum stepping rate and speed; however, the higher power must be reached gradually by ramping up from the lower bidirectional speed—and the same for slowing down prior to stopping.

Unijunction transistor (UJT) A breakover switching device that is used in gate control circuits for SCRs and in many other industrial circuits (timers, oscillators, waveform generators).

Up/down counter A counter or counter instruction that has the capability of incrementing (counting up) or decrementing (counting down), depending on a control command.

User program (of a PLC) The collection of programming instructions that are devised by the human user, which are executed by the PLC's processor to control the industrial system.

Vacuum holder A robotic end-of-arm grabbing device that holds onto an object by applying to it a vacuum cup(s) or a vacuum surface, usually obtained from a venturi passing compressed air.

Variable-reluctance stepper motor A stepper motor that utilizes a toothed nonmagnetized rotor instead of a permanent magnet, so that the rotor always moves to a position that minimizes the magnetic reluctance of the overall flux path (that is, the combined total air gaps in the path).

Venturi A necked-down section of a flow container (pipe) at which the fluid velocity is high and the static pressure is low.

Voltage-controlled oscillator (VCO) An oscillator whose oscillation frequency is variable by varying the voltage applied to its control terminal. A VCO is one of the essential parts of a phased-locked loop.

Voltage-to-current converter An op-amp circuit that delivers a current to the load that is exactly proportional to the input voltage even though the load resistance may vary.

Voltage-to-current ratio *V/f* (for an Ac Motor drive) When variable-frequency speed control of an ac motor is implemented, the motor supply voltage must be adjusted to maintain a constant ratio—the *V/f* ratio.

Weber The unit used to measure the amount of magnetic flux without regard to its density (without regard to the area that contains the flux).

Weld interval In an automatic welding system, the time during which current passes through the electrodes and through the metal-to-metal contact, creating the weld (generally between 2 and 10 seconds), taking into account that this time may alternate between current temporarily flowing (Heat), and current temporarily stopped (Cool).

Welding transformer A transformer whose secondary winding carries current to the electrode power leads.

Winding group, Distributed. See *Distributed winding group.*

Windup reel The final reel on strip-handling apparatus where the coils of the strip are edge-aligned with each other for handling and shipping.

Write (W) To access data for the purpose of modifying it.

Wye Configuration A 3-phase connection method in which all three phase voltages are connected to form a neutral point; shown schematically in the shape of the letter Y.

Zero-speed (plugging) switch A centrifugal switch attached to a motor's shaft; its purpose is to detect when the shaft has stopped turning (zero speed) so that the control circuit can disconnect the motor winding from the reverse-applied source to prevent the motor from taking off in the opposite direction after stopping.

INDEX